W9-BSZ-063

ENCYCLOPEDIA OF ENVIRONMENTAL STUDIES

NEW EDITION

WILLIAM ASHWORTH AND
CHARLES E. LITTLE

JANICE M. FOWLER,
RESEARCH ASSOCIATE

Facts On File, Inc.

Encyclopedia of Environmental Studies, New Edition

Facts On File, Inc.
132 West 31st Street
New York NY 10001

Library of Congress Cataloging-in-Publication Data

Ashworth, William, 1942–
Encyclopedia of environmental studies / William Ashworth and
Charles E. Little.—New Ed.
p. cm.
Includes bibliographical references and index.
ISBN 0-8160-4255-1 (alk. paper)
1. Environmental engineering—Encyclopedias. 2. Environmental protection—Encyclopedias. 3. Pollution—Encyclopedias. 4. Ecology—Encyclopedias.
I. Little, Charles E. II. Title.
TD9 .A84 2001
333.7'03—dc21
00-051379

Text design by Joan M. Toro
Cover design by Cathy Rincon
Illustrations by Dale Williams and Jeremy Eagle, © Facts On File, Inc.

Printed in the United States of America.

VB FOF 10 9 8 7 6 5 4 3 2 1

This book is printed on acid-free paper.

CONTENTS

INTRODUCTION

In the introduction to the first edition (1991) of *The Encyclopedia of Environmental Studies,* William Ashworth observed that he "wrote this book out of frustration." As an author dealing with environmental topics, a librarian, and an active conservationist, he would run across terms that technical reports and government documents defined poorly or not at all, requiring a long search through piles of textbooks, glossaries, and specialized dictionaries that proved to be of little help. "The problem," he observed, "is not just that there is no single source for these terms. Often it is that there no source at all." So it was that he set about to remedy that problem, and he did it brilliantly.

As an author myself, and for 25 years a Washington, D.C., environmental policy analyst (for the Conservation Foundation, the Congressional Research Service of the Library of Congress, and as head of the American Land Forum, an environmental think tank), I know what Bill Ashworth is talking about. And so does every other environmental writer, scientist, economist, historian, theologian, policy maker, policy advocate, teacher, and student. Indeed, we are today even more frustrated about the sheer extent of often arcane information concerning the environment than 10 years ago when Bill Ashworth first published this book.

Clearly "the environment" is no longer, if indeed it ever was, the exclusive concern of slightly crazed nature lovers. Today, seemingly intractable crises confront everyone on the planet: drought and starvation in East Africa; the toll on wildlife and people in the Southern Hemisphere of blindness and cancer from ultraviolet rays pouring through the thinning ozone layer; global warming increasing so rapidly that ecosystems are unable to adapt to the changes; the seemingly unstoppable metastasis of urban sprawl affecting even the most remote countryside areas in developed nations; and onward in a list that itself seems unstoppable.

The argument implicit in publishing an environmental encyclopedia is that such issues are not simply matters of opinion but of factual data and informed analysis. How else can we conduct a civil discourse about the environment? And if civil discourse is impossible, how will it ever be possible for needed public decisions dealing with environmental issues to be made effectively or reliably? The penalties, as we have learned, of making wrong-headed choices about the environment are severe. People actually die. Whole species of our fellow creatures are extinguished forever. Food sources are permanently degraded. The earth's beauty is despoiled. Ignorance of environmental science, economics, politics, and even theology is simply not an option in deciding how we must deal with environmental issues in our individual lives, in our enterprises, and in our governmental policies.

It is for this reason that I have been pleased to contribute to the work that Bill Ashworth began, a work that provides the most up-to-date, most convenient, most comprehensive access to environmental information and analysis compared to any other one-volume encyclopedia in the field. Here are the informational handholds that every American—professional or plain citizen—needs to understand just what "the environment" is all about. Some 1,000 new or revised entries are included in this edition, bringing the grand total to more than 4,000 entries.

One more motive: I am the great-grandson of an encyclopedist whose works are still in print after 100 years, most notably *The Cyclopedia of Classified Dates,* published by Funk & Wagnalls in 1900. His name is the same as mine—Charles E. Little—and it pleases me to be following in his footsteps. A Methodist minister, Reverend Little worked in the attic of his parsonage, a narrow, three-story building in Verona, New Jersey, now occupied by a veterinary clinic. When Reverend Little was engaged with a project, the attic windows were tightly shut and absolutely no visitors were allowed. Not for any scholarly or ascetic reason, mind you, but because his data entries were handwritten on small scraps of paper placed in tidy piles throughout the room. A wayward breeze could require weeks of resorting. My task has been infinitely easier. We have computers these days to keep the facts in their proper piles, and even to access the primary information sources themselves. Moreover we have the Internet to connect us with people whose daily concern is to decide what "environmental studies" should consist of, which is, in fact, the way we—my assistant, Janice Fowler, and I—began our work.

In writing the first edition of the encyclopedia, Bill Ashworth relied on a panel of experts to provide guidance on the selection of topics to cover. They were (with their affiliations at that time) Rodney A. Badger, professor of chemistry at Southern Oregon State College; Edward J. Fritz, a chemical engineer and head of his own firm in Cleveland, Ohio; Vawter Parker, an attorney for the Sierra Club Legal Defense Fund in San Francisco; and Zane G. Smith, a forester formerly with the U.S. Forest Service. It was a temptation to empanel a new set of experts for the new edition, but we finally decided that there might be a better alternative. In the 10 years since the first edition of this book was published, the field of environmental studies and environmental science has burgeoned in North American colleges and universities. Accordingly, we selected 10 of the best, and most representative, programs and requested the syllabi for the survey courses offered. Our assumption, which turned out to be correct, was that perhaps more than anyone else in the highly diversified and fragmented field of environmental studies, survey course instructors, as a group, would have a pretty good grasp on the range of topics to be covered and would have developed a carefully considered list of authoritative textbooks (also a burgeoning field) and supplementary sources for the students to use.

In working with the syllabi and in our conversations with instructors, we discovered that environmental studies, as a discipline, has shifted a bit since the original edition of the encyclopedia was published. While during the 1980s the teaching emphasis was more focused on the hard sciences, the courses of the 1990s also included the social sciences (e.g., economics, politics) and the humanities (e.g., history, philosophy, and religion). The new and revised entries in this edition reflect this shift of emphasis toward a more comprehensive approach.

We then consulted the general textbooks that the instructors required for the courses, and this too was a revelation. Many of them are just flat-out excellent. My personal favorite turned out to be by a full-time environmental textbook author, G. Tyler Miller Jr., a North Carolina chemist and professor of chemistry who, since 1966, has been teaching and writing about environmental studies. Another favorite is an ecology textbook by Manuel C. Molles Jr. of the University of New Mexico. (The titles are listed in the bibliography.) Many instructors, of course, did not use general textbooks such as those by Miller and Molles, but readings from environmental books on various topics. So we acquired those too. (Janice is an ace at buying books secondhand.) In the aggregate, then, our revisions and additions to this encyclopedia reflect what a number of authorities on environmental studies believe the field is presently all about, and what they think are the topics and the terms that need to be covered. In conducting the research for the second edition, we consulted some 300 of the most recently published books, scholarly and scientific papers, reports, and articles. A standout source for us in preparing entries for organizations was the *Conservation Directory* published by the National Wildlife Federation.

So there we are. The first edition took its author a half-decade to prepare. This revision has spanned two years. Somewhere in Methodist heaven my great-grandfather's muttonchop whiskers are waggling in glee: Welcome to the club! So I guess I must thank him, the Reverend Little, for the ability to persevere; and thank Janice Fowler who was not only research assistant but also designer of the new illustrations and indexer for the work as a whole. Thanks as well to Max Gartenberg, friend and literary agent, and Frank K. Darmstadt, senior editor at Facts On File, and especially Bill Ashworth, whose wonderful foundation it was a joy to build on and to be associated with. May this new edition be as useful over the *next* 10 years as Bill's original edition was useful to me.

—Charles E. Little
Placitas, New Mexico

aa a type of LAVA characterized by a rough, jagged surface. Aa flows tend to be high in silicates and low in dissolved gases. They extrude at a lower temperature than do PAHOEHOE flows and they move considerably more slowly (5 to 15 miles per hour). When cool they are dark in color and present the appearance of a fused heap of rubble. The name is Hawaiian, and was apparently derived from the involuntary cries of those who have to walk barefoot across the stuff. *See also* VOLCANO; MAGMA.

Abbey, Edward (1927–1989) American environmental writer, born January 29, 1927, in Home, Pennsylvania, and educated at the University of New Mexico and at the University of Edinburgh in Edinburgh, Scotland (Fulbright fellowship, 1951–52). Edward Abbey was probably America's best-known advocate of radical environmentalism. His most popular work, the novel *The Monkey-Wrench Gang* (1974), features as heroes a group of people who stop the development of WILDLANDS by destroying the machinery (bulldozers, etc.) used to develop them. His 15 years as a park ranger in some of the loneliest parts of the American southwest produced *Desert Solitaire* (1968), a spare, intense, highly personal book that has been ranked with Thoreau's *Walden* and Leopold's *Sand County Almanac* (*see* THOREAU, HENRY DAVID; LEOPOLD, ALDO). Other works by Abbey include *The Brave Cowboy* (1958); *Fire on the Mountain* (1962); *Appalachia* (1970); *Slickrock* (1971); *The Journey Home* (1977); *Down the River* (1982); *Beyond the Wall* (1984); *The Fool's Progress* (1988); and *Hayduke Lives!* (1990), a posthumously published sequel to *The Monkey Wrench Gang. See also* MONKEY WRENCHING. He died March 14, 1989 in Oracle, Arizona. Books concerning Abbey works include *Edward Abbey,* a critical study by Garth McCann (1977); *The New West of Edward Abbey* (1982), by Ann Ronald; *Resist Much, Obey Little: Remembering Ed Abbey* (1989), a festschrift edited by James R. Hepworth and Gregory McNamee; *Epitaph for a Desert Anarchist: The Life and Legacy of Edward Abbey* (1994), a biography by James Bishop; and *Coyote in the Maze: Tracking Edward Abbey in a World of Words* (1998), edited by Peter Quigley.

aboriginal species (indigenous species) a plant or animal SPECIES that inhabited a particular geographic area prior to human disturbance. *Compare* ENDEMIC SPECIES; EXOTIC SPECIES.

absolute age in geology, the actual amount of time that has passed since a geologic event took place, as determined by radiocarbon dating, potassium-argon dating, DENDROCHRONOLOGY or some other means of measuring the passage of real time. *See also* GEOLOGIC TIME: *absolute geologic time.*

absolute humidity the weight of water vapor present in a given volume of air, usually expressed in grams/cubic meter. Knowledge of absolute humidity is of limited use because the volume of an AIR MASS changes with changes in temperature and atmospheric pressure, altering the absolute humidity without any change in the actual moisture content. *Compare* RELATIVE HUMIDITY. *See also* HUMIDITY.

absolute pressure in engineering, the actual pressure applied by a gas or a liquid to the inside walls of a containment vessel, ignoring whatever counter pressures may be applied to the outside walls by the ATMOSPHERE or by another surrounding gas or fluid. *Compare* GAUGE PRESSURE.

absolute temperature the TEMPERATURE of a body measured in relationship to the temperature known as *absolute zero,* where all molecular motion ceases. It is measured in degrees Kelvin (K), which are the same size as degrees Celcius (C) but for which "zero" corresponds to absolute zero rather than to the freezing point of water. Absolute zero has been calculated to lie at –273.16°C (–459.69°F); thus, the freezing point of water has an absolute temperature of 273.16°K.

absorption in chemistry, a process in which a liquid or gas is held within a solid without changing the chemical properties of either of the two substances involved. Absorbed molecules penetrate into the intermolecular spaces beneath the absorbing solid's surface. *Compare* ADSORPTION; SOLUTION.

abyssal plain in geology and oceanography, the relatively level portion of the deep-ocean floor, beginning at the foot of the continental rise (*see under* CONTINENTAL SHELF) and extending outward toward the center of the ocean. It was once thought that the abyssal plain extended in a nearly featureless expanse from one continental rise to the next, but this view is now known to be in error. *See* MID-OCEAN RIDGE; SEAMOUNT; GUYOT.

accretion (1) in law, the slow, imperceptible deposition or erosion of soil by a river over time, as contrasted to *avulsion,* which is the rapid, perceptible deposition or erosion of land by a river that changes course as a result of a flood or similar catastrophic event. The distinction is critical for determining the ownership of land: lands added by accretion belong to the owner of the parcel they are accreted to, while lands added by avulsion remain the property of the original landowner. *See also* RELICTION.

(2) in geology, a SEDIMENTARY ROCK structure that has been built up from a small nucleus by the adherence of a layer or layers of particles around it.

(3) in geology, the process by which land is build up through the deposit of stream sediments in a DELTA or alluvial plain (*see* ALLUVIUM).

ACE *See* ALLOWABLE CUT EFFECT.

ACEC *See* AREA OF CRITICAL ENVIRONMENTAL CONCERN.

acequias an irrigation canal or ditch, traditional in the Hispanic villages of northern New Mexico. *Acequia madre* is the term used for the main (or "mother") ditch, from which smaller ditches deliver water to individual orchards or fields. Acequias are developed by cooperative acequia associations that are made up of landowners and overseen by a *mayordomo* who manages the system and allocates irrigation water by means of opening and closing headgates from the main ditch. The acequia system provides for the efficient use of often scarce water resources and maintains community coherence and Hispanic village traditions that date from the beginning of Spanish agricultural development by colonists in the 17th century. *See also* WATER RIGHTS.

acetone a colorless, highly mobile fluid, chemical formula C_3H_6O (also written CH_3COCH_3 or $(CH_3)_2CO$), obtained from the fermentation of corn or from the oxidation of propane or other petroleum gases. Acetone is the simplest member of the class of compounds known as *ketones* and is thus chemically related to the ALDEHYDES. It is an excellent solvent for fats, oils, plastics, paints, and numerous other compounds, and is sold commercially as fingernail-polish remover and as a solvent for the so-called "super glues." It is only moderately toxic (LD_{50} [rats, orally]: 10.7 ml/kg) and is not known to be carcinogenic (*see* CARCINOGEN), but is classed as a hazardous substance due to its extremely high VOLATILITY and low FLASH POINT (–20°C).

acid in chemistry, originally, any compound that dissociates upon contact with water, forming hydronium ions (H_3O^+; *see* DISSOCIATION; ION): often defined more broadly today as a *proton donor,* that is, any chemical substance that gives up protons when combining with other substances. Since the qualities we think of a "acidic" do not normally show up until the acid is dissolved in water and the dissociation (or proton donation) takes place, most people undoubtedly think of acids as liquids, though many of them are actually solids, and a few are gases, at room temperature. Acid solutions have a pH of less than 7; they have a sour taste, are moderately to strongly corrosive, and conduct electricity. All of these characteristics depend directly upon the presence of the H_3O^+ ions: the more completely a substance dissociates (the more H_3O^+ ions a solution of it contains) the more strongly it will show acidic properties. Acids combine with BASES and METALS to form SALTS.

acid deposition the fallout of acidic compounds in solid, liquid, or gaseous form generated by factories, power plants, internal combustion engines, and other, often distant, sources. ACID RAIN is the best-known form of acid deposition. Others include acid fog, acid snow, acid rime ice, and DRY DEPOSITION.

acid extractable organic compound in ENVIRONMENTAL PROTECTION AGENCY terminology, an ORGANIC COMPOUND that may be separated from wastewater by applying acid to the wastewater. The most hazardous of this group of compounds from an environmental standpoint are the PHENOLS. (Note that most chemists use this term to refer to organic compounds that can be separated *into* water, rather than *out of* water, by acids. Under this more common terminology, phenols are *base extractable* rather than acid extractable.) *Compare* BASE/NEUTRAL EXTRACTABLE ORGANIC COMPOUND.

acid rain rainfall with a lower pH than normal. Since snow, hail, and sleet can also be acidic, the term *acid precipitation* is probably preferable, though it is seldom used in the popular press. Acid rain is the best-known form of AERIAL DEPOSITION; it is found worldwide but is a particular problem in Scandinavia, central Europe, eastern Canada, and the northeastern United States.

CAUSES OF ACID RAIN

Normal rainfall is moderately acidic, with a pH of roughly 5.6 (7 is neutral). This acidity results from the dissolving of CARBON DIOXIDE (CO_2) out of the atmosphere by individual raindrops as they fall to earth. The dissolved carbon dioxide turns the water in the raindrops to weak carbonic acid (H_2CO_3). Chemicals released into the air from volcanoes and other natural sources may also contribute to temporary upswings in rainfall acidity. However, the most significant sources of acid rain are manmade emissions, primarily from the burning of FOSSIL FUELS. The combustion temperatures of these fuels are high enough to cause some of the nitrogen and oxygen molecules in the surrounding air to combine, forming NITROGEN OXIDES (NOx). These compounds escape into the atmosphere, where they enter into a series of chemical reactions that convert them to nitric acid (HNO_3). Impurities in the fuel may also combine with oxygen to form acid-making compounds. The most important of these is sulfur, a common constituent of coal (less common in petroleum), which forms SULFUR OXIDES (SOx). Some of these compounds, when acted on by atmospheric chemistry, yield sulfuric acid

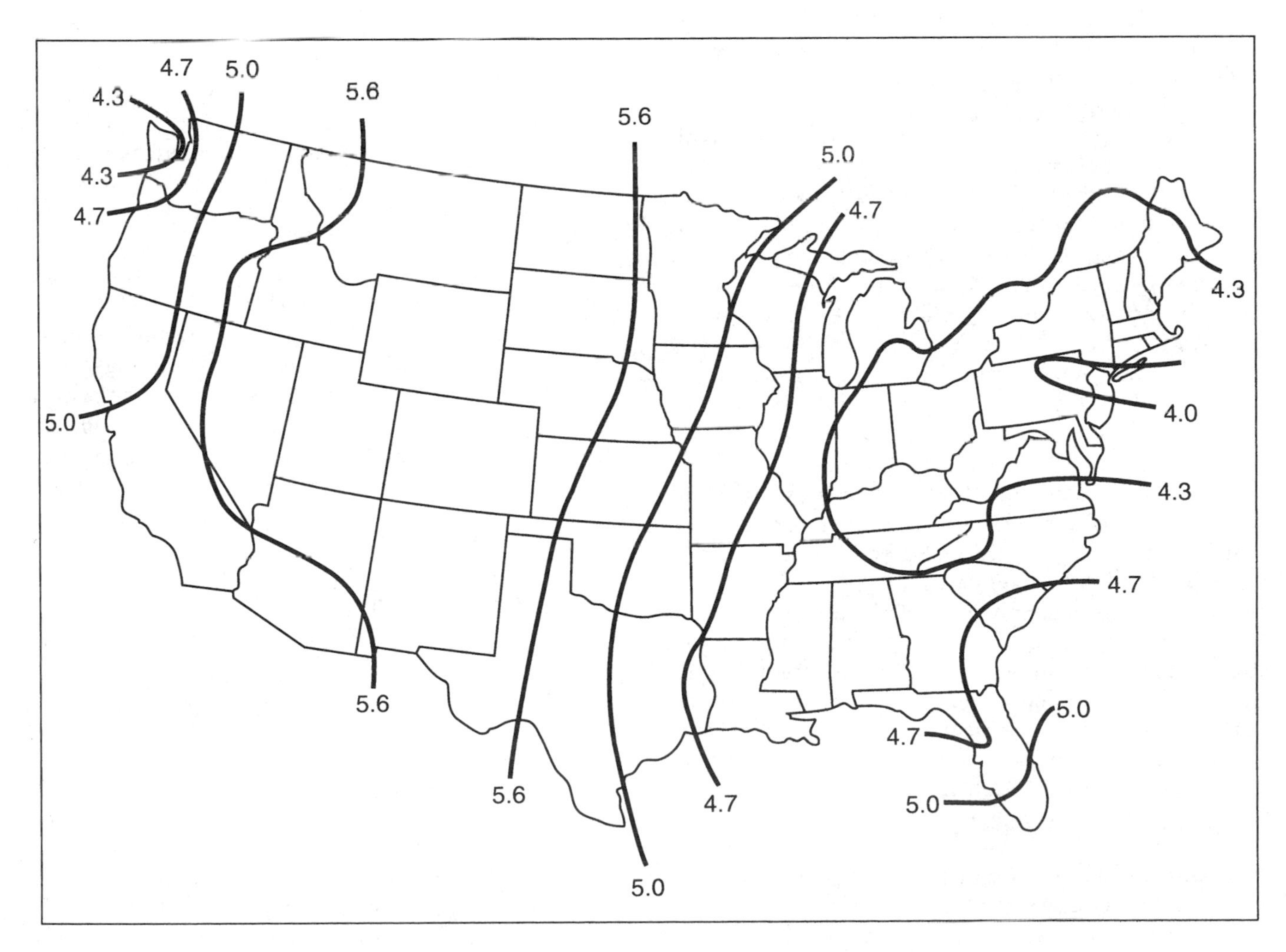

Rainfall acidity

(H_2SO_4). Together, sulfur oxide and nitrogen oxide emissions can profoundly effect the pH of precipitation in areas downwind from an emitting source. During a rainstorm in Wheeling, West Virginia, in 1980, the reading was pH 1.5—about 10,000 times as acidic as normal rainfall.

EFFECTS OF ACID RAIN

The effects of acid rain vary greatly with the type of environment the rain falls upon. Regions of thick soils and basic (alkaline) rock types such as limestone may show little or no effects, while regions of thin soils and acidic rock types such as granite may be significantly altered. This alteration includes, but is not limited to, increases in the acidity of lakes and streams, leading to fish and invertebrate failure; deposition and increased mobility of phytotoxic metals; the increased leaching of essential soil nutrients; decreases of beneficial MICROORGANISMS in the soil; tree death and forest decline in hard-hit areas; and decreases in the growth rate of commercial crops.

HISTORIC BACKGROUND

Acid rain was first noted in Scandinavia, where it was being monitored and studied as early as 1952. Between 1960 and 1980, this monitoring showed average rainfall acidity increasing by roughly 200-fold. Similar results were demonstrated by North American monitoring programs begun in the late 1960s by Cornell University scientists in the Adirondack Mountains of New York and by scientists from the University of Toronto in the La Cloche Mountains of northern Ontario. By 1981 these studies had found approximately 180 lakes in the Adirondacks and 140 in Ontario that were devoid of life due to low pH levels brought about by acid rain. Public concern over these findings led to demands that industrial emissions of SOx and NOx be curbed by law. Accordingly, 1990 amendments to the CLEAN AIR ACT called for the reduction of sulfur dioxide by 10 million tons by 2000 and of nitrogen oxide by 2 million tons by 1995, to be managed by an EPA air pollution "credit" system, whereby the right to pollute can be traded from emissions source to emissions source. Although air pollution laws are expected to reduce acid rain, decades-long deposition in the northeastern United States and Canada have altered the composition of the soils, reducing their nutrient content, which in turn leads to changes in forest cover. *See also* AIR POLLUTION, CLEAN AIR ACT, FOREST DECLINE AND PATHOLOGY, TOXIC PRECIPITATION.

acoustic emissions internal sounds generated by a structural material when placed under stress. Acoustic emissions are a forewarning of structural failure; hence, equipment designed to detect them is an important tool for preventing hazardous materials spills caused by ruptured tanks or pipelines.

acoustic environment *See* NOISE POLLUTION.

acquired lands lands purchased by the federal government from private sources for management as part of a NATIONAL PARK, NATIONAL FOREST, or similar public holding. *Compare* PUBLIC DOMAIN.

acre-foot the amount of water required to cover one acre of land to a depth of one foot. The term is used primarily as a measure of reservoir capacity. One acre-foot is equal to 43,560 cubic feet (325,851 gallons).

acre-yield the amount of water obtainable from one acre of an AQUIFER; also, the amount of anything obtainable from one acre of the land that produces it.

activated carbon (activated charcoal) a form of powdered or granulated CARBON from which most impurities have been removed by treatment with heat and steam. It readily absorbs large quantities of a great variety of substances, principally HYDROCARBONS, and is therefore extremely useful for removing organic pollutants from water and air (*see* ABSORPTION). Most activated carbon is prepared commercially from coconut husks, peach and apricot pits, and other woody BYPRODUCTS of food production.

activated sludge a common form of secondary sewage treatment (*see* SECONDARY TREATMENT), utilizing AEROBIC BACTERIA in a pond or tank that is kept oxygenated either by mechanical mixing or by pumping compressed air through it, both of which agitate or "activate" the WASTEWATER. The bacteria feed on each other and on other organic matter present in the wastewater, forming clumps of a frothy, sticky, gelatinous substance known as *floc,* or *zoogloeal* ("living glue") *masses.* More than 60 SPECIES of bacteria have been found in a single sample of floc; however, the principal species is *Zoogloea ramigera,* which forms multibranched colonies whose sticky, fingerlike protuberances interlock with each other, firmly enmeshing most other microbial life in the water. The floc is separated from the wastewater through the use of a SETTLING BASIN. A small part of it is used to inoculate the wastewater coming into the activated sludge pond with fresh colonies of *Z. ramigera,* while the remainder is disposed of. The clarified wastewater, its BIOCHEMICAL OXYGEN DEMAND reduced by 60% to 95%, is either dumped directly into the RECEIVING WATERS or retained for further purification (*see* TERTIARY TREATMENT).

active recreation outdoor recreation such as team sports, games, and the like that usually require specialized facilities in parks and open spaces, unlike passive recreation, which can be provided by natural, undeveloped areas. *See also* DISPERSED RECREATION.

active solar design any type of SOLAR ENERGY installation requiring pumps, fans, or other mechanical means to transfer heat from the collector to the point of use. *Compare* PASSIVE SOLAR DESIGN.

acute toxicity the ability of a toxic substance to cause death of or serious damage to living ORGANISMS within a short time after their exposure to it. A "short time" is usually defined for this purpose as 96 hours. *Compare* CHRONIC TOXICITY.

Adams, Ansel (1902–1984) American photographer, author, and environmentalist. Born in San Francisco, California, on February 20, 1902, Adams had little formal schooling, being educated partly by his father and largely by his own insatiable curiosity. His first career was as a concert pianist, until he realized at about the age of 28 that what he called "making pictures" (never "taking pictures") was more important to him than his music. Although he continued to play the piano for pleasure to the end of his life, he gave no more concerts after 1930. In 1932 he gave his first important one-man show, at San Francisco's de Young Museum. Four years after that, he became the first young photographer since 1917 to be given a one-man show at Alfred Stieglitz's influential An American Place gallery in New York City. His photographs, primarily of the scenic grandeur of the American west, quickly became known for their clarity of vision and their intense attention to detail, qualities clearly shown in his most famous photograph, *Moonrise, Hernandez, New Mexico* (1941). Called strongly by an urge to protect the world's scenery as well as to capture it on film, Adams became a director of the SIERRA CLUB in 1934, a post he held until 1970. His book *This Is the American Earth* (1960; text by Nancy Newhall) was the progenitor of Sierra Club Books' Exhibit Format Series. An author and lecturer as well as a photographer, Adams published some 30 books and taught numerous seminars and courses in photography in Yosemite National Park and around the country. He received the Sierra Club's John Muir Award in 1963 and the Conservation Service Award of the U.S. Department of the Interior in 1968. A definitive biography, *The Eloquent Light* (Nancy Newhall), was published by Sierra Club Books in 1964. A permanent exhibit of Adams's work may be found at the headquarters of the Wilderness Society, 900 17th Street NW, Washington, DC 20006-2596. His *Ansel Adams: An Autobiography* was reprinted in 1996. Biographies also include *Ansel Adams and the American Landscape: A Biography* (1995), by Jonathan Spaulding, and *Ansel Adams: A Biography* (1998), by Mary Street Alinder. He died in Carmel, California on April 22, 1984.

adaptation in ecology, an alteration in the behavior and/or physical characteristics of an ORGANISM or SPECIES that enables it to adjust to permanent changes in its ENVIRONMENT or to fit into a different NICHE than it has historically occupied.

adequate treatment in sanitary engineering, a term used for a sewage treatment facility that includes at least SECONDARY TREATMENT and provides maximum EFFLUENT concentrations of 30 mg/L SUSPENDED SOLIDS, 30 mg/L BOD (*see* BIOCHEMICAL OXYGEN DEMAND), and 1.0 mg/L total phosphorus (*see* PHOSPHORUS: *phosphorus as a pollutant*).

adhesion, of a fluid *See* WETTING ABILITY.

adiabatic rate in meteorology, the measure of ADIABATIC TEMPERATURE CHANGE caused by the motion of an AIR MASS upward or downward through the ATMOSPHERE. In dry air, the adiabatic rate is about 3°C (5.5°F) for every 1,000 feet of elevation change. *Compare* LAPSE RATE.

adiabatic temperature change in meteorology, the heating or cooling of an AIR MASS due to changes in volume rather than changes in energy content. Generally, the cause of an adiabatic temperature change is the movement of the air mass to a higher or lower elevation. As it moves to a higher elevation, the pressure on it decreases, causing its volume to expand and its temperature to lower; moving to a lower elevation reverses the process. *See also* ADIABATIC RATE.

Adirondack Mountain Club (ADK) hiking and outdoor club founded in 1922 to promote outings and conservation activities in mountain areas, especially the Adirondack Mountains of the State of New York. The ADK maintains portions of various trail systems in the northeastern mountains and operates a pair of lodges; it also has an extensive publications program. There are 26 chapters of the Adirondack Club in New York and New Jersey, with a total membership (in 1999) of 24,000. Address: 814 Goggins Road, Lake George, NY 12845-4117. Phone: (518) 668-4447. Website: www.adk.org.

administrative trail term used historically by both the BUREAU OF LAND MANAGEMENT and the FOREST SERVICE to designate a trail used primarily to help manage a piece of land and only secondarily (if at all) as a route for recreationists. Administrative trails were built to somewhat lower standards of grade and tread width than other types of trail. Most have now been replaced by roads.

adsorption in chemistry, a process in which a liquid or gas adheres to the outside of a solid without penetrating it. The adhering film is normally only one molecule thick. *Compare* ABSORPTION.

advanced treatment *See* TERTIARY TREATMENT.

advection fog *See under* FOG.

adventitious growth in botany, the development of any type of plant tissue in a region of the plant where it is not normally found. A tree branch that bends down and touches the ground, for example, may put out roots (*adventitious roots*) at the point of contact. The ability to produce adventitious growth increases a plant's ability to survive severe damage.

Advisory Council on Historic Preservation independent federal agency which acts as policy adviser to the president, Congress, and federal agencies regarding historic preservation. Created by the National Historic Preservation Act of 1966, the council is comprised of seven federal department heads, a governor, a mayor, and representatives of the National Trust for Historic Preservation and the National Conference of State Historic Preservation Officers. Address: The Old Post Office Building, 1100 Pennsylvania Avenue NW, Suite 809, Washington, DC 20004. Phone: (202) 606-8503. Website: www.achp.gov.

AE *See* ANIMAL EQUIVALENT.

aeolian deposit in geology, soil or sand moved to its current location by wind. Aeolian deposits are particularly important features of the landscape in areas of little vegetation such as deserts (*see* DESERT BIOME) and LITTORAL ZONES. *See also* LOESS; DUNE.

aeration the process of passing air through water, sewage, or some other liquid. Aeration kills ANAEROBIC BACTERIA by exposing them to oxygen; encourages the OXIDATION of POLLUTANTS and growth of aerobic ORGANISMS that remove BIOCHEMICAL OXYGEN DEMAND from the water; and carries off small particles

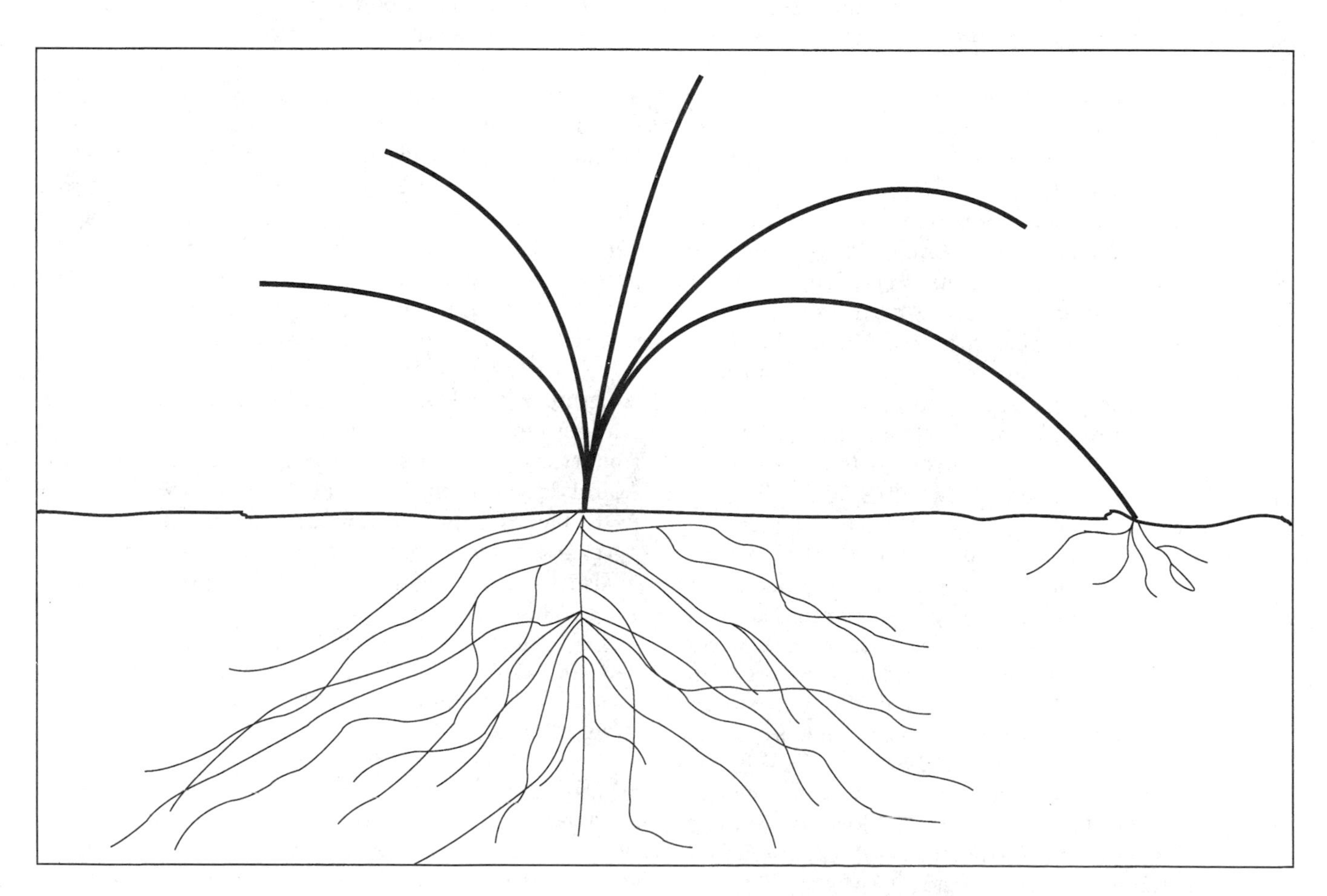

Adventitious roots

in the SURFACE TENSION surrounding the air bubbles. It is performed either by piping air through a screen or other form of diffuser into the bottom of a tank of the liquid to be aerated and allowing it to bubble to the surface, or by mechanically agitating the liquid to a froth. Spraying may also be used as a form of aeration; in this case the liquid passes through the air rather than vice versa. Natural aeration occurs in the waterfalls and rapids of swiftly moving streams (*see* WHITE WATER).

aerial deposition deposition on the land surface or in the water of pollutants that have arrived through the air. Airborne contaminants are usually carried to earth in the rain, though in still air they may simply settle out. *See also* ACID DEPOSITION; ACID RAIN; TOXIC PRECIPITATION; DRY DEPOSITION.

aerial logging any system of logging in which the YARDING of the logs is done through the air. Aerial logging requires fewer roads and avoids the soil compaction and UNDERSTORY damage that results from dragging logs along the ground, and is thus preferable to ground yarding from an environmental standpoint, although it is usually considerably more expensive. The logs may be lifted directly by cable-guided balloons (*balloon logging*) or by helicopters. The most common technique, however, is to use an elevated cable called a *skyline* (*see* SKYLINE LEAD).

aerobic bacteria any species of bacteria (see BACTERIUM) that requires oxygen to live. *Compare* ANAEROBIC BACTERIA.

aerobic stabilizer in sanitary engineering, any device used to remove BIOCHEMICAL OXYGEN DEMAND (BOD) from SEWAGE or other wastes through the use of AEROBIC BACTERIA. The most common aerobic stabilization devices are the TRICKLING FILTER and the ACTIVATED SLUDGE tank, either of which is capable of removing 85%–95% of the BOD in a waste stream in as little as two hours. Aerobic stabilizers are considered to be a form of secondary sewage treatment (*see* SECONDARY TREATMENT).

aerosols droplets of liquid small enough to be suspended in a gaseous medium (usually air). Aerosols can be a particularly dangerous form for POLLUTANTS to take because they can be breathed into the lungs, where they can both damage sensitive lung tissue and be absorbed directly into the bloodstream without the protection afforded by the digestive system. In air pollution control terminology they are sometimes considered to be a form of PARTICULATE (*see* AIR POLLUTION: *types of pollutants*). Note that "aerosol spray" cans are named for the fine mist of droplets they produce, not for the means of propelling the contents from the can. *See also* COLLOID; SUSPENSION.

aestivation *See* ESTIVATION.

age class in ecology or forestry, a geographically related group of individuals of the same species and of roughly the same age. *See also* EVEN-AGED STAND; AGE DISTRIBUTION; COHORT; SURVIVORSHIP.

age distribution the total number of individuals in each AGE CLASS of a POPULATION at a specific instant in time. It is usually represented as a curve on a graph, with the horizontal (x) axis representing ages and the vertical (y) axis representing numbers of individuals. *Compare* SURVIVORSHIP.

Agency for Toxic Substances and Disease Registry (ATSDR) federal agency responsible for coordinating efforts to control or reduce human health problems caused by toxic substances accidentally released into the environment. ATSDR's principal mission is to maintain listings, in as complete and accurate a form as possible, of hazardous substances, their health effects, and the locations within the United States where accidents involving them have caused significant health hazards. The agency also coordinates (and to a certain extent, instigates) scientific studies of the relationships between toxic substances and human health, and acts as a support center for federal and state programs designed to respond to emergencies involving hazardous-substance spills. Established on April 19, 1983 by order of the Secretary of Health and Human Services, as directed by Section 104(i) of the COMPREHENSIVE ENVIRONMENTAL RESPONSE, COMPENSATION AND LIABILITY ACT of 1980, ATSDR is a part of the PUBLIC HEALTH SERVICE of the Department of Health and Human Services (*see* HEALTH AND HUMAN SERVICES, DEPARTMENT OF). Address: 1600 Clifton Road, Atlanta, GA 30333. Phone: (888) 422-8737. Website: www.atsdr.cdc.gov.

Agenda 21 *See* RIO EARTH SUMMIT.

Agent Orange a mixture of two powerful agricultural herbicides, consisting of 2,4-D and 2,4,5-T, sprayed from U.S. aircraft and U.S. Navy vessels during the Vietnam War. The objective of this action, called Operation Ranch Hand, was to defoliate the heavy tree canopy in Vietnam in order to expose troop movements of Vietcong guerrillas and North Vietnamese forces. Between 1962 and 1970, 11.2 million gallons of Agent Orange (named after the orange rings on the barrels) were sprayed, so completely destroying the natural cover that in some areas it has still not returned. In addition to the herbicides responsible for ecosystem damage, the compound contained 2,3,4,7,8-TCDD, an especially dangerous form of DIOXIN which can cause cancer and birth defects. More than 270,000 U.S. veterans registered with a Veterans Administration program to deal with the health conditions brought on by exposure to Agent Orange, with 6,000 (as of 1998) qualified to receive compensation, up to $5,000 a month in severe cases. In addition, Dow Chemical and Monsanto, manufacturers of Agent Orange, settled a class-action lawsuit by providing $180,000 to 20,000 veterans. In Vietnam, of course, the ill effects also persist. The Vietnamese government estimates that 400,000 people were killed or injured by the spraying and 500,000 children have suffered birth defects. In some former storage areas the chemical can still be found in very high concentrations. *See also* TIMES BEACH.

age of recruitment in population biology, the age at which sexual maturity arrives; that is, the age at which an individual animal is "recruited" into the adult population.

aggregate in geology or engineering, gravel, sand, or broken rock quarried for use in making concrete.

aggregation in population biology, a collection (usually temporary) of many individuals of the same species at a single geographic location. An aggregation differs from a herd, flock, etc. by being circumstantial rather than structured. It has little or no social organization and is brought together by environmental forces rather than by any urge of the animals involved to seek their own kind. Examples of aggregations include moths drawn to a light, salmon gathering in the headwaters of rivers to spawn, and birds collecting in one spot before beginning their seasonal migration.

aggressive mimicry in behavioral ecology, a form of MIMICRY in which the mimicking species is a PREDATOR imitating another species or an object in the environment in order to deceive its prey into coming close enough to be caught.

aggressive resemblance *See under* CRYPTIC COLORATION.

agribusiness generally refers to those corporations associated with agriculture as suppliers, such as chemical companies and equipment manufacturers, and those who purchase agricultural commodities for resale or for processing as packaged goods. *See also* CORPORATE FARMING.

agricultural land, U.S. the most productive agricultural resource base in the world, with 968 million acres of land in farms as of 1997. The total amount of agricultural land has declined slightly since the mid-1980s. Harvested cropland constitutes about one-third of the total, with the remainder in pasture or summer fallow, idled in federal programs, and in woodlots or other uses. While prime agricultural land can be found in all regions of the country, the Middle West and Great Plains region constitutes the most important single patch of productive agricultural land in the world, producing basic staples—corn, wheat, soybeans—not only for U.S. consumption but also for export. Increasingly, U.S. agricultural land is being aggregated into larger and larger ownerships, with fewer individual farms and, necessarily, fewer farmers. Some current agricultural land issues include EROSION, BLACK LAND LOSS, and URBAN SPRAWL.

Agricultural Research Service (ARS) branch of the United States Department of Agriculture responsible for overseeing and administering research projects

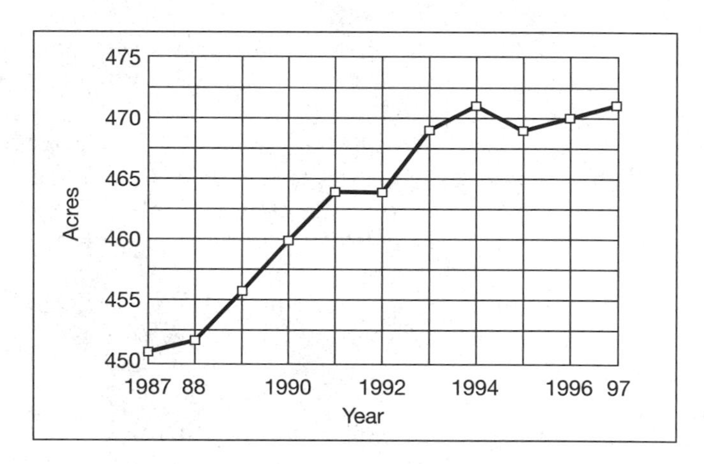

Average U.S. farm size, 1987–1997 ***(U.S. Department of Agriculture, National Agricultural Statistics Service, Farms and Land in Farms)***

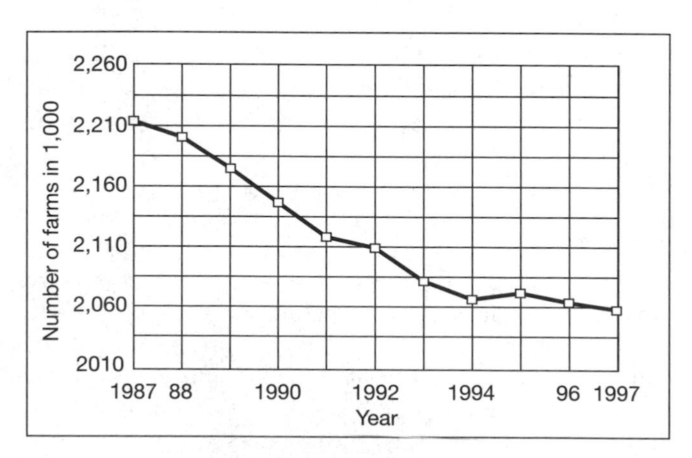

Number of U.S. farms, 1987–1997 ***(U.S. Department of Agriculture, National Agricultural Statistics Service, Farms and Land in Farms)***

relating to all aspects of agriculture. ARS research projects include (but are not limited to) animal and plant protection and production; the use, protection and improvement of soil, water, and air; commodities distribution and storage; and human nutrition. Most of these activities are carried out jointly with state, county, and local agencies through the Extension Service—a separate agency within the Department of Agriculture—and through other, similar partnership agreements. ARS activities take place in 147 sites spread throughout the world and are organized into four administrative regions. Address: Jamie L. Whitten Building, Suite 302-A, 14th Street and Independence Avenue SW, Washington, DC 20250. Phone: (202) 720-3656. Website: www.ars.usda.gov. *See also* AGRICULTURE, DEPARTMENT OF.

Agricultural Stabilization and Conservation Service (ASCS) a branch of the United States Department of Agriculture (USDA), created June 5, 1961 and made responsible for the administration of commodity, land-use, and resource-conservation programs. Among the resource-related programs operated by ASCS are the Agriculture Conservation Program, the Forestry Incentive Program, the Water Bank Program, and the Experimental Rural Clean Water Program. ASCS cooperates closely with the Extension Service—an independent USDA agency—with county and state Extension Agents acting as ex-officio members of the ASCS governing committees on their respective levels. It is now part of the FARM SERVICE AGENCY. *See also* AGRICULTURE, DEPARTMENT OF; FARM SERVICE AGENCY.

Agriculture, Department of United States government agency charged with overseeing all federal programs relating to food and fiber production and distribution, including livestock, vegetables, grains, cotton, forest products, and the conservation of agricultural soil and water. It is the third largest federal agency in terms of work force (after Defense and Health and Human Services) and the second largest in terms of amount of land managed (after Interior). A part of the EXECUTIVE BRANCH of the federal government, the department is headed by a cabinet-level officer (the secretary of agriculture) assisted by a deputy secretary and overseeing seven assistant secretaries, two undersecretaries, and four officers, each of whom is in charge of a major branch of the agency. The most important of these branches from an environmental standpoint is that headed by the assistant secretary for natural resources and the environment, which includes the FOREST SERVICE and the SOIL CONSERVATION SERVICE. Others with environmentally related programs include the assistant secretary for science and education (AGRICULTURAL RESEARCH SERVICE; COOPERATIVE STATE RESEARCH, EDUCATION, AND EXTENSION SERVICE; Extension Service; National Agricultural Library), the undersecretary for international affairs and commodity programs (AGRICULTURAL STABILIZATION AND CONSERVATION SERVICE), the assistant secretary for marketing and inspection

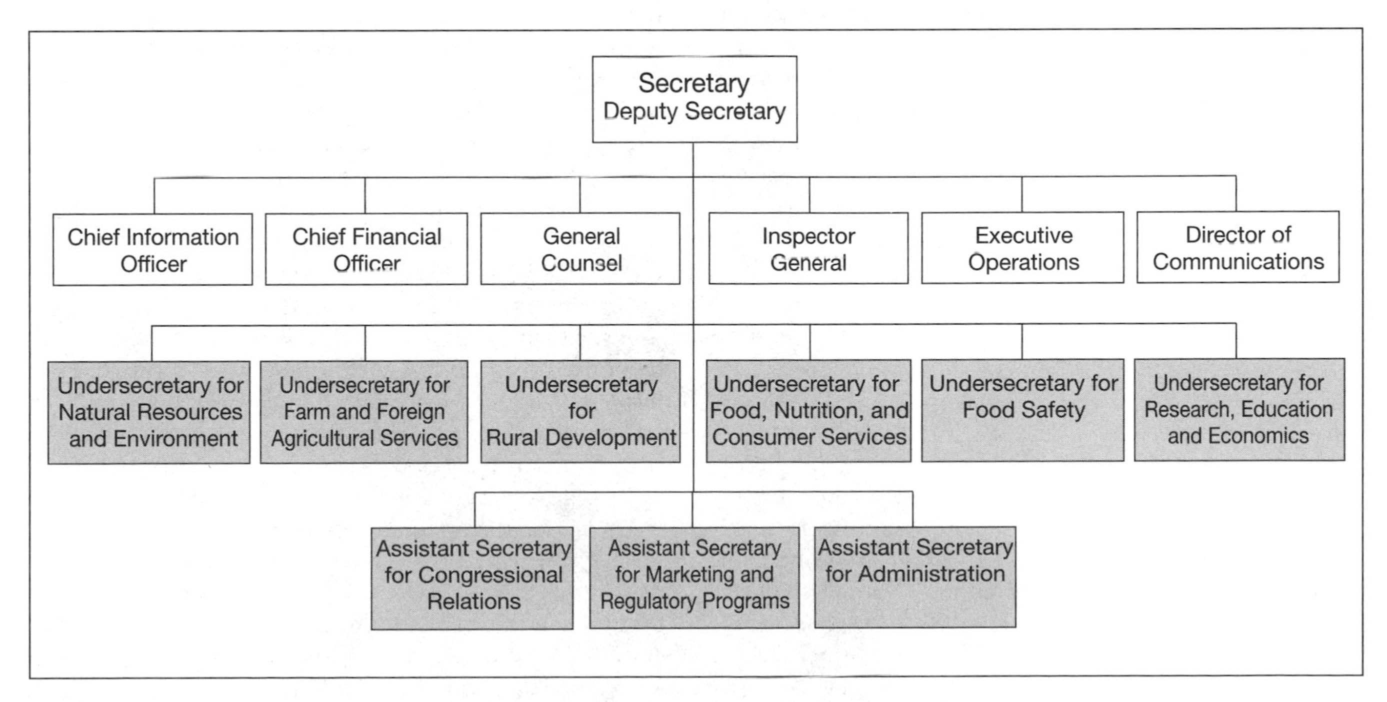

U.S. Department of Agriculture organizational structure (U.S. Department of Agriculture Fact Book, 1998)

services (ANIMAL AND PLANT HEALTH INSPECTION SERVICE, FEDERAL GRAIN INSPECTION SERVICE, FOOD SAFETY AND INSPECTION SERVICE, and the assistant secretary for administration (Board of Contract Appeals).

HISTORY

The set of programs that eventually became the Department of Agriculture was begun about 1836 as a branch of the Office of the Commissioner of Patents (then a part of the State Department). It was charged with distributing seed and commodities to farmers. The agricultural programs moved to the Interior Department with the Patent Office in 1849, became a separate branch of Interior in 1862, and were elevated to cabinet status by act of Congress on February 9, 1889. Since then the department has undergone numerous changes, consolidations, and reorganizations. Among the most important of these are the Transfer Act of 1905, which added the Forest Service to the department; the Soil Conservation Act of 1935, which created the Soil Conservation Service; and the President's Reorganization Plan 1 of 1939, which transferred agricultural surplus programs from the Department of Commerce to the Department of Agriculture. Many food production functions were transferred to a War Foods Administration within the War Department during World War II, but were returned to the Department of Agriculture after the war.

agroforestry a soil conservation measure akin to STRIP CROPPING, but with agricultural crops planted between rows of trees or shrubs rather than alternations such as corn with a grass cover crop. In agroforestry the woody plants, being taller, can provide shade, retain soil moisture, and reduce water and wind erosion by means of their density and root structure. Moreover, the trees or bushes can themselves supply produce such as fruits, nuts, decorative greens, or even firewood or mulch. *See also* CONSERVATION TILLAGE.

AID (USAID) See UNITED STATES AGENCY FOR INTERNATIONAL DEVELOPMENT.

air *See* ATMOSPHERE.

Air and Waste Management Association a nonprofit group founded in 1907 that encourages worldwide interaction on and discussion of environmental issues. The association also promotes technology development, environmental education, and informed environmental management. Publications include a monthly newsletter and the *Journal of the Air and Waste Management Association*. Membership (1999): 13,000. Address: One Gateway Center, 3rd Floor, Pittsburgh, PA 15222. Phone: (412) 232-3444.

air current in meteorology, a wind caused by some environmental factor that is unrelated to a storm. Unlike storm winds, air currents are usually steady and predictable from day to day. *See* ATMOSPHERE: *dynamics of the atmosphere.*

air mass in meteorology, a body of air that tends to act as a coherent, separate whole. Neighboring air masses usually differ from each other in pressure, temperature, and water and POLLUTANT content. They move as units, and they develop internal circulation patterns that are largely independent of one another. The boundaries between air masses are one of the principal determinants of weather and of the distribution patterns of pollutants. *See* FRONT; INVERSION; ATMOSPHERE: *dynamics of the atmosphere.*

air pollution the presence of harmful substances in the earth's ATMOSPHERE. Air pollution from natural causes—dust, smoke from forest fires, gases from volcanic eruptions or decaying matter—has always been an environmental problem. The earliest forms of industrial air pollution, particulate matter from coal-burning factories and residences, posed severe public health problems for centuries in large cities, notably London and New York. The conversion from coal to oil for heating in the mid-20th century reduced particulate matter, but still produced harmful amounts of pollutants—carbon monoxide (CO), oxides of sulfur and nitrogen (SO_x and NO_x), and ozone (O_3), as did the burgeoning number of automobiles used for commuting in large cities. By 1950 photochemical smog became an added pollution hazard in many cities, notably Los Angeles, California, where a temperature INVERSION, a frequent occurrence, can lead to the accumulation of harmful gases at ground level, especially ozone, which affects the lungs and can be deadly to trees and crops, even at many miles distant from the source. In the 1960s tall smokestacks were introduced on factories to reduce AMBIENT AIR pollution in industrialized areas. The idea was that "the solution to pollution is dilution." By injecting polluting gases into the upper atmosphere, it was thought, they would become so dispersed as to be harmless. As it turned out, the gases created ACID RAIN in areas up to 500 miles away. Meanwhile, CHLOROFLUOROCARBONS used as propellants for spray cans and in automobile antifreeze rose into the stratosphere, breaking down the so-called OZONE LAYER that protects the earth from harmful ULTRAVIOLET RADIATION, which can cause cancer and blindness in animals and humans. The ozone hole, acid rain, and excess carbon dioxide from the burning of fossil fuels that has resulted in the GREENHOUSE EFFECT and global warming, are all extremely serious environmental consequences of air pollution. *See also* CLEAN AIR ACT.

air quality classes rating system established by the 1970 revision of the CLEAN AIR ACT to determine allowable levels of AIR POLLUTION in various regions of the United States. The air quality classification system divides the nation into AIRSHEDS and categorizes each airshed into one of three classes:

Class I: no deterioration allowed.
Class II: some deterioration allowed.
Class III: deterioration allowed up to national standards.

Class I airsheds include NATIONAL PARKS, WILDERNESS AREAS, and other pristine or reasonably pristine regions. In these, AMBIENT QUALITY STANDARDS may not be allowed to decrease below December 1974 levels. Class III airsheds are primarily industrial regions and large municipalities where ambient quality standards are largely ignored as long as pollutants do not reach unhealthful levels as currently defined either by Congress or by the ENVIRONMENTAL PROTECTION AGENCY. The definition of Class II airsheds varies from region to region. Here the allowable ambient levels of pollutants are set by the states, as long as they do not exceed national standards.

Air Quality Control Region (AQCR) as defined by the CLEAN AIR ACT, a geographical region consisting of two or more cities, counties, or other government units in which air-pollution problems are relatively uniform. Designated by the secretary of health and human services (*see* HEALTH AND HUMAN SERVICES, DEPARTMENT OF), Air Quality Control Regions are required to adopt consistent pollution-control strategies. They can (and often do) cross state lines. When this happens, the states must cooperate in designing the control strategy, and each state is responsible for enforcing the strategy within its own portion of the AQCR. If enforcement by the states is inadequate, the ENVIRONMENTAL PROTECTION AGENCY is authorized to set and enforce standards that supercede those of the states. Approximately 265 AQCRs have been designated between the passage of the Clean Air Act in 1970, and 1999.

airshed in meteorology, a geographical region that tends, usually because of topographical factors, to have uniform air quality. POLLUTANTS in one part of an airshed are swiftly transferred to other parts of the same airshed but not to neighboring airsheds. Airshed boundaries, like those of WATERSHEDS, are generally along ridge lines.

Alaska Lands Act *See* ALASKA NATIONAL INTEREST LANDS CONSERVATION ACT.

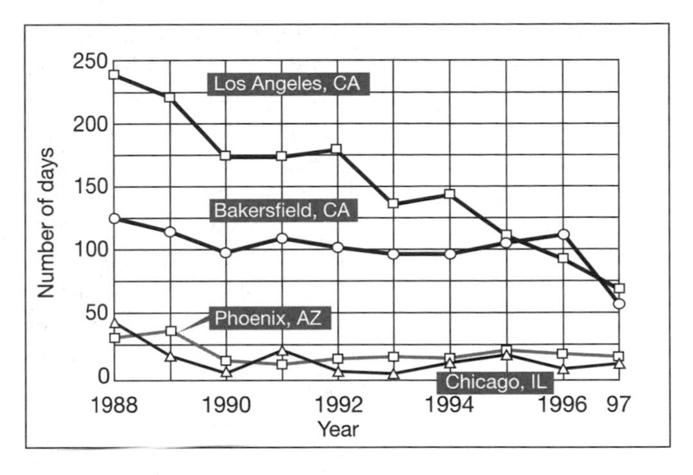

Air quality of selected U.S. cities, 1988–1997 ***(U.S. Environmental Protection Agency, Office of Air Quality Planning and Standards)***

Alaska National Interest Lands Conservation Act (Alaska Lands Act) federal legislation to resolve conflicts between development and preservation interests in Alaska, signed into law by President Jimmy Carter on December 2, 1980. One of the largest single acts of land preservation in U.S. history, the Alaska Lands Act placed 104.3 million acres of land in restricted or preserved categories, including 43.5 million acres of new NATIONAL PARKS, 53.7 million acres of WILDLIFE REFUGES, 56.4 million acres of WILDERNESS AREAS, and 13 new designations under the WILD AND SCENIC RIVERS ACT. (These numbers add up to more than 104.3 million because much of the new wilderness acreage was within the new National Parks and National Wildlife Refuges.) The additions doubled the acreage of the National Park System and the National Wildlife Refuge System, and tripled the acreage of the National Wilderness Preservation System. On the development side, the act (1) mandated an annual ALLOWABLE CUT of 450 million board feet from the Tongass National Forest and created an automatic annual appropriation of $40 million to administer this harvest level; (2) waived the wilderness review provision of the FEDERAL LANDS POLICY AND MANAGEMENT ACT for all BUREAU OF LAND MANAGEMENT lands in Alaska, thereby effectively eliminating wilderness consideration for BLM lands in the state; and (3) contained so-called "hard release" language for National Forest lands, effectively eliminating these lands from further wilderness consideration as well. Finally, the bill tied up the loose ends of the Alaska Native Claims Settlement Act of 1971, making final designations of which lands should belong to the State of Alaska, which should belong to the federal government, and which should belong to the native peoples of the state.

HISTORY

The Alaska Lands Act grew out of section 17(d)(2) of the Alaska Native Claims Settlement Act of 1971, which mandated study of all federal lands in Alaska for designation in the various federal land-management categories, disbursement to the state, or disbursement to the native peoples. The studies were completed by 1977, and hearings were held that year on potential legislation. The House passed its version of the Alaska Lands Act in 1978; when the Senate failed to follow by the end of the 95th Congress, President Carter invoked the ANTIQUITIES ACT, designating most of the House-passed parklands as NATIONAL MONUMENTS by EXECUTIVE ORDER. Thus spurred, the Senate finally acted in the closing days of the 96th Congress; the House accepted the Senate version, and the bill became law with the president's signature on December 2, 1980.

ALASKAN PARKS AND PRESERVES

As a result of the Alaska Lands Act, 10 new parks and preserves were added to three existing national parks (which themselves were redesigned and increased in acreage). The three original parks are (as presently named) *Denali National Park and Preserve,* with North America's highest mountain; *Glacier Bay National Park and Preserve,* with dramatic glacier fronts; and *Katmai National Park and Preserve,* a wildlife and sport fishing area in the Aleutian Range. The new parks and preserves are *Aniakchak National Monument and Preserve,* a 30-square-mile caldera located on the Alaska Peninsula; *Bering Land Bridge National Preserve,* a remnant of the land bridge connecting Siberia with the North America that offered a route for the first human migration to the Western Hemisphere about 12,000 years ago; *Cape Krusenstern National Monument,* archaeological sites dating back 4,000 years near Kotzebue in northwest Alaska; *Gates of the Arctic National Park and Preserve,* in the Brooks Range north of the Arctic Circle; *Kenai Fjords National Park,* a rainforest near Seward; *Kobuk Valley National Park,* north of the Arctic Circle and rich in wildlife and archaeological sites; *Lake Clark National Park and Preserve,* in the Chigmit mountains along the western shore of Cook Inlet; *Naotak National Preserve,* a large, pristine river basin; *Wrangell-St. Elias National Park and Preserve,* the largest unit of the national park system in the "mountain kingdom of North America" east of Anchorage; and *Yukon-Charley Rivers National Preserve,* along the Canadian border in the 1898 gold rush area. *See also* NATIONAL PARK SYSTEM.

Alaska Pipeline oil pipeline running from north to south across the eastern portion of the state of Alaska, from Prudhoe Bay on the Arctic Ocean to the port of Valdez on the Gulf of Alaska. The pipeline is 1.2 meters (4 feet) in diameter and 1,262 kilometers (789 miles) in length, and has a working capacity of 1.2 million barrels of oil per day. It crosses three major mountain ranges and approximately 350 rivers, including the 700-meter-wide (2,300-foot) Yukon. 437 kilometers (364 miles) of the pipeline's path is buried; the remaining 789 kilometers (425 miles) are elevated above ground on some 78,000 towers. Much of the pipeline passes through zones of heavy seismic activity, and as a consequence it has been designed to withstand the theoretical strain from an earthquake measuring up to 8.5 on the RICHTER SCALE. Design considerations include a zigzagging course (for flexibility), teflon-coated supports (to allow slippage), and automatic shutoff valves designed to limit any spill to less than 15,000 gallons. The bases of most of the support towers are refrigerated to avoid damage from PERMAFROST, and parts of the pipeline itself are refrigerated and buried in permafrost to avoid blocking caribou migration routes.

HISTORY

Discovery of the Prudhoe Bay oil reserves was announced on March 13, 1968. By the end of 1969 preliminary plans for the pipeline were completed and a consortium of oil companies, the Alyeska Pipeline Company, had formed to build it and had purchased the required 800 miles of pipe. In what became the first major test of the NATIONAL ENVIRONMENTAL POLICY ACT (NEPA), environmentalists sued to force consideration of a less environmentally sensitive alternate route through Canada and on April 2, 1973, won a U.S. Supreme Court ruling that NEPA had been violated. The Court, however, refused to block construction, and on November 11, 1973, the issue became moot when President Richard Nixon signed a law exempting the pipeline from further NEPA review. Construction began in April 1974: it was halted in September 1975, when evidence came to light that Alyeska had filed falsified x-rays of welds on some of the buried pipe. A subsequent investigation ending in May 1976 revealed that, out of roughly 30,000 welds completed in 1975, 3,995 were defective, including approximately 1,000 at so-called "critical" locations such as river crossings. One hundred forty-eight of these welds were eventually replaced. Work resumed under closer government supervision in the summer of 1976, and the first oil flowed through the pipeline on June 20, 1977.

ENVIRONMENTAL ISSUES

The salient issues raised by the existence of the Alaska pipeline include the migratory routes for caribou, especially the Porcupine herd, thought to number up to 180,000 animals; oil leakage from the pipeline itself, as well as spills at the terminus at Valdes, such as the devastating *EXXON VALDES* spill in 1989, which disrupted the Prince William Sound ecosystem; and maintaining the integrity of the ARCTIC NATIONAL WILDLIFE REFUGE, which energy interests wish to open up for exploration.

Alaska Power Administration originally, one of five federal agencies known as the "federal power administrations." The Alaska Power Administration, formerly within the U.S. Department of Energy, was sold to private Alaskan owners in 1995. Prior to the turnover, the administration had developed two hydroelectric projects: Ekluta, built in 1955, and Snettisham, built in 1975. Unlike other power administration projects, those in Alaska affected only one state and had only one purpose—to provide cheap hydroelectric power. For these reasons, the Alaskan projects could be privatized. *See also* POWER ADMINISTRATION, FEDERAL.

albedo in the physical sciences, the percentage of radiation falling on a surface that is reflected back from it. Earth as a whole has an albedo of about 35%, meaning that for every 100 units of sunlight that strike it, 35 are reflected back into space. The albedo of different surfaces on the Earth varies dramatically with color and texture: new snow, for example, has an albedo of about 80%, while dense conifer forest has an albedo of about 5%. Thus much more of the incoming sunlight is reflected from the Arctic and Antarctic regions than from the temperate-zone forests, an imbalance that helps drive worldwide weather patterns. The clearcutting of vast areas of boreal forest, which removes light-absorbing trees, is thought by some scientists to disrupt regional weather regimes, including the possibility of reducing rainfall in northern wheat-growing areas.

Albright, Horace M. (1890–1987) a founder of the National Park Service. Albright, a lawyer, was assistant director and then director from 1929 to 1933. With STEPHEN T. MATHER, the first director, Albright was responsible for the creation and early progress of the new agency, established in 1916. His principal contributions were to emphasize "interpretation" of the parks, with rangers trained to give authoritative talks and act as guides, and to convince the administration of President Franklin Delano Roosevelt to transfer historic sites, battlefields, and similar facilities in the eastern United States to the park service for administration, thus greatly expanding the agency's range (previously limited to western parks) and prestige. His books include *Oh, Ranger! A Book About the National Parks* (1928), with Frank J. Taylor; *The Birth of the National Park Service* (1985), with Robert Cahn; *The Story Behind the Scenery* (1987), with Russell Dickinson and William P. Mott; *Conservators of Hope* (1988); and *Creating the National Park Service: The Missing Years* (1999), with his daughter, Marian Albright Schenck.

alcohol any of a large group of ORGANIC COMPOUNDS characterized by the replacement of one or more of the hydrogen atoms (H) in the compound by hydroxyl GROUPS (OH^-). Alcohols are generally colorless, flammable liquids with a mildly agreeable odor and a strong taste; they make good SOLVENTS for a wide variety of substances. All are to a greater or lesser degree poisonous. The two most broadly used alcohols are *methyl alcohol* (CH_3OH, also known as *methanol* or *wood alcohol*) and *ethyl alcohol* (C_2H_5OH, also known as *ethanol* or *grain alcohol*). These two compounds have very similar properties, differing principally in their toxicity. Ethyl alcohol is slightly more poisonous than methyl alcohol when taken in very large doses. However, the body is able to metabolize small amounts of ethyl alcohol into water and CARBON DIOXIDE, while small amounts of methyl alcohol metabolize into formic acid, making methyl alcohol far more dangerous in the range of doses normally encountered.

DENATURED ALCOHOL

For industrial purposes, most ethyl alcohol is "denatured" by the addition of small amounts of methyl alcohol and/or other substances that render it toxic, unpalatable, or both. Denaturing is done according to standard formulas: for example, "formula 1" refers to a mix of 5 gallons of methyl alcohol to 100 gallons of ethyl alcohol, while "formula 28A" contains 1 gallon of gasoline to 100 gallons of ethyl alcohol. Denatured alcohol cannot be used for beverages, and thus is not subject to federal excise taxes. However, it can cause severe pollution problems due to the presence of the denaturing agents, many of which themselves rank as PRIORITY POLLUTANTS.

aldehyde in chemistry, any of a large class of ORGANIC COMPOUNDS of the type formula RCHO, where R is any of numerous GROUPS and -CHO (the *aldehyde group*) is a specific group formed from one carbon, one hydrogen, and one oxygen atom. Aldehydes are closely related to *ketones* (type formula: RCOR), and both classes of compounds may be derived from primary ALCOHOLS through oxidation. Unlike ketones, however, most aldehydes may be easily oxidized further to yield carboxylic acids—a property that makes them useful in industrial chemistry as REDUCING AGENTS. They form POLYMERS easily and are widely used in the production of plastics.

aldrin **(octalene, compound 118)** a broad-spectrum chlorinated hydrocarbon pesticide, chemical formula $C_{12}H_8Cl_6$, once widely used as a sheep dip, soil fumigant, seed treatment, and general insecticide. Aldrin is among the most toxic of the cyclic organic compounds, with symptoms of liver and kidney metabolic failure showing up from skin contact with as little as 3 grams; LD_{50} in rats (orally) is 7–15 mg/kg. It is also a powerful CARCINOGEN that is estimated to cause one extra cancer in 10,000 when present in water in concentrations as low as .5 parts per billion. It bioaccumulates easily in aquatic life (*see* BIOLOGICAL MAGNIFICATION). In the soil, it deteriorates rapidly; however, its principal decay product is dieldrin ($C_{12}H_8Cl_6O$), a compound with very similar toxic and carcinogenic properties that is also used as a pesticide. Aldrin has been banned in the United States since 1976.

alewife (***Alosa pseudoharengus***) a small fish of the herring family (Clupidae) native to the North Atlantic Ocean. Since 1960 it has invaded the Great Lakes system, where it has become a major nuisance, suffering major die-offs each spring that clog water intakes and cover beaches with dead fish. By 1967, more than 90% of the tonnage of fish caught in Lake Michigan were alewives, which are commercially useless. Since then, the numbers have fluctuated from year to year but have generally been down, partly because of the widespread planting in the lakes of Pacific salmon, for which alewives serves as an important food source. Length: (North Atlantic) 10–12 inches; (Great Lakes) 3–4 inches.

alfisol any of an important group of soils in the SEVENTH APPROXIMATION soil classification system, characterized by brown to gray-brown surface HORIZONS and an accumulation of clays (an "argillic horizon") beneath the surface. BASE SATURATION is moderate to high (greater than 35%). Most alfisols have at least one dry season of three months or more a year. Developed principally in steppe, shrub-steppe, and deciduous forest environments, and (occasionally) under grasslands, alfisols are important agricultural soils, underlying much of the American midwest and eastern Europe and occurring in western Australia, north-central Canada, and other places. Those developed in dry locations (*xeralfs*) and regions of cold climate (*boralfs*) are used for pasturage. Other varieties (*udalfs, ustalfs*) are used primarily for the production of grains. Aqualfs are alfisols that have developed in areas of poor drainage, thus remaining wet most of the year. If drainage is established they will convert to one of the other types.

algae (singular: **alga**) a group of CHLOROPHYLL-containing, photosynthetic organisms (*see* PHOTOSYNTHESIS; ORGANISM), once classed as a division of the plants but now recognized as belonging to eight divisions in two separate KINGDOMS. This taxonomic scattering means that the term "algae" no longer has any formal descriptive meaning, although it is still in common use. The majority of the approximately 25,000 known SPECIES of algae are one-celled organisms (*see* CELL), although multicellular forms exist, including one form, kelp, which may grow to several hundred feet in length. The simplest algae, the CYANOPHYTA (blue-green algae), belong to the kingdom MONERA. All other forms are classed with the PROTISTA. These include the chlorophytes (green algae), the phaeophytes (brown algae, including kelp and other common "seaweeds"), the Rhodophytes (red algae), the chrysophytes (golden algae, including the DIATOMS), the xanthophytes (yellow algae), the Pyrrophytes (dinoflagellates), and the Euglenophyta (euglenoids). Almost completely aquatic (one exception: those species that combine with fungi to form LICHENS), the algae occupy both fresh and salt water and are found in a broad variety of HABITATS, some of them extremely severe. Some species, for example, have colonized high-mountain snowfields (*see* WATERMELON SNOW), while others live in hot springs at temperatures up to 70°C (160°F); still others are found in highly saline brines, including the evaporation pans of companies producing salt from seawater. *See also* EUTROPHICATION; ALGAL BLOOM; RED TIDE.

algal bloom condition of a body of water in which one or more species of ALGAE are present in such overwhelmingly large numbers that they interfere with the utilization of the water by other forms of life. Algal blooms are characterized by water of low transparency and of a tint the color of the primary species of algae involved in the bloom. Often there will be floating mats of algae on the surface of the water. The odor of decay is generally present, both from masses of dead algae and from the decay of other living things that have been killed by the bloom.

CAUSES AND EFFECTS

Although water temperature and the presence of light can both increase the speed of algal growth—which explains why most problems with nuisance algae take place in shallow water during warm weather—the underlying cause of all algal blooms is the presence in the water of the algae's LIMITING NUTRIENT in more or less unlimited amounts. The source of this nutrient may be either natural (the decay of aquatic organisms; RUNOFF from nutrient-rich lands) or artificial (untreated sewage; industrial EFFLUENT; fertilizer-enriched agricultural runoff). Whatever the source, the excess nutrients feed a cycle of rapid growth, reproduction, and death for the algae. The dead algae fall to the bottom in enough numbers that they have been likened to an underwater rainstorm. Their decay uses up dissolved oxygen and releases TOXINS, making it difficult for other aquatic life to survive. (The living algae may also release toxins: *see* RED TIDE.) Water drawn for domestic use during an algal bloom usually has an unpleasant odor and flavor, and may be unhealthful as well. Under naturally eutrophic conditions there are often two blooms per year: the first as temperatures climb in the early summer, and the second in the fall as the nutrients taken up by the first bloom are re-released into the water through decay and build back up to more than limiting levels. *See also* EUTROPHICATION; BIOCHEMICAL OXYGEN DEMAND.

aliphatic compound any ORGANIC COMPOUND whose molecular structure does not include a BENZENE RING or similar cyclic structure ("aromatic ring") with alternating single and double bonds. *Compare* AROMATIC COMPOUND.

alkali in chemistry, any BASE that dissolves easily in water: specifically, a HYDROXIDE of one of the metals of the so-called "alkaline series" (lithium, sodium, potassium, rubidium, or cesium). *See also* ALKALI FLAT; ALKALINITY; ALKALINIZATION OF SOILS.

alkali flat in geology, the flat floor of a basin, usually an old lakebed from which the water has evaporated, which is covered by deposits of ALKALIS and mineral SALTS. Alkali flats are characteristic of closed DRAINAGE BASINS in very dry climates. During the infrequent rains, these flats often become covered with broad, shallow bodies of water in which the salts and alkalis redissolve. They are then referred to as *alkali lakes.*

alkali lake *See under* ALKALI FLAT.

alkalinity in chemistry, a measure of the ability of a substance to neutralize an ACID. Strongly alkaline solutions contain high concentrations of hydroxyl (OH^-) ions (*see* HYDROXIDE; ION), and are generally extremely caustic, making them extremely dangerous when released into watercourses.

alkalinization of soils occurs when cations of calcium, magnesium, and sodium, and the anions sulfate, chloride, and bicarbonate are introduced into the surface layers. When dissolved, these salts form soda or other alkaline compounds seriously affecting plants, crops, and trees. In arid and semi-arid agricultural regions, salts may accumulate in the soil because the amount of rain is not sufficient to flush the excess down through the soil profile beyond the reach of plant roots. The resulting buildup reduces yield, restricts the kinds of plants that can be grown, and may render some areas virtually unusable for agriculture. The causes of the buildup are sometimes natural—saline seeps from existing subsurface salt beds, for example. More often, alkalinization is the result of irrigation practices, since all irrigation water (unlike rain or other precipitation) contains some salts. In the San Joaquin Valley of California, for example, at least half a million acres have suffered yield reductions of 10% or more. Irrigated lands in other western states have also experienced yield decreases. In some areas fields are flooded to flush out accumulated salts, although this practice can eventually create a more serious problem. Overall, between 55 million and 60 million acres—about 15% of U.S. cropland—are affected by a buildup of salts from the irrigation of poorly drained soils.

Allen's rule in biology, rule of thumb for determining the effect of climate on the body shapes of warm-blooded animals and birds. First proposed by the American ornithologist Joel Asaph Allen (1838–1921), the rule states: *the cooler the climate an animal dwells in, the smaller the relative size of its body appendages.* Thus the arctic hare (*Lepus arcticus*), which dwells on the tundra north of the tree line in Canada and Alaska, has ears only half as long as those of the similarly sized blacktail jackrabbit (*L. californicus*) of the southwestern U.S. deserts. Allen's rule holds most strongly among ALLOPATRIC species. However, the ornithologist J. C. Welty (*The Life of Birds*, 1975) has demonstrated that its effects can be seen within a single species as well, the wings (for example) of the winter wren (*Troglodytes troglodytes*) shrinking by an average of 2% for every degree of latitude one moves northward through its range. *Compare* BERGMANN'S RULE.

allergen a substance in the environment to which a living organism reacts as if it were a MICROORGANISM or other living PATHOGEN, producing antibodies that then react with the allergen, releasing chemicals called *histamines.* The histamines in turn produce the localized tissue swelling that we call an allergic reaction. On the skin, this tissue swelling shows up as a rash; in the respiratory system it constricts the airways, causing ASTHMA or HAY FEVER, and in the digestive tract it generally interferes with the absorption of food and water, causing diarrhea. Most allergens (but by no means all) are PROTEINS. Sensitivity to allergens varies greatly from substance to substance and from individual to individual: for example, nearly everyone is allergic to poison oak and poison ivy, but only a small number of people are sensitive to leather clothing.

allocation type (land allocation type, land allocation) in planning (especially forest planning), the use or mix of uses for which a particular piece of land is supposed to be managed. A map of a forest by allocation type (*a land allocation map* or *resource allocation map*) would show a number of overlapping zones, each one labeled for the type of use appropriate within it—recreation, stream protection, grazing, timber management, etc.

allopatric species in the biological sciences, similar SPECIES that live in separate geographic areas. A pair of species may be allopatric over only part of their RANGE: if one species exists in Georgia and Tennessee and a second exists in Tennessee and Kentucky, the two species are allopatric in Georgia and Kentucky but not in Tennessee. Allopatric species do not interbreed, even when given the opportunity. *Compare* SYMPATRIC.

allotrope *See* ALLOTROPIC FORM.

allotropic form (allotrope) in chemistry, one of two or more different forms taken by a single chemical substance. Each form of the substance has its own distinct set of physical characteristics. Graphite, coal, and diamonds, for example, are all allotropic forms of the element CARBON. Each is composed of the pure element but their physical characteristics differ widely. The particular form an allotropic substance takes depends primarily on the conditions of heat and pressure during its formation.

allowable cut (allowable harvest) in forestry, the maximum amount of timber or other forest products that may be removed from a given area during a given time period and still meet the management objectives for that area. It is usually expressed in board feet (*see* BOARD FOOT). The term "allowable cut" is often associated with even-flow management (*see* NON-DECLINING EVEN FLOW) or other forms of SUSTAINED YIELD but it does not need to be. If the management objective for a piece of timberland is to liquidate the timber growing on it, the allowable cut could conceivably be the total volume of timber on the piece. *See also* ALLOWABLE CUT EFFECT.

allowable cut effect (ACE) in forestry, an increase in the ALLOWABLE CUT made possible by basing harvest calculations on the level of SUSTAINED YIELD expected from a forest after the introduction of improved silvicultural techniques rather than on the current yield from the same forest.

allowable harvest *See* ALLOWABLE CUT.

allowable sale quantity (ASQ) in FOREST SERVICE usage, the maximum amount of timber that may be sold annually from a given piece of land within the MANAGEMENT CONSTRAINTS established by laws and regulations pertaining to that land. The ASQ is averaged over time; it may be exceeded legally in a specific year, but only if compensation is made by cutting less in another year. *Compare* PROGRAMMED ANNUAL HARVEST; ALLOWABLE CUT.

alloy in metallurgy, a mixture of a METAL with one or more other elements, usually other metals, though non-metallic elements such as CARBON may be used in small amounts. The resulting substance has the properties of a metal (ductility, malleability, luster, etc.) but is generally superior to its constituent substances in one or more areas such as hardness, durability, heat resistance, or elasticity. For this reason, most "metals" in use today are actually alloys. Some alloys are simply homogeneous mixtures of tiny crystals of the constituent substances, while others are solutions of one material in the other, formed in the molten state and then allowed to cool. Still others are mixtures of individual molecules that bond to each other and show some of the properties of chemical COMPOUNDS, though without following the rules of VALENCE. Alloys that contain mercury are called AMALGAMS; most of these are liquid at room temperature.

all-terrain vehicle (ATV) a type of OFF-ROAD VEHICLE designed for use in mud, sand, and WETLANDS as well as on dry firm ground. It has an enclosed, watertight drive train and either large balloon tires or caterpillar treads. Most ATVs sold for recreational use are essentially three- or four-wheeled motorcycles. Like all off-road vehicles, ATVs can create serious erosion hazards. However, they are generally less damaging than trail bikes (motorcycles designed for off-road use) due to the greater tire surface—hence, less force per unit area—in contact with the ground.

alluvial soil AZONAL SOIL developed on a recent deposit of ALLUVIUM. Alluvial soils are characterized principally by extreme youth and therefore lack of profile development (*see* SOIL: *the soil profile*). They may be light to dark in color, alkaline to acidic, and poor to moderately rich in organic content. *See* SOIL: *classification of soils.*

alluvium (alluvial deposit) in geology, silt, sand, gravel, or other sediments laid down by the action of a running stream. Specific types of alluvial deposit include the *alluvial plain,* a plain made by the filling in of a river valley with sediment; the *alluvial fan,* a cone-shaped deposit laid down where a stream flows out of a constricted spot such as a canyon or gully and is able to spread out on a plain or down a talus slope or in a lake margin; and the *alluvial dam,* a stream blockage formed by sediment deposited around some obstacle in the streambed, usually in a place where the stream has had to slow down and is ready to drop its sediment load anyway. *See also* FLUVIAL DEPOSIT.

alpha particle **(alpha ray)** a particle consisting of two protons and two neutrons (*see* ATOM: *structure of the atom*) emitted by a decaying radioactive substance. An alpha particle has a double positive charge and an ATOMIC WEIGHT of 4, and may be thought of as a helium atom stripped of its electrons. *Compare* BETA PARTICLE; GAMMA RAY.

alpine zone **(arctic zone, arctic-alpine zone)** the region above or north of the TIMBERLINE, characterized by the absence of trees and the presence of expanses of bare rock, tundra (*see* TUNDRA BIOME), and lingering or permanent deposits of snow and ice. *See also* LIFE ZONES.

alternation of generations the reproductive strategy followed by most living organisms, in which a generation consisting entirely of haploid individuals (individuals in which each CHROMOSOME is represented only once in each cell) gives rise to a second generation consisting entirely of diploid individuals (individuals in which each chromosome is represented twice in each cell), which in turn gives rise to another haploid generation, and so on. In some species the individuals in the haploid and diploid generations are closely similar in size and characteristics and can only be told apart by their reproductive structures (or by microscopic examination of their tissues); in others, the haploid and diploid generations vary widely from each other. In the VASCULAR PLANTS, the haploid generation has been reduced to a few cells that remain enclosed by the tissues of the dominant diploid organism (*see* SPOROPHYTE; GAMETOPHYTE). Animals do not exhibit true alternation of generations, but consist of diploid individuals that produce haploid reproductive cells—the sperm and the egg—which then unite to form a new diploid individual. *See also* MEIOSIS; MITOSIS.

alternative in land-use planning, one of a set of specific alternate plans for managing the same piece of ground. A fully developed alternative—as, for example, one prepared for use in an ENVIRONMENTAL IMPACT STATEMENT or a forest plan—includes a statement of the goals the land is to be managed for, the costs involved in management, and the effects to be expected on surrounding lands and on the larger society, as well as specific prescriptions for managing the resource. Different alternatives usually address different management goals and societal effects, though this is not necessary; two alternatives may be simply two different ways of achieving the same end. *See also* PREFERRED ALTERNATIVE.

alternative energy sources in general, the means of producing energy that does not consume NONRENEWABLE RESOURCES. Usually such energy sources are assumed to be relatively nonpolluting as well. Examples of alternative energy sources include solar power (*see* SOLAR ENERGY), WIND POWER, COGENERATION, HYDROGEN FUEL CELLS, and human muscle power (walking, bicycling, clockwork mechanisms, etc.).

alternative transportation any energy-efficient, low-pollution form of transportation. Alternative transportation preferably uses an ALTERNATIVE ENERGY SOURCE; however, it is often taken to mean any form of transportation other than the private automobile, including everything from walking and bicycling to fossil-fuel powered mass transit.

Amazon rain forest *See* RAIN FOREST BIOME.

ambient air in meteorology, the surrounding air, that is, the air present throughout a vicinity (as opposed to the air at any one location within the vicinity or that wafted in by an incoming breeze). *See also* AMBIENT QUALITY STANDARD.

ambient quality standard in pollution control, the level that pollutants may be allowed to reach in the AMBIENT AIR. Ambient quality standards represent the general air-quality standards for an entire vicinity rather than quality standards for spot concentrations of pollutants in scattered locations. Compliance with the standards is monitored by taking spot samples in numerous locations and averaging the readings. *See also* BACKGROUND LEVEL; AIR QUALITY CLASS; NATIONAL AMBIENT AIR QUALITY STANDARDS.

amenity value **(amenity)** in land-use planning, any unmeasurable or unquantifiable use that must nevertheless be taken into account when planning the management of a particular piece of land. Usually these are things that give pleasure to the senses rather than supplying tangible goods. They include scenic beauty, "naturalness," lack of offensive odors and sounds, and so forth. Amenity values are similar to aesthetic values but cover a broader range, including the spiritual, the religious, and the humanitarian, as well as the merely beautiful.

American Association for the Advancement of Science **(AAAS)** founded in 1848 to promote science with both working scientists and the general public. The association provides for cooperation and responsibility of scientists and promotes scientific freedom. AAAS also works toward furthering the public's interest in and understanding of scientific principles and new developments, especially in the areas involved with human welfare. Membership (1999): 140,000. Address: 1200 New

York Avenue NW, Washington, DC 20005. Phone: (202) 326-6400. Website: www.aaas.org.

American Chestnut Foundation a privately funded organization concerned with preserving and reestablishing the American chestnut tree, greatly reduced in number in the early 1900s due to a fast-spreading blight. The foundation, begun in 1983, conducts breeding research at two farms in Virginia and supports work being done elsewhere in the United States. Membership (1999): 2,800. Address: 469 Main Street, P.O. Box 4044, Bennington, VT 05201-4044. Phone: (802) 447-0110. Website: www.chestnut.acf.org.

American Conservation Association founded in 1958, a nonprofit association established to increase understanding of and interest in conservation. Both scientific and educational in basis, the association seeks to conserve the public's natural resources. Address: 1200 New York Avenue NW, Suite 400, Washington, DC 20005. Phone: (202) 289-2431.

American Farmland Trust (AFT) preserves agricultural land through education, land use planning, and technical guidance. Founded in 1980, the AFT works to help communities develop environmentally friendly farming practices. It publishes many books and booklets designed to educate the public and to provide the necessary information to grassroots activists. Membership (1999): 30,000. Address: 1200 18th Street NW, Suite 800, Washington, DC 20036. Phone: (202) 331-7300. Website: www.farmland.org.

American Fisheries Society (AFS) a professional society comprised of 50 local or regional chapters, plus special sections and divisions devoted to specific areas of fisheries management. The society works to preserve aquatic habitats through the use of proper management and careful conservation. The goal of AFS is the judicious use of U.S. fisheries both for profit and recreation. Membership (1999): 9,200. Address: 5410 Grosvenor Lane, Suite 110, Bethesda, MD 20814. Phone: (301) 897-8616. Website: www.fisheries.org.

American Forest and Paper Association (AFPA) organization created in 1993 by the merger of the American Forest Council, the American Paper Institute, and the National Forest Products Association. The NFPA is the trade association of the wood products industries, representing both multinational and small business. AFPA works to preserve the environment through the sponsorship of conservation programs concerned with paper recycling, sustainable forestry, and environmental health and safety. Address: 111 19th Street NW, Suite 800, Washington, DC 20036. Phone: (202) 463-2700. Website: www.afandpa.org.

American Forest Council (AFC) See AMERICAN FOREST AND PAPER ASSOCIATION.

American Forest Foundation (AFF) national research and educational foundation founded in 1981. AFF supports the American Tree Farm System (71,000 forest landowners, with 95 million acres in tree farms) and the award-winning environmental education program, Project Learning Tree. Address: 111 19th Street NW, Suite 78, Washington, DC 20036. Phone: (202) 463-2462.

American Forestry Association (AFA) See AMERICAN FORESTS.

American Forests (AF) formerly American Forestry Association, an organization established in 1875, the oldest national citizen's conservation organization in the United States. A proponent of forest management and conservation, AF now advances environmental protection through programs such as GLOBAL RELEAF 2000, the Urban Forestry Center, the Forest Policy Center, and the Famous and Historic Tree program, and through the publication of a quarterly magazine, *American Forests*. Address: P.O. Box 2000, Washington, DC 20013. Phone: (202) 955-4500. Website: www.amfor.org.

American Geographical Society founded in 1851. The society sponsors lectures and research and publishes current geographical material in technical and general interest books and journals for readers in the United States and more than 100 other countries. Membership (1999): 8,000. Address: 120 Wall Street, Suite 100, New York, NY 10005-3904. Phone: (212) 422-5456.

American Hiking Society hiking and conservation club, founded in 1977 to protect hiking trails and the interests of hikers in the United States and to promote the building of new hiking trails in natural areas throughout the nation. Membership (1999): 10,000. Address: 1422 Fenwick Lane, Silver Spring, MD 20910. Phone: (301) 565-6704.

American Institute of Biological Sciences an organization of biologists whose goal is to promote biology and the application of biological knowledge and principles to improve human welfare. The institute, founded in 1947, works toward this goal through education, publication, and advisory relationships. Membership (1999): 8,000. Address: 1444 I Street NW, Washington, DC 20005. Phone: (202) 628-1500.

American Land Rights Association (ALRA) formerly the National Inholders Association, a property-rights organization founded in 1978. ALRA's members, including those owning private property within national parks, forests, and other public lands, oppose legislation and regulations affecting what they consider to be their personal rights as landowners. Known for its militant advocacy and generally anti-environmentalist stance, the ALRA has broadened its mission somewhat in recent years to support family recreation and public access to federal lands, although it often opposes park and wilderness designations. Members (2000): 26,000. Address: P.O. Box 400, Battle Ground, WA 98604. Phone: (360) 687-3087. Website: www.landrights.org.

American Littoral Society concerns itself with the examination and conservation of coastal areas such as wetlands, barrier islands, and shorelines and the animals inhabiting these aquatic habitats. The society hosts exploratory trips and publishes materials for both popular and technical audiences. Membership (1999): 8,000. Address: Sandy Hook, Highlands, NJ 07732. Phone: (732) 291-0055. Website: www.alsnyc.org.

American Museum of Natural History a research and educational organization whose museum hosts more than 3 million visitors annually. The museum publishes the well respected magazine *Natural History.* Areas of research and education include all of the natural world, from anthropology and astronomy to mammalogy and paleontology. It is particularly concerned with ecological relationships, evolution, and human cultures. Membership (1999): 515,000. Address: Central Park West at 79th Street, New York, NY 10024. Phone: (212) 769-5000.

American Ornithologists' Union (AOU) founded in 1883. The AOU's goal is to promote ornithological science by means of membership education, yearly conferences, and publications. Membership (1999): 4,500. Address: National Museum of Natural History, MRC-116, Smithsonian Institution, Washington, DC 20560-0116. Phone: (202) 357-2051. Website: http://pica.wru.umt.edu/AOU/AOU.html.

American Planning Association with 46 chapters and 16 planning divisions, an association providing information, research, and education in land use and city planning. The APA sets professional and ethical standards and accredits of college-level degree programs in planning through their American Institute of Certified Planners. Membership (1999): 29,000. Address: 1776 Massachusetts Avenue NW, Washington, DC 20036. Phone: (202) 872-0611.

American Rivers environmental protection organization, founded in 1973 to lobby on behalf of the preservation of America's free-flowing rivers and the lands adjacent to them, including riparian landscapes, wetlands, canyons, and gorges. The organization also sponsors research on rivers issues and hosts an annual conference. Prior to 1986 it was known as the American Rivers Conservation Council. Address: 1025 Vermont Avenue NW, Suite 720, Washington, DC 20005. Phone: (202) 347-7550. Website: www.amrivers.org.

American Society of Landscape Architects a professional association concerned with analysis, planning, design, management, preservation, and rehabilitation of the land. The society, founded in 1899, is involved with both professionals and the general public in land use, open space, transportation, sustainable communities, and urban environment issues. Membership (1999): 12,500. Address: 636 I Street NW, Washington, DC 20001-3736. Phone: (202) 686-2752.

American Water Resources Association a nonprofit association concerned with water resources issues and water resource research, management, planning, and development. The association promotes interaction between engineers, biological scientists, and other scientists through conferences addressing water policy and technology. Publications include conference proceedings and a newsletter. Membership (1999): 3,500. Address: 950 Herndon Parkway, Suite 300, Herndon, VA 20170-5531. Phone: (703) 904-1225.

American Water Works Association (AWWA) professional organization, founded in 1881 to share information and develop standards concerning all aspects of public water-supply systems. The organization includes engineers, chemists, bacteriologists, managers of municipal and investor-owned water systems, manufacturers of water treatment and distribution equipment, government officials, and interested individuals. Membership (1999): 55,000. Address: 6666 West Quincy Avenue, Denver, CO 80235. Phone: (303) 794-7711. Website: www.waterwiser.org.

America the Beautiful Fund promotes local projects designed to improve the quality of the environment through small grants, technical support, and recognition awards for both individuals and organizations. The fund focuses on environmental design; small farm production; land, historical, and cultural preservation; and horticultural therapy. Address: 1730 K Street NW, Suite 1002, Washington, DC 20006. Phone: (202) 638-1649. Website: www.america-the-beautiful.org.

Ames test test for carcinogenity using bacteria instead of human or animal test subjects (*see* CARCINOGEN; BACTERIUM). Developed by Bruce N. Ames of the University of California at Berkeley in about 1979, the test exposes a laboratory-raised strain of the human intestinal bacterium *Salmonella typhimurium* to a concentrated dose of a suspected carcinogen. The bacteria are then tested for their ability to produce the essential AMINO ACID histidine. The Ames test produces results rapidly—in hours or days instead of years—and has approximately a 90% correlation rate with human EPIDEMIOLOGIC TESTS for carcinogenity. This makes it extremely valuable for rapid screening of new substances; however, its results are not accurate enough to be depended on for the permanent classification of a substance as carcinogenic or not.

amine any of a large group of ORGANIC COMPOUNDS derived from AMMONIA by the substitution of an alkyl group (type formula: C_nH_{2n+1}) or an aryl group (type formula: C_nH_{n-1}) for one or more of the ammonia molecule's hydrogen atoms. Primary amines have alkyls or aryls substituting for one hydrogen atom in the ammonia molecule; secondary amines substitute two hydrogen atoms, and tertiary amines substitute all three hydrogen atoms. Amines are fairly widespread in nature (one example: the odor of rotting fish) and are easily made in the laboratory. Their principal commercial use is as hardeners in plastics or synthetic resins. As a group they are not environmentally hazardous, though several individual amines are listed as PRIORITY POLLUTANTS by the Environmental Protection Agency due to their toxic and/or corrosive properties.

amino acid an organic ACID containing an amino group (NH_2), a carboxyl group (COOH), and an attached alkyl or aryl group (*see under* AMINE). Amino acids are the building blocks of PROTEIN molecules, and as such are one of the most important constituents of living tissue. There are roughly 70 amino acids known, 20 of which are used in protein construction. These 20 are divided into two groups: the nonessential amino acids (those that an organism is able to manufacture in its own body), and the essential amino acids (those that cannot be manufactured by the organism but must be obtained from other sources). The essential/nonessential grouping varies from species to species. Generally speaking, the higher an organism lies on the FOOD CHAIN, the fewer amino acids it can manufacture and the more it must obtain by eating other organisms. Humans obtain their 10 essential amino acids most easily from meat, although by choosing with care all of them may be obtained from a vegetarian diet.

ammonia a colorless, painfully pungent gas, chemical formula NH_3, formed as a metabolic BYPRODUCT (*see* METABOLISM) by most animals and excreted in the form of urea (CH_4N_2O) or uric acid ($C_5H_4N_4O_3$), both of which decompose spontaneously in the environment, releasing the ammonia. It is produced commercially from NATURAL GAS or "water gas" (made by passing steam through incandescent COKE) by means of the Haber-Bosch process, in which the gas is heated in the presence of steam and an iron CATALYST. Industrial uses include fertilizer production, pulp and paper processing, explosives manufacture, as a refrigerant, and (in liquid form) as a solvent.

AMMONIA AS A POLLUTANT

Ammonia dissolves extremely easily in water, a saturated solution (*see* SOLUTION) at room temperature consisting of 30% ammonia by weight. It is highly toxic to most animal life, but many aquatic plants—especially ALGAE—can use it as a source of nitrogen. Thus, its presence in water contributes significantly to EUTROPHICATION in those waters where nitrogen is the LIMITING NUTRIENT. Some MICROORGANISMS (*nitrifiers*) can combine it with DISSOLVED OXYGEN, forming nitrates and water and using up 4.44 parts oxygen to each part ammonia as they do so, thus creating a large BIOCHEMICAL OXYGEN DEMAND. It also reacts readily with CHLORINE in water, forming toxic *chloramines* (type formula $NH_{3-n}Cl_n$, where $n = 1$ to 3); thus its presence in public water supplies can make chlorination as a water-purification process ineffective or even dangerous.

Amoco Cadiz The name of the ship involved in the largest European OIL SPILL to date. The *Cadiz,* a VLCC-class ship (*see under* SUPERTANKER) of 229,000 tons DEADWEIGHT CAPACITY sailing under a Liberian flag, lost headway off the coast of Brittany when her steering system failed at 10:45 A.M. on March 16, 1978. Twelve hours later, after towlines had parted three times in heavy seas, the ship ran aground on a reef two miles off the fishing village of Portsall, France, rupturing her hull and ultimately causing her to break in two (March 24). Sixty-nine million gallons of crude oil were released into the English Channel, creating a slick more than 100 miles long that fouled beaches the entire length of the northern Brittany coast.

amoeba any of several thousand species of tiny, one-celled organisms of the order Amoebida, class Sarcodina, phylum Protozoa, kingdom PROTISTA. Other sarcodinae are sometimes also called amoebas (or, in order to differentiate them from the true amoebae, "amebas"). Although each organism consists of only a single CELL, and despite the fact that they are not true animals, amoebas show nearly all of the characteristics we associate with the word "animal." Most are predatory, chasing and engulfing prey, which include algae, other protozoans, and small multicelled animals such as ROTIFERS. They respond to light, chemical stimuli, and mechanical agitation of the water they live in, and exhibit rudimentary learning in the form known as HABITUATION. Reproduction is primarily asexual, by fusion (*see* ASEXUAL REPRODUCTION), but a few forms are known to practice a variety of sexual reproduction that involves briefly merging into a single organism in order to exchange genetic material (*see* GENE; DNA). Most amoebas are harmless to humans; however, several live as parasites in the human digestive tract, and at least one of these can cause a serious, debilitating illness: AMOEBIC DYSENTERY.

amoebiasis *See* AMOEBIC DYSENTERY.

amoebic dysentery (**amoebic colitis, amoebiasis**) human GASTROINTESTINAL DISEASE caused by a parasitic AMOEBA, *Entamoeba histolytica,* which invades the lower intestinal tract. The disease is characterized by violent abdominal cramps and the passage of frequent watery STOOLS, often containing blood and/or mucus. In severe cases the wall of the intestine may become perforated, leading to infection of other organs within the abdominal cavity, particularly the liver and lungs. Death can result.

TRANSMISSION AND OCCURENCE

Although flies can carry the disease by landing first on infected FECES and then on food, the principal means of transmission is through polluted water. Hence outbreaks can and do occur anywhere. Although lack of sanitary facilities in undeveloped countries makes the disease seem more prevalent in the tropics, northern countries are not immune to episodes of it. Four deaths were attributed to amoebic dysentery in South Bend, Indiana, in 1953; 20 years earlier, a similar outbreak in Illinois nearly shut down the Chicago World's Fair. There were 6,632 cases reported in the United States in 1981, up from 2,775 in 1975.

amorphous rock rock that has no crystal structure. Most limestone is amorphous, as are obsidian and pumice. Amorphous rock should not be confused with CRYPTOCRYSTALLINE ROCK, which has a crystalline structure, though it cannot be made out without a high-powered microscope.

amphibian declines and deformities the worldwide phenomenon of massive die-offs of frogs, toads, salamanders, and other amphibian species together with physical deformities such as a leg growing from a frog's mouth. Researchers have found that amphibian populations have been declining since the 1960s. The overall numbers dropped 15% a year from 1960 to 1966, with subsequent population declines at a rate of 20% per decade. By 1993, more than 500 populations of frogs and salamanders on five continents were listed as declining or subject to decline. The common western toad (*Bufo boreas*) is now absent in more than 80% of its original range. The golden toad of Costa Rica (*B. periglenes*) has not been seen since 1988. In Australia 14 frog species have disappeared, including the so-called gastric-brooding frog, which is presumed extinct. In Canada the population of the northern leopard frog, *Rana pipiens,* has crashed. In California both the red-legged frog and the yellow-legged frog have also suffered major declines, the latter having disappeared in two-thirds of its range. In the Pacific Northwest five of 34 species are now candidates for the endangered species list. While there is debate over the extent of amphibian decline, most researchers estimate that, taken together, between one-quarter and one-half of the earth's species could be extinct in the next 30 years if present trends continue. The causes of amphibian declines are multiple, and for the most part can be traced to human actions. Amphibians as a class are especially sensitive to ecosystem changes and modifications of habitat. From eggs and embryos in water to adults on land, amphibians' thin, moist, and sensitive membranes come into close contact with various components of the natural environment, which if modified or made even slightly toxic can have a dire effect on amphibian health. Accordingly, the loss of suitable habitat, agricultural poisons, climate change, and especially increasing doses of UV-B rays from a thinning OZONE LAYER, have taken their toll. A number of important field studies, beginning in 1979, have shown that UV-B can affect the genetic make-up of amphibian eggs and embryos. In general, research has revealed that those species whose eggs are most vulnerable to UV-B radiation show the greatest evidence of population decline. Amphibian deformities may or may not be caused by UV-B radiation or other anthropogenic environmental impacts but they suggest, as do the population declines, that environmental conditions affecting amphibians have been significantly altered. Many scientists see amphibian decline and deformity as an early warning signal of major ecosystem dysfunction. In 1998, the U.S. government established the Task Force on Amphibian Declines and Deformities (TADD). It includes 14 cabinet-level departments and agencies charged with determining why one-third of the 230 native species in the United States were in decline, with reports of deformities on the increase and coming from virtually every state in the union. A TADD website (www.frogweb.gov) has been created to link the public, which is asked to report various phenomena, with the scientific effort.

amphibole any of a group of common rock-forming, silica-rich minerals that are found chiefly in igneous rock. They lie midway in the BOWEN SERIES between PYROXENE and biotite MICA, and are present in diorite but not in BASALT, thus serving as an indicator when doubt exists as to which of these two closely related rock types a specimen should be classed in. Examples: horneblende, nephrite jade.

AMTRAK Alternate name for the NATIONAL RAILWAY PASSENGER CORPORATION.

amygdule in geology, a VESICLE in basalt or similar porous rock that is filled by a mineral deposit. Amygdules are essentially small GEODES; however, they tend to be uniform in composition, while geodes tend to vary in composition from the outside to the center. The average size is around a centimeter across (3/8 inch). The most common filling material is QUARTZ, but many other minerals also amygdulize, including CARBONATES, zeolites (compounds of aluminum silicates with the light metals such as sodium and calcium; a few of the harder ones, especially phrenite [calcium aluminum silicate], are used as gemstones), and on occasion, pure metals (especially copper). *See also* VUG.

anadramous species any SPECIES of fish that lives in the ocean but migrates up rivers to spawn. The life cycle of a typical anadramous fish includes hatching from eggs laid in streambed gravels near the headwaters of a tributary stream (*see* SPAWNING GRAVEL); migrating downstream to the ocean as a FINGERLING; two to four years of life as an ocean fish; and finally joining others of its species in a massive SPAWNING RUN back upstream to leave its own spawn in the headwaters gravels. Almost always, a particular fish will return to the same gravel bed in which it was hatched to lay its own eggs, apparently guided to the correct tributary by small differences in water chemistry and temperature. Anadramous fish ordinarily die immediately following spawning, though some species make two or more runs, returning to the ocean for a year or more in between. Occasionally an anadramous species may be found that lives its entire life in freshwater, passing the "ocean" stage in a large lake (*see* LANDLOCKED SPECIES). The best-known anadramous fish are the various species of salmon; others include smelt, lampreys, shad, steelhead trout, and striped bass. *Compare* CATADRAMOUS SPECIES.

anaerobic bacteria any of several SPECIES of bacteria (*see* BACTERIUM) that do not require OXYGEN to live. *Strict* or *obligative anaerobes* can live only in the absence of oxygen, and are killed by its presence. *Aerotolerant anaerobes* can tolerate the presence of oxygen in their environment but do not use it for metabolic processes (*see* METABOLISM), while *facultative anaerobes* can metabolize either with or without oxygen. In general, aerotolerant bacteria do better without oxygen, while facultative bacteria do better with it. Anaerobic bacteria usually use either NITROGEN or SULFUR as a hydrogen receptor in their metabolic processes instead of oxygen, producing hydrogen sulfide (H_2S), METHANE and AMMONIA as BY-PRODUCTS instead of the CARBON DIOXIDE and water associated with aerobic metabolism. Anaerobic bacteria are found in nature in such oxygen-poor environments as the beds of eutrophic lakes or ponds (*see* EUTROPHICATION). Many are PATHOGENS, which thrive in places such as puncture wounds, the bladder and kidneys, and the lower intestinal tract. They are useful as indicators of oxygen depletion in polluted watercourses (*see* INDICATOR SPECIES). In addition, some species are used in the treatment of sewage and other wastes (*see* ANAEROBIC DIGESTER; FACULTATIVE POND). *Compare* AEROBIC BACTERIA.

anaerobic digester in engineering, a type of waste-treatment facility using ANAEROBIC BACTERIA to decompose the waste material. Anaerobic digesters are used primarily in sewage treatment as a means of deodorizing and improving the disposability of sludge obtained from SETTLING BASINS and from aerobic treatment facilities such as ACTIVATED SLUDGE tanks. The process uses two different types of anaerobic bacteria. Acid-producing anaerobes attack the sludge first. The organic acids these organism produce are then used as food by methane-producing bacteria (Methanobacteria), yielding methane gas and a dark, organically rich, essentially odorless sludge that is high in nitrogen compounds and may be landfilled or used as fertilizer. Anaerobic digesters are somewhat difficult to operate, as they tend to be unstable ECOSYSTEMS. The Methanobacteria are extremely sensitive to small changes in environmental conditions, while the acid-forming bacteria are tolerant of a much broader range, meaning that if conditions shift outside of a narrow "window" the Methanobacteria will die out while the acid producers continue to thrive, resulting in odiferous, highly acidic sludge instead of the desired deodorized material. Hence, anaerobic digesters must be monitored regularly by checking the amount of methane and the pH value of the EFFLUENT they produce. They operate best at a temperature of 35°C (98°F) and thus must be heated, usually by burners fueled by the methane gas produced by the digester itself. The simplest digesters are round concrete-block or steel tanks with covers that float on the sludge in order to seal it away from the air and keep the conditions within it anaerobic. The gases produced by the digestion are collected in a "gas dome" at the center of the cover from which they are drawn off by pipe. More complex digesters may contain mechanically driven paddles to stir the sludge, mixing the bacteria more uniformly through it; this stirring may also be done by bubbling gases through the tank. As the sludge digests, it separates into liquid and solid phases. The liquid, referred to as *supernatant,* is commonly drawn off and combined with the incoming sewage to the treatment plant, to be run through the plant again.

anaerobic stream a stream or portion of a stream in which the level of DISSOLVED OXYGEN has declined to zero. An anaerobic stream is almost always the result of pollution by large amounts of BIOCHEMICAL OXYGEN DEMAND. Normal (oxygen-using) life forms die out in an anaerobic stream, but ANAEROBIC BACTERIA thrive, and their waste gases—principally hydrogen sulfide and ammonia—permeate the water and the surrounding air. Sludge is broken up and carried into the WATER COLUMN by these gases: it either floats on top of the stream or is suspended within it, causing the water to appear black. The only effective treatment is to eliminate the pollution that is the source of the anaerobic conditions. *See also* OXYGEN SAG CURVE.

ancient forest *See* OLD GROWTH.

andesite a form of extrusive volcanic rock (*see* EXTRUSIVE ROCK) midway in composition between RHYOLITE and BASALT. it is fine-grained and light in color, containing numerous small PHENOCRYSTS of feldspar in a plagioclastic matrix (*see* PLAGIOCLASE) whose albite content ($NaAlSi_3O_8$) is between 50 and 70%. A common mountain-building rock, andesite is named for the Andes Mountains of South America.

andesite line a line separating the predominantly andesitic volcanic rocks of the continents from the predominantly basaltic volcanic rocks of the ocean basins. It is especially prominent in the western Pacific.

anemometer any of several commonly used devices for measuring the speed of the wind. The most widespread form of anemometer, the *cup anemometer,* utilizes four rigid cups mounted on short arms that are connected to a shaft at right angles to each other. The cups catch the wind, turning the shaft, whose speed of rotation can then be measured by a device similar to an automobile speedometer. Another form of anemometer, the *Pitot tube,* consists of an L-shaped tube with a colored fluid in it. The long arm of the tube is pointed into the wind with the short arm pointed upwards, and the pressure of the wind drives the fluid into the short arm a distance proportional to the wind speed. The Pitot tube is the most widely used form of airspeed indicator for aircraft. A third, relatively uncommon anemometer contains a length of metal wire whose electrical CONDUCTIVITY varies with temperature. The cooling effect of the wind blowing across the wire varies the amount of current the wire is able to transmit, and this current variation, measured by an ammeter, can be used to calculate the wind speed. *See also* BEAUFORT SCALE.

angiosperm a flowering plant. The term *angiosperm* means literally "enclosed seed," referring to the fact that all plants of this subclass bear their seeds enclosed within an ovary rather than exposed on the surface of a sporophyll, as with the GYMNOSPERMS. There are 300,000 or more species of angiosperms worldwide, including not only those plants with conspicuous flowers such as the lily, the orchid, the rose, and the sunflower, but also the grasses, the cacti, and most broad-leafed trees. Angiosperms appear to have originated in the Permian period, some 200 million years ago; however, they became abundant only about 75 million years ago, in the late Cretaceous. Together with the gymnosperms, they make up the class Pteropsida of the Tracheophyta, or VASCULAR PLANTS. *See also* FLOWER.

angle of repose in geology, the steepest angle at which a slope of unconsolidated material such a sand, soil, or gravel can remain stable. Above the angle of repose, erosion takes place principally through MASS WASTAGE; below it, slower processes such as wind erosion, SHEET EROSION, and gullying (*see* GULLY) tend to take over. The angle of repose varies with the type of material involved: dry sand has an angle of repose of about 33%, while TALUS and SCREE SLOPES—which almost always lie at the angle of repose—may exceed 40%.

angler day in land-use planning and resource management, a measurement of the amount of use of a particular stretch of water by recreational fishermen. One angler day is generally defined as a visit by one fisherman for one hour or more during one 24-hour period. *See also* VISITOR DAY.

Animal and Plant Health Inspection Service (APHIS) agency within the U.S. Department of Agriculture (*see* AGRICULTURE, DEPARTMENT OF) charged with the administration of federal regulations and programs dealing with the health of animals and plants. The most important duties of APHIS from an environmental standpoint are those that deal with pest eradication programs, including the coordination of pest control measures among federal agencies, between federal agencies and private landowners, and on pest infestations that cross state or national boundaries. The agency is committed by law to consider nonchemical as well as chemical means of pest control; however, it is not required to adopt or prohibit any particular method.

animal equivalent (AE) in wildlife management, the amount of land of a specified VEGETATION TYPE required to support a single animal of a specified species. Example: the AE of Douglas fir forest for an adult deer is 20 acres, meaning that it takes 20 acres of forest to support one deer on a continuous year-round basis.

animal rights an emerging ethical proposition and citizen movement. The concept of animal rights in the form of animal "liberation" seeks to link the rights of animals to the ethic of human rights for all oppressed groups, for instance women, minorities, and homosexuals. In a seminal paper on the subject (*New York Review of Books,* 1973) by Peter Singer, an Australian ethicist, the author asserts that it is morally wrong to kill or cause pain to animals for food, medical experimentation, or any other purpose. The animal rights movement is especially vigorous in Great Britain and growing stronger in the United States as well, especially with regard to the use of laboratory animals. The movement springs from a humanitarian sensibility, long held in Europe and America, rather than an ecological one. DEEP ECOLOGY, which seems to coincide with animal rights, deals however with animals in terms of asserting the right of threatened or endangered species to be protected and preserved at all costs. The deep ecologists do not, for the most part, extend this notion to individuals within a species as do those in the animal rights movement.

animal unit month (AUM) in rangeland management, the amount of FORAGE necessary to supply the nutritional needs of one cow and one calf for one month. An animal unit month is usually defined as equivalent to 800 pounds of forage. *See also* GRAZING UNIT.

anion *See under* ION.

annual in botany, a plant which runs through its entire life cycle—sprouting, growth, maturity, reproduction, and death—in a single year. Annuals are found in all climates and BIOMES, but are particularly prevalent in harsh environments such as deserts and tundra due to the much higher survivability of seeds as compared to mature plants. *Compare* PERENNIAL.

annual rings (growth rings) sequence of light and dark layers in the wood of a tree. On a cross-section through the trunk (as, for example, the top of a stump), these layers appear as a series of concentric rings that may be counted to determine the tree's age. The rings are caused by the tree's varying growth rate at different seasons, the lighter wood being the more rapid growth laid down during the warm, wet part of the year (usually spring) and the darker wood representing the near-stagnation of the dry and/or cold seasons. *See also* CAMBIUM; XYLEM; LIGNIN.

anode *See under* ELECTRODE.

anoxia lack of oxygen; especially, lack of sufficient oxygen to sustain life (but *see* ANAEROBIC BACTERIA). The term may refer either to atmospheric conditions or to the amount of DISSOLVED OXYGEN in a body of water.

Antarctica the southernmost of the world's continents, centered roughly on the South Pole: it is the fifth largest continent and the only one without permanent human inhabitants. Antarctica covers about 13.8 million square kilometers (5.4 million square miles), approximately the size of the United States and Mexico combined. Ninety-eight percent of its surface is covered by ice, to an average depth of 2 kilometers (1.2 miles): the total mass of the Antarctic icecap is about 30 million cubic kilometers, which represents just under 70% of the earth's freshwater resources. The portion of the continent toward Australia, known by convention as "East Antarctica," has a semicircular coastline which is almost exactly conterminous with the Antarctic Circle. The portion toward South America ("West Antarctica") has a ragged, deeply indented coastline capped by a long, narrow extension, the Palmer Peninsula, which reaches more than 320 kilometers (200 miles) north of the Antarctic Circle, coming within 1,000 kilometers (600 miles) of Tierra del Fuego. Soundings beneath the ice indicate that West Antarctica is actually an island archipelago separated from East Antarctica by a below-sea-level defile known as the Bentley Subglacial Trench. The Antarctic climate is the harshest on earth: the lowest temperature ever recorded on the planet, –89 degrees C (–128.6 degrees F), was measured near the Soviet Union's Vostok Station in east-central Antarctica on July 21, 1983. Largely because of the extremely cold temperatures, little precipitation falls, and much of the continent, though ice-covered, is technically a desert. Land-based life forms are chiefly LICHENS (about 350 SPECIES). There are only two flowering plants—both found only at the tip of the Palmer Peninsula—and the largest land animals are mosquitoes and midges. By contrast, the surrounding oceans are among the richest ECOSYSTEMS on earth, supporting a wide variety of marine life forms, from krill (small lobsterlike CRUSTACEANS that form the base of the Antarctic FOOD CHAIN: *see* KEY-INDUSTRY ANIMAL) up to large VERTEBRATES such as seals, penguins, and whales. The northern boundary of this ecosystem is the Antarctic Convergence, a zone of turbulence and sharp water temperature and salinity changes, which encircles the continent at between 50° and 60° south latitude and is considered the true boundary of the Antarctic.

ANTARCTIC CONSERVATION

Though the continent contains rich mineral deposits, including CHROMIUM, COPPER, PETROLEUM, and what is probably the largest bed of COAL on the planet, its remoteness and difficult working conditions have kept these resources costly enough that they have remained economically undevelopable until very recently. Within the last decade, economic conditions have shifted so much that current barriers to Antarctic resource exploitation are largely political. Antarctica is considered international territory under the terms of the Antarctica Treaty of 1959, signed by 12 nations including the United States, the Soviet Union, and Japan; 22 others have since added their names to the treaty, with 13 of these becoming full voting members of the governing coalition. Under the treaty's terms, the continent's resources are to be used for peaceful purposes and no territorial claims may be pressed. Seven of the signatory nations, however, continue to hold pre-treaty territorial claims, in some cases overlapping, which they expect to seek recognition for should the treaty be dissolved. In June 1988 the treaty nations signed a Convention on the Regulation of Antarctic Mineral Resources Activity (CRAMRA), which would allow commercial development of Antarctic resources to begin. However, in May 1989, two of the signatories (France and Australia) announced that they had withdrawn support for CRAMRA, effectively blocking its implementation. Environmental groups around the world are seeking to permanently block resource development in the fragile Antarctic environment by having the entire region south of the Antarctic Convergence declared an international wilderness preserve and park. In the meantime, tourism has adversely affected the Antarctic environment. Some 3,000 cruise ships enter Antarctic waters annually, and tourists have been blamed for disturbing penguin breeding colonies. The Antarctic continent suffered its first OIL SPILL on January 28, 1989, when the *Bahia Paraiso,* a 435-foot Argentine supply ship carrying some 25,000 gallons of oil, hit a reef and capsized off the tip of the Palmer Peninsula. Most of the oil was saved, but enough leaked into the sea to destroy a nearby Adélie penguin breeding colony. During the 1990s Antarctica became the focus of attention as indications appeared that GLOBAL CHANGE was causing unprecedented ice-melt on the Antarctic Peninsula. In 1995 a massive slab of ice, 600 feet thick and 1,000 square miles in area, detached from the Larson Ice Shelf and floated free. Since then, thousands more square miles of ice have detached, or are threatening to detach, in the area. The rapid environmental changes due to global warming have affected the Antarctic ecosystem in major ways, with wildlife species such as the elephant seal now taking up residence in the newly warmed waters, as well as fur seals and gentoo and chinstrap penguins, formerly found closer to South America. Meanwhile the local Adélie penguin population is dropping, a sign of major change. Recent studies have shown that increased UV-B radiation, stemming from the thinning of the OZONE LAYER, may be affecting phytoplankton and other organisms at the base of the food chain, with consequences for the entire range of species associated with Antarctica. *See also* ICE SHEET.

antecedent stream in geology, a stream whose course was determined before the formation of current land features. Antecedent streams often cut through entire MOUNTAIN SYSTEMS, maintaining their courses by rapid downcutting as the land rises beneath them (*see* DOWNCUTTING STREAM). *Compare* CONSEQUENT STREAM.

anther in biology, the pollen-producing organ of a flowering plant. *See* FLOWER.

anthracite *See under* COAL.

anthracnose a fungal condition affecting trees and plants that can be identified by dark spots or blotches on the leaves. The term *anthracnose* combines the Greek words for coal (*anthra*) and diseases (*nosos*). For the most part, anthracnose is seasonal. The disease, usually appearing after cold, wet springs, is not carried over the following winter. An exception is a newly identified fungus species causing anthracnose in the eastern flowering dogwood (*Cornus florida*) and the Pacific dogwood (*C. nuttallii*). The fungus, *Discula destructiva,* was discovered in the late 1970s, and within 10 years killed up to 80% of eastern dogwoods throughout parts of their native range in the East. In the West the fungus is not quite so virulent but has caused considerable damage. The fungus, whose origin is unknown, is transmitted from tree to tree by wind and water. Leaves are first affected, then the infection spreads to the branches and the trunk, which eventually becomes so girdled with cankers that the translocation of water and nutrients from the roots is slowed and finally stops, at which time the tree dies.

Regeneration of dogwood groves so afflicted is nearly impossible, since the disease strikes seedlings even more decisively than their parents. So far, the search for reliably resistant genotypes has proved unsuccessful. Although dogwoods used ornamentally in gardens and parks are often free of *D. destructiva,* since they may be at some remove from infected trees and receive ample water and sunlight, most nurseries, landscape architects, and garden designers no longer recommend the use of dogwood.

anthropocentrism describes the view that natural systems should be valued primarily in terms of their utility for human beings. A contrasting view, BIOCENTRISM, suggests that humankind is only a part of natural systems and cannot assert a right to modify natural systems solely for its own benefit. *See also* DEEP ECOLOGY; RELIGION AND THE ENVIRONMENT.

anthropogenic effects the effects of human activities on the natural world, both positive and negative. Examples of anthropogenic effects include POLLUTION, HABITAT DESTRUCTION, habitat restoration, preservation of threatened species, etc.

antibiotic a substance produced by a SPECIES of MICROORGANISM (usually a mold; *see under* FUNGUS) that is harmful to one or more other species of microorganisms. The classic example is penicillin, obtained from the mold *Penicillium chrysogenium,* which is active against numerous species of pathogenic bacteria (*see* PATHOGEN; BACTERIUM). Others include neomycin (obtained from the mold *Streptomyces fradiae,* aureomycin (obtained from the mold *S. aureofaciens*), etc. Antibiotics attack microorganisms in a variety of ways, but the most common methods are by inhibiting cell-wall production or by interfering with growth or reproduction. They are not normally toxic to higher organisms, but may cause allergic reactions (*see* ALLERGEN), especially in large doses. A few may also be CARCINOGENS. They increase growth rates in animals and so are often routinely added to animal feed. The principal hazard of unrestrained antibiotic use is that strains of bacteria will develop that are resistant to them, thus making disease control more difficult. This has happened, for instance, with gonorrhea, which can no longer be controlled by penicillin as it once could.

PRODUCTION OF ANTIBIOTICS

Although penicillin and a few others have been produced synthetically, nearly all commercial production of antibiotics is done by culturing microorganisms and harvesting the product. Antibiotics obtained in this manner may be used directly, or they may be altered slightly in order to improve their usability. Penicillin, which cannot in its natural form be taken orally, is almost always sold in an altered form. Altered antibiotics are known as *semisynthetics.*

anticline in geology, a folded rock structure in which the fold is convex, that is, where the layers of rock involved in the fold have bulged rather than sunk in the middle. Generally (but not always!) the older layers in an anticlinal rock sequence lie toward the center of curvature. *Compare* SYNCLINE.

anticyclone in meteorology, a disturbance in the atmosphere in which the winds flow in a clockwise direction (counterclockwise in the southern hemisphere). Anticyclones have areas of greater than normal air pressure known as *high-pressure cells* at their centers, from which the winds (typically light) spiral outward. They average 1,200–2,500 miles in diameter and their centers move approximately 20–30 miles per hour (ground speed), generally in an easterly direction. The weather within them is usually fair. *Compare* CYCLONE.

anti-environmentalism activities and views by individuals and organizations undertaken in reaction to environmental regulations and laws and designed to protest or obstruct pro-environmental activism and policy reform. Anti-environmentalism, in the form of citizen action, public relations efforts, and lawsuits, has been a feature in efforts to preserve WETLANDS, to reduce logging on public lands, to enforce the ENDANGERED SPECIES ACT, and to set aside land for parks and similar public uses. Some anti-environmental protests are spontaneous and citizen-led, and some are led by political commentators in print and broadcast media. Often they are orchestrated by corporations economically affected by environmental regulation, or by politically oriented organizations ideologically opposed to environmental reform. The most successful efforts are by corporate public relations departments and agencies, sometimes through groups with names that suggest they are citizen organizations, such as Citizens for Sensible Control of Acid Rain, a power-industry lobby which operated between 1983 and 1991 in opposition to the CLEAN AIR ACT. A more recent pseudo-public-interest group, the Global Climate Coalition, has been financed by the oil, auto, and coal industries to lobby against legislation intended to reduce global warming. Another feature of anti-environmental action is the so-called SLAPP suit—"strategic lawsuits against public participation." The lawsuits are brought by corporations, well-financed, politically oriented organizations, and even government agencies to discourage civic activism. The goal is usually not to win settlements, but to intimidate environmental activists into silence.

The vast majority of SLAPP suits are dismissed, but they can still cost defendants thousands of dollars in attorneys' fees and lost work time and can have a decidedly chilling effect on legitimate environmental advocacy. In extreme cases anti-environmentalism has led to crimes against persons and property, including instances of beatings, rape, arson, and even murder. In 1992 New York state ecologist Anne LaBastille's barns were burned to the ground for her role as a commissioner with the Adirondack Park Agency. In 1993 in New Mexico Navajo Leroy Jackson's body was found dead in his truck, thought by some to be murdered for his activism in protecting forest land in the Chuska Mountains. While some of the tactics of the more recent past have been untoward, for the most part the anti-environmental strain in U.S. politics is now expressed within traditional bounds of discourse, even when serious economic impacts are anticipated.

antimony the 51st element in the ATOMIC SERIES; atomic weight 121.75, chemical symbol Sb. Antimony is usually classed with the metals, though it has some nonmetallic properties. Its high melting point and unusual property of expanding slightly upon solidification make it particularly useful in making castings; it is also used in some alloys and medicinal compounds. Though it is an irritant rather than a toxin, causing rashes rather than illness, it is classed as a PRIORITY POLLUTANT by the ENVIRONMENTAL PROTECTION AGENCY due to its tendency to combine with free hydrogen to form stibine (SbH_3), an extremely toxic gas with an LC_{50} (mice) of roughly 100 parts per million.

Antiquities Act federal legislation allowing the creation of NATIONAL MONUMENTS by presidential proclamation, signed into law by President Theodore Roosevelt on June 8, 1906. Prompted originally by the imminent destruction of Arizona's Petrified Forest and of prehistoric Indian dwellings in the American southwest by swarms of souvenir hunters, the Antiquities Act permits the president, by executive order, to set aside areas of federal land to prevent the destruction of "historic landmarks . . . and other objects of historic or scientific interest." Immediately upon passage of the law, President Roosevelt proclaimed the first three national Monuments: Petrified Forest (Arizona), Montezuma's Castle (Arizona), and El Morrow (New Mexico). Two years later, in 1908, Roosevelt considerably broadened the scope of the law by taking advantage of the "scientific interest" clause to set aside the immense tract of land in northern Arizona that eventually became Grand Canyon National Park. This interpretation of the law was upheld by the Supreme Court in 1920. Since then, nearly every president has used the Antiquities Act at least once, and approximately half the acreage of the National Park System is or has been presidentially proclaimed National Monuments. Roosevelt himself proclaimed a total of 18, including the monuments that eventually became Olympic and Lassen National Parks. In 1996 President Bill Clinton used the Antiquities Act to create the 1.7-million-acre Grand Staircase-Escalante National Monument in Utah. Then again, in 1999, President Clinton set aside the 1,500-square-mile Grand Canyon-Parashant National Monument; the Agua Fria National Monument near Phoenix containing Indian ruins; additional lands for Pinnacles National Monument near San Jose, California; and the Coastal National Monument to protect thousands of small, federally owned islands and reefs off the central California coast. In the years since the passage of the Antiquities Act in 1906, only three of 17 presidents—Nixon, Reagan, and Bush—have failed to use the powers conferred by the act to protect nationally significant areas. The single most extensive use of the act has been President Jimmy Carter's Alaska Lands Proclamation of December 1, 1978, creating 17 National Monuments totaling nearly 56 million acres on federal lands in the State of Alaska. *See* ALASKA NATIONAL INTEREST LANDS CONSERVATION ACT.

aphid (plant louse) any of the numerous species of sap-sucking insects belonging to the SUPERFAMILY *Aphidodiae.* Aphids are parasitic on green plants (*see* PARASITE). Their mouthparts are formed into a hard, fine proboscis with which they penetrate the epidermis (outer skin) of the plant, tapping the nutrients passing through the PHLOEM layer. Excess nutrients are passed through the insect and excreted at regular intervals as a drop of sugary nectar that is much desired by ants, which often "farm" aphids, tending them as humans tend livestock. Aphids reproduce by PARTHENOGENESIS during the warm part of the year, producing males and fertilized eggs only in the fall—a REPRODUCTIVE STRATEGY that allows them to expand their population in an extremely rapid manner and still keep sufficient genetic variability for species maintenance. They are among the most damaging of agricultural pests and thus are a prime target of ORGANOPHOSPHATE PESTICIDES, to which they tend to be extremely susceptible. They may also be controlled by biological means, particularly by the introduction of ladybird beetles. *See* BIOLOGICAL CONTROL; INTEGRATED PEST MANAGEMENT.

APHIS *See* ANIMAL AND PLANT HEALTH INSPECTION SERVICE.

apical dominance tendency for the most vigorous growth of a plant to be concentrated in the highest part (the apex). Where present, apical dominance leads to a prominent central stem or trunk and weak or absent side branches. It is strong in CONIFERS, less strong in deciduous trees (*see* DECIDUOUS PLANT), and nearly absent in low spreading SHRUBS such as huckleberry or kinnikinnik.

CAUSES

Apical dominance appears to be controlled by the production of growth hormones called AUXINS in the plant's TERMINAL BUD. The auxins increase the growth rate of the LEADER and thus of the central stem; at the same time, they apparently suppress the growth of side branches (the buds at the ends of the branches also produce auxins, but in lesser amounts). If the leader and its terminal bud are removed from a plant, the next-highest bud becomes the terminal bud, and its branch becomes the new leader (and eventually, the main stem).

Appalachian Mountain Club hiking and conservation organization founded in 1876 to promote protection and enjoyment of the mountains, especially the Appalachian chain of eastern North America, and to educate the public on the values of mountains and other natural areas. The club maintains eight hiking huts and over 1,000 miles of trail in the Presidential Range and elsewhere in the northern Appalachians. It also sponsors research and educational projects. Membership (1999): 74,000. Address: Five Joy Street, Boston, MA 02108. Phone: (617) 523-0636. Website: www.outdoors.org.

Appalachian Trail hiking trail extending for 2,050 miles along the crest of the Appalachian Mountains and neighboring MOUNTAIN SYSTEMS in the eastern United States, from Mount Katahdin in northern Maine to Springer Mountain in northern Georgia. First envisioned in 1921 by regional planner Benton MacKaye, it is one of the longest continuous marked footpaths in the world.

Elevations range from near sea level at Bear Mountain Bridge on the Hudson River near Peekskill, New York, to 6,634 feet at Clingman's Dome in the Great Smokey Mountains on the Tennessee-North Carolina border. Motorized vehicles are forbidden on the trail, which is maintained largely by a consortium of citizens groups coordinated by the APPALACHIAN TRAIL CONFERENCE, established in 1925. Approximately 175 people hike the trail from end to end every year, and many millions enjoy shorter trips. *See also* NATIONAL TRAILS ACT OF 1968.

Appalachian Trail Conference federation of outdoor clubs and individuals, founded in 1925 to manage the APPALACHIAN TRAIL and protect the lands along it. The conference has been given responsibility for the trail by the federal government. Its activities include trail and hut maintenance, the production of maps and guidebooks, and the monitoring of lands near the trail for incompatible developments. Membership (1999): 23,000. Address: PO Box 807, Harpers Ferry, WV 25425. Phone: (303) 535-6331.

appropriate technology technology that achieves its goals through methods that make the least possible impact on the environment. Ideally, appropriate technology should produce no pollution and consume no NONRENEWABLE RESOURCES. An example often given is the sailing ship—built of wood, powered by wind, and leaving no track of its passage across the water. *See also* ALTERNATIVE ENERGY SOURCE; ALTERNATIVE TRANSPORTATION; RENEWABLE RESOURCE.

appropriation a specified amount of money from the federal treasury, granted by Congress to a specified federal agency for a specified project or group of related projects. Appropriations are normally made on an annual basis. They may be earmarked for a specific stage or portion of a project, such as design, construction, maintenance, etc. Money designated for one project or portion of a project cannot be spent in another manner without specific authority from Congress. No money may be appropriated without prior legislation authorizing the money to be spent in that manner; however, this may be a standing authority, which does not need to be renewed annually. Appropriations are handled by the APPROPRIATIONS COMMITTEES of the House and Senate. *See also* AUTHORIZATION; ORGANIC ACT.

appropriations committee congressional committee that handles the direct disbursement of funds from the federal treasury. Although the appropriations committees of the House and Senate cannot act without a prior AUTHORIZATION, their broad powers to give or withhold funds for authorized projects or programs make them among the most powerful and sought-after committee assignments in Congress. *See also* APPROPRIATION.

aquaculture the growing and harvesting of food fish and shellfish for commercial sale. In the past 20 years, coinciding with reduced natural fisheries around the world, aquaculture has become the fastest-growing segment of U.S. agriculture. Worldwide, aquaculture now supplies nearly one-fifth of the fish and shellfish market. Among the environmental problems caused by fish farming are EUTROPHICATION and other forms of pollution from waste, chemicals, and fish food. In southeast Asia natural mangrove swamps, which are cleared for shrimp farming, are threatened—a serious issue since these swamps act as natural nurseries for many species. In northern waters, escapees from salmon farms interbreed with wild salmon, interfering with the ability of wild fish to return to freshwater spawning sites. In the United States, trout, salmon, catfish, shrimp, oysters, and mussels are among the major species grown and marketed. *See also* FISHERIES DECLINE.

aquifer any geologic FORMATION capable of holding water beneath the surface of the ground (*see* GROUNDWATER). Aquifers are generally composed of buried layers of gravel, sand, or ALLUVIUM, though heavily fractured rock beds can serve as well. Water held in aquifers, known as GROUNDWATER, is obtained for human use by digging wells or tapping SPRINGS or SEEPS; it is of prime social and economic importance worldwide. *See also* POROSITY; PERMEABILITY; ARTESIAN WELL; CONE OF DEPRESSION; SALINE INTRUSION; FOSSIL WATER; AQUIFER SENSITIVITY; OGALLALA AQUIFER.

aquifer sensitivity measure of the potential harm to an AQUIFER from chemical contamination. Factors taken into consideration in judging the sensitivity of an aquifer include capacity and extent; depth to the WATER TABLE; location and size of RECHARGE AREAS; current quality of water within the aquifer; POROSITY and PERMEABILITY; and importance for human use (when there is no feasible alternative source of water, an aquifer is automatically classed as highly sensitive). Sensitivity is generally expressed as a number from one to nine, with aquifers classed 1–3 rated as insensitive, 4–6, moderately sensitive, and 7–9, highly sensitive. *See also* LEGRAND RATING SYSTEM.

Aral Sea an inland sea located in central Asia, on the border between Kazakhstan and Uzbekistan. Once the fourth largest inland water body in the world, the Aral Sea has now become an ecological disaster area. In 1960 the USSR diverted water from the Amu Darya and the Syr Darya, two feeder rivers in the semi-arid steppe, in order to irrigate what is now the world's largest cotton belt. Today the Aral Sea is only half its original size in terms of surface area with only a quarter of its original volume. Salinity has tripled. Its fishery is largely demolished, with 20 of its 24 native species now extinct. Some former fishing ports are now 50 miles away from the edge of the water. Moreover, the sea no longer serves to moderate the regional climate, leading to extremes in temperature and a shorter growing season, imperiling the very cotton crop that its waters have been diverted to create. The impoverished population of the Aral Sea region has been afflicted by polluted drinking water and carcinogenic chemicals as well as a harsh climate with frequent winds, laden with toxic salts, sweeping across the treeless flats. In 1993 an appeal by the five states affected by the Aral Sea crisis (Kazakhstan, Kyrgyzstan, Tajikistan, Turkmenistan, and Uzbekistan) appealed to the World Bank for $450 million in aid to meet immediate human needs. A Paris conference organized the following year among potential donor nations yielded disappointing results, ultimately delivering only $15.7 million to the region. No significant relief from either nature or the world community is in sight. The Aral Sea, in the view of scientists, will never recover to its 1960 condition, but will remain little more than a collection of near-lifeless, highly saline lagoons.

arboreal adaptation an alteration in the physique or behavior of an animal or plant that allows it to live wholly or partly in the trees. The long arms of the gibbon, which make it easier for the animal to swing from limb to limb, are an arboreal adaptation; so are the long, flexible toes of the perching birds (order Passeriformes) and the exclusively eucalyptus-leaf diet of the koala. *See also* ARBOREAL SPECIES.

arboreal species any species of animal (occasionally of plant) that lives in the branches or on the trunks of trees, rarely or never touching the ground. Squirrels, sloths, many monkeys, and most birds are arboreal species, as are several reptiles and amphibians, numerous insects, and the class of plants known as EPIPHYTES, or air-roots. *See also* CANOPY DWELLER; ARBOREAL ADAPTATION.

archaeological area in FOREST SERVICE usage, a site or area that has been administratively designated as containing a significant ARCHAEOLOGICAL RESOURCE or a group of archaeological resources. Within an archaeological area, preservation and interpretation of the archaeological resource becomes the DOMINANT USE; other uses may be permitted, but only to the extent that they do not damage the archaeological resource. *See also* SPECIAL INTEREST AREA.

Archaeological Conservancy works to protect and preserve important archaeological sites in the United States in collaboration with the government, museums, and other associations. It publishes *American Archaeology.* Address: 5301 Central Avenue NE, Suite 1218, Albuquerque, NM 87106. Phone: (505) 266-1540.

archaeological resource in land-use planning, any evidence of prior human occupation except written documents, pictures, and maps. Old roadbeds and trail TREADWORK, house foundations, artifacts such as nails, flint chips, or pottery, ashes of past campfires, and many other things would all classify as archaeological resources, which are important principally for the light they can cast on past ways of life and patterns of resource use.

arctic haze a smoglike phenomenon in which a reddish-brown haze thickly blankets the entire arctic region in winter and spring, from the 60th parallel northward. This vast area takes in all of Greenland, Iceland, most of Alaska, Norway, Sweden, Finland, eastern Russia north of Leningrad, and the Siberian uplands. While 19th-century European and U.S. explorers found the arctic air astonishingly brilliant and clear, the 20th century, especially after World War II, brought reports that began to concern environmental scientists. During the 1970s, air chemistry measurements showed clearly that arctic haze was not a natural event, but was caused by the long-range transport of air pollution, with a marked increase after 1956. The haze is heaviest over the Alaska North Slope, where thermal INVERSION causes multiple layers of pollutants to rise as much as five miles above the Earth's surface. At times, haze events in the arctic can produce atmospheric pollution levels equal to those in medium-sized American cities. The pollution is created by acidic sulfur dioxide emissions and other pollutants from factory districts in Russia and eastern Europe and, to a lesser extent, in western Europe. Because of prevailing winds, North American factories produce only about 4% of arctic haze. Since 1980, pollution controls in Europe have decreased the haze somewhat, but there are no formal international agreements dealing with the problem.

Arctic National Wildlife Refuge (ANWR) The federal WILDLIFE REFUGE in northeast Alaska, on the shores of the Beaufort Sea and adjacent to Canada's Northern Yukon National Park. Established in late 1960 as the 9-million-acre Arctic National Wildlife Range, the ANWR was renamed and more than doubled in size by the ALASKA NATIONAL INTEREST LANDS CONSERVATION ACT OF 1980. It now consists of approximately 18.5 million acres of land on the Arctic coastal plain and in the adjacent mountains of the Brooks Range. The wildlife resources of the refuge have been compared to the African Serengeti, with more than 180,000 caribou, as well as large numbers of musk oxen, wolves, foxes, polar bears, and other animals. It also provides nesting grounds for at least 108 species of birds, many of them migratory WATERFOWL. With geology similar to that of the nearby Prudhoe Bay oil fields, the ANWR is suspected to hold large amounts of PETROLEUM, and it has recently become the principal focus of debate between conservationists and the oil industry over the future development of North American oilfields. Conservationists and industry officials agree that energy development would decrease the animal population of the refuge by 40–50 percent. They disagree as to whether or not this impact is acceptable. *See also* ALASKA PIPELINE.

arctic zone *See* ALPINE ZONE.

area guide In FOREST SERVICE usage, historically, a document designed to coordinate planning among two or more neighboring NATIONAL FORESTS with similar management goals and geographical characteristics. The guide normally included a statement of GOALS and MANAGEMENT OBJECTIVES for the forest involved, as well as descriptions of the potential economic and environmental impacts from a broad range of possible ALTERNATIVES. Area guides were largely rendered obsolete by the NATIONAL FOREST MANAGEMENT ACT. *See also* PLANNING AREA.

Area of Critical Environmental Concern (ACEC) an area of federal land within a BUREAU OF LAND MANAGEMENT district that has been determined to present special management problems due to environmental constraints. ACECs may be designated for a number of reasons, including scenic and recreational values; WATERSHED or RIPARIAN ZONE protection; the presence of unusual plant COMMUNITIES, CRITICAL HABITAT for wildlife, or significant archaeological or historical sites; environmental hazards such as unstable slopes (*see* MASS WASTAGE) or FLOODWAYS; or any other factor that is intrinsic to the designated land and is notably different from the conditions of the surrounding environment. Designation of an area as an ACEC does not specifically preclude any other management activity, including timber harvest, off-road vehicle use, road construction, etc.; however, it assures that as these activities are carried out the values for which the ACEC was designated will be protected. The term "Area of Critical Environmental Concern" was first used in the proposed National Land Use Policy and Planning Assistance Act of 1973, where it was applied only to areas on non-federal land. After this act failed to pass, the ACEC concept was attached to the FEDERAL LAND POLICY AND MANAGEMENT ACT OF 1976, as a federal land, rather than a non-federal land, management designation. It applies only to Bureau of Land Management lands.

area rotation in forestry, the practice of determining the amount of timber to be harvested during a given planning period by means of a formula relating the total acreage available for harvest to the ROTATION AGE for the timber on that acreage. The simplest such formula is obtained by dividing the total acreage by the rotation age. Thus, if the rotation age is 80 years, 1/80 of the acreage will be harvested each year. More complex formulas take into account differences in SITE INDEX and planning periods more than one year in length. Area rotation is best suited to EVEN-AGE STANDS and is thus more widely practiced in areas where SECOND GROWTH timber is being harvested than it is in regions of OLD GROWTH.

area source in pollution-control technology, a source that releases pollutants over a broad, diffused area rather than from a single identifiable point such as a smokestack. Area sources are much harder to control than are POINT SOURCES due to the difficulty of tracking and containing the multitude of individual release points involved. They generally require regulatory rather than technological control strategies (e.g., passing laws regulating the times that backyard trash fires may be built rather than supplying clean-burning incinerators to every household). Examples of area sources include quarrying operations, field burning, and the evaporation of volatile substances (*see* VOLATILITY) from numerous spread-out small businesses such as gasoline stations and dry-cleaning establishments. *Mobile point sources* such as automobiles and trucks, taken as a class, create area sources. Thus a heavily traveled highway may be considered an area source for CARBON MONOXIDE and other pollutants associated with motor-vehicle exhausts despite the fact that all of the carbon monoxide arises from individual point sources.

arête in geography and mountaineering, a ridge, specifically a steep-sided ridge with a narrow, usually jagged summit. Arêtes usually form between glacial CIRQUES in the upper reaches of MOUNTAIN SYSTEMS.

arid climate in meteorology, an extremely dry CLIMATE, generally defined as one in which the annual rainfall is less than one-half the local rate of evaporation. *See also* DESERT.

aridisol any soil type characterized by high mineral content and failure to retain moisture. To be classed as an aridisol, a soil must normally be moist no more than three consecutive months in any given year. *See also* SOIL: *classification of soils.*

***Armillaria* fungus** a common fungus in mesic forests. The fruiting body of *Armillaria mellea* is found at the root crown of affected trees, indicating that filaments have girdled major roots, cutting off the transport of water and nutrients. This occurs when trees are young or stressed by drought, cold, insect damage, or environmental pollution. In another form, *A. bulbosa,* ordinary forest mushrooms show above ground, but beneath them is a large body of tendrils that in one place in Michigan have been found to cover a 37-acre area. This massive growth of rhizomorphs—cordlike bundles of filaments—started out as a single microscopic spore but now weighs as much as 10 tons and is at least 1,500 years old. That it is a single organism has been established by DNA tests of filaments from a wide area. *See also* FOREST DECLINE AND PATHOLOGY; MESIC SITES.

armor stone a form of RIPRAP in which the individual stones are particularly large. The lower limit for armor stone is generally taken to be one-half meter in diameter. *See also* EROSION: *control of erosion.*

Army Corps of Engineers (USACE) federal (U.S.) agency charged with the construction and operation of civil works projects. Although the Corps is technically a part of the Army and thus under the control of the Department of Defense, it is primarily a civilian agency, with perhaps 40,000 civilian employees supervised by approximately 200 military officers. Its principal mission concerns navigation and flood control projects, although it also has responsibility for overseeing DREDGE-AND-FILL operations and pollutant discharges into navigable waters, and has historically engaged in some non-water-related construction activity, including building the Alaska Highway, the Washington Monument, and the U.S. Capitol, and constructing and maintaining most Veterans Administration hospitals and domiciliaries around the nation. Dating to 1775, when the first engineers were assigned to the Army by the Continental Congress (their first assignment: fortifications for the Battle of Bunker Hill), the USACE gained corps status in 1802 and took on its first civil works projects in 1824. Its regulatory function was originally assigned in the Rivers and Harbors Act of 1899 (*see* REFUSE ACT) and was greatly expanded by the CLEAN WATER ACT OF 1972. It functions primarily as a contractor, planning and designing projects but generally assigning the actual construction to private firms, which are awarded contracts on a sealed-bid basis. *See also* BENEFIT/COST ANALYSIS; IRON TRIANGLE; PORK BARREL.

aroclor *See* POLYCHLORINATED BIPHENYL.

aromatic compound in chemistry, any ORGANIC COMPOUND whose structure includes an *aromatic ring*—a BENZENE RING or similar cyclical structure with single and double COVALENT BONDS alternating around it. Aromatic compounds tend to be more stable than the non-cyclic organics, and to occur more often as solids. Important aromatics include all vitamins except vitamin C, DDT and many other organic pesticides, PCBs, PBBs and related compounds (*see* POLYCHLORINATED

BIPHENYL; POLYBROMINATED BIPHENYL), and most hormones. Compare ALIPHATIC COMPOUND. *See also* CYCLIC ORGANIC COMPOUND.

ARS *See* AGRICULTURAL RESEARCH SERVICE.

arsenic the 33d ELEMENT in the ATOMIC SERIES, atomic weight 74.91, chemical symbol As. Lying midway in the VA series of the PERIODIC TABLE between antimony and bismuth (which are classed with the metals) and nitrogen and phosphorus (which are not), arsenic is considered a borderline case. One of its three principal ALLOTROPIC FORMS ("gray arsenic") shows moderately strong metallic properties while the others ("yellow arsenic", "black arsenic") do not. Its melting point is higher than its boiling point at atmospheric pressure, so that it does not occur in liquid form (*see* SUBLIMATION). The most common natural occurrence of arsenic is as a substitute for sulfur in metallic ores, from which it is often released into the atmosphere by the heat of smelting. It is used as an insecticide and rodenticide, as an alloying material for lead and other metals, and as an additive in the manufacture of glass to increase the clarity of the finished product—all of which allow it numerous pathways into the environment. This is a problem, because arsenic is virulently toxic; doses as small as 70 milligrams can kill a human. Symptoms of acute arsenic poisoning include severe gastrointestinal distress, nausea, vomiting, and diarrhea, followed by shock and death. Smaller doses lead to chronic poisoning, detected by skin discoloration and peeling and, in severe cases, liver and kidney dysfunction. A reliable test for detecting minute quantities of arsenic in the environment has fortunately existed since the early 19th century (*see* MARSH TEST). Arsenic compounds can be found in groundwater supplies.

arterial in traffic engineering, a major street that carries traffic traveling from one part of a city to another. Arterials are designed to route traffic between neighborhoods rather than to provide access to individual businesses or homes, although such access may be possible from them. They have a tendency to serve as barriers to cross traffic and so should normally be located along neighborhood boundaries rather than being placed through the centers of existing neighborhoods where they will have a divisive effect. Minimum width is generally 18 meters (60 feet), with major arterials often twice that width. Design speed is 60–75 kilometers per hour (35–45 miles per hour). Expected traffic loads can range from 10,000 to 40,000 trips per day. *Compare* COLLECTOR.

artesian well any well in which the STATIC LEVEL lies above the regional WATER TABLE. Artesian wells result from tapping into the lower end of a slanted

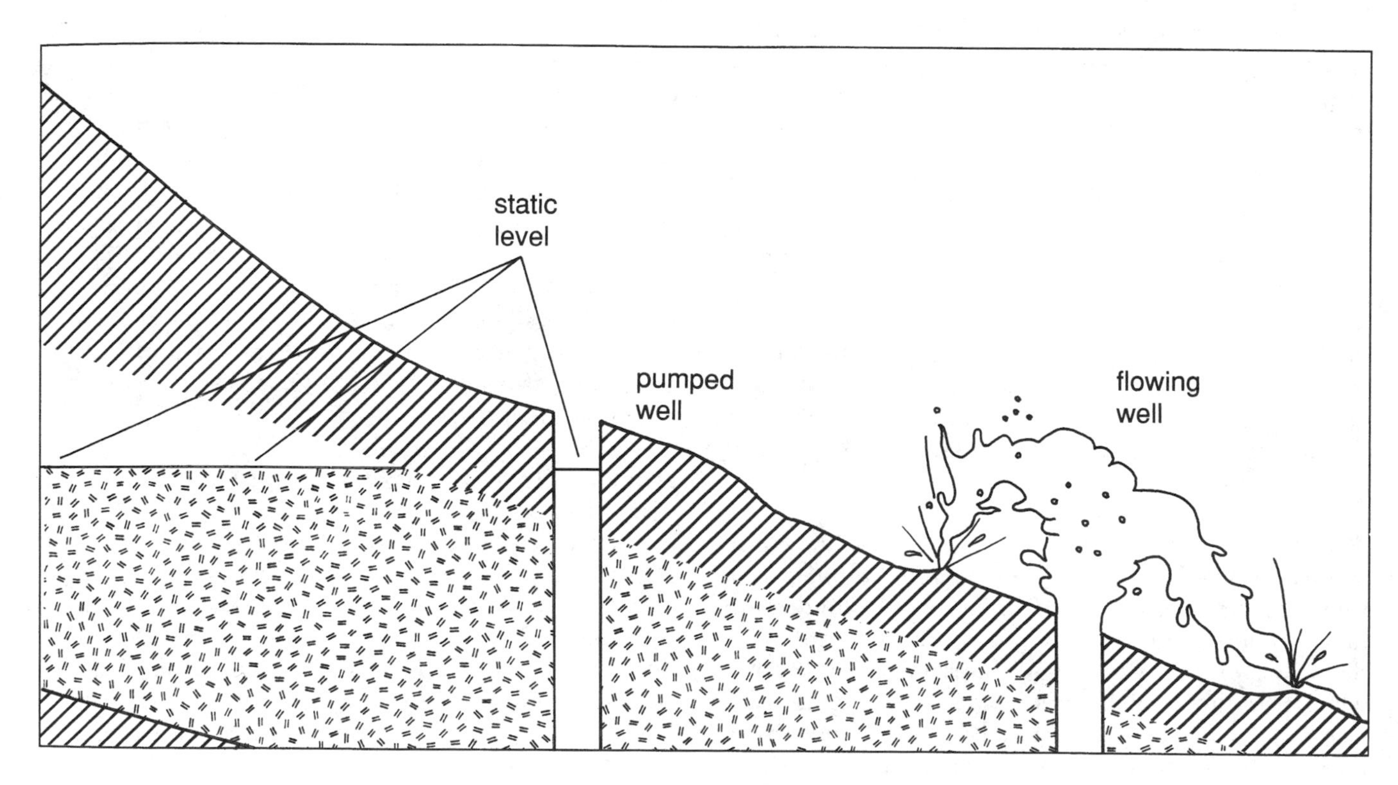

Artesian wells

AQUIFER trapped beneath an impermeable layer of rock (*see* PERMEABILITY). Pressure from the water in the upper end of the aquifer (*artesian pressure*) forces water into the hole made by the well through the impermeable layer. If the pressure in the aquifer is high enough, the water may flow or even spout out onto the surface of the ground, creating a *flowing artesian well.*

asbestos any of several naturally occurring mineral fibers, principally compounds of magnesium, though occasionally iron or calcium minerals may form fibers as well. Nearly all commercial-grade asbestos is magnesium silicate, 95% of it occurring as chrysotile ($(OH)_4 MgSi_2O_5$), a fibrous variety of SERPENTINE. Approximately 3/4 of the known world supply of chrysotile is in the Canadian province of Quebec. Most of the remainder is in South Africa and the Soviet Union.

CHARACTERISTICS AND USES OF ASBESTOS

Asbestos resembles organic fibers such as cotton and wool in its pliability and softness. Unlike cotton and wool, however, it will not burn, and is thus highly useful for weaving flameproof cloth, a practice that is at least 22 centuries old. It also has high heat-absorbing properties, making it an excellent material for brake linings, boiler insulation, and the construction of protective garments for fire fighters and others who must work under conditions of intense heat. Unfortunately, it is also a powerful CARCINOGEN, particularly when inhaled or ingested as microscopic fibers, though its long LATENCY PERIOD hid this fact until comparatively recently. It is classed as a PRIORITY POLLUTANT by the ENVIRONMENTAL PROTECTION AGENCY, and its use is now banned in the United States, where it has largely been replaced by fiberglass.

ASCS *See* AGRICULTURAL STABILIZATION AND CONSERVATION SERVICE.

asexual reproduction any form of organic reproduction in which the parent organism does not exchange genetic material with another organism of the same species. Several varieties of asexual reproduction are possible. *Binary fission*—common in PROTISTA and other one-celled organisms—closely resembles the process of MITOSIS, by which the CELLS of multicellular animals divide. The organism's CHROMOSOMES replicate (duplicate themselves) within the cell nucleus; the nucleus elongates with a group of identical chromosomes in each end; and finally the cell splits down the middle, along the short axis of the elongated nucleus, forming two "daughter cells," which are exact copies of the parent cell. *Schizogony* is identical to binary fission except that several replications of the chromosomes take place before cell division, so that when the organism finally divides it breaks into several daughter cells at once. *Budding* produces a small copy of the parent that begins as a growth on the parent's side and then breaks free and pursues an independent existence. In one-celled organisms this is simply a sort of sack in the cell wall into which replicated genetic material flows, while in higher organisms a complete multicelled copy of the parent usually develops before separation. (Occasionally, separation never takes place. A strawberry plant, for example, buds from RUNNERS, which remain permanently attached to the parent plant.) Finally, in *Parthenogenesis,* eggs are produced and develop into young exactly as in the higher forms of sexual reproduction, but without the crucial step of fertilization.

ADVANTAGES AND DISADVANTAGES OF SEXUAL REPRODUCTION

Asexual reproduction allows extremely rapid population growth that can be controlled entirely by the availability of food and suitable HABITAT, without the need to seek out and mate with a conspecific organism (*see* CONSPECIFIC POPULATIONS). However, since every daughter organism produced asexually is genetically identical to its parent (*see* CLONE), species variation cannot be achieved except by MUTATION and undesirable traits cannot be bred out. For this reason, almost all organisms—even one-celled forms such as AMOEBAS and paramecia (*see* PARAMECIUM)—utilize some form of sexual reproduction at least part of the time. *See* REPRODUCTIVE STRATEGY.

ash in general, the inorganic constituents of a complex organic substance such as wood, coal, or vegetable or meat FIBER. Ash is left behind as a residue when organic substances are burned. If the substance is used as a food, some of the ash may be metabolized (*see* METABOLISM) as a source of minerals while the rest is excreted as a waste product. *See also* FLY ASH; VOLCANIC ASH.

askarel liquid any liquid, such as PCB (*see* POLYCHLORINATED BIPHENYL), used as an insulator for transformers or other electrical equipment.

aspect (1) in geology, the appearance and/or surface composition of a body of rock.

(2) in surveying, forestry, engineering, geology, etc., the compass direction in which a slope faces. Expressed in degrees, the aspect is the compass bearing of a line in the horizontal plane that forms a right angle with the line of intersection of the horizontal plane and the slope. *Compare* DIP; STRIKE; TREND.

(3) in ecology, a division of the year corresponding roughly (but not precisely) to a calendar season (*see* ASPECTION).

aspection in ecology, the annual variation of the ECOSYSTEM in a specific region caused by the change of seasons. Four ecological seasons, or *aspects,* are generally recognized. These are subdivided into nine subseasons, or *sectors*. The *vernal aspect* begins with the appearance of flowers in the spring (*prevernine sector*) and continues through the full foliation of leaves on deciduous trees (*vernine sector*). New growth sprouts; animals come out of hibernation; birds migrate north and begin courtship and nesting activities. The vernal aspect is followed by the *aestival aspect,* consisting of the *cisaestine sector* (the height of nesting activities and vegetal growth) and the *aestine sector* (the end of the rainy season: soils and vegetation dry up; birds enter molt; insects reach maximum concentrations). The *autumnal aspect* finds fruit ripening, birds migrating south, and autumn flowers blooming (*serotinine sector*), followed by the first frosts and the color changes and shedding of leaves from deciduous plants (*autumnine sector*). Finally, the *hiemal aspect* represents life at the year's lowest ebb, commencing with the baring of branches and the dying of annual plants and insects (*hiemine sector*), the hibernation of animals and the disappearance of birds (*hibernine sector*), and at last the first sprouts and buds of the coming spring (*emerginine sector*). It should be noted that the boundaries between the various aspects and sectors are arbitrary and even capricious, varying from place to place, from year to year, and even from vegetal type to vegetal type. The study of aspection is not concerned with fixing temporal boundaries but with making a sort of rhythmic sense of the whole annual ecological cycle and thus improving ecologists' ability to make comparisons between equivalent parts of it in different regions or different years.

asphalt (bitumen) a sticky black substance derived from PETROLEUM through evaporation of some of the lighter HYDROCARBONS and (usually) partial oxidation of the remainder. It occurs naturally throughout the world, in various forms corresponding to those of water: springs, seeps, pools and lakes ("tar pits"), and even the equivalent of AQUIFERS ("rock asphalt"—beds of gravel, sand, or fractured rock impregnated with asphalt instead of water). These natural asphalts have been used since earliest times as building, paving, and caulking materials; however, nearly all the asphalt used today is produced from raw petroleum. Nearly three-fourths of the U.S. annual production of 1.2 million short tons is used for paving; the remainder goes to a broad variety of uses, including principally the manufacture of roofing, paint, and adhesives and caulking compounds. Because it is a residue—meaning that everything that can volatilize or leach from it has already done so—asphalt does not represent much of a threat as a pollutant.

Aspinall amendment loophole in the WILDERNESS ACT OF 1964 allowing mining claims to be staked in wilderness areas for 20 years, and for patented and valid unpatented claims to be worked indefinitely thereafter. The amendment is named for its author, Congressman Wayne Aspinall (D-Colorado), who insisted on its inclusion as a price for allowing the act to pass through the House Committee on Interior and Insular Affairs of which he was then chair. *See* LOGROLLING.

association in ecology, a group of otherwise unrelated plant species that are characteristically found growing together within a particular COMMUNITY. Plant associations serve as dependable guides to soil conditions and MICROCLIMATES, and are therefore closely studied by foresters and botanists. For example, salal (*Gaultheria shallon*) growing in association with Douglas fir (*Pseudotsuga menziezii*) indicates a good-to-excellent growing site for this important U.S. timber species. *See also* ASSOCIES.

Association of American Geographers (AAG) founded in 1904, the AAG strives to stimulate research in geography and to promote its findings in government, commerce, and education. Membership (1999): 7,000. Address: 1710 16th Street NW, Washington, DC 20009-3198. Phone: (202) 234-1450. Website: www.aag.org.

associes in ecology, a term that is sometimes used to differentiate ASSOCIATIONS within seral communities (*see* SERE) from those within CLIMAX COMMUNITIES. In this usage, the term *association* is reserved for climax communities; plant groupings within seral stages of a community are known as *associes*.

asthma a form of allergic reaction (*see* ALLERGEN) in which the bronchial tissues in the lungs become inflamed, swelling up and secreting mucus and thus constricting or blocking the bronchial passages. Asthma attacks cause severe shortness of breath, wheezing, and fits of coughing as the body attempts to clear mucus from the lungs; they can be severely debilitating and even occasionally fatal. The causative allergen may be present in the bloodstream, either from being ingested (chocolate is a common cause) or from being absorbed through the skin. Contrary to popular belief, stress (physical or emotional) will not trigger an asthma attack by itself. However, if an allergen is present, the allergen is much more likely to cause an attack if the individual is under stress. Smoking or the presence of AIR POLLUTION can also increase susceptibility to asthma-causing allergens. *Compare* HAY FEVER; EMPHYSEMA.

Aswan High Dam one of the largest dams in the world, constructed on the Nile in southern Egypt between 1960 and 1970 and financed largely by the USSR. Lake Nasser, rising behind the dam and 2,000 square miles in area, inundated many archaeological treasures although some were removed beforehand. Eight times larger than Arizona's Lake Mead, created by the Hoover Dam on the Colorado River, the Aswan High Dam has been a boon to agriculture along the Nile River for 550 miles southward to Cairo, providing for year-round irrigation with which three crops can be produced annually rather than just one. Nevertheless, severe environmental problems have arisen. Lake Nasser itself is becoming silted up from the vast numbers of distant tributaries of the White Nile and the Blue Nile in the Sudan and Ethiopia. And the absence of fertile silt below the dam has led to excessive dependence on chemical fertilizer, which is too expensive for many farmers. Moreover, the absence of annual flooding and the use of flush-irrigation practices have led to salinization of agricultural soils, further reducing productivity. In the Nile Delta the seawater encroaches, no longer held back by river flows drawn down by the one gallon in five that evaporates from Lake Nasser or is diverted for agricultural use. As a result, coastal settlements in the delta have had to build elaborate sea walls and jetties, and commercial fisheries have been disrupted.

atmosphere the envelope of gases surrounding a planet or other astronomical body. The term is often used interchangeably with the word *air,* but this is incorrect. Air is the substance from which the body known as the atmosphere is made, as water is the substance from which the ocean is made, or rock is the substance from which a mountain is made. Earth is not the only body in the solar system with an atmosphere: Mars, Venus, Jupiter, and indeed most of the other planets have atmospheres, although the structure and composition of them are far different from the Earth's. Even the Moon has an atmosphere, albeit an extremely thin and tenuous one. However, the only atmosphere we are ordinarily concerned with is our own, and therefore it is usually understood that "the atmosphere" and "the Earth's atmosphere" mean the same thing—a convention we will follow for the rest of this article and elsewhere in this encyclopedia. The atmosphere is one of the four divisions into which scientists traditionally separate the environment, the others being the LITHOSPHERE (rock and soil), the HYDROSPHERE (water) and the BIOSPHERE (living things). It is extraordinarily important to life, not only because we breathe it but because it serves as a blanket to hold heat on the Earth that would otherwise escape to space, and as a vast engine to transfer heat from the equator to the poles and to move water vapor from the seas to the land.

COMPOSITION OF THE ATMOSPHERE

The air of which the atmosphere is composed is a MIXTURE of gases rather than a COMPOUND. Therefore, its composition is not constant but varies from place to place, although in the lower atmosphere these variations are ordinarily so small that they can be ignored. By far the largest portion of the lower atmosphere—78%—is NITROGEN; another 21% is OXYGEN, leaving barely 1% to be divided among all other atmospheric constituents. Most of this remaining 1% is argon (0.93%) and CARBON DIOXIDE (0.03%). The final four hundredths of a percent are composed of neon, helium, methane, krypton, hydrogen, and varying amounts of water vapor and PARTICULATES.

STRUCTURE OF THE ATMOSPHERE

The atmosphere is separated by gravity, centrifugal force, and the influence of the Sun into a number of concentric layers, or *spheres.* These spheres may be defined from each other in a number of different ways. The most common method used by meteorologists is the temperature gradient. In the lowest level of the atmosphere, the TROPOSPHERE, temperature decreases gradually with altitude (*see* LAPSE RATE). Approximately 10 kilometers (6 miles) above the surface, at the TROPOPAUSE, the temperature curve suddenly levels off and, a few kilometers later, begins rising. This sphere of rising temperature gradient is called the *stratosphere.* At about 45 kilometers (28 miles) comes a second break in the curve, the *stratopause,* above which temperatures begin falling once more through the *mesosphere.* Finally, at the *mesopause,* 80 kilometers (50 miles) up, a third and final change in direction occurs, and the gradient begins rising again through the *thermosphere,* which extends outward to the limit of the atmosphere, an ill-defined zone of escaping gases that may lie as high as 4,000 kilometers (2,500 miles) above the surface. Because the thermosphere is layered by molecular size, with nitrogen on the bottom, followed by layers of oxygen, helium, and hydrogen, it is sometimes called the *heterosphere.* In this case, the troposphere, stratosphere, and mesosphere are usually lumped together as the *homosphere.* Another commonly recognized layer in the atmosphere is the *chemosphere,* a zone from 20 to 150 kilometers up where chemical reactions commonly take place (it includes both the OZONE LAYER and the IONOSPHERE, a zone of ionized gases [*see* ION] that occupies roughly the lower 300 kilometers of the thermosphere).

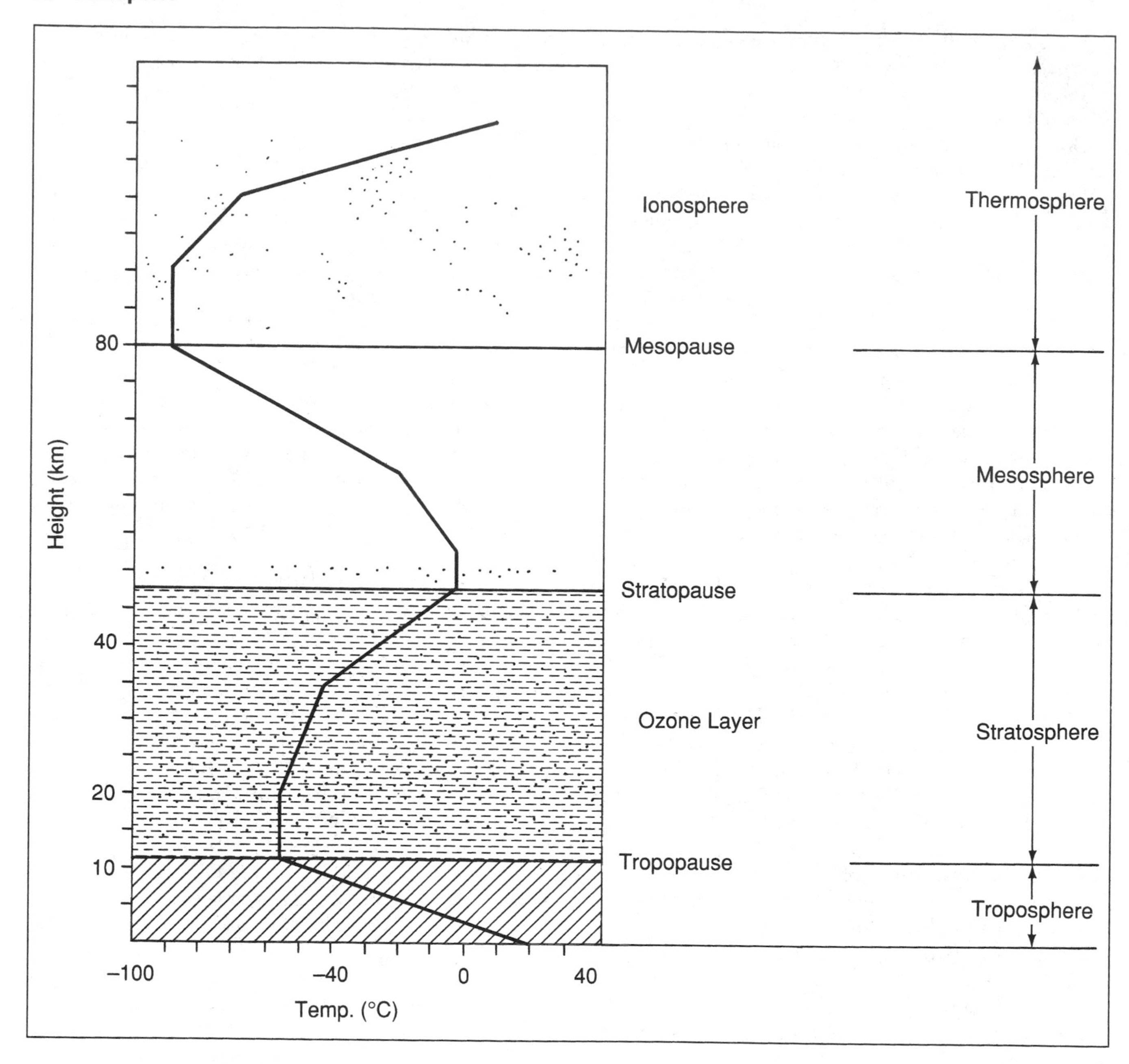

The atmosphere

DYNAMICS OF THE ATMOSPHERE

The stratosphere, the mesosphere and the thermosphere are zones of calm that are rarely or never disturbed by winds and other massive air movement. The troposphere, by contrast, is a region of intensive activity. Two forces account for most of this turbulence: the radiation of heat from the Earth's surface, which sets up large convection currents in the lower atmosphere (*see* HADLEY CELL) as well as causing differential zones of high and low pressure due to differences in the amount of heat radiated from various parts of the surface, and the drag exerted by the Earth's surface as it turns through the atmosphere on its axis, a drag that is most strongly felt in the lowest 5 kilometers or so of the atmosphere, sometimes known as the *friction layer.* The upper-troposphere air-circulation pattern is normally poleward from the equator and equatorward from the poles, thus transferring heat from the Tropics outward and making the planetary temperature average considerably more uniform; in the friction layer, the general trend of air movement is from the seas to the continents. Friction-layer air transport is not totally predictable, however, due to the major influence of local factors such as land and sea breezes (*see* LAND BREEZE; SEA BREEZE) and CYCLONIC STORMS. *See also* WEATHER;

CLIMATE; CYCLONE; ANTICYCLONE; JET STREAM; GREENHOUSE EFFECT; ATMOSPHERIC STABILITY.

atmospheric pressure pressure exerted by the Earth's ATMOSPHERE on objects within it. As with any gas or liquid, this pressure is exerted equally in all directions; that is, the pressure on the underside of an object suspended in the air is the same as the pressure on the top or on either side. Atmospheric pressure is approximately 101 kilopascals (14.7 pounds per square inch) at sea level (slightly more or less, depending upon meteorological conditions). It is cut roughly in half for each 18,500 feet of elevation gain. *See also* MILLIBAR; BAROMETER.

atmospheric stability (a-s) in meteorology and air pollution control, the resistance of an AIR MASS to vertical mixing; that is, the degree to which the air mass either damps out vertical currents (*stable air*) or encourages them to form (*unstable air*). Atmospheric stability depends primarily on the relationship of the LAPSE RATE to the ADIABATIC RATE. Under *superadiabatic conditions* ("strong lapse rate") the lapse rate is greater than the adiabatic rate, meaning that air moving upward will cool more slowly due to adiabatic change than its surroundings are cooling due to the lapse rate, making it warmer and warmer in relation to its surroundings and thus causing it to rise faster and faster. *Subadiabatic conditions* ("weak lapse rate") are the reverse of this. The rising air cools faster than its surroundings, causing it to stop rising and even to sink again. In extreme cases the lapse rate may even reverse, damping out vertical mixing altogether (*see* INVERSION).

atmospheric tank a storage tank designed to operate at or near ATMOSPHERIC PRESSURE, used for storing liquids of low VOLATILITY such as crude oils. Federal standards require that the pressure difference between the inside and outside of an atmospheric tank should be no more than 0.5 pounds per square inch. Most atmospheric tanks maintain equality of pressure through venting systems that allow vapors to escape; hence, they may contribute significantly to air pollution if used for highly volatile materials. *Compare* LOW-PRESSURE TANK.

atom in chemistry and physics, the smallest unit of a substance that cannot be broken down further or changed to another substance by purely chemical means (but *see* ION; ISOTOPE). Atoms are the basic building blocks of matter and the units from which all ELEMENTS and COMPOUNDS are constructed. Ninety-two varieties of them are known in nature, each one corresponding to one of the elements; 14 more (the so-called TRANSURANIC ELEMENTS) have been created in the laboratory.

STRUCTURE OF THE ATOM

Though they cannot be broken down by chemical means, atoms are not indivisible. They are made of tiny, discrete units. There are three basic types of these atomic building blocks, or *subatomic particles,* found in each atom. (The many other known types of subatomic particles are less predictably distributed and do not have the same significance as the three basic units for determining the type of atom being constructed.) The smallest, the *electron,* carries a negative charge and has a mass of about 9.1 x 10^{-28} grams (0.00000000000000000000000000091g): it is essentially a tiny, rapidly moving packet of energy. The *proton* carries a positive charge that balances the negative charge of the electron. It is roughly 5,000 times larger than the smaller particle. The third particle, the *neutron,* carries no charge and is slightly larger than a proton; it is probably itself composed of a tightly bonded proton and electron. Within the atom, the protons and neutrons form a cluster at the center known as the *nucleus,* while the electrons circle them some distance away. The orbit of the closest electron is about 20,000 times as wide as the nucleus. The orbits are not randomly distributed but occur in discrete steps known as *shells.* Electrons may move from lower shells to higher ones only by increasing their energy content by one *quantum*—the basic unit of energy in the universe (this is the origin of the phrase "quantum leap"). Only the electrons in the outer shell, known as the *valence shell,* are available for combining with other atoms to form COMPOUNDS.

Atomic Energy Commission (AEC) obsolete federal agency, created by Congress in 1946 under the terms of the Atomic Energy Act to oversee, regulate, and promote the development of atomic energy for both defense and peacetime purposes. The principal predecessor of the NUCLEAR REGULATORY COMMISSION, the AEC was abolished by the Energy Reorganization Act of 1974.

atomic number *See under* ATOMIC SERIES.

atomic series in chemistry, physics, and related disciplines, an ordering of the various chemical ELEMENTS by the number of protons their atoms contain (*see* ATOM: *structure of the atom*). An element's position in the atomic series is called its *atomic number.* The chemical properties of an element may be predicted with reasonable accuracy from its position in the atomic series (*see* PERIODIC TABLE).

atomic weight the average mass of a single ATOM of an ELEMENT, expressed in terms of a dimensionless unit approximately equal to the mass of one hydrogen atom. (The "standard atom"—the atom that is normally taken to have a unit weight, with the others measured relative to it—is actually not the hydrogen atom but the most common ISOTOPE of the carbon atom, which is defined as having an atomic mass of 12. The unit of atomic weight is therefore precisely one-twelfth the mass of a carbon atom.) Because the most commonly occurring isotopes of most elements have an equal number of protons and neutrons in their nuclei (*see* ATOM: *structure of the atom*), the atomic weight is generally roughly twice the atomic number (*see* ATOMIC SERIES). However, due to the existence of other isotopes of varying masses that must be averaged together with the most common isotope when calculating the atomic weight, this doubling is almost never precise.

ATSDR *See* AGENCY FOR TOXIC SUBSTANCES AND DISEASE REGISTRY.

ATV *See* ALL-TERRAIN VEHICLE.

Audubon, John James (1785–1851) American painter and naturalist, born April 26, 1785, in Les Cayes, in the French colony of St.-Domingue, which later became Haiti, and died January 27, 1851, at his country home on northern Manhattan Island (now part of New York City). The son of a French naval officer, Audubon spent his youth in France, where he may have studied drawing briefly under the well-known painter Jacques Louis David. He had no other formal art training. Audubon came to America in 1803 at the age of 18, settling first on a farm near Philadelphia, where he began to draw birds. After trying several occupations in a number of different locations, including Kentucky, Louisiana, Ohio, and New York City, he decided to make the portraiture of American birds his life work, beginning in 1827 the publication of a series of folio drawings, *The Birds of America,* in which every bird was rendered full-sized. The number of species represented eventually reached 1,065, on 435 hand-colored plates. An excellent marksman, Audubon usually drew from specimens that he shot and then mounted in lifelike poses. By 1833 he had become convinced that American birdlife was endangered by heavy hunting and other pressures of civilization, and he spent the last part of his life advocating bird protection, work that is carried on today in his name by the NATIONAL AUDUBON SOCIETY.

Audubon Society *See* NATIONAL AUDUBON SOCIETY.

AUM *See* ANIMAL UNIT MONTH.

autecology the study of the interaction between a single species, or a single member of a species, and the environment. *See also* ECOLOGY.

authorization as used by the various federal agencies, authority granted by congressional legislation to spend a specified amount of money on a specified project such as a dam, a timber harvest program, a consumer education campaign, etc. An authorization does not grant money for a project, but merely tells the agency where it should spend the money if and when it becomes available. The agency must return to Congress for an APPROPRIATION before it actually has money to spend.

PROCEDURE

Authorization bills are handled by the Senate and House committees designated by the rules of Congress to oversee the agency that will be carrying out the project—the Agriculture Committees for Forest Service and Soil Conservation Service projects, the Environment and Public Works Committees for National Park Service, Bureau of Land Management and Army Corps of Engineers projects, and so forth. Occasionally an authorization may proceed through Congress on its own, but more often a group of related authorizations will be handled in a single piece of legislation known as an OMNIBUS BILL. Most authorization bills carry a form of "sunset clause," which states that if no money has been appropriated for a project for a certain length of time (usually three years) it must be reauthorized. *See* ORGANIC ACT; PORK BARREL. *See also* articles under each agency named above.

auto-oxidation *See* AUTOXIDATION.

autotrophism the ability of an organism to manufacture all of its food from inorganic sources. The most common autotrophic organisms ("autotrophs") are the green plants, which use PHOTOSYNTHESIS to construct CARBOHYDRATE molecules from water and CARBON DIOXIDE. *Compare* heterotrophism (*see under* HETEROTROPH).

autoxidation (auto-oxidation) in chemistry, an OXIDATION reaction that takes place upon the simple exposure of a substance to the environment, without the addition of extra energy or of a synthetic CATALYST. The best known example is rust (hydrated ferric oxide, $Fe_2O_3 \cdot H_2O$), which is an autoxidation of iron in the presence of air and water.

auxin in the biological sciences, one of a small group of growth hormones that are produced in the tips of growing plants. Like all hormones, auxins are cyclical hydrocarbons (*see* AROMATIC COMPOUND). The most common variety is indoleacetic acid ("IAA," chemical formula $C_{10}H_9NO_2$). These hormones cause cells in the SHOOT to elongate, and are responsible for most differential-growth phenomena in plants, including GEOTROPISM, PHOTOTROPISM and APICAL DOMINANCE. In root tissue they become growth inhibitors, decreasing the growth rate instead of speeding it up. As they pass through a plant's tissues, auxins are slowly deactivated by ENZYMES and by other hormones (known as *antiauxins*). It is the complex interactions among these chemicals that determines the actual rate at which differential grown phenomena take place. *Compare* GIBBERELLIN; CYTOKININ.

avoidance in biology, a behavioral trait of living organisms that causes them to draw away from certain stimuli. It is present in all animals, most PROTISTA, and some plants. In simple organisms, avoidance behavior is usually straightforward and predictable: an AMOEBA, for example, will always initially move away from a source of light. In the animals, however, it is complicated by learning. Squirrels will avoid human contact unless they have learned that the humans feed them; cats and dogs will avoid each other unless they live in the same household; and even amoebas will stop avoiding light if, after a while, the light does not go away.

avulsion (in law) *See under* ACCRETION.

A-weighted sound scale a scale used to measure the effects of NOISE POLLUTION in the neighborhood of a sound source. The A-weighted scale is designed to duplicate the idiosyncrasies of the human ear. It deemphasizes low frequencies and slightly emphasizes frequencies between 1,000 and 5,000 Herz (cycles per second).

azonal soil any soil without a B horizon: see SOIL: *the soil profile*. The azonal soils tend to be shallow and young, with little organic content. *See* ALLUVIAL SOIL; LITHOSOL; REGOSOL.

B

bacillus any rod-shaped BACTERIUM.

Bacillus thuringiensis (Bt) a pathogenic BACTERIUM (*see also* PATHOGEN) of the genus *Bacillus,* discovered in Germany in 1911 and widely used since as an alternative to chemical PESTICIDES. It is particularly effective against Lepidopterae (moths and butterflies), including such serious pests as the GYPSY MOTH and the cabbage moth. It does not cause epidemic disease; instead, it produces a crystal of PROTEIN known as a *parasporal body* during the process of sporulation (that is, while the bacterium is entering the spore stage), and this parasporal body happens to be highly toxic to caterpillars. It is apparently harmless to mammals, birds, and reptiles, and to most beneficial insects. The bacillus is easy to culture in the laboratory, stores well, and can be sprayed like a pesticide. It does not reproduce easily in the field, so applications of it tend to be self-limiting and can thus be timed to poison a particular species of caterpillar without harming others that happen to come out of their cocoons at different times. *See also* BIOLOGICAL CONTROL.

background in visual management, the distant part of a visible landscape or VIEWSHED. It is usually defined as that portion of the view in which the individual trees of a STAND can no longer be seen. *Compare* FOREGROUND; MIDDLEGROUND. *See also* VISUAL RESOURCES MANAGEMENT SYSTEM.

background level the amount of radiation, or of a chemical substance, normally present in the environment at a given location. Knowledge of background levels is necessary not only to determine long-term exposure rates to environmental hazards but to establish a local baseline against which measurements following accidental release of hazardous substances or radiation may be compared to find the net effect of the release.

backpackers disease *See* GIARDIASIS.

BACT *See* BEST AVAILABLE CONTROL TECHNOLOGY.

bacteriophage *See* VIRUS.

bacterium (plural: **bacteria**) any of numerous single-celled ORGANISMS belonging to the KINGDOM MONERA. Although not all Monera are bacteria, the line between these organisms and the CYANOPHYTES, or blue-green algae, is not precise, making it impossible to separate them into distinct divisions. The simplest living organisms (but *see* VIRUS), bacteria lack a true nucleus and do not have CELLULOSE in their cell walls. They do not form colonial or multicellular structures, although some SPECIES may not separate completely upon reproduction, creating many-celled masses consisting of independent cells that happen to be stuck together. Many species are motile (*see* MOTILITY), and some have rudimentary sensory organs, such as eyespots and chemical sensors, which are sensitive to minute changes in water chemistry. Most are HETEROTROPHS. As well as being extremely simple, these organisms are extremely old: microfossils of them have been found in sedimentary rocks laid down more than 3.2 billion years ago. Their DNA is identical in basic structure to our own, and it is almost certain that we share a common remote ancestor.

TYPES OF BACTERIA

Bacteria come in three distinct forms: rod-shaped (*bacilli*), spiral-shaped (*spirilli*), and spherical (*cocci*).

Bacilli have flagella (*see* FLAGELLUM) more commonly than the other two varieties, and are thus the most likely to be motile. They sometimes form long filaments called *mycobacteria*. Spirilli include the largest bacteria; they almost never form filaments or other combined structures. Despite their general lack of flagella they are often motile, moving through fluids with a sinuous, snakelike motion. Cocci, the smallest bacteria, are often found stuck together. The type of combined structure they form is representative of the GENUS they belong to, with *Diplococcus* forming pairs, *Streptococcus* forming chains, and *Staphylococcus* forming uneven clusters.

BACTERIA IN THE ENVIRONMENT

Although probably best known as PATHOGENS—the causes of diseases such as syphilis, scarlet fever, and tuberculosis—bacteria are widespread in all parts of the environment and are necessary for the operation of nearly all ECOSYSTEMS. Many act as DECOMPOSERS, breaking down the bodies of dead animals and plants (as well as other bacteria) and re-releasing their component minerals for use by other organisms. Bacteria are resident in the digestive tracts of most animals, including humans, where they aid in the digestion of food. They live in nodules on the roots of some plants, especially the LEGUMES, assisting them in the takeup of NITROGEN from the air and the soil. Some are able to decompose environmentally dangerous COMPOUNDS such as petrochemicals and PCBS. It has been estimated that the weight of all living bacteria, taken together, equals or exceeds the weight of all other organisms on the planet. *See also* GRAM STAIN; ANAEROBIC BACTERIA; AEROBIC BACTERIA.

badlands in geography, arid or semiarid terrain in which poorly developed drainage systems and lack of vegetative cover have allowed extensive erosion to take place. Badlands consist of largely patternless topography characterized by heavily gullied slopes, free-standing walls and pinnacles, and numerous steep-sided buttes and draws. Several national parks and monuments feature badlands topography, including Badlands National Park in South Dakota, Theodore Roosevelt National Park in North Dakota, and El Malpais National Monument in New Mexico.

baghouse in air-pollution control terminology, a means of removing PARTICULATES from stack emissions by filtering them through a fabric bag. The device is similar to a domestic vacuum cleaner, but on a considerably larger scale. Baghouses are among the oldest, most widespread, and most effective means of air-pollution control (*see* AIR POLLUTION), with several hundred thousand currently in operation in the United States and Canada.

DESIGN AND OPERATION

Though there are numerous types of baghouse design, the significant differences among them boil down to two: variations in the type of fabric employed and variations in the method used to clean the bags once they become full. Fabric choice depends largely on the type of particulate being filtered and on the temperature of the gas stream, woven cloth bags being preferable in most cases due to their superior strength, with felt bags employed where the majority of particulates are of a very small size. Felt bags are almost always cleaned by "backflushing," that is, by blowing air backwards through them from the clean-air side. Backflushing may also be used for cleaning woven bags, but more common methods include shaking the bags, rapping the frame holding them, or pulsing jets of compressed air through them. A major factor in baghouse design is the ratio of the volume of stack gases passed (in cubic feet per minute) to the area of fabric in the bags (in square feet), commonly called the gas/cloth (G/C) or air/cloth (A/C) ratio: for most installations, this ratio is set at around 4:1, but it may be as low as 1.5:1 or as high as 12:1 depending upon the composition of the stack gases, the type of filter, and the cleaning method employed.

bag limit in wildlife management, the number of animals of a particular SPECIES which a hunter may legally kill in a given time period, usually either in a day or in a hunting season. A comparable limit on fish is known as a *creel limit*.

Baikal, Lake the world's deepest and oldest lake, located in southeastern Siberia, Russia. Fifty miles wide and nearly 400 miles long, and fed by more than 300 tributaries, Lake Baikal holds 20% of the world's freshwater. Its 2,635 identified species and subspecies comprise the largest number associated with any freshwater body on Earth. Some 1,800 are found nowhere else, including the only freshwater seal species. Logging, agriculture, and industrial development, however, have threatened the delicate ecological balances of Baikal and its surrounding area. Thousands of acres of forest land have been logged off to create land for agriculture, increasing erosion and the runoff of agricultural chemicals into the lake. A gigantic cellulose plant built in the 1960s in the town of Baikalsk on the southern shore has polluted large areas, making the water unsuitable for drinking. In addition, the commercial fish catch in Baikal has been seriously reduced, the death rate in the seal population has risen, and acid rain has increased. In 1993 a consortium of American and Russian organizations and government agencies sponsored a cooperative study directed by the American ecological planner George D. Davis. The study project recommended tough new land use controls in the 115,000-square-mile Lake Baikal region, and called for the designation of the area as a World Heritage Site by the UNITED NATIONS EDUCATIONAL, SCIENTIFIC, AND CULTURAL ORGANIZATION (UNESCO).

bait station in range management, a poison-laced animal carcass designed to lure PREDATORS who will eat it and die from the poison; also, any device or object that combines an attractant with a means of destroying the attracted animals. Bait stations are usually ineffective as a means of PREDATOR CONTROL because they tend to attract SCAVENGERS rather than predators. They are nonselective and can be extremely hazardous to NONTARGET SPECIES, including human children, and therefore their use is generally not looked upon as sound environmental management.

balance of nature the concept that the reproduction rate of a SPECIES has been adjusted by evolution to closely approximate the species' death rate from all natural causes (accidents, illness, PREDATORS, old age, etc.). The stability of this balance varies with the number of different causes of death, because large variations in one cause can be balanced more easily by small variations in many others than by large variations in a few. If too great a variation occurs in one cause of death for the other causes to compensate (the removal of too many predators by human PREDATOR-CONTROL programs, for example), the system is said to be "out of balance," and there will normally be drastic effects on the species' numbers and on the numbers of whatever other species it in turn may effect. While most contemporary ecologists believe that the term "balance of nature" oversimplifies complex natural processes, there is a general view that the ebb and flow of the forces of nature do tend toward a dynamic equilibrium. *See* KAIBAB PLATEAU.

balanoid zone (intertidal zone) the area of the ocean shore between the normal high-tide and normal low-tide lines, characterized by the presence of TIDEPOOLS and MUDFLATS, the prolific growth of ORGANISMS such as crabs, clams, starfish, sea anemones, and ALGAE, and the presence of barnacles on virtually all available solid objects. *Compare* LITTORINA ZONE; SUBTIDAL ALGAL ZONE. *See also* LITTORAL ZONE; SWASH.

Bald Eagle Protection Act federal legislation to prevent the extinction of the bald eagle, signed into law by President Franklin Roosevelt on June 8, 1940. The act makes it illegal to "pursue, shoot, shoot at, poison, wound, kill, capture, trap, collect, molest or disturb" any bald eagle, bald eagle nest, or bald eagle egg without specific permission from the Secretary of the Interior (*see* INTERIOR, DEPARTMENT OF). An exception is made for eagle nests that are interfering with power transmission lines. Possessing and transporting eagles or eagle parts is also made illegal. Penalties of up to $5,000 and a year in jail for each succeeding offense may be levied in either criminal or civil court, and ranchers convicted of violating the law may have their public-land grazing leases canceled. The act's provisions were extended to golden eagles by an amendment signed by President John Kennedy on October 24, 1962. Despite the law, by the 1960s only about 400 nesting pairs of bald eagles remained in the lower 48 states, the species a victim mainly of habitat loss and the effects of DDT on reproduction. In 1973 bald eagles were afforded even more protection with the passage of the ENDANGERED SPECIES ACT. As a result of these laws, together with the Migratory Bird Treaty Act, by the mid-1990s the bald eagle population in the lower 48 states had increased to 4,000 nesting pairs, with an estimated 45,000 more eagles in Alaska. As a consequence, the bald eagle was downgraded from *endangered* status to *threatened* by the Fish and Wildlife Service.

Bankhead-Jones Farm Tenant Act federal legislation that made possible the creation of the National Grassland System (*see* NATIONAL GRASSLAND; NATIONAL FOREST SYSTEM), signed into law by President Franklin Roosevelt on July 22, 1937. Designed primarily as a vehicle to provide loans to tenant farmers during the Depression so that they could purchase the lands they were tilling, the act included language (Title III) authorizing the federal purchase of worn-out, eroded, flood-damaged, or otherwise economically submarginal farmland for the purpose of restoring it for watershed protection, soil protection, and reforestation. More than 7 million acres of these "Title III lands" were purchased under this program, in 28 states, mostly in the Great Plains; in Montana alone the purchases amounted to some 2 million acres. All of these lands remained under the care of the SOIL CONSERVATION SERVICE as so-called "Land Utilization Projects" until 1960, when 3.9 million acres were transferred to the FOREST SERVICE as National Grasslands. The purchasing authority, but not the remainder of the Act, was repealed on September 27, 1962.

barbed tributary in geology, a TRIBUTARY stream that points upstream rather than down as it enters the MAINSTEM; that is, one whose lower course runs in a generally opposite direction to that of the mainstem, entering at an acute angle with the mainstem's downstream limb. Barbed tributaries are characteristic of PIRATED STREAMS, and are usually an indication that the mainstem once ran in the opposite direction.

barber chair in logging, a stump with a large upright sliver of wood and bark on one side, torn from the upper part of the tree as it falls. Barber chairs indicate poor or sloppy felling practices.

bark beetle causes extensive damage in North American forests. In coniferous trees such as the commercially valuable Douglas fir, the bark beetle enters trees weakened by spruce budworm, making galleries beneath the bark that choke off the transport of water and nutrients and introducing a deadly fungus as well. The beetles attract one another to "target" trees by emitting a pheromone. Healthy trees can direct the flow of sap to holes bored by the bark beetle, flushing out the intruder. In severely infested areas, and in trees affected by drought or other stresses, the flushing strategy does not work, however. Another bark beetle carries the *Ceratocystis* and the *Ophiostoma* fungi, both of which kill American elms. The elm has virtually disappeared from its native range, although park and street trees can survive if given large doses of fungicides. *See also* FOREST DECLINE AND PATHOLOGY.

barometer a device for measuring ATMOSPHERIC PRESSURE. Barometers are of two basic types. In the *liquid barometer,* a glass tube closed at one end and containing a heavy liquid, usually MERCURY, is inverted in a bowl containing the same liquid. The level of the liquid in the tube moves up and down in response to changes in the amount of pressure exerted on the liquid in the bowl, with greater pressure forcing more liquid from the bowl into the tube and thus raising the level in the tube. The sides of the tube are usually marked off in standard units such as inches or MILLIBARS. In the *aneroid barometer,* a thin-walled metal chamber has most of the air removed from it. Its sides then act as diaphragms, being forced inward when pressure increases and bowing out again when pressure decreases. A lever attached to one of the sides moves in and out with the side. Its motion is translated through a gear train into rotary motion on a dial, which can be calibrated using the same units as the glass tube of the liquid barometer. Since air pressure varies with height above sea level, barometers may be used to measure altitude as well as to mark passing low- and high-pressure weather systems. When used in this manner, they are called *altimeters* and are marked off in feet or meters above sea level rather than in inches of mercury or millibars of pressure. Adjustments must be made in the altitude readings to compensate for pressure variations caused by the weather. Because of the difficulty of doing this, altimeter readings are never more than approximate and cannot be relied upon for survey data.

barrel a liquid measure, usually defined as 42 gallons, though the steel barrels currently used to transport petroleum and industrial chemicals and wastes have a capacity of 55 gallons rather than 42. *See* DRUM.

barrel of oil equivalent (boe) in energy management, an amount of energy equal to that found in a 42-gallon barrel of CRUDE OIL. One boe = 5,800,000 btu (*see* BRITISH THERMAL UNIT).

barrier, ecological *See* ECOLOGICAL BARRIER.

barrier islands are formed off shallow-water coastlines, such as North America's Atlantic and Gulf coastlines. The most extensive barrier islands are the Outer Banks of North Carolina which guard the entire coastline of the state, from Currituck Sound on the north (North Carolina) to Tubbs Inlet on the south (South Carolina). Barrier islands are usually long, thin landforms made of sedimentary material that run parallel to the shore, creating biologically rich embayments behind them. Barrier islands protect inland areas from storms but can themselves be breached and flattened by hurricanes. The storm damage to barrier islands comes not from the frontal attack of a hurricane but from the sound side. Storm water is pushed landward by the wind up into the estuaries, then rushes back toward the sea and devastatingly across the island when, after the eye of the hurricane is passed, the wind pattern reverses. These reverse surges resemble giant tidal waves that can carry summer cottages and trailers out to sea and undermine larger buildings, creating new inlets that separate the islands in the process. Barrier islands are unstable even without hurricanes, since the normal action of onshore wind and waves creates a kind of "roll-over" effect in which the banks move slowly shoreward year after year. Meanwhile, new shoals are formed offshore which may themselves become barrier islands some centuries hence.

BART *See* BEST AVAILABLE RETROFIT TECHNOLOGY.

Bartram, John and William (1699–1777), (1739–1823) John Bartram, explorer and naturalist, known as the father of American botany. Born March 23, 1699, near Darby, Pennsylvania, and self-educated in botany and the natural sciences, he was a founding member of the American Philosophical Society and in 1765 was appointed botanist to King George III of England. Bartram traveled extensively in the colonies, and his journals and letters influenced George Washington, Thomas Jefferson, and Benjamin Franklin. Bartram is responsible for establishing the first American botanical garden (still maintained today) in 1728 near Philadelphia and for collecting, hybridizing, and exporting plants of the New World. John Bartram's son, William, was the first native-born American naturalist and an author in his own right. Accompanying his father on his explorations into the natural world of the colonies, William wrote his influential and well-received *Travels,* an encyclopedic treatise of American flora and fauna, published in 1791. Both father and son believed in and wrote about BIOCENTRISM, a philosophy that was far ahead of its time. Their works, which articulated the moral equality and intelligence of species other than humans, influenced later writers such as HENRY DAVID THOREAU, RALPH WALDO EMERSON, and JOHN MUIR. John Bartram died September 22, 1777, in Kingsessing (now part of Philadelphia), Pennsylvania, the site of his botanical garden. The lives of the Bartrams have been examined in Ernest Earnest's *John and William Bartram* (1940) and also in Thomas P. Slaughter's *The Natures of John and William Bartram* (1997), which deals with the relationship and emotional lives of the father-son team as well as their scientific, philosophical, and literary contributions.

basalt in geology, a dark-colored, fine-textured IGNEOUS ROCK composed principally of calcic PLAGIOCLASE and PYROXENE, usually extrusive (*see* EXTRUSIVE ROCK), though it commonly forms intrusive DIKES and SILLS as well. It often contains numerous VESICLES, giving its surface the appearance of frozen froth. As it cools from the molten state, basalt usually crystallizes into large hexagonal columns 15–60 centimeters (6–24 inches) across. When these columns break at the face of a basalt outcrop they take on the appearance of a giant staircase, giving the stone the common name of traprock (German *Trappe,* stair). Basalt is sometimes present in massive amounts, as in the upper Great Lakes region and in the Columbia Basin of eastern Washington and northern Idaho. Well-known smaller outcrops include the Giants Causeway in Ireland, Devils Postpile National Monument in the eastern Sierra, and the Palisades on the Hudson River just above New York City. *See also* LAVA.

base in chemistry, originally, any compound that dissociates upon contact with water, releasing *hydroxide ions* (OH^-); defined more broadly today as *a proton acceptor,* that is, any chemical substance that accepts protons during chemical reactions (*see* ATOM: *structure of the atom*). (Because protons are essentially naked hydrogen ions—hydrogen atoms with their single electron stripped away—proton acceptors may also be referred to as *hydrogen acceptors.*) Bases show a pH greater than 7. They are often slippery or "soapy" to the touch, may be corrosive, and usually have a bitter taste. They turn pink LITMUS blue. Common bases include lye, AMMONIA, and nearly all household soaps and cleansers. *Compare* ACID. *See also* ALKALI.

baseflow in hydrology, the amount of water entering a stream from GROUNDWATER sources—therefore, the flow that remains steady from day to day (though it will vary from season to season).

base load in electrical engineering, the lowest demand consistently placed on a transmission line or electrical generator in any 24-hour period. Base load is the minimum current always flowing through a power network. *Compare* PEAK LOAD.

base/neutral extractable organic compound in ENVIRONMENTAL PROTECTION AGENCY terminology, an ORGANIC COMPOUND that may be separated from wastewater by leaching with a liquid of pH 7 or above. Base/neutral extractable organic compounds listed as PRIORITY POLLUTANTS by the Environmental Protection Agency include naphthalene, nitrobenzene, fluorene, chrysene, benzidine, and most HEXACHLORS. *Compare* ACID EXTRACTABLE ORGANIC COMPOUND.

base level in geology, the lowest elevation to which a given piece of terrain may be eroded. Base level is usually represented by the mouth of the region's principal river or stream, though in dry areas it may be the surface of a PENEPLAIN established by wind erosion and MASS WASTAGE rather than the action of water.

baseline the first east-west line of a land survey, established with unusually careful measurements because it will be the line to which all other points and lines in the survey will be referenced and from which ranges and townships will be numbered; hence, any carefully established body of data to which further measurements or observations in the same field may be compared.

basement rock in geology, the rock beneath a sedimentary series (*see* SEDIMENTARY ROCK; SERIES), representing the original surface on which the sediments were deposited. Basement rock is usually igneous or metamorphic (*see* IGNEOUS ROCK; METAMORPHIC ROCK), though it may be an older sedimentary series separated from the newer one by an UNCONFORMITY. The term is also sometimes used to refer to the rock upon which a BASALT flow or other extrusive igneous formation is resting.

base saturation in soil science, the amount of EXCHANGEABLE BASES present in a soil sample, expressed as a percentage of the soil's capacity to hold these bases. This capacity, known as the *cation exchange capacity,* or CEC, is dependent principally on the size and number of CLAY particles present. The more clay particles, and the smaller their size, the larger the total clay surface area, and thus the greater number of ADSORPTION sites, there will be. Thus, the same total amount of exchangeable bases may represent high base saturation in a soil of low clay content and low base saturation in a soil of high clay content. Since plant roots "find" exchangeable bases more easily in a soil in which the base saturation is high, the measurement can be used as a rough index of soil fertility.

basic slag (Thomas phosphate) a form of calcium phosphate, $Ca_3(PO_4)_2CaO$, produced as a slag during steel-making when heated ores containing PHOSPHORUS are slaked with LIMESTONE flux. The resulting product is between 2% and 10% phosphorus and may be used as an agricultural FERTILIZER if finely ground. Since most North American iron ores have a very low phosphate content, basic slag is not commonly produced on this continent, but is produced and used principally in Europe.

basin and range topography a type of landscape characterized by parallel lines of tilted FAULT-BLOCK MOUNTAINS, so that the land rises gently to a relatively level ridge, drops off steeply to a valley, then rises gently again to the next ridge, usually a number of miles distant. HORSTS and GRABENS are also common. Basin and range topography develops where continental plates are breaking up (*see* PLATE TECTONICS), forming RIFT VALLEYS and (eventually) oceans. Its most striking current manifestations are in East Africa and in the GREAT BASIN region of Nevada, northeastern California, and eastern Oregon.

Basel Convention a ban promulgated in 1989 by the Organization for Economic Cooperation and Development (a successor to the Organization for European Economic Cooperation) to restrict rich countries from dumping toxic materials in poor countries. By 2000, 135 countries and the European Union had agreed to the ban.

BAT *See* BEST AVAILABLE TECHNOLOGY.

Batesian mimicry in behavioral ecology, a form of MIMICRY in which a harmless and/or edible SPECIES mimics the appearance and behavior of a toxic, unpalatable or dangerous species in order to be left alone by PREDATORS. Example: the monarch butterfly, which is both toxic and unpalatable, is mimicked by the viceroy butterfly, which is neither—but is left alone by any predator with any experience with monarchs due to the almost exact match in coloration and flying styles of the two species. Batesian mimicry is named for the British naturalist Henry Walter Bates (1825–92), the first scientist to describe the phenomenon in print.

batholith in geology, a large body of INTRUSIVE ROCK of generally uniform composition, usually forming the core of MOUNTAIN SYSTEM. Batholiths are normally shield- or lens-shaped and may be anywhere from a few miles to several hundred miles across. The most common material is GRANITE, which is almost always surrounded by a band of METAMORPHIC ROCK created by the heat and pressure of the intruding batholith. Batholiths exposed by erosion of the overlying rock usually form striking scenery. Examples include the Black Hills batholith in South Dakota, the various batholiths of the Canadian Shield surrounding Lake Superior and extending north to the Arctic Ocean in eastern Canada, and the Sierra Nevada of California, a single enormous batholith 40 to 80 miles across and nearly 400 miles long. *See also* PLUTON.

bathymetric map a map showing the shape of the bed of a lake or other body of water. A bathymetric map is similar to a TOPOGRAPHIC MAP, except that the contours show depths beneath the surface of the water body rather than elevations above sea level.

beach nourishment the process by which beaches are built up through the deposition of sand eroded from upcurrent areas along a coast or upstream along a river (*see* EROSION; DEPOSITION; SHORE DRIFT). Beach nourishment is a continuous process, with the sand deposited by the nourishment process being balanced by sand removed elsewhere on the beach or at other times of the year through erosion; hence, anything that cuts off the flow of sand that is providing the nourishment will rapidly cause shrinkage, "death," and eventual disappearance of the beach. *See* GROIN; DAM.

bearing the straight-line direction from one point on the landscape to another, expressed in terms of a change from some previously established direction. In surveying, the previously established direction is always a north-south line through the observer's location, and the bearing is expressed in degrees and compass directions: N28°W (28 degrees west of north), S36°E (36 degrees east of south), etc. It is important to designate whether the bearing is a *true bearing* ("north" indicates the North Pole) or a *magnetic bearing* ("north" indicates magnetic north), as the two bearings may vary considerably (*see* MAGNETIC DECLINATION). In popular usage, the previously established direction does not need to be oriented north and south, but may be any easily established line of sight. In this case, the bearing is expressed in degrees right or left rather than in compass directions ("face the large rock and take a bearing 45 degrees left").

bearing tree (section tree) in surveying, a tree chosen as close as possible to the corner of a TOWNSHIP or SECTION and marked in some distinctive manner, usually by nailing a SURVEY PLATE to it, although if no survey plate is available the tree may simply be blazed (marked with paint, or with small sections of the bark removed in a distinctive pattern) and a complete description entered in the surveyor's notes, including the bearing and distance from the tree to the actual corner.

Beaufort scale scale of wind velocities, devised in 1806 by the Irish hydrographer and member of the British Admiralty Sir Francis Beaufort (1774–1857). Beaufort developed his scale as a means of measuring the force of winds at sea by their effect on sail canvas. It has since been modified by meteorologists to refer to wind speeds as shown by their effects on common objects and confirmed by clocking on an ANEMOMETER. (*See also* HURRICANE.) The complete scale in its current form is below.

beauty strip a strip of uncut trees along public highways in national forests where intensive logging operations are, or have been, conducted. Beauty strips are found mostly in CLEARCUT areas in western states where the Forest Service wishes to conceal areas devoid of trees from tourists.

bed (1) in hydrology, the land surface lying beneath a body of water. The bed is usually assumed to lie from the normal high-water mark down, so that it is proper, for example, to speak of the "exposed bed" of a river at low-flow stages.

(2) in geology, either (a) a single layer in a stratified rock formation (*see* STRATUM); or (b) any body of rock of uniform composition, usually considerably broader than it is deep.

(3) in mine engineering, a body of ore lying within a single FORMATION and probably formed by a single geologic event.

The Beaufort Scale

Code Number	Wind Speed (mph)	Description
0	<1	calm
1	1–3	light air
2	4–7	light breeze
3	8–12	gentle breeze
4	13–18	moderate breeze
5	19–24	fresh breeze
6	25–31	strong breeze
7	32–38	moderate gale
8	39–46	fresh gale
9	47–54	strong gale
10	55–63	whole gale
11	64–75	storm
12	<75	hurricane

bedding plane in geology, the interface between two neighboring layers in a stratified rock formation; also, any plane lying parallel to this interface.

bed load in hydrology, the sediment and other materials carried along the bed of a moving stream of current (as opposed to those suspended or dissolved in the water). *Compare* DISSOLVED LOAD; SUSPENDED LOAD.

bedrock in geology, the solid rock lying underneath a deposit of soil, sand, gravel, or other UNCONSOLIDATED MATERIALS, or exposed at the surface of the ground; also, the first layer of non-ore- or water-bearing rock beneath an ORE BODY or an AQUIFER. *Compare* BASEMENT ROCK.

beech scale created by a sucking insect (*Cryptococcus fagi*) on the bark located near the base of the American beech, a tree common in forests east of the Mississippi. The scale does not by itself kill beech trees but creates an "infection court" in weakened trees which allows a deadly fungus, *Nectria coccinea* var. *faginata,* to finish the job. The scale, a late-19th-century import from Europe, has become especially virulent since World War II, infesting beech trees broadly from Maine to Ohio and southward to southern West Virginia. *See also* FOREST DECLINE AND PATHOLOGY.

benchmark (1) in surveying, a permanent elevation marker, usually of metal, set into BEDROCK or, where bedrock is not available, attached to a large boulder or concrete block or mounted on top of a steel pipe driven into the ground. The elevation of each benchmark is carefully established and logged so that the benchmark may serve as a starting point for other elevation determinations in the same area.

(2) any carefully established data that may be relied on to compare new data to, as (for example) a complete list of species in an acre of undisturbed forest, to which the species mix in forests with various degrees of disturbance may be compared.

beneficial use in water law, a use of water that confers monetary or other definable advantages on the user. A determination of beneficial use is made by the courts in establishing or vacating a WATER RIGHT. Beneficial use is traditionally defined in broad statutory or constitutional terms as "the basis, the measure and the limit" of an appropriative water right. Actual determination of whether a particular use is beneficial or not is commonly left up to case law, although more and more beneficial uses are being explicitly written into statutes. The two principal criteria for beneficial use are (a) that the water be diverted from its natural source (though the concept of "instream beneficial use" is beginning to be recognized by many states; *see* INSTREAM USE); and (b) that the amount of water used be consistent with the amounts customarily used in the same region for the same purpose, so that, for example, if one farmer is using a water right of 20 cubic feet per second (cfs) to irrigate a certain acreage of tomatoes, a neighboring farmer cannot claim beneficial use of 40 cfs to irrigate the same acreage of the same crop (*see* DUTY OF WATER). All states recognize domestic, municipal, industrial and agricultural use as beneficial. Uses that may or may not class as beneficial, depending upon the state, include stock watering, power generation, mining, recreation, fish and wildlife, groundwater recharge, pollution control, navigation, and aesthetics.

beneficiating (preprocessing) treating a LEAN ORE by milling, crushing, SINTERING, and other consolidation processes, so that it becomes rich enough for conventional smelting processes to extract the metal from it. Beneficiating is usually done at or near the mine so as to reduce the weight that must be carried to the smelter.

benefit in resource management, an effect arising from a particular allocation of land or resources that can be shown to increase the well-being of one or more groups of users. The users need not be human; *see* WILDLIFE BENEFITS. Benefits may be either quantifiable or non-quantifiable ("tangible benefits" or "intangible benefits"); however, resource managers generally attempt to quantify as many as possible, so that they may be compared to the costs of the same action. *See* COST; BENEFIT/COST ANALYSIS.

benefit/cost analysis in economics, analysis of a proposed action to determine whether or not the total benefits exceed the total costs. If benefits exceed costs, the action is valid from an economic standpoint; if costs exceed benefits, the action is economically invalid and should not be undertaken without compelling extra-economic reasons. Federal law has required benefits to exceed costs for water projects (DAMS, LEVEES, canals, etc.) since 1936. Other federal actions also usually undergo benefit/cost analysis before they are undertaken, although it is applied less rigorously and (generally) without the force of law behind it.

METHODOLOGY

The result of benefit/cost analysis is a set of figures called the *benefit/cost ratio* or *b/c ratio,* usually expressed as a function of unity (1.6:1, 0.93:1, etc.). The b/c ratio is obtained by dividing total project benefits by total project costs over the expected life of the project. If benefits exceed costs, the value of the b/c ratio will be greater than 1 and the action may be considered valid. This seemingly straightforward process is complicated by two factors: the need to include the cost of money (i.e., interest rates), and the difficulty in determining dollar values for all costs and all benefits. The cost of money is factored in by assuming an interest rate, called the *social discount rate,* and calculating a running total for future benefits and costs using standard discount formulas: the resulting equation looks like this,

$$BCR = \frac{\sum_{t=1}^{T} [B_t/(1 + i^t]}{K + \sum_{t=1}^{T}[C_t/(1 + i)^t]}$$

where BCR = the benefit/cost ratio; T = the economic life of the project (in years); i = the social discount rate; B_t = annual benefits for year *t*; C_t = annual costs for year *t*; and K = the initial (or startup) costs. The section of the equation above the horizontal line represents the total benefits of the project, corrected for money costs; the section below the line represents the total costs, corrected in the same manner. The presence of the constant factor K in the lower term of the equation, without a corresponding constant in the upper term, means that an increase in the social discount rate will reduce the upper term faster than the lower. Thus the b/c ratio is extremely sensitive to changes in the discount rate, and an action that appears economical at a low discount rate may prove highly uneconomic if the discount rate is raised. This is the basis of commonly heard complaints that an artificially low discount rate has changed an unfavorable benefit/cost ratio (that is, a benefit/cost ratio of less than unity) to a favorable one.

The problem of determining what is a cost and what is a benefit is not quite so tractable as that of factoring in the cost of money. The difficulty lies in the fact that while some costs and benefits are *tangible*—that is, easily quantifiable in dollar terms, such as the cost of concrete or the stumpage price of timber—others are *intangible* and cannot be quantified in any simple way. Most intangible costs and benefits are related to AMENITY VALUES—the loss or gain of recreational benefits, environmental diversity, and so on. Usually an attempt is made to quantify these values by using figures such as projected VISITOR DAYS multiplied by the average amount each visitor would be likely to spend in the community, reduction in wildlife POPULATIONS multiplied by a factor representing the supposed value of each individual animal, etc. These figures by their nature are open to vastly different interpretations, leaving the value of those b/c ratio calculations that include large intangible components largely at the mercy of the proclivities of the economist doing the calculations. *See also* PROJECT ECONOMIC COSTS; INDUCED COSTS; COST-EFFECTIVENESS ANALYSIS; SECONDARY BENEFITS; ECOSYSTEM SERVICES.

Bennett, Hugh Hammond (1881–1960) A farmer's son from Wadesboro, North Carolina, Bennett graduated from the University of North Carolina in 1903 and later worked for the U.S. Department of Agriculture Bureau of Soils. After 30 years as a soil scientist, Bennett was responsible for the formation in 1935 of the Soil Conservation Service (SCS; now the NATURAL RESOURCES CONSERVATION SERVICE) through his efforts in persuading Congress for funding. He served as the director of the SCS until 1951. He also helped establish the SOIL AND WATER CONSERVATION SOCIETY, an organization dedicated to protection and wise use of soil and other natural resources. Recognized as the father of soil conservation, Bennett was author (with William R. Chapline) of *Soil Erosion: A National Menace* (1928), *Soil Conservation* (1939), and *This Land We Defend* (1942). A biography of Bennett, *Big Hugh: The Father of Soil Conservation* (1951), was written by Wellington Brink.

benthic division that portion of the aquatic HABITAT that lies on, in, or just above the bed of a body of water. The benthic division is comprised of a shoreward LITTORAL ZONE and a deep-water *profundal zone*. The division between the two zones in the oceans is generally equivalent to the edge of the CONTINENTAL SHELF. Lakes do not usually have profundals, although Lake BAIKAL (Siberia), Lake Superior, Crater Lake (Oregon), and perhaps a few other are exceptions to that rule. *See also* BENTHIC DRIFT ORGANISM; BENTHOS.

benthic drift organism any tiny bottom-dwelling aquatic organism that is neither motile (*see* MOTILITY) nor fixed in place, but is moved about by the bottom currents of the body of water it inhabits. Benthic drift organisms can be members of any of the five KINGDOMS of living things; however, most are PROTISTA. *See also* BENTHOS.

benthic zones the portion of a river or stream that is associated with the bottom, below the WATER COLUMN. *See also* RIPARIAN ZONE.

benthos a collective term designating those animals and plants that live at the bottoms of bodies of water (*see* BENTHIC DIVISION). Benthic organisms range from the very small (bacteria, one-celled algae) to the very large (carp, sturgeon, giant catfish), and include crabs, clams and mussels, lobsters and crayfish, sponges, starfish, OLIGOCHAETES, insect larvae, and the various microscopic or near-microscopic life forms known collectively as BENTHIC DRIFT ORGANISMS. The types and quantities of benthic life forms in a body of water are influenced by four principal factors: the amount of available light, the amount of DISSOLVED OXYGEN, the amount of available NUTRIENTS, and the texture of the bottom of the water body (that is, whether it is composed of rocks or SEDIMENT DEPOSITS). A fifth factor, water temperature, affects the types of SPECIES that make up the benthos but has little influence on their numbers except indirectly, through its effect on oxygen and nutrient levels. Where the bottom is rocky, SESSILE ANIMALS will tend to predominate: these are mostly "filter feeders" that pass large amounts of water through their digestive systems, filtering out and feeding on the PLANKTON and other small organisms in the water. Sediment-rich bottoms attract burrowing animals such as worms that act as DECOMPOSERS for larger animals and plants whose carcasses fall to the bottom; the sediment is usually also rich in bacteria (*see* BACTERIUM) and other MICROORGANISMS. Bottom-feeding fish and crawling animals such as crabs and starfish are found on both types of bottom; they scavenge dead organisms or act as PREDATORS on each other and on the sessile and burrowing animals. In general, the benthos is richest in shallow, calm tropical waters and poorest in disturbed environments. It is very sensitive to disruption by human activities such as dredging and the dumping of SPOIL. Their feeding habits make benthic organisms strong bioaccumulators (*see* BIOLOGICAL MAGNIFICATION), so that they tend to accumulate toxic substances rapidly in their bodies when exposed to industrial OUTFALLS, contaminated spoil, or other sources of these substances.

bentonite a colloidal clay (*see* COLLOID) with a high aluminum silicate content, formed by the decomposition of VOLCANIC ASH. It is highly hygroscopic (that is, able to take up water vapor from the air), absorbing a minimum of 10 times its own weight in water and swelling to several times its original volume in the process, a property that makes it useful as a sealant for well-drilling. It is also used as an EMULSIFYING AGENT for paints, detergents, and some pharmaceuticals; as a filler in the manufacture of paper and synthetic rubber; and as a decolorizing agent in the refining of petroleum and other oils. *See also* CLAY; DRILLING MUD.

benzene a colorless, mildly odiferous, extremely flammable liquid, chemical formula C_6H_6, found in coal tar and petroleum and first isolated by the English chemist Michael Faraday in 1825. It is only very slightly soluble in water (approximately one part per 1,500) but is an excellent SOLVENT itself for many organic and inorganic compounds, including oils, fats, resins, phosphorous, sulfur, and most of the HALIDES. It is highly volatile (*see* VOLATILITY), with explosive vapors; it is also moderately toxic, both in vaporous and liquid forms, whether ingested, inhaled, or absorbed through the skin, and has recently been demonstrated to be a CARCINOGEN. Notwithstanding these dangers, it is widely used as a laboratory and industrial solvent and in the manufacture of medicines, dyes, floor coverings, artificial leather, varnishes, and many other organic products. Its greatest importance, however, lies in the hexagonal ring structure of its molecule and in its consequent role as the basis for most of the class of chemicals known as AROMATIC COMPOUNDS. *See* BENZENE RING.

benzene ring the benzene molecule (*see* BENZENE) in its role as the structural nucleus for the class of chemical substances known as AROMATIC COMPOUNDS. The benzene ring is a closed structure consisting of six CARBON atoms arranged in a hexagon, with alternating single and double COVALENT BONDS. Each carbon atom has a HYDROGEN atom attached to it. In forming aromatic compounds other than benzene, one or more of the hydrogen atoms will be replaced by a different GROUP. The attached group may be another benzene ring, and it in turn may have other groups attached to it; in this way the more complex aromatics—PCBs, DDT, etc.—are built up.

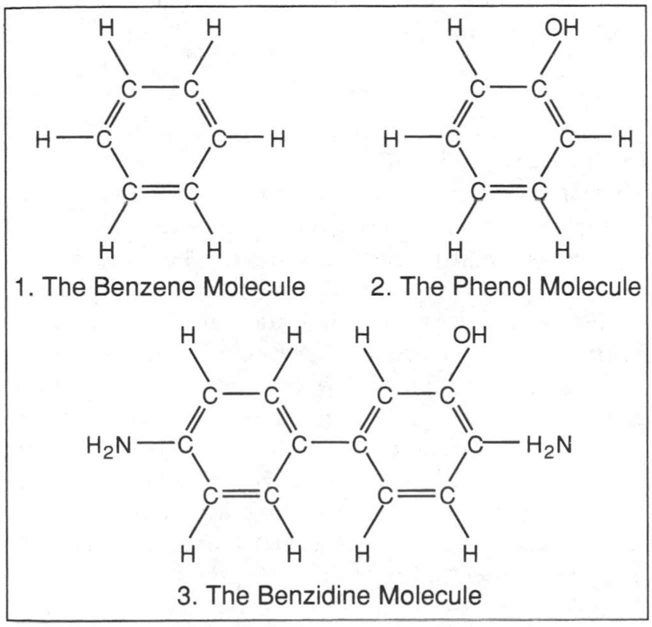

1. The Benzene Molecule

2. The Phenol Molecule

3. The Benzidine Molecule

Benzene ring

Certain groups of aromatics may be composed of the same elements in the same proportions and yet differ from one another in chemical properties. The difference lies in the placement of the various groups around the rim of the benzene ring—whether they are attached to neighboring carbon atoms, opposing carbon atoms, or something in between. For this reason, the names of the aromatic chemicals identify the points of attachment to the benzene ring as well as the elements involved and their proportions. *See* CHEMICAL NOMENCLATURE.

Bergmann's rule in biology, rule of thumb for determining the effect of climate on the body size of warm-blooded animals and birds. In its most succinct form, the rule states: the cooler the climate, the larger the animal. Bergmann's rule holds most strongly for differing RACES within a single SPECIES (the Kodiak bear, a northerly race of *Ursus horribilis,* is considerably larger than the more southerly grizzly), but it also correlates fairly well among closely related species (all races of *U. horribilis* are larger than *U. americanus,* the American black bear, which has a generally more southerly range.) The rule is named for the German zoologist Carl Bergmann, who first proposed it in a paper in 1847. *Compare* ALLEN'S RULE.

berm a long, narrow bank of earth, usually man-made, though the term *natural berm* is sometimes seen. Berms may surround ponds or waste-disposal sites, or they may run along hillsides, forming the outside edges of canals or irrigation ditches. Earth piled on a hillside to build up the outer portion of a trail tread or roadbed is also called a berm.

Berry, Thomas (1914–) a prominent theologian who writes on environmental issues as they relate to spirituality. Born in 1914 in the hills of North Carolina, Berry entered a monastery in 1934 and later earned a doctoral degree from the Catholic University of America in Western intellectual history. He continued studying science and languages, traveling to the East and immersing himself in Asian history, culture, and religion. In 1970 he founded the Riverdale Center for Religious Research in New York City and was its director until he retired in 1995 to a life of writing and lecturing. He has described the center as a place for studying the dynamics of Earth and the role fulfilled by human beings within the universe. Berry served as president of the American Teilhard Association for the Future of Man. A prolific writer, Berry has won the Lannan Literary Award for Nonfiction. His published works include *The Dream of the Earth* (1988), one of the primary influences in the growing interest in ecological theology in the United States and Europe. Later books are

Befriending the Earth: A Theology of Reconciliation Between Humans and the Earth (1991), *Buddhism* (1996), *Creative Energy: Bearing Witness for the Earth* (1996), and *The Great Work: Our Way into the Future* (1999). *See also* RELIGION AND THE ENVIRONMENT.

Berry, Wendell (1934–) American essayist, poet, and novelist. Born August 5, 1934, in Port Royal, Kentucky, Berry received his B.A. and M.A. degrees from the University of Kentucky and studied with noted writer Wallace Stegner at Stanford University. After time spent teaching at Stanford and New York University, as well as traveling to Italy, he returned to teach at the University of Kentucky and live on his family's farm. He retired from teaching in 1977 to devote his time to writing and farming. Berry's writing deals with themes of nature and the necessity of human responsibility to the land. His published works include *The Broken Ground* (1964), his first collection of poetry; *The Memory of Old Jack* (1974), a novel of the reminiscences of a 92-year-old farmer; *The Unsettling of America* (1977), an exploration of current morality and culture; *The Unforeseen Wilderness: An Essay on Kentucky's Red River George* (1971); *The Gift of Good Land* (1981); and *Another Turn of the Crank* (1995).

bessemer ore (direct-shipping ore) any iron ore rich enough to be used directly in a blast furnace without BENEFICIATING. The concentration of iron in a bessemer ore must exceed 50%. *Compare* LEAN ORE.

best available control technology (BACT) in air pollution control, the best means of reducing or eliminating airborne POLLUTANTS that is available for installation during construction of a potentially polluting industry such as a power plant or a steel mill. BACT standards are set on a case-by-case basis, and plant owners are allowed to interpret "best available" partially on economic grounds, so that the term really means "best available at reasonable cost." *Compare* BEST AVAILABLE RETROFIT TECHNOLOGY; LOWEST ACHIEVABLE EMISSION RATE. *See also* CLEAN AIR ACT.

best available retrofit technology (BART) in air pollution control, the best control technology that can be added onto a currently polluting source at reasonable cost. *See also* BEST AVAILABLE CONTROL TECHNOLOGY.

best available technology (BAT) in water pollution control, the best means currently available for reducing or eliminating the discharge of POLLUTANTS into natural waters from industrial POINT SOURCES. The standards do not apply to point-source discharge from municipal sewage plants or other publicly owned treatment works (POTWs) (*see* BEST PRACTICABLE TECHNOLOGY). The CLEAN WATER ACT requires economic considerations to be taken into account when setting BAT standards. However, the courts have ruled that these economic considerations apply to industries as a whole rather than to individual plants, so that a marginally productive plant may not skimp on pollution control simply because it is operating on a less profitable basis than other plants in the same industry. Economic standards are not considered at all when setting effluent limits for PRIORITY POLLUTANTS. *See also* BEST CONVENTIONAL WASTE TREATMENT TECHNOLOGY.

best conventional waste treatment technology (BCT) in water pollution control, a control category established by the 1977 amendments to the CLEAN WATER ACT, which is to be applied in the establishment of standards for CONVENTIONAL POLLUTANTS from industrial sources. Like the BEST AVAILABLE TECHNOLOGY standard, it does not apply to publicly owned treatment works. BCT standards are set on a plant-by-plant basis and must explicitly take into account the BENEFIT/COST RATIO of applying the controls. Note that the term "conventional," as applied here, does not mean conventional technology but conventional pollutants.

best management practice (BMP) in water pollution control, the best means available to control pollution of waterways from nonpoint sources (*see* NONPOINT POLLUTION). BMP standards for nonpoint sources are analogous to BEST AVAILABLE TECHNOLOGY STANDARDS for POINT SOURCES, and the same tests apply for determining them.

best practicable technology (BPT) in water pollution control, the best technology currently available on a practical basis for controlling POINT-SOURCE POLLUTION from industrial sources. Publicly owned treatment works are required to meet a modification of these standards that calls for the best technology practically available "over the life of the project" rather than "currently." BPT standards are less stringent than BAT standards (*see* BEST AVAILABLE TECHNOLOGY), and can take into consideration BENEFIT/COST ANALYSIS, the age and condition of the plant, and other nonenvironmental factors. They are usually set by taking an average of similar plants in the same vicinity. However, the courts and the ENVIRONMENTAL PROTECTION AGENCY have both rejected arguments by industry that "practicable" means "nothing at all" in cases where the receiving waters are already very dirty or are extremely large in relation to the size of the EFFLUENT.

Beston, Henry (1888–1968) American writer and naturalist, born June 1, 1888, in Quincy, Massachusetts, and educated at Harvard University (B.A. 1909; M.A. 1911). After serving in the American Field Service and the United States Navy during World War I, Beston returned to Massachusetts, where he began writing and doing editorial work to support himself. In 1923 he built a tiny two-room cabin just above the beach at Eastham, Cape Cod, intending to use it for weekend retreats. Instead, he moved in and lived a Thoreau-like existence for a year, observing the effects of the seasons on the Cape and its beach and keeping a journal, which became *The Outermost House* (1928), a slim, understated, perfectly observed work that remains a model for good nature writing. Later, and somewhat less successful, books include *Northern Farm* (1948) and *White Pine and Blue Water* (1950). Beston lived long enough to see his Cape Cod cabin declared a National Literary Landmark (October 11, 1964), although the cabin subsequently washed out to sea in 1978. He died at his home in Nobleboro, Maine, on April 15, 1968.

beta particle (beta ray) free electron emitted from an atom in the process of radioactive decay. *Compare* ALPHA PARTICLE; GAMMA RAY. *See also* RADIOACTIVITY; NUCLEAR ENERGY; NUCLEAR WASTE.

bgd abbreviation for *billion gallons per day.*

Bhopal, India site of the world's worst chemical disaster, which took place in 1984 when a methyl isocyanate (MIC) tank at a Union Carbide pesticide manufacturing facility developed a water leak that vaporized 30 to 40 tons of toxic gas. Some 500,000 people in a 30-square-mile area were exposed to MIC, which killed 2,500 within a week, by official count. Unofficial estimates suggested a total of at least 7,000 dead, with 200,000 suffering acute injury. The gas affects the central nervous system; injures the eyes, causing vision loss; and produces extreme difficulty in breathing and pain in the gastrointestinal tract, liver, and kidneys. Studies performed nine years after the event revealed that 50,000 people still suffered from such symptoms. Other long-term studies have shown a 17% increase in mortality. Union Carbide, wishing to limit liability, created a fund capped at $470 million, which is administered by the Indian government. Ten years after the accident, 80% of the claims involving deaths were settled, but only 30% of the injury cases were settled. About half the claims have been rejected. The Bhopal tragedy has raised significant international environmental issues relating to safety regulations that might be required of corporations in their home country but not in the countries where manufacturing facilities are located, often in developing nations. Such matters are central to the concerns about international treaties such as the NORTH AMERICAN FREE TRADE AGREEMENT and the efforts by the WORLD TRADE ORGANIZATION to develop international protocols on tariffs and trade.

BIA *See* BUREAU OF INDIAN AFFAIRS.

biennial in botany, a plant whose life cycle is completed in two years. The first year is spent storing food in a TUBER, BULB, or similar underground structure. During the second year, the plant concentrates on reproduction, producing flowers and seeds. The intervening winter is spent dormant below the ground. Some biennials are geared to cycles of day length, others to cycles of temperature or rainfall. Most can be induced to flower in a single year if the appropriate cycle is artificially produced. *Compare* ANNUAL; PERENNIAL.

big bang theory the widely held assumption by contemporary cosmologists that at the beginning of time the universe began with an explosion of extremely dense matter and energy. The approximate date of this event is most often put at 15 billion years ago, although the estimate is under constant scientific scrutiny. The theory is supported by the discovery of astronomer Edwin Hubble in 1929 that a "red shift" in spectrographic data indicated that the distance between galaxies is constantly increasing. Hubble's Law thus became the first major finding supporting the big bang theory. More recently, the discoveries of quasars and radio background noise have also tended to support the theory. The big bang is significant to the emergence of a creation-oriented "new cosmology" embraced by contemporary theologians and philosophers which connects the human condition with dynamic, ever-evolving natural processes. *See also* RELIGION AND THE ENVIRONMENT.

Big Cut the rapid, systematic and nearly complete logging-off of the forests of the upper Midwest (Minnesota, Wisconsin, and Michigan) during the second half of the 19th century. The term has come to symbolize logging in its most environmentally destructive form.

big game animal in wildlife management, any large animal hunted by humans for sport. The term is generally understood to be limited to mammals as large as or larger than humans (deer, elk, bear, mountain lion, moose, etc.).

bilge water water trapped within the hull of a ship. It must be pumped out at intervals to keep the ship from foundering. However, since it is usually polluted by oil, fuel, human wastes, and the dregs of whatever cargo the ship may have been carrying, pumping should not be done directly into a harbor or the open ocean but should be processed through a waste-treatment facility, either on board the ship or as part of a harbor complex.

bioaccumulation *See* BIOLOGICAL MAGNIFICATION.

bioassay any test that identifies the quantity or activity level of a chemical through its effect on a living organism. *See* e.g., AMES TEST.

bioavailability the ease with which a chemical substance (ELEMENT or COMPOUND) is taken up by living ORGANISMS. The greater a substance's bioavailability, the more likely it is to enter the FOOD CHAIN if it is released into the environment. Thus, bioavailability is as important as TOXICITY in determining the effect of a POLLUTANT. Bioavailability also determines the effects of plant NUTRIENTS. Nitrogen, for example—an important nutrient—is not strongly bioavailable in its pure (atmospheric) form, but must normally be "fixed" as a NITRATE before it can be absorbed by plants. *See* NITROGEN-FIXING BACTERIA; NITROGEN: *the nitrogen cycle.*

biocentrism the view that humankind is only a part of natural systems and cannot assert a right to modify natural systems solely for its own benefit. Also called *ecocentrism,* it is the opposite of ANTHROPOCENTRISM, which asserts that nature should be dominated by humankind and was created solely for the benefit of humans. *See also* DEEP ECOLOGY; RELIGION AND THE ENVIRONMENT.

biochemical oxygen demand (BOD) sometimes called *biological oxygen demand,* the amount of DISSOLVED OXYGEN used during the decomposition of the biodegradable impurities (*see* BIODEGRADABLE SUBSTANCE) in a water sample. Most of the oxygen use measured by BOD is caused by the metabolic demands of MICROORGANISMS (*see* METABOLISM), but a small proportion (in certain samples, a large proportion) is caused by purely chemical oxidation reactions (*see* OXIDATION). Since the time required to decompose the impurities in a water sample varies with the amount of impurities present in the sample, BOD test results are not strictly comparable with each other unless the length of time the test was run is known. This time is generally indicated by a subscript (BOD_{10}=BOD over a 10-day period). Occasionally, BOD testing may extend over a very long period in order to allow the sample to reach equilibrium. The results of such long tests are called "ultimate BOD" (BOD_{ult}).

SIGNIFICANCE OF BOD

BOD is not a pollutant, but a measure of the effect of certain types of pollutants on the RECEIVING WATERS. Generally, a high BOD in a body of water or a WASTE STREAM means that high levels of NUTRIENTS are present, along with a high number of microorganisms feeding on the nutrients. These are the conditions we associate with EUTROPHICATION. BOD is thus an indicator of degraded conditions in a body of water, but it cannot tell us specifically what may be causing the degradation.

TESTING FOR BOD

BOD tests are conducted by filling at least two standard one-liter bottle ("BOD bottles") with the water to be tested. Both bottles are kept in the dark long enough for the temperature to stabilize at 20°C. Then the first one is taken out and tested for dissolved oxygen content. The second sample is tested for dissolved oxygen at the end of the BOD test period, normally five days, and the difference in the two test results is recorded as the BOD. If more than two samples are taken, the extra samples can be tested at regular intervals throughout the test period and the results plotted on a graph (the "BOD curve"). Samples in which the BOD is known or suspected to be high enough that all the oxygen will be used up before the impurities are fully decomposed are diluted with well-oxygenated distilled water before the test series begins. In this case the final results must be multiplied by the dilution factor (for example, if a set of samples is diluted by a 4:1 ratio, the BOD reading for the diluted samples must be multiplied by 4 to obtain the BOD of the undiluted original sample). Similarly, a sample that might be expected to have a high BOD but which contains no microorganisms can be "seeded" with a measured amount of another sample for which the BOD is known. The BOD of the combined sample is then adjusted by subtracting the effects of the seed BOD to obtain the BOD of the unknown sample.

INTERPRETING THE RESULTS

BOD readings are generally given in parts per million (ppm) of oxygen: 10 ppm BOD_5, for example, means that during a five-day test period 10 units of oxygen were used for every million units of water sample present. Since the total amount of oxygen soluble in cold water is only about 10 ppm, this would indicate a fairly strongly degraded stream. "Normal" stream BOD is in the neighborhood of 1–2 ppm. The BOD of untreated SEWAGE is generally around 200 ppm, which is normally reduced to between 10 and 30 ppm by a sewage treatment plant before release into the receiving waters. In measuring the BOD of an EFFLUENT, it is important to keep in mind that a large

BOD in a trickle of effluent will probably have less effect than a small BOD in a massive effluent; that is, that the total nutrient load that an effluent places in the receiving waters is more important than the ratio of nutrients to water in the effluent. For this reason, BOD figures are often converted to "pounds of BOD," which is simply the BOD in milligrams per liter multiplied by the number of liters of effluent entering the receiving waters during the test period (five days for BOD_5) and again by the weight of a liter of water (2.2 pounds). Finally, those using BOD test results should always be aware that the microorganisms responsible for most BOD are highly sensitive to toxic chemicals. This means that BOD readings in streams with high numbers of MICRO CONTAMINANTS present will usually be artificially low, as the microorganisms will be killed by the contaminants before they can decompose the biodegradable substances that accompany the contaminants. *See also* CHEMICAL OXYGEN DEMAND.

biociation in ecology, a characteristic grouping of animals and plants within a CLIMAX COMMUNITY, identified by the dominant animal species present (*see* DOMINANT SPECIES) rather than by the dominant plant species. *Compare* ASSOCIATION. *See also* BIOCIES.

biocide a chemical that kills living things. Any chemical that is toxic to any living thing is technically a biocide; however, the term is generally reserved for chemicals that kill a large number of different SPECIES, both plant and animal. *See* PESTICIDE; HERBICIDE.

biocies in ecology, a BIOCIATION within a seral community (*see* SERE) instead of a CLIMAX COMMUNITY. *Compare* ASSOCIES.

bioconcentration *See* BIOLOGICAL MAGNIFICATION.

biodegradable substance a substance that is broken down into simpler substances by biological action in the environment. A *biodegradable detergent,* for example, is one in which the detergent molecules are converted into ALCOHOLS and other simple HYDROCARBONS through the action of soil and waterborne bacteria and other commonly found MICROORGANISMS. The term should not be used for substances that spontaneously break down (*unstable substances*) nor for those which break down through the action of sunlight, wind, water, or any natural agents other than living things (GEODEGRADABLE SUBSTANCES).

biodiversity a contemporary term referring to the variety of organisms of all kinds in a given habitat. Areas of the greatest biodiversity occur in tropical zones, such as the rain forest biome of the Amazon Basin. Biodiversity implies healthy plant and animal associations resistant to exogenous stresses, such as human-caused ecological disruptions. The best-known exponent of biodiversity is E. O. WILSON. *See also* CONSERVATION BIOLOGY.

Biodiversity Convention *See* RIO EARTH SUMMIT.

bioengineering in the environmental context, a term that refers to the human manipulation of biological materials, as in plants that are engineered to be resistant to pesticides. *See also* GENETIC ENGINEERING.

biofuel liquid or gaseous fuels derived from BIOMASS that can be used directly (such as wood pellets, charcoal, and animal wastes for heat) or processed to power INTERNAL COMBUSTION ENGINES, such as GASOHOL, which is made from corn. *Compare* FOSSIL FUELS.

biogas any gas, such as METHANE, which may be derived from the metabolic activity of living organisms. *See* METABOLISM; ORGANISM.

biogeochemical cycle in ecology, the path taken by a chemical ELEMENT as it passes from inorganic matter to organic matter and back again. PHOSPHOROUS, for example, may exist in water in the form of dissolved phosphates (inorganic); the phosphates may be used as food by ALGAE, which break them down and use the phosphorus to build DNA and other complex living molecules (organic); and when the algae die, the phosphorus may be released by the decomposition of the algal tissue as phosphates once more (inorganic). In particularly short-lived species of algae, a single phosphorus atom may pass through the entire cycle, inorganic to organic and back again, a hundred times or more in a single day. *See* TURNOVER TIME.

TYPES OF CYCLES

There are two principal types of biogeochemical cycles. *Gaseous cycles* are those in which the principal inorganic reservoir of the element is the earth's atmosphere (*see* CARBON: *the carbon cycle;* NITROGEN: *the nitrogen cycle*). *Sedimentary cycles* are those in which the principal inorganic reservoirs are soil, sediment, and sedimentary rock. The two types are not entirely exclusive. Nitrogen, for example, occurs in its inorganic form primarily in the atmosphere, but also exists and is stored as NITROGEN OXIDES in soil and sediments. SULFUR, a constituent of some PROTEINS and ENZYMES, is stored primarily in soil and rock either as sulfates or as free sulfur, but also occurs in small amounts in the atmosphere as sulfur dioxide and sulfur trioxide (*see* SULFUR OXIDE). Inorganic carbon, found principally in the atmosphere as CARBON DIOXIDE, is also present in large quantities in the Earth's crust as coal, graphite, diamonds, and LIMESTONE (calcium carbonate).

COMPLEXITY OF GEOCHEMICAL CYCLES

The phosphorus cycle outlined above is actually just a small side-loop of a much more complex overall cycle, one that has several paths possible for an individual phosphorus atom at any given point. If, for instance, the alga containing the phosphorus atom does not die and decay but is eaten by a fish, the phosphorus will not return directly to the water but will become part of the fish. If the fish is then eaten by a bird, the phosphorus will leave the water altogether, to be deposited as inorganic phosphorus at some distant site on land, in the bird's excrement, in the excrement of another animal which eats the bird, or as a product of the decomposition of the bird's or other animal's body when it dies. Once deposited on the land in this manner, the phosphorus may either be taken up in organic form again by the roots of a plant, or may wash downhill through the processes of EROSION, ending up in the sediment of the sea floor, where it may either be recycled into aquatic life or compressed eventually by geologic processes into sedimentary rock, to be uplifted eons later by TECTONIC ACTIVITY, exposed by erosion and weathering, and made available for plant growth once again. For a complete outline and diagram of this process, *see* PHOSPHORUS: *the phosphorus cycle. See also* PERFECTION.

biological control in agriculture and forestry, the control of pest ORGANISMS through the use of the organisms' natural enemies or through other biological means rather than through the application of chemical HERBICIDES and PESTICIDES. Numerous methods of biological control exist. PREDATORS such as ladybugs or insect-eating birds can be imported to deal with insect outbreaks. Alternatively, PARASITES or PATHOGENS may be released in the infested area. These often have the advantage of being highly specific, attacking only a single GENUS or even a single SPECIES. Pathogens and parasites can also be used against weeds, which can also sometimes be controlled by importing an insect or other animal that feeds selectively on the weed, leaving the crop plants alone. Other means of biological control include the elimination of breeding sites (for example, by draining the stagnant ponds utilized for breeding by mosquitoes); the use of sexual attractants to lure pest insects to traps or to confuse them so that mating is interfered with; the release of sterilized males during the breeding period, so that the females they mate with will lay infertile eggs; the planting of marigolds or other insect-repelling plants among the crop plants so that insects are repelled from the crop plants, too; and so on. Biological control has been practiced at least since the time of the earliest recorded history; the ancient Chinese, for example, placed nests of predatory ants in trees to control insects. More recently, it has had a number of spectacular successes, such as the use of the Vedalia beetle to control scale insects in citrus groves; the importation of the disease myxomatosis to control rabbits in Australia; and the use of the BACILLUS Bt (*see BACILLUS THURINGIENSIS*) against the GYPSY MOTH, a pest of CONIFER forests. Biological control is generally cheap, easily applied, and effective, and usually has the advantage of being highly specific towards the TARGET SPECIES, thus avoiding harm to beneficial insects and plants. Its principal disadvantage is the need to maintain a low level of the pest organism viable in order to keep the control organism from dying out for lack of appropriate hosts or prey, making it impossible to use biological control to totally eliminate a pest population. *See also* INTEGRATED PEST MANAGEMENT.

biological digester *See* ANAEROBIC DIGESTER.

biological magnification (biomagnification, bioconcentration, bioaccumulation) process through which small concentrations of a contaminant in the environment become large concentrations in the bodies of plants and animals. Biological magnification depends on several factors, the most important of which are (1) the affinity of many contaminants for fatty tissue (*see* LIPOPHILIC SUBSTANCE); (2) the extreme longevity and chemical stability of these contaminants (*see* PERSISTENT COMPOUND); and (3) the concentrating effects of the FOOD CHAIN. One-celled aquatic organisms such as PLANKTON "sweep up" contaminants as they drift or swim through the water. Since the contaminants are not metabolized (*see* METABOLISM), they accumulate in the bodies of the organism in greater and greater amounts through time. Larger organisms eat large quantities of the smaller organisms, accumulating the smaller organisms' contaminant loads. Because these contaminants are fat-soluble, they concentrate in the larger organisms' fatty tissues, where they are not excreted, causing their amounts to build even faster. The process is very efficient. Studies of Clear Lake, California, in the mid-1960s demonstrated a 100,000-fold increase in the concentration of the pesticide DDT through biological magnification, from .02 ppm in the water to over 2,000 ppm in the fat of fish-eating birds at the top of the food chain.

biological methylation the addition of the methyl group (chemical formula CH_{3-}) to an ELEMENT or COMPOUND through biological action. The result is always an ORGANIC COMPOUND, regardless of the classification of the original substance. For example, mercury—normally a biologically inert element—may be converted through the action of ANAEROBIC BACTERIA in the bottom sediments of lakes into methyl mercury ($(CH_3)_2Hg$), which is no longer biologically inert but may be metabolized, building up in the FOOD CHAIN (*see* BIOLOGICAL MAGNIFICATION) and leading to mercury poisoning.

biological oxygen demand *See* BIOCHEMICAL OXYGEN DEMAND.

biological potential *See* BIOTIC POTENTIAL.

biological warfare the use of MICROORGANISMS to destroy people, crops, or livestock. Biological warfare was prohibited by the Geneva Convention in 1925; however, some suspect that it was used in the Gulf War (1991) by Iraq. Many fear that biological agents may also be used by state-sponsored or independent terrorists, and could not only cause a toll on human life, but also permanently disrupt life-supporting ECOSYSTEMS.

biomagnification *See* BIOLOGICAL MAGNIFICATION.

biomass the total mass of all living things in a given ENVIRONMENT or sample. The term is also used to indicate the total mass of all living members of a single GENUS or SPECIES in a given environment or sample, and is sometimes extended to include recently living matter that has not yet decayed.

biome a biological environment defined by climate rather than by geographic location. The vegetation and animal life of a given biome is similar in character anywhere in the world that the biome may be found, either because the same SPECIES have found favorable environments there or because convergent evolution has caused different species to take on similar characteristics in response to similar conditions. Four principal factors—two associated with precipitation and two with temperature—control the distribution of biomes on the land. Average annual precipitation—the total amount of rain, snow, and other forms of atmospheric water that falls during an average year—is obviously important. Less obvious, but probably more important, is the pattern of availability of that precipitation to plants and animals through the course of a year (the presence or absence of a pronounced dry season; whether or not most moisture is tied up in snow or ice for a significant part of the year, etc.). Similarly, average annual maximum and minimum temperatures are important. So, less obviously, is the range between them, with plants and animals subject to wide temperature swings requiring a whole different set of adaptations than those whose environmental temperatures remain fairly constant, regardless of whether those constant temperatures are warm or cool.

TERRESTRIAL BIOMES

There is a broad agreement among ecologists that deserts, grasslands, coniferous forests, deciduous forests, rain forests, and tundra each constitute separate biomes. Beyond these, however, opinions vary. Is tropical savanna a separate biome, or a variant of the grassland biome? Should CHAPARRAL brushfields and similar environments elsewhere on the globe be considered a separate biome, or are they part of the deciduous forest—or the coniferous forest? Is CLOUD FOREST separate from rain forest? Is the tropical deciduous forest separate from the temperate deciduous forest? So far, there is little agreement on the answers to these questions. With the exception of the CHAPARRAL BIOME, this book treats all biomes on which there is disagreement as subsets of those on which ecologists generally agree. *See* DESERT BIOME; GRASSLAND BIOME; CONIFEROUS FOREST BIOME; DECIDUOUS FOREST BIOME; RAIN FOREST BIOME; TUNDRA BIOME.

AQUATIC BIOMES

Since temperature ranges are never extreme within bodies of water, and since precipitation has no relevance to aquatic organisms as long as the body of water they are inhabiting does not dry up, climatic criteria cannot be used to separate aquatic biomes from each other. It has therefore become standard practice to refer to only two aquatic biomes, the freshwater biome and the marine biome, although some authorities subdivide the freshwater biome into *lotic* and *lentic* biomes (that is, standing water and running water) and identify as many as six separate marine biomes—the polar seas, the temperate seas, the tropical seas, the sea floor (*see* BENTHIC DIVISION), the shoreline (*see* BALANOID ZONE), and the coral reef. *See also* ECOTONE.

biomonitoring the use of living things to detect changes in environmental conditions. Coal miners, for example, once carried canaries into the mines with them; the birds are extremely sensitive to CARBON MONOXIDE poisoning, and their behavior enabled the miners to detect the presence of this deadly gas at levels far below those which would be dangerous to human health. Other examples of biomonitoring include testing water for the presence of the BACTERIUM *Escherichia coli* as an indicator of the presence of animal FECES and the use of plant INDICATOR SPECIES to determine the condition and trend of grazing land. *Compare* BIOASSAY.

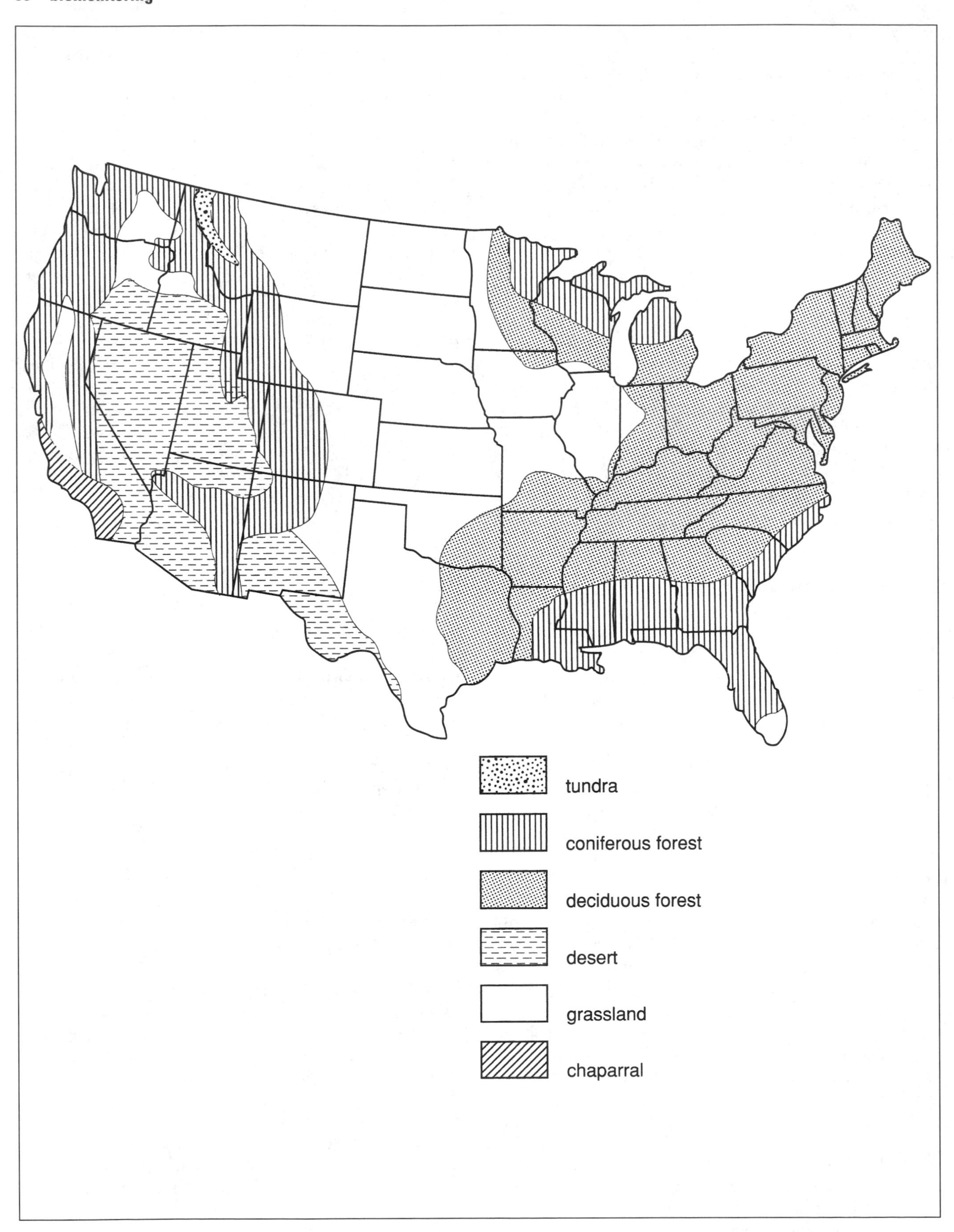

Major biomes of the United States

bioregion a geographical region whose boundaries are defined by ECOTONES or other significant changes in vegetation. A bioregion may contain several different vegetation types or even different BIOMES; however, the mix of these vegetation types or biomes will be consistent within the bioregion and will differ from the mix in neighboring areas. Bioregional boundaries are often determined by soil chemistry; for example, the Sierra Nevada and Cascade mountain ranges in the western United States form an essentially continuous chain with a nearly identical mix of TUNDRA and CONIFEROUS FOREST BIOMES but are considered separate bioregions because the largely granitic soils of the Sierra grow different SPECIES of coniferous trees and tundra plants than do the largely basaltic soils of the Cascades (*see* GRANITE; BASALT).

bioremediation the use of bacteria to clean up sewage, soils, and aquifers made unusable by oil and chemical spills and the pollution caused by mining processes. For example, bacteria that live naturally in horse manure can be isolated and when introduced into sewage sludge at high temperatures can virtually eliminate the organic material. Other bacteria species can actually live on BENZENE, requiring it as their only source of energy, thus permitting their use in treating benzene leaking from underground storage tanks. In mining, bacteria can break down cyanide, used in leaching gold from ore. *See also* BACTERIUM.

biosphere the total realm of all living things, including all plants, animals, PROTISTA, fungi (*see* FUNGUS), and MONERA. The term is not generic, but specific; that is, it is not a collection of categories, but the sum of all individual living ORGANISMS on the planet at any given moment. The biosphere is one of the four principal divisions of the natural world; others are the LITHOSPHERE (rock and soil), the HYDROSPHERE (water), and the ATMOSPHERE (air).

Biosphere II an artificial ecosystem under sealed glass established in the Arizona desert north of Tucson in 1991 that was to provide life support for four women and four men who were sealed inside for a 2-year mission. The structure, which covered 3.15 acres, contained soil, plants, and animals that in an extremely simplified way could mimic the interactions of Biosphere I—Earth—to provide oxygen, water, and food for the "Biospherians." Financed by Texas billionaire Ed Bass with an initial $30 million investment and first led by John Allen, the project intended to demonstrate how humans could live on other planets as well as provide new insights into ecological interactions on earth. It drew on an earlier Russian experiment, Bios 3, in which three scientists lived for three months inside a small, artificially lit structure, breathing air recycled by plants. Biosphere II expanded on this idea, to provide not only air but food and water as well, and to recycle wastes over a long enough period of time to suggest that the system could be self-sustaining. The plan called for the Biospherians to live in small apartments and survive on the produce of a half-acre garden and six BIOMES replicating different environments—tropical rain forest, savanna, marsh, ocean, and desert. Energy for the system was supplied by sunlight shining through the glass. During the first two-year experiment, the Biospherians encountered severe difficulties. Oxygen levels were too low, and nitrous oxide was too high. Eventually a "scrubber" to purify the air had to be installed and external oxygen supplied. Moreover, there was not enough food, for the Arizona weather during 1991–92 was exceedingly cloudy, reducing yields in the garden. Some insects needed to pollinate plant life became confused from the lack of ultraviolet rays they used for navigation, and others were weakened, so much of the plant life requisite to ecosystem balance died. Two insects that did thrive were crazy ants (*Pesatrechina longicornis*) and cockroaches. In 1994 a new crew took the place of the original eight, in the hope that they could better sustain themselves, but the publicity surrounding the project was so negative that the sponsor, Ed Bass, having invested $150 million by this time, fired the director and formed a nonprofit venture with the Lamont-Doherty Earth Observatory of Columbia University to plan and conduct a more academically-oriented program of research. Today, although the facility has become a thriving tourist attraction, complete with its Rainforest Giftshop and Cyber-Cafe, it takes a more rigorous approach to ecological research, without so much emphasis on the stunt-like aspects of the earlier program. *See also* ECOSYSTEM SERVICES.

biosphere reserves *See* INTERNATIONAL BIOSPHERE RESERVES.

biota all living organisms within a particular unit, such as a geographical region. A biota will normally encompass several COMMUNITIES and ASSOCIATIONS; however, the term is also used to describe the living portion of, for example, a soil sample of one cubic centimeter or less.

biotechnology *See* GENETIC ENGINEERING.

biotic potential (biological potential) the maximum rate at which a SPECIES may expand its numbers within the environment. The biotic potential of a given species is defined as its maximum reproductive rate minus its minimum mortality rate; that is, the situation where every fertilized ovum survives until it dies of old age. Since this presupposes a total absence of disease, PREDATORS; and accidents throughout a POPULATION, no species can ever actually reach its biotic potential. However, the closeness with which it approaches biotic potential in a given environment is a useful measure of that environment's ENVIRONMENTAL RESISTANCE. The term *biological potential* can also be used to indicate total growth possible under a given set of management options. For example, the biological potential of a STAND for timber production would be the amount of wood volume the stand could grow under optimum conditions.

biotic province in ecology, a continuous geographic area, usually quite extensive in size, within which the mix of animal SPECIES differs from that in all adjacent areas. A biotic province differs from a BIOME in being determined by geography as well as by ecological ASSOCIATION. The same association occurring in two widely separated areas would be in the same biome but in two different biotic provinces. Biotic provinces in mountainous regions often contain more than one LIFE ZONE. Some of these may contain identical species mixes to the equivalent life zones in adjacent provinces, as long as at least one of the life zones has a distinct FAUNA. *See also* BIOREGION.

birth control in humans refers to the practice of intervening in the reproductive cycle to retard or eliminate pregnancy. Birth control policies in some countries, especially developing nations beset by rampant population growth, are established for environmental, public health, and economic reasons. For women, the approaches to birth control include the use of pills, injections, intrauterine devices, the tying off of fallopian tubes, and, in a few countries, sterilization and abortion. For men, the techniques are mainly surgical vasectomies and the use of condoms. Fierce opposition to birth control policies has been mounted by religious conservatives in the United States and by the Catholic Church on an international basis regardless of environmental conditions and human suffering. Moreover, some policy analysts question the efficacy of population limitation by artificial birth control measures, suggesting that cultural and economic factors affect population growth much more than chemical, surgical, or mechanical means of birth control. Macroeconomic studies suggest that as industrial development takes place, birth rates are reduced, regardless of other factors. In mammalian species, birth control is, for the most part, automatic since most species do not have year-round estrus as humans do and populations are limited, sometimes radically, by population crashes and environmental and competitive factors. In the control of insects and other pests, scientists have created a form of birth control by sterilizing sexually active males of various species. *See also* POPULATION; POPULATION EXPLOSION; REPRODUCTIVE STRATEGY.

birth rate *See* POPULATION; POPULATION EXPLOSION.

bitumen *See* ASPHALT.

black land loss in U.S. agriculture, the downward trend of farmland ownership by African Americans, especially in the South. After the Civil War, Congressman Thaddeus Stevens, a radical Republican, proposed that all former slaves who were heads of household be granted 40 acres and $50. The measure failed, which brought on the practice of sharecropping, under which black farm workers remained constantly indebted, often through dishonest practices by white landowners. Although sharecropping effectively retained the black workforce in a kind of de facto slavery for some decades after emancipation, by the early 1900s, in the "golden age of agriculture," some black farmers were able to acquire land. By 1910 in North Carolina, South Carolina, Mississippi, Alabama, and Georgia, some 240,000 farms were in black ownership, accounting for 16.5% of all Southern landowners. Another 670,000 blacks became tenant farmers, constituting 43.6% of all Southern tenant farmers. Unlike sharecropping, tenant farming could, for some, lead eventually to land ownership. Black ownership of farmland increased until 1920, when it began to decline, owing to a cotton market crash and the general depression in agriculture which preceded the Great Depression of the 1930s. Despite New Deal efforts to have equitable federal agricultural programs, black land loss continued through the 1930s and 1940s. In the post–World War II period, 1945 to 1959, black land ownership declined by 33%, while the number of black tenant farmers, hard hit by mechanization, declined 70%. Then, according to the U.S. Civil Rights Commission, a dramatic shift in black agriculture occurred between 1959 and 1969, a period of rapid farm consolidation into larger and larger units. Where the number of white-operated commercial farms declined by only 26.3%, black commercial farm operations in the South declined by 84.1%. Looking into the reasons for the disparity in 1982, the commission and others concerned with the issue found that the U.S. Department of Agriculture (USDA), lending institutions, agricultural products corporations, local courts, white landowners, and others had been systematically depriving black farmers of the ability to buy farmland and to operate commercially. The USDA was unwilling to provide loans to small, subsistence farmers, thus eliminating

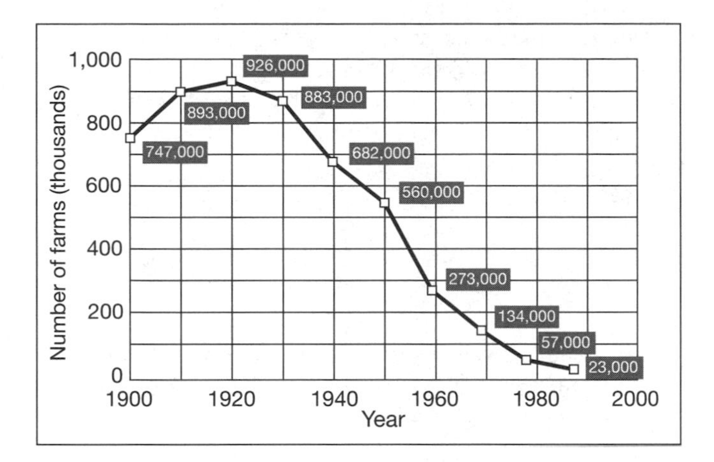

Loss of black-owned farms in the United States ***(Civil Rights Commission, USDA Census of Agriculture)***

many blacks; land owned by blacks was effectively annexed by white neighbors claiming "adverse possession" (acquiring another's land by simply moving onto it or fencing it off for a period of years and then claiming ownership); complicated tax procedures led to acquisition and resale of tax delinquent properties owned by blacks; courts worked in collusion with lawyers and white landowners to force "partition sales" wherein claims of a distant heir would require the sale of a property to honor a presumed undivided interest; banks required two or three times as much down payment for the purchase of land by blacks as they did by whites. These and other findings brought the issue of black land loss to national attention in the 1980s, but by the 1990s it had largely been forgotten. The latest available data (1987) show a total of 23,000 black-owned farms, about 2.5% of the number of farms owned by blacks in 1920.

black smoker in oceanography, a vent on the deep-sea floor that releases a large, rapidly moving plume of extremely hot (350°C) mineral-laden water. Upon contact with the cold water of the oceans, the vent water cools rapidly, releasing its minerals in the form of fine PARTICULATES, which turn the plume black. Some of the metallic minerals are deposited around the vent mouth, building a "chimney" of nearly pure metal, usually manganese or lead. Other minerals are dissolved in the water. Many oceanographers now believe that black smokers (and similar "white smokers"), by circulating ocean water through the earth's crust, are responsible for much of the sea's mineral content. *See also* PLATE TECTONICS; MID-OCEAN RIDGE.

black water the component of household SEWAGE that carries urine, fecal matter (*see* FECES), or other strongly biologically active substances. More specifically, black water is the effluent from toilets and from garbage disposals. *Compare* GRAY WATER.

BLM *See* BUREAU OF LAND MANAGEMENT.

BLM Organic Act *See* FEDERAL LAND POLICY AND MANAGEMENT ACT.

blowdown (1) in forestry, trees have been blown over by winds. Blowdown is most severe where trees are exposed to winds that they have formerly been protected from. The trees along the edge of a CLEARCUT, for example, may suffer severely from blowdown because their root systems were developed when trees in the clearcut blocked the wind from that direction and cannot hold against the stronger winds that occur after the wind-blocking trees have been removed.

(2) in engineering, cooling water or industrial PROCESS WATER removed from a system in order to drain away SALTS that would otherwise accumulate in the system. Most water carries dissolved salts. As part of the water evaporates (for example, in a COOLING TOWER), the salts become concentrated in the water that remains. Some or all of this remaining water must be regularly drained in order to prevent the salts from being deposited in the pipes, corroding them or reducing their effective diameters. Because it carries the concentrated salts, blowdown increases the salinity of the RECEIVING WATERS it enters. Hence, it must sometimes be treated to remove excess salts before release.

blowout in geology, a bowl-shaped depression in a sand DUNE created by wind action. Blowouts occur on already-established dunes, as the result of wind funneling through a small gap on the dune's surface. The increased speed of the wind beyond the funnel excavates the depression. The sand excavated from a blowout often piles up in a separate, smaller dune called a *blowout dune* just beyond. Blowouts are most common in the FOREDUNE area of a beach, but may occur elsewhere.

blue-green algae *See* CYANOPHYTE.

blue stain in logging, a fungal infection of trees that shows up as a blue color on the ends of cut logs and on lumber cut from them. The underlying wood is generally sound; hence, blue stain is not considered a DEFECT unless it is extremely far advanced.

BMP *See* BEST MANAGEMENT PRACTICE.

board foot in forestry, logging, and lumber manufacturing, a measure of wood volume equivalent to that of a board one foot long, one foot wide, and one inch thick.

BOD *See* BIOCHEMICAL OXYGEN DEMAND.

boe *See* BARREL OF OIL EQUIVALENT.

bog a type of WETLAND characterized by the presence of floating mats of vegetation, principally MOSSES, GRASSES, and some reeds. Bog water is dark in color and has a low PH due to the presence of tannic acid from decaying vegetation. The water is poorly oxygenated; there is little circulation. Bogs generally develop in cold climates on poorly drained sites, usually within old lake beds or other environmental depressions. *Compare* SWAMP; MARSH; ESTUARY. *See also* CLIMBING BOG; HANGING BOG; STRING BOG.

bog iron an impure variety of the iron ore LIMONITE, found mixed with sand and plant materials in the floors of dried-up bogs, where it has formed as a PRECIPITATE from the mineral-rich bog water. Bog iron is spongy, soft, and difficult to separate from its impurities, and therefore is of little use as commercial ore.

bolt in logging, a portion of a LOG that has been delimbed and cut ("bucked") to a length short enough for handling on a truck and in the mill.

Bonneville Power Administration *See* POWER ADMINISTRATIONS, FEDERAL.

Boone and Crockett Club outdoorsman's organization, founded in 1887 to promote the conservation and protection of North American wildlife, especially GAME ANIMALS. Most of its members are hunters interested in the preservation of BIG GAME and its environment. Address: Old Milwaukee Depot, 250 Station Drive, Missoula, MT 59801. Phone: (406) 542–1888. Website: www.boone-crockett.org.

BOR *See* BUREAU OF OUTDOOR RECREATION.

borate in chemistry, any of the numerous SALTS of BORIC ACID. Borates as a class are moderately to strongly soluble in water, forming solutions with the properties of BASES. They fuse with metals when heated, forming glassy substances with colors that are characteristic for each metal. This property is the basis of the "borax-bead test" used in chemical labs for identifying metals. The most common of the borates, *borax* (sodium borate, $Na_2B_4O_7{\cdot}10H_2O$) is found naturally in concentrated deposits on dry lake beds, especially those in volcanic regions. It is used as a water softener, as a component of electrical solder, an in the manufacture of glazes and enamels. It is only mildly toxic, has no carcinogenic or mutagenic properties (*see* CARCINOGEN; MUTAGEN), and is considered safe for household use.

borax *See under* BORATE.

boreal forest *See under* CONIFEROUS FOREST BIOME.

boric acid a weak acid obtained by dissolving boric oxide (B_2O_3) in water, forming $B(OH)_3$ (also written H_3BO_3). Deposits of it occur naturally as a result of volcanic action. Boric acid is used as a component of some antiseptics (especially in eye washes), in the manufacture of BORATES, and as a component of heat-resistant glass. Its former use as a food preservative was not appropriate, as it is fairly toxic (less than five grams can cause death in children).

boride in chemistry, a binary COMPOUND in which BORON takes the role of a nonmetal. Borides are important as components of semiconductors because of their high electrical conductivity coupled with high chemical inertness.

boron the fifth element of the ATOMIC SERIES; atomic weight 10.81, chemical symbol B. Free boron does not exist in nature; when isolated in the laboratory, it forms 12-sided crystals (dodecahedrons) that have a HARDNESS approaching that of diamond. Although not classed as a metal, boron has certain properties in common with the metals, including the ability to form ACIDS (*see* BORIC ACID). It is used as a neutron absorber in nuclear reactions and as a component of metal ALLOYS, to which it adds hardness. *See also* BORIDE; BORATE.

botanical area in FOREST SERVICE terminology, a SPECIAL INTEREST AREA set aside to protect an interesting or unique COMMUNITY, SPECIES or group of individual plants. The usual rules pertaining to special interest areas are applied, with special emphasis on keeping the plant community intact. The emphasis in a botanical area is on the recreational value of the plant life. For areas protecting the scientific values of plants, *see* RESEARCH NATURAL AREA.

bottle bill law encouraging or requiring the recycling of beverage containers, usually through the use of a small monetary deposit on each container. The deposit is redeemable when the container is returned to a store. Oregon was the first state to pass a bottle bill: its legislation, which went into effect on October 1, 1972, requires deposits on beer and soft drink containers (both cans and bottles). Other types of beverage containers are exempted. State officials report better than 90 percent compliance with the Oregon bottle bill, which has reduced roadside litter by more than 80 percent over most of the state. By 1987, nine other states (Iowa, Michigan, Massachusetts, Vermont, Maine, California, Connecticut, New York, and Delaware) had passed similar legislation.

bottom dweller an ORGANISM that lives on the bottom of a body of water. The term is usually restricted to animals larger than MICROORGANISMS—shellfish, crustaceans, OLIGOCHAETES, etc. *See also* BENTHOS; BENTHIC DRIFT ORGANISM.

bottom feeder a fish or other free-swimming aquatic animal that feeds primarily on rooted aquatic plants and bottom-dwelling animals (*see* BOTTOM DWELLER). The bottom feeder typically does not spend all its time on the bed of the body of water it inhabits but may rise off the bottom when not feeding. It may also sometimes feed at other levels within the water. Examples of bottom feeders include CARP and pike.

bottom loading the filling of a storage tank through a valve located at the bottom rather than an opening in the top. Bottom loading requires the use of DRY-DISCONNECT COUPLINGS and of a pressurized feeding system powerful enough to overcome the weight of the fluid already stored in the tank. However, it allows loading and unloading through the same valve, eliminates splashing, and greatly simplifies overfill protection, characteristics which often make it an appropriate choice for tanks storing hazardous liquids. *See* HAZARDOUS SUBSTANCE; HAZARDOUS MATERIAL; HAZARDOUS WASTE.

boulder in engineering geology, any loose ROCK larger than about 256 mm in diameter (10.5 inches). Boulders are generally rounded and abraded by weathering. At one time the term implied transport (that is, the composition of the boulder was always different from the COUNTRY ROCK on which it rested) but this distinction is no longer considered essential. *Compare* COBBLE; PEBBLE. *See also* WENTWORTH SCALE.

Boulding, Kenneth (1910–1993) a British-born and -educated economist. Boulding served as an economist with the League of Nations and taught economics at universities in Scotland, Canada, and the United States before joining the faculty of the University of Michigan, where he headed the Center for Research on Conflict Resolution from 1959 to 1966. Boulding's Quaker beliefs led him to work for international peace and explore the interconnections of natural and human systems. As president of the American Economic Association in 1968 and the AMERICAN ASSOCIATION FOR THE ADVANCEMENT OF SCIENCE in 1979, he became an international spokesman for interdisciplinary study of environmental concerns. Boulding lectured widely and was a member of the National Research Council Committee on Nuclear and Alternative Energy Systems in the 1970s. His books include *Economic Analysis* (1941), *The Image: Knowledge in Life and Society* (1956), *The Meaning of the Twentieth Century* (1964), *Peace and the War Industry* (1970), *Collected Papers* (5 volumes, 1971–75), *Stable Peace* (1978), and *Towards a New Economics: Critical Essays on Ecology, Distribution and Other Themes* (1992).

bound water *See* HYGROSCOPIC WATER.

Bowen ratio in climatology, the ratio of SENSIBLE HEAT (H) to LATENT HEAT (LE) in the earth's ATMOSPHERE. The Bowen ratio (H/LE) varies from place to place and time to time. It is lowest where EVAPORATION is actively tying up heat inputs as latent heat. Where the ratio is low (as over the oceans), temperature change is slow and the range between low and high temperatures is narrow. Where the ratio is high (as over the deserts), temperature change is fast and the low and high temperatures tend to be extreme.

Bowen series in geology, the sequence in which different minerals crystallize out of molten basaltic MAGMA as the magma cools. Also known as the *reaction series* or *Bowen's reaction series*, the Bowen series was formulated by the American geologist N. L. Bowen in 1928. It is based on reactions between the newly crystallized minerals and the remaining molten mass of

Discontinuous Series	Continuous Series	Rock Type	Temperature
olivine pyroxine	calcic plagioclass ↓	basalt	hot ↓
amphibole biotite	↓ sodic plagioclass	diorite	↓
potassium feldspar muscovite quartz		rhyolite	↓ cool

The Bowen series

the magma: OLIVINE, for example, reacts with the remaining magma to form PYROXENE, which in turn reacts with the remaining magma to form AMPHIBOLE, and so on. At high temperatures there are two simultaneous series, the *discontinuous series* (in which the different minerals have distinct characteristics) and the *continuous series* (in which a single mineral, plagioclase, simply becomes higher and higher in sodium content and lower and lower in calcium content). At lower temperatures, the two series fuse into one. The Bowen series is responsible for the differentiation of basaltic magmas into BASALT, ANDESITE, and RHYOLITE ROCK.

BP abbreviation for "before present." The term is used by geologists and others to date events by counting backward from today rather than forward from the birth of the Earth. For example, 12,000 BP means 12,000 years ago. *See also* ABSOLUTE AGE; GEOLOGIC TIME.

BPA short for Bonneville Power Administration. *See* POWER ADMINISTRATIONS, FEDERAL.

bract in botany, a modified leaf found at the base of a FLOWER or flower head. Bracts are usually small and scale-like: they may, however, be large and brightly colored, as in the poinsettia, whose "petals" are actually bracts. They lie outside of the calyx (*see* FLOWER: *parts of a flower*) and are not technically considered part of the flower. *Compare* SEPAL.

braided channel in geology, a stream channel in which sediment—usually deposited by the stream itself—has created a series of bars and islands through which the stream passes in a number of dividing, reconnecting, and interweaving channels that appear from above like the strands of a braid. Braided channels are characteristic of glacial outwash plains (*see* GLACIAL LANDFORMS) and other areas of recently deposited ALLUVIUM.

British thermal unit (BTU) unit of measurement used to express heat output. The BTU is defined as the amount of energy required to raise one pound of water one degree Fahrenheit. It is equivalent to 252 CALORIES or 778 FOOT-POUNDS. 3413 BTU are equivalent to one KILOWATT-HOUR.

broadcast burn a CONTROLLED BURN of logging SLASH in which the slash has not been piled but lies scattered over the ground. Broadcast burning requires little or no handling of the slash, and returns NUTRIENTS to the soil over a broad area rather than in a concentrated dose. However, like all controlled burning, it should properly be done under prescribed and rigorously supervised conditions.

broad-spectrum pesticide a PESTICIDE such as DDT, which kills a wide variety of living ORGANISMS. The "shotgun approach" that the broad-spectrum pesticides represent is now discredited as a pest-management tool due to the indiscriminate killing of NONTARGET SPECIES (including bees, ladybugs and other beneficial insects) following their use. *See also* BIOCIDE.

bronchitis a disease of the lungs characterized by an inflammation of the bronchial tubes (the passageways that carry air through the lung tissue). This inflammation causes chest pain and severe coughing, usually accompanied by mucus secretions. *Acute bronchitis* is caused by a VIRUS. It is usually accompanied by fever and is often followed by bacterial complications (*see* BACTERIUM) such as pneumonia. *Chronic bronchitis* is the result of permanent damage to the bronchial tubes, either from acute bronchitis or from the inhalation of irritants such as chemical fumes or cigarette smoke. It is a debilitating, usually progressive, and often fatal disease. *Compare* ASTHMA.

Brower, David R. (1912–2000) American environmental activist, born on July 1, 1912, in Berkeley, California, and educated in the Berkeley public schools and (briefly) at the University of California. An accomplished mountain climber, Brower made a number of first ascents during the 1930s and 1940s—including New Mexico's Ship Rock (1939)—and served as an instructor in the U.S. Army's 10th Mountain Division in Austria during World War II (1943–1945). He also became a good amateur lepidopterist, discovering a SPECIES of butterfly in California that is now named for him. In 1941 he joined the editorial staff of the University of California Press in Berkeley: he returned to that post following World War II and held it until 1952, when he was named executive director of the SIERRA CLUB. His aggressive handling of that role resulted in the club's loss of tax-exempt status, and the resulting furor forced the organization to fire him in 1969. Brower resigned from the club board in May 2000—the third time in 15 years he had done so—to underscore his contention that leaders were not doing enough to save Earth. He then founded two new environmental organizations, the JOHN MUIR INSTITUTE FOR ENVIRONMENTAL STUDIES and the activist FRIENDS OF THE EARTH, which he served as president until 1979. He also helped found or served on the boards of numerous other organizations, including the LEAGUE OF CONSERVATION VOTERS, the EARTH ISLAND INSTITUTE, and the RACHEL CARSON TRUST FOR THE LIVING ENVIRONMENT. Brower was known as a no-holds-barred campaigner whose passion for environmental causes occasionally overruled his judgment, but who was nevertheless highly effective. A personable man, he was well-liked even by his adversaries. Brower was the winner of the 1998 Blue Planet Prize, an award of 50 million yen ($427,600) given by the Asahi Glass Foundation of Japan for significant contributions in solving environmental concerns. He was nominated three times for the Nobel Peace Prize (1978, 1979, and 1998), jointly with professor Paul Ehrlich. The best biography of him is probably John McPhee's *Encounters with the Archdruid* (Farrar, Straus and Giroux, 1971). A prolific writer himself, Brower's works include *Only a Little Planet* (1975); *Headlands* (1976); *Not Man Apart* (1976); *Work in Progress* (1991); *Let the Mountains Talks, Let the Rivers Run: A Call to Those Who Would Save the Earth* (1995); and *For Earth's Sake: The Life and Times of David Brower* (1990), an autobiography. He passed away on November 5, 2000, of cancer at his home in Berkeley, California.

Brown, Lester (1934–) founder and president of WORLDWATCH INSTITUTE, an influential think tank committed to the study of global environmental issues. Brown received a B.S. degree in agricultural science from Rutgers University, an M.S. in agricultural economics from the University of Maryland, and a master's degree in public administration from Harvard University. Before founding Worldwatch in 1974, Brown served as an international agricultural analyst with the U.S. Department of Agriculture and as an adviser to the secretary of agriculture. He is principal author and editor of many Worldwatch publications, including a bimonthly magazine, a series of research papers, and two annual books, *State of the World* (translated into 30 languages) and *Vital Signs*. Brown has received many honorary degrees and awards, including the 1987 United Nations Environment Prize, the 1989 Gold Medal from the World Wide Fund for Nature, and the Blue Planet Prize in 1994 from the Asahi Glass Foundation of Japan.

brownfields land in urban or industrial areas that has been abandoned and where warehouses, factories, and the like stand empty. Brownfield sites can often be toxic and filled with garbage, useless scrap, and other refuse; hence they are not attractive to corporations for redevelopment because of the heavy cost of clean-up as well as for dismantling old structures. Governmental incentives, together with anti-sprawl zoning laws that encourage redevelopment of close-in areas, can result in adaptive reuse of such lands, however. In Portland, Oregon, for example, a brownfield area formerly used by heavy industry was purchased under a public-private partnership scheme, cleaned up, and redeveloped as a mixed-use commercial enterprise called RiverPlace. *See also* URBAN GROWTH BOUNDARY; URBAN RENEWAL; URBAN SPRAWL.

browse any vegetal matter used as food by browsing animals (animals that eat any available palatable material that happens to be within reach rather than looking for specific foods). Deer, sheep, cattle, horses, and other UNGULATES are the principal browsing animals, though squirrels, bears, and even humans will browse occasionally. *See also* BROWSE HEIGHT; BROWSE LINE.

browse height the highest elevation above the ground that a browsing animal can reach to obtain food. Because different species have different browse heights (elk, for example, can reach further up trees than can deer), determination of browse height in a browsed patch is a fairly reliable guide to the type of animal that has been browsing there.

browse line an even, cropped line at the bottom of the CANOPY of a tree or grove of trees, below which all available foliage has been eaten. Because leaves and twigs are normally the last available BROWSE to be eaten (after GRASSES, HERBS, and FORBS), a well-developed browse line generally indicates poor or overgrazed range. *See also* BROWSE HEIGHT.

brush woody SHRUBS or shrublike trees, usually occurring in large masses known as BRUSHFIELDS. Brush has no commercial wood value and little or no grazing value, and is a fire hazard and a nuisance to hikers, horseback riders, and other recreationalists. It is generally disliked by foresters because it often occupies land that might otherwise be used for timber production. However, it provides excellent COVER for small animals and birds, and normally has an extremely well-developed root system that is valuable for holding back SHEETFLOW and reducing erosion after heavy rains. *See also* CHAPARRAL; BRUSH CONVERSION; BRUSHWORK.

brush conversion the process of transforming brushfields (*see* BRUSH) into timberland. Brush conversion involves three steps: removing the brush; planting the trees; and tending the plantation in order to prevent the regrowth of brush before the trees are fully established. The term may also be used to refer to the conversion of brushfields to grasslands for grazing purposes or increased water production.

brushwork the cutting of BRUSH, small tree limbs, etc. from the route of a trail or road so that it is possible for a human or a vehicle to pass along it easily. Doing brushwork is known as *brushing* the trail or road. *Compare* TREADWORK.

bryophyte in botany, any plant belonging to the DIVISION Bryophyta, which includes the liverworts, hornworts, and MOSSES. Bryophytes lack vascular tissue (that is, fluid-conducting veins) and thus cannot move fluids about through their bodies efficiently. This limits their maximum size to about 30 centimeters (12 inches) long, with most being considerably smaller than this. With no vascular tissue, they cannot have true leaves, stems or roots, though most differentiate into leaf- and stem-like structures, probably to gain an advantage in PHOTOSYNTHESIS. Many—especially the mosses—also develop rootlike organs known as *rhizoids,* which serve to anchor them in place. The rhizoids differ from true roots in that they play no part in obtaining water and NUTRIENTS, which are absorbed uniformly through the rest of the plant. Like the VASCULAR PLANTS, the bryophytes reproduce in the pattern known as ALTERNATION of *generations,* and (also like the vascular plants) one of these generations is normally much reduced and is dependent upon the other in what amounts to a parasitic fashion. However, the independent generation in the bryophytes is haploid and the dependent, diploid—the reverse of the vascular reproductive pattern. Because they require moist conditions, bryophytes are somewhat limited in distribution, lying mostly near bodies of water, SPRINGS, or SEEPS, or growing in areas that are often bathed in mists, such as CLOUD FORESTS. Most of the approximately 20,000 known species are tropical; however, some have adapted to cold climates and may be found growing luxuriantly wherever moisture conditions allow in the high mountains and the far north. They are more sensitive than vascular plants are to AIR POLLUTION, especially to the SULFUR OXIDES, and are useful as INDICATOR SPECIES for that reason.

Bt *See* *BACILUS THURINGIENSIS.*

btu *See* BRITISH THERMAL UNIT.

bubble in air-pollution regulation, the treatment of an entire industrial plant or complex of plants as a single POINT SOURCE for regulatory purposes, as if all of its separate stacks, vents and other sources of POLLUTANTS were enclosed in a giant plastic bubble with only one opening from which all emissions emerged. The point of the bubble concept is to allow plant managers to meet overall emissions standards with a choice of alternative strategies that may include decreasing emissions more from some stacks (or entire factories) than from others, rather than forcing them to meet specific emissions goals on each stack. This permits those processes from which emissions are technologically and/or economically easier to control to be reduced below the overall standard, offsetting the application of more costly and difficult controls on other processes, which are then allowed to relax to levels greater than the imposed standards. Variations on the bubble concept include the "new source-old source bubble," which allows firms to install less than state-of-the-art pollution-control equipment in new plants provided that retrofitting of old plants in the same AIRSHED reduce their emissions enough to offset the emissions from the new plant, thus avoiding any increase in total emissions; and the "mobile bubble," which averages the total amount of emissions from all mobile sources (cars, trucks, busses, etc.) in an airshed or other region and allows some vehicles to emit more pollutants than others as long as the total pollutant load meets the bubble standard. Applied to fleets of new cars produced by a single auto maker, the mobile bubble would allow some models to exceed pollution standards as long as other models in the same fleet emitted low enough levels of pollutants to offset the excess. The bubble concept is generally supported by environmentalists as a means of allowing flexibility in meeting pollutant-reduction goals. However, new source-old source and mobile bubbles have been rejected by most environmental groups because they do not reduce pollutant loads but merely maintain them at current, generally unacceptable, levels.

bubble tower *See under* DISTILLATION.

bucking in logging, sawing the trunk of a felled tree into manageable lengths for YARDING and hauling. Bucking is usually done simultaneously with LIMBING. *See also* BOLT.

Buffalo Commons a term describing portions of shortgrass prairie west of the 98th meridian in the Great Plains so beset by drought, erosion, economic depression, declining population, and dependency on farm subsidies that they could be returned to pre-settlement use as a range for buffalo. The notion was presented in a 1987 scholarly article as part prediction, part proposal by Professors Frank and Deborah Popper, a husband and wife who are geographers and land-use planners at Rutgers University and the City University of New York, respectively. In a study of 436 shortgrass prairie counties in Montana, North Dakota, South Dakota, Wyoming, Nebraska, Kansas, Colorado, Oklahoma, Texas, and New Mexico—a sixth of the U.S. land area, and a quarter of the Great Plains, but with only 6.5 million people living in it—the Poppers identified some 110 sparsely populated areas (totaling less than half a million population in 139,000 square miles) that could be included as core areas for such a commons. Here, they suggested, policies should encourage the return of the buffalo and a stable ecosystem, rather than the hopeless efforts to support a failing agriculture and a declining water table. Roundly excoriated by Great Plains politicians, ranchers, and business executives, the Poppers nevertheless took a charts-and-graphs presentation of their idea on the road at the invitation of colleges, universities, and civic organizations in their study area. The story of their adventures is told in a popular book by journalist Anne Matthews (*Where the Buffalo Roam,* 1992), describing death threats and police escorts for the two academics. By the end of the 1990s, however, the Poppers' prediction/proposal was gaining traction not only among geographers and environmentalists, but also with some of the very people who had flung epithets earlier in the decade. An editorial in the Fargo (North Dakota) *Forum* noted in 1999 that "The creators of the Buffalo Commons metaphor, Frank and Deborah Popper, received a polite reception last week in Fargo-Moorhead. The overt anger that greeted them a decade ago was mostly gone, probably because thinking people realized the Poppers were right about the history, heritage and future of the Great Plains." Today the buffalo are coming back to many areas of the Poppers' "commons." The buffa lo market, which few had heard of in 1987, has come into its own with increasing numbers of ranchers, Indian tribes, and investors replacing cattle with buffalo herds.

buffering capacity a measure of the ability to maintain a constant pH by neutralizing excess ACIDS or BASES. In environmental literature, the term is most often used to refer to the ability of a body of water to withstand the effects of ACID RAIN. This ability depends on the alkalinity of the water, which is primarily controlled by two factors: the composition of the BED and the depth and composition of the soil in the surrounding WATERSHED. As a rule, the higher the pH of the bed and watershed soil, the greater the buffering capacity. Small differences can have great effects: studies by the Freshwater Institute of Canada have shown that the ability of a lake to neutralize acid rain can vary by a factor of at least 10,000. The best buffering is shown by lakes developed in LIMESTONE watersheds; the worst, by those developed on GRANITE, especially where the lake is young enough that no sediments have built up in its bed. *See also* ACID RAIN: *effects of acid rain.*

buffer zone in land-use planning or forestry, an area of restricted management activity surrounding an environmentally sensitive core. The purpose of the buffer zone is to protect the environmentally sensitive core from the impacts of the management activity. The core may be a single point location such as a spotted owl nest (*see* SPOTTED OWL MANAGEMENT AREA), or a larger area such as a pond, a meadow, or even an entire WILDERNESS AREA or NATIONAL PARK. The amount of management activity allowed in the buffer varies according to the sensitivity of the core area to be protected, and may or may not include such activities as salvage logging, home building, or road construction. A long, narrow buffer zone along a linear landscape feature such as a stream, trail, or highway is often referred to as a *buffer strip.*

built environment term referring to areas characterized by buildings and public infrastructure as opposed to natural environments such as countryside and wilderness areas.

bulb in botany, an organ for the underground storage of food utilized by some PERENNIALS, primarily of the lily family. It may also serve as a source of vegetative reproduction (*see* ASEXUAL REPRODUCTION). A bulb is essentially a modified leaf bud. It normally consists of a short, conical, downward-pointing stem from which arise several layers of fleshy, tightly packed leaves. Thinner, scale-like leaves form a tight coating over the outside. Adventitious roots (*see* ADVENTITIOUS GROWTH) sprout from the bottom. *Compare* CORM; TUBER.

bulk carrier any cargo carrier (railroad car, ship, etc.) designed to carry GRAIN, SAND, COAL, ORE, or similar material in a loose state rather than confined to containers. Material carried in this manner is known as *bulk cargo.*

bulkhead line in land-use planning, a mapped line marking the legal outer limit of DREDGE-AND-FILL or other development operations in a bay or ESTUARY. Usually (though not always) the bulkhead line is defined as the mean high water mark.

bulk storage storage in large containers of materials that will be dispersed and used in small containers. A gasoline station is one familiar example. The gasoline in its storage tanks is not for use on site, but is parceled out in small amounts for use in such things a automobiles and motorcycles and lawnmowers. Gasoline stations in turn are served by larger bulk-storage facilities known as TANK FARMS. Other than gasoline, materials commonly held in bulk-storage facilities include agricultural and industrial chemicals, propane, crude oil, and water.

bull-of-the-woods (lead feller, bullbuck) in logging, the supervisor of a crew of timber fellers (*see* FELLING). The term is a holdover from the 19th century, when the lead feller was usually the strongest and most aggressive man on the team; however, it is still in use today, even though the position of bull-of-the-woods (or just "bull") is now normally assigned by the front office.

bunker oil *See under* FUEL OIL.

BuRec *See* BUREAU OF RECLAMATION.

Bureau of Biological Survey federal wildlife-management and biological research agency, in existence from 1885 to 1940. Beginning as a department within the Division of Entomology of the Department of Agriculture (*see* AGRICULTURE, DEPARTMENT OF), the bureau became an independent division of the department in 1905 and was moved to the Department of the Interior in 1939, where it was consolidated with the BUREAU OF FISHERIES a year later to form the U.S. Fish and Wildlife Service (*see* FISH AND WILDLIFE SERVICE: *History*). It was while serving as chief of the Bureau of Biological Survey that C. Hart Merriam developed the theory of plant and animal distribution by LIFE ZONES.

Bureau of Commercial Fisheries a defunct branch of the U.S. FISH AND WILDLIFE SERVICE, in existence from 1959 to 1970. The Bureau regulated the biological aspects of commercial fishing operations, including catch limits, population rehabilitation, seasons, and area closures. It also was charged with the management of marine mammals, including whales, seals, and sea lions. It was phased out with the creation of the NATIONAL OCEANIC AND ATMOSPHERIC ADMINISTRATION, which took over its principal duties.

Bureau of Fisheries obsolete federal agency, formed in 1871 as the United States Fish Commission and renamed the Bureau of Fisheries when it became a part of the Department of Commerce and Labor in 1903. Transferred to the Department of the Interior in 1939 (*see* INTERIOR, DEPARTMENT OF), it was merged with the BUREAU OF BIOLOGICAL SURVEY a year later to become the U.S. Fish and Wildlife Service (*see* FISH AND WILDLIFE SERVICE: *History*).

Bureau of Indian Affairs (BIA) agency of the U.S. Department of the Interior (*see* INTERIOR, DEPARTMENT OF) charged with overseeing the relationship of the federal government with the various Indian Nations and the native peoples of Alaska. From an environmental standpoint, the most important role of the BIA is the advice and assistance it gives to native peoples for the development of natural resources on their lands. Created as a branch of the War Department in 1824, the BIA became part of the Department of the Interior when that department was created in 1849. It currently administers, or assists with the administration of, some 56 million acres of land in behalf of American Indian tribes. Indian lands for which the bureau has responsibility include approximately 0.6 million acres owned outright by the federal government; 10.1 million acres held in trust by the federal government for individual Indians; and 45.3 million acres held in trust for distinct Indian tribes.

Bureau of Land Management (BLM) agency of the U.S. Department of the Interior (*see* INTERIOR, DEPARTMENT OF) charged with managing the PUBLIC DOMAIN lands of the United States. BLM lands are generally desert and rangeland, although some are mountainous and/or heavily timbered. Nearly all of these lands are west of the 100th meridian, in Alaska, Arizona, California, Colorado, Idaho, Montana, New Mexico, Nevada, Oregon, Utah, and Washington. However, scattered small parcels exist throughout the nation. Most BLM lands have always been under federal ownership, though some are revested land grants (*see,* e.g., OREGON & CALIFORNIA LANDS) and some have been purchased or traded for by the agency. In addition to the stewardship of its own lands, the BLM is charged with management of subsurface mineral rights on all federal lands (including NATIONAL FORESTS and Wildlife Refuges; *see* FOREST SERVICE; FISH AND WILDLIFE SERVICE), with surveying all federal lands (but *see also* GEOLOGICAL SURVEY), and with granting rights-of-way for roads and utilities across most federal lands. It also manages the nation's wild horse herds under the terms of the Wild Free-Roaming Horse and Burro Act of 1971.

STRUCTURE AND FUNCTION

The BLM is headed by a national director in Washington, D.C., who reports to the assistant secretary of the interior for land and water resources. Beneath the national director are 12 state directors (some of whom serve more than one state) and numerous district managers, who direct the agency's field offices in the *districts,* typically consisting of two to four counties, although they may be much larger (the entire State of Washington is a single district). The Districts may be broken down further into *areas:* if this is done (it is by no means universal) each area will have a separate area manager. Districts and areas are often further subdivided into *range management units* (for rangeland) and *sustained yield units* (for timberland). These serve as the basic units for management planning. In addition to this hierarchic structure, the bureau maintains a Service Center in Denver, which provides support functions to the entire agency and to its minerals, surveying, and other functions on non-agency lands.

HISTORY

The Bureau of Land Management was formed by President Harry Truman on July 16, 1946, by consolidating the GRAZING SERVICE and the GENERAL LAND

OFFICE. It was reorganized and streamlined by the FEDERAL LAND POLICY AND MANAGEMENT ACT of 1976 (sometimes called the "BLM Organic Act"). As the nation's largest land manager—responsible for 264 million acres—the bureau is responsible for one-eighth of the land of the United States. It is charged with subsurface mineral resources management duties for about 300 million additional acres and wildlife suppression and management duties on 388 million acres. Mandated to manage its lands for MULTIPLE USE and SUSTAINED YIELD of RENEWABLE RESOURCES, it is directed to create special management plans for environmentally sensitive lands (*see*, e.g., AREA OF CRITICAL ENVIRONMENTAL CONCERN).

Bureau of Mines agency within the U.S. Department of the Interior (*see* INTERIOR, DEPARTMENT OF THE) created by Congress on July 1, 1910 and charged with conducting research and compiling statistics on all aspects of mineral production, including mineral location and production surveys and working surveys of active mines; improved methods for extracting minerals from ORES; mine safety and health; and methods of protecting environmental values during mining operations. The bureau was closed September 30, 1996, due to a reorganization of the Department of the Interior. Duties of the bureau were reassigned to several other government agencies.

Bureau of Oceans and International Environmental and Scientific Affairs *See under* STATE, DEPARTMENT OF.

Bureau of Outdoor Recreation defunct federal agency, established by Congress in 1963 as a bureau of the Department of the Interior (*see* INTERIOR, DEPARTMENT OF) to coordinate and promote federal activities relating to outdoor recreation. It was replaced by the HERITAGE CONSERVATION AND RECREATION SERVICE in 1978, which was abolished February 19, 1981, and its duties transferred to the NATIONAL PARK SERVICE.

Bureau of Reclamation **(BuRec, BR)** agency within the U.S. Department of Interior (*see* INTERIOR, DEPARTMENT OF) charged with the development of IRRIGATION works, including DAMS, RESERVOIRS, and distribution systems, for the reclamation of arid lands. Most of the BuRec's larger dams are multipurpose projects, with flood control, navigation, recreation, fish and wildlife, and power generation benefits included in the BENEFIT/COST ANALYSIS for the project. The smaller projects, especially those built under the terms of the Small Reclamations Project Act of 1956, are turned over to state or local agencies or to private cooperatives to operate; however, the bureau operates about 50 larger facilities itself. All projects, including the smaller ones, are inspected from time to time, and the bureau retains the responsibility of making certain that they remain safe. Federal funds are used to build the projects, but about 85% of the costs, termed "reimbursable expenditure," are returned to the treasury by the project beneficiaries through the payment of fees established by negotiated contracts. The agency also carries out studies in the water resources field, including watershed planning and the location of new sources of freshwater, and is a major supplier of water-oriented recreation.

HISTORY

The Bureau of Reclamation was created as the Reclamation Service by President Theodore Roosevelt in July 1902, under the authorization of the Reclamation Act of 1902. Originally part of the GEOLOGICAL SURVEY, the Reclamation Service became a separate agency in 1907 and was renamed the Bureau of Reclamation in 1923. The name was changed to the WATER AND POWER RESOURCE SERVICE in 1978, but the new name was abandoned three years later. Environmentalists in the past have objected to its structural approach to water management: this approach, under Commissioner Floyd Dominy, almost succeeded in damming the Grand Canyon in Grand Canyon National Park during the 1960s. However, recently the BuRec has begun to stress conservation in its project designs and has partially succeeded in shifting its emphasis from large MAINSTEM storage projects to smaller, more efficient and more environmentally sound projects on the headwaters of streams. *See also* DAM.

bushel in logging or forestry, a measure of wood volume equivalent to 1,000 BOARD FEET.

bushel rate **(busheling)** in logging, the practice of paying logging crews by the wood volume they produce each day, measured either in BOARD FEET or in BUSHELS, rather than by the time they spend at work. It is a practice widely used by the timber industry. Because of its emphasis on volume produced per unit of time, bushel-rate payment encourages speed over care, and can be both unsafe and environmentally hazardous without very strict oversight.

butt swell the flared-out base of a tree trunk just above the ground. Butt swell is common to all trees, and is really the beginning of the root system, which will continue and increase below the ground the spread represented by the butt swell above the ground. Because the grain in the butt swell is usually heavily contorted—tending to bind saws—and because it often contains large amounts of PITCH, loggers prefer to cut a tree above the butt swell. If their contract requires them to leave short stumps (so that the felling cut must come in the butt swell), the butt swell will usually be bucked off (*see* BUCKING) and left behind as SLASH.

byproduct (by-product) anything obtained from the regular operation of a process that is not the directly desired result of that process. "Process" here is a very general term, ranging from the metabolic functions of a CELL (*see* METABOLISM) to a large industrial operation. Byproducts are not due to malfunctions of these operations, but arise from the nature of the operations themselves, as, for example, smoke from the burning of wood to obtain heat in a stove or heat from the burning of wood to obtain smoke in a smokehouse. If they are thrown away, byproducts may also be called *waste products*. An efficient process, however, minimizes the production of byproducts or produces byproducts that are useful in other contexts as much as possible.

C

Cabinet Committee on the Environment short-lived predecessor of the COUNCIL ON ENVIRONMENTAL QUALITY, in existence from March 5 to July 1, 1970.

cable yarding (cable logging) in logging, any system of YARDING in which the logs are brought to a LANDING by pulling them with a steel cable attached to a fixed power winch rather than being dragged behind a tractor, team of horses, etc.

cadastral survey (land survey) a survey made to determine property lines, land ownership patterns, political boundaries, and the location of existing buildings, highways, and other engineering works. *Compare* TOPOGRAPHIC SURVEY; ENGINEERING SURVEY; CONSTRUCTION SURVEY.

cadmium the 48th element in the ATOMIC SERIES; atomic weight 112.4, chemical symbol Cd. Cadmium is a soft, lustrous, silvery metal, closely allied with MERCURY and ZINC in chemical properties. Its principal natural occurrence is as an impurity in zinc ores, from which it is obtained commercially as a vapor boiled off during the zinc smelting process. Cadmium is used in various metal ALLOYS to make the metal more malleable; it is also an ingredient of solder. Its electrolytic properties make it valuable in plating metals, in storage batteries, and in lamp elements. A PRIORITY POLLUTANT, it is highly toxic in small quantities, either inhaled (cough, headaches, chest pain, vomiting, extreme irritability) or ingested (increased salivation, choking, vomiting, abdominal pain, anemia, diarrhea). It is the probable cause of ITAI-ITAI DISEASE.

CAFE standards *See* CORPORATE AVERAGE FUEL EFFICIENCY STANDARDS.

calcic rock a rock containing a high proportion of calcium, usually in the form of calcium carbonate ($CaCO_3$). The most common variety of calcic rock is LIMESTONE. Calcic rocks are basic (that is, have a low pH: *see* BASE): they dissolve relatively easily in weak acids, neutralizing the acids in the process. A bed made of calcic rocks will help buffer a body of water against ACID RAIN. *See also* KARST TOPOGRAPHY.

caldera in geology, a large basinlike depression in the earth caused by volcanic action. A caldera differs from a CRATER in being considerably broader than the volcanic vent that formed it. There are three main types. A *collapse caldera* is the result of the venting of material from beneath the cone of a VOLCANO, removing the cone's support and causing it to collapse in on itself; an *explosion caldera* is caused by the explosion of a blocked volcanic vent under pressure from rising MAGMA beneath, tearing the surface of the ground off for some distance around the vent; and an *erosion caldera* is an old crater that has been enlarged into a caldera through the action of wind and water erosion.

calefaction *See* THERMAL POLLUTION.

caliche (duricrust) in geology, a thin hard layer on the surface of the ground in arid regions. Caliche is formed by the evaporation of mineralized GROUNDWATER. As it evaporates, the groundwater leaves behind its minerals, causing them to collect on the surface of the ground where they cement together the soil, clay, sand, or gravel particles that made up the original surface. Caliche is often nearly impermeable to water (*see*

PERMEABILITY), and serves to seal off AQUIFERS from their RECHARGE AREAS, creating FOSSIL WATER. It also makes it difficult for plants to grow. *Compare* DESERT PAVEMENT.

California rights in water law, *see* HYBRID RIGHTS.

calorie a unit used in the measurement of energy. It is most often seen as a measure of heat, as opposed to the degree (Fahrenheit or Celsius), which is a measure of temperature. (*Heat* is the energy level of the molecules of a substance; *temperature* is the ability of the substance to transmit heat to its surroundings.) The term "calorie" is somewhat confusing because it has not been fully standardized. Originally, the term *calorie* was defined as the heat required to raise the temperature of one gram of water by one degree C. However, this quantity is not absolute but varies slightly, depending upon the original temperature of the water (*see* TRIPLE POINT). Several attempts have been made over the past century to get around this problem. The *mean calorie* was originally defined as 1/100 of the heat required to raise a gram of water from 0°C to 100°C, that is, from freezing to boiling; the *fifteen-degree calorie* was defined as the quantity of heat required to raise a gram of water from 14.5°C to 15.5°C, and the *international table calorie* abandoned the water standard altogether, being defined as 1/860 of a watt-hour. The mean calorie turned out to be equivalent to 4.184 JOULES (0.004 BTU—*see* BRITISH THERMAL UNIT); renamed the *thermocalorie,* it is defined today in relation to the joule, and is the "calorie" used by chemists (abbreviation: cal). The international table calorie (abbreviation: cal_{IT}) is now defined as 4.1868 joules, and is the "calorie" used by engineers. The fifteen-degree calorie has fallen out of use altogether. The "Calorie" used by nutritionists to measure food energy content is actually a *kilocalorie* (abbreviation: Cal or kcal), equivalent to 1,000 thermocalories, or 4,184 joules (3,968 BTUs). In order to differentiate them from the kcal more fully, the thermocalorie and international table calorie are often referred to as *gram-calories.*

cambium in botany, the thin layer of growth tissue just beneath the bark of a tree, shrub, or other WOODY PLANT where all CELL division in the stem and branches takes place. The cambium consists of three layers, each just one cell thick. The central, or primary, layer consists of so-called "mother" cells; division of these cells produces pairs of "daughter" cells, one of which replaces the original mother cell while the second becomes part of either the outer or inner secondary layer. Each daughter cell in the outer layer splits a second time to form two cells of XYLEM, or inner bark; each daughter cell in the inner layer splits to form two cells of PHLOEM, or outer heartwood. The phloem and xylem cells themselves do not split further.

Canada–United States Environmental Council (CUSEC) coalition of regional, national and international environmental groups, founded in 1974 to work on environmental issues of common concern to Canadian and U.S. citizens. CUSEC's priority project since its founding has been acid precipitation. However, it has also worked extensively on Great Lakes issues and on protection of the Arctic environment. Address: 1101 14th Street NW, Suite 1400, Washington, DC 20005. Phone: (202) 682-9400.

Canadian Long-Range Transport of Air Pollutants and Acid Deposition Assessment (LRTAP) a major and comprehensive study of the effects of ACID DEPOSITION on Canadian forests, undertaken in parallel with the U.S. NATIONAL ACID PRECIPITATION ASSESSMENT PROGRAM during the 1980s. While the U.S. study tended to minimize the effects of acid rain and other polluting agents on forests, the Canadian LRTAP report (1990) did not. The Canadians were particularly concerned with the impacts of cross-border pollutants from the American Middle West on sugar maples, which they judged had been severely and extensively damaged by acid deposition.

Canadian zone the zone of plant and animal life that lies between the HUDSONIAN and TRANSITION ZONES. The Canadian zone corresponds roughly to the CONIFEROUS FOREST BIOME. *See also* LIFE ZONES.

canid in the biological sciences, any animal, wild or domestic, belonging to the dog family *(Canidae)* of the order Carnivora (*see* CARNIVORE). Characteristics of the canids include long muzzles equipped with powerful canine teeth, a scent gland at the base of the tail, and the presence of five toes on the front feet and four on the rear (some breeds of domestic dogs have five toes on all four feet). They possess high intelligence and are usually strongly pack-oriented in behavior. Examples include the wolf, the dog, the coyote, the dingo, and the fox.

canopy in forestry and ecology, the uppermost portion of a forest or a STAND of trees, containing the leaves and upper branches of the trees. *Compare* CROWN.

canopy dweller any animal that spends its life principally or exclusively in the forest CANOPY. Examples include squirrels, monkeys and marmosets, tree frogs, small lizards such as the gecko and the Carolina anole, and numerous insects and spiders.

canopy-nesting bird a bird that builds its nest in the forest CANOPY. It may (and often does) feed elsewhere. Examples include robins, kinglets, and most hummingbirds, vireos, and warblers.

canopy tree refers to tall species, usually in a deciduous forest, that provide a canopy shading and protecting smaller trees and other understory plants. In most forests only a few canopy trees are present—oaks, maples, hickories, tulip poplar, for example. In the MIXED MESOPHYTIC forest, however, up to a dozen canopy trees may be present. *See also* FOREST TYPES.

capability in forestry or ecology, the ability of a unit of land to produce organic growth, usually measured by the amount of total BIOMASS that can be produced in a year. The greater the amount of biomass produced, the higher the land capability.

capillary action (capillarity) the set of forces that cause fluids to rise above or fall below STATIC LEVEL in narrow tubes (*capillary tubes*). Capillary action is principally a product of the difference between the WETTING ABILITY of a fluid and the strength of its SURFACE TENSION. If the wetting ability exceeds the surface tension, the fluid is pulled outward from the center and up the sides of the tube, creating a concave fluid surface and raising the fluid level. If the surface tension exceeds the wetting ability, the pull is toward the center of the tube, creating a convex surface and lowering the fluid level. The actual amount of rise or fall of the fluid depends on a complex interplay of factors, including the size of the tube, the relative degree of difference in the size of the forces of wetting and surface tension, and the force of gravity.

capillary fringe in soil hydrology, a narrow band just above the WATER TABLE where water will rise in spaces between soil particles through CAPILLARY ACTION. The thickness of the capillary fringe depends on the size of the soil particles and the compactness of the soil (that is, how densely or loosely it is packed); however, it is almost always between one and three feet. *See also* SOIL; GROUNDWATER.

capillary water water held by CAPILLARY ACTION in the spaces between soil particles above the water table. It cannot be drained out by gravity, but is available for use by plants. It should not be confused with the CAPILLARY FRINGE. *Compare* FILM WATER; GRAVITATIONAL WATER. *See also* FIELD CAPACITY; SOIL WATER.

caprock (capstone) in geology, a layer of hard, erosion-resistant rock lying above a series of softer layers and preventing them from being eroded. Where the caprock is absent the land surface erodes rapidly, leaving those areas covered by the caprock much higher than their surroundings. *See* MESA; HOODOO.

carbide a COMPOUND of CARBON with one of the METALS or TRANSITION ELEMENTS. As a group, carbides are extremely hard, tough, stable materials, with high melting points. Many of them react strongly with water, releasing acetylene gas. Among the most useful are:

CALCIUM CARBIDE (CaC_2)

Also known as acetylenogen, calcium carbide is the material most commonly used in welding and cutting torches and in cavers' "carbide" lamps for releasing acetylene.

TUNGSTEN CARBIDE (W_2C)

Included in tool steel to decrease brittleness and improve the tool's ability to hold a cutting edge.

BORON CARBIDE (B_4C)

Among the hardest, most chemically resistant substances known, it is used in the manufacture of ceramics and wear-resistant tools.

SILICON CARBIDE (CARBORUNDUM: SIC)

Primarily used as an abrasive, silicon carbide is the principal component of emery boards, emery paper, and emery cloth, and is used on sharpening stones and on saw blades meant for cutting rock or concrete. It is also a component of refractory brick, furnace linings, and antiskid pavement, and is used in producing semiconductors.

carbohydrate any chemical COMPOUND composed of CARBON, HYDROGEN and OXYGEN, with the hydrogen and oxygen in the same proportion as they are found in water, that is, twice as many hydrogen atoms as oxygen atoms. The type formula is usually written $C_x(H_2O)_y$; however, it should probably be written as $C_xH_{2y}O_y$ instead, as these compounds contain H and OH IONS rather than H_2O and are thus not true hydrates. One of the basic structural elements of living tissue, carbohydrates are widespread in nature as sugars, starches, CELLULOSE, and a variety of other compounds. Simple, noncyclic carbohydrates are known as *monosaccharides*. These can be transformed to cyclic compounds (that is, compounds containing a ring structure), which may then combine into POLYMERS called *oligosaccharides* in which each MONOMER is a cyclic monosaccharide. Oligosaccharides composed of two monosaccharides are called *disaccharides*. Those with three or more are called *polysaccharides*. All sugars are mono- or disaccharides. Starches are polysaccharides containing roughly 25 monosaccharide rings each; cellulose molecules contain on the order of 200 monosaccharide rings each. *See also* PHOTOSYNTHESIS; METABOLISM.

carbon the sixth element of the atomic series; atomic weight 12.011, chemical symbol C. The most important element to the existence of life, carbon is a part of every molecule of living tissue, from VIRUSES to human flesh. It is an extremely variable substance, with eight separate ISOTOPES (carbon 9 through carbon 16) occurring in the pure state in three ALLOTROPIC FORMS (coal; amorphous carbon, or graphite; and diamond). Two of the eight isotopes, carbon 12 and carbon 13, are stable; of these, carbon 12 is the more important, accounting for almost 99% of all occurrences of the element on the planet. The most abundant radioactive isotope of carbon, carbon 14, is widely used as a "tag" for tracing the course of specific carbon-based molecules in chemical reactions and environmental contamination and as a means of dating archaeological specimens.

CHEMICAL ACTIVITY

Carbon is extremely versatile chemically, having the ability to form molecular bonds with many if not most of the 105 other known elements (*see,* e.g., CARBIDE; CARBONATE). It also combines readily with itself, creating chains and ring structures that can build to almost unimaginable complexity. It is the size and complex nature of these carbon-based molecules, as well as the exothermic nature of the reaction of many of them with OXYGEN (*see* EXOTHERMIC REACTION), which makes carbon an ideal element on which to build living tissue. Because of the great variety of forms, sizes, and additional constituents available to the carbon-based molecule, there are more carbon compounds known than all other compounds put together.

THE CARBON CYCLE

Because living things die, carbon as a constituent of living matter is not bound in one place but cycles through the environment, passing from organic to inorganic forms and back again (*see* BIOGEOCHEMICAL CYCLE). The most important inorganic reservoir is the ATMOSPHERE, which holds carbon in the form of CARBON DIOXIDE (CO_2). Atmospheric carbon dioxide is taken up by green plants, which convert it to CARBOHYDRATES through the process of PHOTOSYNTHESIS, making the carbon available for use in building the plants' tissues. From this point, a carbon atom may follow any of numerous possible path-

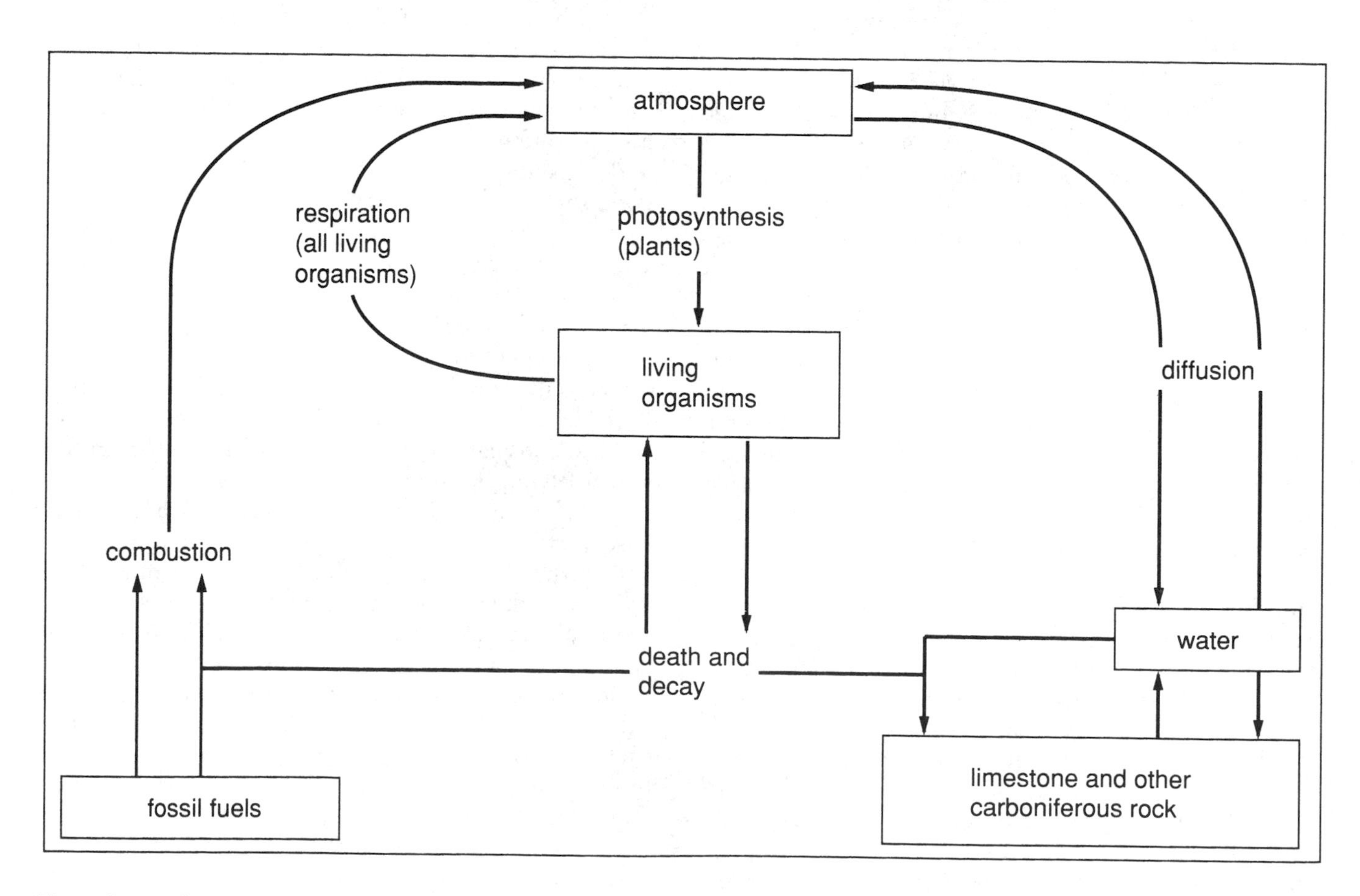

The carbon cycle

ways. It may simply be returned to the atmosphere through the plants' nighttime RESPIRATION (at night, plants, like animals, breathe oxygen and give off carbon dioxide). It may become part of the body tissue of an animal that eats a plant and, later, of a second animal that eats the first animal. From there, it may return to the atmosphere through the animal's respiration. Or it may wait until one of the animals—or the original plant—dies and its tissues are broken down by DECOMPOSER organisms. A small percentage will enter the LITHOSPHERE, deposited as LIMESTONE and other CARBONATES or as FOSSIL FUELS, from which it will eventually be released through weathering or combustion. As it exists in nature, the carbon cycle is close to perfect (*see* PERFECTION). However, human activities have lately been altering that. *See,* e.g., GREENHOUSE EFFECT. *See also* all other entries beginning with *carbo-* (e.g., CARBON MONOXIDE; CARBON DIOXIDE). *See also* COAL; METABOLISM; BENZENE RING; AROMATIC COMPOUNDS; ALIPHATIC COMPOUNDS.

carbonate any compound based on the carbonate GROUP, $-CO_3$. The most common naturally occurring carbonate is calcium carbonate, or *calcite* ($CaCO_3$), the principal constituent of LIMESTONE. Other important natural carbonates include *magnesite* ($MgCO_3$), *siderite* ($FeCO_3$), and *smithsonite* ($ZnCO_3$), ores of (respectively) magnesium, iron, and zinc. Sodium carbonate (*sal soda* or *washing soda,* Na_2CO_3) is used extensively in the manufacture of glass and soap, as a bleaching agent, and as a general cleanser. A separate but closely related group of compounds, the *bicarbonates,* contain a carbonate group with a hydrogen atom attached to it ($-HCO_3$). Of these, the most important from a commercial standpoint is sodium bicarbonate, or *baking soda* ($NaHCO_3$). Carbonates and biocarbonates are generally mildly to moderately caustic, and will effervesce (fizz) when acid is applied to them, releasing CARBON DIOXIDE. They are extremely important in maintaining the BUFFERING CAPACITY of bodies of water against ACID RAIN, and, because they are in equilibrium with dissolved CARBON DIOXIDE (the more dissolved carbonates, the less dissolved CO_2), help to determine the availability of CO_2 to photosynthesizing aquatic plants (*see* PHOTOSYNTHESIS). As a group they are not environmentally hazardous; however, the more caustic members of the group (such as washing soda) will cause damage to living tissue on contact, and are therefore classed as pollutants.

carbon balance the ratio of carbon in the ATMOSPHERE (CO_2) to that contained (as C) in living plants and sequestered in the earth or underwater as, for example, in roots and buried trees, or fossilized as coal, oil, or gas. Should the carbon balance change, as it does from natural causes such as volcanoes, producing an excess of CO_2 in the atmosphere, the growth of forests and plants increases to take up the excess through PHOTOSYNTHESIS, thus maintaining the balance wherein the CO_2 must remain in rough equilibrium with oxygen (O) for the maintenance of plants and animals. Should atmospheric carbon significantly increase relative to oxygen, natural ecosystems as well as human physiology can be damaged by the GREENHOUSE EFFECT and other difficulties. *See also* CARBON CYCLE.

carbon cycle *See* CARBON: *the carbon cycle.*

carbon dioxide a colorless, odorless gas, chemical formula CO_2, produced in nature by the burning of CARBON compounds in forest fires and other natural conflagrations, the weathering of CARBONATE rock (such as LIMESTONE), and the RESPIRATION of plants and animals. It cannot be liquefied at atmospheric pressure, but passes directly from the solid form ("dry ice") to the gaseous form (*see* SUBLIMATION). Carbon dioxide does not normally support combustion—in fact, one of its principal uses is as a fire extinguisher—but certain substances, notably magnesium, will continue to burn in an atmosphere of pure CO_2 if they are first ignited in air. CO_2 is used by industry to "carbonate" fizzy beverages, to manufacture various chemical carbonates (e.g., baking soda; washing soda), and (as dry ice) as a refrigerant.

IMPORTANCE TO LIFE

Carbon dioxide is the principal reservoir of available carbon for all living things (*see* PHOTOSYNTHESIS; CAR-

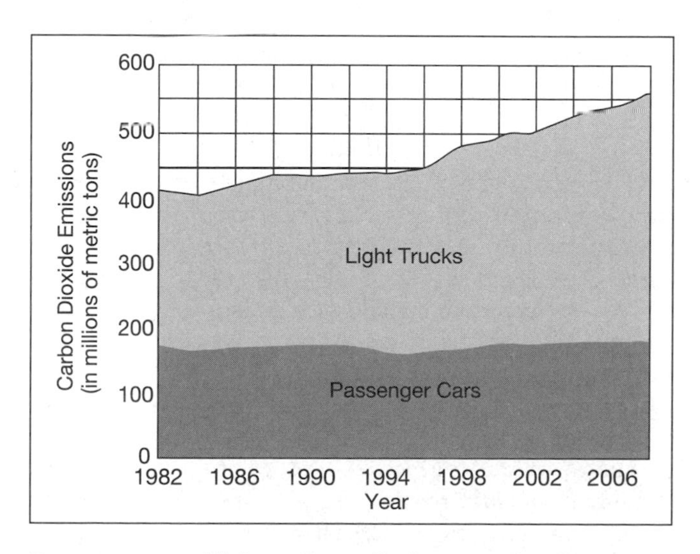

Passenger car and light truck contributions to carbon dioxide emissions *(Western Economic Association, 1999)*

BON: *the carbon cycle*). Green plants in sunlight breathe it directly, and in small amounts it also seems to strengthen the respiration of animals; however, larger amounts of it (concentrations of 10% or more) cause breathing difficulties, loss of consciousness, and eventually, death. Its presence in the atmosphere helps regulate the Earth's temperature (*see* GREENHOUSE EFFECT).

Carbon Dioxide Information Analysis Center the principal global climate change data and information analysis center for the U.S. Department of Energy. The center maintains and analyzes data on carbon dioxide and other gases in the atmosphere to follow long-term climate changes and trends. The data are provided to researchers, educators, policymakers, and others involved with climate change issues. Address: P.O. Box 2008 MS-6335, Oak Ridge, TN 37831-6335. Phone: (865) 574-3645. Website: www.cdiac.esd.ornl.gov.

carbon monoxide a colorless, odorless gas, chemical formula CO, formed by the incomplete combustion of CARBON, either because of insufficient OXYGEN (a fire in an enclosed space) or too-rapid passage of a carbon-based fuel through a combustion chamber (an automobile engine). Carbon monoxide is highly poisonous: a concentration of as little as 01% can bring on symptoms of poisoning, and 2% is usually fatal. It has an affinity for blood HEMOGLOBIN, combining with it in place of oxygen and thus preventing the transport of oxygen by the bloodstream. Its production by automobile engines and faulty furnaces, along with the extremely small amounts required to be dangerous, makes it an important constituent of urban SMOG. The symptoms of carbon monoxide poisoning are deceptively mild, and call little attention to themselves until they are far advanced: they begin with a mild headache and mental fatigue, followed by drowsiness, unconsciousness, and death.

USES

Because it is the product of an incomplete oxidation reaction, carbon monoxide has a strong affinity for oxygen. It burns readily in air, and is used as a constituent of some gaseous industrial fuels; it is also useful as a REDUCING AGENT in certain metallurgical reactions.

carbon tax a proposal to tax carbon emissions rather than (some) carbon-based products. The tax would provide more equitable and comprehensive means to reduce the use of FOSSIL FUELS, derive funds for environmental remediation of the impacts of fossil fuels, and encourage alternative energy sources. *See also* GREENHOUSE EFFECT; ALTERNATIVE ENERGY SOURCES.

carbon tetrachloride a heavy, colorless fluid, chemical formula CCl_4, formerly widely used as cleaning fluid, spot remover and all-purpose industrial SOLVENT. It was also occasionally used as a PESTICIDE and as a fire suppressant (although it is a volatile fluid [*see* VOLATILITY] neither the fluid nor the vapors are flammable). Carbon tetrachloride may be thought of, chemically, as a CHLORINATED HYDROCARBON in which all the hydrogen atoms have been replaced by chlorine. Both the fluid and its vapors are highly toxic. Acute exposure can cause nausea, vomiting, headaches, stupor, and eventually death from kidney and liver failure, while chronic low-level exposure causes severe (and in most cases irreversible) liver and kidney damage. It is easily absorbed through the skin, where it dissolves the underlying fatty tissue on its way to entering the bloodstream. For these reasons it is no longer available as a consumer commodity, although large amounts are still manufactured for use in the production of plastics and other ORGANIC COMPOUNDS.

carbon to nitrogen ratio a means to describe the content of nitrogen, a nutrient, in living tissue. A high C:N ratio indicates low nitrogen content, which is the case in plants where the ratio is 25:1. In animals, bacteria, and fungi, the ratio is about 5:1. Differences in C:N ratios among organisms can influence what they eat, how rapidly they reproduce, and the rate at which they decompose.

carborundum *See* CARBIDE: *silicon carbide.*

carcinogen any substance that increases the likelihood that an individual who comes into contact with it will contract cancer. (It is incorrect to say that carcinogens "cause" cancer, because the cause-and-effect relationship is incomplete; there is always a statistically random group of exposed individuals who do not develop the disease. A carcinogen causes an increase in the cancer rate, rather than directly causing cancer. The rate increase, however, may be extremely large.) Because of the danger they pose to human health, the use of known carcinogens is closely regulated in most parts of the world, and is often banned altogether.

TESTS FOR CARCINOGENITY

Since no one-to-one cause-and-effect correspondence can be established between a carcinogen and a cancer, tests for carcinogenity must rely on the establishment of statistical trends rather than mechanisms. There are two principal ways to do this. One is through EPIDEMIOLOGICAL TESTS on human populations. Though extremely difficult to do (the complexity of the human environment makes isolating the effects of a single substance nearly impossible), these tests have the advantage of applying directly to humans and therefore avoiding the question as to whether data established for one SPECIES can be applied to another. The second method meets this objection by ignoring it; that is, by assuming that the data are transferable (an assumption that is usually accurate). This method—or these methods, because there are more than one—involve studying laboratory animal populations bred to be particularly cancer-prone. The animals (usually mice) are isolated and fed diets high in the suspected carcinogen, or are exposed to it in other ways (inhalation; skin contact). The rate at which they develop cancers is then compared to a control group that has been kept isolated from the substance.

For a third method of testing for carcinogenity, *see* AMES TEST.

carnivore any animal whose principal diet consists of meat. Technically, "meat" in this usage means the flesh of VERTEBRATES; however, the term "carnivore" is often expanded to include animals (and even plants: *see* CARNIVOROUS PLANT) that eat insects, spiders, and other invertebrate animals (*see* INVERTEBRATE). *Compare* OMNIVORE; HERBIVORE; INSECTIVORE. *See also* PREDATOR; SCAVENGER.

carnivorous plant a plant that obtains some of its NUTRIENTS by catching and "eating" living insects and other small animals. Most carnivorous plants live in BOGS, where the acidic waters are unfavorable for the growth of NITROGEN-FIXING BACTERIA. Hence, nitrogen appears to be the principal nutrient sought from animal bodies by the plants, although other elements are usually taken up as well. Carnivorous plants are an example of extreme ADAPTATION, and, like most extreme species, they are particularly sensitive to environmental disruption. Most of them are extremely limited in RANGE and HABITAT. Examples include the Venus's-flytrap, the sundew, the bladderwort (which traps aquatic animals in underwater bladders), and various species of pitcher plant. All except the bladderwort use highly modified leaves to capture their prey.

carp a large bottom-dwelling fish, *Cyprinus carpio,* native to Asia but now resident in all parts of the world. Carp prefer shallow, slow-moving or still waters with muddy bottoms, where they generally feed by uprooting aquatic plants and feeding on the tender roots. They are a nuisance in many North American streams and lakes, where their extreme hardiness and longevity, and their tendency to muddy the waters as they feed, has caused them to displace numerous native species of fish and to damage or destroy much waterfowl HABITAT. *See also* BOTTOM DWELLER; EXOTIC SPECIES.

carriage water in water law, an amount of water over and above a specified DUTY OF WATER that must be diverted into a canal or ditch to make certain that the specified amount reaches its destination. Water is necessarily lost to evaporation and seepage between the diversion works and the point of use. Thus, an amount of water larger than that needed at the point of use must be diverted from the stream in order to assure the desired amount at the point of use. The extra water is needed to "carry" the desired amount to its destination; hence the term "carriage water." In most states, carriage water is made part of a divertor's WATER RIGHT by measuring the amount of water to be appropriated at the point of diversion rather than at the point of use.

carrying capacity in ecology or land-use planning, the number of individuals of a particular SPECIES that a given unit of land or a given ECOSYSTEM can support indefinitely without degrading. Whenever the population of any species—humans included—exceeds the carrying capacity for that species, environmental damage becomes unavoidable. The principal factors affecting carrying capacity are the availability and CAPABILITY of land, the quality and quantity of water, and the circulation patterns of the air.

Carson, Rachel (1907–1964) American scientist and environmental writer, born May 27, 1907, in Springdale, Pennsylvania. Trained as both a writer and a biologist (after initially majoring in creative writing, she received an M.S. in biology from Johns Hopkins University in 1932, with further work at Woods Hole Marine Biological Laboratory between 1932 and 1936), Carson combined the two interests most of her career, serving as editor in chief for the U.S. FISH AND WILDLIFE SERVICE (1941–52) and publishing three best-selling books on the environment: *Under the Sea Wind* (1941); *The Sea Around Us* (1951); and finally *Silent Spring* (1962), a book condemning the indiscriminate use of PESTICIDES, the publication of which is usually credited with starting the modern environmental movement. A reluctant activist ("I may not like what I see," she once wrote, "but it does no good to ignore it"), her scientific credentials and restrained, highly polished writing style helped her work survive a massive campaign by the chemical industry to discredit it. She died April 14, 1964, in Silver Spring, Maryland.

caryopsis *See under* FRUIT.

catadramous species any SPECIES of fish or other aquatic animal that spends most of its life in freshwater, moving into salt water only in order to spawn. An example is the common fresh-water eel of Europe and North America; the eels of the two continents consist of not only a single species but a single POPULATION, part of which meets in the middle of the Atlantic Ocean each year to mingle and spawn. *Compare* ANADRAMOUS SPECIES.

catalyst in chemistry, a substance whose presence causes or speeds up a chemical REACTION between two or more other substance (the *reactants*). The catalyst itself is not changed in the process. Catalysts are of two general kinds. *Homogenous catalysts*—so-called because the catalyst is almost always of the same phase (that is, solid, liquid, or gas) as at least one of the reactants—work by forming an intermediate COMPOUND with one of the reactants, which is then reacted upon by the second reactant, releasing the catalyst in its original form. *Heterogenous catalysts* provide an adsorptive surface (*see* ADSORPTION) on which the reactants are held in contact with each other, allowing time for the reaction to take place. Catalysts are important to many industrial processes and—as ENZYMES—play a critical role in the METABOLISM of all living things. *See also* CATALYTIC CONVERTER.

catalytic converter a device attached to an automobile exhaust system that uses a heterogeneous catalyst (*see* CATALYST) to reduce the amount of POLLUTANTS present in the exhaust by causing some of the more harmful pollutants to react with the OXYGEN in the surrounding air, forming less harmful substances. The catalytic converter is normally placed in the exhaust system just behind the manifold, where the exhaust gases are still very hot. The usual catalysts used are the metals platinum and palladium. These are present as a coating on a bed of refractory beads (small round beads made of a material, such as ceramic, that can stand very high temperatures), in order to maximize both the surface of the catalyst available for reactions and the presence of air spaces for the passage of the exhaust gases. Catalytic converters are capable of eliminating or reducing to low levels both the CARBON MONOXIDE and the unburned HYDROCARBONS resulting from the incomplete combustion of gasoline, converting them to CARBON DIOXIDE and water vapor. However, they may actually add to the content of NITROGEN OXIDES in the exhaust, and they react to the presence of trace amounts of SULFUR in the exhaust by creating hydrogen sulfide (adding a "rotten egg" scent to traffic jams) and sulfur trioxide (creating a fine mist of sulfuric acid droplets: *see* SULFUR OXIDE). Thus, they are probably not the final answer to the control of pollution from automobile exhaust. The catalysts in catalytic converters are easily "poisoned" (that is, their adsorptive surfaces are easily clogged) by impurities in gasoline, and especially by lead. Hence, they require the use of carefully refined unleaded gasoline and even under the best of operating conditions must be periodically replaced or regenerated. *See also* AIR POLLUTION.

catastrophism in geology, the theory that geologic change takes place rapidly, in worldwide "catastrophes," with long periods of quiescence in between, and that it is these catastrophes that render animals extinct. Catastrophism's strongest proponent was Baron Georges Cuvier (1769–1832) of France, an early authority on fossils, who used the theory to explain the differences between fossils in different geologic strata (*see* STRATUM). Long considered obsolete and invalid, catastrophism has had a limited revival in modified form lately in theories connecting the extinction of the dinosaurs (and similar mass extinctions found elsewhere in the geologic record) to the influence of large comets or other celestial bodies colliding with or coming close to Earth.

catchment basin *See* WATERSHED.

cathode *See under* ELECTRODE.

cathodic protection a means of protecting buried steel tanks against the effects of stray current corrosion (*see* CORROSION: *types of corrosion*) by turning the entire tank into a cathode (*see* ELECTRODE). The method employs a so-called "sacrificial" anode, usually made of magnesium, that is connected electrically to the tank and buried near it in the soil. Stray electrical currents in the soil enter the tank (the cathode) and leave at the sacrificial anode, corroding it rather than the tank. *See also* GALVANIC SERIES.

cation *See under* ION.

cation exchange capacity *See* BASE SATURATION.

cat skidding in logging, YARDING logs by hauling them behind a tractor, usually a crawling or "Caterpillar" tractor (hence the name "cat skidding"), although special rubber-tired skidders are increasingly replacing the destructive "cats." Because the only equipment necessary is the tractor itself, some CHOKERS, and a short length of cable to attach the chokers to the tractor, cat skidding is the least expensive form of mechanized yarding. It is also highly flexible, allowing the LEAD to be in nearly any direction when falling the timber, and for these reasons it is very popular, especially for smaller LOGGING SHOWS. However, it can involve massive soil disturbance in the creation of SKID ROADS, which are often routed by the tractor operator ("cat skinner") rather than laid out by an engineer or silviculturist, and inevitably cause serious soil compaction, both from the weight of the tractor and from dragging the logs across the ground rather than elevating them in the air. For these reasons it ranks as the most environmentally damaging form of yarding and is generally regarded as unsuitable for all but the least sensitive sites.

caucus in politics, originally a closed meeting of a particular political faction to discuss policy or choose candidates, now becoming a synonym for the faction itself. A caucus may be thought of as a SPECIAL INTEREST GROUP composed of politicians rather than lobbyists. *See,* e.g., ENVIRONMENTAL CAUCUS.

cavity nesting bird a bird that builds its nest in a hole in the trunk of a tree or in a face of rock or earth, either one it has created (woodpeckers) or one that it finds (owls, wrens, wood ducks).

CBO *See* CONGRESSIONAL BUDGET OFFICE.

CDC *See* CENTERS FOR DISEASE CONTROL.

CDF *See* CONFINED DISPOSAL FACILITY.

cell in biology, the smallest unit of living matter (VIRUSES are not now considered true living ORGANISMS due to their inability to reproduce without assistance). All living organisms are composed of cells. Some (the MONERA and most PROTISTA) have only one cell, while others (plants, animals, and most fungi [*see* FUNGUS]) contain many cells with varying structures and functions. Most cells are microscopic in size, but a few are surprisingly large. Unfertilized birds' eggs each consist of a single cell, and nerve cells in some large mammals (whales, giraffes, elephants) may be as much as 3 meters (10 feet) long.

STRUCTURE OF CELLS

All cells follow the same basic structural plan, with a membrane (and usually, a firmer structure known as the *cell wall*) surrounding a complex fluid called CYTOPLASM within which float CHROMOSOMES (composed of DNA and used in reproduction) and a number of small bodies called *ribosomes* (used in protein synthesis). *Prokaryotic cells* (or *Prokaryotes*)—the bacteria (*see* BACTERIUM) and blue-green algae (*see* CYANOPHYTE)—consist essentially of nothing more than these basic elements, though the blue-green algae also contain CHLOROPHYLL. *Eukaryotic cells* (or *Eukaryotes*)—all other one-celled organisms, and the individual cells of multicelled organisms such as plants, fungi, and humans—are much more complex, protecting their chromosomes within an internal body known as the *nucleus* and containing within their cytoplasm, in addition to the ribosomes, a number of diverse bodies called *organelles,* each with a specific function in regards to the cell's METABOLISM. Among the most important of these are the *mitochondria,* which assist in respiration; the *golgi bodies,* which synthesize a variety of PROTEINS; and, in plants, the *chloroplasts,* which contain the cell's CHLOROPHYLL and thus serve as its organs of PHOTOSYNTHESIS. The mitochondria are the size and general structure of small bacteria, and contain their own DNA, while the chloroplasts are the size and general structure of blue-green algae. Because of this and other evidence, most biologists today believe that these two organelles were once free-living organisms that set up a symbiotic relationship with the ancestral Eukaryotic cells (*see* SYMBIOSIS) and eventually were subsumed within them. A further feature of most Eukaryotic cells is the presence of *vacuoles,* "bubbles" of fluid-filled space within the cytoplasm, which apparently assist in the transport of nutrients and wastes into and out of the cell. In rigid-walled plant cells, the vacuoles line up on top of one another in adjacent cells, forming long multicellular tubes, which may either be separated by a permeable membrane (*see* PERMEABILITY) or may actually connect directly to each other. Finally, nearly all Eukaryotic cells contain *lysosomes,* roughly spherical bodies containing powerful ENZYMES used to destroy cellular invaders. The lysosomes are surrounded by a membrane that is continually replaced by the cell as it is eaten away from the inside by the enzymes. When the cell dies the membrane is no longer replaced, and the enzymes destroy the membrane and, once free in the cytoplasm, the rest of the cell—the principal reason that cells, once dead, decay so very rapidly.

CELLS IN THE ENVIRONMENT

The cell is the basis upon which all life is built, and is the shared heritage of every organism, large and small, in the environment. It is a nearly ideal solution to the problem of protecting the relatively fragile DNA molecule in order to keep it intact long enough to reproduce in a hostile and changing environment, and as such represents the fundamental structure that makes the BIOSPHERE possible. *See also* MEIOSIS; MITOSIS.

cell towers facilities used for the transmission of signals for cellular telephones and similar devices. As the popularity of mobile cell phones has increased, transmissions towers, which are often an aesthetic intrusion and (in the view of some) a health problem, have had to be built in urban, suburban, and rural areas at a rate that has alarmed many land use planners, conservationists, and municipal governments. In 1996 Congress passed the Telecommunications Act (TA) which guarantees the right of municipalities to control tower location. Local governments are, however, often ignored or overpowered by the telecommunications industry, and thoughtless siting has created a strong public reaction to the proliferation of the 50-to-100-foot towers, especially in suburban and rural areas where there are no tall buildings to conceal them. According to the Cellular Telecommunications Industry Association, the number of cell towers increased from 58,000 to 100,000 in just two years—between 1998 and 2000.

cellulose the principal structural material of the cell wall (*see* CELL: *structure of cells*) in plants and ALGAE. Cellulose is a *polysaccharide,* that is, a large POLYMER in which the individual MONOMERS are sugar (in this case, glucose) molecules. It is closely related to the starches. The glucose monomers are attached to each other so that a hydroxide GROUP lies alternately to the left and to the right of the main chain. These groups bind parallel cellulose molecules together through hydrogen bonds, forming the threadlike *microfibrils* that give cellulose its great structural strength (it is stronger, pound for pound, than steel). A major storehouse of energy for the FOOD CHAIN and one of the principal sources of glucose for HERBIVORES, cellulose is nevertheless impossible for most ORGANISMS to digest. Most animals that rely on it depend on the presence of MICROORGANISMS in their digestive tracts that secrete the ENZYMES needed to break down the cellulose into usable glucose units.

Center for Conservation Biology founded in 1984 for the purpose of developing the science of conservation biology. Under the center's current president, PAUL R. EHRLICH, it researches scientific and policy issues relating to the conservation, management, and restoration of species diversity worldwide. The center's intent is to foster methods of preserving Earth's ecosystems while also providing for current and future human welfare. Address: Department of Biological Sciences, Stanford, CA 94305-5020. Phone: (650) 723-5924. Website: stanford.edu/group/CCB.

Center for Environmental Education features an extensive collection of materials for environmental education, including more than 10,000 books, videos, and other materials. The center also publishes a newsletter. Membership (1999): 5,000. Address: c/o Antioch New England, 40 Avon Street, Keene, NY 03431. Phone: (603) 355-3251. Website: www.cee_ane.org.

Center for Marine Conservation involved in the protection of coastal areas and ocean resources through research, education, and analysis of public policy. The center has five program areas: fisheries and wildlife conservation, ecosystem protection, biodiversity conservation, international initiatives, and citizen monitoring and outreach. Founded in 1972, the center has a variety of publications. Membership (1999): 120,000. Address: 1725 DeSales Street NW, Suite 600, Washington, DC 20036. Phone: (202) 429-5609. Website: www.cmc-ocean.org.

Center for Resource Economics (Island Press) seeks to solve local and global environmental problems by developing, publishing, and marketing books concerned with biodiversity and sustainable development. Address: Island Press, 1718 Connecticut Avenue NW, Suite 300, Washington, DC 20009. Phone: (202) 232-7933.

Center for Science in the Public Interest primarily concerned with consumer advocacy, concentrating on issues relating to health, nutrition, and alcohol. Information is disseminated through publications, press releases, and other media. Membership (1999): 1,000,000. Address: 1875 Connecticut Avenue NW, Suite 300, Washington, DC 20009. Phone: (202) 332-9110. Website: www.cspinet.org.

center-pivot system *See* IRRIGATION: *methods of irrigation.*

center rot *See* HEART ROT.

Centers for Disease Control (CDC) agency within the PUBLIC HEALTH SERVICE, created by the Secretary of Health, Education and Welfare on July 1, 1973 (*see* HEALTH AND HUMAN SERVICES, DEPARTMENT OF) and charged with administering public health programs concerned with controlling preventable diseases, especially those transmitted by insects (*see* VECTOR) and those caused by environmental pollution. In addition to administrative and data-gathering duties, the CDC fields "strike teams" to respond to health-threatening environmental emergencies such as HAZARDOUS WASTE spills or nuclear-plant accidents. It is also the federal agency charged with cooperation with other nations on disease prevention and the cleanup of health-threatening environmental problems. Headquarters: 1600 Clifton Road, Atlanta, GA 30333. Phone: (404) 639-3311. Website: www.cdc.gov.

CEQ *See* COUNCIL ON ENVIRONMENTAL QUALITY.

CERCLA *See* COMPREHENSIVE ENVIRONMENTAL, RESPONSE, COMPENSATION AND LIABILITY ACT.

Certified Forest Products Council a nonprofit organization dedicated to promoting responsible forest management by influencing consumer buying practices. The council hopes to reduce and eventually eliminate the harvesting and sale of products originating from endangered forests. It seeks to educate corporate and individual buyers on the benefit of purchasing certified products. Address: 14780 SW Osprey Drive, Suite 285, Beaverton, OR 97007. Phone: (503) 590-6600. Website:www.certifiedwood.org.

cetacean an animal belonging to the order Cetacea, which includes the whales, porpoises, and dolphins. Cetaceans are mammals that have adapted to life in the oceans by taking on the form of fish. They have long, flexible backbones and fused neck vertebrae; their hind legs are absent, and their front legs have become fin-shaped flippers. Dorsal fins are present in most species, and tail fins are present in all. However, a cetacean's tail fins (known as *flukes*) are not vertical (as in the fish) but lie in a horizontal line. The nostrils have fused to form a single "blowhole," and have moved to the top of the head (cetaceans, like all mammals, have lungs rather than gills, and breathe atmospheric air rather than that dissolved in the water). Instead of hair—which would interfere with efficient swimming—these animals have developed a thick layer of fat (*blubber*) directly under the skin to act as an insulator. Cetaceans have no sense of smell, but their sense of hearing, which has been modified to operate more efficiently under water, is extraordinarily sensitive. They have large brains and display a high level of intelligence. The order contains three suborders, six FAMILIES, and more than 35 SPECIES.

CFC *See* CHLOROFLUOROCARBONS.

CFL *See* COMMERCIAL FOREST LAND.

CFM *See* CHLOROFLUOROMETHANE.

CFR *See* CODE OF FEDERAL REGULATIONS.

CFS *See* CUBIC FEET PER SECOND.

chain as a measure, 66 feet. A distance of 80 chains is equal to one mile. The term comes from the actual, physical chain used to measure distances between surveying stations on level ground. It is rarely used outside the field of surveying, and its usage is declining even there.

chaos theory describes change not as an immutably linear series of cause and effect relationships, but as apparently "chaotic." Chaos theorists propose that Newtonian determinism cannot predict outcomes, even if it were possible to identify every single interrelationship at the most minute scale, since the internal dynamics can shift cause and effect relationships at unpredictable moments into entirely new realms, thus creating a set of new relationships that logically cannot be identified in advance. The classic example demonstrating the "nonperiodicity" of cause and effect chains, how sensitive they are to initial causes and how the internal dynamics are beyond ordinary analysis, is that of a butterfly flapping its wings in Beijing resulting in a storm in New York months alter. Such nonpredictability is especially important in understanding that the dynamics of natural systems can be determined by seemingly illogical initial causes, and that such initial causes will not always produce a similar outcome. The insights provided by chaos theorists underscore what ecologists and environmentalists have always maintained, that extreme caution must be used in interfering with natural systems since outcomes can be neither predicted nor controlled. *See also* ECOLOGY.

chaparral a tangled, dry growth of woody brush and shrubs characteristic of sunny slopes in semiarid regions. More than 150 species of plants are classed broadly as chaparral, including buckbrush, ceanothus, manzanita, sagebrush, and various scrub oaks. All of these have in common a many-branched growth form; small, stiff, waxy leaves (usually evergreen); and an extensive, interweaving root system that often contains most of the plant's bulk. Growths of chaparral are almost impenetrable by large animals, making travel difficult through them. They suppress timber growth, even where the rainfall would marginally allow it, and they can pose a significant fire hazard. However, they serve as excellent COVER for small birds and animals, and their root systems are a valuable aid in holding back EROSION. *See also* BRUSH; CHAPARRAL BIOME.

chaparral biome (Mediterranean association) a major semiarid-region BIOME, covering much of the American southwest, southern Australia, parts of Chile and South Africa, and most of the region around the Mediterranean sea. The principal plant life is CHAPARRAL, but there are also scattered trees—usually evergreens or oaks and a variety of GRASSES and flowering HERBS, often adapted to bloom beneath the chaparral. The climate is hot and dry, with the slight rainfall occurring mostly in the winter. Winds are common. Fires sweep through regularly, usually on a 10- to 12-year cycle. These fires destroy the crowns of the chaparral and kill most competing vegetation, but leave the root systems intact, from which the chaparral can regrow. If for some reason these fires do not occur for 25 years or more, FUEL BUILDUP usually leads to a much hotter fire at a later time, killing the roots and allowing erosion and the invasion of other plants to destroy the ECOSYSTEM. *See* FIRE-MAINTAINED CLIMAX.

character displacement in ecology, the evolution of differing characteristics in two otherwise similar

SPECIES enabling them to take advantage of different parts of a shared environment and thus to coexist with each other. The differences may be either behavioral or physical, and are often both.

characteristic landscape in visual management, a view or portion of a view that maintains a natural, unaltered appearance. Native vegetation is present in natural densities; alterations (such as roads, mine pits, HARVEST UNITS, etc.) are either hidden or blended into the general landscape in such a manner as to escape notice.

checkerboard ownership in land-use management, a pattern of land ownership wherein alternate SECTIONS are owned by different entities. In the most common pattern, every other section will be public land (usually managed by the FOREST SERVICE or the BUREAU OF LAND MANAGEMENT), with the intervening sections belonging to private owners. The private sections may be subdivided among numerous small landholders, or a single private landholder may own a number of squares of the checkerboard. Checkerboard ownership derives from the 19th century RAILROAD LAND GRANTS of alternate sections of public land given to private companies as encouragement to build railroads. Where some of these land grants have revested (that is, have come back into public ownership), a checkerboard of alternate Bureau of Land Management and Forest Service sections may develop. Management of public lands involved in checkerboard ownership patterns is complicated by the need to provide access across the public lands to the private sections that they surround, and by the artificiality of single-square-mile grids imposed on dynamic ECOSYSTEMS, requiring cooperation among all owners in the checkerboard in order to properly manage for ecosystem stability and productivity.

chemical nomenclature the scientific naming of chemical compounds. Chemical nomenclature is a complex subject that can only be touched on here. In general, the names of compounds are made up of a series of roots derived from the names of their constituent elements, arranged in order by electronegativity (ability to attract electrons), with the most electropositive first and the most electronegative last. (Relative electronegativity can be empirically determined from the PERIODIC TABLE: the further to the left an element appears in the table, the more electropositive it is). The end of a name is a suffix denoting the compound's general type: *-ide* indicates a BINARY COMPOUND; *-ane* indicates a HYDROCARBON; and *-ite* and *-ate* are SALTS of oxygen-containing acids, with *-ite* referring to salts derived from the lower VALENCE state of bivalent metals and -*ate* referring to salts derived from the higher valence state (example: $NaClO_2$) is *sodium chlorite;* $NaClO_3$ is *sodium chlorate*). Individual roots within the compound name may also carry suffixes and prefixes. The most common root prefixes are derived from Greek numbers and indicate the number of atoms of the root element present in the compound, as : *mono,* 1; *di,* 2; *tri,* 3; *tetra,* 4; *penta,* 5; etc. (example: *pentachlor* indicates a compound containing five chlorine atoms). The two most common root suffixes are *-ic* and *-ous:* like -ite and -ate, these refer to valence states, with the lower valence state of a bivalent element ending in -ous and the higher in -ic (example: $FeCl_2$ is ferr*ous* chloride; $FeCl_3$ is ferr*ic* chloride). Finally, the name of a hydrocarbon usually carries numbers indicating the points of attachment of the various constituent GROUPS to the basic carbon-hydrogen chain or ring of the compound. For aliphatic (chain) hydrocarbons (*see* ALIPHATIC COMPOUND), the numbering begins at the end of the chain nearest the first attached group. For ring hydrocarbons (*see* AROMATIC COMPOUND; CYCLIC ORGANIC COMPOUND), the numbering begins at a carbon atom with an attached group and proceeds clockwise around the ring, with the starting point chosen so as to keep the numbers indicating the points of attachment as low as possible.

chemical oxygen demand (COD) a measurement that is sometimes used in place of BIOCHEMICAL OXYGEN DEMAND (BOD) as a means of determining the amount of NUTRIENTS and ORGANIC MATTER in a water sample. COD is determined by adding a measured amount of an OXIDIZING AGENT, potassium dichromate ($K_2Cr_2O_7$), to the water sample. The sample is boiled until the oxidation REACTIONS cease, and the amount of the oxidizing agent that remains in the sample is measured again. The amount of COD may be calculated from the difference in the "before" and "after" measurements of the oxidizing agent. For "clean" samples (those containing primarily BIODEGRADABLE SUBSTANCES with no toxic materials present), COD is approximately the same as BOD_{ult}, with the advantage of being determinable in a much shorter time.

chemosynthesis in the biological sciences, the production of organic chemicals from inorganic source substances by an autotrophic ORGANISM (*see* AUTOTROPHISM) that utilizes chemical energy rather than the light from the Sun (*see* PHOTOSYNTHESIS). It is highly probable that the earliest life forms on Earth were chemosynthetic rather than photosynthetic; however, the only chemosynthetic organisms remaining today are a few highly specialized bacteria (*see* BACTERIUM) occupying odd NICHES such as geyser pools or anaerobic muck from the bottom of watercourses (*see* ANAEROBIC BACTERIA). In a few places, complete

ECOSYSTEMS have developed around chemosynthesis. The best-developed examples are those surrounding thermal springs in the deep ocean trenches, where chemosynthetic bacteria form the basis of a FOOD CHAIN that includes PROTOZOANS, large OLIGOCHEATES, and even crabs and MOLLUSKS. *See* BLACK SMOKER.

Chernobyl nuclear reactor complex in the northern Ukraine, USSR, which on April 26, 1986, became the site of the world's worst nuclear-energy accident to date. Located on the Pripyat River 100 kilometers (60 miles) north of Kiev and just a few kilometers short of the Belorussian border, the Chernobyl reactor complex was one of the largest in the Soviet Union, with four operating 1,000-megawatt reactors and two more under construction. At 1:00 A.M. on April 25, 1986, plant engineers began an experiment in low-power operation of reactor number 4, during the course of which they assumed manual control of the reactor, turning off the emergency core cooling system and virtually all safety alarms and fail-safe automatic shutdown devices. A computer programmer's miscalculation allowed power levels to drop too low in the reactor's core, and the experimenters attempted to jack power back up quickly by withdrawing the reactor's control rods almost all the way. At 1:23 A.M. on April 26 an uncontrolled chain reaction began that caused the reactor's power output to climb to 100 times its design capacity within approximately four seconds, leading to a core meltdown. The resulting heat caused an immense steam explosion that blew the reactor's containment building apart, spewing tons of burning graphite and other materials over the reactor grounds and releasing a radioactive plume 5 kilometers (3 miles) high. Three people died in the blast, and 28 others died shortly afterwards from its direct effects: another 237 were hospitalized with severe radiation poisoning. A 30-kilometer circle around the plant was evacuated, displacing 135,000 people from their homes. Effects of the accident were felt as far away as the west coast of Wales, 3,200 kilometers (2,000 miles) from the reactor, where fallout raised radioactivity in crops and

Significant Nuclear Accidents

Country	Year	Description
Japan	1981	Tsuruga. Some 100 workers exposed to radiation during repairs to a nuclear plant.
	1999	Tokaimura. Chain reactions resulting from an accidental overload of a uranium container, exposing 300,000 to dangerous radiation that reached 15,000 times accepted levels in areas close to the plant.
Switzerland	1969	Lucien Vad. A coolant malfunction in an underground reactor requiring that the cavern be permanently sealed.
United Kingdom	1957	Liverpool. A fire in the Windscape plutonium production reactor. Released radioactive material blamed for 39 cancer deaths.
United States	1961	Idaho. Reactor accident killed 3 workers, but radiation was contained.
	1966	Michigan. Sodium cooling malfunction caused a partial core meltdown at a demonstration breeder reactor. Radiation contained.
	1971	Minnesota. Some 50,000 gallons of radioactive water spilled into the Mississippi River, with some entering the St. Paul water system.
	1975	Alabama. Reactor fire caused dangerous lowering of cooling water levels.
	1979	Tennessee. Release of highly enriched uranium from a secret nuclear fuel plant contaminated 1,000 people with up to 5 times normal radiation levels.
	1979	Pennsylvania. Partial core meltdown of a reactor on Three Mile Island. Worst disaster in the U.S.
	1981	Tennessee. Workers contaminated when more than 100,000 gallons of radioactive coolant leaked into a containment building.
	1986	Oklahoma. Burst cylinder of nuclear material at the Kerr-McGee plant caused one death and 100 hospitalizations.
USSR	1957	Russian Federation. Explosion of a nuclear waste container at a nuclear plant in the Ural Mountains required 11,000 people to be evacuated.
	1967	Russian Federation. Underwater radioactive waste storage area at Lake Karachy spread radioactive particles when the lake dried up, requiring the evacuation of 9,000.
	1986	Ukraine. Explosion of Chernobyl nuclear plant near Kiev left 31 dead and required the evacuation of 135,000. As a result of radiation released in the atmosphere, tens of thousands of cancer deaths have been projected and large areas sealed off. Worst disaster in history to date.

livestock to levels well beyond health standards for human consumption. Ten years after the accident the Russian government had spent $200 million to clean up the disaster, and put the three remaining reactors at Chernobyl back on line producing electricity. Health impacts over a wide area are still being felt, however, even though official forecasts suggested that continuing health effects would be minor. In fact, researchers have found that the incidence of thyroid cancer has risen in the areas affected, successful pregnancies have declined, and congenital defects in newborns have increased. A comparison of 2,000 newborns in the polluted areas of Belorussia, Russia, and the Ukraine with a like number of children in nearby unpolluted areas showed that more than half the children subject to radiation displayed signs of mental retardation. Experts estimate that as many as 3 million children currently require medical treatment for radiation exposure. Some analysts believe that these findings prove the PETKAU EFFECT, postulated in 1950, that low doses of radiation, even below the thresholds established by public health authorities, can, if continued over an extended period, have an even worse effect on organisms, including humans, than a single high-dose exposure. *See also* NUCLEAR ENERGY and associated entries.

chernozem a grassland soil with a thick, extremely dark A horizon and a B horizon that becomes increasingly alkaline with depth, often containing CARBONATE deposits near or at the bottom (*see* SOIL: *the soil profile*). Chernozem soils develop under cool, semiarid conditions in a STEPPE or SHRUB-STEPPE environment.

chert any of a group of silica-rich rock-forming minerals that result from PRECIPITATION or ORGANIC DEPOSITION of silica compounds in still waters, usually found mixed with LIMESTONE. Cherts are typically fine-grained and extremely hard. The best-known mineral of the group, flint, was the preferred tool-making material of early humans.

Chesapeake Bay the largest and at one time the most productive estuarine ecosystem in the United States. The bay stretches from the northeastern corner of Maryland, north of Baltimore, southward for 200 miles to Norfolk, Virginia. It is 30 miles from east to west at its widest point, at the mouth of the Potomac, and narrows to 4 miles at Annapolis. The Susquehanna, Patuxent, Potomac, Rappahanock, York, and James Rivers are its main freshwater sources, all rivers flowing from the western side of the bay. The rivers of the famed Eastern Shore of Maryland and Virginia, a flattish peninsula separating the bay from the sea, do not contribute much freshwater, although they comprise some of its most biologically rich areas. The combined shorelines of the bay itself measure 4,000 miles. If shoreline measurement were extended

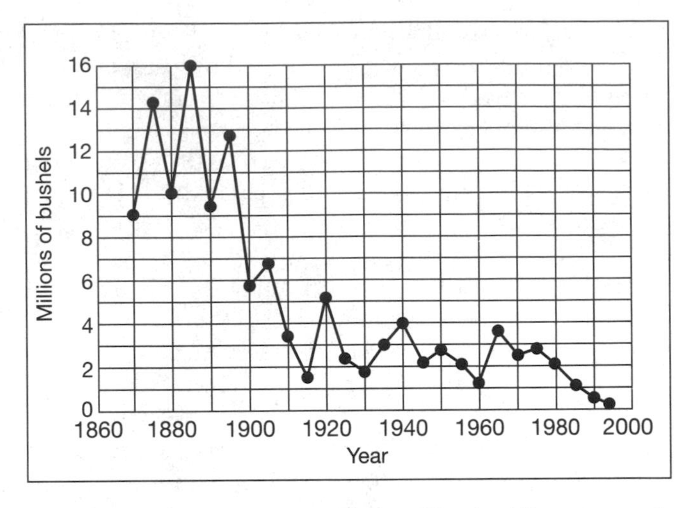

Oyster harvest in Maryland, 1870–1993 ***(Maryland Department of Natural Resources, Fisheries Division, Annapolis, Md.)***

to the high-tide line of the tributaries, the mileage figure would double. Until the 1970s the Chesapeake Bay produced more seafood per acre than any other body of water in the world, although even at that time there were signs that the bay was becoming polluted. Shellfish beds were closed from time to time and fish kills were becoming more frequent. As development increased in cities bordering the bay and as farms within its 64,000-square-mile watershed applied more and more pesticides and chemical fertilizers to their croplands, and as forest cover in the watershed decreased concomitantly, the productivity levels of the bay plummeted. Runoff from city streets entered the rivers and thence flowed into the bay. Oysters, once so plentiful that their shoals were a shipping hazard, began to die out from poisons, over-enrichment, silt, and a change in salinity. When the bay was still a relatively pristine ecosystem, oysters could filter the water, cleansing it of particulates and enhancing the ecosystem for all manner of creatures, including great quantities of blue crab, for which the bay is famous, and striped bass. Both have suffered, along with the oyster. The "dead zone"—waters so depleted of oxygen that productivity is virtually nil—increased 15-fold between 1950 and 1980. In 1983 the states bordering the bay and the U.S. Environmental Protection Agency founded the Chesapeake Bay Program to take the difficult steps to halt, and even reverse, the precipitous decline of the ecosystem. The plan was to reduce nutrients from farm runoff (causing oxygen-depleting algae blooms), protect existing fisheries and associated wildlife, and reduce the effects of toxic substances flowing into the bay, such as pesticides, or dumped into it, such as industrial chemicals and illegal sewage. In many respects, the initial goals have been met and there is a chance for the bay to begin the long road back to stability. The one organism that is the ultimate key to the ecological health of the bay, the oyster, still suffers from a greatly reduced population. Some ecologists recommend a total ban on oyster fishing (it has long been limited) to see whether this crucial component of the natural system can recover and therefore speed the recovery of the entire ecosystem.

Chesapeake Bay Foundation works to protect and restore the Chesapeake Bay and its watershed. Membership (1999): 80,000. Address: Headquarters, 162 Prince George Street, Annapolis, MD 21401. Phone: (410) 268-8816. Website: www.savethebay.cbf.org.

chestnut blight a deadly fungus (*Endothia parasitica*) attacking the American chestnut (*Castanea dentata*), once the dominant tree of the eastern deciduous forest, providing between one-quarter and one-half the CANOPY. The American chestnut is now all but extinct. Inadvertently introduced from Asia when hybridizers wished to combine the board feet (*see* BOARD FOOT) of the big American tree (up to 11 feet in diameter) with the larger nut of its Asian cousin, the fungus chokes the tree to death by cutting off the transport of water and nutrients from the roots. The first evidence of chestnut blight was found on the grounds of the New York Zoological Garden in 1904 by forester Herman Merkel, but despite heroic efforts to stop the blight from spreading, trees rapidly became infected on a broad front throughout their native range, which runs from Maine southward through the Appalachian chain to Alabama and westward into Ohio and southern Indiana. By the 1930s the chestnut had virtually disappeared from eastern forests, replaced by deciduous oaks of various species. Although some small stunted chestnuts sucker up here and there, they usually succumb before they reach great age. Chestnut fanciers have combed the forests to find a resistant genotype that could be cloned, and geneticists have tried, with only limited success, to develop back-crosses with the smaller Chinese chestnut, which is resistant to the fungal blight. Other strategies involve introducing a "hypovirulent" strain of the blight to produce a benign infection in the tree that can "wall off" the spread of disease, and using genetic engineering to create fungus-resistant genes which have been found in a common fungus-eating snail. The AMERICAN CHESTNUT FOUNDATION believes that a successful cultivar may soon emerge and that one day it might be reintroduced into the eastern forest.

China syndrome in connection with nuclear power plants, a term describing a core meltdown so severe that it would go right through the Earth all the way to China. The term was used as the title of a 1979 movie about the cover-up of an accident at a nuclear facility, starring Jane Fonda, Jack Lemmon, and Michael Douglas. The movie was released just two weeks before the accident at THREE MILE ISLAND, which helped make it a hit. While neither the Three Mile Island incident nor the CHERNOBYL explosion seven years later in the Ukraine was a true meltdown, these disasters were serious enough to put the phrase "China syndrome" firmly into the environmental lexicon.

chipping mills in the forest products industry, the facilities that turn logs into chips that then are further processed into pulp for paper and other products. Chipping mills, which are common in the South, are fed by clearcutting surrounding forests within a radius of as much as 60 miles. The mills use not only mature trees but also saplings and branches. *See also* CLEARCUT.

chlordane an ORGANOCHLORIDE pesticide, chemical formula $C_{10}H_6Cl_8$. Chlordane is a thick, amber-colored fluid, formerly applied as an AEROSOL spray either by heating it to 120°F (at which point its viscosity dramatically declines) or by mixing it with an organic SOLVENT such as KEROSENE. Its use is banned in the United States. While only moderately toxic to mammals in single doses, (LD_{50} [rats, orally]: approx. 500 mg/kg), chlordane is a cumulative poison, and long-term exposure may lead to severe liver damage. It absorbs readily through the skin, hence is exceedingly dangerous to farm workers.

chlorinated hydrocarbon a HYDROCARBON compound in which one or more of the hydrogen ATOMS has been replaced by a chlorine atom. As a group, chlorinated hydrocarbon compounds tend to be solids or thick, stable fluids with low electrical CONDUCTIVITY, of low ACUTE TOXICITY to mammals but often of high acute toxicity to invertebrates, and they have found broad use as pesticides, lubricating fluids, ASKAREL LIQUIDS, etc. Unfortunately, their usefulness is severely limited by the fact that most are extremely fat-soluble and bioaccumulative (*see* BIOLOGICAL MAGNIFICATION), and over the long term tend to be carcinogenic and mutagenic (*see* CARCINOGEN; MUTAGEN). For these reasons, the use of most of them is currently banned or highly restricted in the United States. Examples: DDT, PCB (*see* POLYCHLORINATED BIPHENYL), CHLORDANE, HEPTACHLOR. *See also* ORGANOCHLORIDE; HALOGENATED HYDROCARBON.

chlorine the 17th element of the atomic series; atomic weight 35.453, chemical symbol Cl. Chlorine is a dense, yellow-green gas with a choking, suffocating odor, extremely toxic to all forms of life. It is one of the most reactive substances known. Though it is one of the more common of the elements that make up the earth (rank: 14th), chlorine does not exist in nature in the free elemental state but is always found as part of a COMPOUND. An atmosphere of chlorine will support the combustion of CARBON and many other materials, including most METALS. An atmosphere of mixed chlorine and hydrogen, when exposed to light, will explode violently. As a constituent of table salt (sodium chloride; NaCl), chlorine is a necessary element in the human diet. It is also used as a water purifier, as a household bleach (in aqueous solution), and as a chemical REAGENT in many industrial processes. It is prepared commercially by passing an electric current through salt water. U.S. production (1984): 10.8 million short tons.

chlorofluorocarbon (CFC) a family of extremely useful but problematic compounds developed in 1930 by a General Motors chemist. The compounds include CFC-11, trichlorofluoromethane (CCl_3F), used as a coolant in air conditioners and refrigerators and to make the bubbles in plastic foam; and CFC-12, dichlorofluoromethane (CCl_3F_2), a now-banned propellant (Freon) used in AEROSOL cans. In 1974 chemists at the Massachusetts Institute of Technology (MIT) discovered that CFCs, which rise slowly to the stratosphere (10–20 years to make the trip), were breaking down the OZONE LAYER, whose O_3 molecules form a stratospheric blanket keeping harmful ultraviolet rays in the UV-B range from striking the Earth. (In high doses UV-B can cause blindness and cancer in animals and humans and affect the growth of plants.) When the CFC compounds reach the stratosphere, the chlorine atoms break down, which in turn speeds up the breakdown of the protective ozone (O_3) into plain oxygen (O_2). And the effect is cumulative and relentless. The CFC molecule can last between 65 and 110 years in the stratosphere, and each one has the capability to convert 100,000 ozone molecules into oxygen molecules. In 1984, 10 years after the MIT scientists' warning, the thinning of the ozone layer was decisively proven when scientists discovered that 40 to 50% of the ozone layer over Antarctica was being destroyed. Although commonly referred to as the "ozone hole," it was in fact a thinning of the ozone layer, a phenomenon that now takes place in the Northern Hemisphere as well, albeit to a lesser extent. In 1987, 36 nations signed an agreement (the MONTREAL PROTOCOL) to phase out CFCs, and chemists got to work to create alternative coolants for air conditioners and refrigerators. These include a group of hydrochlorofluorocarbons (HCFC) and hydrofluorocarbons (HFC) which eliminate the chlorine atom altogether. These compounds are now in use, and there is some indication that the ozone layer may be responding. Despite the protocol, however, CFCs continue to be manufactured and today are second in dollar value only to cocaine as an illegal substance smuggled into the United States.

chloroform (**trichloromethane**) a colorless, sweet-tasting liquid, chemical formula $CHCl_3$, dense but of extremely low VISCOSITY. Once widely used as a pain reliever during surgery or childbirth, chloroform fell out of use as safer analgesics were discovered. It causes cardiac arrest in roughly one out of 3,000 patients, and when exposed to light tends to decompose, releasing phosgene gas, a deadly poison. Since 1976 it has been banned for drug use in the United States as a suspected CARCINOGEN, but it still finds use in industry as a SOLVENT and cleansing agent and as an ingredient of fire extinguishers.

chlorophyll the green pigment in plants, necessary for PHOTOSYNTHESIS and therefore for the existence of all complex life forms on earth. Chlorophylls are conjugated PROTEINS, chemically and structurally similar to blood HEMOGLOBIN but containing magnesium in place of the hemoglobin's iron. The chlorophyll molecule also sports a long HYDROCARBON "tail" that is absent in hemoglobin. At least four forms are known. Chlorophyll *a* ($C_{55}H_{72}MgN_4O_5$) and chlorophyll *b* ($C_{55}H_{70}MgN_4O_6$) are found, mixed together, in the higher plants and in green algae; blue-green algae contain chlorophyll *a* and chlorophyll *c* ($C_{35}H_{28}MgN_{405}$); red algae mix chlorophyll *a* with chlorophyll *d* ($C_{54}H_{70}MgN_4O_6$). Though it is light-sensitive, chlorophyll will not support photosynthesis in a test tube but must be held in a section of the plant cell known as the *chloroplast* (*see* CELL: *structure of cells*). It therefore has no commercial use as a carbohydrate producer separate from its role in plant life, though it finds use in other ways, as an ingredient in soaps and mouthwashes and in the production of photographic film, vulcanized leather, and some forms of antiknock gasoline. It is produced in commercial quantities by the solvent-extraction process from the leaves of green plants, principally spinach and alfalfa.

Chlorophyta *See under* ALGAE.

choker in logging, a length of cable or chain that is looped around a log (or group of small logs or limbs) to form a "handle" for YARDING equipment to grasp.

choker setter the member of a logging crew who is assigned the job of putting CHOKERS on logs and other materials to be yarded (*see* YARDING). The task of choker setter is usually assigned to the member of the crew with the least seniority.

chordate in biology, any member of the PHYLUM Chordata, characterized primarily by the presence of a spinal chord in at least some stage of the ORGANISM'S life. In most adult chordates, including fish, birds, reptiles, amphibians, and humans and other mammals, the spinal chord is enclosed in a flexible bony sheath called the *backbone*. This sheath and its enclosed spinal chord are formed from a rigid rod of nerve cells and surrounding protective tissue known as the *notochord*, which is present in the embryonic or larval stage of all chordates and which persists into the adult stage in some of the more primitive varieties. (The most primitive, including sea squirts and similar sessile species, lose all traces of the notochord and of the spinal chord itself as adults and come to resemble sea anemones, sponges, and other sessile INVERTEBRATES [*see* SESSILE ANIMAL].) Other characteristics shared by all chordates, at least in the embryonic or larval form, include bilateral symmetry (a body with two similar sides), the presence of gill slits, and the development, however temporary, of a tail. In most members of the phylum, one end of the spinal chord eventually enlarges to form a brain. *See also* VERTEBRATE.

Christmas Bird Count an annual bird census undertaken just before and just after Christmas organized by the NATIONAL AUDUBON SOCIETY. Participants go out in groups to record every bird seen or heard in a 24-hour period, each group operating within a 15-mile circle. The program was started by the famed ornithologist Frank M. Chapman, who suggested the idea in the December 1900 issue of *Bird-Lore*, the Audubon Society magazine (renamed *Audubon* 40 years later). Chapman believed that such a census (originally held on Christmas day itself) would be a more humane way to determine bird populations than taking samples in the field via shotgun. What the bird count lacks in scientific rigor it makes up for in sheer longevity. Long-term trend data in the population of bird species revealed by the count have been extremely useful to ornithological research and in determining the impacts of changes in habitat. Today, the bird count is conducted throughout North America and in certain areas in the Caribbean and Central and South America. Some 45,000 counters join the program each year.

Christmas-tree bill in politics, a piece of legislation designed to provide "gifts" for as many members of the legislature as possible, by including projects that they are particularly interested in or which will bring money into their districts. *See also* OMNIBUS BILL; PORK BARREL.

chromium the 24th element of the atomic series, atomic weight 51.996, chemical symbol Cr. Chromium is a tough, hard, lustrous metal, silvery in color, with an extremely high melting point (1900°C): it will not combine with oxygen under atmospheric conditions, making it very useful in situations where corrosion resistance is important. When alloyed with steel (in so-called "stainless steel") it imparts much of its toughness and corrosion resistance to the alloy in concentrations as low as 10%. Chromium is also an important TRACE ELEMENT in the nutrition of humans and other mammals. Its exact role is not known, but it appears to be necessary to the proper METABOLISM OF CARBOHYDRATES. Notwithstanding the nutritional importance of the element, however, many chromium compounds—particularly the chromate salts—are highly toxic and cause severe tissue damage on contact, and are often CARCINOGENS. Hence, chromium is a serious pollutant, and its mining, smelting, and use in industry must be done under strict controls. Particular care must be taken to see that mine SPOIL and smelter wastes do not reach watercourses. Only about 10% of the chromium used in the United States is minded domestically. The remainder is imported, principally from South Africa, the Philippines, and the Soviet Union.

chromosome the body within a living CELL that holds the cells' genetic material and is responsible for passing on inherited characteristics from generation to generation. In all higher animals and plants, chromosomes come in pairs, with one of each pair provided by each parent, thus guaranteeing genetic diversity and a multitude of individual mixtures of characteristics. The number of pairs varies among species; humans have 23. Chromosomes are large, complex structures that appear under a microscope as long, convoluted strings. They are composed of DNA molecules, RNA molecules, and proteins called *chromatin*. The genetic information is carried on the DNA molecules. Each chromosome carries the information for up to several thousand inherited characteristics. Chromosomes that are altered or damaged cannot pass on these characteristics properly; hence, chemicals that cause chromosome damage are a particularly insidious type of pollutant, whose effects will not normally be felt by the affected individual but by that individual's offspring. *See* MUTAGEN. *See also* GENE; CELL: *structure of cells*.

chronic toxicity a form of TOXICITY in which the symptoms develop only after long-term (*chronic*) exposure to the toxic compound. "Long-term" may be measured in days, weeks, or years, depending on the compound involved. Normally the symptoms caused by chronic toxicity are themselves chronic; that is, they take days, weeks, or years to go away. *Compare* ACUTE TOXICITY.

chrysophyte any SPECIES of ALGAE belonging to the CLASS Chrysophyta of the ORDER Thallophyta. Because the green color of their CHLOROPHYLL is usually partially masked by other pigments, giving them a characteristic golden hue, the chrysophytes are popularly called "golden algae." They have the unusual characteristic of storing food reserves in their bodies as oils rather than as starches. Most of the 17,000 or so known species are DIATOMS.

cinder cone a symmetrical, steep-sided heap of VOLCANIC ASH, cinders, and other pyroclastic materials (*see* PYROCLASTIC ROCK), deposited around the mouth of a volcanic vent as the result of a series of explosive eruptions. Formation of a cone begins with the deposit of a ring of material around the vent. If the eruption ceases at this point, the cone is known as a *tuff ring*. Continued explosive eruptions pile materials both inside and outside the tuff ring, building the cone. Since there is no core of hardened LAVA for structural strength (although the invasion of already formed cinder cones by later lava flows may produce the illusion of a core), cinder cones seldom exceed 100 meters (330 feet) in height. Their slopes almost always lie at a uniform angle of 33°—the ANGLE OF REPOSE for volcanic cinders.

cirque a steep-sided, bowl-shaped depression in a mountain slope, several hundred feet to several miles across, formed by downcutting action at the head of an active GLACIER. It may or may not still contain the glacier. Cirques are often remarkably regular in shape, as if scooped out in a single stroke by a huge shovel or scoop. Those no longer occupied by glaciers usually contain lakes and/ or meadows. *See also* GLACIAL LANDFORMS.

CITES Treaty *See* CONVENTION ON INTERNATIONAL TRADE IN ENDANGERED SPECIES.

city planning *See* LAND USE PLANNING.

Civilian Conservation Corps (CCC) federal agency of the late 1930s and early 1940s, established to provide employment and training for young out-of-work men and to promote the development and conservation of natural resources. Created through EXECUTIVE ORDER by President Franklin D. Roosevelt on April 5, 1933 as a division of the War Department called Emergency Conservation Work, the agency was soon dubbed the "Civilian Conservation Corps" by the press, and when Congress authorized its continuation in June, 1937, the popular name was formally attached to it. It became part of the Federal Security Agency in July 1939 and was abolished shortly after American entry into World War II. To qualify for admission to the CCC, a young man had to be a U.S. citizen, unemployed, and between the ages of 17 and 25. CCC workers received $30/month plus room and board, usually in a work camp run jointly by the War Department and a land-management agency such as the FOREST SERVICE or the NATIONAL PARK SERVICE. At its height around 1940, the agency was finding work for some 3 million men.

LEGACY

The CCC was among the most successful public-works agencies ever created. CCC workers reforested hundreds of thousands of acres of burned and logged-over land, built fire lookouts, rehabilitated streams, and created fish ponds. They constructed hundreds of miles of high-standard forest trails and built trail shelters and campgrounds in National Forests and National and State Parks, many of which remain in use today. CCC-built structures are characterized by solid stonework and heavy, rough-hewn woodwork, and have set standards for most forest architecture since. The agency itself has been the model for other federal and state conservation and employment programs, notably the JOB CORPS.

Cladophora a widespread GENUS of green ALGAE, often associated with CULTURAL EUTROPHICATION of lakes and streams, where it is a common constituent of ALGAL BLOOMS. *Cladophora* are multicelled algae that grow as branched filaments in dense bunches that may reach 15 inches or more in length. They are equipped with tendrils called *holdfasts* to anchor them to solid objects in the streambed or lakebed, though they can also live as free-floating organisms. Unlike most green algae, they reproduce through ALTERNATION OF GENERATIONS. Unlike true plants that practice this form of reproduction, the algae's haploid and diploid generations are the same size and have the same growth form (are "isomorphic"). They are very hardy and adaptable organisms, and the only effective means of controlling them is to control the NUTRIENT load entering the water.

clarifier *See* SETTLING BASIN.

Clarke-McNary Act federal legislation with a major effect on shaping the growth and functions of the United States FOREST SERVICE, signed into law by President Calvin Coolidge on June 7, 1924. Essentially a series of amendments to the WEEKS LAW of 1911, the Clarke-McNary Act had two principal purposes: (1) to allow the Forest Service to provide technical forestry assistance to private landowners, thus laying the basis for today's programs in cooperative forestry; and (2) to expand the authority of the secretary of agriculture (*see* AGRICULTURE, DEPARTMENT OF) to purchase private lands, accept land donations, and seek the transfer of lands from other agencies so that these purchases, donations, and transfers could be made for timber production as well as watershed protection. Passage of this act was pivotal in the evolution of the National Forests from preserves to timber farms. *See also* MULTIPLE USE-SUSTAINED YIELD ACT; NATIONAL FOREST MANAGEMENT ACT.

class in biology, a taxonomic category (*see* TAXONOMY) lying between the PHYLUM (or DIVISION) and the ORDER. The differences between individual classes in the same phylum are equivalent to those between, for example, mammals (class Mammalia) and birds (class Aves), or between conifers (class Angiospermae) and hardwoods (class Gymnospermae).

class, air quality *See* AIR QUALITY CLASS.

class, recreation land *See* RECREATION LAND CLASS.

clay an extremely find-grained mineral constituent of soil that will form a colloidal paste when mixed with water (*see* COLLOID). The definition regarding the largest particles that may still be called clay varies, but a common test is that the particles must be small enough to pass through a No. 200 sieve (0.074 millimeters in diameter, about .003 inches). Clays are produced primarily by the weathering of FELDSPAR and secondarily by the weathering of SHALE and SLATE (which are themselves composed of compacted and hardened clay deposits). Chemically, they are almost exclusively hydrated aluminum silicates, that is, aluminum silicates modified by the addition of a small amount of WATER OF CRYSTALLIZATION; a few may be magnesium silicates. Mixed with water, clay forms a plastic, easily worked paste that holds shapes well and hardens permanently when baked with sufficient heat. It has been used since prehistoric times to make pottery, sculpture, and bricks. It is also an ingredient of cement, a sizing for paper and a filler for paint, among numerous other applications. Not a pollutant in the strict sense of the term (that is, not a TOXIN, CARCINOGEN or MUTAGEN), it can still be a serious problem in lakes or quiet watercourses because it can render water opaque, greatly lowering the water body's PRIMARY PRODUCTIVITY (*see* TURBIDITY; COMPENSATION DEPTH)—a problem compounded by the extraordinarily long time a fine clay will take to settle out in still water. *See also* KAOLIN.

Clean Air Act term applied to several related pieces of federal legislation designed to protect air quality in the United States. The original Clean Air Act was signed into law by President Lyndon Johnson on December 17, 1963. However, little of this act remains intact, and the term "Clean Air Act" as used today generally refers to one or both of the extensive sets of amendments passed in 1970 (date of president's signature: December 31, 1970) and 1977 (date of president's signature: August 7, 1977). A previous, smaller set of amendments was enacted in 1966. As currently constituted, the act has two major, interrelated thrusts: to prevent the further deterioration of general air quality in the United States, and to reduce or eliminate the incidence of a group of specific pollutants (LEAD, OZONE, CARBON MONOXIDE, NITROGEN OXIDES, sulfur dioxide [*see* SULFUR OXIDES], HYDROCARBONS, and PARTICULATES) which have been found to be particularly hazardous to human health or to the environment. Standards are set both for the emission of these pollutants by industrial sources and motor vehicles and for their presence in the AMBIENT AIR (*see* NEW SOURCE PERFORMANCE STANDARDS; NATIONAL EMISSION STANDARDS FOR HAZARDOUS AIR POLLUTANTS; NATIONAL AMBIENT AIR QUALITY STANDARDS). Enforcement of these standards is left up to the states through STATE IMPLEMENTATION PLANS; however the ENVIRONMENTAL PROTECTION AGENCY is given broad powers to compel the states to adopt and enforce adequate plans, including the withholding of federal grant money from other agencies for projects in states that have not adopted plans in compliance with federal guidelines. Areas whose air meets the National Ambient Air Quality Standards are designated *attainment areas* and are required to maintain these standards through a program known as PREVENTION OF SIGNIFICANT DETERIORATION (PSD). Areas whose air does not meet the National Ambient Air Quality Standards are designated NONATTAINMENT AREAS and must take steps to bring themselves into attainment.

THE 1990 AMENDMENTS

On November 15, 1990, after an intense four-year campaign, the Clean Air Act of 1990 was signed into law by President George Bush. The new act reauthorizes the existing act and strengthens it in several key areas. The issues of OZONE LAYER destruction, ACID RAIN and the retraining of workers displaced by regulation-induced plant closures are addressed for the first time. In addition, the act lowers permissible emission rates of HYDROCARBONS and NITROGEN OXIDES by 35 to 60 percent; extends emission standards to light utility vehicles; establishes five new categories of areas suffering from urban SMOG (marginal, moderate, serious, severe, and extreme) within which control programs will have to be instituted; and makes it significantly easier for citizens to sue to ensure enforcement of the act by government agencies. *See also* AMBIENT QUALITY STANDARD; BEST AVAILABLE RETROFIT TECHNOLOGY; BEST AVAILABLE CONTROL TECHNOLOGY; LOWEST ACHIEVABLE EMISSION RATE; BUBBLE.

Clean Air Network an association of local, state, and national organizations dedicated to protecting and preserving air quality to safeguard human health and a maintain a healthy environment. The network sets a national agenda and provides its members with the necessary resources and support to take action. Address: 1200 New York Avenue NW, Suite 400, Washington, DC 20007. Phone: (202) 289-2429. Website: www.cleanair.net.

cleaning cut in forestry, a very early form of PRECOMMERCIAL THINNING, done while the stand to be thinned is still in the SAPLING stage. Ideally, the timing of a cleaning cut should be just after BROWSE HEIGHT is reached and any remaining OVERSTORY has been removed (*see* REMOVAL CUTTING; SEED TREE CUTTING; SHELTERWOOD SYSTEM). At this stage, individual genetic differences in the young trees are already easy to spot, but very little of the site's growth potential has gone into the trees that will be removed.

Clean Sites, Inc. a program to encourage the use of innovative technology in hazardous site cleanup to reduce costs, increase effectiveness, and decrease time involved. The program consists of five basic components: arrangement of partnerships and agreements, assessment of appropriate technologies, management of technology development and reclamation projects, public policy evaluation, and education and communication. Address: 1199 North Fairfax Street, Suite 400, Alexandria, VA 22314. Phone: (703) 683-8522.

Clean Water Act term popularly applied to the federal Water Pollution Control Act Amendments of 1972, signed into law by president Richard Nixon on October 18, 1972. The name was subsequently adopted by Congress as the official title to the 1977 amendments to the act, signed into law by President Jimmy Carter on December 27, 1977, and for all amendments since that time. The goal of the 1972 act—expressly stated in the act itself—was to make all navigable waters of the United States fishable and swimmable by 1983 (*see* NAVIGABLE STREAM), and to eliminate entirely all discharge of pollutants into these waters by 1985. Though the deadlines have been abandoned, these goals remain as guiding principals of the act today. The most important provision of the act from an enforcement standpoint is Section 402, which requires permits for all POINT SOURCE DISCHARGES of pollutants into navigable waters (*see* NATIONAL POLLUTANT DISCHARGE ELIMINATION SYSTEM). Sections 301–307 are also important: these require waste-treatment facilities to reduce pollutants in the treated EFFLUENT by the greatest amount that is technologically feasible and, in some cases, to eliminate them altogether. (*See* BEST AVAILABLE TECHNOLOGY; BEST CONVENTIONAL WASTE TREATMENT TECHNOLOGY; PRIORITY POLLUTANT; CONVENTIONAL POLLUTANT; MICROCONTAMINANT). Other sections set up a nationwide system of planning to reduce or eliminate pollution from non-point sources (*see* 208 PLAN; NONPOINT POLLUTION); require the pretreatment of industrial wastes before they may be discharged into sewers (*see* PRETREATMENT STANDARDS); mandate a study by each state of all freshwater lakes within its borders to determine each lake's condition and the management steps required to bring it into compliance with the "fishable, swimmable" goal; and continue a long-standing series of federal grants to municipalities for the construction of sewage-treatment facilities.

HISTORY

The Clean Water Act, as it now stands, incorporates two separate lines of legislation. The original Federal Water Pollution Control Act, signed into law by President Harry Truman on June 30, 1948, represents the principal line: the current act is officially a set of amendments to this 1948 law. It was concerned primarily with the health effects of water pollution, and originally took the form of federal grants for the construction of sewage-treatment facilities by municipalities that discharged their sewage into navigable waters. Amendments and extensions were made in 1952, 1955, 1960, and 1966. The second line is concerned with environmental effects rather than health effects. It had its origin in the Oil Pollution Act of 1924, which prohibited the dumping of oil by vessels in coastal waters. Numerous amendments to this act over the years greatly broadened and strengthened it, culminating in the Water Quality Act of 1965, which required states to designate stream segments by their dominant use (recreation, industry, water supply, etc.) and set water quality standards appropriate to this use. The two lines were joined for the first time in the Water Quality Improvement Act of 1970, which effectively made the Water Quality Act and its predecessors a part of the Water Pollution Control Act, setting the stage two years later for the complete legislative overhaul that became the Clean Water Act of 1972.

AMENDMENTS TO THE ACT

The Clean Water Act was modified in 1977, 1981, and 1987. The 1977 amendments expanded on state participation in the pollution discharge program and gave the EPA additional powers to regulate the discharge of toxic substances. In 1981, with a change of administrations in Washington (from Carter to Reagan), amendments reduced the level of funding given to states and municipalities for the construction of wastewater treatment facilities. In 1987 the act was amended to provide a revolving fund to states for funding wastewater treatment infrastructure, the protection of estuaries, and nonpoint source pollution controls. Between 1989 and 1997 the fund provided $13 billion for these purposes, but at a steeply declining rate, as the cost of water cleanup shifted from national to state and local governmental levels. In 1995 amendments were proposed by a Republican-dominated Congress that would have given polluters more time to meet mandated standards for cleanup. The bill was opposed by the Clinton administration and died in the House. Instead, the administration proposed in 1998 a "Clean Water Action Plan" that would devote $10.5 billion to clean up the 40% of the nation's waterways that had failed to meet the "fishable, swimmable" goal set forth in the 1972 act. So far, Congress has failed to undertake a major overhaul of the legislation. Congress does continue, however, year after year, to appropriate funds. And if the ambitious goals of the 1972 Clean Water Act and its amendments have not yet been fully met, the act has demonstrably produced cleaner rivers, lakes, streams, and groundwater resources.

Clean Water Fund concerned with environmental and consumer protection especially in the areas of water pollution and solid waste management. The fund fulfills its mission through research, training, education, and publications. Address: 4455 Connecticut Avenue NW, Suite A 300, Washington, DC 20008. Phone: (202) 895-0420.

clearcut in forestry, a method of timber harvesting in which all trees in a HARVEST UNIT are cut down. Clearcuts are generally easier and cheaper to make than SELECTIVE CUTS or shelterwood cuts (*see* SHELTERWOOD SYSTEM), and are necessary for the establishment of EVEN-AGED STANDS and the reproduction of certain SPECIES under some climate conditions (*see,* e.g., DOUGLAS FIR). However, they destroy (at least temporarily) many of the AMENITY VALUES of the forest; leave soils vulnerable to EROSION; and seriously disrupt biological communities over an area reaching some distance beyond their boundaries, often leading to a permanent change in the types of plants and animals which can grow there. *Compare* ECOFORESTRY; ZEROCUT.

clear lumber in the forest products industry, unblemished lumber, particularly that which has no KNOTS. Sound clear lumber brings the highest price of all forest products—a fact that puts extra economic pressure on the industry for the logging of large OLD GROWTH trees, which contain the highest quantities of clear lumber.

clear well in water treatment technology, the reservoir in which FINISHED WATER is stored following treatment and before delivery into the distribution system.

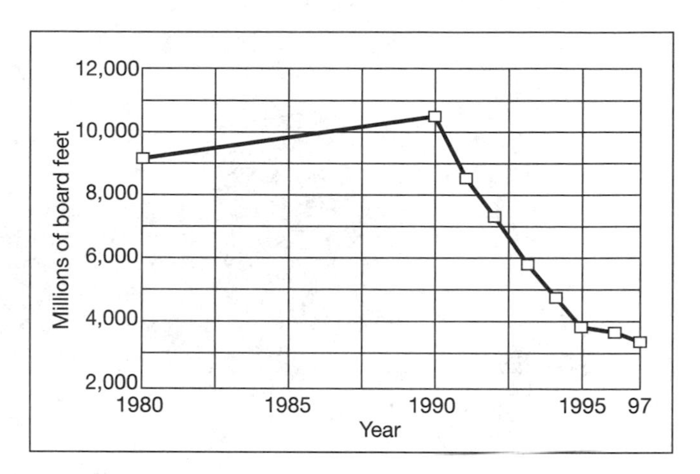

Timber cut in U.S. national forests, 1980–1997 ***(U.S. Forest Service, Timber Demand and Technology Assessment)***

cleavage plane in geology, a flat surface along which a rock of MINERAL will break easily. The angle between adjacent cleavage planes is characteristic for each type of rock or mineral and may be used as a tool for the identification of samples.

cleaver in geology and mountaineering, a long, narrow band of rock, SCREE, or morainal material (*see* MORAINE) separating two sections of a snowfield or GLACIER. It lies up and down a slope rather than across it. A cleaver lies at the center of a face rather than at the angle where two faces meet; hence, it is not a true ridge. *Compare* ARETE.

climate the long-term pattern of changes in atmospheric conditions averaged over a year, a number of years, or many centuries. The description of a region's climate includes factors such as the average annual rainfall; the presence (or absence) of wet and dry seasons; average hours of sunlight; average cloud cover; mean annual temperature, along with the average annual maximum and minimum temperatures and the time of year in which they occur; and so on. *Compare* WEATHER. *See also* ATMOSPHERE: *dynamics of the atmosphere.*

climate change *See* GLOBAL CHANGE.

climate zone a designation describing geographical areas as influenced by weather. In North America, climate zones typically include Polar (at the poles), Subarctic (most of Alaska), Cool (the Northeast and Midwest), Warm (Southeast, southern Pacific coast), Dry (Southwest, Great Plains), Tropical (Caribbean, parts of Mexico) and Highland (Mountains in the West and Northwest). In agriculture, climate zones are used to guide farmers and gardeners on the suitability of certain plants and crops in a given area. These zones (*see*

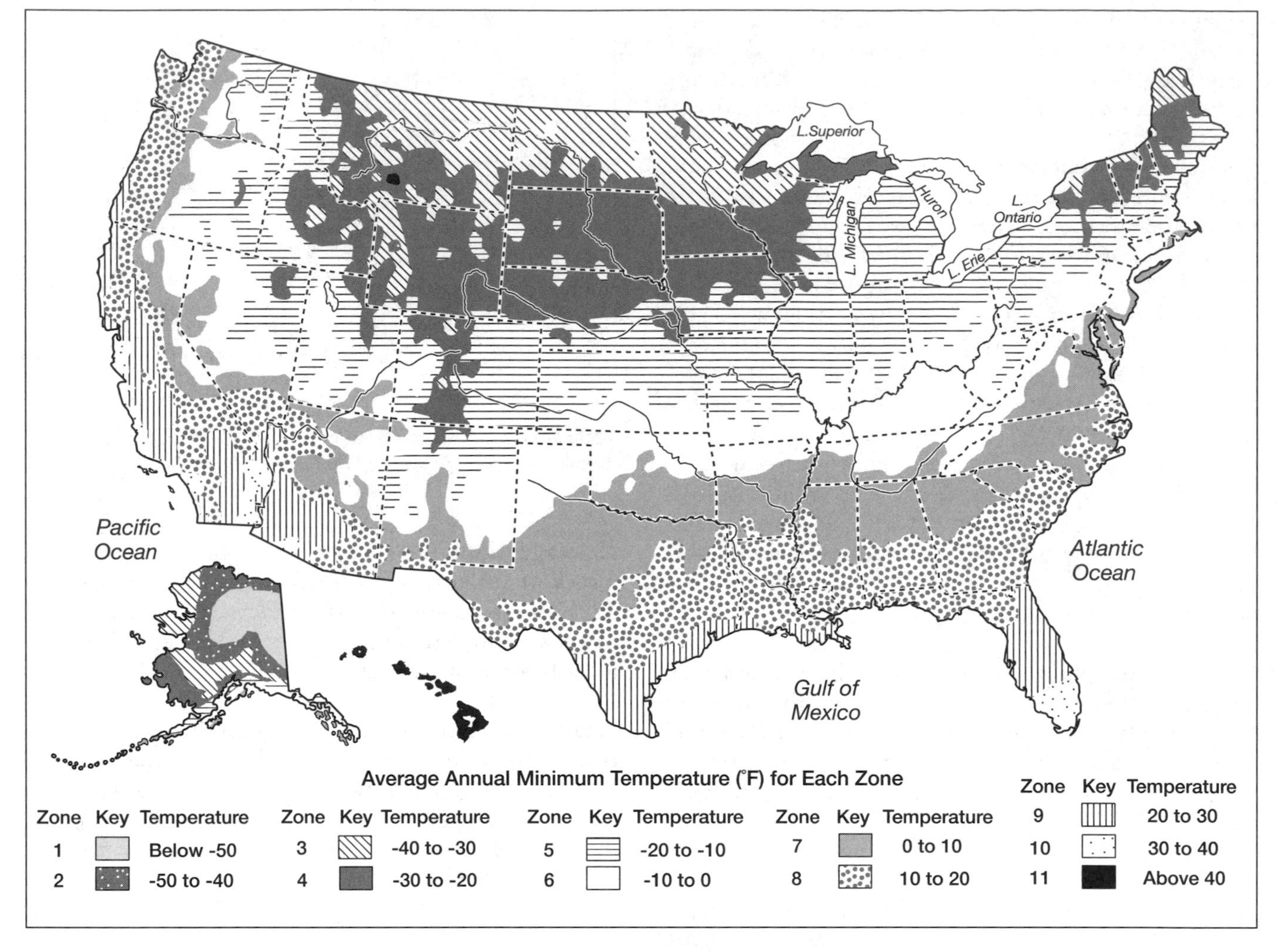

The USDA Plant Hardiness Map (United States Department of Agriculture, 2000)

plant hardiness map above) range from 2 to 11, in the coterminous U.S.

climatic climax In ecology, a CLIMAX COMMUNITY in which the species makeup is determined largely or exclusively by climatic factors such as amount of rainfall, temperature extremes and averages, etc. Most climax communities in nature are climatic climaxes (but *see* FIRE MAINTAINED CLIMAX).

climax (climax stage, climax state) in ecology, the final stage of community SUCCESSION, when stability has been achieved and the ECOSYSTEM is no longer changing. The climax state is characterized by complexity and interdependence; that is, there are ordinarily a large number of plant and animal SPECIES involved, many of which depend on the presence of other species to survive. *Compare* SERE.

climax community in ecology, the plants and animals that characterize the CLIMAX state of a succession.

climax forest in ecology, a forest that has reached the CLIMAX state and is no longer significantly changing in terms of species mix present.

climax vegetation the plants involved in a CLIMAX COMMUNITY.

climbing bog a type of BOG in which some of the characteristic bog vegetation (usually sphagnum moss) extends some distance upward from the bog onto the surrounding "dry" land. Climbing bogs typically develop in areas of heavy rainfall and short, cool summers, often on thin soils with high per-unit water-holding capacity but little volume—all of which means that the bog vegetation will have plenty of water to maintain itself outside the limits of the bog. *Compare* HANGING BOG.

cline in the biological sciences, a gradual gradient in the physical and/or behavioral characteristics of a single SPECIES over geographical distance. Adjacent

POPULATIONS are nearly indistinguishable from one another, but populations at some distance from each other along the cline may be strikingly different in size, color, etc. In extreme cases, populations from opposite ends of the cline will not be able to interbreed, justifying calling them separate species, though both can interbreed with any of a number of common intermediate populations. *See* ALLEN'S RULE; BERGMANN'S RULE; GLOGER'S RULE.

clone in the broadest sense, a group of CELLS which has arisen from a single "parent" cell strictly through the process of cell division, with no fertilization occurring at any point. The group of cells that form the clone is usually a POPULATION of one-celled animals or plants, but it may also be an individual multicelled organism. Since every cell of a multicelled organism (including a human being) carries a complete set of CHROMOSOMES and therefore contains all the genetic information necessary to reconstruct the organism, it is theoretically possible to create a clone of any individual of any species from a randomly chosen cell. In practice, the cloning of complex organisms is limited by the difficulty of creating and maintaining the conditions under which cell division can take place outside the body. It is much easier to create a clone of a plant than of an animal—in fact, many plants use a form of cloning as a principal means of reproduction (strawberry plants reproducing from runners, for example: *see* ASEXUAL REPRODUCTION: *budding*)—and the process is often used in nurseries to maintain a plant's desirable genetic characteristics by growing clones of it from cuttings. Some conifers, including DOUGLAS FIR, may be cloned from their needles, providing a means of producing seedlings for reforestation without having to collect seed. Since each clone is an exact genetic copy of its parents, it is much easier to select for superior stock or stock that is well-adapted to the proposed growing site. However, since each clone carries all of the parent individual's faults as well as its virtues—such as susceptibility to drought, insect attack, and so forth—a plantation of clones is much more easily damaged by environmental disruption than is a plantation of mixed individuals, and is more likely to be completely destroyed if such disruption occurs. The cloning of animals is much more difficult than the cloning of plants, and the cloning of vertebrates has proved to be the most difficult of all, though frogs and other amphibians have been successfully cloned. *See also* MITOSIS; GENETIC ENGINEERING.

closed basin in hydrology, a river basin or WATERSHED that has no outlet to the ocean. Closed basins are characterized by the presence of alkaline lakes into which the rivers run and from which evaporation is the only outlet. All rainfall within the boundaries of the basin is eventually either evaporated or transpirated back into the air (*see* EVAPORATION; TRANSPIRATION; EVAPOTRANSPIRATION), though it may reside for some time in the soil or be moved some distance by rivers first. The watershed surrounding the Great Salt Lake in Utah is an example of a closed basin. *See also* GREAT BASIN.

cloud forest a forest on a mountainside against which clouds often rest. The conditions within a cloud forest are moister and darker than those in forests at similar elevations that are not so often covered by clouds, giving rise to a distinctive FLORA that often include large numbers of BRYOPHYTES and other moisture-loving plants. Trees in these forests are often adapted to obtaining much (or even most) of their water needs by absorbing moisture through their leaves instead of their roots. Cloud forests occur on the seaward side of coastal mountain systems, usually in a band above the rain forests (*see* RAIN FOREST BIOME), to which they are similar. They are normally at fairly high elevations, but where mountains rise directly out of the sea they may exist at near sea-level conditions, as in the so-called "fog forest" which lies in a narrow band along the Pacific coast of North America from San Francisco north to Alaska.

cloud seeding *See* WEATHER MODIFICATION.

Club of Rome Report *See* LIMITS TO GROWTH.

coactor a species that interacts with the DOMINANT SPECIES in such a way as to either increase or decrease the dominant species' role in the community. The red tree vole, for example, is a coactor in DOUGLAS FIR communities because it spreads the spores of the MYCORHIZA that are required to keep the Douglas firs healthy.

coal a type of rock consisting primarily of ORGANIC MATTER that has been highly modified by heat and pressure so that it contains a large proportion of elemental CARBON, that is, carbon that is not part of a COMPOUND. Coal begins as *peat,* a moist organic muck formed as a result of anaerobic decomposition of fallen vegetation on the floors of BOGS and SWAMPS. The peat becomes buried beneath SAND and SILT deposits. When these deposits become thick enough, the pressure exerted by their weight turns their lower layers into SANDSTONE and SHALE, converting the peat to low-rank coal in the process. This lowest rank of coal, called *lignite,* is about 40% water—roughly half the moisture content of peat—and contains large amounts of complex HYDROCARBONS. Its elemental carbon content, known as *fixed carbon* due to the method of analysis used to determine it (*see* FIXED SOLIDS), is roughly 35%. Heat from geothermal sources deep in the earth converts lignite to *bituminous coal,* which is about 50% fixed carbon and 45% hydrocarbons, the moisture content having been reduced to roughly 5%. Further heat, combined with pressure from TECTONIC ACTIVITY, drives most of the remaining hydrocarbons off, forming *anthracite*—still containing about 5% water but with nearly all of the remaining 95% composed of fixed carbon. Coals intermediate in rank between bituminous and lignite are called *subbituminous;* those between bituminous and anthracite are referred to as *semibituminous* or *semianthracite,* depending upon which of the two major ranks their composition is closer to. In addition to these *ranks* (determined by the carbon/hydrocarbon/moisture balance of the coal), coal is assigned to *grades* based on the amount of impurities present. The most serious of these impurities from an environmental standpoint is sulfur, a byproduct of the coal's formation under anaerobic conditions. Low-grade (high-sulfur) coals are one of the main causes of the formation of ACID RAIN. Estimated worldwide reserves of coal are 786 billion tons, sufficient to supply all the energy needs of the planet for about 200 years if it were all mined. About 28% of that total is in the United States. Within the United States, slightly over half the known reserves are low-grade bituminous coals found in the midwest and the Appalachians, easily available but environmentally hazardous due to their high sulfur content. Another 46% is subbituminous coal and lignite, found principally beneath the western High Plains, with a significant amount also known on the Arctic coastal plain of Alaska. These coals are relatively low in energy content but are high-grade (low in sulfur) and are thus preferable environmentally to the high-sulfur midwestern bitumins. Only 2% of the total U.S. coal reserves are high-grade anthracite. These are found primarily in one small area in central Pennsylvania. *See also* SLURRY; STRIP MINE.

coal slurry pipeline a pipeline designed to transport coal by converting it into a SLURRY and using pumps and gravity flow to move it from place to place. Pumping coal in slurry form has the advantage of being a continuous flow rather than a series of discrete loads, and is significantly cheaper than rail or truck transport. However, it requires large amounts of water and significant amounts of real estate for rights of way, and may create a considerable amount of pollution due to leaks or accidentally opened valves. Coal slurry pipelines are not currently considered utilities by the courts, hence cannot use condemnation procedures to acquire rights of way.

coarse solids in pollution control, large solid matter in the FEED WATER to be treated by a water treatment or sewage treatment plant. Coarse solids are generally removed by a screen on the intake to the plant. Anything that will pass through the screen is by definition not a coarse solid.

Coastal States Organization (CSO) organization of government officials, founded in 1970 to promote unified action by its member states to influence federal laws and regulations concerning problems of the coastal and marine environments. Representatives, appointed by the governors of their states, come from states bordering the Atlantic and Pacific Oceans, the Gulf of Mexico, and the Great Lakes. The organization also sponsors coastal conferences and workshops. Membership (1988): 35. Address: 444 N. Capitol Street NW, Suite 312, Washington, DC 20001. Phone: (202) 628-9636.

coastal zone the geographical area on or near the coastlines of ocean waters and (in the United States and Canada) the Great Lakes. The 2000 census is expected to show that 80% of the U.S. population in the coterminous states live within 50 miles of a coastline. Land use in coastal zones is extremely critical owing to the impact of high population and industrialization on fragile estuarine and marine ecosystems. *See also* LITTORAL ZONE.

Coastal Zone Management Act (CZMA) federal legislation to require and encourage the coastal states of the United States to adopt and enforce comprehensive land-use plans for the lands and waters adjacent to their coasts, signed into law by President Richard Nixon on October 27, 1972. "Coastal states," for the purposes of the act, are defined as those states adjacent to the Atlantic, Pacific, or Arctic Oceans, the Gulf of Mexico, or one or more of the Great Lakes. Such states are required to adopt coastal management plans that meet a set of minimum requirements, including the establishment of coastal zone boundaries; a designation of areas of "particular concern" within the coastal zone; an inventory of permissible uses; an enforcement mechanism; and a prioritization of uses. Beach access, energy planning, and erosion control must also be addressed. The act sets up a program of grants that may pay up to 80% of a state's costs to prepare and administer a plan, contingent upon certification of satisfactory progress. CZMA plans are approved, and grants are administered, by the Secretary of Commerce, acting through the NATIONAL OCEANIC AND ATMOSPHERIC ADMINISTRATION. All other federal agencies are required to coordinate their activities on coastal lands with state CZMA plans. The Coastal Zone Management Act was amended and extended in 1976 and again (as the "Coastal Zone Management Improvement Act") in 1980. In November 1990, Congress enacted legislation (the Coastal Zone Act Reauthorization Amendments) to control nonpoint source pollution in the coastal zone. States with approved coastal planning programs are required to show in their plans how NONPOINT POLLUTION can be reduced. Guidelines formulated by the EPA stress the reduction of sediment and discharges from livestock operations into coastal waterways and tributaries. *See also* LAND USE PLANNING; CLEAN WATER ACT.

Coast and Geodetic Survey federal (United States) surveying and mapping agency, established by Congress as the Coast Survey on February 10, 1807 and redesignated the Coast and Geodetic Survey in 1878. Set up originally to create and maintain an accurate survey of the American coastline in regard especially to harbors and shipping hazards, the Survey's duties were gradually broadened to include collecting and collating geodetic data (*see* GEODETIC SURVEY) from all over the world. It was merged into the ENVIRONMENTAL SCIENCE SERVICES ADMINISTRATION on July 13, 1965. Since 1970 its duties have been assumed by the NATIONAL OCEANIC AND ATMOSPHERIC ADMINISTRATION. *Compare* GEOLOGICAL SURVEY.

cobble in engineering geology, a rock midway in size between a PEBBLE and a BOULDER. Cobbles are usually rounded by weathering or the action of water or glaciers; they range in size from roughly 2.5 inches to 10 inches in diameter (64mm to 256mm). *See* WENTWORTH SCALE.

cocarcinogen a substance that significantly increases the likelihood that an individual exposed to a CARCINOGEN will develop cancer. The cocarcinogenic substance is itself not commonly a carcinogen. The action of a cocarcinogen may be compared to that of a CATALYST in a chemical reaction. It does not cause the reaction (in this case, the development of a cancer) but it establishes an environment in which the reaction is better able to take place. An example is oil of croton seeds, used in laboratories to increase the susceptibility of test animals to a broad range of cancers.

Code of Conduct for Responsible Fisheries a voluntary program established by the Food and Agriculture Organization of the United Nations in 1994. The code contains principles and standards meant to conserve food fish resources, threatened marine species, and fragile ecosystems. Included are guidelines covering fisheries management and operations, aquaculture, integration of commercial fisheries into coastal zone management plans, post-harvest practices (dumping unwanted fish and other materials), and commercial and biological research.

Code of Federal Regulations (CFR) master list of all regulations issued by every agency of the U.S. government, published by the National Archives and Records Service of the General Services Administration and updated annually. Regulations published in the CFR have the force of law, and violators—including the issuing agency—may be taken to court to enforce compliance. The source of all regulations in the CFR is the FEDERAL REGISTER.

codominant *See* DOMINANT SPECIES: *codominance.*

coevolution the process by which two or more independent SPECIES develop characteristics that complement each other, often to the point where they are no longer fully independent. Coevolved characteristics begin as a more or less random association of the involved species, which is then specialized through natural selection on both sides. The best-known example of coevolution is that of bees and flowering plants. The ancestors of today's bees were unspecialized insects whose diet included pollen. The ancestors of flowering plants were GYMNOSPERMS whose pollen was stickier than most. The pollen stuck to the insects and was transferred from plant to plant—an evolutionary advantage to both plant and insect (more efficient pollination = more plants = more to eat). Once this link was established, characteristics that aided the process began to appear through natural selection: attractants on the part of the plants (bright colors and sweet scents), locating devices and more efficient pollen carriers on the part of the insects (specialized eyesight and olfactory organs, hairy bodies). Today, bees and flowers are dependent on each other for survival. Numerous other examples of coevolution exist. The most extreme case is probably that of the LICHENS, in which the associated species (an alga [*see* ALGAE] and a FUNGUS) are so closely dependent upon each other that they are treated by biologists as a single organism. *See also* SYMBIOSIS; COMMENSAL SPECIES.

cogeneration a term originally used to designate the production of two different forms of energy (for example, heat and electric power) from the combustion of a single fuel. It is now generally used for any energy generated from what would otherwise be an industrial or domestic waste product. A cogeneration facility connected with a lumber mill, for example, might use wood wastes (bark, trimmings, sawdust, etc.) to fuel a steam boiler; the steam could then be used both to heat the plant and to drive a turbine-powered electrical generator. It has been estimated that cogeneration could provide up to 30% of the United States' energy needs, with production at individual industrial plants or municipalities even higher than that. Properly designed cogeneration facilities using household garbage, for example, might make many cities net exporters of energy. Federal law requires utilities to purchase excess power from cogenerators on their lines, a step designed to encourage the production of cogenerated energy.

cohort in ecology or human studies, a group of individuals of the same SPECIES and roughly the same age within a given POPULATION. A cohort differs from an AGE CLASS in not being strictly geographically related. *See also* AGE DISTRIBUTION; SURVIVORSHIP.

coke a relatively pure form of CARBON produced by heating COAL in the absence of OXYGEN. It is also the ultimate end-product of PETROLEUM cracking (*see* DESTRUCTIVE DISTILLATION), though the cracking process is rarely carried on that long. To make coke, coal is heated to 2,000°F in "coking ovens," which consist of banks of brick-lined chambers surrounded by gas jets that are burned to provide the heat. There are normally 65 ovens to a bank. The impurities in the coal are driven off as gases, which are collected and cooled to produce coal tar and coal gas. The coal gas is fed into the gas jets and used to heat the ovens, while the coal tar is drawn off and sold as a BYPRODUCT. Coke is a hard, gray-black material with a metallic luster, similar in appearance to graphite. It is roughly 92%–95% pure carbon. Most coke produced in the United States is used by the steel industry.

col a high pass in a mountain ridge, typically in the shape of a long, flat parabolic curve, produced where two CIRQUES cutting away from opposite sides of the ridge begin to intersect.

cold deck *See* LOG DECK.

coliform bacteria *See* FECAL COLIFORM BACTERIA.

coliform count a test for the presence of FECAL COLIFORM BACTERIA in a water sample, used as an indicator of the presence of pollution by human or animal wastes in a body of water. The test is generally done in one of two ways. In the *most probable number* (MPN) method, a set of 15 tubes containing lactose solution are inoculated with varying measured amounts of water. If fermentation occurs (indicated by the presence in the tube of gases generated by fermentation), the bacteria are presumed to be present. This "presumed test" is then confirmed (the "confirmed test") by culturing the water from the gas-producing tubes (that is, by placing some of the water from these tubes into an environment that will encourage the growth of bacterial colonies, or *cultures*) and checking the appearance of the colonies that result. If they have the characteristics of typical coliform colonies, they are recultured in a way to make the individual bacteria visible (the "completed test"). The *membrane filtration* (MF) method replaces the first step of the MPN method by filtering the suspected water source and culturing the filter residue directly. It is much quicker than the MPN method but cannot be used if the water is carrying significant amounts of PARTICULATE matter because the filter clogs too easily. If coliform bacteria are found to be present, the final step in either method is to do a *standard plate count,* in which the numbers of colonies of bacteria in a culture medium are physically counted. A standard plate count of 100 colonies or more per milliliter of water is considered grounds for declaring the water contaminated.

collection system in hazardous-materials engineering, any physical structure for intersecting spills, leaks, etc., before they can pose a danger to the environment outside a hazardous-materials storage facility. A typical collection system might consist of an IMPERMEABLE LINER beneath the site; a BERM around it; and one or more lengths of PERMEABLE PIPE buried in the ground to collect the spilled materials for pumpage to a storage tank.

collector in a transportation network, a pipeline, rail line, street, etc. that gathers local traffic and feeds it into a trunk or ARTERIAL line. Collectors serve local neighborhoods rather than entire cities. In a traffic grid, collector streets should be spaced roughly 1/4 mile to 1 mile apart, and should be sized to carry 1,500–10,000 trips per day (40–44 feet in width, including traffic and parking lanes). In sewage systems, a collector is synonymous with a main, and feeds into a trunk line (*see* SEWER).

colloid a material that has been subdivided into extremely small particles or droplets of liquid that are capable of forming a COLLOIDAL SUSPENSION in a proper suspending medium; also (loose usage), the colloidal suspension itself. The particle size small enough to form a colloid varies with the material it is made of, but is generally smaller than .001 millimeter (.00004 inches) in diameter.

colloidal clay *See* CLAY.

colloidal suspension (colloidal dispersion) a form of SUSPENSION in which the suspended material is of a small enough particle size to remain suspended indefinitely (*see* COLLOID). A colloidal suspension has some of the properties of a SOLUTION, notably homogeneity (sameness throughout) and the ability to pass through any form of filter without change. However, colloidal particles will not pass through an osmotic membrane (*see* OSMOSIS) and they do not affect the freezing or boiling point (or most other chemical or physical properties) of the material they are suspended in. Colloidal suspensions have different names depending upon whether the suspended material (*dispersed phase*) and the suspending medium (*dispersion phase* or *dispersion medium*) are solids, liquids, or gases. *See* AEROSOL; SOL; GEL; EMULSION.

Colorado rights in water law, *see* PRIOR APPROPRIATION.

Columbia River Inter-Tribal Fish Commission an organization founded in 1977 comprised of fish and wildlife committees of four Columbia River treaty tribes: Yakima, Umatilla, Warm Springs, and Nez Percé. The mission of the commission is to unify the management of the Columbia River fishery resources regarding research, advocacy, harvest control, and law enforcement. Address: 729 NE Oregon, Suite 200, Portland, OR 97232. Phone: (503) 238-0667. Website: www.critfc.org.

combined sewer system a SEWERAGE SYSTEM in which storm drains are fed into the same set of pipes that carry household sewage and industrial wastes. Combined systems represent an initial cost saving by not requiring a city to lay separate STORM SEWERS. However, the vastly greater volume that the sewers are required to carry during storms frequently exceeds the DESIGN CAPACITY of the sewage treatment plant, causing overflow including a significant amount of untreated household and industrial wastes—to pass around the plant and fall directly into the RECEIVING WATERS in an untreated form, seriously polluting them. For this reason, the building of combined systems is generally not allowed today, and municipalities with existing ones are being encouraged to replace them as rapidly as possible.

combustible liquid any liquid with a FLASH POINT higher than 100°F (38°C) but lower than 200°F (93°C). *Compare* FLAMMABLE LIQUID.

commensal species two SPECIES of animals that normally live together even though neither is directly dependent upon the other. Examples include the New Zealand petrel (a cavity-nesting bird) and the tuatara (a lizard-like reptile), which usually share each other's nesting burrows; the shark and the pilot fish, which often swim and hunt together, rarely if ever attacking each other; and humans and their pets, particularly cats and dogs.

commercial component in forest planning, the portion of a forest's LAND BASE that is designated as COMMERCIAL FOREST LAND. The commercial component always includes the STANDARD COMPONENT; it may also, depending upon the usage desired, include the MARGINAL COMPONENT, the DEFERRED COMPONENT, and/or the RESERVED COMPONENT. *See also* FOREST PLAN; SITE INDEX.

Commercial Fisheries, Bureau of *See* BUREAU OF COMMERCIAL FISHERIES.

commercial forest land (CFL, commercial timber land) land capable of regrowing a salable timber crop after the current timber crop is cut down. To be classed as commercial forest land, a piece of land must be able to grow wood fiber at a rate of at least 20 cubic feet per acre per year; that is, the net volume of all the trees on any acre of the land must increase by at least 20 cubic feet per year. Commercial forest land is divided into numbered categories, or "classes," by its ability to "put on growth," with the faster growing sites receiving the lower numbers (*see* SITE INDEX). From the standpoint of pure forestry, all lands capable of putting on growth at the minimum commercial rate should be classed as commercial forest lands; however, land-management agencies usually remove from the definition those lands that meet the growth criteria but that cannot, for other reasons (such as inclusion within the boundaries of a WILDERNESS AREA or a STREAMSIDE MANAGEMENT unit), be harvested. The term has been replaced by SUITABLE FOREST LAND in official Forest Service terminology: however, the older term is still generally preferred in informal use. *See also* COMMERCIAL LAND BASE.

commercial land base the total amount of immediately harvestable COMMERCIAL FOREST LAND within a NATIONAL FOREST or BUREAU OF LAND MANAGEMENT district. The commercial land base excludes all lands that, though commercial lands by virtue of their growth capability, cannot currently be harvested due to legal, environmental, or economic constraints. *See also* STANDARD COMPONENT.

commercial thinning in forestry, a THINNING in which the trees removed from the thinned stand are large enough to be used in the manufacture of FOREST PRODUCTS such as lumber or chipboard.

Commission on Sustainable Development *See* RIO EARTH SUMMIT.

Commoner, Barry (1917–) American scientist and environmental activist, born in Brooklyn, New York, on May 28, 1917, and educated at Columbia University (B.S. 1937) and Harvard (M.S., 1938; Ph.D., 1941). After teaching briefly at Queens College, New York, he joined the U.S. Navy during World War II, following which he became a member of the faculty of Washington University (St. Louis, Missouri). In 1981 he returned to Queens College, where he expanded the work he was doing in the Center for Biology of Natural Systems, which he founded in 1966. The center's mission is the extensive study of environmental issues: garbage disposal, agricultural pollution, carcinogens; energy issues; conservation, alternative energy sources; conservation issues: organic farming, reduction of waste; and interrelated economic and social issues. Commoner ran for president of the United States on the Citizens Party ticket in the 1980 elections. His books (*Science and Survival,* 1966; *The Closing Circle,* 1971; *Energy and Human Welfare: A Critical Analysis,* 1975; *The Poverty of Power,* 1976; *Making Peace with the Planet,* 1992; *Near-Term Electric Vehicles Costs,* 1993; and *How Government Purchase Programs Can Get Electric Vehicles on the Market,* 1994) are marked by a sense of urgency and a strong concern for the integration of all aspects of life, science, technology, and politics.

community in ecology, the entire mix of plant and animal SPECIES that is present at a particular stage of a SUCCESSION. The term implies not only the numbers of species present but the often complex patterns of dependency and other relationships among them. *Compare* ASSOCIATION; ECOSYSTEM. *See also* CLIMAX; SERE.

compatible use in forest planning, an activity that can take place at the same time as another on the same piece of land, with the land-use objectives being met for both uses. Fishing and hiking are compatible uses because one does not ordinarily interfere with the other. Scenic viewing and timber harvesting may or may not be compatible uses depending on the point the viewing takes place from and the type of harvesting that is going on. Roadbuilding and wilderness are INCOMPATIBLE USES. *See also* CONFLICTING USE; COMPLEMENTARY USE.

compensation depth in limnology, the bottom boundary of the PHOTIC ZONE. Compensation depth is usually defined as the depth to which just enough light penetrates for OXYGEN production by PHOTOSYNTHESIS in green plants and ALGAE to exactly balance oxygen consumption by the same plants and algae. Hence, it is normally the lowest depth at which these organisms may be found. It varies with the transparency of the water, and can be anywhere from a few centimeters to over 100 meters below the surface. *See also* SECCHI DISC.

competition in ecology, conflict arising from the attempted use of a single resource by more than one SPECIES (*interspecific competition*) or by too many individuals of the same species (*intraspecific competition*). Interspecific competition is one of the mechanisms that leads to EXTINCTION. Since two species cannot permanently occupy the same NICHE at the same time (*see* GAUSES'S PRINCIPAL), the species that is better adapted to the niche will eventually crowd the other out. The worse-adapted species often dies out, at least in that locality; however, it may simply move to a similar niche in a neighboring locality where it is slightly less well-adapted but has no competition (*see* COMPETITIVE EXCLUSION), or it may physically adapt to a new niche in the same locality (*see* CHARACTER DISPLACEMENT). Intraspecific competition is a principal driving force behind species change and improvement, and serves also to keep population sizes within the limits of the environment. *See also* CARRYING CAPACITY.

competitive exclusion in ecology, the displacement of a SPECIES from an environmental NICHE it is otherwise capable of filling by another species that is better adapted to that same niche. Competitive exclusion differs from EXTINCTION in that the displaced species is able to survive in the same locality, though it probably will not thrive. For example, a flower that prefers the sun but is capable of blooming in both sun and shade may be crowded out of the sun by a faster-growing plant that cannot bloom in the shade. The first plant will therefore survive as a shade-dwelling species, though it will be less vigorous than if it were allowed to use both niches. The distinctive FLORA that grow on harsh sites such as sand dunes or SERPENTINE soils are almost always a result of competitive exclusion rather than of actual preference for the harsh sites by the flora found there. *See also* COMPETITION.

complementary use in land-use planning, a use for a unit of land that improves the land's ability to serve a second, simultaneous use. Properly conducted timber harvesting is a complementary use to deer hunting because it increases the amount of FOREST-EDGE HABITAT, which deer prefer. Wilderness protection is a complementary use to commercial fishing because it protects SPAWNING BEDS and water quality. *Compare* COMPATIBLE USE; CONFLICTING USE; INCOMPATIBLE USE.

composite in botany, any plant belonging to the family Compositae (also known as Asteraceae), including asters, sunflowers, daisies, dandelions, thistles, marigolds, chrysanthemums, etc. Composites are characterized by the presence of a flower head containing numerous tiny individual flowers (*florets*), so tightly packed that the head has the appearance of a single flower. In most cases, the florets are of two types. The *disc flowers* pack the center of the flower head (the "disc"); they have inconspicuous, often brown petals. The *ray flowers,* which surround the disc flowers, have a distorted, strongly asymmetrical calyx (*see* FLOWER: *parts of a flower*), which makes each flower look like a single petal. A more fundamental difference is that the disc flowers are always perfect, while the ray flowers are imperfect. Despite their difference in appearance, the ray and disc flowers have the same fundamental structure; an inferior ovary, five fused petals (in the ray flowers, three of the five petals generally form the ray while the other two remain tiny), fused ANTHERS (forming a tube around the style), and SEPALS that have been reduced to bristles or hairs. A few composites are grown for food (sunflowers, lettuce, artichokes), and several others (thistles, dandelions, ragweed) are serious pest plants. The vast majority, however, have no economic value one way or the other. Ecologically, they are important for their hardiness and adaptability, which allows them to colonize raw or inhospitable land rapidly. The family contains 900–1,200 genera and 20,000–25,000 species, and is the most widespread, most numerous, most morphologically uniform, and probably most evolutionarily advanced group of flowering plants.

compost organic residues from farming, gardening, or kitchen waste, which are piled, moistened, and allowed to undergo decomposition for later use as a fertilizer or mulch. *See* ORGANIC GARDENING AND FARMING.

Comprehensive Environmental Response, Compensation, and Liability Act (CERCLA) federal legislation to identify and clean up sites where HAZARDOUS WASTES have been improperly or illegally disposed of, signed into law by President Jimmy Carter on December 11, 1980. The most important provision of CERCLA is the establishment of the Hazardous Substances Response Trust Fund, popularly known as "Superfund," earmarked specifically for hazardous waste cleanup. This fund is administered by the ENVIRONMENTAL PROTECTION AGENCY (EPA), which is required to prepare a list of at least 400 sites nationwide that qualify for Superfund cleanup (this list contained over 900 sites in early 1987). Originally set at $1.6 billion, the fund was expanded to $9 billion when the act was renewed on October 17, 1986. This money is to come from taxes on oil, industrial chemicals, motor fuels, and corporate income generally. Congressional dissatisfaction with the rate at which cleanup was proceeding under the 1980 law led to a requirement, included in the 1986 renewal, that work must begin at least 375 sites by 1991. Other provisions include the requirement that all releases of hazardous substances, accidental or otherwise, be reported to the EPA within 24 hours; that all companies operating hazardous-waste disposal facilities report the existence of those facilities to the EPA; and that all deeds to properties containing hazardous wastes specify the locations and types of these wastes. The 1986 revisions added two important provisions: additional funds (not part of Superfund) for research into means of detoxifying, rather than landfilling, hazardous wastes; and a so-called "right-to-know" clause requiring industries

U.S. Superfund Sites (includes both proposed and final sites) 1998

State	*Total sites*
Alabama	12
Alaska	7
Arizona	10
Arkansas	11
California	96
Colorado	17
Connecticut	14
Delaware	17
District of Columbia	1
Florida	53
Georgia	16
Hawaii	4
Idaho	9
Illinois	43
Indiana	30
Iowa	17
Kansas	11
Kentucky	16
Louisiana	15
Maine	12
Maryland	18
Massachusetts	31
Michigan	71
Minnesota	27
Mississippi	3
Missouri	22
Montana	9
Nebraska	10
Nevada	1
New Hampshire	18
New Jersey	111
New Mexico	11
New York	84
North Carolina	24
North Dakota	0
Ohio	36
Oklahoma	12
Oregon	11
Pennsylvania	100
Rhode Island	12
South Carolina	25
South Dakota	2
Tennessee	15
Texas	32
Utah	16
Vermont	9
Virginia	27
Washington	47
West Virginia	7
Wisconsin	40
Wyoming	3
Total U.S. sites	**1,245**

Source: U.S. Environmental Protection Agency, Supplementary Materials: National Priorities List, Proposed Rule, December 1998.

to inform the communities in which they are located concerning the types and amounts of hazardous substances they are handling. While efforts have been made to weaken the provisions of the fund, Congress continues to provide funding at a rate in excess of $1 billion a year.

compound in chemistry, a substance whose MOLECULES are made up of more than one ELEMENT. A compound differs from a MIXTURE in that the molecules or atoms of the separate substances that are brought together to make it up are not simply associated in a more or less random arrangement but have become bound to each other by chemical and physical forces into new molecules with their own distinct properties. The elements in a compound are always in fixed proportions, that is, always occur in the same ratio to each other throughout the compound, wherever and whenever it may be present.

concealing coloration *See* CRYPTIC COLORATION.

condemnation *See* EMINENT DOMAIN.

condensation the conversion of a substance from a gaseous (vapor) state to a nongaseous state due to cooling. Most gases condense into liquids; however, a few materials condense directly from the gaseous to the solid state (*see* SUBLIMATION). During condensation, the heat of vaporization is released and the energy level of the condensing substance is lowered without reducing its temperature. Its primary importance to the environment is its role in the HYDROLOGIC CYCLE: water evaporated or transpirated into a warm AIR MASS as water vapor (*see* EVAPORATION; TRANSPIRATION) condenses when the air mass cools, forming fog, clouds, and dew, and transferring the LATENT HEAT represented by its heat of vaporization from the point on the earth's surface where the evaporation took place to the point where the condensation takes place, thus affecting the atmospheric heat balance and distribution. This is one of the principal methods by which heat is transferred from the equator toward the poles. *See also* DEW POINT; HUMIDITY.

conductivity a measurement of the ability of a substance to conduct electricity. Conductivity is the inverse of RESISTIVITY, and is measured in *mhos* (*ohms* spelled backwards), which are related to ohms by the formula mho=1/ohm.

CONDUCTIVITY AS A MEASURE OF WATER PURITY

In a liquid, conductivity is the result of the presence of free IONS: the greater the number of ions, the greater the conductivity. The number of ions depends in turn on the number of molecules within the liquid that *dissociate* (break apart into two or more independent parts). Pure water dissociates very weakly, and is therefore not a particularly good conductor. It is, however, an excellent SOLVENT, and can carry a large load of dissolved material—much of it fully dissociated by the process of solution. Under these circumstances, the number of free ions becomes very large, and the water becomes a very good conductor indeed. The conductivity of water is thus directly proportional to the amount of dissociable material dissolved in it, allowing conductivity to serve an index to the purity of the water (the lower the conductivity, the purer the sample). For these purposes, conductivity is measured in *micromhos* (μmhos), or thousandths of a mho. Extremely pure water will have a conductivity of less than 10 μmhos/cm, while the conductivity of heavily mineralized waters may run above 40,000 μmhos/cm. *See also* TOTAL ION CONCENTRATION.

cone the FRUITING BODY of a pine, fir, or other GYMNOSPERM. It is the presence of cones that gives the gymnosperms the common names "conifer" or "coniferous tree." Cones are generally composed of woody scales arranged radially around a central shaft, though there are variations: most cypresses (Cupressaceae) have spherical cones, with the scales radiating from a single point, and yews (Taxaceae), junipers (*Juniperus* spp.), and a few others bear fleshy-scaled cones that look very much like berries (*see* FRUIT). The seeds are normally borne at the bases of the scales in groups of one to 20 per scale, though again there are exceptions; yew cones, for example, contain a single seed each, with the scale structure wrapped most of the way around it. Lengths of cones range from 0.5 centimeters (juniper) to 66 centimeters (sugar pine).

cone of depression in hydrology, a localized lowering of the WATER TABLE around a well that is being heavily pumped. The effects of a cone of depression on the individual well (and on others in its vicinity) are essentially the same as those caused by inadequate RECHARGE; however, a cone of depression is not a supply problem, but a transportation problem. Water is simply being drawn out of the well faster than it can flow through the AQUIFER (*see* PERCOLATION). The solution is to drill more wells and pump each of them more slowly so that the percolation rate can keep up with the rate of withdrawal in each well, although the extra wells can, if pumped too rapidly, merely create overlapping cones of depression that lower the entire regional WATER TABLE. *See also* STATIC LEVEL; DRAWDOWN.

confined disposal facility (CDF) a site designed to allow the dumping of contaminated dredge spoil (*see* DREDGE; SPOIL) without allowing the contaminants in it

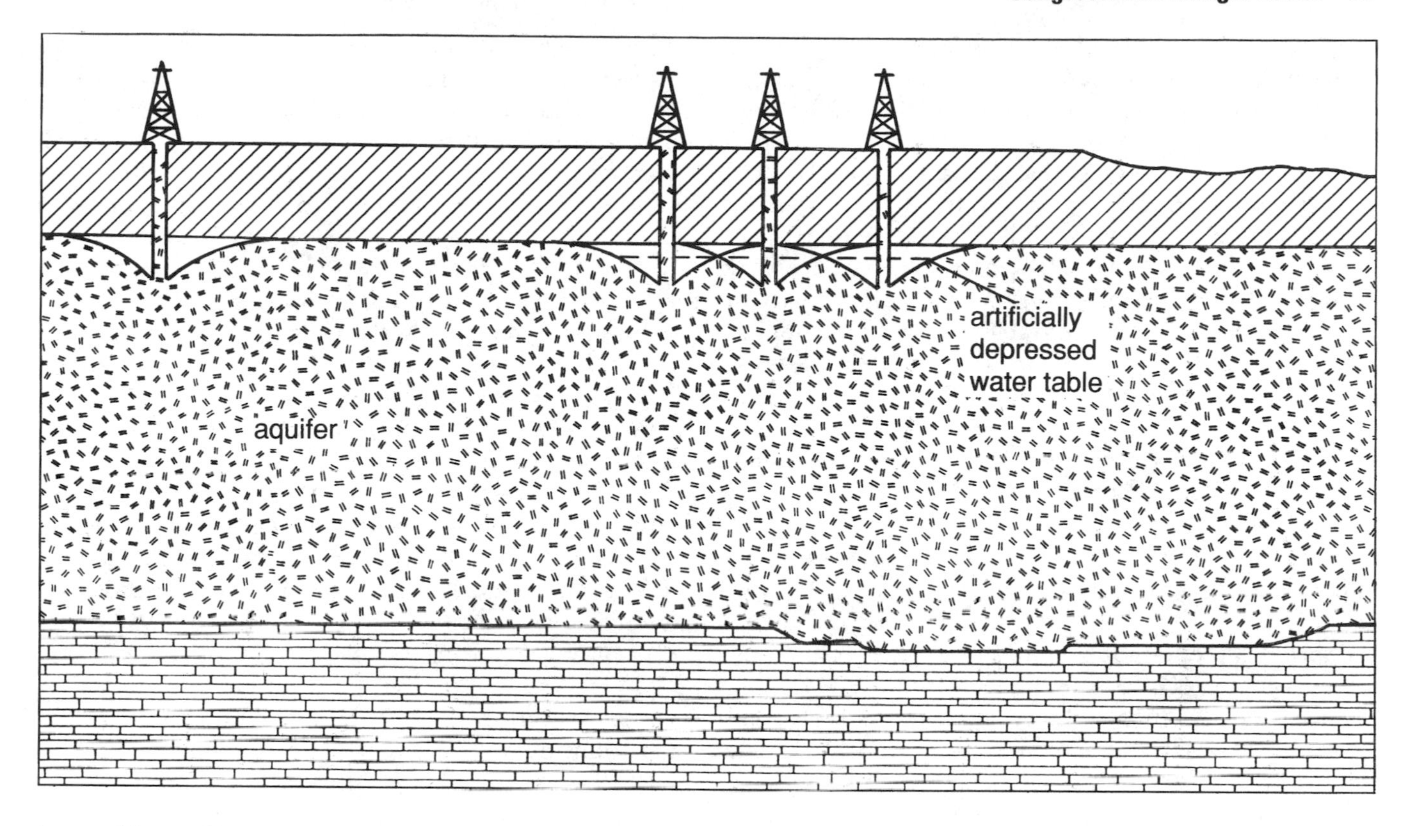

Cones of depression

to return to the water. There are two principal types. In *diked disposal,* a large "filter dike," made of boulders 3–5 feet in diameter and lined internally with a thick layer of sand, is built in the water near the site to be dredged. Contaminated sediments dumped within the space confined by the dike either settle to the bottom of the confined area or are filtered out as the water passes through the sand. The chemical contaminants normally remain bonded to the sediments (*see* ADSORPTION: ABSORPTION), allowing them to be removed by the filtering process as well. In *upland disposal,* a landfill site is prepared, using the same technology and site preparation required for any HAZARDOUS WASTE facility, that is, using a BERM, an IMPERMEABLE LINER, and a COLLECTION SYSTEM. The contaminated dredge spoil is then moved to the site by trucks, railroad cars, or (occasionally) a conveyer belt. Upland sites have the advantage of removing the contaminated material entirely from the dredged waters; however, they are expensive to build and maintain, and if inadequately lined may leach contaminants into nearby watercourses or AQUIFERS. Diked sites contain sediments and their sorbed contaminants extremely well, but fail entirely to contain dissolved contaminants that have not sorbed to the sediments. They may also occasionally be breached by storms. Thus, neither system has a clearcut advantage over the other from an environmental standpoint. *See also* IN-PLACE POLLUTANTS.

confirmed test *See under* COLIFORM COUNT.

conflicting use in land-use planning, a use for a piece of land that diminishes its ability to be used for some other purpose. The two uses may still exist simultaneously, but neither will be able to expand to its full potential. For example, scenic viewing and timber harvest are usually conflicting uses. The choice of SILVICULTURAL PRESCRIPTION is limited by the need to protect the view, while the view is nevertheless unavoidably altered by the presence of road cuts and visible HARVEST units. *Compare* INCOMPATIBLE USE. *See also* COMPATIBLE USE.

conglomerate in geology, a form of SEDIMENTARY ROCK consisting of fairly large, nonuniform pieces of PARENT ROCK contained in a sandstone or siltsone matrix.

Congressional Budget Office (CBO) an agency of the legislative branch of the U.S. federal government, established to provide Congress with independent studies of the federal budget and budget-related issues (such as public-works expenditures and tax bills). Established by the Congressional Budget Act of 1974, the CBO is specifically charged with the tasks of analyzing current economic trends and their effects on current federal fiscal policies and on the federal budget (and vice versa); keeping track of appropriations and other outlays to make sure they remain within budgeted targets or ceilings ("scorekeeping"); making five-year projections of the costs of all major federal actions, including new legislative proposals; and preparing an annual report on the budget to help guide Congress through the budgetary enactment process. In addition, the CBO may be requested by Congress to do special studies on the fiscal impact of specific actions. The comparable agency in the administrative branch of government is the OFFICE OF MANAGEMENT AND BUDGET.

conifer a cone-bearing tree. *See* CONE; GYMNOSPERM; CONIFEROUS FOREST BIOME.

coniferous forest biome a major northern hemisphere BIOME, covering most of northern Europe, Asia, and North America, and existing in broad patches in mountainous regions elsewhere. The dominant vegetation is a forest of mixed CONIFERS (*see* GYMNOSPERM), often with an UNDERSTORY of small HARDWOODS. The climate is cool, but may include very hot days during the short summers. Rainfall may range from moderately light to extremely heavy, but it is invariably concentrated in a single season (usually the winter, though a summer rainy season is common along the northern border of the biome). Animal life must be adapted to extreme swings of temperature, and therefore includes principally those species which migrate, moult, or hibernate. There are few reptiles and amphibians, but many seasonal insects. Water features (lakes, ponds, streams, and bogs) abound, most of them oligotrophic in nature (*see* OLIGOTROPHIC LAKE).

EXTENT AND DIVISIONS

Though there are small isolated patches of it here and there in the southern hemisphere, the coniferous forest biome is almost exclusively a northern hemisphere phenomenon—probably due to the lack of land masses in the proper latitudes of the southern hemisphere. There are four principal divisions of the biome. The *boreal forest,* or *taiga,* forms a broad band across all three northern continents between latitude 50°N and latitude 70°N: the dominant trees are spruce and fir. Many of the animals and plants are Holarctic in distribution (*see* HOLARCTIC SPECIES). Soils are acid; bogs are common. The *montane forest* is an extension of the boreal forest that fingers southward down mountain chains, especially in North America. Spruce is less common in the montane forest; the predominant species are fir, hemlock, and (especially in the lower elevations) pine. The *coastal forest* is predominant along the Pacific coast of North America, rare elsewhere. It is dominated by hemlock, DOUGLAS FIR, spruce, redwood, and American cedar (cypress). The extremely high rainfall of this region allows these trees to reach enormous size. Finally, in eastern North America, a broad band of mixed conifers and hardwoods known as the *north woods* lies along the southern border of the taiga from Minnesota through the Great Lakes region to New York and New England. Though often classed as a separate biological region because of the variety and size of its hardwoods, the north woods probably belongs in the coniferous forest biome due to its dominant species, which are pine and hemlock. In North America, the coniferous forest biome is roughly coextensive with the CANADIAN ZONE (*see* LIFE ZONES).

conjunctive use in water management, the use of surplus surface waters to artificially recharge AQUIFERS during wet years. The aquifers may then be used as a groundwater source to supplement surface water supplies during dry years.

conk in forestry, a DEFECT in CONIFER logs resulting from the presence of the fruiting body of a FUNGUS. Conk invades living trees, almost always entering through a KNOT. The infection normally penetrates several inches to several feet into the tree trunk and spreads from seven to 30 feet on either side of the point of entry; it often spreads around the tree as well, following the ANNUAL RINGS. Its presence may be detected in several ways. The conk itself—the FRUITING BODY of the fungus—may be present on the outside of the tree's trunk, growing out of the knot that was the original site of infection. More commonly, a "punk knot" may be seen, in which the central core of a knot has been replaced by a punky yellow material, the remains of a conk which has fallen off. Finally, there may be "blind conks"—swellings in the wood that indicate the presence of punk knots overgrown by sound wood. Conk is a serious defect and often results in culling the logs in which it is found (*see* CULL MATERIAL). The most common variety of conk, red ring rot, is caused by the invasion of a shelf fungus, *Fomes pini,* which produces a yellow shelflike structure on the outside of the trunk. *See also* LOG SCALING.

consent decree in law, a document drawn up as a means of achieving out-of-court settlement of a lawsuit. The consent decree consists of a statement of facts to which all parties to the suit agree and the outline of a course of action which all parties agree will satisfy their claims. It also almost always includes a specific clause voiding the suit for which it serves as a settlement. Consent decrees are binding on all those who sign them, and are enforceable by court action if they are violated. They are an increasingly common means of settling environmental disputes, especially those between government agencies and individual citizens or corporations.

consequent stream in geology, a stream whose course is determined by current topography. *Compare* ANTECEDENT STREAM. *See also* STRUCTURAL VALLEY.

Conservation and Research Foundation strives to promote conservation, encourage research in biology and related sciences, and discover the depth and breadth of nature/human relationships. Founded in 1953, the foundation advances studies in the areas of air quality, environmental law, human population growth, and wildlife. Address: 24 Schillhammer Road, Jericho, VT 05465.

conservation biology a multidisciplinary specialty in the biological sciences that has recently arisen to undertake research concerning BIODIVERSITY, especially the ways in which natural processes are affected by the loss of diversity caused by the extinction of gene pools, species, and biotic communities through human-caused ecosystem disturbance.

conservation easement a right in land that may be conferred by a landowner to another party, usually a government or quasi-government conservation agency, in order to protect natural values. Conservation easements may be positive, as for hiking trails or bridle paths, or negative, wherein public entrance is not permitted but the land is barred from a change of use that would affect natural values. *See also* DEVELOPMENT RIGHTS.

The Conservation Foundation environmental research and educational center, founded in 1948 to promote conservation and wise use of the earth's natural resources. The foundation was absorbed into the World Wildlife Fund in 1992 and no longer has an independent program.

The Conservation Fund pursues sustainable conservation through an integration of economic and environmental goals. Founded in 1985, the fund now has offices throughout the United States and finances many important programs. These include American Land Conservation Program, American Greenways, Civil War Battlefield Campaign, Conservation Leadership Network, and the Freshwater Institute. The fund also gives out various awards in the field of conservation such as the Alexander Calder Conservation Award and the Cartledge Award for Excellence in Environmental Education. A newsletter, *Common Ground,* is published bimonthly. Address: 1800 North Kent Street, Suite 1120, Arlington, VA 22209. Phone: (703) 525-6300.

Conservation International works with indigenous peoples in tropical and temperate regions to preserve their ecosystems and biological diversity. The organization is active in Bolivia, Botswana, Brazil, Colombia, Costa Rica, Ecuador, Ghana, Guatemala, Guyana, Indonesia, Madagascar, Mexico, Panama, Papua New Guinea, Peru, the Philippines, the Solomon Islands, and Suriname. Membership (1999): 4,000. Address: 2501 M Street NW, Suite 200, Washington, DC 20037. Phone: (202) 429-5660. Website: www.conservation.org.

Conservation Law Foundation a member-supported organization founded in 1996 supporting resource conservation and public health in the New England area. The foundation focuses on water and energy conservation, land preservation, transportation planning, and protection of marine resources. It also publishes documents and reports. Membership (1999): 10,000. Address: 62 Summer Street, Boston, MA 02110. Phone: (617) 350-0990. Website:www.clf.org.

conservation movement the effort, dating approximately from the mid-1800s in the United States, to conserve natural resources such as forests, soils, and water as well as natural, scenic areas. The conservation of nature and aesthetic landscapes was most famously advocated by HENRY DAVID THOREAU and the Concord (Massachusetts) circle of transcendentalists that included RALPH WALDO EMERSON. It was Thoreau's view that nature should be conserved for its own sake and that everyone should have access to wild places. "In wildness is the preservation of the world," he wrote. The necessity for the conservation of natural resources, as a practical economic matter, was earliest and most forcefully expressed in 1864 with the publication of *Man and Nature,* by GEORGE PERKINS MARSH. Marsh, a Vermont lawyer, had been ambassador to Italy, and while there became convinced that the "original balances of nature" could be permanently deranged by human action. In large areas of the Mediterranean basin he adduced that DESERTIFICATION from timber cutting and grazing had long since passed the point of no return, that the natural cover could never be restored. Returning home, he became concerned about the status of his native Vermont, where 80% of the forest had been cleared for crops and pasture. His book, still in print, remains influential. These two authors typify the two main philosophical strains of the conservation movement—the romanticism of Thoreau and the practical natural economy of Marsh. The two strains continued into the early 20th century, with the utilitarian GIFFORD PINCHOT, the father of modern forestry, and his near opposite, JOHN MUIR, whose energetic and passionate leadership promoted the preservation of scenic wonders and the creation of the

SIERRA CLUB. In the 1940s ALDO LEOPOLD proposed a "land ethic" in his book *A Sand County Almanac* that combined economic and aesthetic values as elements of a single, ecologically oriented conservation idea: "A thing is right when it tends to preserve the integrity, stability, and beauty of the biotic community. It is wrong when it tends otherwise." In the 1960s STEWART LEE UDALL, with the publication of *The Quiet Crisis* (1963), also tried to bring the strands together in what he called the "new conservation." The designation did not stick, however. Beginning in 1970, "conservation"—whether utilitarian, romantic or ecological—became absorbed into what is now called the environmental movement. Nevertheless the tenets of the conservation movement still exist: in the practical work of Soil and Water Conservation Districts, say, or, on the other side, in the spiritually based "wildlands" initiatives of activist DAVID FOREMAN. *See* ENVIRONMENTAL MOVEMENT.

Conservation Reserve Program an initiative first authorized in the 1985 Farm Act to induce farmers to stop growing crops on land subject to serious EROSION or that is otherwise environmentally sensitive. In return for planting a protective grass or tree cover on vulnerable cropland, the farmer can receive a "rental" payment for each year of a multiyear contract. In addition, a farmer can receive an additional payment for establishing *permanent* conservation areas that can give shelter and food to wildlife and improve water quality. To date, some 30 million acres of land have been accepted into the program, which is administered by the Farm Service Agency of the U.S. Department of Agriculture. This is only about 10% of total planted cropland, but the program, together with CONSERVATION TILLAGE, an associated soil-saving technique, has reduced soil erosion by 25%. At present, Congress has capped enrollment at 36.4 million acres, although efforts are underway to increase allowable acreage to 45 million, the original statutory limit, since the retirement of marginal lands can reduce erosion on such lands by nearly 90%, with many other benefits realized as well.

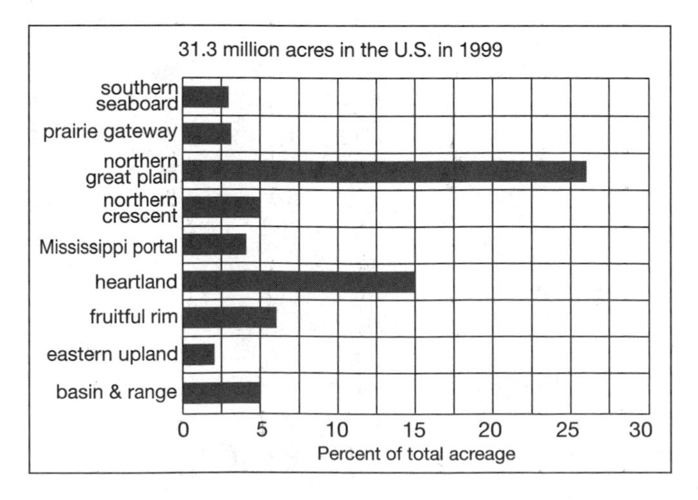

Conservation Reserve Program (in millions of acres) (U.S. Department of Agriculture Fact Book, 1999)

conservation tillage various tillage techniques meant to reduce the EROSION of cropland and the runoff of nonpoint source pollutants into waterways, and to increase yields. Developed in the 1930s and 1940s, conservation tillage was slow to be adopted by farmers. Today it is practiced on about 140 million acres, more than a third of total U.S. cropland. In the corn belt, the percentage is two-thirds. Conservation tillage is formally defined as a tillage system that maintains at least 30% of the soil surface covered by residue after planting, as opposed to the more traditional practice of plowing under the residues. A residue coverage of less than 30% is defined as *clean tillage* and does not qualify for cost-sharing payments sometimes made available by the U.S. Department of Agriculture to establish conservation tillage on a farm. Among the most innovative approaches to conservation tillage are *no-till, ridge-till,* and *strip-till.* No-till involves the use of substantial amounts of HERBICIDE. Instead of plowing under the stubble and weeds in a field after harvest in the fall, herbicides are applied to "burn down" the weeds, after which seeds for a new crop can be planted directly into the stubble of the previous crop, often with special seed-planting machinery adapted for the purpose. A benefit of no-till is that in many areas, double cropping is possible where it was not with conventional tillage. With ridge till, permanent planting beds—ridges—are created, the tops of which are cleared of weeds and residue and then planted. The residue remains on the field between the ridges, however, thereby conserving soil moisture. The herbicide requirement for ridge-till is much lower than for no-till, and in fact, ridge-tillage can be used in organic farming, with no herbicide whatsoever. Strip-till is a variation of ridge-till, but without the ridges: a narrow strip of soil is cleared to create a planting bed, with residue-covered areas between. With strip-till, an in-row planter is often used, a narrow shank that penetrates several feet into the soil to create a break in the hardpan created by conventional plowing so that roots can penetrate to the subsoil for moisture and nutrients. Other than these more inventive approaches, many farmers simply substitute a chisel plow for a moldboard (mulch-till) or use other techniques that leave a residue (reduced-till). Taken together, conservation tillage and the CONSERVATION RESERVE PROGRAM are credited with reducing erosion from cropland by 25% in the United States. *See also* CONTOUR PLOWING; SOIL CONSERVATION.

conspecific populations in the biological sciences, a pair (or group) of POPULATIONS whose members are capable of interbreeding. Individuals in conspecific populations should almost always be classed in the same SPECIES, even though the populations may be widely separated geographically and may have large variations in superficial characteristics such as coloring, ear or beak shape, etc.

construction survey a survey made during construction of a road, dam, building, etc. to make certain that it lies on the landscape in precisely the position indicated by the site plan drawn from the ENGINEERING SURVEY. A properly done construction survey repeats all the steps of the engineering survey that preceded it, to at least the same degree of accuracy.

consumer in ecology, an ORGANISM that consumes other organisms to obtain complex nutrients rather than producing the nutrients through PHOTOSYNTHESIS or CHEMOSYNTHESIS. The term is essentially synonymous with HETEROTROPH, except that "heterotroph" refers to the organism's means of making its living, while "consumer" refers to its function in an ECOSYSTEM. Consumers are classed according to the TROPHIC LEVEL they normally occupy. Thus, consumers that eat photosynthetic organisms directly are called *first-order consumers;* those that eat first-order consumers are called *second-order consumers;* and so on. *Compare* PRODUCER. *See also* FOOD CHAIN; FOOD WEB.

consumerism in popular usage, the tendency for Americans, and to a lesser extent western Europeans, to measure personal achievement by the quantity of goods purchased and consumed, as opposed to a commitment to nonmaterial values. The ethos of consumerism, driven by the profit needs of multinational corporations and fed by the increasing personal income of many American citizens, has led to what most environmentalists believe are disproportionate and unsustainable levels of waste, pollution, and resource depletion. On a per capita basis Americans consume 23 times more goods and services than the average resident of a developing nation. According to one research report, Americans spend nine times more hours per week shopping than they do playing with children. Most of that purchasing is done at shopping malls, which in 1987 exceeded the number of high schools in the U.S. With the rise of the Internet, it is now expected that "e-commerce" will only increase consumerism. If present trends continue, a child born today, by the time he or she reaches 75, will have consumed 43 million gallons of water and 3,375 barrels of oil, and have produced 52 tons of garbage. Analysts of these data conclude that if populations of developing nations were to produce waste and pollution and consume resources at levels even close to that of Americans, life on the planet (including human life) would be threatened. According to the UNION OF CONCERNED SCIENTISTS, a majority of the world's scientific Nobel laureates believe that unless current levels of consumption and pollution are reduced, great human misery will ensue and the natural systems of the planet will be "irretrievably mutilated." *See also* WORLD SCIENTISTS' WARNING TO HUMANITY.

contact poison a TOXICANT that poisons an ORGANISM through direct contact with its epidermis (skin, bark, or other external covering). Contact poisons are generally materials that are absorbed through the epidermis into the capillaries or veins. The toxic reaction associated with them occurs after they have reached the circulatory system and been carried to other parts of the body.

contact rock in geology, a zone of METAMORPHIC ROCK created at the contact line between an intrusive body of MAGMA (such as a DIKE or a PLUTON) and the preexisting body of COUNTRY ROCK. Contact rock is generally hard and fine-grained. Its chemical content and crystalline structure grade outward from the intrusive body, with the most highly altered next to the intrusion. The body of contact rock (the *aureole*) surrounding the intrusive body may be anywhere from less than 1 centimeter to 5 or 10 kilometers thick, depending upon the size of the intrusion, the chemistry of the country rock, and the depth at which the intrusion and aureole were formed. *See also* HORNFELS.

containment structure in nuclear technology, a building constructed around a nuclear reactor to prevent the escape of radiation in case of damage to the reactor. It is designed to contain the *maximum credible accident,* that is, the worst accident likely to occur at the site. Containment structures must be constructed of a material able to withstand intense bombardment by radiation without either weakening or passing any radiation products through. They must also be able to withstand extremely high heat (in case a core meltdown should occur) and at least moderately high explosive force (although a nuclear plant cannot suffer an atomic explosion, a steam explosion or the ignition of explosive gases may occur). The material of choice is normally steel-reinforced concrete at least 2 feet, and preferably 4 feet, thick. *See also* NUCLEAR ENERGY.

contaminant any substance present in an environment where it does not belong. Soil in a field is part of the field: soil on food after it has been removed from the field is a contaminant. Usually (as in the previous example) the term "contaminant" implies that the foreign substance is harmful; however, this is not a necessary part of the definition. The trace ingredients of honey, which vary from batch to batch and even from sample to sample within the same batch, may be properly termed "contaminants" even though it is they that impart to the honey its distinctive flavor and much of its nutrient value. *See also* MICROCONTAMINANT.

continental climate in meteorology, the climate in the interior of a continent, beyond the modifying effects of the oceans. Without the heat sink of the mass of ocean water present to warm the winters and cool the summers, and without the massive amounts of moisture that evaporate from the ocean and blow over the nearby land, continental climates are drier and more severe, with wider swings of temperature. Although the mean annual temperature may be the same at a continental station as that at a similarly situated marine station, the annual temperature range will be much more extreme at the continental station, with hotter summers and colder winters. The Great Plains of the United States and Canada have a typical continental climate.

continental divide the WATERSHED boundary separating the streams that flow to one side of a continent from those that flow to the other side. It is often a mountain range, but does not need to be. The nearly dead-level prairie country of western Minnesota and southeastern North Dakota, for example, is part of the continental divide separating waters flowing to the Arctic Ocean (the Red River du Nord and tributaries) from those flowing to the Gulf of Mexico (the Mississippi system). If the term is capitalized (Continental Divide), it normally refers—at least in North America—to the specific continental divide lying along the crest of the Rocky Mountains that separates Pacific Ocean waters from those flowing to the Atlantic, Gulf, and Arctic Oceans.

continental drift *See* PLATE TECTONICS.

continental glacier (continental ice sheet) in geology, a GLACIER covering a significant portion of a continent or large island. The word "significant" in this definition remains vague. The Antarctic and Greenland ice sheets, the only continental glaciers currently in existence, each cover between 80% and 90% of their respective landmasses. The Pleistocene ice sheets of North America and Eurasia, however, at their greatest extension, covered only between 20% and 30% of their landmasses. Unlike alpine glaciers, which flow downhill, continental glaciers flow outward in all directions from *centers of accumulation* in the upper latitudes. The flow is actually a spreading of the glacier's mass caused by its great weight (continental ice sheets are often several kilometers thick). The sculpting action of continental ice has created—or at least altered—most of the landforms of northern North America and Eurasia, including such major features as Hudsons Bay, the Baltic Sea and the Great Lakes. *See also* ICE AGE.

continental rise *See under* CONTINENTAL SHELF.

continental shelf a broad, gently sloping region of the sea floor extending from the low tide line along the coast outward to roughly the 200-meter depth, where it terminates in an abrupt, precipitous decline (the *continental slope*), dropping several thousand meters to the beginning of the *continental rise,* which slopes gently off to the ABYSSAL PLAIN. Continental shelves surround all the world's continents. They vary in width from less than 10 miles (off northwestern Africa) to approximately 1,000 miles (off northern Scandinavia and the north coast of the Soviet Union), but generally average between 40 and 50 miles wide. In most places they are smooth, but in some (notably off glaciated coastlines) they contain deep canyons and considerable undersea mountains.

IMPORTANCE OF THE CONTINENTAL SHELF

Because most fish prefer to breed in shallow water, and because the continental shelf underlies virtually all of the shallow water of the oceans, the region of the continental shelf produces the vast majority of ocean fish. It is also the region where almost all sedimentary rocks form (from erosion products from the continents and from chemical deposits from microscopic life forms that thrive in the shallow waters), hence is both the primary location of petroleum formation and the prime remaining site for oil-well prospecting and oil production. Since oil production and fisheries are CONFLICTING USES, management of the continental shelves is likely to be a controversial subject for many years.

For the geologic history of the continental shelf, *see* PLATE TECTONICS.

continental slope *See under* CONTINENTAL SHELF.

Contingent Valuation Method (CVM) an elaboration on BENEFIT/COST ANALYSIS that assigns a dollar value to goods and services, including ECOSYSTEM SERVICES, that are difficult to evaluate in dollar terms. Thus the value of, say, clean air—in terms of health costs, safety, ecosystem impacts, and quality of life—would be introduced into an economic analysis of the benefits and costs of mandating a reduction of air pollution. Without the use of such a contingent valuation method, the costs of air pollution effects, because they do not take in the contingencies, can be externalized by corporations, which means they will be borne by the general public and posterity.

contour a line connecting points of equal elevation along a slope; also, the physical act of traveling such a line. For the use of contours in map-making, *see* TOPOGRAPHIC MAP. *See also* CONTOUR PLOWING.

contour interval *See under* TOPOGRAPHIC MAP.

contour map *See* TOPOGRAPHIC MAP.

contour plowing (contour cultivation, contour planting) cultivation method in which furrows are plowed across slopes rather than up and down them. Contour plowing results in level furrows that trap RUNOFF, thereby significantly cutting erosion. It is estimated that approximately 50% of the SILT lost from a field plowed in a straight grid can be saved by plowing contours instead. Crop production is usually improved due to the increase in available moisture for the plants that results when the trapped runoff soaks into the ground. During severe storms, contoured furrows may fill with water and overflow, causing a "cascading effect" that washes out the furrows below and actually increases rather than controls erosion. To prevent this, a technique known as "ridged contouring" is now employed. At periodic intervals down the hillside, the ridge between two furrows is made much higher than normal, thereby acting as a dam to cut off the cascading effect in mid-slope. *See also* CONSERVATION TILLAGE.

controlled burn in forestry or agriculture, a fire set purposely to fulfill a specific management objective. The objective may be to get rid of SLASH, to remove FUEL BUILDUP, to destroy weeds, or even to create a FIREBREAK. Controlled burns are made only when very specific conditions of wind speed, temperatures, and soil and vegetation moisture are met so that the burn will proceed slowly and be easy to keep under control. The normal time for controlled burning is during calm, cool weather just after the rainy season. *See also* FIRE ECOLOGY.

conventional pollutant loosely, any substance found in normal household sewage; more specifically, a substance that pollutes a body of water by altering its TROPHIC LEVEL, affecting its dissolved oxygen balance, or physically changing its character. The category includes NUTRIENTS (particularly PHOSPHORUS and NITROGEN compounds); CHEMICAL OXYGEN DEMAND; BIOCHEMICAL OXYGEN DEMAND; bacteria (*see* BACTERIUM); SILT and other PARTICULATE matter; and heavy, sludge-forming oils and greases. It does not normally include toxic substances. *Compare* CONVENTIONAL TOXIC SUBSTANCE; MICROCONTAMINANT. *See also* CULTURAL EUTROPHICATION.

conventional toxic substance an inorganic or simple organic COMPOUND (*see* INORGANIC COMPOUND; ORGANIC COMPOUND) that is poisonous to animal or plant life. Conventional toxic substances include (among others) ARSENIC, CYANIDE, AMMONIA, BORIC ACID (especially as sassolite: *see* BORON), CHLORINE and other HALOGENS, and PHENOL. *Compare* CONVENTIONAL POLLUTANT; MICROCONTAMINANT.

Convention on International Trade in Endangered Species (CITES) a treaty signed by 145 countries (as of 1999) that prohibits international trade in some 700 endangered or threatened species, either as live specimens or as wildlife products (such as elephant tusks). First promulgated in 1975, enforcement of the treaty provides only a partial answer to controlling the illegal $6 billion international endangered species export trade—in parrots, rare orchids, spotted cat pelts, powdered rhinoceros horn (thought to be an aphrodisiac), and thousands of other creatures, plants, and products. While the agreement is binding, any nation can obviate its terms simply by delisting a species as endangered if it chooses. Moreover, many countries that export exotic species for (mainly) North American and European collectors are not signatories to the treaty and can export exotic species and products as well as serve as entrepôts for smugglers. Some of the centers of illegal trade are Argentina, Indonesia, Singapore, Taiwan, and Thailand. *See also* ENDANGERED SPECIES ACT.

convergent evolution in the biological sciences, the development of similar characteristics in two unrelated species due to similar environmental pressures. The eyes of the octopus and squid are remarkably like the eyes of mammals, for example, despite the great evolutionary gulf separating the two groups. Similarly, the powerful hind legs of the kangaroo and of the rabbit developed completely independently of each other. In both cases, they were a response to the problem of avoiding PREDATORS in open country. *Compare* PARALLEL EVOLUTION; COEVOLUTION.

Cool Communities Program a project of AMERICAN FORESTS with the U.S. Department of Energy to encourage the strategic planting of trees next to houses and along streets, measurably reducing energy consumption and thereby reducing CO_2 and resultant global warming. The program is based on the estimate that growing trees can absorb 48 pounds of CO_2 a year, and that three trees shading a house can cut air conditioning needs by up to 50%. Seven communities around the country participate in the program: Miami, Florida; Frederick, Maryland; Atlanta, Georgia; Springfield, Illinois; Tulsa, Oklahoma; Tucson, Arizona; and Austin, Texas. The Clinton administration's Climate Change Action Plan called for the Cool Communities program to expand to 250 cities and 100 federal facilities. *See also* GLOBAL RELEAF, GLOBAL CHANGE.

cooling tower a structure designed to dissipate waste heat from industrial processes or thermal power generating plants. Cooling towers are most familiar to the public in connection with nuclear plants, but are utilized as well in coal, natural gas, or petroleum-powered electricity-generation plants and in some manufacturing processes. Nearly all use water as a coolant, although a few closed-system, dry process towers utilize other fluids (e.g., AMMONIA or CHLOROFLUOROMETHANE).

DESIGN AND OPERATION

There are two principal methods used in cooling towers to dissipate heat. In the traditional "wet tower" method, the hot water is sprayed or allowed to drip over a series of fine grids or a column of loose packing material, maximizing the water's surface-to-volume ratio and thus allowing rapid EVAPORATION. The evaporation cools the remaining water in the tower in the same way that evaporating perspiration cools human skin, by drawing *heat of vaporization* from it for the conversion from a liquid to a vapor. Air circulated through the tower, either by CONVECTION or through the use of large fans (*forced updraft*) carries the water vapor away, bearing the heat with it. The system works but loses from 2 to 10% of each charge of cooling water to the atmosphere, depending upon such factors as the design of the tower and the temperature and humidity of the surrounding air. Hence, especially where water supply may be a problem, a "dry tower" method is sometimes employed. The dry tower acts on the same principal as the automobile radiator, running the water through a network of fine pipes interspersed with airways through which a forced draft passes. Unfortunately, this method, while far preferable to the wet method for conservation of water, utilizes a considerable amount of energy to force both water and air through fine passages, and as a consequence is not used as widely as it perhaps deserves to be. Both wet towers and dry towers may be operated either as "once-through" or "closed" systems. In the once-through system, water is pumped from the source (usually a lake or river), passed through the plant and the cooling tower, and dumped back into the source. Since this has the effect of warming the source, and therefore potentially damaging it (*see* THERMAL POLLUTION), once-through systems are seldom used today except where the water source is extremely large, as in the Great Lakes region or along the seacoast. Closed systems, by contrast, remove water from the source once to fill ("charge") the cooling system and then recirculate it again and again, adding only enough to make up for that lost to leakage and evaporation or drawn off to prevent mineral buildup in the system (*see* BLOWDOWN). They do not create thermal pollution in the water source, but because the water, once withdrawn, is never returned, they rank as a consumptive use, and may contribute to water scarcity in arid or semiarid regions.

cooperation in ecology, a number of individual animals or plants acting together in a way that benefits all of them. Cooperation may occur either among members of the same SPECIES (*intraspecific cooperation*) or among individuals of different species (*interspecific cooperation*). *See also* COMMENSAL SPECIES; SYMBIOSIS.

Cooperative State Research, Education, and Extension Service (CSREES) provides for a connection between the research and education sources and programs of the U.S. Department of Agriculture and state land-grant institutions. CSREES works in partnership with more than 130 colleges and universities, 9,500 research scientists at 59 agricultural experiment stations, 9,600 local extension agents in 57 cooperative extension services, various federal and state agencies, and many nonprofit organizations. Programs include agribusiness, home economics, disaster education, environment and natural resources, farm safety, plant and animal production, pest management, food safety, rural agricultural improvement, home gardening, and home care. The 4-H program involves over 5.4 million youth. Local county extension offices are listed under local government in the phone book. Headquarters address: CSREES, U.S. Department of Agriculture, Washington, DC 20250-0900. Phone: (202) 720-3029. Website: reeusda.gov. *See also* AGRICULTURE, DEPARTMENT OF.

copepod any of a number of small aquatic crustaceans (subclass Copepoda, class Crustacea) characterized by long, multibranched antennae at the front and prominent feathery appendages (*caudal rami*) at the tail end of the body. Most copepods are tiny free-swimming animals varying from less than a millimeter to a centimeter or so in length; however, there are parasitic forms ("fish lice") that attain a length, including caudal rami, of a foot or more. The free-swimming species form a large portion of the PLANKTON of both fresh and salt water, and as such are an important part of the aquatic FOOD CHAIN. The subclass Copepoda contains seven orders, 21 families, and over 4,500 species.

copper the 29th ELEMENT in the ATOMIC SERIES, ATOMIC WEIGHT 63.54, chemical symbol Cu. An orange, lustrous METAL with high ductility (ability to be drawn into a wire) and excellent electrical CONDUCTIVITY, copper is the second most widely used metal (after IRON). It was almost certainly the first to be extensively used by humans. Copper implements and ornaments are found in prehistoric deposits on all continents: they are especially numerous on Cyprus and the British Isles and in the Lake Superior region of North America. The metal occurs fairly widely in the pure state in nature ("native copper"), but is most commonly found as copper sulfite ore (chalcopyrite, bornite, chalcocite, covellite, and others) from which the metal is extracted through smelting. Copper is nontoxic by itself but most of its COMPOUNDS have toxic properties. The most serious environmental hazards associated with its use are the effects of the smelting process, which releases significant quantities of sulfur and (in some cases) arsenic fumes and results in TAILINGS, which are highly toxic to plant and animal life. Most copper mining and smelting in the United States today takes place in Arizona and Utah, with smaller amounts in Michigan, Nevada, Montana, and New Mexico.

Copper Basin an area near Ducktown, Tennessee, on the Tennessee-Georgia border where the ambient air pollution from copper smelting beginning in the mid-1800s killed virtually all trees and plants in a 56-square-mile region. The tree cover has still not fully recovered.

coprolite a pellet of fecal matter (*see* FECES), especially one which has been preserved in some manner, either by drying out (short-term) or fossilization (long-term). Coprolites are valuable to environmental scientists because they can be used to determine what the animal which passed the feces ate, which in turn gives clues to animal behavior, abundance of plants and of PREY species, relationships between PREDATORS and their prey, and even weather and climate patterns from the time the coprolite was created.

coral reefs garden-like shallow-water formations in coastal areas made up of the exoskeletons of living corals, a small marine animal related to the sea anenome. The best known and the largest reefs, and the most beautiful, are in the Caribbean and Australia, although coral reefs exist in most tropical marine waters. Coral reefs are extremely sensitive to pollutants as well as to changes in temperature, which cause them to bleach to a dead white color. The destruction of the reefs constitutes a vivid advance warning of serious environmental impacts to come from industrialization and GLOBAL CHANGE.

cord the volume of wood in a tightly packed stack 4 feet wide, 4 feet high, and 8 feet long—approximately equivalent to 128 cubic feet, although as a practical matter the full volume is never achieved due to the spaces between stacked wood pieces. The cord is used principally as a firewood measure. Thus, "cordwood," or wood capable of being stacked into a cord, is synonymous with stove-sized pieces. A *cord-foot,* occasionally seen as a volume measure, is the volume of a stack 4 feet by 4 feet by 1 foot, or 16 cubic feet: there are eight cord-feet in a cord.

core sample (1) in geology, a column of rock and/or soil obtained by a *core drill* (a drill with a hollow bit; cutting is done exclusively at the circumference of the bit, freeing the center of the cut circle, which is then pushed up through the hollow center of the bit as a continuous column). Core samples enable geologists to read the sequence of stratigraphic layers (*see* STRATUM; STRATIGRAPHIC UNIT) in the center of a rock or soil FORMATION as well as at its exposed edges; hence they are invaluable as a mapping tool and as a means for determining the geologic history and structure of otherwise inaccessible areas such as the sea floor.

(2) in forestry or dendrochronology, a column of wood taken with a core drill in like manner to a geologist's rock column (see [1]). A core sample of wood is taken across the grain from the outside to the center of a tree or branch; the rings and their spacing are then read to determine the tree's age and the climatic conditions at various stages of its life. *See* ANNUAL RINGS; DENDROCHRONOLOGY.

Coriolis effect in mechanics, the combination of forces that causes an object moving in a straight line across a rotating surface to trace a curved path on the surface. If the surface is rotating counterclockwise, the deflection will be to the right; if it is rotating clockwise, the deflection will be to the left. In a fluid medium such as air or water, the Coriolis effect causes a swirling motion that is responsible for the direction of ocean currents and prevailing winds as well as determining the rotational direction of whirlpools and of cyclonic storms (*see* CYCLONE). The Coriolis effect is named after the French physicist Gaspard Gustave de Coriolis (1792–1843), who first described it mathematically; in the environmental sciences it is sometimes known as *Ferrel's law* after the American meteorologist William Ferrel (1817–91), who was the first to recognize that Coriolis's principle, which the Frenchman had described in relation to a rotating plane surface such as a turntable, applied as well to the rotating earth.

corm in botany, a thick, fleshy underground stem utilized by some plants (example: gladiolus) for food storage and as an organ of vegetative reproduction (*see* ASEXUAL REPRODUCTION). The corm differs from the BULB in being primarily stem tissue rather than leaf tissue. It may, however, have a few rudimentary, scalelike underground leaves forming a coating around it. During vegetative reproduction from a corm, the new plants arise from buds formed at the bases of these reduced leaves. The corm will usually have short adventitious roots (*see* ADVENTITIOUS GROWTH) arising from it.

corn snow snow in which the surface has developed a layer of coarse, rounded crystals, usually frozen together to form a crust. Corn snow is formed by alternating cycles of nighttime freezing and daytime thawing during fair weather.

Corporate Average Fuel Efficiency (CAFE) Standards in force since 1978 and introduced in response to the oil embargoes of the 1970s. The CAFE standards of the federal government require that manufacturers' "light-duty vehicles," meaning, effectively, passenger cars, meet a sales-weighted fuel efficiency minimum for the entire fleet of cars sold by the manufacturer. In 1978 the minimum was 18 miles per gallon (mpg). The purpose of the standard was mainly to conserve FOSSIL FUEL and thus reduce reliance on foreign oil. Secondarily it was to reduce AIR POLLUTION. While other countries increased taxes to reduce the use of gasoline, the U.S. Congress chose instead to enforce a fuel efficiency minimum; the United States remains the only advanced industrial nation to do so. The standard set for passenger cars under the law slowly increased over the years, with the figure as of 1997 set at 27.5 mpg. Despite the CAFE standard increases, fuel efficiency has not actually improved commensurately because of the popularity of light trucks, vans, and sport utility vehicles (SUVs), now comprising half the consumer automobile market. For these vehicles, the CAFE standards have been much lower: 20.7 mpg (as of 1997), thus limiting the potential for the CAFE standards to markedly improve efficiency. In 1999 President Clinton issued an executive order requiring that SUVs and other trucks and vans be included in the "light-duty vehicle" category under the program. Manufacturers whose fleets do not meet the required minimums are fined $55 per mpg for each vehicle sold. The CAFE approach, even with its fines structure, is not, in the view of most environmentalists and economists, an efficient or effective way to encourage fuel efficiency and, concomitantly, to reduce CO_2 emissions—an increasingly significant issue in terms of GLOBAL CHANGE. With the high profit margins earned by SUVs (which encourages manufacturers to promote them heavily) and the historically low cost of gasoline, many believe that direct fuel taxes would be a more effective means of achieving fuel efficiency and would require much less interference with the operations of an important component of the private sector. *See also* CARBON TAX.

corporate farming refers to a type of business organization for farms which is increasing, as opposed to individual operations, often referred to as "family farms," although most incorporated farms are owned by family members related by blood or marriage. Currently (1996 data) corporate farms comprise about 10% of the total number of farms but own 29% of the farmland. Sales volume for corporate farms is nearly four times greater than for individual operations—$247,000 in sales per farm on the average versus $63,000. *See also* AGRIBUSINESS; BLACK LAND LOSS.

Corps of Engineers *See* ARMY CORPS OF ENGINEERS.

corrosion the chemical or electrochemical destruction of a substance, differentiated from means of mechanical destruction such as abrasion and EROSION. Abrasion and erosion break off small pieces of a substance and carry them away without altering their chemical properties; corrosion changes the substance into another, usually weaker substance (though with some materials, notably aluminum in air, corrosion may create a thin film of a much tougher substance, which then protects the original substance from further damage).

TYPES OF CORROSION

There are two principal types of corrosion, each of which presents its own problems for environmental engineering. In *chemical corrosion,* a chemical REACTION takes place between the corroding substance and the agent of corrosion, creating one or more new substances with substantially different properties. A common example is the rusting of iron, which results from the reaction of iron and oxygen in the presence of water to form hydrated iron oxides. In *electrolytic corrosion,* the passage of an electrical current through a metal or other conductor submerged in an ELECTROLYTE (such as salt water or damp soil) removes electrons from the metal at the point where the current exits (the anode), thereby slowly converting the atoms of the metal to free IONS, which migrate away. Electrolytic corrosion is further broken down into two subtypes. *Stray current corrosion* results from the migration of direct current from some external source (such as improperly insulated machinery) through the surrounding electrolyte and in the metal (alternating current is not a problem because it converts the anode into a cathode during every other cycle, redepositing any ions that have migrated off). *Galvanic corrosion* takes place when a difference of electrical potential exists within the metal or the surrounding electrolyte, causing the corroding system to generate its own weak electric current. There are numerous factors that can lead to this self-generating type of electrolytic corrosion. Differences in soil chemistry from one end to the other of a buried piece of metal may do it; so may differences in moisture content, or the presence of certain types of bacteria (*see* BACTERIUM). A major cause is the burial or submergence of different metals with different electrical properties in close proximity to each other within an electrolyte; for example, the burial of a steel tank next to a copper pipeline. *See* GALVANIC SERIES.

PROTECTION AGAINST CORROSION

Chemical corrosion can be avoided by coating susceptible areas (such as the inside of steel storage tanks) with inert materials, and by making sure that seals, fittings, hoses, and so forth are made of materials that are compatible with the chemicals that they will be exposed to. Stray-current corrosion can be eliminated by insulating or otherwise electrically isolating the region of the electrolyte containing the at-risk material. Galvanic corrosion is a knottier problem because differences in electrical potential can be affected by so many factors. Control of soil composition, acidity, and moisture content can help a great deal in the case of buried metal structures, as can making certain that all parts of the structure are made of the same metal and that there are no other metal objects buried nearby. None of this, however, will help very much in the case of objects submerged in an excellent electrolyte such as sea water. Probably the best overall approach to both forms of electrolytic corrosion is to use a so-called "sacrificial anode"—a rod of magnesium or some other material from the low end of the galvanic series, connected electrically to the structure that needs protection and—because it serves as the anode for the entire system—concentrating all the corrosion in itself. When combined with an insulating coating on the protected structure, a sacrificial anode provides nearly 100% protection against both galvanic corrosion and stray-current corrosion. *See also* CATHODIC PROTECTION; CORROSIVE MATERIAL.

corrosive material a material (especially a liquid) that reacts corrosively with normal storage and transportation mediums such as steel, aluminum or rubber; also, any material which damages or destroys human skin tissue on contact. *See also* CORROSION.

cost *See* BENEFIT/COST ANALYSIS; COST-EFFECTIVENESS ANALYSIS.

cost/benefit analysis *See* BENEFIT/COST ANALYSIS.

cost-effectiveness analysis in economics, the evaluation of alternative means to achieve the same end with the goal of finding the least-cost alternative. Cost-effectiveness analysis takes into consideration all of each alternative's costs, both economic and induced (*see* PROJECT ECONOMIC COSTS; INDUCED COSTS), and then examines the relationship between the total cost of each alternative and the effectiveness with which the alternative achieves the goal: for instance, an alternative that only achieves 70% of the goal may be more attractive than one that achieves 90% if the first alternative only costs half as much. Since comparisons between the desirability of different levels of effectiveness usually boil down to opinion, it is usually necessary in practice to specify a minimum effectiveness level. Alternatives that reach this level can then either be compared directly on a cost-for-cost basis, or can be evaluated with a formula that takes into account the ratio between cost and effectiveness. *See also* BENEFIT/COST ANALYSIS.

cotyledon the portion of a plant embryo corresponding roughly to the leaf of a growing or mature plant, sometimes called the "seed leaf." Within the seed, the cotyledon acts as the embryonic plant's digestive and storage system, absorbing food from the ENDOSPERM. When the seed sprouts, the cotyledon is the first part of the infant plant to break the ground surface, and acts as the plant's center of PHOTOSYNTHESIS until its true leaves can form (these do not form from the cotyledon, but from the *epicotyl,* the portion of the embryo above the cotyledon). The number of cotyledons a plant has is diagnostic of its class or subclass within the family of seed-bearing plants. *See* COTYLEDONAE; MONOCOTYLEDON; DICOTYLEDON; ANGIOSPERM; GYMNOSPERM.

Cotyledonae the group of seed-bearing plants; that is, those which form COTYLEDONS. The Cotyledonae include all members of the classes Gymnospermae and Angiospermae (*see* GYMNOSPERM; ANGIOSPERM). The term is not an official classification (the subdivision Pteropsida, which includes the Cotyledonae, also includes the ferns, which are non-seed-bearing), but it can be useful as a term of reference when both conifers and flowering plants are included in a discussion. *Compare* CRYPTOGAMIA.

couloir a broad, shallow, steep gully on a mountain face, usually above TIMBERLINE or in an untimbered area below timberline. The term is French; it derives from the verb *couler* (to slide, flow), probably from the fact that couloirs so often serve as avalanche paths.

Council on Environmental Quality (CEQ) federal (U.S.) agency located within the Executive Office of the president, created by the NATIONAL ENVIRONMENTAL POLICY ACT OF 1969 and charged with overseeing and coordinating all federal policy decisions with respect to the environment. The CEQ sets federal environmental policy, monitors compliance with this policy by other federal agencies, and adjudicates conflicts among other federal agencies and between federal and state agencies over environmental matters. It also prepares guidelines for the creation of ENVIRONMENTAL IMPACT STATEMENTS, reviews all environmental impact statements where the impacts are national in scope or where substantial controversy exists, collects and publishes information regarding problems and trends in national environmental quality, and prepares the annual State of the Environment Report for delivery to Congress by the president. It has no enforcement power in the private sector: for these powers, *see* ENVIRONMENTAL PROTECTION AGENCY. Headquarters (CEQ): 722 Jackson Place NW, Washington, DC 20503. Phone: (202) 456-6224.

countershading *See* OBLITERATIVE COLORATION.

country neighborhood in land-use planning, a rural area of dispersed housing that nevertheless manages to maintain a communal feeling due to the presence of some common social focal point. Sometimes this focal point is a small, centrally located HAMLET within the neighborhood; more often it is an isolated country store, school, church, or grange hall. Occasionally a country neighborhood will develop without a focal point, maintaining its identity and sense of community from some external factor such as similarity of terrain. The households within a particular small WATERSHED, for example, might form a country neighborhood even though no specific social gathering place exists.

country rock in geology, the rock surrounding an intrusive body such as a mineral VEIN or a magmatic DIKE or SILL. *See also* INTRUSIVE ROCK; MAGMA.

Cousteau, Jacques-Yves (1910–1997) French oceanographer and explorer, born June 11, 1910, in Saint-Andre-de-Cubzac, a small town at the tip of the Gironde Estuary a few kilometers north of Bordeaux, France. After attending high school in New York City, Cousteau joined the French navy, graduating from the Brest Naval Academy in 1933. Turning to ocean diving as a sport around 1936, he quickly became fascinated by it, and was one of the small group of French divers who developed the wet suit (ca. 1940) and the aqualung (ca. 1942). Following World War II he turned to oceanography as a career, founding the Underwater Research Group at Toulon, France, on the Mediterranean. In 1950, in partnership with an anonymous English donor, he purchased a 140-foot U.S.-built minesweeper, outfitted it as a floating laboratory, and renamed it *Calypso,* forming a nonprofit foundation, Calypso Oceanographic Expeditions, to support his work. By 1960 he was becoming concerned about worldwide ocean pollution, a concern which led him, in 1975, to form the COUSTEAU SOCIETY. Though Cousteau had little formal oceanographic training, he constantly surrounded himself with trained scientists and learned directly from them, and he was recognized as one of the world's foremost experts on the oceans. His award-winning films, books and television programs on underwater topics have made an immeasurable contribution to popular understanding of the world ocean and its environmental problems. He died June 25, 1997.

The Cousteau Society environmental organization founded in 1973 to support the work of the French oceanographer JACQUES-YVES COUSTEAU. The group is concerned with planet-wide environmental issues, especially the protection and conservation of the marine environment. Its staff of 85 concentrates primarily on scientific research and public education through television programs, lecture series, and publication of the society's journal, the *Calypso Log.* 870 Greenbrier Circle, Suite 402, Chesapeake, VA 23320. Phone: 1-800-441-4395. Website: www.cousteausociety.org.

covalent bond in chemistry, a bond formed between two ATOMS in which a pair of ELECTRONS is shared equally between the two atoms. Atoms joined by covalent bonds do not capture electrons from each other, and thus do not form IONS; instead, the atoms trade the shared pair of electrons back and forth so that as each atom captures one of the electrons it releases the other. The effective result is that the orbit of each electron in the shared pair has been expanded to include both atoms. More than one covalent bond may exist between the same two atoms. Covalent bonds are considerably stronger than ionic (*monovalent*) bonds and create COMPOUNDS (*covalent compounds*) that are chemically more stable than ionic compounds. *See also* VALENCE.

cove a small EMBAYMENT on a body of water, especially one broader than it is deep. In mountainous areas, the term is also applied to small openings ("hollows") where a plain or valley floor extends a short distance into a mountainside, or where a treeless area such as a meadow extends into an otherwise unbroken forest wall.

cover in wildlife management, plant growth that enables an animal to get partially or completely out of sight beneath or within it. *Dense cover* is cover able to hide 90% or more of an animal at a viewing distance of 150 feet; *hiding cover* is cover able to hide 90% of an animal at a viewing distance of 200 feet. *Optimal cover* is dense cover which extends for at least six animal lengths in every horizontal direction. In big-game management, the "animal" in these definitions is usually understood to be the size of a mature elk.

cover crop in agriculture, a crop planted on a field or pasture solely to provide vegetation to cover the soil and thus protect it from dying out or washing away. Cover crops are not normally harvested, but are simply plowed under or burned off as the field or pasture is being prepared for use. *See also* GREEN MANURE.

coyote getter (cyanide gun, M-44) device for killing animals by making them inject themselves with cyanide, once widely used on the rangelands of the American west as a means of PREDATOR CONTROL directed principally against coyotes. The coyote getter consists of a metal tube roughly 1 foot long and 1 inch in diameter containing a spring-loaded dart tipped with cyanide and a triggering mechanism connected to a "bait wick," usually cotton cloth dipped in carrion. The device is buried in the ground with just the bait wick showing. A curious animal tugging on the wick will release the trigger and propel the dart into its mouth. Like most unattended predator-killing devices (*see also* BAIT STATION) the coyote getter kills a large number of NONTARGET SPECIES, including dogs, cats, and even small children, as well as the targeted coyotes. For this reason, it is now outlawed in most areas.

crag and tail in geology, a GLACIAL LANDFORM found where a GLACIER has ridden over relatively unresistant rock with a more resistant INTRUSION in it. The resistant intrusion forms a knob (the "crag") that protects the "downstream" portion of the less resistant rock, causing it to form a long ridge stretching downstream from the crag (the "tail"). Crags and tails are thus a very good indication of a vanished glacier's direction of flow. *Compare* DRUMLIN; ROCHE MOUTONNEE.

crater any generally round, steep-sided depression in the earth, caused by either an impact or an explosion. The orifice at the top of a VOLCANO through which it ejects new material is usually a crater, as is the hole caused by the impact of a meteorite or other extraterrestrial body, or by the explosion of a bomb. A similar hole caused by the collapse of an underground cavity is not a crater but a SINKHOLE. *See also* CALDERA.

creationism the belief, in opposition to CHARLES DARWIN's theory of evolution and the BIG BANG theory, that the universe literally emerged in six days as stated in the Book of Genesis, and that God created living beings in their present forms approximately 10,000 years ago. This belief, especially as proposed by its adherents to be taught in public schools as "creation science," tends in the view of many Catholic and mainstream Protestant theologians to separate God from nature and religion from science, with significant implications for the development of an environmental ethic. "Creationism" should not be confused with "creation spirituality" or "creation theology," which for the most part are based on evolutionary process theology. *See also* RELIGION AND THE ENVIRONMENT.

Creative Act of 1891 *See* FOREST SERVICE: *history.*

creel limit *See under* BAG LIMIT.

crepuscular species any species of animal whose activity peaks during the twilight hours, that is, at dawn and at dusk. Examples include deer, rabbits, house cats, and many species of insects. *Compare* DIURNAL SPECIES; NOCTURNAL SPECIES.

criteria pollutant *See under* NATIONAL AMBIENT AIR QUALITY STANDARDS.

critical habitat the HABITAT necessary to the survival of a SPECIES of plant, animal, or other ORGANISM. Critical habitat includes all air, land, and water space that the species in question requires to carry out its normal living patterns; it also includes other living things utilized by the species for food, shelter, or other necessary activities. (Critical habitat for a PREDATOR, for example, includes its accustomed PREY.) The ENDANGERED SPECIES ACT requires the designation and protection of critical habitat for all ENDANGERED SPECIES. A crucial legal argument, as yet unresolved, revolves around the question of whether this requirement refers to the designation of critical habitat for the species at its current population levels or whether enough extra (currently uninhabited) habitat must be designated to serve the species if and when its population rebounds to nonendangered levels. *Compare* CRUCIAL HABITAT.

Cross-Florida Barge Canal project to connect the Atlantic Ocean to the Gulf of Mexico by a water-level route from Jacksonville to Yankeetown (80 kilometers north of Tampa) through the base of the Florida peninsula, halted by citizen complaints when approximately 1/4 complete. It is considered one of the environmental movement's most significant victories over the Army Corps of Engineers. Authorized on July 23, 1942—partly to provide Texas oil tankers a route that would be protected from Nazi submarines—the Canal remained unfunded until the early 1960s due to its extremely low BENEFIT/COST RATIO. Construction of the 185-mile-long canal finally began on February 26, 1964. On January 15, 1971, canal opponents—led by Florida Defenders of the Environment and the ENVIRONMENTAL DEFENSE FUND—won an injunction in federal court against further work on the canal, citing violations of the NATIONAL ENVIRONMENTAL POLICY ACT, the FISH AND WILDLIFE COORDINATION ACT, and the canal's authorizing legislation itself (which had authorized construction only on the condition that groundwater supplies would not be affected). Four days later, on January 19, President Richard Nixon permanently halted construction of the canal by EXECUTIVE ORDER.

crown of a tree, the portion of the tree above the bare trunk, that is, the branches and trunk from the first major leaf- or needle-bearing branch upward. *Compare* CANOPY.

crown closure in forestry, the percentage of the CANOPY that is without significant openings.

crown fire *See* WILDFIRE: *types of wildfire.*

crucial habitat the portion of the HABITAT of any species (endangered or otherwise) that is required for carrying out some necessary activity. Crucial habitat may be only a small part of a species' total habitat, but it is usually the size of the crucial habitat—not the total habitat—that acts as the principal limiting factor to population growth. Examples of crucial habitat include winter range, breeding and denning sites, COVER for hiding from PREDATORS, and the growing sites of any food that an animal may require in its diet and for which no substitute is possible. *Compare* CRITICAL HABITAT.

crude as a noun, usually denotes CRUDE OIL.

crude oil (crude) PETROLEUM as it is pumped out of the ground, before any process of refinement takes place. Crude oil is highly varied in its properties. Its color may vary from black or dark brown through red to light yellow. It may be so thin that it pours like water ("light crude") or so thick that it barely pours at all ("heavy crude"). In structure it is a complex mix of HYDROCARBONS that varies considerably from sample to sample. Three principal structural classes are recognized. *Asphaltic crudes* contain only hydrocarbons whose molecules are composed of precisely twice as many HYDROGEN atoms as CARBON atoms. *Paraffin crudes* have two extra hydrogen atoms on each molecule (no. of hydrogen atoms = 2 x no. of carbon atoms + 2). *Mixed-based crudes* are composed of a mixture of the other two types. Impurities consisting of non-hydrocarbon-based molecules are often found in all three types: the most common of these is SULFUR, which may compose up to 5% of a sample of crude. Low-sulfur crude is known as "sweet crude"; high-sulfur crude, as "sour crude." *See also* PETROLEUM SUPERTANKER; DESTRUCTIVE DISTILLATION; OIL SPILL.

cruiser in logging, see TIMBER CRUISER.

cryptic coloration (concealing coloration) in zoology, a color pattern on an animal that makes it difficult for other animals to see it. Cryptic coloration is constant throughout a SPECIES. Though there may be minor individual variations, each member of the species will have essentially the same markings. There are four principal types. In *protective resemblance* the animal's color pattern is similar to the generalized background it will most often be resting against; a moth may be the color of tree bark, an aphid may be green, etc. (The spots of leopards and the reticulated color pattern of giraffes are also forms of protective resemblance, in these cases resembling dappled sunlight.) In *obliterative coloration* the color pattern is dark on top and light on the bottom, so that the tone of the part of the animal that is in the shade will be the same as that in the sun and it therefore cannot easily be detected by its shadow. This form of cryptic coloration is used by many birds, reptiles, and amphibians. In *disruptive coloration,* a slash of dark color across an animal's light-colored body makes it difficult to visually join the two light-colored halves and see the animal as a single creature. The killdeer's dark neckband is coloration of this type. When the eye sweeps across it, the bird appears as two separate blocks—head and body—neither of which particularly looks like prey. Finally, in *aggressive resemblance* the animal's coloring combines with body shape to make it look strikingly like some common object in the environment. This is most common among insects (there are insects that look like twigs, leaves, pieces of bark, and even bird droppings) but it can also be found among a few VERTEBRATES, such as the sea dragon (a relative of the sea horse), which looks like a clump of seaweed. *See also* WARNING COLORATION; MIMICRY.

crypto-biotic crust in some western American deserts, a small, hard crust of mosslike growth that serves to trap moisture and provide a purchase for seeds. The crust is grayish, hard to see, and extremely fragile; it can be uprooted by a single step. Disturbed crypto-biotic crust can take 500 years to grow back. *See also* DESERT BIOME.

cryptocrystalline rock rock composed of crystals so tiny that they cannot be seen with a magnifying glass: crystals are, however, present. Cryptocrystalline rocks result from rapid cooling of IGNEOUS ROCK bodies, usually on the surface. Example: CHERT. *Compare* AMORPHOUS ROCK.

cryptogamia the group of plants that reproduce by spores, notably ferns and BRYOPHYTES. Fungi, which are sometimes included in the group, are not members of the plant KINGDOM and thus should not be lumped with the others, although they do reproduce by means of spores (*see* FUNGUS). *Compare* COTYLEDONAE.

cryptophyte any plant that bears part or all of its reproductive structures either underwater or underground, that is, a plant capable of reproducing from a CORM, BULB, or TUBER. The potato is an obvious example.

cubic feet per second (cfs) a measure of the rate of flow of a substance—usually of water, though the term is also applied to other liquids, to gases, and even occasionally to solids (for example, to volcanic EJECTA). One cubic foot per second is a volume of one cubic foot of the substance passing a stationary point in one second's time.

cull material (cull) in logging and forestry, any tree or part of a tree that is large enough for making lumber but is not suitable for this use due to some factor other than its size. Though it is possible to have "standing culls"—that is, culls that are rejected before harvesting—the term generally implies that the tree the cull came from was harvested first. Cull material includes principally deformed or unsound logs ("cull logs": *see* DEFECT) and those that are broken or otherwise damaged during the harvesting operation. Together, cull material and SLASH generally make up 40 to 60% of a normal harvest.

culmination of mean annual increment in forestry, the age at which the MEAN ANNUAL INCREMENT of a tree's growth is at its largest, that is, the age at which the tree is "putting on volume" at the fastest rate. Measured as a percentage of the tree's total volume, mean annual increment decreases regularly throughout the tree's life; however, measured in terms of absolute volume, it increases for the first roughly 20% of the tree's life and then begins decreasing. The age at which the turnaround occurs is the culmination of mean annual increment, and is by common definition the point at which the tree reaches maturity. It is important to note that growth does not stop at this point, but continues throughout the tree's life. In fact, the wood put on by the tree after culmination is denser, has finer grain and fewer knots, and generally makes better lumber than that put on before (*see* OLD GROWTH).

cultural eutrophication the alteration of the TROPHIC LEVEL of a body of water from less to more eutrophic due to human activities. Cultural eutrophication usually results from an increase in the NUTRIENT

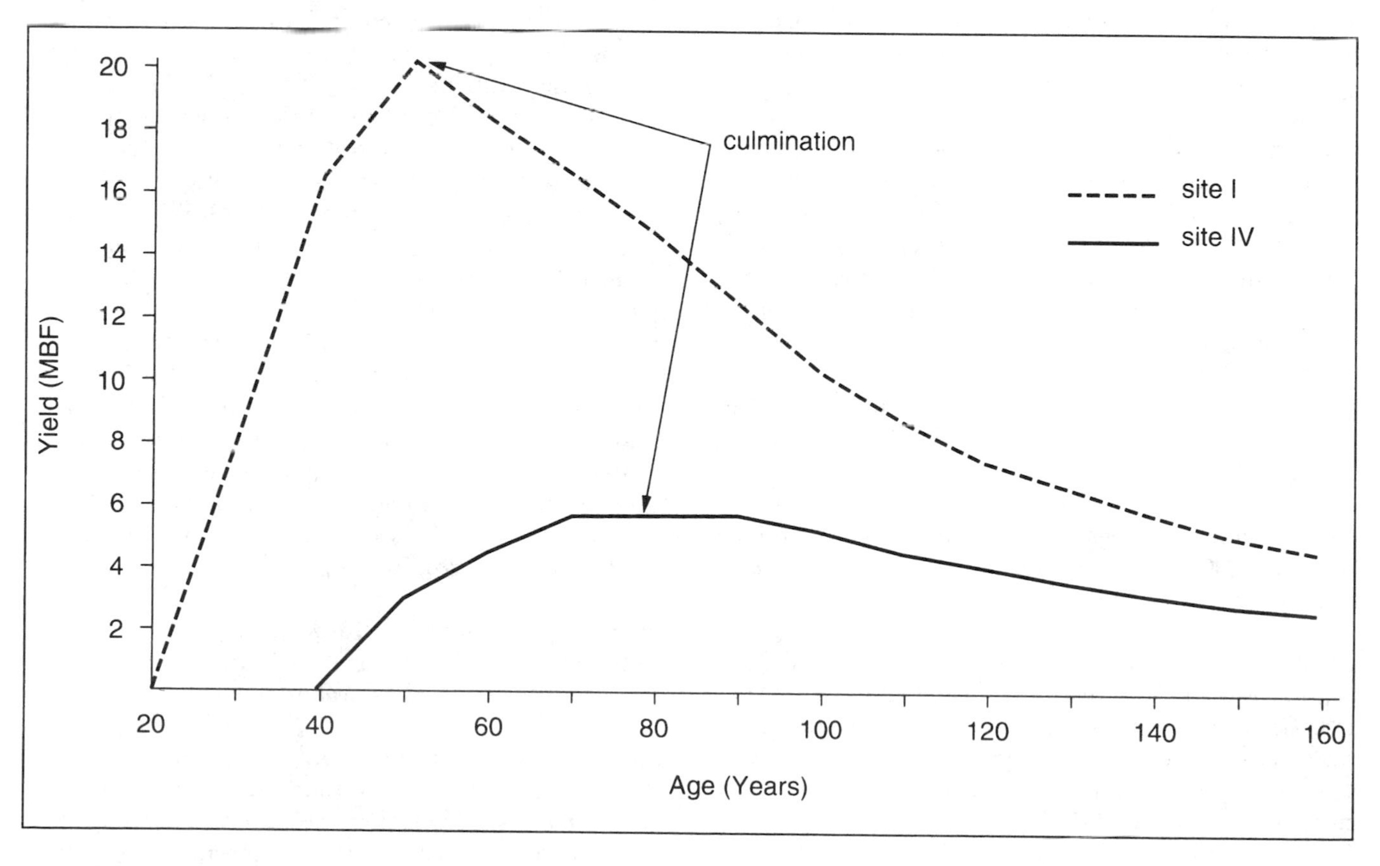

Mean annual increment

load in the water body, but may also be caused by THERMAL POLLUTION or by morphometric (shape) changes resulting from dredge-and-fill operations or increased sedimentation. It creates health hazards and aesthetic problems for all users of the water, including recreationists, shippers, and municipal and industrial waterworks. Typical problems include increased levels of bacterial and algal TOXINS (*see* BACTERIUM; ALGAE); disagreeable taste, odor, and color; and the clogging of waterways, intakes, and equipment by excess algal growth. *See* further discussion under EUTROPHICATION.

cultural resource in land-use planning, an archaeological site, historical site, or other evidence of past use of an area by humans. The term includes everything from abandoned buildings and roadbeds to ancient pottery shards, flint chips, or even buried bits of charcoal from a campfire. Federal agencies are required by law to evaluate cultural resources that may be damaged or destroyed by their actions and to avoid or mitigate the damages wherever possible (*see* ANTIQUITIES ACT).

cultural prescription in forestry, the recommended procedure or set of procedures for carrying out a desired CULTURAL TREATMENT. For example, the cultural treatment called for might be the removal of competing vegetation from a plantation of young CONIFERS. The cultural prescription for achieving this end could be hand removal, the use of HERBICIDES, or even the introduction of domestic goats.

cultural treatment in forestry, any activity designed to modify natural vegetation in order to achieve some management objective. Examples include THINNING, fertilization, pruning, insect control (either with pesticides or through the use of biological controls), and the removal of competing vegetation. *See also* CULTURAL PRESCRIPTION.

cumec in water flow measurements, cubic meters per second (m^3/sec). One cumec is equal to 35.7 cubic feet per second (cfs).

c unit (c-unit) in logging or forestry, a unit of volume equal to 100 cubic feet (1,200 board feet). *See also* BOARD FOOT.

curie a unit for measuring the amount of radiation emitted by a radioactive substance (*see* RADIOACTIVITY). One curie is equivalent to the number of ALPHA PARTICLES emitted by one gram of radium-226 in one second (approximately 37 billion or 3.7×10^{10}). The curie is named for Marie Curie (1867–1934), the French physicist who first isolated radium and described its effects.

curie point (curie temperature) the temperature above which a magnetic substance such as iron or nickel loses its ability to maintain a magnetic field. It is unrelated to the unit of radiation known as the curie. The curie point is named for the French physicist Pierre Curie, husband of Marie Curie (*see* CURIE).

cut in forest management, the amount of timber that is removed from a land-management unit such as a National Forest or Ranger District (*see* FOREST SERVICE: *structure and function*) in a given amount of time, usually one year (the "annual cut"). The cut is usually expressed in BOARD FEET. *See also* ALLOWABLE CUT.

cutbank a nearly vertical soil slope that has been exposed by cutting action, either natural (through rapid stream or wave EROSION) or manmade (in the course of building a road or trail or leveling a foundation). Cutbanks are generally unstable; they erode easily, and may simply collapse under the influence of gravity. For this reason, cutbank stabilization is an important part of construction engineering, and can also play an important role in stream or shoreline management. Several stabilization methods are commonly practiced. The most important, at least for artificial cutbanks, is to make certain that the PITCH of the cutbank is less than the ANGLE OF REPOSE for the type of soil the bank is composed of. However, this must usually be coupled with other methods. In *vegetal stabilization* the cutbank is sown with a plant, such as grass, which tends to create a thick, interlocking set of roots; these help keep the soil in place. In *terracing,* the cutbank is interrupted at intervals by narrow horizontal terraces, ensuring that the length of any column of earth up the face of the bank can only be as long as the distance from one terrace to the next. This limits the weight of each column and reduces the chances that gravity will cause mechanical bank failure, as well as provides a stopping point partway up the bank for any soil that does happen to slide. Finally—and especially in areas where the angle of a cutback must be at or greater than the angle of repose—the bank may be faced off by RIPRAP, a retaining wall, or a set of GABION BASKETS, covering the soil so that it cannot slide. *See also* EROSION: *control of erosion.*

cut block in logging or forest management, a unit of timber that is to be clearcut; also, the CLEARCUT unit itself, after the timber has been removed from it.

cutting unit in logging or forest management, a clearly defined area of forest land that is to be harvested in a single set of logging operations, following a single SILVICULTURAL PRESCRIPTION. Several cutting units will usually be associated with each other in a single harvest plan; these cutting units may or may not be contiguous to each other and may or may not follow the same silvicultural prescription. On federally managed forest land, a cutting unit is generally the same as a *sale unit*.

Cuyahoga River a river passing through the city of Cleveland. During the 1960s the Cuyahoga was so polluted that it rendered parts of Lake Erie, into which it enters, nearly lifeless. Finally, in 1969, a buildup of volatile POLLUTANTS on the surface of the water caused the Cuyahoga to catch fire, an event so vivid that it called national attention to the problems of WATER POLLUTION, especially in urban areas. The Water Pollution Control Act Amendments of 1972—now called the CLEAN WATER ACT—followed, and in time the waterway was cleaned up. Lake Erie, which many thought might be permanently dead, has recovered. Today the Cuyahoga is well known as the location of the Cuyahoga Valley National Recreation Area. A unit of the NATIONAL PARK SYSTEM created by Congress in 1974, the recreation area preserves 33,000 acres of pastoral valley along a 22-mile segment of the river between Cleveland and Akron.

cyanide any of numerous chemical SALTS containing the cyanogen GROUP (a COMPOUND of carbon and NITROGEN, chemical formula CN^-). The most common commercial industrial forms are sodium cyanide (NaCN) and potassium cyanide (KCN), both of which are white granular powders that tend to fuse into needlelike crystals. Odorless and tasteless themselves, they combine readily with the moisture in air to form hydrogen cyanide (HCN), a gas with the pronounced odor and flavor of bitter almonds. The cyanides are extremely poisonous (LD_{50} [rats, orally]: NaCN, 15 mg/kg; KCN, 10 mg/kg), as is the associated hydrogen cyanide gas (LC_{50} [rats, mice, dogs]: 300 ppm for 3 minutes). Symptoms of cyanide poisoning include headache, vertigo, nausea, vomiting, paralysis, unconsciousness, convulsions, and respiratory arrest leading rapidly to death. HCN also causes pronounced gasping for breath, which increases the victim's intake of the poison.

CYANIDE AS A POLLUTANT

The cyanides are used occasionally as RODENTICIDES and insecticides (*see under* PESTICIDE) and in the control of livestock PREDATORS (*see* COYOTE GETTER); however, they enter the environment more commonly as a result of their use in metallurgy. A solution of NaCN or KCN in water is one of the very few liquids that will dissolve gold and silver, and is therefore widely used to extract those metals from their ores (*see* LIXIFICATION). Cyanide salts are also an important part of the solution ("pickling water") used to case-harden steel and to clean it prior to electroplating. Waste waters from gold and silver extraction and from steelmaking may thus both contribute cyanide to watercourses, where it is a PRIORITY POLLUTANT.

cyanide gun *See* COYOTE GETTER.

cyanobacteria *See* CYANOPHYTE.

cyanophyte any member of the PHYLUM (or DIVISION) Cyanophyta, commonly known as the "blue-green algae." The cyanophytes contain CHLOROPHYLL, and share many characteristics with the true ALGAE, with which they were originally classed. However, they lack a complete nucleus and have no CELLULOSE in their cell walls (*see* CELL: *structure of cells*), and most authorities today class them with the bacteria (*see* BACTERIUM), even going so far in some cases as to call them "cyanobacteria" or "blue-green bacteria." (Questions as to whether they are plants or animals are neatly avoided by classifying both bacteria and cyanophytes into a separate KINGDOM, the MONERA: *see* TAXONOMY.) Cyanophytes are unicellular organisms, though some species form filamentous colonies. It is the filamentous varieties, notably those of the genus *Cladophora,* that are principally responsible for the phenomenon known as the ALGAL BLOOM (sometimes called the "bacterial bloom") in which, in response to optimal temperatures and nutrient supplies, cyanophyte populations explode to the point where they clog waterways, dye water green, and kill other forms of aquatic life (*see* EUTROPHICATION). The division contains approximately 7,500 known SPECIES.

cycad any member of the family Cycadaceae of the class Gymnospermae (the coniferous plants; *see* GYMNOSPERM). Cycads are small, shrublike trees, normally less than 10 feet tall, though one Australian SPECIES reaches 50 feet. They look very like palms, and often carry popular names that reflect this appearance (sago palm; fern palm): however, they bear cones, and are more closely related to the pines and firs than to the true palms. They are rather primitive in structure, and many botanists believe that they are not too far removed, in an evolutionary sense, from the common ancestor of all GYMNOSPERMS and ANGIOSPERMS. Extremely widespread in the fossil record, this once numerous family is today reduced to about 100 species in nine genera, almost all in the tropics.

cyclic organic compound an ORGANIC COMPOUND that contains a ring of CARBON atoms somewhere in its structure. The carbon ring may contain as few as three atoms, but the usual number is five or six: occasionally, a cyclic compound may be found with more than six atoms in its carbon ring. Atoms of other ELEMENTS (most commonly OXYGEN, NITROGEN, or SULFUR) are sometimes substituted for one or more of the carbons in the ring: these are known as *heteroatoms*. Examples of cyclic compounds include some sugars; most vitamins; and BENZENE and its derivatives, including PCBs (*see* POLYCHLORINATED BIPHENYL) and DDT. *See also* AROMATIC COMPOUND; BENZENE RING.

cyclone (low-pressure system) in meteorology, an area of low air pressure ("low-pressure cell") surrounded by a circular pattern of winds blowing in a counterclockwise direction (northern hemisphere: in the southern hemisphere, cyclones blow clockwise—*see* CORIOLIS EFFECT). From a scientific standpoint, all such low-pressure systems are cyclones; popular terminology usually restricts the term to the winds accompanying CYCLONIC STORMS, one of the few areas in which popular usage is more restrictive than scientific usage. (Funnel-cloud storms, often erroneously called "cyclones," arise from a significantly different set of circumstances: *see* TORNADO.)

cyclonic storm in meteorology, a well-developed CYCLONE with particularly strong encircling winds, usually accompanied by precipitation. Cyclonic storms create stormy weather conditions over large areas, although generally speaking the larger the area the weaker the storm. Small, intense cyclonic storms are called *typhoons* or *hurricanes: see* HURRICANE.

cytokinin in botany, any of a group of closely related plant growth hormones that are derived from nucleic acids (*see* DNA; RNA). The cytokinins are the functional opposite of the AUXINS. They are formed in the roots and migrate upward through the plant, and are responsible primarily for promoting CELL division. It is believed that the interactions between auxins and cytokinins are responsible for most tissue differentiation in plants. For example, it has been demonstrated in the tobacco plant that an auxin-cytokinin mixture will cause roots to form if auxins are the predominant hormones, but will merely cause undifferentiated tissue growth if the two classes of hormone are present in equal amounts.

cytotoxin a TOXIN that destroys cell cytoplasm (*see* CELL: *structure of cells*). Generalized cytotoxins are relatively rare: most toxins are directed toward specific types of cell, and do not usually destroy them but merely interfere with their function. *See,* e.g., NEUROTOXIN; ENTEROTOXIN.

Daly, Herman (1938–) an ecological economist. Born in 1938, Daly received a B.A. degree in economics from Rice University and a doctorate in economics from Vanderbilt University. He served as senior economist in the environment department of the World Bank from 1988 to 1994 and, since 1994, has been senior research scholar with the School of Public Affairs at the University of Maryland. A founding member of the International Society for Ecological Economics, he currently serves as associate editor of *Ecological Economics,* the society's journal. His concerns regarding connections between the economy, the environment, and ethics have been the topic of numerous books, articles, and lectures. His books include *Towards a Steady-State Economy* (1973); *Economics, Ecology, Ethics* (1980); *For the Common Good: Redirecting the Economy Toward Community, the Environment, and a Sustainable Future* (with theologian John Cobb, 1989); and *Beyond Growth: The Economics of Sustainable Development* (1996).

dam a barrier to a flow, especially a structure built to act as a barrier to the flow of a stream or a river. Dams may be built for a variety of purposes, including the reduction of downstream floods; the impoundment of water for IRRIGATION, domestic, or industrial water supply; the creation of a recreational lake; the diversion of part or all of a stream's flow into canals for transport to other areas; the deepening of streams in order to improve navigation; and the creation of a HEAD for the generation of HYDROELECTRIC POWER. Dams built to accomplish several of these goals at once are known as *multipurpose dams,* and are the principal focus of environmental concern due to their generally larger size and proportionately greater impact on their surroundings.

TYPES OF DAMS

Structurally, dams are of four main types. *Embankment dams,* the oldest and still most widespread type, consist of a mass of soil (*earthfill*) or crushed rock (*rockfill*) placed in the path of the river. Because such a mass is usually pervious (that is, allows water to seep through), these dams usually have an impervious core of clay, masonry, or concrete. The base of an embankment dam is four to seven times as broad as the height of the dam, with most of the extra material placed upstream, so that the structure's slope is considerably gentler on the upstream side. *Gravity dams* are massive concrete or masonry structures that, like *embankment dams,* are prevented from washing out by their sheer bulk. They are wedge-shaped and steeper upstream than down, a shape which tends to transfer the press of the water against the back of the dam into a rotational force pushing the front of the dam down into the riverbed and thus assisting gravity to keep it in place. *Arch dams* are constructed as an upstream-pointing arch: the water pushing against the center of the arch tends to flex the dam, pressing its ends more strongly into the banks on either side and thus preventing it from moving. Arch dams can only be built where the banks of the stream at the damsite are structurally solid rock. Finally, *buttress dams* consist of a relatively thin sheet of impervious material braced on the downstream side by triangular buttresses, giving it the shape of a gravity dam without the gravity dam's bulk. Buttress dams depend principally on the rotational portion of the gravity dam's system of forces, without the assistance of the latter type's great weight. A buttress dam may be built as a series of upstream-facing curves mimicking the single curve of the arch dam; in this case the arch shape transfers the river's force to the buttresses instead of to the banks of the stream. All four types of dams have an *outlet works* through which the water normally runs, and a *spillway* to release excess flows in times of flood. The downstream ends of the outlet works and the spillway, called in each case the *apron,* must be shaped to move the water rapidly away from the base of the dam to avoid undermining its foundations. Spillways on gravity and buttress dams are usually just a rounded-off section of the dam's crest that is slightly lower than the rest. Spillways on embankment dams are similar, except that they normally lead to chutes lined with cement to prevent erosion of the embankment. Spillways in arch dams cannot be notches in the rim due to the structural weakness this would impart to the arch. They lead around the dam instead, usually by tunneling through one or both banks of the damsite.

ENVIRONMENTAL EFFECTS

Dams can have a number of environmental benefits, including an improvement of summer streamflow conditions downstream; a reduction in the amount of BIOCHEMICAL OXYGEN DEMAND and bacterial growth (*see* BACTERIUM) carried by the river; oxygenation of the river water at the apron; and the creation of bodies of standing water, together with their resident organisms, in regions where they may be otherwise rare or absent. These benefits, however, must be balanced against some severe environmental costs. Dams and their associated RESERVOIRS inundate and destroy riparian habitat (*see* RIPARIAN ZONE) and block access to the river's headwaters by ANADRAMOUS SPECIES. (Though many dams include fish ladders, these have proved generally ineffective in passing fish upstream around the barriers the dams create, which, for fish that orient themselves according to the direction the water is running, include the quiet waters of the reservoir as well as the dam itself.) Waterfalls, rapids, gorges, beaches, and other scenic areas may be destroyed. In fact, beaches are often eliminated downstream from dams as well as upstream, due to the dam's tendency to block the passage of sand and gravel down the river in times of flood (*see* BEACH NOURISHMENT). The large fluctuations in water level required in the operation of most reservoirs—for example, the daily fluctuations in releases through hydroelectric turbines to accommodate fluctuations in electric demand, or the need to keep the water level low in a flood-control reservoir during flood season, so that there will be room to trap the excess water of a flood as it comes down the river—mean that reservoirs do not operate at all like lakes in an ecological sense, with no bankside vegetation and with a constantly expanding and contracting volume of water that prevents the stabilization of populations of fish and INVERTEBRATES (*see* DRAWDOWN). Finally, the tendency of dams to "silt in" as the sediments normally carried by the running stream are deposited in the still waters of the reservoir means that the useful life of these structures must usually be measured in decades rather than centuries—at which point most of the benefits will be lost, while most of the costs will still be present. Accordingly, because of silting up, the need for structural modification to meet safety standards, and environmental impacts, an increasing number of older dams in the United States are being removed and their rivers restored to a free-flowing condition. The rule of thumb among civil engineers is that the expected life of a dam is 50 years. Of the 87,500 dams in the U.S., nearly all of them (85%) will be that age by the year 2020. Currently, several hundred removal projects are under way. Typically these projects involve older and smaller structures on secondary rivers, not major multiuse or hydroelectric projects, although a power company dam was removed on the Clyde River in Vermont. In this case there was a likelihood that the dam would not be relicensed by the FEDERAL ENERGY REGULATORY COMMISSION for environmental reasons. It had been seriously affecting salmon runs up the river. The dam was demolished in 1996. It may be safely anticipated that the next decades will see more dam removal than dam building. *See also* ARMY CORPS OF ENGINEERS; BUREAU OF RECLAMATION; FLOOD.

Darling effect in behavioral ecology, a correlation between the onset of the breeding period and the density of colonies (or of population AGGREGATIONS) in social animals. Generally, the larger and denser the colony, the earlier breeding begins and the more rapidly it spreads throughout the colony. First noticed by the ornithologist F. F. Darling in the 1930s, the Darling effect seems to be the result of the greater number of individuals that a single courting display can be seen by and, hence, can stimulate, in dense colonies. It is particularly prominent among birds.

Darwin, Charles (1809–1882) English scientist and naturalist noted for his theory of evolution. Darwin started but never finished studying medicine at the University of Edinburgh and theology at the University of Cambridge. In 1831, when he was 22, he set sail for South America as the naturalist on the H.M.S. *Beagle.* During the voyage, Darwin observed that many of the islands, especially the Galápagos Islands off the coast of Ecuador, supported closely related finches, mockingbirds, and tortoises which exhibited different forms of behavior. Upon returning to England in 1836, he formulated his ideas in his *Notebooks on the Transmutation of Species.* His popular and much discussed book, *On the Origin of Species,* was published in 1859. In this book he theorizes that a species evolves by natural selection as a result of the pressure for survival. The young of a species that survive to reproduce carry with them any positive variations that may have helped them succeed. These variations are passed on to their offspring, which may in turn have positive variations, and through the generations the species evolves, embodying these new traits. Darwin also introduced the idea that related species descended from a common ancestor. Contemporary scientists had trouble accepting Darwin's work owing to its inability to be tested and proven. He also received much criticism from religious leaders for opposing biblical creation and placing humankind on the same biological level as animals. Darwin continued with his work, however, and published three other books: *The Variation of Animals and Plants Under Domestication* (1868), *The Descent of Man* (1871), and *The Expression of the Emotions in Animals and Man* (1872). He was elected to the Royal Society in 1839 and to the French Academy of Sciences in 1878. He was also honored with burial in Westminster Abbey.

day-neutral plant a plant whose flowering time does not depend on the length of the PHOTOPERIOD. Day-neutral plants appear to respond to clues such as temperature, rainfall, and their own maturation rate (that is, the time since sprouting) to determine flowering time. Most grow, or at least originated, near the equator, where day length is close to uniform throughout the year. *Compare* SHORT-DAY PLANT; LONG-DAY PLANT. *See also* PHOTOPERIODISM.

Dayny's virus a BACTERIUM, *Salmonella enteritidis* var. *daynysii,* isolated by the French pathologist Jean Daynys from a dying vole population near Charny, France, in 1893, and used in France for many years as a BIOLOGICAL CONTROL for reducing vole IRRUPTIONS. Mixed success coupled with human danger (*S. enteritidis* is the PATHOGEN associated with typhoid fever) have caused its virtual abandonment since World War II.

DBCP (1,2-Dibromo-3-chloropropane) a CYCLIC ORGANIC COMPOUND used in agriculture as a soil FUMIGANT, principally against NEMATODES. A brown, pungently scented liquid, DBCP has relatively high toxicity to mammals (LD_{50} [rats, orally]: 173 mg/kg) and acts as a skin and mucous membrane irritant and as a CARCINOGEN.

DBH *See* DIAMETER, BREAST HEIGHT.

DDT in full *dichlorodiphenyltrichloroethane,* a CHLORINATED HYDROCARBON pesticide, heavily used in the United States from the end of World War II until December 1972, when it was banned by the ENVIRONMENTAL PROTECTION AGENCY. It is still in wide use elsewhere in the world. Structurally, the compound consists of a pair of BENZENE RINGS connected to each other by a trichloroethyl GROUP (C_2HCl_3): each benzene ring has a CHLORINE atom attached to it at the CARBON atom directly opposite its attachment to the group. Highly toxic to most insects, DDT has a low TOXICITY to humans and other VERTEBRATES, a pair of properties that initially made it extremely popular as a "safe and effective" pesticide. It was used on food crops and on cotton, broadcast-sprayed in residential areas to kill mosquitoes, and placed directly on the skin of humans and their pets and livestock to control ECTOPARASITES. However, several dangerous properties of the chemical became apparent after a decade or so of heavy use. It is strongly lipophilic (*see* LIPOPHILIC SUBSTANCE), allowing it to build up rapidly through the FOOD CHAIN (*see* BIOLOGICAL MAGNIFICATION). It is also extremely persistent, breaking down very slowly in the environment and thus capable of accumulating in soil, SEDIMENTS, and vegetation. This persistence also means that it can be transported great distances without change, allowing it to build up in the Arctic and Antarctic regions and other areas far from its point of use. Finally, it has proved to be a powerful MUTAGEN and TERATOGEN, especially among birds, in which it causes reproductive failure by thinning eggshells so that eggs are crushed by the weight of the incubating parent bird. These properties, together with evidence that the target insects were becoming resistant through natural selection (only those resistant to DDT were surviving to breed the next generation) caused the 1972 EPA ban. Canada has also banned use of the compound. However, it is still widely used in Mexico, India and other parts of the third world, and in the Soviet-bloc countries—a total of about 6,000 tons per year. Some of this material is currently showing up in U.S. and Canadian waters and wildlife due to distribution by global wind patterns (*see* TOXIC PRECIPITATION).

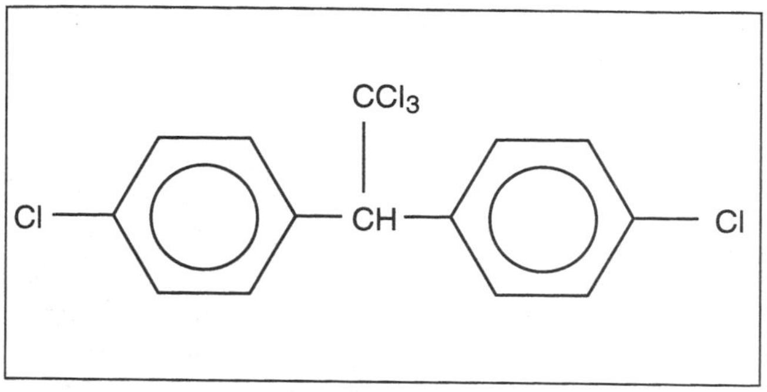

DDT structure

dead weight capacity ([dwc] dead weight tonnage [dwt]) in shipping, the total amount of weight that a ship carries when full, including not only cargo but fuel, crew and passengers, and stores for the voyage.

The dwc of a vessel is found by calculating the weight of the water volume displaced by the hull of the ship when it is riding at its PLIMSOLL LINE.

debt-for-nature swap an approach to protecting rapidly disappearing tropical forests in developing nations. The idea, first proposed in 1984 by Thomas Lovejoy of the Smithsonian Institution, is to provide debt relief to participating governments—often financially beset—in return for their commitment to protect selected forest areas. At the outset, debt-for-nature agreements have taken place under the aegis of nonprofit conservation organizations, such as THE NATURE CONSERVANCY. By the mid-1990s the conservancy had converted some $100 million in debt in 18 countries to protect more than 20 million acres of tropical forest land in Latin America. Often the organization could buy part of a country's debt from commercial banks for as little as 10 cents on the dollar. CONSERVATION INTERNATIONAL has also been involved in debt-for-nature swaps. In some cases the swaps have worked well, but in others the terms have been abrogated by participating governments. In recent years the range of debt-for-nature swapping has been significantly increased beyond the private purchase of commercial debt by conservation groups. The Lugar-Biden Tropical Forest Conservation Act, signed into law in 1998, provides for developing countries to reduce their debt to the U.S. government in return for setting up trust funds to finance the protection of threatened rain forest areas. Spain has also established a debt-for-nature swap program. In 1998, under its *Araucaria* project, it forgave $70 million in debt in Latin America. *See also* DEFORESTATION; RAIN FOREST, BIOME.

decay product in chemistry or physics, one chemical COMPOUND or ELEMENT derived from another through a process of spontaneous or environmentally induced decay ("environmentally induced decay" is taken here to mean decay caused by sunlight or by a chemical REACTION between the decaying compound and some element or compound present in the environment). The decay process may be either chemical (the failure of weak molecular bonds, as when OZONE decays to OXYGEN) or radioactive (the change of one ELEMENT to another of a lower atomic number through the emission of the specialized decay products known as ALPHA PARTICLES and BETA PARTICLES, as when uranium decays to lead). Decay products are almost always both simpler and more stable than the original substance.

decibel (dB) in acoustical engineering, the measurement most commonly used to express the loudness of sounds. The decibel is not a unit, but a logarithmic ratio: it is defined as $10 \log (I/I_0)$, where I = the sound pressure of the source being measured and I_0 = the sound pressure of a sound at the threshold of hearing (that is, a sound just loud enough to be heard). ("Sound pressure" means the small differences in air pressure caused by the sound wave. It is these differences that are perceived as "loudness" or "softness" by the ear.) Since the human ear is itself logarithmic in character, the average person can just make out the difference in loudness between two sounds one decibel apart, even though the actual differences in sound pressure are much greater at the upper end of the scale than at the lower end. Zero dB is the threshold of hearing: 120 dB is the "threshold of pain," where the sound being measured becomes physically painful to the listener. Damage to hearing actually begins below the threshold of pain, at about 85 dB. Some representative decibel levels: quiet room (bedroom or library), 30 dB; whisper, 35 dB; normal conversation, 60 dB; traffic, 70–90 dB; jet aircraft 500 feet away, 110 dB.

deciduous forest biome in ecology, a widespread BIOME in which the DOMINANT SPECIES are trees that lose and regrow their leaves in an annual cycle. The best-known expression of the deciduous forest biome is the *temperate deciduous forest,* found in eastern North America, central Europe, west-central Asia, New Zealand, and parts of China and Japan, but there is also a *tropical deciduous forest* found over a fairly large expanse of Africa, South America, and Southeast Asia. Precipitation in the temperate deciduous forest averages 30–50 inches per year, and is spread nearly evenly throughout the four seasons; there are pronounced temperature shifts from winter to summer. Leaf fall occurs at the onset of winter. The tropical deciduous forest has no pronounced temperature swings, but has distinct wet and dry seasons, with leaf fall occurring at the beginning of the dry season. Study of these tropical forests has led some ecologists to the conclusion that it is probably lack of *available water* in the temperate-zone winter—not the low temperatures—that causes the temperate deciduous forest to lose its leaves. Forests in the deciduous forest biome typically have much more profuse undergrowth than is found in either the rain forest or the coniferous forest. Most of this undergrowth is PERENNIAL herbs and low shrubs that flower profusely in the brief span of time between the onset of available moisture (either at the end of the winter or at the beginning of the rainy season) and the time at which the trees, which respond more slowly, "leaf out" and block the sunlight, preventing it from reaching the undergrowth.

deciduous plant a plant that drops all of its leaves during one or more seasons of the year. Tropical deciduous plants drop their leaves during the dry season or seasons, apparently as a means of reducing TRANSPIRATION and thus keeping moisture loss to a minimum.

Temperate-zone deciduous plants drop their leaves during the winter; this may also be moisture-rather than temperature-related, however, as the BIOAVAILABILITY of water usually declines during the winter due to the fact that it is bound up as ice and snow for much of the time. *See also* DECIDUOUS FOREST BIOME.

declaration of taking in law, a statement filed with a court by a public agency as part of a condemnation proceeding in order to obtain private property for public use (*see* EMINENT DOMAIN). A legal description of the specific property to be taken must be included in the declaration. The time of filing of the declaration of taking is considered to be the point at which legal ownership of the land changes hands, although it normally precedes the determination of fair market value and the awarding of compensation to the former landowner.

declining flow in forest management, the harvest of timber from a NATIONAL FOREST or other management unit at a rate faster than it can be regrown, so that the flow of FOREST PRODUCTS from the unit declines over the long run. Declining flow management is not permitted on public lands in the United States except under certain extremely limited conditions: *see* SUSTAINED YIELD. *Compare* NON-DECLINING EVEN FLOW.

decommissioning with reference to nuclear reactors, *decommissioning* means taking power plants off line, dismantling the structures and machinery, and decontaminating the sites. Today only a small number of nuclear plants have been decommissioned (although quite a few stand idle), but that situation is expected to change as some of the early plants get older. The most difficult aspect of decommissioning is site cleanup and disposal of NUCLEAR WASTE.

decomposers in ecology, living ORGANISMS that break down the MOLECULES of dead organic matter (*see* ORGANIC COMPOUND) into smaller organic molecules and/or into INORGANIC COMPOUNDS and ELEMENTS. Most decomposers are bacteria (*see* BACTERIUM) and other MICROORGANISMS, though insects and other INVERTEBRATES can often be classed as decomposers for at least part of their lives. Decomposers are an extremely important part of all GEOCHEMICAL CYCLES, because they free up and make available for re-use the elements and simple compounds such as NITROGEN and PHOSPHORUS that are necessary for the existence of life. *Compare* CONSUMERS; PRODUCERS.

deep ecology an environmental philosophy that takes as its fundamental tenet the concept of *biocentrism,* that is, that life as a whole is valuable, and that no one species—including humans—can be singled out as more important (or less important) than any other. Deep ecology is contrasted by its proponents with what they call "shallow ecology," which they define by its emphasis on stewardship rather than preservation of natural systems—an emphasis that they say continues to see the Earth as a collection of resources assembled for human benefit and which should therefore be seen as an extension of resource "rape" rather than a rejection of it. Biocentrism emphasizes the preservation of natural systems for their own benefit rather than for any benefit such preservation might bring to humans. To this end, it rejects industrialism, whether communist, socialist, or capitalist; calls for a reduction in human population to something approaching natural CARRYING CAPACITY, probably around 1 billion persons worldwide; and supports a return to small-scale "sustainable communities," which draw RENEWABLE RESOURCES only from their own regions and do not use nonrenewable resources at all. First articulated by the Norwegian philosopher and mountain climber Arne Naess (1912–) in a 1973 article, "The Shallow and the Deep, Long-Range Ecology Movements: A Summary," published in Oslo in an annual collection called *Inquiry 16,* deep ecology achieved movement status around 1980 through the work of several American philosophers and ecologists, most notably Bill Devall and George Sessions. It has been one of the building blocks of the so-called GREEN MOVEMENT, and has been the principal impetus behind bioregionalism (*see* BIOREGION) and related developments in the politics of ecology. *See also* EARTH FIRST!; GAIA HYPOTHESIS; PERMACULTURE.

deepwell disposal a means of disposing of liquid wastes by injecting them into permeable rock formations (*see* PERMEABILITY; FORMATION) at least 1,000 feet, and occasionally as deep as 20,000 feet, below the earth's surface. Originally used by the petroleum industry as a means of disposing of brines brought to the surface during oil drilling—estimates are that as many as 40,000 brine injection wells have been used since the practice first became widespread around 1900—the process was adopted by the chemical industry and others beginning about 1950 as a means of disposing of a variety of liquid hazardous wastes. Currently, some 600 injection wells are in use. These dispose of roughly 60% of all liquid hazardous wastes produced in the United States. In order to avoid contaminating drinking-water supplies, ENVIRONMENTAL PROTECTION AGENCY regulations require that the formation in which the wastes are deposited (the "depository formation") be at least one-fourth of a mile below any AQUIFER used as a water supply (*see* GROUNDWATER) and be separated from it by a layer of impermeable rock (*see* IMPERMEABLE LAYER). It is best if there is a layer of impermeable rock beneath it also. In choosing a depository formation, geologists generally look for formations that contain highly saline water—greater than 10^4 parts per million total dissolved solids (*see under* TOTAL SOLIDS)—because this usually indicates a high degree of isolation from other permeable formations. The well used to inject the wastes, known as an *injection well* or a *reinjection well,* must be lined with impermeable material down to its interface with the impermeable layer above the depository formation. Its cap must be equipped with a pressure-sensing device to warn of sudden drops in pressure within the depository formation, such as might result from breaching the impermeable barriers above or below it.

SAFETY OF DEEPWELL DISPOSAL

Though it is generally considered to be safer than surface disposal, deepwell disposal nevertheless presents some serious environmental hazards. Since the waste materials are injected into the depository formation under pressure, even a small fracture in the impermeable rock cap above it will allow wastes to leak upward toward drinking-water aquifers that may lie above. If such a fracture does not exist, the pressure of the injected wastes may cause one to develop; or the wastes may react with the impermeable cap or with the lining of the injection well, corroding them away and forming a pathway into a drinking-water aquifer. Wells accidentally drilled into a pressurized depository formation may erupt as geysers of waste; old, uncapped wells—many of them abandoned and unmapped—may provide pathways for the wastes to return to the surface; or the injection well itself may "blow its cap" if the back pressure from the injected wastes becomes too high. In one case near Denver, Colorado, in the 1960s, injection of liquid wastes into a 12,000-foot-deep well caused a series of minor earthquakes, apparently by lubricating a buried FAULT so that it could slip more easily. Because of these and similar problems, several states, including Alabama, Florida, New York, and New Jersey, currently ban deepwell disposal.

defect in logging, any condition of a log that reduces its value as a source of lumber. Five classes of defect are recognized. *Interior defect,* which lies within the tree and is surrounded on all sides by sound wood, usually results from a fungal infection such as HEART ROT, although it may also be caused by wood-boring insects or by the growth patterns of the tree itself, such as spiral GRAIN or overgrown wounds. *Side defect* is on the outside of the log and results from CONK, fire scars, beetle damage, damage suffered during FELLING, and similar circumstances. *Sweep* (also known as *curve*) consists of a bend in the log as a result of its own growth; it is often seen in the lower logs taken from trees grown on steep slopes. *Crotch* is a splitting of the log into two separate branches of approximately the same size. Finally, *knots* may count as a defect if they are excessively large or are present in unusually large numbers. Allowance must be made for the presence of defect in order to arrive at the true monetary value of a log (*see* LOG SCALING).

Defenders of Wildlife environmental activist organization, founded in 1947 to protect wildlife and its HABITAT by all legal means, including lobbying, litigation, and public advocacy. Growing out of an anti-trapping organization known as the Anti-Steel-Trap League (formed in 1925), the Defenders have since broadened their scope to all wildlife. They have a strong ENDANGERED SPECIES program as well as PREDATOR and fur-bearer protection sections. Membership (1999): 250,000. Address: 1101 14th Street NW, Suite 1400, Washington, DC 20005. Phone: (202) 682-9400. Website: www.defenders.org.

Defense Nuclear Agency agency within the U.S. Department of Defense charged with the management of nuclear weapons research, development, testing, and deployment. It is required by law to coordinate its activities with the Defense Programs Office of the Department of Energy (*see* ENERGY, DEPARTMENT OF).

deferred component (deferred forest land, unavailable forest land) in public forest management, commercial timberland that is not currently available for harvest due to administrative decisions on the part of the forest manager. Lands that fit into the deferred component are almost always there for environmental reasons. They include WILDERNESS STUDY AREAS, RESEARCH NATURAL AREAS, SPOTTED OWL MANAGEMENT AREAS, and other locally or regionally established SPECIAL INTEREST AREAS. Because the decision to place land in the deferred component is administrative rather than being directed by Congress, deferred lands are subject to administrative reallocation as STANDARD COMPONENT lands at any time; hence, they remain in the TIMBER BASE and are included in calculations of the ALLOWABLE CUT.

deforestation refers to the destruction of natural forests around the world that has taken place since the dawn of civilization but at current rates has become a problem with dire environmental consequences. The causes of deforestation are international timber harvesting, land clearing for agriculture, the requirements for fuel wood by growing populations in developing nations, and drought, fire, and disease, often exacerbated by pollution or other human-caused ecosystem disturbances. Worldwide, half the original forests have been eliminated over the course of human history. In

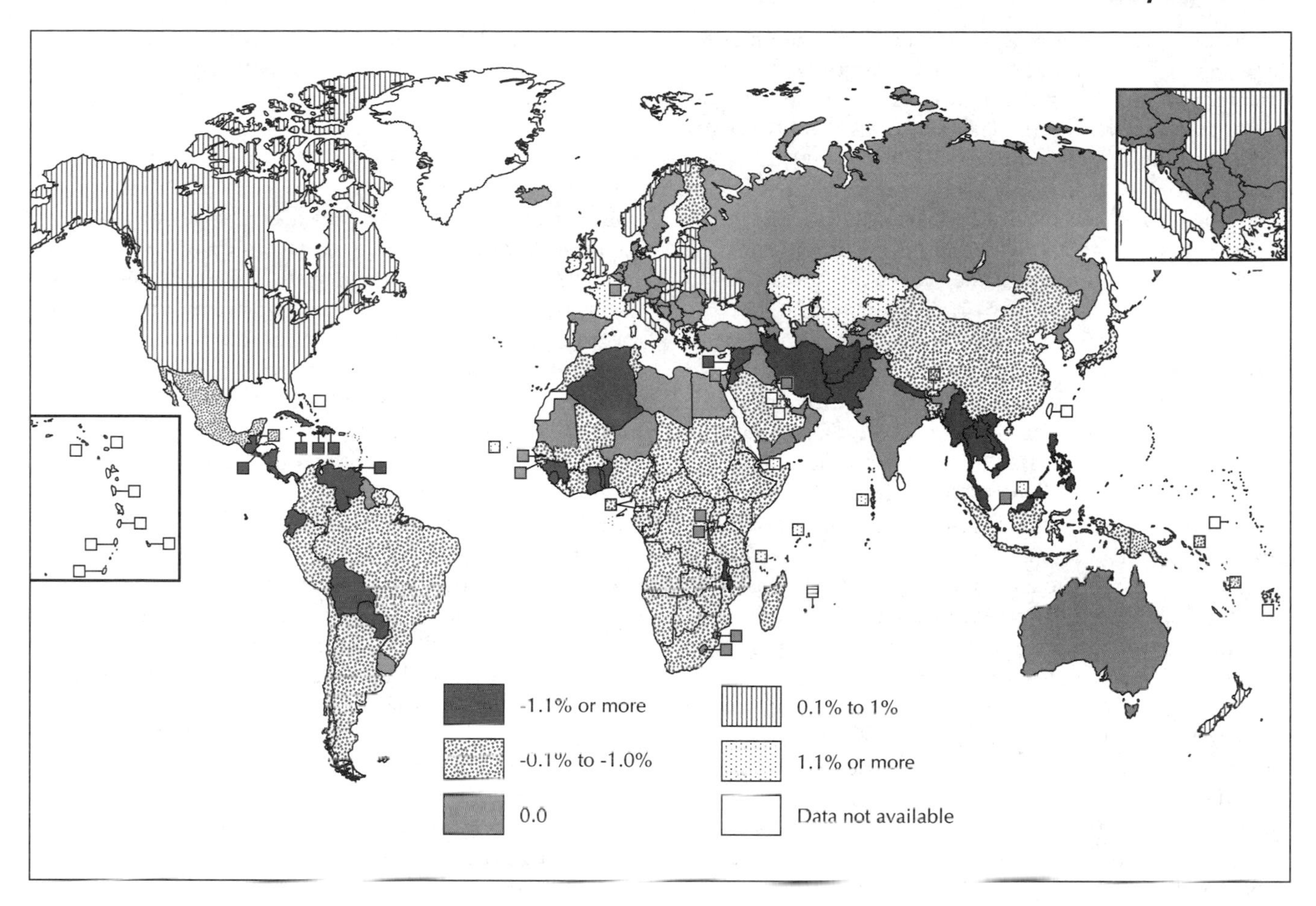

Deforestation

western Europe virtually all the original forest land has long been cleared, as it has in much of the eastern United States, where the primeval forest once stretched unbroken from the Atlantic shore to the edge of the Great Plains. In the tropics over the past two decades, deforestation has taken place at an annual rate of more than 50,000 square miles—an area seven times the size of New Jersey. Deforestation not only eliminates a valued resource and disrupts the indigenous human communities dependent on the forest, but also has manifold environmental impacts at the global scale. These include the loss of species at such a rate that whole ecosystems can be threatened. Since forests provide a "sink" for the absorption of CO_2, deforestation becomes a crucial matter in view of the increase of CO_2 in the atmosphere as a result of industrialization and the burning of fossil fuel. While efforts have been made to create international agreements to stem deforestation, nothing significant has taken place so far. Indeed, in the view of many, the development-oriented approach of the WORLD TRADE ORGANIZATION (which replaced the General Agreement on Tariffs and Trade) could intensify, rather than lessen, the worldwide deforestation crisis.

degree-day in climatology and in heating, ventilating and air-conditioning engineering (HVAC), a measure of the difference between some specified *threshold temperature* and the observed temperature in a region. *Heating degree-days* are calculated by taking the average temperature of each day and comparing it to the threshold, usually 65°F. If the average temperature is at or above the threshold the number of degree days is 0, but if it is below the threshold the number of degree-days is equal to the difference between the average temperature and the threshold. For example, if the average temperature on a given day is 48°F, the number of degree-days recorded is 17(65–48). The number of heating degree-days during a year is the sum of the numbers recorded for each day. *Cooling degree-days* are the opposite of heating degree-days. They are calculated on days when the temperature lies *above* a threshold value, usually 72°F. The number of heating and cooling degree-days in a region is a fairly accurate predictor of the amount of energy that will be required for household heating and air-conditioning in that region.

Delaney Clause a provision enacted in 1958 that prohibited the use of food additives causing or suspected to cause cancer in humans or animals. The unambiguous language of the Delaney Clause (written by James Delaney, a Democratic congressman from New York City), administered by the U.S. FOOD AND DRUG ADMINISTRATION (and later jointly with the ENVIRONMENTAL PROTECTION AGENCY [EPA]), became a problem as many new food additives were developed in subsequent years, ostensibly to improve the nutrition and safety of food products. Whereas at the outset the application of the law allowed for zero tolerance for a relatively few known carcinogens, the concept of "negligible risk" arose as the number and chemical complexity of additives and pesticide residues grew. In practice this meant that the cancer risk could be no more than one in 12 million in additives meant to increase nutrition and safety. For cosmetic additives, such as coloring agents, the courts in a 1987 decision required the continuation of zero tolerance. Despite some relaxation of Delaney Clause standards, the food industry had long pointed out that modern measuring equipment, unknown in 1958, can pick up traces of pesticides down to 1 part per billion, or even trillion. Thus, they argued, the clause was unreasonable if strictly interpreted as written. Finally in 1996, in an overhaul of laws governing the use of pesticides, Congress repealed the Delaney Clause in a trade-off arrangement whereby the EPA would require of pesticide residues on food a "reasonable certainty that no harm will result from aggregate exposure." As a practical matter, the new law established a cancer-risk threshold no greater than one in 1 million. For the most part, environmental and consumer-advocate organizations believed that the new laws would result, on balance, in safer foods and safer pesticides than would a continuation of the Delaney Clause. *See also* RISK ASSESSMENT AND MANAGEMENT.

Delaware River Basin Commission founded in 1961 by Delaware, New Jersey, New York, Pennsylvania, and the United States (the parties of the Delaware River Basin Compact). The commission, which has the leadership role in the management and preservation of the Delaware River Basin water resources, strives to provide regional support in water pollution prevention, drought management, ground and surface water conservation, public education and involvement, and resolution of disputes. Address: P.O. Box 7360, West Trenton, NJ 08628-0360. Phone: (609) 883-9500. Website: www.state.nj.us/drbc/.

delta in geology, a deposit of ALLUVIUM laid down by a river at its mouth. It is normally triangular in shape, with the apex of the triangle upstream. Deltas are characterized by low-lying land, easily inundated at times of high water. They are crisscrossed by BRAIDED CHANNELS, and contain a considerable amount of WETLANDS. They appear flat, but actually slope slightly upstream.

FORMATION OF DELTAS

Deltas owe their existence to the fact that moving water can hold considerably more suspended material than still water can, causing rivers to drop their SEDIMENT loads when they enter still water. A delta grows from both its downstream and upstream ends; however, the principal growth is at the downstream end, at the mounts of the channels that wind through the already-deposited material. The BED LOAD of the stream is deposited early, as the stream slows down abruptly upon encountering still water. The suspended materials then sort themselves out, the heavier ones first, the lighter ones further out as the water becomes progressively quieter. At the outer edge of the delta, deposition is countered by sediment removal through wave and current action in the receiving water, making the rate of growth dependent on a balance between river deposition and shoreline erosion. This tends to keep the "toe" of the delta steep, so that deep water often exists remarkably close to the delta's end. Since the materials suspended in a river usually include a considerable amount of ORGANIC MATTER as well as soil particles, delta sediments tend to be extremely fertile farmlands, a condition that encourages developments that are then threatened by the floods that created the farmlands in the first place—a conflict that is not at all easy to resolve.

demographic transition in population biology, the adjustment of a SPECIES' birth rate to parallel a change in its death rate and therefore keep the population size relatively constant. Among students of human biology, "the demographic transition" usually refers to the specific transition that occurs when a society's death rate is dramatically lowered by advances in health care and nutrition. If the birth rate does not also drop dramatically (that is, if the demographic transition does not occur), a POPULATION EXPLOSION will take place, depleting the society's resources and ultimately leading to a DEMOGRAPHIC TRANSPOSITION.

demographic transposition in population biology, a condition where a SPECIES' death rate exceeds its birth rate, leading to population decline. The greater the transposition, the more pronounced the decline. The term "transposition" is used here because the normal condition of a POPULATION is assumed to be that births will moderately exceed deaths, leading to a slow population growth that is stabilized by outmigration of the excess ORGANISMS and the establishment of new populations. Demographic transpositions are caused by severe adverse conditions such as epidemics or exhaustion of food supplies, and are almost always abrupt. *See also* DEMOGRAPHIC TRANSITION.

demulsibility rate the speed at which an EMULSION separates into its two constituent liquids.

dendritic drainage pattern in geology or hydrology, the branching, treelike pattern (Greek *dendros,* tree) normally formed by a stream and its tributaries, with many small streams coming together to form fewer large ones that all eventually converge in a single MAINSTEM. *Compare* BRAIDED CHANNEL.

dendrochronology the dating of past events through the study of tree rings (*see* ANNUAL RINGS). Dendrochronology can precisely date any environmental changes that affected the growing tree, including especially droughts and fires. It can also be used to date prehistoric buildings if the buildings include well-preserved wooden structural members.

denitrification (1) in soil science, the removal of NITROGEN from soil through bacterial action (*see* BACTERIUM). Denitrification takes place when certain normally aerobic (air-using) bacteria (principally of the genera *Bacillus* and *Pseudomonas*) are grown under anaerobic (airless) conditions, bringing about metabolic changes (*see* METABOLISM) that cause them to break down nitrogen compounds such as AMMONIA that are present in the soil. The freed nitrogen is released into the air. *See also* NITROGEN-FIXING BACTERIA; NITROGEN: *the nitrogen cycle;* ANAEROBIC BACTERIA; FACULTATIVE ORGANISM.

(2) in wastewater treatment, the removal of ammonia from wastewater through the use of denitrifying bacteria similar to those found in soil. Bacteriological denitrification of wastewater is a two-stage process. Nitrogen-fixing bacteria (*nitrifiers*) are first employed to change the ammonia to nitrates. The nitrates are then stripped of their nitrogen by denitrifying bacteria (*denitrifiers*). The nitrogen is released into the air. Denitrifying tanks are similar to ANAEROBIC DIGESTERS, but the sludge is stirred less aggressively and the cover does not need to be completely airtight. Wastewater denitrification is still largely experimental, although a few full-scale installations have been made.

density in the life sciences, the number of individuals of a given SPECIES or group of species present in a given unit of living space. The unit used varies according to the size and habits of the species in question. It can be anything from a square centimeter for MICROORGANISMS to a SECTION or TOWNSHIP for large PREDATORS such as grizzly bears. It is usually a unit of area, but for aquatic or aerial species it may be a unit of volume.

MEASUREMENT OF DENSITY

It is usually impossible (or at least impractical) to count each individual member of a species within a study area, so some form of estimation of density must normally be used. Often it is sufficient merely to estimate the *relative density,* that is, to establish the ratio between numbers of the study species in two different HABITATS, or in two different parts of the same habitat. The best way to do this is through the use of "quadrats"—sample plots of known size within which individuals may be counted. Several randomly chosen sample plots within each habitat type are counted and averaged, and the results for the different types are compared to each other. Quadrats are also the best way to estimate *absolute density,* which requires an accurate estimate of the total number of individuals of the study species present in the study area. The method is the same as that for estimating relative density, but many more quadrats should be used, and care must be taken to make sure that all habitat types within the study area are represented. When the use of quadrats is difficult or impossible (for example, in estimating the absolute density of a widely dispersed and highly mobile population such as that of a TOP PREDATOR), a second method, known as "catch-and-release," may be used. In this method a number of individuals of the study species are caught, marked, and released back into the population. A second set of catches is then done, and the proportion of marked to unmarked individuals is measured. This proportion is assumed to be the same as the proportion of total marked individuals to total population. This method is less accurate than the use of quadrats, and considerable care must be taken in interpreting and using its results.

CONTROLS ON POPULATION DENSITY

Since populations cannot continue to grow indefinitely, but must remain within the CARRYING CAPACITY of their environments, some factors within these environments must act as controls over the maximum densities that populations can attain. The most universal of these factors is simply the individual ORGANISM's life span—the time between the organism's birth and the point at which its bodily systems begin to wear out and fail ("old age"). Other controls on population density are environmental, and are known collectively as ENVIRONMENTAL RESISTANCE. These include COMPETITION, predation, WEATHER and CLIMATE, and the availability of sufficient food and light for all organisms present. *See* DENSITY-DEPENDENT FACTOR; DENSITY-INDEPENDENT FACTOR.

density-dependent factor in population biology, a factor influencing the size or growth rate of a POPULATION whose strength is directly related to the density of the population; that is, the higher the density, the more the density-dependent factor comes into effect. Density-dependent factors include intraspecific COMPETITION, disease, predation and parasitism (*see* PREDATOR; PARASITE), and population DISPERSAL. *Compare* DENSITY-INDEPENDENT FACTOR. *See also* DENSITY: *controls on population density.*

density-independent factor in population biology, a factor influencing the size or growth rate of a POPULATION whose effect is largely unrelated to the density of the population (it is probable that no size- or growth-determining factor is wholly independent of population size). Density-independent factors include WEATHER and CLIMATE, interspecific COMPETITION, RELIEF and other topographic factors, and seasonal changes such as the length of the PHOTOPERIOD. *Compare* DENSITY-DEPENDENT FACTOR. *See also* DENSITY: *controls on population density.*

deoxyribonucleic acid *See* DNA.

deposition in geology, the setting-down of materials picked up elsewhere by a transporting medium such as wind or water. Deposition is the primary force responsible for forming and shaping DUNES, DELTAS, bars and SPITS, and other alluvial and aeolian landforms. *See* ALLUVIUM; AEOLIAN DEPOSIT.

CAUSES OF DEPOSITION

All deposition is a result of a reduction in the capability of the transporting medium to carry things. Sometimes this diminishment is due to an actual change in volume (for example, evaporation of standing or slowly moving water will result in the deposition of whatever materials the water was carrying); more commonly, however, it is due to a change in speed. The slower a stream of air or water is moving, the less material it can carry, and the smaller the pieces of that material must be. Consequently, every time a stream slows down, some of the material it is carrying drops out. When the GRADIENT of a river is reduced, the river slows down, and deposition takes place. When a current in an ocean or large lake runs off a point into deep water, the current slows down, and deposition takes place. When a wind runs up against a ridge or other geographical barrier, the wind slows down, and deposition takes place. The more abrupt the slowdown, the faster the material is released and the more rapidly the depositional landforms build up.

ENVIRONMENTAL IMPORTANCE OF DEPOSITION

Deposition is the other side of EROSION, that is, it is the process through which the eroded materials come to rest again—and must therefore be taken into account every time erosion-control measures are designed. Care must be taken to assure that erosion-control measures in one area do not result in the elimination of materials necessary for deposition in another area, as, for example, on a popular beach or a valuable WETLAND (*see* BEACH NOURISHMENT; DELTA). *See also* ALLUVIUM; LOESS.

desalination (desalinization) the removal of SALTS from saline water in order to make the water available for industrial, agricultural, or domestic use. For details *see* DISTILLATION; FREEZE DISTILLATION; OSMOSIS; REVERSE OSMOSIS.

desert an area of land that receives very little rainfall. A region is generally considered to be a desert when its rainfall is less than 9 inches per year; however, *see* ARID CLIMATE. *See also* DESERT BIOME.

desert biome in ecology, a BIOME in which the principal factor shaping the structure of the plant COMMUNITIES is aridity (lack of moisture). In deserts, the rate of EVAPO-TRANSPIRATION is greater than the rate of PRECIPITATION, placing great stress on all living things. Days are typically hot; however, lack of moisture in the air prevents it from holding much heat, so that nights are often cold. Vegetation is sparse and composed primarily of ANNUALS, which can exist as seeds between rainstorms. There are also some PERENNIAL herbs that survive dry periods as BULBS or TUBERS; some WOODY PLANTS with small, succulent leaves whose surfaces are coated with a thick, waxy coating (the *cuticle*) to prevent moisture loss; and cacti and similar SUCCULENTS that retain moisture in their stems. Nearly all of these plants are ANGIOSPERMS; GYMNOSPERMS are uncommon to nonexistent in desert areas. Soils are typically rocky and sandy, although areas of fertile soil exist that may be farmed with proper irrigation. Most deserts, including the Sahara, the Sonoran, and the Great Australian deserts, are situated in a pair of broad bands lying roughly 30 degrees north and south of the equator (*see* HORSE LATITUDES), but a few lie further north or south, especially on the lee side of mountain ranges (*see* RAIN SHADOW).

desertification as currently used, the alteration or elimination of natural vegetation in a region due to a decrease in the amount of available SOIL WATER which has been brought about by human activities rather than by climate change. The cause may be a lowering of the WATER TABLE resulting from the overpumping of wells, or it may be overgrazing or other inappropriate agricultural techniques that place too heavy a direct demand on the soil moisture. As the soil water available for plant use decreases, the species composition of the plants that grow in the soil alters in favor of drought-resistant types such as thistles and cactus. If the soil continues to dry out, even these plants will disappear, and the land will be lifeless except for EPHEMERAL SPECIES. Desertification currently affects about 225 million acres of land in the United States, mostly in the southwest. *See also* GROUNDWATER; OVERDRAFT.

desert pavement in geology, a loose covering of pebbles found on top of the ground in DESERTS. Desert pavement occurs where winds have blown away sand, silt, and other small particles, leaving behind the pebbles, which are too big for the winds to move. *See also* LAG GRAVEL.

Desert Protective Council environmental organization founded in 1954 to promote the preservation of the deserts of the American southwest, including their indigenous plants and wildlife. Address: PO Box 3653, San Diego, CA 92163-1635. Phone: (714) 397-4264. Website: www.dpcinc.org.

desiccation the process of drying all the water out of something. Thus "desiccated soil" is soil that has lost its SOIL WATER, either on a temporary or a permanent basis (*see* DESERTIFICATION); "desiccated plants" are plants whose tissues have dried out, killing and destroying the plant's CELLS, and so forth.

design capacity in engineering, the maximum flow of goods, products, or other items that a manufacturing plant or transportation system has been set up to reliably handle. The key word here is "reliably"; it is often possible to force higher levels of items through the system, but not without substantially increasing the associated costs and/or the risk of breakdowns, accidents, or other system failures. Such a system is said to be operating *above* (or *beyond*) *design capacity.*

destination resort in land-use planning, a resort development or other large-scale DEVELOPED RECREATION SITE, usually including a hotel, at which it is expected that people will plan to spend their vacations (as opposed to a site at which people merely stop on their way to something else).

destructive distillation the heating of a complex substance in the absence of OXYGEN and other reactive substances so as to break it down into simpler substances through the action of heat alone. These simpler substances are usually driven off as vapors as they are formed, and may be separated out and collected by means of a fractionating tower (*see* DISTILLATION: *methods of distillation*). Destructive distillation of PETROLEUM to obtain GASOLINE, FUEL OILS and other products is known as *cracking,* and the accompanying fractionating tower is called a *cracking tower. Pyrolization,* which takes place in the first stages of burning of a piece of wood and yields a cleaner-burning fuel, is also a form of destructive distillation. Wood stoves designed to pyrolize efficiently can contribute significantly to the solution of air-pollution problems caused by overreliance on wood heat, a common problem in the American West and New England.

detergent in chemistry, any of a moderately large group of COMPOUNDS that act as EMULSIFYING AGENTS for oil and grease, breaking them down into microdroplets small enough to be carried away by water. Because most "dirt" found on clothing, furniture, woodwork, etc. has a large amount of oil and grease in it, detergents function well as cleansers. This is their primary commercial use, although they may also be used for other purposes, including the breaking up and flushing away of oil slicks. Soap, the oldest known detergent and still probably the most widely used, consists of a "head" of a sodium or potassium SALT formed from a fatty acid (an acid containing the carboxyl GROUP, COOH; these acids are normally found in animal and plant fats, hence the name) attached to long "tail" of CARBOHYDRATES. The head dissolves in water, while the tail dissolves in oil or grease. Soap works well under most conditions, but tends to form a scum of insoluble salts in water containing magnesium or calcium IONS ("hard" water), and to convert back to sodium or potassium ions and fatty acids in water of low PH. So-called "artificial detergents" avoid these problems by substituting salts of inorganic acids, most commonly sulfuric acid, for the salts of the fatty acids in the head end of the molecule. These form soluble rather than insoluble salts in hard water, and do not release fats if broken down under conditions of low PH.

THE DETERGENT PROBLEM(S)

Detergents pose two unrelated problems in the environment. Some (notably the so-called "ABS" detergents, those made from alkylbenzenesulfonate) have branched hydrocarbon "tails" instead of straight ones. These are not broken down by bacterial action (*see* BACTERIUM), making them nonbiodegradable (*see* BIODEGRADABLE COMPOUND) and allowing them to build up in the environment, collecting as nearly permanent masses of foam on lakes and streams. This problem can be solved by using straight-tailed molecules instead of branched ones, the procedure used by virtually all detergent manufacturers since about 1965. The second problem is the presence, in most commercial detergents, of water softeners and other compounds not directly related to the detergent action itself. These substances, known as "builders," can be used as NUTRIENTS by many nuisance aquatic plants and ALGAE, contributing to CULTURAL EUTROPHICATION of the waters the detergent residues end up in. The most common of the builders, trisodium phosphate, is a ready source of PHOSPHORUS—often a LIMITING NUTRIENT in lakes and streams—and its presence even in small quantities can lead to ALGAL BLOOMS and other serious water-degradation conditions. For this reason, phosphates as a component of detergents have been limited or banned in many areas. Most of the substitutes for these compounds, however, have also been found to cause environmental problems: nitrilotriacetic acid (NTA), for example, once the principal replacement for detergent phosphates, had to be abandoned when it proved to be a potent MUTAGEN that also mobilized HEAVY METALS (that is, formed bioavailable compounds with them) and released another common group of limiting nutrients, nitrates, upon decomposition.

deuterium (heavy hydrogen) a naturally occurring ISOTOPE of HYDROGEN that contains, in addition to the single proton and single ELECTRON of "normal" hydrogen, a single neutron (see ATOM: *structure of the atom*). Chemically, it is virtually identical to normal hydrogen; however, its MOLECULAR WEIGHT is twice that of hydrogen, slightly altering its physical properties and those of COMPOUNDS made with it. For this reason it is given a chemical symbol of its own (D: also occasionally ^2H). It is nontoxic and nonradioactive and no more harmful to the environment than is normal hydrogen, although its associations with nuclear reactors (as a radiation shield) and with the hydrogen bomb (as a component of lithium deuteride (LiD), which reacts with plutonium to free the neutrons used in the fusion reaction) have given it a bad reputation. About one atom of hydrogen in every 5,000 is actually deuterium. *See also* HEAVY WATER.

deuterium oxide *See* HEAVY WATER.

developed recreation in land-use planning, any form of recreation which includes modification of the natural environment and construction of facilities before it can be carried out. Examples include camping in developed campgrounds, driving for pleasure, golf, tennis, etc. *Compare* DISPERSED RECREATION. *See also* DEVELOPED RECREATION SITE.

developed recreation site in land-use planning, any piece of land on which the natural environment has been significantly modified in order to serve recreational purposes. Examples include developed campgrounds and picnic areas; golf courses; parks with lawns and flower beds; viewpoints with parking areas, railings, and walkways; etc. *See also* DEVELOPED RECREATION.

development rights a right in land that can be separated from the "bundle of rights" that constitutes fee simple land ownership. The right to develop a given parcel of land—structures, roads, etc.—can be purchased from an owner by a government of quasi-government agency for the benefit of the general public, or it can be donated to such agencies by a landowner, or in some jurisdictions, traded, bought, or sold. In the latter case the transfer of development rights (TDR), the right to develop at a greater density than might be allowed under existing zoning law, can be secured by transferring development rights from another parcel in an area where municipal or state land-use plans call for a reduction in density. The transfer can take place between properties in a single ownership, or can be "brokered" by a planning and zoning authority. The acquisition and transfer of development rights has become an increasingly important means for governments to control the use of land in areas beset by URBAN SPRAWL, since it does not require the purchase or management of land that may not be needed to carry out a planning goal. *See also* CONSERVATION EASEMENT.

dewatering in sewage treatment, the removal of water from residual sludge. This was traditionally accomplished through air drying of residual sewage material, but more recently through centrifuges and presses, accompanied by chemical conditioning. *See also* WASTEWATER; SEWAGE TREATMENT PLANT.

dew point in meteorology, the temperature at which dew begins to form in a cooling AIR MASS. At the dew point the RELATIVE HUMIDITY of the air mass is 100%; that is, it is holding as much moisture as it can at that temperature, a condition known as *saturation*. Further cooling reduces the moisture-holding capacity of the air further, forcing some of the moisture present to condense out.

diabase a dark, iron-rich IGNEOUS ROCK with a visible crystalline structure. The crystals are smaller than those found in GABBRO, into which it grades. Diabase forms under conditions of medium pressure, and is commonly found as a component of DIKES and SILLS; it is also occasionally found at the base of a thick LAVA flow, or as all or part of a BATHOLITH. *See also* INTRUSIVE ROCK.

diameter breast height (DBH) in forestry, the diameter of a tree measured at the height of the human breast, which is defined as a point 4 1/2 feet above the ground. If the tree is on a slope, the measurement is taken at a point 4 1/2 feet above the upslope side. A DBH measurement is generally above the tree's BUTT SWELL and well into the area where the trunk has established its standard TAPER, making it a far more useful gauge of a tree's size than the diameter at ground level would be. When foresters refer to a tree's "diameter," they almost always mean DBH.

diameter class in forestry, a grouping consisting of all trees of a given diameter (*see* DIAMETER BREAST HEIGHT). In practice, diameter classes normally cover 2-inch intervals. That is, the 8-inch diameter class consists of all trees within a stand having diameters from 7 to 9 inches; the 10-inch diameter class contains all trees within the stand that have diameters from 9 to 11 inches, and so on.

diameter tape in forestry, a flexible tape measure carried by TIMBER CRUISERS in order to measure the diameter of trees. Its scale has been multiplied by a factor of *pi* (3.14159"=1"), allowing the diameter to be read directly when the tape is placed around the tree at breast height (*see* DIAMETER BREAST HEIGHT).

diastrophism in geology, the large-scale deformation of the earth's crust that results in mountain chains, continents and ocean basins, GRABENS, and similar large-scale geographic features. Diastrophism is one of three principal forces involved in shaping the earth's surface, the others being EROSION and DEPOSITION. *See also* OROGENY; FAULT; VOLCANO; PLATE TECTONICS.

diatom any of numerous SPECIES and ALGAE belonging to the class Chrysophyta (*see* CHRYSOPHYTE) and possessing a hard outer shell of silica. Diatoms are the most populous class of PHYTOPLANKTON, and as such are a principal food source for marine animal life. One-celled ORGANISMS ranging in size of .005 centimeters in diameter (slightly smaller than the breadth of a human hair) downward, they are mostly solitary, though some species form large colonies. Many of the solitary species are motile (*see* MOTILITY). The most striking thing about them is their shells. These are highly varied in form, but are always regular (equilateral triangles, spheres, ellipsoids, and diverse other geometrical shapes). They are covered with fine networks of holes and lines that serve, along with the overall shape, to identify the species (the ability to resolve these markings is a traditional test of the quality of microscope lenses). The shells are composed of pure silica, and are made in two symmetrical halves that are often likened to the halves of a pillbox. During reproduction by CELL fission, the shell of the mother cell splits open, one half remaining with each daughter cell: the daughter cells then each grow new "lids." Estimates of the number of species of diatoms range from 9,200 to 17,000. *See also* DIATOMACEOUS EARTH.

diatomaceous earth a soft, extremely friable (*see* FRIABILITY) white rock made of the fossil shells of DIATOMS. Though it sometimes contains small amounts of organic matter and/or clay, diatomaceous earth is usually remarkably pure, consisting of almost 100% silica. Deposits occur worldwide, mostly dating from the early Cenozoic (100 million years BP). These are quarried for use as abrasives (common applications: silver polish and toothpaste), as filtering and insulating materials, and as an inert filler for explosives. In its pure form, diatomaceous earth is known as *diatomite*.

diatomite *See under* DIATOMACEOUS EARTH.

dibenzofuran *See* FURAN.

dicotyledon (dicot) any member of the subclass Dicotolydonae of the class Angiospermae, or flowering plants (*see* ANGIOSPERM). The dicotolydons are the principal subclass of the angiosperms, containing roughly 250,000 SPECIES, or approximately five-sixths of all known species of flowering plants. They are the only angiosperms that produce true woody tissue (*see* XYLEM; PHLOEM; *see also* GYMNOSPERM). The veins of their leaves almost always exhibit a dendritic pattern, that is, they branch off from each other rather than lying parallel through most of the leaf. The veins in their stems are arranged in cylindrical groups known as *vascular bundles*. Their roots are woody rather than fibrous and tend to form TAPROOTS rather than TUBERS or RHIZOMES. Their seeds each contain the two COTYLEDONS that are the source of their name. Typical species: oak, rose, pea, buttercup, sunflower. *Compare* MONOCOTYLEDON.

dieback in wildlife biology, a reduction in population of species that have grown too numerous for a food supply or have experienced an ecological change in HABITAT. In forestry, dieback or "branch dieback," refers to branches in the canopy of trees dying back because of lack of water or nutrients or pests, or because of pollution. *See also* FOREST DECLINE AND PATHOLOGY.

dieldrin *See under* ALDRIN.

dielectric any substance that can be used as an electrical insulator. In a dielectric, the electrons remain bound to the individual atoms of the material rather than flowing through it, as in a conductor. However, they tend to concentrate at the ends of their orbits nearest a positive charge, or opposite a negative charge, creating an electrical polarization within the dielectric material that is known as a *bound charge*. The greater this polarization, the more efficient the insulating properties of the dielectric. The strength of the insulation is expressed by a number called the *dielectric constant*, which is defined as the ratio of the insulating ability of the material in question to the insulating ability of a vacuum (the dielectric constant of a vacuum is by definition 1). The dielectric constant of air is 1.0005; that of soda glass (a relatively good insulator) is 7; and some ceramic materials—the best insulators known—have dielectric constants of around 100.

diel movement in limnology and marine biology, the daily vertical migration of PLANKTON in a body of water, from near the surface at night to as much as 60 meters below the surface during the day. Diel movement is usually explained as a phototrophic adjustment; that is, the plankton move in response to light levels, up during darker conditions and down during lighter periods. Not all species of plankton exhibit diel movement, and among those that do the movement is usually only a few meters (or even only a few centimeters) in each cycle. The phenomenon is less pronounced in eutrophic than in mesotrophic or oligotrophic lakes, presumably because of variations in light penetration (*see* TROPHIC LEVELS; EUTROPHICATION).

diffusion pressure in chemistry, name given to the tendency of SOLUTIONS to become homogeneous, that is, for the MOLECULES or IONS of the SOLUTE to spread equally throughout the SOLVENT. If one part of a solution has less molecules of solute than another, solute will flow "down the diffusion gradient" from the area of greater concentration to the area of lesser concentration until the concentrations in the two areas become equal. Diffusion pressure is the principal driving force behind OSMOSIS and is thus a chief factor in the ability of living CELLS to absorb NUTRIENTS. *See also* DIFFUSION RATE.

diffusion rate the rate at which a SOLUTE spreads through a SOLVENT, for example, the rate at which a glass of water into which a drop of ink has been dropped becomes uniformly colored by the ink. The diffusion rate of a substance depends primarily on its DIFFUSION PRESSURE, but may also be influenced by temperature, pressure, and other physical variables. Diffusion rates are important in a wide variety of environmental situations, from the control of METABOLISM in the individual CELL to the prediction of the behavior of POLLUTANTS in a body of water (*see* PLUME).

dike (1) in engineering, *see* LEVEE.

(2) in geology, a tabular rock body—that is, one which is considerably thinner than it is broad or tall—that cuts across the structure of a larger rock body. Dikes are formed by MAGMA forced into cracks in a preexisting rock formation, and take the shape of the crack. Thus, they may range in thickness from less than an inch to several thousand feet and in length from a few feet to many miles. They are *discordant* (i.e., they cut across layers rather than lying parallel to them), and, because they cool rapidly due to contact with the surrounding COUNTRY ROCK, are almost always hard and finely crystalline in structure, causing them to sometimes remain standing in formations known as *dike walls* after the surrounding rock has eroded away. They often occur in parallel groups known as *dike sets* or *dike swarms*. *Compare* SILL.

diked disposal *See* CONFINED DISPOSAL FACILITY.

Dillard, Annie (1945–) American author, born April 30, 1945, in Pittsburgh, Pennsylvania, and educated at Hollins College in Roanoke, Virginia (B.A., Phi Beta Kappa, 1967; M.A., 1968). From 1973 to 1975 she wrote a regular column for the Wilderness Society journal *Living Wilderness*. In 1975 she began a teaching career at Western Washington State University in Bellingham and (since 1979) at Wesleyan University in Middletown, Connecticut. Her environmental reputation rests primarily on a single book, *Pilgrim at Tinker Creek* (1974), a lyrical exposition of nature-in-the-small that won its author the Pulitzer Prize for nonfiction in 1975. She is also the author of many other books, including her memoir, *An American Childhood* (1998); a novel, *The Living* (1993); *Mornings Like This: Found Poems* (1996); *Holy the Firm* (1999); and her most recent, *For the Time Being* (1999), a discourse on human birth defects. Dillard has received fellowship grants from the John Simon Guggenheim Foundation and the National Endowment for the Arts. She has also received the New York Press Club Award, the Washington Governor's Award, the Connecticut Governor's Award, and the Campion Medal from the Catholic Book Club in 1994.

dimethyl sulfide (DMS) a compound found in nature that is produced mainly in the world's oceans by means of the algae-caused breakdown of the covering plates of coccolithospores, a microscopic single-celled organism. The resultant DMS molecules rising into the atmosphere attract water droplets, thus forming clouds. This phenomenon has led some researchers to speculate that increasing amounts of DMS, through the action on the coccolithospores of an increasing amount of algae from the warming of the world's oceans due to GLOBAL CHANGE, may then increase cloud formations that can, in turn, help cool the planet so that DMS operates as a kind of global "thermostat," possibly counteracting the GREENHOUSE EFFECT.

dioxin in full, polychlorinated dibenzo-*para*-dioxin (PCDD); any of a group of numerous closely related AROMATIC COMPOUNDS consisting of two BENZENE RINGS joined together by a pair of OXYGEN atoms, with CHLORINE atoms attached to two or more of the free corners of the rings. The type formula is $C_{12}O_2Cl_n$, where *n* can be any number from two to eight; the structural formula is shown in the accompanying diagram.

A total of 75 dioxins are theoretically possible, depending upon the number and position of the attached chlorine atoms. They vary widely in TOXICITY, although they are similar in other characteristics. The most dangerous form, 2,3,7,8-tetrachloro-dibenzo-*para*-dioxin (2,3,7,8-TCDD, or just TCDD), found as a contaminant in the PESTICIDES Silvex and 2,4,5-T, is the most toxic chemical ever synthesized by humans. A grayish-white crystalline substance with a physical resemblance to baking soda, it has an LD_{50} (in guinea pigs) of 0.0006 mg/kg. Less than 3/100 of a gram is enough to kill 500 people. Other forms of dioxin are much less dangerous. The dichloro forms (two chlorine atoms), for example, are approximately 1 million times less toxic than the tetrachloro forms, while the hexachloro forms (six chlorine atoms), found as contaminants of 2,4-D, are probably 15,000 to 50,000 times less toxic than the tetrachloro forms. No dioxin, however, is completely safe. Symptoms of exposure are similar to those of the closely related PCBs (*see* POLYCHLORINATED BIPHENYL) and dibenzofurans (*see* FURAN). They include chloracne, nerve and liver damage, weakness and pain in the hands and feet, and fetal abnormalities and miscarriages. The chemicals are strongly lipophilic (*see* LIPOPHILIC SUBSTANCE) and build up rapidly in the FOOD CHAIN. Trace amounts of dioxins have been found as contaminants in incinerator smoke, automotive exhaust, cigar and cigarette smoke, and indeed virtually anywhere where combustion of aromatic hydrocarbons has taken place. Most of these, however, have not been the more toxic forms.

Dioxin structure

dip in geology, the angle at which sedimentary STRATA, layers of BASALT, or other planar geologic features (that is, those whose surfaces lie approximately all in one plane) slope away from the horizontal. The dip is measured at right angles to the STRIKE. Knowledge of the dip of underground strata is important to help locate oil deposits and AQUIFERS and to determine the stability and strength of rock formations for engineering purposes such as roadbuilding and dam construction. *See* DIP SHOOTING.

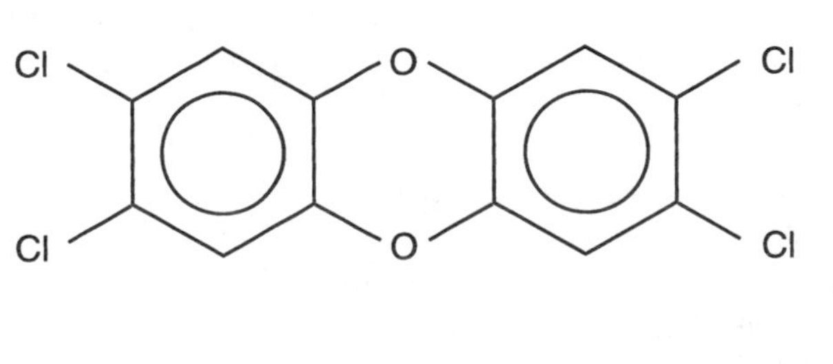

dip compass a magnetized needle pivoted from a central balance point that allows it to move only up and down, used to locate deposits of iron ore, buried metal pipelines, or any other underground object which will attract a magnet.

dip shooting in engineering geology, the process of determining the DIP of underground STRATA. There are three principal methods of doing this. The easiest, but least reliable, is simply to extrapolate from surrounding formations: determine the dip of strata on an outcrop and extend it as an imaginary line beneath the surface. This method may be accurate enough if there is reason to believe that the strata are continuous and the dip does not change (for example, if another outcrop further along shows the same dip, strike, and general content). When more accuracy is required, *seismic shooting* may be used. An array of SEISMOGRAPHS is set up on the ground surface; a series of explosive charges is set off; and differences in the underlying strata are read from the seismographs as differences in the behavior of the sound waves passing through the earth. Readings from the array as a whole may then be correlated to calculate the dip. For still greater accuracy, a series of CORE SAMPLES may be taken and the dips of the various strata may then be determined from their actual physical locations in the cores.

direct loading in environmental technology, the amount of pollutants released directly into a receiving body such as a lake (for example, from shipping spills or from sewage OUTFALLS directly into lake waters) as opposed to those that arrive in a roundabout manner (for example, from sewage outfalls into tributary streams, from ground RUNOFF, or from AERIAL DEPOSITION). *Compare* INDIRECT LOADING.

direct-shipping ore *See* BESSEMER ORE.

disclimax in ecology, a COMMUNITY maintained in an artificial CLIMAX stage due to continuous massive disturbance. A disclimax differs from a SERE in that natural SUCCESSION has been halted; no new species are invading the community, and the proportions of SPECIES within the community are remaining relatively constant. It differs from a true climax in that once the disturbance has been removed, succession will resume. Heavily grazed rangeland, for example, may be kept in a state of disclimax by the systematic removal, by the grazing animals, of the most succulent grasses and

FORBS. When the grazing pressure diminishes, these GRASSES and forbs will reappear and the disclimax will alter toward a true climax.

discontinuous distribution in population biology, the occurrence of a plant or animal SPECIES in two or more distinct clumps of individuals (*see* POPULATION), with the spaces between being empty of the species or (at the most) traversed by only a few transient individuals. A discontinuous distribution is usually the remnants of a formerly continuous distribution that has been interrupted by changes of HABITAT, and so almost always represents a population in decline. *See also* DISPERSAL; DISTRIBUTION OF PLANTS AND ANIMALS.

dispersal in ecology, the means by which animals and plants expand their RANGES to colonize new areas; also, the means by which a single POPULATION of a SPECIES expands and mixes with other populations in order to maintain the genetic stability of the species as a whole. Dispersal is extremely important from an environmental standpoint. It allows populations to expand their numbers without increasing their density, and provides the means for a species that has depressed or exhausted the food supply in one area to find another area where food exists, thus guaranteeing the species' survival. It is also the primary (and almost the only) way for plants and animals to recolonize regions that have been laid bare by earthquakes, vulcanism, floods, or other large-scale natural or man-made environmental disturbances, and thus provides the principal means for making certain that life itself remains on the planet.

MEANS OF DISPERSAL

Dispersal can take a wide variety of forms, but all of them can be included under two general headings: *active dispersal,* in which the ORGANISM deliberately seeks new areas to live, and *passive dispersal,* in which the organism depends on the chance activity of natural forces such as winds, river currents, and transportation by active dispersers. Of the two, passive dispersal is the more important. It is the form used by almost all plants, most MICROORGANISMS, and many small insects and arachnids (spiders and their relatives), and is occasionally resorted to by larger animals as well. In the water, most passive dispersers simply drift with the currents, although a few may hitch rides on other organisms (for example, the parasitic COPEPODS known as "fish lice" disperse along with their hosts). On land, passive dispersers usually depend on the wind. Small flying insects such as APHIDS are carried before the wind as they fly aimlessly about; spiders spin "traveling webs," launch them into the wind, and ride them for great distances; and even relatively heavy creatures such as snails and small frogs may occasionally ride windblown leaves to new territory. A few passive dispersers, on land as on sea, use active hosts. These include ticks, mites, and other ECTOPARASITES; plants with burrs such as thistles; and plants such as strawberries, whose seeds are eaten along with the fruit by active dispersers, pass through the active dispersers' digestive tracts, and are deposited—well fertilized—some distance away. Active dispersers, on the other hand, deliberately seek out new areas, and walk, crawl, swim, or fly to get there. Grazing animals, for example, will look for new RANGE when their own begins to be overgrazed; PREDATORS seek out areas where more PREY may be found. Young beavers leave their home stream and may travel some distance in search of another with a suitable—and untaken—dam site.

MEASURING RATES OF DISPERSAL

Means of measuring rates of dispersal are very similar to those used for measuring population DENSITIES. In this case, however, the measurements are made not of an overall population but of the population's expanding margins. Traps are set up in the path of the dispersing animals, at different distances from the center of population, and these are checked over a period of time to determine the changes in total catch at each distance from the center. These changes may then be used to calculate the speed at which the population is advancing. Note that it is the *rate of population advancement* that is sought—not the travel speed of each individual. For this reason, there is no need to mark individuals, but simply to count numbers. *See also* DISPERSAL CENTER; DISPERSAL PATHWAY; DISTRIBUTION OF PLANTS AND ANIMALS.

dispersal center in population biology, the point at which the DISPERSAL of a POPULATION begins. It is usually coincident either with the point of origin of the SPECIES, or with the point of introduction of the species into a region in which it has not previously been found. The location of a historic dispersal center must often by inferred from the location and density of currently surviving populations, especially in the case of a DISCONTINUOUS DISTRIBUTION. Densities get smaller, and islands of population get further apart, the farther one moves away from the dispersal center. *See also* DISPERSAL PATHWAY; DISTRIBUTION OF PLANTS AND ANIMALS.

dispersal pathway in population biology, any route along which a population DISPERSAL takes place. A dispersal pathway is often a distinct corridor, such as the path of the prevailing wind, a stream, or a valley—any place in which conditions are favorable (if only marginally so) for the survival of the dispersing species. Manmade features are sometimes used as dispersal pathways: a railroad embankment whose shaded north bank allows the spread of plants through an otherwise inhospitable desert; a canal which connects two previously unconnected bodies of water; etc. The Welland Canal in southern Ontario, for example, has become a dispersal pathway for lamprey eels, allowing them to bypass Niagara Falls and enter the upper Great Lakes, where they have caused widespread damage to the fisheries. *See also* DISPERSAL CENTER; DISTRIBUTION OF PLANTS AND ANIMALS.

dispersal unit in ecology, the form in which an ORGANISM usually spreads to new areas (*see* DISPERSAL). Plants, for example, usually use seeds or spores as dispersal units. Animal dispersal units are generally young at or near the AGE OF RECRUITMENT. The term may be used for organisms at the dispersal developmental stage even if they are not dispersing, allowing easier comparison of dispersal rates (by determining the ratio between the number of dispersal units present and the number that actually disperse).

dispersed activity in land-use planning, any activity that takes place over a broad reach of land rather than being concentrated in one spot. Grazing, for example, is a dispersed activity; a cattle feedlot is not. *See also* DISPERSED RECREATION.

dispersed recreation in land-use planning, any form of recreation that utilizes a broad reach of land without depending upon concentrated facilities. Hiking and backpacking, off-road vehicle use, and hunting are all forms of dispersed recreation. Fishing, river rafting, and other forms of river- or lakeshore-based recreation represent a special form of dispersed recreation known as *linear dispersal* due to their tendency to disperse recreationists along a line (the rivercourse or the lakeshore) rather than over a broad area. *Compare* DEVELOPED RECREATION.

dispersing agent in chemistry, a compound that speeds up or otherwise assists in the formation of a COLLOID. *See also* EMULSIFYING AGENT; DETERGENT.

dispersion in chemistry, see COLLOID.

disposable income in economics, the money an individual (or family) receives and has a choice on how to spend. In most cases it is equivalent to take-home pay; that is, an individual's wages or salary after taxes and other workplace deductions have been removed.

disruptive coloration *See* CRYPTIC COLORATION.

disseminule in the biological sciences, an individual member of a dispersing population (*see* DISPERSAL). Broadly speaking, any dispersing individual is a disseminule; however, the word is usually restricted to refer only to those individuals who are in the life stage where their species usually disperses. Thus, for plants, the "disseminules" would be seeds or spores; for birds, fully fledged juveniles (*see* FLEDGLING); for most species of deer, yearling fawns (*see* YEARLING); etc.

dissociation in chemistry, (1) the separation of a dissolved compound into positive and negative IONS, which then move about in the solution independently of each other (see SOLUTION).

(2) the decomposition of a solid compound into two separate compounds, one a solid, the other a gas (for example, calcium carbonate [$CaCo_3$], when heated, dissociates into calcium oxide [CaO—a solid] and carbon dioxide [CO_2—a gas]). The rate at which this *dissociation reaction* takes place is constant for a given compound at a given temperature, and is expressed in terms of the vapor pressure of the gas produced, called in this case the *dissociation pressure* (*see* EVAPORATION).

dissociation pressure *See under* DISSOCIATION.

dissolved load in geology, the MINERALS, SALTS, and other chemical ELEMENTS and COMPOUNDS carried in SOLUTION by a stream. *Compare* BED LOAD; SUSPENDED LOAD. *See also* EROSION.

dissolved oxygen (DO) the OXYGEN carried in SOLUTION in a body of water. Dissolved oxygen is necessary for the survival of all aquatic life except ANAEROBIC BACTERIA. It is separated from the water by osmotic action (*see* OSMOSIS) in the gills of fish and other water-dwelling ORGANISMS. Oxygen enters into solution from air at the water surface, so the rate of solution can be increased either by increasing the surface area of a body of water or by mixing the water so that as much of the volume as possible spends enough time at the surface to pick up new oxygen. The most effective natural means of doing these things are waterfalls and breaking waves, both of which both mix and *aereate* (pass air bubbles through) the water. Such *reoxygenation* is necessary because dissolved oxygen is constantly being used up by living things and by chemical reactions involving substances suspended or dissolved in the water. The maximum amount of dissolved oxygen able to be held in a body of water decreases as the temperature of the water increases. It is also reduced by increases in water salinity and by altitude (the higher a body of water lies above sea level, the less dissolved oxygen it can hold). Sample maximum DO concentrations for pure water at sea level: 12.8 parts per million (ppm) at 5°C (59°F); 8.4 ppm at 25°C (77°F). *See*

BIOCHEMICAL OXYGEN DEMAND; CHEMICAL OXYGEN DEMAND; OXYGEN SAG CURVE.

dissolved solids in pollution control, minerals and other materials that are carried in SOLUTION in a body of water, as opposed to those that are carried in SUSPENSION (*see* SUSPENDED SOLIDS) and those that are bounced along the bottom in a moving watercourse (*see* BED LOAD). Dissolved solids are divided into two fractions—a *volatile fraction* that evaporates when the sample is heated to 650°C (1200°F) for 30 minutes and a *fixed fraction* that remains behind when the volatiles have been driven off (*see* VOLATILE SOLIDS; FIXED SOLIDS). *See also* TOTAL SOLIDS.

distillate *See under* DISTILLATION.

distillation the process of purifying a liquid MIXTURE or SOLUTION by heating it to drive off one or more of its components as vapors and then catching and condensing the vapors. The condensed vapors, known as *distillates,* are drained off as the primary product of the distillation process. The residue of unevaporated materials is either discarded as waste or tapped as a source of BYPRODUCTS. Distillation is a primary method of purifying saline water (*see* DESALINATION) and is also used for separating crude PETROLEUM into GASOLINE, KEROSENE, and other products; for producing turpentine and other materials from wood wastes; for manufacturing industrial ALCOHOL and alcoholic beverages; and for certain other separation tasks. It is of little or no use for purifying water containing MICROCONTAMINANTS because of the volatile nature of many of these compounds (*see* VOLATILITY).

METHODS OF DISTILLATION

In its simplest form, distillation consists of introducing the mixture to be distilled into the bottom of an enclosed chamber (known as a *retort,* an *alembic,* or simply a *still*), heating it to the boiling point of the COMPOUND being purified so that it is driven off as a vapor, drawing off this vapor from the top of the still, cooling it until it condenses as a liquid, and collecting the liquid. This process is adequate when only small amounts are desired, when only one substance is to be purified, and when the substance to be purified has a boiling point substantially different from that of any of its impurities. If any of these three conditions is not met, however, modified methods must be used. In *flash distillation,* used principally to desalinate water, large quantities of saline water are heated in pressure chambers to beyond the boiling point. The superheated water is then introduced into a low-pressure vessel, where it "flashes" into steam. This steam—still superheated—helps to heat the water in the pressure chambers on its way to being condensed as pure FINISHED WATER. In *fractional distillation* or *fractionation,* a device called a "fractionating column" is employed to separate materials with similar boiling points. The mixed vapors (with their slightly different boiling points) are passed up through a tower containing a series of baffles, or "plates." Some of the vapors condense on the plates and flow back down through the tower, forcing the rising vapors to bubble through them; hence the popular name "bubble tower." As they pass through the descending condensate, the vapors redistill some of it, losing heat in the process, until eventually there is no longer enough heat to revaporize the material with the higher boiling point. Condensate drawn off at that level will be a pure concentration of the higher-boiling-point material; condensate drawn off higher up will be a pure concentration of the lower-boiling-point material. If there are several different materials involved, tapping the tower at appropriate levels can draw off each one of them as they "drop out" of the vapor. The more plates used in the tower, the more pure the final products will be. A 10% concentration of alcohol in water may be increased to 97% with 100 plates (due to a quirk in the behavior of alcohol, no greater concentration can be achieved; *see* ALCOHOL), and separation of different ISOTOPES of the same substance has been achieved with 500 plates. For very complex mixtures such as petroleum, the number of plates necessary to achieve pure products is too large to be commercially practical; hence, the commercial fractions of petroleum (gasoline; kerosene; etc.) are mixtures of several compounds with nearly identical boiling points rather than single purified compounds. *See also* DESTRUCTIVE DISTILLATION; FREEZE DISTILLATION.

distribution of plants and animals the patterns in which the various SPECIES of living things are distributed over the surface of the earth. Many things account for distribution patterns, which are rarely random. The major factors that govern distribution are (1) the distribution of the species' PREFERRED HABITAT; (2) the presence or absence of COMPETITION for the species' NICHE within that habitat as it is found in various locations; (3) the presence of a CLIMATE that the species can survive in, including a favorable temperature regime, sufficient rainfall and availability of water, and (especially for plants) an annual cycle of day lengths that the species is, or can be, adapted to (*see* LONG-DAY PLANT; SHORT-DAY PLANT; DAY-NEUTRAL PLANT); and (4) the existence of DISPERSAL PATHWAYS connecting favorable areas with each other. The most common pattern of distribution that arises from the interaction of these four factors is *clumped dispersion,* in which the species occurs in more or less discontinuous groups. The location of the groups is determined primarily by the factors mentioned above, while distribution within the groups is largely uniform due to the pressures of territoriality (*see* TERRITORY). *See also* DENSITY; DISPERSAL; DISPERSAL CENTER; MICROHABITAT; POPULATION; DISCONTINUOUS DISTRIBUTION; LIFE ZONE; BIOME.

disturbance in ecology, an event or action that disrupts a state of natural equilibrium within an ecosystem. Ecological studies have suggested that higher levels of disturbance decrease species diversity in an ecosystem but that low levels of disturbance do not have the opposite effect. In fact, low levels can lead to increased competition which in turn reduces diversity. The highest levels of diversity, which produce the greatest resistance to disruption by exogenous forces, may obtain with disturbance at an "intermediate" level. Disturbance can be caused by human actions as well as natural forces. *See also* LANDSCAPE ECOLOGY.

diurnal species any SPECIES of animal or plant whose activity peaks during the daylight hours. Examples include bees, houseflies, wolves (and most other CANIDS), and most birds. *Compare* NOCTURNAL SPECIES; CREPUSCULAR SPECIES.

diversion of water, the removal of part or all of the flow of a stream for use outside its natural bed. Possible uses include irrigated agriculture (*see* IRRIGATION), HYDROELECTRIC POWER, municipal or industrial water supply, the filling of farm ponds, the creation of wildlife HABITAT, etc. *See also* DIVERSION WORKS; DAM; RIPARIAN DOCTRINE; WATER RIGHT; BENEFICIAL USE; OUT-OF-BASIN TRANSFER.

diversion works a structure designed to divert all or part of the flow of a stream for use outside its natural bed (*see* DIVERSION). A diversion works usually consists of a low barrier stretching partway or completely across the stream to create an IMPOUNDMENT, a canal beginning directly upstream of the barrier, and a moveable barrier known as a *headgate* that can be used to control the amount of water entering the canal. Headgates usually open and close on vertical tracks, moving up and down along a threaded rod (usually turned by a large wheel) to adjust the size of the opening between the gate and the bottom of the canal. They may also include a grating (*trash rack*) to keep floating debris out of the canal. *See also* DAM.

division in botany, a major taxonomic category (*see* TAXONOMY), falling between the KINGDOM and the CLASS; that is, kingdoms are divided into divisions, which are in turn divided into classes. The term "division" is used only for plants. The corresponding term for the other kingdoms is *phylum.*

DNA (deoxyribonucleic acid) any of a large group of closely related complex organic acids (*see* ORGANIC COMPOUND; ACID) found in the nuclei of living CELLS. DNA MOLECULES carry the genetic information of all living things, determining both how ORGANISMS will grow and what characteristics they will pass on to their descendants. The molecules are self-replicating, which means that they are able to manufacture exact copies of themselves—a characteristic that is the driving force behind all multicellular organic growth and reproduction. It is also the driving force behind the diversification of life into the wide variety of forms we see today, because once a change has been made in the structure of a particular molecule of DNA (*see* MUTATION) this change is faithfully copied by every molecule replicated by that DNA and its descendants. Thus the protection of DNA is of supreme importance for the preservation and orderly development of life, and those environmental factors that attack it are among the most dangerous of all POLLUTANTS (*see* CARCINOGEN; MUTAGEN).

STRUCTURE OF DNA

DNA molecules are extremely large, with MOLECULAR WEIGHTS measured in the millions. Each DNA molecule is composed of a very great many smaller structures known as *nucleotides,* each of which consists of three parts: (1) a molecule of a type of sugar known as D-2-deoxyribose; (2) a molecule of one of four "nitrogenous bases" (*see* BASE) that fall into two broad types, the *purines* (*adenine, guanosine*) and the *pyrimidines* (*thymine, cytosine*); and (3) a molecule of phosphoric acid. In the DNA molecule, the nucleotides are arranged in two long, identical chains that lie side by side so that the head of each chain is next to the tail of the other ("antiparallel"), with the nucleotides bound to their neighbors in their own chain by COVALENT BONDS between the sugar and phosphoric acid molecules and to their neighbors in the opposing chain by HYDROGEN BONDS between the purines and pyrimidines. The paired chains are then twisted into a corkscrew shape (the so-called "double helix"). The whole thing resembles a twisted rope ladder, with the phosphate-sugar chains as the ropes and the organic bases stretched between them as the rungs. The genetic information carried by the molecule is encoded in the pattern made by the differing "rungs"—the bases—from one end of the "ladder" to the other. To reproduce, the molecule unzips itself down the center of the helix, breaking each rung at the purine/pyrimidine bond and creating two identical single chains, each of which then attracts the materials to construct for itself a new antiparallel chain from the organic "soup" within the cell nucleus in which it resides.

OCCURRENCE OF DNA

DNA is found in all living things. Complex organisms such as redwood trees or human beings contain a number of molecules, or "strands," of DNA in the nucleus of each cell, where they are found as the principal ingredients of the CHROMOSOMES. Simpler organisms such as bacteria (*see* BACTERIUM) may contain only one or two strands. At the lowest extreme lie VIRUSES, most of which consist entirely of a single strand of DNA covered by a thin protective sheet of PROTEIN. *See also* RNA; GENE.

DO *See* DISSOLVED OXYGEN.

DOD abbreviation for Department of Defense. In environmental literature, this abbreviation will be most often seen in documents dealing with the ARMY CORPS OF ENGINEERS.

DOE abbreviation for Department of Energy (*see* ENERGY, DEPARTMENT OF).

DOI abbreviation for Department of the Interior (*see* INTERIOR, DEPARTMENT OF THE).

dolphin die-off refers to events taking place between 1987 and 1990, when large numbers of dead dolphins were washed ashore along the Mediterranean and both the Pacific and Atlantic coastlines in North America. Scientists estimate that in 1987 half of the bottlenosed dolphin population along the U.S. East Coast, from New Jersey to Florida, were killed. Massive die-offs also occurred in the Gulf of Mexico that year and in subsequent years. Hundred of striped dolphins were killed in the Mediterranean as well. In each of these die-offs, researchers found evidence of morbillivirus, which, though not necessarily deadly in itself, can be fatal in animals with weakened immune systems. The die-offs took place in heavily polluted coastal areas, and tissue samples have shown that PCBs (DIOXIN) and the persistent pesticides DDT, dieldrin, and all organochlorines were implicated in the dolphins' inability to resist the viral infections. The presence of these compounds, researchers suspect, also reduces testosterone in males and therefore reproductive rates. *See also* OCEAN POLLUTION.

dolos segmented, interlocking structures placed just offshore in high-energy wave zones within bodies of water to absorb the force of the waves and prevent shore EROSION. Known technically as "offset asymmetric tetrapods," dolos are composed of diamond-shaped, four-sided structures, usually made of concrete, which are placed side to side in a weaving pattern. They are anchored both to the bottom and to each other, allowing the barrier as a whole to bend with the force of the waves so that more of each wave's energy is dissipated. *See also* EROSION.

dominant overstory in forestry, a SPECIES or GENUS of tree that makes up more than 50% of the OVERSTORY at a specified forest location. For example, if the overstory at the location being examined is composed entirely of oak and chestnut, then if there are more oak trees than chestnut trees, oak is the dominant overstory. If there are more hickory trees than oak trees, hickory is the dominant overstory.

dominant species in ecology, a SPECIES which, through size, shape, activity level, or just sheer numbers, controls the character of the ecological COMMUNITY of which it is a part. Usually (but not always) the dominant species comprises more than 50% of the BIOMASS of the community. It always controls the community ENERGY BUDGET, often substantially altering the nearby environment as it does so and in the process creating the conditions that allow the rest of the community to exist. For example, a tree decreases the light and heat and increases the humidity in the area directly around its base. These changes allow animals and plants that could not otherwise survive in the area due to heat, dryness, or lack of shade to live (or even thrive) there. In terrestrial communities, the dominant species is usually a plant; in aquatic communities dominance is less clearly defined, but the dominant species, when present, is likely to be an animal.

CODOMINANCE

In some communities, two or more species may share dominance; that is, two or more species may account as a group for more than 50% of the community biomass and/or energy flow, with no single species in this group enjoying clear superiority over the others. In this case, the species are said to be *codominant*. In general, the more favorable the growing site, the more likely that codominance will occur in the community occupying the site. *See also* DOMINANT OVERSTORY; ASSOCIATION.

dominant use in land-use planning, the principal activity—in human terms—that is taking place on a piece of land. The dominant use does not exclude other uses, but it places them in a subordinate role and may make them more difficult to carry out. For example, if cattle grazing is the dominant use of a piece of grassland, it will be difficult or impossible to use the same piece of land as a golf course. *Compare* MULTIPLE USE. *See also* DOMINANT USE MANAGEMENT.

dominant use management the management of a piece of land in order to maximize the benefits derived from the DOMINANT USE, with little or no regard to the effect of this management on other uses. Dominant use management is the opposite of *multiple use management,* in which the land is managed so that, even though the dominant use takes precedence (as it usually must), other uses are not precluded but are taken into account as much as possible in planning for the dominant use (*see* MULTIPLE USE).

DOT short for Department of Transportation (*see* TRANSPORTATION, DEPARTMENT OF).

Douglas, David (1799–1834) Scottish botanical explorer, born sometime in 1799 in Scone, Perthshire, Scotland: the day and month of his birth are unknown. Douglas was almost completely self-taught, but his knowledge of plants—gained as a gardener's apprentice—was sufficient to win him a post as the assistant to the well-known botanist William Jackson Hooker at Glasgow University in 1820. In the spring of 1823, Hooker recommended Douglas for the position of plant collector with the Horticultural Society of London. Douglas spent the rest of his life traveling and collecting for the Horticultural Society in North America and Hawaii. He discovered, named, and collected seeds and foliage from over 1,000 plants—650 in California alone—including the Monterey cypress, the DOUGLAS FIR, the California poppy, and seven of the 17 native pines of the western United States. The number of individual SPECIES named for him still stands as a record. Dour, puritanical, and probably mentally unstable, he nevertheless left a journal full of cogent observations on nature and its crass and needless destruction by humans. He was trampled to death by a wild bullock in Hawaii on July 12, 1834, under circumstances that strongly suggest suicide. The best biography of David Douglas is William Morewood's *Traveler in a Vanished Landscape* (Clarkson N. Potter, 1973).

Douglas, William O. (1898–1980) Supreme Court justice and conservationist. After completing his studies at Whitman College and Columbia University, Douglas was admitted to the New York bar in 1926, then taught at Columbia and Yale University Law School until his appointment to the U.S. Supreme Court in 1939. A strong supporter of Frankin Delano Roosevelt's New Deal, Douglas became known as a champion of civil liberties and minorities' rights during his 36 years on the Supreme Court. He was also an activist on behalf of landscape preservation, and wrote on conservation, history, and politics. His more than 30 books include *Of Men and Mountains* (1950), *The Bible and the Schools* (1966), *Towards a Global Federalism* (1968), and *Points of Rebellion* (1970). He was instrumental in the designation of the C&O Canal as a National Historical Park in Washington, D.C., and Maryland. In Washington state a 168,000-acre wilderness area—the William O. Douglas Wilderness—bears his name.

Douglas fir (*Pseudotsuga menziesii*) important timber tree of the western United States and Canada and one of the largest and most distinctive trees in the world. Despite its popular name, it is not a true fir; although it has sometimes been called "Oregon pine" and "Douglas's spruce," it is neither a pine nor a spruce. It is instead one of five species classed by botanists as the genus Pseudotsuga, or "false hemlock." One other Pseudotsuga, *P. macrocarpa* ("bigcone Douglas fir"), grows in western North America; three others are found in eastern Asia.

CHARACTERISTICS

Douglas firs are extremely large trees, up to 14 feet through and 320 feet tall; specimens 250 feet tall and 8 feet through are relatively common in untouched STANDS. They are long-lived, commonly reaching an age of 700 years and occasionally doubling that. Their dark green "bottle-brush" needles, roughly an inch long, grow all around the branchlets that bear them; their CONES are small (3–4 inches) with three-pointed BRACTS protruding from beneath the scales. (The Rocky Mountain variety, *P. Menziesii var. glauca,* has blue-green needles and cones with bracts that curve backward, and is sometimes considered a separate species.) Douglas fir wood is soft and easily worked, but tough; it makes excellent structural lumber. When grown in continuous stands the species becomes "self-pruning," meaning that the lower branches drop off as the tree ages, leaving as much as 200 feet of trunk from which to make "clear" (i.e., unknotted) lumber. It often forms vast, nearly pure EVEN-AGE STANDS, many of great age, with a profusion of well-developed undergrowth; however, it does not reproduce well in its own shade, and most botanists feel that these Douglas-fir forests—stable as they seem—are not climactic (*see* CLIMAX) but are a long-lived SERE leading to a climax stage where the dominant species will be hemlock.

downcutting stream a stream that is deepening its valley, particularly one that is doing so rapidly. *Compare* GRADED STREAM.

downwelling the seasonal downward movement of flowing and standing water bodies. *See also* UPWELLING.

dowser (water witch) a person who purports to be able to find water (or sometimes other substances) beneath the ground through psychic means, usually by holding a forked stick called a "dowsing wand" in his or her hands and walking over the ground until a location is found where the tip of the stick seems to pull downward. Occasionally a simple pendulum or some other tool may be used instead of the wand. Dowsing to find water almost always works, not because it is really possible to locate "veins" of water in this way but because water underlies virtually all land if one digs deep enough, making it nearly impossible for the dowser to fail. *See* GROUNDWATER; AQUIFER.

draft statement *See under* ENVIRONMENTAL IMPACT STATEMENT.

drainage basin *See* WATERSHED.

drainage density in geology, the ratio of the total length of stream channel (including the MAINSTEM and all tributaries) within a given WATERSHED, to the land area of that watershed. It is used by geologists and landscape managers as a measure of the texture of a piece of land—land with a higher drainage density generally being considerably rougher than that with a lower density. It is also useful as an indicator of the vulnerability of a stream system to surface contamination, especially from nonpoint sources (*see* NONPOINT POLLUTION), because a stream with a higher drainage density will have more opportunities to be polluted than a stream with the same volume but with a lower drainage density.

drainage liquor in waste-management technology, the liquid that leaks from a SECURE LANDFILL or other waste-disposal site. Drainage liquor consists of a combination of LEACHATE and liquids leaking from corroded drums or other broken containers within the site.

drainfield (tilefield) in sanitary engineering, the part of a septic system or similar waste-disposal facility through which WASTEWATER is infiltrated into the soil (*see* SEPTIC TANK). A typical drainfield consists of a series of short lengths of clay pipe ("tiles") laid end-to-end in a slightly sloped 4-foot-deep trench, with the ends of the tiles separated from each other by roughly 1/2 to 1 centimeter. Wastewater is introduced into one end of this line of tiles and flows down it, seeping out into the soil through the gaps between the tiles. The number of tiles needed for a drainfield depends on the speed at which water will infiltrate into the surrounding soil (*see* PERC TEST).

drawdown the vertical distance between the full-pool stage of a reservoir or lake (*see* FULL POOL) and any given water level lower than full pool. Drawdown increases as the reservoir is drained. Operating reservoirs generally reach full pool in late spring and begin to be drawn down about the middle of the summer, reaching their lowest level ("maximum drawdown") around mid-November. They are maintained at that level until the end of the flood season, generally around mid-April (*see* FLOOD), when they begin filling again. Because of this yearly cycle of inundation and drying out, few plants will grow in the shore area exposed by drawdown, giving it a naked aspect. This naked shore area is usually traversed by horizontal lines marking levels where the pool elevation stabilized long enough to deposit some minerals and/or do some erosion—the so-called "bathtub ring effect." *See also* DAM; RESERVOIR; MINIMUM FLOOD CONTROL POOL; MINIMUM CONSERVATION POOL.

dredge in engineering, a mechanical device, usually massive in scale, for removing loose materials such as SILTS and GRAVELS from the beds of rivers, lakes, and other bodies of water. Several different types of dredge are in common use; these may be organized into three broad categories, depending upon how they remove the dredged material (called SPOIL) from the site.

Bucket dredges use a large scoop, or "bucket," to physically dig the spoil out and lift it up. In a "dipper dredge," the bucket is on the end of a long arm, or "boom," operated in the same manner as the familiar power shovel; in a "grab-bucket dredge," the bucket is suspended from cables, which drag it along the bottom. An "elevator dredge" uses a conveyer belt with a whole series of buckets fastened to it; these ride up the belt full, dump their loads into the dredge as they turn around the end of the conveyer, and ride down empty.

Suction dredges use a pump to suck the spoil up through a hose onto the dredge and generally on over to the shore for disposal. The spoil is usually loosened from the bottom of the body of water either by a jet of water ("hydraulic dredging") or a jet of compressed air ("pneumatic dredging").

Stirring dredges do not lift the spoil at all; they merely agitate the bottom to suspend the silts and other small particles in the water, depending on the currents to carry them away. Most stirring dredges, like most suction dredges, use either hydraulic or pneumatic dredging to loosen and suspend the spoil particles.

DREDGING AND THE ENVIRONMENT

Dredging is widely used to deepen and maintain harbors and shipping channels; to clean accumulating silt from water-storage reservoirs to keep their capacity from shrinking; to mine the beds of bodies of water for gold and other precious materials; and to obtain materials for filling WETLANDS and shallow waters in order to create new lands (*see* DREDGE-AND-FILL). There are usually no substitutes for these uses. Dredging is also one of the most environmentally destructive of all human activities. The entire benthic community in the dredged area is eliminated (*see* BENTHIC DIVISION; BENTHIC DRIFT ORGANISM). Silts suspended in the water by stirring dredges or through incidental agitation by buckets or suction nozzles drastically increase TURBIDITY and decrease light penetration. These silts often settle on SPAWNING GRAVELS, making it difficult or impossible for fish to breed. They also smother shellfish and small crustaceans (crabs, lobsters, crayfish) that cannot move out of their way. The spoil, or SLUDGE, dredged from shipping channels and harbors is often heavily polluted, and disposing of it can pose serious problems (*see* IN-PLACE POLLUTANTS). *See also* STREAM CHANNELIZATION; ARMY CORPS OF ENGINEERS.

dredge-and-fill operation designed to increase the usable land surface of a shoreline by dredging materials from the floor of a body of water and using them to fill shallows and wetlands. (Some of the fill may also come from landward sources. Chicago's shoreline has been extended into Lake Michigan with sand trucked in from the nearby Indiana Dunes, and San Francisco's land area was drastically increased following the 1906 earthquake by dumping the rubble from destroyed buildings into the bay.) Along with the environmental hazards posed by dredging (*see* DREDGE), dredge-and-fill operations destroy wetlands—thus seriously damaging the PRIMARY PRODUCTIVITY of the body of water—and, if carried to extremes, may change the climate of coastal cities. *See also* ARMY CORPS OF ENGINEERS; BULKHEAD LINE.

drift (1) in geology, *see* GLACIAL DRIFT.

(2) in mining engineering, a horizontal tunnel in a mine; specifically, one that follows a mineral vein rather than cutting across it.

drift ice ICEBERGS, FLOES or other free-floating pieces of ice, formed by the breakup of PACK ICE or by "calving" from a GLACIER whose terminus happens to be in the water. Drift ice is usually carried far from its point of origin by ocean currents, often penetrating deep into the middle latitudes, a circumstance that not only interferes with shipping but lowers water and air temperatures and thus may significantly affect world CLIMATE.

drift-net fishing involves the use of a gargantuan net shaped like a tennis net but 50 feet from top to bottom and up to 40 miles long. Pulled by trawlers that can cover large areas of ocean, these deadly and very efficient nets trap any and all creatures that come near, whether a desirable commercial species, TRASH FISH, ENDANGERED SPECIES, marine mammal, or any other organism. Because drift nets make such a clean sweep of large areas, the U.N. General Assembly, recognizing the threat to aquatic biodiversity, declared a moratorium on larger drift nets in 1992. The ban is voluntary, however, and there is no way to enforce compliance in any case. *See also* CODE OF CONDUCT FOR RESPONSIBLE FISHERIES; FISHERIES DECLINE; TRAWLING.

drilling mud a heavy, viscous fluid (*see* VISCOSITY) used as an aid in drilling deep wells. Drilling mud is composed of a COLLOIDAL SUSPENSION of fine mineral particles in water (less commonly, in oil). The mineral used is normally BENTONITE, though barite (barium sulfate, $BaSO_4$) may also be employed. In use, the mud is pumped down through the hollow center of the pipe string (that is, the set of pipes connecting the drill bit to the surface) and flows back to the surface in the space between the pipe string and the sides of the well, carrying with it the rock waste created by the drilling process. As it rises, the mud fills cracks and seals the well against GROUNDWATER intrusion, thus largely eliminating the need for a well casing. On reaching the surface, the mud is usually sieved to remove the rock fragments and recycled through the pipe string once more.

drinking water *See* POTABLE WATER.

drip irrigation a method of IRRIGATION in which water is supplied in small amounts directly to the roots of plants. A typical drip system involves the use of flexible 1/2- to 3/5-inch plastic distribution tubing from the water source into the area (or row) to be irrigated, to which 1/4- or 3/8-inch emitter tubing can be attached. Finally an emitter or other distribution device, such as a tiny sprinkler or bubbler, is attached to the end of the emitter tubing. Used in orchards, plant nurseries, and sometimes for row-crops in arid areas, drip systems can use 50% to 80% less water than comparable sprinkler or flood irrigation systems, generally increase plant health, keep down weeds (because they are not watered), and have been shown in tests to raise productivity by as much as 80%. They are extensively used in Israel and other parts of the Middle East and increasingly in the United States.

drip line in general, a line on the ground marking the outer limit of a rainproof (or relatively rainproof) overhang, that is, the line along which rain will drip from the edge of the overhang. More particularly, in ecology, the term refers to a line on the ground marking the outer limits of the CROWN of a tree or the CANOPY of a dense STAND of trees. The UNDERSTORY vegetation often changes fairly radically at the drip line

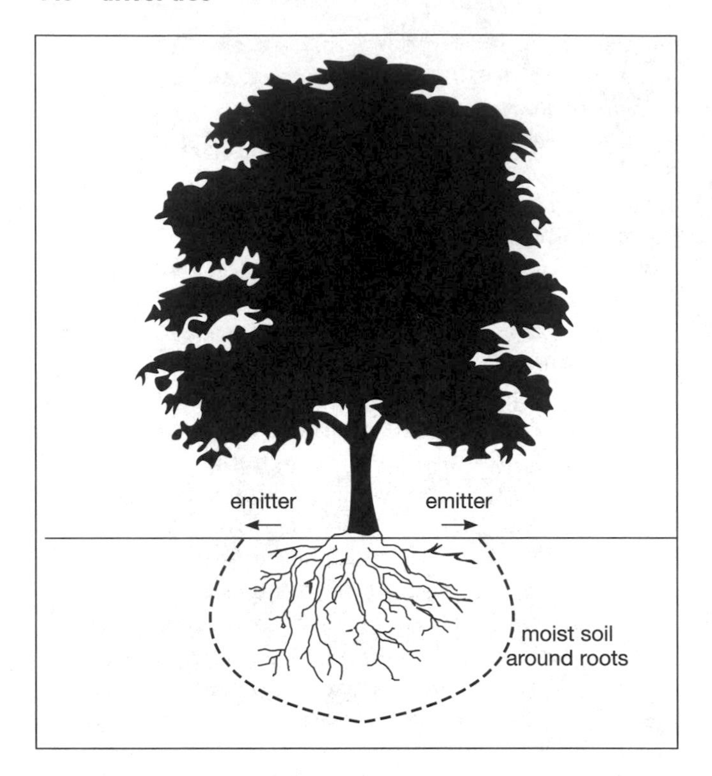

Drip irrigation ***(Rain Bird Sprinkler Manufacturing Corp.)***

due to the change in moisture availability; also, the tree's (or stand's) roots very rarely extend beyond that point.

driver tree in logging, either a windfall or a previously cut tree that is leaning against a second tree and which, when the second tree is felled, will "drive" it so that it falls in a predictable direction.

drought a time of lower-than-normal rainfall. Meteorologists technically define *drought* as a period of 15 or more consecutive days during which the mean daily rainfall does not exceed 0.01 inches. If the total rainfall on any given day during the period never exceeds 0.01 inch, the period is an *absolute drought;* if the total rainfall on one or more days exceeds 0.01 inch but remains under 0.04 inch, the period is no longer a drought but is technically a *dry spell.*

drum a metal container for storing and transporting liquids. The drum most commonly employed in the United States is made of rolled steel and has a capacity of 55 gallons. *See also* BARREL.

drumlin in geology, a GLACIAL LANDFORM found where a resistant knob of rock has caused a GLACIER to slow down, depositing drift materials (*see* GLACIAL DRIFT) "downstream" of the rock knob. Drumlins are characteristically long and narrow, steep on the "upstream" face and tapering off gradually downstream. Their axes line up with the direction of flow of the glacier. *Compare* CRAG AND TAIL; ROCHE MOUTONNEE.

drupe *See under* FRUIT.

dry deposition in pollution control terminology, the deposit of airborne pollutants directly onto the land or water surface, as opposed to TOXIC PRECIPITATION, in which the pollutants are washed from the air by raindrops. Dry deposition is not always "dry"; that is, the pollutants may be liquids as well as solids or gases, the important distinction being that they are not carried down in the rain. The process accounts for a significant proportion of the organic pollutants (such as DDT and POLYCHLORINATED BIPHENYLS) reaching the Great Lakes and other large bodies of water each year. *See also* AERIAL DEPOSITION.

dry-disconnect coupling a connector between two hoses, or between a hose and a tank, in which both parts of the connector automatically close when they are separated, thus reducing or preventing spills during connecting and disconnecting. One end of a dry- disconnect coupling is usually equipped with a spring-loaded valve that must be closed by a lever before the two halves of the coupling can be separated. The other end may have a similar valve, or it may simply contain a one-way device, such as a ball valve, which allows liquid to flow into it but not out of it. *Compare* QUICK-DISCONNECT COUPLING.

drylands in farming, arid lands that receive less than 20 inches of rainfall per year. In dryland farming areas, part of the land will lie fallow in alternate years to conserve moisture. Many dryland farmers in those parts of the Great Plains suitable only for wheat and similar crops have installed center-pivot irrigation systems to grow crops such as corn and soybeans. *See also* IRRIGATION.

dry spell *See under* DROUGHT.

Dubos, René (1901–1982) a famed bacteriologist best known for his philosophical writings on the relationship of humans to the Earth and his work as an international environmental leader. Dubos, who was born in 1901 in Saint Brice, France, and educated in France, came to the United States in 1924 to study bacteriology at Rutgers University. His pioneering work in bacteriology while on the faculty at Rockefeller University led to the development of streptomycin and the tetracycline antibiotics. Through his books Dubos shared his insights on human adaptability and the effects of environmental degradation on human life. In 1972 he was appointed chairman of the committee responsible for the first United Nations Conference on the Human Environment, and in 1975 he founded the Center for Human Environments to encourage the public and decision makers to develop environmentally supportive policies and to foster new environmental values. His books include *Man Adapting* (1965); the Pulitzer Prize winner, *So Human an Animal* (1968); *A God Within* (1972); and *Only One Earth: The Care and Maintenance of a Small Planet* (with Barbara Ward, 1972). Dubos died in 1982 on his 81st birthday.

duck stamp federal migratory WATERFOWL hunting permit in the form of a stamp, issued annually on or about July 1 and sold at post offices. All duck and goose hunters over the age of 16 are required to purchase a duck stamp in addition to any required state and local hunting permits. Non-hunters who are interested in waterfowl conservation are also encouraged to purchase them. Ninety per cent of the $17–20 million raised annually through the sale of duck stamps goes to preserving and improving waterfowl HABITAT, with the remainder being used to administer the stamp program.

HISTORY

The duck stamp program was authorized by Congress on March 16, 1934, through passage of the Federal Migratory Bird Hunting Stamp Act. The first stamps went on sale that same year. By the 1990s, sale of the stamps had raised more than $500 million, used to purchase and maintain waterfowl habitat. The annual contest to design the stamps is the only regularly scheduled federally sponsored competition for graphic artists. The stamps themselves are the longest-running continuous series of postage or revenue stamps in U.S. history.

Ducks Unlimited wildlife conservation organization, founded in 1937 to restore, maintain, and create HABITAT for migratory WATERFOWL on the North American continent. Begun by and still largely oriented toward duck hunters, the group has also developed a large following among non-hunting wildlife enthusiasts for its active and effective WETLANDS conservation program. Membership (1999) 600,000. Address: Headquarters, One Waterfowl Way, Memphis, TN 38120. Phone: (901) 758-3825. Website: www.ducks.org.

duckweed any of several SPECIES of small, free-floating aquatic plants of the family *Lemneacae,* resembling ALGAE but possessing flowers and vascular tissue (*see* VASCULAR PLANT) and therefore classed as ANGIOSPERMS. Depending upon the species and the age, an individual duckweed plant may range from the size of a pinhead to the size of a dime. They are roughly disc-shaped and are leafless, but with a modified stem (the *thallus*) that serves as a pseudo-leaf. Most species produce rootlets that dangle below the floating plant, apparently serving as a conduit for water into the plant's circulatory system. Duckweed produces seed, but its principal means of reproduction is through budding at the edge of the thallus. The process is quite rapid, allowing the plant to cover an entire pond surface with a myriad of floating plants in a few weeks' time (*see* ASEXUAL REPRODUCTION). It prefers shallow, still water with no current or wave action, usually over a muck or peat bottom. A preferred food of most WATERFOWL, it is considered a nuisance by humans because of its tendency to clog filtration equipment and IRRIGATION canals.

duff the partially decayed or reduced ORGANIC MATTER that makes up the upper surface of the soil beneath trees, especially CONIFERS. Duff is usually light-colored and retains the texture of the LITTER from which it derives. It has not yet reached the dark, uniform color and consistency of HUMUS. It often forms a mat half an inch or more in thickness, effectively preventing the growth of most HERBS.

dune a hill composed of windblown sand. Though dunes may form in DESERTS and other locations with steady winds and an adequate supply of sand, they are most common along the shores of large bodies of water. They have a characteristically asymmetric shape, with the slope toward the wind lying at about 10° and the backslope at about 33° (the ANGLE OF REPOSE for sand grains). The dunes closest to the water, known as the *foredunes,* are formed when winds blowing off the water carry, roll, or bounce ("saltate") sand grains inland, piling them on top of each other. As the pile grows, it forms a barrier to the wind, so that more sand grains are deposited on it and growth becomes more rapid. The dune moves slowly away from the beach as the wind picks up sand grains from the face of the dune and blows them over the top, where they are deposited on the back side. Eventually it proceeds far enough from the beach that a new foredune is built in front of it, blocking the wind, at which point the older dune will normally slow down and become stabilized with vegetation. Foredunes are almost always *transverse dunes,* that is, with their crests crossways to the wind. As they move away from the beach, however, they may develop into a parabola (*parabolic dunes,* or *crescent dunes*) with the two ends of the parabola pointing upwind. In extreme cases, the center of the parabola may be breached. In this case, the two arms will eventually straighten out to form ridges (*longitudinal dunes*) parallel to the direction of the wind. *See also* BLOWOUT; AEOLIAN DEPOSIT.

dune stabilization in the management of shoreline areas, beaches, and barrier islands, the use of plantings or artificial structures, such as seawalls, GROIN, and RIPRAP, to retard EROSION from wind and water.

duricrust *See* CALICHE.

Dust Bowl an area of the southern Great Plains that experienced the greatest ecological calamity in U.S. history. During most of the 1930s, western Kansas, the Oklahoma and Texas Panhandles, and parts of eastern Colorado and New Mexico—the epicenter of the Dust Bowl—experienced prolonged drought in an area that had been extensively plowed out, first to capitalize on high commodity prices during and after World War I and then to cope with the mounting agricultural depression of the 1920s, which preceded the Great Depression that affected everyone else. To cope with falling commodity prices, the newly mechanized farmers took their tractors deeper and deeper into virgin perennial grasslands under the mistaken notion that "rain follows the plow." The first warnings came in 1931 and 1932 in Kansas and Colorado, and by 1935 wind was stripping the topsoil—which had no perennial grass roots to hold it down—off a four-state area. Thousands and thousands of tons of dust swirled up in storms that reduced visibility to near-zero at midday. Animals died of asphyxiation, as did babies and small children whose mouths became clogged with dust. Many died from "dust pneumonia." When it was over, the topsoil was virtually all gone in many areas, and in all, 100 million acres of land had been ravaged. Some 3.5 million people pulled up stakes and went west in the decade of the 1930s, a migration that included not only farm families but those who were part of the farm economy as well: shopkeepers, veterinarians, farm equipment salesmen, schoolteachers. The disaster produced a permanent demographic, political, and cultural change in the U.S. and spawned a new understanding of agronomy, farm economics, and the governmental relationship with the farm sector. *See also* EROSION; SOIL CONSERVATION; SOIL CONSERVATION SERVICE; CONSERVATION RESERVE PROGRAM; BUFFALO COMMONS.

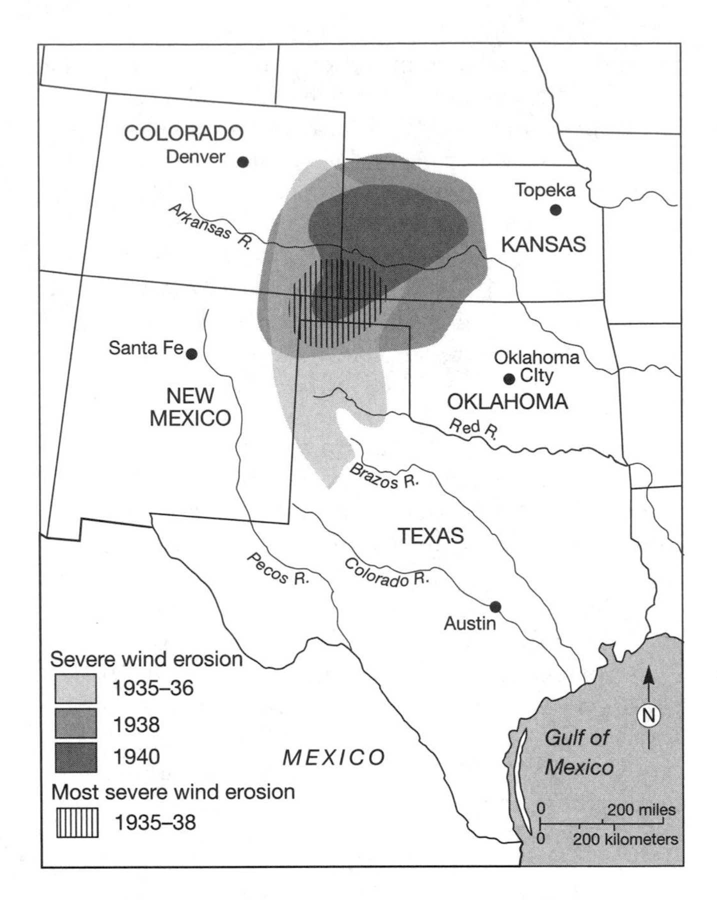

Extent of area subject to severe wind erosion, 1935–1940

Dutch elm disease a fungus originating in Asia that reached Europe in 1919 and devastated the elm trees planted to stabilize the soil of the dikes in the Netherlands. The fungus, *Ceratocystis ulmi,* was first discovered in American elms in Ohio in 1930. Vectors for the fungus are two tiny bark beetles, *Scolytus multistratus,* the European elm bark beetle, and a beetle native to the United States, *Hylurgopinus rufipes.* The fungus came into the United States via the European beetle, which was embedded in some English elm logs that American cabinetmakers had imported for veneers. Although arborists made a heroic effort to contain Dutch elm disease during the 1930s, it soon spread throughout the native range of the American elm. Today half or more of the elms have been killed, with most of the survivors ornamental trees that must be heavily dosed with fungicide to survive. Only a very few elms remain in natural forests. An effort to develop a disease-resistant cultivar has been going on for decades, a program that yielded the so-called Liberty elm in 1983. Clones from this one tree are available at low cost from the Elm Research Institute, which holds a plant patent. The Liberty elm does not reproduce true to type, but among the thousands of cloned saplings distributed, one only has succumbed to disease. *See also* FOREST DECLINE AND PATHOLOGY.

dutchman in logging, an extension of the undercut around the side of a tree trunk in order to compensate for LEAN and to gun the tree into the desired LAY (*see* FELLING).

duty of water in water law, the maximum amount of water that may be used by the holder of a WATER RIGHT under the doctrine of PRIOR APPROPRIATION.

Duty of water limitations are set to prevent the waste of water by irrigators (*see* IRRIGATION) who withdraw more water from a stream than the plants they are growing can use. They are usually set at what is felt to be the maximum amount of water that crop plants can use, delivered at the maximum rate at which it can be assimilated, considering the climate of the region the laws apply to, the soils of the growing site, and other factors. Occasionally in areas of extreme water scarcity they may be set lower than the maximum utilizable rates and amounts in order to encourage conservation and the growing of less water-hungry crops. Once set, duty of water limitations are rarely subject to judicial or administrative review; hence, they are most useful in setting the amounts of new rights or transferred rights rather than cutting back on existing abuse. Common duty-of-water amounts are 1 cubic foot per second per acre, or 3 acre-feet per acre per year. *See also* BENEFICIAL USE.

dystrophic lake *See under* TROPHIC LEVELS.

EA *See* ENVIRONMENTAL ACTION. The term is also used by the Forest Service as an abbreviation for Environmental Assessment (*see* ENVIRONMENTAL ANALYSIS REPORT).

EAR *See* ENVIRONMENTAL ANALYSIS REPORT.

Earth Charter a project of an informal group of nongovernmental organizations (NGOs) which, frustrated by the lack of forthrightness in the pronouncements of the RIO EARTH SUMMIT, developed its own manifesto. The Earth Charter Commission was convened by Maurice Strong, a Canadian businessman and de facto founder of the U.N. Environment Programme, and former Soviet leader Mikhail Gorbachev. The original drafting committee was headed by Stephen A. Rockefeller, a professor of religion, and has been under revision since then. The charter commission hopes that some version of its statement will be adopted by the U.N. General Assembly in 2002, the 10th anniversary of the Rio Earth Summit, and that it will become as influential in the conduct of U.N. affairs as the Universal Declaration of Human Rights, one of the U.N.'s core documents.

Earth Day an annual observance of the need to protect the natural environment, held on April 22. Conceived by Senator GAYLORD NELSON of Wisconsin, the first Earth Day was held in 1970, organized by environmentalist Dennis Hayes, who has continued in this role ever since. The concept caught the imagination of many Americans, especially students, and some 20 million people participated in Earth Day demonstrations that first year, including 10,000 grade and high schools, 2,000 colleges, and 1,000 communities. Congress adjourned for the day so that members could speak at environmental rallies in their districts. At the 30th anniversary of Earth Day, in 2000, the theme was clean energy and the need to deal with GLOBAL CHANGE stemming from the use of FOSSIL FUEL. Many magazines and television programs were devoted to the event, including a special issue of *Time* (April–May 2000). In an obvious appeal to American youth, Leonardo DiCaprio, the heartthrob young actor, was chair of Earth Day 2000. Although many thought Earth Day would die out, it obviously did not, and is now observed in 140 countries. Some churches and synagogues also celebrate the occasion with an "Earth Day Sabbath" on the Saturday or Sunday nearest to April 22.

Earth First! (EF!) radical environmental organization, founded in 1980 to support demonstrations, direct action, and civil disobedience "in defense of Mother Earth." Essentially an anarchic organization, Earth First! has little central organization or control over its member groups. Its philosophy is biocentric, emphasizing the precedence of natural processes over human needs (*see* DEEP ECOLOGY). It does not believe in compromise. Address: P.O. Box 1415, Eugene, OR 97440. Phone: (541) 344-8004.

earthflow in geology, the movement of water-soaked earth downhill in response to gravity. Earthflows differ from *mudflows* in that they are too slow to be perceptible to the eye and are not confined to channels. They differ from SLUMPS in that there is no backward rotation of the land surface involved, the mass of earth in the flow coming to rest in roughly the same orientation as it had when it began a few feet to a few hundred feet away, although it is usually distorted lengthwise in the direction of the flow. *See also* SOLIFLUCTION.

Earth Island Institute strives to alleviate threats to the Earth's cultural and biological diversity through education and activism. The institute, founded in 1982 by DAVID BROWER, supports and creates projects that advance preservation, conservation, and restoration of the environment. Earth Island Institute publishes the *Earth Island Journal*. Membership (1999): 20,000. Address: 300 Broadway, Suite 28, San Francisco, CA 94133. Phone: (415) 788-3666. Website: www.earthisland.org.

Earthjustice Legal Defense Fund a public interest law firm founded in 1971 as the Sierra Club Legal Defense Fund to provide legal services to Sierra Club chapters. In 1997 the name was changed to reflect the organization's expanded role as a legal advocate for the environment on a wide range of issues and for various organizations, not just the Sierra Club. The fund concentrates exclusively on using environmental law to promote natural resource protections, water and air quality, and similar goals. Since its founding, the fund has (as of 2000) provided some 500 organizational clients legal representation on environmental cases at no cost. Address: 180 Montgomery Street, Suite 1400, San Francisco, CA 94104–4209. Phone: (415) 627–7600. Website: www.earthjustice.org.

earthquake a sudden, violent movement of the earth's surface. Most earthquakes result from motion along FAULTS in the underlying BEDROCK; others are related to the motion of bodies of MAGMA beneath the surface. The point within the earth where the actual motion of the earthquake takes place (that is, the point where the rocks slip or fracture from the forces working on them) is called the *seismic focus*. The area on the surface directly over the seismic focus is known as the *epicenter*. A *deep-focus earthquake* has its seismic focus 300 kilometers (180 miles) or more beneath the epicenter. A *shallow-focus earthquake* (the most common variety) has its seismic focus less than 70 kilometers (42 miles) beneath the epicenter. Earthquakes with their foci falling between these two extremes are known as *intermediate-focus earthquakes*. Earthquakes are usually felt for a wide radius around the epicenter. This is due to the presence of *seismic waves*, rhythmic disturbances of the earth's surface that radiate out from the epicenter like waves in a pond after a rock is tossed into it (*see* P-WAVE; S-WAVE). At least 20,000 earthquakes are recorded by SEISMOGRAPHS in the United States each year: most of these are too small to do any damage, but a few are large enough to create massive disruption. The most severe earthquake to occur in the United States in historic times was the New Madrid, Missouri, earthquake of 1812. Its intensity has been estimated at 8.7 on the *Richter scale*. *See also* PLATE TECTONICS; MERCALLI SCALE; TSUNAMI.

Earth Share a nonprofit fund-raising coalition of 44 nonprofit environmental organizations. Founded in 1988, Earth Share seeks to raise funds through payroll deduction campaigns. These funds support the member organizations and provide for public service announcements for the education of the public. Address: 3400 International Drive NW, Suite 2K, Washington, DC 20008. Phone: (202) 537–7100. Website: www. earthshare.org.

Earthwatch Institute sponsors scientific field research worldwide. Founded in 1971, it funds research projects by recruiting paying volunteers to accompany scholars in their field research. Earthwatch has sponsored projects in 50 countries and 25 states in various areas of study. Publications include *Earthwatch Magazine*, *Earthwatch Expedition*, and *Solutions*. Membership (1999): 23,000 in the U.S., 35,000 internationally. Address: 680 Mt. Auburn Street, Box 9104, Watertown, MA 02471. Phone: (617) 926–8200. Website: www.earthwatch.org.

ECCS short for *emergency core cooling system* (*see under* NUCLEAR ENERGY).

ecesis in ecology or population biology, the permanent establishment of a SPECIES or COMMUNITY in an area where it was not previously found. Ecesis represents the final stage of a process that begins with the invasion of a species or community into a new environment and continues with adjustment of the conditions in the new environment so that the invading organism or organisms can be self-sustaining and can maintain a balanced population (*see* BALANCE OF NATURE). The process is not considered complete until the invading species have lived through a complete range of climatic extremes in the new environment. A palm tree, for example, might make it through several mild winters in Utah, but ecesis could not be considered to have taken place until it had survived a severe winter as well—and reproduced afterward. *See also* DISTRIBUTION OF PLANTS AND ANIMALS; DISPERSAL.

ecofeminism a term coined in 1974 by Françoise D'Eaubonne describing the affinities between feminism and environmentalism politically, philosophically, and psychologically. Among the several aspects of ecofeminism: the view that both women and nature are exploited by a patriarchal culture and therefore share a kind of common cause; that women, as nurturers and homemakers, are instinctively more in tune with the preservation of the natural environment; and that women, because of their tendency to seek consensus, are better suited to provide political leadership for environmentalism. While sociologists and political scientists agree that there is a strong connection between feminism and environmental concerns, many contend that neither a sensitivity to nature, a commitment to radical reform, nor a sense of solidarity with nature arising from oppression suggests that women are "better environmentalists" than men, or that environmental politics should be feminist in its approach. At the same time, the history of the environmental movement, and the conservation movement that preceded it, indicates that women have, compared to their role in other movements, been very much at the forefront as writers, scientists, and political activists.

ecoforestry a method of timber harvesting developed by Canadian forest owner Merve Wilkinson. Ecoforestry leaves the biggest and oldest trees in place, chooses trees to cut that are shading out an area where seedlings have sprung up and that are widely separated, and skids them out of the forest one at a time to reduce damage to the forest floor. Wilkinson has reported that his resulting production is only a bit smaller than that of more commercially-oriented tree farms, and that the total amount of wood on his farm is roughly the same as when he started managing it in the 1950s. *See also* CLEARCUT; SELECTIVE CUT; HIGH-GRADING; NEW FORESTRY.

ecojustice a term used mostly in a religious or philosophical context that refers to the application of justice as an ethical proposition, not to human beings only but to all beings and to all creation. The associated term used in connection with political action on behalf of human rights is *environmental justice*. *See also* RELIGION AND THE ENVIRONMENT; DEEP ECOLOGY.

ecolabeling refers to the growing practice of providing environmental and associated health information on the labels of consumer products. Such information might include whether foods are organically grown, whether wood products have not been taken from old growth forests, or whether products are biodegradable and recyclable.

E. coli *See* ESCHERICHIA COLI.

ecological barrier an ENVIRONMENTAL FACTOR that serves as a blockade to the DISPERSAL of a POPULATION. An ecological barrier may be a physical feature such as a river or a mountain range, or it may be a more subtle factor such as a difference in climate or in soil chemistry. Human-made ecological barriers include freeways, canals, fences (for some species), and areas of urban development.

ecological economics an emerging discipline that tries to reconcile economic system thinking, based on labor, capital, resources, and technical innovation, with natural system thinking, based on dynamic equilibrium, diversity, interdependence, and the recycling of energy and matter. Both disciplines are based on discerning the chains of cause and effect—indeed, prior to the general use of the term *ecology*, the field of study was often referred to as *natural economy* (as opposed to *political economy*). Today, where the disconnection between natural economy (ecology) and political economy (economics) has led many to the general view that their goals are mutually exclusive, ecological economics assumes the opposite—that good economics is good ecology, and vice versa, even (if not especially) in a free-market system in which massive amounts of capital are concentrated in relatively few hands. In the United States ecological economics came into its own with the oil embargoes of the 1970s, when "energetics" was proposed as a kind of medium of exchange instead of the dollar and the economic benefits of "closed-loop living" were espoused, wherein nature's "free goods," such as sun, wind, and growing plants and trees for food, shelter, and clothing, could provide the basis for a satisfying way of life. The alternative ecological economic theories of the 1970s later developed into new industrial technologies that today are commonplace: solar and wind energy, the use of industrial and consumer waste to create profitable new products, and the development of promising new technologies such as hydrogen fuel cells. For the merging of natural and political economic systems to work well, however, a much more complex method of economic and ecological analysis must be perfected in which "costs" and "benefits" are evaluated over a long time horizon and at the global scale. *See* BENEFIT/COST ANALYSIS; ECOSYSTEM SERVICES.

ecological efficiency the percentage of energy used by an ORGANISM or group of organisms out of that available in the environment. Ecological efficiency is most often calculated between TROPHIC LEVELS in a FOOD WEB; that is, in a given ECOSYSTEM, the ecological efficiency of a particular organism is found by measuring its energy consumption and dividing that figure by the energy consumption of its PREY. In general, efficiency decreases as a food web is ascended; CARNIVORES are less efficient than HERBIVORES and top carnivores are less efficient than mid- or low-level carnivores. The most efficient organisms in any ecosystem, however, are the DECOMPOSERS, and the least efficient are the PRODUCERS. In general, more efficient organisms will tend to replace less efficient ones at the same trophic level or in the same NICHE (*see* COMPETITION; SUCCESSION).

ecological footprint a term that refers to the amount of land required to provide for the needs of a given population under a given rate of consumption on a sustainable basis. A major city, for example, would have a much larger ecological footprint than the square miles of land within the city limits.

ecological pyramid **(Eltonian pyramid)** in the biological sciences, a means of describing the relationships among TROPHIC LEVELS in a COMMUNITY of organisms, developed by the population biologist Charles Elton around 1940 and widely used since. The pyramid shape is a result of the loss of energy between each pair of trophic levels, so that a large number of plants is necessary to support a small number of HERBIVORES, which supports a still smaller number of CARNIVORES (*see* ECOLOGICAL EFFICIENCY). Three kinds of ecological pyramids are generally recognized. In the *pyramid of numbers,* the actual number of organisms in each trophic level is counted or closely estimated. In a classic study of a bluegrass field by Eugene P. Odum in 1959, for example, 5,842,424 PRODUCERS (green plants) were found per acre. The number of primary consumers (herbivorous insects) was found to be 707,624; that of the secondary consumers (ants, spiders and carnivorous beetles), 354,904; and that of the tertiary consumers (birds and moles), 3. The *pyramid of biomass* is slightly more meaningful than the pyramid of numbers: it looks at the total mass of the organisms at each level rather than at the numbers, thus adjusting for the fact, for example, that each of the three birds and moles in the previous example is many times larger than the insects they feed upon. The pyramid shape of the community is still present, but the proportions between levels decline in a more regular fashion. Most meaningful of all is the *pyramid of energy,* in which the energy consumed by the organisms at each trophic level is calculated. In another classic Odum study (Silver Springs, Florida, 1957), it was found that the plants in an aquatic ecosystem were using 20,810 kilocalories of energy per square meter per year; the herbivores that fed on them, 3,368 kcal/m^2/yr; the carnivores that fed on the herbivores, 383 kcal/m^2/yr; and the TOP PREDATORS, only 21 kcal/m^2/yr. The concept of the ecological pyramid is important in understanding the effects of environmental alterations on ECOSYSTEMS, especially those alterations that will affect one trophic level more severely than others. *See also* FOOD CHAIN; FOOD WEB.

Ecological Society of America **(ESA)** the principal professional association of ecologists of the United States. The society encourages responsible employment of ecological principles to solve environmental problems. ESA convenes an annual conference, provides expert testimony to Congress, and publishes numerous reports and journals to distribute current ecological research findings. Membership (1999): 7,200. Address: 2010 Massachusetts Avenue NW, Suite 400, Washington, DC 20036. Phone: (202) 833-8773. Website: www.esa.org.

ecological succession *See* SUCCESSION.

ecology the study of the relationships of living things to each other and to their surroundings. The term was coined by the German naturalist Ernst Haeckel in 1869. It is made up of the two Greek root words *oikos* ("household," "dwelling place") and *logy* ("study of"). Its use as a substitute for the word ENVIRONMENT or ECOSYSTEM is improper; however, using it to refer to the complex, interconnected web of relationships among organisms within an environment or ecosystem has become accepted usage, mainly because no other word exists that has precisely the same meaning.

ecology, deep *See* DEEP ECOLOGY.

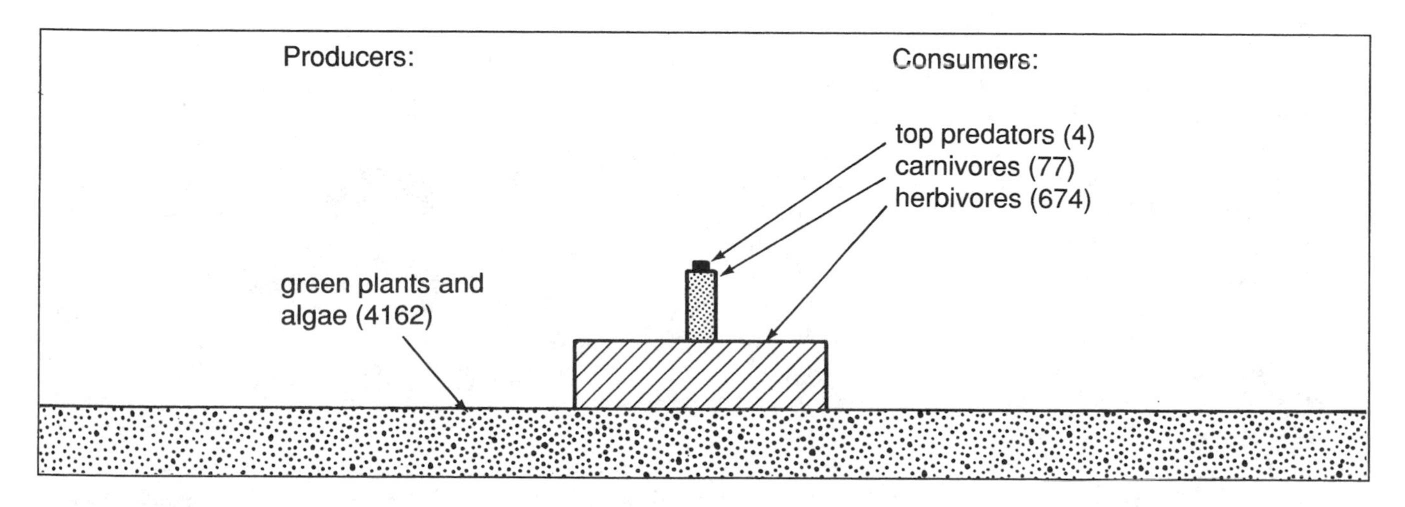

Pyramid of numbers

ecology, evolutionary *See* EVOLUTIONARY ECOLOGY.

ecology, restoration *See* RESTORATION ECOLOGY.

economic externalities *See* EXTERNALITIES.

economic poison any toxic substance (*see* TOXICANT) whose toxic properties are used as a management tool in a manner that contributes to the economic well-being of the user. The term encompasses HERBICIDES, FUNGICIDES, and insecticides and other PESTICIDES; the specialized poisons, such as ROTENONE, used in fisheries management; CYANIDE, 1080, and other toxic materials used in PREDATOR CONTROL; and all similar materials.

economic selection system in logging, *see* HIGH-GRADING.

ecosystem the total physical and biological environment in a given area, with an emphasis on the relationships and connections among the various parts. As the term "-system" implies, an ecosystem tends to function as a synergistic unit (*see* SYNERGY): changes in any one component of the system will cause changes in others as the system adjusts to the new conditions created by the first change. In a complete ecosystem, the flow of material will always be circular, requiring no net input of matter, only an input of energy from solar radiation or geothermal heat; hence, the area involved must be large enough to include all the necessary participants in this circular flow. (Energy flows *through* an ecosystem in the same balanced manner that matter flows *around* it. The net energy input is always equal to the work done within the system plus the energy lost to ENTROPY. *See* ENERGY BUDGET; THERMODYNAMICS, LAWS OF.) Because of the importance of relationships in determining whether a system will survive or fail, ecologists tend to think of ecosystems as sets of NICHES rather than as sets of specific plants and animals. *Compare* COMMUNITY.

ecosystem services the tangible and monetarily quantifiable benefits to human society of natural systems. In 1997 Robert Costanza, a University of Maryland ecological economist, attempted, with his colleagues, to quantify the "value of the world's ecosystem services and natural capital." Costanza's controversial project, and the like-minded efforts of other ecological economists, attempts to overcome the difficulty in valuing ecosystem services via normative economic analysis. For example, the tangible economic benefits—in terms of jobs, taxes, retail sales, property values—of building a marina in a marshland area would appear to overwhelm the monetary benefits of preserving the marsh. Perhaps only the value of shellfish harvest would be counted in a typical analysis. And yet the marsh, if left alone, could produce many valued ecosystem services over the years relating to offshore fisheries, aesthetics, the cleansing of water pollutants, or social stability in nearby human communities. What if, in fact, the marsh were a kind of ecological keystone, the destruction of which could disrupt the functioning of a wide area of land and water and create a public cost several times greater than any advantages that a marina might bring? Thus, if ecological services are factored into the economic calculus, an objective view of the monetary benefits of a major change in ecosystem function might well be entirely different. Costanza's project, at the macro scale, proposes that ecosystem services produce $33.3 trillion worth of benefits annually. Significantly, the services provided by

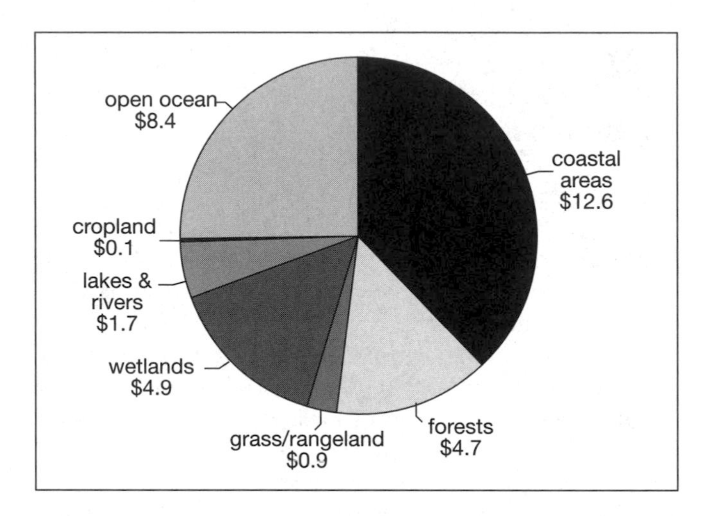

Ecosystem services by biome, trillions of dollars/year
(R. Costanza et al. "The Value of the World's Ecosystem Services and National Conservation."* Nature, *May 15, 1997)

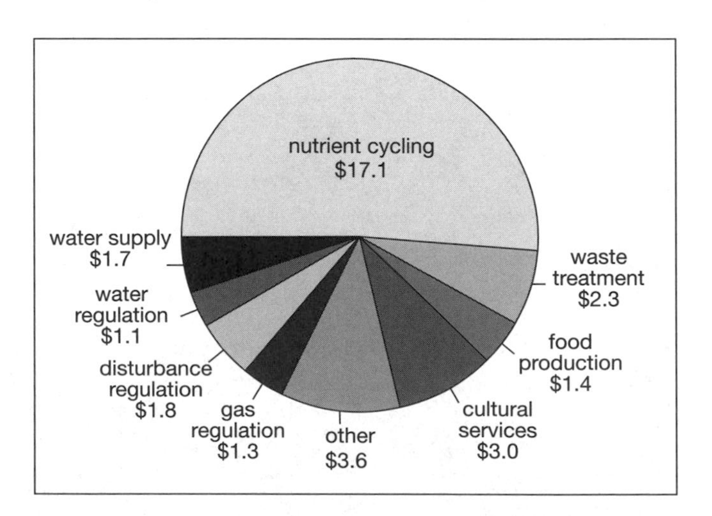

Ecosystem services by service, trillions of dollars/year
(R. Costanza et al. "The Value of the World's Ecosystem Services and National Conservation."* Nature, *May 15, 1997)

forests and grasslands are worth only about $100 an acre in Costanza's model, but coastal ecosystems can produce as much as $8,000 to $9,000 an acre in services. Even though these ecosystems constitute only 6.3% of the surface of the planet, they produce in monetary terms 43% of its ecosystem services. *See also* ECOLOGICAL ECONOMICS.

ecotage refers to various acts of sabotage, although with no intention to harm individuals, meant to disrupt development projects taking place in wildernesses or other natural areas that the ecoteur feels should be protected. Sometimes called "MONKEY WRENCHING" (after EDWARD ABBEY'S novel *The Monkey Wrench Gang*), ecotage most often consists of pulling up survey stakes, putting sugar or abrasives in the fuel tanks of earth-moving equipment, or engaging in other forms of civil disobedience to publicize environmental abuses. Although "tree spiking" to discourage logging was once a popular form of ecotage, this tactic has largely been discontinued because of the danger to those operating chain saws and mill saws without knowledge of the spikes. Ecotage should not be confused with ECOTERRORISM. It is, instead, an extreme form of direct-action environmental protest.

ecoterrorism a term properly used in reference to terrorist acts involving the disruption or destruction of natural systems, such as the burning of oil fields, the poisoning of water supplies, and the like. For political purposes anti-environmentalists often categorize environmental protests involving ECOTAGE as ecoterrorism.

ecotheology reconciles normative theologies with the principles of ecology and contemporary environmental issues. *See* RELIGION AND THE ENVIRONMENT.

ecotone the boundary between two BIOREGIONS, BIOTIC PROVINCES, or ECOSYSTEMS. An ecotone is normally a band or zone rather than a sharp line; however, the width of this band can vary enormously, from a few inches (the zone of lapping wavelets at the waterline of a pond) to many miles (the transition between the TUNDRA and CONIFEROUS FOREST BIOMES on the arctic plains). Generally, *climatic ecotones* (those caused by changes in average rainfall, temperature, or length of PHOTOPERIOD) are more gradual than those caused by physical factors such as changes in soil chemistry, availability of soil water, or elevation differences along mountain slopes.

ecotourism refers to vacation travel to largely undisturbed places of high ecological interest for educational as well as recreational reasons. Ecotourism facilities and services seek to have a minimal impact on nature, operate at a small scale, employ local people as guides, foster respect for different cultures, and contribute to local economies. In its ideal form ecotourism should advance conservation goals in terms of funding and publicity in the areas visited. Ecotourism began in the 1970s in Latin American rain forests and in East African wildlife areas. Since then it has become the fastest-growing segment of the $4 trillion worldwide tourist industry. Estimates of ecotourism's share have been as high as $30 billion annually, with an annual growth rate of up to 30%. Critics of ecotourism, which is often sponsored by nonprofit conservation organizations and government agencies in many developing nations, is that its very success might serve to degrade the ecosystems it seeks to preserve. Most experts believe that the solution to this dilemma lies in educating ecotourists not to expect luxury services and to limit the size of tourist facilities.

ecotype in ecology, a genetic variant of a SPECIES that is adapted to the conditions in a particular physical or geographical environment. Ecotypes can breed with one another but exhibit differences in physical characteristics (size, color, breeding time, ability to survive low temperatures or dry conditions, etc.) that make it easier for them to fit the peculiarities of their particular environment. Often all of the ecotypes in a particular geographical area or HABITAT type (such as a swamp or grassland) will show the same sort of adaptations. For example, all CONIFERS at timberline, regardless of the species, tend to grow as KRUMMHOLZ, a tendency that is probably at least partially genetically determined.

ectoparasite any PARASITE that lives on the outside of its host, penetrating the skin only to obtain food. Examples include fleas, lice, mites, and chiggers. Because they often transfer from host to host, ectoparasites are one of the most important means of transmission for many diseases (*see* VECTOR). *Compare* ENDOPARASITE.

edaphic factors in the biological sciences, the characteristics of soils as they relate to the growth of plants. Edaphic factors include both soil chemistry (NUTRIENT content and the presence, in certain soils, of potentially harmful chemicals such as ARSENIC or copper) and soil mechanics (particle size, compactness, FIELD CAPACITY, and so forth). Generally, the more severe the climate (drier, colder), the more important edaphic factors become. They are highly site-specific, and are usually the factors to look for when the composition of a plant community changes noticeably over a very short distance (but *see* MICROENVIRONMENT).

edge cities term used to describe satellite cities in large conurbations that have become, essentially, full-service cities in themselves. Examples are White Plains,

New York, north of New York City; Rockville, Maryland, northwest of Washington, D.C.; Kirkwood, Missouri, west of St. Louis; and Santa Ana, California, west of Los Angeles. Edge cities tend to exacerbate a sprawl pattern of development by creating suburbs of their own in an ever-widening metropolitan land area. *See also* URBAN SPRAWL.

effluent any waste product in the process of being discharged into the environment. The term applies to biological and household wastes and to waste heat as well as to industrial discharges and sewage-plant outfalls. Since not all waste products are environmentally harmful, the discharge of an effluent does not automatically create POLLUTION; however, nearly all pollution results, in one form or another, from effluent discharges, and therefore reduction or elimination of effluents is one of the most important forms of pollution control.

EFFLUENT LIMITATION

Since effluents are a classic form of external cost (*see* EXTERNALITIES), they will not normally be limited by their producers without external coercion in the form of legal strictures of some type. There are two principal approaches to this external coercion. In *regulation,* the most common approach, standards are set concerning the types of wastes that may be discharged and the amount of these wastes permissible in the effluent. This amount may be expressed either as a concentration, such as parts per million (ppm) or parts per billion (ppb)—meaning the number of molecules of the waste present for each million (or billion) MOLECULES of water—or it may be expressed as a rate of flow, that is, so many pounds of the waste per day, or per week, or per unit of manufactured product. The effluent is tested on a regular basis, and the firm or agency responsible for the effluent is subject to criminal penalties if the standards are not met. A considerably less common but promising alternative to regulation is the development of a system of *effluent charges.* In this approach, effluent-creating firms or agencies are charged fees for the privilege of discharging their wastes into the environment. These costs then become part of the economic climate in which the firm or agency operates, to be handled on a normal economic balance sheet. If the fees are set high enough, there will be strong economic encouragement to invest in pollution-control technology or to modify the processes used in the plant in order to reduce or eliminate the wastes they produce.

effluent standard in pollution control, a standard applied to the amount of material actually released by a pollution source (*see* EFFLUENT) rather than to the amount of material (relative or actual) present in the water or the air. Effluent standards may be stated either as concentrations (percentage, weight per unit volume, etc.) or as total permissible loading (weight or volume per unit of time). Relatively easy to enforce due to their lack of ambiguity (a particular effluent either is or is not exceeding the standards, with no middle ground), effluent standards nevertheless have two serious drawbacks: they cannot be used to solve NONPOINT POLLUTION problems where the effluent is diffused over many unknown sources, and they make it more difficult to take into account factors of scale (a pollutant loading of 5 pounds per day, for example, has a significantly different effect on a stream flowing at 100 cubic feet per second than it does on one flowing at 100,000 cubic feet per second, and the same standard should not necessarily be applied to both). *Compare* AMBIENT QUALITY STANDARD.

Ehrlich, Paul (1932–) American population biologist, born May 29, 1932, in Philadelphia, Pennsylvania, and educated at the University of Pennsylvania (B.A., 1953) and the University of Kansas at Lawrence (M.A., 1955; Ph.D., 1957). After two more years as a research associate at Kansas (1958–59), he joined the faculty of Stanford University, where he has taught ever since. His numerous books on biology, ecology, and the environment—many co-authored with his wife Anne (Howland) Ehrlich, also a biologist—include the best-selling *The Population Bomb* (1968) and *The End of Affluence* (1974). Ehrlich's other books include *Population, Resources, Environment* (1970), *Biology and Society* (1976), *Ecoscience* (1977), *Extinction* (1981), *The Population Explosion* (1990), and *Betrayal of Science and Reason* (1996). The intensity and passion of Ehrlich's crusade for human population controls has caused some critics to overlook the firm scientific basis of his work, which has always remained impeccably grounded in his own discipline.

EIS *See* ENVIRONMENTAL IMPACT STATEMENT.

ejecta in geology, any material thrown from a VOLCANO during the course of an eruption. Ejecta includes dust and volcanic ash particles, LAVA BOMBS, and anything else that is thrown through the air from the mouth of the volcanic vent. It does not include LAVA, which is not thrown but flows over the ground.

electric vehicles (EVs) cars, trucks, bicycles, buses, trams, and other conveyances powered solely by electricity. Automotive EVs (i.e., vehicles not powered by direct means, such as pantographs, the power-collecting booms used by electric streetcars and buses to make connection with overhead lines) are powered by storage batteries, and were introduced early in the 20th century. Like automotive steam engines, electric battery motors soon gave way to INTERNAL COMBUSTION

ENGINES powered by FOSSIL FUELS (including natural gas, gasified coal, gasoline, kerosene, and diesel fuel). While some automotive electric vehicles (especially delivery vans) have commonly been seen on city streets, storage battery technology did not advance significantly until the oil embargo of the 1970s added to the already increasing concern about SMOG and other forms of AIR POLLUTION from automobiles and trucks, and, most recently, global warming resulting from CO_2—the primary byproduct of the combustion of carbon-based fuels. Recent improvements in rechargeable batteries have been significant, so that by the end of the 20th century there were about 500,000 alternative-fuel vehicles in the United States. Where once electric vehicles could not travel much farther than 50 miles between charges, new battery technology—specifically the nickel-metal-hydride battery (NiMH)—now permits EVs to go 170 miles on a single charge. At Sandia National Laboratory in Albuquerque, New Mexico, further improvements are under development through a lithium-ion battery with a specially formulated cathode that can reduce costs while maintaining high capacity. One incentive to produce a better battery is a California state law mandating that 10% of vehicles sold in the state must meet zero-pollution standards, a requirement that only EVs can meet. Electric vehicles are not, however, pollution-free, since the electricity needed to recharge the batteries comes from power plants that use fossil fuel for electrical generation. Even so, the reduction in pollution can be substantial, since even if a power plant were to use 50% coal (the most polluting energy source) for generation, the reduction of greenhouse gas emissions would be reduced by half in city driving conditions. In the longer range, it may be that cars powered by HYDROGEN FUEL CELLS will become a power source soon enough (some say as early as 2010) to make further work on EV technology impractical, since the fuel cells would produce power with less pollution per vehicle mile than storage batteries. Thus, in the view of many, the transition to hydrogen-powered vehicles is more likely not through EVs but through HYBRID-ELECTRIC CARS, first available in 2000, with low-horsepower internal combustion engines that recharge the batteries during operation. Some automobile experts, notably Lee Iacocca, former chairman of Chrysler Corporation, and Robert Stempel, former chairman of General Motors, believe that in the end smaller vehicles may offer the best use of EV technology. These two industry leaders have joined forces to produce an electric bicycle for the world market that can retail for less than $1,000.

electrode the physical structure ("pole") through which an electrical current enters or leaves a vacuum or a nonmetallic conductor such as a fluid, a quartz or germanium crystal, damp soil, etc. Usually (but not always) the electrode is made of metal. Electrodes always come in pairs, one of which emits ELECTRONS while the other attracts them. The electron-emitting pole is called the *cathode* and the electron-attracting pole is called the *anode,* names given to them by the English chemist Michael Faraday around 1845. *See also* CORROSION: *types of corrosion;* ELECTROPLATING.

electrodialysis method of removing dissolved materials from a SOLUTION through the use of electrical attraction and diffusing membranes, used to purify water and, in medicine, to purify blood in the so-called "artificial kidney." An electrodialysis cell consists of a closed container divided into three sections by the use of permeable membranes (*see* PERMEABILITY). One side section contains a cathode and the other an anode (*see under* ELECTRODE), while the center section contains no electrodes at all. The fluid to be purified is fed into all three sections and a current is sent through the electrodes, causing the positive IONS to migrate through the membrane into the side section containing the cathode and the negative ions to migrate into the side section containing the anode. Purified (ion-free) fluid may then be drawn from the center section. The first desalination plant to produce POTABLE WATER from brackish sources through electrodialysis went into operation in Buckeye, Arizona, in 1962.

electrolyte a fluid that has the ability to conduct electricity. The best electrolytes are usually SOLUTIONS in which a high degree of DISSOCIATION has taken place, creating numerous positively and negatively charged IONS. These ions then become carriers of the electrical current through the fluid. Some solids also become electrolytes upon being melted by heat. An example is sodium chloride (common table salt), which is an electrolyte in both its molten and dissolved forms.

electrolyte corrosion *See* CORROSION: *types of corrosion.*

electron a tiny charged particle, smallest of the three principal constituents of the ATOM, with a mass of 9.1×10^{-28} gram (1/837 that of a HYDROGEN atom) and a negative charge of 1.6×10^{-19} coulomb. Electrons are arguably the most important units of matter. Their number and arrangement determine many characteristics of the ELEMENTS, including chemical activity, electrical conductivity, magnetism, and VALENCE. All COMPOUNDS are formed by electron bonding among the elements. Electrons are the carriers of electricity through circuits. Under the name *cathode rays* they are responsible for the operation of vacuum tubes and for the creation of the pictures on television screens; under the name *beta rays* they are one of the three principal forms of RADIOACTIVITY. *See also* ATOM: *structure of the atom.*

electroplating process through which a thin, even coating of metal is electrically bonded to the surface of an object made of another type of metal or (less commonly) of a non-metal. The object to be plated is attached to an electrical circuit in such a way that it will form a cathode (*see* ELECTRODE). It is then immersed in a solution of one or more SALTS of the plating metal ("the bath"), and a small electric current, on the order of 1–6 volts, is passed through the system, causing the free metallic IONS in the bath to be deposited on the object. In some cases, particularly those involving the plating of gold and silver, the presence of CYANIDE ions in the bath improves the uniformity of the plating; hence, either complex salts of the plating metal that include cyanide are used, or other cyanide compounds are added to the bath. In either case the spent bath becomes a significant hazard if released into the environment.

electrostatic precipitator device for removing PARTICULATES from air, flue gases, or other gaseous mediums by inducing a negative electrical charge on the particulates, allowing them to be attracted to a positively charged "collector plate" from which they can be removed and collected by mechanical means. The negative charge is provided by a set of ELECTRODES (*discharge electrodes*) that ionize the gases around them (*see* ION). The ions in the gases then collide with the particulates, passing their electrical charges on to them and causing them to migrate to the large collector plates, where they cling in much the same way that clothing clings together after a static charge has been induced in it in a hot-air dryer. Periodically, the clinging particulates are removed from the collector plates by washing, brushing, or striking the plates with specially designed "rapper plates" that cause the particles to fall off into a bin beneath the collectors. In a high-voltage or "single-stage" precipitator, ionization of the gases and collection of the charged particles takes place within a single unit. In a low-voltage or "two-stage" precipitator ionization is accomplished by one set of electrodes, with collection taking place at a second set. The efficiency of either type of precipitator depends largely on the electrical properties of the particulates to be collected. Particles whose electrical resistance is too low will not hold a charge long enough to be attracted to the collector, while those whose resistance is too high may coat the collector so thickly that its charge cannot be maintained even with frequent cleaning. Despite these limitations, electrostatic precipitators have found broad use both as emission-control equipment for industry and as so-called "electronic air filters" installed in air-conditioning systems or sold as free-standing units for home or office use.

element in chemistry, a substance made up entirely of the same kind of ATOM; hence, one that cannot be broken down into other substances through purely chemical means. The differing properties shown by the elements are a result of the different numbers of protons and neutrons and ELECTRONS that make them up and of the different arrangements of their electrons (*see* VALENCE). These differences occur in a regular, periodic manner, allowing the elements to be arranged according to their properties in a repeating table (*see* PERIODIC TABLE). Eighty-seven elements occur naturally on the earth and in its surrounding atmosphere. As of 1988 another 21 had been synthesized by atomic physicists, for a total of 108 (*see* TRANSURANIC ELEMENTS). Sample natural elements: oxygen, hydrogen, sodium, carbon, iron, copper, gold. Sample man-made ("transuranic") elements: plutonium, curium, californium, einsteinium. *Compare* COMPOUND. *See also* individual articles on most elements.

El Niño, La Niña El Niño is a large-scale disturbance in a normal weather pattern due to increases in water temperature in the eastern Pacific Ocean and lower than average barometric pressure. Properly termed the El Niño Southern Oscillation, such climate events favor the formation of winter storms with heavy precipitation in much of North and South America. (The term *El Niño,* Spanish for the Christ child, refers to the incidence of storms at the Christmas season.) The 1997–98 El Niño resulted in storms that killed an estimated 2,100 people and caused some $33 billion in property damage. Sometimes an El Niño is followed by a climate event called *La Niña,* as happened in 1999–2000. These events occur when the water temperature in the eastern Pacific decreases and barometric pressure is higher than usual. As may be expected, La Niña can bring periods of drought. In the past century, according to the NATIONAL OCEANIC AND ATMOSPHERIC ADMINISTRATION, there have been 23 El Niño events and 15 La Niña events. The four strongest of these have occurred since the 1980s. Some environmental scientists believe that these extreme weather conditions are exacerbated by GLOBAL CHANGE due to CO_2 emissions and the GREENHOUSE EFFECT.

Eltonian pyramid *See* ECOLOGICAL PYRAMID.

embayment a bay, especially one that has a wide mouth or is formed by a broad indentation of the shoreline. Geologists also use the term for any of the several processes by which bays are formed, and, by extension, for any material or landform of one sort that is extended into another. For example, an indentation at the base of a hill could be called an embayment of the adjacent prairie; an area where sedimentary rocks extend into an indentation in a body of granite or basalt would be an embayment of the sedimentary rocks; etc. *See also* COVE.

emergent macrophyte an aquatic MACROPHYTE that is rooted in the bed of a body of water but extends most of its stem and leaves above the water's surface into the free air above. Examples: cattail, bulrush, wild rice.

Emerson, Ralph Waldo (1803–1882) American poet, lecturer, essayist, and philosopher of religion and nature. Emerson entered Harvard College when he was 14, graduated when he was 18, and continued his studies at Harvard Divinity School after three years of teaching. After ordination in 1829, he was appointed minister of a large Boston Unitarian church. After the death of his wife and brothers in 1832, he broke away from church ritual, resigning from his pastorate. He traveled to England where he met with English romantic poets Samuel Taylor Coleridge and William Wordsworth, and with social reformer Thomas Carlyle, who became a lifelong friend. Upon his return to the United States, Emerson actively lectured in Boston on his philosophy of nature-oriented TRANSCENDENTALISM. He held that truth could be revealed through emotion and reason instead of the widely held viewpoint that truth was discovered through observation and experience. Emerson helped in the founding, writing, and editing of *The Dial,* the journal of transcendentalism in New England. His major published works include *Nature* (1836), *Essays, First Series* (1841), *Essays, Second Series* (1844), *Poems* (1847 and 1865), *Representative Men* (1850), *English Traits* (1860), *Society and Solitude* (1870), *Natural History of Intellect* (1893), and *Journals* (1909).

eminent domain in law, the right of a government to transfer property from private to public ownership on demand. All nations throughout history have had some form of eminent domain. In most countries today, however, it is tightly constricted by law and may be used only under certain circumstances. In the United States, the taking of private land for public use must be demonstrated to increase public welfare in some form, and the private landowner must be awarded just compensation for the loss of his land. "Just compensation" generally means fair market value; the actual amount is normally determined by a court of law. As sovereign units, the federal government and the various state governments all have the right of eminent domain; city and county governments do not, but may be granted it by the state for specific purposes. States (and the federal government) may also grant rights of eminent domain to private corporations if it can be shown that to do so will increase the public welfare. For example, a power company may be granted a right of limited eminent domain in order to obtain corridors to string its transmission lines. *See also* DECLARATION OF TAKING.

emission standard a legally determined limit on the amount of a particular POLLUTANT that may be released into the air from a single source. Emission standards are normally set as concentrations, that is, the proportion of total emissions that may be composed of each pollutant. These concentrations are usually expressed in terms of parts per million (ppm) or parts per billion (ppb), meaning the number of molecules of the pollutant present for each million—or billion—molecules of air. Alternatively, especially for STATIONARY SOURCES, the standard may be set as a flow rate, that is, as the total weight of the pollutant that may be released per unit of time (day, week, year) or per unit of production. *See also* AIR POLLUTION; AMBIENT QUALITY STANDARD; CLEAN AIR ACT.

emphysema a chronic, progressive disease of the lungs, involving a hardening and thickening of the tissues of the terminal *bronchioles,* the *air sacs,* and the *alveoli* (the smallest branches of the lung passageways, the air chambers that terminate them, and the cuplike organs off these chambers that serve as the exchangers of OXYGEN and CARBON DIOXIDE into and out of the blood). In the early stages, the symptoms of emphysema are similar to those of ASTHMA, including coughing, wheezing, and shortness of breath. As the disease progresses and the lungs become less and less able to expel air (that is, as the air sacs become less and less flexible and the passageways leading to them become more and more constricted), they expand in size, both in an effort to create more exchange tissue and as a consequence of filling up like a balloon with stale, unexpelled air. The patient develops a characteristic "barrel chest"; the skin turns blue as a result of the inability to exhale carbon dioxide. The enlarged lungs put physical pressure on the heart which, together with the extra load placed on the circulatory system by the increased energy needed to breathe and by the poisons the body is unable to excrete through the damaged lungs, often leads to heart attacks. Death usually results. There is no cure: what treatment there is involves purification of the air being breathed and the administration of drugs that can temporarily expand the bronchioles and allow for a slightly greater exchange of air.

CAUSES

Emphysema is strictly an environmentally caused disease, with little or no involvement of BACTERIA or VIRUSES. The highest incidence of the disease is among cigarette smokers; however, air pollution can aggravate the condition and cause it as well. Among the common air pollutants, the presence of sulfur dioxide (SO_2) and ozone (O_3) appears to correlate most strongly with the incidence of emphysema, with a secondary, carrier role played by PARTICULATES. Particulates are probably the prime cause of the aggravation of cases of emphysema during pollution incidents, as they increase the flow of mucus and block the already-constricted lung passages. Hydrogen cyanide also appears to play a role, and may be the principal causative agent in cigarette smoke.

INCIDENCE

Emphysema-related deaths in the United States increased more than 14-fold between 1950 and 1970, peaking at 11.2 deaths per 100,000 people in 1970. Since then the death rate has declined to approximately one-half that amount, closely paralleling a decline in the amount of SO_2 released into the atmosphere as a result of clean air legislation. *See* CLEAN AIR ACT; ASTHMA.

emulsifying agent (emulsifier) in chemistry, a COMPOUND whose presence stabilizes an EMULSION and prevents it from separating out into its components. Emulsifying agents act in two ways. One type consists of asymmetrical molecular chains: one end of each chain is soluble in one of the two liquids that make up the emulsion, while the other end dissolves readily in the second liquid. When an emulsion forms, these asymmetrical MOLECULES form a thin layer over each of the dispersed microdroplets, with the ends soluble in the microdroplets pointing inward and the ends soluble in the emulsifying medium pointing outward. These layers prevent the droplets from coalescing and separating out. It is this type of emulsifying action that gives DETERGENTS and soaps their cleaning power. The second type of emulsifying agent apparently works by altering the SURFACE TENSION of one of the two liquids of the emulsion while either not affecting the second liquid or altering its surface tension in a different manner. The difference in the relative surface tension between the two liquids allows them to remain mixed. This is the type of emulsifying agent most commonly found in foods. Examples include the lecithin (provided by egg yolks) that stabilizes the emulsion of oil and vinegar we know as mayonnaise, and the casein (found naturally in milk) that prevents homogenized milk from settling out. *See also* WETTING AGENT.

emulsion in chemistry, a form of COLLOIDAL SUSPENSION in which both the suspended material (*dispersed medium*) and the material it is suspended in (*dispersion medium*) are liquids. The dispersed medium is scattered through the dispersion medium in the form of "microdroplets" of extremely small size that are capable of remaining in suspension indefinitely. Since these microdroplets are liquid, however, they have a tendency to coalesce when they run into each other, forming larger droplets that eventually reach the size where they begin settling out, breaking the emulsion down. Therefore a small amount of a third substance must usually be added to prevent this coalescence and keep the emulsion intact. *See* EMULSIFYING AGENT.

enabling legislation a law allowing a person, corporation, or government agency to act in a way which it would not otherwise have authority to. Legislation that gives a utility company the right to condemn land for transmission-line corridors (*see* EMINENT DOMAIN) is enabling legislation, as is legislation granting asylum to aliens who may have entered the country illegally. From an environmental standpoint, the most important enabling legislation is that granting federal and state agencies the authority to make regulations in specific policy areas, such as water resources, pollution control, and mine reclamation. *Compare* ORGANIC ACT.

endangered species as defined in the ENDANGERED SPECIES ACT, a SPECIES of plant or animal that is in imminent danger of EXTINCTION ". . . over all or a significant portion of its range . . ." Insect pests that present an "overwhelming and overriding risk to man" are excluded. Tests to determine whether or not a species is in imminent danger are not defined by the act; however, the criteria developed by the California Fish and Game Commission in 1974 have been generally accepted by other land managers and by the judiciary. According to the commission, a species should be listed as endangered if any of a set of five specific conditions are met. These include (a) the death rate of the species exceeds its birthrate; (b) the species is highly specialized and incapable of adapting to environmental change; (c) the species' HABITAT is seriously depleted or disturbed; (d) the introduction of one or more EXOTIC SPECIES poses a threat due to COMPETITION, predation (*see* PREDATOR), parasitism (*see* PARASITE), or disease; or (e) environmental pollution threatens its viability or survival. *Compare* THREATENED SPECIES; RARE SPECIES.

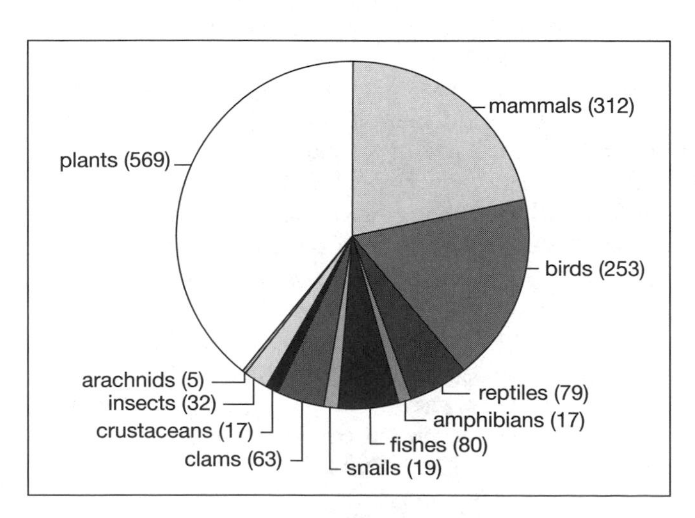

U.S. endangered species as of 1999 ***(U.S. Fish and Wildlife Service,*** **Endangered Species Technical Bulletin,** ***quarterly)***

Endangered Species Act federal legislation to prevent the extinction of rare and endangered animals and plants, signed into law by President RICHARD NIXON on December 28, 1973. The act requires the secretary of the interior (*see* INTERIOR, DEPARTMENT OF), in cooperation with the secretary of commerce, to prepare regularly updated lists of ENDANGERED and THREATENED SPECIES for publication in the FEDERAL REGISTER. These lists are known collectively as the *Endangered Species List.* Private possession of animals or plants on the Endangered Species List, their body parts, or any products made from them is expressly forbidden without a permit, and commercial trade is not allowed. Federal funds may not be used for any purpose that will decrease the HABITAT of any species on the list or otherwise endanger it further, and federal agencies are required to make certain that all of their actions are consonant with the purposes of the act. Finally, the secretaries of commerce and the interior are required to designate areas of CRITICAL HABITAT for species on the list, and to take steps to make certain that this habitat is preserved, including where necessary the purchase of private lands for the purpose of consolidating habitat and preventing its destruction by commercial development.

HISTORY

Although U.S. legislative concern with endangered species goes back at least as far as 1875, when the Buffalo Protection Act was vetoed by President Ulysses Grant, the first comprehensive legislation to directly address endangered species was the Endangered Species Preservation Act of 1966. This act and its amended version, the Endangered Species Conservation Act of 1969, established the Endangered Species List and prohibited commerce in species on that list or in products made from them. The 1973 act is not an amendment to these earlier laws, but a replacement that specifically repeals them. It has itself been amended twice, once (1978) to establish an Endangered Species Committee to review federal projects for possible exemptions from the act, and again (1982) to require that additions to and deletions from the Endangered Species list be based solely on biological considerations. The 1982 amendments also allowed for the reintroduction of locally extinct species into their former ranges on an experimental basis. All told, from 1973 to 1999 the Fish and Wildlife Service listed 925 U.S. species, including 569 plants, as *endangered* (likely to become extinct throughout all or a significant part of their natural range), and 258 species (including 136 plants) as *threatened* (likely to become endangered in the foreseeable future). *See also* REINTRODUCTION OF SPECIES; EXTINCTION.

Endangered Species Coalition founded in 1982, an alliance of environmental, scientific, and religious organizations joined together to expand and energize public support for the preservation and protection of endangered species. Membership (1999): 450 organizations. Address: 1101 14th Street NW, Suite 1400, Washington, DC 20005. Phone: (202) 682-9400.

endemic species a plant or animal SPECIES that occurs naturally only in a particular limited geographic area. *Compare* ABORIGINAL SPECIES; EXOTIC SPECIES.

endocrin disrupters environmental contaminants, such as POLYCHLORINATED BIPHENYL (PCB), DIOXIN, and breakdown products of DDT, that disrupt hormones and other chemical messengers in humans and terrestrial and marine wildlife. The disruption from these compounds can result in reproductive failure and abnormality, impairment of immune systems, and difficulties relating to sexual differentiation, growth, and development.

endoparasite an internal PARASITE, that is, a plant or animal which lives within the body of its HOST species. Examples include tapeworms, liver flukes, and the amoeba *Entamoeba histolytica* (*see* AMOEBIC DYSENTERY). *Compare* ECTOPARASITE.

endosperm in botany, the fleshy material within a ripening seed that serves as food for the developing plant embryo. In ANGIOSPERMS (flowering plants), the endosperm develops as a result of so-called "triple fusion" between one of the two sperm cells in a grain of POLLEN and two POLAR BODIES (incomplete cells) thrown off by the female GAMETOPHYTE as it develops; hence the endosperm tissue is triploid, that is, each of its cells contains three sets of CHROMOSOMESS. In most DICOTYLEDONS, the developing plant embryo uses up the entire store of endosperm tissue before GERMINATION. In most MONOCOTYLEDONS, some endosperm tissue remains to nourish the sprouting plant. GYMNOSPERM seeds do not contain true endosperm, although the term is sometimes used for the tissue formed by the thickening of the gametophyte wall that precedes fertilization and that serves the same purpose for the gymnosperms as endosperm does for the angiosperms.

endothermic reaction a chemical REACTION that requires the addition of energy in order to take place. Endothermic reactions slow down and stop if the source of the energy feeding them is removed. The source of energy may simply be the heat present in the reacting substances. In such cases, if no energy is provided to the system during the reaction, the temperature of the reaction product will be lower than the temperatures of the reacting substances before they were placed together. *Compare* EXOTHERMIC REACTION.

endrin a CHLORINATED HYDROCARBON PESTICIDE, chemical formula $C_{12}H_8Cl_6O$. A so-called "hard" or BROAD-SPECTRUM PESTICIDE, endrin is an isomeric form of dieldrin (*see* ISOMER) that differs only in the means of attachment of its chlorine atoms. It is, however, about two to three times as toxic as dieldrin, and about five times as toxic as ALDRIN, another closely related COMPOUND. (An aldrin MOLECULE is essentially an endrin molecule without its OXYGEN atom.) Like its relatives, endrin bioaccumulates readily (*see* BIOLOGICAL MAGNIFICATION) and does not break down easily in the environment. Widely used from its introduction in 1959 to about 1970, it is now banned by the ENVIRONMENTAL PROTECTION AGENCY. Symptoms of acute overexposure are similar to those of aldrin, and include liver and kidney damage, convulsions, and respiratory failure. Chronic overexposure leads to symptoms similar to HEPATITIS. The chemical is equally dangerous when inhaled, ingested, or absorbed through the skin.

Energy, Department of United States government agency charged with overseeing federal activities related to energy production and conservation, including policy-making; research and development; regulation and siting of hydroelectric facilities (*see* HYDROELECTRIC POWER), nuclear generation facilities (*see* NUCLEAR ENERGY), electric transmission lines and NATURAL GAS pipelines, and other major power generation and transmission works; and the coordination and oversight of the energy-related activities of other government agencies. In addition, the agency has been given responsibility for nearly all energy-related national defense activities, including the management of the STRATEGIC PETROLEUM RESERVES and the control of atomic weapons production and waste disposal. It is the fourth largest federal agency in terms of land ownership but among the smallest of all cabinet-level agencies in terms of work force. A part of the EXECUTIVE BRANCH of the federal government, the department is headed by a cabinet-level officer (the secretary of energy) assisted by a deputy secretary, an under secretary, eight assistant secretaries, a general counsel, and an inspector general. Of these, the most important from an environmental standpoint are the under secretary and the five so-called *program assistant secretaries* (nuclear energy; fossil energy; conservation and renewable energy; defense programs; and environmental protection, safety, and emergency preparedness) who report directly to the under secretary. In addition, the department encompasses the FEDERAL ENERGY REGULATORY COMMISSION, a quasi-independent agency whose chairman reports directly to the secretary of energy.

HISTORY

Among the youngest of the major federal agencies, the Department of Energy was formed on October 1, 1977 under the terms of the Department of Energy Organization Act passed on August 4 of that year. Most of the offices, programs, and employees for the new department were transferred from the Interior Department, including the Federal Energy Administration, the Federal Power Commission, the Energy Research and Development Administration, the five regional power administrations (*see* POWER ADMINISTRATIONS FEDERAL), and portions of the BUREAU OF RECLAMATION. The Departments of Commerce, Housing and Urban Development, and Defense also gave up certain functions to the Department of Energy. *See also* ENERGY INFORMATION ADMINISTRATION; OFFICE OF CIVILIAN RADIOACTIVE WASTE MANAGEMENT; OFFICE OF ENERGY RESEARCH; and separate articles under most agency or office titles listed above.

energy budget an accounting of the energy flowing into and out of a system. The system may be living (an animal, a plant, a POPULATION of animals or plants), nonliving (a building, a motor, a manufacturing process), or a combination of the two (a nation, an ECOSYSTEM, the entire planet). In any case, the system must follow the first law of thermodynamics (*see* THERMODYNAMICS, LAWS OF), which states that the flow of energy into the system must exactly equal the flow of energy out. The preparation of an energy budget is useful for planning energy-conservation measures such as building weatherization, as well as being a necessary step toward the assignment of TROPHIC LEVELS and the determination of the needs of ORGANISMS within an ecosystem.

energy crisis *See* OIL EMBARGO.

energy flow in ecosystems, *see* ENERGY BUDGET.

Energy Information Administration agency within the U.S. Department of Energy (*see* ENERGY, DEPARTMENT OF) charged with collecting, collating, and distributing data on energy research. The administration provides a centralized location for government officials and the public to find up-to-date information on trends in energy technology, usage, conservation techniques, and similar matters. The agency's mission is directed toward a general readership rather than scientists or engineers. It does not initiate research or publish technical papers.

energy resources principally FOSSIL FUEL resources, such as coal, natural gas, and SYNFUELS (derived mainly from coal); nuclear power; geothermal energy; hydroelectric power (from dams and tidal movement); solar and wind energy; biomass and biofuels; and hydrogen gas. For the most part, RENEWABLE ENERGY resources, such as solar, wind, geothermal, biomass and biofuels, and hydrogen, have less widespread environmental impact than nonrenewable resources—fossil fuels and nuclear energy.

Energy Star Standards an ENVIRONMENTAL PROTECTION AGENCY program begun in 1995 to reduce the amount of electricity used in common home and office appliances, including computers and printers. For those products meeting the standards, a green "Energy Star" is awarded, which manufacturers can use in their advertising and on their packaging. Energy Stars may also be awarded to home builders whose houses and apartments are 30% more energy efficient than those meeting regular building code energy standards. The program has had modest success in garnering commercial cooperators.

energy tax *See* CARBON TAX.

engineering survey a survey made for the purpose of planning the construction of a specific structure on a specific site. Engineering surveys begin by establishing physical control points on the site and determining their precise relationship in space. These, combined with the data established by a TOPOGRAPHIC SURVEY, allow planners to know precisely where on the site the structure will be placed and thus determine the constraints that the topography, soils, etc. will make on the design. *Compare* CONSTRUCTION SURVEY.

Engineers, Army Corps of *See* ARMY CORPS OF ENGINEERS.

enrichment of ore, *see* BENEFICIATING.

enterotoxin a TOXIN that disrupts digestive processes, usually leading to diarrhea. In extreme cases, the disruption can be so severe that it is fatal. Common enterotoxins include those produced by SALMONELLA bacteria (*see* BACTERIUM) and by the Entamoebae (*see* AMOEBIC DYSENTERY).

entisol a soil in which the HORIZONS are largely undeveloped, that is, one in which little or no vertical differentiation has taken place. Generally, entisols are young soils in which development has not had time to proceed; however, some are simply too shallow for development, and in others the development has been arrested by climatic conditions. In the latter case, there may be an accumulation of SALTS or other minerals low in the soil.

entrainment the capture of a nonmoving substance by a moving one; also, the capture of a slowly moving substance or one with little mass by a more rapidly moving substance or one with considerably larger mass moving in another direction. Wind passing over a windbreak, for example, entrains some of the still air behind the windbreak, carrying it away; thus, the air pressure directly behind the windbreak is lower than that in the surrounding area.

entropy *See* THERMODYNAMICS, LAWS OF: *the second law.*

entry in logging or forestry, any stage in the management of a timber STAND that requires the removal of living material. Cruising would not be considered a stand entry (*see* TIMBER CRUISER); THINNING, RELEASE CUTTING, or (in the extreme case) the creation of a CLEARCUT would. Because each entry into a stand has certain unavoidable costs connected with it, both economic (labor, machine time) and environmental (soil compaction, wildlife disturbance), it is generally better management to plan as few entries per ROTATION as possible, although frequent entries on lands with a high SITE INDEX may be worthwhile if they will significantly increase the growth rate of the timber on the site.

entry cut in logging, the first free-cutting operation in a timber STAND during a given ROTATION.

environment the total setting in which a given object rests or a given action takes place, including all physical, chemical, biological, physiological, and psychological factors. The term can be modified and restricted by specifying the object or the action. Thus the *work environment* is the environment immediately surrounding a job to be done, including not only the walls of the room (or the lack of them) but lighting, temperature, humidity and noise levels, interactions with colleagues and other persons, air quality, presence or absence of superiors, and so forth; the *environment of a logging operation* would include not only the trees to be cut but the soil beneath them, the air above them, all other vegetation on the site, all affected wildlife, any recreational sites within earshot or line of sight, and so forth. The *immediate environment* is that directly surrounding the object or action; the *total affected environment* includes everything, however remote, that could be affected by the object or action. Used without a modifier, the term "environment" generally means the *natural environment,* that is, all things in a given area or region that are not human-made. *Compare* ECOLOGY. *See also* ENVIRONMENT, DIVISIONS OF.

environment, divisions of in ecology, a set of general categories into which those environmental factors affecting an animal or POPULATION of animals may be placed. Four divisions of the environment are generally recognized; these include FOOD, WEATHER, SHELTER, and *other animals*. *See* articles dealing with the various aspects of these topics, e.g., NUTRIENT; CLIMATE; HABITAT; COVER; COMPETITION.

Environmental Analysis Report ([EAR] Environmental Assessment [EA]) FOREST SERVICE document summarizing the findings of a study of the impact of a proposed action on the natural environment. The principal purpose of an EAR is to determine whether or not a formal ENVIRONMENTAL IMPACT STATEMENT will be required and to provide the documentation for that decision. It is prepared primarily for internal use by the agency itself, and is not generally released for public comment. (Exceptions: Forest Service policy requires a 30-day public comment period for EARs of three specific types: [1] those covering actions for which no precedent has been established; [2] those covering actions closely similar to previous actions that have required Environmental Impact Statements; and [3] those covering actions involving or affecting FLOODPLAINS or WETLANDS.) Depending on the action covered, the EAR may vary from a one- or two-page photocopied document to a printed and bound volume of 100 pages or more. The most important part of the document from a legal standpoint is the *finding*, specified either as a *finding of significant impact* (in which case an Environmental Impact Statement will be required) or as a *finding of no significant impact* (in which case an Environmental Impact Statement will not be required). It is only on the basis of these findings that an EAR, which is essentially an "in-house" document, may be successfully challenged in court.

Environmental Assessment *See* ENVIRONMENTAL ANALYSIS REPORT.

Environmental Career Center a service organization founded in 1980 to help people find careers related to the environment. Through its Environmental Partnership Program, it offers paid apprenticeships in natural resources management and environmental protection. The center works in partnership with universities and employers and provides career seminars for professional societies and agencies. Address: 100 Bridge Street, Suite A1, Hampton, VA 23669. Phone: (757) 727-7891. Website: ecc.cybros.net/ECC.

Environmental Careers Organization serves to promote, develop, and motivate professionals working in environmental vocations through placement, advisement, career research, and consulting. Founded in 1972, the organization has five regional offices in the United States and an extensive community of over 6,000 alumni. Address: 179 South Street, 3rd Floor, Boston, MA 02111. Phone: (617) 426-4375. Website: www.eco.org.

Environmental Caucus a loosely knit CAUCUS of members of Congress who cooperate with each other on environmental issues. The Environmental Caucus has its own WHIP system to alert members to environmentally related votes that may be coming up.

environmental corridor a narrow strip of relatively undeveloped land running through a developed region. *See* LANDSCAPE ECOLOGY.

Environmental Defense Fund (EDF) environmental activist organization, founded in 1967 to defend environmental quality through legal action, lobbying, and regulatory reform. EDF's primary focus is on toxic chemicals, air quality, energy, wildlife, and global environmental problems. Its membership consists largely of attorneys, economists and research scientists. Membership (1988): 50,000. Address: 444 Park Avenue South, New York, NY 10010. Phone: (212) 505-2100.

environmental economics *See* ECOLOGICAL ECONOMICS; ECOSYSTEM SERVICES.

environmental education in an academic context, the formal study of the environment in terms of the many disciplines that pertain to it. There is an emerging distinction between *environmental science* and *environmental studies* as an academic specialty, albeit with a great deal of overlap. Environmental science deals with the environment largely in terms of biology, ecology, geology and other earth and life sciences, and can include physics, chemistry, even astronomy and cosmology. Environmental studies programs in schools and colleges are more likely to approach the subject in terms of the social sciences and, often, the humanities. In both cases, the primary topics of environmental inquiry are covered, for example (in no particular order): the dynamics of natural systems, evolution, human population, energy sources and use, renewable natural resources, minerals and soils, food production, human health, air and climate, water resources and pollution, solid waste and hazardous materials, environmental economics, biodiversity. In the end, the basic discipline for an *environmental sciences* approach is, broadly speaking, ecology, and the basic discipline for an *environmental studies* approach is, broadly speaking, economics. In a Brown University project to provide a website listing for all college and university environmental studies and environmental sciences programs, more than 100 institutions of higher learning asked that their programs be included. The website address for the listing is www.brown.edu/Departments/Environmental_Studies/espgm.html.

Programs in Environmental Science and Environmental Studies in the United States (colleges and universities)

Institution	Department or Program
Alfred University	Environmental Studies
Allegheny College	Department of Environmental Science
Antioch College	Department of Environmental and Biological Sciences
Antioch New England Graduate School	Department of Environmental Studies
Arkansas State University	Environmental Sciences Ph.D. Program
Assumption College	Interdisciplinary Environmental Association
Bethel College	Natural Sciences—Environmental Science
Boston University	Center for Energy and Environmental Studies
Brown University	Center for Environmental Studies
Bucknell University	Environmental Studies
California State University at Hayward	Environmental Studies
California State University: San Francisco State University	Department of Environmental Studies
Carnegie Mellon	Civil and Environmental Engineering
Carroll College	Environmental Studies Program
Claremont McKenna College	Roberts Environmental Center Environment, Economics, and Politics Program
Clark University	Environmental School
Colby College	Environmental Research/Science
Columbia University	Center for Environmental Research and Conservation (CERC) Department of Earth and Environmental Sciences/ Biosphere2 Center
Colorado College	Environmental Science
Cornell University	Center for the Environment
Dartmouth College	The Environmental Studies Program
Davis and Elkins College	Department of Biology and Environmental Science
Delaware Valley College	Agronomy and Environmental Science
Denison University	Environmental Studies
Dickinson College	Environmental Studies
Drake University	Environmental Science and Policy Program
Drexel University	School of Environmental Science, Engineering and Policy
Duke University	School of the Environment Duke Law School: Law and the Environment
Duquesne University	School of Natural and Environmental Studies
Elizabethtown College	Environmental Science
Florida International University	Department of Environmental Studies
Georgia Tech	School of Civil and Environmental Engineering School of Public Policy: Environmental Policy
Gettysburg College	Environmental Studies Program
Grinnell College	Environmental Studies Concentration
Harvard University	Environment at Harvard
Hampshire College	Environmental Studies
Indiana University	School of Public and Environmental Affairs School of Law
John Carroll University	Environmental Studies
Johns Hopkins University	Department of Geography and Environmental Engineering
Juniata College	Environmental Science and Studies
Keene State College	Environmental Studies Program
Lake Superior State University	Environmental Science Environmental Chemistry
Lehigh University	Earth and Environmental Science
Louisiana State University	Agronomy
Louisiana Tech University	Environmental Sciences Program
Manchester College	Environmental Studies
Massachusetts Institute of Technology	Center for Environmental Health Sciences Center for Environmental Initiatives Civil and Environmental Engineering Department of Earth, Atmospheric, and Planetary Sciences
Massachusetts Maritime Academy	Marine Safety and Environmental Protection
McNeese State University	Biological and Environmental Sciences
Michigan Technological University	School of Forestry and Wood Products
Middlebury College	Program in Environmental Studies

Northern Arizona University	Center for Environmental Science and Education Program in Environmental Politics and Policy
New York University (NYU)	Environmental Health Sciences
Oberlin College	Environmental Studies
Ohio State University	Environmental Science Graduate Program
Oklahoma State University	Environmental Science Graduate Program
Oregon Graduate Institute	Department of Environmental Science and Engineering
Oregon State University	Environmental Sciences
Undergraduate Program	Environmental Sciences
Graduate Program	Environmental Soil Science, Masters of Science Program
Pace University Law School	Environmental Law Program
Pennsylvania State University	The Earth System Science Center] Environmental Pollution Control
Portland State University	Environmental Programs
Princeton University	Princeton Environment Institute
Rensselaer Polytechnic Institute	Environmental Studies Program
Rice University	The Environmental at Rice
Sam Houston State University	Environmental Science
San Jose State University	Environmental Studies
Slippery Rock University of Pennsylvania	Environmental Geosciences Geography/Environmental Studies
Sonoma State University	Department of Environmental Studies and Planning
Southampton College, Long Island University	Environmental Studies
Southern Illinois University at Edwardsville	Environmental Science Program
Southeast Missouri State University	Environmental Science Program
State University of New York (SUNY)	College of Environmental Science and Forestry
Stephen F. Austin State University	Environmental Science Division
Temple University	Environmental Studies
Texas A&M University	College of Agriculture and Life Sciences Plant Pathology and Microbiology
Towson State University	Department of Geography and Environmental Planning
Tufts University	Department of Urban and Environmental Policy
Tulane University	Environmental Studies Program
University of Alaska, Anchorage	Environmental Studies Program
University of Arizona, Tucson	The Institute for the Study of Planet Earth
University of California, Davis	Department of Environmental Studies Department of Environmental Horticulture
University of California, Los Angeles	Environmental Science and Engineering Doctoral Program
University of California, Santa Barbara	Donald Bren School of Environmental Science and Management Environmental Studies Program
University of California, Santa Cruz	Environmental Studies
University of Charleston	Environmental Studies Department
University of Cincinnati	Civil and Environmental Engineering College of Medicine, Department of Environmental Health
University of Colorado, Boulder	Cooperative Institute for Research in Environmental Sciences Center for Environmental Journalism
University of Delaware	Center for Energy and Environmental Policy
University of Denver	Environmental Science
University of Georgia	School for Environmental Design
University of Hawaii, Manoa	School of Ocean and Earth Science and Technology
University of Idaho	Environmental Science Program
University of Illinois, Springfield	Department of Environmental Studies
University of Iowa	Center for Global and Regional Environmental Research
University of Kansas	Environmental Studies
University of Kentucky	Natural Resource Conservation and Management
University of Maine	Marine Law Institute
University of Massachusetts, Amherst	Environmental Sciences Program Department of Forestry and Wildlife Management
University of Massachusetts, Boston	Environmental Sciences Program

University of Miami	Environmental Science Program
University of Michigan	School of Natural Resources and Environment
University of Montana, Missoula	Environmental Studies Environmental Organizing Semester
University of North Carolina at Chapel Hill	Carolina Environmental Program Institute for Environmental Studies
University of North Texas	Environmental Science
University of Oklahoma	Civil Engineering and Environmental Science
University of Oregon	Environmental Studies
University of Pennsylvania	Penn Environmental Group Master of Environmental Studies
University of Rochester	Department of Earth and Environmental Sciences Environmental Health Science Center
University of South Carolina	Baruch Institute for Marine Biology and Coastal Research School of the Environment
University of Southern California	Environmental Studies
University of Southern Maine	Environmental Science and Policy Program
University of Vermont	Environmental Program
University of Virginia	Department of Environmental Sciences
University of Washington	Program of the Environment (POE)
University of Wisconsin, Madison	Institute for Environmental Studies
University of Wyoming	Institute for Environment and Natural Resource Research and Policy
Virginia Commonwealth University	Center for Environmental Studies
Virginia Tech	Environmental Engineering and Science Environmental Programs at Virginia Tech
Western Michigan University	Environmental Studies
Williams College	Center for Environmental Studies
Wilson College	Environmental Studies
Yale University	Yale Forestry School

environmental engineering those engineering procedures that are specifically designed to minimize or mitigate the effects of a structure on its ENVIRONMENT, or to create an optimal environment for a specific action to take place. *Compare* ENVIRONMENTAL TECHNOLOGY.

environmental ethics the application of ethical considerations to nonhuman creatures, natural forms, and natural systems. Traditional ethics, as a philosophical study, deals with the way human individuals and societies interact with each other in terms of moral principles. Environmental ethics applies moral principles to the ways in which humans interact with nature, rather than with each other. Thus, just as there are inherent human rights—such as the Lockean life, liberty, and property—there are also inherent rights of nature. These would include the right of species to survive (i.e., not be extinguished by human agency), the right of natural systems to remain in balance, and even in extreme forms the right of individual rocks, trees, plants, landforms, and creatures not to be extinguished, harmed, or dislocated. *See also* DEEP ECOLOGY; LAND ETHIC.

environmental factor any factor in the ENVIRONMENT of an object or activity, especially one that will affect or be affected by the object or activity. Common environmental factors associated with the natural environment include weather, soil type, exposure to sunlight, drainage, slope, vegetal cover, and the size and SPECIES composition of resident animal populations. The costs and benefits associated with environmental factors may usually be classed as EXTERNALITIES.

Environmental Impact Statement ([EIS] Environmental Statement [ES]) a written document prepared by a federal agency, or by a private firm under contract to a federal agency, detailing the effects that an action proposed by the agency will have on the environment. Under the terms of the NATIONAL ENVIRONMENTAL POLICY ACT OF 1969 (NEPA), all "major Federal actions significantly affecting the quality of the human environment" require the preparation of an EIS, which *must* cover (in the words of the act):

> (i) the environmental impact of the proposed action, (ii) any adverse environmental effects which cannot be avoided should the proposal be implemented, (iii) alternatives to the proposed action, (iv) the relationship between local short-term uses of man's environment and the maintenance and enhancement of long-term productivity, and (v) any irreversible and irretrievable commitments of resources which would be involved in the proposed action should it be implemented.

The document must also follow guidelines established by the COUNCIL ON ENVIRONMENTAL QUALITY

(CEQ) as published in the FEDERAL REGISTER (40 CFR 1500–8), and any additional requirements imposed by the responsible agency.

FORMAT OF AN EIS

During the first decade of NEPA, Environmental Impact Statements were generally organized according to the five areas of coverage required by the act (listed above); however, revisions to the CEQ guidelines in mid-1979 established a new format that all agencies preparing EISs were strongly urged to follow, and nearly all EISs published since that time have conformed to these recommendations. The document begins with a *Summary* of the findings and recommendations, organized in the same manner as the EIS as a whole. This is followed by a *Table of Contents* for the document and a *Statement of Purpose and Need* indicating why the proposed action is being undertaken. The meat of the document is contained in the next section—*Alternatives, Including the Proposed Action.* This is a full description of several alternative courses of action, including the action proposed by the agency (the "preferred alternative"); a "no-action alternative" in which nothing is done; and a range of other possible actions that generally includes a "no-change alternative" that represents a straight continuation of present policies, an "amenities alternative" that emphasizes the protection of environmental values, and a "commodities alternative" that is designed for maximum economic production. The discussion of the alternatives is followed by a description of the impacted environment; a discussion of the expected environmental consequences of the various alternatives (this must fulfill the NEPA requirements for accountings of irreversible commitments of resources that cannot be avoided, adverse environmental impacts that cannot be avoided, and the relationship of the proposed action to both the short-term and long-term productivity and environmental health of the area the action impacts); a list of the preparers of the EIS, annotated with their qualifications; a list of those receiving the document, including other agencies, private organizations, and individuals; an index; and any required appendices, which usually will include a glossary, brief accounts of special studies undertaken during the preparation of the EIS, and discussions of agency policies that are particularly relevant to the action in question.

PREPARATION OF AN EIS

Preparation of an Environmental Impact Statement begins with the preparation of an ENVIRONMENTAL ANALYSIS REPORT (EAR) to determine if an EIS will be necessary. If the EA determines that an EIS must be prepared, a process called *scoping* takes place, in which the agency gathers input from other agencies and from the public concerning the potential impacts of the proposed action and uses this input (often referred to as "issues, concerns and opportunities," or ICOs) to determine what should be discussed in the EIS and how deep the discussion should be on each point. Data-gathering follows, usually through the use of an INTERDISCIPLINARY TEAM; this data is then sorted, analyzed, and written up as a *Draft Environmental Impact Statement (DEIS),* which is circulated to other affected agencies, to organizations, and to the public in general. Disagreements among agencies often show up at this point (for example, the FISH AND WILDLIFE SERVICE may object to plans for a dam proposed by the ARMY CORPS OF ENGINEERS). Finally, comments received on the DEIS are analyzed, and a *Final Environmental Impact Statement (FEIS)* is prepared. This final statement is required to contain responses to all substantive comments received by the agency on the DEIS.

VALUE OF THE EIS PROCESS

NEPA merely requires the preparation of an EIS; there is no "threshold value" of environmental damage that will require cancellation of a project if the EIS demonstrates that this damage will occur. An agency may be challenged in court on the adequacy of a particular EIS, but not on actions based on that EIS, as long as the actions stay reasonably close to the preferred alternative. The value of the EIS process thus lies not in the legal shackles it places on an agency, but in the close examination it forces the agency to make of the results of its decisions and the public light that is shed on the decision process, often resulting in a preferred alternative that is far sounder from an environmental standpoint than the action that might have be taken had the EIS process not been undertaken.

Environmental Law Institute an internationally known education and research organization in the area of pollution abatement and resource conservation. Founded in 1969, ELI researches important environmental issues and then educates professionals and the general public concerning these issues via positive change and constructive solutions. Address: 1616 P Street NW, Suite 200, Washington, DC 20036. Phone: (202) 939-3800. Website: www.eli.org.

environmental movement a post–World War II outgrowth of the CONSERVATION MOVEMENT, which expanded concerns about nature, natural history, and the management of natural resources to include air and water pollution, toxic substances (e.g., DDT) and toxic waste, population increases, nuclear energy, and URBAN SPRAWL, among other issues. Arguably, a major shift from a prewar emphasis on nature preservation and resource conservation came with postwar housing expansion, highway building, and the baby boom, all of which happened in the 1950s. Suddenly, "the environment" had radically changed, to the detriment of an urbanizing population. The growing alarm over postwar excesses was chronicled in a number of books published between the late 1950s and the late 1960s, including *The Exploding Metropolis* (1958), by the editors of *Fortune;* RACHEL CARSON'S *Silent Spring* (1962); STEWART LEE UDALL'S *The Quiet Crisis* (1963); and Robert and Leona Rienow's *Moment in the Sun* (1967). These books, among many others, gave rise to new laws governing pesticide use, water and air pollution, and most importantly, the NATIONAL ENVIRONMENTAL POLICY ACT of 1969. Then, in 1970, the first EARTH DAY took place. If before that date the words "environmental" and "movement" were never uttered in the same sentence, after it the conjoining of the terms became more and more frequent. Since that time the movement has gained great strength, with major organizations like the NATIONAL AUDUBON SOCIETY and the SIERRA CLUB growing exponentially, and newcomers such as the NATURAL RESOURCES DEFENSE COUNCIL and the ENVIRONMENTAL DEFENSE FUND becoming major forces in the field. By the 1990s the movement had reached so thoroughly into American life and thought that nearly three-quarters of the adult population considered themselves to be "environmentalists." The 1990s also saw a renascence of the earlier "grass roots" aspects of the movement, wherein independent, local organizations were formed at an increasing rate, in contrast to the continued growth of what came to be called "the majors." This trend is thought to be a healthy one, even by the larger organizations. Moreover, as environmental problems are becoming more and more international, the environmental movement is beginning to shift its emphasis from lobbying the U.S. Congress with practical analyses of the issues to a worldwide concern with ENVIRONMENTAL ETHICS, particularly as ethical perceptions are informed by religious doctrine. During the early part of the 21st century, many of those in the environmental movement believe, their greatest opportunity will lie in making common cause between RELIGION AND THE ENVIRONMENT.

Environmental Policy Institute (EPI) environmental research and lobbying center, founded in 1974 to influence federal laws and regulations concerning water-resource development, energy conservation, toxics regulation, ground-water protection, and nuclear power. It also works to see that laws passed by Congress in these fields are being properly implemented. Originally the research arm of the Environmental Policy Center, it has taken over and expanded the center's role. Address: 218 D Street SE, Washington, DC 20003. Phone: (202) 544–2600.

Environmental Protection Agency (EPA) independent federal (U.S.) agency charged with administering federal laws dealing with pollution control and cleanup, especially those regarding air pollution, HAZARDOUS WASTE, MICROCONTAMINANTS, and the protection of surface waters and GROUNDWATER. The administrator of the EPA reports directly to the president: s/he is assisted by a deputy administrator, a chief of staff, and a number of assistant administrators, of which the most important are those in charge of the four program offices: Air, Noise and Radiation; Water; Solid Waste and Emergency Response; and Pesticides and Toxic Substances. An assistant administrator for research and development coordinates all research programs within the agency, as well as overseeing research grants to universities, corporations, and individuals. An associate administrator for legal and enforcement counsel and general counsel coordinates all legal actions undertaken by the agency, including those initiated by the program offices, and an associate director for policy and resource management is responsible, among other things, for setting standards and regulations. The EPA operates 11 regional offices, one for each of the 10 STANDARD FEDERAL REGIONS and a separate Great Lakes Program Office for the Great Lakes Basin. Each of these is headed by a regional director.

HISTORY

The Environmental Protection Agency was created on December 2, 1970, by President Richard Nixon, who combined the NATIONAL AIR POLLUTION CONTROL ADMINISTRATION, the FEDERAL WATER QUALITY ADMINISTRATION, the Bureau of Solid Waste Management, and various programs relating to radiation control, pesticide regulation, and environmental research from throughout the federal government to form the new agency. This structure was subsequently ratified by Congress, which has since specifically charged the agency with the administration of a number of pollution-control laws, among them the CLEAN WATER ACT; the SAFE DRINKING WATER ACT; the CLEAN AIR ACT; the COMPREHENSIVE ENVIRONMENTAL RESPONSE, COMPENSATION AND LIABILITY ACT; and the TOXIC SUBSTANCES CONTROL ACT. Originally a highly activist agency—it is required by law to analyze all federal proposals that impact the environment, and to publish its analyses of these proposals when it finds that the impact will be negative—the EPA became scandal-ridden and moribund under Administrator Anne Gorsuch Burford in the early years of the Reagan Administration, but has since recovered its original purpose. Headquarters:

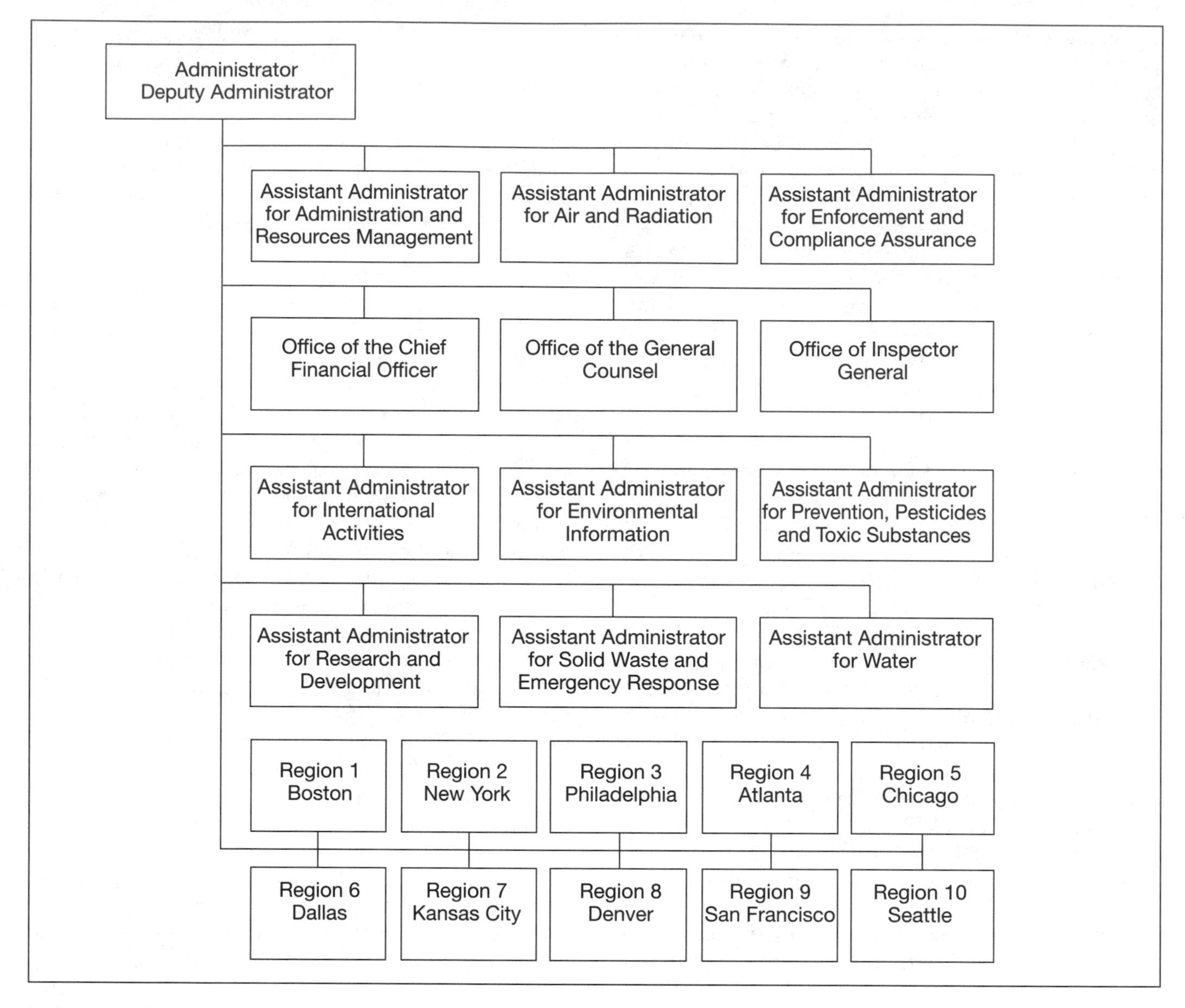

Environmental Protection Agency organizational structure

401 M Street SW, Washington, DC 20460. Phone: (202) 564–1657.

environmental resistance the set of factors that act to restrict the growth of a POPULATION of animals or plants or of a biotic COMMUNITY. Environmental resistance is measured indirectly, by calculating the rate of unrestrained growth of the population or community from birthrate and old-age-mortality data (*see* BIOTIC POTENTIAL) and subtracting from it the actual observed rate of population growth. *See also* CARRYING CAPACITY.

Environmental Statement *See* ENVIRONMENTAL IMPACT STATEMENT.

environmental stewardship *See* STEWARDSHIP.

environmental technology the body of laboratory testing procedures involved with determining the impacts of an object or action on the environment. *Compare* ENVIRONMENTAL ENGINEERING.

enzyme in biology, an ORGANIC COMPOUND that serves as a CATALYST in metabolic reactions (*see* METABOLISM). Enzymes are complex PROTEINS that are highly specific to particular reactions. They have little or no effect on others. Most apparently operate by temporarily binding one or more of the MOLECULES involved in the reaction, a process that takes place at specific locations on the enzyme molecule known as *active sites*. Interference with enzyme production or operation is one of the principal means by which POLLUTANTS and PATHOGENS cause illness and death in living ORGANISMS.

EO *See* EXECUTIVE ORDER.

EPA *See* ENVIRONMENTAL PROTECTION AGENCY.

epeirogeny in geology, a movement upward or downward of a broad portion of the earth's crust, with little or no deformation of rock STRATA. It was once thought that periods of epeirogeny alternated with periods of OROGENY in the uplift of continents, and that they were the result of different processes (*see* DIASTROPHISM). Today, it is believed that epeirogeny and orogeny go on simultaneously as a result of a single process, the collision of continental plates (*see* PLATE TECTONICS); however, the term remains useful as a means of distinguishing between grades of uplift.

ephemeral species (ephemeral) a SPECIES of plant or animal with an extremely short active life span, especially one that appears only during or immediately after rains in a DESERT or other dry region. Desert ephemerals generally survive long periods of drought as dormant seeds or eggs that sprout or hatch only when moistened. Some ephemeral plants, however, survive as rootstocks (TUBERS, BULBS or CORMS) that put forth shoots only in wet weather, and some ephemeral animals bury themselves in the mud of drying-up ponds and drastically slow their life processes down, ready to burst into a frenzy of activity when the pond fills up again. *See also* ESTIVATION.

ephemeral stream in geology, a stream that flows only during or immediately after a rainstorm. The bed of an ephemeral stream lies entirely above the WATER TABLE, and serves only as a conduit for RUNOFF. *Compare* INTERMITTENT STREAM; PERENNIAL STREAM.

epidemiologic test an attempt to determine the causative agent of a disease by a statistical matching of outbreaks of the disease with potential causative agents. Epidemiologic tests do not establish a direct link between the agent and the disease, but they can demonstrate strong circumstantial proof that such a link exists. Example: the statistical correlation between outbreaks of EMPHYSEMA and the amount of PARTICULATES in the air the emphysema victims are breathing.

epilimnion the zone of well-mixed, well-oxygenated and relatively warm water extending downward from the surface of a lake or other large body of water. It lies above the HYPOLIMNION and is separated from it by the THERMOCLINE. Mixing within the epilimnion is done principally by the wind and secondarily by convection currents set up through the different rates of heating of shallow and deep waters. Beginning as a narrow band just below the surface early in the spring, the epilimnion grows downward through the warm months. If the lake is shallow enough, it will reach the bottom by autumn, in which case the thermocline and the hypolimnion will vanish (*see* THERMAL STRATIFICATION). Because of mixing downward from the surface, the epilimnion contains more DISSOLVED OXYGEN than the hypolimnion; however, since dissolved NUTRIENTS recycled from the sediments in the lake bottom cannot cross the barrier of the thermocline, the nutrient content of the epilimnion is generally less than that of the hypolimnion, especially in large, deep lakes. *See also* LAKE; FIRST-ORDER LAKE; SECOND-ORDER LAKE; THIRD-ORDER LAKE.

epiphyll an EPIPHYTE that grows on the leaves of other plants.

epiphyte a plant that grows on the outside of another plant, utilizing it for support but not parasitising it (*see* PARASITE). Some simple plants such as LICHENS and MOSSES are incidental epiphytes, growing equally well on a tree trunk, a stone, or the forest floor. More complex plants may be specifically adapted to life as epiphytes. Tropical orchids, for example, often grow high in the rain forest CANOPY in order to obtain sufficient light. Their roots ("air-roots") dangle in the air, taking advantage of the humidity produced by the TRANSPIRATION of the rain forest trees. Since they are not in contact with the soil, obtaining minerals can be a problem for epiphytes. Many have solved it by becoming carnivorous, developing specialized leaves that trap and digest small insects (*see* CARNIVOROUS PLANT).

epizootic referring to the simultaneous death of a large number of individuals within a species.

equilibrium in ecology, a relatively steady state achieved when environmental conditions are stable. The population of a given species, for example, may be said to be in equilibrium when the reproductive rate is balanced by the death rate due to age, predation, or natural disaster. *See* DISTURBANCE.

erosion the removal of sand, soil, rock fragments, and other mineral materials from the surface of the land by the forces of wind, water, and gravity. Erosion that takes place under natural conditions is called *geological erosion* and is responsible for the shaping (and ultimate removal) of hills and mountains, the forming of valleys and canyons, and the sculpting of shorelines and river channels. *Accelerated erosion* takes place when humans alter the environment in such a way as to increase the rate at which the land surface may be removed. In addition to these two broad categories, erosion can be classified in a number of different ways. *Shoreline erosion* is the removal of sand and stone from lakeshores and ocean coastlines. It generally takes place laterally along the shore (*see* LONGSHORE CURRENT; SHORE DRIFT), although material may also be carried away from the shore perpendicularly toward deep water. *Bank erosion* is similar to shoreline erosion, but takes place along running water. *Glacial scouring* is erosion caused by glacial ice: it often remodels the planet on a very large scale (*see* GLACIER: *the work of glaciers;* GLACIAL LANDFORMS; CONTINENTAL GLACIER). *Hillslope processes* is the collective name for erosion that moves surface materials from the upper part of a slope to the lower part. It may include both grain-by-grain movement and large-scale transposition (*see* MASS WASTAGE). The removal of soil from cultivated fields and pastures is called *soil erosion,* and is always considered to be artificially accelerated.

CAUSES OF EROSION

Erosion may be caused by wind, water, or mechanical disturbance. Wind erosion tends to selectively remove sand-through silt-sized particles; particles smaller than silt adhere to each other and resist erosion, while materials larger than sand cannot be moved by the wind except under rare circumstances. Most wind-eroded materials move by *saltation*—bouncing along the surface—although the finer material will be suspended in the air, and coarser material may simply roll along the ground, a process known as *surface creep.* Shore drift, bank erosion, and the downcutting of stream channels are all forms of water erosion. In addition, there are four types of water erosion that contribute to soil erosion and hillslope processes, including *sheet erosion* (the removal of material evenly from the surface of a slope); *rill erosion* (the creation of a multitude of tiny channels, called *rills,* which constantly change course); *gully erosion,* in which a number of rills coalesce into a semipermanent ditch or GULLY down the hillside; and *splashing,* which takes place when raindrops hit a bare soil surface, scattering the particles downslope. Mechanical erosion is a result of the displacement of surface materials by vehicles or by the feet of animals, including people (*see* TRAMPLING DISPLACEMENT).

COSTS OF EROSION

Erosion has some obvious costs, including the destruction of roads and trails, the removal of TOPSOIL from fields and pastures, the undercutting of shoreline developments (homes, parking lots, businesses, recreation facilities, etc.), and the filling of reservoirs with silt deposits, rendering them unable to perform water-storage functions. Less obvious costs include the transport of pollutants that become attached to the eroding particles (*see* ADSORPTION; ABSORPTION), the added costs to water-treatment facilities that must filter the eroded materials out of water they provide for drinking and industrial uses, and the reduction of fertility that occurs due to the selective transport of the soil particles with the most NUTRIENTS within them, which results from their size range and position in the upper stratum of the soil. (Studies by soil scientists have indicated that the first materials eroded from a patch of recently bared soil include from two to 20 times the nutrients of those that remain behind.) These costs are much greater under conditions of accelerated erosion than under geological erosion; cultivated soil erodes at rates up to 15,000 times faster than soil that retains its natural plant cover.

CONTROL OF EROSION

Geological erosion is largely uncontrollable, and should be "controlled" by choosing land uses that will not be damaged by it, such as locating developments back far enough from shoreline bluffs that normal shoreline processes do not cause the collapse of the bluffs directly under the developments. Accelerated erosion, on the other hand, can be greatly reduced or even eliminated through proper land-use practices. These include CONSERVATION TILLAGE; the construction of windbreaks on the windward side of fields; the retention of as much vegetation (ground cover) as possible on erodible sites (and the planting of vegetation on CUTBANKS and other spots laid bare by necessity); and, where other means are impractical or impossible, the stabilization of streambanks, cutbanks, and shorelines with concrete or rock structures (*see* DOLO; GROIN; RIPRAP; GABION BASKET). Such structural means of erosion control should be limited to use only when all else fails, as they usually have large and unpredictable effects on erosion patterns downslope, downshore, or downstream. *See also* SOIL CONSERVATION.

erosion pavement *See* DESERT PAVEMENT.

erratic in geology, a loose rock—often of fairly large size—that differs in type and/or composition from the BEDROCK on which it rests or the rock of nearby outcrops. Erratics result from transport by GLACIERS or ICEBERGS. If the erratics carried by a vanished glacier can be traced to the outcrops or mountainsides from which they were derived (by comparison of chemical composition and stratigraphic structure; *see* STRATUM), it may allow the direction of flow of the glacier to be mapped with some precision. Erratics may be carried some distance: some have been traced to sources as far as 500 miles from their current position.

Escherichia coli (E. coli) a gram-negative (*see* GRAM STAIN), anaerobic (*see* ANAEROBIC BACTERIA) rod BACTERIUM commonly found living as a COMMENSAL SPECIES in the intestinal tract of mammals, including humans. Though exposure to unfamiliar strains of it has been implicated as the principal cause of "travelers' diarrhea," *E. coli* is normally harmless and may even make a positive contribution as a DECOMPOSER of waste products. Its principal environmental interest is as an INDICATOR SPECIES for polluted water. Since it is found in almost all mammalian FECES, and since it does not survive well outside the gut, its presence in water correlates extremely well with the presence of recent fecal contamination. It is also used as the principal vehicle for a common screening test for CARCINOGENS. *See* COLIFORM COUNT; AMES TEST.

esker a long, sinuous ridge of sand and gravel, deposited within a tunnel formed by a stream flowing beneath the surface of a glacier and remaining standing as a mold of the stream tunnel after the glacier melts away. Eskers are a relatively common feature of recently glaciated landscapes. *See also* GLACIAL LANDFORMS; CONTINENTAL GLACIER.

ester in chemistry, any ORGANIC COMPOUND created by a reaction between an ACID and an ALCOHOL. Esters are roughly analagous to inorganic SALTS. A large and extremely important group of chemicals, esters include the chemicals responsible for the flavors and fragrances of fruits and flowers; the LIPIDS (fats and oils); and many drugs and related compounds, including aspirin, oil of wintergreen, and nitroglycerin. POLYMERS formed from esters, called *polyesters,* are used to make fabrics, plastics, and artificial rubbers. Most esters can be recognized from their chemical names, which are normally formed by adding -yl to the root of the alcohol's name and -ate to the root of the acid's name. Thus, for example, oil of wintergreen, formed of methanol and salicylic acid, has the chemical name methyl salicylate.

estivation (1) in zoology, a period of dormancy or extreme torpor during hot, dry weather, used by many insects and some amphibians and reptiles as a means of surviving the time between desert rainfalls in the summer. *Compare* HIBERNATION. *See also* EPHEMERAL SPECIES.

(2) in botany, the arrangement of floral parts within an immature bud.

estrus the period when a female animal is sexually receptive ("in heat"). Estrus corresponds to RUT in the male. The *estrus cycle* is the hormone-driven cycle of reproduction in the female, consisting of preparation for pregnancy, ovulation (formation of the egg or eggs), sexual receptivity, and, if pregnancy does not occur, reabsorption or sloughing off of the egg or eggs and the uterine lining—events that occur on a cyclical basis, usually either approximately once a month or once or twice a year.

estuarine zone *See under* ESTUARY.

estuary broadly speaking, the zone at the mouth of a river where the river waters mix with the RECEIVING WATERS. In the restrictive sense more generally used, the term *estuary* means the complex of WETLANDS that usually develops within this zone of mixing, often referred to as the *estuarine zone.*

DEVELOPMENT AND CHARACTERISTICS OF ESTUARINE WETLANDS

Because moving water can carry more sediments than still water can, rivers deposit sediments at their mouths (see DELTA), and as these sediment deposits build up, the river mouths become broad and shallow. Sheltered within the mouths of the rivers, these shallows are protected from wave action, encouraging the establishment of emergent plants (*see* EMERGENT MACROPHYTE). These plants form a physical barrier to the flow of the water, slowing it down further and causing even more sediment deposition; thus the wetland becomes self-perpetuating, healing itself when breached and even growing upstream. The presence of the nutrient-rich sediments and the shallowness of the waters (meaning that PHOTOSYNTHESIS can take place all the way to the bottom and that no THERMOCLINE can ever develop to block nutrient transport) encourages the growth of PLANKTON in these estuarine waters, and the high plankton numbers draw high numbers of fish. Fish also find the estuary a good breeding ground due to the shelter from waves and (perhaps to an even greater extent) to the subtle shifts in water chemistry as the river water mixes with the receiving water. (This shift in chemistry is obviously greatest in salt-water estuaries, with the gradation of salinity through the estuary from low at the landward end to high at the seaward end, and the constant daily shifts of salt water in and out with the tides. However, there is also generally a shift in chemistry from river water to lake water, especially in terms of the types and numbers of dissolved IONS and SUSPENDED SOLIDS.) The high numbers of fish in turn support high numbers of predatory mammals, reptiles, and birds, making the estuary the most productive of all ECOSYSTEMS. It is not uncommon for a temperate-zone salt-grass estuary to produce 20,000 pounds of BIOMASS per acre per year—roughly 33 times the productivity of a typical dry grassland.

ethanol *See under* ALCOHOL.

ethylene in chemistry and biology, a light, colorless, flammable gas, chemical formula C_2H_4, found in ripening fruit, where it apparently functions as one of the hormones that speed the ripening process. It is also manufactured commercially by a number of processes, the most important of which is the catalytic cracking of PETROLEUM (*see* CATALYST; DESTRUCTIVE DISTILLATION). It has a multitude of industrial uses, including fueling welding torches and serving as a raw material for the manufacture of plastics ("polyethylene") and other organic products. Small amounts are produced by the burning of KEROSENE; hence, a kerosene stove burning in a room where fruit is ripening will speed the ripening of the fruit. It is nontoxic, although large amounts cause drowsiness and, eventually, death through asphyxiation. Lethal concentration (in mice): 950,000 ppm.

etiolation in botany, a condition of green plants marked by long, spindly stems, poorly-developed leaves, and paleness from lack of CHLOROPHYLL. Etiolation is caused by incomplete PHOTOSYNTHESIS, and results from attempts by the plant to grow in darkness or semi-darkness.

etiologic agent the agent causing a disease or disease-like condition. An etiologic agent can be a MICROORGANISM (a BACTERIUM or VIRUS), a chemical POLLUTANT, or an environmental condition (such as either too much or not enough sunlight). Etiologic agents that cause specific conditions often have specific names (*see,* e.g., CARCINOGEN; MUTAGEN; PATHOGEN).

ET rate *See* EVAPOTRANSPIRATION.

eukaryote in biology, an ORGANISM made up of eukaryotic CELLS, that is, one in which each cell making up the organism contains a true nucleus. Eukaryotes make up four of the five KINGDOMS of living things currently recognized by biologists. Only the MONERA (bacteria and blue-green algae) are not eukaryotic. (VIRUSES have no cell structure at all, and there is some question as to whether they can qualify as living organisms.) *See* CELL: *structure of cells.*

European elm bark beetle *See* DUTCH ELM DISEASE.

eutrophication in limnology, an increase in the PRIMARY PRODUCTIVITY of a lake, pond, bay, or other body of water, especially an increase that is sufficient to shift the trophic state of the body of water from oligotrophic or mesotrophic to eutrophic (*see* TROPHIC CLASSIFICATION SYSTEM). Under natural conditions, eutrophication is an extremely slow process, normally requiring a decrease in the depth of the lake due to infilling by SEDIMENTS or erosion of the outlet, or both. The decreased depth allows sunlight to penetrate to more of the bottom, warming the water and allowing the development of rooted aquatic plants (*see* MACROPHYTE). It also means that the oxygen-rich layer of water near the surface is a higher proportion of the total water mass. The increase in sediments also usually means an increase in available NUTRIENTS, as these are primarily carried into the water bound to the sediment particles. Man-caused eutrophication (*cultural eutrophication*), on the other hand, can be quite rapid. Some of the major causes of cultural eutrophication include the dumping of SEWAGE and other nutrient-rich substances into the lake waters; increases in sedimentation due to logging, road building, or agricultural activities; the use of chemical FERTILIZERS on fields, lawns, or forests from which the RUNOFF will enter the lake, either directly or through its inlet streams; concentrated livestock grazing, causing a buildup of animal wastes, which are high in nutrients, in and on the soil of the lake's WATERSHED; and increases in water temperature due to power-plant cooling, reservoir construction, deforestation, or other activities which either heat the water directly or increase its exposure to the sun. *See also* LIMITING NUTRIENT; WATER POLLUTION.

evaporation the conversion of a substance from the liquid state to the gaseous state. Evaporation takes place when a MOLECULE of liquid gains enough energy (in the form of heat) to escape from the liquid into the atmosphere above it. The amount of heat required, called the *heat of vaporization,* varies according to the composition of the liquid (examples: water, 2,258 joules/gram [15,354 calories/ounce]; ammonia, 1,368 joules/gram [9,302 calories/ounce]). Since the heat energy absorbed by an evaporating molecule is used to change its state rather than raise its temperature, it is called *latent heat.* When evaporation is taking place rapidly enough, significant amounts of heat may be tied up as latent heat, causing the surrounding temperature to drop. This is why the air is usually cooler after a rainstorm or beneath a tree (*see* EVAPOTRANSPIRATION).

ENVIRONMENTAL EFFECTS OF EVAPORATION

The evaporation of toxic liquids is a minor but important pathway for these substances into the environment. It is one of the principal sources of the chemicals in TOXIC PRECIPITATION and may cause health problems for those close enough to the source to breath the vapors in concentrated amounts. The most important environmental role of evaporation, however, is the part it plays in the distribution of water and heat through the atmosphere and over the surface of the earth. Water molecules evaporated from one location are carried by the winds to another, where they condense and precipitate, not only increasing the amount of water at the new location but releasing the latent heat absorbed when they evaporated. The presence of the winds that do this carrying is in itself largely due to the differential air temperatures and pressures caused by evaporation. *Compare* SUBLIMATION. *See also* HYDROLOGIC CYCLE; WEATHER; CLIMATE; FOG.

evaporite basin in geology, a dry lakebed from which the water has evaporated rather than being drained away. Evaporite basins develop in closed WATERSHEDS with no outlet to the sea, such as those of the GREAT BASIN of the western United States. Their soils are high in SALTS and MINERALS that have been deposited out of SOLUTION by the evaporating water, and will not grow plants well (*see* SHADSCALE DESERT). *See also* ALKALI FLAT.

evapotranspiration the sum of EVAPORATION and TRANSPIRATION, that is, the total amount of water vapor released to the atmosphere from all sources at a specific site. The rate of evapotranspiration for a given area at a given time is known as the *ET rate;* it varies greatly with temperature, wind speed, type of vegetation, and relative humidity (*see under* HUMIDITY).

even-age management in forestry, any combination of management techniques that is designed to lead to a forest composed of EVEN-AGE STANDS. The principal tool of even-age management is the CLEARCUT (or PATCH CUT), although SEED TREE CUTTING or the SHELTERWOOD SYSTEM may be used as well. In any case, an aggressive program of reforestation needs to be followed in order to make certain that the area cleared for REGENERATION does indeed regenerate. Even-age management is the preferred method for DOUGLAS FIR and other SHADE-INTOLERANT SPECIES—which grow back poorly under other systems—and it has some economic benefits for other species as well due to its area-intensive nature, meaning that management practices can be concentrated in a small area and practiced intensively, increasing the amount of management that can be done for the same amount of money. The mosaic of even-age stands that results, however, is usually not ecologically comparable to a natural forest.

even-age stand a STAND of timber in which all the trees are approximately the same age. In the case of natural stands, "approximately the same age" means that all ages fall within 20 years of each other; in the case of managed stands, it means that the age difference between the oldest and youngest trees in the stand is less than one-fourth the ROTATION AGE. Even-age stands are characteristic of SECOND GROWTH forests and of forests that grow up following natural catastrophes such as fires and avalanches. Because all the trees in an even-age stand have experienced the same environmental stresses at the same points in their lives, they are more alike in their response to new stresses than are trees in a multi-age stand and are thus more susceptible to epidemics and other causes of mass death; however, because they are of a nearly uniform size and condition, they are considerably simpler to harvest and mill. *See also* MONOCULTURE.

even-flow management *See* NON-DECLINING EVEN FLOW.

Everglades *See* FLORIDA EVERGLADES.

evolutionary ecology, evolutionary biology the study of the evolutionary process, including genetic mutation, natural selection, and adaptation as they contribute to speciation. Because the rate of extinction of species, particularly in the tropics, has been so rapidly increasing, evolutionary ecology and biology has become essential to the understanding of biodiversity and the maintenance of stable ecosystems.

exchangeable base in soil science, any of the four light metallic ELEMENTS, calcium, magnesium, sodium, or potassium, which, though technically acidic (that is, which form cations rather than anions when included in ionizing COMPOUNDS; *see* ION), commonly form compounds which, when dissolved in water, increase the number of hydroxide (OH^-) ions present and thus raise the PH. All four are necessary to plant growth. They are called "exchangeable" because their cations, when adsorbed to CLAY particles (*see* ADSORPTION), go through a continuous exchange process, with some cations being released while others are adsorbed, thus assuring a steady supply of ions to be taken up by the roots of plants. *See* also BASE SATURATION.

exclusive farm use zoning a land regulation strategy aimed at maintaining operating farms and farmland in the face of URBAN SPRAWL and other changes in land-use patterns. Originating in Oregon in 1973, exclusive farm use zoning requires that lands so designated can be sold only for continued farm use and that, further, they cannot be subdivided into parcels any smaller than the area of land needed to carry on the kind of commercial farming operations that are prevalent in a given area. Of greatest use in or near rapidly expanding metropolitan areas, the exclusive farm use approach permits the preservation of traditional land uses and local economies and is specially useful in protecting specialized and relatively small-scale farming—grape vineyards, orchards, flower farms, and the like—as opposed to commodity-crop agriculture (wheat, soybeans, corn) that ordinarily requires quite large acreages at some remove from the path of urban development.

executive branch the branch of government charged with putting laws into action, as compared to the *legislative branch* (Congress), which passes the laws, and the *judicial branch* (the courts), which oversees their enforcement and rules on their Constitutional validity. The executive branch is headed on the federal level by the president and on the state level by the governor, and consists principally of regulatory and management agencies that are organized according to the activity regulated or managed into a relatively few large departments. The heads of these departments (generally termed *secretaries*) form the president's or governor's *cabinet*. Executive-branch agencies have no power to make laws; however, the regulations they design in order to carry out their legislatively set duties generally have the force of law, considerably blurring the distinction between the executive and legislative branches. Major executive-branch agencies on the federal level that deal with environmental regulation and management include the Department of the Interior (including the BUREAU OF LAND MANAGEMENT and the BUREAU OF RECLAMATION; *see* INTERIOR, DEPARTMENT OF); the Department of Agriculture (including the FOREST SERVICE and the SOIL CONSERVATION SERVICE; *see* AGRICULTURE, DEPARTMENT OF); the Department of Energy (*see* ENERGY, DEPARTMENT OF); the ENVIRONMENTAL PROTECTION AGENCY; and the Department of Defense, including especially the ARMY CORPS OF ENGINEERS.

executive order (EO) a regulation or policy directive issued by the president or by a state governor rather than by a federal or state agency or legislative body. Unless otherwise specified, executive orders are binding on all agencies of the EXECUTIVE BRANCH; however, they do not apply to the legislative branch (Congress) or the judicial branch (the courts), and they have no force in the PRIVATE SECTOR (although they may regulate private behavior on public land). Executive orders have often been used for environmental purposes, including the establishment of NATIONAL MONUMENTS, the regulation of the use of OFF-ROAD VEHICLES in parks and on other public lands, and the restriction of PREDATOR CONTROL on federal and state rangelands.

exhaustible resource *See* NONRENEWABLE RESOURCE.

exothermic reaction a chemical REACTION in which energy is released, usually in the form of heat. The energy released is known as the *heat of reaction* and is indicated by the symbol ΔH. Exothermic reactions are usually self-perpetuating (that is, once started, they will continue as long as the reacting compounds are present) and result in more stable COMPOUNDS than those with which the reaction began. The burning of CARBON in OXYGEN, as in a wood fire, is a common example of an exothermic reaction. *Compare* ENDOTHERMIC REACTION.

exotic species a SPECIES of plant or animal not native to the region in which it is found. Exotic species are often brought into a region by human activity, either accidental or purposeful; however, they may also be carried in by natural forces such as storms or floods. *Compare* ENDEMIC SPECIES; ABORIGINAL SPECIES.

exponential growth in population studies, an increase at an increasing rate due to the multiplication of the whole by a fixed percentage in a given time. Thus, if a population is growing by 10% per year, a population of 1,000 would, after one year, total 1,100. But because the 10% is applied to the new total, the growth does not rise in a straight line (linear growth). The next year the population would total 1,210; the year after 1,331, and so forth. Shortly after the seventh year, the original figure would be doubled. In 20 years, the population (assuming no deaths or other subtractions) would be 6,728. An oft-used example to illustrate the power of exponential growth (albeit at a much higher factor) is that of a lily pond, which on the first day has one lily pad, on the second day, two, on the third day four, and so forth until on the 29th day, half the surface of the pond is covered by lily pads. The question is, how much of the surface will be covered on the next day? The answer, of course, is all of it. *See also* POPULATION EXPLOSION.

exposure (1) in mountaineering, a term used to indicate the risk of a fall, especially a fall with fatal consequences. It is a combination of the likelihood of slipping and the probable distance of a fall, and increases with both the height and steepness of the climbing face.

(2) in the environmental sciences, the direction a slope faces; e.g., *northern exposure* indicates a north-facing slope. *See also* ASPECT.

(3) in medicine, an obsolete term for HYPOTHERMIA.

external benefit *See under* EXTERNALITIES.

external cost *See under* EXTERNALITIES.

externalities (**economic externalities**) in economics, those COSTS and BENEFITS of an economic transaction that do not accrue directly to the producer or consumer of the good or service being transacted but are borne instead by those who are not parties to the transaction. Externalities do not affect supply and demand curves or profit-and-loss calculations, and thus are not accounted for by classical economic theory. Their importance has been fully recognized only since about 1960. *See* MAXIMUM SOCIAL WELFARE; MATERIALS BALANCE APPROACH.

SIGNIFICANCE OF EXTERNALITIES

Many—if not most—environmental problems are the result of the tendency of economic systems to ignore externalities. Because they do not show up on a balance sheet, externalities seldom effect corporate decision making. Environmental damages such as pollution, EROSION and soil depletion on public land, and the disappearance of wildlife HABITAT and of opportunities for DISPERSED RECREATION are almost exclusively external costs. Dumping industrial EFFLUENT in a river, for example, does not represent a cost to the plant producing the effluent. The costs of this dumping are borne by the fish and other living things that inhabit the river water, by the animals and people who drink it, and by the recreationists who can no longer eat the fish they catch or enjoy the spectacle of a clean, unpolluted river. External benefits arise in the same way. A small CLEARCUT, for example, may be made strictly to get the most wood production out of a timber STAND, but may nevertheless benefit deer (and deer hunters) by increasing the amount of available FOREST-EDGE HABITAT. However, external costs usually far outweigh external benefits.

EXTERNALITIES AND ENVIRONMENTAL PROTECTION

The problem of environmental protection often boils down to finding ways to internalize as many external costs as possible, that is, to make certain that the costs associated with pollution, environmental degradation, etc., become costs of doing business for the firm. There are two principal approaches to this. *Regulation* sets standards for pollutant release levels, plant and waste dump design and siting, and so forth. As long as the standards are met the externalities being regulated remain externalities, but they become sharply internalized in the form of plant closures, fines, and possible jail terms if the standards are exceeded. *Taxation* sets up free structures—effluent charges, siting fees, and so on—which appear directly on a company balance sheet and can be dealt with like any other cost. Attempts to maximize profits by the firm will then lead it to minimize these costs, that is, to reduce or eliminate effluents and to locate facilities where siting fees are low. Critics of taxation as a means of controlling externalities contend that the fees involved give firms the right to purchase a "license to pollute," while critics of regulation point out that the fact that the costs of pollution remain external until the regulations are exceeded makes it difficult for firms to see them as internal costs, leading to a "game" of defying the regulations and keeping the costs external in the hope that they will not be caught. Probably the best approach is a combination of regulation and taxation, coupling (for instance) maximum allowed effluent concentrations with a sliding fee structure for effluent levels below the maximum. *See also* BENEFIT/COST ANALYSIS; BENEFITS; COSTS; SECONDARY BENEFITS.

extinction the eradication of all members of a specific group of animals or plants. If the group involved is merely a local POPULATION, the process is referred to as *local extinction.* When the word is used with no qualifier, total extinction of a SPECIES is generally what is meant.

CAUSES OF EXTINCTION

The ultimate cause of all extinctions is the same: a change or series of changes in the group's ENVIRONMENT to which it cannot adapt. These changes take many forms. In nature, the most common cause of extinction is probably COMPETITION from better-adapted species utilizing the same NICHE, either because a new species has evolved or because a DISPERSAL PATHWAY has been opened to allow the new species to invade the RANGE of the less-well-adapted one. Predation is less of a factor in natural extinctions, because PREDATORS and PREY usually evolve together and are adapted to each other's needs: however, the introduction of a new predator into an area can doom a species or population if it cannot adapt fast enough. Predation is a major factor in human-caused extinction. It was the principal cause of the demise of the great auk and the dodo, certainly affected the passenger pigeon, and was at least partly to blame for the disappearance of the mastodon and the mammoth. The most common cause of extinction, however—natural or man-caused—is the destruction or alteration of HABITAT. This is especially true of plants, which are rooted to the ground and cannot move around in search of new habitat; however, it is also true to a surprising degree of animals. A species of bird that requires the COVER provided by OLD-GROWTH FOREST will die out if the forest is cut off. A species of fish that breeds only in running water cannot survive the building of a dam. A contributing factor is often an animal's own susceptibility to extinction, either because it has developed requirements for a strictly limited set of environmental conditions such as a special food or a limited temperature range (*niche overspecialization*) or because particularly beneficent living conditions have created in it a very low breeding rate that is not able to recreate the population fast enough to make up for newly introduced predation or competition pressures. This undoubtedly helped the dodo to extinction and was a major contributor to the death of the passenger pigeon. Finally, POLLUTION presents as serious hazard to many species today. DDT, for example, nearly eradicated the peregrine falcon and numerous insect-eating songbirds (including the familiar American robin), and complete aquatic communities in the mountain lakes of eastern North America and northern Europe have become extinct in recent years due to contamination of the lake water by ACID RAIN.

COSTS OF EXTINCTION

The costs of extinction to the extinct species itself are obvious and are the same as the costs of death to an individual member of the species: disappearance without possibility of return. The costs to the world at large are a little less apparent but are very real: they include loss of genetic diversity (*see* GENE POOL), loss of possible utilization of the species for food, fiber, or medical value, and—especially in human-caused extinctions—the vacation of the species' niche, with a consequent decrease in the stability of the ecosystem (*see* the discussion under MONOCULTURE). *See also* ENDANGERED SPECIES; THREATENED SPECIES; ENDANGERED SPECIES ACT; BIODIVERSITY; MASS EXTINCTION.

extinct species a SPECIES of animal or plant that has no living members. *See* EXTINCTION; *compare* THREATENED SPECIES; ENDANGERED SPECIES.

extrusive rock any IGNEOUS ROCK that flows or is ejected onto the Earth's surface in a still-molten state. Examples include BASALT, RHYOLITE, and ANDESITE. *Compare* INTRUSIVE ROCK. *See also* LAVA; VOLCANO.

extreme weather event (EWE) the phenomenon, first arising in the 1990s, of an unusual number of droughts, severe storms, and record high and low temperatures. Extreme weather has been predicted by climatologists since the 1980s as a result of global warming. *See* GLOBAL CHANGE.

Exxon Valdez the name of the ship involved in the greatest OIL SPILL to date in American waters. Outbound from the port of Valdez, Alaska—the southern terminus of the ALASKA PIPELINE—the 987-foot-long *Valdez* was carrying approximately 53 million gallons of CRUDE OIL when she ran aground on Bligh Reef, at the entrance to Prince William Sound from Valdez Narrows, at four minutes past midnight on March 24, 1989, cutting a 600-foot-long gash in her bottom and spilling 10.9 million gallons of crude into the sound. Near-hurricane-force winds in the next few days complicated the cleanup, keeping work crews on shore and breaking up and spreading the slick. Twelve hundred miles of coastline were eventually affected, including 90% of the coast of Katmai National Park. At least 100,000 birds—including 150 bald eagles—were killed, along with 1,000 sea otters and uncountable numbers of fish and marine INVERTEBRATES. Although much of the fault was placed on the captain of the ship, the ship itself and its builders and owners were also to blame. Despite a great deal of pressure mounted by environmentalists and government agencies to use double-hulled tankers for the transport of oil, the *Exxon Valdez* was single-hulled. Had Exxon spent the $22.5 million necessary to fit the tanker with a double hull, the corporation could have avoided an accident that is estimated to have cost $8.5 billion. Exxon spent $2.2 billion directly on the cleanup, and in a 1994 class-action suit a jury ordered the corporation to pay affected fishermen, landowners, and others $5 billion in individual damages. Ten years after the spill, the parts of Prince William Sound that are subject to strong tidal action have seemingly recovered. Tarry deposits, drifts of oil-soaked vegetation, and surface oil slicks may still be found in the calmer backwater parts of the sound, however. Accordingly, the permanent ecological effect of the disaster still cannot be assessed, but it is likely that the original natural balances of the ecosystem will never fully return to their pre-spill condition. *See also* OIL SPILLS.

faciation in ecology, a regional division within a plant COMMUNITY that differs noticeably from the rest in terms of SPECIES composition but is not different enough to warrant naming as a separate community. The jack pine (*Pinus banksiana*) community of the upper midwest, for example, contains black spruce (*Picea mariana*) in its northern reaches but not elsewhere. The portions with black spruce are a separate faciation within the jack pine community. *See also* ASSOCIATION; FIFTY PER CENT RULE.

fac pond *See* FACULTATIVE POND.

facultative organism an ORGANISM that can live in two or more different lifestyles. When used alone, the term usually refers to *facultative anaerobes,* that is, organisms that can live either with or without the presence of free OXYGEN. (This term is somewhat of a misnomer, in that most facultative anaerobes actually survive and grow more readily when exposed to oxygen. They are able to survive without it, but do not thrive. A better name for these might thus be *facultative aerobe,* with "facultative anaerobe" reserved for those organisms that ignore the presence or absence of oxygen altogether.) The term "facultative" can also be used to refer to organisms that may or may not function as PARASITES. These *facultative parasites* can survive on their own, but prefer to live on or in a HOST SPECIES. *Compare* OBLIGATIVE ORGANISM.

facultative pond in sewage treatment technology, a sewage lagoon that operates as an OXYGENATION TANK part of the time and as an ANAEROBIC DIGESTER the rest of that time. It generally has a larger number of FACULTATIVE ORGANISMS in it than either of the two "pure" treatment facilities it mimics, but the name comes from the operation of the pond rather than from the type of organisms present. Facultative ponds work as oxygenation tanks during the warm part of the year, when the water surface is open and breezes can keep it stirred up. They work as anaerobic digesters in cold weather when the pond surface is frozen and covered with snow, cutting off both light and oxygen to the organisms in the pond. Sludge is normally left in the pond through a full 12-month cycle, thus gaining the benefits of both aerobic and anaerobic treatment without having to be pumped. Facultative ponds (or "fac ponds") are obviously best suited to cold climates, but can be used in warm climates if there is a means to artificially cover the surface of the pond.

fallen tree succession *See under* SUCCESSION.

family in biology, a group of genera (*see* GENUS) that share enough characteristics for it to be reasonable to assume that they are *monophyletic* (evolved from a common ancestor). As a level of organization, the family lies between the genus and the ORDER. Family divisions are generally based on the morphology (overall shape) of such features as (among the plants) flowers and leaves or (among the animals) skeletal, dental, and reproductive systems. They are often named according to a *type genus* that shows all or nearly all of the family characteristics in obvious form. For example, the pine family (*Pinaceae*) is named for the pine genus (*Pinus*) but also contains the spruces, larches, hemlocks, firs, and Douglas firs. *See also* TAXONOMY.

family farms a term, greatly celebrated by farm-belt politicians, that suggests a 19th century mixed farm—with crops; livestock, poultry, and vegetables—operated solely by the work of a single family that provides a whole living. Family farms still predominate: nine out 10 farms are individual operations but are usually specialized and fully mechanized. Some are incorporated even though family-owned. And many of the family farms are multimillion-dollar enterprises. The general view that family farming has a less adverse impact on the environment than agricultural enterprises owned by nonfamily corporations sounds logical but is difficult to prove.

family planning the effort deliberately to limit the number of children born in a family in order to reduce impacts on the environment and the consumption of natural resources, and to increase quality of life, especially in developing nations. Family planning programs in various nations include encouraging the use (sometimes with free distribution) of contraceptive devices such as condoms and IUDs (intrauterine devices); sterilization of both women and men; antifertility pills and subdural implants; pills to induce miscarriage early in pregnancy; and abortion. Although family planning programs are underfunded throughout the world, the birth rate has declined markedly in many countries. *See also* POPULATION; POPULATION EXPLOSION.

FAO *See* FOOD AND AGRICULTURE ORGANIZATION OF THE UNITED NATIONS.

farm bill comprehensive farm-assistance legislation enacted biennially by the U.S. Congress between 1933 and 1996. The farm bills provided price-support payments and other benefits to farmers, some of them for conservation practices. In 1996 the so-called Freedom to Farm bill was enacted which radically changed the agriculture support structure and the need for a new price-support farm bill every other year. *See also* FEDERAL AGRICULTURE IMPROVEMENT AND REFORM ACT OF 1996.

farmland *See* AGRICULTURAL LAND, U.S.

Farm Service Agency part of the U.S. Department of Agriculture that since 1994 has administered the Agricultural Stabilization and Conservation Service, the Federal Crop Insurance Corporation, and the Farmers Home Administration. The FSA operates via local offices at the county level, with a farmer-selected review committee for each office. *See also* AGRICULTURE, U.S. DEPARTMENT OF.

Farm Tenant Act *See* BANKHEAD-JONES FARM TENANT ACT.

fascicle in botany, a cluster of needles arising from a single node on the twig of a CONIFER and thus forming a single leaf structure. The number of needles in each fascicle is characteristic of the SPECIES of conifer bearing the fascicles and may be used for taxonomic purposes (*see* TAXONOMY).

fat-soluble substance a chemical COMPOUND that dissolves in fat or, more generally, in any of the LIPIDS. They are generally insoluble or only slightly soluble in water. Many organic TOXINS are fat-soluble. Examples include DDT, TOXAPHENE, MIREX, the PCBs (*see* POLYCHLORINATED BIPHENYL), and the DIOXINS. Because these chemicals go into solution easily in the fat of living animals, they are not readily flushed from the body by metabolic processes (*see* METABOLISM) but, once ingested, tend to accumulate to higher and higher levels, a process known as BIOLOGICAL MAGNIFICATION or bioaccumulation. If the fat is later used by the animal (drawn upon for energy during a famine, for example), the fat-soluble toxins may be released rapidly into the bloodstream, leading to sickness or even death.

fault in geology, a break in a rock structure, caused by an EARTHQUAKE, in which the parts of the structure on opposite sides of the break have been displaced relative to each other. The displacement may be anywhere from an inch or two to several miles. In the case of larger displacements, more than one earthquake has generally occurred along the same fault. Motion along the fault may be horizontal (*strike-slip fault*), vertical (*dip-slip fault*), or a combination of the two (*oblique-slip fault*). The plane along which this motion occurs is the *fault plane.* The two opposing rock surfaces that move past each other along the fault plane are known as the *walls.* Usually the fault plane is not strictly vertical but is sloped so that one wall (the *footwall*) lies beneath the other (the *hanging wall*). In a *normal fault,* the hanging wall slips downward along the footwall. In a *reverse fault,* the hanging wall slips upward. If the fault plane intersects the ground surface a *fault line* is formed, generally marked by a *fault scarp* where either the hanging wall or the footwall has broken the surface and lies exposed as a cliff from a few inches to several thousand feet in height.

SIGNIFICANCE OF FAULTS

Faults are a major creator of landforms, forming the faces of mountain ranges such as the Sierra Nevada and the Rockies and determining the alignment of valleys such as the St. Lawrence and the Hudson. Most earthquake activity takes place along established faults, both because the earth's structure has already been weakened there and because the underlying earth strain that caused the fault in the first place ordinarily will not be relieved by a single quake but will require a whole series of them; hence, any structure that might be particularly sensitive to earthquake damage (such as a dam or a nuclear power plant) should not be built on or near an existing fault line. *See also* FAULT-BLOCK MOUNTAIN; FAULT-LINE SCARP; FAULT ZONE; HORST; GRABEN; PLATE TECTONICS.

fault-block mountain in geology, a mountain consisting of a portion of a rock FORMATION that has been raised above the rest of the formation by faulting (*see* FAULT). The STRATA in a fault-block mountain are displaced from the rest of the formation (raised and/or tilted) but are otherwise unchanged. *See also* HORST; BASIN AND RANGE TOPOGRAPHY.

fault-line scarp a scarp created along a FAULT line by differing erosion rates along the two sides of the fault rather than by motion along the fault plane.

fault zone a band of rock in which numerous FAULTS have occurred, so weakening the rock structure that movement caused by an earthquake does not tend to follow established fault lines but may create a new fault line anywhere within the zone.

fauna animals: specifically, an assemblage of animals found within a particular set of delimiting boundaries. The boundaries may be geographical (the fauna of Arkansas, the fauna of North America, the fauna of the upper Hudson River), temporal (the fauna of the Pleistocene, the summer fauna) or biological (the fauna of the rain forest, the fauna of the Douglas fir-salal association, the mammalian fauna). A small segment of a larger fauna is sometimes known as a *faunule.* Examples of faunules include the fauna of small, segregated sample plots within a large study area or the fossilized fauna found in a particular rock STRATUM.

faunule *See under* FAUNA.

FDA *See* FOOD AND DRUG ADMINISTRATION.

fecal coliform bacteria any of a group of related gram-negative rod bacteria (*see* GRAM STAIN; BACTERIUM) that are normally found in the lower intestinal tract of mammals, including humans. Fecal coliform bacteria are important environmentally because of their contribution to water pollution and their use as indicators of recent contamination by animal or human FECES. Important members of the group include *ESCHERICHIA COLI, Enterobacter aerogenes,* and the various species of *Klebsiella. See also* COLIFORM COUNT.

fecal matter *See* FECES.

feces (fecal matter) animal excrement. About one-third of the dry weight of normal feces consists of living and dead bacteria (*see* BACTERIUM; FECAL COLIFORM BACTERIA; *ESCHERICHIA COLI*). The remainder is undigested food residues.

Federal Agriculture Improvement and Reform Act of 1996 legislation that over a seven-year period intends to move U.S. agriculture toward a "free market" footing as opposed to continued dependency on price supports. Since the 1933 Agricultural Adjustment Act, the legislative centerpiece of Franklin Delano Roosevelt's New Deal, the U.S. government had set fair target prices for a wide range of agricultural products on an annual basis. If prices fell below that level, the government would make up the difference by buying up the surplus. U.S. agriculture, beset by mounting economic depression even before the Great Depression began in 1929, could not have survived without price supports. And clearly the emergence of a strong agriculture industry served the nation well during World War II, when U.S. farmers were called upon a feed a good part of the world, as well as our own troops and those remaining on the home front. After the war, however, the price supports became more and more difficult to administer fairly, and more and more expensive as the agricultural economy grew. By 1985 agricultural farm subsidies peaked at $26 billion, at which point many wondered if a technologically sophisticated agriculture with fewer and bigger farms operating in a global marketplace really should be supported by a program invented half a century before when farms were small, numerous, and, in most cases, still reliant on mules. Now, with the Freedom to Farm Act, support payments are phased out over a seven-year period, with a series of fixed annual transition payments to farmers based neither on prices nor on production of specific crops. The government no longer requires that land be taken out of production in return for the payments, nor does it prohibit farmers from switching to new kinds of crops regardless of presumed scarcity or surplus. While the 1996 act maintains various conservation programs, including the CONSERVATION RESERVE PROGRAM, many are now voluntary. In any case, the financial incentives to participate in conservation programs—for soil, water, wildlife, and the like—are no longer as powerful, since the 1996 act tends to encourage farmers to plant every available acre in the most profitable crop.

Federal Energy Regulatory Commission (FERC) a quasi-independent federal regulatory agency housed within the Department of Energy (*see* ENERGY, DEPARTMENT OF) and charged with overseeing siting and rate-setting activities for all energy production and distribution facilities that fall within federal jurisdiction. These include interstate oil and natural gas pipelines, interstate electrical transmission lines, and all hydroelectric generation plants built on NAVIGABLE

STREAMS. Essentially identical to the old FEDERAL POWER COMMISSION, FERC is governed by a five-member board whose chairman reports directly to the secretary of energy. It is considered a judicial body, and hearings before it on permit applications follow strict legal procedures. Address: 888 First Street NE, Room 2-A, Washington, DC, 20426. Phone: (202) 208-1371. Website: www.ferc.fed.us.

Federal Highway Administration (**FHWA**) agency within the U.S. Department of Transportation (*see* TRANSPORTATION, DEPARTMENT OF) charged with overseeing the federal government's role in all aspects of highway travel, including federal aid to highways, construction of the interstate highway system and of other federal highways, research in highway and vehicle engineering, traffic safety, and highway beautification. The FHWA is responsible for licensing highway carriers of hazardous materials and for promulgating regulations concerning the transport of these materials over the public roads (but *see also* MATERIALS TRANSPORTATION BUREAU). Created in 1894 within the Department of Agriculture (*see* AGRICULTURE, DEPARTMENT OF) as the Office of Road Inquiry, the FHWA became the Bureau of Public Roads in 1918; it was transferred to the Department of Commerce in 1949 and was renamed the Federal Highway Administration and made a part of the Department of Transportation when that Department was created on October 15, 1966. Headquarters: 400 Seventh Street SW, Washington, DC 20590. Phone: (202) 366-0537. Website: www.fhwa.dot.gov.

Federal Insecticide, Fungicide, and Rodenticide Act (**FIFRA**) federal legislation establishing registration and labeling requirements for PESTICIDES, HERBICIDES and other ECONOMIC POISONS, signed into law by President Harry Truman on June 25, 1947, and since amended numerous times. In its current form, FIFRA prohibits the sale of any economic poison that has not been registered by the ENVIRONMENTAL PROTECTION AGENCY (EPA). Registration is not automatic, but requires documentation, in which the burden is on the applicant to prove that the material being registered will not damage human health or the environment if used as intended. The EPA may suspend a registration at any time, and may cancel a registration following a hearing in which the registrant is allowed the opportunity to rebut evidence in favor of the cancellation. All materials must be reregistered every five years. Each registered poison receives a registration number, which must be displayed on the label of any product containing the poison. Other labeling requirements include an ingredients list and clear instructions for proper use and for disposal of any unused residues.

Federal Land Policy and Management Act (**FLPMA**) federal legislation to provide authority and guidelines for the management of lands administered by the BUREAU OF LAND MANAGEMENT (BLM), signed into law by President Gerald Ford on October 21, 1976. Often called the "BLM Organic Act," FLPMA states for the first time the legislative intent of Congress to maintain the remaining PUBLIC DOMAIN lands in public ownership and gives the BLM authority to make and enforce regulations for the use of these lands and others under its jurisdiction. The act also directs the agency to prepare land management plans based on the principals of MULTIPLE USE and SUSTAINED YIELD; makes the protection of AREAS OF CRITICAL ENVIRONMENTAL CONCERN a management priority; and subjects BLM lands to the provisions of the WILDERNESS ACT OF 1964. A critical provision, Sec. (401)(b), significantly modifies the TAYLOR GRAZING ACT OF 1934. Modeled on the Knutson-Vandenberg Act requirements for reforestation, this section, which applies to the FOREST SERVICE and the FISH AND WILDLIFE SERVICE as well as the BLM, sets up a special fund in the U.S. Treasury into which one-half of all grazing fees collected on federal lands are to be deposited and from which monies are to be drawn for rehabilitation of depleted RANGELANDS.

Federal Maritime Commission independent federal (U.S.) agency created by President John Kennedy on August 12, 1961, and charged with overseeing federal laws and regulations relating to waterborne commerce, including those dealing with financial responsibility for water pollution caused by shipping activity and accidents and by offshore oil drilling. The five-member commission is headed by a chairman who reports directly to the president. Headquarters: 1100 L Street NW, Washington, DC 20573.

Federal Power Commission obsolete federal agency, established in 1920 and given broad powers to regulate the interstate aspects of natural gas transmission and electrical generation and transmission, including the siting of power dams, transmission lines, and pipelines. Some of the most important environmental battles of the 20th century were waged largely before the Federal Power Commission, including the successful efforts to keep dams out of Hells Canyon, the Grand Canyon, and Dinosaur National Monument. The commission was abolished in 1977 by the act that set up the Department of Energy (*see* ENERGY, DEPARTMENT OF), and most of its powers were transferred to the FEDERAL ENERGY REGULATORY COMMISSION.

Federal Railroad Administration (**FRA**) agency within the U.S. Department of Transportation (*see* TRANSPORTATION, DEPARTMENT OF) created by Congress on October 15, 1966, and charged with overseeing all federal activities in regard to railroad construction and safety, including the regulation of rates, the promulgation of safety regulations, and the promotion of rail travel as part of an integrated national transportation policy. The FRA also operates the federally owned Alaska Railroad. Headquarters: 400 Seventh Street SW, Washington, DC 20590.

Federal Register a daily publication of the National Archives and Records Service (NARS) of the General Services Administration (GSA), a part of the EXECUTIVE BRANCH of the U.S. federal government. The *Federal Register* contains all rules, both proposed and final, promulgated by federal agencies; presidential proclamations and EXECUTIVE ORDERS; and other documents of regulatory law that Congress or the president may require to be printed. Since passage of the Federal Register Act in 1935, no regulation produced by any federal agency is legally binding until it has appeared in final form in the *Federal Register.* The law also requires a 30-day comment period before regulations become final in order to allow them to be challenged. Regulations that are published in final form in the *Federal Register* are codified each year as the CODE OF FEDERAL REGULATIONS. The *Register* is available at all federal depository libraries and most larger public and university libraries, as well as at Federal Information Centers and the regional headquarters of most federal agencies.

Federal Water Pollution Control Act *See under* CLEAN WATER ACT.

feedback loop The self-modification of a system whereby a given aspect of the system is fed back thus leading to a change in the dynamic of the system. A feedback loop in an environmental context can make a system self-regulating, or can alter it significantly.

feedlot in agriculture, a facility for the concentrated feeding of livestock or poultry, usually with a fenced compound surrounding it. Feed is brought to the animals in a feedlot rather than turning them loose to forage; thus the animals do not have to move about and are able to put on higher quality (i.e., fatter) meat more rapidly. At least 40%, and probably well over half, of all beef grown in the United States is feedlot-raised, most of it in the midwest, where there is easy access to corn and wheat grown on irrigated acreage. The principal disadvantage of feedlots lies in their high contribution to water pollution. At least half a billion tons of manure, and probably more, are produced by cattle in feedlots each year in the United States alone. Since it is concentrated in the feedlot rather than scattered over the range, it overwhelms the capacity of the soil to absorb and buffer it, and therefore creates polluted RUNOFF into nearby watercourses. It may also migrate downward through the soil into AQUIFERS, polluting GROUNDWATER reserves over broad areas. Most states now require farmers to collect the runoff from feedlots; however, protection of groundwater has proved much more difficult from both a regulatory and technical standpoint, and the problem remains unsolved.

feed water water entering a situation or procedure in which it will be manipulated and used in some manner by humans. Water diverted (*see* DIVERSION) for IRRIGATION, stock watering, etc., is *agricultural feed water;* water diverted for industrial uses such as cooling or the manufacture of goods is *industrial feed water;* saline water running into a DESALINATION plant is a *saline feed;* and so forth. *Compare* PROCESS WATER; FINISHED WATER.

feldspar any of several closely related rock-forming MINERALS composed of aluminum, silica, and one of a small group of light metallic elements that includes sodium, potassium, calcium and (occasionally) barium. The various forms are difficult or impossible to tell apart in the field. As a group, the feldspars are among the most common of minerals. They are a major constituent of GRANITE, and are found to a greater or lesser degree in nearly all IGNEOUS rocks. They are usually white, but may be pink, yellow, green or transparent. Their luster is similar to that of QUARTZ, but they are slightly softer, and they cleave much more easily, showing a 90° angle between CLEAVAGE PLANES. Hardness on the MOHS SCALE: 6.

feldspathoid a FELDSPAR with a lower-than-normal silica content. The feldspathoids show poorer cleavage (*see* CLEAVAGE PLANE), and generally have less LUSTER, than the true feldspars.

felling severing a standing tree from its stump. Felling is the first, and in many ways the most crucial, step in the harvest of timber. It creates the most safety hazards and the most opportunities for economic loss, and runs a close second to YARDING in its potential for damage to the forest environment. Improper felling causes at least 80% of the amount of breakage that takes place between standing wood and finished lumber, and can raise costs and lower profits from a logging operation by as much as one-fourth.

FELLING PRACTICES AND PROCEDURES: THE LEAD

The first step in felling is to determine the *lead,* that is, to decide which direction the tree should be made to fall. The most crucial factor in choosing the lead is the *length of ground*—the presence or absence of features such as rocks, gullies, stumps, and previously-felled logs, all of which can cause breakage. The feller should avoid "short ground" (rough terrain), and should wherever possible fell uphill so that the tree will not have to fall as far. The tree's natural lean is also a factor and may in some cases be the only determinant, although there are a number of methods available to fell a tree away from its lean (*see,* e.g., DRIVER TREE). Finally, all other things considered, the lead should be toward the direction the tree will be yarded (*see* YARDING) if at all possible, in order to minimize damage to soils and LEAVE TREES. This is especially crucial when cable yarding is being used.

FELLING PRACTICES AND PROCEDURES: THE CUT

Once the lead is determined, the feller will clear himself a pathway through the brush 20 feet or so to the side for an escape route once the tree starts toppling; he may also prepare a bed to receive the tree, clearing away obstacles and smoothing the ground along the path of the lead (this is especially important for very large trees such as old-growth redwood and Douglas fir, and for very breakable trees such as cedar). The actual cut begins with a wedge-shaped *undercut* or *face cut* on the side of the trunk toward the lead, one-fourth to one-third of the trunk diameter deep and with an opening at the front of the wedge of roughly one-fifth the trunk diameter. The *backcut* follows. Directly opposite the undercut on the trunk, it is a little deeper and an inch or two higher. Finally, wedges are placed in the backcut and driven in, "wedging the tree over" in the direction of the lead. If the tree is especially large and heavy, or if the lean is severe away from the lead, a jack may be employed in place of the wedges.

SPECIAL PRACTICES

Normally, all the trees on a STRIP will be led in the same direction. On steep or badly broken ground, however, this is not always possible. In such cases the feller must "lead to terrain," walking the entire strip and choosing the lead for each tree before felling any in order to avoid crossing leads, with the concurrent increased risk of breakage. An experienced feller can also sometimes bunch leads, felling several trees so that their tips are close together to decrease yarding time. With timber small enough for breakage not to be a problem, he may even practice "group felling," dropping several trees across one another in such a manner that a single CHOKER can be placed around them and they can be yarded together. These practices must be followed with extreme care, however, as the so-called "jackstrawing" that results from trees lying randomly across one another makes yarding more, not less difficult. *See also* DUTCHMAN; BARBER CHAIR; BUCKING; LIMBING; LOGGING SHOW.

felsenmeer in geology, a field of broken rock, rather like a TALUS slope but without a cliff above it. Felsenmeers generally lie at or above TIMBERLINE on mountain peaks; they are caused by FROST-WEDGING, with little or no motion of the resulting rubble taking place after formation. The name is German and means, literally, "boulder-sea."

felsic rock (felsite) rock composed of FELDSPARS, FELDSPATHOIDS, QUARTZ and similar silica-rich minerals. Felsic rocks are generally fine-grained in texture and light in color, and have a dense, heavy feel to them. Crystals are generally present but are too small to be seen without a magnifying glass or a microscope.

feral animal a wild animal descended from domestic stock, or an animal that was once domesticated but has returned to the wild. Examples include the feral burros of the Grand Canyon, the feral camels of Australia, and the feral cats of city alleyways. *Compare* EXOTIC SPECIES.

Ferrel's law *See* CORIOLIS EFFECT.

fertilizer in agriculture, a substance added to the soil/plant system to aid plant growth or increase productivity by providing extra NUTRIENTS for the plants' use. Fertilizers may be sprayed directly onto a plant's leaves (*foliar application*), but they are more commonly added to the soil so that they may be taken up by the plant's roots (*soil application*). Some are natural ("organic) materials such as manure, plant mulch, bone meal, or PHOSPHATE ROCK; others are synthetic ("inorganic") substances specifically manufactured to be sources of plant nutrients; and a few, such as BASIC SLAG, are BYPRODUCTS of other industries.

THE NEED FOR FERTILIZERS

Plants do not need nutrients in the same sense that animals do; that is, they do not need to consume complex ORGANIC COMPOUNDS such as vitamins and PROTEINS, but can manufacture the compounds they need through PHOTOSYNTHESIS and related processes. They do, however, require certain chemical ELEMENTS to supply the raw materials for these manufacturing processes, and these elements must be provided in an assimilable form. Of the 16 elements currently considered necessary for the growth of VASCULAR PLANTS, three—OXYGEN, HYDROGEN, and CARBON—are readily available from air and water. The remainder must be obtained from the soil. Soils, however, are not uniform in structure and material, but in many cases are deficient in one or more nutrient elements. These deficiencies are increased by agricultural cropping, which removes the mature plants for use elsewhere rather than allowing them to die and decay on site, releasing their nutrients back to the soil. If healthy plants are to be grown in such nutrient-deficient soils, some means must be found to supply the missing nutrients. Fertilizers provide that means.

THE CHEMISTRY OF FERTILIZERS

In order to be assimilated by plants, most nutrient elements must be present in ionic form (*see* ION) rather than as complete, undissociated MOLECULES. NITROGEN, for example, is present in vast amounts in the air (*see* ATMOSPHERE) but only a relatively few plants, nearly all of them LEGUMES, can use it directly. Fertilizers therefore need to be made up, not simply of the required elements, but of COMPOUNDS of those elements that dissolve and dissociate readily in water (*see* SOLUTION; DISSOCIATION). In theory, any number of compounds could be used. In practice, all fertilizers, both organic and inorganic, are chosen from the same small group of compounds ("carriers") whose rate of dissociation, soil MOTILITY, and BIOAVAILABILITY are particularly appropriate for use as plant food. Of the three nutrient elements commonly provided by fertilizers, nitrogen is normally carried by AMMONIA (NH_3) or a related compound such as urea ($CO(NH_2)_2$) or ammonium sulfate ($(NH_4)_2SO_4$); PHOSPHOROUS, by orthophosphoric acid (H_3PO_4); and potassium, by potassium carbonate (K_2CO_3) or potassium chloride (KCl). Ammonia and its relatives may be derived naturally from manure or GREEN MANURE, or may be manufactured from NATURAL GAS. Orthophosphoric acid is provided by bone meal or by PHOSPHATE ROCK, either in natural form or one of several processed varieties (monoammonium phosphate, diammonium phosphate, TRIPLE SUPERPHOSPHATE) designed to make the phosphate more available. Potassium carbonate is supplied by wood ashes ("potash"); potassium chloride, by sylvite or sylvinite ores, usually processed to yield pure KCl ("muriate of potash").

N-P-K FERTILIZERS

Because it is simpler, cheaper, more convenient, and generally more accurate to apply fertilizer in a single application rather than many, most commercial fertilizers are sold as mixes of nitrogen, phosphorus, and potassium compounds. Such mixes are known as N-P-K fertilizers, from the chemical symbols of the nutrient elements included. N-P-K fertilizers are mixed in various proportions, which are chosen among by farmers or gardeners on the basis of the nutritional needs of the plants to be fed and the known deficiencies in the soil these plants are growing in. These proportions are indicated by a set of three numbers known as the *grade*. The numbers represent percentages, and are always listed in the N-P-K order. Thus a 10-20-10 grade indicates a fertilizer that is 10% nitrogen, 20% phosphorous and 10% potassium; the remaining 60% is made up of the non-nutrient parts of the carriers plus a (generally small) amount of filler, often PEAT or DIATOMACEOUS EARTH. Sometimes micronutrient carriers, such as BORAX (for BORON) or ferrous sulfate (for iron) may be added to the filler.

THE USE AND ABUSE OF FERTILIZER

Fertilizers are an essential part of all agricultural practice, both primitive and modern. Without them, all the nutrients in the soil would rapidly be carried away in the harvested crops, and the soil would not longer be able to support plant life. The environmental questions concerning fertilizers, therefore, are not whether to fertilize, but when, how much, and with what. Timing is crucial because fertilizer applied at the wrong time is not absorbed by the plants but washes off into watercourses or migrates through the soil to pollute the GROUNDWATER, often removing other nutrients from the soil as it goes. Fertilizers in excess of what can be used by plants also wash off into watercourses or migrate to the water table, leading to increases in nutrient levels and TOTAL DISSOLVED SOLIDS (*see* EUTROPHICATION; WATER POLLUTION). The choice of which fertilizer to use depends on the chemistry of the soil and the type of plants one is trying to grow, and must be carefully made to avoid over-fertilization of one or more nutrients. The question of organic versus inorganic fertilizers is relatively unimportant to the plants, since all fertilizers—organic and inorganic alike—provide the same nutrient elements carried on the same set of carriers. Organic fertilizers usually contain more micronutrients, and are applied in a less mobile form, so that they release more slowly, making timing of the application less of a problem. They are considerably more difficult to apply in amounts figured precisely for soil conditions, however.

FES short for FINAL ENVIRONMENTAL STATEMENT (*see* ENVIRONMENTAL IMPACT STATEMENT).

FHWA *See* FEDERAL HIGHWAY ADMINISTRATION.

fiber (1) in the agricultural sciences, a crop such as cotton, flax, or timber that is grown and harvested for the production of nonfood items (clothing, paper, building products, etc.).

(2) in food science, the undigestible portion of a food (*see also* ASH). Fiber is important in the diet not only because it helps stimulate the digestive tract to flow properly, but because it has the ability to absorb or adsorb excess cholesterol and other potentially harmful food substances and carry them out of the body.

fibrous root system a plant root system consisting of a thick mat containing many short, slender, interwoven roots, without the presence of a TAPROOT. The plant family with the best-developed fibrous root systems is the GRASSES. Fibrous root systems have excellent soil-holding properties and are able to take full advantage of the nutrients in the TOPSOIL, but do not penetrate far enough into the soil to hold a tall plant against the wind or to follow a declining WATER TABLE into the subsoil; hence, plants with fibrous root systems can exist in dry climates only as ANNUALS, and in windy climates only as low-growing plants with little wind resistance. They are useful in environmental engineering for stabilizing CUTBANKS and other unstable slopes, and for holding soil against the wind.

field capacity in hydrology or soil science, the amount of water a soil can hold against gravity, that is, the amount of water remaining in the pores of the soil after saturation by a storm and completion of drainage to the WATER TABLE following the storm's end. *See also* CAPILLARY WATER.

FIFRA *See* FEDERAL INSECTICIDE, FUNGICIDE, AND RODENTICIDE ACT.

50% rule in ecology, the requirement that at least 50% of the predominant SPECIES in one of a pair of geographical aggregations of plants and animals must be different from those in the other aggregation in order for the two aggregations to be considered separate COMMUNITIES. The rule works both ways: two aggregations *cannot* be considered separate communities if less than 50% of the predominant species are different, and they *must* be considered different if more than 50% of the predominant species differ from each other (less than 50% are shared). Notice the emphasis on *predominant species*. A predominant species from one community may be present as a minor species in another community without violating the 50% rule.

film water (bound water, hygroscopic water) in soil science, the thin film of water that clings tightly to soil particles in all moist soils. It is not available to plants and cannot be drained out by gravity, but it can be driven off by heat. The "film" is only a few MOLECULES thick and is held to the soil particles by the electrostatic attraction that results from the highly polarized nature of the water molecule (*see* WATER: *structure of water*). Film water is highly immobilized, to the point where the molecules slow down dramatically. The energy lost in this slowing-down process is released as heat ("heat of wetting"). Because it adheres to the soil particles, film water is sometimes referred to as "water of adhesion" in order to differentiate it from water held in place by surface tension ("water of cohesion") (*see* CAPILLARY WATER). *Compare* GRAVITATIONAL WATER. *See also* SOIL WATER.

filterable virus *See under* VIRUS.

filter feeder in ecology, an ORGANISM (usually an animal) that feeds by passing sea water through an internal filter. Organisms that are too large to pass through the filter pores are collected and used as food. Most filter feeders are SESSILE ANIMALS such as mussels, which utilize PLANKTON; however, a few large, mobile animals are filter feeders, including the blue whale—the largest animal of all—which feeds primarily on krill (tiny lobsterlike crustaceans), obtained by passing massive quantities of sea water through bony filters (*baleen* or "whalebone") in the whale's mouth.

final cutting in logging, the removal of all the remaining OVERSTORY in a STAND after REGENERATION has become firmly established. *See* SHELTERWOOD SYSTEM.

fine a very small solid particle, especially one in a MIXTURE that also contains larger particles. The upper boundary size varies with the use of the term. In air pollution terminology, a fine is a PARTICULATE smaller than 2.5 microns (p.0025 mm, or about 0.0001 inches) in diameter. Such particulates are inhaled into the lungs rather than being caught by the mucus membranes of the nose and throat, and are thus responsible for most of the respiratory-system damage caused by air pollution. Fines can be controlled by the use of BAGHOUSES, ELECTROSTATIC PRECIPITATORS, or SCRUBBERS. In soil science and engineering geology, the term *fine* refers to a particle of SILT or CLAY, generally less than 0.02 mm in diameter. Engineers occasionally extend the term to fine sands (0.075 mm or less), especially when speaking of AGGREGATE or crushed rock.

fingerling a young fish that has reached a stage of growth where it is approximately the size of a human finger (2–4 inches in length). Most anadramous fish (*see* ANADRAMOUS SPECIES) migrate from their spawning beds to the sea as fingerlings. *Compare* FRY.

finished water in water-supply terminology, water that has been through a water treatment plant and is ready for delivery to users.

firebreak a linear opening in a forest or grassland that has a minimal amount of combustible materials in it and which therefore serves as an effective barrier to the spread of WILDFIRE. A firebreak may be a natural feature such as a river or a band of cliffs, a man-made feature created for another purpose such as a highway, or an opening ("fire lane") created specifically to halt fires. The minimum effective width is about 10 feet, and to prevent the spread of crown fires (*see* WILDFIRE: *types of wildfire*) the break must be considerably wider than that.

FUELBREAKS

Firebreaks are almost always thrown around wildfires in an effort to stop them; however, good forestry practice dictates that they also be created in advance wherever substantial FUEL buildup makes the rapid spread of fires likely, or where fire would be particularly harmful, such as in a municipal WATERSHED. A break created for this purpose is called a *fuelbreak*. A *partial fuelbreak*—the removal of downed fuel without cutting down many trees—can often be as effective as a full fuelbreak, and is considerably better from both ecological and recreational standpoints. *See also* WILDFIRE: *wildfire management;* PRESCRIBED BURN.

fire-dependent species a SPECIES of plant that reproduces poorly, or does not reproduce at all, without the aid of fire. Many species of pines are fire-dependent: the knobcone pine (*Pinus attenuata*), for example, bears resinous cones that open and expel their seeds only in the heat of a fire, while the slash pine (*Pinus elliottii*) depends on periodic fires to eliminate brown-spot disease and allow the young trees to remain healthy. *See also* FIRE-MAINTAINED CLIMAX.

fire ecology the study of the role of wildfire in natural COMMUNITIES and ECOSYSTEMS. *See* FIRE-DEPENDENT SPECIES; FIRE-MAINTAINED CLIMAX.

fire-maintained climax a COMMUNITY that is maintained in an apparent CLIMAX state by having periodic WILDFIRES sweep through it. If fire is eliminated, the so-called "climax" alters, often quite rapidly, to something else. Grasslands, for example, which may persist unchanged for thousands of years if swept by fire every few years, are quickly invaded by trees and shrubs when fire is eliminated. Other examples of fire-maintained climaxes include the oak woodlands of the northeast, the DOUGLAS-FIR forests of the Pacific northwest, and the eucalyptus forests of Australia. *See also* FIRE-DEPENDENT SPECIES.

firn (neve) old, compacted SNOW; technically, snow that has reached a density of at least 0.4 grams per cubic centimeter. For convenience's sake, it is usually more loosely defined as snow that has survived through at least one summer without melting or sublimating away (*see* SUBLIMATION). In firn, the crystalline structure that characterizes snow has broken down completely, but the pores between crystals have not yet become entirely filled. Firn that survives through several seasons gradually becomes *ice*. *See* GLACIER: *requirements for glacier formation. See also* FIRN LINE.

firn line the lowest elevation at which SNOW that falls on a GLACIER'S surface will survive through the summer and into the next winter. Above the firn line, ICE is formed and the glacier grows; below the firn line, it is always in the process of melting back, though ice flowing down from the zone above the firn line may make it seem to grow. *See also* FIRN; GLACIER: *requirements for glacier formation.*

first law of thermodynamics *See* THERMODYNAMICS, LAWS OF.

first-order lake in limnology, a lake in which overturn never occurs; that is, one with a well-developed THERMOCLINE that never reaches the lake bed. Temperatures at the bottom of a first-order lake remain at 4°C throughout the year, and the lake is usually oligotrophic (*see* TROPHIC LEVELS). In general, first-order lakes are those that are at least 300 feet deep, though northerly or high-elevation lakes may achieve first-order status at shallower depths, and tropical lakes may require considerably more depth in order to reach it at all.

Fish and Wildlife Coordination Act (FWCA) federal legislation requiring mitigation for the loss of wildlife HABITAT due to the construction of federal water resources projects, signed into law by President Franklin Roosevelt on March 10, 1934, and amended numerous times since. The act requires designers of federal dams, reservoirs, and irrigation works to include the costs and benefits to fish and wildlife when determining the BENEFIT/COST RATIO of a project; requires consultation with state and federal wildlife and fisheries agencies during the design of a project in order to minimize the project's effects on fish and wildlife; and authorizes the purchase of private lands to replace habitat lost due to the construction of a project. A separate section calls for ongoing studies by the Interior Department (*see* INTERIOR, DEPARTMENT OF) on the effects of waterborne SEWAGE and industrial wastes on fish and wildlife.

Fish and Wildlife Service (FWS) agency within the U.S. Department of the Interior (*see* INTERIOR, DEPARTMENT OF) charged with conserving and managing the nation's wildlife resources, including birds, game and nongame animals, freshwater and anadramous fish (*see* ANADRAMOUS SPECIES), and some marine mammals. The agency's principal role is the management of the National Wildlife Refuge System (*see* WILDLIFE REFUGE); however, it also operates fish hatcheries, enforces federal wildlife laws (including the ENDANGERED SPECIES ACT), administers wildlife research programs, and provides technical expertise and field assistance for wildlife management programs carried out by other federal agencies or by state and local governments.

STRUCTURE AND FUNCTION

The FWS is headed by a national director who reports to the assistant secretary of the interior for fish and wildlife and parks. Seven regional offices, each run by a regional director, administer 508 national wildlife refuges.

HISTORY

The Fish and Wildlife Service was created in 1940 by President Franklin Roosevelt, through the combination of the existing BUREAU OF FISHERIES and BUREAU OF BIOLOGICAL SURVEY into a single agency under the authority of Presidential Reorganization Plan III. The Fish and Wildlife Act of 1956 gave the agency legislative status and separated it into two major divisions, the BUREAU OF COMMERCIAL FISHERIES and the BUREAU OF SPORT FISHERIES AND WILDLIFE. In 1970 the Fish and Wildlife Service was temporarily dissolved, with the Bureau of Commercial Fisheries moving to the Department of Commerce and the Bureau of Sport Fisheries and Wildlife becoming an independent bureau within the Department of the Interior. This Bureau was redesignated the United States Fish and Wildlife Service by Congress in 1974. Often castigated by environmentalists for its emphasis on management by HABITAT manipulation instead of habitat preservation, and for its longtime cooperation with state and local Animal Damage Control units (*see* PREDATOR CONTROL), the Fish and Wildlife Service is nevertheless the only major agency of the federal government whose entire mission revolves around BIOSPHERE conservation, and its success may be viewed as an important touchstone for gauging the success of building what the naturalist and environmental writer Aldo Leopold (1887–1948) referred to as the "environmental ethic" into the fabric of American society. Address: 1849 C Street NW, Washington, DC 20240. Phone: (202) 208-3100. Website: www.fws.gov.

Fisheries, Bureau of *See* BUREAU OF FISHERIES.

fisheries decline the recent worldwide phenomenon of a reduction in the commercial catch of many favored food-fish species. Species with seriously declining stocks in ocean fishing areas include salmon, cod, yellowtail flounder, pollock, redfish, red snapper, tuna, swordfish, yellow croaker, and even shark. The causes include overfishing to meet increasing demand (see chart for U.S. consumption), wasteful factory-ship practices, the destruction of fish-spawning habitats pollution, and changes in climate regimes. According to the United Nations Food and Agriculture Organization (FAO), almost 70% of the world's marine fish stocks are either fished to capacity or are already overfished or depleted. In U.S. waters, the National Marine Fisheries Service has classified more than 82% of commercial stocks as being overfished. The population of some species, such as the Atlantic bluefin tuna, has been reduced to one-fifth its former size. Worldwide, in the 20-year period 1970–90, the number of large fishing vessels doubled, to nearly 1 million. Many of these vessels use lines of hooks up to 80 miles long or DRIFT NETS that ensnare virtually all marine life in their path. Such practices produce a huge "bycatch"—species that are not wanted and are tossed back, usually dead. The United Nations estimates that the bycatch amounts to a third of the volume of the regular catch. In shrimp trawling, an estimated 15 tons of fish are dumped for every ton of shrimp landed. Many fisheries scientists believe that if fishing pressure can be reduced, many of the species in serious decline can come back. Others are concerned that global warming may permanently disrupt ocean currents and temperatures leading to a permanent deficit of food fish for a growing world population.

fisherman's trail *See under* WAY TRAIL.

Fishery Conservation and Management Act federal legislation regulating ocean fishing, signed into law by President Gerald Ford on April 13, 1976. The act established a 200-mile limit for U.S. territorial waters in the oceans and required foreign fishing vessels operating within that limit to have a permit issued by the United States. It also set up a federal fisheries management program to supersede those of the various states, and created eight regional fisheries councils to design and enforce the management program within their respective territories, subject to a set of conditions included in the act which were meant to guarantee sustained harvest of fishery resources and protection of marine ECOSYSTEMS.

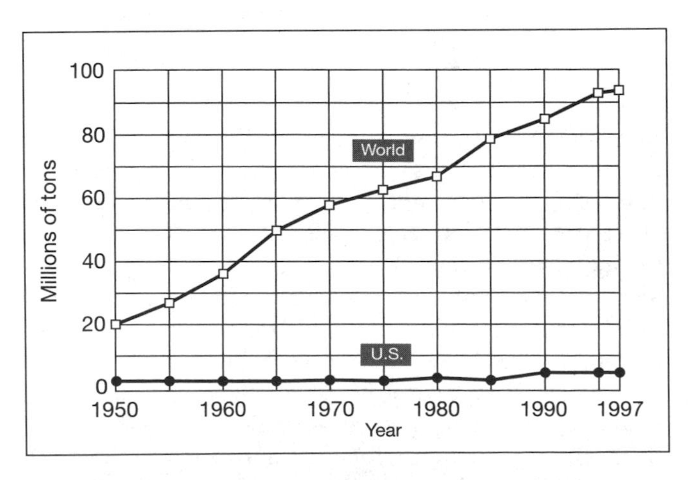

Fish catch, 1950–1997 (world catch and U.S. catch) ***(FAO,*** **Yearbook of Fishery Statistics: Catches and Landings,** ***and U.S. Department of Agriculture, National Agricultural Statistics Service,*** **Agricultural Statistics,** ***annual)***

fish farming *See* AQUACULTURE.

fish louse *See* COPEPOD.

fixed solids in pollution-control technology, those SOLIDS that are not vaporized by high heat, and thus remain after the VOLATILE SOLIDS are driven off. Fixed solids are generally MINERALS and other inorganic materials (*see* INORGANIC COMPOUND).

flagellate any of numerous species of protozoa (see PROTISTA) of the subphylum Mastigophora, characterized by the presence of one or more flagella (*see* FLAGELLUM) and therefore motile in nature (*see* MOTILITY). The flagellates are divided into two groups, the *phytoflagellates,* which contain CHLOROPHYLL and are often classified with the one-celled ALGAE, and the *zooflagellates,* which are unpigmented, and are generally thought to be phytoflagellates that have lost their pigment through MUTATION. Some, especially among the phytoflagellates, form large colonies that act like rudimentary multicelled organisms, with the individual cells taking on specialized duties such as food-catching, reproduction, etc. There is a considerable body of evidence that suggests that the phytoflagellates are the ancestral stock from which all so-called "higher organisms," both plant and animal, evolved. Their principal importance from an environmental standpoint lies in their great abundance in bodies of water, where as the main component of phytoplankton (*see* PLANKTON) they form the basis of the aquatic FOOD CHAIN. Certain species are also important as PATHOGENS, among them *Trypanosoma brucei gambiense,* which causes sleeping sickness, and *Giardia (lamblia) intestinalis,* which causes GIARDIASIS, or "hikers disease."

flagellum a long, whiplike appendage found on numerous one-celled ORGANISMS and on the motile reproductive cells (such as spores and sperm) of more complicated organisms, used principally as a means of propulsion (*see* MOTILITY) but also occasionally as a means of catching food and as an organ of touch. The flagella of all eukaryotic cells (*see* EUKARYOTE) are identical in structure and consist of a bundle of 11 hollow filaments called "microtubules" in so-called "9+2" arrangement, that is, with two central filaments surrounded by a circle of nine others. The flagella of PROKARYOTES are much finer, and appear to be composed of three long PROTEIN molecules braided together. *See also* FLAGELLATE; CELL: *structure of cells.*

flag line a line marked through the woods by the use of twists of colored plastic tape ("flagging tape" or "flags") tied to branches along the line in such a way that each flag is visible from both the flag in front of it and the flag behind it. The flag line may mark a proposed trail or road route ("flagged trail" or "flagged road") or the boundary of a proposed CUT BLOCK ("flagged boundary"), and is generally color-coded. Typical colors include yellow for roads, blue and white for trails, and red for cut blocks.

flammable liquid any liquid with a FLASH POINT lower than 100°F.

flammable solid any solid (other than an explosive) for which special care must be taken to prevent accidental combustion. The category includes those solids that sublime easily and for which the vapors are flammable (*see* SUBLIMATION; FLASH POINT), those with a low IGNITION POINT, and those which undergo spontaneous chemical change into flammable forms, especially when the chemical change takes place through an EXOTHERMIC REACTION.

flark *See* STRING BOG.

flash distillation *See under* DISTILLATION.

flash point the temperature at which a substance begins giving off flammable vapors in a heavy enough concentration to be ignitable by ordinary means such as a spark or a match. The flash point is determined by one of several standardized tests. The test used should normally be specified along with the temperature.

flat water water without rapids in it. The term may refer to a lake or a reservoir, or to a quiet section of a river with little or no perceptible current. *Compare* WHITE WATER.

flavanoid in organic chemistry, any of a group of chemical COMPOUNDS composed of two BENZENE RINGS connected by a HYDROCARBON GROUP containing a chain of three CARBON atoms. Flavanoids appear in nature primarily in the ANGIOSPERMS, where they are found as flower pigments—they are responsible for the colors of nearly all red, blue, and white flowers—and as protective pigments in the leaves, used to screen out ultraviolet radiation while letting through the greens and blues needed for PHOTOSYNTHESIS.

fledgling a young bird that has grown feathers and left the nest but still requires feeding and protection by the parent birds in order to survive.

floater a CONTAMINANT that floats on the surface of the water without either dissolving or sinking. To be classed as a floater in analysis, a contaminant must have a solubility ratio of less than 1% (*see* SOLUTION) and a SPECIFIC GRAVITY of less than 1.0. The term is often used in sanitary engineering as a synonym for FECES.

floating macrophyte an aquatic MACROPHYTE that is not rooted to the bottom of a body of water but floats on the surface instead. A common example is DUCKWEED (*Lemna* spp).

floc *See under* ACTIVATED SLUDGE.

floe a free-drifting ICE platform, unattached to land, formed by the breakup of the frozen surface of a body of water. Floes are considerably thinner and flatter than ICEBERGS, and tend to drift about in large, dense groups (*see* PACK ICE). Particularly large, thick floes, known as *ice islands,* occasionally develop in the polar seas, where they may drift about for years or even centuries, developing topographic features akin to those of solid land, including hills, streams, and ponds. *See also* FLOEBERG.

floeberg an ICEBERG created by the collision and piling-up of two or more FLOES rather than by glacial calving.

flood the inundation of normally dry land by water. Floods may result from several different causes. Along large bodies of water, storm waves and tidal waves (*see* TSUNAMI) may inundate portions of the coastline. A rise in the water level of a lake, either due to the construction of a dam at the outlet or to increased RUNOFF in the WATERSHED above the lake, may flood shoreline properties, as may subsistence (sinking) of the land itself from such causes as GROUNDWATER withdrawal or the construction of heavy buildings. *Inflooding*—the temporary ponding of runoff on the ground surface during and immediately after a storm—may flood basements and damage crops and roadways. The most damaging floods, however, are generally those resulting from *outflooding*—the discharge of excess water down a streambed, so that the stream's banks cannot contain it. Outflooding normally occurs as a result of rainfall in a stream's watershed, although there are other causes (earthquakes, for example, may cause outflooding by triggering landslides into lakes, displacing the water of the lake down the outlet stream). They cause severe, recurrent damage to structures in the stream's FLOODPLAIN (highways, bridges, homes, commercial buildings, and so on). The social costs of these damages, as well as their money costs, make it imperative that some sort of effort be made to eliminate (or at least reduce) them. Attempts to do this are referred to as *flood control.*

METHODS OF FLOOD CONTROL

There are two principal approaches to controlling floods. The most common, known as the *structural approach,* consists of building protective structures, including lining streambanks with LEVEES and concrete or masonry walls (*flood walls*) to physically prevent the flood waters from leaving the channel, constructing dry channels known as FLOODWAYS to carry the extra water during times of flood, and building upstream DAMS to catch the floodwaters and release them slowly so that the stream never exceeds bankfull stage. STREAM CHANNELIZATION is also considered a structural means of flood control, as is so-called "floodproofing"—building structures on stilts or on high, sturdy foundations to prevent the flood waters from damaging them. The alternative, the *nonstructural approach,* is usually considered sounder from an environmental standpoint. Recognizing that floods are part of a river's natural cycle and that the floodplain is just the dry portion of the river's bed, nonstructural methods focus on zoning and other methods to discourage the construction of buildings on the floodplain. They also may aim at reducing flood crests through watershed controls such as the planting of trees and other vegetation to increase the INFILTRATION rate during storms (runoff from bare soil can be more than 20 times as great as runoff from forested land). A mixture of structural and nonstructural approaches is probably the ideal, especially if the structural approaches used are those that are environmentally less damaging, such as substituting several small headwaters dams for one large dam on the MAINSTEM, and operating dams as so-called "dry reservoirs" (so that normal streamflow passes under the dam, and only floods back up in the reservoir area). *See also* NATIONAL FLOOD INSURANCE PROGRAM.

flood insurance *See* NATIONAL FLOOD INSURANCE PROGRAM.

flood irrigation *See* IRRIGATION: *methods of irrigation.*

floodplain (1) in geology, an alluvial plain, that is, the plain of alluvial sediments (*see* ALLUVIUM) deposited and being currently maintained by a stream.

(2) in land-use planning and engineering, the portion of a stream's valley likely to be covered by water during a FLOOD. Floodplains are mapped by using past rainfall and RUNOFF data, with extrapolations based on changes in the character of the stream's WATERSHED (such as increased urbanization). Several different floodplains are usually designated in a given area, differentiated on the basis of the expected frequency of flooding. For example, the *50-year floodplain* consists of the area that would be flooded, on the average, once every 50 years (meaning that it has a 2% chance of being flooded in any given year). Commonly designated areas for zoning and other

regulatory purposes include the 10-, 30-, 50-, 100-, and 500-year flood plains. *See also* NATIONAL FLOOD INSURANCE PROGRAM.

floodway a channel or course used by a river only during floods. It remains dry when the river is not flooding. Natural floodways can pose a problem if they are not recognized as such, because structures may be built in them that are then inundated during floods, always with inconvenience and often with loss of property. Artificial floodways are sometimes constructed to help direct and control a river's flooding pattern (*see* FLOOD: *methods of flood control*).

flora plants; specifically, an assemblage of plants found within a particular set of delimiting boundaries. The boundaries may be geographical (the flora of Arkansas, the flora of North America, the flora of the upper Hudson River), temporal (the flora of the Pleistocene, the summer flora) or biological (the flora of the rain forest, the flora of the Douglas fir-salal association, the vascular flora). *Compare* FAUNA.

Florida Everglades a unique ecosystem in south Florida consisting of the largest saw grass marshland in the world, some 10 million acres. Water from torrential rainfall in the area and the outflow from Lake Okeechobee (which collects much of it) at the northern edge of the Everglades create a "river of grass" running 100 miles southward to a mangrove swamp that provides an essential nursery for myriad species of fish, amphibians, and birds. The American alligator (*Alligator mississippiensis*) thrives in the Everglades and contributes to the diversity of species by digging "gator holes" in the dry season, which provide refuge for fish, turtles, snails, and other creatures that would otherwise die in the drying sun. Many species in the Florida Everglades are on the brink of extinction, including the snail kite, wood stork, Florida panther, American crocodile, loggerhead turtle, manatee, and indigo snake. About one-seventh of the Everglades is protected as part of Everglades National Park. The remainder is threatened by the diversion of the freshwater flowing through the saw grass for agriculture and housing.

flower the reproductive structure of an ANGIOSPERM. Flowers are layered structures that appear to have been derived from TERMINAL BUDS. Their sudden massive appearance in the late Cretaceous (75 million years ago) marks one of the major advances in the evolutionary development of life on Earth.

PARTS OF A FLOWER

A complete flower consists of four concentric layers. At the center are the *carpels,* the female reproductive organs, consisting of an *ovary* surmounted by a narrow shaft (the *style*) with a swollen tip (the *stigma*) equipped with POLLEN receptors. If the carpels are fused into a single structure they are sometimes known collectively as a *pistil,* although this term is passing out of use. Surrounding the carpels is a circle of *stamens,* the male reproductive organs, each one a slender, flexible *filament* topped by a small baglike organ called the *anther,* which produces and stores the pollen. Enclosing the stamens are the *petals,* brightly colored, often flamboyantly shaped leaflike structures whose purpose seems exclusively to be to attract pollinating insects and birds. The circle of petals is known collectively as the *corolla.* The outermost layer of the flower consists of *sepals.* These too are leaflike in form. Their primary purpose is to protect the flower while it is "in bud"—developing prior to opening—and hence they are commonly green, though once the flower is open they may also serve as insect attractants, in which case they are apt to be colored like the petals and are often indistinguishable from them except by their position in the outside ring of the flower's structure. The sepals are known collectively as the *calyx;* the calyx and corolla together form the *perianth.* All four layers grow from a single cuplike collar of tissue at the base of the flower, the *receptacle,* which forms the tip of the flower stalk or PEDICEL. The receptacle itself may be enclosed by one or more BRACTS; these are not technically part of the flower, but of the pedicel.

THE CLASSIFICATION OF FLOWERS

Flowers are classified structurally in two ways: as *complete* and *incomplete,* or as *perfect* and *imperfect.* In a complete flower, all four of the structural layers are present; in an incomplete flower, one or more entire layers are missing, as in the meadow rue (*Thalictrum*), which has colored sepals but no petals, or the poinsettia, which has no perianth at all, the entire role of sepals and petals alike having been taken over by colored bracts. A perfect flower contains both male and female sex organs, that is, both stamens and carpels; an imperfect flower is missing one or the other. Plants with imperfect flowers may mix male and female flowers on a single plant (*monoecious*) or grow them on separate male and female plants (*dioecious*). The order in which the layers of the flower grow out of the receptacle is also important for classification purposes. If the ovaries (at the base of the carpels) lie above the perianth, they are said to be *superior* and the flower is *hypogynous.* If the ovaries lie below the perianth, they are *inferior,* and the flower is *epigynous.* In a few flowers, the ovaries are neither inferior nor superior but lie on the same level as the perianth, which encircles them like a collar. These flowers are said to be *perigynous.* The classification of flowers into families, genera and species depends primarily on these structural variations

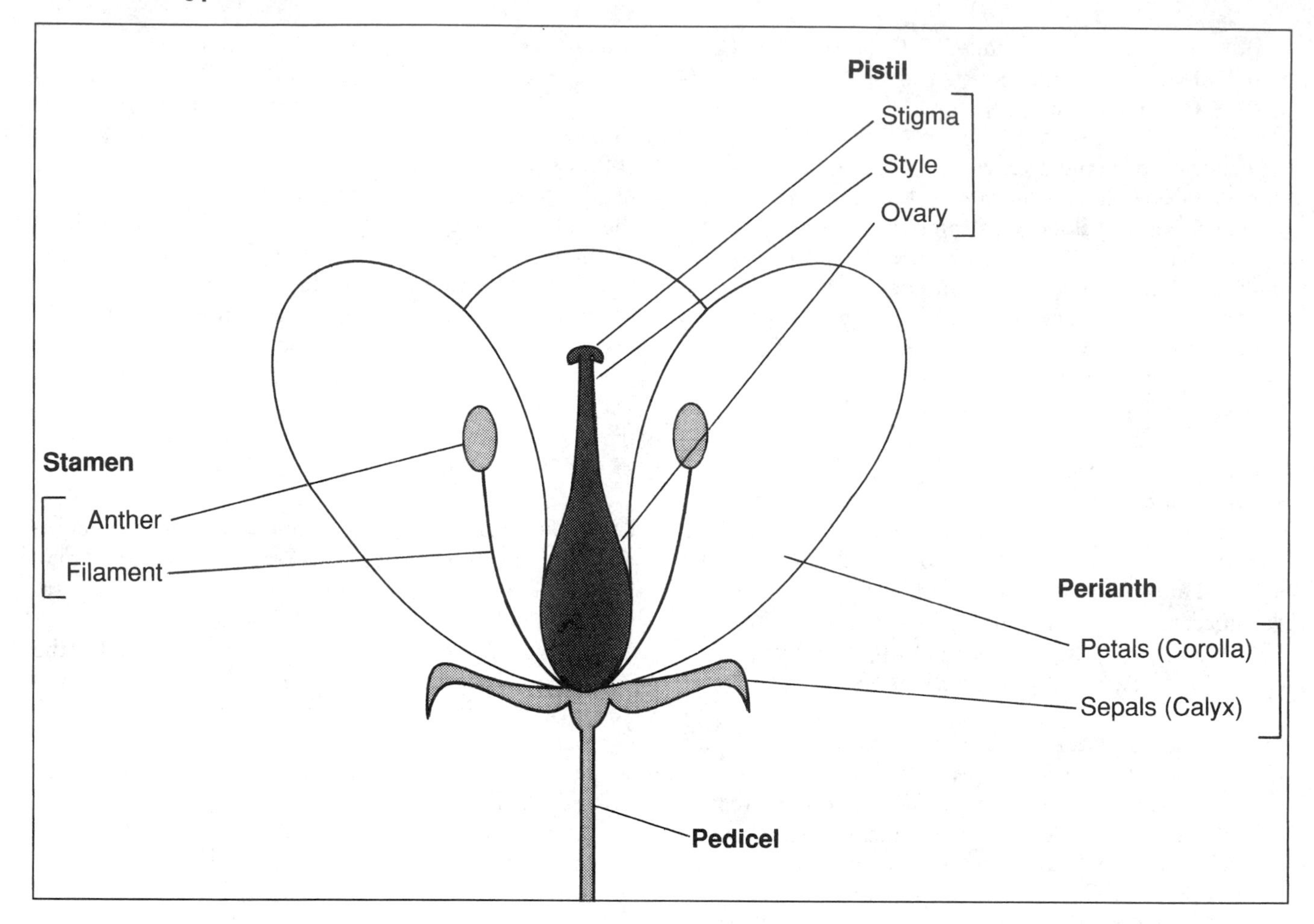

The parts of a flower

and only secondarily on the many variations of color and shape within each structural layer. *See also* ANGIOSPERM; INFLORESCENCE.

flowering plant *See* ANGIOSPERM.

FLPMA *See* FEDERAL LAND POLICY AND MANAGEMENT ACT.

flue gases the gaseous waste products of combustion that are released into the atmosphere through a chimney.

fluoridation of water the addition of small amounts of a FLUORINE compound, usually sodium fluoride (NaF), to public water supplies in order to reduce or prevent tooth decay in the children drinking the water. The fluoride IONS provided by the dissolved compound are incorporated into the tooth enamel as it is formed, hardening it and making it considerably more resistant to bacterial action (*see* BACTERIUM). The process has little or no effect on adult teeth in which the enamel has already been laid down. Because sodium fluoride is a virulent poison (ingestion of as little as 5 grams can kill an adult human; ingestion of less than a gram can lead to severe abdominal pain, nausea, vomiting, and muscular weakness; chronic long-term ingestion of as little as two parts per million can cause FLUOROSIS), fluoridation is a controversial process, though the extremely small amounts used (0.7–1.0 ppm) appear to pose no health hazard, and naturally fluoridated waters—which are fairly widespread in the world—have been consumed for centuries without demonstrable harm to the populations consuming them.

fluorine the ninth element in the ATOMIC SERIES, atomic weight 19.0, chemical symbol F. The lightest and most reactive of the HALOGENS, fluorine is the most chemically active of all the nonmetallic elements, combining (or combinable with) all other elements except helium, neon, and argon. The binary compounds thus formed are known as *fluorides*. Free fluorine can only be obtained in the laboratory due to its high reactivity. It is thick, greenish-yellow gas with a choking, suffocating odor, highly corrosive and toxic to all life forms (many organic molecules simply fall apart in its presence). Water burns in an atmosphere of free fluorine, igniting by spontaneous combustion. Wood and paper burst into flame in its presence, and glass and most metals will burn in it if their temperature is slightly elevated. Fluorine compounds, on the other hand, are usually remarkably stable, and though they are also often toxic, their toxicity is commonly much lower than that of the pure element. Many are widely used. They include hydrofluoric acid (HF), which etches glass for decorative glassware and scientific instruments; uranium hexafluoride (UF_6), used in the refining of fuel for atomic reactors (*see* NUCLEAR ENERGY); sodium fluoride (NaF), added to public water supplies as a preventative of tooth decay (*see* FLUORIDATION OF WATER); and a wide variety of FLUOROCARBONS. *See also* FLUOROSIS.

fluorocarbon any HYDROCARBON COMPOUND in which one or more of the HYDROGEN ATOMS has been replaced by an ATOM of FLUORINE. Fluorocarbons are generally chemically inert and extremely stable, with high DIELECTRIC properties. They are used as hydraulic fluids, refrigerants, lubricants, insulators, and in a wide variety of other industrial applications. Teflon, a fluorocarbon POLYMER, is used in the manufacture of nonstick cookware, artificial heart valves, automotive plastics and "breathable" rain wear. *See also* CHLOROFLUOROMETHANE.

fluorosis disease caused by chronic, long-term overexposure to FLUORINE compounds. In animals (including humans), the symptoms of fluorosis include mottled teeth and brittle bones. In plants, the disease shows up as a loss of growth vigor and a blue tinge to the foliage. *See also* FLUORIDATION OF WATER.

fluvial dam a type of FLUVIAL DEPOSIT that forms where a small, swift stream enters a larger, slower one. The smaller stream slows down, forcing it to drop part of its sediment load as it enters the larger stream. This sediment is deposited in the bed of the larger stream, creating a barrier. *See also* ALLUVIUM.

fluvial deposit in geology, material laid down by running water. Fluvial deposits are always sorted with GRAVELS and other heavier materials on the bottom, forming what are known as *graded beds*. They are made of coarser particles than are LACUSTRINE DEPOSITS, and generally occur in a less continuous fashion. Examples include GRAVEL BARS, FLUVIAL DAMS and natural LEVEES. *See also* ALLUVIUM.

fly ash finely divided particles or ASH produced by the combustion of fuels, principally powdered COAL, and carried into the atmosphere by FLUE GASES. The size range of fly ash particles is generally defined as between one micron and 200 microns in diameters. *See also* PARTICULATE.

flyway in ornithology, a pathway taken by migrating birds, especially WATERFOWL, from their nesting grounds in northern North America to their overwintering grounds in Central America and South America. (The term could conceivably be applied just as well to migratory bird pathways in Europe, Asia and Africa, but this usage is uncommon among ornithologists.) Flyways are broad but well-defined clumpings of MIGRATION CORRIDORS. Their heavy use by waterfowl means that it is especially important to preserve WETLANDS along them (*see* WILDLIFE REFUGE). Four major flyways have been mapped across the United States. These are (from west to east) the Pacific Flyway, which uses the Pacific Coast and the Cascade and Sierra Nevada mountains as north-south guidelines; the Central Flyway, which crosses the Great Plains in a wide band along the east slope of the Rockies; the Mississippi Flyway, which follows the Mississippi River; and the Eastern or Atlantic Flyway, along the Atlantic coast and the Appalachian Mountains.

F/M ratio (loading) in sewage-treatment technology, the food-to-microorganisms ratio, that is, the ratio of the mass of BIOCHEMICAL OXYGEN DEMAND (BOD) in the incoming WASTE STREAM to the mass of MICROORGANISMS in an ACTIVATED SLUDGE tank or similar SECONDARY TREATMENT facility. The mass of the microorganisms is generally estimated by measuring the mass of the MIXED LIQUOR SUSPENDED SOLIDS (MLSS); thus, the F/M ratio, as measured, is actually the BOD/MLSS ratio. It is generally in the neighborhood of 0.2 to 0.5, but may be as low as 0.05 (in a so-called "extended aeration" facility) or as high as 2 (in a "high-rate" facility) (see SEWAGE TREATMENT PLANT). Monitoring the F/M ratio to make certain that it is at or near the design rate is one of the best means of keeping track of the health of a secondary-treatment facility.

fog in meteorology, a cloud resting on the ground; in more technical language, a collection of water droplets in the air at or very near ground level. In order to be reportable in a weather summary, a fog must reduce straight-line visibility to a kilometer or less (0.6 miles). Meteorologists recognize five separate types of fog, depending upon the means of formation of each. *Radiation fog* occurs during cold weather when a ground surface that has been heated by the sun during the day radiates that heat away at night, causing it to cool dramatically. The air directly in contact with the ground cools as well, dropping its temperature below the DEW POINT and causing fog to condense. This type of fog is rarely more than a few meters thick. *Advection fog,* by contrast, may be several hundred meters thick. It results when a warm, wet AIR MASS moves over a cold surface, as when oceanic air moves inland over a cool coastline. *Upslope fog* results from adiabatic cooling as a mass of air is forced upward by a mountainside or a slanting plain (*see* ADIABATIC TEMPERATURE CHANGE). *Frontal fog,* or *precipitation fog,* results from the intersection of a warm air mass with a cold one, and is essentially the same as advection fog except that the incoming air mass is cooled by contact with another air mass rather than with the land. Finally, *steam fog* results when moisture evaporating from a water body or an extremely wet surface (such as a street following a heavy rain) meets air that is substantially colder than the surface from which it is evaporating and immediately condenses out again. Steam fogs are rarely more than a meter or two thick.

fold in geology, a rock structure in which the layers of rock have been bent by forces acting on them after they were laid down. Folds may be small, occupying only a few cubic meters of rock, or they may be large, underlying or forming entire mountain ranges. They are often, but not always, accompanied by faulting (*see* FAULT).

FONSI abbreviation for Finding of No Significant Impact (*see* ENVIRONMENTAL ANALYSIS REPORT).

Food and Agriculture Organization of the United Nations (FAO) works to improve living conditions for rural citizens, to enhance yields and the distribution of agricultural products, and to enrich the nutrition and standard of living for populations worldwide. Founded in 1945, the FAO promotes sustainable development and the satisfying of human needs without jeopardizing the prospects of future generations. Membership (1999): 174 nations. Address: Viale delle Terme di Caracalla, Rome 00100, Italy. Website: www.fao.org.

Food and Drug Administration (FDA) agency within the U.S. Department of Health and Human Services (see HEALTH AND HUMAN SERVICES, DEPARTMENT OF) created by Congress on May 27, 1930, and charged with assuring the purity of foods, drugs, and other substances intended for human consumption and with enforcing truth-in-labeling laws. The most important part of the agency from an environmental standpoint is the National Center for Toxicological Research, which is devoted to initiating and supporting studies on the biological effects of toxic chemicals in the environment, especially as these affect human health. The Center maintains permanent research programs on the chronic long-term effects of very small doses of toxicants (*see* MICROCONTAMINANT) and on the development and improvement of tests for chemical contaminants and for their health effects on the human organism. Headquarters: 5600 Fishers Lane, Rockville, MD 20857. Phone: (888) 463-6332. Website: www.fda.gov.

food chain in ecology, a set of predator-prey relationships (*see* PREDATOR; PREY) through which the food energy of an ECOSYSTEM flows from the bottom to the top, that is, from the PRODUCERS to the top CONSUMERS. A food chain always begins with a green plant or other autotroph (*see* AUTOTROPHISM) and ends with a TOP PREDATOR. For example, a vole might eat grass and in turn be eaten by a coyote; this would form the food chain:

grass → vole → coyote

where the symbol "→" stands for "eaten by." Similarly, a small algae-eater such as a COPEPOD might be eaten by a tadpole, which might in turn be eaten by a largemouth bass, forming the food chain:

algae → copepod → tadpole → bass

Because some energy gets lost at each link of the chain, no food chain can ever be very long. Aquatic food chains tend to average about four links, land-based chains about three. Almost no straight predator-prey chain is over five links long. It is possible to lengthen a chain somewhat by including SAPROBES and/or PARASITES, but even then the number of links will seldom exceed five:

maple leaf → mushroom → mouse →
owl → bobcat

See also FOOD WEB; ECOLOGICAL PYRAMID; TROPHIC LEVEL.

Food Safety and Inspection Service (FSIS) agency within the U.S. Department of Agriculture (*see* AGRICULTURE, DEPARTMENT OF) charged with making certain that foods shipped across state lines meet federal standards for safety and wholesomeness. The agency has no authority over food shipments within individual states. Its primary environmental responsibility is to police foods for chemical CONTAMINANTS such as pesticide residues.

food web in ecology, the set of interconnecting FOOD CHAINS that represents the total pattern of pathways along which food energy may flow through an ECOSYSTEM. The food web concept recognizes that each

PREDATOR may have several different types of PREY, and that these various types of prey may exist on several TROPHIC LEVELS. A frog may eat flies, for example, and in turn be eaten by either a water snake or a bass:

flies → frog → snake
flies → frog → bass

However, the bass may also eat the snake, and either the snake or the bass may sidestep the frog stage and eat flies directly, considerably complicating the ecosystem structure (see diagram on the following page). The food web concept is among the most important tools of the science of ecology. By tracing food webs, it is possible to get a sense of the stability of an ecosystem (the more complex the food web, the more stable the system) and to identify those points where the system is most vulnerable to change or destruction. (The fewer prey SPECIES an animal uses, the more vulnerable it is to the disappearance of any one of them; hence, the least stable points in any ecosystem are those points where the strands of the food web are the least dense.) It is also possible to see where energy is being lost from the system, and to determine from that the CARRYING CAPACITY of the system for each species active in it. For example, in the food web

corn → hogs → humans
↑________________↑

the chain → hogs → humans has one more link than the chain corn → humans. Since each extra link means an additional loss of energy, the carrying capacity of this system for humans is less than that of one in which the human stopped eating pork. *See also* ECOLOGICAL PYRAMID.

foot-pound the amount of energy required to lift an object weighing 1 pound a distance of 1 foot. There are approximately 778 foot-pounds in one BRITISH THERMAL UNIT.

forage food obtained by animals through browsing or grazing. Forage is generally vegetable matter—GRASSES, FORBS, and berries—but the term is occasionally extended to include grubs and other INVERTEBRATE animals that may be thought of as being found rather than hunted down.

foraging acre in range management, an amount of FORAGE equivalent to that which would be found on 1 acre of densely packed GRASSES and FORBS.

forage unit *See* GRAZING UNIT.

foraminiferan (foram) any of numerous SPECIES of one-celled heterotrophic protozoans (*see* HETEROTROPH; PROTISTA) of the ORDER Foraminifera, CLASS Sarconida. The foraminifera are closely related to the amoebae (*see* AMOEBA); they are animal-like in appearance and behavior, and are undoubtably among the strongest reasons why it took so long for the protista to be recognized as a separate KINGDOM. They are also among the most massive of unicellular ORGANISMS, some being large enough to be seen by the naked eye. Almost all species have shells. A few accrete them from sand grains cemented together with mucous, but most secrete them in calcareous form. These calcareous varieties look remarkably like tiny shellfish. The shells ("tests") are usually multi-chambered, the organism beginning with a single globular shell and adding new, larger chambers, as necessary, in an overall pattern (straight, curved, spiral, etc.) characteristic of its species. It occupies all chambers at once, the protoplasm in each chamber being connected to the others through holes between chambers and remaining part of a single cell. Feeding and locomotion are accomplished by means of pseudopods extended through ports in the shell. Nearly all forams are marine. Most are BOTTOM DWELLERS, but a few species have buoyant shells and float in the middle or upper layers of the ocean, where they form a major part of the PLANKTON and are thus extremely important as a basic link in the marine FOOD WEB. Foraminifera shells are one of the principal components of the ocean-floor sediments, which, over the course of geologic time, become petroleum deposits, thus undergirding much of the modern world economy. In fossil form, they also make a good tool for tracing geologic STRATA. *Compare* RADIOLARIAN.

forb any herbaceous plant (*see* HERB) that is not a GRASS; especially any non-grass herb, such as clover, that is used for FORAGE by livestock or wildlife.

forced main in water or sewage treatment technology, a MAIN in which the water or sewage is pumped uphill instead of flowing downhill. Forced mains normally connect two separate gravity-flow grids. Flow from the first grid is collected in a single main at the bottom of the grid and directed into a pump house (*lift station*) that lifts it through the forced main to the top of the second grid. Other uses include conveying water from a treatment plant to an elevated reservoir and conveying sewage from the bottom of a SEWERAGE SYSTEM to a treatment plant located at a higher elevation or beyond an intervening ridge.

foredune in geology, one of the DUNES in the dune line lying closest to the water along a beach. Foredunes are the most recently formed dunes and the most actively migrating.

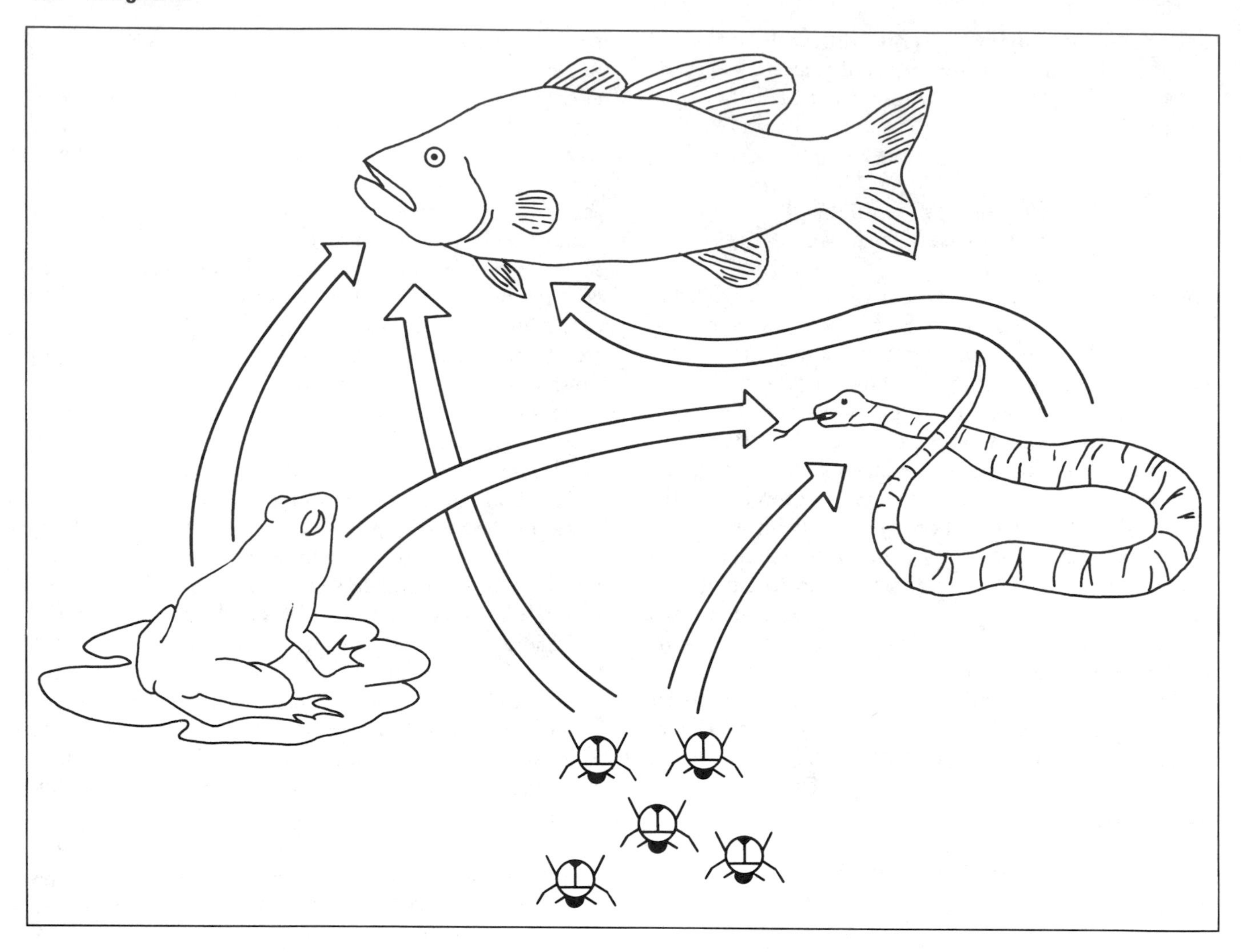

The food web

foreground in visual management, that portion of a view nearest to the observer. The foreground includes all visible objects on which textural features and patterns the size of tree branches or smaller can be seen. It is generally definable as those parts of the landscape lying within one-fourth mile of the point on which the observer stands. *Compare* MIDDLEGROUND; BACKGROUND. *See also* VIEWSHED; VISUAL RESOURCE MANAGEMENT SYSTEM.

Foreman, David (1947–) environmentalist and founder of the radical environmental group EARTH FIRST!. Born in 1947, Foreman was chair of Young Americans for Freedom, a conservative organization, and then worked for the Wilderness Society in the 1970s. He became convinced, however, that large environmental organizations had overcompromised on major environmental issues, particularly those concerning western land resources, and did not have the ability to stand up to the environmental destruction of corporations and government agencies. Inspired by EDWARD ABBEY's fictional "monkey wrench gang," Foreman founded Earth First! in 1980 to take direct action to protest logging, road-building, power lines, dams, and other developments threatening the western wilderness. While anti-environmentalists characterized Foreman and his group as ecoterrorists, great care was taken to cause no physical harm to individuals in their actions. (*See* ECOTAGE and ECOTERRORISM.) Even so, the group was put under surveillance by the FBI's counterterrorist unit for encouraging and participating in the disabling of equipment and structures and, most famously, for tree spiking. (Although all such trees were marked, the practice was abandoned when an undiscovered spike, placed by someone outside the group, was hit by a sawmill blade causing severe injury to a worker.) At length, after a two-and-a-half year, $2 million investigation, an FBI SWAT team in full body armor burst into Foreman's home in Tucson in 1989 and arrested him for disabling electrical transmission towers associated with an Arizona power project and for damaging a ski lift. Although Foreman neither directed this action nor participated in it, he was nevertheless convicted of conspiracy in 1991 after a lengthy trial. In return for his pledge to cease advocating MONKEYWRENCHING, his sentence was suspended. He soon organized a new group, the Wildlands Project, based on his book (written with Howie Wolke) *The Big Outside: A Descriptive Inventory of the Big Wilderness Areas of the United States* (1989). The project advocates the protection of large, interconnected wilderness areas to maintain a functional habitat for wildlife. Other books include *Confessions of an Eco-Warrior* (1991); *Defending the Earth: A Dialogue Between Murray Bookchin and Dave Foreman* (1991); and *Ecodefense: A Field Guide to Monkeywrenching* (1985). Susan Zakin's *Coyotes and Town Dogs* (1993) provides a fascinating account of the activities of Foreman and Earth First! during the 1980s.

Forest and Rangeland Renewable Resources Planning Act (RPA) federal legislation to create and implement a coordinated national management plan for utilizing the RENEWABLE RESOURCES of the NATIONAL FOREST SYSTEM, signed into law by President Gerald Ford on August 17, 1974. RPA requires the FOREST SERVICE to prepare for submission to the president a series of National Assessments of the condition and trend of renewable resources in the United States in as detailed a manner as possible, including both supply and demand factors, and to accompany its submission with a National Program for the utilizing of these resources on the National Forests and for improving their yields in a manner consistent with the principles of MULTIPLE USE and SUSTAINED YIELD. The first assessment and program were completed in 1975, and the second in 1979. After that, the legislation required revisions every five years for the program, and every 10 years for the assessment. Annual reports on compliance with the program were to be prepared for use by the president in preparing his budgetary requests to the Congress. In addition, the act legislatively defines the National Forest System ("all . . . lands, waters and interests therein administered by the Forest Service. . . ."), mandates the use of INTERDISCIPLINARY TEAMS for preparing resource management plans on individual units of the system, and requires the Forest Service to maintain offices in locations convenient to the public it serves, with special emphasis on rural areas near or within the forests themselves. *See* also NATIONAL FOREST MANAGEMENT ACT OF 1976.

forest coordinating requirements term used by the United States FOREST SERVICE to designate the pattern in which the various elements of a FOREST PLAN must be coordinated with each other. For the management and planning of any specific resource (water, timber, recreation, etc.), the forest coordinating requirements spell out the web of consultations among other management teams that must be undertaken before their own planning can be finalized.

forest decline and pathology in an environmental context, the degeneration of forests through natural and human-caused impacts such as pollution, climate change, and the inadvertent introduction of exotic pests and pathogens. Human-caused forest decline and concomitant tree death came to broad public attention in the 1970s and 1980s with WALDSTERBEN, the widespread dying of trees in Germany due to air pollution and acid rain. The same phenomenon also took place in the United States with a massive DIEBACK of native species on Mount Mitchell, North Carolina, Camels Hump, Vermont, and other high elevations of the Appalachian chain. On Mount Mitchell, 80% of mature spruce and Fraser fir trees died. Rime ice collected on Mount Mitchell was found to have a pH of 2.1, somewhere between battery acid and lemon juice, as a result of acid deposition from factory emissions many hundreds of miles away. On Camels Hump, 75% of the red spruce were similarly afflicted. Subsequent research, even at lower elevations, confirmed not only that trees were dying from acid deposition and tropospheric ozone (O_3) from factories and internal combustion engines, but also that over the years this deposition had leached essential nutrients from forest soils so that regeneration of many tree species was severely affected. In California researchers discovered as early as the mid-1950s that ozone-laden smog had decimated stands for ponderosa and Jeffrey pines, species that are susceptible to O_3. AIR POLLUTION is rarely the proximate cause of tree death and forest decline. But it does weaken trees sufficiently for them to be vulnerable to adventitious pests and diseases, drought, and cold. Forest decline can also result from mismanagement of natural forests, such as clearcutting and high-grading. In Michigan the removal of the pine forests in the northern part of the state turned out to be permanent, since pines cannot regenerate from stumps. The replacement forest, largely oak and aspen is imperfectly suited to this ecosystem and is therefore vulnerable to insect attack, especially gypsy moths. Elsewhere the removal of certain desirable species from a forest, such as ponderosa pine in the West, has resulted in an increased number of species that are also susceptible to insect attack and, subsequently, to forest fire due to a buildup of flammable material. In CLEARCUT areas of the Northwest, remaining forests are susceptible to BLOWDOWN and are weakened because of exposure to sun and cold weather. Scientists estimate that for every clearcut tree in this area, another dies from the ecological impact of clearcutting. In recent years improvements in air pollution laws and improved forestry practices have reduced the incidence of forest decline, but it still remains a significant environmental issue. In the 1980s the U.S. government funded the NATIONAL ACID PRECIPITATION ASSESSMENT PROGRAM to measure the effects of acid deposition and other pollutants on freshwater and forest ecosystems. The report on the forest research downplayed the effects of pollution to such a degree that the $50 million program was discredited by independent scientists and the environmental community as being tainted by political considerations. A Canadian counterpart study on the Long Range Transport of Air Pollution was more forthright, suggesting serious consequences for sugar maples in eastern Canada unless steps were taken to curb air pollution. Amendments to the CLEAN AIR ACT in 1990 call for further reductions in oxides of nitrogen and in ozone.

forest-edge habitat in ecology, the interface between forests and open space such as grasslands or watercourses, characterized by the growth of thick shrubbery and small trees. The forest edge is intermediate in humidity and light availability between the forest and the open space, and attracts a distinctive COMMUNITY of animals, including songbirds, deer, chipmunks, and other species (*see* FOREST-EDGE SPECIES). Properly conducted logging operations can increase the amount of forest-edge habitat in a region and thus increase the population of forest-edge animals, including particularly deer: this is the principal basis of the claim that logging "helps" wildlife.

forest-edge species in wildlife biology, an animal or plant SPECIES whose exclusive or principal HABITAT is the ECOTONE separating a forest from an adjacent open space such as a meadow, grassland, or body of water (*see* FOREST-EDGE HABITAT). Forest-edge plants are typically those which require open shade. They include dogwood, vine maple, rhododendron, and other woody shrubs and small trees, along with herbs such as the delphinium, the tiger lily, and the shooting-star. Forest-edge animals are generally those which prefer food found in open spaces but which also require a substantial amount of protective COVER. These include deer, rabbits, bob-whites, pheasants, and most perching birds (order *Passeriformes*). Humans should also probably be classified in this category.

forest fire *See* WILDFIRE.

Forest Handbook *See FOREST SERVICE HANDBOOK.*

Forest History Society nonprofit education organization founded in 1946 to increase the public knowledge of the historical significance of human interaction with forests. The society is concerned with forest industries, conservation, and recreation. It supports research, publications (*Forest History Today* and *Environmental History*), and an archival library. Membership (1999): 1,500. Address: 701 Vickers Avenue, Durham, NC 27701. Phone: (919) 682-9319. Website: www.lib.duke.educ/forest.

Forest Plan (in full: **National Forest Land and Resource Management Plan**) a document prepared by the FOREST SERVICE under the terms of the NATIONAL FOREST MANAGEMENT ACT OF 1976 (NFMA) to guide the management of a single NATIONAL FOREST over a 10–15-year period. It is supposed to be revised and reissued at intervals not to exceed 15 years. Replacing the earlier concept of the UNIT PLAN, the Forest Plan is prepared by an INTERDISCIPLINARY TEAM appointed by the forest supervisor. The team also prepares an ENVIRONMENTAL IMPACT STATEMENT to accompany the plan. The plan must conform to conditions spelled out in the NFMA as well as those in Forest Service planning regulations based on the NFMA and finalized in September 1979. Among these conditions are requirements for the protection of streams and their RIPARIAN ZONES, the maintenance of wildlife diversity, the evaluation of lands for biological and economic suitability for timber production and the conformance of the plan to this evaluation, and the management of the forest for SUSTAINED YIELD and full MULTIPLE USE of all its resources. Twelve so-called Program Elements must be considered at every stage of the process: these include recreation, wilderness, fish and wildlife resources, livestock range, timber production, water protection, mining and minerals development, community development (that is, a consideration of the impact of the plan on local communities), fire protection and pest control, land ownership and use patterns, soils, and facilities (campgrounds, corrals, fire lookouts, ranger stations, etc.).

THE PLANNING PROCESS

There are 10 steps to the Forest Plan process. These steps are normally seen as a circular array, with the 10th step leading directly back into the first one so that the next planning cycle can begin directly after the release of the plan. The process begins with the IDENTIFICATION OF ISSUES, CONCERNS AND OPPORTUNITIES (ICOs)—a collection of opinions as to how the forest should be managed, taken from both Forest Service employees and forest users (recreationists, loggers, ranchers, local community officials, and so on). This is followed by *selection of planning criteria* of two types: "process criteria," which will be used to evaluate the progress of the plan through its various stages, and "decision criteria," which will control the actual resource allocations made by the plan. Step 3 is an *inventory* and collection of as much information about the forest that is relevant to the preparation of the plan as is possible. Data is collected on the amounts and ages of timber, soil types and conditions, streamflow, wildlife, recreational use, the economic condition of local communities, and a variety of other things. The fourth step in the process is an *analysis of the management situation,* in which the data collected in step three is evaluated in light of the ICOs determined by step 1 to see how many of these ICOs the forest is capable of meeting. *Formulation of alternatives* follows as step 5. This involves determining a range of possible actions that meet one or more of the ICOs and which analysis of the management situation determines can be accomplished with the available resources of the forest. Once the alternatives are formulated, their effects on the various resources and on the forest users are estimated (step 6) and they are evaluated in terms of these effects for ranking in apparent value (step 7). In step 8 a *preferred alternative* is chosen based on the evaluation in step 7. Once it is finalized, this alternative becomes the actual Forest Plan. (It is at this stage that the Environmental Impact Statement is usually issued, allowing an opportunity for public review and comment before the preferred alternative is implemented.) Step 9, *implementation,* puts the plan into effect on a forest-wide basis. Step 10, *monitoring and evaluation,* determines how the plan is functioning over the next few years—feeding directly into step 1, the *formulation of issues, concerns and opportunities,* for the next planning cycle.

See also REGIONAL PLAN.

forest products manufactured products made from wood. In its broadest sense, the category includes lumber; particleboard, chipboard, and plywood; paper made from wood pulp; wood-based plastics; turpentine and wood resins; bark MULCH; fuel wood; and virtually anything else that can be made from a tree and sold. In some uses, the term may be limited to building materials, furniture, veneer, and other manufactured materials that are still recognizable as wood.

forest receipts any moneys received by the FOREST SERVICE as payment for the use of a resource. Forest receipts include funds received from the sale of timber; camp-ground fees; Christmas-tree permit fees; vacation-home leases; special-use permit fees, such as those paid by the operators of ski areas or commercial lodges on Forest Service land; and all other moneys paid directly to the National Forest System by consumers, groups of consumers, or commercial operators. *Compare* APPROPRIATIONS.

Forest Response Program *See* NATIONAL ACID PRECIPITATION PROGRAM.

forest roads the system of (largely) unpaved roads constructed in the National Forests by the U.S. Forest Service to facilitate logging. The road-building program, which is financed by the government, not the private loggers that it benefits, has long been a matter of controversy. The roads now total 378,000 miles in the nation's 155 national forests, eight times the length of the interstate highway system. In 1999 President Bill Clinton directed the Forest Service to cease building roads on more than 40 million acres of remote forest land. While members of Congress from logging states objected, fearing a loss of jobs, the ban is expected to survive a congressional challenge since very little of the U.S. softwood timber harvest—about 3%—now comes from federal forest land. Instead of logging, the most important road use is now recreational. Camper-back pickup trucks and other four-wheel-drive vehicles now outnumber logging vehicles by more than 100 to 1 on forest roads. Accordingly, in the 1990s Forest Service policy changed markedly, tending to emphasize recreation and ecosystem management more than logging, which peaked in 1987. In fact, the Forest Service is removing forest roads in some areas where they are no longer needed and pose an erosion threat.

Forest Service agency within the U.S. Department of Agriculture (*see* AGRICULTURE, DEPARTMENT OF) charged with management of the NATIONAL FOREST SYSTEM. The agency also provides advice and technical assistance to help manage forests on state and private lands, and conducts research on forest management, forest ecosystems, wood products manufacture, recreation, wilderness, fire management, and other forest-related topics.

STRUCTURE AND FUNCTION

The National Forest System consists of 155 NATIONAL FORESTS and 19 NATIONAL GRASSLANDS, totaling some 191 million acres, mostly west of the Mississippi River. The forests are organized into nine Regions, each of which operates under a regional forester; these nine regions are separate from the STANDARD FEDERAL REGIONS and are recognized only by the Forest Service itself. Overall authority for the system resides in the chief forester in Washington, D.C., who is in turn under the authority of the assistant secretary of agriculture for natural resources and the environment. Within the regions, each forest is broken down into Ranger Districts, each one headed by a district ranger. The structure of each level from the district to the office of the chief forester is loosely parallel, with separate functional staffs (or on the smaller Ranger Districts, separate individuals) assigned on each level to timber management, wildlife management, range management, recreation management, water and minerals management, forest engineering (road and trail construction), fire management, land management (land use planning), and administration. The regulations and overall policies for the agency are coordinated in the FOREST SERVICE MANUAL.

HISTORY

Under the Forest Reserve Act of March 3, 1891 (also known as the "Creative Act"), the president was given authority to declare certain areas of the public lands as Forest Reserves, to be managed by the secretary of the interior. Six years later, management of these reserves was codified by the ORGANIC ACT OF 1897, which established the outlines of what was to become the Forest Service. Management was transferred to the Department of Agriculture in 1905; the Transfer Act also named the managing agency the Forest Service. In 1907 the president's right to set aside Forest Reserves by proclamation was repealed. By that time, however, the National Forest System totaled more than 132 million acres. Another 60 million acres has been added since, mostly lands purchased from private ownership under the terms of the WEEKS LAW OF 1911. Until the early 1950s the National Forests were operated primarily as timber reserves, with private timber companies opposing the sale of timber from these reserves on the grounds that it provided unfair competition for privately owned timber. However, as private timber reserves were used up, National Forest timber became the major supply source for private sawmills, and timber sale activities were expanded greatly, leading to a flurry of protective legislation, including the MULTIPLE USE-SUSTAINED YIELD ACT OF 1960, the WILDERNESS ACT OF 1964, the FOREST AND RANGELAND RENEWABLE RESOURCES PLANNING ACT OF 1974, and the NATIONAL FOREST MANAGEMENT ACT OF 1976. Since the mid-1990s, however, logging has been de-emphasized in favor of recreation and ecosystem management. This is partly due to public demand for recreational facilities and aesthetic concerns, partly due to challenges under the ENDANGERED SPECIES ACT (especially to protect the spotted owl), and partly because much of the "easy timber" has already been logged off. *See also* ROADLESS AREA REVIEW AND EVALUATION and various entries beginning with the words *forest* and *forestry.*

Forest Service Land Areas

The U.S. Forest Service manages 191.6 million acres of national forest land—8.3% of the total land area of the U.S.

National Grasslands	3.9 million acres
National Primitive Areas	173,762 acres
National Scenic-Research Areas	6.630 acres
National Wild and Scenic Rivers	4,385 miles, 95 rivers
National Recreation Areas	2.7 million acres
National Game Refuges and Wildlife Preserves	1.2 million acres
National Monument Areas	3.3 million acres
National Historic Areas	6,540 acres
Congressionally Designated Wilderness	34.6 million acres

Source: U.S. Department of Agriculture Fact Book, 1998.

Forest Service Employees for Environmental Ethics Founded in 1989, this national organization unites Forest Service employees and other concerned persons who are interested in changing the Forest Service management philosophy by working from the inside to provide for a more ecologically and economically sustainable system. Membership (1999): 12,000. Address: P.O. Box 11615, Eugene, OR 97440. Phone: (541) 484-2692. Website: afseee.org.

Forest Service Handbook one of a series of detailed instruction manuals relied upon by FOREST SERVICE personnel for assistance in performing individual forest-management tasks such as timber sale preparation, road construction, and wildlife enhancement. The handbooks differ from the FOREST SERVICE MANUAL in that they offer principally technical guidelines rather than legal and administrative rules. They are, however, keyed to individual sections of the *Manual* and may be thought of as supplemental to it.

Forest Service Manual a detailed set of policy statements and guidelines, administrative rules, and organizational directives, used by the FOREST SERVICE to guide its day-to-day operations. It is not considered legally binding, although a clear violation of a *Manual* directive may be used as evidence of capricious decision making by the agency and will generally provide grounds for winning an appeal. Originating during the Theodore Roosevelt administration as a small pocket volume called *The Use of the National Forest Reserves* (The "Use Book") written by Chief Forester Gifford Pinchot, the *Manual* today runs to over 20 thick loose-leaf volumes organized into eight series, with each series covering a broad policy area such as land-use planning, timber planning, or agency organization. The series are broken down further into numbered titles, sections, subsections, and paragraphs. Regional Supplements augment the *Manual* for each Forest Service administrative region; Forest Supplements provide further guidance at the individual National Forest level (the FOREST PLAN is considered a supplement to the *Manual*). Continual upgrades are produced at each level to reflect new laws, judicial decisions, EXECUTIVE ORDERS, administrative rule changes, and other alterations in the administrative climate. These are

Interpreting Paragraph Numbers in the *Forest Service Manual*

Each paragraph in the *Manual* is identified by a number consisting of a four-digit whole number followed by a two-digit decimal fraction (example: 1234.56). In these numbers, the thousands place represents the series; the hundreds place, the title within the series; the tens, the chapter within the title; the ones, the section within the chapter; and the decimal fraction, the specific paragraph within the section. Our hypothetical example would break down this way:

Series	1000	(first series)
Title	1200	(second title of series 1000)
Chapter	1230	(third chapter of title 1200)
Section	1234	(fourth section of chapter 1230)
Paragraph	1234.56	(paragraph 56 of section 1234)

generally issued first as Emergency Directives and are later integrated into the *Manual* as replacement pages. Because of the bulk of the document, complete copies are not generally available outside of regional offices; however, all individual forests, and most Ranger Districts, have partial copies (including all relevant Emergency Directives), and some complete sets are available at these levels and in large local libraries.

forest types a designation system developed by the U.S. FOREST SERVICE that describes different forests by their dominant species.

formation in geology, a distinctive regional grouping of rock and soil types that may be easily identified wherever it crops up, whether or not parts of other formations intervene. Formations are generally genetically related; that is, they were formed by the same set of forces at roughly the same time. Specific elements within the formation such as sedimentary layers or batholithic bodies (*see* BATHOLITH) are usually recognizable from one outcrop of the formation to the next, allowing the formation to be traced even when it seems discontinuous. The Niagaran Dolomites, for example, are recognizable as a single formation due to their consistent composition and bedding structure, even though outcrops of them are scattered discontinuously from Rochester, New York, to Green Bay, Wisconsin. *Compare* SEQUENCE.

formation water in hydrology, water included within a geological FORMATION at the time it was laid down. *See also* FOSSIL WATER.

forty in surveying, one-sixteenth of a SECTION. A forty is a square one-quarter mile on each side, containing exactly 40 acres of land. *See also* ROUND FORTY.

fossil in geology, the preserved physical evidence of an ancient ORGANISM or of an ephemeral natural phenomenon such as a raindrop, a watercourse, a beach, or a lake bed. (By "ancient" geologists generally mean greater than 10,000 years of age.) The most common type of fossil is probably the *cast,* an impression in mud, clay, or sand preserved unchanged as the material hardened into stone. Typical casts include impressions of ripple marks, splash marks, footprints, leaves, and soft-bodied animals, often remarkably detailed. Other types of fossils include *permineralized* ("petrified") tissues,

U.S. Forest Types

Eastern Forests	Western Forests and Hawaii	Alaska Forests
White-red-jack pine	Douglas fir	Hemlock-Sitka spruce
Spruce-fir	Hemlock-Sitka spruce	Closed spruce-hardwoods
Longleaf-slash pine	Ponderosa pine	Open, low growing spruce
Loblolly-shortleaf pine	White pine	
Oak-pine	Lodgepole pine	
Oak-hickory	Larch	
Oak-gum-cypress	Fir-spruce	
Elm-ash-cottonwood	Redwood	
Maple-beech-birch	Chaparral	
Aspen-birch	Pinyon-juniper	
	hardwoods	
	Ohia	

Source: U.S. Geological Survey

where mineral deposits have invaded the soft body parts of an organism and gradually replaced them, resulting in a mineral deposit that is an exact replica of the organism; and carbon residues (often in the shape of the original organism) left behind as the body parts of the organism are broken down by DESTRUCTIVE DISTILLATION in the absence of decomposing organisms (*see* DECOMPOSER), as beneath flows of LAVA. Occasionally, hard (and even soft) body parts of ancient organisms may be found unaltered. The most dramatic examples of this have been mammoth bodies, preserved in PERMAFROST, which appear fresh enough to cook and serve (this was reportedly actually done by a Russian army regiment in the late 19th century) and fossil POLLEN grains found in ancient PEAT BOGS that have been successfully sprouted. Fossils show an increase in complexity and diversity over geologic time, and are among the best forms of evidence for the existence of organic evolution. *See also* FOSSIL FUEL; FOSSIL WATER.

fossil fuel fuel made from the organic remains of fossilized ORGANISMS (*see* FOSSIL) and recovered from the rock STRATA in which the fossilized organisms are found. The two principal types of fossil fuels are COAL (made from fossilized SWAMP and BOG plants and composed largely of elemental CARBON) and PETROLEUM (made primarily of fossilized MICROORGANISMS deposited in the beds of ancient seas and composed of a mixture of complex ORGANIC COMPOUNDS: NATURAL GAS, often classed as a separate type of fossil fuel, is almost certainly formed as a BYPRODUCT of petroleum formation and thus should properly be placed with the petroleums in classification schemes). Fossil fuels are the primary energy source for modern society, and exist in vast stores; however, these stores are finite and, once used up, will not be replaced for many millions of years, making modern society's near-complete reliance on them for such important functions as agricultural production and household heat a foolish and, ultimately, a dangerous practice. However, as public understanding of the environmental impact of fossil fuel use has slowly developed, conservation practices and alternative fuel sources, such as SOLAR ENERGY, HYDROGEN FUEL CELLS, and the like have been gaining ground. In 2000 the HYBRID-ELECTRIC CAR was introduced to the mass market, perhaps signaling a basic change in public attitudes.

fossil water water trapped in ancient AQUIFERS that are not being recharged, that is, aquifers that have no means of replacing water after it is drawn out. Generally, fossil water lies within aquifers that are overlain by impermeable rock (*see* PERMEABILITY). They once had connections to the surface, allowing water to enter and fill them up, but these connections have eroded away or have been blocked off, leaving the water within them. Fossil water usually has a high mineral content due to the amount of time it has spent underground. Wells can bring it up for human consumption; however, since it is not being replaced, its use should be looked upon as a form of mining, which—like any other mining—will eventually reach the limits of the mineral LODE being mined. Thus, no water-dependent developments should be based on fossil water unless a plan exists to convert to non-fossil sources before the fossil water runs out. Huge lodes of fossil water underlie much of North America's High Plains region, where they have served as the principal water source for much of the continent's food production. Most of these lodes are currently very close to being "mined out," with potentially disastrous results for the world's food supply. *See* OGALLALA AQUIFER.

FRA *See* FEDERAL RAILROAD ADMINISTRATION.

fractal geometry developed by Benoit Mandelbrot in 1982 to provide a means to measure the dimensions of complex shapes, such as coastlines. Fractal geometry uses different scales to measure the same shape, the scale to be determined by the purposes of the measurement. Thus the number of miles in the coastline, of, say, Chesapeake Bay, for navigation would be many fewer than the miles that might be measured for a study of commercial shellfishing beds.

fractional distillation *See under* DISTILLATION.

fractionation *See under* DISTILLATION.

fracture in geology, (1) a break in a body of rock such as a BATHOLITH or a sedimentary STRATUM in which the separate pieces formed by the break have not moved apart from each other any significant distance. *Compare* FAULT.

(2) the shape taken by the face of a rock sample when it is broken. It is usually characteristic of the type of rock or mineral found in the sample, and can be used for identification purposes. *See also* CLEAVAGE PLANE.

Framework Convention on Global Climate Change an international agreement, developed at the 1992 United Nations Conference on the Human Environment (RIO EARTH SUMMIT), to negotiate a protocol to reduce greenhouse gases. The agreement (the "Rio Treaty") endorsed voluntary, nonbinding measures to stabilize greenhouse gases "at a level that would prevent dangerous anthropogenic interference with the climate system." After several false starts, the parties to the Rio treaty met in Kyoto, Japan, to create a binding protocol which took at least a faltering step toward international efforts to deal with climate change. *See also* KYOTO PROTOCOL.

Francis of Assisi (1182–1226) Italian mystic, poet, and founder of the Franciscan order. After a pleasure-seeking youth, Francis had a total change of heart during his imprisonment in Perugia. While in prison, he became severely ill and vowed to change his way of life. He sold all he had, giving the money to the church, and devoted himself to serving the poor and sick. Francis is now considered the patron saint of ecology, based on his mystical sensibilities toward the natural world as expressed in his poem, "Canticle of Brother Sun." *See also* RELIGION AND THE ENVIRONMENT.

"Frankenfoods" See GENETIC ENGINEERING.

Freedom to Farm Act *See* FEDERAL AGRICULTURE IMPROVEMENT AND REFORM ACT OF 1996.

freeze distillation a method of obtaining freshwater from saline or heavily mineralized sources by freezing it instead of boiling it. (Since no evaporation occurs, the term "distillation" for this process is technically a misnomer; hence, some authorities prefer to call it "freeze separation" or "crystallization.") Freeze distillation depends on the fact that water with salt in it freezes at a lower temperature than does freshwater; hence, ice that forms on top of a body of salt water is largely freshwater ice, and may be skimmed off and melted as a source of water for domestic use or for irrigating plants. The process may be repeated several times, with purer water obtained each time. Efficiency declines rapidly, however, with only marginally purer water being obtained after the second or third ice "crop" is melted and refrozen.

Freon *See* CHLOROFLUOROCARBON.

friability in geology and soil science, crumbliness, that is, the ease with which a sample of rock or soil breaks apart. A rock sample is considered friable if it can be crumbled in the hand. A friable soil is a moist soil that neither falls apart nor clumps stiffly together.

Friends of the Earth (FOE) environmental activist organization, founded in 1969 to foster the development of a planet-wide environmental ethic. Although FOE engages in lobbying and litigation on what it considers to be particularly offensive projects, its prime efforts are devoted to the promotion of environmental values and to the worldwide acceptance of the precedence of these values over national and economic interests. It produces a journal, *Not Man Apart.* Address: The Global Building, 1025 Vermont Avenue NW, Suite 300, Washington, DC 20005. Phone: (202) 783-7400. Website: www.foe.org.

Friends of the River (FOR) environmental activist organization, founded in 1975 to engage in lobbying, litigation, and advocacy on behalf of wild rivers. Originally formed to oppose the New Melones Dam on the Stanislaus River in California, FOR has expanded its focus to river protection nationwide, although most of its projects are still in California. It has occasionally engaged in civil disobedience. Membership (1999): 10,000. Address: 128 J Street, 2nd Floor, Sacramento, CA 95814. Phone: (916) 442-3155.

front in meteorology, the boundary between a pair of AIR MASSES. Fronts move ("advance") across the landscape as the air masses move. If the air mass following the front is cooler than the air mass it is replacing, the front is called a COLD FRONT. If the new air mass is warmer than the air it is replacing, the front is called a WARM FRONT. The boundary between the air masses always lies in a slanting line in relation to the ground, with the warmer air above the line and the cooler air below. Thus cold fronts lift the edge of the warmer air they are replacing and move in beneath it, while warm fronts ride up over the edge of the resident air mass. When a fast-moving front overtakes a slower one, the result is an *occluded front,* with the older front generally riding on top of the overtaking one, a condition that leads to extremely unstable weather. Fronts are the locations of most storms ("frontal weather patterns") because of the instability caused by the steep temperature and pressure gradient from the leading edge to the trailing edge of the front. Generally, the steeper this gradient, the more severe the accompanying storms.

frontal fog *See under* FOG.

frost heave an upward movement of the soil due to the freezing of water within it. The expansion of the water as it freezes pushes the soil upward. If the soil is on a slope, it will settle slightly downslope upon thawing, causing a slow "flowing" of the soil surface called *soil creep.* Frost heave is a powerful force that can throw building and bridge foundations askew, cause waving and bumps in highway surfaces, and move large stones upward through the soil. It is the chief cause of PATTERNED GROUND features in cold climates. *See also* FROST-WEDGING.

frost-wedging in geology, the breaking apart of a body of rock through the freezing and thawing action of water. Frost-wedging depends on the fact that water expands slightly upon freezing (*see* WATER;

ICE). Water flows into cracks in a rock body while it is above 0°C (32°F); when the temperature of the rock drops below 0°C, the water in the cracks freezes, and the force of its expansion breaks the rock apart. The broken-off piece is usually held in place by the ice until the ice thaws, at which time the rock falls away. Frost-wedging is one of the major erosive forces in the high mountains, where the nights are generally below freezing and the days generally above, so that each 24-hour period brings a complete cycle of freezing and thawing.

fruit in biology, the seed-bearing body of an ANGIOSPERM, including the seed itself and the fleshy tissue that develops around it in the ripened state. (The horticultural definition is somewhat more limited. To class as a fruit in horticulture, the fleshy tissue surrounding the seed or seeds must be edible and must be produced by a tree, or at least by a PERENNIAL.) The fruit develops from the ovary of the flower following fertilization (*see* FLOWER: *parts of a flower*), sometimes with other parts of the flower included as secondary structures. In the simplest kind of fruit, the *berry* (grape, tomato) the mature seed is simply embedded in a more or less swollen, usually water-engorged, ovary. The *drupe* (peach, plum) resembles a berry but surrounds the seed with a hard coating (the "pit") derived from the inner wall of the ovary. The *pome* (apple, pear) is like a drupe but contains many small pits ("carpels") instead of a single large one, and the *nut* (oak, walnut) is like a drupe but with the flesh reduced to a thin covering (the "husk"), which often does not entirely surround the pit. *Capsules* (maple, pea) are thin, usually soft, one- to many-chambered shells in which the seeds are suspended; they split open to scatter the seeds when they become ripe (single-chambered capsules are sometimes known as *legumes,* after the principal family of plants that bears them). *Aggregate fruits* (raspberry, blackberry) consist of a number of berries fused into a single structure; *multiple fruits* (strawberry, pineapple) are aggregate fruits that fuse other parts of the flower into the fruit along with the ovary. Other important types of fruit include the *hesperidium* (orange, grapefruit; like a berry but with a hard, oily rind and with the flesh divided into multiple chambers), the *pepo* (squash, banana; a pome with a skin or rind formed from the receptacle of the flower rather than from the ovary), and the *caryopsis* (corn, wheat; an unfused aggregate fruit in which the ovary wall fuses to the seed and becomes the seed wall, producing a fleshy structure, known as a *kernel,* that is essentially all seed).

fruiting body in biology, the specialized spore-producing organs of the higher fungi (class Basidiomycetes: *see* FUNGUS), including the mushrooms, toadstools, puffballs, shelf fungi, and related forms. The fruiting body is usually the only visible (above-ground or out-side-of-tree-trunk) portion of the ORGANISM. The term "fruiting body" is also sometimes used loosely for any spore-producing body (*sporangium*), including those of ferns, mosses, club mosses, and slime molds.

fry in wildlife biology, a juvenile fish, especially one that is less than one year old. *Compare* FINGERLING.

FSIS *See* FOOD SAFETY AND INSPECTION SERVICE.

F-scale *See* FUJITA INTENSITY SCALE.

fuel in forestry, all potentially burnable natural materials, living and dead, in a forest. The term refers especially to the woody parts of shrubs and to dead wood materials such as SLASH and WINDTHROW. An accumulation of such materials over a number of years' time is referred to as a *fuel buildup. See also* FUEL HAZARD.

fuel break *See under* FIREBREAK.

fuel cells *See* HYDROGEN FUEL CELL.

fuel hazard in forestry, the amount of fuel buildup present (*see* FUEL), especially that which is dry enough to be easily ignited and massive enough to allow a fire to spread quickly and to make suppression difficult once spreading has begun. *See* WILDFIRE.

fuel oil the petroleum fraction midway in VOLATILITY between lubricating oil and GASOLINE (*see* PETROLEUM; DISTILLATION). Fuel oil, as the name implies, is used principally as a fuel for heating and for driving diesel engines. It comes in grades numbered 1 through 6 according to VISCOSITY, with the more viscous ("heavier") grades receiving the higher numbers: nos. 1 and 2 ("light oils") include kerosene, range oil, diesel oil, and similar fuels, while no. 6 ("residual oil") is used primarily in industrial burners. Nos. 4 and 5 are blends of no. 2 with no. 6 in varying amounts to meet viscosity or sulfur-content requirements. Grades nos. 5 and 6, the so-called bunker oils, are semisolid at ordinary temperatures and must be preheated to be used. Though only moderately toxic, fuel oils are highly flammable and explosive and thus are properly listed as hazardous materials. The heavier grades pose particular problems during OIL SPILLS due to their tendency to coat all solid matter, including the bodies of animals and plants, with a thick film that interferes with natural functions such as air exchange, grooming, and the flight of WATERFOWL. It also makes foods unpalatable and is extremely difficult to remove once it has formed.

fuelwood consumption an issue in developing countries that has reached crisis proportions. In a village economy, fuelwood is gathered from nearby woods and forests for cooking and heat, in contrast to the delivery of fuel, such as coal, oil, or gas, in industrial societies. According to the United Nations Food and Agricultural Organization (FOA), the fuelwood shortage will affect 2.7 billion people in 77 countries in the near future. The environmental impacts of fuelwood gathering can be extremely severe, denuding hillsides and large wooded areas and thus leading to disastrous flooding.

Fujita Intensity Scale (F-scale) scale used by meteorologists to measure the intensity of TORNADOES. The F-scale correlates wind speed with the damage experienced along the tornado's path. It runs from 0 to 5, with F0 ("light damage") indicating winds of less than 116 kilometers/hour (70 miles per hour), while F5 ("incredible damage") indicates winds of over 419 km/hr (250 mph). In use, the scale is read backwards. The amount of damage is assessed; an F-number is assigned; and the corresponding wind speed can then be estimated.

full pool in civil engineering, the highest water level that a RESERVOIR has been designed to contain. In times of particularly high runoff, a reservoir may sometimes operate "above full pool"; in such cases it will sometimes flood out of its banks.

Fuller, Buckminster (1895–1983) American philosopher, writer, inventor, and environmentalist, born on July 12, 1895, in Milton, Massachusetts, and largely self-educated (he was twice expelled from Harvard, in 1914 and again in 1915). Best known as the inventor of the geodesic dome (patented 1953), Fuller was a man of extremely broad interests and aptitudes who also developed the Dymaxion map (a means of representing a three-dimensional world on a two-dimensional surface without distortion) and the prefabricated house, and who served as an international lecturer and consultant on the application of technology to problem-solving in an immense variety of fields. His philosophy, with its emphasis on APPROPRIATE TECHNOLOGY, synergistic relationship (*see* SYNERGY) and the need to concentrate on the use of RENEWABLE RESOURCES, had a profound impact on what might be called "applied environmentalism," that is, the practical application of environmental idealism. He is credited with coining the term "Spaceship Earth" as an analogy of the need for technology to be self-contained and to avoid waste. His infectious enthusiasm for these concepts was probably as important as their content. Fuller died in Los Angeles, California, on July 1, 1983.

Fujita Intensity Scale

Rank	Wind Speed (km/hr)	Expected Damage
F0	<116	light
F1	116–180	moderate
F2	181–253	considerable
F3	254–332	severe
F4	333–419	devastating
F5	>419	incredible

fullers earth a nonplastic aluminum magnesium silicate CLAY of the same chemical composition as KAOLIN but with a higher HYGROSCOPIC WATER content. It appears to be derived from the weathering of augite and HORNBLENDE rather than from FELDSPAR. Fullers earth has the ability to maintain its colloidal structure (*see* COLLOID) through a wide range of conditions, which makes it an excellent adsorber (*see* ADSORPTION). Once used primarily to clean or "full" the natural oils from wool fibers before spinning (hence the name), fullers earth is used today principally as a decolorizing and purifying agent during the refinement of PETROLEUM and as a substitute for ACTIVATED CARBON in water-treatment filters. It is also used as a filler in rubber products, and as an adsorbent material to help clean up OIL SPILLS.

fumigant any PESTICIDE that is active in the gaseous, rather than the liquid or solid, state. Being gases, fumigants usually affect the respiratory tract, although some function as CONTACT POISONS. To be effective, they must be applied in an enclosed space such as a building, a tent erected especially for crop fumigation, etc. One class, the *soil fumigants,* is applied by burying volatile chemicals (*see* VOLATILITY) in the soil. As the volatilized gases rise through the soil they fill the soil pores, killing NEMATODES and other pest ORGANISMS.

functional response in ecology, the tendency for animals to increase the rate of feeding as more food is made available, but only to a certain point, after which the feeding rate levels off.

Fund for Animals animal welfare organization, founded in 1967 to promote the rights of all animals, both domestic and wild. In addition to its work in preventing cruelty to animals, the fund does a significant amount of work in ENDANGERED SPECIES protection, HABITAT preservation, and PREDATOR-CONTROL reform. It is among the most activist-oriented of all animal-rights groups. Membership (1999): 250,000. Address:

200 West 57th Street, New York, NY 10019. Phone (212) 246-2096.

fungicide a chemical that can kill a FUNGUS. Most of those used commercially are COMPOUNDS of ZINC, NITROGEN, and/or SULFUR. The most common are probably those used on ringworm, athletes' foot, and other fungal infections of humans and domestic animals, including zinc undecylenate ($C_{22}H_{38}O_4Zn$: a SALT of undecylenic acid), Tolnaftate ($C_{19}H_{17}NOS$: an ESTER of one of the carboxylic acids), and similar compounds. *Bordeaux mixture*—a mixture of copper sulfate and calcium hydroxide—was once widely used as a fungicide on agricultural crops and landscape plants, but is no longer recommended due to the dangers it poses to animals and to the plants themselves. It has been replaced by safer compounds, primarily carboxylic acids and their esters. Fungicides are classed as ECONOMIC POISONS and are regulated by the federal government under the FEDERAL INSECTICIDE, FUNGICIDE, AND RODENTICIDE ACT.

fungus (plural: **fungi**) any of numerous plantlike ORGANISMS belonging to the KINGDOM Fungi, including the molds, mushrooms and toadstools, mildews, yeasts, and numerous plant PATHOGENS such as rusts, blights, and rots. Once considered part of the plant kingdom, fungi are now classed separately due to significant differences in their structure, METABOLISM, and feeding and reproductive strategies. All are HETEROTROPHS; most are saprobic (*see* SAPROBE), although some are PARASITES and a few, especially among the so-called Imperfecta, are predatory (*see* PREDATOR). Ecologically, they are classed as DECOMPOSERS, responsible (together with the bacteria; *see* BACTERIUM) for the breakdown of large organic MOLECULES (*see* ORGANIC COMPOUND) and the release of NUTRIENTS back into the environment.

STRUCTURE AND FUNCTION

Most fungi are multicellular organisms, although a number of one-celled SPECIES exist. All share a cell-wall structure made of chitin (the same material found in insect exoskeletons) instead of CELLULOSE and a feeding habit of excreting ENZYMES to digest the food outside the organism, after which the freed nutrients are absorbed through the cell wall. All reproduce by means of spores. The multicellular forms are composed entirely of long filaments known as *hyphae* (singular: hypha). Specialized structures such as mushrooms are formed of numerous hyphae compressed together side by side. Collectively, the hyphae form a body known as the *mycelium*. This is primarily a widespread net of hyphae, with the visible reproductive structures (mushrooms, puffballs, etc.) forming only a small portion of the total mass. Parasitic fungi have specialized hyphae, known as *haustoria*, that are able to penetrate the host species; saprobic varieties often have similar structures, known as *rhizoids*, that anchor them to the soil or to the dead plant and animal matter they feed upon.

CLASSIFICATION OF FUNGI

Because the fungi have only recently been separated into their own kingdom, their taxonomic classification remains somewhat unsettled. The growing tendency among mycologists (students of fungi) is to recognize five major divisions (note that the plant term, *division*, is retained in place of the animal term *phylum*). The Basidomycota, or "club fungi," produce their spores on tiny clublike structures known as *basidia*. This division includes the mushrooms, the puffballs, the shelf fungi, and many rusts and smuts (plant diseases). The similar-appearing Ascomycota (the "sac fungi") produce their spores in sacklike structures, the *asca*, instead of on clubs. These include the morel "mushroom," the truffle, the yeasts, most molds and mildews, and some blights. Zygomycota, including bread mold and similar fungi, develop a "resting stage" called *zygospore*, which amounts essentially to a fertilized but unreleased spore. The zygospores form a hairlike mass of filaments, usually black, on the surface of the substance invaded by the fungus. They may remain dormant in this stage for several months. Oomycota ("egg fungi") differ substantially from other fungi in that they release motile reproductive cells similar to plant sperm. Their CHROMOSOME structure is also different, leading some mycologists to conclude that they evolved separately from the other fungi. Some go so far as to include them with the PROTISTA instead of the Fungi. The Oomycota include the so-called "water molds" and some diseases of plants, notably the notorious "late blight" that caused the great potato famine in Ireland in the middle of the 19th century. Finally, the Deuteromycota, or "fungi imperfecta," are a heterogeneous group with one major similarity: they reproduce asexually, without going through a sexual stage at all. Most of these fungi appear to be degenerate Ascomycota that have lost their sexual stage. They include the fungi that produce penicillin and other ANTIBIOTICS; ringworm and athlete's foot; and the yeastlike organisms that create most cheeses, soy sauce, and the first-stage fermentation of saki (Japanese rice wine). (Note that not all authorities accept these divisions: some place all fungi in a single division, the Mycota, with the divisions suggested above being treated as separate CLASSES instead.) *See also* LICHEN; MYCORRHYZA.

furan in chemistry, any of a large group of COMPOUNDS containing the *furan ring*, which consists of

The furan ring

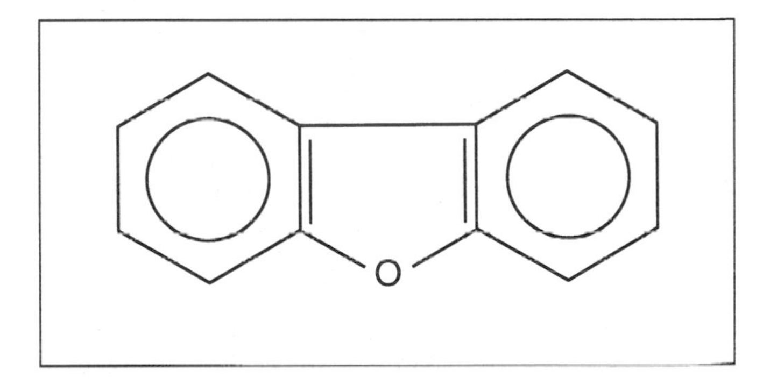

Dibenzofuran (generalized structure)

four CARBON atoms and one OXYGEN atom bound together in a cyclic form. The furans include a very large number of compounds, including some sugars (the *furanoses*), a number of ALCOHOLS, and others; many of these are beneficial. In the popular press, however, the term "furan" has come to refer to a relatively small subgroup of the furans, the *dibenzofurans*. These compounds consists of a furan ring integrated into a pair of BENZENE RINGS, so that each benzene ring shares two carbon atoms with the furan ring. When CHLORINE atoms are attached to one or more of the outer carbon atoms of the benzene rings, the result is a *polychlorinated dibenzofuran* (PCDF): PCDFs are closely related in structure and characteristics to the PCBs (*see* POLYCHLORINATED BIPHENYL) and the polychlorinated dibenzo-*para*-dioxins (*see* DIOXIN). Dibenzofurans are found as trace CONTAMINANTS in most combustion products, including incinerator smoke, cigarette and cigar smoke, and automobile exhaust.

fusion point in the physical sciences, the melting point; that is, the temperature at which a substance is transformed from the solid phase to the liquid phase (such as the transformation of ice to water at 0°C, or 32°F).

FWS *See* FISH AND WILDLIFE SERVICE.

G

gabbro in geology, a dark-colored, coarse-grained IGNEOUS ROCK in which the interwoven mineral grains are large enough to be identified as to type. Gabbro is formed as part of a PLUTON. It consists largely of PLAGIOCLASE FELDSPAR and PYROXENE, and usually has a very high iron content, giving soils decomposed from it a reddish tint. The stone holds its shape well, and is sometimes used in place of GRANITE for monuments and building stones.

gabion basket (gabion) an EROSION-control device consisting of a rectangular basket made of heavy steel mesh and filled with broken rock or river stones. In use, a number of gabions are wired together to form a retaining wall. Gabions are relatively inexpensive and have the great advantage, from an environmental standpoint, of being filled with native stone and thus not introducing any foreign element into the site except the mesh itself. The spaces between stones collect soil and eventually support plant life, helping to keep the stones in place after the steel mesh rusts away. When first installed, however—and for a number of years afterward—gabions look distinctly messy, and they never stop looking artificial. They are used for river-bank erosion control, as retaining walls in highway cuts and other permanent excavations, and for the improvement of fish HABITAT in streams.

Gaia hypothesis the idea that the BIOSPHERE functions as a single huge self-regulating ORGANISM; in other words, that life itself creates the conditions in which life can flourish, maintaining the Earth's atmospheric composition, planetary temperature balance, and other parameters within the narrow ranges that living organisms can tolerate. The idea exists in two forms, the so-called "weak hypothesis," which merely states that life significantly alters atmospheric conditions and other planet-wide phenomena, and the "strong hypothesis," which states that these alterations tend to operate within negative-feedback loops that stabilize themselves within the beneficial life-supporting ranges. A common model used by Gaia proponents to explain the concept is "Daisyworld": a hypothetical planet on which the only living things are two species of daisies, a light-colored one that tolerates warm conditions and a dark-colored one that tolerates cold. Assume the planet is covered mostly by dark-colored daisies, indicating cool conditions. Because they are dark, these daisies will absorb extra solar radiation, warming the planet. As it warms, conditions worsen for the dark-colored daisies but improve for the light-colored ones, so that eventually light-colored daisies cover most of the planet. They reflect more solar radiation into space, thereby cooling the planet and allowing the dark-colored daisies to take over again. Thus the overall planetary temperature is kept within a narrow range. Gaia proponents state that a large number of similar, but much more complex, negative-feedback loops maintain the complex conditions of Earth in a state that allows life to flourish. There is a considerable amount of circumstantial evidence to support the Gaia hypothesis, including the remarkable stability of the atmosphere's composition over geologic time and the recent discovery that oceanic PLANKTON communities play a major role in determining the amount of cloud cover over the oceans and thereby significantly alter the Earth's climate.

HISTORY

The belief that the earth is a living organism is strong among so-called primitive peoples, and the term "Gaia" used to describe the hypothesis is the ancient

Greek name used for the "Earth-mother," or living Earth. As a scientific idea, the Gaia hypothesis may be traced back to the 18th-century Scottish geologist James Hutton, who referred to the earth as a "superorganism." In its modern form, it dates from the early 1970s, when it was stated as a formal hypothesis by the British ecologist JAMES LOVELOCK (1919–) working in collaboration with U.S. biologist Lynn Margulis. Lovelock's first book on the topic, *Gaia: A New Look at Life on Earth,* was published in 1979. In 1988 he published *The Ages of Gaia: A Biography of Our Living Earth,* which more resolutely connected the hypothesis with the manifold global environmental issues that were gaining greater scientific attention in the 1980s. Since then, the Gaia hypothesis, like the new cosmology (BIG BANG THEORY) and CHAOS THEORY, has become a fixture in the effort to understand the dynamics of global ecosystems and the implications of intractable human-caused environmental problems such as global warming and species extinction. *See also* DEEP ECOLOGY.

Galápagos Islands located off the coast of Ecuador and made significant by CHARLES DARWIN whose findings there (in 1835) led to his theory of natural selection. The Galápagos archipelago, made up of 13 large islands, is named after the huge, 500-pound tortoises found there. Many unique species are indigenous to these sparsely vegetated volcanic islands, including extremely tame large land and marine iguanas, a flightless cormorant, and unusual marine life in the lagoons and tidal shallows. Especially telling to Darwin, however, was his finding that of the 26 species of land birds on the island, almost all of them (21 or 23, he was not quite sure) were "peculiar" (his word), while of the 11 species of marine birds, only 2 were peculiar. This indicated evolutionary adaptations based on environmental factors through natural selection. (He did not use the phrase "survival of the fittest.") The case for natural selection was most persuasively made by the so-called Darwin's finches, a group of 13 species that are found nowhere else but the Galápagos. The different species developed distinct characteristics, such as bill size and thickness, based on the food supplies and other factors on the particular island to which they were confined. Darwin postulated that all 13 of the finches evolved from a single species originating on the South American mainland. Darwin published his field findings in 1837 in *The Voyage of the H.M.S. Beagle* but waited for more than 20 years before he published *The Origin of Species* (1859), whose theory of evolution created the discipline of modern biology and revolutionized scientific thought. The Galápagos, which helped trigger Darwin's great insight, are still remarkable, with most of the species he found there still ensconced. Protected by Ecuador as a wildlife sanctuary, the Galápagos enjoy limited tourism, which is nevertheless increasing (to about 50,000 visitors a year in the late 1990s), and the living creatures of the islands—survivals of species arrested at various stages of evolutionary adaptation—are still studied by biologists from all over the world.

gall in botany, a fruitlike growth on a plant caused by the plant's reaction to the presence of ENZYMES released by a FUNGUS, a colony of bacteria (*see* BACTERIUM), or an insect LARVA. Most especially prominent galls ("oak apples," "gallnuts," etc.) are caused by a relatively few SPECIES of parasitic wasps of the family Cynipidae (*see* PARASITE). Their shape and location on the plant, and the species of plant they infest, is characteristic of the particular wasp species involved in the infestation. The enzymes released by the wasp larvae cause extremely rapid cell division and growth in the plant, assuring the developing larvae both a food supply and a protective covering. Like all parasites, gall-forming ORGANISMS sap vitality from the host plant; however, in most cases this loss of vitality is not very great. Galls that form on oaks are often high in TANNIC ACID, and are an important commercial source of this compound.

gallinaceous bird any bird belonging to the ORDER Galliformes, including the domestic chicken, the turkey, pheasant, quail, peafowl, grouse, and others. There are about 256 SPECIES worldwide, organized into seven FAMILIES. The Galliformes are typically medium-sized to large ground-dwelling birds with short, rounded wings that enable them to fly swiftly but not far. They are usually fast runners. Their short beaks curve downward: their long legs generally have spurs on them and end in large, clawed feet. They have small heads. Normally omniverous (*see* OMNIVORE), their primary foods are seeds, grains, and small insects. Most GAME BIRDS belong to this order.

galvanic corrosion *See* CORROSION: *types of corrosion.*

galvanic series a ranking of METALS and metallic ALLOYS according to their tendency to give up ELECTRONS. If a pair of dissimilar metals are connected with a wire and placed in an ELECTROLYTE, an electron flow will be produced in the wire away from the metal that gives up electrons more easily (is lower in the galvanic series, or "less noble"). This produces CORROSION of the less noble metal, and the deposition of a thin coating of ATOMS from the electrolyte on the metal higher up the series ("more noble"). The rate of CORROSION and of plating, and the strength of the electrical current through the wire, depends upon the distance the two metals lie apart on the series; hence, to protect tanks

Galvanic Series

Less noble

magnesium
zinc
galvanized steel
aluminum
cadmium
mild steel
wrought iron
cast iron
stainless steel
lead
tin
naval brass
nickel (active)
yellow brass
aluminum bronze
red brass
copper
silicon bronze
nickel (passive)
silver
graphite
gold
platinum

More noble

used to store hazardous materials against corrosion and eventual leakage, the metals used in their construction should be chosen as close together on the galvanic scale, and as far toward the noble end, as possible. *See also* CORROSION: types of corrosion.

game animal any animal for which hunting is legal. Game animals are generally divided into *big game* (deer, antelope, bear, etc.) and *small game* (rabbits, squirrels, raccoons, etc.). POPULATIONS of game animals are generally managed through the use of GAME TAGS, BAG LIMITS, CLOSED SEASONS, and similar means. *Compare* NONGAME WILDLIFE. *See also* GAME BIRD.

game bird any bird for which hunting is legal. Game birds are generally divided into *upland game birds* (those that inhabit fields and pastures, such as pheasant and quail) and *waterfowl* (ducks, geese, and other large birds that spend most of their time on the water). POPULATIONS of game birds are managed in the same manner as those of GAME ANIMALS. *See also* NONGAME WILDLIFE.

game tag in wildlife management, a permit to kill a GAME ANIMAL. A separate tag must be purchased for each individual animal to be killed, over and above the purchase of a general hunting license. Usually both the total number of tags available for a particular game SPECIES and the number of tags each individual hunter may buy for that species is limited. Game tags are used primarily to manage big game (*see* BIG GAME ANIMAL), though they are occasionally used for small game as well. *Compare* BAG LIMIT; DUCK STAMP.

gamete in biology, a haploid reproductive cell (that is, a reproductive cell containing only a single set of CHROMOSOMES) that must fuse with another haploid cell, forming a *zygote,* before cell division can begin. The male gamete is known as the *sperm;* the female gamete, as the *egg. See also* MEIOSIS; MITOSIS; GAMETOPHYTE; ALTERNATION OF GENERATIONS.

gametophyte in biology, a plant that produces GAMETES rather than spores. In the VASCULAR PLANTS and in most ALGAE, gametophytes alternate every other generation with SPOROPHYTES (*see* ALTERNATION OF GENERATIONS). Among primitive plants, the gametophyte and the sporophyte may be the same size; however, evolution has gradually reduced the size of the gametophyte until, in the flowering plants (*see* ANGIOSPERM), the female gametophyte has been reduced to a seven-celled body, the *ovule,* which dwells within and is parasitic upon the ovary of the sporophyte flower. The male gametophyte has been reduced even further, to a three-celled pollen grain (*see* POLLEN).

gamma ray in physics, electromagnetic radiation with extremely short wavelengths. Gamma rays are technically defined as that part of the electromagnetic spectrum with wavelengths of less than 1×10^{-10} meters (by contrast, visible light ranges from 4×10^{-7} meters to 7×10^{-10} meters, and the radio waves used for communication may be up to 10 meters or more in length). Gamma rays are produced by the radioactive decay of ATOMS. They are very high in energy, and are the most dangerous form of atomic radiation, capable of causing MUTATIONS, cancer, and general cellular damage in living ORGANISMS. They are highly penetrating, and are the principal reason that nuclear reactors must be so heavily shielded. Current evidence suggests that they may lie on the boundary between matter and energy, not being completely characterizable as either. *Compare* ALPHA PARTICLE; BETA PARTICLE.

gangue in geology and mining, the nonmetallic—hence, economically worthless—portion of an ORE.

ganister *See* QUARTZITE.

gap phase replacement in forest ecology, the reproduction of relatively shade-intolerant tree species within a mature forest by taking advantage of small gaps in the forest CANOPY created by windthrow, lightning strikes, or the death of old trees due to disease. Gap phase replacement is the principal means of reproduction in mixed-hardwood forests such as those of the north-central United States, where WILDFIRE is relatively rare and many of the CLIMAX SPECIES cannot reproduce in their own shade.

garbage *See* SOLID WASTE.

gasohol a mixture of 90% gasoline and 10% ethanol derived from plant material, such as corn. Now marketed widely, especially in the Middle West, the fuel reduces carbon monoxide (CO) from INTERNAL COMBUSTION ENGINES and is somewhat more efficient as an energy source. Ethanol-gasoline mixtures were introduced in the 1920s to increase octane, thus permitting higher-compression engines and reducing "knock." Under great pressure from those industries that felt threatened by the potential of ethanol to outsell petroleum-based fuels, ethanol was replaced by TETRAETHYL LEAD (TEL), sold under the brand name *Ethyl,* as an anti-knock compound. In 1986, however, after long study, TEL was banished as a constituent of motor fuel. An extremely poisonous element, LEAD, including the lead in TEL, is a neurotoxin that does not break down in the environment. This concern rekindled enthusiasm in the 1970s for biomass-derived alternative fuels, as did the earlier oil embargoes of that decade. In the late 1990s interest in gasohol—with ethanol as a constituent—was once again revived because of serious pollution effects from yet another oxygenate, MTBE, a petroleum derivative which if leaked into aquifers can render groundwater undrinkable. All along, agricultural interests have promoted the development of ethanol fuels for cars and trucks, but since the 1990s greater emphasis has been placed on the development of other energy sources, such as HYDROGEN FUEL CELL. By the mid-1990s gasohol represented 8% of the light duty vehicle fuel market. *See also* ALTERNATIVE ENERGY SOURCES, METHYL TERTIARY BUTYL ETHER (MTBE).

gasoline the fourth fraction of PETROLEUM (*see* DISTILLATION), ranking just above KEROSENE in VOLATILITY. It is sold primarily for use as a motor fuel, but also sees a small amount of use as a SOLVENT and (occasionally) a PESTICIDE. Gasoline is a mixture of several complex HYDROCARBONS with boiling points ranging from about 40° to 200°C (100° to 400°F). As it is sold at service stations it is a manufactured product that has been carefully blended to achieve a desired overall boiling point and has had a number of additives mixed into it to control its ignition point and speed of burning, as well as to keep motor parts clean. The blend differs from place to place and even from winter to summer, with winter gasoline generally having a lower boiling point than that sold in the summer. The most common measurement of gasoline performance is the *octane rating,* which is a comparison of the gasoline's ignition characteristics with those of pure iso-octane ($C_3H_3(CH_3)_5$), one of its normal constituents. If (for example) its performance is 90% as good as that of iso-octane, it is given an octane rating of 90; 85% as good, an octane rating of 85; etc. Aside from problems associated with its use, including air pollution and the proliferation of pavement, gasoline is a serious environmental contaminant due to its TOXICITY. Leakage from underground storage tanks, such as those found at nearly every service station or other dispensing facility, have made it a major contributor to the pollution of GROUNDWATER.

gastrointestinal disease (GI) any disease in which the principal symptoms are disruptions of digestive processes (nausea, vomiting, stomach cramps, diarrhea or constipation, etc.). Most GI diseases are waterborne, and are often caused by PARASITES. Examples include GIARDIASIS and AMOEBIC DYSENTERY.

GATT *See* GENERAL AGREEMENT ON TARIFFS AND TRADE.

gauge pressure the pressure within a tank, pipeline, or other enclosed space, as measured by a standard pressure gauge. Since such a gauge normally reads 0 at rest, that is, when vented to the atmosphere, the gauge pressure is the difference between the ABSOLUTE PRESSURE within the tank and the ATMOSPHERIC PRESSURE around it.

Gause's principle in ecology, the rule that two SPECIES cannot indefinitely occupy the same NICHE at the same time. Gause's principal holds true because, given two species with identical ecological requirements, the more efficient species (*see* ECOLOGICAL EFFICIENCY) will always outcompete the less efficient species for the food and HABITAT they both require. The difference in efficiency does not have to be very large to be effective.

gcd in sewage disposal, short for gallons per capita per day.

gel a colloidal dispersion of a solid in a liquid (*see* COLLOID) in which the individual particles in the dispersed phase have become connected to each other in a more or less firm three-dimensional matrix, imparting to the dispersion as a whole many or all of the characteristics of a solid, such as rigidity, resistance to deformation, etc. Gels contain a great deal of water, and are considered liquids despite their apparently solid characteristics. Most will flow slowly over a long period of time (*see* VISCOSITY). Jellies and gelatin desserts are familiar examples of gels, which also comprise the greater part of all living tissue. *See also* SOL.

gem a MINERAL or (occasionally) a ROCK that is valued for its beauty and its uniqueness rather than its utility. Gems are generally either highly purified forms of minerals or minerals that contain specific, usually rare, impurities. Often they have been partially modified by heat and pressure. They are found most often in igneous intrusions (*see under* IGNEOUS ROCK) or in CONTACT ROCK, where conditions favor their formation. Three classes of gems are recognized. *Precious stones* are hard, transparent gems that refract light in patterns. The term is generally restricted to diamonds, emeralds, and the corundum minerals (rubies and sapphires). *Semiprecious stones* are those that are less hard, less transparent and/or less refractory. They include the QUARTZ minerals (quartz, opal, amethyst, agate, etc.), topaz, garnet, zircon, jade, and others, and grade off at the bottom into *ornamental stones,* a category that includes jaspar, onyx, obsidion, turquoise, malachite, hematite, and even occasionally hard COAL (as "jet"). Since gems generally occur in small, scattered lodes, the mining of them does not normally—with rare exceptions, such as the South African diamond mines—have a significant effect on the environment.

gendarme in mountaineering, a rock pillar on top of a ridge or ARETE, especially an isolated pillar large enough to make passage around it difficult.

gene the smallest unit of heredity, that is, the smallest chemical group that effects the characteristics that an ORGANISM inherits from its parents. Genes are said to "lie on the CHROMOSOME," meaning that they are identifiable parts of the long body of the chromosome. Their location on the chromosome, as well as their composition, determines their function. The chemical differences may be very small—the reversal of a pair of bases on the DNA molecule, for example, or the isomeric transformation of a side group (*see* ISOMER). Each gene exists in two forms called *alleles,* one inherited from each parent, with the characteristics of the organism being determined by which allele is dominant. Modification of an allele by chemical or radiological MUTAGENS is the cause of MUTATIONS.

gene pool in the biological sciences, the total genetic content, including all alleles, of a POPULATION. Larger gene pools tend not only to be more varied, but more stable, because INBREEDING is less likely to occur, leading to a smaller chance that recessive genes (*see* GENE) will be expressed. Preservation of more than one gene pool is generally necessary for the preservation of a SPECIES, because different gene pools develop slightly different responses to environmental conditions, thereby increasing the number of responses to changes in those conditions and improving the chance that these changes will be survivable by at least some of the ORGANISMS. *See also* GENETIC DRIFT.

General Agreement on Tariffs and Trade (GATT) predecessor to the WORLD TRADE ORGANIZATION. Established in 1948 as a postwar economic measure, GATT sought to eliminate import quotas among member nations and reduce tariffs, thereby helping improve the war-ravaged world economy. Over the years the agreement effectively enhanced the increasing hegemony of transnational corporations through the worldwide marketing of goods and services. Many American environmental and labor leaders believe this trend—also encouraged by the INTERNATIONAL MONETARY FUND and the WORLD BANK—to be antithetical to the public interest if left uncontrolled. Strict environmental and labor laws in one country can easily be avoided by transnational corporations relocating their manufacturing facilities to countries with the least restrictive environmental and labor constraints, thus eliminating jobs at home, encouraging labor exploitation abroad, and increasing global pollution.

General Land Office (GLO) obsolete federal land-management agency, in existence from 1812 to 1946. Created originally within the Department of the Treasury, the GLO functioned primarily as a real-estate agency whose chief function was to liquidate the PUBLIC LANDS. Even when it was given management responsibilities, beginning in 1836, the agency continued to view itself largely in the land-liquidation role. Chronically understaffed (at one time, 145 employees were managing lands that totaled over one-half of the United States), the GLO was the focus of numerous scandals in the late 19th century as it attempted to administer such far-reaching laws as the HOMESTEAD ACT and the RAILROAD LAND GRANTS. It was combined with the GRAZING SERVICE in 1946 to form the Bureau of Land Management (*see* BUREAU OF LAND MANAGEMENT: *History*).

genetic diversity as a principle of BIODIVERSITY, the presence of a variety of species in a given biotic community. Without such variability, as in agricultural monocultures of, for example, corn, species are vulnerable to decline from pests, diseases, or changing environmental conditions.

genetic drift the gradual change in a GENE POOL that results from the random loss of alleles from the population due to the failure of individuals carrying those alleles to breed. Genetic drift is faster in small populations than in large ones because fewer individuals carry each allele, increasing the chance that rare alleles will be lost. An allele with an incidence of 1%, for example, will be present in 50 individuals in a population of 5,000 but in only five individuals in a group of 500, and is unlikely to be present at all in a group of less than 100.

genetic engineering involves gene splicing with DNA to restructure the genetic makeup of plants and animals. Genetic engineering generally involves removing molecules from the DNA of a donor organism, "splicing" them into the genetic material of a virus, which can then be introduced, via a bacterium (such as *Escherichia coli*), into a host organism. Using such techniques, genetically modified organisms (GMOs) have been created that can greatly increase world food supply while decreasing the cost of production, thus promising an entirely new kind of GREEN REVOLUTION to feed a growing population, especially in developing nations. Some genetically engineered plants can create their own pesticides, thus eliminating the need for intensive spraying. Others have been created to resist herbicides, so that weeds can be eliminated by direct herbicide application without harming the crop. Cows can be engineered to produce more milk, sheep to produce more wool, and fish to grow faster and bigger for the table. Concern that GMOs will escape and modify the genetic material of natural plants and animals through reproductive processes, or have other unintended consequences, is high among many ecologists and environmentalists, although government agencies (such as the USDA) and agribusiness corporations (such as Archer-Daniels-Midland) seek to reassure the general public that genetic engineering can be controlled and can create safe products. Europeans, as well as Americans, have become especially suspicious of genetically engineered foods, dubbing them "Frankenfoods" (after Frankenstein's monster). And some countries have required that imported foods be labeled as to their genetic make-up.

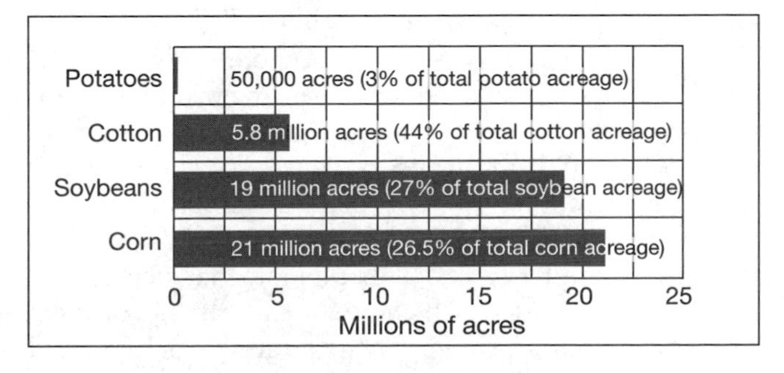

U.S. genetically altered crops, 1998 *(Earth Island Journal, winter 1999–2000)*

genus (plural: genera) in the biological sciences, a group of closely related SPECIES that have enough features in common that they may be easily distinguished from all other groups. As a taxonomic classification (*see* TAXONOMY), the genus falls between the species and the FAMILY. The genus name forms the first half of the Latin binomial used as the scientific name for the species. It is by tradition always capitalized, and is usually presented in italic type.

CLASSIFICATION BY GENERA

Since the genus, like all biological names, is an artificial classification, there are no hard and fast rules for grouping species into genera, and the practice is fairly difficult to define, though it is usually easy enough to recognize. Generally, genera are designated on the sum of characteristics of the organisms involved looked at on a worldwide basis rather than on the shapes of specific parts or on point-to-point variations in color, size, etc. There are no set limits to genus size, which may vary from a single species (a *monotypic* genus) to as many as 3,000 species. Since the division is somewhat arbitrary, the number of species involved in a particular genus often depends upon the predilections of the speaker (*see* LUMPER; SPLITTER). Generic names are often paralleled in popular nomenclature; for example, a scientist referring to a *Quercus* and a layman referring to an oak would be talking about the same broad grouping, with finer divisions possible to each (layman: "white oak"; scientist: *Quercus alba*).

geochemical cycle the path (or set of paths) followed by an *element* through the inorganic environment (*see* INORGANIC COMPOUND). If the cycle includes organic compounds, it is called a BIOGEOCHEMICAL CYCLE. (Another way to put this is that the geochemical cycle may involve the ATMOSPHERE, the LITHOSPHERE and the HYDROSPHERE, but not the BIOSPHERE.) A typical geochemical cycle might begin with free silica within a molten MAGMA combining with oxygen to form silicon dioxide, SiO_2 (QUARTZ). The quartz could become part of a granite BATHOLITH that was later exposed by EROSION, allowing the quartz to be carried away grain by grain to become sand, which might eventually be compressed into SANDSTONE and then metamorphosed into QUARTZITE. The cycle would eventually end millions of years later with the subduction of the crustal plate carrying the quartzite (*see* PLATE TECTONICS), carrying it below the crust where it would be reconverted into magma, melting the quartz and (possibly) releasing the silica and oxygen again as separate elements.

geode a cavity within a ROCK that has been filled with MINERAL deposits, either completely or partially. In the latter case, the minerals often "crystallize out," forming a coating of crystals over the inside of the cavity. Such geodes are valued for their beauty when cut open. Since during the formation of the cavity the stone directly around the cavity usually becomes hardened, geodes often "erode out" of the formation they are in, remaining behind on the erosional surface as separate stones.

geodegradable substance a COMPOUND or MIXTURE that can be broken down into smaller compounds or into its constituent ELEMENTS by the physical and chemical forces of the environment, such as weathering, SOLUTION, the action of sunlight, etc. *Compare* BIODEGRADABLE SUBSTANCE; NONDEGRADABLE POLLUTANT.

geodetic survey a survey utilizing spherical coordinates, instead of planar coordinates, in order to account for the curvature of the Earth. Generally, a geodetic survey is done by making a PLANE SURVEY and then correcting it for curvature using a set of standard "geodetic tables" which give the curvature for each point on the globe (since the Earth is not quite a true sphere, this curvature differs significantly from place to place). Geodetic surveys are necessary for very long or very large features such as highways, coastlines, mountains and mountain systems, and state boundaries.

Geographic Information System (GIS) a computer-driven technology capable of producing complex data sets and maps useful in geographic and ecological analysis and land-use planning. Among the uses of GIS systems is the "layering" of information on maps to show how certain features relate to one another: plant associations related to hydrological data, for example, or landownership patterns with land uses. Also, GIS systems can create predictive models in "what if" scenarios that, for instance, combine soil-type data with weather patterns to determine potential erosion in a given agricultural area, which can then be mapped as an aid in determining optimum soil conservation practices. REMOTE SENSING, such as Landsat data collected from satellite imagery, adds another dimension to GIS analysis, so that large regions, as well as confined areas, can be interpreted.

geographic province an area or region of the Earth's surface set off from surrounding regions by similarities in landforms, vegetation, and (usually) underlying rock types. A province will often contain parts of more than one BIOME. In such cases, the mix of biomes will be characteristic and will be different from the mix in neighboring provinces. Example: the basin-and-range province, which covers most of Nevada and parts of Oregon, Idaho, Utah, and California, and consists of FAULT-BLOCK MOUNTAINS made mostly of IGNEOUS ROCKS and covered largely by desert vegetation, with small patches of coniferous forest and alpine tundra near the tops of the higher peaks.

geological area in FOREST SERVICE terminology, an area on a NATIONAL FOREST that contains interesting or unique geologic features and is set aside to be managed for the preservation of these features and the interpretation of them to forest visitors. The emphasis in a geological area is on recreation. For the preservation of geologic features for scientific study, *see* RESEARCH NATURAL AREA.

Geological Survey (USGS) agency within the U.S. Department of the Interior (*see* INTERIOR, DEPARTMENT OF) created by Congress in 1879 and charged with inventorying and mapping the nation's soil, water, land, and mineral resources. Since 1962 it has extended these duties to areas outside the boundaries of the United States as well as within them. The agency is best known for its detailed, highly accurate TOPOGRAPHIC MAPS, which cover the entire United States on a scale of 1 inch to the mile (the *15-minute series*) and much of it at 2 inches to the mile (the *seven-and-one-half-minute series*). It is also responsible for reporting streamflow data on the nation's rivers.

geologic time the time scale used to date rock formations and geologic events. A special scale is required both because of the great age of the Earth (4.5 billion years) and because it is often impossible to assign precise ages to geologic phenomena, so that some sort of relative scale—rather than the absolute scale of written history—becomes necessary. If geologic periods are dated in relation to each other rather than in relation to some abstract scale of years, it is not necessary to revise the entire system merely because the rocks in one formation are found to be several million years older than they were previously thought to be. One merely has to revise the dates assigned to the periods.

METHODS OF GEOLOGIC DATING

The basic principal used in geologic dating is the *principal of superposition,* which states that younger sedimentary beds are always on top of older sedimentary beds. "On top" here means, of course, "on top when it was deposited," so the original position of a rock SERIES must be determined before it can be dated. This is usually done by looking for phenomena such as fossilized ripple marks (which would mark the original surface), graded particle size (the larger particles would have originally been on bottom), and so forth. To extend the dating beyond a single outcrop, the

The Geologic Time Scale

Time (B.P.) (millions of years)	Epoch	Period	Era	Eon
0.01	Holocene	Quaternary	Cenozoic	Phanerozoic
4	Pleistocene			
10	Pliocene	Tertiary		
25	Miocene			
40	Oligocene			
55	Eocene			
65	Paleocene			
135		Cretaceous	Mesozoic	
190		Jurassic		
225		Triassic		
280		Permian	Paleozoic	
315		Pennsylvanian		
345		Mississippian		
400		Devonian		
440		Silurian		
500		Ordovician		
570		Cambrian		
700		Precambrian	Proterozoid	Cryptozoic
3,400			Archean	
4,500			Azoic	

principal of horizontal continuity is used, which states that two identical rock series in neighboring rock outcrops were once continuous and are actually the same series. Using this principal, relative dating can be established over a broad area by following individual layers until they disappear and then picking another layer higher or lower in the series to continue from that point. Strata in different formations can be dated in relation to each other by comparing their fossils. If they contain remnants of the same or similar life forms, they were almost certainly formed in the same era. Finally, IGNEOUS ROCKS may be dated by the *principal of cross-cutting*. If an igneous body cuts through a sedimentary formation, it may be presumed to have formed after the sediments were laid down.

DIVISIONS OF GEOLOGIC TIME

As historical time is divided into two large periods—A.D. and B.C.—so geologic time is also divided into two large periods, Phanerozoic time and Precambrian time, the division between the two being the point where fossils of living things suddenly become prolific in the rock record, roughly 570 million years ago. Phanerozoic time and Precambrian time are further subdivided into *eons, eras, periods,* and *epochs.* Phanerozoic time consists of a single eon, the Phanerozoic eon. It is divided into three eras: the Cenozoic (most recent), Mesozoic (middle), and Paleozoic (most ancient). There are two periods in the Cenozoic era (Neogene and Paleogene), three in the Mesozoic (Cretaceous, Jurassic, and Triassic), and seven in the Paleozoic (Permian, Pennsylvanian, Mississippian, Devonian, Silurian, Ordovician, and Cambrian). Only the Cenozoic is usually subdivided further. It contains six epochs, three in the Neogene (Holocene, Pleistocene, and Pliocene) and three in the Paleogene (Oligocene, Eocene, and Paleocene). Precambrian time is divided into two eons, the Proterozoic and the Archean. Further divisions have often been proposed, but there is little or no agreement on where these divisions should be placed. (For details, see the accompanying chart.)

ABSOLUTE GEOLOGIC TIME

Radiometric dating techniques such as radiocarbon dating and potassium-argon dating have allowed the establishment of approximate ages in years for many rock formations, thus allowing rough age boundaries to be set on the varying divisions of geologic time. These are constantly being revised as new data is available. The currently recognized ages are given in the accompanying chart. Absolute geologic time is usually given in terms of the number of years before the Δpresent (abbreviated B.P.): thus, the Mesozoic (for example) runs from 225 million years B.P. to 65 million years B.P., for a duration of 160 million years.

geomagnetic pole one of the two poles of the Earth's magnetic field, located roughly 4,000 miles above the planetary surface at opposite ends of the globe. It does not correspond precisely with either the MAGNETIC POLE or the GEOPHYSICAL POLE. The south geomagnetic pole lies above a point near the Soviet Union's Vostok Antarctic Station, roughly halfway between the south geophysical pole and the coast of Antarctica on a bearing toward the west coast of Australia; the north geomagnetic pole lies above the center of the Hayes Peninsula of Greenland, north of Baffin Bay. The geomagnetic poles may be thought of as points where the lines of force of the earth's magnetic field converge. They are of interest primarily for their effects on the Van Allen Belts and similar upper-atmosphere phenomena.

geometric population growth *See* POPULATION EXPLOSION.

geophysical pole (**geographical pole**) the point where the earth's axis of rotation pierces the surface at the north or south extreme of the planet: it differs from both the MAGNETIC POLE and the GEOMAGNETIC POLE. The geophysical poles are the "true" poles—the points referred to when the terms "north pole" and "south pole" are used without qualification.

geosyncline *See* SYNCLINE.

geothermal energy energy derived from the internal heat of the Earth and utilized for human purposes. Available principally in regions of recent or ongoing volcanic activity (*see* VOLCANO) or along active earthquake FAULTS, geothermal energy nevertheless represents—at least potentially—a vast and continually renewable form of energy due to its tapping of heat constantly being generated at the Earth's core and by the forces of TECTONIC ACTIVITY in the Earth's crust. At least 2.3 million acres of known geothermal resource areas (KGRAs) exist in the United States on federal lands alone. The potential power generation represented from these areas is in the neighborhood of 20 QUADS each year (annual U.S. energy consumption in the early 1980s was about 75 quads). Currently among U.S. states, only California uses significant amounts of geothermally generated electricity, including an 85 megawatt plant at the Geysers just north of San Francisco and 200 megawatts from a group of plants in the Imperial Valley, although geothermally heated water is used for space heating in several other locations, including Boise, Idaho, and Klamath Falls, Oregon. Elsewhere in the world, the principal users of geothermal energy are Italy, Iceland, and New Zealand.

DEVELOPMENT OF GEOTHERMAL ENERGY

Four types of geothermal resources are generally recognized. *Drysteam fields* are the most productive. These are areas where GROUNDWATER in contact with a body of MAGMA or other geothermal heat source has been converted to steam beneath the Earth's surface. Piped to the surface, this steam can run electrical turbines directly. *Wet-steam fields* are regions where groundwater in contact with a geothermal heat source has been heated to above the boiling point (100°C, or 212°F) but remains in a liquid state due to pressures on it within the AQUIFER. This *superheated water,* which may have a temperature as high as 350°C (700°F), partially flashes into steam when brought to the surface, creating a product that averages about 20% steam and 80% hot water. As in the drysteam field, the steam can be used to drive electrical generators. *Low-temperature fields* consist of water at or below the boiling point. Such fields cannot currently be used for electrical generation, although they are sufficient for heating buildings. Finally, some *hot-rock fields* exist in which little or no groundwater is present to transfer the heat to the surface. These cannot currently be utilized at all, although experimental designs exist in which water may be injected into the ground next to the hot rock to create a body of artificial groundwater.

ENVIRONMENTAL EFFECTS

Developments of geothermal fields can cause significant changes in groundwater flow patterns, which may dry up GEYSERS and HOT SPRINGS that serve as tourist attractions. The extraction of the hot water may also cause subsistence of the ground over the geothermal source. Water in contact with magma generally has a high mineral content, including a variety of SALTS and HEAVY METALS, and must be disposed of after use in such a manner as not to pollute freshwater aquifers or surface supplies. It may also carry a fair amount of radioactivity. A common constituent of geothermic steam is hydrogen sulfide gas, which creates a "rotten-egg" smell that may carry for miles from the plant, as well as contributing to the acidification of rainfall downwind (*see* ACID RAIN). Finally, most geothermal plants are extremely noisy, and may drive away wildlife from a region as well as disturbing visitors to nearby national parks and other recreation areas (most KGRAs in the United States are associated with significant recreation features, including Yellowstone, Lassen, and Crater Lake National Parks). They are, nevertheless, cleaner and generally less hazardous than many other forms of energy production, including coal-fired and nuclear power plants, and if carefully developed will represent a major source of ALTERNATIVE ENERGY to substitute for dwindling worldwide reserves of FOSSIL FUELS.

geotropism the orientation of plants in response to gravity, causing the roots to grow downward and the SHOOTS to grow upward. Geotropism in the shoot appears to be a result of differential concentrations of AUXINS. If a shoot is trained horizontally, the auxins collect in the lower side, which as a result grows more rapidly, causing an upward bend of the shoot tip. Geotropism in the root is less well understood, but seems to have something to do with the growth of the *root cap,* a thimble-like protective covering over the tip of the root. If the root cap is removed, a root that would otherwise grow downward—the typical geotrophic response—will grow instead in whatever direction it is pointed by the experimenter. *Compare* PHOTOTROPHISM.

germination in the biological sciences, the resumption of growth of a mature seed or other plant DISPERSAL UNIT (such as a spore) following dispersal from the parent plant. In some cases germination takes place immediately upon separation from the parent; however, in most plants—especially those of cold and temperate climates—a period of quiescence, known as *dormancy,* is necessary before germination can successfully take place. In such cases, germination depends upon the removal of *inhibitors.* The precise nature of these inhibitors varies from SPECIES to species, but all have essentially the same function: to make certain that conditions are proper for the plant's growth before that growth begins. Some inhibitors are mechanical: a resinous coating that must be melted in the heat of a WILDFIRE, a hard shell that must be abraded by being dragged along the bottom of a running stream, a substance of extremely low POROSITY through which water soaks at a controlled rate. Others are chemical. These include ENZYMES or similar growth-controllers that must be removed by leaching in damp soil or by the digestive juices of animals, substances that undergo chemical change during long periods of cold (allowing germination to take place as soon as the weather warms up again), and substances that are destroyed upon exposure to light (releasing the seed's growth as the PHOTOPERIOD lengthens). Some of these dormancy mechanisms are remarkably effective: 10,000-year-old seeds thawed from arctic PERMAFROST have been successfully germinated under laboratory conditions. Germination following dormancy serves the ecological function of making certain that a plant will not grow until conditions are right for it, thus allowing species to survive harsh and rigorous conditions. It is generally bred out of crop plants in order to allow the farmer rather than natural processes to control the rate of germination, which is one of the principal reasons that domesticated plants often do not survive well in the wild.

geyser a spout of water emanating from the earth. The water is usually hot and is accompanied by clouds of steam. Geysers are caused by GROUNDWATER coming into contact with bodies of MAGMA and being heated until they escape upward. A few geysers are continuous, but most are periodic, requiring the accumulation of a certain amount of hot water and steam next to the magma before the pressure is sufficient to "blow" through the cooler water that has accumulated in the channel leading to the surface. Geysers are most common in areas of recent or ongoing vulcanism (VOLCANO formation) such as Iceland, the Aleutian Islands, and the Yellowstone CALDERA. *See also* HOT SPRING; GEOTHERMAL ENERGY.

GFC *See* GIRARD FORM CLASS.

GI *See* GASTROINTESTINAL DISEASE.

giardiasis (backpacker's disease, hiker's disease) a GASTROINTESTINAL DISEASE caused by a parasitic PROTOZOAN, *Giardia (lamblia) intestinalis,* and characterized by abdominal cramps, flatulence, and alternate diarrhea and constipation. Giardiasis is spread principally by the contamination of water supplies with infected FECES. Since the protozoan is not particularly specific with regard to its host, but can live in the intestinal tract of any warm-blooded animal, it is not easy to eliminate this contamination, and even wilderness streams can carry high concentrations of it. Though *Giardia* cannot survive low temperatures, it overwinters well in deer and beaver—its principal hosts among wild populations—and is thus almost certainly permanent once it has infected an area. It passes from host to host in the form of cysts (hard capsules within which the organism exists in a dormant state). These cysts, which can remain viable outside the host for long periods, are best destroyed through boiling (20 minutes or more) or removed by filtering through an ultrafilter (*see* REVERSE OSMOSIS). Once established in the intestinal tract, *Giardia* becomes a permanent resident, and can only be killed by the ingestion of certain potent antibacterial agents (Quinacrine hydrochloride; furizolidone) whose side effects are serious enough that the cure is often worse than the disease, which may be mild or even asymptomatic (i.e., with no visible symptoms) in many individuals. *Giardia* is of European origin, and only became a problem in North America after about 1975. INCUBATION TIME: one to three weeks.

gibberellin (gibberellic acid) any of a group of about 40 closely related plant growth hormones, active in the promotion of stem elongation, flowering, and seed GERMINATION. The gibberellins are designated by the letters GA (for gibberellic acid) with a numerical subscript. The most biologically active member of the group is GA_3, first isolated in 1926 from the fungus *Gibberella fujikuroi* but since found in many species of VASCULAR PLANTS. Application of the GAs to dwarf plants causes them to exhibit normal growth rates. Application of small amounts to normal plants causes an increase in size, while application of larger amounts causes rapid, spindly stem growth and a distinct paling of the plant's pigmentation. Application to buds in some species stimulates flowering, and application to dormant seeds can end the dormancy period and begin germination. Gibberellins are widely available commercially for use in nurseries and greenhouses. They appear to have no affect on human or other animal life; however, care must be taken with them because pollution of water with large amounts could lead to damage to plants. *Compare* AUXIN.

Girard form class (GFC) in forestry, a classification system for CONIFERS based on the amount of TAPER present in the trunk, developed by the Oregon forester James W. Girard around 1947 and widely adopted since. As taper tends to be uniform from the top of a tree to its base, knowing both the GFC and the height allows a fairly accurate estimate of the volume of lumber the tree holds. Since taper also tends to be uniform within any given STAND, determining the GFC and the average height of the stand gives a good estimate of total stand volume. In fieldwork, the GFC is determined by dividing the DIAMETER BREAST HEIGHT (DBH) of a tree by the SCALING DIAMETER of the first standard (16-foot) log. The volume is then read from a table. A major advantage of the GFC method is that, with practice, a forester can estimate taper quite accurately without measurement, allowing extremely rapid preliminary cruises (*see* TIMBER CRUISER).

glacial drift in geology, any loose material laid down on a landscape by a GLACIER or by the meltwaters associated with a glacier. The term includes both *till,* unsorted material dropped where it was when the ice melted, and *stratified drift,* sorted material deposited in the beds of lakes and watercourses by each spring's surge of meltwater. Drift is often carried long distances by the glacier before it is put down, and its composition and orientation can tell a great deal about the size and general course of the vanished ice. It is also important economically in some areas as sources of sand and gravel and as a major factor in determining the productivity and tillability of soils (drift generally makes very poor agricultural soils). *See also* KAME; ESKER; MORAINE; OUTWASH DEPOSIT; ERRATIC; GLACIAL LANDFORMS.

glacial landforms in geology, the characteristic features of a landscape shaped by a GLACIER or by its meltwaters. They are often visually striking. Glacial landforms are of two different types. *Drift deposits* are landforms created by materials carried into an area by a glacier and left behind when the glacier melts. These include KAMES, ESKERS, MORAINES, DRUMLINS, and outwash plains (*see* OUTWASH DEPOSIT). The drift deposits may surround KETTLES left behind when blocks of ice became separated from the glacier, were surrounded by drift, and then melted. The other type of glacial landform is created by the removal of material. Such landforms typically show much exposed rock, ground smooth and often carrying parallel scratches in the direction of the glacier's travel (*see* GLACIAL POLISH; STRIATIONS). Valleys are U-shaped and usually terminate in CIRQUES at their upper ends; their sides often show *truncated spurs,* steep triangular faces of rock that mark where the ends of side ridges have been worn away. Ridges are ground into ARETES and COLS; mountain peaks become HORNS. Lakes are common in both types of terrain, either dammed up behind the drift deposits or lying in depressions in the stone floors of the cirques and U-shaped valleys. *See also* PATERNOSTER LAKES; CRAG AND TAIL; ROCHE MOUTONNEE; HANGING VALLEY; GLACIAL DRIFT.

glacial polish a smooth, flat, generally somewhat shiny surface characteristic of rock outcrops that have been recently covered by GLACIERS. Glacially polished surfaces are generally marked by STRIATIONS and often have ERRATICS scattered about on them. They show up best on hard, light-colored rock such as GRANITE. *See also* GLACIAL LANDFORMS.

glacier a moving body of ICE on the landscape, typical of polar and high mountain regions. The ice derives from compacted SNOW that has failed to melt over a number of years' time, and thus contains more air than ice derived directly from water. Glaciers actually do flow, although the flow is many orders of magnitude slower than the flow of water, averaging only about 1 meter per year (some glaciers, known as "galloping glaciers," may for short periods move as much as 100 times this rate). There are two mechanisms involved in this flow. The more important method, *outward deformation,* results from the lack of structural rigidity ("competence") of glacial ice. When a great enough thickness accumulates, its weight begins to deform it so that it sags toward its edges, thus flowing outward from the highest point, or *center of accumulation.* The critical thickness is normally around 20 meters (65 feet). The secondary method, *basal flow,* also results from the glacier's weight; it is enough in most cases to reliquify the bottom of the glacier, creating a thin film of water that lubricates the bed and allows the glacier to physically slide along the path of least resistance.

TYPES OF GLACIERS

Glaciers are of three general types. *Alpine glaciers* form in mountainous areas. They consist of two subtypes: *valley glaciers,* which flow down the principal valleys of a mountain range, and *hanging glaciers,* which occupy the CIRQUES and short valleys among the peaks that line the troughs occupied by the valley glaciers. Since hanging glaciers may contribute much of the valley glacier's ice, they are sometimes referred to as "feeder glaciers." *Piedmont glaciers* are alpine glaciers that have flowed beyond the confines of their mountain valleys and spread out on the plain at the mountain's foot. One piedmont glacier may be fed by several valley glaciers. Finally, *continental glaciers* ("ice caps" or "ice sheets") form from the accumulation of snow in polar regions. They often cover vast areas, moving solely by outward deformation from centers of accumulation that may be several miles thick.

THE WORK OF GLACIERS

Glaciers are among the most active shapers of the landscape, excavating great masses of rock from their beds and moving it long distances across the earth (*see* GLACIAL LANDFORMS and all references under that heading). They perform this carving action in two ways. *Plucking,* probably the more important of the two, results when the thin layer of water at the base of the glacier trickles down into cracks in the underlying rock and then freezes, wedging the rock apart (*see* FROST WEDGING) and incorporating it into the body of the glacier, which then carries it off. *Abrasion* results from the presence of these plucked rocks in the underbody of the glacier. As they are dragged across the land, they act precisely like the teeth of a file, cutting away the rock beneath them. Abrasion often results in very fine particles that may then be carried in the glacier's meltwaters, creating accumulations of powdered stone known as *glacial flour* (or, since it turns the meltwaters a characteristic opaque white color, *glacial milk*).

REQUIREMENTS FOR GLACIER FORMATION

In order for a glacier to form, snow must accumulate without melting for a number of years. This requires relatively cold conditions, but, even more importantly, it also requires a relatively high incidence of snowfall. For this reason, glaciers are actually less likely to form at the poles (where precipitation tends to be light) than in the zone of heavier snowfall near the Arctic Circle, and in high mountains in the temperate and tropical zones. The center of accumulation for the continental glacier that covered the northern half of the North American continent during the last ice age, for example, was not at the North Pole but in the region just south of Hudson Bay. *See also* GLACIAL DRIFT; GLACIAL POLISH; STRIATIONS; KAME; ESKER; MORAINE; OUTWASH DEPOSIT; HANGING VALLEY; CRAG AND TAIL; ROCHE MOUTONNEE; DRUMLIN; KETTLE.

glei *See* GLEY.

Glen Canyon Dam major dam on the Colorado River near Page, Arizona, approximately 10 kilometers (6 miles) south of the Utah border. Its reservoir, Lake Powell, lies primarily in Utah. Glen Canyon Dam was authorized as part of the Upper Colorado River Basin Storage and Flood Control Act of April 11, 1956, and construction began in October of that same year: the floodgates were closed in January of 1963. The dam, an arch type (*see under* DAM), stands 216 meters (710 feet) from buttress to buttress at its greatest width. The 27 million ACRE-FOOT reservoir has a surface area of 161,390 acres and stretches 298 kilometers (186 miles) upstream from the dam. Considered one of the environmental movement's greatest defeats, the dam inundated a spectacular portion of the Colorado River Canyon and backed water to within a few hundred feet of Rainbow Bridge—the world's largest natural bridge—potentially threatening its foundations. Worse than these things, however, have been its downstream effects, which include lowering the silt load in the Colorado and controlling its level through the Grand Canyon so that beaches and gravel bars in the canyon are no longer renewed by flooding: as a consequence, most of these are disappearing, and the entire ecology of the canyon bottom has been dramatically altered. Mismanagement of the dam was responsible for the so-called "Fool's Flood of 1983" during which sudden increases in the river's flow caused by emergency releases to keep the dam from overflowing flipped at least three rafts and required the evacuation of 140 people from the canyon by helicopter: the sudden releases also caused $50 million in damage to the dam's spillways. In 1996, however, an experimental planned release was conducted in an effort to restore, at least partially, the original Glen Canyon ecosystem (which includes GRAND CANYON NATIONAL PARK) that had been radically altered by the dam. Prior to the building of the dam, the river was slow-moving and muddy, but with extraordinary variations throughout the year, with high water flows in the summer 10 times greater than those in winter. This pattern of variation gave rise to a continuing shifting of sandbars, the stirring up of nutrients, the seasonal removal of invasive plant species, the creation of backwater areas for fish breeding, and other unique characteristics that created the ecosystem. After the dam was built, these characteristics were curtailed, resulting in several warm-water fish species becoming endangered, the decline of nutrient levels, and the elimination of the sandbars and backwater breeding areas. The planned release, meant to mimic a spring flood, took place from March 26 through April 2, 1996. The results, while not spectacular, were encouraging, with the restoration of sandbars and backwaters, and an increase in nutrient levels. Perhaps the larger implications of what might be called "restoration flooding" were political, in that government agencies, heretofore concerned only with electric power generation and delivering water to agriculture and cities, could now operate dams with the ecological viability of downstream rivers in mind.

gley (glei, gleyed layer) in soil science, a heavy, somewhat sticky, mottled, gray to blue-gray layer of soil that replaces the B horizon (*see* SOIL: *the soil profile*) in poorly drained soils such as those beneath low-lying meadows and bog woodlands. Gleys are rich in iron and ORGANIC MATTER; however, these materials have been chemically reduced rather than oxidized (*see* REDUCTION; OXIDATION) due to the poor aeration in these constantly damp soils. The term "gley soil" is now considered obsolete (*see* SOIL: *classification of soils*).

global change, global climate change, global warming all these terms refer to increasing temperatures of the atmosphere and of ocean waters, the effects of increased bombardment of UV-B rays through a thinning ozone layer, and the ecosystem degradation from the long-range transport of air pollutants. These interrelated aspects of global change, deriving from industrialization and resulting alterations in the CARBON BALANCE from excess carbon dioxide in the atmosphere, have produced the GREENHOUSE EFFECT and many other potentially harmful impacts on ecosystems and people. Among the impacts are the weakening of forests and plant communities not only because of pollution, but also because plants cannot easily adapt to global warming; weather extremes—tornadoes, hurricanes, floods, heat waves, cold spells, and the like; and tropical diseases, such as dengue and malaria, that can now reach populated areas which had heretofore been too cold for them to get a foothold. Because of the extremely serious effects of global change, the RIO EARTH SUMMIT of 1992, under the auspices of the United Nations, called for international agreements to reduce CO_2 emissions in order to slow the rate of global warming and climate change. *See also* KYOTO PROTOCOL.

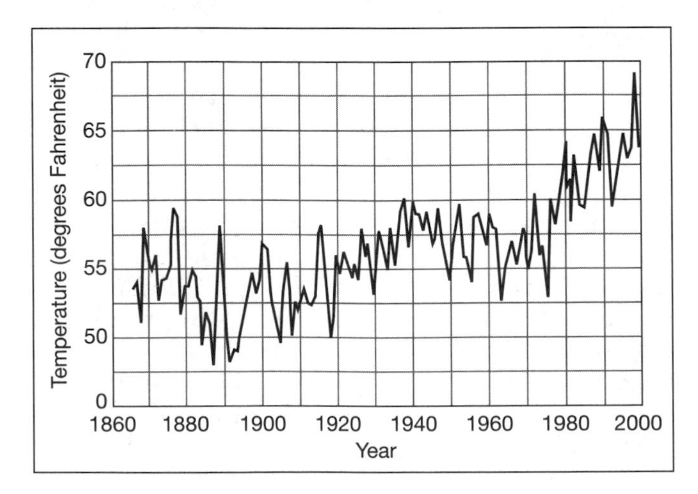

Global temperature, 1866–1999 ***(NASA)***

Global Climate Coalition an industry lobbying group, including energy, automotive, and allied corporations, that wishes to downplay the potential impacts of global warming and to discourage the development of meaningful international agreements to reduce CO_2 emissions. In 1999, the Ford Motor Company, convinced that global warming was a serious worldwide concern, resigned from the group. *See also* ANTI-ENVIRONMENTALISM.

Global Environment Facility (GEF) an international fund administered by the WORLD BANK, the UNITED NATIONS ENVIRONMENT PROGRAMME, and the U.N. Development Programme, to assist developing countries in protecting biodiversity, saving energy, conserving soil and water, phasing out the use of CHLOROFLUOROCARBONs, and similar efforts. Since its creation in 1991, GEF has invested more than $2 billion in 500 projects in 120 countries. *See also* RIO EARTH SUMMIT; UNITED STATES AGENCY FOR INTERNATIONAL DEVELOPMENT (USAID).

Global Environment Monitoring System (GEMS/AIR) a project of two United Nations agencies, the World Health Organization and the UNITED NATIONS ENVIRONMENT PROGRAMME, to monitor air pollution in the world's cities and provide technical assistance to cities in developing countries. Over the life of the program (1976–96), some 80 cities were monitored. Among the most polluted were Mexico City, Beijing, Shanghai, Teheran, and Calcutta. For children living in such cities, according to GEMS/AIR, the simple act of breathing is the equivalent of inhaling two packs of cigarettes a day.

Global Forum *See* RIO EARTH SUMMIT.

global positioning system via a relatively inexpensive hand-held device that can interpret data broadcast from satellites, a means to determine the exact longitude and latitude of any point on Earth.

Global ReLeaf a program sponsored by AMERICAN FORESTS to encourage the planting and care of trees as a means of combating global warming caused by the buildup of CARBON DIOXIDE. The program, conceived by R. Neil Sampson, a former executive vice president of AMERICAN FORESTS, was begun in 1989. The concept is based on the finding that a growing tree can absorb 48 pounds of CO_2 a year, and that three trees shading a house can cut air conditioning needs by up to 50%. Since then, hundreds of municipal cooperators and corporate sponsors have joined in the effort, along with a number of foreign countries. After 10 years of operations, Global ReLeaf projects had seen 7 million trees planted, which according to the U.S. Department of Energy have removed 27,286 tons of carbon dioxide from the atmosphere. In 2000, the program encouraged the planting of more than 1 million trees as part of the "Global ReLeaf 2000" program, which has the goal of planting 20 million trees in the next millennium.

global warming *See* GREENHOUSE EFFECT; GLOBAL CHANGE.

Gloger's rule in ecology, the observation that the pigmentation of warm-blooded animals varies with latitude, becoming darker as the center of the animals' genetic stock moves toward the equator. Gloger's rule holds most completely true among different POPULATIONS of a single SPECIES. It is also largely true among closely related species and even among related genera (*see* GENUS), though it breaks down among distantly related genera. Differences in human skin color are a prime example of Gloger's rule in action. Similar color shifts are found among brown bears, red-tailed hawks, black-capped chickadees, and numerous other species. *Compare* BERGMANN'S RULE; ALLEN'S RULE.

GL-SLS abbreviation for Great Lakes–St. Lawrence Seaway.

GMT abbreviation for Greenwich Mean Time (*see* GREENWICH MERIDIAN).

GNP *See* GROSS NATIONAL PRODUCT.

goal in land-use planning, a desirable end to be sought over the long term for a management unit. Goals ordinarily cover broad areas of policy and are idealistic in nature. They are often unobtainable in practical terms, but serve as a direction toward which the actual day-to-day management of the unit will be aimed. *Compare* OBJECTIVE; MANAGEMENT CONSTRAINT.

golden algae *See* CHRYSOPHYTE.

goods in economics, items of commerce. *Trade* or *retail goods* are items offered to prospective purchasers; *manufactured goods* are items produced for sale; *raw goods* are raw materials offered for sale to manufacturers; *personal goods* are the personal property an individual has accumulated, either through trade or through his own manufacture. Goods are important as an environmental concept because the more an economy depends upon the exchange of goods, the more raw materials it requires to manufacture the goods and the more pressure it puts on its natural resources to obtain those materials. *Compare* SERVICES. *See also* CARRYING CAPACITY.

Gore, Albert, Jr. (1948–) politician, environmentalist, author, and vice president of the United States (1992–2000). Born to the son of a congressman from Tennessee, Gore earned a degree in government from Harvard University, and served in the army as a military reporter in Vietnam. He worked as a reporter and writer for *The Tennessean* in Nashville for five years and earned a law degree from Vanderbilt University in 1976. Gore served four terms in the U.S. House of Representatives and one term in the Senate. During his time in Congress, he initiated efforts to clean up hazardous waste sites and brought attention to environmental issues such as the depletion of the ozone layer. As vice president, Gore took the initiative to stabilize climate change, encourage environmental technologies, boost sustainable development, and develop energy-efficient vehicles. Gore has also promoted nationwide access to the Internet by working to ensure classrooms and libraries have the necessary equipment, and was instrumental in organizing a religious response to the environmental crisis. He was the Democratic Party candidate for president in 2000. Although Gore won the popular vote, he failed to win a majority in the electoral college. His books include *Earth in the Balance: Ecology and the Human Spirit* (1992).

graben in geology, a block of land or section of a geologic FORMATION bounded by FAULTS and lower in elevation than the corresponding blocks or sections on either side. Grabens are usually considerably longer than they are wide. Large grabens are one of the principle means by which STRUCTURAL VALLEYS are formed. *Compare* HORST.

grade (graded profile) in geology and hydrology, the angle at which DEPOSITION and *erosion* exactly balance in a streambed so that downcutting ceases. The precise angle involved varies with the size of the stream and the nature of its bed. A stream that has reached this equilibrium condition is called a graded stream and is said to have *eroded to grade*.

graded stream in geology, a stream in which the erosion of material from any place along the bed is exactly balanced by the deposition of new material being carried in from upstream. Such streams are said to be "running at grade." They are no longer deepening their valleys, which tend to assume a parabolic shape, steep at the headwaters and gradually flattening out toward the mouth, known as a *graded profile*. *Compare* DOWNCUTTING STREAM. *See also* GRADE.

grain (1) in lumber, the orientation of the wood fibers. This orientation is always longitudinally toward what was once the top of the tree, and is consistent down to the level of the individual CELLS of XYLEM. It can be seen in the alternate lines of dark and light wood that represent the tree's ANNUAL RINGS.

(2) in geology, the size of the individual mineral particles in a rock sample. "Fine-grained rock"—that is, rock in which the mineral particles are small and uniform in size—is generally stronger than that with larger grains. Grain size and uniformity is usually consistent among rocks of the same type.

(3) in horticulture, a plant that produces edible kernels (*see under* FRUIT). Almost all grains are members of the GRASS family. The group includes the earliest and most widely cultivated domestic food plants, among them rice, corn, and wheat.

gram a unit of mass in the METRIC SYSTEM, equivalent to the mass of one cubic centimeter of water at maximum density, that is, at 4°C. There are 453.6 grams in an English (avoirdupois) pound.

gram-molecular weight *See* MOLECULAR WEIGHT.

Gram-negative bacteria *See* GRAM STAIN.

Gram-positive bacteria *See* GRAM STAIN.

Gram stain procedure for the identification and classification of strains of bacteria (*see* BACTERIUM), developed by the Danish physician Hans Christian Gram around 1883 and since standardized as one of the primary tools of microbiology. The procedure begins with the addition of a solution of a dye, *gentian violet* or *crystal violet,* to a fixed culture of the unknown bacteria, staining them purple. A solution of ACETONE or a mixture of acetone and ethanol (*see under* ALCOHOL) is then applied, washing the dye from some species (*gram-negative bacteria*) while leaving others (*gram-positive bacteria*) stained brightly purple. A second dye, known as the *counterstain,* is usually added after the washing stage to make the gram-negative bacteria stand out in a distinct color (usually red). Gram staining is important because the difference in the staining reaction correlates well with differences in the sensitivity of bacteria to ANTIBIOTICS, apparently due to variations in cell-wall chemistry between the two groups. The walls of the gram-negatives contain a group of chemicals called *lipopolysaccharides* ("fatty-multiple-sugars") that make them thinner and less permeable (*see* PERMEABILITY) to those antibiotics with particularly large molecules such as penicillin and actinomycin.

GRAND Canal acronym for the Great Recycling And Development Canal, a proposed Canadian water-development project designed to transfer water from the James Bay watershed of Quebec to the Great Plains farming regions of Canada and the United States via the Great Lakes, first suggested in 1933 by Montreal engineer Thomas Kierans and actively promoted by Kierans and others since. GRAND Canal proponents envision a 160-kilometer (100-mile) dike across the mouth of James Bay where it connects to Hudson's Bay. Inflow from the region's rivers would then turn the bay into a 30,000-square-mile freshwater lake, from which the water could be pumped to the 285-meter-high (960-foot) divide between the James Bay and Great Lakes watersheds, probably near Amos, Quebec. From there it would run by gravity into Lake Huron. Other canals and pumps would distribute the water from the western side of the Great Lakes to the Great Plains and perhaps beyond to Arizona and southern California. Total project costs, including the western distribution system, are estimated at $100 billion (Canadian). On the positive side, the project would replace depleted GROUNDWATER supplies on the Great Plains (*see,* e.g., OGALLALA AQUIFER) and would stabilize water levels and improve water quality in the Great Lakes–St. Lawrence system. Negative impacts include elimination of major wildlife HABITAT areas (including the almost certain loss of the James Bay polar bear POPULATIONS), large added power-generation needs, and probably climatic changes due to shifts in the continental water balance, as well as the normal environmental impacts of OUT-OF-BASIN TRANSFERS, here magnified to an immense scale. *Compare* NAWAPA.

Grand Canyon National Park one of the "crown jewels" of the NATIONAL PARK SYSTEM. Located in northern Arizona, the park encompasses 178 linear miles of the Colorado River with the canyon itself about 1 vertical mile deep and 18 miles wide at its widest point. Rock formations are exposed by the erosive power of the river, cutting into the earth as the river slowly rises from tectonic uplift. The earliest formations are 1.7 billion years old, and the youngest laid down quite recently in the Holocene era. The canyon is, according to naturalist and author Joseph Wood Krutch, "the most revealing single page of Earth's history anywhere open on the face of the globe." Even so, the Grand Canyon is beset with several significant environmental challenges, including 5 million visitors a year; visible pollution from the Four Corners power plant nearby; overflights by light planes; and the alteration of the riverine ecosystem by upstream dams. While park authorities cannot control the pollution, the overflights have been severely restricted in recent years, reducing noise pollution, and plans are afoot to radically reduce auto traffic in the park. Moreover, an experimental release of waters from the upstream GLEN CANYON DAM is expected to help restore the original canyon ecosystem.

granite a fine-grained igneous intrusive rock (*see* IGNEOUS ROCK; INTRUSIVE ROCK) composed largely of FELDSPAR and QUARTZ. It is usually light in color and has a speckled appearance due to the dark grains of OLIVINE, MICA and HORNBLENDE that are normally scattered through it. It usually occurs in large lens-shaped bodies known as BATHOLITHS, which are highly resistant to erosion, and because of this are often sculpted into spectacular mountain scenery (e.g., the Sierra Nevada Mountains of California). *See also* GRANODIORITE.

granodiorite an igneous intrusive rock (*see* IGNEOUS ROCK; INTRUSIVE ROCK) similar in appearance to GRANITE, but with larger grains that are less well cemented together. The difference appears to relate to both the temperature of formation and to the chemical constituents of the FELDSPAR, which, along with QUARTZ, makes up the primary rock material (granite is composed primarily of plagioclase feldspar; granodiorite, of oligoclase feldspar). Granodiorite is sometimes called "rotten granite" or "decomposed granite."

grass any of approximately 6,000 SPECIES of plants belonging to the FAMILY Gramineae (also known as Poaceae) of the CLASS Monocotoledonae (*see* MONOCOTOLEDON). Grasses are among the most widespread and environmentally important of all plants. They are grouped into two loose types, the *bunchgrasses* (reproducing entirely by seed) and the *sod-forming grasses* (reproducing both by seed and by RUNNER): the bunchgrasses are mostly ANNUALS, while the sod-forming grasses are almost all PERENNIALS. All share a round, hollow or pith-filled stem consisting of short segments separated by nodes, a FIBROUS ROOT SYSTEM that forms extensive, thick mats beneath the soil surface, and a specialized leaf consisting of two parts: a *sheath* that wraps around the stem (always beginning at a node) and a long, narrow, ribbonlike *blade*. Often the connection of the sheath to the stem node develops a ring of harder tissue called a *ligule*. The flowers are wind-pollinated rather than insect-pollinated, and tend therefore to be small and inconspicuous. Grasses include numerous species of great economic value, including wheat, corn, sugarcane, and bamboo as well as the typical lawn and pasture grasses such as fescue, bluegrass, bermuda grass, bentgrass, and timothy. Environmentally, they are important for two principal reasons. Their fibrous root systems do an excellent job of holding soil in place and reducing EROSION; and their unusual habit of growing from the base, rather than the tip, allows them to grow continually while their tips are being grazed off, thus supporting far larger numbers of grazing animals under a far greater range of climatic conditions than top-growing plants could without suffering harm. *See also* GRASSLAND BIOME.

grassland biome an extensive BIOME in which the dominant species are GRASSES. Other PERENNIALS, such as Compositae (daisies and sunflowers; *see* COMPOSITE) and Ranunculae (buttercups), are usually present; however, ANNUALS are almost nonexistent due to the difficulty they have becoming established among the grasses. Trees are usually present only along watercourses, and are limited to willows, cottonwood, and similar hardy species of HARDWOODS. Rainfall is from 10 to 30 inches a year. Soils are thick and well-developed. Large herds of grazing animals are typically present. Fire is a common occurrence. There is some evidence that most grasslands are actually maintained by the presence of fire, and that suppression of these "prairie fires" leads to encroachment of forests into the grassland areas (*see* FIRE ECOLOGY; FIRE-MAINTAINED CLIMAX). Temperate-zone grasslands, also known as STEPPES, occur in large areas of central Asia, in central North and South America, and in Australia. They are sometimes divided into two zones, the *tallgrass prairie* and the *shortgrass prairie,* on the basis of vegetal differences caused by differences in rainfall. Tropical grasslands, also called *savannas,* occur widely in Africa, with smaller expanses in northern South America and in east-central Mexico. They differ from the temperate grasslands in having broadly spaced trees scattered relatively evenly through them, and are sometimes treated as a separate biome.

Grassland Heritage Foundation *See under* TALLGRASS PRAIRIE ALLIANCE.

grassroots organizations *See* ENVIRONMENTAL MOVEMENT.

graupel in meteorology, a form of PRECIPITATION consisting of snowflakes that have been modified into pellets, either by the melting and refreezing of the snowflakes' arms or by the addition of a raindrop that has frozen around the snowflake's nucleus and filled in the interarm spaces. It does not show the characteristic layered structure of HAIL.

gravel in geology, an accumulation of rock fragments of PEBBLE size or greater (more than 2mm in diameter; *see* WENTWORTH SCALE). In engineering, and often in popular usage, the term is restricted somewhat farther, to materials less than 78mm in diameter (roughly hens-egg size) that will not fall through a number 4 screen (4.76mm square holes). *See also* GRAVEL BAR.

gravel bar a deposit of GRAVEL in the bed of a stream, usually partially exposed at low water. Three kinds of gravel bars are generally recognized. *Point bars* (also known as *meander bars*) lie on the insides of turns in the stream channel. They result from the fact that the water on the inside of a turn has less distance to travel than water on the outside, and therefore moves more slowly and can carry less material. The material it deposits forms the point bar. *Channel bars* lie in relatively straight portions of the stream channel. They result from various causes, chiefly either a change in the stream's gradient that causes it to slow down and drop some of its load (*see* BRAIDED CHANNEL) or the uncovering of a previously formed alluvial deposit from which the smaller materials are then removed by the stream, leaving the larger materials behind as an aquatic version of LAG GRAVEL. Finally, *delta bars* are deposits laid down where a small, active stream enters a large, slower one, depositing its BED LOAD just outside its mouth. Gravel bars trap smaller particles in the spaces between their individual stones, thus serving as a natural filter for the stream. They also create riffles—helping to aerate the water—and serve as spawning beds for many forms of aquatic life, including salmon and trout (*see* SPAWNING GRAVEL). They should not be removed from a stream unless there is clear evidence that the effects of the removal can be mitigated.

gravitational water in hydrology and soil science, the water in a saturated soil that will drain out in response to gravity. Gravitational water is held in the large pore spaces of the soil. It drains to the WATER TABLE within two to three days following a rainfall, hence is only briefly available for the use of shallow-rooted plants. *Compare* CAPILLARY WATER; HYGROSCOPIC WATER. *See also* FIELD CAPACITY; SOIL MOISTURE TENSION.

gray water (sullage) water that has been used for bathing, laundry, dishwashing, cooking, mopping, etc. It is not potable (*see* POTABLE WATER), but is generally satisfactory for watering plants, although sanitation laws prohibit its use in some areas and it may harm some plant species that are particularly sensitive to DETERGENTS. The category does not include wastewater from toilet flushing or from garbage disposals (*see* BLACK WATER). *See also* SEWAGE.

Grazing Service obsolete federal land-management agency, in existence from 1934 to 1946. The Grazing Service was charged with the management and improvement of livestock range on the PUBLIC DOMAIN lands of the West (*see* TAYLOR GRAZING ACT; BUREAU OF LAND MANAGEMENT: *History*).

grazing system the mix of management techniques chosen to meet the GOALS and OBJECTIVES relating to livestock on a management unit of the NATIONAL FOREST SYSTEM or the BUREAU OF LAND MANAGEMENT. The term is also applied by livestock managers on private lands. *See* RANGE MANAGEMENT.

grazing treatment *See* RANGE TREATMENT.

grazing unit in range management, (1) 1,000 pounds of grazing animals(s), or one cow plus one calf. The grazing unit is also known as the *animal unit month conversion factor* (*see* ANIMAL UNIT MONTH).

(2) a conterminous area of land on which a herd of animals is grazed. The term is used most commonly in relation to grazing permits issued by federal agencies to grant ranchers permission to use specific areas of public land.

Great Barrier Reef at 1,250 linear miles, the world's largest coral reef, in effect a living biological entity created by the accumulated calcareous secretions of billions of small organisms. Located 10 to 100 miles off the northeast coast of Australia, the reef provides millions of visitors a look at colorful coral gardens, together with associated plants and wildlife in a complex of cays, islands, and lagoons. As one of the most easily damaged of all ecosystems, the Great Barrier Reef has declined in ecological stability and beauty in recent years as a result of coral-collecting, siltation, water pollution, and a population outbreak of the crown-of-thorns starfish, which consumes the living coral organisms.

Great Basin large intermontane region of the western United States from which there is no external drainage, all waters being evaporated or transpirated away (*see* EVAPORATION; TRANSPIRATION; EVAPOTRANSPIRATION) rather than flowing to the sea. The rivers of the basin generally flow into lakes with no outlets. All water leaves the lakes through evaporation, a process that leaves the lake water heavily mineralized. The climate of the region is dry, with hot summers and generally cold winters. Vegetation is sparse. Principle land use activities are grazing and mining. The Great Basin covers nearly all of Nevada, a large part of Utah, and smaller parts of Oregon, Idaho, Arizona, and California, and is roughly contiguous with the basin-and-range province (*see* GEOGRAPHIC PROVINCE; BASIN AND RANGE TOPOGRAPHY).

Great Lakes Fishery Commission Founded in 1955 as an outgrowth of the Canada–U.S. Convention on Great Lakes Fisheries, the commission's duty is to act as an adviser to both governments in areas of fishery improvements, fishery research, management of sea lamprey, and the improvement and perpetuation of fishery resources. Address: 2100 Commonwealth Boulevard, Suite 209, Ann Arbor, MI 49105-1563. Phone: (734) 662-3209. Website: www.glfc.org.

Great Lakes Indian Fish and Wildlife Commission established in 1983, provides support to its member tribes in Michigan, Minnesota, and Wisconsin on matters regarding off-reservation fishing and hunting rights. The commission issues technical reports, assists in treaty enforcement, and provides legal services. Address: P.O. Box 9, Odanah, WI 57861. Phone: (715) 682-6619.

Great Lakes United (GLU) coalition of environmental, labor and other groups, founded in 1982 to coordinate efforts to preserve and protect the Great Lakes Basin and the St. Lawrence River. GLU takes an ECOSYSTEM approach, emphasizing WATERSHED and coastline protection as well as pollution control within the lakes themselves. Its principal efforts are put into the control of toxics (including AERIAL DEPOSITION) and monitoring U.S. and Canadian compliance with the GREAT LAKES WATER QUALITY AGREEMENT OF 1978. Address: State University College at Buffalo, Cassety Hall, 1300 Elmwood Avenue, Buffalo, NY 14222. Phone: (716) 886-0142.

Great Lakes Water Quality Agreement of 1978 (GLWQA) agreement between the federal governments of the United States and Canada to restore and protect the water quality of the Great Lakes system, signed at Ottawa, Canada, on November 22, 1978. The GLWQA contains two unique provisions that make it an important milestone in international environmental law. First, it specifically addresses, not just the boundary waters between the two nations, but the entire ECOSYSTEM of which they are a part. This "ecosystem approach" means that pollution of the lakes can be addressed by land-use management and other source-specific techniques, without regard to international boundaries, as well as by cleaning up the damage after it occurs. Second, the agreement calls for "virtual elimination" of whole classes of MICROCONTAMINANTS from the water of the lakes. Known as "zero tolerance," this provision means that there is to be no permissible level of these materials in wastewater EFFLUENTS around the lakes—in other words, that risk analysis is not to be used as an excuse for inaction. Other portions of the agreement designate the INTERNATIONAL JOINT COMMISSION to enforce the agreement, and set up a Science Advisory Board and a Great Lakes Program Office to assist them in this task; mandate the cleanup of 41 specific "areas of concern" (mostly harbors and shipping channels with high pollutant loads); and call for the establishment of strategies dealing with AERIAL DEPOSITION, NONPOINT POLLUTION, and the release of PHOSPHATES into Great Lakes waters.

Great Plains a vast grassland region of North America lying west of the Mississippi River and east of the Rocky Mountains, and running from the prairie provinces of Canada (Alberta, Saskatchewan, and Manitoba) southward nearly to the Rio Grande. Agriculture in the Great Plains consists of staple crops such as wheat, soybeans, and corn, some of it irrigated courtesy of the OGALLALA AQUIFER, plus cattle and sheep ranching. Increasingly, parts of the western plains are being returned to the buffalo, especially where cattle overgrazing has degraded the grassland ecosystem. *See also* BUFFALO COMMONS; GRASSLAND BIOME.

Great Recycling And Northern Development Canal *See* GRAND CANAL.

Great Smoky Mountains National Park offers a substantial area of relatively undisturbed old-growth Eastern forest as well as some of the highest mountain summits east of the Mississippi River. Located in North Carolina and Tennessee, the park has more than 130 native tree species and 1,300 varieties of flowering plants. Authorized by Congress in 1926, the park was established for full development in 1934. At 520,000 acres, it has been designated an INTERNATIONAL BIOSPHERE RESERVE (1967) and a WORLD HERITAGE SITE (1983).

great soil groups in the earth sciences, a soil-classification scheme developed by the United States Department of Agriculture (USDA; *see* AGRICULTURE, DEPARTMENT OF) during the 1930s and 1940s. Based on the work of the Russian soil scientist Vasily Dokuchaev (1846–1903) in the late 19th century, the great soil groups classification system was elaborated and codified by the American C. F. Marbut beginning about 1927 and reached its highest development in USDA publications in 1949. Classification under the great soil group system is done by presumed genetic relationships; that is, soils of one type are presumed to evolve into soils of other, related types over time and under the influence of climate, vegetation, and local rock type, and the classifications reflect these lines of evolution. Three orders are recognized: *zonal soils* ("normal" soils), with fully developed HORIZONS, formed in areas where unimpeded water movement in the soil allows leaching to occur and where little or no EROSION and DEPOSITION is taking place on the soil surface; *azonal soils,* with no horizons, formed in areas where physical processes keep the soil continually mixed up, such as alluvial plains (*see* ALLUVIUM), landslides, and regions of ongoing LOESS DEPOSIT; and *intrazonal soils,* with weakly developed horizons, formed in areas where impeded drainage inhibits leaching action but allows some vertical differentiation on the basis of particle size. Within each order, the soils are divided into *great soil groups* on the basis of the development processes that formed them. Typical great soil groups include the PODZOLS (zonal soils formed under forests in humid, temperate climates); CHERNOZEMS (zonal soils formed under semiarid grasslands); HUMIC GLEYS (intrazonal soils formed under damp meadows); and REGOSOLS (azonal soils formed on talus slopes). Zonal great groups, and some intrazonal and azonal great groups, are also sometimes grouped together by the type of leaching that occurs in them: those that develop a calcic horizon (the *pedocals,* formed in temperate arid and semiarid regions), those that accumulate iron and aluminum in the B horizon (the *pedalfers,* formed in temperate humid regions), and those that accumulate iron and other metals in the A horizon (the *laterites,* formed primarily in the tropics). Finally, the great groups are broken down further into *associations, series, types* and *phases.* These designations are the same under the great soil group system as they are under the newer USDA scheme, the so-called SEVENTH APPROXIMATION. The great soil groups scheme has been properly criticized for devoting too much attention to genetic process while ignoring or playing down the large influence of differing PARENT MATERIALS on soil profile and structure. However, it is considerably easier to grasp than the overly detailed technical language of the seventh approximation, and has therefore remained in some use, particularly among geomorphologists and others whose work inclines them toward a genetic view anyway. *Compare* UNIFIED SOIL CLASSIFICATION SYSTEM.

Greater Yellowstone Coalition A nonprofit organization dedicated to the preservation and protection of the unique Great Yellowstone Ecosystem through increased public awareness and political action. The coalition works through its individual members and through some 120 national and regional member organizations. Its publications include *Greater Yellowstone Report, Greater Yellowstone Today,* and *EcoAction Alerts.* Membership (1999): 7,500. Address: P.O. Box 1874, 13 South Willson, Bozeman, MT 59771. Phone: (406) 586-1593. Website: www.greatyellowstone.org.

greenbelt in land-use planning, a region of open space surrounding a city and maintained in an undeveloped condition either through the use of zoning laws or by designation as parkland. Besides providing a division between the city and its suburbs and thus helping contribute a sense of civic identity, the greenbelt improves urban air quality, moderates temperatures (*see* URBAN HEAT ISLAND), and provides an opportunity for city residents to experience a natural, undeveloped environment close at hand.

green building the use of sustainable, nontoxic, energy-efficient materials and designs for housing and other structures. The ultimate in green building would feature compact passive-solar architecture, the use of rammed earth or straw-bale construction, an earthen floor stabilized with cement, and a roof of recycled materials. Green building is generally more expensive on a square-foot basis than standard stick-built homes made out of two-by-fours on a concrete slab.

greenhouse effect a warming of the Earth's surface due to the trapping of heat by the ATMOSPHERE. The greenhouse effect arises from the fact that certain gases—notably water vapor and CARBON DIOXIDE (CO_2)—are transparent to radiation with short wave lengths but are opaque to radiation with long wave lengths. Energy reaching the Earth from the Sun is predominantly short-wave radiation, so it passes easily through the atmosphere. It is absorbed by the earth and reradiated as long-wave radiation, which is absorbed by the CO_2 and water vapor in the atmosphere and reradiated back to Earth rather than being lost to space. It has been estimated that the greenhouse effect has raised the Earth's average temperature by 35°C (63°F) above what it would be in the absence of greenhouse trapping, thus making life possible on the planet.

GREENHOUSE WARMING AS AN ENVIRONMENTAL PROBLEM

Carbon dioxide makes up only about 0.03% of the atmosphere, yet this tiny percentage is principally responsible for the greenhouse effect. Hence, it follows that even a very small increase in the amount of atmospheric CO_2 would be likely to create a corresponding increase in global average temperatures. Such an increase has been going on in this century, largely as a result of two human activities: the burning of FOSSIL FUELS to obtain energy, and the destruction of forests. (Since trees utilize CO_2 and release oxygen during daytime respiration, forest destruction increases CO_2 levels in two ways: by releasing CO_2 stored in the trees, and by eliminating a principal user of atmospheric CO_2.) Since 1860, atmospheric carbon dioxide has increased by about 15%, and it is expected to double in the next 50 to 75 years. Climatic models predict that this increase will lead to a 2°C (3.6°F) rise in global temperatures by the year 2040, with as much as 9°C (16°F) possible by the year 2100. The effects of such a change in the atmospheric heat balance could include dramatic changes in rainfall patterns and up to a 2-meter (6.5-foot) rise in sea level, inundating broad areas of the world's seacoasts and rendering many

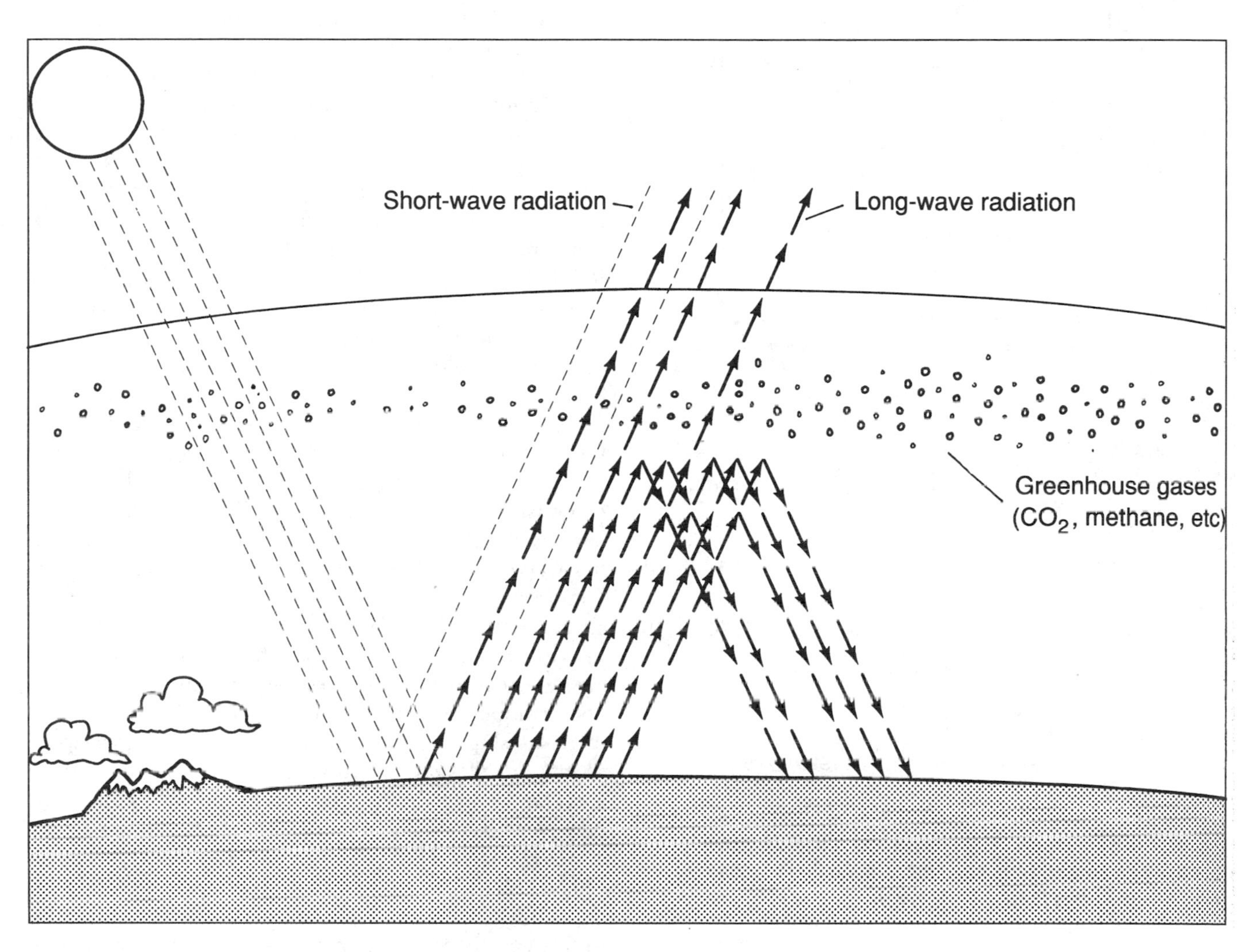

The greenhouse effect

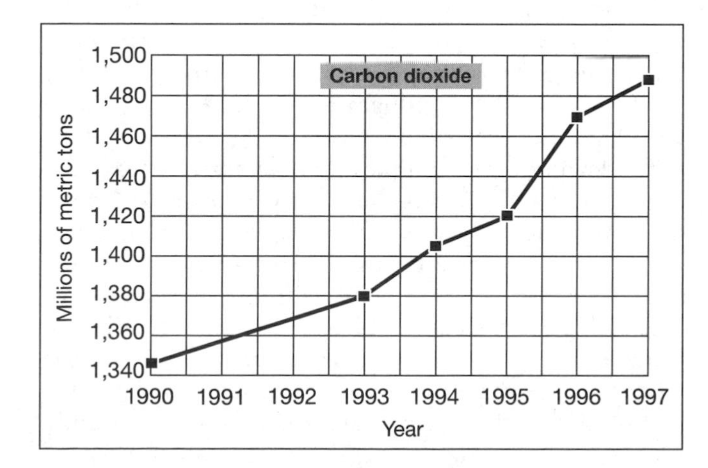

U.S. greenhouse gas emissions from human activities, 1990–1997 (carbon dioxide) ***(U.S. Environmental Protection Agency)***

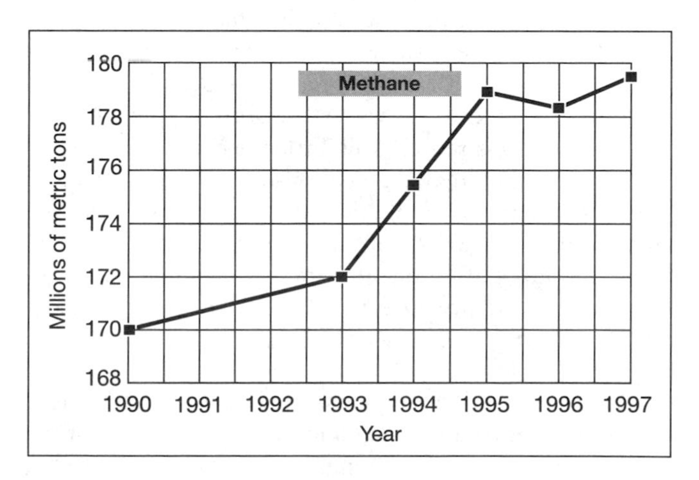

U.S. greenhouse gas emissions from human activities, 1990–1997 (methane) ***(U.S. Environmental Protection Agency)***

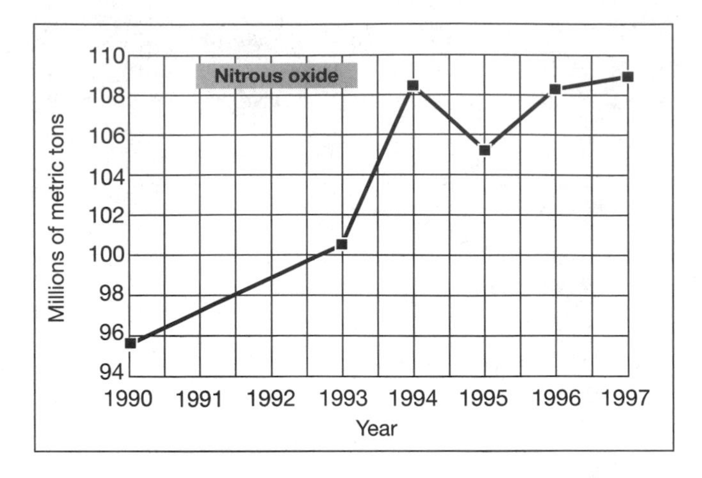

U.S. greenhouse gas emissions from human activities, 1990–1997 (nitrous oxide) ***(U.S. Environmental Protection Agency)***

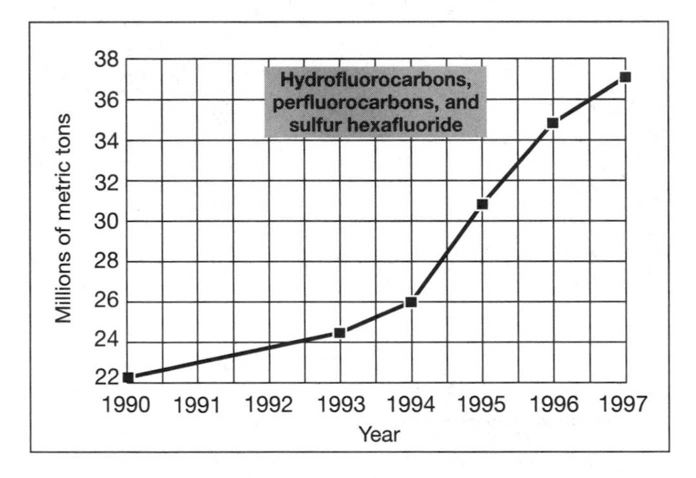

U.S. greenhouse gas emissions from human activities, 1990–1997 (hydrofluorocarbons, perfluorocarbons, and sulfur hexafluoride) ***(U.S. Environmental Protection Agency)***

coastal AQUIFERS unfit to drink due to SALINE INTRUSION. Adding to the potential problem is the fact that many of the man-made gases that have been released into the atmosphere since about 1960—including especially the CHLOROFLUOROCARBONS—are also highly effective greenhouse gases. Most studies to date indicate that there is remarkably little we can do to slow or reverse this manmade greenhouse warming, although the increased atmospheric ALBEDO caused by the greater amounts of PARTICULATES in the air—also largely a human-caused phenomenon—may counter it to some extent. *See also* GLOBAL CHANGE; KYOTO PROTOCOL.

green manure crop plants that are plowed under upon maturation in order to increase the fertility of the soil. Plants used for green manure are usually LEGUMES, chosen because of their ability to fix NITROGEN and thus enrich concentrations of this necessary NUTRIENT, which will then be available for other crops grown later in the same field. *See also* FERTILIZER.

Green Party a political party begun in Germany in 1983 that now includes organizations in 75 countries. The U.S. Green Party, started in the early 1990s, has elected 68 Greens to local political office in 17 states. The platform of the U.S. Green Party includes not only "ecological wisdom," but also social justice and personal responsibility, nonviolence, and community-based economics. In many states Greens have fielded candidates for state and national office, including president and vice president. In 2000, consumer advocate Ralph Nader was the Green Party presidential candidate, and Winona LaDuke, a Native American activist, was candidate for vice president. Nader and LaDuke also ran for these offices in 1996.

Greenpeace radical environmental organization, founded in 1979 to confront and attempt to halt environmentally destructive activities through the use of demonstrations and nonviolent direct action. It also engages in lobbying and educational work. Best known for its whale and fur-seal protection work, in which volunteers interpose their bodies between the animals and commercial whalers or sealers, Greenpeace is also heavily involved in pollution abatement and protests against nuclear weapons testing. It gained a considerable amount of international sympathy and support when its ship, the *Rainbow Warrior,* was blown up by agents of the French government in the harbor at Auckland, New Zealand on July 10, 1985, killing one crew member. Membership (1999): 420,000. Address: 1436 U Street NW, Washington, DC 20009. Phone: (202) 462-1177.

green revolution term applied to the development and deployment of crop plants that have been genetically altered to increase their yield. This increase can be enormous. In one documented instance, the introduction of a strain of rice known as IR-8 to parts of the Philippines raised harvests from approximately 1,500 to over 9,000 kilograms per hectare (1,350 to 8,100 pounds per acre). Other plants improved by green-revolution techniques include corn, wheat, and rye. Wheat and rye have also been crossed to form a new high-yield species called *triticale.* In general, green-revolution strains are shorter and thicker than their "normal" counterparts, with larger seed heads. The stiff, short stems, which are less susceptible to blowdown, give them the common names "dwarf wheat," "dwarf rice," etc. Once thought to have the potential to totally eliminate world food shortages, green-revolution techniques began to fall into disfavor during the mid-1970s due to their intensive cultivation requirements, including large amounts of water (often requiring the construction of large-scale IRRIGATION works) and massive applications of FERTILIZERS and PESTICIDES, often leading to severe pollution problems. *Compare* PERMACULTURE.

greenway a linear open space established along either a natural corridor, such as a riverfront, stream valley, or ridgeline, or overland along a railroad right-of-way converted to recreational use, a canal, a scenic road or other route. The greenway concept was developed in the 1950s and popularized by urbanist William H. Whyte in the 1960s. In the 1970s greenways began to be developed in urban areas: a trailway and nodal park system, for example, along the South Platte River in Denver, Colorado; streamside trails in Raleigh, North Carolina; a 140-mile connected system of linear parks around Portland, Oregon. The greenway movement has largely been a grassroots effort rather than a state or federal government-sponsored initiative. Today greenways can be found in every state and in most U.S. cities and surrounding metropolitan areas. A few long-distance greenways have also been established outside of urban and urbanizing areas. The American Greenways Program, a division of THE CONSERVATION FUND, estimates that by the late 1990s there were at least 3,000 greenway projects in the United States.

Greenwich meridian the LONGITUDE line passing directly through Greenwich, England. It is the vertical baseline for the global system of geographical coordinates (longitude 0°), and is the location of Greenwich Mean Time (GMT), the reference time upon which all time zones around the world are based.

grit chamber in sanitary engineering, a means of removing sand, metal fragments, and other inorganic solids (*see* INORGANIC COMPOUND) from the WASTE STREAM in a SEWAGE TREATMENT PLANT. A grit chamber is similar to a SETTLING BASIN, except that the flow of the waste stream is slowed down to a considerably smaller degree so that only the inorganic material, which averages more than twice the density of the organic wastes, settles out of the stream. This inorganic material, or "grit"—unlike SLUDGE—normally needs no further treatment beyond washing. It is usually disposed of by landfilling (*see* SANITARY LANDFILL). The flow rate in a grit chamber is usually maintained at 0.25 to 0.3 cubic meters per second. *See also* PRIMARY TREATMENT.

groin (groyne) a structure built to control shoreline EROSION, consisting of a low wall built perpendicular to the shore and extending from the shore out into the body of water a distance of several feet to a quarter of a mile or more. The most common construction material is stone, though wood and concrete are occasionally used instead. Groins operate by slowing down LONGSHORE currents and interrupting SHORE DRIFT. They have a tendency to collect sand on the upcurrent side, "starving" beaches beyond them (*see* BEACH NOURISHMENT), and thus may actually increase erosional problems rather than curing them unless they are very carefully sited and designed. Groins built on either side of the mouth of a river in order to prevent bar formation and keep the channel clear are called *jetties*. *See also* GROIN FIELD; EROSION: *control of erosion*.

groin field a set of several small GROINS that are designed and built as a unit. The groins of a groin field are generally less than 100 feet long and about the same distance apart. Groin fields are slightly more effective than single large groins at controlling erosion without interrupting BEACH NOURISHMENT in the area downdrift from the structure. However, the advantage is not great enough to make them an environmentally attractive means of controlling erosion. *See* EROSION: *control of erosion*.

gross erosion in geology, the total amount of material removed from a given site or by a given erosional agent (such as a specific stream or the LONGSHORE CURRENT in a large body of water) in a specified length of time, usually one year. In calculating gross erosion, the effects of DEPOSITION are ignored. *Compare* NET EROSION.

Gross National Product (GNP) in economics, the total sum of money expended in a nation within a given period of time. The GNP includes expenditures for both GOODS and SERVICES, and is calculated by adding together all personal expenditures by individuals, all government expenditures, and all capital-goods investment by corporations (that is, all corporate investment in machinery, plant space, and inventory). The concept of the GNP was developed in the 1930s by the American economist Simon Smith Kuznets as a means of measuring national economic health. At that time, U.S. GNP was running between $50 and $100 billion per year. By 1999, U.S. GNP was up to $9.5 trillion per year.

ENVIRONMENTAL SIGNIFICANCE OF THE GNP

Economists generally agree that GNP must show steady annual growth for an economy to remain healthy: it must at least match population growth in order to avoid a decline in per capita income. The problem with this from an environmental standpoint is that traditional means of increasing the GNP require massive infusions of natural resources and increases in manufacturing, leading to the destruction of wildlands and the release of higher and higher levels of pollutants into the air and water. There is also a solid-waste disposal problem caused by the "flow-through" of consumer goods from manufacturer to trash heap, which increases as personal spending increases—traditionally the best place to "prime" the GNP. The problem for environmental economists, therefore, is to find ways to maintain growth in the GNP without increasing manufacturing and natural-resource consumption. This can be done by increasing the proportion of the GNP that goes to investment in services over that which goes to investment in goods—a trend that is being seen in the United States today. In 1960, personal investment in services was approximately one-third of total personal expenditure. Today, it is closer to one-half.

ground lead in logging, any system of YARDING in which the logs are dragged over the ground, including horse logging, CAT SKIDDING, and some forms of CABLE YARDING. *Compare* HIGH-LEAD LOGGING.

ground-nesting bird any bird that builds its nest on the ground. Ground-nesting birds are more vulnerable to PREDATORS than any that build their nests in bushes or trees, and therefore have a greater need for COVER. It is also more difficult to avoid disturbance of nesting sites by human land-use activities, including especially construction, logging, and recreation. Principal ground-nesting SPECIES include the GALLINACEOUS BIRDS (quail, pheasant, grouse, turkey, etc.), the nightjars (whippoorwills, nighthawks, etc.), most arctic species, and many species of WATERFOWL.

groundwater water held beneath the surface of the ground. The term is generally reserved for water lying at or below the WATER TABLE. Water above the water table is called SOIL WATER. Groundwater is held in porous rock or soil bodies called AQUIFERS; it is the source of the water which flows from SPRINGS and SEEPS, and provides BASEFLOW for PERMANENT STREAMS. It may be withdrawn for human use by drilling wells: such withdrawals account for approximately 20% of the water used in the United States.

PROBLEMS OF GROUNDWATER

As with surface water, groundwater problems may be divided into two types, those of supply and those of quality. Supply problems result from OVERDRAFT (*see* CONE OF DEPRESSION; FOSSIL WATER; OGALLALA AQUIFER), and are addressable through conservation measures and (in some cases) through CONJUNCTIVE USE. Groundwater quality problems are complex and not easily addressable by any readily available current means. Some are caused by natural mineralization of the waters (*see,* e.g., SALINE INTRUSION): others result from pollution of groundwater RECHARGE areas by toxic wastes, leaking underground storage tanks, septic drainfields, and similar sources. Once in the aquifer, the pollutants spread through DIFFUSION and CAPILLARY ACTION, often in unpredictable ways; see PLUME. They are difficult or impossible to eliminate. ENVIRONMENTAL PROTECTION AGENCY researchers have found more than 200 manmade toxic chemicals in groundwater supplies at some 7,000 sites around the nation, affecting up to 28% of U.S. groundwater reserves. More than 1,100 wells were closed due to contamination in the decade between 1975 and 1985. *See also* AQUIFER SENSITIVITY; LEGRAND RATING SYSTEM.

groundwater dam a rock body or artificial structure beneath the surface of the ground that creates a barrier to the flow of GROUNDWATER, so that the WATER TABLE is higher on one side than on the other. Groundwater dams can be used together with INTERCEPT WELLS to form a barrier to PLUMES of polluted groundwater.

group (functional group) in chemistry, broadly speaking, any subgroup of ATOMS within a MOLECULE (as opposed to the molecule as a whole). More specifically, the term *group* is usually used for a characteristic grouping of atoms that recurs in several different COMPOUNDS, unifying them as a chemical family. *See also* RADICAL.

group selection system *See* SELECTIVE CUT.

growth curve a set of measurements of some specific aspect of the growth of an ORGANISM or a POPULATION or of some specific aspect of economic growth such as resource consumption, individual wealth, etc. A growth curve is usually plotted graphically with the growth measurement on the vertical axis over time on the horizontal axis. Biological growth curves are almost always *sigmoid,* that is, S-shaped, with an initial phase of acceleration (an upward curve) followed by a period of constant growth (a straight slanting line) and closed off by a period of deceleration (a flattening curve). The slope of the constant phase and the proportions of the three phases in relation to each other are generally characteristic of the SPECIES involved, as are the locations of any temporary increases or decreases ("inflections") such as human puberty. Economic growth curves are also often sigmoid, but less reliably so. They have a disturbing tendency to become *asymptotic,* that is, to curve upward more and more rapidly until they approach a straight vertical line. Such growth curves, of course, cannot be maintained to their conclusion, which would require infinite growth in an instantaneous time period.

growth form in botany, the overall shape taken by a mature plant. It is generally characteristic for each SPECIES, though there can be broad individual variations within a species due to variations in moisture conditions, sunlight, exposure to winds, soil mineralization, and so forth.

growth rings *See* ANNUAL RINGS.

groyne *See* GROIN.

grumusol a soil of semiarid grasslands, characterized by high clay content, lack of a B HORIZON, a thin brown A horizon, and a thick C horizon with vertical cracks in it that often extend into and through the A horizon, especially in dry weather. The term is now generally considered obsolete, and has been replaced by the more general term VERTISOL. *See also* SOIL; GREAT SOIL GROUPS.

Gulf States Marine Fisheries Commission (GSMFC) founded in 1949, a compact (approved by the U.S. Congress) between Alabama, Florida, Mississippi, and Texas. The 15-member commission's mission is the promotion and conscientious use of the Gulf Coast resources and marine fisheries through cooperative programs and waste prevention. Address: P.O. Box 726, Ocean Springs, MS 39566-0726. Phone: (228) 875-5912. Website: www.gsmfc.org.

guttation in the biological sciences, the exuding of water as liquid droplets through the leaves (and occasionally, the stems) of plants. The roots of plants are constantly taking up moisture from the soil through OSMOSIS, thus increasing the water pressure within their XYLEM. If this "root pressure" is not balanced by TRANSPIRATION from the leaves, the excess water must be forced out at the upper end of the plant's vascular system. Guttation generally takes place when transpiration is low or nonexistent, as on extremely humid nights. Most plants guttate through specialized STOMATA called *hydathodes*, which are able to release water without altering the plant's CARBON DIOXIDE balance.

GUY botanist's nickname for any of numerous extremely similar yellow flowers of the COMPOSITE family. The letters stand for Goddamned Unidentifiable Yellow.

guyot (tablemount) in oceanography, a SEAMOUNT with a broad, flat top. Guyots are former islands, generally of volcanic origin, whose tops were planed off by wave action and which have subsequently sunk below sea level. Their tops often bear wave-rounded COBBLES and the fossilized remains of shallow-water organisms. Their sinking is generally the result of seafloor subsistence associated with continental drift (*see* PLATE TECTONICS). Because they represent a region of relatively shallow water surrounded by much deeper waters, guyots can be important producers of fish. They are also sometimes heavily mineralized, and are close enough to the surface to make mining potentially attractive. Guyots approaching closer than 100 meters to the ocean surface are called *banks*.

gymnosperm a seed-bearing VASCULAR PLANT that produces its seeds in CONES or conelike structures rather than in FLOWERS. They are pollinated by wind action rather than by insects. Nearly all gymnosperms are trees or large shrubs. More "primitive" in structure than the ANGIOSPERMS, and more ancient in the fossil record, the gymnosperms are nevertheless better adapted to certain conditions, notably cold climates, and thus remain widespread today. Many, especially the pines, true firs, hemlocks, and DOUGLAS FIRS, are important economically as sources of lumber. Because the gymnosperms do not appear to be monophyletic (that is, they do not appear to have evolved from a single ancestor), disagreement remains over their taxonomic classification; however, most botanists place them in a single class, the Gymnospermae, with four subclasses: the Cycadinae (cycads), Ginkgoinae (ginkgos, or maidenhair trees), Coniferinae (conifers) and Gnetinae (a diverse group that includes several gymnosperms with angiosperm-like vascular tissue and reproductive structures). Together, these four subclasses contain 14 families, 62 genera (*see* GENUS) and about 500 SPECIES.

gyppo a small-scale, independent logging operator. Gyppo logging operations generally employ fewer than 10 workers. They almost never own their own timber, but must either purchase government timber or work under contract to a larger logging firm or other private landowner. Though there is often an unsavory, cut-and-run aura associated with the word "gyppo," there is no hard evidence that small firms are any less careful of the forest resource than large ones. The origin of the term probably comes from the itinerant ("gypsy") nature of small logging operations rather than from any inference of rampant cheating.

gypsy moth a large moth, *Lymantria dispar,* introduced from Europe and now a serious insect pest in North America. Gypsy moths lay their eggs on the leaves of oaks, birches, maples, and other HARDWOODS. The caterpillars that hatch out of the eggs then eat the leaves, often temporarily defoliating the tree and sometimes killing it. The adult moths are white and black (females) or brown and black (males), with a wingspread of 2–3 inches. The caterpillars are yellow and hairy, with red and blue knobs in a line down either side. They hatch from the eggs at a length of about 1/8 inch and grow to about 2 inches before pupating (that is, before spinning a pupa and metamorphosing into an adult moth). They were accidentally introduced into North America in 1869 in Medford, Massachusetts, by the French scientist Leopod Truvelot while trying to obtain a disease-resistant cross between the gypsy moth and the silkworm moth, and have since spread throughout the northeastern and mid-Atlantic states, and into parts of the Middle West. So far, some 60 million acres of forest have been defoliated by the gypsy moth, the worst period being 1981, when 12 million acres were defoliated. An Asian strain of the same species (called the AGM) appeared in scattered locations in the Northwest in the early 1990s and may be more of a threat to U.S. forests than the European strain. While AGMs look the same as the European moths (albeit a bit larger), they do not share two characteristics that have kept the European moths from doing even more damage. First, the female AGM can fly a fair distance to lay her eggs, which the European female cannot, suggesting a much more rapid dispersion of the AGM. Second, the AGM will attack coniferous trees as well as deciduous trees, suggesting problems for the economically important coniferous forests of the Northwest. Attempts during the 1950s and 1960s to eradicate gypsy moths with massive aerial sprayings of DDT (now banned) and other hard pesticides were singularly unsuccessful. Since then, efforts to control the moths have been more highly targeted, with Bt (*Bacillus thuringiensis*), a relatively benign (to humans) biological agent, the preferred weapon. The active ingredient in Bt is a bacterium that exists in nature and causes the gypsy moth to lose its appetite. Natural controls of gypsy moth infestation have also been observed, which provides some hope. In certain situations, "gypsy moth wilt" takes place, the result of a naturally occurring virus familiarly called NPV (for nucleopolyhedrosis virus) which can cause a massive gypsy moth population crash. In the late 1980s another natural regulator, a fungus named *Entomophaga maimauiga,* appeared, which was in fact imported from Japan as a biological control in 1909. It was long thought that the fungus had simply died out. Gypsy moth populations killed by *Entomophaga* have a similar appearance to those killed by NPV, which may account for the mystery. The problem with all these controls—Bt, NPV, and the *Entomophaga* fungus—is that they work well only under quite specialized circumstances. Application must be at exactly the right stage in the moth's life cycle, humidity and temperature are crucial factors, and for the most part the controls are effective only on the most severe infestations. It would appear that the gypsy moth will be a problem in U.S. forests and gardens for a while longer.

habitat in ecology, the type of ENVIRONMENT in which an ORGANISM of a particular SPECIES is likely to be found. Organisms are said to "prefer" their proper habitat. That is, sparrows prefer shrubs and hedgerows; trilliums prefer well-drained soils in moist, heavily shaded woods; spotted owls prefer old-growth forests; etc. The term includes vegetation, soil and rock types, amount of SHELTER and COVER, ASPECT (and exposure to sunlight), and numerous other factors. Access to the proper habitat is critical for the preservation of most species; hence, destruction of a particular habitat type can led to declines of species and even local extinction. The fragmentation of habitat can also have a profound effect on indigenous species, some of which require large undisturbed areas. Fragmentation occurs through changes in land use, such as for housing, or altering natural systems, such as draining marshlands for agriculture or converting natural forests to tree farms. Most notably, fragmented and modified habitat has led to a decline of migrating warblers, which require large areas of undisturbed woodlands for breeding. *Compare* NICHE. *See also* BIOME.

habitat conservation area *See under* THOMAS REPORT.

habituation in the biological sciences, a type of learning process in which an ORGANISM gets used to a particular stimulus and responds more sluggishly to it, or no longer responds at all. Habituation is among the simplest forms of learning, and is exhibited even by one-celled organisms. Amoebae (*see* AMOEBA), for example, will contract into as small a space as possible and stop all activity when a light is shone on them. If the light continues to shine, however, they will eventually become habituated to it and will resume their normal behavior.

Hadley cell in meteorology, one of a pair of convection currents in the ATMOSPHERE that circulate between the equator and the regions 30° north and south of it (the HORSE LATITUDES). The Hadley cells are caused by the heating of the air over the equator. The heated air rises and flows poleward; this flow of air is deflected by the CORIOLIS EFFECT so that, by the time it reaches the horse latitudes, it is flowing east-west. Here a combination of cooler conditions and drag from the rotating earth cause it to slow down and sink to the surface, where it flows back to the equator. The name *Hadley cell* honors the 18th-century English physicist George Hadley, who proposed the first model for global atmospheric circulation (1735). It is sometimes also applied to the similar convection cells between 30° and 60° north and south of the Equator, and between 60° north and south and the North and South Poles. *See also* ATMOSPHERE: *dynamics of the atmosphere.*

hail in meteorology, a form of PRECIPITATION consisting of roughly round pellets of ICE. Hail is formed in turbulent AIR MASSES containing strong updrafts within which temperatures range from above to below freezing (0°C, or 32°F). A raindrop in the lower, warmer portion of the air mass is carried aloft on an updraft until it freezes. The resulting ice granule grows by collecting water vapor out of the air through SUBLIMATION until it becomes heavy enough to fall. Other updrafts may lift it aloft again for a new cycle of freezing, giving it a concentric, layered structure. As they fall to earth, hailstones average about a centimeter across. The largest on record (Coffeyville, Kansas, September 3, 1970) was 14 centimeters (about 5 1/2 inches) through and weighed 3/4 of a kilogram (1.6 pounds).

half-life in chemistry and physics, the time it takes half of an ELEMENT undergoing radioactive decay to transmute to the new element that is the end product of the decay process. Uranium-238, for example, decays to lead, with a half-life of about 4.5 billion years. Thus, roughly half the U_{238} present at the creation of the earth has now been converted into lead. Half-lives of other materials may be considerably less than this. The half-life of radium-226, for instance, is about 1,600 years, while strontium-90 has a half-life of 28 years and that of curium-242, one of the TRANSURANIC ELEMENTS, is less than half a year. Generally, 10 half-lives are considered sufficient for the volume of a radioactive substance to decline to insignificance.

half-wave depth in geology and hydrology, a depth of water equal to one-half the distance between the crests (or the troughs) of a successive pair of WAVES in a body of standing water such as a lake or an ocean. The half-wave depth represents the distance below the surface that water is normally disturbed by wave action. Thus, as a wave approaches a shoreline, it first begins to affect the bed of the body of water ("feel bottom") at the half-wave depth, making that depth the point where shoreline EROSION begins.

halide in chemistry, a BINARY COMPOUND in which one of the two constituents is a member of the halogen family of elements (*see* HALOGEN). The other constituent is commonly a METAL. Most halides are crystalline solids that dissociate well in water (*see* DISSOCIATION) and are good conductors of electricity. *See also* SALT.

halogen in chemistry, one of the five ELEMENTS that form the halogen series in the PERIODIC TABLE. They include (in order of increasing atomic size) flourine, chlorine, bromine, iodine, and astatine. The halogens all have seven ELECTRONS in their outer electron shells (*see* ATOM: *structure of the atom*), giving them a VALENCE of –1. They are all chemically active, and are very similar in their chemical properties, though they show a general gradation from flourine (most active; least metallic in character) to astatine (least active; shares some properties with the metals, including LUSTER and electrical CONDUCTIVITY). *See* HALIDES; SALT; HALOGENATED HYDROCARBON. *See also* articles on each of the individual halogen elements.

halogenated hydrocarbon in chemistry, a HYDROCARBON compound with one or more HALOGEN atoms attached to it. Most halogenated hydrocarbons are biologically active. As a group, they dissolve easily in fats, allowing them to bioaccumulate rapidly (*see* BIOLOGICAL MAGNIFICATION). Many are carcinogenic (*see* CARCINOGEN). Representative halogenated hydrocarbons include POLYCHLORINATED BIPHENYLS (PCBs), POLYBROMINATED BIPHENYLS (PBBs), DDT, 2,4-D and 2,4,5-T.

hamlet in land-use planning, a settlement with a population of less than 250 persons. Hamlets are generally named, are often incorporated (*see* MUNICIPALITY), and almost always include two or more commercial or public buildings (stores, schools, churches, etc.). *Compare* COUNTRY NEIGHBORHOOD; VILLAGE.

Handbook, Forest Service *See* *FOREST SERVICE HANDBOOK*.

hanging bog a BOG developed on a slope rather than in a depression. Hanging bogs result in areas where AQUIFERS reach the surface beneath shallow soil cover on a hillside, resulting in a broad area of SEEPS that keep the soil permanently soaked. Generally there is also a considerable amount of atmospheric moisture present as well, as for example on the upper slopes of the windward side of a mountain range. The excess moisture both increases plant growth and slows its decay, allowing the development of typical bog acid-water conditions and encouraging the growth of bog vegetation. Small hanging bogs which develop around hillside springs are sometimes called "spring bogs."

hanging valley in geology, a small tributary valley whose floor is elevated some distance above the valley floor of the MAINSTEM. Hanging valleys result when the mainstem valley is downcut much faster than the tributaries (*see* DOWNCUTTING), usually as a result of glacial action, with the GLACIER occupying the main valley not only rapidly increasing its depth but also acting as a temporary BASE LEVEL beneath which erosion in the tributary cannot take place. Streams occupying hanging valleys almost always end in waterfalls.

hardness of rocks and minerals, *see* MOHS SCALE.

hard pesticide unofficial term for any PESTICIDE that is not easily biodegradable (*see* BIODEGRADABLE SUBSTANCE) and therefore tends to persist in the environment, allowing it to bioaccumulate readily up the FOOD CHAIN (*see* BIOLOGICAL MAGNIFICATION). Generally included in the group are DDT, ALDRIN, Dieldrin, HEPTACHLOR, and other ORGANOCHLORIDES. *See also* BROAD-SPECTRUM PESTICIDE.

hard rock mine a mine cut into an igneous or metamorphic FORMATION (*see* IGNEOUS ROCK; METAMORPHIC ROCK) in which the mineral is removed through tunnels excavated along the mineral VEINS. Hard rock mines are generally less environmentally damaging than other forms of mining, and if carefully handled will not disturb the scenery. However, they can still cause a significant amount of stream pollution due to mineralized RUNOFF from within the mine and the leaching of rainwater through the tailings (*see* LEACHATE). *Compare* STRIP MINE; PLACER MINE.

Hardin, Garrett (1915–) biologist, educator, and author. Professor emeritus of human ecology at the University of California, Santa Barbara, Hardin is an advocate of population control to avert environmental disaster. Born in 1915, he is known for his article "THE TRAGEDY OF THE COMMONS" (1967) and for the concept of "lifeboat ethics," concerning the ethical issues surrounding survival in the face of excess population. A prolific writer and lecturer, his books include *Nature and Man's Fate* (1959), *Birth Control* (1970), *Exploring New Ethics for Survival: The Voyage of the Spaceship Beagle* (1972), *Mandatory Motherhood: The True Meaning of Right to Life* (1974), *The Limits of Altruism: An Ecologist's View of Survival* (1977), *Biology, Its Principles and Implications* (1978), *Promethean Ethics: Living with Death, Competition, and Triage* (1980), *Naked Emperors: Essays of a Taboo-Stalker* (1983), *Living Within Limits: Ecology, Economics, and Population Taboos* (1993), *The Immigration Dilemma: Avoiding the Tragedy of the Commons* (1995), and *The Ostrich Factor: Our Population Myopia* (1999).

hardwood any tree belonging to the class Angiospermae (*see* CLASS; ANGIOSPERM). Hardwoods are flowering plants; most are deciduous (lose their leaves in winter; *see* DECIDUOUS PLANT). Their GROWTH FORMS tend to be much broader than those of conifers, and not so centrally oriented. Representative hardwoods include the maple, the beech, the oak, the alder, the madrone, the walnut, and the trees of the rose family (apple, apricot, plum, cherry, pear, almond) that are commonly grown for their fruit. *Compare* SOFTWOOD.

harvest prescription in forestry, the specifications laid down when planning for the harvest of a particular STAND of trees. The harvest prescription covers the harvest method to be used (CLEARCUT, SHELTERWOOD SYSTEM, etc.); FELLING practices; YARDING methods; number and location of LEAVE TREES; and, in most cases, the location or roads and LANDINGS. It does not directly cover REGENERATION or other treatments of the site that take place outside the scope of the current stand ENTRY, although these things must be taken into account. For example, felling and yarding must be planned so that the soil will not be damaged to the point where regeneration will not occur. *Compare* SILVICULTURAL PRESCRIPTION.

harvest unit in forestry, an area of forest land on which the timber is to be cut in a single operation or set of operations, following a single HARVEST PRESCRIPTION. Harvest units vary in size from a few acres to 1,000 acres or more. Silvicultural conditions on the entire unit should be essentially uniform; that is, such factors as the slope of the land, the ASPECT, the soil type, and the vegetation mix should be similar throughout the unit. The boundaries should be mapped, surveyed, and flagged so that they can be easily found. *See also* CUTTING UNIT.

Hawk Mountain Sanctuary Association founded in 1934 to protect a noted site in central Pennsylvania for the observation of migrating hawks. The association is committed to the conservation of RAPTORS and to improved knowledge of the central Appalachian environment. It employs a full-time staff assisted by interns and volunteers to provide programs in education and research and to maintain the 2,380-acre sanctuary, which is open to the public year-round. It publishes *Hawk Mountain News; The Mountain and the Migration;* and *Hawks Aloft.* Membership (1999): 10,000. Address: 1700 Hawk Mountain Road, Kempton, PA 19529. Phone: (610) 756-6961. Website: www.hawkmountain.org.

Hawken, Paul (1946–) businessman-environmentalist and author. As principal of Smith & Hawken, a catalog sales concern, Hawken has become a persuasive proponent of corporate reform, with writings encouraging cost-effective ecological practices. He founded Erewhon Trading Company, a natural foods wholesaler, in the 1960s; cofounded Smith & Hawken in 1979; and later founded Datafusion, a software company, in 1995. Hawken drew on his experience in a best-selling book, *Growing a Business* (1987), with a companion PBS series on the challenges and opportunities involved in starting and running a socially responsible corporation. He has served as cochair of The Natural Step, an educational foundation concerned with environmental sustainability. Currently a consultant on sustainability issues, he works with corporations, foundations, and educational institutions. His honors include the Small Business Administration's Entrepreneur of the Year (1990), the Council on Economic Priorities Environmental Stewardship Award (1991), and the Green Cross Millennium Award for International Environmental Leadership (1999). Among his other books are *The Next Economy* (1984), *The Ecology of Commerce: A Declaration of Sustainability* (1994) and *Natural Capitalism,* with AMORY LOVINS and L. Hunter Lovins (1999).

hay fever a form of allergy in which the ALLERGEN is POLLEN grains and the reaction involves the swelling of nose, throat and other upper respiratory system tissues; usually the eyes become involved as well. People with hay fever are often also allergic to other items, and may show hay-fever-like symptoms after exposure to airborne bacteria (*see* BACTERIUM), fungi (*see* FUNGUS), dust particles, bits of hair or feathers, etc. As with other allergies, control of the symptoms is usually possible with antihistamines. *Compare* ASTHMA.

hazardous material loosely, a HAZARDOUS SUBSTANCE; specifically, in regulatory law, a substance regulated by the Department of Transportation (DOT: *see* TRANSPORTATION, DEPARTMENT OF) under their mandate to make certain that highways, railroads and other transportation corridors remain safe. The DOT currently lists about 1,700 of these substances, grouped into seven categories: explosives, flammables (*see* FLAMMABLE LIQUID; FLAMMABLE SOLID), corrosives (*see* CORROSION), compressed gases, poisons (*see* TOXICANT), ETIOLOGIC AGENTS, radioactive materials, and miscellaneous ("other regulated materials," or ORM). DOT regulation of these substances includes package design, handling techniques, routing requirements, prohibitions on the co-shipment of certain materials, and paperwork requirements, including the reporting of materials shipped, amounts they are shipped in, routes they are shipped over, and "incidents" (spills and leaks) that may happen either on route or during transfer to and from storage at either end of the shipment. The lists and regulations are published in the CODE OF FEDERAL REGULATIONS (49 CFR 171–177).

Hazardous Materials Transportation Act federal legislation to decrease the dangers to the public resulting from the transportation of HAZARDOUS MATERIALS, signed into law by President Gerald Ford on January 3, 1975. The act authorizes the secretary of transportation to issue and enforce regulations concerning the shipping, handling, packaging, labeling and storage of hazardous materials being transported across state lines, or of hazardous materials being transported within individual states if such transportation is related to an interstate shipment. Carriers of hazardous materials are required to register with the government, to keep complete records, and to make periodic reports of their activities. Other sections of the act mandate a "continuing" study of means to upgrade the safety of hazardous materials transport, and forbid the transportation of non-medically related radioactive materials by passenger aircraft.

hazardous substance loosely, any substance that may cause harm to human beings or to the natural environment. Specifically, the term "hazardous substance" is used in regulatory law to describe a substance regulated by the ENVIRONMENTAL PROTECTION AGENCY (EPA) under the terms of Section 311 of the CLEAN WATER ACT. The EPA classifies these substances into four categories. *Flammables* (also known as *ignitables; see* FLAMMABLE LIQUID; FLAMMABLE SOLID) are materials that burn easily and thus constitute a fire hazard. *Toxics* (*see* TOXICANT) are materials that make living things ill. *Corrosives* (*see* CORROSION) are materials that react readily with other substances, including those commonly used as containers, "eating them away." These materials are hazardous because they can harm living tissue, and they escape readily into the environment if care is not taken to make sure that they are placed in containers that they are unable to corrode. Finally, *reactives* are unstable materials that may undergo spontaneous reactions, either alone or in combination with air or water, resulting either in explosions or the release of substances fitting into one or more of the other three categories. The EPA publishes a list of hazardous substances in the FEDERAL REGISTER (40 CFR 116). There are currently approximately 300 chemicals on the list. Spills or other accidental discharges of any substance appearing on this list must be reported to the EPA if they exceed designated size threshholds. These threshholds are also listed in the Federal Register (40 CFR 117). *See also* HAZARDOUS MATERIAL; HAZARDOUS WASTE; PRIORITY POLLUTANT.

Hazardous Substances Act federal legislation to identify and limit public exposure to HAZARDOUS SUBSTANCES, signed into law by President Dwight Eisenhower on July 12, 1960, and amended numerous times since. The act defined several classes of hazards and allowed the secretary of health, education and welfare to list any substance whose characteristics fit into any of these classes as a hazardous substance. Such substances were subject to strict labeling requirements, and in extreme cases could be banned from interstate commerce. Weak enforcement provisions and numerous loopholes made the act nearly useless, and it has been almost completely superseded by later legislation, although it is technically still in force. *See* HAZARDOUS MATERIALS TRANSPORTATION ACT; RESOURCE CONSERVATION AND RECOVERY ACT; TOXIC SUBSTANCES CONTROL ACT; COMPREHENSIVE ENVIRONMENTAL RESPONSE, COMPENSATION AND LIABILITY ACT.

hazardous waste loosely, any material that, when thrown away, constitutes a hazard to the health or safety of humans or the natural environment. Specifically, in regulatory law, the term "hazardous waste" refers to substances regulated by the ENVIRONMENTAL PROTECTION AGENCY (EPA) under the terms of the RESOURCE CONSERVATION AND RECOVERY ACT OF 1976 (RCRA) and listed in the CODE OF FEDERAL

REGULATIONS under 40 CFR 261. There are approximately 500 substances covered by RCRA regulations, which pertain only to the disposal of materials, not their handling or storage prior to disposal. Among hazardous household wastes, just to take one category as an example, oven cleaners, septic tank cleaners, paints and wood preservatives, roofing tar, pesticides, gasoline and motor oil, and dry cell batteries are all considered hazardous and must be disposed of properly. *See also* HAZARDOUS SUBSTANCE; HAZARDOUS MATERIAL; PRIORITY POLLUTANT.

HCA in forestry, abbreviation for habitat conservation area: *see under* THOMAS REPORT.

head in hydrology, the vertical height of a column of water. The amount of head is used as a measure of the ability of water to do work; the greater the head, the higher the water pressure and the more work can be obtained from an equivalent amount of water. One hundred feet of head is equivalent to 43 pounds per square inch (psi) of GAUGE PRESSURE. *See* HYDROELECTRIC POWER.

headframe the structure at the top of a well or vertical mine shaft that holds the pulleys over which cables are run to operate an elevator, a hoist bucket, or some other form of lifting device.

headgate *See under* DIVERSION.

head lean *See under* LEAN.

headsaw (headrig) in lumbering, the first lumber-forming saw a log meets when it enters a sawmill. The headsaw is generally a form of bandsaw. Its purpose is to cut the log lengthwise into flat slabs. For most (but not all) logs, the headsaw will be preceded by a *cutoff saw,* which cuts the logs to lumber-forming length.

Health and Human Services, Department of branch of the federal (U.S.) government charged with maintaining the health and well-being of U.S. citizens. Created by Congress on April 11, 1953, as the Department of Health, Education, and Welfare, the agency was renamed the Department of Health and Human Services upon the creation of the Department of Education on October 17, 1979. It is headed by a cabinet-level official, the secretary of health and human services, who reports directly to the president. The department has four major program divisions; all environmental health programs are located within a single division, the PUBLIC HEALTH SERVICE. *See also* AGENCY FOR TOXIC SUBSTANCES AND DISEASE REGISTRY; CENTERS FOR DISEASE CONTROL; FOOD AND DRUG ADMINISTRATION.

heap leaching in mining, a method for concentrating and removing traces of metals from broken pieces of lower-grade ORES and TAILINGS. Developed by copper miners, the process involves stacking the materials in alternate foot-thick layers of fine and coarse pieces until a heap about 20 feet high is obtained. Water is then poured over the heap several times, at long intervals. OXIDATION takes place in the intervals between pourings, and the LEACHATE is therefore rich in copper oxides, from which the copper can be chemically precipitated. The same process can be used, with modifications, for other metals. To obtain gold, for example, the water poured on the heap is first mixed with potassium cyanide (*see under* CYANIDE), which is capable of dissolving the gold; the leachate is then treated with ZINC to precipitate the gold. The resulting cyanide- and zinc-laced leachate is a severe environmental hazard if allowed to escape.

heart rot (center rot) a disease of both HARDWOOD and SOFTWOOD TREES, caused by a FUNGUS, usually *Fomus annosus.* Acting in its role as a DECOMPOSER, the fungus attacks the HEARTWOOD of the tree, usually working from the center out. It does not damage the XYLEM or any other living layer of the tree. The tree is, however, seriously damaged structurally, and may break easily, in addition to being rendered commercially valueless by the destruction of its wood. For methods of estimating the loss of value, *see* LOG SCALING.

heartwood the dead inner wood of a tree or other WOODY PLANT. Heartwood consists of XYLEM in which the water-carrying tubes have been filled with LIGNIN and have solidified. It serves as the tree's skeleton, allowing it to stand erect and support its CROWN. The great majority of wood in a living tree is heartwood and is thus technically dead.

heat low *See* THERMAL LOW.

heaviside layer the lower section of the ionosphere, consisting of the D and E layers. *See* ATMOSPHERE: *structure of the atmosphere.*

heavy hydrogen *See* DEUTERIUM.

heavy metal any of a large group of metallic elements (*see* METAL; ELEMENT) with relatively high ATOMIC WEIGHTS and similar health effects. The group is somewhat poorly defined. As generally understood, it ranges down to an atomic weight of 24 (CHROMIUM) and includes at least two NON-METALS (ARSENIC, SELENIUM) as well as the true metals. Some (iron, cobalt, ZINC) are important TRACE ELEMENTS. Others interfere with or inhibit METABOLISM. Most are toxic in relatively small amounts. They have in common an ability to interfere with each other's metabolic actions (lead, for example, may interfere with the body's production of HEMOGLOBIN, thus inhibiting oxygen transfer in the bloodstream) and a sensitivity to diet (a high-protein diet will generally slow down their absorption by the body). Large amounts of them are often present in industrial and mining wastes and can be carried as microscopic particles in factory and power-plant emissions over long distances, polluting lakes and other water bodies and entering the tissue of plants and trees, sometimes severely weakening them.

heavy water in chemistry, water in which the normal hydrogen ISOTOPE has been replaced by heavy hydrogen (*see* DEUTERIUM), increasing its MOLECULAR WEIGHT by two, from 18 to 20. Heavy water occurs in nature in the proportion of about 1:6900 (heavy water: normal water). It is very slightly more viscous than normal water (*see* VISCOSITY), and has slightly higher boiling and freezing points. Its REACTION time is slightly longer, making chemical reactions involving it slightly slower than those involving normal water, though they are otherwise identical. It is not radioactive, but is an excellent absorber of ALPHA PARTICLES, making it useful to moderate nuclear reactions (*see* NUCLEAR ENERGY).

heavy water reactor *See under* NUCLEAR ENERGY.

hedged shrub in range or wildlife management, a SHRUB that has a dense, pruned ("hedged") appearance, like a suburban hedge that is kept regularly trimmed. Though it may occasionally be caused by severe weather conditions, especially in exposed places such as ridgetops, hedging is generally a result of overgrazing by animals and is thus an indicator that the condition of the RANGE is deteriorating. The amount of deterioration can be judged fairly well by the type of shrub being hedged. The less palatable the hedged SPECIES is to the animals using the range, the farther advanced the deterioration should be assumed to be.

helispot any place in which a helicopter may safely be set down. To be mapped as a helispot, a location should have a level clear area at least 50 feet in diameter that can be approached at an angle no steeper than 45°. If clearing, leveling, or other engineering work has been done on the site, the helispot is usually referred to as a *helicopter pad.*

Hells Canyon popular name for the Grand Canyon of the Snake River and the central feature of the Hells Canyon National Recreation Area, on the boundary between Oregon and Idaho south of the city of Lewiston, Idaho. The canyon is 160 kilometers (100 miles) long and has an average depth of more than 1,675 meters (5,500 feet), with a maximum depth of 2,407 meters, (7,901 feet), making it the deepest canyon in North America and one of the deepest in the world. From an environmental standpoint, Hells Canyon is most important as the subject of a U.S. Supreme Court opinion handed down on June 5, 1967, in a case known as *WPPSS* vs. *FPC.* The case had been argued as a question of whether dams should be built in the canyon by public or private power entities. The decision, written by Justice William O. Douglas, made this issue subordinate to the larger question of whether any dams should be built at all. The determination of "in the public interest," Douglas wrote, could be decided "only after an exploration of all issues relevant to the 'public interest,' including . . . the public interest in preserving reaches of wild rivers and wilderness areas, the preservation of anadramous fish for commercial and recreational purposes, and the protection of wildlife." The Court's decision launched an eight-year battle for the preservation of the canyon, culminating in the signing of the Hells Canyon National Recreation Area Preservation Act by President Gerald Ford on December 31, 1975.

hematite a rock-forming MINERAL, chemical formula Fe_2O_3, containing about 70% iron. It is the most important commercial iron ore. Hematite occurs occasionally as rhombohedral crystals of gemstone quality, usually found in groups ("iron roses") that radiate outward from a central point. More often it is massive and granular ("compact") or composed of MICA-like flakes ("specular"). It may also be soft and earthy ("red ocher"). In any form, it makes a dark cherry-red STREAK when rubbed on a test surface. Large quantities of it occur in the Lake Superior region, in the Appalachians, in central Europe, in eastern Canada, and in Venezuela. *Compare* LIMONITE; TACONITE.

hemoglobin the material that carries OXYGEN in the bloodstream. Hemoglobin is a red pigment composed of a PROTEIN base known as *globin* carrying an iron-containing chemical GROUP known as *heme.* A MOLECULE of oxygen binds to the heme under conditions of oxygen surplus (e.g., in the lungs) and releases from it under conditions of oxygen debt (e.g., in tissues in need of oxygen). There are four heme groups in each hemoglobin molecule, hence each is capable of carrying four molecules of oxygen. The hemoglobin molecules themselves are carried in the red blood cells, composing about 33% of the bulk of each cell. A very large molecule (molecular weight: 64,458), hemoglobin contains 574 units of AMINO ACID formed into four chains, one for each heme group. It is structurally similar to CHLOROPHYLL, and may be thought of as the animal equivalent of this compound. The blood of all VERTEBRATES, and of nearly all of the so-called "lower animals," contains hemoglobin, making materials that interfere with its function (such as CARBON MONOXIDE), or with its manufacture by the bone marrow (such as uranium and other radioactive substances; *see* RADIOACTIVITY) among the POLLUTANTS that are the most broadly damaging to the BIOSPHERE. *See also* RESPIRATION.

hemp a herbaceous plant whose fibers are useful in making cordage, cloth, and paper—in the last case as an alternative to the use of wood pulp. Because hemp, a member of the Cannabaceae family of herbs, is also a source of marijuana, its cultivation is banned in the United States. Sisal hemp and manila hemp, also used for cordage, are not true hemps.

hepatitis disease of the liver, characterized by jaundice (a sallow, yellow skin tone), fever, weakness, loss of appetite, and abdominal and muscle pains. Two principal types have been identified. Hepatitis A (also known as *infectious hepatitis*) is caused by a filterable VIRUS called HAV. It is carried primarily by contaminated water supplies and food, although it may also be spread by injection by contaminated needles. Hepatitis B (*serum hepatitis*), caused by a closely related virus called HBV, is spread primarily through injection by contaminated needles, although mosquitoes may also occasionally carry it. Both types lead to death in approximately 10% of all cases. Those who survive may suffer permanent liver damage. The acute phase in both types lasts approximately two weeks, followed by a convalescence of roughly six months. Incubation time: HAV, 2–6 weeks; HBV, 6 weeks–6 months.

heptachlor a CHLORINATED HYDROCARBON PESTICIDE, chemical formula $C_{10}H_5Cl_7$, extensively used in the 1950s to control root borers, termites, boll weevils, and other insect pests. Heptachlor is a central nervous system stimulant that can cause paralysis and death in relatively small amounts (LD_{50} [rats, orally]: 90 mg/kg). Long-term exposure can cause liver damage and increased risk of cancer. The half-life of the chemical in soil is two to four years. One of its decay products, heptachlor epoxide, is even more toxic than heptachlor itself, and may persist in dangerous amounts for up to 11 years. Both heptachlor and heptachlor epoxide bioaccumulate extremely well (*see* BIOLOGICAL MAGNIFICATION) and are thus dangerous at any point that they enter the FOOD CHAIN. Restriction of the use of heptachlor began in 1959 with a ban on its use on hops. By 1975 restriction was virtually complete.

herb (herbaceous plant) any flowering plant (*see* ANGIOSPERM) that does not develop woody tissue. Herbs have no form of SECONDARY GROWTH; therefore the entire SHOOT (or the entire plant) must die if PRIMARY GROWTH ceases. For this reason, all herbs of temperate or cold climates remain low, relatively small plants—those generally thought of as "wildflowers"—which either die completely each year (*see* ANNUAL) or overwinter as buried roots and sprout a new shoot each spring (*see* PERENNIAL), though in the tropics a few may reach large size by continuing their primary growth for a number of years.

herbicide a chemical COMPOUND or mixture of compounds used to kill unwanted vegetation. The most common use of herbicides is as weed killers in landscape management and roadside vegetation control. However, there are other uses, notably to release greater growth rates in commercial CONIFER plantations by thinning the OVERSTORY and eliminating competing vegetation; to destroy crops and eliminate forest cover for enemy troops in wartime; and even to assist in harvesting a commercial crop, such as cotton, which is treated with a chemical, usually calcium cyanamide (CaNCN) to make its leaves fall off before mechanical harvesting so that the cotton will not be stained green by the leaves. Herbicides work in a variety of ways. *General herbicides,* such as sodium arsenite and sodium chlorate, simply poison the plant. Since they usually poison nontarget plants (*see* NONTARGET SPECIES), insects, small animals, and even large animals and humans as well as the pest species they are aimed at, these are normally only used under highly controlled conditions. *Contact poisons,* such as paraquat ($C_{12}H_{14}N_2C_{12}$ or $C_{12}H_{14}N_2(CH_3CSO_4)_2$) kill only those parts of the plant they come into contact with. *Defoliants,* such as 2,4,5,-T and 2,4-D, cause a plant to drop its leaves; most work by interfering with AUXIN production, causing excess auxin to accumulate in the leaf nodes so that the leaves are pinched off and starved for nutrients from the stem by growing their veins shut. Defoliants are generally most effective against DICOTYLEDONS. Since most of these have broader leaves than the MONOCOTYLEDONS and the GYMNOSPERMS, the defoliants are said to be "selective" for broadleafed plants, and for that reason are used to control weeds in lawns and to destroy HARDWOODS ("weed trees") without affecting nearby conifers in commercial timber plantations.

APPLICATION OF HERBICIDES

Herbicides may be applied either by *broadcast spraying*—spraying a fine mist over the target vegetation, usually from an aircraft or a land vehicle—or by TOPICAL APPLICATION, using a nozzle to place the herbicide directly on specific plants, usually accomplished by means of backpack spray gear carried by field workers walking through the target area. Broadcast spraying is by far the more dangerous of the two, primarily due to the tendency of the sprayed pesticide to drift into adjacent nontarget areas. Regulations tightly restrict its use in most cases. For example, FOREST SERVICE regulations require that wind velocities be no more than 5 miles per hour, that RELATIVE HUMIDITY be below 50%, and that temperatures be above freezing but below 70°F (21°C); that buffer strips at least 100 feet wide be left along watercourses; and that all landowners or those who have water intakes within one mile of the spraying site be directly notified in advance.

HAZARDS OF HERBICIDE USE

The principal hazard of herbicide use is the potential damage to nontarget species, including animals and humans, as even the so-called "selective herbicides" are toxic, carcinogenic and/or mutagenic (*see* TOXIC COMPOUND; CARCINOGEN; MUTAGEN), at least to some degree. (*See,* e.g., the discussions under 2,4,5-T and 2,4-D.) Drift from broadcast spraying into nontarget areas, especially watercourses, poses the greatest problems, but even topical application can present hazards, chief among them the tendency for wildlife and livestock to selectively browse on the treated foliage. Even plants not normally used for BROWSE may be eaten in the wilted stage after pesticide treatment. The reason for this appears to be alterations in the plant's metabolism brought on by the pesticide that cause it to produce more sugars and other CARBOHYDRATES while in the process of dying from the spray. The dangers are that the animals too will die, either from the effects of the herbicide itself or from eating toxic plants that they normally ignore but that have become attractive to them due to the spraying.

herbivore any animal whose diet consists principally or exclusively of plants. The favored food of terrestrial herbivores is generally HERBS—hence the name—but most will also eat the leaves of WOODY PLANTS, and occasionally the twigs and bark as well. Aquatic herbivores feed principally on ALGAE, though those that live in shallow water also eat reeds, mosses, cattails, and other water-loving BRYOPHYTES and VASCULAR PLANTS. In ecological terms (*see* ECOLOGY), herbivores are *first-level consumers,* that is, the immediate consumers of the producers (*see* FOOD CHAIN; ECOLOGICAL PYRAMID). Since this is not a particularly efficient role, herbivores generally must eat more often, and consume larger quantities of food, than CARNIVORES or OMNIVORES.

CLASSIFICATION OF HERBIVORES

Herbivores are generally classified according to the range of their diet. Those restricted to a single plant species (certain APHIDS, for example) are called *monophagous;* those that eat only a few well-defined, usually closely related species are *oligophagous;* and those that eat any plant matter available are *polyphagous.*

Heritage Conservation and Recreation Service obsolete federal agency charged with coordinating federal activities dealing with the recreational aspects of the nation's natural, historical, and cultural heritage. Formed in 1978 by reorganizing and expanding the BUREAU OF OUTDOOR RECREATION, the service was abolished by Interior Secretary James Watt in 1981.

hesperidium *See under* FRUIT.

Hetch Hetchy valley in Yosemite National Park, California, that was the site of a crucial early battle of the environmental movement. Located on the Tolumne River approximately 65 kilometers (40 miles) northwest of Yosemite Valley, Hetch Hetchy was called "another Yosemite" by JOHN MUIR and others. Included in the park when it was established on October 1, 1890, the valley was sought by the city of San Francisco as a domestic water supply beginning in 1901. The city's original application, dated October 16, 1901, was turned down three times (January 20, 1903; December 22, 1903; and February 20, 1905). A second application, filed on December 27, 1905, was granted on June 11, 1908. The SIERRA CLUB, in its first major conservation battle, launched a campaign against the project on the grounds that a dam and reservoir were incompatible with National Park goals and that the valley was needed more as an alternate scenic destination of Yosemite Valley than as a water supply. A city election in San Francisco on November 11, 1908, revealed public sentiment 6 to 1 in favor of the dam; nevertheless, the Sierra Club continued to fight it, over the objections of some of its own members. A potential split in the club was averted when an advisory measure on the club's internal election ballot of January 29, 1910, revealed that the membership opposed damming Hetch Hetchy by a ratio of nearly 4 to 1. Subsequent appeals to the courts and Congress were unsuccessful, however, and on December 19, 1913, President Woodrow Wilson signed the Raker Act, authorizing San Francisco to construct O'Shaughnessy Dam and Hetch Hetchy Reservoir. Construction began early in 1914. The dam was completed in 1923, and the 149-mile-long San Francisco Aqueduct was opened to the Crystal Lake Reservoir on the San Francisco Peninsula on October 24, 1934. In 1938 the gravity-type dam (*see under* DAM) was raised an extra 85 feet: it currently stands 131 meters (430 feet) high, with a crest 277 meters (910 feet) long. The capacity of Hetch Hetchy reservoir is 360,360 acre-feet (*see* ACRE-FOOT).

heterotroph any ORGANISM that cannot synthesize the full range of organic molecules it needs (*see* MOLECULE; ORGANIC COMPOUND) but must obtain some or all of them by consuming other organisms. Heterotrophic organisms may be predaceous (*see* PREDATOR), herbivorous (*see* HERBIVORE), parasitic (*see* PARASITE), or saprobic (*see* SAPROBE). They include all animals, all fungi (*see* FUNGUS), nearly all bacteria (*see* BACTERIUM), about half of the PROTISTA, and a few parasitic, saprobic or carnivorous plants. *Compare* AUTOTROPHISM.

hexachlor in chemistry, loosely, any ORGANIC COMPOUND containing six CHLORINE atoms. More specifically, the term "hexachlor" standing alone usually refers to hexachlorobenzene (HCB: chemical formula H_6Cl_6), the simplest of the six-chlorine aromatics, sold commercially as a FUNGICIDE under the name Bunt-no-more. HCB is a suspected CARCINOGEN that is known to cause skin rashes, headaches, and nausea if taken internally. It is listed as a PRIORITY POLLUTANT by the ENVIRONMENTAL PROTECTION AGENCY. Other hexachlor compounds include Endosulfan, Alodan, Lindane, ENDRIN, Dieldrin and ALDRIN. Hexachlorophene, a biphenyl hexachlor (that is, a hexachlor containing two PHENYL groups), was once widely used as an over-the-counter germicidal soap but is now closely regulated due to its suspected carcinogenity and its demonstrated ability to cause nerve and brain damage in humans.

hibernation a time of extreme torpor during cold weather, used by many animals to survive the winter in the temperate and polar zones. In the hibernating animal, metabolism slows to a crawl. The heart may beat less than 10 times per minute (about 1% of its active rate in a small animal such as a ground squirrel), and breathing may slow to as little as four times per minute. Body temperature may drop below 10°C (50°F). Most hibernators are rodents, although many bats, and a few birds (primarily of the poor-will family), also hibernate. Reptiles in cold weather enter a state called *diapause,* which is similar to hibernation. The "hibernation" of bears is incomplete and is properly termed a state of DORMANCY rather than of hibernation. *Compare* ESTIVATION.

hiding cover *See under* COVER.

high-grading in logging, the practice of harvesting only the best (that is, the largest and highest quality) timber in a STAND, leaving the remainder growing on the site. High-grading is the precise opposite of selective logging (*see* SELECTIVE CUT) in that selective logging purposely leaves the best trees behind for seeding purposes and takes the rest, while high-grading takes the best and leaves the rest. It is generally considered poor silvicultural practice, especially in MIXED STANDS, where the tendency is to leave a stand composed entirely of unmerchantable species (*see under* UNMERCHANTABLE TIMBER); however, it leaves a relatively intact CANOPY, minimizes disturbance of the ground cover and UNDERSTORY, and generally does less harm to the forest ECOSYSTEM than most other forms of logging. Widely practiced in the 19th century, especially in the western forests, high-grading fell out of favor in the early 20th century. It has recently seen a mild resurgence under the name "economic selection system."

high-lead logging a system of CABLE YARDING in which the felled trees are brought to the LANDING by means of cables strung through pullies at the top of a SPAR TREE or portable tower and wound onto drums on a *yarder* (stationary donkey engine). The tips of the logs drag along the ground, while the butt ends are suspended in the air. High-lead logging centralizes the yarding process, reducing the number of roads and landings that must be built. The soil is lightly scarified (*see* SOIL SCARIFICATION) rather than furrowed, and the scarification is in a radial rather than a parallel pattern, which reduces EROSION and helps prepare the site for REGENERATION. The system also makes it easier and cheaper to practice YUM yarding, as the entire yarding system is already in place and simply needs to be utilized, not set up, for the YUM task. However, it is rarely used with any type of harvest system other than clearcutting (*see* CLEARCUT) due to the damage it causes to LEAVE TREES and UNDERSTORY vegetation. *See also* SKYLINE LEAD.

high-pressure cell in meteorology, *see* ANTICYCLONE.

high volume sampler in pollution control, a device to measure PARTICULATE levels in air, consisting primarily of a suction pump and a filter. The pump pulls the air through the filter, which collects the particulates. Weighing the filter before and after a test run gives a gross measurement of the amount of particulates collected. Since the amount of air moved by the pump can be precisely calculated from flow rates and elapsed time, this *gravimetric analysis* can be easily converted into an accurate reading of the total amount of particulates present in the air. Sampling time is normally 24 hours, during which a typical high-volume sampler will pass 70,000 cubic feet of air through its filter. Particulate standards in most regulatory law, including the CLEAN AIR ACT, are based on readings taken by high-volume samplers.

Highway Trust Fund created as part of the Federal-Aid Highway Act of 1976 to provide a continuous and growing source of funding for constructing and maintaining the interstate highway system. The fund derives its revenues from federal motor fuel taxes. While a 1983 amendment diverted a fraction of fund income (about 10%) to nonautomotive programs, the environmental impact of the interstate highway system that has burgeoned because of the fund has been striking. Planners and environmentalists blame the interstate highway system for creating extensive URBAN SPRAWL, rural air pollution, aesthetic degradation, and the loss of fuel-efficient rail transportation systems, which could not compete with trucking via publicly funded highways. In 1991 the U.S. Congress enacted the INTERMODAL SURFACE TRANSPORTATION EFFICIENCY ACT to provide some balance to an overemphasis on highways in transportation policy, but there is no remedy for the vast land-use changes that have already taken place.

hiker's disease *See* GIARDIASIS.

hinge line in geology, *see* ISOSTATIC UPLIFT.

"Historical Roots of Our Ecological Crisis" title of an extremely influential and controversial article by historian LYNN WHITE, published in the journal *Science* in 1967 (March 10, 1967, pp. 1203–1207). The article argued that the Judeo-Christian traditions of dualism (humans apart from nature) and the associated biblical injunction that humans "subdue the earth" and "multiply" gave license for the despoliation of natural processes by industry. "We shall continue to have a worsening ecologic crisis," White wrote, "until we reject the Christian axiom that nature has no reason for existence save to serve man." The article was a foundation for an emerging emphasis on ecological theology and environmental justice on the part of churches and synagogues. *See also* RELIGION AND THE ENVIRONMENT.

historic resource lands in land-management, lands with significant cultural remains, such as buildings or foundations, ditch BERMS, roads, etc., dating from times following the beginning of written records for the lands in question. Lands with cultural materials dating to before written records are *archaeological resource lands* (*see* ARCHAEOLOGICAL RESOURCE; ARCHAEOLOGICAL AREA). The "historic resource lands" designation may be given to lands that show no current historical artifacts if written records point to an important historical use. For example, the site of a fur-trading post may be termed historic resource land if its location may be positively identified from the record, even though no trace of the post may currently exist.

hogback in geology and mountaineering, a narrow, steep-sided ridge with a convex crest, that is, humped in the middle and descending at either end. Hogbacks are most often formed where sedimentary or volcanic STRATA dip very steeply (*see* DIP), but they can also result from the erosion of small ANTICLINES or from the action of a pair of parallel GLACIERS on a ridge intervening between them. *Compare* ARETE.

Holarctic region in the biological sciences, the BIOTIC PROVINCE surrounding the North Pole and including the northern portions of both the North American and Eurasian continents. The southern border is somewhat indistinct, but the region includes all of the arctic portions of the TUNDRA BIOME and at least a part of the boreal forest (*see under* CONIFEROUS FOREST BIOME). The mix of genera (*see* GENUS)—and even the mix of SPECIES—is constant throughout the region, with most of the species involved having circumpolar distribution, that is, being found all around the pole (*see* HOLARCTIC SPECIES). The region is customarily divided into two subregions (FACIATIONS) based on the slight differences between Eurasian and American flora and fauna. The Eurasion faciation is known as the *Palearctic region,* while the North American faciation is called the *Nearctic region.*

Holarctic species in the biological sciences, any SPECIES of plant or animal whose natural RANGE includes the northern portions of both North America and the Eurasian continent. Holarctic species are generally cold-tolerant ORGANISMS that spread laterally along the southern margin of the northern ice cap during the most recent ICE AGES while North America and Eurasia were connected by a land bridge across the Bering Strait, though some are aquatic species that inhabit the Arctic Ocean, and a few—especially birds, their PARASITES, and those plants whose seeds might be carried in the birds' digestive tracts—may have spread in modern times. The term is also sometimes used loosely to indicate any species found within the HOLARCTIC REGION.

holistic approach an approach to study and problem-solving that takes as broad and integrated a view as possible of the object, ORGANISM, or system being studied. To look at an organism holistically is to look at it as a whole, observing the integration of its structure and function rather than trying to isolate each part. Holistic observers are more likely to ask *why* something happens than *how* it happens. The opposite of the holistic approach, the *mechanistic approach,* breaks large things down into smaller and smaller components in an attempt to reduce complicated organisms, activities or forces to a few simple building blocks that may then be built up in a variety of ways to achieve the observed diversity of nature. *See also* SYNERGY; GAIA HYPOTHESIS.

home range in behavioral ecology, the RANGE of an individual ORGANISM (as opposed to the range of the SPECIES to which the organism belongs). An animal's home range consists of the territory the animal will traverse regularly and get to know well during its lifetime. It will include a core TERRITORY that the animal will protect against intruders and a surrounding buffer zone that it will share with others of its species whose home ranges overlap its own. Animals normally feel uncomfortable outside of their home range, and will attempt to return if displaced. The size of home ranges varies with the requirements of the animal and the ability of the terrain to provide those requirements. Some lizards require only a few dozen square feet, while large PREDATORS such as bears and mountain lions sometimes need 100 square miles or more. Since an animal's home range must supply all its needs, including foods of the proper species, water, shelter and COVER, and such idiosyncratic details as high places for lookouts, wallows for bathing, and dead logs for claw-sharpening, mapping and preservation of home ranges is one of the most important tools for the preservation of ENDANGERED SPECIES. *See also* HABITAT.

Homestead Act federal (U.S.) law granting inexpensive ownership of publicly owned lands to citizens willing to settle and improve them, signed into law by President Abraham Lincoln on May 20, 1862. It was severely restricted by the TAYLOR GRAZING ACT OF 1934, and repealed altogether by the FEDERAL LAND POLICY AND MANAGEMENT ACT OF 1976. Although commonly thought to provide free land, the act actually required payment of a small fee (typically $25) in addition to five years' occupancy and the construction of a house at least 12 by 16 feet in size before title could be granted to the settler—who had to be over 21, a veteran, or the head of a family. Lands were granted in 160-acre parcels (quarter sections, or "quarters"; see SECTION), one to a claimant. A married couple could thus claim 320 acres. If a claimant did not want to wait the full five years before taking possession, he or she could obtain title after six months at a cost of $1.25/acre.

IMPACTS OF THE HOMESTEAD ACT

The Homestead Act was designed to lure settlers to the empty lands of the American West, and in this it largely succeeded. Some 975,000 homestead claims were filed under the terms of the act between 1862 and 1900. However, abuses of the law allowed certain individuals to amass huge quantities of land, either through outright fraudulent claims or by paying others to claim it and then turn over ownership. These "land barons," as they came to be called, were responsible for numerous resource abuses (*see,* e.g., BIG CUT). The law also encouraged the settlement of lands in deserts and other arid regions that were impossible to develop without IRRIGATION, thus helping to shape the United States' current overdependence on irrigated agriculture. Finally, many homesteaders failed to "prove up" their claims, which then reverted to federal ownership, creating a difficult-to-manage checkerboard pattern of interspersed public and private ownership that persists to this day (*see* CHECKERBOARD OWNERSHIP). *See also* PUBLIC LAND; BUREAU OF LAND MANAGEMENT: *History.*

homothermalism the ability of an ORGANISM to maintain a constant body temperature when the temperature of its environment is shifting. Homothermalism is popularly called "warm-bloodedness," but this term is inaccurate. In the first place, it is the metabolic processes within the body's cells—not the blood—that maintain the body's temperature (*see* CELL; METABOLISM); in the second place, the body may just as well be cooler than, as warmer than, the environment. All organisms are homothermic to some degree. The phenomenon is most advanced among the birds and mammals, which have developed insulating body coverings as well as highly efficient metabolic regulators, but reptiles, fish, insects, and even plants, fungi, PROTISTA and MONERA show some homothermic response. Generally speaking, the larger the organism, the better it is able to maintain a constant body temperature—one of the reasons why so many small animals hibernate (*see* HIBERNATION) and so many small plants are ANNUALS. *Compare* POIKILOTHERMIC ORGANISM.

hoodoo in geology, a vertical stone pillar that has been eroded into a strange or fantastic shape. Hoodoos are usually found in deserts or on high mountain ridges. In both places, they are a product of wind EROSION acting differentially on horizontal STRATA of varying degrees of hardness. *See also* GENDARME.

horizon in soil science, a layer within the soil, distinguishable from the layers above and below it through differences in color, texture, structure, and/or mineral composition. *See* SOIL: *the soil profile.*

hormone any of a number of different ORGANIC COMPOUNDS used by living ORGANISMS to modulate biological activities such as growth, development, digestion and METABOLISM, reproduction, and some reflexes (such as the "fight or flight response"). Hormones are essentially chemical neurons, and form a chemical complement to the nervous system called the *endocrine system.* They act as messengers, dispersing through the bloodstream as a result of specific stimuli to promote responses in parts of the organism that may be distant from the stimulus. At the affected "target cells" they usually function as CATALYSTS for specific chemical REACTIONS, matching up to specific "receptor sites." Only those compounds that fit the receptor sites properly will affect the cell's activities, thus ensuring highly specific responses to hormone release in the bloodstream. The chemistry of hormones is highly complex; however, most are PROTEINS or other polypeptide-based compounds (one important exception to this is the *steroids,* whose structures are built on LIPIDS). Because of this complexity, hormone production and use must be closely controlled by the body, and unregulated exposure to hormones or compounds that mimic hormonal behavior can be exceedingly dangerous to living systems, causing deformities, cell abnormalities, reproductive and other system malfunctions, and in some cases cancer.

horn in geology, a mountain that has been eroded by glacial action (*see* GLACIER) into a sharp-pointed, steep-faced, generally three-sided pyramidal peak. Horns are surrounded by a radiating system of CIRQUES whose headwalls make up the sides (*faces*) of the peak. The type example is the Matterhorn, on the Swiss-Italian border. *See also* GLACIAL LANDFORMS.

hornblende a dark (black, brown or dark green), shiny silica-containing MINERAL that is an important constituent of both METAMORPHIC and IGNEOUS ROCKS. It occurs often in gneiss and schist, where it usually appears as glossy six-sided crystals. Hornblende is the most common member of the AMPHIBOLE group of minerals. *See* BOWEN SERIES.

hornfels in geology, a hard, tough, fine-grained CONTACT ROCK formed from SHALE or SLATE by the heat and pressure of igneous intrusions (*see under* IGNEOUS ROCK). It is dark in color and often has lost its sedimentary layered structure, making it easy to confuse with BASALT, especially along the contact line between the intrusion and the COUNTRY ROCK.

horse latitudes in meteorology, bands of high atmospheric pressure, low precipitation, and weak and variable winds at approximately latitude 30°N and latitude 30°S, also known as the *subtropical highs.* The horse latitudes are apparently a result both of (1) a convergence or "piling up" of the air aloft due to the CORIOLIS EFFECT; and (2) a general cooling of upper-atmosphere air that has moved northward and southward from the equator due to heat convection. The combined result of these forces creates a descending mass of hot, dry air. The horse latitudes are the location of the earth's tropical deserts (*see* DESERT BIOME), and were a major barrier to exploration in the days of sail due to the almost constant becalment of ships in them. The name apparently derives from the Spanish practice of throwing horses overboard from becalmed ships when it was no longer possible to feed and water them. *See also* ATMOSPHERE: *dynamics of the atmosphere.*

horst in geology, a block of land or section of a geologic FORMATION bounded by FAULTS and higher in elevation than the corresponding blocks or sections on either side. Horsts are usually considerably longer than they are wide. They are a major mountain-building feature in some parts of the world. *Compare* GRABEN. *See also* FAULT-BLOCK MOUNTAIN; BASIN AND RANGE TOPOGRAPHY.

host species in the biological sciences, a living ORGANISM that is being parasitized (*see* PARASITE) by another organism. The variety of host species that a given parasite can utilize is generally highly limited due to metabolic restrictions (*see* METABOLISM) on the host and/or the parasite. The term "host species" is also sometimes used for one member of a symbiotic relationship if the relationship is clearly more beneficial to one partner than to the other: *see* SYMBIOSIS.

hot pond in logging, a LOG POND that is artificially heated to keep it free of ice and allow the associated sawmill to keep operating all winter. The source of heat is usually waste heat from the milling operation itself, though some mills use steam furnaces powered by the burning of wood wastes. Hot ponds were developed around Muskegon, Michigan, in the late 19th century.

hot spring any SPRING in which the water temperature noticeably exceeds the temperature of the surrounding environment, or of other nearby surface waters. Hot springs are caused by contact between GROUNDWATER and a body of MAGMA, and are an indication that the region they are in is tectonically active (*see* TECTONIC ACTIVITY). They usually form along FAULT LINES or on the sides of VOLCANOES. Because hot water is a better SOLVENT than cold, and because MINERALS percolate out of magma more readily than out of solid rock, hot springs are often highly mineralized. The minerals are deposited where the water cools down upon emerging from the ground, forming terraces and coating stream bed stones. *See also* GEYSER.

Housing and Urban Development, Department of established in 1965 to provide assistance for housing and community development. HUD works with state and local governments, the housing industry, and mortgage lenders to coordinate housing programs. Address: HUD Building, 451 7th Street SW, Washington, DC 20410. Phone: (202) 755–5111. Website: www.hud.gov.

Hudsonian zone in the Merriam life zones system (*see* LIFE ZONES), the zone between the UPPER SONORAN zone and the ALPINE ZONE, characterized by small groves of CONIFERS interspersed with lakes, ponds, and open meadows. The Hudsonian lies at TIMBERLINE on mountain peaks in temperate regions and at the northern tree line in the arctic. It is now considered to be an ECOTONE between the TUNDRA BIOME and the CONIFEROUS FOREST BIOME.

human ecology the study of the interaction of human beings and human society with nature and natural processes. More than simply a subset of ecology as a biological science, human ecology necessarily encompasses anthropology, sociology, economics, philosophy, politics, and religion. Issues of carrying capacity, habitat restoration, population dynamics, environmental ethics, and the like are of current interest in the field.

human population trends *See* POPULATION EXPLOSION.

Humane Society of the United States established in 1954 to protect both domestic and wild animals. Programs of the society include animal control, cruelty investigation, humane and environmental education, farm animal projects, federal and state legislative efforts, habitat and wildlife protection, and laboratory animal welfare concerns. Its Center for Respect of Life and Environment has been a major voice in support of organized religion taking an active role in environmental concerns (*see* RELIGION AND THE ENVIRONMENT) and has sponsored major national and international conferences. The center also publishes a quarterly journal, *Earth Ethics*. The Humane Society offers resources to the general public, local organizations, the media, and governmental agencies. Membership (1999): 6,200,000. Address: 2100 L Street NW, Washington, DC 20037. Phone: (202) 452–1100.

Humboldt, Alexander von (1769–1859) German naturalist and explorer, born September 14, 1769, in Berlin, and educated at the universities of Frankfurt and Gottingen and at the School of Mines in Freiberg, Saxony. Trained primarily as a mining engineer, Humboldt worked for the Prussian government from 1892 to 1896, when the death of his mother left him with a substantial legacy that enabled him to devote himself to his passion for exploration. After four years of intense preparation, including extensive private studies in botany, geography, and meteorology, Humboldt sailed for the Americas, arriving at Cumana, Venezuela, near Caracas, on July 16, 1799. He spent the next five years exploring and making scientific observations in South America, primarily in the Andes, where he located the Earth's magnetic equator and became the first observer to formally state the close relationship between the effects of altitude and latitude on living things (*see* HUMBOLDT'S RULE): he also made the first oceanic temperature studies of what became known as the Humboldt Current, off the continent's west coast. After a brief visit to North America, he returned to Europe in 1804, settling in Paris. Outside of a transect of Siberia in 1829 to do a minerals survey for the Russian government, he made no other extensive trips. In 1832 he moved back to Berlin, where he became a court adviser to the king of Prussia, a position he retained for the rest of his life. A superb observer whose stated purpose was "to investigate the interaction of all the forces of nature," Humboldt was an ecologist before the term was invented, and his work remains a model of accurate, dispassionate scientific study. He died in Berlin on May 6, 1859.

Humboldt's rule in meteorology and ecology, a rule, formulated in 1850 by the pioneering German naturalist Alexander von Humboldt (1769–1859), which states that the average temperature of a region rises 1°F for every degree of latitude one travels toward the equator. Humboldt also observed that the same roughly 1°F change in average temperature can be obtained by a change in elevation of 300 feet. Therefore, to climb 300 feet up the side of a mountain is the same, ecologically, as traveling one degree north—a phenomenon that has a profound effect on the distribution of plants and animals on a mountain slope.

humic gley a GLEY with a strongly developed A horizon that is rich in organic matter. *See* SOIL: *classification of soils;* HUMUS.

humic matter *See* HUMUS.

humic site a growing site with a high level of soil moisture, such as a WETLAND or a rain forest (*see* RAIN FOREST BIOME). *Compare* XERIC SITE; MESIC SITE. *See also* MOISTURE GRADIENT; SOIL WATER.

humidity in meteorology, the amount of water vapor in the air. *Absolute humidity* is the weight of the water vapor present in a given volume of air, usually expressed in grams per cubic meter (g/m^3). *Specific humidity* is the weight (or mass) of water vapor present in a given weight (or mass) or air, usually expressed in grams per kilogram (g/kg). It is more useful than absolute humidity for most purposes because the weight of a given volume of a gas (including both air and water vapor) varies so radically under varying conditions of temperature and pressure that a ratio of weight to volume is essentially meaningless. The same weight of water can give a wide range of values for absolute humidity, but it always gives the same value for specific humidity. Finally, *relative humidity* is the ratio of the air's actual water content to the total amount of water vapor the same volume of air could theoretically hold under its current conditions of temperature and pressure. Relative humidity is the most useful of the three methods of expressing humidity because it has the most immediate predictive value. It tells how close the air is to saturation and therefore how rapidly EVAPORATION and CONDENSATION will take place, which in turn affects the rate of formation of FOG, clouds, and PRECIPITATION. Air with a high relative humidity also has an increased capacity to hold heat, due to the high SPECIFIC HEAT of the water vapor. This means that humid air at a given temperature actually holds more heat than dry air at the same temperature, with profound effects on atmospheric heat transfer and the formation of storms. Relative humidity is measured with a hygrometer or a *psychrometer* (wet-and-dry-bulb thermometer). *See also* ATMOSPHERE; DEW POINT.

humus (humic matter) in soil science, organic materials (*see* ORGANIC MATTER) within the soil that have been partially or largely decomposed, so that they may still be identified as organic from the chemical COMPOUNDS present but cannot be identified as to precise source. Humus is formed by the feeding activity of MICROORGANISMS and by the leaching action of water passing downward through the soil. It consists primarily of PROTEINS and LIGNIN, with small amounts of LIPIDS, CELLULOSE and other materials. Two principal types are recognized. *Mor humus,* which forms under acidic conditions, is developed primarily by leaching. It is generally thin and light-colored, with the outlines of the parent organic materials often recognizable. Mor humis is commonly found under CONIFERS and in the dry beds of old BOGS. *Mull humus* forms under neutral or alkaline conditions, and is developed principally by microbial action. It is dark in color and up to a foot thick, and is the common humus of HARDWOOD forests and grasslands. A layer of pure (or nearly pure) humus at the top of a soil is equivalent to the O_2 HORIZON. It is usually covered by leaf and twig LITTER (the O_1 horizon).

hunter day in wildlife management, a measurement of hunting pressure in an area, defined as 12 hours of use by hunters. It may consist of use by one hunter for 12 hours, 12 hunters for one hour each, or any intermediate combination between these two extremes. The figures are rounded up to the nearest hour; thus, hunter visits of less than one hour are classed as one-hour visits. *Compare* ANGLER DAY. *See also* VISITOR DAY.

hunting with reference to sport hunting, a troublesome issue for some environmentalists and ecologists. While many wildlife experts believe that hunting is needed to manage populations of game species (deer, for example), ecologists point out that when hunting involves predatory species, such as mountain lions and wolves, harmful population imbalances in prey species throughout an ecosystem can ensue. Moreover, to the extent that governmental funds are used to manage ecosystems, many wildlife ecologists believe that to manage them (marshland areas, for example) exclusively for game (e.g., ducks) also introduces distortions in wildlife populations. Commercial hunting—especially of charismatic megafauna (e.g. elephants, Bengal tigers) in places such as Africa, Asia, South America, the Arctic, and Polynesia—has become a quite serious ecological matter in that both legal and illegal hunting have seriously depleted populations of rare and endangered species. *See also* ENDANGERED SPECIES ACT.

hurricane in meteorology, an intense cyclonic storm (*see* CYCLONE) with high winds that move in a circular pattern around a zone of calm air in the center. By international agreement, the winds of a storm must be moving at least 120 kilometers per hour (75 miles per hour) to qualify as a hurricane. At speeds less than this, they are termed "tropical storms." In popular terminology, the term "hurricane" is used only for those storms that develop over the Atlantic Ocean. Those that develop over the Pacific are called "typhoons" and those that develop over the Indian Ocean, simply "cyclones." Meteorologists classify all of these storms as hurricanes.

CHARACTERISTICS OF HURRICANES

Hurricanes are among the most powerful natural forces known, the energy developed by two medium sized storms of this type being roughly equivalent to that consumed by the entire population of the United States for a full year. They derive this energy from the CONDENSATION of water vapor; hence, they can only develop over tropical oceans, where there is a large supply of warm, moist air. The typical hurricane is roughly 600 kilometers in diameter (360 miles). At the center is a zone of extremely low pressure about 20 kilometers across (12 miles) known as the *eye.* Surrounding it is the *eye wall,* a solid mass of cumulonimbus clouds ("thunderheads") 10–12 kilometers high (6–7 miles), rotating at an extremely high speed and surrounded in turn by inward-spiraling winds heavily laden with clouds and producing torrential rains. The whole thing moves forward (in the northern hemisphere, northwestward), driven by the *easterlies* or "trade winds"—the band of westward-flowing air just south of the HORSE LATITUDES—at a rate of

10–20 kilometers an hour until it encounters cooler, dryer air where, deprived of its source of energy, it quickly dissipates. Generally this happens over land. Often a hurricane will appear to "bounce off" a continental mass and move back out to sea. These are storms that have moved far enough north to encounter the westerlies (the winds flowing predominantly west to east that are characteristic of middle and upper latitudes), changing their direction of forward motion. Although best known for their destructive force, they are actually beneficial in many parts of the world, where they bring dependable sources of rain to areas that might not otherwise get enough (*see* MONSOON FOREST). The destruction they bring principally affects seafront property, and is mostly caused not by the winds but by the accompanying *storm surge,* a wall of water 10–15 meters high driven by the high winds and low air pressure in the eye wall and capable of rapidly inundating low-lying seacoast areas.

CLASSIFICATION AND NAMING OF HURRICANES

Hurricanes are classified according to the *Saffir/Simpson scale*, which assigns numbers from 1–5 depending on the storm's intensity. This scale is given in the accompanying chart. Before 1978, the tropical storms and hurricanes of the Atlantic were named from a list of female names chosen at the beginning of each hurricane season. Since 1978, the names have come from a sequence of five lists on which male and female names alternate. *See also* MONSOON; TORNADO.

hybrid in the biological sciences, an ORGANISM whose male and female parents are members of two separate and distinct genetic lines (*see* GENE). Hybridization takes place most easily between two varieties within the same SPECIES (as, for example, hybrid corn, hybrid roses), but it also can take place between two SPECIES (the mule, for example—a hybrid between a horse and a donkey—or the tangelo, a hybrid between a tangerine and a grapefruit). If the hybrid is *fertile* (has the capacity to bear young), it may give rise to a *hybrid species* with characteristics that are a cross between those of the two parent species. An example is the London plane tree, *Platanus acerifolia,* a widely-used ornamental species which resulted from the accidental hybridization of the Oriental plane tree (*P. orientalis*) with the North American species *P. occidentalis* in the parks of London. Hybridization is considerably more common among the plants, where it is occasionally possible even to cross genera (*see* GENUS), than it is among the animals, which makes classification of plants by species and genus more difficult than classification of animals (*see* TAXONOMY).

Saffir/Simpson Scale

Rank	Wind Speed (km/hr)	Eye Pressure (mb)	Damage
1	119–153	>980	minimal
2	154–177	965–979	moderate
3	178–209	945–964	extensive
4	210–250	920–944	extreme
5	>250	<920	catastrophic

hybrid-electric car passenger auto combining a small, efficient INTERNAL COMBUSTION ENGINE with an electric motor to produce a self-charging, low-pollution vehicle. The hybrid combines the high torque of the electric motor at low speeds with the high torque of a gasoline engine at high speeds using a computer to switch between them to provide optimum power and fuel efficiency up to 70 miles per gallon. Japanese manufacturers Honda and Toyota made hybrid-electrics available in the United States in 2000. American automakers are planning to introduce their models in 2003 or 2004. *See also* ELECTRIC VEHICLES, HYDROGEN FUEL CELL.

hybrid rights in water law, any system of state WATER RIGHTS laws that encompasses portions of both PRIOR APPROPRIATION doctrine and RIPARIAN DOCTRINE. Hybrid rights systems have usually come into existence because states have initially recognized only riparian rights, and have later converted to the prior-appropriation system, making allowance for established riparian rights as they did so. In most cases this means that new riparian rights cannot be established, and that all new rights, whether or not they are held by riparian landowners, are subject to prior-appropriation restrictions. Only those riparian rights that were actually being put to BENEFICIAL USE at the time of passage of the prior-appropriation laws are normally recognized. The hybrid-rights system seems to function best in regions of moderate rainfall. All states with hybrid-rights systems are currently found either along the Pacific Coast or on the 100th meridian, regions of considerably more rainfall than the pure prior-appropriations states but considerably less than those which practice pure riparian rights.

hydrocarbon in chemistry, any COMPOUND that contains only HYDROGEN and CARBON in its molecular structure (*see* MOLECULE). Hydrocarbons range from the very simple (METHANE, chemical formula CH_4) to the very complex (PETROLEUM and its derivatives). They are a numerous and important class, and probably formed the basic material from which life evolved on the earth. Hydrocarbons are divided into two broad categories based on their molecular structure. *Aromatic hydrocarbons* (*see* AROMATIC COMPOUND) contain one or more BENZENE RINGS or similar cyclic structures with alternating single and double bonds. *Aliphatic hydrocarbons* (*see* ALIPHATIC COMPOUND) have linear, branched, or non-alternating cyclic molecules. Most exist in several ISOMERS with slightly differing properties.

hydroelectric power electricity generated utilizing the energy of falling water. A hydroelectric generating facility has four principal parts: a RESERVOIR or IMPOUNDMENT (usually created by a DAM) to draw the water from; a *turbine* (also known as an *impeller*) consisting of a number of blades connected to a central shaft that rotates when the water strikes the blades; a PENSTOCK to carry the water from the reservoir to the turbine; and a *generator*, turned by the turbine shaft, which does the actual work of generating the electricity. The amount of power obtained is dependent primarily upon the volume of water passing through the penstock and upon the HEAD (the height differential between the top and bottom of the penstock). It has no relationship to the size of the reservoir as long as the reservoir is sufficiently large to maintain the required volume flow. Hydroelectric power has two principal advantages from an environmental standpoint. It is renewable (as long as the HYDROLOGIC CYCLE keeps the river running), and it is clean (it generates no air pollutants and no radiation.) Its hazards are chiefly those associated with all types of dams, reservoirs, and penstocks (*see* under these headings), although the rapidly spinning turbine blades pose a physical hazard to fish accidentally entering the penstock, and the transmission lines required to carry the power away from the damsite pose problems of their own that may be accentuated in the remote, scenic settings characteristic of many hydroelectric facilities. *See also* PUMPED STORAGE.

hydrofluorocarbon (HFC) an alternative to CHLOROFLUOROCARBONS (CFCs) as a refrigerant and for other uses. HFCs break down faster in the atmosphere than CFCs and do not contain ozone-destroying chlorine. They are, nevertheless, a greenhouse gas and in one form (HFC-134A) produce methyl chloroform, which does break down the O_3 molecule. *See also* GREENHOUSE EFFECT, OZONE LAYER.

hydrogen the first ELEMENT in the ATOMIC SERIES; ATOMIC WEIGHT 1.01, chemical symbol H. Hydrogen is by far the most common of the elements—it makes up approximately 90% of the known universe—and it is probable that all other elements are derived from it through nuclear fusion, primarily within stars. On the earth it occurs primarily as a component of WATER and HYDROCARBONS. A colorless, odorless, extremely lightweight gas—a liter of it weights less than a tenth of a gram (0.004 ounces) at room temperature—hydrogen burns explosively in the presence of OXYGEN, making it useful as a fuel, although the heavy tanks and tight seals required to contain it have limited its application primarily to stationary installations. (There is also some fear of it as a mobile power source due to its explosive nature, although it is probably safer for that use than is GASOLINE.) Nontoxic, noncarcogenic and nonmutagenic, it is nevertheless classed as a HAZARDOUS SUBSTANCE due to its flammability. *See also* HYDROGEN BOND.

hydrogen bond an electrostatic bond between an ATOM of HYDROGEN in one MOLECULE and an atom of another ELEMENT in a neighboring molecule. Much weaker than the internal bonds that hold molecules together (*see* COVALENT BOND; ION), hydrogen bonds are nevertheless a significant attractive force, giving water its high boiling point and making possible such phenomena as SURFACE TENSION and CAPILLARY ACTION.

hydrogen fuel cell an energy-producing device that uses a kind of reverse electrolysis combining oxygen and hydrogen into water which generates electricity in the process. Fuel cells can provide electrical power to motor vehicles and to buildings, plus hot water, a by-product, in the case of buildings for heating. Emissions are negligible.

hydrologic cycle the cyclical path followed by water through the atmosphere, responsible for all freshwater features of the Earth, including LAKES, STREAMS, RIVERS and GROUNDWATER features such as SPRINGS and SEEPS. The hydrologic cycle depends on the ability of water to exist in both gaseous and liquid forms at Earth-normal temperatures and pressures. The cycle has four principal phases. In the *evaporation phase*, heat from the sun converts liquid water in the sea to water vapor, which then becomes part of the atmosphere, rising and moving about with the other atmospheric gases. In the *condensation* phase, the water vapor condenses back to liquid water. The droplets formed, however, are light enough to remain aloft on the wind. *Rainfall* results when the droplets of liquid water, massed together into clouds, coalesce into drops large enough for their mass to overcome the bouyancy of the wind. Finally, *runoff* carries the rainfall that has reached the Earth's surface downhill into the sea again. This picture is complicated by numerous side trips and short cuts. Rain may, of course, fall directly into the

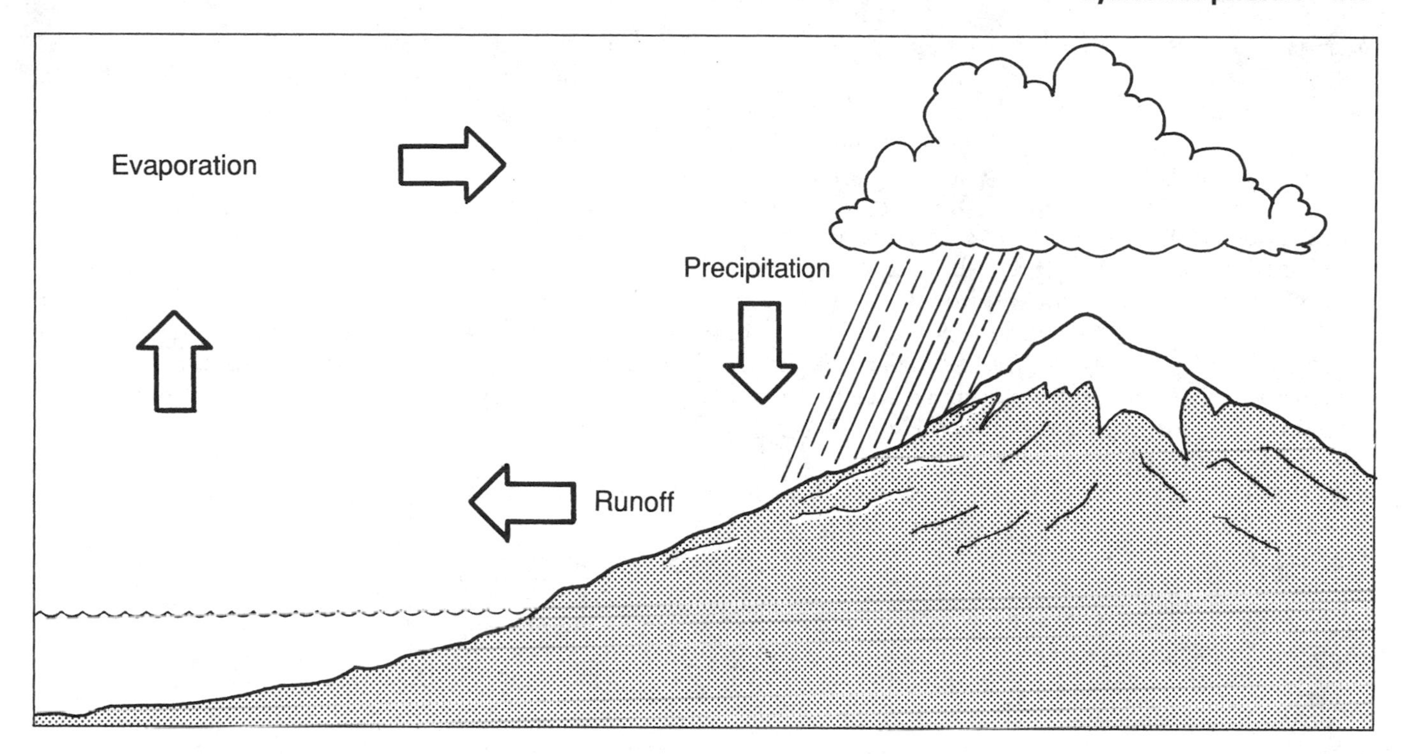

The hydrologic cycle

sea, eliminating the runoff phase; or it may evaporate before reaching the ground, again eliminating runoff. Water may evaporate from the damp ground surface or from rivers and lakes as well as from the sea. Some water is taken up by plants, which return it to the air through EVAPOTRANSPIRATION instead of evaporation. Some falls as a solid—snow or hail—instead of as a liquid. In colder climates, it may remain a solid for many centuries, pausing the cycle indefinitely between rainfall and runoff. Finally, some may soak into the earth, entering AQUIFERS, where it may reside for considerable periods of time (up to several million years) before flowing out on the surface again and reentering the cycle. *See also* WATER; ICE; and separate entries on each phase of the hydrologic cycle (RAINFALL; RUNOFF; EVAPORATION; CONDENSATION).

hydrologic equation in hydrology, a generalized equation used to indicate the relationships of the factors controlling RUNOFF. The equation is generally written

$$R = P - E \pm C$$

where R = runoff, P = the total amount of PRECIPITATION, E = EVAPOTRANSPIRATION, and C = the change in so-called "storage water," that is, GROUNDWATER reserves, SOIL WATER, and water in lakes and reservoirs. The C term represents a form of governor on streamflow. During precipitation events (storms), lakes and groundwater reservoirs tend to fill up, removing excess water from the stream and keeping the flow rate fairly constant, while during dry spells these reservoirs tend to release water to the stream, again keeping its flow fairly constant. *See* STREAMFLOW.

hydroponics use of nutrients in solution to grow plants, instead of soil. Roots are immersed, misted, or held in a sterile medium, such as pure sand or gravel. With hydroponic techniques, plants can be grown closer together than in soil, yields can be increased, and multiple-cropping is possible.

hydropower generation of electrical power from flowing water, usually a dam, but also from tidal movement, or by a waterwheel driven by a flowing stream or river.

hydrosphere in ecology, the waters of the earth. The hydrosphere includes all free WATER, in whatever form and wherever it occurs: oceans, lakes and rivers, SNOW and ICE, atmospheric water (water vapor and rain), and GROUNDWATER. It is one of the four main divisions of the planetary environment, the others being the ATMOSPHERE (air), the LITHOSPHERE (soil and rock) and the BIOSPHERE (living organisms).

hydrostatic pressure in hydrology, the amount of pressure at any specific point within a body of water that is at rest. It is zero at the surface of a free-standing water body (such as a lake or pond), and increases by 0.433 lbs/in^2 for every foot of depth below the surface. It is independent of direction, being applied equally to all sides of a body immersed to that depth. It is also independent of the absolute size of the body of water, the pressure per square inch being the same at the bottom of a pipe 1 inch in diameter and 50 feet high as it is at the bottom of a lake 10 miles across and 50 feet deep. In GROUNDWATER the situation is complicated somewhat by having to factor in the POROSITY and PERMEABILITY of the AQUIFER, and by the possibility of having constrictive pressure on the top of the aquifer as well as on the sides and bottom due to the possible presence of an IMPERMEABLE LAYER above it. However, once a well is drilled into the aquifer, the water in the well becomes free-standing and the complications no longer apply. *See also* HEAD; STATIC LEVEL.

hydroxide in chemistry, any INORGANIC COMPOUND containing HYDROGEN and OXYGEN bound together in the form of a negatively charged ION (OH^-). Hydroxides are generally BASES; they turn pink LITMUS blue, have a bitter taste, and react with ACIDS to form SALTS and WATER.

hygroscopic water (bound water, film water) in soil science, water bound so tightly to the outside of soil particles that it cannot be moved by either gravity or CAPILLARY ACTION but can only be evaporated away (*see* EVAPORATION). Hygroscopic water exists as a film several MOLECULES thick on the outside of the soil particles. The innermost layers (*adhesion water*) are bound to the soil particle by molecular adhesion of the water molecules to the soil surface, while the outer layers (*cohesion water*) are bound to the inner layers and to each other by the cohesive force of HYDROGEN BONDS—the same force that creates SURFACE TENSION. All but the outermost layers of cohesion water are unavailable for use by plants. The amount of hygroscopic water in a sample of soil varies with the evaporation rate, the depth and texture of the soil, and the time since the last rainstorm. *Hygroscopic capacity*—the amount of water a soil is capable of holding in bound form—is generally measured by allowing a sample to reach equilibrium in an atmosphere of 98% humidity at a temperature of 25°C. *Compare* CAPILLARY WATER; GRAVITATIONAL WATER. *See also* FIELD CAPACITY.

Hymenoptera important ORDER of insects that includes the ants, bees, and wasps (termites, often confused with ants, belong to a separate order, the Isoptera). Hymenoptera are characterized by high intelligence and complex behavior. Physically, nearly all show the so-called wasp waist, an extreme narrowing of the connection between the abdomen and the thorax. The young are legless, wingless, and usually eyeless grubs. The order contains more than 15,000 named and described SPECIES. Undescribed species probably total well over 100,000.

hypercar passenger auto powered by efficient ALTERNATIVE ENERGY SOURCES, especially one using a HYDROGEN FUEL CELL. Such cars, using easily available hydrogen gas (derived from "splitting" the water molecule, H_2O, or "stripping" it from natural gas), would be almost pollution-free, emitting mainly water vapor and no CO_2. Early hypercars are expected to be on the market by 2010.

hypereutrophic lake *See under* TROPHIC LEVELS.

hypolimnion in limnology, that portion of a body of water that lies below the THERMOCLINE. The hypolimnion receives no inmixing of water from the surface, hence has no way of obtaining fresh stores of DISSOLVED OXYGEN. However, it contains a high level of NUTRIENTS due to its contact with the SILTS and decomposing ORGANIC MATTER of the bottom. Its temperature is generally cold—often a constant 4°C—allowing it to act as a refuge for cold-water ORGANISMS during the warm part of the year. The depth of the hypolimnion below the surface depends on the season of the year, the altitude and LATITUDE of the water body, and numerous other factors. *Compare* EPILIMNION. *See also* TURNOVER TIME.

hyporheic zone in a river or stream, the area beneath the bed of the watercourse that serves as a transitional zone between surface water flow and groundwater.

hypothermia in medicine, the cooling of the body of a warm-blooded animal to the point where normal bodily functions can no longer be maintained. In humans, the critical point is reached at 92°F (33°C). If the temperature of the internal organs (the *core temperature*) drops below this point, the body can no longer generate enough heat on its own to recover but must be warmed externally to avoid death. The principal causes of hypothermia are not cold temperatures and high elevations but wetness, wind, and exhaustion. Water next to the skin evaporates, taking its heat of vaporization (*see* EVAPORATION) from the body. Wind lowers the VAPOR PRESSURE in the body's vicinity, thus increasing the rate of evaporation. Exhaustion robs the body of the supplies of energy it needs to keep METABOLISM going at a fast enough rate to counter the loss of heat to the environment. Symptoms of the first stages of hypothermia include a general dulling of the senses, slurring of speech, and loss of motor control, especially in the hands. As the core temperature approaches the critical point, violent shivering takes place as the body tries to supply through mechanical means the heat it can no longer supply through metabolism. Stupor and a glassy stare come next, followed by shallow breathing, slow, irregular heartbeat, and death. Treatment consists of removing all wet garments and warming the individual. Mild hypothermia (*exposure*) can be countered by eating CARBOHYDRATE-rich foods (especially hot sweet liquids) and by vigorous exercise. More severe cases require the application of external heat sources such as hot blankets, fires built on both sides of the body, and/or skin-to-skin contact with a healthy person or persons.

I

IAA *See* AUXIN.

ice the solid phase of WATER. Ice is a compact, transparent MINERAL that forms hexagonal crystals. Due to the strongly polar nature of the water MOLECULE (*see* POLAR LIQUID), ice is about 9% less dense than liquid water at its FREEZING POINT (0°C or 32°F). This allows it to float on the surface of the water, a peculiarity without which all bodies of water would long since have frozen from the bottom up, extinguishing life on the planet. When first formed, it breaks easily; however, once compressed into its most compact form (*pressure ice*) it can be amazingly strong and tough. It is a major agent of geologic change (*see* FROST-WEDGING; GLACIER; GLACIAL LANDFORMS) and an important storage medium for water within the HYDROLOGIC CYCLE. *See also* SNOW; FIRN; HAIL; SLEET; ICE AGE; ICE CAP; ICEBERG; ICE SHELF; FLOE; PACK ICE.

ice age in geology, a period of GEOLOGIC TIME during which GLACIERS cover substantial portions of the earth's surface. The term is most commonly used to refer to the period from 2 million to 12,000 years before present (BP), called by geologists the *Pleistocene,* during which glaciers expanded to cover roughly 30% of the Earth (roughly three times today's ice cover); however, it is now known that ice ages have occurred on a cyclical basis, roughly ever 225 million years, for at least the last billion years. The reasons behind the advance of continental ice sheets are not known, but they appear to result from a number of minor "wobbles" in the Earth's orbit that affect the amount of energy the planet receives from the sun. These occur with different periods, so that at times they cancel each other and at other times they have a cumulative effect. Ice ages are not uniformly cold. The Pleistocene saw at least four major advances and retreats of the glaciers, separated by relatively warm periods called *interstadials.* There is some evidence that the ice age has not really ended, and that we are simply in a fifth interstadial at the present time. The ice age was a major shaper of the landscape (*see* CONTINENTAL GLACIER; GLACIAL LANDFORMS), and probably of animal and plant SPECIES as well. There is strong circumstantial evidence, for example, that it helped trigger our own evolutionary advance from early to modern humans.

iceberg a large, irregularly shaped mass of ice, broken from the toe of a tidal glacier (*see* GLACIER) or from the edge of an ICE SHELF—a process called *calving*—and floating in the ocean. Icebergs may extend up to 250 or 300 feet above the water's surface and as much as 1,500 feet below; 90% of their mass, and roughly 80% of their height, is below the water line. They may drift 2,000 miles or more from their point of origin before melting. Numerous schemes have been proposed for harnessing icebergs and towing them to warm-water ports, where they could presumably be tapped as a source of freshwater; however, the large energy requirements and complicated logistics involved to obtain more than a minimal amount of water in this manner make such schemes highly unrealistic. *Compare* FLOE; PACK ICE.

ice cap a body of ICE lying on a flat surface such as a PLATEAU or the summit of a mountain, generally feeding active GLACIERS from two or more sides. Isolated ROCHES MOUTONNÉES may stick up through its surface, which will often be scored by crevasses. The ice in an ice cap moves extremely slowly, but it does move. This movement is radial, away from the thickest part of the ice cap (the *center of accumulation*) toward the edges, and is caused by pressure differences between the thick and thin portions. *See also* CONTINENTAL GLACIER.

ice island *See under* FLOE.

icepack *See* PACK ICE.

ice sheet, ice shelf vast polar areas covered with ice that are of environmental concern in connection with global warming. If temperatures continue to warm, the ice sheets in Greenland and Antarctica will melt, leading to further warming because they will no longer reflect the Sun's rays back into the atmosphere. Once the ice sheets have disappeared, the darker waters or land areas will absorb heat from the Sun rather than reflecting it, thus accelerating the warming trend. In Antarctica, ice shelves form around edges of the continent and are 600 to 4,000 feet thick. They are disappearing rapidly in several areas, especially along the Antarctic Peninsula in the Northwest, probably due to global warming, although some of the melting may be a natural cyclical phenomenon. Since the ice shelves already displace water, some researchers believe that sea levels will not be greatly affected by melting, although others predict a major sea level rise over the next century. In either case, under the pressure of the ice, extremely cold and extremely dense Antarctic bottom waters are created that affect major ocean currents. The currents, which interact with one another and influence world weather patterns, may be radically altered if the Antarctic bottom waters are warmed. *See also* ANTARCTICA; GREENHOUSE EFFECT.

Ickes, Harold L. (1874–1952) conservationist, lawyer, and U.S. Secretary of the Interior during the administration of President Franklin D. Roosevelt and part of the Truman administration. Ickes studied at the University of Chicago and worked as a newspaper reporter in Chicago. He became active in politics and in 1932 encouraged progressive Republican support for Democratic presidential nominee Franklin D. Roosevelt. He served as Secretary of the Interior from 1933 to 1946, longer than any other secretary in this department, and as head of Roosevelt's Public Works Administration from 1933 to 1939. The Public Works Administration undertook the construction of bridges, dams, public buildings, and housing developments. Throughout his life Ickes was known for his dedication to social reform and especially for championing the conservation of natural resources. Arguably the most effective Interior Secretary ever, with the preservation of millions of acres of Western land to his credit, he is acknowledged as a major figure in the conservation history of the 20th century. His books include *New Democracy* (1934), *The Autobiography of a Curmudgeon* (1943), and *Secret Diary* (three volumes, 1953–54). A prizewinning authoritative biography, *Righteous Pilgrim: The Life and Times of Harold L. Ickes, 1874–1952*, was published by historian T. H. Watkins in 1990.

igneous rock rock that is formed by the cooling of part or all of a molten body of MAGMA. Igneous rocks may be either *intrusive* (cooled underground; *see* INTRUSIVE ROCK) or *extrusive* (cooled on the surface; *see* EXTRUSIVE ROCK). An *igneous intrusion* is an igneous rock body that has invaded another rock body along a line of weakness such as a FAULT or FRACTURE. It is generally tabular in form (long and wide but not very thick). Examples of igneous rocks include GRANITE, BASALT, ANDESITE and RHYOLITE; examples of igneous intrusions include DIKES and SILLS. *Compare* SEDIMENTARY ROCK; METAMORPHIC ROCK. *See also* PLUTON; VOLCANO; BATHOLITH.

ignition point (kindling temperature) for any substance, the temperature at which it catches fire. More precisely, the ignition point of a substance is that temperature at which the substance begins undergoing an exothermic oxidation reaction with the surrounding ATMOSPHERE. *See* EXOTHERMIC REACTION; OXIDATION.

IJC *See* INTERNATIONAL JOINT COMMISSION ON BOUNDARY WATERS.

imbiber beads small plastic beads that are capable of absorbing as much as 27 times their own volume of a broad range of ORGANIC COMPOUNDS. They do not absorb water. Imbiber beads are used in a number of pollution-control applications. Woven or sewn into a blanket, they can be placed on OIL SPILLS, absorbing the oil and leaving the water behind. Made into a filter (usually by being loosely packed between a pair of screens), they can be placed in a water line, where they will allow water to pass as long as it is free of organic pollutants but will swell up and block off the flow if pollutants enter the system. They are cheap, easily obtainable, and one of the best tools available for the control and management of HAZARDOUS SUBSTANCES in the water supply.

impermeable layer in groundwater hydrology, a horizontal layer of ROCK, CLAY, CALICHE, or other natural material through which water cannot pass. Impermeable layers define the vertical limits of an AQUIFER. They also may affect the aquifer's rate of RECHARGE, an impermeable layer close to the ground surface effectively cutting off recharge for a deeply-buried aquifer. An impermeable layer tilted on its side may also prevent the horizontal movement of groundwater (*see* GROUNDWATER DAM).

impermeable liner a layer of watertight material used to line a storage area for HAZARDOUS SUBSTANCES in order to prevent the contamination of GROUNDWATER by the hazardous substances. The two principal types of liner materials are CLAYS (especially BENTONITE) and sheets of synthetic organic POLYMERS such as polyethylene, polyvinyl chloride (PVC), or neoprene. Clays have the advantage of being relatively cheap and easily applied; however, they dry and crack easily, may break down due to ION exchange with the hazardous substance, and cannot be used in areas of high groundwater due to their lack of resistance to water pressure. Polymer linings avoid these problems and are highly resistant to most chemicals, but must be installed extremely carefully due to the ease with which they can rip, tear, or have holes punched in them by underlying rocks or woody growth. They also deteriorate upon exposure to sunlight or OZONE and may be dissolved by certain organic SOLVENTS. No liner can be used beneath the WATER TABLE. Lined ponds should be considered temporary expedients. They should not be used for permanent storage or disposal. They are best used in climates where net EVAPORATION exceeds net PRECIPITATION.

implementation plan a specific plan of action designed to make certain that a particular set of regulatory standards are met. For example, an implementation plan for a TURBIDITY standard for the water in a particular stream might include EROSION-control activities in the stream's WATERSHED, restrictions on construction or on the removal of ground cover, stream bank protection, the installation of filters on sewage OUTFALLS, or all of the above. The term is also used by the FOREST SERVICE and other land-management agencies to designate a plan of action to carry out a land or resource management plan.

imponderable value *See* INTANGIBLE VALUE.

impoundment in environmental science and engineering, the backing up of water or another fluid behind a barrier. An impoundment is essentially the same as a RESERVOIR, except that the impoundment may be either temporary or permanent and either manmade or natural, while the term "reservoir" almost always is applied to a man-made impoundment that is meant to be permanent. *See also* DAM.

improvement cutting in silviculture, the removal of specific trees from an immature STAND in order to increase the volume and/or quality of timber that will be obtained from the stand after it matures. The removed trees may be noncommercial species such as HARDWOODS, or they may be individual trees that would be of low commercial value due to disease, injury, or genetic defects. *Compare* PRECOMMERCIAL THINNING. *See also* TIMBER STAND IMPROVEMENT.

inbreeding in the biological sciences, the production of offspring by the sexual union of two organisms that lie close together on the same genetic line (*see* GENE), such as a sister and brother, a father and daughter, a mother and son, or a flowering plant and its own POLLEN or that of its first-generation offspring. Inbreeding can be used to emphasize and stabilize desirable traits, such as overall size, length of coat (in animals), or size and color of blossoms (in plants). However, it also runs the risk of emphasizing and stabilizing undesirable traits, such as susceptibility to disease. *See also* INBREEDING DEPRESSION.

inbreeding depression in wildlife biology, a reduction in the viability of a POPULATION of wild animals due to INBREEDING, usually caused by a reduction of the population size through overhunting or the destruction of HABITAT. Three principal factors contribute to inbreeding depression. These are (1) a reduction in the viability of offspring (that is, an increase in the percentage of offspring that die before reaching sexual maturity); (2) an increase in sterility rates and miscarriages due to inbred genetic defects (breeding failure or "fecundity depression"); and (3) a change in the sex ratio, with the number of males increasing in proportion to the number of females ("sex ratio depression").

inceptisol in geology and soil science, any of a group of young soil types with low organic content and present but poorly developed HORIZONS. Inceptisols are most commonly found beneath arctic and alpine tundra (*see* TUNDRA BIOME), where they are the principal soil type. They are also found on the slopes of VOLCANOES, where they exist as a stage in the development of better-differentiated soils such as ULTISOLS. They are acidic in character, are usually built on a CLAY base, and typically show poor drainage, leaving them moist much or all of the year. *See also* SOIL: *classification of soils.*

inclusion in geology, a random piece of rock or crystal of foreign composition embedded in an IGNEOUS ROCK. Inclusions are almost always from one of two sources: either they are crystals that have formed in cavities within the igneous body or they are chunks of rock torn from the edges of a MAGMA CHAMBER, the throat of a volcanic vent, the COUNTRY ROCK with which a DIKE or SILL has formed, etc., which did not melt before the magma around them cooled.

income distribution effect in economics, a change in the pattern in which monetary income is distributed within the population of an area as the result of a change in land-use patterns, the imposition of a new regulation, or some similar action. Income distribution effects are called *positive* if they narrow the gap between high and low income groups and are called *negative* if they increase it. *See* BENEFIT/COST ANALYSIS.

incompatible uses in land-use planning, activities that cannot take place on the same piece of land at the same time. Examples include timber harvest and wilderness preservation, urban development and agriculture, and reservoir construction and white-water rafting. *Compare* CONFLICTING USE; COMPATIBLE USE; COMPLEMENTARY USE.

increment (growth increment, volume increment) in forestry, the volume of new growth added by a tree or a STAND in a given time period, usually one year. *Mean annual increment* (MAI) is the average volume put on by the tree or stand each year during the course of its life to date, and is computed by measuring the total volume and dividing by the age (of the tree) or the average age (of the stand). *See also* CULMINATION OF MEAN ANNUAL INCREMENT.

incubation time in epidemiology, the length of time between exposure to a disease and the first appearance of symptoms. Each separate PATHOGEN has a characteristic range of incubation times. Within that range, the shorter the incubation time the more virulent the disease will generally be. *Compare* LATENCY PERIOD.

Indian Affairs, Bureau of *See* BUREAU OF INDIAN AFFAIRS.

indicator species a SPECIES that can be used to assess the condition of an ECOSYSTEM, or a portion of an ecosystem; to point to the presence of some other, less obvious species or other component of the ecosystem (such as a soil mineral); or to indicate if current conditions in the ecosystem are getting better or worse. The northern spotted owl, for example, is an indicator of the amount and condition of OLD GROWTH forest in the Pacific Northwest (*see* SPOTTED OWL MANAGEMENT AREA). *See also* RANGE MANAGEMENT.

indigenous species *See* ABORIGINAL SPECIES.

indirect loading in pollution control, the sum of all POLLUTANTS present in a body of water that have not entered it directly, including pollution from NONPOINT SOURCES and from all sources, both point and nonpoint, which flow into TRIBUTARIES rather than entering the body of water directly. *Compare* DIRECT LOADING. *See also* INTERNAL LOADING.

indoor pollution pollution occurring in homes, factories, stores, schools, and offices. Indoor pollutants include ASBESTOS, an extremely dangerous insulating material, now banned, leading to lung disease; LEAD, also banned, a constituent of paint and solder (for water piping) that if ingested by children can lead to mental retardation and other maladies; RADON, a naturally occurring, cancer-causing gas emitted from subsoils and rocks beneath foundations that can accumulate in closed buildings; and a number of gases (such as carbon monoxide from space heaters, and cigarette smoke) and volatile compounds (such as cleaning fluids) that can be harmful. Factories, where toxic materials are routinely used, are a special case, with procedures established by the OCCUPATIONAL SAFETY AND HEALTH ADMINISTRATION (OSHA) to reduce health risks. Some modern public and commercial buildings, and sometimes even residences, are sealed with unopenable windows and limited venting in order to reduce heating and air conditioning costs. In such cases, when airborne toxic materials are introduced into the air circulation system, a condition known as "sick building syndrome" can occur. In extreme situations, "sick" office buildings and schools have had to be radically remodeled or even abandoned.

induced costs in BENEFIT/COST ANALYSIS, the costs to a community that are indirectly attributable to the construction and operation of a project, such as street-repair costs due to extra traffic; the need to build new schools and sewage-treatment facilities to take care of the needs of construction workers and their families; etc. *Compare* PROJECT ECONOMIC COSTS.

industrial chemical any substance used as a raw material or CATALYST by industry, whether or not it is also used in the home. The term "industrial chemical" generally implies that the substance is purchased and used in bulk quantities.

inert material (inert substance) any substance that does not react chemically with other substances that it is likely to come into contact with at the conditions of temperature and pressure that the contact is likely to occur. *Compare* STABLE SUBSTANCE.

infill term used to describe an urban development strategy that encourages building on vacant land close to urban centers in preference to allowing a sprawl pattern of urban growth. *See also* URBAN SPRAWL.

infiltration the flow of a liquid into a porous substance or container by passing through small holes. Infiltration implies that the liquid will stay in the substance or container once it gets there, as opposed to

PERCOLATION, which implies that the liquid will flow through the substance or container on the way to something else. For example, water *infiltrates* an unsealed clay vessel by *percolating* through the walls of the vessel. Surface water infiltrating into an AQUIFER recharges it; rainfall infiltrating soil provides water for the growth of plants. Groundwater infiltrating into sewage lines poses problems for designers of sewage treatment plants, who must size the plants large enough to handle not only the sewage flow itself but the extra volume generated by the infiltration, which will increase as the sewage system ages. *See also* INFILTRATION CAPACITY.

infiltration capacity in soil science, the maximum rate at which a soil can absorb rainfall. Infiltration capacity is highest at the beginning of a storm and decreases as the storm continues due to the shrinkage of soil pore sizes. It is much lower for bare soils than for soils on which plants are growing because raindrops falling on bare soils tend to flatten out and pulverize the soil surface, clogging its pores (on vegetated soils, the energy which would otherwise go into pulverizing the soil is absorbed by the plants, allowing the soil pores to remain open). Soils with a high CLAY content have a high dry infiltration capacity because they crack easily when dry; however, their wet infiltration capacity approaches zero due to their rapid absorption rate, which swells the cracks shut. Infiltration capacity is inversely related to RUNOFF—the lower the infiltration, the higher the runoff—and thus plays a large part in determining EROSION rates and how apt a stream is to flood out of its banks during a heavy rain. *See also* INFILTRATION.

inflorescence in botany, a group of flowers growing together on a single stalk. The type of inflorescence a plant shows is characteristic of its SPECIES, and is generally a clue to its ecological role and to its evolutionary history. An inflorescence may be *simple,* with unbranched flower stems (*pedicels*) attached directly to an unbranched central stalk (*peduncle*), or it may be *compound,* with either the pedicels or the peduncle—or both—branching one or more times. It may be *determinate,* with flowering beginning at the tip of the inflorescence and proceeding toward the base, or INDETERMINATE, with flowering beginning at the base and proceeding toward the tip. The simplest form of inflorescence, the *solitary inflorescence*—a single flower—is classed by definition as determinate.

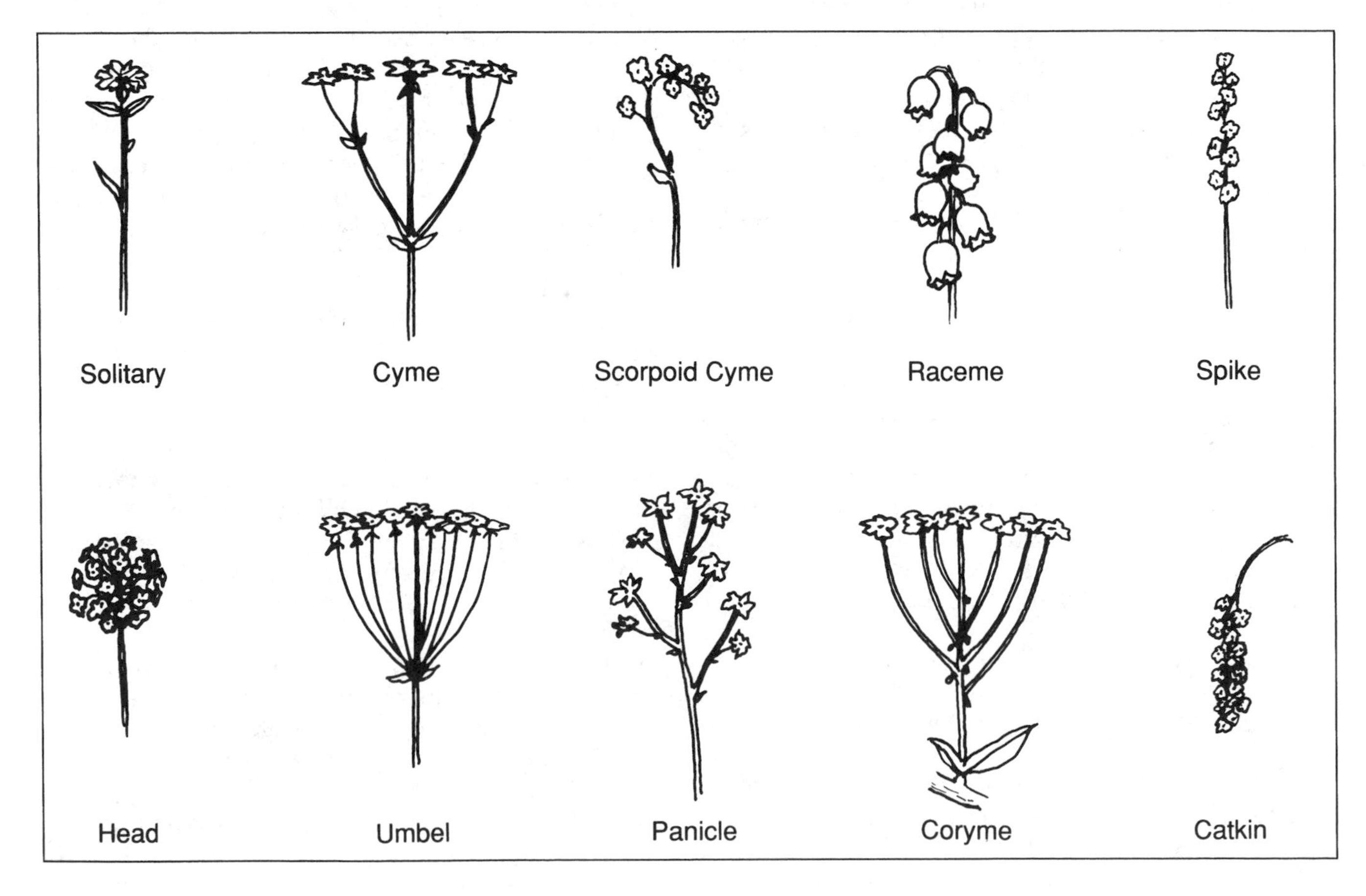

Types of inflorescence

The only other major determinate type is the *cyme,* in which the solitary flower at the tip of the stalk is flanked by two or more clearly subsidiary flowers with their pedicels attached to the stalk's sides (a variant, the *scorpioid cyme,* has subsidiary flowers on one side only, causing the tip of the stalk to curl up like the tail of a scorpion). Among indeterminate forms of inflorescence, the most common form is the *raceme,* with the individual flowers on horizontal pedicels that grow all around the vertical peduncle. The pedicels are generally longer, and the flowers larger, at the base of the inflorescence, and grow progressively smaller toward the tip. There is no flower on the tip itself. In one variant of the raceme, the *spike,* the flowers have no pedicels, growing directly out of the sides of the peduncle. In another variant, the *corymb,* the pedicels curve upward and are much longer at the base of the inflorescence than at the top, producing a flat-topped effect. A *panicle* is a compound raceme. A *catkin* is a spike with tiny, densely packed, usually incomplete flowers, usually found hanging downward on a short, weak peduncle. Besides the raceme and its variants, the major type of simple indeterminate inflorescence is the *head* (sometimes called the *capitulum*), in which the pedicels all arise close together near the tip of the peduncle, forming a dense, globular cluster of flowers. The principal variant in this line, the *umbel,* is a head in which the pedicels—like those of the corymb—get progressively much shorter from the outside in, creating a flat-topped effect. *See also* COMPOSITE.

Inform, Inc. an environmental education and research group founded in 1973 which identifies and describes practical, business-oriented solutions for dealing with such environmental problems as solid waste disposal, hazardous material handling, unhealthy air quality, and alternative fuels for vehicles. Inform publishes a newsletter, *INFORM Reports,* as well as various topical works such as *Gearing up for Hydrogen; Tracking Toxic Chemicals;* and *Rethinking Resources.* Membership (1999): 1,000. Address: 120 Wall Street, 16th Floor, New York, NY 10005. Phone: (212) 361-2400. Website: www.informinc.org.

informal recreation in land-use planning and management, any casual form of recreation that takes place outside of a developed recreation site. Informal recreation includes sliding down roadside snowbanks (as opposed to skiing on marked trails); swimming at a random spot along a riverbank (as opposed to swimming at a designated swimming beach); picnicking beneath an attractive tree (rather than in a state park, roadside rest area, Forest Service picnic area, or other developed site); etc. *Compare* DISPERSED RECREATION.

injection well a method of disposing of HAZARDOUS WASTES by pumping them into deep permeable strata (*see* STRATUM; PERMEABILITY), usually 3,000 feet or more beneath the surface of the earth. The strata are carefully mapped beforehand to make certain that they are not part of a known AQUIFER system, and that they are completely capped by impermeable rock. There are numerous hazards associated with injection-well disposal, the most serious being the need for extremely tight well casings: since the contents of the well are usually under pressure, even a small leak in the casing as it passes through an upper-level aquifer can lead to massive contamination of groundwater supplies. Chemicals disposed of by injection well can dissolve the strata they are injected into, creating underground cavities that may then collapse, breaking the impermeable layers above the disposal area and allowing the contaminants to migrate upward into water supplies. In some cases the caps of injection wells have failed and the wells have literally erupted, spewing masses of hazardous chemicals into nearby watercourses (and in at least one case, into Lake Erie). For all these reasons, injection wells have come under increasing attack as a means of hazardous-waste disposal, and have been made illegal in many areas. However, roughly 60% of all U.S. hazardous wastes continue to be disposed of in this manner. *See* DEEPWELL DISPOSAL.

inland delta an area along a river that resembles a true DELTA in patterns of sediment deposition, vegetation and wildlife, the presence of BRAIDED CHANNELS, and so forth, but which does not lie at the river's mouth. Inland deltas almost always derive from the filling in of shallow lakes (or broad, shallow river segments) with sediments to the point where there is no longer any standing water in them. However, they may also occasionally develop on flat places caused by other geomorphic (landform-shaping) forces, such as the floor of a GRABEN, a wind-flattened area among desert sand dunes, or a deposit of LOESS.

inorganic carbon in chemistry, CARBON as it exists alone or in simple COMPOUNDS (compounds without a carbon-carbon bond or a carbon-hydrogen bond) such as CARBON DIOXIDE (CO_2), CARBON MONOXIDE (CO), and the CYANIDES. The ratio of inorganic to organic carbon in a particular environment can be used as a measure of that environment's ability to support life (*see* PRIMARY PRODUCTIVITY). *Compare* ORGANIC CARBON.

inorganic compound in chemistry, any COMPOUND that does not contain CARBON. The class is usually also taken to include the most simple carbon-containing compounds, including CARBON DIOXIDE, CARBON MONOXIDE, the CARBONATES, and the CYANIDES. Due to the particularly active nature and chemical versatility of the carbon ATOM, there are only about one-tenth as many inorganic compounds (by number) as there are ORGANIC COMPOUNDS. These inorganic compounds, however, make up most of the LITHOSPHERE and virtually all of the ATMOSPHERE and HYDROSPHERE, and are thus extremely important to life.

in-place pollutant a POLLUTANT that has sunk to the bed of a body of water and has become incorporated into the bottom sediments. Nearly any material denser than water has the potential of becoming an in-place pollutant. Lighter materials may also become in-place pollutants if they have the ability to sorb (*see* ADSORPTION; ABSORPTION) to PARTICULATES in the water, which then settle out, carrying the pollutants with them to the bottom. In these ways, bottom sediments in particularly heavy-use areas such as industrial harbors and shipping channels may actually come to contain more pollutants than sediments. In-place pollutants adversely affect rooted aquatic vegetation (cattails, water lilies, etc.) and bottom-dwelling INVERTEBRATES such as mayfly larvae and OLIGOCHAETES, easily entering the FOOD CHAIN through these creatures, where they become subject to BIOLOGICAL MAGNIFICATION in fish, birds, and mammals, including humans. They also serve as a reservoir from which pollutants may be resuspended in the water. Evidence suggests, for instance, that PHOSPHORUS may cycle repeatedly between the sediments and the water in a lake before it "flushes away" through the outlet stream. Once laid down, these pollutants are extremely difficult to get rid of. Dredging them out suspends large quantities of them in the water, endangering the aquatic ECOSYSTEM, and presents the major problem of what to do with the contaminated SPOIL, which is an extremely bulky and often very potent form of HAZARDOUS WASTE. Where it can be done, the best method of dealing with them is probably to cover them with a protective layer of unpolluted sediments and simply leave them alone. This, however, means that particular stretch of harbor or channel can never again be dredged, which often means it can never again be used. *See also* WATER POLLUTION; SLUDGE.

input-output analysis *See* LEONTIEF ANALYSIS.

insecticide *See* PESTICIDE.

Insecticide, Fungicide and Rodenticide Act *See* FEDERAL INSECTICIDE, FUNGICIDE AND RODENTICIDE ACT.

insectivore loosely, any animal (or plant: *see* CARNIVOROUS PLANT) that eats insects, arachnids (spiders, scorpions, and their relatives), and related creatures. More specifically, an insectivore is an animal belonging to the ORDER Insectivora, a group of small, primitive placental mammals (that is, mammals that develop a placenta during the gestation of their young) that have derived relatively unchanged from the earliest placental stock. Insectivores are characterized by small, unlobed brains; five-toed, plantigrade feet (that is, the animal walks on its entire foot rather than just its toes); long, pointed noses; an overshot upper jaw (that is, the upper jaw longer than the lower); and numerous small three-pointed teeth. They are nocturnal in habit (*see* NOCTURNAL SPECIES), and are present on every continent except Australia. The most widespread members of the order are the shrews (Soricidae), which live throughout the world. Others are the moles (Talpidae) of North America, Europe, and Asia; the hedgehogs (Erinaceidae) of Europe, Africa and Asia; the tenrecs (Tenrecidae) of Madagascar; the golden moles (Chrysochloridae), otter shrews (Potomogalidae) and elephant shrews (Macroscelididae) of Africa; and the solenodon (Solenodontidae), which lives only on the islands of Cuba and Haiti. Most are voracious eaters with extremely high metabolic rates and are valuable as a means of insect population control, as well as being an important food source themselves for larger animals and birds, including especially RAPTORS.

insectivorous plant *See* CARNIVOROUS PLANT.

insolation the amount of total solar radiation received by a specified area or ORGANISM over a specified time period. Insolation is generally expressed in langleys per minute, where one langley = one gram-calorie per square centimeter. It may also be expressed in watts per square meter. The portion of the insolation lying within the visible spectrum is known as *illuminance* and is measured in lumens per square foot; one langley yields approximately 10,000 lumens/ft^2 of illuminance under standard conditions, that is, full midday sunlight at moderate elevations in the middle latitudes. The insolation on a particular piece of ground varies with its ASPECT, steepness, elevation above sea level, and latitude above or below the equator, as well as with the time of day, the season of the year, and the cloud cover. The maximum obtainable at sea level (clear air conditions, no clouds, the Sun directly overhead) is approximately 1.42 langleys/min. Variations in annual average and annual extreme insolation figures have a profound impact on the type and quantity of vegetation that will grow on a particular site and the type of animal life that will inhabit it, as well as determining the suitability of the site for SOLAR ENERGY installations. *See also* SOLAR CONSTANT.

Institute for Environmental Studies A multidisciplinary program of the University of Wisconsin created in 1970 to provide a centralized setting for students, professors, and professionals of different backgrounds to gain a better understanding and overall view of environmental problems. The program offers more than 90 courses and eight interdisciplinary degree and certificate programs. IES also has an outreach program to share its expertise through the award-winning Earthwatch radio series; a newsletter, *In Common;* and an e-mail newsletter, *InterView.* Its research program deals with such topics as local and regional water management, climate change, recycling, and the restoration of disturbed ecosystems. Address: Institute for Environmental Studies, University of Wisconsin-Madison, Science Hall, 550 North Park Street, Madison, WI 53706. Phone: (608) 263-1796. Website: www.ies.wisc.edu.

instream minimum flow in water law, the reservation by a state of a specified amount of streamflow for INSTREAM USES. Once the flow of a stream is reduced to the instream minimum requirement no further water can be withdrawn even if rights to it exist (*see* WATER RIGHT). *Compare* INSTREAM RIGHT.

instream right in water law, a WATER RIGHT held by a government body under the doctrine of PRIOR APPROPRIATION to maintain water in a stream for fish and wildlife, recreation, and other INSTREAM USES. The normal prior-appropriation requirements of diversion and beneficial use (as usually defined) are waived. Instream rights are similar to INSTREAM MINIMUM FLOWS, except that they are treated legally in the same manner as other rights in that any right senior in time to the instream right has priority of use (meaning, for example, that an irrigator with senior rights could legally dry up the stream despite the existence of the instream right, but an irrigator with junior rights could not).

instream use in water law, a BENEFICIAL USE of the water in a stream that requires it to be left in the streambed as natural flow rather than diverted into pipelines or ditches. Instream uses include wildlife and fisheries preservation, recreation and aesthetics, LIVESTOCK watering (where the livestock can drink directly from the stream), HYDROELECTRIC POWER generation, the maintenance of enough streamflow to carry away SEWAGE or industrial EFFLUENT, and any other uses for which at least a minimum streamflow level is required. (At least one state, Montana, also recognizes the reservation of minimum streamflows for possible future diversion as an instream use.) The concept of instream use is not recognized by all states. Where it is, it is generally incorporated as part of the doctrine of PRIOR APPROPRIATION. Under this doctrine, there are two general approaches to protecting instream flows. One method is to withdraw the stream from appropriation, recognizing all currently existing diversion rights but not granting any further ones. A second method is for a state agency such as the Fish and Game Department to apply through normal channels for an appropriative right, which it then leaves in the stream. Such "instream minimum rights" are usually considered to take precedence over private rights no matter in what order they were granted, although the courts have sometimes held that strict prior appropriation requires that instream minimum rights must take their turn behind whatever private rights preceded them in time of establishment and have not yet been vacated. As of 1984, the states allowing establishment of instream minimum rights were Alaska, California, Colorado, Idaho, Montana, Kansas, Oklahoma, Oregon, and Washington. *See also* RIPARIAN DOCTRINE; WATER RIGHT; INSTREAM RIGHT.

insular ecosystem (island ecosystem) an ECOSYSTEM that is highly self-contained and strongly differentiated from its surroundings and has little or no direct influence from similar ecosystems elsewhere. Insular ecosystems tend to be simpler and more easily disrupted than larger ecosystems which have no clear boundaries. Examples of insular ecosystems include lakes (especially those without outlet streams), islands, isolated mountain peaks, desert MESAS, city parks, and all other situations where one type of ecosystem is completely surrounded by another.

intangible value (imponderable value) in land-use planning, a BENEFIT, use, or value (of a piece of land, a resource, or a proposed activity) that cannot be quantified, and which therefore cannot be directly compared with other benefits, uses, or values. Since no monetary value or other measure can conveniently be placed upon them, intangible values pose a substantial problem in the use of BENEFIT/COST ANALYSIS to make decisions concerning land allocations. The category includes aesthetic pleasure (scenic viewing), religious significance, historical significance, importance for scientific study, and the values of SOLITUDE and DISPERSED RECREATION.

integrated pest management (IPM) the control of pest ORGANISMS through the use of a mix of biological, mechanical and chemical methods. In general, integrated pest management uses mechanical and biological means (*see* BIOLOGICAL CONTROL) to maintain pest populations at a low level, resorting to chemicals (*see* PESTICIDE) only to control outbreaks. Chemical pesticides are not applied at a predesignated time and a predetermined rate, but only when crop damage reaches some threshold value (usually 5%) and then only in the amounts necessary to knock the pest population back under the threshold level rather than to eliminate it completely. The pesticide chosen is as specific to the TARGET SPECIES as possible, and as short-lived as possible, to avoid harming the beneficial insects, birds, and small mammals that are necessary to maintain biological control. Where IPM is applied, pesticide use (and costs) can be reduced by up to 90%, with crop yield losses cut in half. Moreover, according to one estimate, if IPM were practiced by all U.S. farmers, the health impacts from pesticides would drop by 75%.

Inter-American Tropical Tuna Commission established in 1949, the commission is charged with the conservation of tuna and dolphin resources in the eastern Pacific Ocean. Participating nations are the United States, Costa Rica, El Salvador, Ecuador, France, Japan, Nicaragua, Panama, Vanuatu, and Venezuela. Address: c/o Scripps Institution of Oceanography, 8604 La Jolla Shores Drive, La Jolla, CA 92037. Phone: (619) 546-7100.

interceptor *See under* SEWER.

intercept well a well drilled into a contaminated AQUIFER in order to intercept the flow of contaminants and pump them out, thus protecting the downgrade portions of the aquifer. Intercept wells are generally drilled in groups, forming a curved line around the downgrade perimeter of the contaminant PLUME. They are commonly used to guard against seawater intrusion in coastal aquifers as well as to limit pollution of aquifers from SEWAGE or HAZARDOUS WASTES.

interdisciplinary team a management or problem-solving group composed of individuals whose areas of expertise differ from one another: for example, a forester, a wildlife biologist, a soil scientist, and an economist. Interdisciplinary teams are widely used by the FOREST SERVICE, the BUREAU OF LAND MANAGEMENT, the ARMY CORPS OF ENGINEERS, and other federal land-management agencies, where they are generally assembled to work on a specific problem or set of problems such as preparing a timber sale or designing a recreation plan. The members of the team are expected to meet regularly and pool their knowledge and skills in an effort to work out a unified approach, rather than having each individual work separately in only that area in which his or her own expertise lies (a team practicing this latter approach is known as a *multidisciplinary team* rather than an interdisciplinary team).

interflow in hydrology, water that flows down a slope within the upper layers of the soil rather than on top of them. Interflow may remain in the soil as SOIL WATER, but more often it either flows back to the surface further down the slope or enters a streambed, contributing to the flow of the stream. It is a major portion of RUNOFF following rainstorms anywhere that the soil is deep enough to serve as a conduit.

interflow zone a layer of soil, sediments, streambed gravels, VOLCANIC ASH, or similar material found between two discrete BASALT flows of differing ages. An interflow zone develops whenever sufficient time ensues for EROSION and DEPOSITION to take place between two flows in the same SERIES. Each zone represents an old land surface that has developed on one flow and been covered by a later flow. Interflow zones may be from less than an inch to several dozen feet thick. The thicker zones are likely to contain ORGANIC MATTER, including the trunks of trees engulfed by the onflowing lava. They are usually highly porous and permeable (*see* POROSITY; PERMEABILITY), and make excellent AQUIFERS.

Intergovernmental Panel on Climate Change (IPCC) established in 1988 under the auspices of the UNITED NATIONS ENVIRONMENT PROGRAMME and the World Meteorological organization to examine global warming trends. The IPCC, representing more than 60 nations, started out with 300 eminent climatologists. Their initial findings, suggesting that global warming might have anthropogenic causes, were published in 1990 and led to the U.N.'s RIO EARTH SUMMIT in 1992 where 154 parties signed the U.N. Framework Convention on Climate Change. A 1995 report, this time representing a consensus of more than 1,500 IPCC atmospheric scientists, confirmed that the balance of evidence suggested "a discernible human influence on the climate system." The scientists projected that over the next 100 years a temperature rise of 2 to 6.5 degrees Fahrenheit could be expected, an average increase greater than any seen in the last 10,000 years. *See also* GLOBAL CHANGE; KYOTO PROTOCOL.

intergrade organism an animal, plant, or other living ORGANISM that represents an intermediate physical type between two distinct RACES of a SPECIES, or between two separate POPULATIONS that can be clearly differentiated from each other on the basis of physical characteristics.

Interior, Department of the United States government agency charged with overseeing most federal natural resource activities, including those relating to minerals, water, parks, fish and wildlife, and management of the remaining PUBLIC DOMAIN LANDS. (The department does not oversee the NATIONAL FOREST SYSTEM, which is managed instead by the Department of Agriculture; *see* FOREST SERVICE; AGRICULTURE, DEPARTMENT OF). It is the largest federal agency in terms of land jurisdiction and the fifth largest in terms of work force. A part of the EXECUTIVE BRANCH of the federal government, the department is headed by a cabinet-level officer (the secretary of the interior) assisted by an under secretary and overseeing six assistant secretaries, seven directors, an inspector general, and a solicitor, each of whom is in charge of a major branch of the agency. Nearly all of these officials have important environmental roles; however, the most important from an environmental standpoint are the assistant secretary for fish and wildlife and parks (National Park Service, Fish and Wildlife Service), the assistant secretary for energy and minerals (Geologic Survey), and the assistant secretary for land and water resources (Bureau of Land Management, Bureau of Reclamation, Office of Water Policy).

HISTORY

One of the oldest of the current cabinet-level agencies of the United States, the Department of the Interior was created by Congress on March 3, 1849 by lumping together the existing Office of Indian Affairs, General Land Office, Pension Office, and Patent Office. Conceived as a sort of general housekeeping agency for the nation's internal affairs, the department slowly altered its focus to concentrate more and more on natural resource management. The Department of Agriculture was split off in 1889, the Department of Commerce and Labor in 1903, and the Veterans Administration in 1930. A few agencies dealing with resource conservation have been transferred out as well—most notably, the Forest Service (to the Department of Agriculture, 1905), the Soil Conservation Service (to the Department of Agriculture, 1935), and the Bureau of Commercial Fisheries (to the Department of Commerce, 1970: *see* NATIONAL OCEANIC AND ATMOSPHERIC ADMINISTRATION). Nevertheless the Department of the Interior remains today the chief steward of U.S. natural resources and the nation's leading environmental agency. The Department of Interior has jurisdiction over about 450 million acre of federal lands and 3 billion

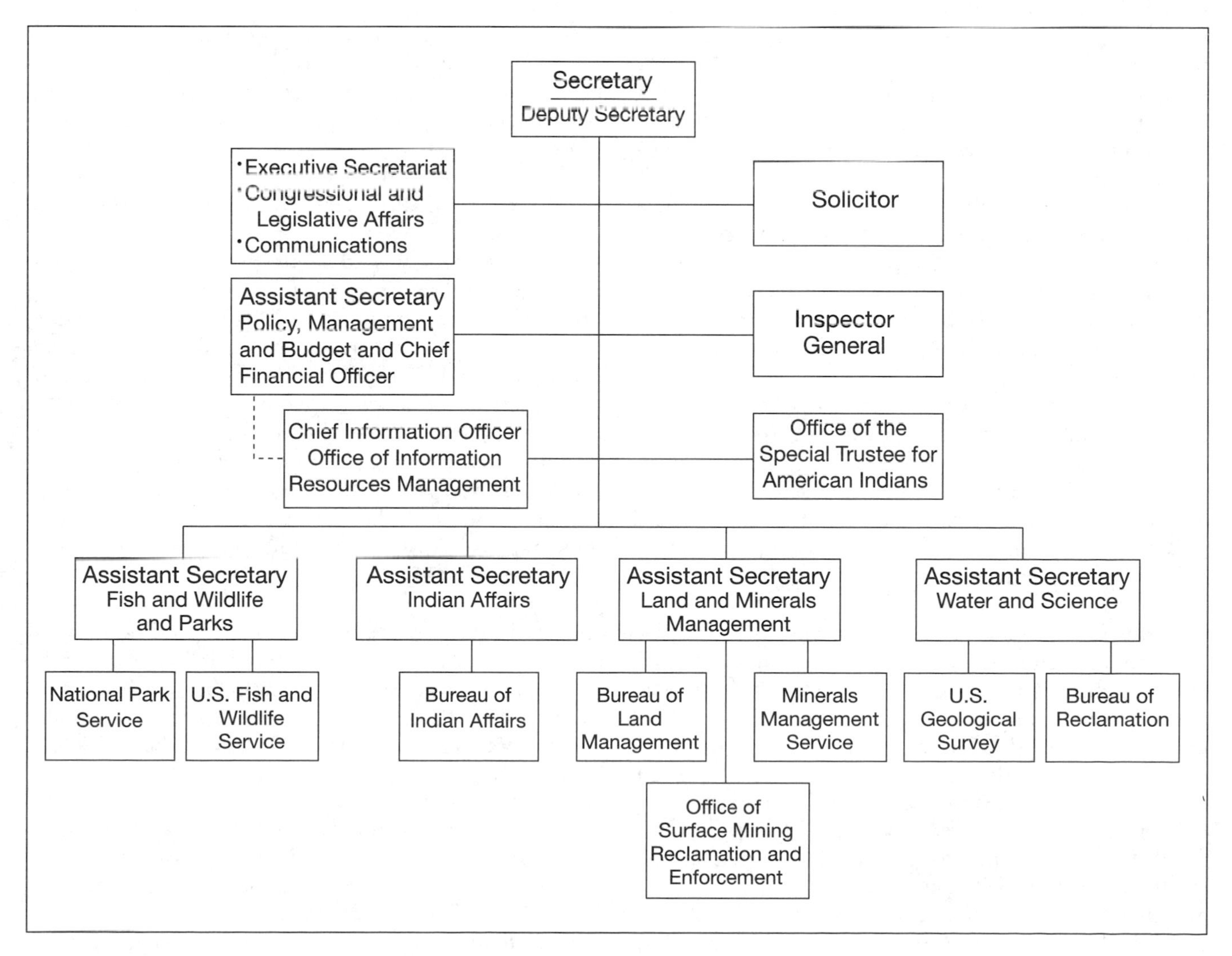

U.S. Department of the Interior organizational structure

acres of the Outer Continental Shelf, and manages more than 57,000 buildings. The department employs some 66,000 employees nationwide. Address: 1849 C Street NW, Washington, DC 20240. Phone: (202) 208-3100. Website: www.doi.gov.

intermediate cut in forestry, any removal of MERCHANTABLE TIMBER from a STAND before the time of the FINAL CUTTING. The term is sometimes generalized to include PRECOMMERCIAL THINNING and other cuts of nonmerchantable timber as well.

intermittent stream any stream that stops flowing during dry seasons or periods of DROUGHT. Intermittent streams intersect only the upper, fluctuating portion of the WATER TABLE. They contain BASEFLOW, but only so long as the water table remains above the level of their beds. *Compare* PERENNIAL STREAM; EPHEMERAL STREAM.

Intermodal Surface Transportation Efficiency Act of 1991 (ISTEA) allocates federal funding for surface transportation programs, under a matching grant program with states, normally 80% federal, 20% state. Unlike earlier federal funding programs, which were based on a single transportation mode (e.g., highways, mass transit, etc.), ISTEA encourages comprehensive solutions of transportation needs. Although ISTEA funds are derived from fuel taxes collected via the federal HIGHWAY TRUST FUND, money can be used under the act for purposes other than highway building and maintenance, including a significant amount for mass transit and a 10% reserve for bicycle paths, landscaping along highways, historical preservation, and a "Scenic Byway" program to preserve low-use "country" roads. In 1998 Congress enacted "TEA-21," a reauthorization providing $217 billion over a five-year period to the program, including $41.4 billion for mass transit. Among the environmental provisions were funds for reducing air pollution from cars and for encouraging carpooling and bicycle commuting.

internal combustion engine the power source for virtually all vehicles on land, plus boats, ships, and some aircraft. The principal feature of the engine is the controlled explosion of gasified liquid fuel delivered to combustion chambers. Moving pistons alternately compress the gas and are in turn pushed downward by the explosion, set off by a spark plug, delivering a twisting energy, called *torque,* to a crankshaft and thence to a transmission that moves the wheels (or propellor in the case of ships and some aircraft) that moves the vehicle. After each cycle the residue of the gas explosions is exhausted through a valve that opens alternately with the intake of fuel. This four-stroke operating cycle is the most common and was first conceived by Alphonse Beau de Rochas of France in 1862 and manufactured commercially by a German firm, Otto and Langen, in 1867. A two-stroke engine was later developed (in 1878) by Dugald Clerk of Scotland. An engine that did not require a spark for ignition and could run on kerosene and other low-cost fuel was manufactured by Rudolf Diesel in 1897. Automobiles began using internal combustion engines in 1895, beginning with a three-wheeled vehicle fabricated by Karl Benz in Germany with an engine designed by Gottlieb Daimler. Wilbur and Orville Wright installed a four-stroke engine in a biplane in 1903, marking the first self-powered, heavier-than-air flying machine. In the early 1900s, steam-powered vehicles were cheaper to manufacture and operate and electric-motor-powered vehicles were much more efficient, especially at lower speed where their torque is greatest and the internal combustion engine's least. However, given the availability of inexpensive fuel that could be carried in substantial amounts by the vehicle itself, internal combustion engines quickly overcame electric motors and steam engines to become the dominant power source for automobiles, trucks, buses, and farm equipment. Despite the inefficiency of internal combustion engines (they require most of the power developed to drive the engine itself) and the toxicity of the exhaust, 20th-century highway building has helped petroleum-powered vehicles proliferate, thus creating dispersed settlement patterns that require more and more personal transportation that only internal-combustion-engine vehicles could provide, which in turn has led to serious health problems from increasing pollution. Now that the sprawl pattern of development and the concomitant AIR POLLUTION must be greatly reduced for reasons of public health, economic efficiency, and environmental quality, alternatives to the internal combustion engine are being taken seriously as the power source for future vehicles. *See also* HYBRID-ELECTRIC CARS; ELECTRIC VEHICLES; URBAN SPRAWL.

internal loading in pollution control, the total amount of POLLUTANTS entering a body of water by recycling from the SEDIMENTS in its bed. Internal loading tends to continue for some time after other pollutant sources are controlled. *See also* IN-PLACE POLLUTANTS.

International Association of Fish and Wildlife Agencies (IAFWA) professional organization founded in 1902 to coordinate the activities of state and provincial fish and wildlife agencies and officials in the United States and Canada to share ideas for wildlife management and to educate the public and national, state, and provincial legislatures on wildlife-management issues and the need for sound wildlife legislation. Address: 444 North Capitol Street NW, Suite 544, Washington, DC 20001. Phone: (202) 624-7890.

International Atomic Energy Agency (IAEA) agency of the United Nations established January 24, 1957, to promote the peaceful use of NUCLEAR ENERGY and the sharing of non-military nuclear technology among nations. One hundred eleven nations currently belong to the IAEA, which has its headquarters in Vienna, Austria. The United States has been a member since July 29, 1957.

International Biosphere Reserves protected areas designated under the United Nations' Man and the Biosphere Program (MAB) as having unique natural characteristics of worldwide importance. MAB was inaugurated in 1971, with the first listing of reserves made in 1976 based on nominations submitted by participating countries. By the 1990s, there were 337 reserves totaling 500 million acres in 85 countries. Early reserves were considered mainly research sites, but in 1995, at the second world conference on biosphere reserves, held in Seville, Spain, a plan termed the Seville Strategy was adopted. Under this plan, each reserve should fulfill three complementary functions: (1) preservation of important ecosystems and landscapes; (2) fostering sustainable development in the reserve area of benefit to local people; and (3) providing a well-supported research and education capability. Biosphere reserves typically consist of a *core area* with a relatively undisturbed ecosystem characteristic of a regional ecosystem; a *buffer zone* where land uses are managed in ways that help protect the core; and a *transition area* which is a "zone of cooperation" with land managed cooperatively with local authorities. Most reserves are national parks or preserves, such as YELLOWSTONE NATIONAL PARK in the United States, the GALÁPAGOS ISLANDS off the coast of Ecuador, or LAKE BAIKAL in Russia. The protection of core areas and the management of surrounding lands is the responsibility of the cooperating country. The U.N.'s MAB Programme has no authority to determine land uses or activities taking place in the reserves. *See also* BAIKAL, LAKE; WORLD HERITAGE SITES.

International Boundary and Water Commission established by treaty between the United States and Mexico to provide timely, environmentally sensitive, and fiscally responsible water and boundary services in the border region. All work of the commission, pursuant to the Convention of 1889 and the 1944 Water Treaty, is to benefit the social and economic welfare of people on both sides of the international boundary. Of primary concern are projects affecting the flow and quality of the waters of the Rio Grande and the Colorado River, including storage reservoirs, dams for hydroelectric energy, levee and floodway projects, and water sanitation. Address: U.S. Headquarters: 4171 North Mesa, C-310, El Paso, TX 79902.

International Crane Foundation endeavors to preserve cranes through conservation, research, captive breeding, field ecology, and education programs. Established in 1973, the foundation publishes a quarterly magazine, *ICF Bugle,* and distributes a book entitled *Reflections: The Story of Cranes.* Membership (1999): 7,000. Address: P.O. Box 447, E-11376 Shady Lane Road, Baraboo, WI 53913-0447. Phone: (608) 356-9462.

International Geophysical Year (IGY) a landmark effort in international scientific cooperation in the geophysical disciplines which took place over an 18-month period from July 1, 1957, to December 31, 1958, a period of maximum sunspot activity. Some 60,000 scientists from 67 nations took part in research projects involving climatology, the measurement of radiation fields, geology, vulcanology, and the like. Among the discoveries were the Van Allen Radiation Belts which produce the aurora borealis, the nature of deep-ocean rifts, and an understanding of the interrelated dynamics of weather patterns and ocean currents. One result of the IGY was an international agreement to set aside ANTARCTICA as a nonmilitary region to foster continued international scientific study. The IGY was also predecessor to many cooperative international scientific programs such as the INTERNATIONAL BIOSPHERE RESERVES and the INTERGOVERNMENTAL PANEL ON CLIMATE CHANGE.

International Joint Commission on Boundary Waters (IJC) bilateral agency of the United States and Canada devoted to managing jointly owned waters, especially those of the Great Lakes and the St. Lawrence River. Established by treaty on January 11, 1909, the IJC adjudicates disputes between the two nations over uses of waters that form or flow across the boundary, and is responsible for monitoring compliance with the GREAT LAKES WATER QUALITY AGREEMENT OF 1978. It is assisted in this work by two advisory boards: the Great Lakes Water Quality Board, consisting of ranking officials of the pollution-control agencies of both nations, and the Science Advisory Board, consisting of scientists from government agencies, universities, and the private sector. Headquarters: 1250 23rd Street NW, Suite 100, Washington, DC 20440, Phone: (202) 736-9000. Great Lakes Regional Office: 100 Ouellette Avenue, 8th Floor, Windsor, Ontario Canada N9A 6T3.

U.S. Biosphere Reserves, July 1999

Aleutian Islands National Wildlife Refuge
Beaver Creek Experimental Watershed
Big Bend National Park
Big Thicket National Preserve
California Coast Ranges Biosphere Reserve (8 Units)
  Heath and Marjorie Angelo Coast Range Preserve
  Elder Creek Area of Critical Environmental Concern
  Jackson Demonstration State Forest
  Landels-Hill Big Creek Reserve
  Redwood Experimental Forest
  Redwood National Park
  North Coast Redwoods District State Parks (3 Units)
  —Del Norte Coast Redwoods State Park
  —Jebediah Smith Redwoods State Park
  —Prairie Creek Redwoods State Park
  Western Slopes of Cone Peak
Carolinian-South Atlantic Biosphere Reserve (11 Units)
  Blackbeard Island and Wolf Island National Wildlife Refuges
  Cape Lookout National Seashore
  Cape Romain National Wildlife Refuge
  Capers Island Heritage Preserve
  Cumberland Island National Seashore
  Gray's Reef National Marine Sanctuary
  Hobcaw Barony (North Inlet)
  Little St. Simons Island
  Santee Coastal Reserve and Washoo Reserve
  Tom Yawkey Wildlife Center
  Cascade Head Experimental Forest & Scenic Research Area
Apalachicola National Estuarine Research Reserve
Central California Coast Biosphere Reserve (See Golden Gate Biosphere Reserve)
Central Gulf Coastal Plain Biosphere Reserve
Central Plains Experimental Range
Champlain-Adirondack Biosphere Reserve (3 Units)
  Adirondack Park Agency
  Green Mountain National Forest
  Mount Mansfield State Natural Area
Channel Islands Biosphere Reserve (2 Units)
  Channel Islands National Marine Sanctuary, NOAA
  Channel Islands National Park
Coram Experimental Forest
Denali National Park and Biosphere Reserve
Desert Experimental Range
Everglades National Park (with Dry Tortugas N.P.)
Fraser Experimental Forest
Glacier Bay-Admiralty Island Biosphere Reserve (2 Units)
  Admiralty Island National Monument
  Glacier Bay National Park
Glacier National Park
Golden Gate Biosphere Reserve (13 Units)
  Bolinas Lagoon Preserve, Cypress Grove Preserve
  Bodega Marine Reserve
  Cordell Bank National Marine Sanctuary
  Farallon Islands National Wildlife Refuge
  Golden Gate National Recreation Area
  Gulf of the Farallones National Marine Sanctuary
  Jasper Ridge Biological Preserve (Coordinator)
  Marin Municipal Water District
  Mount Tamalpais State Park
  Point Reyes National Seashore
  San Francisco Peninsular Watershed
  Tomales Bay State Park
  Samuel P. Taylor State Park
Guanica Commonwealth Forest Reserve
H. J. Andrews Experimental Forest
Hawaiian Islands Biosphere Reserve (2 Units)
  Haleakala National Park
  Hawaiian Volcanoes National Park
Hubbard Brook Experimental Forest
Isle Royale National Park
Jornada Experimental Range
Konza Prairie Research Natural Area
Land between the Lakes
Luquillo Experimental Forest
Mammoth Cave Area
Mojave and Colorado Deserts Biosphere Reserve (5 Units)
  Anza-Borrego Desert State Park
  Death Valley National Monument
  Joshua Tree National Monument
  Philip L. Boyd Deep Canyon Desert Center
  Santa Rosa Wildlife Management Area, San Bernardino National Forest
New Jersey Pinelands Biosphere Reserve
Niwot Ridge Biosphere Reserve
Noatak National Preserve (2 Units)
  Gates of the Arctic National Park
  Noatak National Preserve
Olympic National Park
Organ Pipe Cactus National Monument
Rocky Mountain National Park
San Dimas Experimental Forest
San Joaquin Experimental Range
Sequoia-Kings Canyon National Parks
South Atlantic Coastal Plain Biosphere Reserve
Southern Appalachian Biosphere Reserve (5 Units)
  Coweeta Hydrologic Laboratory
  Grandfather Mountain
  Great Smoky Mountains National Park
  Mt. Mitchell State Park
  Oak Ridge National Environmental Research Park
Stanislaus-Tuolumne Experimental Forest
Three Sisters Wilderness
University of Michigan Biological Station
Virgin Islands National Park and Biosphere Reserve
Virginia Coast Reserve
Yellowstone National Park

International Monetary Fund (IMF) one of several international organizations with great influence over national economic policies, and therefore of significance to global environmental concerns. An agency of the United Nations established in 1945, the IMF is headquartered in Washington, D.C. Utilizing a fund of approximately $300 billion from the member nations, the organization buys foreign currencies to stabilize exchange rates and pay international debt. The fund also acts as a financial adviser and provides technical assistance to 182 member nations around the globe. Some believe that the advice and the money lead developing nations into pattern of excessive resource and labor exploitation, financial indebtedness, and environmentally unsound industrial development. *See also* GENERAL AGREEMENT ON TARIFFS AND TRADE; NORTH AMERICAN FREE TRADE AGREEMENT; WORLD BANK; WORLD TRADE ORGANIZATION.

International Pacific Halibut Commission coordinates scientific research and management of the Pacific halibut resources. The commission was established in 1923 by a treaty convention between the United States and Canada. Address: P.O. Box 95009, Seattle, WA 98145. Phone: (206) 634-1838.

International Whaling Commission established in 1946 by the International Convention for the Regulation of Whaling and concerned with the conservation of whale populations. Member nations are Antigua and Barbuda, Argentina, Australia, Austria, Brazil, Chile, People's Republic of China, Costa Rica, Denmark, Dominica, Finland, France, Germany, Grenada, India, Ireland, Italy, Japan, Kenya, Republic of Korea, Mexico, Monaco, Netherlands, New Zealand, Norway, Oman, Peru, Russian Federation, Saint Kitts and Nevis, Saint Lucia, Saint Vincent and the Grenadines, Senegal, Solomon Islands, South Africa, Spain, Sweden, Switzerland, United Kingdom, United States, and Venezuela. Address: U.S. Commissioner, U.S. Department of Commerce, Room 5128, Herbert C. Hoover Building, 14th Street and Constitution Avenue NW, Washington, DC 20230.

interstadial in the earth sciences, a time of warm climate between two periods of glacial advance (*see* GLACIER). Interstadials are characterized in the geologic record by wind- and water-formed erosional deposits found on top of one set of glacial deposits and beneath another set. Recent interstadial deposits may also contain ORGANIC MATTER. They are comparable to INTERFLOW ZONES in a BASALT SERIES, with glacial deposits taking the place of the basalt flows. At least four interstadials are clearly represented in the records of Pleistocene glaciation from northern North America and Eurasia (*see* ICE AGE; CONTINENTAL GLACIER), and many glaciologists feel that we are currently living in what is termed the "fifth interstadial"—in other words, that it is only a matter of time before the glaciers return once more. *See also* STADE.

Interstate Commission on the Potomac River Basin interstate compact that collects historical data, coordinates water programs in the Potomac Basin states, and cooperates in ongoing studies regarding water supply, water quality, and land resources associated with the watershed. The commission was established in 1940 by the District of Columbia, Maryland, Pennsylvania, Virginia, and West Virginia. Address: 6110 Executive Boulevard, Suite 300, Rockville, MD 20852. Phone: (301) 984-1908.

intertidal zone *See* BALANOID ZONE.

intrazonal soil a soil with poorly to moderately well-differentiated HORIZONS, whose development has been strongly influenced by local topography, drainage patterns, and climate. Intrazonal soils typically occur as small "islands" in a surrounding zonal or azonal soil; they include the soils of alpine meadows, dry south-facing slopes, stabilized sand DUNES, and similar situations. The term is now generally considered obsolete. *See also* SOILS: *classification of soils*.

introduced species *See* EXOTIC SPECIES.

intrusive rock in geology, an IGNEOUS ROCK that has cooled and solidified entirely beneath the surface of the ground. Intrusive rocks cool much more slowly than EXTRUSIVE ROCKS, commonly giving them a hard, fine-textured, extremely erosion-resistant crystalline structure. The most common intrusive rock is GRANITE and its relatives. *See also* DIKE; SILL; BATHOLITH.

inventory in silviculture, the number of trees in each DIAMETER CLASS found within a given area. Inventory is usually taken only of MERCHANTABLE TIMBER. Two separate types of inventory are possible. *Standing inventory* is an actual count of standing timber currently "on the stump" in the inventoried area; *future inventory* or *potential inventory* takes growth INCREMENT into account and thus gives a sense of how rapidly wood is being added to a stand as well as how much is currently present.

inversion (**thermal inversion**) in meteorology, an atmospheric condition in which a layer of relatively cool air is overlain by a layer of relatively warmer air, thus inverting the normal tendency of air to decrease in temperature with altitude (*see* LAPSE RATE). Inversion result from the nighttime loss of heat from the ground surface through radiation. If there is a lack of wind mixing (that is, if the air is still), the air in contact with the ground will cool faster than the air higher up, causing the reverse lapse rate characteristic of an inversion to develop. If the ground is warmed sufficiently by the sun the next day, the inversion will "burn off." Inversions that persist for a number of days usually develop only in mountain-ringed valleys or basins, where they are the result of drainage of cold air down the valley slopes due to the tendency of cold air to settle. The blanket of cold air thus formed is too thick for solar heating to disperse, and will persist until broken up by winds. Because cool air settles and warm air rises, there is little or no mixing across the interface between the cold and warm bodies of air during an inversion. This tends to trap pollutants next to the ground and cause them to build up without dispersing. In the Los Angeles area, for example, cool air blowing in from the Pacific Ocean displaces warmer air, forcing it to a higher altitude. Because the cool air is trapped by the San Gabriel Mountains on the east and cannot rise through the warmer inversion layer above it, it takes very little time for the trapped cool air to fill up with highly toxic pollutants—ozone, oxides of sulfur and of nitrogen, volatile hydrocarbons, and particulate matter. *See* AIR POLLUTION.

invertebrate in the biological sciences, any animal without a backbone. The invertebrates include insects, arachnids, mollusks, crustaceans, starfish, sea anemones, and virtually all other animals except for fish, amphibians, reptiles, birds, and mammals. *Compare* VERTEBRATE.

ion in chemistry, an ATOM that has either lost or gained one or more ELECTRONS and has thus acquired a small positive or negative electrical charge. The properties of ions are often quite different from those of the atoms they were formed from. SODIUM and CHLORINE atoms, for instance, are both highly poisonous, but sodium and chlorine ions—as found in common table SALT—are necessary for human health. Metallic ELEMENTS tend to oxidize (lose electrons) during ionization, forming positive ions (*cations*), while nonmetals tend to be reduced (gain electrons), forming negative ions (*anions; see* OXIDATION; REDUCTION). Metals and nonmetals usually react with each other to form *ionic compounds,* in which the metals lose electrons to the nonmetals: the resulting ions are then held together by the attraction of their opposite electrical charges. When an ionic compound is dropped into a POLAR LIQUID such as water, the positive and negative ions dissociate from each other ("ionize"), each one becoming surrounded by several molecules of the liquid oriented in such a way that the pole on the liquid molecule is pointed toward an ion of the opposite charge: such a solution conducts electricity easily, and is known as an ELECTROLYTE (*see* DISSOCIATION; SOLUTION). Ions play a fundamental role in the chemistry of life: the liquid contents of the living CELL are mostly electrolytes whose ion content is very carefully controlled by the cell membrane. *See also* COVALENT BOND.

ion exchange in pollution control, the removal of dissolved POLLUTANTS (*see* SOLUTION) by means of a chemical REACTION that exchanges the pollutant's IONS for less harmful and/or more easily controllable ones. The earliest use of ion exchange was to soften "hard" water by replacing the calcium and magnesium ions that gave the water its hardness with sodium ions. This was accomplished by bubbling the water through a bed of crushed zeolite (sodium aluminosilicate, $NaAlSi_2O_6$). The calcium and magnesium in the hard water reacted with the zeolite, combining with the aluminosilicate to release the sodium. More recently, groups of synthetic RESINS have been developed that will exchange HYDROGEN ions for virtually all positively charged ions in a solution. The dydrogen ions may then be removed through reaction with a second resin that binds them into an insoluble compound. The result is completely demineralized water. Ion-exchange resins are normally packed in long columns through which the water is passed. The zone of reaction moves slowly up the column until all the resins have been reacted with, at which point the column may be removed and backflushed with a solution that reverses the ion-exchange reaction, recreating the original resins—a process known as *regeneration.*

iron the 26th ELEMENT in the ATOMIC SERIES, ATOMIC WEIGHT 55.847, chemical symbol Fe. Iron is highly reactive and forms COMPOUNDS easily, so it is rarely found in the pure state in nature. However, its compounds are widespread and it ranks overall as the fourth most abundant element on the planet. It is important as a TRACE ELEMENT in life processes, especially those of animals, in which it forms an essential part of the HEMOGLOBIN molecule and is thus responsible for the transport of OXYGEN within the bloodstream. It is also critically important as an item of commerce and manufacture: humans use 14 times more iron than they do all other metals combined.

IRON AND STEEL MANUFACTURING

The basic process for refining iron from its ORES has been known since at least 3000 B.C.: it consists of crushing the ores, heating them to beyond iron's melting point (in the impure form, roughly 1,000°C [1,832°F]), and drawing off the impurities (slag), which generally float on the surface of the melted iron. In its pure form, iron is too soft and brittle to be of much use, so CARBON and other impurities are generally added to the mix. The most common process in use today begins with the production of so-called "pig iron" in a *Bessemer converter,* a tall (27.5-meter [90-foot]) cylindrical retort which is charged with a mixture of iron ore and LIMESTONE and heated by burning COKE: the coke produces CARBON MONOXIDE, which provides the carbon for strengthening the iron, while the limestone serves as a flux, combining with impurities in the ore so that they do not recombine with the iron upon cooling. The pig iron is then further refined in an open-hearth furnace heated to approximately 3,000°C (5,400°F), lowering its carbon content slightly and distributing the carbon uniformly through the iron to form a true ALLOY. The molten material—now properly called steel—is usually cooled quickly by "quenching" it with water in order to form a hard crystalline structure. Other refinements that may be used, depending upon the product desired, include the addition of MAGNESIUM, CHROMIUM or NICKEL to the mix to form stainless steel and other alloy steels (*ferroalloys*) and the use of a so-called "pickling solution" of CYANIDE or other SALTS to harden the steel's surface through the formation of a surface film of nitrides or similar compounds ("case-hardening").

ENVIRONMENTAL CONSIDERATIONS

Steelmaking uses a great deal of water (about 25,000 gallons for each ton of finished product), mostly in the quenching process. This water is polluted by the impurities in the iron and by cyanide and other pickling salts, and can be a serious POLLUTANT if released into watercourses. If it is not released, however, it must be counted as "used up" and may badly deplete water reserves in the region where the steel mill is located. Cleaning and recycling the quenching water within the plant can help solve both of these problems. Release of excess carbon monoxide from the Bessemer converter can also be a hazard, although this is usually minimal because the gas is collected for use as a fuel for the open-hearth furnaces that make up the next stage of the steelmaking process. *See also* TACONITE; LIMONITE; HEMATITE; DIRECT-SHIPPING ORE; BENEFICIATING.

iron bacteria (iron-fixing bacteria) any of several genera of bacteria (*see* GENUS; BACTERIUM) that, through their metabolic processes, cause iron compounds to precipitate from sea water and become deposited in seafloor sediments. It is thought that this process is at least partially responsible for the concentration of iron into ORE BODIES suitable for mining in many parts of the world.

iron triangle in politics, a term used to describe the lines of mutual support that develop among backers of a pork-barrel project (*see* PORK BARREL). The three apexes of the triangle are formed by the congressman in whose district the project will lie, the federal agency that will build the project, and the local lobbying group in favor of the project.

irreversible effect in land-use planning, any effect of a proposed planning action or resource commitment that cannot be changed, or is highly unlikely to be changed, by future generations. Irreversible effects include the EXTINCTION of one or more plant or animal SPECIES; the destruction of a CLIMAX COMMUNITY; the physical removal of large amounts of soil and rock, either through accelerated EROSION or through purposeful activities such as quarrying or strip-mining; and so forth. They also include *irretrievable commitments of resources;* that is, changes in the use patterns of resources that are technically possible to reverse but are so costly to establish that they are unlikely to be abandoned in the future, such as the replacement of a free-flowing stream with a RESERVOIR or the building of a shopping center or housing development on land previously used for agriculture.

irrigation in agriculture, the process of supplying water to a plot of land for the purpose of growing plants with higher water requirements than those that can grow on the plot under natural conditions. Almost all human cultures living in dry climates have practiced some form of irrigation. In North America, the practice dates back at least to the time of the Hohokam Indians of what is now central Arizona, who irrigated crops from the Salt River as early as the first century A.D. with canals that are in some cases still in use today. There are currently (1988) some 51 million acres of irrigated cropland and pastureland in the United States. Most of it is in the 17 western states; however, the practice is increasingly spreading in the midwestern and eastern states as a means of increasing crop yields and removing the uncertainty of fluctuations in rainfall.

METHODS OF IRRIGATION

Four principal methods are in use to apply irrigation water to the soil. The *ditch-and-furrow* method utilizes irrigation ditches ("lateral canals," or just "laterals") to move the water from the main irrigation canals to individual fields, where it is allowed to run across the fields in plowed furrows between the rows of crops. *Sheet irrigation* also uses ditches to move the water to the fields, but the water is allowed to run out onto the level surface of the field and soak into the soil rather than being led across it in furrows. This method is used most commonly for pastures and for crops that need irrigating only once or twice a year. *Sprinkler irrigation* uses sprinklers to apply the water to the field, sometimes as rotating jets from a so-called "water gun" but more commonly in the form of lengths of pipe up to a quarter mile long with holes in them like a common lawn or garden sprinkler. These pipes are mounted on large wheels to allow them to move across a field or pasture. They come in two forms, either as "crawling systems" that move slowly across a field from one side to the other or as "center-pivot systems" that rotate around a fixed pipe in the center of the field, creating a circular irrigated area as much as half a mile across. Sprinkler irrigation normally uses considerably less water than sheet or ditch-and-furrow irrigation. Finally, *subsurface* or *drip irrigation* uses perforated tubing buried in the soil to apply water directly to the roots of these plants being irrigated. Drip irrigation uses even less water than sprinkler irrigation, and up to 60% less than is used in either the sheet or ditch-and-furrow methods.

PROBLEMS ASSOCIATED WITH IRRIGATION

Although irrigation has substantially increased food production in arid parts of the world, and is thus undoubtedly a net social good, it is not without problems. The most serious of these is the problem of SALT accumulation in irrigated soils. All natural waters carry dissolved mineral salts. When irrigation water is spread out over a field, much of it evaporates, leaving the salts behind to accumulate on and in the soil. Excess water that runs off the field at the far end has an increased salt load, both from concentration through evaporation and from picking up salts deposited on the soil by previous irrigation. This increased salt load then enters watercourses as "return flow," increasing the salinity of the watercourses and compounding the salt-deposition problem for irrigators who withdraw their water from these watercourses further downstream. Salt deposition took less than 40 years to convert much of California's Imperial Valley from an irrigated "paradise" to a salt desert where not even the arid-land vegetation that grew there formerly could survive. Other problems associated with irrigation include water wastage (only about 25% of the water withdrawn for irrigation purposes is actually used by the plants); high consumption rate (about 95% of the water used for irrigation is consumed rather than being returned to its source; this contrasts to a consumption rate of 25% for domestic water use and less than 10% for industrial use); land subsistence and other problems caused by the depletion of groundwater reserves; and, in some regions, the inundation of large areas of formerly productive land due to rising WATER TABLES as "lost" irrigation water builds up in the soils.

irritability in the biological sciences, the ability of an ORGANISM to respond to a change in its environment (*stimulus*) with a change in its behavior (*response*). Irritability is a property of all living organisms, including even VIRUS particles (which do not activate unless their environment is changed by the approach of a living CELL with the appropriate genetic chemistry for their use in replicating themselves), and is one of the principal tests for determining if an organism is living or not (a dead amoeba will not respond to a beam of light by contracting, as a live one will; a dead plant will not respond to watering by rebuilding the TURGOR in wilted tissues; and so forth). *See also* HABITUATION; AVOIDANCE.

irruption in population biology, a sudden and spectacular increase in the POPULATION of a particular SPECIES of animal of plant to the point where environmental damage occurs. Irruptions are distinct from the normal density peaks of animals with regular POPULATION CYCLES such as rabbits, voles (meadow mice), and lemmings, although they almost always occur in conjunction with these peaks if a cyclic species is involved in the irruption. Their exact cause remains unknown, and almost certainly varies from irruption to irruption. However, a disturbance of more than one natural control on the population, such as disease, predation (*see* PREDATOR), or food-supply limits, can normally be found. Irruptions almost always take place on land that has been cultivated or otherwise disturbed by human activities, and should thus be classed as a symptom of environmental degradation. They are usually brought to an end by epidemic disease, often coupled with malnutrition as the food supply give out (animals) or the soil becomes depleted (plants). Poisoning programs rarely make much of a dent in an irruption, and may actually be counterproductive: poison is not transferred from individual to individual and so does not spread through a population as disease does, and poisons often kill PREDATORS and VECTORS as well as the irrupting species, thus relaxing environmental control further and permitting increased growth of the irruption.

island ecosystem *See* INSULAR ECOSYSTEM.

Island Press *See* CENTER FOR RESOURCE ECONOMICS.

isobar in meteorology, a line connecting points of equal ATMOSPHERIC PRESSURE on a weather map. An isobar represents a *locus,* that is, a line on which all points meet a particular criteria, in this case equality of air pressure. Thus it does not connect two points of the same pressure across an intervening zone of different pressure. Like CONTOURS on a TOPOGRAPHIC MAP, isobars delineate "hills" and "valleys" of greater or lesser air pressure, allowing the meteorologists to locate CYCLONES and ANTICYCLONES, predict the direction of winds, and follow the course of storms across the earth's face.

isohyet in meteorology, a line connecting points of equal PRECIPITATION on a weather map. An isohyet represents a *locus,* that is, a line on which all points meet a particular criteria, in this case equality of precipitation. Thus it does not connect two points of the same precipitation across an intervening zone of different precipitation. Isohytes help trace storm fronts and show how fast storm systems are advancing across the landscape; however, they are considerably more valuable if drawn to average annual figures rather than on a day-to-day basis, since average annual rainfall is such an important determinant of water availability and hence of ECOSYSTEM structure. Average annual isohytes, along with average monthly *isotherms,* are among the most important delineators of ECOTONES.

isomer in chemistry, a COMPOUND that has the same chemical formula as a second compound but which differs from it in structure. In other words, a compound and its isomer will contain the same ELEMENTS, in the same proportions, but they will be attached to each other in different ways. Isomers often differ radically in physical properties. Acetic acid, for example—the ACID that gives vinegar its bite—is the compound $C_2H_4O_2$, arranged as CH_3COOH. It has a boiling point of 118.1°C and a freezing point of 16.6°C, and in its pure form is a strongly caustic acid with a pungent odor. Its isomer, methyl formate, is also the compound $C_2H_4O_2$, but arranged as $HCOOCH_3$: reversing the attachment points of the hydrogen ion and methane groups in this manner creates a pleasant-smelling, less caustic (but more flammable) compound that boils at 31.5°C and freezes at about –100°C. Most complex compounds have at least one isomer, and many—especially the large MOLECULES characteristic of ORGANIC COMPOUNDS—have more than one. It may be important from an environmental standpoint to know which isomer of a compound one is dealing with (*see*, e.g., DIOXIN).

isostatic uplift in geology, a rebounding of a broad reach of the earth's surface and all the features on it following the removal of a great weight, such as a CONTINENTAL GLACIER. The boundary defining an area of isostatic uplift is known as the HINGE LINE. Isostatic uplift is caused by the tendency of all mechanical systems to try to reach stasis, that is, to equalize all pressures throughout the system. A glacier's weight substantially increases the pressure on the ground surface beneath it, compressing it until its internal pressures balance those pressing down on it. Once the glacier has melted, the ground beneath must move upward until the pressures within it are once more in balance with those in its surroundings. Isostatic uplift resulting from the disappearance of the Pleistocene glaciers is still proceeding in northern North America. At its center near Hudson Bay the rate of uplift is more than 2 meters (7 feet) per century.

isotherm in meteorology, a line connecting points of equal temperature on a weather map. An isotherm represents a *locus*, that is, a line on which all points meet a particular criteria, in this case equality of temperature. Thus it does not connect two points of the same temperature across an intervening zone of different temperature. Like CONTOURS on a TOPOGRAPHIC MAP, isotherms delineate "hills" and "valleys" of greater and lesser temperature, allowing meteorologists to locate and follow air masses and their accompanying weather patterns across the Earth's face. Isotherms representing average monthly temperatures show overall continental climatic patterns and, together with average annual ISOHYETS, are one of the principal means of delineating ECOTONES.

isotope in chemistry, one of a set of two or more ATOMS that have the same number of protons but differing numbers of neutrons (*see* ATOM: *structure of the atom*). Since it is the number of protons, not neutrons, that determines an atom's number of ELECTRONS and therefore establishes its chemical properties, differing isotopes of the same element will have identical chemical properties; however, since their ATOMIC WEIGHTS are different, their physical properties will differ slightly from each other. Two or more isotopes are known for nearly all elements in the ATOMIC SERIES; some of these are of considerable importance to modern technology. *See,* e.g., HEAVY WATER.

itai-itai disease literally "ouch-ouch disease," a health condition caused by chronic CADMIUM poisoning, first described in 1970 in the city of Toyama, Japan, approximately 150 miles northeast of Tokyo on the east coast of the island of Honshu. The city was located near a ZINC smelter, and cadmium released by the smelter had poisoned the soil of nearby rice paddies, elevating the levels of cadmium in the rice consumed by the local residents. Symptoms include severe pain in bones and joints, softening of the bones (leading to multiple fractures), kidney damage, and eventually death.

Izaak Walton League of America (IWLA) sportsmen's conservation organization, founded in 1922 to protect freshwater sports fisheries. It has since expanded its role to the protection of other wildlife habitat and the conservation of soil, water, and forest resources. It operates the Izaak Walton League Endowment, which purchases parcels of endangered private land for later conveyance to the federal government. The IWLA is among the most conservative of major conservation groups. Membership (1999): 50,000. Address: 707 Conservation Lane, Gaithersburg, MD 20878-2983. Phone: (301) 548-0150. Website: www.iwla.org.

J

Jack Ward Thomas Report *See* THOMAS REPORT.

Jackson, Wes (1936–) geneticist, author, and leader in sustainable agriculture. Currently president of THE LAND INSTITUTE, which he founded in 1976, Jackson earned a B.A. degree in biology from Kansas Wesleyan University, an M.A. in botany from the University of Kansas, and a Ph.D. in genetics from North Carolina State University. He is best known for his research in establishing perennial food crops in the Great Plains as an alternative to MONOCULTURES such as corn and soybeans which lead to erosion, require substantial irrigation, and deplete the soil. He was a 1990 Pew Conservation Scholar and in 1992 received a MacArthur Fellowship. Jackson is the author of *New Roots for Agriculture* (1980), *Altars of Unhewn Stone* (1987), *Becoming Native to This Place* (1994), and other books.

Jackson Turbidity Unit (JTU) unit for measuring the amount of TURBIDITY in a water sample, developed around 1900 by the American hydrologist and chemical engineer Daniel Dana Jackson (1870–1941). Measurement of JTUs was originally done using the so-called "Jackson candle," a device consisting of a candle burning beneath a flat-bottomed glass tube. The sample to be tested was poured into the tube until the outline of the candle flame could no longer be seen; its depth was then compared to a table based on a standard SUSPENSION of 1 milligram/liter DIATOMACEOUS EARTH in water, 1 centimeter's depth of the standard solution being one JTU. Turbidity testers today use a SPECTROPHOTOMETER calibrated directly in JTUs, but the standard unit remains the same.

Japanese nuclear disaster *See* TOKAIMURA.

jet stream in meteorology, a powerful upper-atmosphere wind moving from west to east at an average speed of 125 kilometers per hour (75 mph). Occasional clockings have reached as high as 300 kilometers per hour (180 mph). (These are winter speeds; summer speeds drop to approximately half these values.) Jet streams develop at or near the TROPOPAUSE over very strong surface FRONTS, especially those connected with polar AIR MASSES.

jetty *See* GROIN.

Job Corps U.S. government program that provides training and jobs for disadvantaged young people ages 16 through 24 in conservation and development, outdoor recreation, and community projects. Created by the Job Training Partnership Act in 1964, the Job Corps has provided employment skills to more than 1.7 million youth. The program is managed regionally through the Civilian Conservation Centers and more than 110 Job Corps campuses nationwide. Address: Department of Labor, Employment and Training Administration, Frances Perkins Building, 200 Constitution Avenue NW, Washington, DC 20210. Phone: (202) 219-8550. Website: www.jobcorps.org.

Johnson, Lady Bird (1912–) environmentalist and wife of President Lyndon Johnson. Born in Karnack, Texas, as Claudia Alta Taylor, she was given the nickname "Lady Bird" by a nursemaid. Lady Bird married Lyndon Johnson in 1934 and took an active role in his political life. As first lady from 1963 to 1969, she promoted national beautification projects and other environmental causes. An accomplished businesswoman as well, she built up a multimillion-dollar broadcasting company starting in 1943 with the purchase of a single Texas radio station. In 1983 Mrs. Johnson endowed a 63-acre LADY BIRD JOHNSON WILDFLOWER RESEARCH CENTER near Austin Texas. She is the author of *A White House Diary* (1970). Lewis I. Gould's book *Lady Bird Johnson and the Environment* (1988) details her environmental contributions.

Jordan's rule in ichthyology, the observation, first made by ichthyologist David Starr Jordan (1851–1931), that species of fish that live in colder waters tend to have more vertebrae in their backbones than do species that live in warmer waters. More recently, the same correlation has been observed for increases in salinity; that is, the more saline the environment, the more vertebrae in the backbones of the species which live there.

joule a measurement of energy, defined as one newton of force acting over 1 meter of distance. A newton is the force required to accelerate a 1-kilogram weight at a rate of 1 meter/sec. Thus a joule can also be defined as the energy required to accelerate a 1-kilogram weight from a speed of 0 meters/sec to a speed of 1 meter/sec in a distance of 1 meter. It is equivalent to approximately .7376 FOOT-POUNDS.

JTU *See* JACKSON TURBIDITY UNIT.

jungle *See under* RAIN FOREST BIOME.

juvenile water water formed from chemical reactions within the Earth, which has not yet existed as surface water or as atmospheric water vapor. The amount of juvenile water produced each year by the Earth is not known, but it is thought to be extremely small.

Kaibab Plateau a large, forested plateau in northern Arizona. It forms part of the North Rim of the Grand Canyon of the Colorado River and is the location of the Grand Canyon Lodge. The Kaibab Plateau is best known to environmental science as the scene of one of history's most disastrous experiments in PREDATOR CONTROL. In the mid-19th century, the Kaibab supported a herd of approximately 30,000 deer, in balance with a free predator POPULATION made up principally of cougars, coyotes, and wolves. Beginning about 1885, competition from domestic livestock began to shrink the deer herd, which was down to about 5,000 by 1906. In that year the domestic stock was moved out and an aggressive predator shooting and trapping program was begun that virtually eliminated all predation on deer on the plateau. Within 18 years, the deer population had increased more than 20 times, to over 100,000, and the amount of FORAGE produced by the plateau was not adequate to support it. Sixty thousand deer died in the winter of 1924. The population crash continued over the next seven years, until the herd was down under 20,000. In 1931 the predator-control program was suspended. By 1939 the deer herd had stabilized at 10,000 animals and was back in balance with its predators. The Kaibab example is probably the single best-known incident of wildlife mismanagement in North America, and has nearly become a textbook cliche.

kame in geology, a steep-sided, conical hill of glacial debris. Kames are formed in two principal manners. In the most common means of formation, a depression on a GLACIER'S surface will collect windblown sand, dirt, and small gravels. When the glacier melts, this collection of materials is left behind as a kame. Alternatively, a kame may be formed from materials transported across the surface of a glacier by a stream and deposited at the edge of the glacier where the stream passes onto solid ground. *Compare* MORAINE; ESKER; DRUMLIN. *See also* GLACIAL LANDFORMS.

kaolin (kaolinite) an extremely soft, friable (*see* FRIABILITY) stone composed principally of compressed CLAY. The term is sometimes also used as a synonym for clay, especially the fine, pure anhydrous clay required for pottery manufacture. *See also* FULLER'S EARTH.

karst topography a type of landform resulting from the solution of CALCIC ROCKS such as LIMESTONE, MARBLE, or MARL by GROUNDWATER. The phenomenon is named for the Karst region of southern Yugoslavia, but is widespread throughout the world wherever large deposits of calcic rocks exist. Typical features of karst topography include the existence of solutional caverns (caves) beneath the ground surface; the presence of numerous SINKHOLES, caused either by the collapse of cavern roofs or by solutional widening of vertical cracks in the rock; *blind valleys,* where streams flow up to the base of vertical walls of stone and disappear under them; *Lapies,* a knobby, gnarled-appearing surface rounded and hollowed by solution, probably beneath a soil OVERBURDEN that was later stripped off by erosion; and *terra rossa,* remnant soil deposits that have the appearance and consistency of LATERITIC SOILS and probably are caused by excessive leaching of minerals due to the porosity of the underlying rock. *Compare* PSEUDOKARST; THERMOKARST.

Keep America Beautiful, Inc. an industry-sponsored public relations organization founded in 1953 by container and packaging corporations to educate the public about litter prevention. Keep America Beautiful also works with communities and provides training in the Keep America Beautiful System, a method for improving waste handling practices within these communities. Address: 1010 Washington Boulevard, 7th Floor, Stamford, CT 06901. Phone: (203) 323-8987.

kenaf a tall, fibrous, reedlike annual plant related to okra that can be processed to make paper. A "tree-free" substitute for book-text and stationery, kenaf paper is nevertheless expensive and has not captured an appreciable share of market from wood-pulp paper. Some environmental publications use kenaf not only because it saves trees and forests from the ravages of CHIPPING MILLS, but also because kenaf needs little or no pesticide to maintain high yields, requires no soil nitrogen, uses fewer chemicals and energy in manufacturing, and has no major water pollution impacts as do pulp and paper facilities.

kerosene (kerosine) the fifth fraction of PETROLEUM (*see* DISTILLATION), ranking between the petroleum ethers and the oils. It is two cracking steps heavier than GASOLINE. Kerosene is a light, volatile, yellow-to-white, extremely mobile liquid with a characteristic sweetish odor. It is a mixture of several COMPOUNDS, among them *n*-dodecane, BENZENE, and naphthalene. It has a boiling point of 175°–375°F and a FLASH POINT of 150°–185°F. Kerosene is used primarily as a fuel for lanterns and for space heaters. The mixture is highly toxic (LD_{50} [orally, rabbits]: 28mg/kg), causing vomiting and diarrhea when ingested and headaches, drowsiness and coma when inhaled. Incomplete combustion produces CARBON MONOXIDE, making the use of kerosene-burning equipment hazardous in closed spaces. Incomplete combustion also produces small amount of ETHYLENE, which can speed the ripening of fruits; hence, kerosene stoves are often used in commercial fruit-ripening rooms.

kettle in geology, a large, roughly circular depression in a glacial OUTWASH DEPOSIT, often filled with water to form a *kettle pond* or a *kettle lake*. Kettles are formed when blocks of ice fallen from the face of a GLACIER become surrounded by outwash deposits. When the ice melts, the space it occupied is left behind as the kettle. Lakes formed in this manner are extremely common in Michigan, Wisconsin, Minnesota, and other areas along the southern margin of the last advance of the North American CONTINENTAL GLACIER. Some can be quite large. Albert Lea Lake, a kettle lake in southern Minnesota, covers 26,000 acres and has approximately 70 miles of shoreline. *See also* GLACIAL LANDFORMS.

key-industry animal in ecology, a SPECIES (or one of a group of species) of HERBIVORE that is so numerous that it forms the basis for all or most of the FOOD CHAINS in an ECOSYSTEM. Key-industry animals are generally small animals such as mice or insects, although the role is occasionally filled by large UNGULATES. In the sea, the key-industry animals almost always turn out to be COPEPODS. Birth rates among key-industry animals are always high in order to maintain the species' numbers despite high predation (*see* PREDATOR). *See also* R-SELECTED SPECIES.

keystone species those species that interact with and support a variety of other species by playing a key role in the natural dynamics of an ecosystem. Alligators, for example, are a keystone species in the FLORIDA EVERGLADES because they dig "gator holes" in the dry season which, filled with water, provide protection to many species that might otherwise not be able to survive.

kilowatt-hour (KWH) a measurement of electrical power, equivalent to 1,000 watts applied for one hour (1 watt = 1 JOULE applied for one second). One kilowatt-hour is equivalent to 3413 BRITISH THERMAL UNITS. The kilowatt-hour has become the standard unit for measuring domestic electrical usage, and forms the basis for nearly all billing systems for electricity.

kinesis in the biological sciences, movement of an ORGANISM that is random in direction but whose intensity is controlled by some ENVIRONMENTAL FACTOR such as light intensity, chemical concentration, etc. Kinesis is common among MICROORGANISMS, especially those in which POPULATION densities are generally high enough that randomly directed motions will bring a fairly high number of individuals into a favorable position with respect to the stimulus, even if that number represents only a small percentage of the actual population. *Compare* TAXIS.

kingdom in the biological sciences, any of the five major classifications of living things. The kingdom is the most inclusive category in TAXONOMY. Until about 1970, all ORGANISMS were classed into one of two kingdoms; the plants (Plantae) or the animals (Animalae). It is now thought that this classification presents too many borderline cases where it is difficult to tell which kingdom an organism should be classed in. Current biological thinking assumes, in addition to the plants and animals, three more kingdoms: the Fungae (mushrooms, molds, etc.; *see* FUNGUS); the PROTISTAE (ALGAE, PROTOZOANS, and other complex single-celled, colonial or [occasionally] multicellular but poorly differentiated organisms); and the MONERA (bacteria and blue-green algae; *see* BACTERIUM; CYANOPHYTE).

kinin *See* CYTOKININ.

knot in the lumber industry, an interruption in the GRAIN of a piece of wood caused by the inclusion of the base of one of the tree's branches. Knots are generally round and dark in color. The grain divides around them in a curving pattern, and the knots themselves are extremely hard and often contain large amounts of PITCH, making them difficult to saw through. They also weaken the wood structurally. For all these reasons, they are avoided as much as possible in lumbermaking (*see* CLEARLUMBER), although in some woods (e.g., knotty pine, birdseye maple) they are sought for their decorative effect.

Knutson-Vandenberg Act federal legislation to provide for the restoration of cutover federal land, signed into law by President Herbert Hoover on June 9, 1930. The act authorizes the FOREST SERVICE, at its discretion, to require the purchaser of a NATIONAL FOREST timber sale to deposit into the federal treasury funds sufficient to reforest the sale area, including both replanting and TIMBER STAND IMPROVEMENT after the new STAND is established. Recent revisions allow the agency to require the deposit of funds for other environmental restoration work, such as wildlife mitigation, as well. These "K-V funds" cannot be levied in an amount greater than the average cost of restoring similar land on a per-acre basis. They are not considered part of the sale receipts, and hence are not subject to revenue sharing with the county in which the sale is located. Surplus K-V funds go into the general Forest Service budget. The Knutson-Vandenberg Act also authorized the appropriation of money from the Treasury to build and operate tree nurseries and allowed the use of National Forest SEEDLINGS to reforest NATIONAL PARK lands following forest fires.

Koeppen climate system a system for classifying world climates, devised around 1918 by the German meteorologist Wladimir Koeppen (1846–1940) and widely adopted since. Born out of an attempt to correlate differences in climate with observable differences in vegetation type (*see* BIOME), the Koeppen system utilizes variations in temperature and precipitation to classify the world into five broad climatic regions, each designated by a capital letter. Type A climates are tropical, while type E climates are polar. Both types exhibit little temperature variation during the year, although A is warm and E is cold. Types B, C, and Dare the mid-latitude climates. These are strongly seasonal, with type B being arid (DESERT and STEPPE), type C humid with mild winters (coastal), and type D humid with harsh winters (continental). These five basic types are then broken down further by the presence or absence of such factors as wet summers, wet winters, high or low maximum and minimum temperatures, and so on. The subtype is indicated by the presence of a lower-case letter (or sometimes two) next to the capital letter indicating the basic climatic type. For example, the letter s indicates a dry summer (wettest winter month at least three times the precipitation of the driest summer month); w indicates a dry winter; f indicates roughly evenly spaced precipitation around the year; and a, b, c, and d indicate temperature regimes, with a being the warmest and d being the coldest. The complete system is given in the table on the previous page.

Koeppen Climate System

Major Climate Types and Subtypes

A:	Tropical Humid	
B:	Tropical and Mid-Latitude Dry	
		BS: Steppe
		BW: Desert
C:	Coastal (Mid-Latitude Humid, mild winters)	
D:	Continental (Mid-Latitude Humid, cold winters)	
E:	Polar	
		EF: Ice Cap
		ET: Tundra

Descriptive Modifiers

a:	average temperature of warmest month>22°C
b:	warmest month <22°C, but at least four months >10°C
c:	warmest month <22°C, but at least one month >10°C
d:	average temperature of coldest month below –38°C
f:	rainfall approximately even year-round
g:	warmest month precedes summer solstice
h:	hot (average annual temperature >18°C)
i:	uniform annual temperature (variation <5°C)
k:	cool (average annual temperature <18°C)
k':	cold (average temperature of warmest month <18°C)
m:	monsoon (rainfall year-round but with strong peaks)
n:	foggy
s:	dry season in the summer
s':	maximum rainfall in the autumn
t':	warmest month in the autumn
w:	dry season in the winter
w':	dry season in the autumn
w":	two distinct dry seasons
x:	maximum rainfall late spring or early summer

Samples of Koeppen climate designations from various parts of the United States include Cfa (coastal climate, precipitation roughly evenly spaced, generally warm summers: Atlanta, Georgia), Cs (coastal climate, dry summer: Seattle, Washington), Dfb (continental climate, precipitation roughly evenly spaced, cool but not cold summers: Duluth, Minnesota), and Aw (tropical climate, dry winters: Miami, Florida).

Krakatoa a small volcanic island in the Sunda Strait between the islands of Java and Sumatra in Indonesia. During the night of August 26–27, 1883, pressures built up behind a plug of MAGMA in the throat of the Krakatoa VOLCANO caused the island to explode, reducing its area from 18 square miles to 6 and producing a tidal wave (*see* TSUNAMI) that killed an estimated 36,000 people and destroyed millions of dollars' worth of property in the western Pacific basin. The sound of the explosion was heard 3,000 miles away, and dust blown into the stratosphere dimmed the sun and raddened sunsets for three years. The island is less important to environmental science for the violence of its explosion, however, than it is for the near-laboratory conditions that explosion produced for the study of animal and plant DISPERSAL and ecological SUCCESSION. Three years after the explosion, the only life forms present—aside from an occasional spider or insect—were MONERA (bacteria and blue-green algae). Ten years later, scientists found the shores ringed by typical-beach vegetation, including coconut palms. Inland there were only algae, mosses, and a few hardy seed plants such as grasses. By 1906, typical rain forest vegetation and animal life were beginning to be established, and by 1930—less than 50 years after the explosion—the island once again contained a complete rain-forest ECOSYSTEM.

krummholtz a growth form taken by CONIFERS at or near TIMBERLINE in the Arctic and on mountain slopes. Trees growing as krummholtz are dwarfed, with a stem only a few inches through often representing several hundred years of growth. They develop in a prostrate manner rather than vertically, often forming dense mats of vegetation 1–2 feet high over large areas of land. The needles are shortened and lie close together on the twigs. The krummholtz form appears to be a result of several environmental factors common at the treeline, including high winds, short growing seasons, thick winter snowpacks, and poorly developed soils. The most common krummholtz genera are the firs (*Abies*), spruces (*Picea*), and hemlocks (*Tsuga*). It has occasionally been suggested that some SPECIES in these genera have evolved separate RACES for the krummholtz form and the erect form, but this concept has never been proved. *See also* TUNDRA BIOME.

K-selected species in population biology, any animal SPECIES whose POPULATION size is limited primarily by DENSITY-DEPENDENT FACTORS and which therefore tends to expand its numbers up to the CARRYING CAPACITY of the environment (represented in population equations by the letter *K*). K-selected species tend to be large, mobile, adaptable animals with the ability to exploit more than one environmental NICHE. Since such adaptability requires learning, these animals also tend to be intelligent and to have a long juvenile training period, which means (since the young must stay with their mothers for several years) that the birth rate tends to be low and a high proportion of young must survive in order to perpetuate the species, a reproductive pattern known as the *K strategy.* Bears are probably the best-known K-selected animals. *Compare* R-SELECTED SPECIES.

K strategy *See* K-SELECTED SPECIES.

K-V funds *See under* KNUTSON-VANDENBERG ACT.

Kyoto Protocol potentially the most significant and far-reaching international environmental policy initiative to date. The Kyoto Protocol was an outgrowth of the Framework Convention on Climate Change developed at the RIO EARTH SUMMIT in 1992 and subsequently ratified by 175 countries. The Framework Convention called for voluntary reductions of greenhouse gases (primarily carbon dioxide, methane, and nitrous oxide) to 1990 levels to alleviate global warming. It soon became clear, however, that voluntary reductions would not be effective. Accordingly, in 1997 representatives from some 160 countries gathered in Kyoto, Japan, to work out mandatory reductions, although developing nations, while encouraged to comply, would not be required to meet the same standards as developed nations—for reasons of economic hardship and because their contribution to global warming was minor. The United States' requirement under the protocol was to reduce average greenhouse gas emissions 7% below the 1990 level for the period 2008–12. Before the protocol can take effect, however, at least 55 countries must ratify the agreement. As of 1999 only seven had done so, with most nations waiting for the United States to take the lead, which the U.S. Senate has thus far refused to do. In fact, its first vote was 95-0 against the protocol. The reasons given for the senators' reluctance ranged from concerns about the emissions reduction requirement (too low for some, too high for others) to an objection to the flexibility in reduction rates allowed to Third World nations. The protocol hoped to overcome this last problem by allowing developed nations to meet their targets by "buying" offsetting reductions in less energy-efficient countries; and in response to this feature the WORLD BANK created a "market" for the buying and selling of emissions credits. Another feature of the protocol was a provision allowing for emission-level targets to be reduced if changes in land use created or improved so-called "sinks." A new carbon sink, for example, would be a reforestation project whereby new trees would absorb and sequester some of the CO_2. Although the protocol has extremely complex procedures, coupled with quite minimal requirements for limiting greenhouse gas emissions, it is expected to be ratified eventually, as global warming becomes more and more evident. When that takes place, a beginning point in the international control of global warming will have been achieved, with some hope of reducing it significantly in the long-term future. *See also* GREENHOUSE EFFECT; GLOBAL CHANGE.

lacustrine deposits in geology, SEDIMENTS, SANDS, and GRAVELS deposited in the beds of LAKES. The lakes may or may not be present any longer. Lacustrine deposits tend to be composed of fine, rounded particles, evenly bedded. Ripple marks, if present, are symmetrical. Any gravels are well sorted and rounded, as would be found on a rocky beach. *Compare* FLUVIAL DEPOSITS; GLACIAL DRIFT.

Lady Bird Johnson Wildflower Center promotes the preservation and use of native plants as landscape materials by educating the public about the natural beauty, environmental benefits, and economic value of America's native wildflowers. The center was founded in 1982 by former first lady LADY BIRD JOHNSON. Membership (1999): 19,000. Address: 4801 La Crosse Avenue, Austin, TX 78739. Phone: (512) 292-4200.

LAER *See* LOWEST ACHIEVABLE EMISSION RATE.

lag gravel in geology, pieces of GRAVEL left behind on a wind-scoured surface (*see* WINDSCOUR) because they were too big for the wind to carry off. Lag gravel often occurs as widely scattered stones from roughly the size of a nickel upward, each sitting on a low pedestal of smaller particles it has protected from the wind. It is most characteristic of the surface of unsorted deposits such as glacial MORAINES. *See also* VENTIFACT; DESERT PAVEMENT.

lagoon (1) any relatively small, shallow body of water, especially one that borders a sea or a large lake and is connected to it by a restricted passage. Lagoons are common behind sand dunes and within coral atolls. Because they are warm and protected from winds and wave action, they are usually richer in life than the larger body of water nearby.

(2) in sanitary engineering, a POLISHING POND.

lake any body of standing water on the land. Usually, a lake is larger than a POND, although the distinction between them is by no means hard and fast, and it is probably better to think of ponds as a special class of lakes rather than as something separate (but see the discussion under *pond*). The depressions occupied by lakes are formed in a variety of ways. They may be basins scooped out by GLACIERS or blocked by glacial deposits such as MORAINES or KAMES; they may be valleys blocked by landslides or by alluvial dams (*see* ALLUVIUM); they may be GRABENS or other STRUCTURAL BASINS; or they may be the result of any of a large number of other hollow-forming geologic processes, such as volcanic activity, meteor impact, KETTLE formation, windhollowing, etc. From a geologic standpoint, most lakes are extremely short-lived, succumbing rapidly to the combined processes of EROSION and DEPOSITION, although very large and deep lakes, such as Siberia's Lake BAIKAL, may last for several million years (the American Great Lakes, by comparison, are less than 10,000 years old). Though they appear static, lakes are actually dynamic features. The water in them is constantly being exchanged. Inflow takes place from streams, overland RUNOFF, and infilling from upslope AQUIFERS (most lakes are a fairly accurate representation of the local GROUNDWATER level, representing those spots where the land surface dips below the WATER TABLE). Outflow results from a combination of EVAPORATION, loss to downslope aquifers, and (usually) overflow through overland channels. If evaporation dominates, the lake will usually become brackish (that is, contain a high proportion of dissolved MINERALS and SALTS; *see* ALKALI FLAT), but if overflow and inground flow predominate, the lake water will usually remain fresh. Lakes are important in the environment, both because of the large COMMUNITIES of living things they support and because of their effect on watercourses. Besides helping to regulate streamflow (*see* HYDROLOGIC EQUATION), they serve as SETTLING BASINS and as oxygenation devices, so that the water leaving a lake is almost always cleaner and less turbid (*see* TURBIDITY) than that which entered it. *Compare* LAGOON. *See also* FIRST-ORDER LAKE; SECOND-ORDER LAKE; THIRD-ORDER LAKE; EUTROPHICATION; MESOTROPHIC LAKE; OLIGOTROPHIC LAKE; TROPHIC CLASSIFICATION SYSTEM; PLUVIAL LAKE; PATERNOSTER LAKES.

lake-effect snow in meteorology, the additional snowfall on the leeward side of a large lake such as the Great Lakes, Lake Winnipeg, Lake BAIKAL, and others, caused by the presence of the lake. As an AIR MASS passes over the (relatively) warm water, it picks up both heat and moisture. Upon reaching the far side, it is cooled again as it passes over the land, causing the moisture picked up over the water to condense out and fall as snow. Lake-effect snow has only about half the water content of ordinary snow ("storm snow"), and is therefore considerably lighter and dryer, making it excellent as a base for skiing. The phenomenon is especially pronounced in areas where the air must rise to cross hilly or mountainous land immediately after crossing the water, thus making it subject to adiabatic cooling (*see* ADIABATIC TEMPERATURE CHANGE). The 700-foot-high ridge of the Keweenaw Peninsula of Michigan, which juts northward into Lake Superior, has been known to receive over 300 inches of (largely lake-effect) snow in a single winter.

land in the context of environmental planning and ecology, a quite inclusive term. "The land" is often used to stand for all the natural constituents of land, including soils, waters, plants, and animals and their interactions with one another and with humans and human settlements. *See also* ALDO LEOPOLD.

Land and Water Conservation Fund (LWCF) a funding source, now administered by the NATIONAL PARK SERVICE, for the acquisition and maintenance of park and other conservation lands by the federal government and, through a matching grant program, state and local entities. Established by Congress in 1964, the LWCF is charged with preserving "irreplaceable lands of natural beauty and unique recreation value" with funding derived from royalties from offshore oil and gas leases. By the late 1990s the LWCF had been used to acquire more than 7 million acres of park and open space land. Originally authorized at $300 million a year, it was later limited to $900 million, although this amount was rarely appropriated. In the 1980s and most of the 1990s, the LWCF program was severely cut back, with the 1996 and 1997 appropriations only $140 million for each of those years. In 1998, however, Congress approved a total of $699 million, and in 1999, under a Clinton Administration "Lands Legacy" initiative, a total of $1 billion was requested. Of this amount, $558 million would go to state and local governments, private land trusts, and nonprofit groups for GREENWAYS, wetlands, and wildlife sanctuaries; for open space and "smart growth" planning; for CONSERVATION EASEMENTS; and for the revitalization of urban parks. This approach was designed to empower local officials and civic leaders to be imaginative and flexible in developing creative, socially useful ways to provide for park and recreation needs and for the preservation of natural areas.

land base in land-use planning, the area of land available for planning purposes, either for all resources uses (*total land base*) or for a specific resource use such as logging (*commercial land base*), recreation (*recreation land base*), etc.

land breeze in meteorology, a local wind that blows from the land toward the sea. Land breezes are the opposite of SEA BREEZES. They result from the fact that water cools much more slowly than does land, hence during shifts from warmer to cooler temperatures the air over water remains warm longer than that over the adjacent land. The warm air over the water then rises, pulling the cool air over the land in to replace it. Land breezes usually occur at night during the warm season, beginning shortly after sundown. *See also* LOCAL LOW.

land ethic a proposition first advanced by ALDO LEOPOLD in the early 1930s and refined in his book *A Sand County Almanac* (1949) that ethical behavior should not be confined to the social interactions of humans, but should also include those between humans and "the LAND." "That land is a community," wrote Leopold, "is the basic concept of ecology, but that land is to be loved and respected is an extension of ethics." Leopold suggested that the use of land should be governed by ethics, not merely economics. He saw the land ethic as part of a historic evolution of ethical behavior, beginning with relationships within a family, then tribe, then race, then the nation-state, and finally to all humanity. The extension of ethics into "man's relationship to land" should, he thought, be part of a normal progression given an ecological understanding of humans' place in nature. During the 1960s the land ethic became very much a part of a new conservation consciousness and a new appreciation of ecology. The land ethic (or *environmental ethic,* as some prefer) is now taught in schools and colleges, has scholarly journals devoted to it, and is the subject of many important books—for example, Roderick Frazier Nash's *The Rights of Nature: A History of Environmental Ethics* (1989). Although Leopold was not religious in any traditional sense, the land ethic has been exhaustively treated in ecotheological books as well. *See also* RELIGION AND THE ENVIRONMENT.

landfill *See* SANITARY LANDFILL; SECURE LANDFILL; DREDGE-AND-FILL.

landform any distinctive geomorphological feature, such as mountains, stream valleys, deserts, etc.

landing in logging, the area where logs are loaded onto a truck for transport to the mill. Landings must be large enough for a logging truck to turn around and for the loading equipment to maneuver, and they must be relatively flat. In flat terrain they will be located in the center of the STRIP to be logged. In hilly terrain it is generally best to locate them near the top of the strip, as YARDING uphill is generally easier and less environmentally disruptive than yarding downhill.

Land Institute, The a research and education organization founded in 1976 by WES JACKSON to "develop an agricultural system with the ecological stability of the prairie and a grain yield comparable to that from annual crops." The institute has several ongoing programs at its research sites in Kansas, including the Natural Systems Agriculture program, graduate and undergraduate study opportunities, and a rural community studies program. Address: 2440 E. Water Well Road, Salina, KS 67401. Phone: (785) 823-5376. Website: www.landinstitute.org.

landlocked species a SPECIES, RACE or POPULATION of a normally marine or anadramous GENUS (*see* ANADRAMOUS SPECIES; MARINE SPECIES) that lives its entire life cycle in freshwater. Landlocking of marine species usually results from the isolation of an arm of the sea by DUNE building or TECTONIC ACTIVITY, and the subsequent conversion of the isolated impoundment from saltwater to freshwater due to the inflow of rivers. ORGANISMS trapped by the isolation event either adapt to a freshwater life-style or die. Anadramous species may become landlocked in the same manner, but a more common cause is genetic suppression of the urge to migrate; since these organisms can live in either fresh or salt water, individuals whose drive to reach the sea is reduced may find the sea-like conditions in a large lake adequate to fulfill their needs. If a large enough breeding population of these individuals becomes established, it may eventually become reproductively isolated, forming a new species. Examples of landlocked species include the kokanee, a species of salmon that lives in the large lakes of the northwestern United States and western Canada; *Carcharhinus nicaraguensis,* a landlocked shark that inhabits Lake Nicaragua in Central America; and the Great Lakes populations of the lamprey.

Land Management, Bureau of *See* BUREAU OF LAND MANAGEMENT.

land reform traditionally, the redistribution, often by insurrection, of large tracts of farm and forest land in feudal-like ownership of the wealthy to peasants and others actually working the land. In the United States no large-scale violent land-reform movement has taken place (as it has in South America), although the distribution of public land to individual farmers, called for by Thomas Jefferson, was finally achieved with the HOMESTEAD ACT of 1862.

Landsat *See* REMOTE SENSING.

landscape architecture originally the "laying out of grounds" and later "landscape gardening." Prior to the 19th century, these terms were used to refer to the rearrangement of natural features and the introduction of constructed facilities, usually in the private gardens and estates of the wealthy. During the 18th and early 19th centuries, landscape design was considered a "sister art" to poetry and painting. The term *landscape architecture* came into vogue in the mid-1800s in the United States with FREDERICK LAW OLMSTED, who in effect professionalized garden design and, while not forsaking the commissions of the wealthy, championed the creation of public parks and other spaces for civic improvement, most famously Central Park in New York City, which he designed with his partner Calvin Vaux. Since then landscape architecture has become concerned with a wide range of land use planning and ecological issues as well as horticulture and the building of garden structures.

landscape ecology a relatively new scientific discipline concerned with the study of natural systems, structures, and processes at the larger, heterogeneous "landscape" scale as opposed to the more traditional natural "community" or ecosystem scale of analysis. A vocabulary for landscape ecology is already well developed, which deals with "patches," "corridors," "matrices," and other terms used to describe the elements of a landscape that are structurally interconnected. Patches are remnant natural areas, such as a woodland surrounded by farm fields and housing, for example. Corridors are streamways, hedgerows, and other linear elements that connect patches. The whole of the landscape is a matrix. A "disturbed" matrix would be, say, a semirural subdivision; an "undisturbed matrix" would be a pristine wilderness area. Landscape ecology has great utility in land-use and conservation planning in large regions with a wide variation of elements in their structure which might include shopping malls, highways, housing, and a scattering of intact natural areas. To conserve indigenous wildlife in such a developing area, a landscape structure should include corridors needed to link remnant patches of habitat in a disturbed matrix so that certain species do not become isolated in small, inbred populations that would soon die out.

landscape management zone in land-use planning, an area of land whose scenery is recognized as its most important value. Classifying an area as a landscape management zone does not preclude other resource activities such as logging or mining. It does, however, restrict these activities to forms that will not significantly alter the area's scenic character. *See also* VISUAL RESOURCE MANAGEMENT SYSTEM.

landslide *See* MASS WASTAGE.

land stewardship the preservation and protection of land resources for posterity under the biblical concept of stewardship. *See also* STEWARDSHIP; RELIGION AND THE ENVIRONMENT.

land survey *See* CADASTRAL SURVEY.

land trust in the context of land use planning, a nonprofit civic organization that acquires private land or rights in land for conservation purposes. The largest land trusts in the United States are THE NATURE CONSERVANCY and the TRUST FOR PUBLIC LAND. However, there are many local and regional land trusts as well. Land trusts can manage the land they purchase themselves, resell it while maintaining a CONSERVATION EASEMENT to assure that the land remains protected, or turn it over to a government agency for park and open space purposes. *See also* LAND TRUST ALLIANCE.

Land Trust Alliance, The provides services and resources to expand the skills and effectiveness of both regional and local trusts. Through public education and policy changes, the alliance, founded in 1982, works to increase awareness and conservation of diminishing land resources and to promote the role of land trusts in saving these lands. Publications include *Exchange: Conservation Easement Handbook; Conservation Easement Stewardship Guide; Starting a Land Trust;* and *National Directory of Conservation Land Trusts.* Membership (1999): 2,100. Address: 1319 F Street NW, Suite 501, Washington, DC 20004. Phone: (202) 638-4725. Website: www.lta.org.

land use constraints in land use planning, any factor that limits options for the planning or management of a piece of land. Land use constraints fall into two broad categories. *Intrinsic constraints* are physical constraints arising from the character of the site, such as steep topography, poor soils, or unfavorable ASPECTS. *External constraints* are constraints arising from laws or regulations, such as WILDERNESS designation, riparian management laws as (*see* RIPARIAN LAND), designation as a LANDSCAPE MANAGEMENT ZONE or other special use area, or the presence of legally protected resources such as archaeological remains or ENDANGERED SPECIES.

land use planning the physical planning of municipalities, metropolitan areas, and large regions to provide for transportation, public health, aesthetics, environmental quality, and other needs of the population. City planning is as old as civilization, with cities such as Nineveh, Rome, and others given plans for the layout of streets. In more recent times, land use planning has extended its concerns to the natural environment, with major contributions in the United States made by Ian McHarg, Philip Lewis, and others who have developed planning techniques based on natural processes. A major concern of planners today is URBAN SPRAWL.

La Niña The opposite of EL NIÑO.

lapse rate in meteorology, the rate at which the temperature of the ATMOSPHERE changes in the vertical direction. It differs from the ADIABATIC RATE in being a measure of different parts of a still AIR MASS rather than the same part of a moving one. In dry air, the lapse rate is about 3.6°F for every 1,000 feet of elevation change. In general, the higher the lapse rate of an air mass, the less stable the air mass will be (*see* ATMOSPHERIC STABILITY).

larva (plural: larvae) the young of an ORGANISM in which the juvenile form differs substantially from the adult form. The presence of a larval stage is characteristic of insects, shellfish, crustaceans, and other INVERTEBRATES. Examples include caterpillars (butterfly or moth larvae), grubs (beetle larvae), and maggots (fly larvae).

latency period the time between exposure to a TOXIN, CARCINOGEN, or other environmental stress, and the development of symptoms resulting from that exposure. *Compare* INCUBATION TIME.

latent heat in the physical sciences, the heat required to change a solid to a liquid, or a liquid to a vapor. Since the temperature of a substance does not change as its state is changed, the heat is thought of as hidden ("latent") within the substance. It will be released as "real" (temperature-affecting) heat when the state change reverses. Latent heat is important in weather and climate patterns, and is the prime method by which heat is transferred from the tropics toward the poles. Large quantities of water are evaporated from the tropical oceans, taking up latent heat in the process, and the resultant mass of water vapor is driven north by convection (*see* HADLEY CELLS). Condensation then releases the heat into the atmosphere at a new, more northerly location.

lateral *See* SEWER.

lateral bud in the biological sciences, a bud that is produced on the side of a plant stem instead of at its tip. *Compare* TERMINAL BUD.

laterific soils the heavy, reddish, nutrient-poor soils found commonly under tropical rain forests (see RAIN FOREST BIOME). Most of the usual soil-supplied NUTRIENTS are tied up in the abundant plant and animal life in the rain forest ECOSYSTEM or have been leached from the soil by the heavy-rains. Organic HORIZONS are almost non-existent. When exposed to the sun, these soils bake in a few years into a hard, rock-like substance known as *laterite* that will not support plant life. Hence, tropical forests cannot be cleared to any large degree for modern agricultural practices, and the much-derided SLASH-AND-BURN system actually turns out to be about the best possible agricultural system for the laterite regions. Lateritic soils are widespread in most of the Earth's tropic regions, including Southeast Asia, Central Africa, Central America, and the Amazon Basin. *See also* TROPICAL FOREST BIOME.

latitude the distance a point lies above or below the equator, measured in degrees, minutes, or seconds of arc. There are 90 degrees of latitude between the equator and either pole. Each degree is divided into 60 minutes and each minute, into 60 seconds. One degree of latitude is equal to approximately 110 kilometers (67 miles). *See also* PARALLEL. *Compare* LONGITUDE.

lava in geology, molten rock flowing across the surface of the earth. The term is also used for the rock that results when the flow cools and solidifies. Nearly all lava flows consist of MAGMA that has reached the earth's surface through volcanic vents or fissures (*see* VOLCANO). The few exceptions to this include rock melted by the impact of a large body (such as a meteorite) and rock melted by intense heat created by human activities (such as metal smelting or the explosion of a nuclear weapon). The forms taken by lava are similar to the forms taken by other flowing liquids. Thus there are *lava fountains* where lava spurts from a vent under pressure; *lava lakes* where the material pools in a depression on the surface it is flowing across; *lava falls* where the flow makes an abrupt descent, as over a cliff; and so on. *See also* LAVA TUBE; LAVA BOMB; PILLOW LAVA; BASALT; RHYOLITE; ANDESITE; AA; PAHOEHOE; PSEUDOKARST.

lava bomb a discrete "drop" of LAVA that is thrown through the air by the eruption of a VOLCANO and has become either partially or completely solidified before it hits the ground. Lava bombs are smooth, pear-shaped structures up to 2 feet or more through. Very small lava bombs are known as *lapilli* (singular *lapillus*). These are about the size of drops of water. If made of obsidian (volcanic glass), they are sometimes called *Pele's tears,* after the Hawaiian goddess of volcanoes.

lava tube a linear cavity in a LAVA flow, formed when the surface of the flow cools and solidifies while the interior is still fluid. If there is an opening at the lower end of the flow, the fluid interior can run out, leaving the crust behind as a hollow shell. Lava tubes are usually sinuous, following the direction of flow of the stream of lava. They show a number of interesting features, including *flow lines* in the floors, walls and ceilings; *flow ledges* along the walls, where the flow level stabilized long enough to begin to solidify at its edges, creating a narrow flat shelf with (usually) an upraised rim; and small stalactites called *lavacicles,* which form as hot gases pass through a newly formed tube, partially remelting the ceiling so that it begins to drip toward the floor. Lava tubes may be a few centimeters to 30 meters (95 feet) or more in diameter. The larger ones often form complex, many-chambered caverns with a mile or more of traversable passageway (Ape Cave, in Mt. St. Helens National Monument, Washington, is slightly over 2 miles long. Rumors of longer ones persist in Hawaii). *See also* PSEUDOKARST.

lay in logging, the position of a felled tree after it has come to rest on the ground. Lays are classified as "good" or "bad" depending upon how much damage has occurred to the tree; how easy it is to buck and limb (*see* BUCKING; LIMBING); and how accessible it is to YARDING equipment. *Compare* LEAD; LENGTH OF GROUND.

LC_{50} *See* LD_{50}, LC_{50}.

LD_{50}, LC_{50} in toxicology, the amount of a substance that will kill half (50%) of a POPULATION of test animals within a given time after they are exposed to it.

Depending upon the test and the type of data sought, this "given time" may be anywhere from four to 96 hours. The time is sometimes designated by a preface to the abbreviation: "$96LD_{50}$," for example, would refer to a 96-hour test. LD_{50} ("lethal dose 50") refers to the amount of the material actually ingested, breathed in, absorbed through the skin, or otherwise taken within the animal's body, while LC_{50} ("lethal concentration 50") refers to the concentration of the material present in the animal's environment, that is, in the air it breathes or the water it swims in. For the LD_{50} and aerial LC_{50} tests, the test animals are generally white rats, although mice, guinea pigs and other animals are used on occasion; in water, the test animals are usually trout. Both the test animal and the means of transmission of the chemical (orally, topically, etc.) should normally be given along with results of the test. The animals to be used and the length of time to run the test may be specified by law or regulation. For instance, the Department of Transportation category of "class B poisons" are separated from the less toxic "class A poisons" by the requirement that the poison must have a $48LD_{50}$ of 50 milligrams or less per kilogram of body weight when administered to a group of 10 or more white rats with weights of between 200 and 300 grams each.

leachate a liquid containing dissolved materials that it has picked up by percolating through ("leaching through") a substance that contains the materials in concentrated form (*see* PERCOLATION). Two forms of leachate are particularly hazardous from an environmental standpoint. The first is the leachate formed when rainwater leaches through a HAZARDOUS WASTE dump or landfill (*see* SANITARY LANDFILL). It contains high levels of any soluble materials found in the dump, and may carry those materials down to the WATER TABLE, leading to GROUNDWATER contamination (*see* HAZARDOUS WASTES). The second type of environmentally dangerous leachate comes from the processing of mineral ORES, which often involves leaching the mineral out by passing a SOLVENT through piles of pulverized ore. These solvents, which remain in the leachate after it has passed through the ore, are often hazardous materials such as hydrochloric acid or CYANIDE. *See* HEAP LEACHING; LIXIFICATION.

lead the 82nd element in the ATOMIC SERIES; atomic weight 207.2, chemical symbol Pb. Lead is the most common of the so-called HEAVY METALS and was among the few metals known to the ancient world, where it was widely used for piping and containment vessels of various kinds, though it could not be used for cookware due to its low melting point (327.4°F, well below the IGNITION POINT of paper). It is highly malleable and easily worked, and may be sliced with a steel knife blade. The freshly cut surface has a silvery luster that quickly tarnishes to a characteristic dull blue-gray. Because of its high ability to absorb GAMMA RAYS, lead is among the best materials available to serve as a shield against atomic radiation, and is widely used in the manufacture of containers for radioactive materials (*see* RADIOACTIVITY). It is also used to make corrosion-resistant industrial and laboratory vessels. Alloyed with tin (*see* ALLOY), it is used to make solder. A number of its COMPOUNDS are used as paint pigments, including white lead ($(PbCO_3)_2\ Pb(OH)_2$), red lead (Pb_3O_4), and chrome yellow ($PbCrO_4$). *Tetraethyl lead,* $Pb(C_2H_5)_4$, was widely used until recently as an antiknock component of GASOLINE. This use has been almost completely discontinued due to its role as a major air pollutant (*see* AIR POLLUTION).

LEAD IN THE ENVIRONMENT

Lead has little or no effect on plant life except in extremely high concentrations; however, it and its compounds are highly toxic to humans and to most other animals, either when ingested or when inhaled. It interferes with nerve activity and with the formation of red blood cells (*see* HEMOGLOBIN), thereby inhibiting the transfer of OXYGEN through the body. Its effect are cumulative, making long-term (chronic) exposure to low concentrations as dangerous as short-term (acute) exposure to high concentrations. In either case, the symptoms include lethargy, muscle weakness, gastrointestinal (digestive system) distress, and eventually convulsions, paralysis, and permanent brain damage. Although lead is not soluble in water, it reacts with water in the presence of oxygen to form lead hydroxide ($Pb(OH)_2$), which is soluble in weak acids. Since nearly all public water supplies are slightly acidic, lead cannot be used in water piping or containers, even as a solder, without endangering public health, nor can it be used as an additive to gasoline or paints. *See also* TETRAETHYL LEAD.

lead in logging, the direction in which a tree falls after it has been severed from the stump (*see* FELLING). Lead is largely under the control of the feller (but *see* LEAN). It is generally chosen primarily according to the LENGTH OF GROUND and secondarily by its potential effect on YARDING difficulty and costs. Experienced fellers avoid "crossing the lead"—that is, felling one tree across another, which leads to breakage—but will often "bunch the lead" by felling several trees so that their tips touch, allowing one CHOKER to be set around all of them together. The old saying "always lead toward the mill" (or "toward the LANDING") has little relevance today in light of advancements in yarding practices.

leader the portion of the SHOOT of a WOODY PLANT in which active PRIMARY GROWTH is going on. In forestry, the term is most commonly used to refer to the terminal portion of the main stem of a CONIFER, above the highest whorl of branches. It may also be used to designate the terminal portion of a branch. *See also* RUNNER; TERMINAL BUD.

Leadership in Energy and Environmental Design (LEED) *See* UNITED STATES GREEN BUILDING COUNCIL.

League of Conservation Voters (LCV) political activist organization founded in 1970 to elect candidates to public office who are sympathetic to environmental causes. The group offers monetary and manpower support to candidates with good environmental records at all levels, concentrating on close races. It also publishes voting records and analyses of the environmental positions of selected candidates. Membership (1999): 25,000. Address: 1707 L Street NW, Suite 750, Washington, DC 20036. Phone: (202) 785-8683. Website: www.lcv.org.

League of Women Voters nonpartisan organization whose mission is to encourage political responsibility through the informed and active involvement of the citizens in government. Founded in 1920 as part of the suffragist movement, the league has membership groups in all 50 states, the District of Columbia, Hong Kong, and the Virgin Islands. Through its education fund, it provides educational activities, hosts conferences, and publishes materials. Long concerned with environmental issues, the league focuses on water and air quality, waste management, land use, and energy conservation. Membership (1999): 100,000. Address: 1730 M Street NW, Washington, DC 20036. Phone: (202) 429-1965.

lean in logging, the direction and amount that a tree leans away from the vertical. There are two types of lean: *head lean,* which is the angle the tree makes with the ground in the direction of its principal lean, and *side lean,* which is at right angles to the head lean and may be thought of as the angle the plane defined by the head lean makes with the ground. Lean is almost always caused either by wind ("windlean") or by slippage of the ground the trees stand on ("ground lean"). In either case, the direction and angle of the lean tend to be uniform, or nearly uniform, throughout a STAND. Lean has a strong effect on determining the LEAD during logging, a heavily leaning tree ("leaner") being nearly impossible to pull (or "gun") into a LAY against the lean. This in turn affects YARDING costs and breakage rates. A number of techniques have been developed to "turn the lead" away from the lean, with varying rates of success (*see,* e.g. DUTCHMAN; DRIVER TREE). In extreme cases, a cable is sometimes attached to a mobile YARDER and the tree is physically dragged over against the lean.

lean ore in mining, an ORE in which the ELEMENT or MINERAL being sought is present in such small quantities that it cannot be profitably worked; that is, the cost of obtaining the material from the ore is higher than the value of the material. Whether or not an ore is lean depends both on the cost of working it and the market price of the element or mineral worked out of it, and is therefore subject to change over time as extractive technologies and markets improve. Thus, no ore can ever be classified permanently as lean, and TAILINGS should be thought of as stock piles as much as waste dumps.

leapfrog development term coined during the post–World War II building boom. Leapfrog development describes the practice of large-scale residential developers leaping over smaller, closer-in vacant lands around a city, preferring instead to develop larger parcels (farms, for example) that are cheaper albeit at a greater remove for those commuting to the city core. Leapfrog development, made possible by public investment in the interstate system and other commuting highways, has created discontinuous, economically inefficient, seriously polluted, and aesthetically unattractive metropolitan regions in the United States. *See also* URBAN SPRAWL.

leasable material in mining law, any of several nonmetallic MINERALS covered under the MINERAL LEASING ACT of 1920. Deposits of leasable minerals on federal lands may not be patented under the MINING LAWS. Instead, rights to mine the deposit are leased from the government. Leasable minerals include PETROLEUM, OIL SHALE, gas, COAL, PHOSPHATE, potassium, sodium, and SULFUR. *Compare* LOCATABLE MINERAL.

leave tree in logging, any tree that is not cut during a harvesting or thinning operation. The species, condition, and number of leave trees ("leavers") depend upon the type of operation being conducted. They may be a few outstanding specimens left as SEED TREES, a small STAND left as wildlife COVER, the inferior and unmerchantable trees (*see* UNMERCHANTABLE TIMBER) selected against during HIGH-GRADING, and so on. Leave trees are generally chosen by the silviculturist rather than the feller, and are marked as leavers before the harvest or thinning operation begins.

legislation by chainsaw battle phrase of the wilderness movement, referring to the purposeful siting of timber sales within ROADLESS AREAS in order to make them unsuitable for addition to the wilderness system.

Since wilderness designation is a legislative act (*see* WILDERNESS ACT OF 1964; WILDERNESS AREA), cutting timber on an area specifically to halt such legislation is "legislation by chainsaw."

LeGrand rating system (LeGrand analysis) method of determining the dangers to GROUNDWATER posed by a HAZARDOUS-SUBSTANCES storage site, developed by the American hydrogeologist Harry E. LeGrand (1917–) of the United States Geologic Survey's Groundwater Bureau in the mid-1970s and widely applied since. LeGrand analysis is a relatively simple eight-step process; little or no fieldwork is usually required, as the analysis uses data that are generally already available. Factors considered include (1) a "hydrogeologic rating," which combines ratings for the distance to the nearest well or other water-supply source utilizing groundwater, the depth to the WATER TABLE, the groundwater gradient (that is, the slope of the water table), and the PERMEABILITY and sorption (*see* ABSORPTION; ADSORPTION) ratings of the materials comprising the aquifer; (2) an AQUIFER SENSITIVITY rating; and (3) a hazard potential rating for the substance or substances to be stored at the site. These ratings are arrived at using a standard set of tables. The three figures resulting are then combined into a three-statement expression in the order *hydrogeology-aquifer sensitivity-hazard potential,* which is then compared to a graph to establish a letter grade of A through F, with *A* indicating a good storage site and *F* indicating a poor one. A "confidence rating" estimating the reliability of the data is given for each stage of the process.

legume any member of the FAMILY Leguminosae, including peas, beans, lentils, locust trees, mesquite, acacias, and related plants. The legumes are among the few plants able to directly utilize ("fix") elemental NITROGEN from the air, and are thus extremely important components of any ECOSYSTEM in which they are found. The family contains approximately 500 genera (*see* GENUS) and 13,000 SPECIES.

length of ground in logging, the roughness of the ground surface (the LAY) where a felled tree will be landing. "Roughness" here refers to the presence of rocks, gullies, downed logs, and so forth—anything that might increase breakage when the falling tree lands. A rough, broken lay is referred to as "short ground."

Leontief analysis (input-output analysis) in economics, the study of the interrelationship of the various units of an economy in terms of their inputs and outputs to each other. The analyst constructs a matrix rather like a highway mileage table, with each unit of the economy (individual manufacturers, raw-materials suppliers, consumers, etc.) represented on both the horizontal and the vertical axes. The cell at the intersection of each row and column is filled in with the output of the unit at the head of the row attributable to input from the unit at the head of the column. The sums of the rows then represent total output and the sums of the columns, total input. Changes in technology that alter the amount of one unit's output that another unit requires (such as the replacement of wood chips with recycled paper) can placed in the appropriate cells and their effects calculated on all other cells that they may connect to. A major advantage of Leontief analysis is its ability to incorporate EXTERNALITIES such as pollution costs into economic equations. They are simply given a row and column of their own in the matrix, and their effect is calculated into the totals along with the rest of the rows and columns. *See also* MATERIALS BALANCE APPROACH.

Leopold, Aldo (1887–1948) American writer and pioneer environmental activist, born in Burlington, Iowa, on January 11, 1887, and educated at Yale University (B.S., 1908; Master of Forestry, 1909). Upon graduation from forestry school he joined the United States FOREST SERVICE in the southwest, becoming supervisor of the Carson National Forest in New Mexico in 1912 and assistant regional supervisor in 1917. While working on the Carson he became convinced that the Forest Service should set aside large areas of roadless land to be kept in a wild state, and his internal lobbying for this concept culminated in the formation of the Gila Wilderness in New Mexico—the world's first—in 1924. In 1925 he was appointed associate director of the Forest Service's new Forest Products Laboratory in Madison, Wisconsin. It was not work he enjoyed, and after two uncomfortable years he left government service in 1927. For the next six years he worked as a private consultant in forestry and wildlife, eventually producing a book, *Game Management* (Scribners, 1933), which today is considered the foundation document of the science of wildlife management. That same year he began teaching wildlife management at the University of Wisconsin, a position he would hold for the rest of his life. In 1935 he was named by BOB MARSHALL to the organizing committee of the WILDERNESS SOCIETY. Leopold had a spare, polished, poetic writing style, which he honed through many drafts of each published work. His best-known work, *A Sand County Almanac,* published posthumously in 1949, defined and explored what he termed a "land ethic" in which land was to be treated not as "a commodity which belongs to us," but as "a community to which we belong." He died on April 21, 1948, of a heart attack suffered while fighting a brush fire on a neighbor's property. Other books by Leopold include his essays collected under the title *Round River* (available in a paperback edition along with *A Sand County Almanac*); *The River of the Mother of God and Other Essays* (1991); and most recently *For the Health of the Land: Previously Unpublished Essays and Other Writings* (Island, 1999). Curt Meine's biography, *Aldo Leopold: His Life and Work* (1988), is authoritative and complete. Criticism and commentary include *Companion to A Sand County Almanac* (1987); *Aldo Leopold: The Man and His Legacy* (1987); *Thinking Like a Mountain* (1994); and *Aldo Leopold: A Fierce Green Fire* (1999).

Leopold Report officially *Wildlife Management in the National Parks,* a report presented in 1963 to Secretary of the Interior STEWART LEE UDALL by a committee of scientists headed by University of California zoologist A. Starker Leopold (son of ALDO LEOPOLD). The report was also published as an article in *American Forests* magazine. The chief recommendation of the Leopold Committee was that the ecology of the parks should "be maintained, or where necessary recreated, as nearly as possible in the condition that prevailed when the area was first visited by the white man." A four-step procedure was suggested, including historical research to determine the park's characteristics prior to European contact; ecological research to determine the functional ECOSYSTEM that best fit those characteristics; a series of pilot projects to discover the best way to restore the ecosystem to its pre-contact condition; and adoption of the most successful of the pilot projects as a management strategy for the park as a whole. Most contemporary management practices in the parks, including the "let-it-burn" WILDFIRE policy, the elimination of PREDATOR CONTROL practices, the closing of garbage dumps utilized by bears as food sources, and similar means of returning natural controls to park ecosystems, can be directly traced to the influence of the Leopold Report.

levee an artificial embankment, usually made of rock and earth, built along the bank of a river in order to prevent flood waters from spreading out to cover the FLOODPLAIN, thus protecting buildings, agricultural fields, and other floodplain development from flood damage. Levee-like FLUVIAL DEPOSITS along natural streambanks are also sometimes referred to as "levees" ("natural levees"). Levees have two principal drawbacks as flood-control devices. In times of high water, levees force a stream to flow more rapidly, thus increasing damages downstream of the levees where the greater-velocity floodwaters finally are allowed to spread out into the floodplain; and in times of low water, a river with a large SILT load will deposit silts between the levees, eventually raising its bed above the level of the surrounding countryside and drastically increasing the potential for damage should the levee be breached. Levees also destroy riparian habitat (*see* RIPARIAN LAND; RIPARIAN VEGETATION) and WETLANDS, and interfere with the natural renewal of floodplain soil fertility through silt deposition during floods. *See also* FLOOD: *methods of flood control.*

lichen a "plant" that is actually a symbiotic association (*see* SYMBIOSIS) between two separate ORGANISMS, an ALGA and a FUNGUS, growing together in such a way that they appear to be a single entity. (The symbiosis is skewed, apparently benefiting the fungus more than the alga, as indicated by the fact that lichen algae are often found as free-living species, while lichen fungi never are. Because of this, some biologists prefer to think of the association as a form of controlled parasitism rather than as a true symbiosis [*see* PARASITE].) The fungus's resistance to extremes of temperature and to lack of moisture, coupled with the algae's ability to photosynthesize (*see* PHOTOSYNTHESIS), makes lichens among the hardiest and most adaptable organisms on the planet. They occur in nearly all conceivable HABITATS except the open ocean, and are the principal living things found in the arctic, above timberline on temperate and tropical mountain ranges, and on the Antarctic continent. They grow on any available hard SUBSTRATE, including rocks, tree trunks, compacted soil, and other lichens, and they grow very slowly, a few square millimeters per year. On the basis of growth rates and current size, some individual lichen organisms in the Arctic have been calculated to be as much as 7,000 years old. Their importance to the BIOSPHERE lies principally in their role in soil making. The acids they secrete help break down the rocks they cling to, and their bodies add ORGANIC MATTER to young soils. They also serve as a source for several commercially useful chemicals, including the indicator die LITMUS, and have occasionally been used as human food. One GENUS, *Cladonia* (reindeer moss), is the principal food of caribou, thus supporting much of the Arctic ECOSYSTEM.

TYPES OF LICHEN

Lichens are divided into three principal types. *Crustose lichens* are those that form thin layers ("crusts") on rocks and tree trunks. They are usually black or gray. *Foliose lichens* look like overlapping leaves, and may be nearly any color. *Fruticose lichens*—the most complex varieties—form branched, linear structures that look very plantlike. They are often light green in color, although grays, reds, and blues are commonly found as well. Lichens are classed in the KINGDOM Fungae; most are clearly related to the Ascomycetes. Classification by SPECIES is somewhat difficult. It usually is made by referring exclusively to the fungal partner, although the same fungus associated with different algae may produce widely differing growth forms. *See also* LICHEN LINE.

lichen line in range management, an even, regular line on an exposed vertical rock surface, parallel to the soil level and a few inches above it, with LICHENS growing above the line but not below it. Lichen lines indicate recent EROSION, with the rock surface below the original soil level (represented by the line) not having had time to develop a lichen coating. They are most commonly found on or near ridgetops or other elevated land features.

life expectancy an issue with respect to global population growth. With advances in medical treatment of diseases, better nutrition, and control of natural disasters, a longer life expectancy has contributed to the geometric increase of human population. Thus, as the birth rate has greatly increased, the death rate has decreased almost everywhere in the world. Life expectancy, which on a worldwide basis was only 46.5 years in 1950, is now 66 years. As of 2000, the lowest life expectancy is in Malawi, at 36.0 years; the highest in Australia, at 80.4. The U.S. life expectancy for 2000 is 76.3, up from 68.2 in 1950.

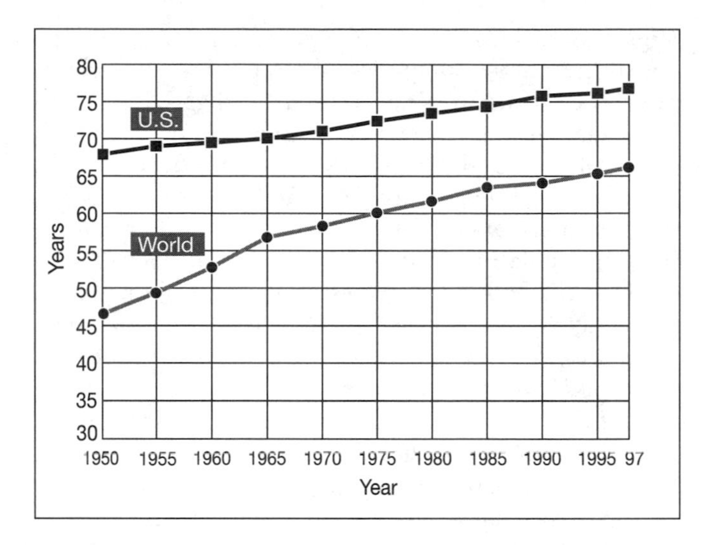

Average life expectancy (United States and world) ***(United Nations,*** **World Population Prospects** ***and U.S. National Center for Health Statistics,*** **Vital Statistics of the United States*****)***

life zone a continental region characterized by roughly similar climate and life forms, used as a guide to classifying plant and animal SPECIES; now considered largely obsolete. The life zone system was first proposed around 1898 by the pioneering ecologist C. Hart Merriam (1855–1942), who divided the North American continent into six broad east-west zones separated from each other by ISOTHERMS based on average annual temperature. From north to south, these zones were the Arctic, Hudsonian, Canadian, Transition, Upper Austral, and Lower Austral. Later ecologists, noting the extreme differences in life forms in the western and eastern parts of Merriam's two Austral zones—differences cause principally by the much greater rainfall in the eastern portions, which Merriam had largely ignored—divided these zones further, into the Upper and Lower Sonoran (west) and the Carolinian and Australriparian (east). They also pointed out (as Merriam had recognized but not used) that elevation made a significant difference—that, for example, climbing a mountain peak in the Upper Sonoran might take the climber through successive bands of Transition, Canadian, Hudsonian, and Arctic vegetation as the climb passed through cooler and cooler layers of the earth's atmosphere (*see* LAPSE RATE; HUMBOLDT'S RULE). For this reason, the Arctic zone was renamed the Arctic-Alpine zone. This and other developments convinced ecologists that the life-zone system was inadequate, and in the 1950s and 1960s it was largely replaced by the ecological distribution system of BIOME and BIOTIC PROVINCE. The life-zone names have lingered in the literature, however, as rough synonyms for the various biomes; the Canadian life zone, for example, is approximately the same as the CONIFEROUS FOREST BIOME. See separate articles under each life zone. *See also* DISTRIBUTION OF PLANTS AND ANIMALS.

Lighthawk an international organization that provides low-altitude flights for journalists, photographers, policy makers, and other influential people in small planes over important environmental sites as a means of expanding the public's knowledge of environmental impacts, such as clearcutting. In partnership with environmental groups, conservation information projects have been developed in Alaska, British Columbia, Washington, Oregon, the Rocky Mountains, Central America, and Florida. Address: 230 California Street, Suite 207, San Francisco, CA 94111. Phone: (415) 715-6400. Website: www.sni.net/lighthawk.

light pollution the interference by "waste" lighting from buildings, streetlights, and commercial sources, of clear views of the night sky in settled areas. Where there is no light pollution, some 2,500 stars can be seen, with a clear and almost startling view of the Milky Way as well. At present, however, only 10% of Americans are able to enjoy a natural sky view because of light pollution. In a suburban setting, only 200–300 stars are visible; in a large city perhaps fewer than 50. Not only is light pollution an aesthetic problem for metropolitan residents, but it also interferes with migration of birds, the food-gathering patterns of bats and other nocturnal species, the reproductive cycle of sea turtles which will not approach a lighted beach to lay eggs, and the light-influenced seasonal growth cycles of trees and plants. Physiological effects of light pollution on humans are suspected as well, including the possibility of decreased melatonin, a known cancer-fighting regulator of estrogen production. Policies governing light pollution are beginning to be formulated, especially for the national parks. Parks and wilderness areas are, by law (*see* CLEAN AIR ACT), to be protected from pollution that hampers visibility. Based on the assumption that congressional intent to provide visual protection can be extended to the night sky, some park administrators are reducing ambient waste light. For example, at Chaco Culture National Historic Park in New Mexico, mercury-vapor lamps (which can throw light up to 100 miles) were changed to shielded incandescent bulbs fitted with timers. The cost of retrofitting was minimal and the park saved 30% on its total electric bill. Some cities have recognized light pollution as well. In Tucson, for example, outdoor lighting ordinances have reduced waste light so that night skies are only twice as bright as they would be without any pollution whatever, compared to Los Angeles, where night skies are 100 times brighter. With regard to "crime lighting," there is growing evidence that the harsh, yellow mercury-vapor lamps are not as effective as relatively dim, well-spaced, shielded halide lamps. According to the International Dark-Sky Association, eliminating outdoor lighting shining above the horizontal plane could save U.S. cities an aggregate of $1 billion annually in illumination costs.

lignin a complex ORGANIC COMPOUND found as a constituent of the walls of plant CELLS, especially in WOODY PLANTS, where it may form between 25 and 30% of the cell wall. The lignin MOLECULE is a massive, many branched POLYMER. The molecular branches interlock with each other, contributing to the hardness and rigidity of wood. The lignin content—hence, the rigidity—of plant cells generally increases as they age, a process known as *lignification.* Extracted from wood during paper making, plywood manufacture, and similar procedures, lignin is an important source of vanillin for food flavoring, and is used as a hardener for plastics and as a SUBSTRATE for the growing of yeasts and other MICROORGANISMS.

limbing in logging, the removal of limbs from the trunk of a felled tree to prepare it for transport to the mill. *See also* BUCKING.

limestone a common SEDIMENTARY ROCK formed largely or completely of calcium carbonate ($CaCO_3$). Limestone is white to gray or blue-gray in color, depending upon the amount and color of any impurities that might be present; it is moderately soluble in water (*see* KARST TOPOGRAPHY) and fizzes in mild ACIDS such as vinegar. It is formed principally in warm, shallow tropical seas, far from the mouths of rivers, though small amounts may be formed in freshwater. The stone is used as a building material and in the production of steel and FERTILIZER.

TYPES OF LIMESTONE

Limestone occurs in numerous varieties, depending upon the means of formation. These varieties differ widely in hardness and FRIABILITY. Pure calcium carbonate crystals precipitated from solution (*see* PRECIPITATE) are called *calcite.* Limestone made largely or completely of calcite crystals is called *calcerous limestone. Dolomitic limestone,* or *dolomite,* is similar to calcerous limestone except that the crystals are made of calcium-magnesium carbonate ($CaMg(CO_3)_2$) instead of pure $CaCo_3$. It is harder and less soluble than regular limestone, and will only fizz in strong acids. *Travertine* is a soft, friable limestone deposited in hot springs. It is often spectacularly colored and patterned by impurities. *Clastic limestones* are formed of the shells and skeletons of aquatic animals and the cell plates of DIATOMS, foraminifera (*see* FORAMINIFERAN), and other PROTISTA; these vary from *coquina,* composed of large shell fragments cemented together by calcerous limestone, to *chalk,* made up completely of the cell plates of foraminifera. MARL is an impure limestone (less than 80% calcium carbonate) that includes large amounts of CLAY. It usually indicates freshwater deposition.

limited use zone in forest management, an area of a NATIONAL FOREST or BUREAU OF LAND MANAGEMENT district on which use restrictions of some sort apply, especially those in which the restriction is on some resource extraction activity such as logging or mining. Limited use zones include STREAMSIDE MANAGEMENT UNITS, scenic management zones, RESEARCH NATURAL AREAS, and similar designations. The locations and boundaries of these areas are determined by the managing agency through administrative rules rather than by acts of Congress; hence, WILDERNESS AREAS, NATIONAL PARKS, and other regions set aside by law are not, technically speaking, limited use zones. *See also* SPECIAL INTEREST AREA.

limiting factor in ecology, any ENVIRONMENTAL FACTOR whose availability or presence serves as a limit to the growth of an ORGANISM, a COMMUNITY or a POPULATION. In all these cases, growth will stop when the limits of the limiting factor are exceeded, even if all other factors contributing to growth are still underutilized. For example, photosynthetic plants (*see* PHOTOSYNTHESIS) can grow on a forest floor only to the extent that light penetrates through the forest CANOPY; those plants that need more light than is available cannot grow there, even if soil conditions, temperature, moisture, and all the other factors are highly favorable to their growth. In this case, light is the limiting factor. Other common limiting factors are temperatures (minimum, maximum, and average), rainfall, and NUTRIENT availability. *See also* LIMITING NUTRIENT.

limiting nutrient in ecology, a chemical ELEMENT necessary for organic growth that is present in an ECOSYSTEM in such small amounts that its availability partly or largely determines the ecosystem's makeup. ORGANISMS will grow and multiply in an ecosystem until all of the limiting nutrient is tied up in living tissue, or being recycled through living tissue as rapidly as possible. At that point, further multiplication will cease, even though there may be sufficient quantities of everything else the organisms need still available to them. The identity of the limiting nutrient varies from place to place and from SPECIES to species at a particular location. The most common limiting nutrients are PHOSPHOROUS and NITROGEN. *See also* LIMITING FACTOR; CULTURAL, EUTROPHICATION.

Limits to Growth popularly known as the "Club of Rome Report," a book (Universe Books, New York, 1972) produced by a team of scientists from Massachusetts Institute of Technology headed by Dennis L. and Donella H. Meadows, analyzing the effects of increasing resource scarcities on human civilization. The book was an outgrowth of the Project on the Predicament of Mankind, a set of studies undertaken in 1970 for the Club of Rome, a group of some 75 private citizens from a number of different nations founded in Rome in 1968 to promote solutions to world problems based on the interrelatedness of complex disciplines such as economics, sociology, ecology, and engineering. The MIT team used a computer to analyze a number of different models of economic growth and found that they all became asymptotic (that is, the curves started going straight up) within the next 100 years, leading to crashes as the exponential growth curves ran into the barrier of limited resources. As a solution, they proposed a conversion to a so-called "steady-state economy" in which growth in resource use no longer took place, which they determined could be accomplished by conversion to renewable resources and services, rather than extractive resources and goods, as the basis of economic systems. The book had a great deal of influence on the environmental movement, and is probably the single most important cause behind the maturation of environmentalism from a preoccupation with local problems such as individual wilderness areas or polluted rivers to a concern for the planet as a whole.

limnetic division (**limnosphere**) in ecology, the freshwater portion of the HYDROSPHERE, that is, all fresh surface water on the planet, including lakes, rivers, ponds, streams and springs. Though it is only a tiny portion of the hydrosphere, accounting for barely 0.015% of the total, the limnetic division is critical for all terrestrial life forms, including humans. With the exceptions of ocean transport, marine fisheries and tidal power, all our uses of water are actually uses of the limnetic division, though we are able to augment it to a degree by using GROUNDWATER in its less saline forms.

limnology one of the biological sciences, concerned with the study of freshwater life forms and their interactions with the environment in lakes, ponds, and streams.

limonite a hydrated oxide of iron, chemical formula $2Fe_2O_33H_2O$. Limonite is a soft, earthy, yellowish mineral with a characteristic yellow-brown STREAK. It is occasionally used as a commercial iron ore, but is more useful as an indicator of the presence of HEMATITE, with which it is usually associated. A relatively pure, powdery form of limonite, known as *yellow ochre,* is used as a paint pigment.

Lindbergh, Anne Spencer Morrow (1906–2001) American author, born June 22, 1906, in New York City and educated at Smith College, Northampton, Massachusetts (B.A., English, 1928). In 1929 she met and married the aviator Charles A. Lindbergh and subsequently became his copilot and navigator, logging more than 40,000 miles of aerial survey work around the world between 1929 and 1934, when they jointly received the Hubbard Medal from the National Geographic Society for their contributions to aviation. In 1955 she published *A Gift from the Sea,* a lyrical, intensely personal narrative of the healing power of nature on the human psyche, based largely on her own recovery from depression following the kidnaping and murder of her infant son in 1932. The work is considered seminal by both the environmental movement and the women's movement.

lipid in biology and chemistry, any of the large group of COMPOUNDS that includes the fats and fat-like substances (oils, waxes, etc.): though the term is often used as a synonym for *fat,* the category is actually much broader than that. Lipids are ORGANIC COMPOUNDS that contain considerably more HYDROGEN ATOMS than OXYGEN atoms. They are insoluble in water, but dissolve well in organic SOLVENTS such as CHLOROFORM. They have a high percentage of carbon-hydrogen bonding within their MOLECULES, hence store a great deal of energy per unit weight. Principal types of lipids include the triglycerides, the phospholipids and associated compounds, and the terpenoids.

Triglycerides, or *triacylglycerides* (fats and oils), consist of glycerol molecules (an alcohol, $C_3H_5(OH)_3$) to which are attached three long fatty-acid chains. The fatty-acid chains may be *saturated* (contain as many hydrogen atoms as possible) or *unsaturated* (contain places where two or more extra hydrogen molecules might be attached. In such places, the molecule has generally formed a carbon-carbon bond). Unsaturated fats, which are characteristically found in fish, birds, and temperate-zone plants, are generally liquids; saturated fats, which are characteristically found in mammals and tropical plants, are generally solids. The triglycerides are used in nature principally as storage mediums for reserve energy, as insulation (as in the whales and walruses), or as waterproof coatings.

Phospholipids have a structure similar to that of the fats and oils, but one or more of the fatty-acid chains is replaced by a complex functional GROUP consisting of a PHOSPHATE attached to an ALCOHOL. The portion of the atom containing the phosphate-alcohol group is soluble in water while the remainder of the molecule is not, causing phospholipids poured into water to form a film on the surface with the phosphorous "heads" of the molecules immersed in the water. Phospholipids are found in nature principally as the inner layer of the three-layer structure of cell membranes (*see* CELL; *structure of cells*), where their unique properties play a large part in the transport of materials into and out of the cell.

Terpenoids (terpenes and sterols) are POLYMERS consisting principally of repeating units of isoprene (C_5H_8), usually with the addition of one or more "foreign" functional GROUPS. This extremely varied set of compounds includes natural rubber, vitamin D, and the group of hormones known as *steroids. See also* LIPOPHILIC SUBSTANCE; LIPOPROTEIN.

lipophilic substance any substance that shows an affinity for LIPIDS; especially, those which are soluble in lipids. Lipophilic substances tend to concentrate in the fatty tissues of plants and animals rather than metabolizing through and being excreted; hence, they are particularly prone to BIOLOGICAL MAGNIFICATION. Examples: PCBs (*see* POLYCHLORINATED BIPHENYL), DDT.

lipoprotein any of a group of important chemical COMPOUNDS composed of a LIPID joined to a PROTEIN. Lipoproteins are used by the body principally in energy transport systems. They are found in the bloodstream, in nerve sheathes, and, within the cell, in the membranes of mitochondria (*see* CELL: *structure of cells*).

lithology the study of the physical characteristics of rocks and minerals, including their shape, color, hardness, texture, FRIABILITY, and crystalline structure.

lithosol a young, shallow, freshly weathered soil with little ORGANIC MATTER present, the soil consisting principally of rock fragments of varying sizes. The term is now generally considered obsolete, and the lithosols are classed with the ENTISOLS. *See* SOIL: *classification of soils.*

lithosphere the solid portion of the earth, including all solid inorganic materials such as ROCK, SAND, CLAY, inorganic soils, etc. The ice of GLACIERS and ICECAPS, though a solid, is not generally included in the lithosphere but is properly part of the HYDROSPHERE. *See also* ATMOSPHERE; BIOSPHERE.

litmus a substance found in various SPECIES of LICHENS (*Variolaria, Roccella* and others) that is used in chemical analysis as an indicator of the acidity or alkalinity of liquids (*see* ACID; BASE). As prepared and sold, litmus is a blue, powdery material that is a mixture of a number of chemicals. The active principal appears to be *azolitmin,* a dark-red, scaly powder that has proved difficult to analyze and may itself be a MIXTURE. Litmus (azolitmin) turns red in acidic liquids and blue in basic ones. It has an active range of about PH4.5 to 8.3. Paper strips impregnated with litmus (*litmus paper*) are a convenient field test kit for water acidity, and are widely used for that purpose.

litter waste materials strewn randomly over a surface. The term can refer to natural as well as human wastes (leaf litter, rock litter, etc.), and to large materials as well as small (a litter of auto parts, a litter of worn-out appliances). However, when standing alone the word "litter" generally refers to the small bits of rubbish tossed aside after use by consumers along roads and sidewalks and in public facilities such as parks or bus terminals. Besides being unsightly, litter of this type can cause significant environmental damage. Food residues in discarded containers attract rats and cockroaches and form a breeding ground for disease; wildlife can be injured by ingesting litter in the belief that it is food (turtles have been known to mistake plastic bags for jellyfish; birds have swallowed—and have choked on—pop-tops from beverage containers, apparently in the belief that they were shiny insects) or by becoming trapped in string, wire, plastic six-pack holders, etc. The best overall means of controlling litter has been found to set out plenty of conveniently placed trash containers and to keep them emptied. People will generally deposit waste in trash cans if they can easily do so. Other, more specific means of litter control include the so-called "litter tax," imposed on manufacturers of items likely to be littered (such as disposable diapers, gum wrappers, fast-food containers, and so on) and used to fund "litter patrols" to clean up littered areas, and container-recycling laws that place a deposit on bottles, cans, and other containers to encourage them to be saved and turned in for cash rather than thrown away (*see* BOTTLE BILL; RECYCLING).

littoral a coastline; that is, the region of intersection between the land and a standing body of water such as a lake or an ocean. *See also* LITTORAL ZONE; LITTORAL BENTHIC ZONE.

littoral benthic zone in ecology, the sea floor from the extreme high tide line to the edge of the CONTINENTAL SHELF, including the LITTORAL ZONE and the shoreward portion of the BENTHIC DIVISION. The littoral benthic zone includes all portions of the sea floor that sunlight can reach; hence it is sometimes referred to as the "lighted bottom."

littoral zone in ecology, the zone between extreme high tide (including the swash line; *see* SWASH) and extreme low tide on an ocean shore. The littoral zone includes the LITTORINA ZONE, the BALANOID ZONE, and the SUBTIDAL ALGAL ZONE. Extremely productive (*see,* e.g., SALT MARSH; TIDE FLAT; TIDEPOOL), the littoral zone is also among the most highly stressed of all environments due to the twice-daily cycle from wet to dry conditions caused by the tides, which also means that there are wide daily fluctuations in light, oxygen, and temperature conditions. The effects of oil and chemical spills tend to concentrate in the littoral zone as this is where the waves wash them ashore. *See also* LITTORAL BENTHIC ZONE.

littorina zone (littorina fringe) the upper edge of the LITTORAL ZONE, wet only by the highest tides and by the splashing of waves. Rocks in the littorina zone are generally black in color due to the presence of black LICHENS; the principal animal inhabitants are small snails.

livestock collectively, any group of domestic animals, especially grazing UNGULATES such as sheep, goats, cows, or horses. *See also* RANGE MANAGEMENT.

lixification leaching; especially, in mining and metallurgical terminology, any technique for removing a metal from its ORE by washing the ore with a SOLVENT and recovering the metal from SOLUTION in the LEACHATE. *See,* e.g., HEAP LEACHING.

LNG acronym for liquid natural gas (*see* NATURAL GAS).

loading (1) the rate at which a POLLUTANT is being added to the water or the air. Loading is usually expressed in one of two ways: either as a ratio of the weight of pollutant released overtime (e.g., pounds per day; tons per year) or as a ratio of the volume (or weight) of the pollutant per unit volume (or weight) of the total EFFLUENT (e.g., pounds per thousand ft^3; milligrams/liter; parts per million).

(2) in sewage-treatment technology, *see* F/M RATIO.

loam a soil with a texture consisting of a mixture of SAND, SILT and CLAY. A good loam has between 25 and 50% sand, between 30 and 50% silt and between 10 and 25% clay. The "ideal" loam is defined as a mixture of 40% sand, 40% silt and 20% clay. Loams form a broken ribbon when wetted to putty texture and smeared on a hard surface. They are found in all soil groups, and will generally be the best agricultural soils in each group. *See* SOIL: *classification of soils.*

local low in meteorology, a region of low air pressure created by local conditions. The main cause of local lows is differential heating. For example, the air over barren land will heat up more rapidly than the air over a nearby forested region, and this warmer air, by expanding and rising, will create a low-pressure area relative to the air over the forest. A second cause of local lows is wind friction caused by rough ground, or by the movement of the wind from water to land. Greater friction causes the wind to slow down and "pile up," raising its pressure in comparison to the region down-wind of the roughness, which then develops a local low.

locatable mineral in mining law, any MINERAL for which a mining claim may be filed and patented under the MINING LAWS OF 1866, 1870, AND 1872. The locatable minerals generally include METALS and GEMS. They do not include so-called "common-variety minerals" such as GRAVEL, building stone or CONGLOMERATE, nor do they include OIL, gas PHOSPHATE ROCK, SULFUR, or other valuable but non-metallic and non-precious materials. *Compare* LEASABLE MINERAL.

lode in mining, an ORE BODY consisting of several VEINS close enough together to be worked in a single operation. Lodes are, properly speaking, always *primary ore deposits;* that is, the metal-rich portions of the deposit exist in the PARENT ROCK where they were formed. Compare placer (*see under* PLACER MINE).

loess a soil that has been transported to its present location by wind. Loess is principally composed of fine silt, with a small proportion of CLAY and SAND. It is fairly well sorted, with the larger materials on the bottom. It often forms thick deposits, with soil depths of 50 feet being common and over 200 feet not unknown. The surface is generally wavy and may form DUNE-like hills. Loess deposits are generally mechanically compact and form near-vertical stream banks and road cuts; however, they are highly erosive due to their small particle size, gullying easily (*see* EROSION). They are often very fertile. Extensive deposits of loess occur in the Palouse country of eastern Washington and northern Idaho, in the Missouri and Mississippi valleys, in the Rhine valley of Europe, and in the Hwang-Ho and Yellow River valleys of China. Most of these deposits appear to have derive from glacial silts created during the last ICE AGE (*see* PLEISTOCENE; CONTINENTAL GLACIER).

log (1) a tree after it has been felled and bucked (*see* FELLING; BUCKING) but before it has been cut into lumber or other products.

(2) a record of measurements or observations taken at specified intervals (hourly, daily, weekly, etc.) over a period of time. The process of recording these measurements or observations is called LOGGING ("logging the data," "logging the observations").

log deck a pile or stack of logs, either at a sawmill or at a LANDING. A deck intended for long-term storage rather than being part of an active transport or milling operation is known as a *cold deck*. A cold deck that is to be maintained for any length of time must be sprayed with water regularly to prevent the logs from drying out and cracking. *See also* LOG POND.

logging chance the physical site of a logging operation. *Compare* LOGGING SHOW; STRIP.

logging show a logging operation while it is under way. The show is the work being done—FELLING, BUCKING, YARDING, etc.—rather than the timber being logged or the land it stands on. *Compare* LOGGING CHANCE; STRIP.

logging strip *See* STRIP.

logistic population growth with reference to exponential growth of animals and other organisms, the limitation on such growth dictated by environmental factors such as food supply, predation, disease outbreak, and the like. Logistic population growth curves tend to flatten out when environmental limits are reached, a phenomenon that can apply to human population growth curves as well. *See also* POPULATION EXPLOSION.

The Scribner Scale

diam (in)	Length (feet)										
	8	**9**	**10**	**12**	**14**	**16**	**18**	**20**	**24**	**28**	**32**
6	0.5	0.5	1	1	1	2	2	2	3	4	5
8	1	1	2	2	2	3	3	3	4	6	7
10	3	3	3	3	4	6	6	7	9	10	12
12	4	4	5	6	7	8	9	10	12	14	16
16	8	9	10	12	14	16	18	20	24	28	32
20	14	16	17	21	24	28	31	35	42	49	56
24	21	23	25	30	35	40	45	50	61	71	81
30	33	37	41	49	57	66	74	82	99	115	131
36	46	52	58	69	81	92	104	115	138	161	185
42	67	76	84	101	117	134	151	168	201	235	269
48	86	97	108	130	151	173	194	216	260	302	346
60	135	152	169	203	237	270	304	338	406	473	541
72	197	222	247	296	345	395	444	493	592	691	789
84	275	309	343	412	481	549	618	687	824	961	1099
96	357	402	446	536	625	715	804	893	1072	1251	1429

To scale a log using this table, find the column containing the log's length; find the row containing its scaling diameter (small end); locate the point within the table where this column and row intersect; and multiply the figure found there by 10. The result is the estimated log volume in board feet.

log pond a pond next to a sawmill, used for storing logs. The stored logs are floated on the surface of the pond. Log ponds are generally preferred to LOG DECKS for long-term storage of logs ("cold storing") because logs stored in a pond do not dry out and therefore will not crack or check. The ponds are almost always manmade.

logrolling in politics, a process in which one legislator trades votes with another in order to advance a bill in which he has a strong personal interest. A senator from Pennsylvania, for example, might agree to vote for price supports on corn if a senator from Kansas votes for funding the construction of a dam in Pennsylvania. Logrolling generally has a bad reputation because it is seldom used except when no other way can be found to pass a particular bill—often interpreted as a sign that the bill has little overall value. It is, however, one of the best ways of building majorities for controversial measures, including reform measures and wilderness legislation, on which legislators might otherwise fear to take a stand. A variant, *negative logrolling,* trades votes against bills rather than for them. *See also* PORK BARREL.

log scaling in logging and forestry, any of various methods of determining the volume of wood in a LOG by measuring its length and diameter. Scaling is usually done "on the truck"; that is, the logs are measured as they are being carried on a logging truck, either at a booth ("scaling station" or "scaling shack") located on the road out of the forest, or as they arrive at a mill's LOG DECK. Occasionally they may be scaled as they float in a river or a LOG POND. Tools for such "water scaling" vary slightly from those used in truck scaling, but the methodology is the same. To scale a log, the scaler first measures the diameter of the small end of the log inside the bark (the "scaling diameter"), using an average of two measurements through the log's true center (not the center of the growth rings) to account for logs that are out of round. Next he or she measures the length of the log, rounded to the nearest 2-foot length above the measured length (usually a few inches will be subtracted from the measured length before rounding up to allow for "trimming,") or squaring the ends of the log, at the mill; (*see* MINIMUM TRIM ALLOWANCE). Third, the scaler determines the gross volume of the log by applying any of several standard scales, known as *rules*. Fourth, a certain percentage of the log's volume is subtracted for DEFECT. This percentage varies from log to log depending upon the type and amount of defect present, and must be estimated by the scaler. A 50-foot measuring tape and a knowledge of scaling rules is all that is absolutely necessary for scaling; however, most scalers use a "scaling stick," a 4- to 6-foot-long stick carrying measurements that have already been calculated according to one or more of the standard scaling rules so that the scaled volume may be read directly, without the need for tables or calculations.

SCALING RULES

Over 100 different scaling rules have been utilized in the United States at one time or another during the past 150 years; only a few of these remain in use today. The choice of which rule to use depends in part on the unit the volume is to be expressed in (e.g., BOARD FEET; cubic feet; CUNITS) and the use the log will be put to (a scale that allows for saw kerf, for example—that is, deducts a certain amount of volume for the width of the cut a sawblade makes while cutting boards from a log—would be inappropriate if the log is going to be converted to chips for wood pulp or particleboard, as chipping does not produce kerfs). The most common rule is probably the *Scribner,* devised by the forester J.M. Scribner around 1825. It is a so-called "diagram rule," in which a series of tables is used to determine volume based on a diagrammatic view of the scaling end of the log. In the absence of the tables a rule-of-thumb formula may be used to determine the approximate scribner volume. The most common of these, Knouf's rule, may be stated as:

$$V=L/2 \text{ x } (D^2-3D)/10$$

where V=volume, L=length and D=scaling diameter. A modification of the Scribner rule, called Scribner Decimal C, has been adopted by the FOREST SERVICE and the BUREAU OF LAND MANAGEMENT to measure all timber coming from federal lands in the United States. This rule is simply the regular Scribner rule rounded up to the nearest 10 board feet and expressed without the final digit. For example, a log that scaled out at 533 board feet (Scribner rule) would be recorded as 54 (Scribner decimal C). Another commonly used rule is the *International 1/4 inch rule,* a "formula" rule that is calculated using a standard formula that includes deductions for saw kerf, slabs (the rounded exterior portions of the log which cannot be used for lumber) and edges (the trim taken from the individual boards after they are cut). Considered the most accurate of all scaling rules, but somewhat difficult to apply because of the complex mathematics involved, it gives results in board feet and may be expressed in decimal C form like the Scribners. Less common but still important rules include the *Brereton rule,* which gives the volume of the entire log with no deductions for saw kerf, slabs, edging, or defect, and is used primarily to measure logs for export; the *Humboldt rule,* used for redwood lumber, which is a modified Scribner rule that assumes a wider saw kerf and allows a standard 30% deduction for defect; and several so-called "cubic foot rules" that give scaled volume in cubic feet rather than in board feet, including the Huber rule, the Smalian rule, the Sorenson rule (or "one in ten" rule), and the Rapraeger rule (or "one in eight" rule). The Huber and Smalian rules require measuring both ends of the log and are used primarily for scientific work, while the Sorenson and Rapraeger rules assume a standard taper (1 inch in 10 feet for the Sorenson; 1 inch in 8 feet for the Rapraeger) and so require measuring only the scaling diameter. The *Doyle Rule,* once widely used for POLETIMBER and other small-diameter logs, has been largely abandoned due to its tendency to underscale small logs and overscale large ones. It may still be found occasionally, especially in the southern pine timber region of the United States.

long-day plant in botany, a plant that will not flower until the length of daylight during a day is at or above some threshold value. *Compare* SHORT-DAY PLANT; DAY-NEUTRAL PLANT. *See also* PHOTOPERIODISM.

longitude the distance a point on the Earth's surface lies east or west of the GREENWICH MERIDIAN. Longitude is measured in degrees, with 360 degrees constituting an entire circle of the globe. The actual distance in miles or kilometers will vary according to how far north or south of the equator the point is located (*see* LATITUDE).

longshore current a current in a body of water that moves parallel to the shore. Longshore currents result primarily from WAVES that approach a shore at an angle. Part of the energy of such a wave will be expended on the shoreline, while the rest will be deflected parallel to it. The direction and strength of longshore currents are determined by the angle the waves make with the shore, which in turn depends upon the PREVAILING WIND; thus, longshore currents may reverse from season to season as the prevailing wind changes. These currents are responsible for the transport of large amounts of SAND, SILT, and other materials along shorelines, and shape many common shoreline features (*see,* e.g., BAR; SPIT; BEACH NOURISHMENT). A longshore current may often be detected visually as a band of "dirty-looking" water close to the shore, with "cleaner" water appearing abruptly at its outer edge. (For the effects of attempts to control longshore currents, *see* GROIN.)

Long Term Ecological Research (LTER) Network a group of 24 sites located throughout the United States and its territories and in Antarctica where detailed data on long-term ecological trends are collected, shared, and analyzed. The LTER program was begun in 1980 by the National Science Foundation, resulting in a collaborative effort among some 1,400 scientists and students who share information via increasingly comprehensive bibliographies and data sets, all available on the Internet. Sites include tropical forests, temperate forests, savanna, deserts, and arctic tundra. A special interest of the LTER network is the effect of climate change in different ecosystems. Efforts are underway to link the U.S. LTER sites with similar networks developed by other nations so that ecological effects of climate change can be studied at the global scale. Address: Network Office, University of New Mexico, Department of Biology, Albuquerque, NM 87106. Phone: (505) 272–7316. Website: www.lternet.edu.

long-term turbidity TURBIDITY that does not significantly decrease when a water sample is left undisturbed for a specified period of time, usually seven days. Long-term turbidity is the result of the presence of colloidal clays (*see* COLLOIDAL SUSPENSION; CLAY). It is much more important than peak turbidity values in determining water quality and its effect on water-supply systems and aquatic life.

long ton 2,240 pounds. Both the SHORT TON and the long ton are defined as 20 hundredweight; however, in determining the long ton, the Imperial or "long" hundredweight is used, which is equivalent to 112 pounds rather than 100 pounds. The long ton and the metric TONNE are close to the same size (1 long ton = 1.016 tonnes).

longwall mine a type of underground mining in which the seam of COAL or ORE is removed entirely, allowing the roof to collapse. The results are similar to STRIP MINING, except that the OVERBURDEN is not removed but merely settles in place. There is somewhat less EROSION and ACID drainage resulting from this method than from strip mining, but the disturbance of GROUNDWATER flow and the requirements for (and difficulties of) surface reclamation are virtually identical in the two methods.

Lopez, Barry (1945–) American environmental writer, born January 6, 1945, in Port Chester, New York, and educated at Notre Dame University (B.A., 1966; M.A., 1968) and at the University of Oregon. Since 1970 he has devoted full time to writing, producing an extended series of books on the interrelationship of humanity and nature, the best-known of which are *Of Wolves and Men* (1978; John Burroughs medal, 1979); *Arctic Dreams* (1986; National Book Award); *Crossing Open Ground* (1988); and *About This Life* (1998). Lopez's work is characterized by a lucid balance between the spiritual and the scientific approaches to nature which has led him to be compared to JOHN MUIR. He lives in Eugene, Oregon.

Love Canal housing development in Niagara Falls, New York that has become a national symbol for improper hazardous-waste disposal technology. The original "Love Canal" was an actual canal, dug around 1900 to supply water and hydroelectric power to the planned community of Model City. It was abandoned after construction of approximately 1 mile. From 1942 to 1952, the abandoned canal was used as a dumpsite for approximately 21,800 tons of hazardous wastes by the Hooker Chemical Company (now Occidental Chemical Corporation). The company then sold the site to the Niagara Falls Board of Education (April 1953), which proceeded to build an elementary school (completed 1957) and sell homesites, despite Hooker officials' warnings that the wastes buried in the area were a potential health hazard. By the mid-1970s, chemical sludge was migrating into basements and appearing in back yards, and epidemiological studies had begun to demonstrate a strong correlation between living at Love Canal and the development of a complex of serious health problems, including cancer, birth defects, and liver and kidney failure. The miscarriage rate was calculated to be 250 times the national average. On August 7, 1978, President Jimmy Carter declared the central Love Canal area a national disaster; on May 21, 1980, the area covered by the disaster designation was increased to include a strip 1,500 feet wide on either side of the filled-in canal. Eight hundred and fifty families were eventually evacuated from the area, with state and federal funds used to compensate them for the loss of their homes, and a $17 million cleanup of the site was begun. Love Canal helped focus national attention on the inadequacies of current hazardous-waste disposal practices and was, along with the VALLEY OF THE DRUMS, one of the two primary moving forces behind the passage of the so-called "superfund" act of 1980; *see* COMPREHENSIVE ENVIRONMENTAL RESPONSE, COMPENSATION AND LIABILITY ACT.

Lovelock, James (1919–) originator of the GAIA HYPOTHESIS, which he and Lynn Margulis first defined in 1972, proposing that the planetary ecosystem behaves like a single living organism. Lovelock, a British scientist, now works independently from his home laboratory in Cornwall, England. After studying at the University of Manchester and the University of London (he has a Ph.D. in medicine), he was on the staff of the National Institute for Medical Research and taught at Yale, Baylor, and Harvard. He has been president of the Marine Biology Association and is a fellow of the Royal Society, London. His books include *Gaia: A New Look at Life on Earth* (1979), *The Ages of Gaia* (1988), and *Healing Gaia: Practical Medicine for the Planet* (1991).

Lovins, Amory (1947–) noted proponent of alternative energy strategies and the more effective use of available energy resources, which he calls "soft energy paths." A physicist, Lovins and his wife, Hunter, established the Rocky Mountain Institute in Aspen, Colorado, in 1982 to develop new approaches to energy conservation. The institute, now with a full-time staff of 45 and a $4.2 million annual operating budget, conducts research in various fields related to energy, including transportation, green development, climate change, water, economic renewal, corporate sustainability, and forest health. In 1993 Lovins received a MacArthur fellowship for his contribution to finding practical ways to deal with issues of energy and the environment in an economy dominated by large transnational corporations. He is author of 22 books, including *World Energy Strategies: Facts, Issues, and Options* (1975), *Non-Nuclear Futures: The Case for an Ethical Energy Strategy* (1975), *Soft Energy Paths* (1977), *Energy and War: Breaking the Nuclear Link* (1980), *Least-Cost Energy: Solving the Carbon Dioxide Problem* (1981), *Energy Unbound: A Fable for America's Future* (1986); and the coauthor of other books, including most recently the influential *Natural Capitalism* (1999), with PAUL HAWKEN and L. Hunter Lovins.

low-density development in land use planning, residential, commercial, or other development in which the total number of buildings is kept low in relation to the amount of open space. A common maximum density for low-density development is two units (homes, stores, etc.) per acre. Low-density development may be achieved either through *large-lot zoning,* in which the minimum lot size is designated, or through *cluster zoning,* in which a developer is allowed to group developmental units close together provided that enough open space is incorporated elsewhere in the development so that the average density is at or below the maximum limit. Under a two-unit-per-acre limit, for example, large-lot zoning would simply designate a minimum lot size of 1/2 acre, whereas cluster zoning might allow 10 units on a single acre provided that four adjacent acres were left undeveloped.

lowest achievable emission rate (LAER) in air pollution law, the highest degree of control possible over the emission of POLLUTANTS from industrial sources. The LAER for each class of pollutants is defined by the CLEAN AIR ACT to be the most stringent emission standard set for that pollutant by any state IMPLEMENTATION PLAN, or the lowest emission rate for that pollutant achieved in practice anywhere in the country, whichever is lower. LAER standards may be waived only if the industry can prove that it is impossible to comply with them. *See also* BEST AVAILABLE CONTROL TECHNOLOGY.

low-pressure system *See* CYCLONE.

low-pressure tank a containment vessel designed to operate at a GAUGE PRESSURE between 0.5 and 15 pounds per square inch. *Compare* ATMOSPHERIC TANK.

loxodrome *See under* MERCATOR PROJECTION.

lumper in the biological sciences, one who tends to classify ORGANISMS with similar but slightly differing characteristics into a single SPECIES rather than dividing them into separate species on the basis of the small differences among them. *Compare* SPLITTER.

lunar cycle (metonic cycle) in astronomy, the 19-year cycle required to match the Moon's period of revolution around the Earth to the Earth's period of revolution around the Sun so that the Moon appears in the same phase on the same day of the year. The lunar cycle is important to environmental science because the net gravitational attraction of the Moon and Sun varies considerably according to their alignment with the earth, affecting the height of the tides and the elevation of sea level (*see* MEAN SEA LEVEL).

lunar rhythm a biological or physical cycle that is at least partially controlled by the passage of the Moon around the Earth. The most obvious lunar rhythm is the cycle of ocean tides. This cycle has a strong effect on the life cycles of littoral ORGANISMS (*see* LITTORAL ZONE), most of which have evolved inner biological rhythms that match the tides. Barnacles, for example, will continue to feed at the time of daily high tide even when placed in a tank of still water in the laboratory. Some lunar rhythms are exceedingly complex. For example, the grunion (*Leuresthes tenuis*), a small fish of the California and Baja California coasts, lays its eggs just below the high tide mark on sandy beaches on the first four nights following a full moon or a new moon. The eggs are timed to hatch with the next lunar peak tide, two weeks later. The existence of lunar rhythms in land organisms is less certain, although many biological cycles—for example, human menstruation—are approximately one lunar cycle long and may have originated as lunar-timed events.

LUST an acronym for LEAKING UNDERGROUND STORAGE TANK.

luster the way in which light is reflected from an object, especially when that object appears to glow as the light strikes it. Luster is a principal means of identifying rocks and minerals, and geologists have developed a complete set of terms to describe it. The most common of these terms are *adamantine* (diamondlike), *metallic, pearly, resinous, silky,* and *vitreous* (like broken glass).

M

maar in geology, a CRATER formed by a volcanic explosion that is not accompanied by lava flows (*see* LAVA) or by cone-building activity. Maars form circular pits with slightly elevated rims. They often fill with water to form small, deep lakes. Their principal cause appears to be steam explosions caused by the contact of GROUNDWATER with a rising body of MAGMA.

McCall, Thomas (1913–1983) conservationist and politician. As governor of Oregon from 1967 to 1975, McCall worked to create within the state government a balance between economic growth and environmental conservation. During his time as governor, Oregon became a national model for environmental reforms, including stringent pollution standards on pulp mills and sewage treatment, protection of beaches from development, recycling programs including the first "bottle bill" requiring deposits on beverage containers, and the implementation of sophisticated state-level land use regulations to control urban sprawl and protect farms and farmland. McCall served on the boards of the Nature Conservancy, the Center for Growth Alternatives, and the Conservation Foundation. His autobiography, *Tom McCall: Maverick,* was published in 1978, and a biography, *Fire at Eden's Gate: Tom McCall and the Oregon Story,* by Brent Walth, was published in 1994. *See also* OREGON LAND USE LAWS.

McCloskey, Michael (1934–) American environmentalist, born on April 26, 1934 in Eugene, Oregon, and educated at Harvard University (B.A., 1956) and at the University of Oregon School of Law (J.D., 1961). After graduation from law school he was hired by the SIERRA CLUB to serve as its Northwest Field Representative, becoming the first Sierra Club staff member to serve outside of San Francisco. At first he operated out of his parents' Eugene home, but in 1963 he was able to open an office in Seattle, Washington. In 1965 he resigned as northwest representative to move to San Francisco, where he held a variety of positions on the Sierra Club staff, eventually (1969) becoming the club's second executive director upon the departure of DAVID R. BROWER. In 1985 he became chairman of the Sierra Club, a position created for him. Quiet, conscientious, and dedicated, McCloskey has brought a high level of managerial professionalism to the environmental movement.

McPhee, John (1931–) American environmental writer, born March 8, 1931, in Princeton, New Jersey, and educated at Princeton (B.A., 1953) and at Magdalene College of the University of Cambridge, England. His first professional appointment was as a scriptwriter for a New York–based television show, *Robert Montgomery Presents* (1955–56); in 1957 he joined the staff of *Time* magazine, a position he held until 1965, when he left to become a staff writer for the *New Yorker.* That same year he published his first book, *A Sense of Where You Are.* McPhee's work is characterized by an immediacy of style and an ability to get inside a landscape and the people it has shaped, allowing the reader to empathize with many separate points of view. His outlook on his subjects is probably best summarized by noting that he is a Fellow of both the Geological Society of America and the American Academy of Arts and Letters. Since 1975 he has served as Ferris Professor of Journalism at Princeton. McPhee won the Pulitzer Prize for general nonfiction in 1999 for his book *Annals of the Former World,* on the geology of North America.

macrophyte in the biological sciences, a large plant. "Large" in this case means that the individual plant ORGANISMS are multicellular and can be seen by the naked eye; thus, the term includes everything from *Lemna* (DUCKWEED) to *Sequoia sempervirens* (giant sequoia). Since the so-called "true" plants—those belonging to the KINGDOM Plantae—are all now understood to be macrophytes, the term is probably obsolete, although it lingers strongly in limnology (the study of the biology of freshwater), where it is used to differentiate reeds, rushes, pond lilies and other plants that belong to terrestrial ORDERS but that happen to have taken up an aquatic life-style from the ALGAE and other plantlike, photosynthesizing MICROORGANISMS that float in the water and make up the PHYTOPLANKTON. *See* EMERGENT MACROPHYTE; FLOATING MACROPHYTE, SUBMERGED MACROPHYTE.

mafic rock rock composed primarily of dark-colored MINERALS such as HORNBLENDE, OLIVINE, PYROXENE, or biotite MICA. The color of almost all mafic rocks is the result of the inclusion of some form of magnesium silicate; thus, soils weathered from mafic rock are high in magnesium, a necessary plant TRACE ELEMENT. *Compare* FELSIC ROCK.

magma in geology, molten (hot liquid) rock beneath the Earth's surface. The heat to melt rock into magma appears to derive principally from the decay of radioactive materials deep within the earth, though some may be left over from the formation of the planet and some may result from physical forces such as compression and friction as the plates of the Earth's surface slide over one another (*see* PLATE TECTONICS). Magma may cool beneath the planetary surface, forming INTRUSIVE ROCKS, or it may flow out onto the surface as LAVA, forming EXTRUSIVE ROCKS. The actual rock formed will depend on both the chemical composition of the magma and the rate of cooling. Several different rock types may be formed from one body of magma, a process called *magmatic differentiation* (*see* BOWEN SERIES). *See also* MAGMA CHAMBER; VOLCANO.

magma chamber a large underground chamber filled with molten MAGMA. Magma chambers usually develop through a process known as *magmatic stoping* in which an area of fractured rock beneath the earth's surface, created either by EARTHQUAKE activity or by the pressure and heat of the magma itself, is invaded by a body of magma. The magma plucks pieces of the rock away and carries them off much as a GLACIER would do to rock on the earth's surface. The space that results is filled with the magma. Magma chambers form the active reservoirs for nearly all volcanic eruptions and are responsible, through their evacuation and collapse, for most caldera formation (*see* VOLCANO; CALDERA).

magnetic declination at any given location on the Earth's surface, the angle between a line drawn from that location to the North Pole ("true north") and the direction indicated by the needle of a magnetic compass ("magnetic north"). Magnetic north lies generally in the direction of the north MAGNETIC POLE, with local variations due to anomalies in the earth's magnetic field ("magnetic anomalies"). It varies slightly over time due to migration of the magnetic pole and to the effects of solar radiation. The line of zero declination—that is, the line upon which a magnetic compass will point to true north—lies near Sheboygan, Wisconsin. The declination on either the Atlantic or the Pacific Coasts of the continental United States varies from about 15° in the south to about 20° in the north. It is westerly on the Atlantic Coast and easterly on the Pacific Coast.

magnetic pole either of two points on the Earth's surface—one in the Northern Hemisphere and one in the Southern Hemisphere—where the lines of magnetic force from the Earth's magnetic field converge and become vertical to the ground. They lie approximately 1,000 miles from the "true" or GEOGRAPHIC POLES—that is, the points where the Earth's axis of rotation intersects the surface. The magnetic poles are the points on the Earth's surface toward which magnetic compasses approximately point (*see* MAGNETIC DECLINATION). They migrate a few miles each year as the earth's magnetic field changes. The north magnetic pole lies among the northern Queen Elizabeth Islands of Canada, on or near Ellef Ringnes Island. The south magnetic pole lies just off the Adêlie Coast of Antarctica, on a line extending between the true South Pole and the city of Adelaide, Australia. *See also* GEOMAGNETIC POLE.

MAI *See* MEAN ANNUAL INCREMENT.

main *See under* SEWER.

mainstem The principal or central watercourse of a river system; that is, a river from its mouth to its principal source, excluding all TRIBUTARIES. Generally, at a fork in a stream, the identity of the mainstem is determined by which branch carries the most water, however, it may at times be determined by which branch's bed seems to be a physical continuation of the bed of the combined branches downstream from the fork, or by which branch is longer.

Malthus, Thomas (1766–1834) English clergyman, economist, and educator, born near Guilford,

Surrey, England, on February 17, 1766, and educated at Jesus College, Cambridge, where he was elected a Fellow in 1793. He took religious orders in 1798 and was appointed curate of the parish of Albury, Surrey, but held the post only a short time. In 1805 he was appointed professor of modern history and political economy at the East India Company College at Haileybury, thus becoming the first professional economist in English history. The appointment was based in large part on the fame of his *Essay on the Principle of Population as It Affects the Future Improvement of Society,* published in pamphlet form in 1798 and, in revised form, as a book in 1803. In this work he pointed out that populations tend to increase geometrically while resources increase arithmetically, thereby leading to a constant outstripping of resources by population and a consequent need for some form of population control. Though the details of Malthusian theory remain controversial today—and though his writing style is universally conceded to be that of a barely controlled fanatic—the main thrust of his argument is now generally accepted in broad outline. It led in his own day to significant reforms in the English poor laws, and has been a major influence in the work of scientists from Charles Darwin to PAUL EHRLICH. Malthus died at Haileybury on December 23, 1834.

management concern in Forest Service usage, a specific problem in resource use that is to be addressed by a set of GOALS or MANAGEMENT OBJECTIVES as expressed in a FOREST PLAN.

management constraint in Forest Service and Bureau of Land Management usage, anything that limits a manager's scope of action in planning the management of a resource. *Legal constraints* are laws and administrative rules covering the resource and any proposed actions; *physical constraints* are the physical parameters of the site where the action will take place, such as SITE INDEX, ASPECT, SLOPE, soil and rock type, average annual temperature, and so on.

management direction in Forest Service usage, the broad, overall policy framework that guides management planning, consisting of the laws and regulations governing the agency together with general directives from the various supervisory levels. Management directions do not set target levels for any resource use. They point out the direction the agency is to take rather than telling it how far to go. Management directions guide the formulation of GOALS, which in turn guide the formulation of MANAGEMENT OBJECTIVES.

management framework plan (MFP) in Bureau of Land Management usage, the management plan for a specific area, including all MANAGEMENT DIRECTIONS, GOALS, and OBJECTIVES, together with the proposed means of carrying them out and their predicted effects on the various resources involved. It is directly comparable to the Forest Service FOREST PLAN or UNIT PLAN.

management intensity in Forest Service usage, the degree of management activity and manipulation of a resource involved in a management plan for that resource. It is usually expressed as the cost of a particular management plan relative to the costs of alternative plans for managing the same resource.

management objective in Forest Service usage, a specific, clear target for the management of a piece of land or of a resource on that land. Management objectives are limited and obtainable over the life of a plan. *Compare* GOAL; MANAGEMENT DIRECTION.

management prescription in Forest Service usage, the specific set of activities that will take place on a piece of land in order to meet the MANAGEMENT OBJECTIVES for that land and its resources, including the tasks to be done, the methods to be used, and standards to be met while using them.

management situation in Forest Service usage, the collection of data concerning conditions on a MANAGEMENT UNIT or MANAGEMENT ZONE together with the MANAGEMENT DIRECTIONS, GOALS, and MANAGEMENT OBJECTIVES that pertain to the unit or zone. In other words, the management situation is the bank of all information on the land use of a particular area or type of area that is available to those who are formulating management plans.

management unit in Forest Service usage, a specific area with fixed boundaries upon which a single set of GOALS, OBJECTIVES, and directions is to be applied. A management unit is the smallest unit of a FOREST PLAN; examples include individual WATERSHEDS; SPECIAL INTEREST AREAS, and STREAMSIDE MANAGEMENT UNITS. *Compare* PLANNING UNIT; MANAGEMENT ZONE.

management zone in Forest Service usage, a particular type of area that requires similar management techniques no matter where it appears. Examples include the RIPARIAN ZONE, the TRAVEL INFLUENCE ZONE, and the SPOTTED OWL HABITAT AREA. *Compare* MANAGEMENT UNIT.

mangrove swamp a crucial ecosystem on tropical coastlines made up of salt-tolerant mangrove trees. The exposed root systems, which are intricate and very sturdy, create a thicket that can shelter many species and, by trapping sediment washed from the land, provide a rich nursery and feeding ground for thousands of fish, invertebrates, and plant species. In addition, the mangrove swamps protect coastlines from storm damage and erosion, a special concern in Southeast Asia where mangroves are being removed to make way for shrimp AQUACULTURE and in Florida where they are removed to make way for coastal development.

mantle (1) in geology, the region of the Earth's interior between the crust and the core. The mantle begins at the MOHOROVICIK DISCONTINUITY, 10 to 35 km below the surface, and continues to the so-called *Gutenberg discontinuity* at a depth of about 2,900 km. It appears to be composed primarily of molten PERIDOTITE. Convection currents in the mantle are almost certainly the driving force behind the motion of the Earth's crustal plates (*see* PLATE TECTONICS). It is also the principal source of heat for the formation of MAGMAS and, indirectly, GEOTHERMAL POWER.

(2) in biology, a large flap of skin that covers the midportion of the body of a MOLLUSK. The space between the mantle and the body, known as the *mantle cavity,* contains the lungs and/or gills and serves (in aquatic mollusks) as a chamber for the absorption of food, the excretion of waste products, and the ejection of reproductive bodies (eggs and sperm). The outer surface of the mantle excretes the shell.

Manual, Forest Service *See* FOREST SERVICE MANUAL.

maple-beech-birch forest type *See* FOREST TYPES.

maple decline a long-term phenomenon in the Northeast afflicting sugar maples (*Acer saccharum*). The decline is manifested by crown and branch DIEBACK, premature leaf abscission, and mortality. First observed in Pennsylvania in 1912, the decline reached northern New England and southern Canada by the 1930s. In 1988, in Quebec's Appalachian Mountain region, researchers found that the number of maple trees showing symptoms of decline had reached 91.3%. ACID RAIN, which directly affects maple trees as well as leaching the nutrients in forest soils, is thought to be a contributing factor to maple decline in recent years. *See also* FOREST DECLINE AND PATHOLOGY.

marble metamorphosed LIMESTONE (*see* METAMORPHIC ROCK). Marble is harder than limestone, and often shows no bedding. When bedding exists, it is usually contorted. Long a sought-after material for buildings and sculpture because of its strength and durability under normal conditions, marble is now less in favor because it is a CALCIC ROCK that is highly vulnerable to damage by ACID RAIN and other forms of AIR POLLUTION.

marginal component in Forest Service usage, historically, land that is classed as COMMERCIAL FOREST LAND but that cannot currently be harvested due to technological limitations, economic costs, or environmental constraints. Land in the marginal component ("marginal land") is usually on very steep hillsides or on highly erodible soils, or contains commercial-sized timber only in scattered, isolated stands. Despite the improbability or impossibility of harvesting it, the timber of the marginal component usually remains in the TIMBER BASE and may be figured into the ALLOWABLE CUT. The term is no longer officially used by the Forest Service. *Compare* DEFERRED COMPONENT; STANDARD COMPONENT.

Marine Mammal Center rescues and rehabilitates ill or orphaned marine mammals that are beached along the central and northern California coast. Founded in 1995, the center also promotes public awareness of the oceanic environment through educational programs for its more than 100,000 visitors per year. Membership (1999): 35,000. Address: Marin Headlands, Golden Gate National Recreation Area, Sausalito, CA 94965. Phone: (415) 289-7325.

Marine Mammal Commission established in 1972 by the MARINE MAMMAL PROTECTION ACT. The commission periodically reviews the population status of marine mammals, researches aspects of marine mammal conservation, and makes recommendations on federal policies and activities which may affect marine mammals. Address: 4340 East-West Highway, Room 905, Bethesda, MD 20814. Phone: (301) 504-0087.

Marine Mammal Protection Act federal legislation to control the taking of marine mammals and to prevent their extinction, signed into law by President Richard Nixon on October 21, 1972. The act, which applies to U.S. citizens anywhere in the world and to citizens of all nations within U.S. territorial waters, forbids the taking of any marine mammal (broadly defined to include whales and dolphins, seals, sea lions, sea otters and walruses, polar bears, and any other mammal which is "morphologically adapted to the marine environment" or "primarily inhabits the marine environment") without a permit; forbids the importation or ownership of marine mammals, parts of marine mammals, or products made from marine mammals that have been taken without a permit; and sets up stringent requirements for the issuing of permits, including a finding that the taking of animals pursuant to the terms of the permit will not endanger their POPULATIONS and a requirement that the permit must specify precise numbers to be taken, acceptable methods of taking, and a specific period of time for which the permit is valid. No permit may be issued for an ENDANGERED SPECIES, except that subsistence hunting by Native Americans living in traditional cultures may be permitted as long as it does not further endanger the species' existence. The incidental killing of dolphins and other marine mammals during commercial fishing operations must be "reduced to the maximum extent feasible." Before the act, the accepted method of netting tuna—encircling them—had been killing as many as 500,000 dolphins a year. After passage of the act in 1972, U.S. fishermen greatly reduced the number of dolphins killed, although the kill rate by foreign fishing boats increased to the point that a public outcry was raised anew over the slaughter. In 1990, environmentalists successfully sued the administration of President George H. W. Bush to bar U.S. import of foreign-caught tuna, and European nations soon joined in with similar actions. Following the 1995 "Panama Declaration," prompted by the embargo, in which countries with major tuna-fishing fleets agreed to stop using the dolphin-drowning encircling nets, the U.S. Congress amended the Marine Mammal Protection Act to conform to the agreement. The amendments further required the National Marine Fisheries Service (NMFS) to find ways to reduce accidental marine mammal killings to a practical minimum, which though called "zero mortality" actually means that optimum populations of mammal species must be maintained at a sustainable level.

marine pollution *See* OCEAN POLLUTION.

Marine Protection, Research and Sanctuaries Act federal legislation to implement the International Convention on the Prevention of Marine Pollution by the Dumping of Wastes and Other Matters, signed into law by President Richard Nixon on October 23, 1972, and to establish a system of NATIONAL MARINE SANCTUARIES. The act requires a permit for the dumping of any materials whatsoever into oceans, bays, ESTUARIES, or any other tidal waters (waters whose levels are determined by and fluctuate with the ocean tides) controlled by the United States. Permits for dredged materials (*see* DREDGE; DREDGE-AND-FILL; SPOIL) are issued by the ARMY CORPS OF ENGINEERS (Corps), except that the ENVIRONMENTAL PROTECTION AGENCY (EPA) may veto such permits within 30 days of their issue if they are found to pose a threat to human health or the marine environment. Permits for all other materials are issued directly by the EPA. High-level radioactive wastes and biological and chemical warfare wastes may not be dumped at all. The EPA and the Corps may impose strict conditions on the permits, including establishing dumping sites, times, and amounts, and may withdraw any part of the CONTINENTAL SHELF from dumping altogether. In general, permittees are encouraged to dump beyond the edge of the continental shelf, in the deep ocean. Other provisions of the act set up a research and monitoring program on the effects of ocean dumping, and of dumping in the Great Lakes, to be administered by the Coast Guard, and require annual reports to Congress on compliance with the act.

marine sanctuaries *See* NATIONAL MARINE SANCTUARIES.

marine species any ORGANISM that lives its entire life in the ocean or on its immediate margins. Marine species are adapted to a saltwater environment and generally cannot survive in freshwater.

marl in geology, an impure LIMESTONE containing large amounts of CLAY. The clay content may range from around 10% ("marly limestone") to over 70% ("calcerous clay"). Marl usually indicates freshwater deposition. It generally occurs in lens-shaped bodies that probably represent the beds of ancient lakes. It is structurally weak and has little or no economic value.

marsh in ecology, a WETLAND characterized by the presence of large quantities of emergent vegetation (*see* EMERGENT MACROPHYTE), chiefly grasses, sedges and cattails, and reeds. Marshes develop along the edges of lakes and ponds, in shallow, slow-moving sections of streams, and in areas where the WATER TABLE is high enough to intersect the land surface part or all of the year. They differ from BOGS in that circulation is good so that there is a continuous supply of fresh oxygenated water. They are among the most productive of all ECOSYSTEMS, both in terms of total BIOMASS and in the great variety of ORGANISMS that inhabit them. They are also excellent natural water treatment devices, removing virtually all PARTICULATES and a high proportion of the dissolved solids (*see* DISSOLVED LOAD; TOTAL DISSOLVED SOLIDS) in the incoming water. *Compare* SWAMP. *See also* ESTUARY; SALT MARSH.

Marsh, George Perkins (1801–1882) pioneering American naturalist, born at Woodstock, Vermont, on March 15, 1801, and educated at Dartmouth College. After graduating from Dartmouth in 1820, Marsh "read law" in Burlington, Vermont, where he was admitted to the bar in 1825, allowing him to set up his own practice. He was elected to the Vermont legislature in 1834 and to the U.S. House of Representatives in 1842. From 1849 to 1854 he served as U.S. ambassador to Turkey, and in 1861 he was appointed ambassador to Italy, a position he held for the rest of his life. All of this was almost incidental to his primary interest, however, which was the study of the natural world. Marsh was among the first to recognize that nature consisted of a web of relationships rather than a group of unrelated objects, and that careless use by humans could harm those relationships. He believed that the principal cause of the fall of Rome was the Romans' degradation of their environment to the point that it could no longer support them—"destroying the principal instead of living off the interest," as he put it—and he saw the same thing beginning to happen in the United States. In 1864 he published *Man and Nature*, a study of the relationship between humans and their environment that revolutionized the study of geography and had a profound influence on the thinking of the generation of early conservationists that included GIFFORD PINCHOT, THEODORE ROOSEVELT and JOHN MUIR. Marsh died at Vallombrosa, Italy, on July 23, 1882.

Marsh test a means of detecting small amounts of ARSENIC, developed around 1840 by the English chemist James Marsh (1794–1846) and still widely used today. The substance suspected to contain the arsenic is introduced into a flask in which HYDROGEN is being generated, either by electrolysis or by a chemical reaction. If arsenic is present, it will combine with some of the hydrogen to form arsine (A_sH_3), a gas that is carried out of the flask with the hydrogen. If the hydrogen is passed through a heated glass tube, the heat and light in the tube will cause the arsine to decompose, releasing the arsenic, which precipitates onto the inside of the tube as a highly visible fine black powder. As little as 0.1 mg of arsenic—less than a hundred-thousandth of an ounce—may be detected in this manner.

Marshall, Bob (1901–1939) American forester and environmental leader, born on January 2, 1901, in New York City, and educated at Syracuse (Bachelor of Forestry, 1924), Harvard (Master of Forestry, 1925) and Johns Hopkins (Ph.D., 1930). Bob Marshall spent most of his short, active life in government service, as an employee of the FOREST SERVICE and the BUREAU OF INDIAN AFFAIRS. He eventually rose to become the Forest Service's first chief recreation officer (May 1937). A boyhood spent in the Adirondacks had given him a strong attraction to wild places, and as an adult he spent every spare minute hiking, traveling and exploring, from the Adirondacks—where he became the first person to climb all 46 peaks over 4,000 feet (1,218 meters) high, giving rise to a club known today as the "46'ers"—to the Brooks Range of Alaska, where he discovered and named the Gates of the Arctic. As a college student, he was a charter member of the ADIRONDACK MOUNTAIN CLUB; as an adult, he became a founder and a guiding spirit of the WILDERNESS SOCIETY, preparing most of its early policy statements and paying nearly all of its expenses out of his own pocket. His article, "The Problem of Wilderness" (*Scientific Monthly*, February 1930) is considered one of the most important position papers of the early environmental movement. An indefatigable hiker, Marshall often walked 50 to 65 kilometers a day (30 to 40 miles), and once climbed 14 of the highest Adirondack peaks in a single 18-hour period. Marshall Peak in the Adirondacks, Marshall Lake in the Brooks Range, and the Bob Marshall Wilderness in Montana are all named for him. He died of heart failure on November 10, 1939, on the train between Washington, D.C., and New York City.

mass extinction the dying out of species at a rate substantially higher than the ordinary "background" extinction rates. In the process of biological evolution, new species are formed continuously and others die out. In recent years, however, especially since the destruction of tropical rain forests, which are the richest places on Earth in terms of species diversity, biologists have determined that a major planetary "mass extinction" event is now taking place. The causes include disturbance of ecosystem integrity in the tropics as well as temperate areas, global warming, and intensified bombardment by ultraviolet rays in the "B" range (UVB) striking the Earth through a thinning ozone layer. In the view of biologist E. O. WILSON, this is the "sixth great extinction" and the first created by the activities of humankind. The last mass extinction took place 66 million years ago, when the dinosaurs disappeared along with 75% of all other species, triggered by a massive meteorite striking the Earth off the Yucatán Peninsula. The present event is characterized by between 18,000 and 73,000 species becoming extinct every year, a loss somewhere between 1,000 and 10,000 times the normal background rate. The implications of the current mass extinction are unclear to conservation biologists. Most do believe, however, that there is probably a threshold level in the rate of extinction that, once exceeded, may lead to a kind of ecological unraveling since species within ecosystems are interdependent. One writer has likened the integrity of natural systems at the global level to a pyramidal display of canned vegetables in a supermarket. A number of cans can be removed from the display without affecting it, but at some point, when the very next (threshold) can is removed, the display will necessarily come tumbling down. *See also* EXTINCTION.

massif a large mountain or mountain mass, usually with several summits, which is set off from neighboring mountains by valleys and deep passes and is clearly definable as a single unit. It is generally of a uniform geologic structure.

mass transit any systemized means for moving people from one place to another within urban areas and their suburbs without using private vehicles. Some authorities allow use of the word for any such system, while others limit it to systems running fixed routes on fixed schedules, with other forms of public or semipublic transportation (taxis, minivans with flexible routes that are able to pick people up at their doors, sliding sidewalks, etc.) being referred to as *paratransit.* Within the stricter definition many varieties of transit systems remain possible, including buses, rail (subways and commuter trains), light rail (trollies), and so-called "people movers" (automated, ultra-light rail vehicles, usually suspended above ground level, with short, numerous runs, often as little as three minutes apart, instead of long, widely spaced runs). Besides providing transportation for the poor and others without cars, either through choice or through necessity, mass transit systems reduce fuel use and pollution and dramatically increase the carrying capacity of city streets. Automobile traffic on a typical city street can move a total of about 1,400 people per hour, while buses can move 7,500 people per hour and light rail, as much as 30,000 people per hour.

mass wastage in geology and engineering, the movement of masses of soil or rock under the influence of gravity. Water may be involved as a lubricant, especially in areas underlain by clay soils, but it does not actually carry the earth material. Slow mass wastage is known as *creep.* More rapid mass wastage is termed a *landslide.* Special types of landslide include the SLUMP, the EARTHFLOW, and the mudflow (*see under* EARTHFLOW). Special types of creep include rock creep, TALUS creep and the *rock glacier,* a slowly moving mass of large and small rocks which acts in most respects like an ice GLACIER and may in some cases have an ice component. *See also* ANGLE OF REPOSE.

materials balance approach in economics, the application of the law of conservation of matter (*matter can neither be created nor destroyed*) to economic theory. The materials balance approach recognizes that all physical inputs to an economic system eventually show up as physical outputs, either as consumer products or as waste materials. The consumer products become waste materials when they reach the end of their useful life. Materials-balance economists use flow charts to analyze the proportion of consumer goods to wastes at any given point in an economy; the time lag between the introduction of a consumer good into the system and its exit from the system as waste; and the amount of waste materials which can flow back to the beginning of the cycle and become inputs. The ideal system, under this approach, allows all wastes eventually to become inputs; other systems can be evaluated by the proportion of their wastes that they have to absorb into the environment rather than placing back into production. *See also* LEONTIEF ANALYSIS.

materials recovery *See* SOLID WASTE; SOLID WASTE DISPOSAL ACT.

Materials Transportation Bureau (MTB) agency within the U.S. Department of Transportation (*see* TRANSPORTATION, DEPARTMENT OF) created by the Secretary of Transportation in July 1975 and charged with the coordination of federal policy regarding the transportation of HAZARDOUS MATERIALS by all methods, including highway, railroad, pipeline, and air (barge transport and other bulk transport by water is excluded from the bureau's coverage, however). The bureau issues regulations concerning hazardous materials transport; coordinates and approves hazardous-transport regulations issued by other agencies (such as the FEDERAL HIGHWAY ADMINISTRATION); and enforces pipeline safety regulations and all hazardous-transport regulations involving more than one mode of transport (e.g., truck and rail). Headquarters: 400 Seventh Street SW, Washington, DC 20590.

Mather, Stephen T. (1867–1930) founder and first director of the NATIONAL PARK SERVICE, born in San Francisco, California, on July 4, 1867, and educated at the University of California at Berkeley (B.A., 1887). From 1887 to 1893 he served as a reporter for the New York *Sun.* He then went into business with the U.S. Borax company in Chicago, rapidly amassing a personal fortune of several million dollars but wearing himself out through overwork in the process, so that he suffered a nervous breakdown in 1903. For therapy he joined the SIERRA CLUB and began participating in hiking and mountain climbing, meeting JOHN MUIR in 1906 and becoming profoundly influenced by Muir's environmental philosophy: from that time forward he used his considerable energy and much of his money promoting the cause of scenic preservation. In December 1914 he was asked to join the Department of the Interior (*see* INTERIOR, DEPARTMENT OF) as an assistant to the secretary, to promote and expand the National Parks, which at that time had no centralized management. He joined the department in January 1915, and by August 1916 had shepherded through Congress the bill creating the Park Service. As the agency's first director, he set extremely high standards of professionalism for its employees, paying many Park Service expenses out of his own pocket to avoid any chance that politics and Congressional cronyism would determine National Park policies. Driving himself at a relentless pace, Mather suffered two more episodes of serious depression as Park Service director (1916; 1918) and was finally forced to retire by a series of strokes (1928–29). He died in Darien, Connecticut, on January 22, 1930.

mature timber in forestry, a STAND of timber in which growth has slowed but the trees remain healthy and vigorous. The boundary between "young" and "mature" is highly subjective. Some authorities use the average number of ANNUAL RINGS per inch in the trunks of individual trees within the stand, six or more being usually considered mature. Others use the termination (or drastic slowdown) of height growth, while others relate maturity to the CULMINATION OF MEAN ANNUAL INCREMENT. Lately, the FOREST SERVICE has been moving toward a definition based on the presence of a number of stand characteristics, including trunks with an average DBH of 20 inches or more (*see* DIAMETER BREAST HEIGHT); CROWN CLOSURE of 50% or greater; the presence of at least two SNAGS of greater than 12 inches DBH per acre, with at least 7% of the snags in the stand exceeding 20 inches DBH; and the presence of six more downed logs (fallen trees) per acre, with an average diameter of at least 12 inches and an average length of at least 20 feet.

maximum modification in visual management, a term denoting a zone where the landscape may be heavily modified by human activity. The area may appear substantially or completely modified when viewed as part of the FOREGROUND or MIDDLEGROUND; however, when viewed as part of the BACKGROUND, a natural appearance should still be substantially present.

maximum social welfare (MSW, MXSW, Pareto optimum) in economics, strictly speaking, a state of the economy in which no household can have its welfare improved without decreasing the welfare of one or more other households. In a state of maximum social welfare all resources are fully allocated and price equilibrium has been reached, with the cost of producing each additional unit of any good (the *marginal cost*) precisely equal to the price individuals are willing to pay to purchase the good. The concept is valuable for analysis; however, it does not take into account the equity of distribution among households (the poor cannot get richer without decreasing the welfare of the rich) or the presence of EXTERNALITIES (the "welfare" spoken of is strictly monetary and cannot be used to measure AMENITY VALUES such as fresh air or scenic beauty), and thus should not be confused with a maximization of the quality of life or the ability to maintain current economic conditions into the future.

MBF abbreviation for 1,000 board feet (*see* BOARD FOOT).

MBM abbreviation for 1,000 feet, board measure. The term is exactly equivalent to 1,000 BOARD FEET. The abbreviations MBF and MBM may be used interchangeably.

m-discontinuity *See* MOHOROVICIK DISCONTINUITY.

mean annual increment (MAI) in forestry, the average yearly growth rate for a tree or a STAND, calculated by measuring the total volume and dividing by the age (for an individual tree) or the average stand age (for a stand). For most SPECIES, the mean annual increment increases each year during a tree's youth (that is, the young tree "puts on wood" faster and faster each year, with NET ANNUAL INCREMENT exceeding mean annual increment) and decreases each year during its maturity (the aging tree puts on wood more and more slowly, with net annual increment being less than mean annual increment). Where this curve changes direction—where mean annual increment and net annual increment are equal—is called the CULMINATION OF MEAN ANNUAL INCREMENT, and is defined as the age at which the tree reaches maturity. From an economic standpoint, the culmination of mean annual increment is usually specified as the best time to cut a tree down (*see* ROTATION AGE), although the tree will continue to put on volume after that point, and the wood produced after culmination—with its tighter growth rings—is generally of higher quality than the wood produced before.

meander (oxbow curve) in geology, a bend in a WATERCOURSE in which the watercourse changes direction by 180 degrees or more over a short distance. Meanders usually occur in groups and lie roughly at a right angle to the general course of direction of the stream, which appears to wander back and forth in a regular pattern. They are characteristic of old, established streams on flat terrain such as an alluvial plain (*see* ALLUVIUM), where they are caused originally by some event, such as a bank SLUMP or the entrance of a tributary, which causes the streamflow to be swifter on one side of the stream than the other, creating a difference in erosion rates. Once established, they tend to enlarge and to propagate downstream because the current always runs faster on the outside of a curve than it does on the inside (thus maintaining the differential erosion rate) and because the current tends to "bounce" from bank to bank once it has been deflected, creating differential erosion rates that switch from bank to bank at uniform intervals downstream. *See also* OXBOW LAKE.

mean sea level the average height of the ocean in relation to the land—not simply the midpoint of the range between extreme high and extreme low tides, but a weighted average calculated by taking hourly measurements over the entire 19 year period of the LUNAR CYCLE (waves, of course, are ignored). Mean sea level is the reference standard for all other elevation measurements on the Earth's surface, and is what is meant when one speaks, for instance, of Mt. Everest being 29,028 feet "above sea level." It rises and falls over geologic time due to factors such as the size of the Arctic and Antarctic ice caps and the slow motion of the Earth's crustal plates (*see* PLATE TECTONICS).

median threshold limit (MTL, 96HRTLM) in toxicology, the concentration of a TOXIN that will be lethal to 1/2 of the individuals in a test POPULATION within 96 hours of exposure. *See also* LD_{50}, LC_{50}.

medical waste a serious pollution issue that first came to public attention in 1988 when discarded hospital syringes, dumped at sea, washed up on eastern seaboard beaches. The next year, the Medical Waste Tracking Act was passed whereby dangerous wastes had to be "red bagged" and disposed of by incineration or (sometimes) other means rather than by ocean dumping or depositing in landfill sites. The result of the law was a vastly increased use of onsite incineration of medical waste, which itself caused serious pollution problems near hospitals, especially from DIOXIN, an extremely carcinogenic compound. In 1994 the ENVIRONMENTAL PROTECTION AGENCY declared that medical incinerators were the largest source of dioxin emissions in the country and by 1998 had forced many incinerators to close. Since then many hospitals have developed quite sophisticated disposal policies, involving the recycling of some wastes, the use of disposal techniques other than incineration, and the overall reduction of the waste stream through reuse of some formerly disposable items. According to the American Hospital Association, 15% of hospital waste is considered infectious. *See also* SOLID WASTE.

Mediterranean association *See* CHAPARRAL BIOME.

megalopolis term coined by city planner and geographer Jean Gottman in 1957 to describe complex, multinodal conurbations such as the rapidly developing metropolitan strips between Boston and Washington, Chicago and Pittsburgh, and San Francisco and San Diego. *See also* EDGE CITIES.

meiosis (meiotic division) the process of CELL division utilized prior to sexual reproduction, in multicellular ORGANISMS. Meiosis is a complex, multistage process in which the CHROMOSOMES are duplicated once but cellular division takes place twice, resulting in four daughter cells, each of which contains half the chromosomes of the original mother cell. The daughter cells then become *gametes* (eggs and sperm), which unite during sexual reproduction to form a new organism with a full set of chromosomes, half of which come from each parent. *Compare* MITOSIS. *See also* ALTERNATION OF GENERATIONS.

meltdown the most serious kind of accident at a nuclear power plant. A meltdown, or "core meltdown," occurs when a runaway nuclear chain reaction takes place in the emergency core cooling system. Accidents at CHERNOBYL, Ukraine; THREE MILE ISLAND, Pennsylvania; and TOKAIMURA, Japan, involved partial core meltdowns. *See also* CHINA SYNDROME.

membrane diffusion *See* OSMOSIS.

Mendel, Gregor (1822–1884) Austrian biologist and botanist, father of plant genetics. After studying science in Vienna, Mendel became an abbot at Brno where he conducted research on the inherited characteristics of plants, especially peas. He crossed varieties that displayed certain characteristics to observe the genetic effects. His observations of more than 28,000 plants showed that traits are not directly passed down from generation to generation, but that discrete factors (genes) are responsible for the expression (dominance) or lack of expression (recessiveness) of various traits. Mendel presented his findings to the local Natural Science Society in 1865. The importance of his work was not recognized until the early 1900s, but Mendel's principles of inheritance have since provided the basis for an understanding of genetics. *See also* GENE; GENETIC ENGINEERING.

Mercalli scale in geology, an EARTHQUAKE scale, developed by the Italian seismologist Giuseppe Mercalli around 1902 and slightly modified since. Unlike the RICHTER SCALE, which measures the *magnitude* of an earthquake (that is, the energy it releases) objectively through the use of seismographs, the Mercalli scale measures the earthquake's *intensity*—a subjective factor that depends not only on the amount of energy released but also the manner in which it is released, the location of the epicenter, the structure of the underlying bedrock, and various other factors. The Mercalli scale is expressed in Roman numerals. I is detectable only by instruments; IV may awaken sleepers and rattle doors; VII damages buildings and affects wells; and XII (the highest Mercalli rating) throws boulders in the air, changes the courses of rivers, and generally causes complete destruction of buildings. During a Mercalli XII earthquake, the ground may be seen to wave up and down several feet. (See following table.)

The Mercalli Scale

- I detectable only by instruments
- II felt on upper floors of buildings; suspended objects may swing
- III felt noticeably indoors; sensation like a heavy truck passing
- IV dishes, windows, doors disturbed; some sleepers wakened
- V dishes, windows broken; unstable objects overturned; trees and poles sway slightly
- VI many people run outdoors; heavy furniture moved; slight damage to plaster and brickwork
- VII felt in moving vehicles; slight damage to ordinary structures; everyone runs outdoors
- VIII considerable damage to ordinary structures; some chimneys, columns, smokestacks, etc., fall; sand and mud ejected in small amounts
- IX most masonry and many wooden structures destroyed; rails bent; riverbanks and steep slopes slide; water splashes over riverbanks and lakeshores
- X most masonry and many wooden structures destroyed; rails bent; riverbanks and steep slopes slide; water splashes over riverbanks and lakeshores
- XI all masonry and most wooden structures destroyed; bridges destroyed; broad fissures in ground
- XII damage total; objects thrown upward; waves seen on ground surface.

Mercator projection a means of representing the surface of a sphere (such as the earth) on a flat plane(such as a map), developed by the Flemish geographer Gerhardus Mercator (1512–1594) and first used by him on a map made in 1568. The Mercator projection treats the North and South Poles as if they were an infinite distance from the equator, thus allowing all north-south lines to be drawn parallel to each other (*see* LONGITUDE). It has a great advantage for navigation because compass directions always correspond to map directions (that is, north-south is always straight up and down and east-west is always exactly horizontal), so that a line drawn from one place to another on the map always represents the exact compass heading necessary to reach that spot (a mapped line with these characteristics is known as a *rhumb* or *loxodrome;* all lines on a mercator projection are rhumbs). Areas near the poles are greatly distorted in size and appearance, however, and the poles themselves cannot be represented without (literally) an infinitely long piece of paper; hence the Mercator projection cannot be used to represent the actual shape of the features of the earth's surface except on a very small scale where the difference in distortion between the top and the bottom of the map is small enough to be insignificant. A variant of the Mercator projection, the *transverse Mercator,* treats the longitude line that happens to run down the center of the map as a vertical "equator" and projects east and west to an imaginary "east pole" and "west pole" 90° away. Because distortion is always smallest near the equator, this projection is more accurate than the standard Mercator for those areas which appear near the map's center, and it has become the standard projection for TOPOGRAPHIC MAPS.

merchantable timber generally, any tree large enough and free enough from DEFECT to be sold for use in manufacturing lumber products. In Forest Service usage, the minimum SCALING DIAMETER for merchantable timber is 6 inches (over a 16-foot "standard log") and the maximum allowable defect is 75%. In private forestry, the minimum scale is often set at 8 inches and the maximum defect at 66%. In practice, this means that the tree must have a minimum DBH of 12 inches (*see* DIAMETER BREAST HEIGHT) and contain at least one sound standard log.

mercury the 80th element in the atomic series; atomic weight 200.59, chemical symbol Hg. Mercury, one of the HEAVY METALS, is a silvery, mobile fluid with an extremely high SURFACE TENSION and good electrical CONDUCTIVITY. It is the only metal that exists in the liquid state at room temperature. Mercury is nearly inert at normal temperatures and conditions but will form a variety of COMPOUNDS when heated or exposed to certain metabolic processes (*see* METABOLISM). The element is used in electrical switches and fluorescent lamps, in thermometers and thermocouples, as a FUMIGANT, and in numerous industrial processes, including pulp and paper manufacturing, gold and silver extraction, and as a CATALYST in the manufacture of a variety of ORGANIC COMPOUNDS.

MERCURY AS A POLLUTANT

Mercury is highly toxic in all forms, both as an element and in virtually every known compound, whether inhaled, ingested, or absorbed through the skin. Exposure to small amounts of elemental mercury over short periods of time is not normally a problem due to the substance's high chemical immobility. However, exposure to high amounts or chronic, long-term exposure to small amounts can cause nervousness, irritability, timidness, and tremors (the so-called "hatter's shakes," so-named because hatters were often afflicted by the condition as a result of their exposure to mercury during the curing of felt; hence, the term "mad as a hatter"). At very high concentrations it can cause nerve damage and severe birth defects (*see* MINAMATA DISEASE). Although it is only very slightly soluble in water, it is a severe hazard to water bodies because it concentrates in the bottom sediments, where ANAEROBIC BACTERIA convert it

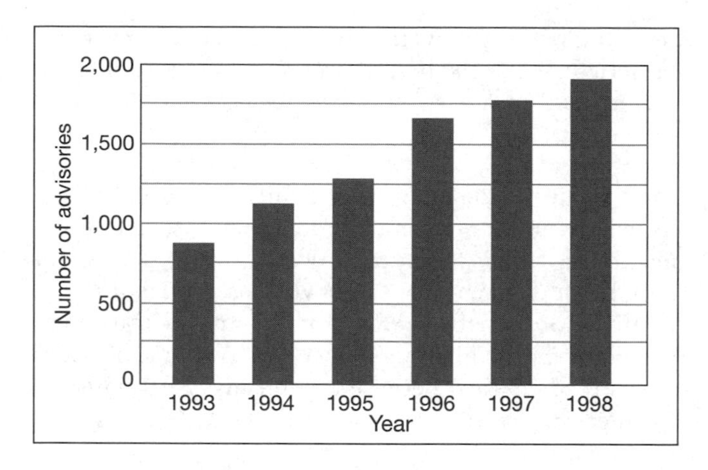

Mercury fish advisories ***(U.S. Environmental Protection Agency, Office of Water, Fish and Wildlife, 1999)***

through metabolic processes to *methyl mercury* (C_2H_6Hg), a moderately to strongly fat-soluble compound which bioconcentrates readily up the food chain (*see* BIOLOGICAL MAGNIFICATION), making fish living in mercury-contaminated waters a clear health hazard to the humans and animals that eat them. Current U.S. drinking water standards for mercury allow a maximum concentration of just 0.002 mg/liter (Code of Federal Regulations 40 CFR 141.11).

Merriam life zone *See* LIFE ZONE.

mesa a small, isolated, elevated PLATEAU with a flat, table-like top and at least one steep side. Mesas are caused by the presence of a horizontal surface layer of erosion-resistant CAPROCK overlying a weaker STRATUM or series of strata. Where the caprock breaks down or ends, winds and water are able to get at the weaker materials and carry them away, leaving only the covered and protected portion standing. Mesas are most common in dry climates such as the American Southwest, evidently because dry soils are more easily wind-erodible than are damp soils.

mesic site in the biological sciences, a moderately well-watered, moderately warm growing site. The soil is moist over half the year but is rarely saturated. Mean soil temperature ranges from 8°C to 15°C (45°F–60°F). Mesic sites are normally the best growing sites for a wide variety of plants. *Compare* HUMIC SITE; XERIC SITE. *See also* MOISTURE GRADIENT; SOIL WATER.

mesotrophic lake a lake that contains moderate levels of dissolved NUTRIENTS in its water. *See* TROPHIC CLASSIFICATION SYSTEM.

metabolism in the biological sciences, the set of chemical REACTIONS taking place within the living CELL that allow energy to be used and transferred and tissues to be constructed and repaired—in other words, the life process. There are two basic types of metabolism. *Anabolic metabolism,* or *anabolism,* consumes energy in the process of building up complex MOLECULES, and is the process by which tissues, ENZYMES, and indeed all molecules of a living ORGANISM are constructed. *Catabolic metabolism,* or catabolism, is the reverse of anabolism. It breaks complex molecules down into simpler ones, releasing energy. The speed at which these reactions take place is known as the *metabolic rate.* The minimum metabolic rate for a given organism—that is, the rate its metabolism proceeds at when it is totally at rest—is called its *basal metabolism.*

THE METABOLIC PROCESS

All metabolism—plant, animal, FUNGUS, BACTERIUM or PROTISTA—proceeds in roughly the same manner. At its heart is a COMPOUND called adenosine triphosphate, or ATP. ATP contains three PHOSPHATE molecules, the third of which is bonded rather weakly to the remainder of the molecule. It can be easily removed (liberating energy) or replaced (consuming energy). Thus, ATP can consume energy in one part of the metabolic cycle and release it in another, moderating the energy flow and keeping it constant. The "fuel", which provides the energy for the ATP reactions is primarily sugar, either built up through PHOTOSYNTHESIS (as in the plants) or consumed from external sources (as in the animals). Other compounds and ELEMENTS are utilized as needed for the various processes the cell undergoes. The process as a whole is extraordinarily complex. There may, for example, be anywhere from 12 to 140 steps required to complete the oxidation of a single molecule of sugar within the cell, a process which takes only one step in the atmosphere. *See also* RESPIRATION.

metal broadly speaking, any ELEMENT that has a positive VALENCE, that is, one that gives up ELECTRONS when forming ionic compounds (*see* COMPOUND; ION). More narrowly defined, metals are those elements that give up their electrons easily enough to form *metallic bonds,* in which a single electron may be shared by three or more atoms (*compare* COVALENT BOND). Metals in their pure state do not consist of whole atoms, but of IONS embedded in a so-called "sea of electrons" that are shared among the ions and move freely between them, giving rise to the excellent electrical CONDUCTIVITY characteristics of most metals. Other characteristics of metals (also generally attributable to the "sea of electrons" structure) are *ductility* (the ability to be drawn into a wire), *malleability* (the ability to be pounded into a thin sheet that conforms to a surface without breaking), good heat conductivity, the characteristic sheen known as *metallic luster* (*see* LUSTER), and the ability to form ALLOYS. *See also* HEAVY METAL.

metalimnion *See* THERMOCLINE.

metamorphic rock rock whose physical structure and (usually) MINERAL content have been changed by heat and pressure, sometimes with the addition of new chemicals from GROUNDWATER, MAGMA, or neighboring rock masses. The conditions necessary to metamorphose one rock type into another are generally found only in regions of vulcanism (*see* VOLCANO) or intense TECTONIC ACTIVITY, though sometimes the weight of a very thick body of rock may generate enough pressure to metamorphose the bottommost layers. Metamorphic rock may form from either IGNEOUS ROCK or SEDIMENTARY ROCK. Modified igneous rocks are called *metavolcanics* and modified sedimentary rocks are known as *metasediments*. In either case the modified rock is usually denser and tougher ("more competent") than the parent rock it came from. The crystals are packed more tightly together and are usually oriented in the direction that the pressure was applied. Examples of metamorphic rocks include MARBLE (metamorphosed LIMESTONE), SLATE (metamorphosed SHALE), and QUARTZITE (metamorphosed SANDSTONE). *See also* CONTACT ROCK.

metasediment *See* METAMORPHIC ROCK.

metastable system a physical, chemical, or biological system that has temporarily stabilized at a higher than normal energy level, usually due to some outside inducement. A metastable system may be anything from an atom in which one or more electrons have skipped up to higher level shells (*see* ATOM: *structure of the atom;* PHOTOSYNTHESIS) to an entire ECOSYSTEM such as a grassland (*see* GRASSLAND BIOME) in which the SPECIES composition has been changed by intense grazing pressure. Metastable systems are always in danger of collapse.

metavolcanic *See under* METAMORPHIC ROCK.

methane a colorless, odorless, flammable gas, chemical formula CH_4, formed during the anaerobic decomposition of ORGANIC MATTER (*see* ANAEROBIC BACTERIA). It can also be formed by the direct union of CARBON and HYDROGEN at very high temperatures. Methane is the principal component of NATURAL GAS. It is the prime constituent of the atmospheres of the outer planets (Jupiter, Saturn, Uranus, and Neptune) and appears to have been present in large amounts in the early atmosphere of the earth as well. As *firedamp* it is often present in coal mines, where it is regarded as a danger due to its explosive and asphyxiant qualities. As *marsh gas* or SWAMP GAS it bubbles to the surface of stagnant bodies of water from colonies of anaerobic bacteria living in the bottom muck. It burns readily and cleanly, and makes an excellent fuel (*see* ALTERNATIVE ENERGY SOURCE).

methylation the addition of a methyl GROUP (CH_3–) to an element or compound, usually through the metabolic activity (*see* METABOLISM) of a class of ANAEROBIC BACTERIA known as *methyltropes*. Methylation is particularly acute as an environmental problem when it involves HEAVY METALS, as it provides a means for these otherwise organically inert materials to enter the FOOD CHAIN (*see,* e.g., MERCURY: *mercury as a pollutant*).

methyl mercury *See under* MERCURY.

methyl tertiary butyl ether (MTBE) an oxygenating compound that added in small quantities to gasoline boosts octane and reduces harmful hydrocarbon emissions. Under the 1990 Clean Air Act, the ENVIRONMENTAL PROTECTION AGENCY (EPA) required that the oxygen content of gasoline be increased 2%. To achieve this goal, either ethanol, made from corn (or BIOMASS), or MTBE, made from methanol derived from natural gas, could be used to reduce hydrocarbon emissions. Gasoline reformulated under EPA guidelines required a blend containing either 11% MTBE or 5.7% ethanol in areas of high AIR POLLUTION. Hydrocarbon emissions are reduced by about 15% with either compound, but MTBE was the additive of choice for most petroleum companies since they could supply it to themselves at a lower cost. Moreover, it has less volatility than ethanol, thus reducing evaporation. By the late 1990s, about one-third of all gasoline used in the United States contained MTBE as an additive. A serious problem with the compound arose in 1996 when residents of Santa Monica, California, began complaining of a turpentine-like taste and odor in their drinking water, requiring the city to shut down half of its water supply wells. Analysis of the tainted water revealed that the cause was MBTE, which had escaped from leaky service station storage tanks. Since then, storage-tank leakage has contaminated water in 16 states. The compound is extremely water-soluble and can travel long distances very quickly in GROUNDWATER plumes. MTBE is, in addition, quite persistent, not biodegrading, and even at trace levels remains discernible. At high concentrations, which have been obtained in several cases in California and elsewhere, MTBE is toxic, having been shown to cause cancer in laboratory animals. Among human symptoms of MTBE poisoning are dizziness, rashes, swelling, respiratory problems, and diarrhea. In 1999 the EPA told gasoline producers to curtail the use of MTBE because of its potential to contaminate drinking water supplies, and the U.S. Congress was asked by a coalition of Northeastern states, where MTBE has been found in 15% of the drinking water (albeit at low levels), to repeal the 2% oxygen requirement altogether. In 2000, in response to EPA requirements for additives, gasoline producers raised prices radically in several areas, especially the Middle West.

metonic cycle *See* LUNAR CYCLE.

metric system system of weights and measures based on measurable physical standards and related to each other exclusively by powers of 10. When first proposed by French scientists in 1790, the metric system was based exclusively on the METER, which was defined as 1/10,000,000 of the distance from the equator to the North Pole. The unit of mass, the *kilogram,* was defined as the mass of a cube of pure water 1/10 of a meter on each side at a temperature of 4°C (Centigrade, or Celsius, a temperature scale which uses the freezing point of water as 0° and the boiling point of water as 100°). In the late 19th century it was discovered that these "standards" were not invariable (the earth is not perfectly round; water, an extremely broad-spectrum SOLVENT, cannot be made perfectly pure), causing the meter to be redefined as the length of a specially made platinum-iridium bar kept in a vault in Washington, D.C. and the kilogram as the mass of a platinum-iridium cylinder kept in a vault in Paris. In the early 20th century, this so-called "mks system" (for meter-kilogram-second, the second—a unit of time—being defined in terms of the Earth's period of revolution about the Sun) was supplemented for scientific work requiring small measurements by the "cgs system" based on the centimeter and gram (1/100 of a meter: 1/1000 of a kilogram) instead of the meter and the kilogram. In 1960, these systems were both replaced by the International Standard System (Systeme International, abbreviated SI) at an international conference held in Paris. This system has now been adopted worldwide. The SI system defines six basic units: the meter, the kilogram, the second, the degree Celsius, the *candela* (a unit of light intensity), and the *ampere* (a unit of electrical current). All other units are derived from these. The meter and the second are defined in terms of atomic properties. The degree Celsius is defined in relation to the so-called "triple point" of water (the point, just above freezing, where the solid, liquid, and gaseous states of water are in perfect equilibrium with each other), and the kilogram is still defined in terms of that platinum-iridium cylinder in Paris. The candela is defined as the light emitted by 1/60th of one square meter of the surface of a perfect radiator ("black body") heated to the melting point of platinum, and the ampere is defined in terms of the force generated by current flowing in wires exactly 1 meter apart in a perfect vacuum.

The Metric System

	SI Unit	U.S./English Unit	Conversion Factors: SI to U.S.	Conversion Factors: U.S. to SI
length	centimeter (cm)	inch (in)	0.3937	2.540
	meter (m)	foot (ft)	3.2808	0.3048
area	square centimeter (cm^2)	square inch (in^2)	0.155	6.4516
	square meter (m^2)	square foot (ft^2)	10.764	0.0929
volume	cubic centimeter (cm^3)	cubic inch (in^3)	0.061	16.3872
	cubic meter (m^3)	cubic foot (ft^3)	35.3133	0.0283
capacity	liter (l)	quart (qt)	1.0567	0.946
weight	gram (g)	ounce (oz)	0.0353	28.3495
	kilogram (kg)	pound (lb)	2.2046	0.4536

Multiply by the SI to U.S. factor to convert a measurement in metric system units to its English system equivalent; multiply by the U.S. to SI factor to convert a measurement in English system units to metric.

M-44 *See* COYOTE GETTER.

MF method *See* COLIFORM COUNT.

MFP *See* MANAGEMENT FRAMEWORK PLAN.

MGD abbreviation for *million gallons per day,* a rate of flow equivalent to 1.55 cubic feet per second (cfs). *See also* MIGD.

mica in geology, any of a group of closely related rock-forming MINERALS characterized by numerous prominent parallel CLEAVAGE PLANES that are evenly spaced and extremely close together so that the mineral splits easily into thin, uniform, elastic sheets. Micas are silicate minerals that are closely related to QUARTZ. They lie close to the bottom of the BOWEN SERIES, crystallizing out of MAGMA at relatively low temperatures, and are found commonly as tiny flecks scattered through granite and similar IGNEOUS ROCKS. They also form in some metamorphic rocks, notably marbles and schists. Crystals of pure mica, known as "books" due to the resemblance of the individual layers to pages, are hexagonal, and may be several feet across. They have been used as substitutes for glass under the name *isinglass*. The four principal types of mica are *muscovite,* a white or transparent variety that is a compound of aluminum; *phlogopite*, a brown variety in which the aluminum has been replaced by magnesium; *biotite,* a dark brown or black variety containing both magnesium and aluminum; and *lepidolite,* containing lithium, and appearing in a variety of pastel purples and pinks.

microbe *See* MICROORGANISM.

microclimate a small, localized CLIMATE pattern that differs in some significant way from the overall climate of the region in which it is found. The air beneath a tree, for example, is always cooler and moister than the air above an adjacent open patch of grass. One way to express this is to say that these two spots have separate microclimates. Likewise, the air among the stalks of wheat in a wheat field will have a different microclimate from the air above the stalks, and the north-facing slope of a gully will have a different microclimate (and grow different vegetation) from the south-facing slope. The term is also used to differentiate the climate of cities from that of the surrounding countryside (*see* URBAN HEAT ISLAND), the climate of mountains from that of the surrounding prairies, the damp "fog belt" along a coast from the drier areas just inland, and so forth.

microcontaminant any POLLUTANT that poses a significant health hazard even when present in the water or air at a very low concentration. Microcontaminants may be dangerous at levels measureable in parts per million (ppm), parts per billion (ppb), or even parts per trillion (ppt). A few (such as MIREX) are so hazardous that they are considered a significant danger if they can be measured at all. Most substances classed as microcontaminants are complex ORGANIC COMPOUNDS such as DDT, PCBs (*see* POLYCHLORINATED BIPHENYL) or DIOXINS.

microenvironment a small area within which the environmental conditions differ significantly from those of the larger area around it. These differences may exist in terms of climate (*see* MICROCLIMATE); availability of light, heat, or moisture; soil chemistry; or any of a number of other factors or combinations of factors. *See also* MICROHABITAT.

microgram (µg, mcg) one millionth of a gram.

microhabitat a small portion of a HABITAT—often only a few square inches to a few square feet in extent—within which conditions are right for the survival of one or more living SPECIES that cannot live elsewhere in the same habitat. The shade of a small boulder, for example, might lower temperatures enough that a group of plants could grow there that might be found nowhere else in the vicinity; these plants might in turn support a small community of insects and even one or two VERTEBRATES such as lizards or mice. The study of microhabitats is an important tool toward the understanding of the distribution of POPULATIONS of living things. *See also* NICHE; MICROCLIMATE; MICROENVIRONMENT.

micron unit of length in the METRIC SYSTEM, equivalent to one-millionth of a meter, or roughly 0.00004 inches.

micronutrient *See* TRACE ELEMENT.

microorganism (microbe) any living ORGANISM that is too small to be seen without the aid of a magnifying glass or a microscope. There are well over a million known species of microorganism, including all MONERA (bacteria and blue-green algae), most PROTISTA (algae and protozoa), many fungi (including yeasts, smuts, and related organisms; *see* FUNGUS), and some animals (rotifers, hydra, and similar creatures, along with the LARVAE of many marine invertebrates). No true plants are microscopic. VIRUSES are a special case. Since they are really submicroscopic rather than microscopic in size (they are too small to be seen by any optical instrument and can only be viewed with a scanning electron microscope), and since there remains some question as to whether they are truly living or not, they are generally referred to as PARTICLES rather than as microorganisms.

ENVIRONMENTAL EFFECTS OF MICROORGANISMS

Microorganisms are present everywhere: in the air, in and on our bodies and those of other animals, in the water, and in the ground. Roughly 30% of the bulk of animal and human FECES is made up of microorganisms. Over a billion of them may be present in a single gram of soil. A few of these cause diseases (*see* PATHOGEN); however, the overwhelming majority are necessary components of the ECOSYSTEM. Microorganisms form the base of all aquatic FOOD CHAINS. In the soil and in the bottom sediments of lakes, streams and seas, they act as DECOMPOSERS, breaking down dead ORGANIC MATTER into its constituent MOLECULES so that they may be used over again by new living things. With very few exceptions they are the only organisms that can fix elemental NITROGEN into organically usable form, and so are depended on by all other life forms for access to this necessary NUTRIENT. The RESPIRATION of photosynthesizing microorganisms such as algae is probably responsible as well for the presence of most of the free OXYGEN in the air. *See also* PLANKTON; BENTHOS.

microtopography the immediate surface TOPOGRAPHY of a local region. Where "topography" refers to the presence of hills and valleys, "microtopography" refers to the location of individual stones, soil PEDONS, etc., on the hills or in the valleys, or even to the pattern of RILLS or other small erosive features that may develop ephemerally after a rainstorm. *See also* MICROHABITAT.

middleground in visual management, the region of a VIEWSHED lying roughly from 1/4 mile to 5 miles from the observer, in which trees and similarly sized objects are individually visible but do not stand out from their surroundings. *Compare* FOREGROUND, BACKGROUND.

mid-ocean ridge a long, linear MOUNTAIN RANGE stretching down the center of the earth's oceans and forming, with its offshoots, a single MOUNTAIN SYSTEM 46,000 miles long. It is the largest geophysical feature on the planet. Beginning in the Arctic Ocean very near the North Pole, the Mid-Ocean Ridge passes down between Greenland and Europe into the Atlantic, becoming the *Midatlantic Ridge;* Iceland, the Azores, and Ascension Island are above-water portions of it. Rounding the southern tip of Africa, it enters the Indian Ocean, where it joins with the *Carlsberg Ridge* coming down from the Arabian Sea to form the *Mid-Indian Ocean Ridge,* passes into the Pacific between Australia and Antarctica as the *Pacific Atlantic Ridge,* and turns northward through the Pacific as the *East Pacific Rise,* giving birth to Easter Island before ending against North America at the Gulf of California.

ORIGIN AND SIGNIFICANCE OF THE RIDGE

The mid-ocean ridge lies along the active, growing edges of most of the Earth's crustal plates (*see* PLATE TECTONICS). As the plates pull apart, MAGMA wells up from the Earth's interior. It is deposited on the edges of the plates as a twin row of basalt peaks that slowly separate as the plates move away from each other—a phenomenon known as *seafloor spreading*—forming a twinned mountain range with a rift down the center that is floored by the youngest rock in the ocean basins. As the rock cools and compresses, huge fractures develop in it at right angles to the ridge. Slow spreading, as in the Atlantic, forms high, steep mountains with prominent fractures; rapid spreading, as in the Pacific, results in a smooth "rise" rather than a mountain cordillera and makes the cross-fractures much less distinct. Rapid spreading also gives rise to thermal features such as HOT SPRINGS and BLACK SMOKERS, which recirculate ocean water and charge it with minerals. It is estimated that the entire volume of the oceans passes through the mid-ocean ridge approximately once every 200 million years, and it is now thought that this circulation—rather than the erosion of minerals from the continents by rivers—is primarily responsible for the "saltiness" of the sea. The hot springs ("vents") are also the locations of the only known complex ECOSYSTEMS that do not depend on energy from the Sun. Using a process called *chemosynthesis* instead of PHOTOSYNTHESIS, bacteria (*see* BACTERIUM) utilize the hot SULFUR in the vent water to provide energy for fixing the CARBON in dissolved CARBON DIOXIDE into ORGANIC COMPOUNDS, forming the basis for a FOOD CHAIN that includes giant tube worms, crabs, clams, and even a few fish—all living in permanent darkness at the bottom of the deepest portions of the ocean.

MIGD abbreviation for *million imperial gallons per day,* a flow rate of 1.86 cubic feet per second (cfs). *See also* MGD.

migration corridor in ornithology, a localized, specific route used for travel by migrating birds. Wherever possible, migration corridors follow some linear ground feature, such as a river or a range of hills, that can be used for visual guidance. Where these features do not exist, migration corridors tend to be straight-line courses between easily identifiable features such as lakes and prominent hills or mountains. The term is also used by biologists to describe the pathways used by migrating land animals such as caribou. *See also* FLYWAY.

Migratory Bird Conservation Commission founded under the Migratory Bird Conservation Act of February 18, 1929, the commission considers and fixes the price for land recommended by the Secretary of the Interior for purchase or lease as migratory bird refuges in the National Wildlife Refuge System. Address: Interior Building, Washington, DC 20240. Phone: (703) 358-1716.

millibar in meteorology, a measurement of air pressure in the METRIC SYSTEM. One millibar is equal to 100 pascals, or about .015 pounds per square inch.

milligram one thousandth of a gram (*see* METRIC SYSTEM).

mimicry in the biological sciences, a complex of coloration, scent, shape, and/or behavior traits exhibited by a SPECIES of plant or animal that makes it closely resemble some portion of its ENVIRONMENT—commonly another species, though it may be something else (a rock, a dead twig, even, in at least one case, a bird dropping). Mimicry goes beyond CRYPTIC COLORATION ("camouflage") in that the mimic usually acts like the model and smells like the model as well as looking like the model. A species will usually mimic another to gain some sort of protection by association; for example, the harmless Louisiana snake, *Lampropeltis triangulum,* looks and behaves almost identically to the venomous coral snakes (*Micruroides*) that share its habitat. The two principal types of mimicry are MULLERIAN MIMICRY, in which a number of species that share some harmful characteristics advertise that fact by similar coloration and behavior (as in the bees, wasps, and hornets), and BATESIAN MIMICRY, in which a harmless species mimics a harmful one (as in the snakes mentioned above, or in the various flys and beetles that copy the behavior and coloration of bees, wasps, and hornets). A variant of Batesian mimicry called *speed mimicry* is sometimes seen in which a slow-moving mimic patterns itself on a model that is rarely caught due to its fast reaction times and great speed, and is thus no longer chased by the majority of predators. *See also* AGRESSIVE MIMICRY.

Minamata disease chronic MERCURY poisoning. The disease is named for Minamata Bay, Japan, where mercury-containing wastes from a local factory in the 1950s contaminated shellfish and other seafoods taken from the bay, causing local residents who depended on these foods to develop severe symptoms of the disease, including tremors, loss of coordination, memory loss, personality disorders, and severe learning disabilities. See the more complete description of the symptoms under MERCURY.

mineral any naturally occurring chemical COMPOUND or ELEMENT found as a constituent of ROCK. Minerals are generally inorganic (*see* INORGANIC COMPOUND), though pure CARBON in its various forms is usually classed as a mineral, and FOSSIL FUELS sometimes are. There are approximately 2,000 known minerals, the most common of which are FELDSPAR (30%–50% of the Earth's crust) and QUARTZ (28%–30%). Other widely occurring minerals include the calcite group (*see* LIMESTONE), the halide group (*see* SALT), the MICAS, and the various iron oxide compounds (LIMONITE, HEMATITE, TACONITE, etc.) that serve as ORES of iron. Minerals almost always occur in the form of crystals. Normally these are too small to be seen with the naked eye (*microcrystals*), but larger ones are fairly common as well, especially in areas where a mineral-rich MAGMA has cooled slowly or where mineral-laden GROUNDWATER has percolated through a rock body containing cavities (*see* GEODE). Particularly well-formed, attractive and/or rare mineral crystals may be economically valuable (*see* GEM). For the role of minerals in the diet, *see* NUTRIENT; TRACE ELEMENT.

Mineral Leasing Act of 1920 federal (U.S.) legislation dealing with ownership of MINERAL resources on federal lands, signed into law by President Woodrow Wilson on February 25, 1920. The law withdraws most nonmetallic mineral deposits on federal land from the jurisdiction of the MINING LAW OF 1872, explicitly declaring both these minerals and the lands beneath which they are found to be the property of the federal government. It provides for leasing exploration and development rights to private individuals and corporations. If mineralization has been shown to be present, these leases can be auctioned to the highest bidder. A percentage of the developer's profits are returned to the federal treasury as royalties. A loophole in the act provides for the patenting of COAL, OIL, SHALE, and other lands for which claims were filed before 1920, provided that the requirements of the mining act concerning annual upkeep, etc., have been met. Originally applied only to the PUBLIC DOMAIN LANDS of the continent, the Mineral Leasing Act was extended to the outer continental shelf in 1953 via the Outer Continental Shelf Leasing Act. *See also* LEASABLE MINERAL.

Mineral Policy Center a research, advocacy, and education organization committed to the prevention and cleanup of mining pollution. Founded in 1988, the center publishes educational materials on the impact of mining and works with local organizations in areas affected by mining damage. Membership (1999): 2,500. Address: 1612 K Street NW, Suite 808, Washington, DC 20006. Phone: (202) 887-1872.

mineral soil a SOIL composed primarily or exclusively of MINERALS, with little or no ORGANIC MATTER present.

miner's inch in water management, a unit of water flow, originally defined as the rate of flow of water escaping through a hole 1 inch square in a 2-inch-thick board. Since the amount of this flow varies rather widely with conditions such as the shape and roughness of the hole, the HEAD, and the actual board thickness (a "2-inch" board today being defined as 1 1/2 inches thick), the term has been standardized in regulatory law to a value of 1.5 cubic feet per minute (cfm), or 2,160 cubic feet per (24-hour) day.

Mine Safety and Health Administration a division of the U.S. Department of Labor that is to promote safety and health of miners through health screening and training assistance and to prevent disasters in the mining industry. MSHA operates the National Mine Health and Safety Academy, an 80-acre facility in Beckley, West Virginia, that is responsible for training mine safety and health inspectors and providing technical support for personnel. Address: Department of Labor, Ballston Tower 3, 4015 Wilson Boulevard, Arlington, VA 22203. Phone: (703) 235-1452. Website: www.msha.gov.

Mines, Bureau of *See* BUREAU OF MINES.

mine spoil *See* SPOIL.

minimum conservation pool in hydrology, the lowest level to which a RESERVOIR can be drawn down (*see* DRAWDOWN). The reservoir level is not lowered to the minimum conservation pool during normal operation of a dam. *Compare* MINIMUM FLOOD CONTROL POOL.

minimum flood control pool in hydrology, the lowest level to which a RESERVOIR is normally drawn down in order to catch the excess waters from a flood (*see* DRAWDOWN). The reservoir level is lowered to the minimum flood control pool at the beginning of the flooding season (the end of November for most United States climates) and kept there, except for temporary filling during floods, throughout the season (generally, until the end of April). The minimum flood control pool is generally significantly higher than the MINIMUM CONSERVATION POOL in order to provide a "cushion" of water that can be released to maintain streamflow if a drought should hit during what is normally the flooding season.

minimum pool *See* MINIMUM CONSERVATION POOL; MINIMUM FLOOD CONTROL POOL.

minimum management requirements (MMRs) in Forest Service usage, the legal constraints on managing a piece of land, particularly those imposed by the NATIONAL FOREST MANAGEMENT ACT and the regulations designed to implement this act. A management plan that does not meet MMRs cannot be legally implemented. Some of the more important of the MMRs for forest land are: (a) timber may be harvested only from SUITABLE LANDS; (b) harvest levels must be able to be sustained in perpetuity; (c) the ROTATION AGE for SECOND GROWTH must be at least 95% of the age of CULMINATION of MEAN ANNUAL INCREMENT; (d) harvest levels must be compatible with NON-DECLINING EVEN FLOW management on a decade by decade basis; (e) CLEARCUTS may not exceed 60 acres in size; (f) harvest of RIPARIAN LANDS, where allowed at all, must be limited to a 300-year rotation cycle; and (g) sufficient OLD GROWTH and MATURE TIMBER must be left to provide HABITAT protection for ORGANISMS that require these habitats.

minimum trim allowance in logging, the extra length added when measuring a log prior to BUCKING it, in order to allow room to compensate for mistakes in measurement, squaring off slanted or broken cuts, etc. The minimum trim allowance is generally defined as 8 inches for each log 40 feet long or under, plus 2 inches for each 10-foot length (or portion thereof) over 40 feet. A 20-foot log, for example, must actually be at least 20 feet 8 inches in length; if it is under that, it will be scaled as an 18-foot log (*see* LOG SCALING).

Mining Law of 1872 a now-controversial statute, still in force, that opens all public land in the United States to mining except lands specifically withdrawn by Congress. The law spells out the maximum dimensions of mining claims and states the conditions under which claims can be *patented,* that is, become private property of the miner. The laws have proved surprisingly robust: though large areas have been withdrawn from entry (including all NATIONAL PARKS; NATIONAL MONUMENTS; NATIONAL WILDLIFE REFUGES; WILDERNESS AREAS [but *see* ASPINALL AMENDMENT]; military reservations; and the entire states of Wisconsin, Minnesota, Michigan, Alabama, Missouri, and Kansas), and though the definition of a "locatable mineral" has been somewhat restricted (*see* LOCATABLE MINERAL; LEASABLE MINERAL), the basic structure and procedures outlined in the law of 1872 have remained unchanged ever since. At the same time, opposing arguments are becoming better known to the general public, including the estimate that since the law's enactment, $385 billion worth of publicly-owned mineral resources has been given to private mining interests for a $500 filing fee and a payment of between $2.50 and $5.00 an acre. In a famous case in 1994, Secretary of the Interior Bruce Babbitt was required under the law to sell the rights to mine some $10–15 billion worth of gold to a Canadian company for $10,000. Moreover, establishing a patent does not require that the land be actively mined. In fact, patent holders can sell mineral rights at a great profit, or even develop the land for nonmining uses, such as a ski resort or vacation-home development. Those who advocate reforming the law suggest that the patent system be abandoned and replaced with nonrenewable 20-year mining leases; that miners meet strict pollution control standards during the leasing period; that they clean up and restore sites, under bond; and that they pay a royalty of 12.5% on the gross value of minerals removed. While mining companies object to these restrictions, environmentalists argue that companies can still make a significant profit if they are imposed.

Minnesota-Wisconsin Boundary Area Commission conducts research, forms recommendations, and coordinates planning for the protection, use, and development of lands and waters that form the boundary between Minnesota and Wisconsin. Address: 619 Second Street, Hudson, WI 54016. Phone: (651) 436-7131.

Mirex a chlorinated organic PESTICIDE, chemical formula $C_{10}Cl_{12}$, introduced in 1955 and widely used for the next two decades, especially against the fire ant in the southern United States. Small amounts of it were also used as a fire retardant in plastics, synthetic rubbers, paper, and paint. Mirex is a white, crystalline powder that dissolves poorly in water but is easily soluble in LIPIDS. Although of low TOXICITY to vertebrates (LD_{50} [female rats, orally]: 600 mg/kg), it is a highly potent MUTAGEN and CARCINOGEN that can cause birth defects in concentrations as low as 12.5 mg/kg. Extremely persistent in the environment, Mirex bioaccumulates readily (*see* BIOLOGICAL MAGNIFICATION), apparently without reaching a SATURATION THRESHOLD. The acceptable level of Mirex in foods meant for human consumption has been set by the ENVIRONMENTAL PROTECTION AGENCY as 0.01 parts per million (ppm). The compound is exceptional, if not unique, among regulated chemicals in that its acceptable level in water is defined by international treaty between the United States and Canada as none at all ("less than detection levels"). The use of Mirex has been banned in the United States since 1975.

mistletoe a small, shrubby, evergreen plant, FAMILY Loranthaceae, that grows as a PARASITE on the branches of various trees. Worldwide, the family has approximately 30 genera (*see* GENUS) and 1,000 SPECIES, mostly limited to the tropics. Two genera and about 100 species grow in the United States. *Phoradendron* spp. ("common mistletoe"), the mistletoe of HARDWOODS, is a dark chalkgreen, many-branched plant with small leaves. The leaves and stems are the same color. The white, semitransparent berries are toxic to mammals but are a favorite food of birds. *Phorodendron* is particularly common on oaks, and is the mistletoe sold in stores at Christmas. *Arceuthobium* spp. ("dwarf mistletoe"), the mistletoe of CONIFERS, is smaller than common mistletoe and is yellow-green in color. The leaves, where present at all, have been reduced to scales. The berry is similar to that of the common mistletoe. *Arceuthobium* creates a condition in conifers known as "witch's broom" in which small, weak branches proliferate from the point of attachment of the parasite, forming a brushy, hanging mass. It has a highly specific host/parasite relationship, so that a given species of *Arceuthobium* will ordinarily infest only a single species of tree. Mistletoe-infected trees are weakened and slow-growing, although the quality of the wood is not normally damaged.

mitosis (mitotic division) the normal process of cellular reproduction (*see* CELL), in which the CHROMOSOMES reproduce and divide so that the number of chromosomes in each daughter cell is identical to that in the mother cell. As mitosis begins, each chromosome replicates itself so that there are twice as many chromosomes as the cell would normally contain. The nuclear envelope then breaks down and the pairs of chromosomes separate, with one of each pair migrating to each end of the elongating cell. Finally, the cell splits down the middle between the two groups of chromosomes, and each new daughter cell forms a new nuclear envelope. Mitosis is the method of cellular division used primarily in the ASEXUAL REPRODUCTION of one-celled ORGANISMS and in the duplication of cells for growth within a multicelled organism. *Compare* MEIOSIS.

mixed liquor in sewage-treatment technology, the contents of an ACTIVATED SLUDGE tank or similar SECONDARY TREATMENT FACILITY. Mixed liquor consists of the influent wastes (that is, the incoming waste stream) plus a small amount of sludge that is recycled from the SETTLING BASINS in order to keep a sufficient supply of MICROORGANISMS in the tank. *See also* MIXED LIQUOR SUSPENDED SOLIDS; F/M RATIO.

mixed liquor suspended solids (MLSS) in sewage-treatment technology, the solid matter suspended in the MIXED LIQUOR of an ACTIVATED SLUDGE tank or similar SECONDARY TREATMENT facility. *See* F/M RATIO.

mixed mesophytic forest term coined by pioneering botanist E. (Emma) Lucy Braun in 1916 to describe the biologically rich woodlands of the west-central Appalachians in West Virginia, eastern Kentucky, eastern Ohio, and parts of Tennessee and Pennsylvania. Alone among all other forest regions in the temperate zone of the United States, this area was not flooded by rising seas in geological time, nor was it glaciated. This created a "mother forest" of as many as 80 woody species in all, with perhaps dozen or more "canopy" trees rather than the usual two or three. Growing to 100 or more feet, the mixed mesophytic canopy includes American beech, tulip poplar, four species of basswood, three species of sugar maple, sweet buckeye, species of white and black oak, a straight-growing form of black locust, and (once) the American chestnut, now disappeared except for remnant suckers from blighted trees. Because of its location on the western slope of the Appalachians, the mixed mesophytic has been subject to serious damage from AIR POLLUTION from factories and power plants in the Ohio and Tennessee Valleys. Recent studies have shown tree mortality at three to five times historic norms. *See also* CHESTNUT BLIGHT; FOREST DECLINE AND PATHOLOGY.

mixed stand a STAND in which more than one SPECIES of tree is present. Most naturally occuring stands are mixed. *Compare* PURE STAND.

mixture in chemistry, a substance containing two or more ELEMENTS or COMPOUNDS that are not chemically bound to each other but retain their own chemical and physical identities. Because the materials in a mixture remain discrete and do not bind in fixed proportions, the mixture itself does not have fixed proportions. A mixture of salt and sugar, for example, can be anything from 99.9% sugar to 99.9% salt, whereas the ratio of sodium to chlorine in the salt itself—a compound, rather than a mixture—will always be exactly 1:1. Mixtures may be *homogenous*—that is, have the same proportions throughout a given sample—but two homogenous mixtures of the same pair of substances may differ widely from each other. You can, for example, stir either a small or a large amount of sugar into solution in a cup of coffee (a solution being a special type of homogenous mixture; *see* SOLUTION). Mixtures are always separable by mechanical means (heat, filtration, gravitational sorting, etc.), though the separation may at times be hard to effect. Besides the solution, special types of mixture include the ALLOY, the SUSPENSION, and the COLLOIDAL SUSPENSION. *Compare* COMPOUND.

MLSS *See* MIXED LIQUOR SUSPENDED SOLIDS.

MMRs *See* MINIMUM MANAGEMENT REQUIREMENTS.

mobile compound a COMPOUND that spreads readily through the environment. *See* MOTILITY.

mobile source in air pollution control, an automobile, aircraft, ship, train, or other moving source of air pollution. Mobile sources are POINT SOURCES in the sense that the emissions of each individual source are theoretically controllable, but the fact that there are very many of them and that they move around often makes such control practically impossible. Clean air legislation usually treats them as point sources (for example, by mandating emission controls on automobiles); however, at times it is easier to ignore individual emissions and simply treat an entire highway, parking facility, airport, etc., as an AREA SOURCE. Mobile sources are the single largest contributor to air pollution in most urban areas, and are particularly important as sources of HYDROCARBONS, NITROGEN OXIDES, and especially CARBON MONOXIDE. According to the National Commission on Air Quality, 84% of all carbon monoxide in the atmosphere derives from mobile source pollution.

modification of character in visual management, a state in which the details of a landscape have been partially or completely altered from the characteristic state (*see* CHARACTERISTIC LANDSCAPE), but in which the overall form, line, and design remains substantially natural. Roads, HARVEST UNITS, and so on must blend into the natural landscape to the point where they appear from a short distance away as if they could have occured naturally. Structures must be inconspicuous and designed to blend into their surroundings as thoroughly as possible. *Compare* MAXIMUM MODIFICATION; PARTIAL RETENTION OF CHARACTER. *See also* VISUAL RESOURCE MANAGEMENT SYSTEM.

moho, Mohole Project *See* MOHOROVICIC DISCONTINUITY.

Mohorovicic discontinuity (moho) the boundary between the Earth's crust and MANTLE, marked by an abrupt change (*discontinuity*) in the way in which seismic waves (waves created by EARTHQUAKES or other seismic events; *see* P-WAVE; S-WAVE) travel through the ground. The depth of the Mohorovicic discontinuity varies from place to place, but it is generally about 35 kilometers (21 miles) below the continents, and about 10 kilometers (6 miles) below the ocean basins. In 1957 an informal scientific organization called the American Miscellaneous Society proposed the *Mohole Project,* which would take a core drill through the Earth's crust to and beyond the Mohorovicic Discontinuity into the mantle to determine the properties of the boundary and to provide an understanding of the gaps in the geological record. The project, backed by the National Sciences Foundation and the National Academy of Sciences, was finally unsuccessful, but a small experimental shaft was dug off the coast of California to a depth of 122 miles below the ocean floor. The project was abandoned as too costly in 1966.

Mohs scale in geology, a scale devised by the German mineralogist Karl Friederich Mohs (1773–1839) to classify the hardness of ROCKS and MINERALS in the field. It is still in universal use today. The Mohs scale ranks hardness in numerical order from 1 to 10, with 1 being the softest and 10 the hardest. The minerals in each numerical class can scratch those in the class below them, and are scratched by those in the class above them. One characteristic mineral is designated for each class, and a good field kit will contain a small sample of each of these characteristic minerals for use in hardness tests. They are:

1. talc	6. feldspar
2. gypsum	7. quartz
3. calcite	8. topaz
4. fluorite	9. corundum
5. apatite	10. diamond

For quick approximations without a kit, the hardness of a fingernail is 2.5; of a copper penny, 3; of a knife blade or piece of window glass, 5.5; and of a steel file, 6.5.

moisture gradient in ecology, a gradual shift over distance from relatively wet conditions in one location to relatively dry conditions at another. The distance under discussion may be quite long (for example, the shift in rainfall totals and thus general climatic moisture on a TRANSECT west from Ohio to Utah), or it may be quite short (for example, the shift in ambient air moisture in the few inches separating the dry air over bare soil from the damper air among the stems of a patch of grass). Short gradients with relatively great differences between their end points (as, for instance, the soil moisture gradient from the muck of a pond's edge to the dry DUFF under a tree a few feet away) are referred to as "steep". The study of moisture gradients is particularly important to plant ecologists because plant SPECIES can often only live under a very narrow range of moisture conditions and will therefore cluster at specific points along a gradient.

mol *See under* MOLECULAR WEIGHT.

molecular weight (formula weight) in chemistry, the weight of one MOLECULE, ION, GROUP, or other formula unit. The molecular weight is the sum of the ATOMIC WEIGHTS of the atoms that combine to make up the formula unit. The unit of measurement, known as the *atomic mass unit (amu),* is 1/12 the weight of an atom of carbon-12 (*see* CARBON). A related concept, the *gram-molecular weight* (*mole* or *mol*) is the amount of a COMPOUND or ELEMENT whose weight in grams (*see* METRIC SYSTEM) is equal to its molecular weight. Thus, one mol of carbon-12 would weigh 12 grams. The concept of the mol is particularly useful in the study of gases, because one mol of any substance always contains the same number of molecules ($6.22 x 10^{23}$, a number known as *Avogadro's number*) and, in the gaseous state, occupies the same volume (22.4 liters at STANDARD TEMPERATURE AND PRESSURE).

molecule in chemistry, the smallest unit that a COMPOUND may be divided into and still retain its physical and chemical characteristics. The composition of a molecule is directly represented by its chemical formula; thus water, chemical formula H_2O, has a molecule composed of two HYDROGEN ATOMS (H) and one OXYGEN atom (O). Most elements are *monatomic,* that is, exist in the pure state as individual atoms; however, several, including hydrogen, oxygen, CHLORINE, and a few others, form molecules containing two atoms each (oxygen also exists in a triatomic form, O_3; *see* OZONE). Molecules are bound together by shared ELECTRONS. In the most common form of bonding, the *covalent bond,* two atoms share a pair of electrons, with the shared pair orbiting around both atomic nuclei. *Multivalent bonds,* in which electrons are shared among three or more atoms, are also known (*see* METAL). A third type of bonding also exists, in which one atom "steals" an electron from another, creating a pair of charged IONS that then cling together by electrostatic force. These *ionic bonds,* however, are considerably weaker than those created by covalent or multivalent bonding, and the compounds that result cannot be said to contain true molecules.

mollisol any of a group of soils characterized by both high ORGANIC MATTER content and high pH, normally found in semiarid to semihumid climates. Mollisols are the soils of grasslands, such as the American Great Plains and the Asian steppes. They have enough rainfall to develop good plant cover, but not enough to leach out the base MINERALS (especially calcium). They generally have well-developed HORIZONS. A buried CALICHE LAYER is often present; the depth to this layer is a good index of the average soil moisture. Mollisols usually border (and grade into) ARIDISOLS. They are well adapted to growing wheat and corn (which are both grasses), and are among the best agricultural soils in the world.

mollusk in the biological sciences, any member of the PHYLUM Mollusca, including the shellfish, snails and slugs, nudibranchs (sea slugs), and cephalopods (octopuses and squid). Mollusks are a highly successful, evolutionarily advanced group of INVERTEBRATES that are second only to the Arthropoda (insects and spiders) in number of SPECIES. In size they range from tiny (*Neopilina,* a limpet-like shellfish of the CLASS Monoplacophora, is less than 2 cm [3/4 inch] across as an adult) to huge (the giant squid, the largest and most complex invertebrate, may be over 18 meters [60 feet] in length). They inhabit virtually all HABITATS on earth, from the deep ocean trenches to mountainsides at over 7,000 meters (23,000) above sea level. Most are marine, but there are numerous freshwater species, and some snails and slugs have developed lungs and invaded the land. (Even these terrestrial forms, however, require moist habitats, and most require calcium-rich soils, such as those derived from LIMESTONE, to obtain the materials to build their shells.) Structurally, all are similar. They are characterized by a soft body that is differentiated into a head, a foot, and a visceral hump that contains most of the organs and gives rise to the MANTLE, a highly specialized organ that aids the animal in digestion and respiration and secretes the characteristic shell (absent in octopi, slugs, nudibranchs, and a few others). Most are slow-moving and rather dim-witted, and some are completely sessile (*see* SESSILE ANIMAL), but the octopi and squids are highly mobile predators that have an intelligence level on the order of that of mammals. The larvae of all mollusks are microscopic or near-microscopic in size. Released from eggs in great numbers, they form an important part of the

PLANKTON community. Oysters, mussels, clams, and other shellfish are important food sources for humans. they are also highly susceptible to pollution because of their habit of filter feeding (straining MICROORGANISMS from the water) and their presence on the beds of watercourses, where pollutants tend to concentrate. Oysters, for example, are known to be able to bioaccumulate pollutants at up to 100,000 times the level that they are found in the surrounding waters (*see* BIOLOGICAL MAGNIFICATION). The phylum contains seven classes and over 100,000 recognized species.

Monera in the biological sciences, the KINGDOM containing the bacteria (*see* BACTERIUM) and blue-green algae (*see* CYANOPHYTE). The Monera are the only living ORGANISMS composed of prokaryotic cells (*see* PROKARYOTE). They are considerably simpler than eukaryotic organisms (*see* EUKARYOTE) and show up much earlier in the fossil record, and for this reason are considered primitive; however, they are extraordinarily successful and adaptable, and thrive under a much broader range of environmental conditions than organisms of the other four kingdoms can tolerate, including excessive heat, hazardous chemical concentrations, and lack of OXYGEN (the kingdom Monera is the only class of organisms that include true anaerobes; *see* ANAEROBIC BACTERIA). Ecologically, they function primarily as DECOMPOSERS, and the breeding of SPECIES that can decompose hazardous chemicals such as PETROLEUM and PCBs (*see* POLYCHLORINATED BIPHENYL) shows great promise as a means of dealing with pollution. It has been estimated that over half of all living organisms on the Earth, by weight, are Monera.

monitoring well a well drilled to study the composition, extent, and spreading pattern of GROUNDWATER pollution. Monitoring wells provide the only direct means of studying pollution patterns in AQUIFERS and are therefore an essential tool in pollution control efforts. However, the results obtained from them must always be used with great care due to the fact that each well is only a single-point sample within a highly complex and unpredictable environment. Considerable study of such factors as water-table gradient, rate of flow, and POROSITY and PERMEABILITY of soil and rock layers must be made before monitoring wells may be sited and before the results from them may be properly interpreted.

TYPES OF WELLS

There are two principal types of monitoring wells. The *vapor well* or "sniff well" is simply a hole extending to the WATER TABLE, either uncased or with a permeable casing (*see* PERMEABILITY), capped by a device designed to detect the presence of organic vapors. It can reveal the presence of CONTAMINANTS in the water or migrating downward through the soil, but it cannot be used to measure their quantities, pinpoint their precise locations, or predict very accurately their course through the aquifer. The *wet well* is used to take actual samples of the groundwater. It is more expensive to install than a vapor well but allows actual measurements rather than mere indications of pollution. The simplest form of wet well, the *single interval well,* consists simply of a well with a casing that is impermeable above the water table and permeable below. It can help determine the type, amount, and horizontal extent of a PLUME of pollutants, but provides no information about vertical distribution. The *well cluster* overcomes this by providing several closely spaced wells, each drilled to a different depth. It often consists of a number of well casings occupying the same large-bore hole, with the sampling screens at different depths and separated from each other by layers of impermeable material such as BENTONITE. A variant on the well cluster, the *nested well,* places several sampling screens in the same hole, connecting them to the surface with piping or flexible hosing instead of well casings.

monkey wrenching a form of sabotage by radical activists against industries, land developers, loggers, and others they see as environmentally destructive forces by vandalizing equipment (smashing distributors and spark plugs, pouring sugar into gas tanks, driving spikes into trees in order to destroy chain saws that encounter them, and so forth). The term monkey-wrenching derives from the name of a novel, *The Monkey Wrench Gang,* by EDWARD ABBEY (1927–1989). *See also* ECOTAGE.

monocotyledon (monocot) any plant that is a member of the subclass Monocotolydonae of the CLASS Angiospermae (*see* ANGIOSPERM). The monocots are the smaller of the two subclasses of angiosperms, containing roughly 50,000 SPECIES, or approximately one-sixth of all known species of flowering plants. They do not produce woody tissue. Their leaves are almost always parallel-veined, that is, with the veins running parallel to each other for most of the length of the leaf, coming together in a single node at the leaf's base. The veins in their stems are scattered randomly through the stem tissue rather than being bundled into cylinders. Their roots are fibrous, and tend to form TUBERS or RHIZOMES instead of TAPROOTS. Their seeds contain a single COTYLEDON. Typical species include GRASS, iris, and lily. *Compare* DICOTYLEDON.

monoculture the growth of a single type of crop plant over a relatively large reach of land. A wheat field or a corn field is a monoculture; so is a SECOND GROWTH forest planted to a single SPECIES of tree such as white pine or DOUGLAS FIR. A suburban lawn qualifies as a monoculture under most definitions. Because they are highly simplified ECOSYSTEMS, monocultures lack most of the built-in feedback loops common in natural systems, and must have the functions normally served by these loops provided instead by the crop grower. Competitors must be kept out by weeding, spraying, and other means. Insect pests multiply profusely, helped along by the abundance of their preferred food, and must likewise be sprayed. Since the plants are often not only of the same species but of the same age and genetic stock, vulnerability to disease is similar to identical throughout a crop stand, allowing any diseases that catch hold in the stand to spread rapidly. *See also* GENETIC ENGINEERING.

Mono Lake large saline lake in east central California, near the town of Lee Vining and less than 20 kilometers (12 miles) from the eastern border of Yosemite National Park. Thought by geologists to be the oldest lake in North America, Mono Lake is the remnant of a much larger body of water that once spread eastward from the Sierra Nevada mountains well into the present state of Nevada. The lake has a current elevation of 6,378 feet and a surface area of approximately 38,400 acres. It is the breeding ground for approximately 25% of the known world population of the California gull, *Larus californicus* (95% of the California population).

HISTORY

When first measured in 1857, Mono Lake stood at a surface elevation of 6,376 feet, or approximately the same as its current level. The level slowly rose until 1919, when it was measured at 6,427.7 feet, giving the lake a surface area of roughly 57,600 acres. In 1920, the city of Los Angeles entered into an agreement with the U.S. BUREAU OF RECLAMATION to produce a study of the Mono Lake watershed as a possible adjunct to its water supply project in OWENS VALLEY. Los Angeles voters approved bonds to build the Mono Basin project in 1930, and the United States Congress passed enabling legislation on March 4, 1931. Diversions began from Rush, Parker, Walker, and Lee Vining Creeks (four of the five major Mono Lake tributaries) in 1941, and the lake began to shrink. Its surface area dropped approximately .3 meters (1 foot) per year from 1941 to 1970, when the so-called "second barrel" aqueduct was completed (*see under* OWENS VALLEY), allowing Mono Basin diversions to double, from 55,000 acre-feet/year to 110,000 af/y. Thereafter the lake dropped approximately .45 meters (1.5 feet) per year, and environmental groups began an intensive campaign to reduce diversions and stabilize the lake's water level. In 1978, a land bridge to the major gull breeding ground on Negit Island was exposed, allowing predators access to the island: this and a crash in the population of the lake's brine shrimp caused gull nesting success to decline to near zero (it improved again in following years). On September 25, 1984, President Ronald Reagan signed the California Wilderness Act, which included provisions for a Mono Lake National Scenic Area as part of the Inyo NATIONAL FOREST. While explicitly protecting Los Angeles water rights, the legislation also directed the Forest Service to protect the ecology of Mono Lake and authorized the agency to enter into agreements with the City of Los Angeles toward that end. In 1989 the City of Los Angeles was ordered by a California judge, in a lawsuit brought by the Mono Lake Committee, a conservation organization, to maintain the lake at a level of 6,377 feet. While still about 50 feet below the 1919 level, the order put a stop to diversions. Then, in 1994, the California Water Resources Board increased the minimum level to 6,392 feet, high enough in its view to protect the ecosystem (including shrimp and migrant birds as well as resubmerging the Negit Island land bridge) but still keep the famed "tufa" above water. The tufa, now a tourist attraction, are white limestone columns that formed after water levels declined, deriving from the interaction between calcium-rich spring water from lakebed seeps and carbonates in the lake water. After some deliberation Los Angeles officials agreed not to appeal the Water Board's decision since a new tertiary sewage treatment plant serving the city could provide enough irrigation and industrial cooling water to offset the loss from the Mono Lake diversions. Geologists have estimated that the required 6,392-foot level will be reached by about 2015. When achieved, it will bring the lake to within 35 feet of its 1919 elevation.

monomer in chemistry, a chemical COMPOUND, or portion of a compound, composed of nonrepeating parts. ELEMENTS (or even GROUPS) may appear more than once, but the pattern of their relationship to other parts of the molecule will be different each place that they occur in the molecular structure. *Compare* POLYMER.

Monongahela decision court decision banning CLEARCUTS on U.S. NATIONAL FORESTS, handed down by the 4th Circuit Court of Appeals (Maryland, Virginia, West Virginia, North Carolina, and South Carolina) on August 21, 1975. Known officially as *Izaak Walton League* vs. *Butz,* the Monongahela case was brought by attorneys for the IZAAK WALTON LEAGUE and several other national conservation organizations against the U.S. Department of Agriculture (*see* AGRICULTURE, DEPARTMENT OF) and its secretary, Earl Butz, as a means of stopping management abuses on West Virginia's Monongahela National Forest. The forest, which had traditionally managed its timber with SELECTIVE CUTS, had suddenly begun clearcutting heavily in 1964, turning a deaf ear to protests from local conservationists, chambers of commerce, tourist organizations, and even the two U.S. senators from West Virginia. After failing to gain concessions through negotiations with the FOREST SERVICE, the Izaak Walton League and others filed suit in District Court in May 1973 under an obscure provision of the ORGANIC ACT OF 1897, which stated that all trees harvested by the Forest Service had to be individually marked. On November 6, 1973, the District Court found for the appelants, banning clearcutting on the Monongahela. The Forest Service appealed to the 4th Circuit Court, which upheld the decision of the lower court on August 21, 1975, thus applying the clearcutting ban to the entire five-state jurisdiction of the Circuit Court. The agency then asked Congress to modify the law. Congress responded by passing the NATIONAL FOREST MANAGEMENT ACT OF 1976, which may be thought of as a direct result of the Monongahela decision, though it addresses considerably more forest management practices than just clearcutting. In fact, the 1976 legislation permitted clearcutting as part of a long-range management plan for national forests. Despite the Monongahela decision, the language of the new law was so vague that it allowed the Forest Service such wide discretion in matters of harvesting that in many areas, especially the West, clearcuts, while of smaller dimension than earlier ones (usually between 25–40 acres), are so close to one another that many wonder whether the Monongahela decision had any effect at all.

monsoon the strong seasonal wind of Southeast Asia that blows from the sea to the land during the summer (the *summer monsoon*) and from the land to the sea during the winter (the *winter monsoon*). The monsoons are essentially a land- and sea-breeze system (*see* LAND BREEZE; SEA BREEZE), magnified in intensity by the great size of the Asian continent and of the Pacific Ocean. The summer monsoon, born over the ocean, is wet, and breeds torrential rains and relatively frequent CYCLONIC STORMS (but the cyclonic storms are not identical to the monsoon); the winter monsoon is dry. This alternate wet/dry cycle breeds a special type of tropical forest, the so-called MONSOON FOREST, containing many plants that are deciduous in response to lack of moisture instead of to low temperatures (as is common in the temperate forest; *see* DECIDUOUS PLANT).

monsoon forest in ecology, a form of tropical deciduous forest (*see under* DECIDUOUS FOREST BIOME) existing in areas of the tropics where MONSOON winds cause a distinct cycle of wet and dry seasons.

montane forest *See under* CONIFEROUS FOREST BIOME.

montane species any SPECIES of plant or animal whose natural RANGE is a mountain or MOUNTAIN SYSTEM. The term is particularly useful to differentiate mountain-dwelling species of a particular GENUS from lowland species of the same genus. Montane species are adapted to lower temperatures than their lowland cousins. They also have greater protection against ultraviolet radiation. Montane animals have larger lungs in order to compensate for the rarified air. Their blood contains more HEMOGLOBIN for the same reason. Montane plants generally have small leaves, and often grow as low mats rather than as upright stems (*see* KRUMMHOLTZ).

Montreal Protocol a 1987 international agreement made in Montreal, Canada, among 54 nations to cut the emissions of CHLOROFLUOROCARBONS (CFCS) into the atmosphere in order to protect the OZONE LAYER. The protocol originally called for a 35% reduction of CFCs by 2000 and did not cover other compounds that break down the ozone molecule. Subsequent agreements with a larger group of nations improved the standards and added other ozone-depleting compounds to the proscribed list. By 1996 developed countries had phased out the use of ozone-depleting compounds; however, developing countries were unable to follow suit. In 1999 revisions to the protocol provided for grants to those countries seriously impacted economically by CFC cutbacks, including China, the world's largest CFC producer and consumer. Scientists have predicted that under the protocol, ozone depletion will increase slightly during the decade of 2000–10 and then decrease until the ozone layer returns roughly to normal by 2050. However, these calculations do not take into account the effect of global warming on the ozone layer. As greenhouse gases cause the warming of the lower atmosphere, by trapping heat and not allowing it to escape into the upper atmosphere, the upper atmosphere becomes colder than before. (*See* GREENHOUSE EFFECT.) Because the ozone molecule is more easily broken down in extreme cold, global warming could affect the rate of ozone recovery. An additional problem is the emergence of a black market in CFCs. In the 1990s CFCs were second only to cocaine in street value as a contraband substance smuggled into the United States. Despite these issues, which many hope will not prove to vitiate the effort to reduce ozone depletion, political scientists agree that the Montreal Protocol as an extremely important demonstration that countries around the world can work together to solve a global environmental problem.

moraine in geology, a mound of ROCKS, SAND, SILT, and other MINERAL debris deposited along the margins of a GLACIER. Those deposited at the end of the glacier are called *terminal moraines,* while those that are deposited along the sides are called *lateral moraines*. As a glacier retreats, its moraines are left behind as long, linear hills, sometimes of a considerable height. Many of the hills of Michigan, Wisconsin, and other parts of the American midwest originated as moraines deposited by the CONTINENTAL GLACIER that extended into these areas in the Pleistocene. Moraines are often waterproof, and lakes may form behind them. The Great Lakes were all created at least partially by morainal dams. *See also* GLACIAL LANDFORMS.

mortality in the biological sciences, the death rate within a given population (or, in forestry, a given STAND), usually expressed in percent over time, that is, the percentage of the population that has died (or that may be expected to die) in a standard time period, usually one year.

mor humus *See under* HUMUS.

moss any of numerous primitive plant SPECIES belonging to the CLASS Musci. They are members of the DIVISION Bryophyta (*see* BRYOPHYTE), which they share with the liverworts and hornworts. Mosses are true plants, but unlike the ANGIOSPERMS and GYMNOSPERMS they do not contain a vascular system (a specialized system for the circulation of fluids; *see* VASCULAR PLANT), although a few genera (*see* GENUS) contain rudimentary vascular tissue in the form of elongated water-conducting cells or regions where rows of dead, hollow cells alternate with rows of living cells. The lack of a vascular system limits their overall size; most individual moss plants are less than 15 centimeters (6 inches) in length, and the maximum length (found only in waterborne SPECIES) is about 4 meters (13 feet). Mosses reproduce by means of spores rather than seeds. Individual plants are differentiated into leaves and stems that are similar in structure and function to those found in the vascular plants, though they are generally much smaller. The plants are anchored to the soil or rock they grow on by rootlike organs known as *rhizoids*. These differ from true roots in being much less complex structurally, and in serving only as anchors, with no part in the absorption of NUTRIENTS or water for the plant. Three ORDERS are recognized. The Sphagnales, or sphagnum mosses, are adapted to acidic conditions. They flourish particularly in BOGS, where they form thick mats known as PEAT. The Andreales, or granite mosses, are small black plants that bear a similarity to LICHENS. They too are adapted to acid conditions, in this case acid rocks such as GRANITE, and are found primarily in high altitudes or in the polar regions. All other mosses belong to the order Bryales, sometimes called the "true" mosses. Bryales are found worldwide (except in the sea), and include the common mosses found on the trunks of trees and as carpets on rocks near streams and in damp places on the forest floor. Generally hardy and often found as PIONEER SPECIES, mosses are nevertheless extremely sensitive to pollution and are among the first organisms to die when pollutants build up in water-courses or in the air. The class contains about 650 genera and nearly 14,000 species.

motility the ability to move from one place to another. When said of ORGANISMS, it implies self-propulsion. A *motile protozoan* moves through the water by vibrating its flagellae (*see* FLAGELLUM) or by similar means, while a *nonmotile protozoan* simply drifts with the current. When said of non-living materials, it implies some means of rapid distribution: a *motile compound* (also: *mobile compound*) spreads rapidly through soil and is taken up easily by plants after a spill, whereas a *nonmotile compound* will tend to stay in the immediate vicinity of the spill.

motor vehicle emissions *See* ACID RAIN; AIR POLLUTION; SMOG.

mountain any portion of the earth's surface that is significantly elevated above its surroundings. "Significantly" in this context is generally taken by geographers to mean 2,000 feet or more. *See also* MOUNTAIN RANGE; MOUNTAIN SYSTEM; MASSIF.

Mountaineers hiking and conservation organization founded in 1906 to explore, enjoy, and protect mountain scenery, especially that of the Pacific Northwest and adjacent regions. The Mountaineers conduct mountain-climbing courses, hold outings of various kinds, and lobby for mountain conservation measures. They own four climbing huts and a 180-acre scenic preserve in the state of Washington. The group was formerly known as the Seattle Mountaineers. Membership (1999): 15,000. Address: 300 Third Avenue W, Seattle, WA 98119. Phone (206) 284-6310.

mountain range usually, a group of MOUNTAINS whose bases are connected to each other. Mountain ranges are generally monogenetic (created at nearly the same time by the same set of tectonic forces; *see* TECTONIC ACTIVITY; PLATE TECTONICS) and are usually linear in nature, that is, are considerably longer than they are broad. *Compare* MOUNTAIN SYSTEM.

mountain system a group of MOUNTAINS or MOUNTAIN RANGES that are related in some manner. Mountain systems usually cover a broader area than mountain ranges; they are not often linear; and their bases are not necessarily connected. They are less likely than ranges to be monogenetic, although they are usually related structurally in some manner. A mountain system will usually contain similar FLORA and FAUNA throughout its length and breadth and from range to range within it; that is, it will belong to a single BIOTIC PROVINCE.

mountain-top removal a strip-mining technique to remove low-sulfur coal lying below the surface of mountaintops in Appalachia, especially in West Virginia. With the requirement of the 1990 Clean Air Act amendments that low-sulfur coal be used for fuel in power plants (to reduce emissions of oxides of sulfur which produce acid rain), strip-mining mountaintops for coal lying in thin horizontal veins in southern West Virginia became economically feasible. The practice has led to severe environmental damage to the region, affecting water quality, wildlife, recreation, landscape aesthetics, and community stability in the valleys where much of the spoil is deposited. Under the SURFACE MINING CONTROL AND RECLAMATION ACT OF 1997 (SMCRA), miners are required to restore strip-mined areas to the original contour and to replant with soil-holding vegetation so that eventually the site will return to roughly the same condition as before. Enforcement of these and other provisions of SMCRA has been uneven and lax over the years, however. Moreover, miners have tried to use an exception to the restoration requirement that allows strip-mined sites to be put to new "commercial or recreational" use. In West Virginia those removing mountaintops for coal have not been able to return the land to its original contour since the mountain had been substantially lowered, and they have found that native hardwoods will not return to the sites. Accordingly, miners have designated areas as potential golf courses and hunting and fishing areas, although there is little demand for the former and no possibility of the latter on the disturbed sites. In addition, the miners have done little or no remediation with respect to interrupting stream flow, filling valleys with SPOIL, and similar problems. While a move in Congress to exempt the mountaintop strip mines from the provisions of SMCRA failed in 1999, the mining continues. Environmental organizations have brought suit in federal court to force a proper interpretation of the SMCRA provision calling for a comprehensive study of the effects of strip mining on water systems, and to show that strip mines are in violation of the Clean Water Act in that they bury streams with mine waste. If the litigants prevail, valley filling, a necessary part of mountaintop removal, will become illegal. But even if the practice is abandoned soon, mountaintop removal will leave a long-term legacy of severe environmental degradation in West Virginia. *See also* STRIP MINE.

mountain zonation the concept that the vertical distribution of plans and animals on a mountain slope is likely to fall into broad bands determined largely by elevation above sea level, and that these bands may be correlated from mountain to mountain within the same LATITUDE. *See also* LIFE ZONES; BIOMES; HUMBOLDT'S RULE.

MPN method *See under* COLIFORM COUNT.

MTB *See* MATERIALS TRANSPORTATION BUREAU.

MTBE *See* METHYL TERTIARY BUTYL ETHER.

Muir, John (1838–1914) pioneering American naturalist and environmentalist, born in Dunbar, Scotland, April 21, 1838; died in San Francisco, California, December 24, 1914. Muir's family emigrated to a farm near Portage, Wisconsin, in 1849, when Muir was 11. As a youth he won prizes for his inventions, and in 1860 he entered the University of Wisconsin at Madison with the intention of becoming an inventor. Temporary blindness caused by a slipping tool which pierced his eye turned his interests from technology to natural history, and in 1864 he traveled to Canada on the first of the many extensive "rambles" (his term) he would engage in for the rest of his life. In 1867 he set out from Louisville, Kentucky, intending to walk to Brazil. He made 1,000 miles, to the Gulf Coast of Florida, where a bout of malaria caused him to alter his destination to California. From March 28, 1868, to the end of his life he made the San Francisco area his home, though he continued to "ramble" all over the world, to Alaska, Arizona, Hawaii, Australia, and New Zealand, and numerous other places. His name is strongly linked to the Sierra Nevada Mountains of California, where he was instrumental in early preservation battles, especially those surrounding the creation of Yosemite National Park. He was the founder of the SIERRA CLUB and directed most of its early conservation efforts, and was one of the lobbyists chiefly responsible for the passage of the ANTIQUITIES ACT. As a scientist, Muir was an early proponent of the theory of continental glaciation (*see* CONTINENTAL GLACIER) and was among the first to recognize that the high Sierra scenery was glacially sculpted. He was also among the first to study ECOSYSTEMS as complex webs of relationships rather than as individual ORGANISMS. The Muir Glacier in Alaska; Muir Woods National Monument and the John Muir Trail in California; and Camp Muir on Mt. Rainier, Washington, are among the many sites named for him. His writing, often quite spiritual, perhaps reflected a religious upbringing by his stern father, a Presbyterian minister. Muir rejected organized religion, however, finding pantheistic solace and inspiration in nature. In recent years his literary works have become of great interest to those concerned with ecological theology and the sacredness of nature. The

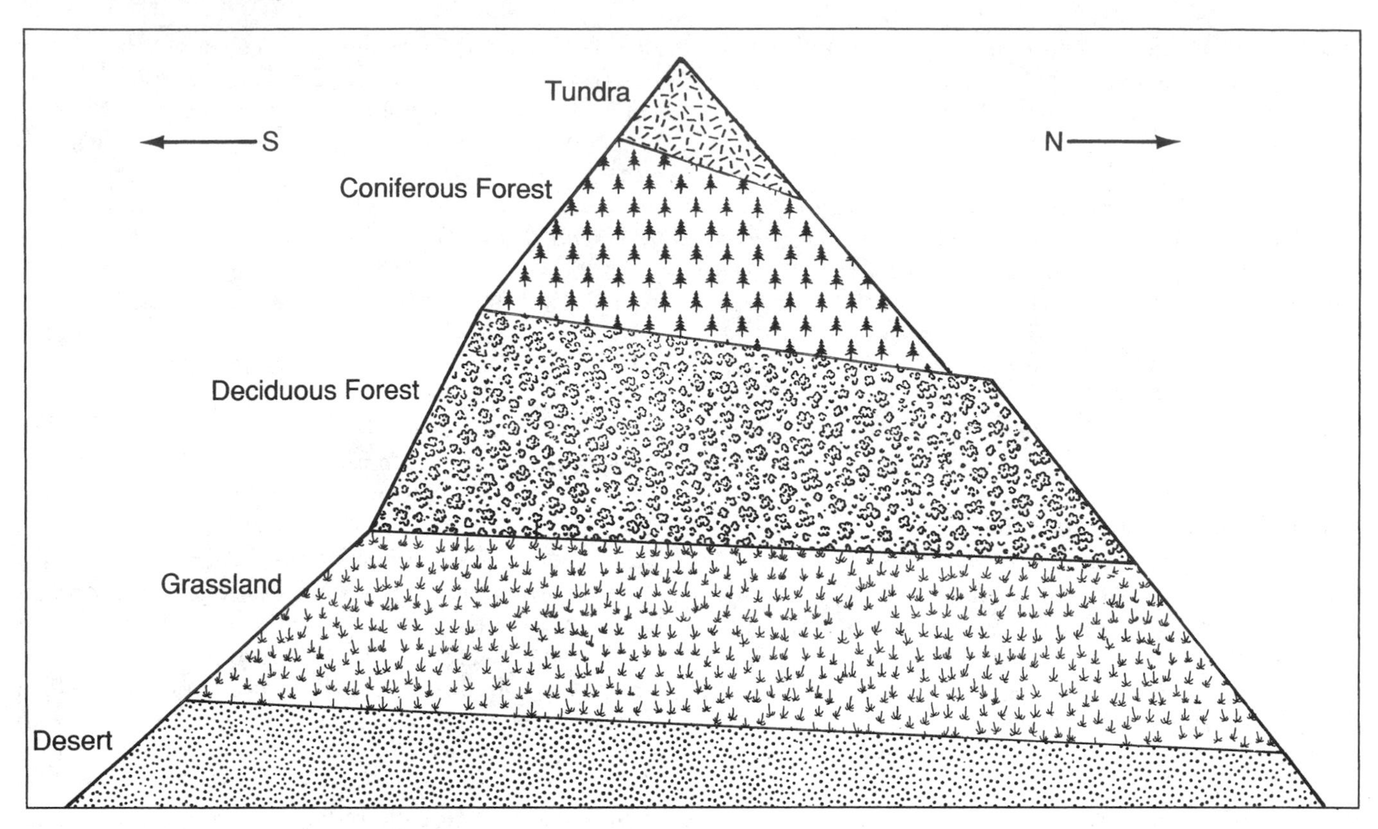

Mountain zonation

books include *The Mountains of California* (1894), *Stickeen* (1909), *The Yosemite* (1912), *The Story of My Boyhood and Youth* (1913), and *A Thousand Mile Walk to the Gulf* (1916). Among many biographies of Muir is Frederick W. Turner's *Rediscovering America: John Muir in His Time and Ours* (1985), which deals effectively with Muir's spirituality. *See also* RELIGION AND THE ENVIRONMENT.

Mullerian mimicry in ecology, a form of MIMICRY in which both the mimic and the model are dangerous to other ORGANISMS. The yellow and black stripes on the abdomens of bees, wasps and hornets are a form of Mullerian mimicry. All these insects, which are only loosely related in other ways, deliver painful stings, and their similar appearance serves to warn other animals of this danger. Mullerian mimicry may be thought of as an extension of warning coloration (*see under* CRYPTIC COLORATION) to pattern and behavior as well as color. *Compare* BATESIAN MIMICRY.

mull humus *See under* HUMUS.

multiple use in land management, the management of a single piece of land for a variety of different uses. The term is most commonly applied to the lands of the NATIONAL FOREST SYSTEM, which are required to be managed for multiple use by the MULTIPLE USE-SUSTAINED YIELD ACT OF 1960. The principal uses referred to in this context are timber management, recreation, wildlife, grazing, and watershed protection. While efforts are made to manage the National Forests for some of these uses simultaneously, others are conflicting or incompatible with each other (*see* CONFLICTING USE; INCOMPATIBLE USE) and must be separated either in space or time, or sometimes both. Lands that are managed for these uses, however, may still meet the definition of multiple use lands if the areas devoted to them are kept in balance with lands on which they are not being carried out, and if the activity that causes the conflicting use is done in such a way as to have little or no permanent impact on the land's ability to provide other uses. *Compare* DOMINANT USE.

Multiple Use-Sustained Yield Act of 1960 (MUSYA) federal legislation mandating the general management direction for the NATIONAL FOREST SYSTEM, signed into law by President Dwight Eisenhower on June 12, 1960. Essentially a codification into law of management practices that the FOREST SERVICE had previously established but which it was under increasing pressure to change, MUSYA directs the agency to manage the RENEWABLE RESOURCES of the National Forests for "multiple use and sustained yield of the several products and services obtained therefrom." "Renewable resource" is specified as "outdoor recreation, range, timber, watershed, and fish." The key concepts of "multiple use" and "sustained yield" are legislatively defined:

> (4)(a) "Multiple use" means the management of all the various renewable surface resources of the national forests so that they are utilized in the combination that will best meet the needs of the American people; making the most judicious use of the land for some or all of these resources or related services over areas large enough to provide sufficient latitude for periodic adjustments in use to conform to changing needs and conditions; that some land will be used for less than all of the resources; and harmonious and coordinated management of the various resources, each with the other, without impairment of the productivity of the land, with consideration being given to the relative values of the various resources, and not necessarily the combination of uses that will give the greatest dollar return or the greatest unit output.
>
> (4)(b) "Sustained yield" means the achievement and maintenance in perpetuity of a high-level annual or regular periodic output of the various renewable resources of the national forests without impairment of the productivity of the land.

The act specifically excludes MINERALS management and the management of non–National Forest federal land from its jurisdiction (Section 1) and states that WILDERNESS AREAS are "consistent with the purposes and provisions of this Act." (Section 2). Beginning in the mid-1990s, the interpretation of the act by the U.S. Forest Service began a slow but decided shift away from seeing national forests in terms of their commercial timbering potential, which has been greatly diminished through decades of logging pressure, and toward "ecosystem management" and a greater emphasis on outdoor recreation as the economic base. *See also* ORGANIC ADMINISTRATION ACT OF 1897; NATIONAL FOREST MANAGEMENT ACT; FOREST AND RANGELAND RENEWABLE RESOURCES PLANNING ACT.

Mumford, Lewis (1895–1990) educator, author, historian, social philosopher. Mumford, a professor of city and regional planning at Stanford University, the Massachusetts Institute of Technology, and other institutions, became well known through his insightful books. His works centered on the effects of the city environment on the human condition and emphasized the negative effects of technology on society. He was honored with the U.S. Medal of Freedom and was a knight of the Order of the British Empire. His books include *Technics and Civilization* (1934), *The Culture of Cities* (1938), *The Condition of Man* (1944), *The Human Prospect* (1955), *The City in History* (1961), *The Highway and the City* (1963), and *The Urban Prospect* (1968).

municipality in civil law, a city, town or other localized area that has incorporated for self-government. Incorporation is an important part of the definition. An unincorporated area is not a municipality and may not make its own laws, no matter how dense its development has become, but must operate under the laws of the county or state in which it is located. Municipalities are treated legally as a special form of corporation and are granted a number of powers by the state, including the right to raise money through taxes, bond sales, and other means; the right to control commerce within their boundaries; the right to own property; and the right to direct and control their development and provide for the welfare of their residents through zoning, nuisance-abatement ordinances, and similar means.

Munsell chart in forestry, a color chart used to determine the NITROGEN content of CONIFER needles. The chart consists of a graduated sequence of colors varying from light blue-green to deep green. The tree's needles are compared to the colors on the chart to find the best match, and the nitrogen content is then read directly from the chart.

mussel a bivalve shellfish found in fresh and salt waters. The blue mussel, which lives in colonies in shallow coastal seawaters of the north Atlantic coast in the United States and Canada, is edible and has a major commercial fishery. Two exotic mussel species from Europe have created a serious environmental problem in the Great Lakes and more recently in Chesapeake Bay and other Atlantic coastal areas. The freshwater zebra mussel which has no natural enemies, was first discovered in 1986 in western Lake Erie (near Detroit). It has clogged water intake pipes for irrigation and cooling, grown into huge masses on boat hulls, piers, and other marine structures, and depleted food supply for other species in the lake. The annual costs for cleanup are expected to reach $5 billion unless the zebra mussel population begins to decline from some natural cause. Another mussel, the quagga, appeared in 1991 in the Great Lakes and in Chesapeake Bay and is thought to be potentially even more destructive and difficult to control since it is larger than the zebra mussel and can survive at greater depths and under a greater range of temperatures and salinity.

mutagen any physical or chemical agent capable of causing a MUTATION. A mutagen may be radiation such as cosmic rays or the rays given off by a radioactive substance (*see* RADIOACTIVITY; ALPHA PARTICLE; BETA PARTICLE; GAMMA RAY); alternatively, it may be a chemical substance capable of passing through the barrier of the cell wall and reaching the nucleus (*see* CELL: *structure of cells*), where it reacts chemically with the DNA molecule to alter its structure, replacing part of it or breaking it into two parts that then recombine in a different pattern. The effects of a mutagen do not show up in the individual who was exposed, but in that individual's offspring, where they usually take the form of birth defects. Since cancer is also a form of cellular abnormality, the same chemical is often both a mutagen and a CARCINOGEN (*see* AMES TEST).

mutation in the biological sciences, a change in the structure of the reproductive DNA of an ORGANISM, resulting in a change in the characteristics of the organism's offspring. The change is *heritable,* meaning that it will be passed on to the offspring's offspring, and all future generations, as well. Mutations are of two general types. *Chromosomal mutations* (also known as *cytological mutations*) occur when damage is done on the chromosomal level. Genes may be left out (*deletion*); a segment of the chromosome may be spliced in backwards (*inversion*); extra chromosome sets may be present (*polyploidy*); etc. *Gene mutations* result from direct damage to an individual gene. One or more of the base pairs (the "rungs" of the DNA ladder) may be either replaced by other pairs (*base substitution*) or absent altogether (*frameshift*). Most mutations are harmful, because they disrupt the "fine-tuning" between organisms and their environment that has taken place over the millennia since life began. The rare beneficial mutations (those that cause a change that confers a selective advantage on the offspring) are one of the chief driving forces of evolution. *See also* MUTAGEN.

mutualism *See* SYMBIOSIS.

mycorrhyza in botany and forestry, a FUNGUS living in a symbiotic association (*see* SYMBIOSIS) with the roots of a VASCULAR PLANT. Approximately 90% of all known vascular plants form mycorrhyzal relationships, and most plants that form these relationships cannot live in soil that has not been innoculated with spores of one or more of the particular fungi they associate with. Often the relationship is highly specific, with a given SPECIES of fungus never found without its plant associate and vice versa. In most of these relationships the fungal partner appears to obtain sugar and other CARBOHYDRATES from the plant partner, while the plant obtains bioavailable minerals—particularly PHOSPHORUS—from the fungus (*see* TRACE ELEMENT; BIOAVAILABILITY). The soft fungal material that encases rootlets of trees and woody plants substantially increases the ability of the roots to take up water. There is fairly strong evidence that mycorrhyzal relationships have been around ever since the evolution of the vascular plants, and that they are the principal means by which the plants were enabled to move onto dry land.

Myers, Norman (1936–) environmental scientist and authority on tropical rain forests. Myers has taught at Oxford University and Cornell University and served as an adviser at the WORLD BANK's GLOBAL ENVIRONMENT FACILITY. He is particularly concerned about the loss of species through DEFORESTATION of the RAIN FOREST BIOME and has won many awards including the Volvo Environment Prize and the United Nations' Sasakawa Prize for his advocacy. Among his books are *A Wealth of Wild Species: Storehouse for Human Welfare* (1983), *Not Far Afield: U.S. Interests and the Global Environment* (1987), *Population, Resources, and Environment: The Critical Challenges* (1991), *Ultimate Security: The Environmental Basis of Political Instability* (1993), *Scarcity or Abundance? A Debate on the Environment* (1994), and *Environmental Exodus: An Emergent Crisis in the Global Arena* (1995).

Nader, Ralph (1934–) consumer advocate, lawyer, lecturer, presidential candidate. After being educated at Princeton and Harvard Universities and admitted to the bar in 1958, Nader practiced law in Connecticut. His best-selling book, *Unsafe at Any Speed* (1965), exposed the auto industry's poor safety standards and influenced Congress in its passage of the National Traffic and Motor Vehicle Safety Act of 1966. Nader's devotion to public safety and to corporate and governmental ethics led him to create several public-interest organizations including the Center for the Study of Responsive Law, Public Citizen, the Clean Water Action Group, and the U.S. Public Interest Research Group. Nader ran as the Green Party candidate in the 1996 U.S. presidential election and again in 2000. Among his many books are *Action for a Change* (1972), *Government Regulation: What Kind of Reform?* (1976), *The Menace of Atomic Energy* (1977), *Who's Poisoning America?: Corporate Polluters and Their Victims in the Chemical Age* (1981), *Case Against "Free Trade": GATT, NAFTA, and the Globalization of Corporate Power* (1993), and *Collision Course: The Truth About Airline Safety* (1995).

Naess, Arne (1912–) Norwegian philosopher and ecologist. Naess is best known for his concept of DEEP ECOLOGY, first expressed in his 1973 article "The Shallow and the Deep, Long-Range Ecology Movements: A Summary." Deep ecology, unlike scientific ecology, which deals with the objective observation of nature, is concerned with personal connectedness to the environment around us and with the right of nature to pursue its own processes without undue human interference. Naess has been published widely and his work translated into several languages including English, French, and German. Professor emeritus at Oslo University in Norway, he has been visiting professor at the University of California, the University of Hong Kong, and the Academy of Social Science of China. He is the recipient of the Mahatma Gandhi Prize for Peace (1994) and the Sonning Prize (1977) for contribution to European culture. He is founder and former editor of the journal *Inquiry,* former editor of the journal *Filosofiske Problemer,* and continuing editor of the journal *Synthese: An International Journal of Epistemology, Methodology, and Philosophy of Science.* His books include *Scepticism* (1968), *Four Modern Philosophers: Carnap, Wittgenstein, Heidegger, and Sartre* (1969), and *Ecology, Community, and Lifestyle: Outline of an Ecosophy* (1989).

NAFTA *See* NORTH AMERICAN FREE TRADE AGREEMENT.

nannoplankton in limnology, very small PLANKTON, generally defined as those that are small enough to pass through a No. 20 mesh net, that is, those which are smaller than 60 MICRONS in diameter. They may be isolated from a water sample either by centrifuging them against the bottom of the sample jar or by treating the sample with a fixative (usually a mixture of iodine and potassium iodide) that kills them, dyes them, and precipitates them to the bottom of the jar. Once collected on the bottom (by either method), they may be counted and identified through the use of a microscope. *Compare* NET PLANKTON.

natality in population biology, the birth rate, that is, the number of births taking place in a given POPULATION within a given period of time. *Maximum natality* is the highest number of births per unit time that could possibly take place in the population, given the population's size, age range, and sex ratio. *Actual natality* (or simply "natality") is the birth rate that is actually observed for the population. It is always lower than maximum natality, often considerably so. In the study of human populations, natality is often given in terms of number of births per thousand people per unit of time. In mathematical terms

$$N = (\Delta P/P \, x\Delta t)m1000$$

where ΔP stands for the change in population; Δt stands for the change in time; and P stands for the total population size.

National Academy of Sciences (NAS) created by the U.S. Congress in 1863 to advise the government in scientific and technical matters as they may relate to policy decisions. The NAS receives 85% of its funding from the federal government in the form of contracts and grants, not directly from congressional appropriations. This form of funding helps ensure independent research and investigation by distancing the researchers from political agendas. The National Research Council, the active arm of what are now three national academies (National Academy of Sciences, National Academy of Engineering, and the Institute of Medicine), performs much of the actual research for the NAS. The NAS works to disseminate the findings of the Research Council through briefings, reports to the media, presentations at professional meetings or private groups, and publication. Address: 2101 Constitution Avenue NW, Washington, DC 20418. Phone:(202) 334-2000. Website: www.nas.edu.

National Acid Precipitation Assessment Program (NAPAP) an ultimately controversial 10-year government study initiated in 1980 to assess the impact of ACID RAIN on lakes and streams and (later) on forests in the eastern United States. Alarmed by reports from Scandinavia of the disappearance of fish from lakes and rivers, scientists from the United States and Canada discovered that the same thing was happening in the Adirondocks in New York State, New England, and southern Quebec. In the late 1970s President Jimmy Carter and Canadian prime minister Pierre Trudeau began a series of talks that led to an international agreement to reduce emissions from midwestern power plants and factories, which were widely acknowledged to be the prime factor in the acid rain that was so acidifying lakes that fish could no longer survive. In 1980 the U.S. Congress, meaning to provide a scientific basis for making the major policy reforms, voted to establish NAPAP. Initially the study concentrated on the acidification of lakes and streams but was later expanded to look into the impact of acid rain on trees and forests, as had happened in Germany (WALDSTERBEN) and which had been detected as well in New England. As the study progressed, however, environmentalists and many of the scientists involved complained that NAPAP was being used to delay necessary AIR POLLUTION legislation and that, worse, an interim report issued in 1987 had seriously distorted the scientific findings by minimizing the seriousness of acid rain. The conclusions of the 28-volume final report, issued in 1991, continued to downplay acid rain, despite a great deal of scientific evidence found in individual NAPAP scientific studies themselves (some of which were ignored). While the electric power industry and its allies in and out of government seized on the report as a reason to delay reforms in air pollution laws, Congress in its deliberations on the CLEAN AIR ACT amendments of 1990 paid little attention to the NAPAP study, which had cost in excess of $500 million. Subsequent research by independent scientists has shown that the worst fears regarding the impacts of acid rain were justified. *See also* FOREST DECLINE AND PATHOLOGY.

National Ambient Air Quality Standards (NAAQS) federal (U.S.) standards for allowable levels of pollutants in the AMBIENT AIR, set by the ENVIRONMENTAL PROTECTION AGENCY under the terms of the CLEAN AIR ACT AMENDMENTS OF 1970. NAAQS levels are set for individual pollutants or classes of pollutants on the basis of criteria established and published by the agency; hence pollutants covered under these standards are known as "criteria pollutants." There are currently seven criteria pollutants: TOTAL SUSPENDED PARTICULATES, sulfur dioxide (*see under* SULFUR OXIDE), CARBON MONOXIDE, NITROGEN DIOXIDE, OZONE, HYDROCARBONS, and LEAD. Two classes of standards are set. *Primary standards* are set on the basis of human health effects: their levels theoretically reflect the concentrations at which a significant number of those exposed to them will suffer impaired health. The law requires the standards to take into account infants and the infirm as well as healthy adults. *Secondary standards,* which are stricter than primary standards, are set on the basis of economic and environmental damage. These take into account damage to buildings, clothing, and automobiles; health effects on wildlife, plants, and other living ORGANISMS; visibility; and other effects that have only an indirect impact on human lives. Amendments to the Clean Air Act in 1977 and 1990 provided that the NAAQS could be administered by each state; and in 1997 the new standards for SMOG (ozone) and soot (particulate matter) were promulgated by the Environmental Protection Agency. *See* STATE IMPLEMENTATION PLAN; AIR QUALITY CONTROL REGION; AIR QUALITY CLASS.

National Association of Conservation Districts the national organization serving official conservation districts. Founded in 1946, the association consists of 54 state and territorial groups and 3,000 local districts. Through research, publication, educational programs, and conferences, it encourages the prudent use and carefully planned development of natural resources including soil, water, and forests. Address: 509 Capitol Court NE, Washington, DC 20002. Phone: (202) 547-6223. Website: www.nacdnet.org.

National Audubon Society (NAS) environmental organization, founded in 1905 to preserve and protect wildlife HABITAT. It has since expanded its interests to natural area conservation, pollution control, and the promotion of environmental awareness. Not a lobbying organization (although it maintains a Washington office), the NAS operates primarily through educational programs and media outreach. Membership (1999): 550,000. Address: 700 Broadway, New York, NY 10003. Phone: (212) 979-3000. Website: www.audubon.org.

National Cancer Institute *See under* PUBLIC HEALTH SERVICE.

National Center for Atmospheric Research (NCAR) established in 1960 to research atmospheric and related science problems. NCAR is operated under an agreement between the National Science Foundation and the University Corporation for Atmospheric Research. Areas of research include climate change, atmospheric make-up, interactions between the Earth and Sun, weather development and forecasting. In the 1980s and 1990s NCAR has focused on global warming and related issues, providing expert and authoritative scientific data and opinion on the controversial aspects of this issue. Address: 1850 Table Mesa Drive, Boulder, CO 80303. Phone: (303) 497-1000. Website: www.ncar.ucar.edu/ncar.

National Center for Health Statistics *See under* PUBLIC HEALTH SERVICE.

National Center for Toxicological Research *See under* FOOD AND DRUG ADMINISTRATION.

National Coalition against the Misuse of Pesticides dedicated to providing communities, organizations, and individuals with valuable information on pesticides and their alternatives. The coalition, founded in 1981, is an information source on alternatives to pesticides, food safety, agricultural worker safety, groundwater pollution, protection of farm children, and related legislative efforts. Address: 701 E Street SE, Suite 200, Washington, DC 20003. Phone: (202) 543-5450. Website: www.ncamp.org.

National Cooperative Soil Survey *See* SEVENTH APPROXIMATION.

National Emission Standards for Hazardous Air Pollutants (NESHAP) federal (U.S.) standards for the maximum emission of hazardous air pollutants from STATIONARY SOURCES, set by the ENVIRONMENTAL PROTECTION AGENCY (EPA) under the terms of Section 112 of the CLEAN AIR ACT. To qualify for NESHAP standards, a pollutant must be (a) demonstrably hazardous to human health, and (b) not already covered by ambient quality standards (*see* NATIONAL AMBIENT AIR QUALITY STANDARDS). The EPA is responsible for setting permissible emission levels for NESHAPs but met with difficulty in applying them, especially after the 1990 amendments to the Clean Air Act which provided for the control of 189 hazardous air pollutants. After many years of contention between the EPA and the chemical industry, a new "HON" rule (hazardous organic NESHAP) was promulgated in 1996, intended to cut AIR POLLUTION from chemical plants—the most serious hazardous air polluters—by 88%. The measure has substantially reduced VOLATILE ORGANIC COMPOUNDS (VOCs), a primary source of urban SMOG, an outcome estimated by the EPA to be the equivalent of eliminating 25% of all the cars in America. The HON rule controls the emissions of 111 of the 189 hazardous air pollutants listed in the 1990 amendments, affecting some 370 chemical plants in 38 states, although most are located in New Jersey, Texas, and Louisiana.

National Environmental Policy Act (NEPA) federal (U.S.) law establishing a framework for policy decisions regarding actions that will have a significant effect on the environment. The act's most important provision is section 102(2)(c), which requires federal agencies to file an ENVIRONMENTAL IMPACT STATEMENT on all "proposals for legislation and other major Federal actions significantly affecting the quality of the human environment". Other sections of the act define broad federal policy with respect to the environment; establish a framework for cooperation between the United States and other nations, and among federal, state, and local governments within the United States, on environmental matters; and mandate the formation of a COUNCIL ON ENVIRONMENTAL QUALITY.

EFFECTS

A great deal of litigation has taken place as a result of the National Environmental Policy Act, nearly all of it based on section 102(2)(c). Most of this litigation has turned on the question of what constitutes a "significant" federal action and thus requires an Environmental Impact Statement. A smaller but still significant number of NEPA cases have revolved around the adequacy of Environmental Impact Statements as filed by various agencies. The courts have generally interpreted the law strictly, resulting in numerous slowdowns and changes in design for the projects under litigation; however, few federal activities have been permanently halted by suits brought under NEPA. Occasionally, Congress has felt it necessary to abrogate NEPA suits by declaring certain actions outside the jurisdiction of the act, or by declaring specific Environmental Impact Statements to be adequate and not subject to further judicial review. *See*, e.g., ROADLESS AREA REVIEW AND EVALUATION.

HISTORY

The National Environmental Policy Act was first introduced in the 90th Congress by Senator Henry M. Jackson (D-WA) on December 15, 1967. This bill failed to pass. The act was reintroduced in the 91st Congress by Jackson in the Senate on February 18, 1969, and by Rep. John Dingell (D-MI) in the House four months later (July 1). The bill passed both houses of Congress in the week before Christmas, 1969, and was signed into law by President Richard M. Nixon on January 1, 1970.

National Flood Insurance Program (NFIP) federal (U.S.) program to provide insurance for homes and businesses in certain flood-prone communities. To qualify for the program, a community must (a) lie within or partially within a flood-prone area as identified on a federally prepared Flood Hazard Boundary Map; and (b) have adopted a land-use plan requiring new structures within the floodplain to either be floodproofed or elevated above the level of the 30-year flood. Communities may choose not to participate in the NFIP; however, federally assisted financing for development within flood-prone areas is not available in communities that have not joined the program, and these communities are not eligible for federal disaster-relief funds. Federal banking regulations also forbid the use of real estate in uninsured floodplains as collateral to secure a loan. Enacted by Congress as part of the Housing and Urban Development Act of 1968, the NFIP administered by the Federal Insurance Administration (FIA), a part of the Federal Emergency Management Administration (FEMA), was intended to provide a nonstructural, more environmentally sound alternative to DAMS and LEVEES as flood-control devices. However, it has proved to be more an adjunct than a replacement. Indeed, beginning in the late 1980s, environmentalists, along with floodplain managers, persuaded Congress to determine why so few people purchase flood insurance under the program. The reason was that while lenders required homeowners to have flood insurance to qualify for mortgages, once the mortgages were secured owners allowed the insurance to lapse, assuming that disaster relief funds and the Federal Emergency Management Agency "buy-out" program would take care of any catastrophes—which sometimes occurred every few years in flood-prone areas. To meet this problem, Congress enacted the National Flood Insurance Reform Act in 1994 requiring mortgage lenders to establish escrow accounts from which flood insurance premiums would be paid. In addition, the 1994 act mandated that as of 1996 mitigation insurance would be required whereby structures suffering damage would have to be modified with flood-resistant structural changes or simply moved to another location. In the latter case, repeatedly flooded land can be converted to public open space or GREENWAYS when the buildings are removed.

National Forest a unit of the NATIONAL FOREST SYSTEM composed primarily of forest lands, or of mountain lands above TIMBERLINE, and managed by the FOREST SERVICE (units of the system composed primarily of prairie are termed *grasslands; see* NATIONAL GRASSLAND). National Forests are required by law to be managed for the MULTIPLE USES of wood, water, wildlife, forage, and recreation (*see* MULTIPLE USE-SUSTAINED YIELD ACT OF 1960). Most lands in the National Forests were originally set aside as part of the "Forest Reserves" withdrawn by presidential proclamation under the Forest Reserve Act of 1891. Other lands have since been added by purchase (*see* WEEKS ACT), by congressional mandate, and by trades with other federal agencies, notably the BUREAU OF LAND MANAGEMENT. Within lands declared by law to be part of the National Forest System, boundaries between individual forests are determined administratively. These usually follow state lines or watershed boundaries, and are designed to create easily manageable units. As storehouses of much of the nation's remaining timber as well as much of its scenic beauty and most of its remaining wild (i.e., ecologically "pure") areas, the National Forests have become a major battle ground for federal land-use policy, and their management has become an increasingly tightly controlled tightrope act. *See also* NATIONAL FOREST MANAGEMENT ACT OF 1976.

National Forest Management Act of 1976 (NFMA) federal legislation setting management policy for the NATIONAL FOREST SYSTEM, signed into law by President Gerald Ford on October 22, 1976. Couched as a set of amendments to the FOREST AND RANGELAND RENEWABLE RESOURCES PLANNING ACT (RPA), and therefore sometimes referred to as the "Forest and Rangeland

Renewable Resources Amendments", the NFMA goes far beyond its parent legislation in a number of significant respects. The most important portion of the legislation is Section 6, subtitled *National Forest System Resource Planning.* This section spells out specific principles that the FOREST SERVICE is required to adhere to while doing land-use planning. The agency must: (1) provide for full public participation in all aspects of the planning process; (2) follow principles of MULTIPLE USE and SUSTAINED YIELD for all renewable resources obtainable from the NATIONAL FORESTS, including recreation, range, timber, water, wildlife and fish, and wilderness; (3) create integrated plans for each unit of the National Forest System (that is, each National Forest and NATIONAL GRASSLAND) that conform to the requirements of the NATIONAL ENVIRONMENTAL POLICY ACT; (4) provide for diversity of wildlife and plants rather than MONOCULTURES, while balancing economic and environmental demands equally insofar as possible; (5) allow timber harvest only where soils, watersheds and bodies of water will not be damaged, and where RESTOCKING is possible within five years; (6) choose harvest systems according to some other criteria than the greatest dollar return or the greatest unit volume of timber; (7) identify lands unsuitable for timber production; and (8) set the ROTATION AGE no lower than CULMINATION of MEAN ANNUAL INCREMENT. Other sections of the legislation require research on improved efficiency in the utilization of wood fiber, including recycling used materials and utilizing SLASH; increased emphasis on reforestation, including a regular five-year inventory of all lands not meeting stocking standards (*see* STOCKING LEVEL); increased cooperation with small woodlot owners to improve private forestry; and detailed rules for timber sale administration, including public notification, competitive bidding, requirements for harvest completion within a specific time period, and—at the sale administrator's option—the establishment of a fund similar to the KNUTSON-VANDENBERG ACT reforestation account into which the sale purchaser must deposit funds to cover the costs of road building and sale administration. *See also* ORGANIC ACT; MONONGAHELA DECISION; MULTIPLE USE-SUSTAINED YIELD ACT; WEEKS LAW.

National Forest Products Association coalition of forest products manufacturers, founded in 1902 to represent the forest products industry before Congress. The association also publishes technical booklets on the use of forest products and promotional materials representing the industry viewpoint on forest management issues. The NFPA is now part of the AMERICAN FOREST AND PAPER ASSOCIATION, which was created on January 1, 1993, by the merging of the American Forest Council, the American Paper Institute, and the National Forest Products Association.

National Forest Reservation Commission obsolete federal commission charged with evaluating and purchasing lands for addition to the NATIONAL FOREST SYSTEM. Created under the terms of the Weeks Forest Purchase Act of 1911 (*see* WEEKS LAW), the National Forest Reservation Commission consisted of the secretaries of war, agriculture, interior, and two members from each House of Congress. Over 20 million acres of land were added to the National Forest System by the commission between 1911 and 1976, when it was terminated by Congress and its duties transferred to the secretary of agriculture.

National Forest System nationwide system of federally owned lands held under the jurisdiction of the Department of Agriculture (*see* AGRICULTURE, DEPARTMENT OF) and managed by the United States FOREST SERVICE. The National Forest System is composed of the NATIONAL FORESTS, the NATIONAL GRASSLANDS, and a number of scattered individual parcels known variously as Purchase Units, Land Utilization Projects, Experimental Forests, and Experimental Research Areas. (These are generally managed as part of the closest National Forest.) Originally a system of so-called "Forest Reserves" created under the auspices of the FOREST RESERVE ACT OF 1891 and managed by the Department of the Interior (*see* INTERIOR, DEPARTMENT OF THE), the National Forest System was reorganized and transferred to the Department of Agriculture by the Transfer Act of 1905. The System's 155 National Forests, 19 National Grasslands, and numerous scattered jurisdictional units total 191,000 acres in 44 states and the Territory of Puerto Rico.

National Grassland a unit of the NATIONAL FOREST SYSTEM comprised primarily of lands belonging to the GRASSLAND BIOME and therefore officially termed a grassland rather than a forest. All currently existing National Grasslands consist of lands acquired by the federal government in the late 1930s under the terms of the BANKHEAD-JONES FARM TENANT ACT. Originally termed *land utilization projects,* they were consolidated and renamed as National Grasslands in 1960, making them a permanent part of the National Forest System. (Some 160,000 acres of widely scattered land utilization projects remain in existence under the old title and are not considered permanent federal lands, though they are classed as part of the National Forest System.) The National Grasslands are managed primarily as rangeland, for stock grazing and water and soil retention, though some contain areas utilized for irrigated agriculture under private leases, and most contain recreational features. For the most part they are treated administratively as parts of nearby NATIONAL FORESTS, having the administrative status of Ranger Districts. Seventeen of the 19 existing National Grasslands are clustered on or just west of the hundredth meridian, in Texas, Oklahoma, New Mexico, Kansas, Colorado, Nebraska, Wyoming, and North and South Dakota. The other two are in the Pacific Northwest, one each in Oregon and Idaho. Total land area: 3.9 million acres. *See also* FOREST SERVICE.

National Inholders Association (NIA) See AMERICAN LAND RIGHTS ASSOCIATION.

National Institute of Environmental Health Sciences *See under* PUBLIC HEALTH SERVICE.

national interest lands *See under* BUREAU OF LAND MANAGEMENT.

National Lakeshore a unit of the NATIONAL PARK SYSTEM that protects lands along the shore of a lake. It is identical in all but name to a NATIONAL PARK. All four currently existing National Lakeshores lie along the shores of the Great Lakes.

National Marine Sanctuaries Underwater coastal nature reserves established under the National Marine Sanctuaries Act of 1972 and administered by the National Oceanic and Atmospheric Administration (NOAA). After a serious oil spill off the coast of Santa Barbara, California, in 1969, legislation was enacted to identify and provide protection to nationally significant underwater natural areas so that they would not be spoiled by such catastrophes. The act, however, was bitterly opposed by offshore drilling interests and the U.S. Department of Defense and remained unfunded for seven years. Finally, in 1979, the first marine sanctuary was established off the North Carolina coast to protect the wreckage of the USS *Monitor*, an ironclad Union warship that had sunk in 230 feet of water during the Civil War. Since then, 11 more marine sanctuaries have been established as underwater preserves containing historical, cultural, ecological, recreational, and aesthetic resources.

National Marine Sanctuaries

Channel Islands. Pacific Ocean 20 miles off Santa Barbara, California. 1,252 square nautical miles. Protects many species including gray whales, harbor and elephant seals, shellfish and corals, and sea birds.

Cordell Bank. Pacific Ocean waters north of San Francisco from Point Reyes to Bodega Head. 397 square nautical miles. Most northerly temperate zone reef in the United States with green, leatherback, and Pacific Ridley turtles.

Fagetele Bay. Off Tutuila Island, American Samoa. 0.25 square nautical miles. Preserves traditional fishing in a area surrounded by a submerged volcanic crater.

Florida Keys. Waters surrounding the Keys to the Dry Tortugas. 2,800 square nautical miles. Protects 6,000 species including corals, fish and bird life, and historic shipwrecks including a Spanish treasure ship and a slave ship.

Flower Garden Banks. In the Gulf of Mexico, 105 nautical miles southeast of the Texas-Louisiana border. 41.7 square nautical miles. The northernmost coral reef in the United States.

Gray's Reef. 17.5 nautical miles east of Sapelo Island, Georgia. 17 square nautical miles. Calving grounds for right whales.

Gulf of the Farallones. Pacific Ocean from Half Moon Bay south of San Francisco to Bodega Head, north of the city. 948 square nautical miles. A western estuarine ecosystem protecting the largest breeding population of marine mammals and seabirds in the continental United States plus more than 100 shipwrecks.

Hawaiian Islands Humpback Whale. In a four-island area of Maui Island, Hawaii. 1,300 square nautical miles. Principal breeding ground of the humpback whale in the northern Pacific.

Monitor. Southeast of Cape Hatteras, North Carolina. 1 nautical mile in diameter. Wreckage of USS *Monitor,* a Union ironclad vessel sunk in 1862.

Monterey Bay. A stretch of the Pacific Ocean between Cambria, California, north of Santa Barbara, and the Marin Headlands north of San Francisco. 4,024 square nautical miles. A near-shore deep-ocean environment that includes one of the deepest underwater canyons off the continental shelf.

Olympic Coast. Off Washington State's Olympic Peninsula, between Cape Flattery and Copalis Beach. 2,500 square nautical miles. Protects the large population of bald eagles in the United States.

Stellwagen Bank. Massachusetts Bay, between Cape Cod and Cape Ann, Massachusetts. 638 square nautical miles. Summer feeding ground for endangered right, humpback, and fin whales.

Data from NOAA, National Audubon Society

National Monument a unit of the NATIONAL PARK SYSTEM or (in a few cases) the NATIONAL FOREST SYSTEM, managed in the same manner as a NATIONAL PARK but generally smaller and less diverse in character and designed to protect a specific feature of particular historic or scientific significance. Although National Monuments may be designated by Congress, most have come into being by presidential proclamation under the terms of the ANTIQUITIES ACT of 1906. The power to declare National Monuments has sometimes been used by presidents in order to break legislative deadlocks over the preservation of National Park-quality land (*see,* e.g., ALASKA NATIONAL INTEREST LANDS CONSERVATION ACT).

National Oceanic and Atmospheric Administration (NOAA) agency of the U.S. Department of Commerce, created by President Richard Nixon on October 3, 1970, and charged with coordinating federal policy and administrating most federal laws in regard to the oceans and ocean resources; the ATMOSPHERE, especially as it relates to WEATHER and CLIMATE (but not air pollution); and those solar and outer-space phenomena that impact the earth. The agency's important environmental roles include weather prediction and atmospheric research; enforcement of the COASTAL ZONE MANAGEMENT Act, the MARINE PROTECTION, RESEARCH AND SANCTUARIES ACT, and the MARINE MAMMALS PROTECTION ACT; regulation of ocean fisheries, minerals exploration, and mining; preparation of nautical charts and maps; research into alternatives to ocean dumping of HAZARDOUS WASTES; and the operation of artificial satellites for environmental data gathering. As presently constituted, NOAA has four principal program branches: the National Ocean Survey, the National Marine Fisheries Service, the National Weather Service, the Office of Global Programs, and the Office of Oceanic and Atmospheric Research. The agency's chief administrator holds the rank of Under Secretary of Commerce for Oceans and Atmosphere. Address: U.S. Department of Commerce, 14th Street and Constitution Avenue NW, Room 6013, Washington, DC 20230. Phone: (202) 482-6090. Website: www.noaa.gov.

National Ocean Survey *See under* NATIONAL OCEANIC AND ATMOSPHERIC ADMINISTRATION.

National Park generally, an area of land set aside by a national government to protect a natural feature or group of features of outstanding scenic, scientific, or recreational value. In the United States, the term means further an area of federal land designated by Congress through the legislative process to be preserved and protected for the benefit and enjoyment of all citizens. National Parks are not WILDERNESS AREAS (although many National Parks contain designated wilderness). They are generally managed with as much access as is possible while maintaining the integrity of the natural features they were set aside to protect. The largest U.S. National Park is Wrangell-St. Elias in Alaska (13.2 million acres), created by the ALASKA NATIONAL INTEREST LANDS ACT of 1980. The oldest is Yellowstone in Wyoming, Montana, and Idaho (2.2 million acres), created by act of Congress on March 1, 1872. *See also* NATIONAL PARK SERVICE; NATIONAL PARK SYSTEM; NATIONAL PARK FOUNDATION; NATIONAL PARKS AND CONSERVATION ASSOCIATION; NATIONAL PARK TRUST FUND BOARD.

National Park Foundation quasi-public agency, founded in 1967 to channel financial support from corporations, labor unions, foundations, and individual donors to projects of the NATIONAL PARK SERVICE. It also funds grants for publications and other projects relating to NATIONAL PARKS conservation, and publishes a biennial guide to the National Parks. The foundation holds a charter from the U.S. Congress. Address: 1101 17th Street NW, Suite 1102, Washington, DC 20036. Phone: (202) 785-4500. Website:http://nationalparks.org.

National Parks and Conservation Association environmental organization founded in 1919 to protect existing NATIONAL PARKS and similar preserves and to promote the formation of others in appropriate locations. The association has an active publications program and a large library of park and conservation-related material. Prior to 1970 it was known as the National Parks Association. Membership (1999): 500,000. Address: 1776 Massachusetts Avenue NW, Washington, DC 20036. Phone: (202) 223-6722. Website: www.npca.org.

National Park Service (NPS) agency within the U.S. Department of the Interior (*see* INTERIOR, DEPARTMENT OF THE) created by Congress on August 25, 1916, and charged with the management of the nation's NATIONAL PARKS and NATIONAL MONUMENTS. Over the years the agency's mission has grown, and in addition to the National Parks and Monuments it is now in charge of most NATIONAL RECREATION AREAS, NATIONAL SEASHORES, NATIONAL LAKESHORES, National Rivers, National Historic Sites, and National Parkways (some areas in these categories are operated by other agencies, notably the FOREST SERVICE and the BUREAU OF LAND MANAGEMENT). It also operates the municipal park system of Washington, D.C. Since the demise of the HERITAGE CONSERVATION AND RECREATION SERVICE in 1981, the National Park Service has served as the principal coordinating body for federal studies of outdoor recreation and historic preservation, including studies for the establishment of National Scenic Trails and National Recreation Trails (*see* NATIONAL TRAILS ACT); Wild and Scenic Rivers (*see* WILD RIVER; SCENIC RIVER; RECREATIONAL RIVER); the National Historic Register; and the Historic American Buildings Survey. It administers part of the LAND AND WATER CONSERVATION FUND and coordinates most interagency planning efforts in natural and historic preservation, recreation management, and archaeological investigation. The National Park Service is headed by a national director in Washington, D.C., who reports to the assistant secretary of the interior for fish and wildlife and parks. It operates a national support center for technical services (architecture, building construction, etc.) in Denver and another for research services in Harpers Ferry, West Virginia. Otherwise, the structure of the agency conforms to the STANDARD FEDERAL REGIONS.

National Park System nationwide system of parks and other preserves held by the federal government and operated by the NATIONAL PARK SERVICE, a branch of the Department of the Interior (*see* INTERIOR, DEPARTMENT OF THE). Besides the NATIONAL PARKS themselves, the National Park System contains the NATIONAL LAKESHORES, the NATIONAL SEASHORES, and most NATIONAL MONUMENTS, NATIONAL RECREATION AREAS, and National Historic Sites and similar areas (a few areas among these latter groups are operated by other federal agencies, notably the FOREST SERVICE). The city park system of Washington, D.C. is also operated by the National Park Service, and is sometimes included in the National Park System. As of 2000, the National Park System included 378 individual areas containing a total of 83.3 million acres, located in all 50 states except Delaware, plus the District of Columbia and the territories.

National Pollutant Discharge Elimination System (NPDES) a nationwide system of permits for the discharge of EFFLUENT into waterways, mandated by Section 402 of the CLEAN WATER ACT and administered by the ENVIRONMENTAL PROTECTION AGENCY (EPA) or by the various states under the oversight of the EPA, provided that the states have adopted statewide water-quality plans in accordance with EPA guidelines. NPDES permits are required for any POINT SOURCE DISCHARGE of effluent, including sewage and industrial OUTFALLS, dredge SPOIL, power-plant cooling system discharge, BILGE WATER and other shipping wastes, drainage ditches, FEEDLOT runoff, and the outflow from some DAMS. To qualify for a permit, an effluent discharger must prove that pollutants in the effluent and controlled using BEST AVAILABLE TECHNOLOGY, and that the effluent will not degrade the water quality in the RECEIVING WATERS below the AMBIENT QUALITY STANDARD set for them. In addition, specific limitations are set on the total quantity of certain pollutants that may be discharged. NPDES permits are generally for a term of five years. Renewal is not automatic, but requires certification that the permit terms have been adhered to and that water quality standards, as set forth in Sections 301–307 of the Clean Water Act, are being met.

National Railway Passenger Corporation (AMTRAK) quasi-official federal (U.S.) agency charged with improving, maintaining, and operating railroad passenger service in the United States. Part public agency and part private corporation, AMTRAK is governed by a nine-person board of directors of whom three are appointed by the president (subject to confirmation by the Senate); two are chosen by passenger-advocate organizations; and two are chosen by the corporation's stockholders (the remaining two are the chairman, chosen by the board itself, and the U.S. secretary of transportation, who serves in an *ex officio* capacity; *see* TRANSPORTATION, DEPARTMENT OF). Created by Congress in 1970 to counter the increasing elimination of passenger service by private railroads, the National Railway Passenger Corporation currently operates a system of approximately 22,000 route miles serving some 500 stations in 45 states. Approximately 730 miles of track, mostly in the Washington-to-Boston corridor, are owned by the agency; the remaining miles are leased from private railroads. Amtrak employs more than 24,000 people operating up to 265 trains per weekday. In 1998 it served more than 21 million passengers. Address: 60 Massachusetts Avenue NE, Washington, DC 20002. Phone: (202) 906-3860. Website: www.amtrak.com.

National Recreation and Park Association works to promote public understanding and appreciation of outdoor recreation and to improve the programs, leadership, and facilities of public parks. Its publications include *Parks and Recreation Magazine, Journal of Leisure Research, Recreation and Parks Law Reporter,* and *Park Practice Program.* Membership (1999): 23,500. Address: 22377 Belmont Ridge Road, Ashburn, VA 20148. Phone: (703) 858-0794.

National Recreation Area (NRA) an area of federal land containing outstanding recreational opportunities and set aside by Congress for that purpose. It is generally not predominantly natural in character, although it may be so. The first National Recreation Areas were created out of the lands surrounding federal DAMS and RESERVOIRS and were devoted to boating and other water-oriented sports. More recently, the concept has been expanded to include non-reservoir-based recreation, although NRAs are still almost always focused on bodies of water (Delaware Water Gap NRA, Pennsylvania/New Jersey; Golden Gate NRA, California; Cuyahoga Valley NRA, Ohio). Occasionally the designation has been used for NATIONAL PARK–caliber lands that happen to be managed by another federal agency such as the BUREAU OF LAND MANAGEMENT or the FOREST SERVICE; in these cases, the National Recreation Area name allows national-park style management while permitting the Forest Service or the BLM to retain jurisdiction over the land (example: Hells Canyon NRA, Idaho and Oregon). The first National Recreation Area, Coulee Dam, was created in 1946 around Grand Coulee Dam and Lake Roosevelt in the State of Washington. *See also* NATIONAL PARK SYSTEM.

National Recreation Trail *See under* NATIONAL TRAILS ACT OF 1968.

National Religious Partnership for the Environment a coalition of religious denominations established in 1993 to encourage religious bodies to address environmental issues from a faith-based perspective. Comprised of the U.S. Catholic Conference, the National Council of Churches of Christ, the Coalition on Environmental and Jewish Life, and the Evangelical Environmental Network, the partnership seeks to integrate environmental awareness and caring into religious thought and practice through five moral precepts: the sanctity of creation; the intrinsic value of all species and habitat; the ethical duty of stewardship; the inseparability of social justice and environmental sustainability; and the responsibilities of private property as measured against the greater good of the commonweal. Through grant programs and the publication of instructional materials, the partnership encourages local congregations to become involved in environmental issues, projects, and policies. Address: 1047 Amsterdam Avenue, New York, NY 10025. Phone: (212) 316-7441. Website:nrpe.org.

National Resource Lands lands in the western states administered by the BUREAU OF LAND MANAGEMENT. These public lands provide energy and mineral resources, rangeland, and recreational sites. BLM also manages environmentally sensitive lands under the designation AREA OF CRITICAL ENVIRONMENTAL CONCERN.

National Resources Planning Board obsolete federal resource planning and coordinating agency, created by Congress in 1939 by merging the Federal Employment Stabilization Office and the National Resources Committee. The board was one of the first serious attempts to do large-scale resource and land-use planning in the United States, and included both data-gathering and coordination activities covering a large range of federal and state actions. It was legislated out of existence by Congress in 1943.

national sacrifice area a term somewhat loosely applied in legislative policy-making to identify areas surrounding, say, a nuclear waste facility, that are "sacrificed" for the greater good of society as a whole. Other national sacrifice areas are associated with power generation, liquefied natural gas storage, chemical manufacturing plants, oil fields, coal mining, solid-waste dump sites, airports, and the like. Too often, national sacrifice areas are home to citizens who are less well off economically and not politically powerful. The notion of asking some people to sacrifice a quality environment for the good of others has given rise to NIMBY ("Not In My Back Yard") as a form of political resistance and provided an impetus for the advocacy of environmental justice for the poor among religious organizations. *See also* RELIGION AND THE ENVIRONMENT.

National Scenic Trail *See under* NATIONAL TRAILS ACT OF 1968.

National Science Foundation (NSF) independent agency of the U.S. government responsible for the support of science and engineering research and the development of science education programs. The NSF was established in 1950. The National Science Board, composed of 24 presidential appointees, sets policy for the foundation. Address: 4201 Wilson Boulevard, Arlington, VA 22230. Phone: (703) 306-1234. Website: www.nsf.gov.

National Seashore a unit of the NATIONAL PARK SYSTEM that protects lands along a seacoast. It is identical in all but name to a NATIONAL PARK. The first National Seashore, Cape Hatteras (North Carolina), was designated by Congress in 1953. *See also* NATIONAL LAKESHORE.

National Trails Act of 1968 federal (U.S.) legislation to promote the establishment and maintenance of hiking and horseback trails for recreational use, signed into law by President Lyndon Johnson on October 2, 1968. The law establishes three classes of trails. *National Scenic Trails* are designated only by Congress. They are long (generally several hundred miles) and built to high standards, maintaining a maximum grade of no more than 10%. Motor vehicles and mountain bikes are banned from them. Two National

Scenic Trails, the APPALACHIAN TRAIL and the PACIFIC CREST TRAIL, were designated by the original act; a third, the Continental Divide Trail, was added to the system in 1978. NATIONAL RECREATION TRAILS are designated by the secretary of the interior or (on NATIONAL FOREST lands) the secretary of agriculture. They are built to a somewhat lower standard than National Scenic Trails, and may be used by motor vehicles and mountain bikes unless these are expressly forbidden. The designating secretary must certify that the trails will remain available for recreational use without impairment for at least 10 years following their inclusion in the system. Finally, *Side* and *Connecting Trails* are short trails that provide access points to the longer National Scenic and National Recreation Trails. They are designated by the secretaries of interior and agriculture, and must meet the same standards as national recreation trails. All three classes of trails may pass over non-federal (i.e., state and private), as well as federal, lands. In the case of National Recreation Trails and Side and Connecting Trails, however, agreement must be reached with all affected landowners before the designation can be finalized. Taken together, there are now more than 820 recreational trails, along with eight National Scenic Trails and 12 National Historic Trails.

National Transportation Safety Board (NTSB) independent federal (U.S.) agency created by Congress on April 1, 1975, and charged with research into the causes of transportation accidents and methods of making transportation safer. The NTSB investigates all accidents to commercial carriers in the United States, including airlines, railroads, bus lines, truck lines, and shipping firms, especially those where one or more fatalities have occurred or where HAZARDOUS SUBSTANCES have been released into the environment; attempts to pinpoint the cause or causes of these accidents; and makes recommendations for avoiding similar problems in the future. The five members of the board are appointed by the president and confirmed by the Senate. Headquarters: 800 Independence Avenue SW, Washington, DC 20594.

National Trust for Historic Preservation private organization chartered by Congress in 1949 to stimulate public participation in the preservation of America's historic and cultural history through education, advocacy, technical support, grants to nonprofit groups, and demonstration programs. The National Trust publishes *Historic Preservation News. Historic Preservation Magazine, Preservation Law Reporter,* and *Historic Preservation Forum.* Membership (1999): 250,000. Address: 1785 Massachusetts Avenue NW, Washington, DC 20036. Phone: (202) 588-6000.

National Wild and Scenic Rivers Act of 1968 *See* WILD AND SCENIC RIVERS ACT.

National Wildflower Research Center *See* LADY BIRD JOHNSON WILDFLOWER CENTER.

National Wildlife Federation (NWF) conservation organization, founded in 1936 to promote the protection of wildlife resources and the conservation of the life-supporting systems of the planet. Formed originally as an offshoot of the National Rifle Association, the NWF has grown to become the largest citizen conservation organization in the world. It does little lobbying, but has strong educational and litigation programs and cooperates extensively with other groups. The largest citizen conservation organization in the United States, NWF maintains nine regional and project offices across the country, including the Alaska Project Office, specializing in wetlands protection, and the Everglades Project Office, specializing in protecting the Big Cypress Watershed region. The federation, which consists of conservation groups operating at the state level, publishes a variety of magazines including *International Wildlife, National Wildlife,* and two magazines for children, *Ranger Rick* and *Your Big Backyard.* NWF's *Conservation Directory,* published annually, provides an authoritative listing of environmental programs, organizations, and individual leaders. Membership (1999): 4,000,000. Address: 8925 Leesburg Pike, Vienna, VA 22184. Phone: (703) 790-4000. Website: www.nwf.org.

National Wildlife Refuges areas administered by the U.S. Fish and Wildlife Service of the Department of the Interior to protect habitats of game and of threatened or endangered species. The first refuge was established by President Theodore Roosevelt at Pelican Island, Florida, in 1903, and the refuge system was established by President Franklin Delano Roosevelt in 1940. Some of the 508 refuges today are extremely critical in carrying out the ENDANGERED SPECIES ACT, such as the ARCTIC NATIONAL WILDLIFE REFUGE in Alaska. *See also* FISH AND WILDLIFE SERVICE.

native species *See* ABORIGINAL SPECIES.

native stone in architecture or city planning, building stone quarried from a FORMATION underlying the region in which it is to be used. GRANITE quarried from the Sierra, for example, would be considered native stone if used for a building in the Sierra but not if used for a building in Kansas. The use of native stone helps a building fit into its surroundings better and is therefore usually preferable from an environmental standpoint.

natural capital the natural resources that can be used to create goods and services (e.g., copper for telephone wires, oil for energy) producing monetary wealth in market-oriented economic systems.

natural enemy a PREDATOR, PARASITE, or PATHOGEN that is indigenous to the same region as another SPECIES of living thing, and which will attack the other species under natural conditions. A species and its natural enemies have generally coevolved (*see* COEVOLUTION) so that their numbers remain in rough balance with each other and with the environment; thus, species introduced into regions where their natural enemies are absent are likely to overbreed, leading to an IRRUPTION, or population explosion. Such explosions are usually best controlled by introducing (or reintroducing) one or more of the species' natural enemies. DDT spraying in the California citrus groves in the late 1960s, for example, led to an explosion of cottony-cushion scale disease due to destruction of the scale insect's chief natural enemy, the vedalia beetle. The disease outbreak was brought under control by reintroduction of vedalia beetles from elsewhere in the west. *See also* BIOLOGICAL CONTROL.

natural gas a type of FOSSIL FUEL consisting of a mixture of organic gases (*see* ORGANIC COMPOUND; GAS), formed as a byproduct of the decomposition of MICROORGANISMS and trapped in rock STRATA as geologic forces convert the decomposing microorganisms to PETROLEUM. All petroleum deposits contain some quantity of natural gas; some contain large amounts, and a few are composed entirely of gas that has leaked away from petroleum deposits below and migrated upward to form "pockets" of gas on the underside of layers of impermeable rock (*see* IMPERMEABLE LAYER; PERMEABILITY). As found in nature, natural gas consists largely of METHANE (50%–80%) and ethane (5%–15%), with some larger alkanes (mostly butane and propane) mixed into it. During processing, nearly all of the propane and butane are removed, leaving a product known as "dry gas," which is composed almost entirely of methane and ethane. The dry gas is then pumped to consumers through a network of natural gas pipelines, while the butane and propane are packaged in steel cylinders ("bottled gas") for sale to consumers who are not served by pipelines or who require the gas in a transportable form.

ENVIRONMENTAL EFFECTS

Natural gas is an extremely "clean" fuel, burning with a very hot flame (1930°C, or about 3500°F) and producing almost no waste products beyond water vapor and CARBON DIOXIDE. Though high concentrations are lethal through simple asphyxiation, it does not cause a toxic reaction in living things. From the user's standpoint, natural gas is thus a nearly "perfect" fuel; however, that perfection does not extend to gas production and refining. Hazardous and/or offensive substances (hydrogen sulfide; the higher alkanes) are often present in the natural product, and must be removed before shipment to consumers. Gas pipelines often require large areas of land, and serve as barriers to wildlife migration. Gas wells, like all wells, are noisy to drill; gas is also noisy in production. As conventional ("casing-head") gas sources become increasingly difficult to find, production is shifting more and more to so-called "unconventional sources" such as COAL SHALES, whose exploitation can seriously disrupt GROUNDWATER supplies. Finally, the carbon dioxide produced as the fuel is burned, like that produced by other fossil fuels, contributes significantly to the GREENHOUSE EFFECT. Nevertheless, natural gas is probably the least environmentally harmful of the so-called "conventional" fuel sources. World reserves from known sources (1998): about 5,100 trillion cubic feet (TCF). U.S. annual use: 22 TCF (23% of the nation's total energy budget).

natural hazards hurricanes, earthquakes, volcanoes, tsunamis, drought, and other natural phenomena that put human society at risk. The prediction of natural hazards is a function of several government agencies, including the NATIONAL OCEANIC AND ATMOSPHERIC ADMINISTRATION, the Department of the Interior, and the Department of Agriculture. Many statutes, notably the NATIONAL FLOOD INSURANCE PROGRAM, govern disaster relief and mitigation. *See also* AGRICULTURE, DEPARTMENT OF; INTERIOR, DEPARTMENT OF.

natural reproduction (natural regeneration) in forestry, the reestablishment of forests on logged-over, burntover, or otherwise denuded lands by windblown and animal-carried seed, without the intervention of humans. The success of natural reproduction depends on the presence of nearby SEED TREES and on the ability of the site to provide adequate shade and moisture for the establishment and survival of SEEDLINGS. It is often surprisingly rapid and complete; however, it cannot be counted on for the adequate stocking of sites on which it is desired to maintain forest cover. *See also* REGENERATION; STOCKING LEVEL.

natural resource in economics, a substance found in nature that can be used to produce GOODS and SERVICES desired by consumers. Inclusion of the term "services" is necessary from an environmental standpoint; otherwise, the term "natural resource" becomes synonymous with "raw materials." Natural resources as a source of *goods* implies that trees, for example, are only valuable if harvested and made into something.

Natural resources as a source of *goods and services* recognizes that trees may have economic value as part of the living forest, even though this economic value may be hard to quantify. *See* BENEFIT/COST ANALYSIS; LEONTIEF ANALYSIS. Examples of natural resources include TIMBER, MINERALS, HYDROELECTRONIC POWER sites, and such intangibles as scenic attractions, swimming beaches, and trout streams.

Natural Resources Conservation Service consisting of the SOIL CONSERVATION SERVICE and other resource conservation programs, the agency of the U.S. Department of Agriculture (USDA) responsible for working with ranchers, farmers, and other landowners to conserve natural resources, especially soil and water. With more than 50 offices nationwide, NRCS provides technical assistance for landowners in cooperation with local soil and water conservation districts. Headquarters address: USDA, 14th Street and Independence Avenue SW, P.O. Box 2890, Washington, DC 20013. Phone: (202) 720-3210. Website:www.nrcs.usda.gov. *See also* AGRICULTURE, DEPARTMENT OF.

Natural Resources Council of America an association of environmental organizations interested in the conservation, protection, and management of American natural resources. The council provides for coordination between its members, governmental agencies, public groups, and the business community. Membership (1999): 69 organizations. Address: 1025 Thomas Jefferson Street NW, Suite 109, Washington, DC 20007. Phone: (202) 333-0411.

Natural Resources Defense Council (NRDC) public interest law firm and environmental organization, founded in 1970 to use litigation, appeals and similar legal actions to protect environmental values in the United States. In addition to its own legal work, the NRDC provides legal advice and support services to other environmental groups, especially in cases that will set legal precedents or protect significant natural resources. Its staff consists of principally of attorneys, scientists and land-use planners. The NRDC program focuses on air and water pollution, nuclear safety, global warming, ocean and fisheries protection, energy efficiency, land use management, protection of wilderness and wildlife, and urban environmental planning. Membership (1999): 400,000. Address: 40 West 20th Street, New York, NY 10011. Phone: (212) 727-2700. Website; www.nrdc.org.

natural selection the theory, developed by CHARLES DARWIN, that species evolve through the reproduction of those individuals in a given population that are most successfully adapted to their environment. Improved adaptability (speed, visual acuity, an opposed thumb) emerges via random genetic mutation which in turn produces unusual and superior characteristics that give such individuals an advantage in the struggle for existence and make them most successful in mating and breeding.

natural succession *See* SUCCESSION.

The Nature Conservancy (TNC) land-preservation organization, founded in 1917 to find ways of preserving ecologically significant lands in private ownership. TNC often purchases lands outright; it owns and operates more than 1,600 preserves, some of fairly large size, scattered throughout the United States. At other times it will negotiate sales or gifts to federal or state agencies. Prior to 1946 it was a committee of the Ecological Society of America. Membership (1999): 832,000. Address: 4245 North Fairfax Drive, Arlington, VA 22208. Phone: (703) 841-5300. Website: www.tnc.org.

navigable stream any WATERCOURSE that meets minimum standards of navigability (ability to sustain waterborne travel) under state of federal law. Three different levels of definition may apply, depending upon the purpose to which the law is directed. Ownership of streambeds is determined by the stream's historic "navigability in fact." If the stream could sustain waterborne commerce by normal means at the time of statehood, the streambed below the mean high-water mark belongs to the state; if it could not, the streambed belongs to the riparian landowners (*see* RIPARIAN LANDS). A broader definition of "navigability" is allowed under the commerce clause of the Constitution for determining federal rights to regulate the water of a stream (as distinct from regulating the land beneath the water). The federal government has regulatory powers if the stream is currently "navigable in fact" to waterborne commerce, if it has ever been navigable in fact, or if it could be made navigable in fact through improvements such as DAMS and STREAM CHANNELIZATION. These regulatory powers may be extended to the non-navigable TRIBUTARIES of a stream declared navigable under the commerce-clause definition if it can be shown that developments on these tributaries will hinder the navigability of the MAINSTEM. Finally, a state's right to regulate the use of stream water, the building of stream-obstructing structures, and the public's right of access to streambanks may depend on the state's own definition of "navigability". It may if it so desires designate as "navigable" all streams that can float canoes, rubber rafts, or even simply logs on their way to a mill. Declaring a stream navigable under such loose standards, however, will not automatically allow the commerce clause to be applied to it or determine the ownership of its bed. For these purposes, the stricter definitions must still be met.

NAWAPA *See* NORTH AMERICAN WATER AND POWER ALLIANCE.

NCI acronym for National Cancer Institute. *See under* PUBLIC HEALTH SERVICE.

Nearctic region *See under* HOLARCTIC REGION.

negative declaration statement the portion of an ENVIRONMENTAL ANALYSIS REPORT that states formally that an ENVIRONMENTAL IMPACT STATEMENT will not be required. The negative declaration statement ("finding of no significant impact" or FONSI) may simply be a sentence declaring that the proposed action is not a major federal action having a significant impact on the environment. However, separate documentation and a 30-day public comment period are required if the proposed action will set a precedent, if the proposed action will affect WETLANDS or FLOODPLAIN lands, or if the proposed action is closely similar to an action that the courts have previously held to require an Environmental Impact Statement.

Negative Population Growth (NPG) population-control organization, founded in 1972 to promote a reduction in world population to roughly half the present level. NPG believes that such an approach is the only one consistent with the survival of both human life and the planetary environment. The organization seeks financial and tax incentives to lower the birth rate, rather than government coercion. Membership (1998): 18,000. Address: 1717 Massachusetts Avenue NW, Suite 101, Washington, DC 20036. Phone: (202) 667-8950. Website: www.npg.org.

negentropy the opposite of entropy; that is, the tendency of a system to become more organized, rather than less organized, over time (*see* THERMODYNAMICS, LAWS OF: *the second law*). Negentropy is a characteristic of life: it occurs only rarely in non-living systems (exceptions include the formation of MINERAL crystals from molten MAGMA and the sorting of SEDIMENTS that occurs as flowing streams enter still water). It is probable that every example of negentropy that occurs can be shown to exist at the expense of greater entropy elsewhere; hence the term is only useful in a highly localized frame of reference.

nekton in the biological sciences, the general term for free-swimming aquatic organisms. The term encompasses fish, squid, lobsters, CETACEANS, and all other organisms that propel themselves through the water rather than drifting with the currents or crawling along the bottom. *Compare* PLANKTON; BENTHOS.

Nelson, Gaylord (1916–) founder of EARTH DAY, former U.S. senator from Wisconsin (1963–81), environmental leader. Born June 4, 1916, in Clear Lake, Wisconsin, Nelson received a B.A. degree from San Jose State College and a J.D. degree from the University of Wisconsin. He served as a first lieutenant during the Okinawa campaign in World War II, and began his public life shortly after the war, in 1948, as state senator from Dane County. In 1958 he was elected governor of Wisconsin and after two terms was elected to the U.S. Senate, where he served until 1981. Since organizing Earth Day in 1970, he has been an active environmental educator and speaker. Often with little support from his fellow senators, Nelson wrote, sponsored, or championed such important environmental legislation as the WILDERNESS ACT OF 1964, the NATIONAL TRAILS ACT OF 1968, the national WILD AND SCENIC RIVERS ACT (1968), the NATIONAL ENVIRONMENTAL POLICY ACT (1970), the CLEAN WATER ACT (1972), the ENDANGERED SPECIES ACT (1973), the ban on DDT, and strip mining controls. After retiring from the Senate, Nelson has involved himself as a counselor to the WILDERNESS SOCIETY, where he is engaged in a variety of land preservation issues such as the elimination of logging subsidies, the protection of national parks, and the expansion of the National Wilderness Preservation system.

nematode (roundworm) any member of the PHYLUM Nematoda, especially those that are parasitic on plants and animals (*see* PARASITE). Most are microscopic, though those that inhabit animal intestines may be up to a meter (3.26 feet) in length. Nematodes are typically elongated, unsegmented worms covered by a smooth, tough skin, or *cuticle*. Their bodies are unusual in having only longitudinal muscle fibers. With no transverse fibers encircling them, these longitudinal fibers must be held in place by hydrostatic pressure within the cuticle, requiring the animal to live in an extremely moist environment. Most can, however, encyst (that is, enclose themselves in hardened shells, or *cysts*) during dry conditions, allowing them to survive in many places that might otherwise seem inhospitable. They are extremely common in the soil, where they live largely in the film of HYGROSCOPIC WATER surrounding the soil particles. A single gram of topsoil may contain half a billion of these tiny organisms. Those that are parasitic on plants generally cause abnormal root growth; those parasitic on animals include trichinosis, hookworms, whipworms, and filarial worms (elephantiasis). Soil nematodes are generally controlled through fumigation (*see* FUMIGANT), though they can also be combated by planting marigolds (which secrete a substance from their roots that is toxic to nematodes), or by seeding the soil with fungi (*see* FUNGUS) such as *Arthrobotrys dacteloids* that are PREDATORS of nematodes, trapping them with their hyphae and literally devouring them. Distribution: worldwide except among the PLANKTON. Number of SPECIES: at least 10,000 and perhaps as many as 500,000.

NEPA *See* NATIONAL ENVIRONMENTAL POLICY ACT.

neritic zone in oceanography, the coastal waters of the ocean, defined either as (a) the zone of water lying between extreme low-water line and the edge of the CONTINENTAL SHELF, or (b) the zone of water beginning at the extreme low-water line and extending outward until the ocean reaches a depth of 200 meters (650 feet). The two definitions are essentially synonymous. Because it is well supplied with both light (due to its shallowness) and NUTRIENTS (due to RUNOFF from nearby landmasses and proximity to the nutrient-rich bottom ooze of the ocean itself), the neritic zone is the most hospitable portion of the ocean for life, producing roughly 75 times more fish (by weight) than the open ocean does, though it contains only about a 1% as much volume. *Compare* BENTHIC DIVISION: PELAGIC ZONE; LIMNETIC DIVISION.

nested well *See under* MONITORING WELL.

net annual increment the increase in usable wood volume in a STAND of trees during the course of a year. Net annual increment is equal to total new growth (*gross increment*) minus wood volume lost to decay, WINDTHROW, disease, breakage, and other natural hazards. *Compare* MEAN ANNUAL INCREMENT.

net erosion in geology, the difference between the amount of material eroded from a site during a given time period and the amount of material deposited on the same site during the same time period. *See* EROSION; DEPOSITION.

net plankton in oceanography and limnology, any PLANKTON large enough to be caught by a #20-mesh plankton net; that is, any which are 60 MICRONS or more in diameter. *Compare* NANNOPLANKTON.

neurotoxin in medicine, any substance that damages or destroys nerve tissue. In mild doses, neurotoxins cause itching, lethargy, and a sensation of pain generally described by its victims as heavy and dull. In larger doses they cause convulsions, paralysis, and death. Many insecticides—especially those classed as ORGANOPHOSPHATES—function as neurotoxins, killing insects by interfering with the conduction of nerve impulses through their nervous systems. The general structural similarity of nerve tissue throughout the animal KINGDOM, unfortunately, means that these highly effective pesticides are also the most dangerous to NONTARGET SPECIES, including humans.

neuston in limnology, the MICROORGANISMS inhabiting the surface film of a body of water (*see* SURFACE TENSION). Most neuston are members of the PROTISTA kingdom. *Hyponeuston* are aquatic organisms that live on the underside of the surface film; they generally differ little from those inhabiting the rest of the body of water, and are often neuston only by virtue of the fact that they happen to have drifted up and lodged against the film. *Epineuston,* or "true" neuston, live on the upper surface of the film. They are generally aerial organisms that have adapted so that they can use the surface film as a SUBSTRATE, usually by developing a broad "foot" of lipid tissue (*see* LIPID). They are much more specialized than the hyponeuston, and will not normally be found in other environments.

neutron *See* ATOM: *structure of the atom.*

neve *See* FIRN.

New England Interstate Water Pollution Control Commission facilitates interstate communication on water pollution issues, provides training opportunities for state employees and educational and informational materials for the public on environmental issues. Address: Boott Mills South, 100 Foot of John Street, Lowell, MA 01852. Phone: (978) 323-7929. Website: www.neiwpcc.org.

new forestry popular name for a set of silvicultural practices designed to obtain SUSTAINED YIELD of timber from a forest by mimicking the forest's natural processes as closely as possible. New forestry advocates speak of "sustainable forests" rather than "sustained yield," and have summarized their position by stating "the important thing is not what you take out of a forest, but what you leave behind." Harvests are designed to mimic natural catastrophic events such as WILDFIRE and WINDTHROW as closely as possible. The soil and water on a growing site are considered basic resources which should be protected as thoroughly as possible. Specific practices include so-called "sloppy CLEARCUTS" in which large downed logs, snags, and UNDERSTORY vegetation are left on the site; clustering HARVEST UNITS to mimic the concentrated effect of natural events; delayed CANOPY closure, to allow understory vegetation time to flourish for a long enough period to fix NUTRIENTS into the soil; and the maintenance of significant reservoirs of OLD GROWTH as seed sources, WATERSHED protection, and wildlife HABITAT (note that these old-growth reservoirs are not "preserves" in the usual sense of the word: as harvested lands return to old-growth conditions, they may be added to the reservoirs and other lands removed). Critics of new forestry deride it as being more concerned with public relations and profitable timber extraction than with maintaining ecosystem integrity. Nevertheless, the approach is clearly preferable to forestry methods that do not take the natural processes of regeneration into account.

New Source Performance Standards (NSPS) term used in both air and water pollution control to refer to emission standards for new industrial plans and other major sources of pollutants. For air pollutants, the authority for NSPS is found in section 111 of the CLEAN AIR ACT. It applies to modifications of older plants as well as the construction of new ones, and mandates the use of BEST AVAILABLE TECHNOLOGY to control the emission of pollutants. For pollutants emitted by FOSSIL FUELS (SULFUR OXIDES, NITROGEN OXIDES, and PARTICULATES), the act requires reduction of emissions by a certain percentage below those that would be present without controls. For the remaining criteria pollutants (*see* NATIONAL AMBIENT AIR QUALITY STANDARDS), standards are set in terms of pounds of pollutant allowed per unit weight of emissions. The cutoff point for application of the law is 100 tons of pollutants per year; plants emitting less than that amount of any given pollutant do not need to meet standards for it. NSPS standards for water pollution are spelled out in Section 306 of the CLEAN WATER ACT. They mandate the use of BEST AVAILABLE CONTROL TECHNOLOGY, and apply only to new plants, without reference to modification of older ones. The law specifically provides that the standards may be set to zero (that is, no discharge of the pollutant in question) if such a standard is both economically and technologically feasible.

new urbanism a set of principles adopted by city planners to counter the adverse effects of URBAN SPRAWL and the decay of center-city areas in large, growing metropolitan regions. New urbanists champion infill development, involving the use of smaller acreages of close-in land rather than allowing builders to leapfrog into the countryside; revitalization of downtown commercial and cultural centers; major mass transportation facilities including light rail, subways, and monorails as well as improved bus service; mixed-use development as opposed to segregating housing from commercial facilities; open space networks, greenways, and bikepaths; and similar stratagems to allow urban life to operate at human scale, reduce AIR POLLUTION and the waste of water resources, preserve agriculture in and around metropolitan areas, and increase social stability and amenity.

NFIP *See* NATIONAL FLOOD INSURANCE PROGRAM.

NFMA *See* NATIONAL FOREST MANAGEMENT ACT.

NFPA *See* NATIONAL FOREST PRODUCTS ASSOCIATION.

NGO *See* NONGOVERNMENTAL ORGANIZATIONS.

niche in ecology, the overall place of an ORGANISM in the structure of a COMMUNITY, including both the space it occupies (*see* HABITAT) and the role it plays in maintaining the community's balance. In other words, the niche is both an organism's home and its job. The idea of the niche is one of the most important concepts in the biological sciences, allowing easy comparisons among communities and providing one of the principal driving forces for the differentiation of SPECIES. When a community is looked at as a group of niches rather than a group of species, two facts rapidly emerge: (1) similar habitats produce similar sets of niches; and (2) in a stable community, no niche is left unfilled. The niche of the coyote, for example, is to roam the grasslands of North America eating rodents, insects, carrion, and occasionally larger animals. There are no coyotes on the grasslands of Africa; there, the same niche is filled by the jackal. The wolf has been largely exterminated in America, leaving an empty niche, and the coyote has gradually been expanding its role to fill that of the wolf as well. Species filling the same niche generally evolve the same characteristics. Hedgehogs (European) look very much like porcupines (American); swallows (which catch insects and eat them in flight by day) have strikingly similar silhouettes and flight patterns to bats (which fill the same niche at night). Striking examples of the drive of nature to fill every available niche are found in Australia, where marsupial "mice," "wolves," "bears," etc. (that is, animals that resemble these creatures but that do not develop a placenta during the reproductive process, developing their young from the embryonic stage onward in external pockets rather than in a womb) have evolved in the absence of their placental counterparts elsewhere in the world; and in the Galápagos Islands, where a group of 13 closely-related finch species has evolved, probably from a single ancestor, to fill a diverse group of empty niches that would normally be occupied by warblers, sparrows, thrushes, and woodpeckers. *See also* GAUSE'S PRINCIPAL.

night soil a euphemism for human excrement. *See* FECES.

NIMBY an acronym for "Not In My Back Yard," used to describe the common phenomenon of public protest of a waste-disposal dump or other environmentally disruptive facility imposed on an area—usually poor or minority—without a voice in the decision. *See* NATIONAL SACRIFICE AREA.

96HRTLM *See* MEDIAN THRESHOLD LIMIT.

nitrate in chemistry, any COMPOUND containing the nitrate GROUP (NO_3). Technically, all nitrates are either SALTS or ESTERS formed from the action of nitric acid (HNO_3) on a METAL or an ALCOHOL. Only sodium nitrate ($NaNO_3$) and potassium nitrate (KNO_3) are commonly found in nature. Nitrates are the principal—almost the only—means by which nitrogen is absorbed by plants (*see* NITROGEN: *the nitrogen cycle*), and are therefore essential to plant growth. For this reason, they are a common constituent of FERTILIZERS. The compounds are generally not toxic, but can be environmentally hazardous because of their tendency to undergo conversion to NITRITES through bacterial action. Excessive nitrates from overfertilization and from concentrated livestock operations such as FEEDLOTS are not excreted by plants but are concentrated in their leaves, where they are converted to nitrites by intestinal bacteria after ingestion by LIVESTOCK and other RUMINANTS, an often-lethal condition known as *nitrate poisoning*. Excessive nitrates also enter WATERCOURSES, where they contribute to EUTROPHICATION (*see* LIMITING NUTRIENT).

nitrite in chemistry, any COMPOUND containing the nitrite GROUP, NO_2–. Technically, nitrites are SALTS and ESTERS formed by the reaction of METALS or ALCOHOLS with nitrous acid (HNO_2). They are also commonly formed through the action of bacteria on NITRATES. Though they are an important part of the nitrogen cycle (*see* NITROGEN: *the nitrogen cycle*), nitrites are extremely hazardous compounds due to their affinity for blood HEMOGLOBIN, which they convert to methemoglobin, disabling its OXYGEN-carrying abilities. Some nitrites are also known or suspected

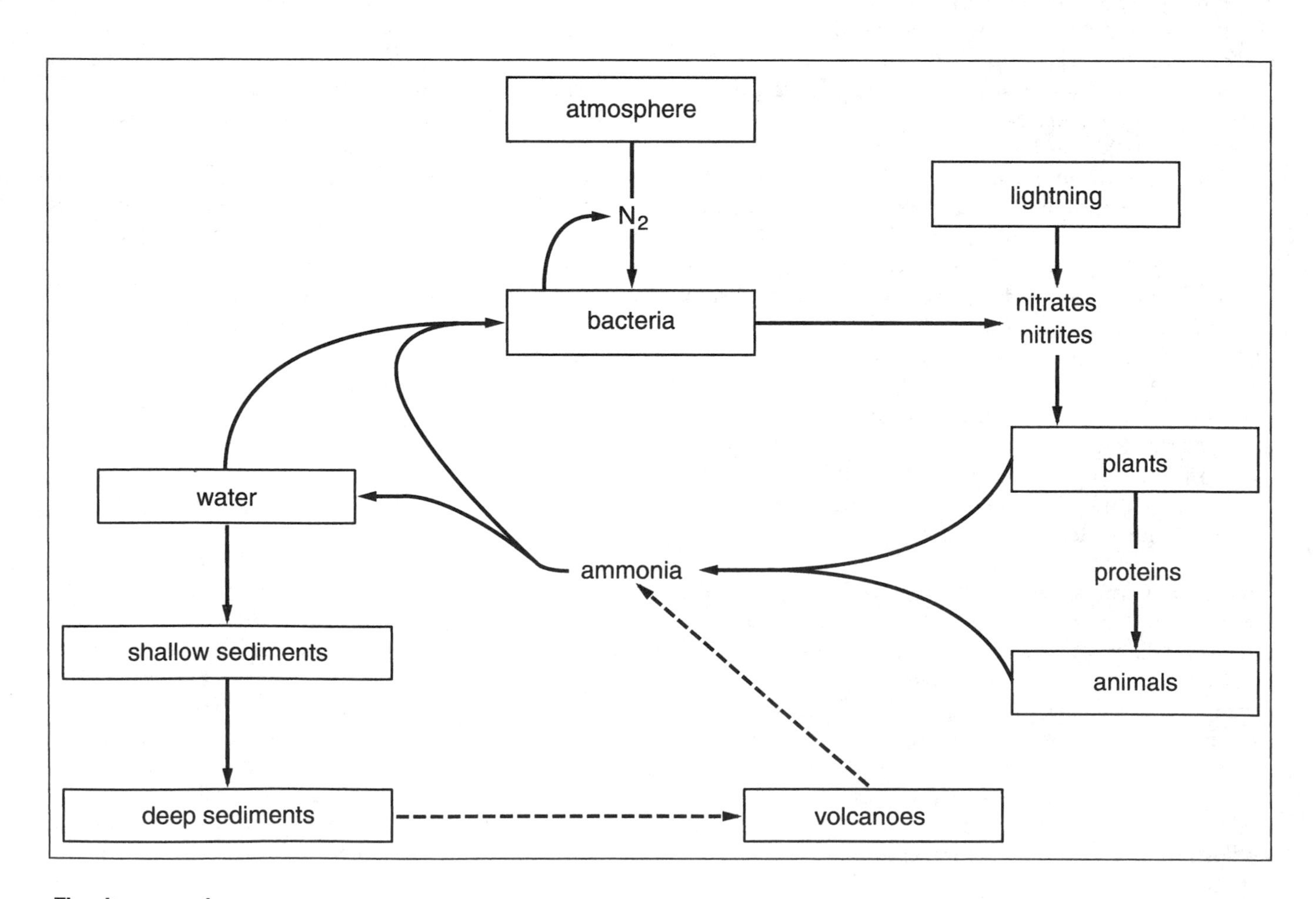

The nitrogen cycle

CARCINOGENS or precursor materials to other carcinogens such as nitrosamines. Nevertheless, sodium nitrite and potassium nitrite are both common food additives, partly because of their ability to prevent the growth of MICROORGANISMS (notably *Clostridium botulinum,* which causes botulism) and partly because the methemoglobin which they produce in meats looks just like hemoglobin but does not oxidize, allowing the meats to maintain their red color and thus appear fresh for a long time. LD_{50} (orally): sodium nitrite, 180 mg/kg (rats); potassium nitrite, 108 mg/kg (rabbits).

nitrogen the seventh ELEMENT of the ATOMIC SERIES; atomic weight 14.0067, chemical symbol N. Nitrogen is a colorless, odorless gas that comprises approximately 78% of the Earth's atmosphere by volume (about 75% by weight). The nitrogen MOLECULE is formed of two nitrogen ATOMS with an extremely strong triple bond between them, making the material nearly inert in its ordinary gaseous state. However, if this bond is severed, the individual nitrogen atoms become extremely active, with the ability to bond ("fix") to nearly all other elements. Large electrical discharges, such as lightning bolts, will free nitrogen atoms and cause nitrogen fixation; so will the metabolic action of some species of bacteria (*see* NITROGEN-FIXING BACTERIA). Nitrogen is one of the principal components of PROTEINS and of DNA; hence, the phenomenon of nitrogen fixation is critical to the survival of life.

THE NITROGEN CYCLE

Like all other nutrient elements (*see* NUTRIENT), nitrogen follows a circular pathway known as a BIOGEOCHEMICAL CYCLE through the environment. The nitrogen cycle begins with atmospheric nitrogen, N_2, which is converted to NITRATES through the action of lightning or nitrogen-fixing bacteria. The nitrates are taken up by plants, which in turn are eaten by animals, which excrete excess nitrogen in their body wastes, mostly in the form of AMMONIA and related compounds. These body wastes—and the bodies of dead animals and plants—are attacked by denitrifying bacteria and other DECOMPOSERS, converting the proteins, ammonia, and other nitrogen compounds back to nitrates and to elemental nitrogen, which is released to the atmosphere again. *See* DENITRIFICATION. *See also* NITRITE; NITROGEN OXIDE.

nitrogen-fixing bacteria any of several genera of bacteria (*see* BACTERIUM; GENUS) that have the ability to convert atmospheric NITROGEN into AMMONIA and NITRATES, which can then be used by other ORGANISMS. Two basic types of nitrogen fixation exist. In the first type, free-living soil bacteria (*Azobacter, Bacillus, Clostrodium, Pseudomonas,* and a few others) take up atmospheric nitrogen and convert it to ammonia ("ammonification") through the use of a special enzyme called *nitrogenase;* the ammonia is then converted to NITRITES ("nitrification") by a second group of bacteria (*Nitrosomonas, Nitrosococcus*), while a third group (*Nitrobacter*) convert the nitrites into nitrates, which are then taken up by the roots of plants. In the second type of nitrogen fixation, a close SYMBIOSIS exists between certain bacteria (*Frankia, Rhizobium*) and certain ANGIOSPERMS (alders, ceanothus, LEGUMES). The bacteria, which form nodules on the roots of the angiosperms, use nitrogenase to convert atmospheric nitrogen to ammonia, which is then taken up directly by the host angiosperms through tissue-to-tissue contact. Excess ammonia is dumped into the soil, becoming available to nitrification organisms for conversion into nitrites and nitrates. The relationship between the bacteria and their host angiosperms is extremely specific; a given species of angiosperm will only associate with a single SPECIES (sometimes, only with a single genetic strain) of bacteria. A few other MICROORGANISMS besides bacteria are capable of nitrogen fixation, including several genera of CYANOPHYTES and at least two of fungi (*see* FUNGUS); and at least one genus of bacterium, *Klebsiella,* performs nitrogen fixation for an animal host, the termite. *See also* NITROGEN: *the nitrogen cycle.*

nitrogen oxide (NOx) in chemistry, any of several COMPOUNDS consisting of varying combinations of NITROGEN and OXYGEN. From an environmental standpoint, the most important varieties of NOx are nitric oxide (NO), nitric oxide dimer (N_2O_2) nitrogen dioxide (NO_2), and dinitrogen tetraoxide (nitrogen dioxide dimer, N_2O_4). Nitric oxide and its dimer are formed in air by the direct union of nitrogen and oxygen at high temperatures or in the vicinity of an electrical discharge; these two compounds then slowly combine with more oxygen to form nitrogen dioxide and dinitrogen tetraoxide. All four compounds dissolve in raindrops, forming nitrous and nitric acids, which then fall to earth as ACID RAIN. The principal sources of atmospheric NOx are lighting storms, high-temperature combustion for power generation or industrial processes, and the spark plug of the internal combustion engine. The deep-brown color of nitrogen dioxide is a principal reason for the "brown air" we associate with SMOG. The other common atmospheric NOx are colorless. Other forms of NOx include nitrous oxide ("laughing gas," N_2O), dinitrogen trioxide (N_2O_3), and dinitrogen pentaoxide (N_2O_6).

Nixon, Richard Milhous (1913–1994) American politician. Nixon was the 37th president of the United States, from 1969 to 1974. Born in Yorba Linda, California, into a Quaker family, he graduated second in his class from Whittier College in 1934 and finished third in his graduating class at Duke University Law School in 1937. He practiced law in Whittier, California, until he enlisted in the U.S. Navy in 1942, where he served as a supply officer in the South Pacific. Running as an outspoken anticommunist, he was elected to the U.S. Congress in 1946 and the U.S. Senate in 1950. In 1952 he was nominated as the running mate of Dwight D. Eisenhower and served as an active vice president for two terms. In 1960 he easily won the Republican presidential nomination and lost a close race to John F. Kennedy. After an unsuccessful gubernatorial race in California, Nixon angrily announced his withdrawal from politics in 1962. In 1968 he ran again for the presidency and was elected with a narrow popular vote majority. As president, he focused on foreign affairs, traveling widely. He worked effectively to lessen tension between the United States and the People's Republic of China and attended a summit meeting there in 1972. Domestic policy did not escape his attention and, although he never made much of it, he could arguably be called the most important postwar "environmental president." During his tenure foundational environmental legislation was enacted, including the NATIONAL ENVIRONMENTAL POLICY ACT, air and water pollution legislation, the ENDANGERED SPECIES ACT, and much else. Although Nixon was reelected by a landslide in 1972, he was soon mired in the Watergate scandal and resigned on August 8, 1974. Pardoned of all criminal charges by his successor, former vice president Gerald R. Ford, Nixon kept a low profile in his retirement. Gradually he rebuilt his public image through a series of interviews, memoirs, and writings on world affairs. He died in New York City on April 22, 1994, and is buried in Yorba Linda. Published writings include *Memoirs* (1978), *The Real War* (1980), *Six Crises* (1981), *Real Peace* (1984), *No More Vietnams* (1985), and *Seize the Moment: America's Challenge in a One-Superpower World* (1992).

NOAA *See* NATIONAL OCEANIC AND ATMOSPHERIC ADMINISTRATION.

nocturnal species any SPECIES of living ORGANISM that is primarily active at night. The principal reasons for nocturnalism are to utilize the cover provided by darkness and to take advantage of the fact that the DIURNAL SPECIES are asleep, thus temporarily freeing their NICHES. Common nocturnal species include owls, mice, bats, and crickets. *See also* CREPUSCULAR SPECIES.

noise and number index (NNI) in NOISE-POLLUTION control, a number used to express the irritability of airport noise. The noise and number index is calculated according to the empirical formula

$$NNI = PNdB + 15 \log N - 80$$

where PNdB is the PERCEIVED NOISE LEVEL and N is the average number of flights in a 24-hour period. The NNI is a unitless number.

noise pollution the presence of sound in the environment at levels that may be injurious to human health or property. The term "injurious" may be defined in several ways. Loud, abrupt sounds may affect hearing by damaging the eardrum; sustained sounds at a somewhat lower volume may affect hearing by damaging the middle ear. Both types of sound may also cause psychological damage, either by triggering the "fight-or-flee" response or through simple irritation. Persons regularly exposed to loud sounds develop heart and respiratory problems and neurological disorders at rates significantly exceeding those in quieter environments. Property damage is generally economic (noise pollution lowers property values); however, it may also be direct, in the form of structural damage from vibrations induced by sound waves or the loss of LIVESTOCK or pets that may be traceable to their reaction to noise. The most widespread source of noise pollution in modern society is vehicular traffic; however, airports and industrial plants are also common noise sources, and both are generally louder than traffic noises, though they affect a much smaller area.

TYPES OF NOISE

Noise may be divided into two broad types: *ambient noise* (or *background noise*), which is constantly present (the rumble of traffic, the sound of flowing water), and *peak noise,* which is of shorter duration but may be considerably louder than the ambient noise. Peak noise is further subdivided into *steady-state noise* (machinery, sports events, the neighbors' stereo) and *impact sound* (gunshots, drumbeats, or any other sudden, sharp sound). The effects of any of these types of noise depend both on the actual loudness level (usually measured in DECIBELS) and on the range of frequencies comprising the sound (that is, the sound's "highness" or "lowness"). "White noise" (that is, noise that covers the entire frequency spectrum) is generally less damaging than noise of a specific pitch. High-pitched noises and low-pitched noises are more damaging than those in the middle of the frequency range. Impact sound is more damaging than steady-state sound of the same loudness and frequency characteristics.

NOISE REGULATION AND CONTROL

Noise is regulated on the federal level through the Noise Control Act of 1972, which requires the ENVIRONMENTAL PROTECTION AGENCY to set standards for noise emissions from a wide variety of sources and authorizes all federal agencies to establish rules for human exposure to noise from sources within their jurisdiction (for instance, the Federal Aviation Administration sets airport noise standards); and by the Quiet Communities Act of 1978, which assists community planning for noise control through grants and research programs. In 1982, however, the Noise Abatement and Control Office, which administered the act, was defunded, and hundreds of state and local programs languished. In 1996 the World Health Organization revealed that 20 million Americans are exposed to noise at such high levels that physiological and psychological damage may result. In 1998 an effort to reestablish the Noise Abatement and Control Office was mounted in the form of a proposed "Quiet Communities Act." Although the program called for only a $5 million budget, it was not enacted. *See also* TRAFFIC NOISE INDEX; PERCEIVED NOISE LEVEL; NOISE POLLUTION LEVEL; NOISE AND NUMBER INDEX; A-WEIGHTED SOUND SCALE.

noise pollution level (NPL) in noise pollution control, an index used to measure noise irritability, calculated by an empirical formula that takes into account the greater irritability of impact sound as opposed to steady-state sound (*see* NOISE POLLUTION: *types of noise*). The formula is

$$NPL=L_{50+(L}10_{-L}90_{)+(L}10-L_{90)}^{2} / 60$$

where L_n = the level of sound exceeded n% of the time; that is, n% of the sounds heard during the period of measurement will be louder than L_n. The unit of measurement is normally dB(A) (*see* DECIBEL; A-WEIGHTED SOUND SCALE.)

nonapatite inorganic phosphorus any INORGANIC COMPOUND of PHOSPHORUS that does not contain the apatite GROUP ($CA_5(PO_4)_3$). Nonapatite phosphorus tends to weather more rapidly than apatite phosphorus; hence, nonapatite deposits indicate soils with increased phosphorus availability over the short run but with a tendency to become rapidly depleted. *See also* FERTILIZER.

nonattainment area in air pollution control, any area of the country that does not meet one or more of the primary AMBIENT QUALITY STANDARDS as defined in the CLEAN AIR ACT. Permits for new STATIONARY SOURCES of pollutants, or for significant modifications of existing ones, may not be issued in nonattainment areas unless (1) an implementation plan exists with a timetable for reaching attainment; (2) the new plant or modification installs pollution-control equipment that meets LOWEST ACHIEVABLE EMISSION RATE standards; and (3) pollution from existing sources is reduced sufficiently that the overall air quality does not diminish. If one of the standards not met is that for CARBON MONOXIDE, a motor vehicle emissions inspections program must be implemented. Failure to meet these requirements can result in the withdrawal of all federal funding from the area, including funding (such as grants to the arts and education) that is unrelated to the pollution problem.

noncalcic soil any soil that is low in calcium carbonate; specifically, a soil that does not form a CALICHE layer or a calcic HORIZON. Noncalcic soils are generally moist and high in ORGANIC MATTER. *See* SOIL: *classification of soils.*

noncommercial forest land (noncommercial component) in forestry, that portion of a forest on which the trees are not likely to regenerate properly after they are harvested. In FOREST SERVICE usage, the noncommercial component includes all lands that produce wood volume at a rate of less than 20 cubic feet per acre per year. *See also* SITE INDEX.

non-declining even flow in FOREST SERVICE usage, the policy of keeping timber harvest levels in a region at or below the growth rate of the remaining timber in the same region, so that the harvest can be maintained at the same volume in perpetuity. Even-flow management is accomplished by making sure that the PROGRAMMED ANNUAL HARVEST never exceeds the POTENTIAL YIELD. For example, if a particular planning area consisted of 100,000 acres of timber growing at an average rate of 80 cubic feet per acre per year, the total sawlog volume taken from that area cannot exceed 8 million cubic feet per year (80 x 100,000=8,000,000). In practice, non-declining even flow is generally calculated on a 10-year cycle. Volume in excess of what will grow back may be harvested in one or more years of the cycle provided that this excess is balanced by harvesting less than what will grow back in other years so that the average harvest for the decade does not exceed 10 years' worth of average growth. Forest Service policy calls for such *departures from even flow* to be carefully documented, however, normally allowing them only when the departure is justified by circumstances such as high mortality due to fire or bug kill, temporary economic dislocation in a community due to a timber flow problem, etc. Non-declining even flow was adopted as official Forest Service policy in 1973 and written into the NATIONAL

FOREST MANAGEMENT ACT OF 1976. The basic even-flow unit is usually a single NATIONAL FOREST. Programmed harvest may exceed potential yield on a given ranger district without violating even-flow, as long as one or more of the other districts balance the excess by harvesting less. *See also* ALLOWABLE CUT; ALLOWABLE CUT EFFECT; SUSTAINED YIELD.

nondegradable pollutant any POLLUTANT that is not readily broken down or otherwise chemically altered by exposure to the natural environment. Nondegradeable pollutants remain potentially harmful as long as they are present. *Compare* BIODEGRADABLE SUBSTANCE; GEODEGRADABLE SUBSTANCE.

nongame wildlife any animal for which a hunting season is not officially declared, including rodents, songbirds, raptors, reptiles, etc. The management of nongame wildlife poses a problem because most wildlife management funds come from the sale of hunting licenses and are earmarked for game management. Recently, a number of states have been experimenting with alternate means of funding nongame management: two of the more successful approaches have been the sale of specialized license plates and the so-called *wildlife checkoff,* where the individual taxpayer is allowed to mark a box on his or her tax return that dedicates a certain portion of his or her taxes to nongame management.

nongovernmental organization (NGO) term used most often to describe public-interest citizen, professional, charitable, religious, and scholarly organizations working with the United Nations and other international bodies to influence their deliberations and improve their effectiveness. NGOs played a significant part in the RIO EARTH SUMMIT, for example.

nonnative species *See* EXOTIC SPECIES.

nonpoint pollution the pollution of water bodies by diffuse sources such as agricultural or urban RUNOFF or AERIAL DEPOSITION. Conveyances for runoff, such as drainage ditches and storm sewers, are generally classed as nonpoint sources; so are concentrated, but largely uncontrollable, sources such as livestock FEEDLOTS, parking lots, and individual SEPTIC TANK systems. Nonpoint pollution is usually the most important overall source of contaminants in a water body; 90% or more of the total pollutant load may come from nonpoint sources, especially if treatment plants have been installed for the point sources. It is much more difficult to deal with than is pollution from point sources because the pollutants come from a large number of small, often unidentifiable sources, making treatment of the WASTE STREAM impractical or impossible. Control can, however, be approached through controls on land use (*see* BEST MANAGEMENT PRACTICE). Assessment for the need for controls is made by measuring the *unit area load;* that is, the total amount of pollutants contributed from a WATERSHED or other definable region divided by the area of the source region and expressed as weight or mass over area, as pounds/acre, tons/square mile, kilograms/hectare, etc. Some factors affecting the unit area load include the amount of rainfall; the slope of the land; the rate of INFILTRATION; the amount of impervious surface, such as pavement or rock; the amount of bare soil surface, particularly in regard to SEDIMENT LOADINGS to a water body, which are considered a form of nonpoint pollution; and the amount of pollutants actually placed on the land, such as FERTILIZERS, PESTICIDES, animal wastes (urine and FECES), and the particles of lead, rubber, ASBESTOS, etc. placed on streets and highways by wear and tear on automobiles. The comparable term for air pollution is AREA SOURCE.

nonrenewable resource any NATURAL RESOURCE that is not replenishable by natural forces such as organic growth or the HYDROLOGIC CYCLE within the lifetime of a human culture. Industries built on a base of nonrenewable resources are not sustainable beyond the life span of the resource. Nonrenewable resources include ORE BODIES, PETROLEUM deposits, building stone, and deposits of FOSSIL WATER. *Compare* RENEWABLE RESOURCE.

nonselected roadless area in FOREST SERVICE usage, any ROADLESS AREA on NATIONAL FOREST land that was not recommended for either wilderness protection or wilderness study during the ROADLESS AREA REVIEW AND EVALUATION process or the forest planning process (*see* FOREST PLAN). *Compare* WILDERNESS AREA; WILDERNESS STUDY AREA.

nontarget species any SPECIES of living ORGANISM other than the particular species or group of species that a given application of PESTICIDE or HERBICIDE is supposed to kill. The ideal pesticide or herbicide is one that is narrowly focused on a particular TARGET SPECIES, with little or no effect on nontarget species in the same area; however, this ideal is rarely (perhaps never) met, and the death of nontarget species continues to be a major concern, particularly when these nontarget species have considerable economic value (ladybugs, honeybees).

North American Free Trade Agreement (NAFTA) a tariff-reducing pact between Canada, Mexico, and the United States intended to create new markets for manufactured goods via a growing Mexican population and to help Mexico emerge as a fully industrialized nation. Despite concerns that jobs would be lost to Mexico and that environmental conditions along the border between the United States and Mexico would deteriorate due to lax standards under Mexican law, the U.S. Congress narrowly ratified the agreement in 1993. Early indications concerning the success of the pact suggest that the predictions of both opponents and proponents have come to pass. The Mexican economy has improved, especially along the border regions, with new *maquiladoras* (factories) springing up suddenly and now beginning to move to the interior of the country. While the new transnational factories took some assembly-line jobs away from American and Canadian industries, they added others because of their increased requirements for sophisticated materials and components that only the industrial partners to the north could supply. At the same time, the population explosion on the U.S.–Mexican border has led to serious difficulties in controlling illegal immigration. Presumably this has reduced the number of jobs for some American citizens, although the low unemployment rate in the United States that emerged in the late 1990s has eliminated earlier fears concerning the economic impact of illegal, low-cost labor. Indeed, the use of immigrants is now openly accepted in many areas of the United States since the largely menial jobs taken by illegal immigrants are not often coveted by American citizens. On the environmental front, the results of NAFTA are also mixed. While the agreement created a new "North American Development Bank" to finance the cleanup of areas along the U.S.–Mexican border, the Rio Grande is still severely polluted. Some of the new *maquiladoras* are built to U.S. standards of pollution controls, but not all. In addition, increasing quantities of imported Mexican fruits and vegetables often escape pesticide residue restrictions. However, international governmental and citizen organizations along the border areas are influencing the state and national governments of Mexico to reduce environmental problems. Moreover, the increased affluence of Mexicans in border cities such as Mexicali, Nogales, and Juarez is leading to citizen demands for improved environmental conditions. Although NAFTA has benefited transnational corporations more than the civic sector of society, the dire predictions of economic and environmental catastrophe have not come to pass. Even so, citizen concern over the empowerment of the transnationals in terms of their environmental and social impact has not abated, as was revealed by protests at meetings of the WORLD TRADE ORGANIZATION, which can be seen as potentially a kind of worldwide NAFTA, in Seattle in 1999 and of the INTERNATIONAL MONETARY FUND and the WORLD BANK meetings in Washington, D.C., in 2000.

North American Water and Power Alliance (NAWAPA) continent-wide water redistribution scheme for North America, first advocated by Los Angeles hydrological engineer Donald McCord Baker around 1960. Essentially a master plan for re-plumbing the continent, NAWAPA calls for major dams on the Yukon, MacKenzie, Yellowknife, Peace, and Coppermine Rivers that would divert virtually all of their combined flow southward. Pumps would be utilized to raise the water to the Rocky Mountain Trench, a natural defile in the eastern part of the Canadian Rockies, where a 500-mile-long reservoir would be created from here, water would flow, primarily by gravity, to the Great Lakes, the Missouri River, the American southwest (via Montana and Idaho) and eventually into Mexico. Barge canals connecting Hudson's Bay and southern Laborador to Lake Huron would complete the project, which carried an estimated price tag of $100 billion in 1964. One hundred ten million acre-feet of water per year would be redistributed through the system, including 22 million to Canada, 78 million to the United States, and 10 million to Mexico. Strongly advocated by a broad coalition of engineers, hydrologists, and Congressmen in the mid-1960s, NAWAPA received Congressional committee hearings but was never funded. It is today largely defunct. *Compare* GRAND CANAL.

Northeast Atlantic Fisheries Commission encourages international cooperation in the management and conservation of the fishery resources of the northeast Atlantic. Address: Room 425, Nobel House, 17 Smith Square, London, SW1P 3JR United Kingdom. Phone: 071-238-5920.

Northeastern Forest Fire Protection Commission an international organization for forest fire protection, suppression, planning, and training. The organization is composed of three commissioners each from the states of Connecticut, Maine, Massachusetts, New Hampshire, New York, Rhode Island, and Vermont; the Canadian provinces of New Brunswick, Nova Scotia, and Quebec; and representatives of the Green Mountain National Forest and White Mountain National Forest. Address: 36 Roslyn Avenue, Warner, NH 03278. Phone: (603) 456-3474.

North Pacific Anadromous Fish Commission dedicated to conserving the anadromous fish resources (species, such as salmon, that migrate from salt water to freshwater to spawn) in the North Pacific Ocean. The commission was established in 1993 by a convention between Canada, Japan, Russia, and the United States. Address: 889 West Pender Street, Suite 502, Vancouver, British Columbia V6C 3B2 Canada. Phone: (604) 775-5550.

no-till agriculture *See* CONSERVATION TILLAGE.

NOx *See* NITROGEN OXIDE.

NPDES *See* NATIONAL POLLUTANT DISCHARGE ELIMINATION SYSTEM.

NRC *See* NUCLEAR REGULATORY COMMISSION.

NRDC *See* NATURAL RESOURCES DEFENSE COUNCIL.

nuclear energy as generally used, energy produced by using the radioactive decay of heavy elements, particularly uranium and plutonium, as a heat source. All nuclear energy installations function in approximately the same way. The heart of the installation is the chamber where the radioactive decay takes place, known as the *reactor.* A fissile material (that is, a material that decays readily, such as uranium-235 [U_{235}: the ISOTOPE of uranium with an ATOMIC WEIGHT of 235] or plutonium-239 [Pu_{239}]) is processed into *fuel rods* composed of cylindrical pellets, each roughly half an inch in diameter and an inch high. These fuel rods are bundled with *control rods* made of a material, such as cobalt or graphite, which is capable of absorbing the extra neutrons. The combined rod assembly is placed in the reactor, and the control rods are drawn out far enough to allow a *chain reaction* to begin within the fuel as the neutrons released by the decay of one atom collide with other atoms, causing them to decay more rapidly. The chain reaction generates heat, which is used to drive an ordinary steam turbine, producing electricity. The amount of heat produced is proportional to the speed of the reaction, which is controlled by the position of the control rods. Usually the bundle of fuel and control rods—also known as the *reactor core*—is surrounded by a neutron-absorbing material called the *moderator.* In the design most commonly used in the United States, the moderator is a bath of HEAVY WATER, which is also used to drive the turbine. This type of installation is known as a *heavy water reactor.* Other variants on the basic design include the *fast reactor* (which dispenses with the moderator); the *breeder reactor* (in which some of the neutrons from the chain reaction are used to convert U238—a nonfissile material—into fissile Pu239, thus creating more fuel than they use); the *pressurized water reactor* (PWR), in which ordinary ("light") water is used to drive the turbines by superheating it under pressure rather than by simply converting it to steam; and the *advanced gas reactor* (AGR), in which the coolant and propellant for the turbines is a pressurized inert gas such as argon. Whatever its design, the reactor is usually housed in a *containment structure* to prevent the spread of stray radiation from the facility. The containment structure is built of radiation-absorbing material, such as concrete, and is usually operated at negative pressure (that is, at less than atmospheric pressure) so that any leakage that occurs will be inward rather than outward. A COOLING TOWER to cool the water that has passed through the turbine so that it can be either released into the environment or sent back to the reactor to be reheated, and an *emergency core cooling system* designed to be able to quickly cool the reactor core by flooding it with massive amounts of cold water in the case of a threatened core meltdown (see below), completes the standard nuclear power plant.

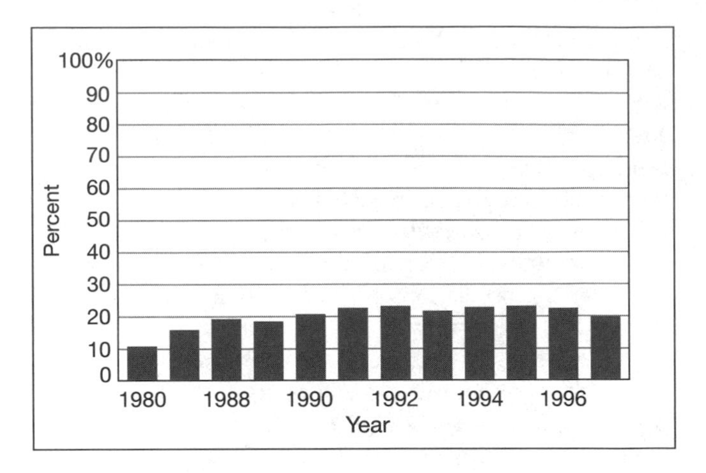

U.S. nuclear power as a percent of all electric power ***(U.S. Energy Information Administration)***

BENEFITS AND HAZARDS OF NUCLEAR ENERGY

Pound for pound, nuclear fuel contains roughly 2.3 million times as much energy as coal, and its costs per kilowatt-hour of electricity generated are roughly 1/4 those of conventional sources. It is quiet, compact, and "clean," in the sense that it does not release chemical pollutants or PARTICULATES into the air or water. It does, however, release small amounts of radiation at all times, and the threat is always present of massive releases in the case of accidents, as at THREE MILE ISLAND (Pennsylvania) in 1979, CHERNOBYL (USSR) in 1986, and TOKAIMURA, Japan, in 1999. The worst type of accident, the *core meltdown,* results from withdrawing the control rods too far, causing the fission reactions in the fuel rods to speed up to where enough heat is created that the core can literally melt its way out of the containment structure, spreading large amounts of radiation into the atmosphere (the design of the fuel rods prevents a totally uncontrolled chain reaction—a nuclear explosion—from occurring, even without the control rods present). Under normal operation all parts of a nuclear reactor slowly become radioactive, and in approximately 25 years the installation must be shut down and dismantled and its pieces somehow safely disposed of. Together with the disposal of the spent but still highly radioactive nuclear fuel, this creates a problem of no small dimensions (*see* NUCLEAR WASTE). Finally, some critics charge that the amount of FOSSIL FUELS required to mine, process, transport, and store uranium is so great that nuclear power plants may be net users, rather than producers, of energy. Despite these hazards, nuclear power is widely used. In the United States, there were 107 operating units as of 1998. In other countries with large numbers of reactors, there are (as of 1998) 56 in France, 53 in Japan, 35 in Great Britain, 29 in Russia, and 21 in Canada. Many countries—notably France, Japan, and Great Britain—are utterly dependent on nuclear power for their electricity. In the United States nuclear power provides about 20% of electricity demand. *See also* PRICE-ANDERSON ACT.

Nuclear Regulatory Commission (NRC) independent agency of the federal (U.S.) government, created by President Richard Nixon on January 15, 1975, as directed by Congress in the Energy Reorganization Act of 1974. Essentially a reworked and expanded version of the old ATOMIC ENERGY COMMISSION, the NRC is organized into four Program Offices, each responsible for a specific portion of the agency's mission. The Office of Nuclear Material and Safeguards regulates the mining of nuclear fuel, its processing and transport, and the disposal of spent fuel and other radioactive waste materials (*see* NUCLEAR WASTE AND CLEANUP). The Office of Nuclear Reactor Regulation oversees the construction and operation of nuclear reactors. It includes divisions for engineering, safety devices, and human safety (evacuation plans), and contains a special Program Office for dealing with the aftermath of the 1979 Three Mile Island nuclear accident in Pennsylvania. The Office of Nuclear Regulatory Research investigates all nuclear "incidents," no matter how small, in connection with reactor operation or the handling of radioactive materials, and coordinates programs of advanced research on reactor and nuclear materials safety, much of which is done through grants to universities and other research facilities. Finally, the Office of Inspection and Enforcement coordinates the actual on-site inspection of nuclear plants and oversees the actions of state regulatory agencies to make certain that they comply with federal law concerning radioactive materials. The commission has a resident inspector stationed at most major nuclear plants. These are coordinated and backed up by field staff operating out of five District Offices. Address: Washington, DC 20555. Phone (301) 415-7000.

nuclear waste and cleanup any waste material that emits detectable levels of ALPHA PARTICLES, BETA PARTICLES, or GAMMA RADIATION, requiring management and a cleanup program. Nuclear wastes are classed in four separate categories. *Spent fuels* are fuel rods from nuclear reactors (*see* NUCLEAR ENERGY) that no longer contain enough concentrated fissile material to sustain an efficient reaction. They continue to emit large amounts of radiation, and are physically very hot. Nuclear warheads that no longer contain enough fissile material for an uncontrolled chain reaction (explosion) may also be classed as spent fuels. *High-level wastes* (HLW) are wastes generated by the reprocessing of spent fuels or warhead materials. They are normally liquid, and are also extremely radioactive and extremely hot (since federal law currently forbids the reprocessing of spent fuels, virtually all HLW in the United States comes from military sources). Low-level wastes (LLW) are primarily materials that have become radioactive through contact with HLWs, operating fuels, or spent fuels, including tools used to work with them, protective clothing worn while handling them, cleaning solutions, coolant, wiping rags, structural parts of disassembled reactors, and so forth. The category also includes uranium mine tailings, radioactive wastes generated by medical facilities that use radiation treatments on cancer patients, and wastes generated by industries (such as the pharmaceuticals industry) that use radioactive materials in manufacturing processes. Finally, *transuranic wastes* (TRU) are low-level wastes that contain trace levels (10 nanocuries or more; *see* CURIE) of elements with higher ATOMIC WEIGHTS than uranium, such as plutonium.

STORAGE OF NUCLEAR WASTES

Since radioactive emissions can cause cancer, genetic MUTATIONS, and other problems, nuclear wastes must be isolated from the environment as long as they exhibit significant levels of radioactivity. The key problem here is the definition of the term "significant." All parties to the debate agree that 10 HALF-LIVES are a sufficient time to isolate any material, but for the materials comprising nuclear waste, 10 half-lives can range anywhere from 5 years to 45 billion years. Fortunately, the more dangerous materials tend to be those with shorter half-lives. It has been calculated that the radioactivity emitted by virtually all forms of nuclear waste materials will decline to or below the level emitted by uranium ores within 1,000 years (the principal exception to this is plutonium, an extremely dangerous material that has a half-life of 24,000 years and thus requires 2.4 million years to decline to negligible amounts). Nevertheless, it has been generally accepted that the only genuinely safe method of disposal is to create a *geologic depository*—an underground storage site in a stable geologic formation that is not likely to be fractured or faulted by TECTONIC ACTIVITY while the

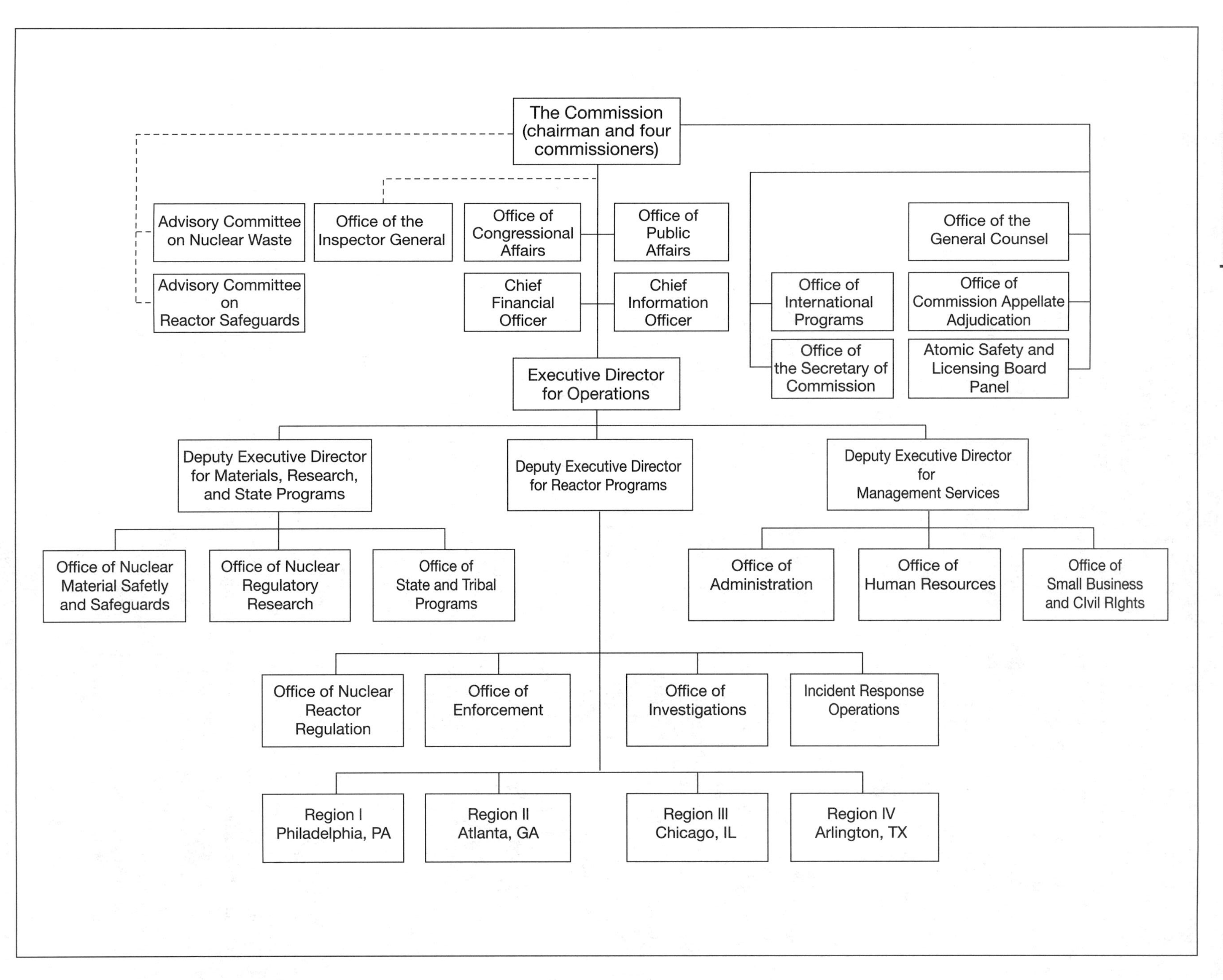

Nuclear Regulatory Commission organizational structure

wastes stored in it remain dangerous. Three types of formations—salt domes, deeply buried basalt INTERFLOW ZONES (that is, sands, gravels and similar materials lying between layers of basalt), and accumulations of so-called "red clays" on the floors of the deep ocean basins—have been suggested as valid sites for geologic depositories, but all exhibit potential problems, and the question about what to do permanently with these materials remains unanswered (the decision of the federal government to bury them in Nevada, under the terms of the Nuclear Waste Repository Act of 1982, notwithstanding). In the meantime the materials accumulate, and must be placed in temporary storage. Most spent fuels are currently stored in pools of water at the sites of the reactors in which they were used. Low-level wastes are accumulating in three licensed "dump sites" in various parts of the nation (they are no longer jettisoned at sea, as they were before 1970), where federal law requires them to be monitored for 100 years and kept from environmental contact for 400 years after that. High-level wastes and TRUs have been stored in underground steel tanks, principally in the states of Washington and Idaho, a practice that critics charged has contaminated nearby AQUIFERS with radioactivity as the often caustic wastes corrode the tanks and leak into the ground. Liquid wastes may be encapsulated in ceramic or glass ("vitrified") to immobilize them; however, the resulting blocks of ceramic or glass still need to be handled and stored, and the heat generated by the radioactivity within them may in some cases be enough to melt them. As of 1999 there were about 2,000 metric tons of wastes in temporary storage in various parts of the United States. After many years of study, the U.S. government has begun the storage of nuclear wastes deep underground in geological depositories. The first of these to open is the highly controversial 2,150-foot-deep WASTE ISOLATION PILOT PROJECT (WIPP) in a salt deposit near Carlsbad, New Mexico. The facility is used to provide permanent disposal of TRUs left over from the research and production of nuclear weapons. Another controversial facility, YUCCA MOUNTAIN, Nevada, has been approved as a permanent disposal facility but is not yet operating. And a third, WARD VALLEY in Southern California, has been fought to a standstill by a coalition of Indian tribes and environmentalists.

nuée ardente in geology, a highly mobile, extremely hot mass of volcanic gases, ash, pulverized rock, and other EJECTA, formed when part of the side of a VOLCANO is blown off by internal pressures. A triggering event such as an earthquake must usually occur before the explosion. Nuées ardentes (French, "glowing clouds") may be thought of in simplest terms as horizontal eruptions. They move at roughly 330 kilometers/hour (200 miles an hour) and may reach temperatures of as much as 700°C. Well-known nuées ardentes include the eruption of Mt. St. Helens (Washington state) on May 18, 1980, which killed 64 people and destroyed 595 square kilometers of forest, and the explosion of Mt. Pelée (Martinique) on May 8, 1902, which killed all but one of the 28,000 inhabitants of the city of St. Pierre.

nuisance in law, the use of a piece of property in such a way as to substantially interfere with the use of neighboring properties. Two types of nuisance are recognized: *public nuisances,* which cause damages to a large number of people or to the public in general; and *private nuisances,* which cause damages to one person or to a few people in significantly greater measure than to others. Individuals may sue for relief from private nuisances in the form of awards for monetary damages, or injunctions against further operation of the nuisance, or both. A state, municipality, or other public body must generally file charges in the case of public nuisances, although there are a few states which allow individuals to bring suit in public-nuisance cases as well. Relief in these cases is generally in the form of fines, or injunctions against further operation, or both. Nuisance law has been used successfully in numerous environmental cases, particularly those involving AIR POLLUTION, NOISE POLLUTION, and the accumulation of garbage and eyesores. Its use is limited, however, by (a) the need to prove that damages to neighboring properties from continued operation of the nuisance are greater than the costs to the proprietor of abating it; (b) the requirement that the uses damaged by the nuisance be established before operation of the nuisance begins (so that, for example, one cannot build a house next to an operating stone quarry and then seek damages for the noise and dust from the quarry, a course of events known legally as "moving to the nuisance"); and (c) the difficulty of proving standing (that is, legal right to sue), especially in public-nuisance cases.

nurse log in forestry, a fallen tree in the process of decaying that has one or more SEEDLINGS established on its upper surface. The seedlings utilize the NUTRIENTS released by the decay of the dead tree. Nurse logs are most common in rain forests (*see* RAIN FOREST BIOME), both because nutrients are scarce in rain-forest soil due to leaching (*see* LATERITIC SOIL) and because the POROSITY of decaying wood makes adequate water retention for seedling establishment difficult in dry climates. The location of nurse logs can be determined even after they have rotted away by the presence of straight lines of closely-spaced SAPLINGS or young trees, all the same age.

nut *See* FRUIT.

nutrient in the biological sciences, an ELEMENT or a COMPOUND required by an ORGANISM to build living tissue and maintain the chemical reactions of the life process. All life is composed primarily of the elements CARBON, HYDROGEN, and OXYGEN, with smaller amounts of NITROGEN, PHOSPHORUS, potassium, sodium, calcium, magnesium, iron, and a few others. Nutrient requirements differ considerably from SPECIES to species, however, due to the different proportions of these elements needed by different organisms and the different forms in which assimilation of them into the body takes place. Plants obtain most of the nutrients they need directly from elements or simple INORGANIC COMPOUNDS, using PHOTOSYNTHESIS and other processes to build the complex PROTEINS and CARBOHYDRATES that form their bodies and the ENZYMES that regulate living functions. Animals have considerably less ability to build complex chemicals from elements, and must obtain most of their nutrients in the form of preconstructed proteins and carbohydrates, which they do by eating plants and other animals. (The proteins and carbohydrates obtained in this manner are generally not used unaltered, but are broken down by the animal's digestive process into their constituent AMINO ACIDS and sugars, which are then reassembled into the proteins and carbohydrates the animal requires.) The nutrient requirements of fungi (*see* FUNGUS) and of MONERA and PROTISTA fall somewhere between the extremes represented by plants and animals: that is, the organisms in these KINGDOMS use more elements directly as nutrients than animals do, but require more complex compounds as nutrients than plants do. *See also* METABOLISM.

nutrient cycle the movement from the nonliving environment of nutritive elements such as carbon, sulfur, and nitrogen into biological systems, from which they are then returned to the nonliving systems through decay, burning, leaching, or other mechanisms.

O&C lands *See* OREGON & CALIFORNIA LANDS.

objective in land-use planning, a specific, obtainable end toward which management should be directed. Objectives are subordinate to GOALS and should be in harmony with them. They differ from goals primarily in being obtainable through realistically available means within a specifiable time frame. A goal, for example, might be to restore a waterway to natural conditions. An objective in harmony with that goal would be to eliminate the discharge of raw SEWAGE into the waterway within 10 years. *See also* MANAGEMENT CONSTRAINT.

obligative organism (obligate) an ORGANISM that must live in a certain way. When it is used by itself, the term usually refers to *obligative anaerobe;* that is, an organism—usually a BACTERIUM—which must live in an OXYGEN-free environment and is killed by exposure to free oxygen. Obligative anaerobes lack the ENZYME necessary to convert hydrogen peroxide (H_2O_2)—a bactericide that is also a common product of bacterial respiration in oxygen—to water and oxygen (H_2O and O_2). Thus they are poisoned by their own wastes. The category includes numerous PATHOGENS. The term "obligate" may also be used to describe PARASITES that cannot live outside their host organisms (*see* HOST SPECIES). *Compare* FACULTATIVE ORGANISMS. *See also* ANAEROBIC STREAM.

obliterative coloration (countershading) a form of CRYPTIC COLORATION in which an animal's back is dark and its underparts light in color, reversing the normal pattern caused by shadows and thus making it more difficult for the animal to be seen. Obliterative coloration is one of the most common forms of protective coloration, and is seen in everything from fish and reptiles to birds, mice, deer, and antelope.

occupational hazard any job- or workplace-related condition that can lead to injury, illness, or death. Occupational hazards range from repetitive motion, such as that required for data entry in offices, which can result in carpal tunnel syndrome, to inherently dangerous industrial operations, such as logging. Hazardous substances and fumes from the use of complex chemical compounds in the workplace environment are of special concern with regard to cancer and respiratory diseases. *See also* INDOOR POLLUTION; RISK ASSESSMENT AND MANAGEMENT.

Occupational Safety and Health Administration (OSHA) agency within the U.S. Department of Labor, created by Congress in 1970 and charged with promoting safe, healthful conditions in the workplace. OSHA's principal environmental responsibility lies in regulating the use of toxic, carcinogenic or otherwise hazardous materials (*see* TOXICANT; CARCINOGEN; HAZARDOUS SUBSTANCE) so that they do not pose a hazard to workers. OSHA regulations apply to all private businesses with one or more employees who are not family members of the employer. Headquarters: (OSHA) 200 Constitution Avenue NW, Washington, DC 20210. Phone: (202) 693-1999. Website: www.osha.gov.

Occupational Safety and Health Review Commission independent, quasi-judicial agency of the federal (U.S.) government created to oversee the operation of the OCCUPATIONAL SAFETY AND HEALTH ADMINISTRATION (OSHA) and to act as an appeal body for decisions of OSHA involving workplace safety and worker health issues, including those relating to the handling of HAZARDOUS SUBSTANCES and other aspects of environmental health. Headquarters: 1825 K Street NW, Washington, DC 20006.

ocean dumping the disposal of solid and liquid waste materials in the ocean. Serious environmental problems have resulted and will continue to ensue from the dumping of some 200,000 tons of chemical weapons after World War II; the disposal of nuclear wastes at sea; crude oil from the illegal cleaning of transport ship tanks at sea; and near-shore garbage dumping by cities. In 1988 the U.S. Congress passed the Ocean Dumping Ban Act, which prohibits the dumping of sewage sludge, industrial waste, and other contaminants in U.S. coastal waters.

ocean pollution perhaps the most difficult of all international environmental issues to deal with. Whereas the world's oceans were largely pristine 50 years ago, ocean pollution now threatens many of the world's fisheries, the integrity of critical estuarine habitats, recreational shorelines, and even the basic circulation patterns that determine climate regimes. Pollutants include solid and liquid materials dumped at sea (*see* OCEAN DUMPING), a practice that produces about 15–20% of ocean pollution. An estimated 44% of the pollution comes from sewage, animal waste, industrial and medical waste, and agricultural chemicals entering ocean waters either from outfall pipes or through polluted rivers and streams. Another 33% of ocean pollution is caused by atmospheric deposition of acidic compounds, heavy metals, and particulate matter and from the disturbance of benthic zones from trawling and mining. The immediate and most observable result of pollution has been its effect on fish and shellfish used for food, especially along coastlines, where 80 to 90% of the global fish catch is taken. According to the United Nations, 70% of global fish stocks are in serious decline. National and international efforts to reduce ocean pollution, in view of the gravity of the issue, have been quite modest. A few countries have established marine sanctuaries; and some provisions of the U.N.'s Law of the Sea can be applied to the pollution problem, as can the Convention on Biological Diversity and the Commission on Sustainable Development. In addition, the UNITED NATIONS ENVIRONMENT PROGRAMME has set up the broadly-based Global Program of Action for the Protection of the Marine Environment. These international efforts do not, however, address ocean pollution comprehensively, nor are any of their provisions enforceable internationally. *See also* FISHERIES DECLINE.

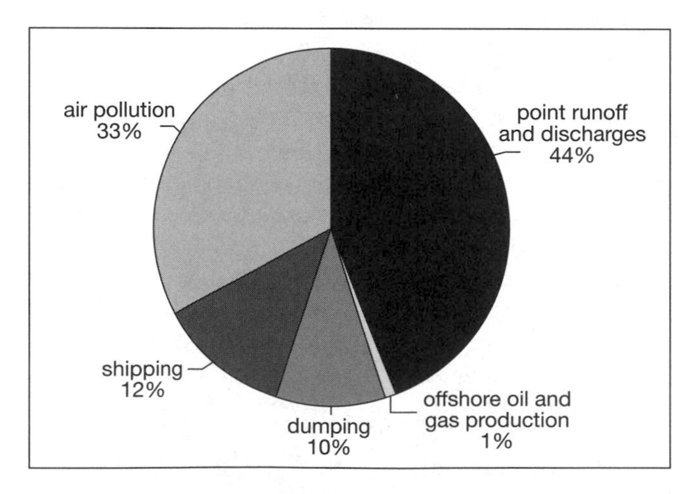

Major sources of marine pollution, 1990 ***(United Nations Environmental Programme [UNEP], "Gesamp: The State of Marine Environment,"*** **UNEP Regional Seas Reports and Studies,** ***No. 115 [UNEP, Nairobi, Kenya, 1990], p. 88)***

Oceans and International Environmental and Scientific Affairs, Bureau of *See under* STATE, DEPARTMENT OF.

octalene *See* ALDRIN.

Octalox a trade name for dieldrin (*see under* ALDRIN).

Office of Civilian Radioactive Waste Management agency within the U.S. Department of Energy (*see* ENERGY, DEPARTMENT OF) created in 1982 and charged with overseeing the handling and disposal of spent nuclear fuel and other high-level radioactive waste (*see* NUCLEAR WASTE) produced by the private sector and by non-defense-related federal activities. The agency is also incharge of the ongoing federal efforts to find ways of safely storing radioactive waste and to establish a permanent high-level waste depository. The director of the Office of Civilian Radioactive Waste Management is ranked within the department at the same level as an assistant secretary.

Office of Energy office within the U.S. Department of Agriculture (*see* AGRICULTURE, DEPARTMENT OF) charged with the coordination of energy policies and the promotion of energy efficiency throughout all other agencies of the department. The office also serves as a liaison between the Department of Agriculture and the Department of Energy (*see* ENERGY, DEPARTMENT OF) and acts as a point of contact for private parties concerned about department energy policies. It operates under the authority of the assistant secretary for economics.

Office of Energy Research agency within the U.S. Department of Energy (*see* ENERGY, DEPARTMENT OF) charged with coordinating federal research dealing with energy production, transmission, and conservation, including the environmental effects of these activities. The office also coordinates federal grant programs that assign money to colleges and universities for research on energy issues, and engages in some basic research in its own laboratories, including especially research in fusion power and high-energy physics. The director of the Office of Energy Research is ranked within the department at the same level as an assistant secretary.

Office of Management and the Budget (OMB) agency within the Executive Office of the president of the United States, created by President Richard Nixon on July 1, 1970, and charged with assisting the president in the preparation of budget recommendations to Congress and with coordinating policies and regulations among the various agencies of the EXECUTIVE BRANCH. Its importance from an environmental perspective lies in its ability to block or hinder the implementation of an agency's proposed policies and regulations if it feels that these conflict with overall federal policy or with the policies and regulations of other federal agencies. Headquarters: Executive Office Building, Washington, DC 20503.

Office of Naval Petroleum and Shale Reserves *See under* ENERGY, DEPARTMENT OF.

Office of Surface Mining Reclamation and Enforcement agency within the U.S. Department of the Interior (*see* INTERIOR, DEPARTMENT OF THE) created by Congress through the terms of the SURFACE MINING CONTROL AND RECLAMATION ACT OF 1977 and charged with the task of regulating the STRIP MINING OF COAL to meet the twin goals of environmental protection and an adequate coal supply. The office acts primarily by assisting in the development of plans by state governments for the regulation of surface mining within the states, approving those plans, and monitoring compliance with them. It also takes a direct regulatory role in those states that have not yet developed approved plans. The Office of Surface Mining Reclamation and Enforcement is headed by a national director in Washington, D.C., who reports to the assistant secretary of the interior for energy and minerals. He is assisted by three assistant directors: an assistant director for program operations and inspection, responsible for the regulatory functions of the agency, including the approval and monitoring of state plans and the designation of areas unsuitable for mining; an assistant director for technical standards and research responsible for the development of environmental standards and the exploration of new techniques for mining and reclamation; and an assistant director for management and budget. Address: Department of the Interior, Interior South Building, 1951 Constitution Avenue NW, Washington, DC 20240. Phone: (202) 208-2565. Website: www.osmre.gov.

Office of Water Policy (OWP) office within the U.S. Department of the Interior (*see* INTERIOR, DEPARTMENT OF THE) charged with the development and coordination of water policies for lands under the jurisdiction of the department. The OWP functions primarily as an internal body, relating the water policies of the various branches of Interior to each other; however, it also coordinates the water activities of the department with the development of water resources by state and local governments and by private individuals to the extent that these developments affect or are affected by activities on Interior Department lands.

off-road vehicle (ORV) any motor vehicle designed for part-time or full-time use on lands where roads either don't exist or exist but have been made undrivable by deep snow, mud, lack of maintenance, etc. The category includes motorcycles ("dirt bikes," "trail bikes"), four-wheel drive vehicles, snowmobiles, dune buggies, ALL-TERRAIN VEHICLES, and any conventional vehicle (such as a pickup or van) sturdy enough to be driven for short distances off the road. Off-road vehicles are useful for providing emergency service in rural areas and for delivering mail and supplies to snowbound communities or homes. Their principal attraction, however, is as recreational vehicles. Approximately 25% of the United States population engages in off-road vehicle recreation each year—nearly as high a proportion as those who engage in hiking and backpacking (28%).

ENVIRONMENTAL CONCERNS

The two principal environmental problems associated with off-road vehicle use are NOISE POLLUTION and damage to soils and plant life. Noise levels are often kept purposefully high on ORVs, partly because an unmuffled or lightly muffled engine is marginally more powerful and responsive, but mostly because high volumes of sound directly under the operator's control contribute to the sense of power over the environment that is a principal part of the ORV experience. These high noise levels are profoundly disturbing to wildlife—they have been known to cause stillbirths and death from shock in wild POPULATIONS—and to those recreationists whose goal is union with nature, one unmuffled ORV is enough to pretty well destroy any sense of this union for several miles in any direction. Problems of soil and plant destruction are at least as serious as, and considerably longer-term than, problems of noise. ORV operators often push their machines to the edge of their ability to maintain traction through mud and up steep hills, destroying ground cover and grinding up the upper few inches of soil so that it is much more easily eroded. In the process, they may muddy streams and destroy CRITICAL HABITAT for plants and small animals. ORVs also make it easier for a statistically few "slob outdoorsmen" to spread vandalism and litter in the backcountry, to poach game, and to engage in such pursuits as running deer and coyotes to death by chasing them with snowmobiles till they die of exhaustion.

CONTROLS ON ORV USE

Since most off-road vehicle use takes place on federal lands—over half of it on lands managed by a single agency, the BUREAU OF LAND MANAGEMENT—the abuses detailed above are subject to federal control. Off-road vehicles are not allowed in WILDERNESS AREAS or on National Scenic Trails (*see* NATIONAL TRAILS ACT); in addition, federal land managers may, at their discretion, ban vehicles from other lands under their care. National Park officials, especially, have been encouraged by environmental organizations to restrict the use of ORVs. In 1999 a coalition of 68 groups filed a petition with the National Park Service asking for a ban of off-road vehicles in all the parks. A survey conducted by the coalition revealed that ORVs were in use in 56 of the 108 large national parks and that 38 of these parks reported serious damage by ORVs. Regulations governing ORV use are based on two EXECUTIVE ORDERS, EO 11644 (signed by Richard M. Nixon, February 8, 1972) and EO 11989 (signed by Jimmy Carter, May 24, 1977), and on section 219.12(i)(7) of the NATIONAL FOREST MANAGEMENT ACT. In summary, these regulations state that all public lands and trails must be designated either "open" or "closed" to ORV use and that ORV use of those areas designated as "open" must be shown to have minimal impact on soil, water, vegetation, and wildlife habitat, and present minimal conflicts with non-motorized recreational use.

Ogallala Aquifer a very large AQUIFER underlying a large part of the American High Plains and supplying most of that region's IRRIGATION water and much of its drinking water. Currently it is in serious danger of exhaustion. The Ogallala underlies all of Nebraska, most of Kansas, and portions of South Dakota, Colorado, Oklahoma, Texas, and New Mexico. Its greatest length (north-south) is approximately 700 miles, and its greatest width (east-west) is about 300 miles. Total surface area is about 220,000 square miles and total holding capacity, roughly 3 billion acre-feet (*see* ACRE-FOOT).

THE PROBLEM

Because of the High Plains region's low rainfall (10–20 inches/yr) and the existence of a CALICHE layer overlying most of the aquifer, the Ogallala has only negligible RECHARGE and thus principally contains FOSSIL WATER, the bulk of it approximately 3 million years old. Current withdrawals amount to about 21 million acre-feet/year, while recharge is less than 100,000 acre-feet/year, a discrepancy that results in an approximately 10-foot decline in the regional WATER TABLE each year. This is an OVERDRAFT of roughly 99.6%, with potentially dire consequences for U.S. agricultural production, which is centered in the High Plains region and built on a base of Ogallala water. Parts of the aquifer, mostly in the south, are already exhausted. Water experts predict that 25% of the aquifer will be depleted by 2020. No alternate source of water has been found locally, and to import enough from the Mississippi/Missouri River system or from the Great Lakes to meet current use levels would cost roughly $2,000/acre-foot, which probably makes such imports impossible. *See also* DESERTIFICATION.

Ohio River Valley Water Sanitation Commission interstate agency founded in 1948 by Illinois, Indiana, Kentucky, New York, Ohio, Pennsylvania, Virginia, and West Virginia to provide a coordinated effort to monitor and control water pollution in the Ohio River Valley Compact District. Address: 5735 Kellogg Avenue, Cincinnati, OH 45228. Phone: (513) 231-7719.

oil technically, any substance that is soluble in ether, CHLOROFORM, or other nonpolar solvent but is insoluble in water (*see* POLAR LIQUID; SOLVENT; SOLUBILITY). Oils are typically either liquids or soft solids that have a slippery, "greasy" feel and are capable of being spread into extremely thin films (*see* OIL FILM). They are classed into three groups. *Mineral oils* are natural oils derived from PETROLEUM and thus consisting of fossilized organic remains. *Fixed oils* are non-mineral oils that do not volatilize easily (*see* VOLATILITY). This group consists principally of LIPIDS. *Essential oils* are non-mineral oils that volatilize easily, including substances such as oil of wintergreen and turpentine. *See also* OIL SHALE; OIL SPILL.

oil embargo the curtailment of oil sales by members of the ORGANIZATION OF PETROLEUM EXPORTING COUNTRIES (OPEC) to the United States and other industrialized nations in 1973–74. The Arab nations

The Ogallala aquifer

in OPEC (Venezuela, a non-Arab member, did not participate) decided in 1973 to embargo oil shipments to protest the support of Israel by Western nations during the Yom Kippur War. At once, the world price of crude oil quadrupled, with the price of gasoline at the pump rising from about $.35 to nearly $2.00 in some areas. (Another slowdown in Arab oil supplies occurred in 1978–79, when Iranian revolutionaries called a general strike against the Shah of Iran, who was supported by the United States.) The result of the embargo was a per-barrel price of $34, more than 10 times its previous level. The embargo and the Iranian action produced new federal policy initiatives to reduce U.S. dependence on foreign oil, many of them quite significant environmentally. Among them were programs to exploit OIL SHALE resources in the overthrust belt of the Rocky Mountains, the development of the ALASKA PIPELINE, the establishment of the STRATEGIC PETROLEUM RESERVE, and government support of ALTERNATIVE ENERGY SOURCES such as NATURAL GAS (on which restrictions were lifted). The federal government also supported initiatives to develop wind and solar energy; low-head (i.e., small dams) hydropower; and increased use of mass transit. Since energy conservation was the most effective way to reduce reliance on imported oil, the federal government imposed a national speed limit of 55 miles per hour and encouraged setting thermostats at 65 degrees or lower. While economic dislocations stemming from the embargo and the Iran crisis were severe, leading to double-digit inflation, the experimental work on alternative energy sources, although largely suspended by the Reagan and Bush administrations, will nevertheless contribute to reducing dependence on foreign oil in the future. An important incentive for continued interest in alternative energy sources is not so much to counter the power of OPEC, which in fact was greatly weakened by the embargo, as it is to curtail the use of FOSSIL FUEL in order to stem global warming, reduce SMOG, and repair the thinning of the ozone layer. *See also* GREENHOUSE EFFECT.

oil film a thin layer of OIL floating on the surface of a body of water. Oil films range in thickness from a single MOLECULE to 2.5 MICRONS (.0001 inch). If they are thicker than 2.5 microns they are called *slicks*. They refract light at a considerably different angle from the refraction caused by water, making them appear iridescent. The appearance of an oil film is a good guide to the amount of oil it contains, and standard terminology has been developed to reflect this (see the accompanying table). *See also* OIL SPILL.

Classification of Oil Films

Term (and characteristics)	Gallons (per sq. mi)
barely visible (barely visible under favorable light)	25
silvery (silvery sheen on surface)	50
slightly colored (trace of color visible)	100
brightly colored (bands and swirls)	200
dull (colors contain some browns)	700
dark (browns predominate: little color)	1300

oil reserves the quantity of oil reasonably recoverable from identified deposits given current extraction technologies, or technologies fairly certain to be developed in the near future. Worldwide, the 13 OPEC countries control two-thirds of all oil reserves. The United States has only 2.3% but uses 30% of the world's oil. American oil reserves decreased by about 10% between 1990 and 1997, from 26 billion barrels to 23 billion barrels. Some experts estimate that known world reserves will be virtually depleted by the mid-2000s, and that newly discovered fields can extend the supply for only another 20 to 40 years beyond that. However, if prices rise, additional fields will become economical to bring on line even as demand is reduced, which suggests to many that oil will never be fully depleted as an energy source or for use in plastics and other products. Since it seems likely that alternative energy sources (including natural gas) will become dominant before oil reserves are depleted, the environmental argument that oil use should be limited to avoid such depletion is not now so persuasive. Instead, environmentalists now argue that oil use must be curtailed because of the dire effects of global warming caused by smokestack pollution and internal combustion engine emissions. *See also* GREENHOUSE EFFECT; ORGANIZATION OF PETROLEUM EXPORTING COUNTRIES; STRATEGIC PETROLEUM RESERVE.

oil shale SHALE containing rich organic remains, mostly in the form of *kerogen*, an oily solid that is a precursor to PETROLEUM. It can be converted into petroleum for refining purposes. Some oil shale is rich enough in kerogen to burn when lit, but most contains considerably less than that. The average is the equivalent of about 25 gallons of petroleum per ton of rock. The refining process is considerably more complex than that for liquid petroleum, and has not yet proved economically feasible, although vast tracts of oil-shale lands remain under lease to oil companies (and, in some cases, under patent; *see* LEASABLE MINERAL, LOCATABLE MINERAL; MINERALS LEASING ACT OF 1920) in the Rocky Mountain states of the American west, and development may resume there at any time. Oil shale deposits are developed by STRIP MINING. The waste rock from the refining process takes up more room than the original rock, and so poses a considerable disposable problem. U.S. reserves: about 1.43 trillion barrels, 80 trillion of which are considered recoverable.

oil spills the accidental discharge of PETROLEUM or of a petroleum product (such as GASOLINE) into a body of water. The body of water involved is frequently the ocean, but it may also be a lake, a river, or a small stream or pond. Large oil spills take place as a result of shipping accidents, BLOWOUTS or other drilling accidents (especially in offshore oil fields), and pipeline ruptures. While these large spills are spectacular and attention-grabbing, the cumulative effects of small spills are probably worse. Small spills result from such routine circumstances as leaky pipeline gaskets, discharge of ship BILGE WATER, transient amounts of oil escaping during coupling and uncoupling of pipelines, and so on. Estimates indicate that as much as 90% of all oil spills—by volume—are of this small, routine variety.

EFFECTS OF OIL SPILLS

Because oil is lighter and less dense than water, and because neither oil nor water dissolves in the other ("oil and water do not mix"; *see* SOLUBILITY; SOLUTION), oil tends to form a film of "slick" on the surface of the water it is spilled into (*see* OIL FILM). This film cuts off the circulation of OXYGEN from the air into the water, kills all NEUSTON and many other PLANKTON, and covers larger animals that blunder into it with a coating of oil, which can kill them, especially if the oil is ingested in the process of grooming or when a PREDATOR eats oil-covered PREY. When they wash ashore, oil spills foul beaches, boats, docks and pilings, coat the feathers of WATERFOWL, and smother barnacles, sea anenomes, and other living things in the intertidal area (*see* BALANOID ZONE). Broken up by wave action, the oil may emulsify in the sea water (*see* EMULSION), where it may damage the gills of fish and other NEKTON as it passes through them. It may also sorb to SEDIMENT particles (*see* ADSORPTION; ABSORPTION), which carry it to the bottom and damage or destroy the benthic community (*see* BENTHIC DIVISION; BENTHIC DRIFT ORGANISM; BENTHOS). Most of these harmful effects are short-lived, and disappear after a few weeks or months. Long-term effects (as measured in years) are generally minimal or nonexistent, though this should not minimize the importance of short-term effects.

CLEANING UP OIL SPILLS

Oil slicks on surface waters may be burned, but this is usually impractical, both because the slicks are often broken up by wave action and therefore must be ignited in numerous places, and because the burning of a large slick may send burning material against the shore. (Some instances have also been known of oil, driven into the air by winds caused by burning slicks that escalate into firestorms [*see* WILDFIRE: *types of wildfire*], falling as "rain" on coastal lands). Chalk or other heavy SORBENTS can be spread on a slick to soak it up and carry it to the bottom; however, this is harmful to the benthic community. Some success is being shown in the breeding of bacteria (*see* BACTERIUM) that can "eat" the oil and break it down metabolically. When the oil is gone, the bacteria die. This technique is still experimental. The best practical approach is probably to use light sorbents such as straw or IMBIBER BEADS, which remain floating and may be easily scooped up after they have taken up the oil. Floating booms should be employed to confine the spill and to keep it away from sensitive shoreline areas wherever possible. Fouled sands are usually best dug up and carted away; oil may be cleaned from rocks, pilings, etc., with high-pressure water or steam jets. Waterfowl, otters, and other aquatic wildlife may have their feathers or fur cleaned with DETERGENT; however, care must be taken to allow time for the plumage's natural oils to replenish themselves before the animals go back in the water, in order to prevent the plumage from absorbing water, which may cause the animal to drown or to die of HYPOTHERMIA. There were 82 major spills (of more than 1,000 gallons) in 1997, totaling 943 million gallons of oil contaminating U.S. waters.

old growth (ancient forest) generally, an area of forest characterized by a high proportion of large, old trees. It may also be described as a "forest that has never been logged" or "a forest where the majority of trees are past the ROTATION AGE." The U.S. FOREST SERVICE has a fairly precise definition. To qualify as old growth on Forest Service lands, an area must (1) be 10 acres or more in size; (2) have an OVERSTORY consisting largely or exclusively of MATURE and OVERMATURE TIMBER; (3) have a multilayered CANOPY and UNDERSTORY; (4) contain standing SNAGS and dead and down timber; and (5) show few or no signs of human intrusion. Under this definition, about 25% of forests in the Pacific Northwest are old growth. Taken together, these areas are called "ancient forests" by environmentalists. Elsewhere in the United States, where logging has been continuous for several hundred years, old growth stands are widely scattered in quite small tracts.

THE VALUE OF OLD GROWTH

Many SPECIES of plants and animals (examples: elk, pileated woodpecker, indian pipe, spotted owl) are adapted to old-growth conditions and require old-growth HABITAT to survive. Such *old-growth dependent species* become extinct if areas of old growth no longer exist for them to live in. The air in old-growth forests is calmer, cooler, and moister than the air in SECOND GROWTH, helping to suppress forest fires and creating an "air-conditioning" effect for nearby open spaces. Old growth forests are usually preferred for recreation because of the cooler, moister air, the profusion of life forms, and the sense of conditions in balance, as well as the impressive size of the trees (these are the stands usually described as "cathedral-like").

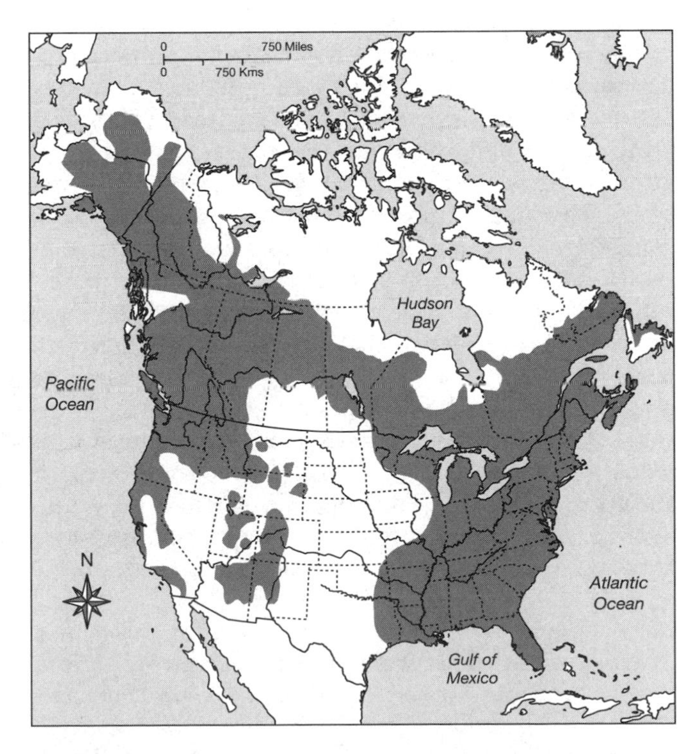

Old growth forests ***(Virgin forests, circa 1600, The Wilderness Society, the U.S. Forest Service and* Atlas Historique du Canada, *Vol. 1)***

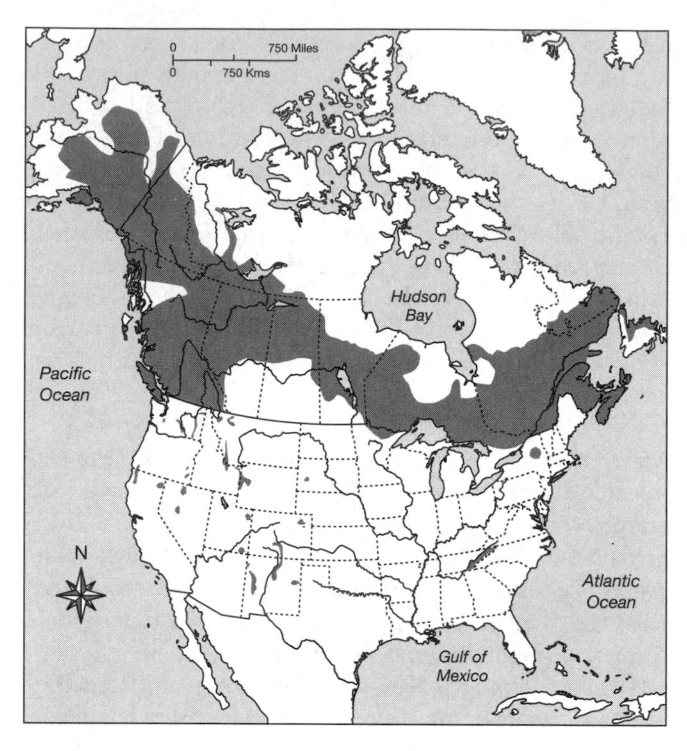

Old growth forests ***(Virgin forests, circa 1993, The Wilderness Society, the U.S. Forest Service and*** **Atlas Historique du Canada,** ***Vol. 1)***

THE "OLD GROWTH PROBLEM"

Because most of the trees in an old-growth stand have passed the CULMINATION OF MEAN ANNUAL INCREMENT, old-growth forests are not efficient lumber producers; that is, they do not "put on wood" as rapidly as do second-growth forests. To increase the timber growth rate and therefore the timber harvest, old-growth stands must be cut down and replanted to second growth, a process known as *old-growth conversion.* If extended to all forests, however, old-growth conversion conflicts with recreation, the preservation of old-growth dependent species, and the other values of old growth listed above. The "old-growth problem," therefore, may be defined as the need to decide when and where to practice old-growth conversion so as to maximize timber production while maintaining enough old growth to serve the other needs it answers. This is probably the single most important issue in contemporary forest planning. *See also* SPOTTED OWL MANAGEMENT AREA; THOMAS REPORT.

oligochaete a member of a CLASS of segmented worms that includes the terrestrial earthworm and a variety of aquatic species. Oligocheates are characterized by a tubular structure made up of short, uniform segments; highly simplified digestive and circulatory systems; and a much reduced or absent head. Nearly all species are blind. Reproduction is hermaphroditic, that is, with each individual possessing both male and female sex organs. Some aquatic species live in the bottom sludge of lakes and streams and surround themselves with tubes made of silt particles cemented together with mucus; hence the common names "tube worm" or "sludge worm." Terrestrial species are economically valuable as soil aeraetors and mixers (*see* SOIL). Most aquatic species are highly pollution-tolerant and are able to survive well in OXYGEN-depleted environments, so the proportion of oligocheates to other bottom-dwelling INVERTEBRATES is a useful index of the quality of bottom waters (*see* INDICATOR SPECIES). The class contains two ORDERS, 12 FAMILIES, and numerous genera and species.

oligotrophic lake a lake containing low levels of available NUTRIENTS in its water and SEDIMENTS. Normally it will contain high levels of DISSOLVED OXYGEN, especially in its deeper waters. Oligotrophic lakes are characteristically deep and cold, and occur principally in northern LATITUDES or at high altitude. They contain few MICROORGANISMS and little plant life, but may have a considerable population of cold-adapted first such as trout, white-fish, and grayling. *Compare* MESOTROPHIC LAKE. *See also* TROPHIC LEVEL; EUTROPHICATION.

olivine (chrysolite) a common rock-forming MINERAL, green to olive-brown in color, with a glassy LUSTER, a HARDNESS of 6.5 to 7, and a SPECIFIC GRAVITY of 3.3. Translucent, gem-quality olivine is known as *peridot.* Olivine is primarily composed of magnesium silicate (*forsterite,* Mg_2SiO_4) with a small percentage of iron silicate (*fayalite,* Fe_2SiO_4) mixed in. The balance between the two constituents is known as the *Fo:Fa ratio.* The mineral is a common constituent of IGNEOUS ROCKS formed deep in the earth. *Compare* PYROXENE. *See also* PERIDOTITE; BOWEN SERIES.

Olmsted, Frederick Law (1822–1903) American landscape architect. Olmsted, acknowledged to be the father of American LANDSCAPE ARCHITECTURE and a foremost apologist for landscape aesthetics, had a varied career before he found his calling. After studying engineering and science at Yale University, he was, in turn, an apprentice civil engineer, a seaman, a Connecticut farmer, a Staten Island nurseryman, and a journalist. In this last pursuit, he was indeed gifted. His book on the pre–Civil War South, *Cotton Kingdom,* first published in 1861, is still in print. In between jobs he traveled extensively in Europe studying landscaping, gardening, and agriculture. Finally, at the age of 35, Olmsted landed an administrative position as the superintendent of the unimproved site that had been designated to be a "central" park for New York City. During this period he befriended British architect Calvert Vaux, who had immigrated to America and was working as a garden designer. When a design competition was announced for developing Central Park, he and Vaux entered a design conceiving of the park as a pastoral retreat for city dwellers. They won the commission. Olmsted's work as a landscape architect (a term he favored as describing a combination of garden design and architecture) was interrupted by the Civil War, during which he served briefly as secretary of the Washington, D.C.–based Sanitary Commission, devoted to the "sick and wounded of the Army." Returning to work on Central Park in 1863, he and Vaux soon resigned after a dispute with city officials. It was during this period that he founded *The Nation,* with a reform-minded journal that is still published, with Edwin L. Godkin. The partnership with Vaux was reinstated by park officials in 1865, but Olmsted, who needed money, took a leave of absence in favor of a job managing the California landholdings and gold-mining operations of John Charles Frémont, the famed explorer, army general, governor, and presidential candidate. The mining operations soon went broke, and Olmsted returned to landscape architecture—for Yosemite Park (then state-owned), and for several commissions in the San Francisco Bay area. He finally returned to the East to resume his partnership with Vaux on a new commission, Prospect Park in Brooklyn. After that Olmsted never again wandered from his profession. His firm undertook hundreds of commissions for public parks and gardens throughout the United States, including Riverside Park in New York City; Jackson Park and Washington Park in Chicago; and the grounds of the Capitol in Washington, D.C. His design approach—to preserve where possible, or recreate where preservation is not possible, a naturalistic landscape—has defined the practice of landscape architecture in America ever since. Laura Wood Roper's two-volume *FLO: A Biography of Frederick Law Olmsted* (1973) is a comprehensive academic study of his life. Elizabeth Stevenson's *Park Maker: A Life of Frederick Law Olmsted* provides a first-rate account of Olmsted's career. *Frederick Law Olmsted: Designing the American Landscape* (1995), with text by Charles E. Beveridge and photographs by Paul Rocheleau, is a lavishly illustrated book on Olmsted's designs. And Witold Rybczynski's book, *A Clearing in the Distance: Frederick Law Olmsted* (1999), provides the most recent exploration of Olmsted's works.

Olson, Sigurd F. (1899–1982) American ecologist, activist, and writer, born April 4, 1899, in Chicago and educated at Northland College (Ashland, Wisconsin) (B.S., 1920) and the University of Illinois (M.S., 1931). In 1922, Olson took a job teaching biology at Ely Junior College in northern Minnesota, on the edge of the Quetico-Superior canoe country; he devoted most of his spare time for the rest of his life exploring by canoe from Ely north to the Arctic Ocean and working actively for the preservation of the opportunity for others to continue these explorations. Though not a member of the organizing committee of the WILDERNESS SOCIETY, he became a member of the society during its first year of existence and made it the principal focus of his activities, serving on its executive committee for many years and as its president from 1968 to 1971: he also served as president of the National Parks Association (1954–60) and as a fellow of the Association of Interpretive Naturalists. He retired from Ely in 1945 to pursue writing and activism full time. Olson's writing style was intense and vivid, and all of his work reflected his passion for wild places in general and for wilderness lakes in particular; his many books all play like variations on a single theme, as if they were one seamless work spread over a number of volumes. The best of them, *The Singing Wilderness* (1956), ranks among the finest pieces of nature writing of the 20th century. Sigurd Olson died on January 13, 1982, of a heart attack suffered while snowshoeing near his home in Ely. His will provided a bequest to Northland College to establish what is now the Sigurd F. Olson Environment Institute.

OMB *See* OFFICE OF MANAGEMENT AND THE BUDGET.

omnibus bill in politics, a legislative bill containing numerous separate, unrelated or loosely related provisions, each of which could be an individual bill, but which are lumped together under one title and voted on as a whole with a single yes or no vote. Omnibus bills are most commonly used in the areas of APPROPRIATIONS and project AUTHORIZATION. *See also* CHRISTMAS-TREE BILL; LOGROLLING.

omnivore an ORGANISM that eats both plant and animal food; that is, one that is both a primary and a secondary consumer (*see* TROPHIC LEVELS). Omnivores have teeth adapted for both grinding (as in the HERBIVORES) and gnawing (as in the CARNIVORES), and have digestive tracts that are no longer than those of the carnivores but typically lack the multichambered stomachs and other specialized digestic apparatus of the herbivores. Examples of omnivores include humans, bears, pigs, and most Corvidae (jays, magpies and crows).

oncogenic substance in medicine, a substance that causes the formation of tumors. Since not all tumors are cancerous, not all oncogenic substances are CARCINOGENS, though there is a high correlation between the two. Nearly all viral particles (*see* VIRUS) that are oncogenic cause cancerous tumors, so the term *oncogenic virus* has come to mean a virus that causes cancer. The term comes from the theory, only partially substantiated, that all living cells contain a repressed tumor-growing GENE, the *oncogene,* which is "derepressed" by the oncogenic agent, causing the growth of the tumor.

One Thousand Friends (1,000 Friends) citizen organizations in several states (notably Oregon, Florida, and New Mexico) formed to discourage URBAN SPRAWL and the destruction of natural and aesthetic resources by indiscriminate development. More than simply "antisprawl," the 1,000 Friends organizations encourage what they call "smart growth" planning, which is based on the principles of the NEW URBANISM—infill development, open space systems, mass transit, and the like. The first group, 1,000 Friends of Oregon, was started in the 1970s by lawyers and planners in support of Oregon's pioneering land use legislation. *See also* OREGON LAND USE LAWS.

on-site disposal detoxifying, lagooning (*see* LAGOON) or otherwise permanently disposing of HAZARDOUS WASTES on the site where they were created, rather than transporting them to a remote disposal site. On-site disposal avoids spills, leaks, and other environmental dangers involved in transporting hazardous materials, and prevents the formation of "superdumps" filled with the wastes of numerous industries, which may become unmanageable due to sheer size; however, it complicates oversight of waste-disposal practices, encourages cost-cutting disposal methods, and poses a greater threat to GROUNDWATER because the limited size of factory grounds generally preclude the consideration of AQUIFER SENSITIVITY in locating disposal sites.

OPEC *See* ORGANIZATION OF PETROLEUM EXPORTING COUNTRIES.

open-pit mine *See* STRIPMINE.

open space preservation a major focus of a large number of environmental organizations and of suburban municipal governments beset with rapid population growth and development. Preserving open space (alternatively *open lands, green space*) became an urgent priority in the immediate post–World War II years and has remained so to this day. After the end of that war, large numbers of returning servicemen began to form new families and needed a place to live. The war, and the economic depression that preceded it, had curtailed residential building, resulting in a severe housing shortage. Accordingly, developers bought up vast tracts of open farmland outside of major cities to create massive developments of identical single-family houses. In 1949, for example, a returning veteran, using the G.I. Bill, could purchase a three- or four-bedroom house just slightly beyond the pre-existing suburban fringe for less than $10,000 with no down payment required and a mortgage rate of 4%. From that point forward, the metropolitan "race for open space" was on, with the developers on one side acquiring land at low prices and conservationists on the other encouraging local, state, and even the federal government to purchase land for park and open space use before it was all gone. Given low-cost mortgages via the G.I. Bill, and (beginning in the mid-1950s) the creation of the Interstate Highway System, the federal government effectively subsidized the open-space-consuming sprawl pattern of development which continues to this day. Because municipalities and nature preservation organizations could not compete with developers for increasingly expensive land, a number of stratagems were devised to protect land without actually having to pay the full "fee-simple" (that is, "all rights in land") price for it. The earliest and most obvious technique was for suburban municipalities to revise zoning codes to require large lots on undeveloped land. One- or two-acre residential zoning became common, and some localities required a 10-acre lot for a single family house. This approach backfired, however, leading to the development in the boom years of the 1960s of greater and greater acreages for fewer and fewer families, thus resulting in even more sprawl and consequent loss of open space. Other techniques created by land conservationists included the acquisition or transfer of DEVELOPMENT RIGHTS, as opposed to purchasing the land in fee simple. Also, programs were initiated to induce wealthy landowners to donate their estates, or parts of them, for nature sanctuaries or similar uses. For land that had to be developed, some—but not all—conservationists encouraged so-called cluster development, whereby houses would be clustered in 1/4-acre lots on, say, a one-acre-zoned tract which could then produce unbuilt-upon land dedicated as open space. These same techniques, perfected somewhat, are in use today for open space preservation. But they were then, and are now, of limited utility in the absence of affirmative urban development policies to encourage a more centralized and compact form of metropolitan growth. In the 1990s a new round of sprawl development began—a third major iteration, after the crises of the late 1940s and mid-1960s, since World War II—consuming more open space for less housing than ever before. *See also* URBAN SPRAWL; NEW URBANISM.

opportunistic organism in medicine, a bacteria (*see* BACTERIUM) or other MICROORGANISM that is harmless to its host (*see* HOST SPECIES) under normal conditions, but which may cause disease or infection if the host is weakened in any way: in other words, a microorganism that does not usually cause disease but can do so when the opportunity arises. An opportunistic organism is often—but not always—a normal resident on or in the host's body, and may in fact be beneficial as long as the host is healthy. Conditions allowing opportunistic organisms to become pathogenic (*see* PATHOGEN) include (but are not limited to) the presence of other diseases; genetic disorders or other inherited conditions; alcoholism; exhaustion or HYPOTHERMIA; suppression of the immune system due to chemotherapy, AIDS, or other reasons; and chronic exposure to hazardous chemicals or to radiation. The term is also used in ecology to refer to animals with unspecialized food and HABITAT needs (such as jays, crows and ravens; gulls; pigeons; and rats and mice) that can fill a variety of NICHES, depending upon what is available, and because of this adaptability are able to thrive around human habitations.

opportunity cost in economics, the true cost of a GOOD or SERVICE, measured in terms of foregone opportunities to put the resources tied up by the good or service to use in other ways. To spend $100 on theatre tickets means to forego not only the opportunity to spend the $100 on dinner at an expensive restaurant but also the opportunity to spend the evening of the performance at home reading a book. To build a road into a previously unroaded area is to forego not only the opportunity to spend the money for other purposes but also the opportunity to use the land for unroaded recreation. *See* BENEFIT/COST ANALYSIS.

optimal foraging theory as part of evolutionary ecology, describes how organisms optimize energy sources (forage) to provide for their success in adapting to any given environment. *See also* NATURAL SELECTION; EVOLUTIONARY ECOLOGY.

order in biology, a level of classification lying between the CLASS and the FAMILY. The difference between orders is the difference, for example, between perching birds such as sparrows, finches, and jays (order Passeriformes) and WATERFOWL such as ducks, geese, and swans (order Anseriformes). *See also* TAXONOMY.

order of magnitude in the sciences, a size difference expressed in powers of 10:100 is an order of magnitude larger than 10; 1,000 is an order of magnitude larger than 100; 0.1 is two orders of magnitude larger than 0.001; and so on. The term is usually used loosely as an indication of relative sizes by simply counting decimal places, a stone (for example) anywhere between 10 and 100 centimeters across being considered an order of magnitude larger than one anywhere between 1 and 10 centimeters across.

ore in geology or engineering, mineral-bearing rock (*see* MINERAL), especially that in which the concentration of minerals is high enough to be mined commercially. The *grade* of an ore depends on the ratio between the amount of mineral and the amount of GANGUE present in each sample. "Low-grade ore" is barely commercially viable while "high-grade ore" may in extreme cases be pure mineral. *See also* ORE BODY.

ore body in geology or engineering, a continuous rock body consisting of ORE of a more or less uniform grade, surrounded by COUNTRY ROCK of a different composition. The size, shape, and extent of ore bodies determine the economic potential of a mine nearly as much as does the grade of the ore itself, because they determine the mining methods. Ore bodies often occur in groups called *ore clusters* of similar age and composition, funneling outward from an originating body called the *root*. Determining the amount and accessibility of an ore body (the *ore expectant*) is the first step in determining whether the ore will be mined or not.

Oregon & California Lands (**O&C lands**) lands originally granted to the Oregon & California Railroad to build a railway from Portland, Oregon, to the California border. They were returned to the PUBLIC DOMAIN (*revested*) in 1916 by an act of Congress after the Supreme Court determined that the company had failed to construct the railroad and had disposed of some of the lands illegally. A second act of Congress, the Oregon & California Act of 1937, withdrew them from homesteading and sale and placed them permanently under the jurisdiction of the federal government. The O&C lands are historically important for being the first federal lands specifically designated (in the 1937 act) to be managed for SUSTAINED YIELD, and for the unique revenue-sharing formula under which they continue to be managed: 50% of revenue derived from them is returned to the counties in which they lie; 25% is returned to the State of Oregon; and 25% is returned to the federal treasury earmarked for management of the lands themselves. There are 2.3 million acres of O&C lands, all in Oregon. About 1.9 million acres of these land are administered by the BUREAU OF LAND MANAGEMENT, with the remainder under the jurisdiction of the FOREST SERVICE. *See also* RAILROAD LAND GRANTS.

Oregon land use laws a series of laws enacted during the early 1970s that pioneered the establishment of statewide standards for growth management and regional planning. The major objectives were to control major land use changes of greater than local concern (such as the conversion to farmland to a shopping center) and to control land uses affecting landscape resources of greater than local concern (such as an ocean beach). Although the laws developed to meet these objectives were complex, the key features were the establishment of statewide "exclusive farm use zoning" to be administered by local (usually county) authorities under the law; and the establishment of "urban growth boundaries," whereby urban services, such as transportation, sewage disposal, and the like could not be extended into countryside areas, thus limiting sprawl development. The Oregon experience helped planners and conservationists throughout the United States assess the potential for limiting sprawl by maintaining the viability of desirable land uses, which for a state like Oregon were important to its tourist economy. The leadership of Oregon governor THOMAS MCCALL was essential in the development and passage of the "land use package," as it was called, and the laws also spawned the emergence of the ONE THOUSAND FRIENDS civic organizations in several states that support such efforts. *See also* URBAN SPRAWL.

organic act a law passed by Congress to create a federal agency and define its powers and the scope of its operations. *See,* e.g., ORGANIC ADMINISTRATION ACT OF 1897; FEDERAL LAND POLICY AND MANAGEMENT ACT OF 1976.

Organic Administration Act of 1897 federal legislation establishing the authority of the FOREST SERVICE to manage the NATIONAL FOREST SYSTEM, signed into law by President William McKinley on June 4, 1897. Passed as a rider to the Sundry Civil Expenses Appropriation Act for Fiscal Year 1898, the Organic Administration Act confirmed the legality of the reserves set aside under the Creative Act of 1891 (*see* FOREST SERVICE: *History*); stated the purposes for which the NATIONAL FORESTS were to be managed ("to improve and protect the forest within the boundaries, or for the purpose of securing favorable conditions of water flows, and to furnish a continuous supply of timber for the use and necessities of citizens of the United States"); confirmed the right of public access to the National Forests; and gave the secretary of agriculture authority to "make such rules and regulations and establish such service as will insure the objects of such reservations [i.e., the National Forests], namely, to regulate their occupancy and use and to preserve the forests thereon from destruction". Civil and criminal penalties were provided for those who failed to follow the secretary's rules. Other provisions confirmed the president's ability to create new National Forests (repealed in 1907) and established a procedure for returning to the PUBLIC DOMAIN lands within the National Forest System that could not be used for the purposes of timber or water supply management (repealed in 1976; *see* NATIONAL FOREST MANAGEMENT ACT OF 1976). *See also* WEEKS LAW; CLARKE-MCNARY ACT; MULTIPLE USE-SUSTAINED YIELD ACT OF 1960.

organic carbon in chemistry, CARBON as it exists in ORGANIC COMPOUNDS (COMPOUNDS with a carbon-carbon bond or a carbon-HYDROGEN bond; *compare* INORGANIC CARBON). The ratio of inorganic to organic carbon in a particular environment can be used as a measure of that environment's ability to support life (*see* PRIMARY PRODUCTIVITY).

organic compound in chemistry, originally, any COMPOUND found as a constituent of living matter. Since approximately 1830 the term has been redefined as any compound that contains CARBON. More specifically, an organic compound is a compound containing a carbon-carbon bond, a carbon-hydrogen bond, or both—a definition that excludes such simple carbon compounds as CARBON MONOXIDE, CARBON DIOXIDE, and the CYANIDES. There are currently more than 10 million organic compounds known—roughly 50 times as many as the number of all other known compounds—with up to 500,000 more being created each year. *Compare* INORGANIC COMPOUND. *See also* AROMATIC COMPOUND; ALIPHATIC COMPOUND; HYDROCARBON.

organic deposition in geology, the formation of ROCK or MINERAL bodies by biological action. Organic deposition is responsible not only for such obvious organically derived materials as COAL and PETROLEUM, but also for chalk, LIMESTONE, and some metallic ORES (*see,* e.g., IRON BACTERIA).

organic gardening and farming agricultural and horticultural practices that feature the use of organic rather than synthetic materials for fertilization and pest control, and tillage methods that maintain the integrity of the soil. The modern organic gardening and farming movement dates from the 1940s and was championed by Sir Albert Howard in England (*The Soil and Health: A Study of Organic Farming*) and J. I. Rodale in the United States. Their books and, later, Rodale's magazines (notably *Organic Gardening*) created a small coterie of enthusiasts concerned with health issues as well as horticulture. The impetus was the introduction of mechanized agriculture in the 1920s, and then the increasing use of artificial fertilizers and chemical pesticides during the 1930s and 1940s, which became especially prevalent in the years following World War II when wartime pesticides (especially DDT) and herbicides were adapted for agricultural use and shown to increase yields significantly. While organic gardening at the hobby level presented no difficulties, commercial organic farming remained very much a fringe enterprise through the mid-1980s, at which time the environmental impacts of agricultural chemicals (even though DDT and other persistent pesticides had been banned) became better known, and when alternative tillage techniques (*see* CONSERVATION TILLAGE) were introduced to reduce the loss of topsoil and build up organic matter. In 1990 the U.S. Congress passed the federal Organic Foods Production Act to set national standards for the production of organic foods. At length, the USDA proposed draft standards under the act in 1997, but these were deemed so permissive (allowing irradiation, sewage sludge, and genetic engineering to be used in organic agriculture) that they had to be entirely reworked. Some 280,000 public comments were received, most of them by advocates of stricter standards. Meanwhile, certification for organic growers has been handled by 15 state agencies and several dozen private organizations. Typically, certification requires that no prohibited pesticides or fertilizers are used and that a farm undergo a three-year transition period in which strict organic methods are used so that agricultural chemicals have been worked out of the farm's soils and waters through several cropping cycles. By the end of the 1990s, there were approximately 10,000 certified organic farms in the United States, constituting somewhat less that 1% of farmland. Organic farms are usually small, averaging 140 acres compared to the 487 acres of the average conventional farm. Even so, the USDA predicts a $6 billion market for organic products by 2001, and though that is still a tiny percentage of U.S. agriculture (total food expenditures exceeded $709 billion in 1997), the USDA believes the organic sector will grow by 20 to 25% annually. Organic foods cost, on average, 57% more than their conventional counterparts. This factor has led those who are skeptical of organic agriculture to claim that it will never succeed on an industrial basis because it cannot "feed the world" with low-cost products. Organic farmers counter that food production is not the problem, since there are grain surpluses in many countries but hundreds of millions of malnourished people nevertheless. What starving populations need, they maintain, is access to tools and techniques that would allow people to feed themselves. They point out that the tools and techniques of organic agriculture are much less costly in terms of capital and operating requirements than those required by heavily mechanized chemical agriculture. *See also* PERMACULTURE; ROBERT RODALE.

organic matter material that was formed by the bodily processes of an ORGANISM. The category includes leaves, twigs, stems, flowers, pollen, and other plant parts; FECES; MICROORGANISMS and small organisms such as DUCKWEED or LICHENS, living and dead; the bodies of animals, or body parts, such as insect wings or exoskeletons; etc. Whole live animals and large live plants are not generally included. Organic matter is a principal source of BIOCHEMICAL OXYGEN DEMAND, and of NUTRIENTS found in the water and on the soil. It also can physically clog waterworks and navigation corridors. It contains numerous complex ORGANIC COMPOUNDS which, though not usually hazardous themselves, can combine with other chemicals in the environment to form hazardous substances (*see*, e.g., TRIHALOMETHANE). *See also* POLYCYCLIC ORGANIC MATTER.

organism a living thing; a specific, individual member of a SPECIES. All organisms share four characteristics: (1) they are discrete units, easily separable from the environment around them; (2) they respond in some form to stimuli (by AVOIDANCE, growth, pursuit, etc.); (3) they have the ability to maintain themselves, by taking matter and energy from outside themselves and using it to build and repair their bodily structure and to keep a continuous series of chemical reactions going (the "life process"); and (4) they have the ability to reproduce themselves with exact or closely similar copies that also exhibit these four defining characteristics. Note that by this definition VIRUSES may not qualify as organisms, since they do not exhibit characteristic (3).

Organization of Petroleum Exporting Countries (OPEC) an international cartel of oil-producing nations established in 1960 to fix production quotas and prices for the world petroleum market. Member countries in 2000 were Algeria, Indonesia, Iran, Iraq, Kuwait, Libya, Nigeria, Qatar, Saudi Arabia, United Arab Emirates, and Venezuela. These countries hold 67% of the world's OIL RESERVES. Saudi Arabia has the largest reserve, at 26% of the total, followed by Iraq with 10%. Until 1973 OPEC members set the price of oil in consultation with petroleum companies. This practice was changed with the so-called Teheran Agreement of 1971, which effectively created the cartel and made the OIL EMBARGO of 1973–74 possible. By 1981, owing to the Iran-Iraq war, OPEC had raised prices to more than 10 times the pre-1973 level. Since then, however, the OPEC nations have had difficulty in agreeing on pricing, especially since the Persian Gulf

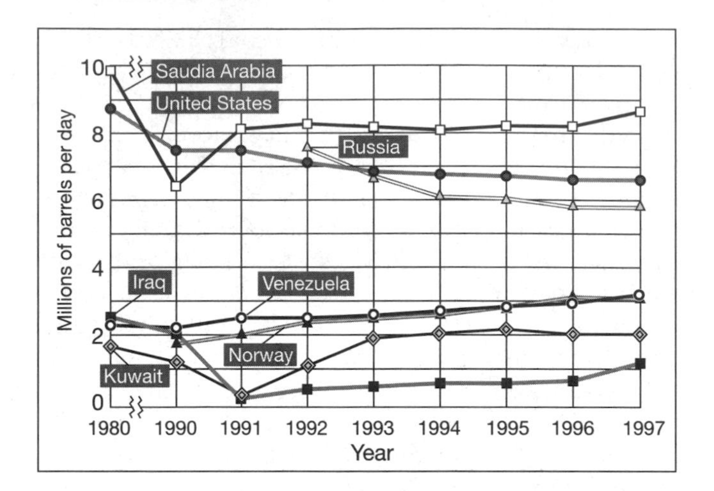

World oil production by major producing country ***(U.S. Energy Information Administration, International Energy Annual, 1997)***

War (1991), after which many members produced as much oil as they could, leading to a rapid decline in oil prices to all-time lows toward the end of the 1990s.

organochloride any ORGANIC COMPOUND that contains CHLORINE. Examples: DDT, VINYL CHLORIDE, the POLYCHLORINATED BIPHENYLS.

organophosphate any ORGANIC COMPOUND that contains at least one phosphate group (*see* GROUP; PHOSPHATE). Examples: malathion, parathion.

orogeny in geology, the building of mountains, especially those created by earth movement rather than vulcanism (*see* VOLCANO); also, the time period during which a particular mountain range was being formed (as: the *Appalachian orogeny;* the *Rocky Mountain orogeny*). It was once thought that the Earth went through periods of widespread orogeny alternating with periods of quiescence (*geomorphological cycles*), but this theory has been largely discredited (*see* PLATE TECTONICS).

ORV *See* OFF-ROAD VEHICLE.

OSHA *See* OCCUPATIONAL SAFETY AND HEALTH ADMINISTRATION.

OSHRC *See* OCCUPATIONAL SAFETY AND HEALTH REVIEW COMMISSION.

osmosis in chemistry and biology, the passage of water or another SOLVENT through a membrane in response to differences in the concentrations of dissolved materials on opposite sides of the membrane. In order for osmosis to take place, the membrane must be either semipermeable or differentially permeable; that is, it must contain a large number of microscopic holes ("micropores") that are big enough to pass small MOLECULES but not large ones. DIFFUSION PRESSURE causes the concentration ratio of SOLUTE to solvent to try to equalize on both sides of the membrane. Since the large molecules of the solute cannot pass through the holes in the membrane, the smaller molecules of the solvent pass through instead. This process continues until the diffusion pressure at the membrane's surface (the *osmotic pressure*) is equalized by hydrostatic pressure caused by the greater number of molecules that end up on the side of the membrane with the higher solute concentration. Osmosis is extremely important in living systems, where it provides the principal means of transporting water through CELL walls, which are differentially permeable membranes. The rate of water transport through the cell wall (and hence, the concentration of the NUTRIENT solution within the cell) is controlled by the difference between the osmotic pressure on the outside of the cell and the turgor pressure within it (*see* TURGOR), a difference that is known as *osmotic potential. See also* REVERSE OSMOSIS.

out-of-basin transfer in water law, the collection and transport of water from one WATERSHED ("river basin") for use in another, usually unrelated watershed. Out-of-basin transfer is widely used as a means of obtaining extra water. Delaware River water is transferred to the Hudson Basin for use by New York City; Colorado River water is transported to the Platte Basin for use by Denver and to the Los Angeles Basin for use by metropolitan Los Angeles; Arkansas River water is transported to the Colorado Basin (and hence ultimately to the Platte and Los Angeles Basins as well) to augment the Colorado's flow; and so on. However, the procedure is environmentally questionable on several grounds. The temperature and chemistry of the incoming water is often sufficiently different from the local water to change the species makeup of the PLANKTON and BENTHOS, thus altering the FOOD CHAIN all the way to the top. New SPECIES may also be imported with the water and may drive local species to extinction, especially if the local POPULATIONS have been weakened by physical changes in the water. Finally, the source watershed from which the imported water is taken suffers the loss of that water, which usually leads to a lower WATER TABLE and to the death of vegetation whose roots can no longer reach the water and of animals that have depended on SPRINGS that the lowered water table causes to dry up. Hence, out-of-basin transport for water-supply purposes has been outlawed in some states, and should be used everywhere only as a last resort.

Outdoor Recreation, Bureau of *See* BUREAU OF OUTDOOR RECREATION.

Outdoor Recreation Resources Review Commission federal study commission created by Congress in 1958 to compile and review data concerning outdoor recreation needs and opportunities in the United States and to make recommendations on the roles that should be played by federal, state, and local governments and by private businesses in providing outdoor recreation. The commission's final report, submitted on September 1, 1962, led directly to the formation of the BUREAU OF OUTDOOR RECREATION.

outfall the point where SEWAGE, industrial wastes, or other piped waterborne waste matter is released into the environment, usually after treatment in a SEWAGE TREATMENT PLANT or other waste-treatment facility. Most outfalls release their wastes into water (*see* RECEIVING WATERS).

outwash deposit in geology, SEDIMENTS, GRAVELS, etc. deposited by the meltwaters of a GLACIER. Unlike TILL, which is deposited directly by the glacier in an unsorted form, outwash deposits are sorted by size, with the smaller materials higher in the deposit and farther from the source (in this case, the ice). Typical outwash deposits include *outwash plains,* formed by sheet RUNOFF from the glacier; *valley trains,* which are outwash plains confined between the walls of a previously-formed valley; KAMES; and ESKERS. *See also* KETTLE.

overburden in engineering, the rock, soil, and vegetation overlying an ORE BODY, especially that which must be removed in order to mine the ore; also, the soil and vegetation that must be removed to reach BEDROCK in order to use the rock to anchor the foundation of a building or a bridge, stabilize a roadbed, etc.

overdraft in hydrology, withdrawal of water from an AQUIFER at a faster rate than the aquifer can supply. *Local overdraft* results when the rate of withdrawal from a well exceeds the flow rate in the surrounding aquifer, creating a CONE OF DEPRESSION around the well. *Regional overdraft* takes place when the STATIC LEVEL in all wells of a region declines. It may be caused by overlapping cones of depression, or by the withdrawal of more water from an aquifer than is flowing back in through RECHARGE. Overdraft is a serious environmental problem, not only because it indicates that those dependent on the overdrawn supply are running out of water, but because it lowers the HYDROSTATIC PRESSURE in the aquifer, which may cause its quality to be degraded. *See* SALINE INTRUSION.

overgrazing occurs when the number of grazing animals exceeds the ability of rangeland to sustain itself. Overgrazing of domestic animals—sheep, goats, and cattle—has led to desertification of large areas of northern Africa and permanent changes around the entire Mediterranean basin as well as in the rangelands of the western United States. *See also* RANGE MANAGEMENT.

overmature timber in forestry, trees that are long past the age of CULMINATION of MEAN ANNUAL INCREMENT and are thus growing new volume ("putting on wood") slowly or not at all. The volume of the STAND as a whole may even be declining due to disease, insect infestations, etc. From the standpoint of environmental forestry, however, overmature timber should often be retained, as it represents a series of HABITAT types that are otherwise unavailable, increasing SPECIES diversity and therefore helping to stabilize the forest ECOSYSTEM. *See* OLD GROWTH.

overstory in forestry, the trees (or occasionally, shrubs) whose CROWNS form the uppermost layer of foliage of a STAND. The term is roughly synonymous with CANOPY, except that "canopy" refers just to the upper foliage layer, while "overstory" generally includes the whole tree. Overstory trees shade the younger trees of the UNDERSTORY, drastically slowing their growth; hence, removal of a mature overstory is one of the best silvicultural methods for speeding the overall growth rate of a stand. *See* RELEASE CUTTING.

overturn in limnology, the total mixing of a lake that occurs when the THERMOCLINE is either at the water's surface (*spring overturn*) or has been driven all the way to the lakebed (*fall overturn*). Between the spring and fall overturns, the lake is colder at the bottom than at the top: after the fall overturn, but before the spring overturn, it is colder at the top than at the bottom. The presence of overturns is important to keep OXYGEN and NUTRIENTS thoroughly mixed throughout the body of the lake's water. *See also* TROPHIC CLASSIFICATION SYSTEM; HYPOLIMNION; EPILIMNION; FIRST-ORDER LAKE; SECOND-ORDER LAKE; THIRD-ORDER LAKE.

ovipositor the organ with which a female insect lays her eggs. The ovipositor is located at the rear of the abdomen. It is often long, slender, and needlelike so that it can pierce vegetable or animal tissue to allow the eggs to be placed in a protected spot with plenty of NUTRIENTS for the growing larvae (*see* LARVA). The stinger of bees and wasps is morphologically a modified ovipositor.

Owens Valley valley in southeastern California that serves as the source of approximately 60% of the water used by residents of the city of Los Angeles. The valley is bordered on the west by the Sierra Nevada and on the east by the Inyo and White Mountains. It is approximately 130 kilometers (80 miles) long and from 30 to 60 kilometers wide (20 to 40 miles). Geologically it is a STRUCTURAL VALLEY, with SCARPS on both sides reaching more than 4,265 meters (14,000 feet) above SEA LEVEL, while the valley floor stands at about 1,200 meters (4,000 feet). The Owens Valley is drained by the Owens River, which has its headwaters near Mammoth Lakes, California, and flows generally south, emptying historically into Owens Lake, a large, shallow, saline body of water with no outlet other than EVAPORATION. Nearly all of the Owens River now flows into the 355-kilometer (222-mile) Los Angeles Aqueduct, which exports 666 CUBIC FEET PER SECOND (CFS) of Owens Valley water to the Los Angeles Basin. The river provides anywhere from 350 to 480 cfs: the difference is made up by pumping GROUNDWATER reserves, mostly from the upper end of the valley. Completed in 1913, the Los Angeles Aqueduct has been controversial from the beginning; it has been bombed at least 17 times, and in November 1924 it was briefly captured by a group of farmers who diverted the water back into Owens Lake for five days. Modern controversy has resolved primarily around the so-called "second barrel," a second aqueduct, parallel to the first one, which has increased the system's capacity by 50 percent. The second barrel was dedicated on June 26, 1970, and, critics charged, led to an immediately noticeable decrease in groundwater levels throughout the valley, drying up wells and springs and killing groundwater-dependent vegetation. Owens Valley residents filed suit on November 15, 1972, seeking to have Los Angeles enjoined from further groundwater withdrawals in the valley unless it could be proved that these withdrawals would not harm the valley's environment, and on June 27, 1973, the Third District Court of Appeals ruled in the valley residents' favor. Subsequent legal decisions fixed the maximum groundwater withdrawal rate at 140 cfs. Arguments continue to rage over precisely what constitutes "harm" to the valley's environment. In the late 1990s state and federal regulatory agencies pressed Los Angeles to allow some water to flow into the now-dry Owens lakebed, once plied by steamboats that carried silver ore across its 112-square-mile expanse. Today, windstorms whipping across the dry lakebed can carry salts and toxic materials at levels up to 20 times the particle pollution allowed by law, making Owens Lake one of the worst sources of AIR POLLUTION in the United States. *See also* MONO LAKE.

owl pellet the fur, feathers, bones, shells, and other indigestible parts of an owl's PREY, formed into a compact mass in the bird's stomach and regurgitated through the gullet. (This regurgitation should not be thought of as vomiting; although the two procedures are mechanically identical, vomiting is the result of disease or other stress on the system, while the regurgitation of pellets is a routine activity.) The examination of owl pellets is useful not only to determine the owl's diet but to help identify the small animals present in a region and to get some rough idea of their numbers. Nocturnal rodents, for example, may be extremely adept at hiding from humans but may nevertheless be captured by owls, their presence in the owl's hunting RANGE being confirmed by their presence in owl pellets.

OWP *See* OFFICE OF WATER POLICY.

oxbow lake in geology, a lake formed when a river cuts through the neck of a MEANDER ("oxbow curve") in its course, leaving the former meander closed off from the river but still full of water. Since meanders almost always occur in NUTRIENT-rich alluvial soil (*see* ALLUVIUM) rather than in rock, oxbow lakes tend to be shallow and eutrophic (*see* EUTROPHICATION) and to fill in rapidly. The fish, INVERTEBRATES, and vegetation that inhabit these lakes are largely those that were present in the meander when it was cut off from the rest of the river, hence tend to be those characteristic of slow-moving rivers rather than those characteristic of lakes.

oxidation in chemistry, originally, the combination of OXYGEN with another ELEMENT or a COMPOUND; today defined as the loss of ELECTRONS by an ELEMENT during the course of a chemical REACTION, regardless of whether or not oxygen is involved. Oxidation is usually an EXOTHERMIC REACTION. The oxidation of CARBON and its compounds by oxygen during the process of RESPIRATION is responsible not only for the construction of many of the compounds found in living tissue but also for the release of much of the energy used by living organisms and is thus the main basis for the continuation of life on Earth. *Compare* REDUCTION. *See also* REDOX; OXIDIZING AGENT.

oxide in chemistry, any BINARY COMPOUND containing OXYGEN as one of its two constituent ELEMENTS. Almost all elements form or can be made to form oxides, the only exceptions being the so-called "noble gases": helium, argon, radon, neon, and krypton. Most oxides of METALS are BASES and most oxides of nonmetals are ACIDS, though a few of each class are amphoteric (that is, show characteristics of both acids and bases). Common oxides include WATER (H_2O); CARBON DIOXIDE (CO_2); and iron rust (Fe_2O_3). *See also* OXIDATION.

oxidizing agent in chemistry, an ELEMENT or COMPOUND that receives ELECTRONS during a chemical REACTION, thus causing the OXIDATION of the element or compound from which the electrons are moved. A "good" oxidizing agent is one that will easily oxidize numerous other materials. The primary example of such a material is OXYGEN; hence, the name "oxidizing agent." The ability of a material to act as an oxidizing agent is related to its *electronegativity,* that is, its ability to attract electrons to itself. *Compare* REDUCING AGENT.

oxygen the eighth ELEMENT of the ATOMIC SERIES, atomic weight 15.9994, chemical symbol O. Oxygen is the most abundant element on the Earth, forming roughly 24% of the ATMOSPHERE, 48% of the LITHOSPHERE, and 88% of the HYDROSPHERE. Its extremely high electronegativity (*see* OXIDIZING AGENT) means that it combines well with other substances, forming stable COMPOUNDS with nearly all other elements (*see* OXIDE). It also combines well with itself. Ordinary molecular oxygen is formed of two oxygen ATOMS joined together (chemical symbol O_2), and OZONE, a common ALLOTROPIC FORM of oxygen, consists of three atoms of oxygen joined together (chemical symbol O_3). Oxygen is necessary for all life. Even ANAEROBIC BACTERIA, which are said to live "without oxygen," are made up primarily of compounds containing oxygen, obtained in this case by breaking down other oxygen-containing compounds (obtained primarily from the bodies of dead air-breathing ORGANISMS) through the process of anaerobic RESPIRATION. *See also* DISSOLVED OXYGEN; OXYGEN SAG CURVE.

oxygen sag curve in water quality technology, a graph showing changes in the amount of DISSOLVED OXYGEN in a stream as the distance below a pollution source increases. Pollution typically raises a stream's oxygen demand (*see* BIOCHEMICAL OXYGEN DEMAND; CHEMICAL OXYGEN DEMAND), leading to a decrease in the dissolved oxygen. This oxygen deficiency then increases the rate at which oxygen from the air dissolves in the stream, which slowly brings the oxygen level back up to normal again at some point downstream from the

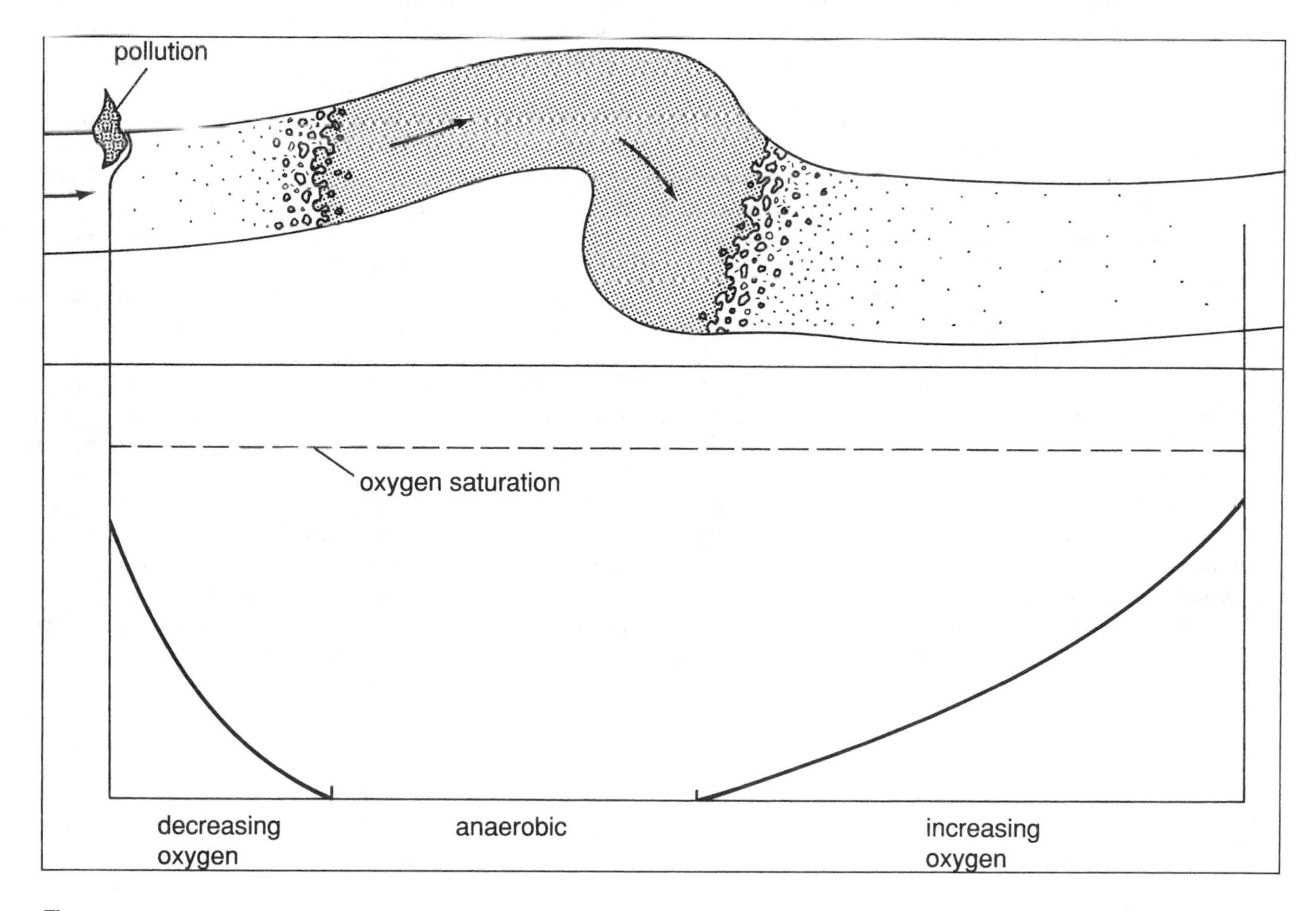

The oxygen sag curve

pollution source. The oxygen sag curve is thus generally assymmetrical in shape, plunging steeply downward at the source of the pollution and slowly climbing again downstream. In extreme cases the curve may be discontinuous, plunging all the way to zero and remaining at that level for some distance, until the oxygen demand of the pollution being carried downstream sinks low enough for the stream to begin reaerating (*see* ANAEROBIC STREAM).

ozone (O_3) an ALLTROPIC FORM of OXYGEN in which the MOLECULES contain three ATOMS each rather than the two atoms per molecule that the gas normally has. Ozone is a bluish, unstable gas with a characteristic "electric spark" odor (electrical discharges in air are one way of creating the gas). Because it is a powerful oxidant (*see* OXIDATION), ozone is a serious air POLLUTANT, irritating respiratory passageways and altering cellular METABOLISM. Even a few parts per million can cause tissue damage in living ORGANISMS (symptoms in humans: dry throat, headache, disorientation, difficulty in breathing). It is a minor but important component of photochemical smog (*see under* SMOG). It is, however, also necessary for the existence of life on Earth due to its ability to absorb ultraviolet radiation and its tendency to collect in the upper atmosphere (*see* OZONE LAYER).

Ozone Action nonprofit civic organization that in 2000 merged with GREENPEACE USA focused on global warming and stratospheric ozone depletion. Founded in 1992, Ozone Action served to educate the public and provided current information on these topics to the media, with a special emphasis on countering attempts by industry to distort facts. Its publications included *Ozone Action News; Ties That Bind,* and *Climate Change Current Effects Summaries.*

ozone layer, ozone shield a band of increased OZONE in the Earth's ATMOSPHERE, centered roughly 25 kilometers above the surface and extending about 15 kilometers above and below the area of highest concentration. At its peak, the ratio of ozone to normal diatomic OXYGEN is about 1:100,000. The ozone layer is created by the bombardment of the ATMOSPHERE by ultraviolet rays (sunlight in the band immediately above the visible part of the spectrum). The radiation rips O_2 molecules apart, each separate O atom then combining with another molecule of O_2 to form O_3. Ozone is transparent to normal wavelength radiation but is nearly opaque to ultraviolet; hence the presence of the ozone layer is crucial to protect the Earth's surface from excess ultraviolet radiation. Recent evidence suggests strongly that some human activities, including especially the production and release of CHLOROFLUOROCARBONS (CFCs), is damaging the ozone layer, the strongest evidence to date being an ozone "hole" that has developed over Antarctica each spring as the so-called "polar stratospheric clouds" (PSCs) disperse. The size of the hole waxes and wanes in near-perfect correspondence with the concentration of CFCs in the polar atmosphere. It is thought that these wholly man-made products reach the Antarctic through atmospheric circulation and become trapped on the surfaces of the ice crystals that make up the PSCs, where they apparently undergo reactions with NITROGEN OXIDES trapped on the same crystals to create ozone-destroying chlorine monoxide (ClO). In 1999, according to the World Meteorological Organization, the Antarctic's seasonal ozone hole covered an area of 16 million square miles, about 2.5 times the size of Europe. Because of the curtailment of chlorofluorocarbon production under the MONTREAL PROTOCOL, however, scientists hope that within the first decade of 2000 the Antarctic ozone layer will begin to heal, returning to normal at about midcentury. Nevertheless, in recent years significant ozone loss has also been recorded over the North Pole, possibly the result of the buildup of greenhouse gasses over the Northern Hemisphere. While the GREENHOUSE EFFECT warms up the lower atmosphere by trapping heat, the upper atmosphere, where the ozone layer is located, is cooled. The cooling increases the rate of decay of the ozone molecule, which becomes unstable in extremely low temperatures. If the stratospheric cooling trend continues, Arctic ozone depletion may be twice as high as it might have been without increases in greenhouse gas emissions. In fact, some scientists fear that, given increasing carbon dioxide emissions from fossil fuel use and the burning of tropical rain forests, the upper atmosphere could become cold enough that ozone molecules will decay on their own, in a continual feedback loop, even without the presence of additional CFCs.

Pacific Crest Trail hiking and horseback-riding trail along the crest of the Cascade, Klamath, and Sierra Nevada Mountains in the western United States, stretching approximately 2,600 miles from the Mexican border near Campo, California, to the Canadian border at Manning Provincial Park, British Columbia. Elevations along the trail range from near sea level at the Columbia River between Oregon and Washington to 13,200 feet near Mt. Whitney, California. Although the trail was actively advocated as early as 1926 by Catherine Montgomery, a student at Western Washington State College in Bellingham, Washington, and was mapped in 1935 by Clinton Clarke of Pasadena, California, it did not officially exist until passage of the NATIONAL SCENIC TRAILS ACT in 1968—and then only as a concept. The final trail segment was not completed until 1988 (near Mt. Ashland, Oregon). The Pacific Crest Trail lies almost completely on lands managed by the BUREAU OF LAND MANAGEMENT and the FOREST SERVICE, and trail construction and maintenance has been done primarily by those two agencies. *Compare* APPALACHIAN TRAIL.

Pacific Salmon Commission bilateral body created in 1985 to implement the Pacific Salmon Treaty between the United States and Canada. The commission provides recommendations for managing the Pacific salmon stocks and works to resolve problems related to this management. It works to provide equity in salmon harvest between the two countries based on the relative numbers of salmon produced in their respective waters. Membership consists of four appointed commissioners and four alternates from both the United States and Canada. Address: 1155 Robson Street, Suite 600, Vancouver, British Columbia V6E 1B5 Canada. Phone: (604) 684–8081.

Pacific States Marine Fisheries Commission promotes coordinated conservation, development, and management of marine and fisheries resources in the region which includes Alaska, California, Idaho, Oregon, and Washington. Founded in 1947, the commission conducts research and monitors fishery practice with a special concentration on marine debris, preserving fisheries habitat, and marine mammal/fishery relationships. Address: 45 SE 82nd Drive, Suite 100, Gladstone, OR 97027. Phone: (503) 650–5400.

Pacific Whale Foundation nonprofit organization founded in 1980 and dedicated to preserving whales, dolphins, and their aquatic habitats generally; also conducts research and promotes education and conservation specifically regarding Pacific whales. With its award-winning Ocean Outreach Program, the foundation has reached more than 700,000 people with programs on marine conservation, educational materials for schools, and research internships, plus "Adopt-a-Whale" and "Adopt-a-Dolphin" programs. Address: 101 North Kihei Road, Kihei, HI 96753. Phone: (808) 879–8860.

pack ice flat, consolidated masses of floating ICE spreading over the surface of a sea or large lake. Pack ice results from the freezing of the surface of a body of water under cold-weather conditions, and is usually broken up into numerous ice plates, or *floes,* which shift with the winds and currents, pulling apart to create narrow passageways of open water of driving against each other to form long linear ridges of humped-up ice called PRESSURE RIDGES. The permanent ice packs of the Arctic and Antarctic can grow to as much as four meters in thickness. They significantly increase the earth's ALBEDO at the poles, thus playing a major role in the determination of climate and weather patterns over the entire globe. *Compare* ICEBERG; ICE SHELF.

pahoehoe a type of LAVA characterized by a ropy, smooth surface. Pahoehoe flows tends to be low in silicates and high in dissolved gases, and they extrude at a higher temperature than AA flows. They move rapidly downslope, often forming caves as the molten interior of the flow runs out from under the already-hardened skin (*see* LAVA TUBE). *See also* MAGMA; VOLCANO.

palatable species in range or wildlife management, a SPECIES of plant that is preferred over other species by browsing or grazing animals, and which will be eaten first if it is available. Palatability is determined by a complex set of factors, including the general chemistry of the plant, the effects of soil chemistry and water supply, the GROWTH FORM (especially as it relates to the proportion of edible to inedible plant parts and to the presence or absence of protective devices such as thorns), and the nutritional needs of the grazing or browsing species. In general, it appears to correlate most strongly with succulence (water content: *see* SUCCULENT), PROTEIN content, and percentage of ASH. The animal's preferences as determined by heredity, of course, also play a large part: FORBS are preferred over GRASSES by sheep, and grasses over forbs by cattle, though the two species have similar nutritional needs. Palatable species are always the first to disappear from an overgrazed range (*see* RANGE MANAGEMENT).

Palearctic region *See under* HOLARCTIC REGION.

paleo- a prefix or combining form meaning *ancient.* It is generally used to refer to study fields or types of vanished phenomena whose characteristics have to be inferred from a study of FOSSILS or other geological evidence.

palmate form in botany, a GROWTH FORM of plants that is characterized by deeply cleft leaf lobes radiating outward from a center. The lobes may be either portions of a single leaf (as in the maple) or individual leaves whose stems radiate from a single node (as in the buckeye or horsechestnut).

pampas *See* GRASSLAND BIOME.

pantheism, panentheism theological beliefs that divinity is to be found in nature—the creation—here on Earth and throughout the universe. Pantheism stands in opposition to both theism (a personal God) and deism (an impersonal "clockmaker" God)—beliefs that generally suggest a deity separate from, and above, creation rather than part of it. Pan*en*theism, a term coined by 19th-century German philosopher K. C. F. Krause, combines pantheism with theism wherein God is not a personality but an all-inclusive essence, *Wesen,* which contains the universe within itself. Theologian Matthew Fox, a former Dominican and now Episcopal priest, says that panentheism, especially as practiced by medieval mystics (such as Mechtild of Magdeburg, Hildegard of Bingen, Julian of Norwich, and FRANCIS OF ASSISI), is essentially ecological, since it emphasizes mutuality and interdependence among species, including humans, within the natural environment. Theism, on the other hand, as traditionally promulgated by most Western religions, places the human in an exclusive, privileged relationship with God, which in the view of some critics of organized religion has led to the wanton destruction of nature and the overexploitation of natural resources. *See also* RELIGION AND THE ENVIRONMENT.

PAOT *See* PEOPLE AT ONE TIME.

paper recycling the collection of waste paper and the remanufacturing process for making usable paper products. The collected waste paper is first soaked, which in effect returns it to a pulp SLURRY, and ink, glue, and coatings are removed. Then new pulp is added to give the product sufficient tensile strength, and the slurry is pressed into new paper. In 1998 43% of waste paper was recycled in the United States, reducing pressure on both landfills and forests. Moreover, recycled paper manufacturing is more energy efficient, reduces AIR POLLUTION, and saves water. *See also* RECYCLING; SOLID WASTE.

parallel on a map, a line connecting points of equal LATITUDE, so called because all lines constructed in this manner lie parallel to each other and to the equator.

parallel evolution in the biological sciences, the development of similar characteristics in related but distinctly separate SPECIES as a result of similar environmental pressures, such as a change of conditions that sets in after the two species have diverged. Several similar species of moth in the English Midlands, for example, underwent melanism (darkening of pigment) during the industrial revolution, enabling them to hide better on soot-covered surfaces. *Compare* CONVERGENT EVOLUTION; COEVOLUTION.

paramecium (plural **paramecia**) a large GENUS of one-celled ORGANISMS, KINGDOM PROTISTA, PHYLUM Ciliophora, CLASS Oligohymenophorea, ORDER Hymenostomatida (*see* TAXONOMY). Paramecia are common inhabitants of freshwater lakes and ponds, where the form a major portion of the PLANKTON. They feed on bacteria (*see* BACTERIUM) and are an important means of keeping bacterial numbers under control, hence are a valuable presence in SEWAGE TREATMENT PLANTS and polluted waters in general. They are oval in form and covered with beating filaments called *cilia* that drive them through the water with a spiral motion. Their AVOIDANCE behavior is the simplest known. It consists, whatever the stimulus, of backing up, turning 15° to the left, and moving forward again.

parasite in the biological sciences, any ORGANISM that depends on the metabolic processes of another living organism to maintain itself. The supporting organism (the "host") is usually weakened and is sometimes killed by the parasite. Parasitism differs from predation (*see* PREDATOR), however, in that the parasite does not consume the host but instead diverts some of the products of the host's life systems and uses them to its own ends. (Exceptions to the "does not consume" rule are the so-called *parasitoids* such as the larvae of the ichneumon wasp, which consume a living host from within.) Parasites are generally highly specialized organisms and are often adapted to life on a specific host (*see,* e.g., MISTLETOE). Some (*heteroecious parasites*) utilize two hosts, living most of their lives in one (the *intermediate host*) but reproducing in another (the *definitive host*). Such parasites are usually far less specific in their choice of intermediate host than they are in their choice of definitive host, an adaptation that allows them to survive in a greater range of conditions than those (*autoecious parasites*) that utilize a single host for their entire life cycle. Parasites form an important means of population control for many organisms and are thus a necessary part of a balanced ECOSYSTEM. *Compare* SAPROBE. *See also* ECTOPARASITE; ENDOPARASITE.

parent material in soil science, the material from which a soil, or a particular HORIZON of a complex soil, has been derived. Parent material is classified both according to its chemical and physical makeup and according to the means by which it was transported to the site where the soil developed; thus the parent material of a soil may include BEDROCK of various types as well as ALLUVIUM, LOESS, or glacial deposits (*see* TILL; DRIFT; OUTWASH DEPOSIT) derived from either nearby or distant locations. It may also include ORGANIC MATTER such as tree limbs, leaves and needles, animal and insect bodies, etc. Different parts of the same soil may derive from different parent materials: for example, an A horizon derived from loess may overlay a B horizon developed from glacial till and a C horizon developed from bedrock. The properties of the parent materials largely determine the properties of the soil, though this influence declines with time and may be negligible or non-existent in very old, well-developed soils. *See also* PARENT ROCK.

parent rock in geology and soil science, the rock from which the MINERAL materials were derived to make either a soil or another rock. The properties of the parent rock are reflected in the soil or the secondary rock in such aspects as grain size, degree of consolidation, and chemical makeup. FELDSPAR, for example, weathers into CLAY and reconsolidates into SHALE, while QUARTZ weathers into SAND and reconsolidates into SANDSTONE. Often chemical "fingerprints" enable the parent rock of a soil or a sedimentary SERIES to be traced to a particular "parent outcrop," providing useful clues to prehistoric drainage patterns and climates. *See also* PARENT MATERIAL.

Pareto optimum *See* MAXIMUM SOCIAL WELFARE.

Park Grass Experiment a famous long-term study that suggests the essential stability of ecosystems. The original study, which ran from 1856 to 1872 in Hertfordshire, England, evaluated various fertilizer treatments on a hay meadow to determine their effects on plant growth. After the study scientists continued to monitor the meadow ecosystem, recording changes in the composition of plants, soil structure, microorganisms and the like. Because of the period of time during which the meadow has been carefully monitored—now nearly 150 years—ecologists have been able to draw conclusions about the inherent stability of terrestrial biotic communities. The consensus is that at the macro scale ecosystems do tend toward stability, in that over time the proportion of grasses to other plants has remained constant. The hay meadow has remained a hay meadow. At the micro scale, however, there has been quite a lot of variation in the *species* of grass and other plant families represented. In effect, then, the Park Grass Experiment confirms that ecosystems do not reach stasis as a CLIMAX COMMUNITY but can achieve an overall "dynamic equilibrium." *See also* BALANCE OF NATURE.

parthenogenesis in the biological sciences, the development of an unfertilized egg into a complete individual. Parthenogenesis has the advantage of being both rapid and independent of contact with other members of the reproducing organism's SPECIES. It is therefore useful for rapid POPULATION expansion in new areas, and is employed for this purpose by a number of insects and plants, especially PIONEER

SPECIES (such as the dandelion) and social species (such as many members of the HYMENOPTERA). APHIDS employ it in order to spread their population rapidly in advantageous environments, utilizing sexual reproduction only in the fall, apparently to create hardier eggs that can live through the winter before hatching (parthenogenesis, like all forms of INBREEDING, tends to accentuate weak traits). Since no genetic material is exchanged in parthenogenesis, the individuals that result are exact genetic copies of the individual that laid the eggs; thus they are almost always females, although when parthenogenesis has been forced in birds in the laboratory it has resulted not in females but in sterile males.

partial retention of character in FOREST SERVICE usage, a class of landscape in which management activities such as roads and logging may be visible but must not dominate the view, which remains substantially natural in character. It is the highest level of modification allowed for class A landscapes and for all but the least sensitive regions of class B landscapes (*see* VISUAL RESOURCE MANAGEMENT SYSTEM).

particulate in air pollution control, a particle of solid matter or droplet of liquid small enough to remain suspended in the air for a long period. Particulates range in diameter from roughly 0.01 MICRON upwards to just under 1,000 microns (1.0 millimeters). They are classed according to both size and composition. *Dust particles* are solids ranging from 1 micron upwards, including such materials as road dust, coal dust, sawdust, etc. *Smoke* and *fumes* are smaller solid particles, ranging in size from .01 to 1 micron. *Smoke* includes all solid CARBON compounds in this size range, while *fumes* includes all non-carbonaceous solids. Both smoke and fumes are generally the result of combustion, either incomplete burning (smoke) or the condensation of hot metallic vapors into tiny solid droplets (fumes). Liquid particulates include *mists,* formed by condensation and ranging roughly from .5 to 10 microns in size, and *sprays,* formed by the mechanical breaking-up of larger drops (as in an atomizer) and ranging in size from 10 microns upward.

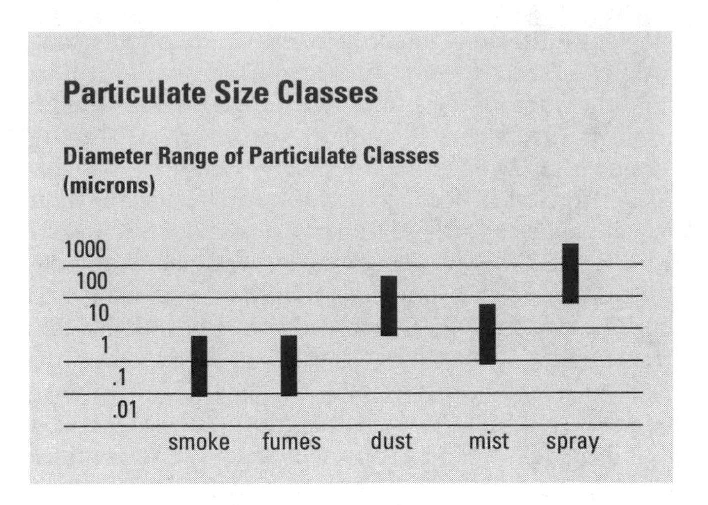

EFFECTS OF PARTICULATE POLLUTION

Particulates are among the most harmful air pollutants from an economic standpoint. They irritate mucus membranes and eye tissue, coat lung linings (*see* EMPHYSEMA) and cut visibility. Drawn into air intakes, they damage heating equipment and internal-combustion engines; settling out of the sky, they coat and corrode surfaces; and swept up by rainfall, they contribute to ACID RAIN. More than 40 million tons of particulates enter the air of the United States each year from all sources. Of this total, nearly 30 million is dust thrown up from unpaved roads; 6.1 million is vehicular emissions (fuel combustion products, tire and brake wear, etc.); 5.8 million results from industrial processes, including fuel combustion, coal processing, milling and manufacturing, and similar sources; and the remainder is split among various minor sources, including agriculture (plowing; grain storage; etc.), home heating (wood stoves and oil burners), electrical generation, solid waste disposal, and so on. *See also* AIR POLLUTION: *types of pollutants;* also various articles on pollution measurement and control technology, e.g. BAGHOUSE; SCRUBBER; HIGH VOLUME SAMPLER; TAPE SAMPLER; etc.

passenger pigeon a now-extinct dove-sized species (*Ectopisted migratorius*) that was once the most abundant of all North American land birds. In the early 1800s ornithologist Alexander Wilson estimated one flock at 2.23 billion birds, which took four hours to fly overhead. Routinely darkening the skies, such giant flocks migrated between the southeastern United States to nesting areas stretching from New York and Pennsylvania to Middle Western and Great Lakes states. A primary food source for American Indians, the bird also became a favorite of colonists, especially in pigeon pie. With the rapid clearing of the virgin eastern forest, the bird was deprived of nesting sites, and its flocking habit made it an easy kill by 19th-century fowlers, who used shotguns, nets, snares—even dynamite—in a wanton market-hunting slaughter that is decried to this day. As passenger pigeon populations diminished, the per-bird retail price in New York City increased from $.50 each in 1855 to $3.50 in 1884, making the pigeon a true luxury item. One of the last flocks of passenger pigeons was sighted by naturalist John Burroughs in 1907, but by then the bird was assumed to be virtually extinct. The last surviving individual, named Martha, died at the Cincinnati Zoological Garden in 1914. *See also* EXTINCTION.

passeriformes *See* PASSERINE.

passerine in ornithology, any bird belonging to the ORDER Passeriformes, or "perching birds." Passerines range in size from the kinglet (wingspread 16–18 centimeters) to the raven (wingspread 1–1.5 meters). Their most outstanding common characteristic is the shape of the foot, which is adapted for perching on tree branches, twigs, etc. It has three toes in front and one somewhat longer toe behind, and contains a trigger reflex that causes the toes to curl around and grasp an object when pressure is put on the underside of the foot. Sparrows, warblers, jays, thrushes, finches, larks, and sandpipers are all passerines, which are generally considered the most intelligent, adaptive, and highly evolved of all birds. The order contains 59 FAMILIES and about 5,100 SPECIES, or roughly 3/5 of all known bird species.

passive recreation *See* DISPERSED RECREATION.

passive solar design any SOLAR ENERGY installation in which solar heat is used directly for space heating rather than being converted into another form (steam; electricity; hot water) for transportation to the point where it will be used. Passive solar installations are generally simply large south-facing windows, combined with a thermally dense storage device such as a concrete floor or a water-filled wall (*see* TROMBE WALL) to store excess heat and re-release it when the sun is not shining in the windows. They are much simpler and cheaper to install than are ACTIVE SOLAR DESIGNS, and have no moving parts to break down; however, they cannot be used to heat rooms away from the sun or to generate other forms of energy such as electricity. Most solar heating installations today are passive solar designs.

pastureland land used for grazing livestock as opposed to land used for crops within a farm ownership. Usually the term is distinguished in the United States from RANGELAND, which implies large areas of land, often in public ownership. *See also* AGRICULTURAL LAND, U.S.

patch in LANDSCAPE ECOLOGY any unitary element of a landscape such as a forest, marsh, farm, or townscape that with other elements forms a mosaic making up the landscape structure of a region as a whole.

patch cut in forestry, a small CLEARCUT, usually with the borders left purposefully ragged to make the cutover unit blend with the topography and allow it to act ecologically as if it were a natural opening in the forest. Patch cuts do not include the entire STAND of timber they are taken from. The timber surrounding a patch cut will typically be left unlogged at least until REGENERATION is firmly enough established to provide shade cover for the forest floor. Patch cuts are typically 2 to 10 acres in size.

paternoster lakes in geology, a series of two or more lakes occupying stairstepped CIRQUES at the head of a glacially carved mountain valley (*see* GLACIER; GLACIAL LANDFORM).

pathogen an ORGANISM or a viral particle (*see* VIRUS) that causes disease in other organisms. Most pathogens are either viruses or MICROORGANISMS (bacteria [*see* BACTERIUM] or PROTISTA). To class as a pathogen, an organism or virus must (a) have some means of invading another organism (the *host*); (b) have some means of temporarily overcoming (or not triggering) the host's defensive systems; (c) be able to reproduce and spread within the host; and (d) cause the host to display an infirmity of some sort. The damage-causing mechanisms used by pathogens are extremely varied, ranging from the production of TOXINS and ALLERGENS to the disruption of CELL METABOLISM and reproductive mechanisms. *Pathogenicity* refers to the likelihood that a pathogen will cause disease. *Virulence* refers to the strength of the symptoms caused, a virulent pathogen being one that causes relatively great damage to the host with relatively few pathogenic organisms present. *Compare* OPPORTUNISTIC ORGANISM.

patterned ground in geology, soil deposits showing regular, rhythmic geometrical patterns that have been created by natural forces. On level surfaces, patterned ground usually takes the form of circles or polygons whose boundaries touch one another; on slopes, alternating stripes are usually formed. The features may be from less than a meter to about 10 meters across (3–35 feet): all patterned-ground features on a given slope or plain are usually roughly the same size. Those outlined with stones are called *sorted patterns,* while those outlined by vegetation and water features are called *non-sorted.* Patterned ground is characteristic of polar regions, high mountains, and other areas where heavy frosts are common. Its formation is not fully understood, but it appears to depend upon FROST HEAVE and related phenomena involving repeated freezing and thawing of the soil.

PBB *See* POLYBROMINATED BIPHENYL.

PCB *See* POLYCHLORINATED BIPHENYL.

PCDF *See* FURAN.

PCDD *See* DIOXIN.

PCP *See* PENTACHLOROPHENOL.

peak load (peak demand) in utility management, the maximum demand placed on a utility system during a given time period. Peak load is generally fairly predictable, both as to the time it will occur and the intensity it will reach. Residential water use, for example, reaches its annual peak during the summer and its daily peak during the period 4:00–6:00 P.M., both due to the increase in residential lawn watering that occurs at those times. Peak load is often twice as high as average use, and is three to four times as high as BASE LOAD. Utility systems must have enough capacity to meet peak load requirements; however, much of this capacity sits idle much of the time. Shifting demand away from the peak and into these periods that have excess supply capacity available is therefore a means of promoting growth without requiring new utilities. *See also* PUMPED STORAGE.

peat the layer of ORGANIC MATTER found at the bottom of old, well-developed BOGS. Peat is composed of the compacted remains of bog plants, primarily sphagnum mosses (*see* MOSS). Dried, it is soft and absorbent, and is used as garden mulch. Bricks of it also may serve as fuel or (in wood- and rock-poor regions) as building materials. Buried and compacted peat deposits metamorphose into COAL.

pebble in engineering geology, a rock fragment smaller than a COBBLE but larger than a grain of SAND. The generally accepted size range is from 2 to 64 millimeters in diameter (roughly 1/16 inch to 2 1/2 inches). *See also* WENTWORTH SCALE.

ped in soil science, a clump of soil particles that adhere naturally to each other. The shape of a ped is characteristic of the soil from which it is formed. Terms used by soil scientists to describe the shape of peds include *granular* or *crumbly* (small angular or spherical units); *blocky* (larger units which are roughly equal-sided), and which may be either angular (*angular blocky*) or rounded (*subangular blocky*); *prismatic* (vertically stretched blocks with flat tops and sides); *columnar* (prismatic peds with rounded tops instead of flat ones); and *platy* (peds arranged as flat, overlapping flakes). Clumps of soil particles that result from human land-use activity such as plowing or shoveling are termed *clods,* and usually consist of several peds (or parts of peds) lumped together.

pedicel in botany, the stalk of an individual FLOWER within an INFLORESCENCE. Pedicels branch off from PEDUNCLES and end in receptacles. *See* FLOWER: *parts of a flower.*

pedon in soil science, the smallest structural unit in which all of the characteristics of a soil may be recognized, that is, the smallest unit in which all HORIZONS are present and within which all characteristic surface relief features and vertical structures such as cracks, SLICKENSIDES, etc. can develop. Pedons are three-dimensional, with a surface area of 1–10 square meters (depending on the soil type) and a vertical profile reaching from the surface to BEDROCK. A soil body as a whole may be thought of as being put together from a number of separate pedons, and is therefore termed by soil scientists a POLYPEDON (*poly*="many"+*pedon*).

peduncle in botany, the stalk of an INFLORESCENCE or of a solitary FLOWER. *Compare* PEDICEL.

pegmatite in geology and mineralogy, a ROCK composed of particularly large MINERAL crystals, usually an inch or more across, but always significantly larger than other rocks of the same mineral composition. Pegmatites are usually chemically identical to GRANITE ("granite pegmatites"), but occasionally may be akin to other rock types such as syenite or diorite instead. They are intrusive in origin (or occasionally metamorphic; *see* INTRUSIVE ROCK; METAMORPHIC ROCK), and form near the surface at lower temperatures and pressures, hence are common constituents of DIKES and SILLS. Among the larger-than-normal crystalline constituents of pegmatites are found garnets, rubies, sapphires, tourmaline, and other gemstones; hence, pegmatite bodies are actively sought for commercial development by gem miners.

pelagic sediment an ocean-floor SEDIMENT that does not have a direct origin from shore or dryland processes. Pelagic sediments include the shells of DIATOMS and other small organisms; organic debris such as the bodies of PLANKTON that reach the bottom and become incorporated into the sediments before they are consumed; the products of the chemical and physical erosion of underwater rock structures such as the MIDOCEAN RIDGE, the continental slope (*see under* CONTINENTAL SHELF), and SEAMOUNTS; and SILTS, SANDS, etc. that may have had a continental origin but have been transported long distances and extensively modified since first entering the sea.

pelagic zone (pelagic division) the open water of the ocean. The term is generally taken to mean the entire body of the ocean water from low tide outward and from the floor upward, although some authorities restrict it to the upper portion of the deep-sea waters, that is, the top 200 meters or so (*see* PHOTIC ZONE) of that portion of the ocean that lies beyond the margins of the CONTINENTAL SHELF. The principal life forms of the pelagic zone are the PLANKTON and NEKTON. *Compare* LITTORAL ZONE; BENTHIC DIVISION; NERITIC ZONE.

peneplain in geology, a plain created by a long period of EROSION that is not interrupted by uplift, faulting, or other TECTONIC ACTIVITY. Peneplains result from the wearing down of mountains and the filling in of valleys; they often have isolated stumps of mountains, known as *monadnocks,* rising from them. Rivers, where present, meander broadly (*see* MEANDER). *See also* BASE LEVEL.

penstock a closed pipe that carries water from a source such as a lake, reservoir, river, or canal to a power-generating device such as a turbine or waterwheel. The HEAD available to a powerplant depends primarily on the elevation difference between the bottom end of the penstock and the surface level of the reservoir or other body that the water is drawn from (minus losses to friction within the penstock) rather than being directly related to the depth of the reservoir. Hence, small reservoirs with long penstocks can normally generate about the same amount of power as can large reservoirs with short penstocks, usually with less harm to the environment, although when a large proportion of a river's flow is diverted through a penstock instead of flowing through the bed of the river it can cause serious damage to aquatic and riparian life. This is especially true if the amount diverted through the penstock tends to fluctuate broadly over the course of a day, a situation typical of the operation of hydropower plants that must meet PEAK LOAD requirements. An additional benefit of long penstocks is the ability they provide to site powerhouses separate from the dams they are associated with, allowing much greater planning flexibility.

penta *See* PENTACHLOROPHENOL.

pentachlorophenol (penta, PCP) a CHLORINATED HYDROCARBON BIOCIDE, chemical formula C_6HCl_5O, once used widely as a defoliant, a general HERBICIDE, and an ingredient of paints and stains designed to protect wooden structures against termites, woodboring bees, etc. It is a powerful human poison as well, either ingested or absorbed through the skin, causing (in either case) a preliminary increase in respiration, blood pressure, and urination, followed by a suppression of all these functions, loss of bowel control, weakness, convulsions, and death. It is highly soluble in organic SOLVENTS such as ether or ALCOHOL, which also increase its level of activity. Its solubility in water is extremely low, but it is nevertheless listed as a PRIORITY POLLUTANT by the ENVIRONMENTAL PROTECTION AGENCY due to its high potency. LD_{50} (rats, orally): 146–175 mg/kg.

people at one time (PAOT, persons at one time) in recreational management, the number of individuals present in a given area at the same time. *Maximum PAOT* represents the capacity of an area, that is, the total number of persons who can use the area without degrading either the physical environment or the recreational experience of other users. *Compare* VISITOR DAY.

pepo *See under* FRUIT.

perched water in GROUNDWATER management, a body of groundwater sitting on top of a layer of impermeable material (*see* PERMEABILITY; IMPERMEABLE LAYER) that raises it above the regional WATER TABLE. Perched water is discontinuous from the main body of groundwater, and often comes from a different source. It may result from the local downflow of GRAVITATIONAL WATER through the soil, for example, while the general regional water table lies in an AQUIFER that is recharged a dozen miles or more away. Perched water bodies are generally small, subject to seasonal fluctuation, and particularly vulnerable to infiltration by POLLUTANTS, and should not normally be trusted as a steady water source.

perched water table the upper limit of the saturated zone of a body of PERCHED WATER. *See* WATER TABLE.

perceived noise level (PNdB) in NOISE-POLLUTION control, the logarithmic average of noise over a 24-hour period, as measured at the observer's position. The unit of measurement is dB(A) (*see* DECIBEL; A-WEIGHTED SOUND SCALE). *See also* NOISE AND NUMBER INDEX.

percolation in general, the passage of a liquid through a permeable solid (*see* PERMEABILITY) or through a mass of unconsolidated material such as SOIL, SAND, GRAVEL, etc. Percolation is usually slow, and includes an element of CAPILLARY ACTION as well as gravitational flow from pore to pore of the material. From an environmental standpoint, the most important use of the word is in relation to the speed of GROUNDWATER movement through a soil body or an AQUIFER. This speed, known as the *percolation* *rate* (or *perc rate*), is a principal factor involved in predicting the effects of CONES OF DEPRESSION and pollutant PLUMES on groundwater supply, as well as determining the suitability of soils for septic systems (*see* SEPTIC TANK). *See also* PERC TEST.

perc test in sanitary engineering, a test to determine the percolation rate of a soil (*see under* PERCOLATION) and therefore to assess its ability to support a SEPTIC TANK. The test is usually run in the following manner. Picking a site typical of the soils in the area of the proposed septic DRAINFIELD, dig a hole 20–40 cm across and as deep as the drainfield will extend, lining the bottom of the pit with gravel. Fill the pit with water, let it stand overnight, and then fill it again to a level approximately 15 cm above the gravel. The percolation rate may now be measured directly as the rate at which the water level drops in the hole, expressed in any convenient unit, such as centimeters/minute or inches/minute. Since the size and location of the hole, depth of the gravel, etc., can vary considerably from test to test, perc test results should always be considered approximate.

perennial in botany, a plant that lives for a number of years. In contrast to ANNUALS and BIENNIALS, whose life span in genetically predetermined, perennials—like animals—live either until they are killed or until they "wear out" metabolically (*see* METABOLISM). Two classes of perennials are generally recognized. *Herbaceous perennials* die back each year, losing their entire SHOOT and surviving the winter as a BULB, TUBER, or other underground structure; they ordinarily flower during each year of their lives (examples: tulip, iris, camas). *Woody perennials* maintain a permanent above-ground structure, though they generally reduce metabolic activity during the winter. They ordinarily flower only upon reaching maturity, which may take a number of years (examples: oak, pine, hazelnut, rhododendron). Reproductively, perennials tend to be K-SELECTED SPECIES that colonize new areas slowly but hang on tenaciously once they have become established (*see* SUCCESSION).

perennial stream a stream that contains water all year round, from its source to its mouth. Perennial streams are more technically defined as those streams whose beds lie below the local WATER TABLE throughout the year and which are thus able to maintain a permanent BASEFLOW. *Compare* EPHEMERAL STREAM; INTERMITTENT STREAM.

perfection in ecology, a state of a BIOGEOCHEMICAL CYCLE in which the ELEMENT involved in the cycle progresses steadily from one stage to the next, without being sidetracked or "trapped" for long periods in one particular form. The degree of perfection of a cycle tends to be characteristic for each element. The OXYGEN cycle, for example, is normally very nearly perfect due to the steady rate at which most ORGANISMS respirate and metabolize (*see* RESPIRATION; METABOLISM), while the calcium cycle is highly imperfect due to the tendency of calcium to get tied up in skeletal structures, shells, and CALCIC ROCK (LIMESTONE and MARBLE). Generally speaking, the more perfect a cycle, the higher the percentage of that element's total planetary supply is available for use by living organisms at any one time.

peridotite in geology, an INTRUSIVE ROCK formed largely of OLIVINE, with small amounts of other minerals present (AMPHIBOLE, PYROXENE, etc.). It contains no FELDSPAR. Peridotite is a massive, heavy rock, dark green to black in color, which may weather to a dull buff-red. It grades into SERPENTINE, which is its chief metamorphic product (*see* METAMORPHIC ROCK). It is formed very deep in the earth, and may be the principal constituent of the MANTLE. Soils weathered from peridotite pose special problems for plants; hence, peridotite outcrops often show unusual or rare plant SPECIES and ASSOCIATIONS. The rock itself is often rich in nickel and/or chromium, and is (in the variety known as *kimberlite*) the only known natural source of diamonds.

periodic table in chemistry, a table displaying the ATOMIC SERIES as a two-dimensional matrix in which each horizontal row ("period") represents ELEMENTS with the same number of ELECTRON shells (that is, the *nth* row contains all the elements with *n* electron shells; *see* ATOM: *structure of the atom*) and each vertical column ("group") represents a set of elements with similar chemical properties, progressing from the elements with incomplete outer shells on the left to those with relatively complete outer shells on the right. The rows vary in length due to the increased number of electrons possible in each shell as one moves outward from the nucleus. (The number of electrons possible is equal to to $2n^2$, where n = the shell number, yielding the progression 2, 8, 18, 32, 50, 72, 98. The actual number of electrons in each row beyond the first two is less than this due to various other constraints on the precise number of electrons that can circle an atomic nucleus. These constraints yield rows 2, 8, 8, 16, 16, 32, and 32 elements long.) The chemical and physical activity of an element can be predicted with some certainty from its position in the periodic table.

peripheral species in land-use planning, a SPECIES of animal or plant whose natural RANGE barely overlaps the boundary of the planning area, so that it is rare or endangered *within the planning area.* Whether or not it is an ENDANGERED SPECIES in absolute terms is not relevant to the designation. Peripheral species should generally be protected in the planning process, both because they represent an extreme (and therefore, possibly variant) POPULATION of the species itself and because they demonstrate the presence of an unusual ECOTYPE for the planning area that should be preserved in the interest of variety if nothing else.

permaculture a method of agriculture in which the object is to utilize ecological principals to produce a SUSTAINED YIELD of renewable GOODS without affecting the environment or decreasing its productivity in any way. Developed by Australian ecologist Bill Mollison—who based many of his ideas on those of the Australian aborigines—and by Masanobu Fukuoka in Japan, permaculture involves planning a farm to operate as an integrated system of humans, domestic animals and plants, and wild animals and plants, emphasizing *diversity* (a wide variety of different SPECIES) and *complexity* (as many functional relationships between species as possible). Permaculture farms are designed in concentric *zones,* with the plants and animals needing the most attention, such as herb gardens, nearest the house (zone 1) and those needing little attention, such as large livestock and woodlots, farthest out (zone 4). If the farm is large enough to contain areas that can be left completely wild, these are placed in an external ring known as zone 5. Each zone is divided into *sectors* based on TOPOGRAPHY, SOILS, PREVAILING WINDS, and similar factors, and attempts are made to locate pastures, woodlots, RESERVOIRS, and so on in the most appropriate sector of the zone they fall into. Farming proceeds with as little interference with the natural cycle as possible, with crops intermingled and sown at times and in manners that will require little or no tilling or weeding. The idea is for the crops and livestock to help each other rather than be dependent upon human intervention. *See also* ORGANIC GARDENING AND FARMING.

permafrost permanently frozen ground, that is, soil or subsoil in which the soil moisture exists in the form of ice throughout the year. The depth at which the frozen soil lies during the hottest part of the year is known as the *permafrost table.* Permafrost underlies most of the world's TUNDRA, usually less than a meter down (about 3 feet): it ranges in thickness from a few centimeters to over 1,000 meters. It forms an impermeable barrier as effective as stone, confining liquid water to the thin layer above it and thus accounting for the extreme bogginess of much of the Arctic. When disturbed, it melts, adding extra water to the surrounding soil and increasing its bogginess. (Permafrost in some areas of the high Arctic is able to refreeze after being disturbed. This is referred to as *active permafrost.*) Building foundations cannot be laid in it because it flows under pressure, like glacial ice. It has been estimated that permafrost underlies approximately 20% of the Earth's dry land. It has also been found beneath part of the seafloor off the north coast of Alaska.

permanent stream in geology, *see* PERENNIAL STREAM.

permanent wilting percentage *See under* WILT POINT.

permeability in hydrology, the ability of a material to pass a liquid, such as water, from one part to another. Materials within which a liquid moves rapidly are termed *highly permeable.* Permeability depends upon the size and number of the connections between the openings in the material rather than on the size and number of the openings themselves. *Compare* POROSITY. *See also* PERCOLATION; PERMEABLE LAYER.

permeable layer in geology or hydrology, a STRATUM of rock, soil, sand, gravel, etc., which allows water to flow within it from one part to another (*see* PERMEABILITY). Solid rock that happens to be permeable (such as SANDSTONE) forms a *consolidated permeable layer:* broken rock, sand, soil, gravel, and so on form *unconsolidated permeable layers. Compare* IMPERMEABLE LAYER. *See also* AQUIFER; INTERFLOW ZONE.

permeable pipe pipe made of a material such as perforated metal, perforated plastic, unfired clay, etc., through which water can easily percolate (*see* PERCOLATION). Water will "leak into" permeable pipes buried in saturated soil; hence they are often buried around the outside of hazardous waste dumps to collect and drain off any LEACHATE that escapes from the confines of the dump.

permissible yield in forestry, *see* ALLOWABLE CUT.

persistent compound a COMPOUND that does not break down when exposed to biological or meteorological action, but remains chemically intact for many years after it has been released into the environment. *Compare* BIODEGRADABLE SUBSTANCE; GEODEGRADABLE SUBSTANCE.

persons at one time *See* PEOPLE AT ONE TIME.

pest control the elimination or management of insect pests considered harmful to humans and human activities such as agriculture. Pest control has been going on for the entire course of human history, but only in modern times has the practice had a widespread impact on the environment and human health. *See also* PESTICIDES.

pesticide technically, any substance used to kill crop pests, including weeds, insects, rodents, and fungi (*see* FUNGUS). Pesticides are classified according to the type of pest ORGANISMS they kill most easily, as *fungicides, rodenticides, insecticides,* and *herbicides.* They may be further classified according to their persistence in the environment ("hard," or persistent, vs. "soft," or easily broken down; *see* HARD PESTICIDE) or according to their chemical content. Principal categories include: *arsenicals,* those that contain ARSENIC or one of its compounds; *organophosphates,* ORGANIC COMPOUNDS containing PHOSPHORUS; *organochlorides,* organic compounds containing CHLORINE; and *biological pesticides,* those containing PATHOGENS of the pests one wishes to kill (*see* BIOLOGICAL CONTROL). (Some pesticides fall outside these chemical categories. Many of these are compounds of ZINC; zinc phosphate, for instance, is commonly used as a rodentcide, while zinc chloride is an effective fungicide.)

THE "PESTICIDE TREADMILL"

While there is some purpose in moderate use of soft pesticides to control IRRUPTIONS and to combat widespread populations of hard-to-eradicate pests, indiscriminate use of any of these materials usually proves to be self-defeating. The pests become resistant to the pesticide through natural selection, requiring heavier and heavier applications, which destroy the pest's natural enemies and thus increase dependence on the ever-less-effective pesticide. Nevertheless, while attention has been called to the dangers of this "pesticide treadmill" by Rachel Carson (*Silent Spring,* 1962) and others, and while some of the most biologically dangerous pesticides have been banned (MIREX, DDT, TOXAPHENE) in the United States, Canada, and a few other countries, use of these compounds continues to increase annually. Current estimates indicate that a billion pounds of pesticides are applied each year in the United States alone, and many other nations, especially in the Third World, use proportionately much more. *See also* BIOCIDE; DDT; and separate articles on HERBICIDE; ORGANOCHLORIDE; and ORGANOPHOSPHATE; ORGANIC GARDENING AND FARMING; GENETIC ENGINEERING.

pesticide resistant species *See* GENETIC ENGINEERING.

petal *See* FLOWER: *parts of a flower.*

Petkau Effect the phenomenon, discovered in 1972 by Abram Petkau of the Canadian Atomic Energy Commission, that continuous low doses of radiation, presumably below the danger level, can cause more damage to cell membranes in humans as well as plants than a single high dose of radiation. The finding, which has been largely ignored, calls into question the long-term safety of nuclear reactors for those living downwind, even when no accidents occur, and explains the observed decline of forests in areas near nuclear power plants. *See also* NUCLEAR ENERGY and associated entries; FOREST DECLINE AND PATHOLOGY.

petiole in botany, the stem of a leaf. The petiole connects the base of the leaf blade to the stem, branch, or twig of the plant. It differs from normal stem tissue in that it is triangular rather than round, with the vascular (liquid-conducting) tissue arranged in a V instead of a circle.

petrification the replacement of organic tissue by MINERAL deposits in such a way that the shape and structure of a vanished ORGANISM can be seen in the resulting rock. Petrification usually takes place when a dead organism is deeply buried by SILT or SAND that is then left undisturbed for several thousand years, during which time the tissues are slowly dissolved by GROUNDWATER and minerals are deposited in the spaces that result. *Permineralization* is a form of incomplete petrification in which the pores and softer tissues of an organism's body have been replaced by mineral deposits but the harder tissues remain. *See also* FOSSIL.

petroleum a mixture of complex ORGANIC COMPOUNDS found as a fluid beneath the surface of the Earth, where it apparently has formed from the organic-rich muck on the bottom of ancient seafloors that have been buried by TECTONIC ACTIVITY. Deposits of petroleum, often accompanied by SALTS and brine from the vanished seas, are found primarily in deep SANDSTONE and LIMESTONE formations, where they tend to move upward until becoming trapped by overlying impermeable rock (*see* IMPERMEABLE LAYER). *Structural traps* are ANTICLINES or inverted SYNCLINES in which the petroleum is trapped in the dome formed by the distorted rock layers. *Stratigraphic traps* are regions where two slanting impermeable layers enclosing an oil-bearing formation come closer and closer together until they pinch off the oil-bearing strata. *Fault traps* occur when a FAULT develops in an impermeable layer, trapping the upward-migrating petroleum in the offset.

PRODUCTION OF PETROLEUM

To remove petroleum from the earth, wells are drilled. These wells go through three stages of production. *Flush wells* are the equivalent of ARTESIAN WELLS for water production; the petroleum flows or gushes out of the well under hydraulic pressure. In *settled wells* the hydraulic pressure (*reservoir pressure*) is inadequate to force the oil to the surface, and pumping is required. In *stripper wells* (over two thirds of those now producing in North America) the reservoir is nearly empty, and the flow rate of oil into the well has decreased to less than 10 barrels per day. A fourth type of well, known as a *secondary recovery well,* is increasingly employed to force greater production from stripper deposits. In these wells, brine or NATURAL GAS is forced into the reservoir through one or more subsidiary wells (often old producing wells), creating enough reservoir pressure to bring the yield back up to an economically productive level. Before it can be used, petroleum must be refined; this is done chiefly by fractionation (*see* DISTILLATION), although cracking (*see* DESTRUCTIVE DISTILLATION), is commonly employed in conjunction with fractionation to break down some of the more complex fractions so that the refining process yields a higher proportion of GASOLINE.

PETROLEUM AND SOCIETY

Petroleum is one of the most widely used materials of modern society, not only as a fuel but as a raw material for numerous chemical substances (*petrochemicals*) including plastics, pharmaceuticals, and fertilizers, and the supply of it is decreasing rapidly. At current use rates, 50–90 years of oil reserves (discovered and undiscovered) remain. The United States controls 2.3% of these reserves. Exploration for new reserves continues, but—even with increasingly sophisticated technology being employed in the search—the average number of reserves discovered each year continues to decrease. With many of the remaining reserves lying in environmentally sensitive areas such as the Arctic and the Outer Continental Shelf, development has become increasingly controversial, and the outlook for society's continued reliance on petroleum is not promising. For the pollution hazards associated with oil, see OIL SPILL; for types of petroleum, see CRUDE OIL. *See also* FUEL OIL; KEROSENE; OIL SHALE; OIL RESERVES.

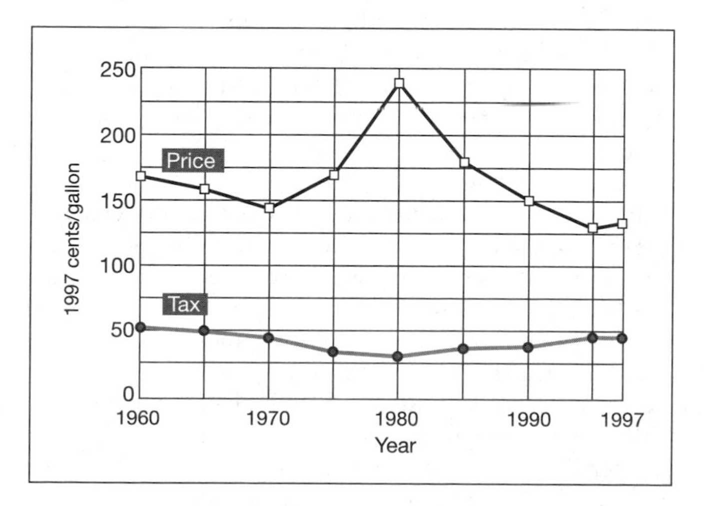

U.S. gasoline prices and taxes (federal and state)

pH literally potential of hydrogen, measuring the acidity or basicity of a liquid (*see* ACID; BASE). Technically, the pH (or "pH value") of a liquid is a measure of the number of HYDROGEN IONS present in the liquid (in mols/liter; *see* MOLECULAR WEIGHT) expressed as a negative logarithm. Thus pure water, with a concentration of 10^{-7} mols/liter of hydrogen ions, has a pH of 7, which is considered neutral. Higher hydrogen ion concentrations give lower pH readings (10^{-6} for example, is greater than 10^{-7} so a pH of 6 represents a higher hydrogen ion concentration than does a pH of 7) and also create more strongly acidic conditions, so pH readings below 7 indicate acidity and pH readings above 7 indicate basicity. Since pH is a logarithmic measure, each step in the scale indicates an increase or decrease by a factor of 10: pH 5 is 10 times as acidic as pH 6, pH 4 is 100 times as acidic as pH 6, and so on. The most accurate pH readings are taken with an electronic device, the *pH meter,* which measures hydrogen ion concentration in terms of the difference in electric potential between a pair of ELECTRODES dipped into the liquid whose pH is to be determined. However, relatively good readings can be obtained with *pH indicators,* such as LITMUS and similar materials, which change color in response to changing pH. This is especially true if several indicators with overlapping color ranges are used, allowing a "fix" within the overlapping ranges. Numerous biological and physical processes are closely dependent on pH, including the breeding cycles of fish and other aquatic life, the metabolic rates and viability of bacteria (*see* METABOLISM; BACTERIUM), the rate of formation of LIMESTONE, and so on. Hence, determination of the pH of watercourses and of the wastes which enter them is of utmost importance in pollution control.

phase-gap replacement *See* GAP PHASE REPLACEMENT.

phenocryst a large, brightly colored or otherwise conspicuous crystal that is structurally part of a rock (as opposed to a crystal formed within a pre-existing rock, such as a GEODE). Phenocrysts occur most often in INTRUSIVE ROCKS that have cooled rapidly, such as GRANITES found near the edges of BATHOLITHS. *See also* PORPHYRITIC ROCK.

phenol in chemistry, (1) a simple AROMATIC COMPOUND, chemical formula C_6H_5OH, consisting of a BENZENE RING with a hydroxyl (OH–) GROUP attached to it. Also known as *carbolic acid,* phenol is a white-to-clear crystalline solid that turns slightly reddish upon exposure to air. It liquefies readily upon the addition of water, so that a mixture of as little as 8% water to 92% phenol will be a liquid rather than a solid. The hydroxyl group gives it some of the properties of an alcohol. Phenol is an extremely toxic chemical; ingestion of as little as one gram can cause death. It absorbs easily through the skin, and must always be handled with gloves. Symptoms off phenol poisoning include nausea, vomiting, circulatory-system collapse, convulsions, and coma, followed by death from either respiratory or circulatory failure. In those cases where death does not occur, necrosis (tissue death) is common in areas exposed to the chemical. The victim's urine turns a smoky green color. Despite these hazards, the material is widely used, principally as a raw material in the manufacture of plastics, as a constituent of industrial disinfectants, and as a chemical REAGENT in industrial processes.

(2) a class of compounds formed by adding one or more extra functional GROUPS to the basic phenol MOLECULE. Examples include cresol ($C_6H_4OHCH_3$) and resorcinol ($C_6H_4(OH)_2$). Phenols may be combined as building blocks into larger organic molecules, in which case they are known as *phenolic rings* or *phenolics.*

phenyl in chemistry, a functional GROUP consisting of a BENZENE RING that has been stripped of one HYDROGEN ATOM (chemical formula: C_6H_5–). The phenyl group attaches to other groups or MOLECULES at the location of the missing hydrogen atom.

pheromones aromatic substances, often undetectable to humans, secreted by various animals, insects, and other creatures that can alter the behavior of individuals of the same species that come into contact with them. Pheromones are secreted by bark beetles, for example, to attract other beetles to a weakened tree that can be invaded for nesting. Also, pheromones (musk, for example) are secreted by many animals in an effort to attract a mate for breeding.

phloem in botany, the tissue system of a VASCULAR PLANT through which the sugar and other food substances (*assimilates*) formed in the leaves are conducted to other parts of the plant. The direction of flow of the assimilates is downward, from the leaves to the roots. The driving force of this flow appears to be the osmotic pressure differences (*see* OSMOSIS) that arise as the assimilates in the sap stream are added by the leaves and used up by the other living portions of the plant. However, the movement itself is massive; that is, the assimilates are carried in a flowing liquid rather than transported by diffusion from CELL to cell. In ANGIOSPERMS, phloem consists of *sieve tubes* composed of elongated cells (*sieve-tube members*) with hollow interiors and perforated end walls (*sieve plates*) through which flowing liquid can pass. Although the sieve-tube member are living cells, they lack a complete nucleus and other important functioning cell parts, and each must be aided by a *companion cell,* a complete cell that has been derived by division from the same "mother cell" as the sieve-tube member and is intimately connected with it. In other vascular plants, phloem consists of *sieve cells* that are perforated throughout rather than just on the end walls, and which overlap each other rather than being placed end to end. Like the sieve-tube members, they are incomplete, and are aided by associated complete cells, the *albuminous cells,* which function like companion cells but are derived from independent mother cells and are therefore not as closely connected to the sieve elements. In WOODY PLANTS the phloem forms the inner back and is separated from the XYLEM by the CAMBIUM LAYER. Most of the bark is formed from dead phloem in which the sieve elements have collapsed (*see* SECONDARY GROWTH). In other plants—and in the growing root and shoot tips of woody plants, where PRIMARY growth takes place—the phloem lies in bundles along the outside of the xylem, generally separated by ridges rising up from the xylem.

phosphate in chemistry, any COMPOUND containing the phosphate GROUP PO_4. Most phosphates are SALTS or ESTERS of phosphoric ACID (acids containing the anhydride group P_4O_{10}). They are critical to the life processes of all ORGANISMS, including VIRUSES and bacteria (*see* BACTERIUM); however, they can also act as POLLUTANTS due to their high level of biological activity. Phosphates are an important component of fertilizers, most commonly as *superphosphate* (monocalcium phosphate, $Ca(H_2PO_4)_2$), formed by treating PHOSPHATE ROCK with sulphuric acid.

See also PHOSPHORUS; EUTROPHICATION; LIMITING NUTRIENT.

phosphate rock (**rock phosphate**) a SEDIMENTARY ROCK composed principally of fluorapatite ($3[Ca_3(PO_4)_2]CaF_2$) mixed with SAND and CLAY, which is mined extensively as a source of phosphate fertilizers (*see* PHOSPHATE; FERTILIZER). Phosphate rock is a light, friable rock (*see* FRIABILITY) that normally occurs as thin layers interbedded with sand and clay above a limestone basement (*see* BASEMENT ROCK). Its origin is obscure, but it seems to be organic (*see* ORGANIC DEPOSITION). It is usually slightly radioactive (80–100 PICOCURIES/GRAM), and contributes some of this radioactivity to GROUNDWATER held in it, making the drilling of wells into phosphate AQUIFERS unwise. Some phosphate rock is ground up and used directly as fertilizer, but most is first chemically converted to superphosphate (*see under* PHOSPHATE). The waste from this conversion process, known as *phosphate slag,* is a low-level radioactive waste, making its common use as road ballast, cement filler, etc., unwise. Most phosphate rock in the United States is located in Florida. Annual U.S. production: about 45 metric tons.

phosphorus the 15th ELEMENT in the ATOMIC SERIES, ATOMIC WEIGHT 31, chemical symbol P. One of the elements necessary for life, phosphorus forms part of the DNA and RNA MOLECULES and is a component of teeth, bones, nerve tissue, and cell walls (*see* CELL). In the form of PHOSPHATES it acts as a buffer within the cell, controlling the cell's pH by accepting and releasing HYDROGEN IONS as needed. Highly chemically active, it does not occur in the free state in nature. When prepared in the free state, it exists in three ALLOTROPIC FORMS, white (or "yellow"), red, and black, differing from each other in crystalline structure, boiling point, melting point, and chemical activity. White phosphorus, the most active form, is among the most dangerous materials known due both to its biological activity (which causes severe "burns" on contact with organic tissue) and its chemical instability (it must be kept under water to prevent it from igniting spontaneously and explosively with the OXYGEN in the atmosphere.

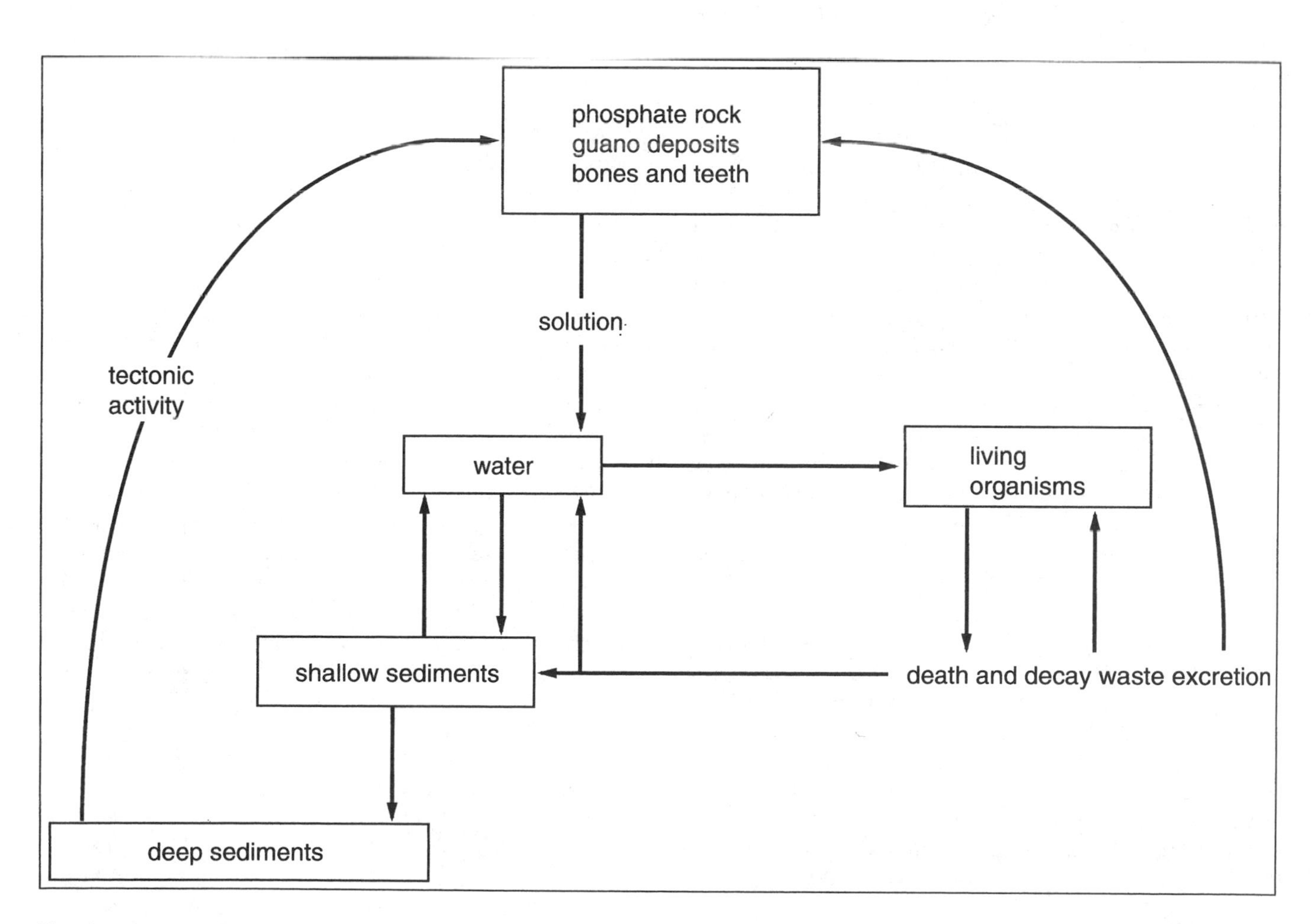

The phosphorus cycle

THE PHOSPHORUS CYCLE

Like all elements utilized in living processes, phosphorus cycles between organic and inorganic forms; that is, any given phosphorus ATOM will spend part of its time as a component of living tissue and part in an inorganic substance such as PHOSPHATE ROCK (*see* BIOGEOCHEMICAL CYCLE). Unlike the cycles of CARBON and other elements, however, the cycle of phosphorus is highly imperfect (*see* PERFECTION). Much of it is incorporated into SEDIMENTARY ROCKS in the deep ocean basins, where it cannot be recycled without geological upheavals, resulting in a continual increase of inorganic phosphorus and a decrease in the organic forms. Some ecologists fear that human activities have accentuated this imbalance by the destruction of seabird colonies (which have in the past been prime phosphate recyclers) and the overapplication of artificial FERTILIZERS, causing them to wash off into the rivers and thence to the sea instead of soaking into the soil for uptake by plants, which would maintain them in the biological phase of the cycle.

PHOSPHORUS AS A POLLUTANT

Because of its high biological activity and generally limited availability, phosphorus is the most common LIMITING NUTRIENT for aquatic ECOSYSTEMS; hence, any process that adds significant amounts of PHOSPHORUS compounds to a water body will generally dramatically increase the growth of plants, ALGAE and other photosynthetic organisms (*see* PHOTOSYNTHESIS). The amount of phosphorus in the water may be expressed in several different ways. The most common measurement is *total phosphorus*—the total amount of all phosphorus compounds present in the water. *Soluble phosphorus* is the percentage of the total phosphorus that will pass through a 0.45 MICRON filter, while *particulate phosphorus* is the percentage of the total phosphorus that the same filter will not pass. *Soluble reactive phosphorus* refers to phosphates; *soluble organic phosphorus* refers to the portion of the soluble phosphorus that is not composed of phosphates. The most common sources of increased phosphorus in water bodies are DETERGENT additives and agricultural RUNOFF.

photic zone in limnology, the region from the surface of a body of water downward to the lowest depth that enough light can penetrate to allow PHOTOSYNTHESIS to take place. The depth to which the photic zone extends varies according to the clarity of the water, and may be anywhere from a few centimeters to nearly 200 meters; however, the most common depth is between 3 and 5 meters. Most aquatic life is limited to the photic zone. *See also* COMPENSATION DEPTH.

photochemical smog *See* SMOG.

photoperiod in the biological sciences, the portion of a regular dark/light cycle during which light is available to living ORGANISMS. In nature, the photoperiod generally corresponds to the daylight hours, although it may be affected by other things; for example, plants growing within the mouth of a cave will have a photoperiod corresponding to the length of time that daylight is able to penetrate the cave. *See also* PHOTOPERIODISM.

photoperiodism in the biological sciences, the triggering of daily or seasonal changes in an ORGANISM's life cycle by changes in the length of the PHOTOPERIOD. Waking and sleeping are simple examples of photoperiodism, as are the daily opening and closing of certain flowers (e.g., poppies) and the DIEL MOVEMENT exhibited by many species of PLANKTON. More complex changes may be triggered by seasonal changes in the photoperiod, including bird migration, HIBERNATION, color changes in the leaves of DECIDUOUS PLANTS, etc. The most thoroughly studied form of complex photoperiodism is probably the flowering response in plants (*see* LONG-DAY PLANT; SHORT-DAY PLANT; DAY-NEUTRAL PLANT).

photosynthesis in biology, the synthesis of CARBOHYDRATES from CARBON DIOXIDE and WATER by living ORGANISMS utilizing light as an energy source. Photosynthesis is catalyzed by chlorophyll (*see* CATALYST; CHLOROPHYLL) and thus is limited to plants, green ALGAE, and those few other organisms (CYANOPHYTES; a few BACTERIA) that contain this complex MOLECULE. With a few very minor exceptions such as ANAEROBIC BACTERIA, all organisms that do not contain chlorophyll obtain their carbohydrates by consuming those who do, either at first or second hand (*see* FOOD WEB; FOOD CHAIN; PRODUCER; CONSUMER; TROPHIC LEVEL). Thus, photosynthesis is ultimately responsible for animal as well as plant life. It is also the principal source of atmospheric OXYGEN, which is released as a BYPRODUCT of the photosynthetic process. Photosynthesis is actually a complex group of chemical reactions divided into two phases, the "light reactions" in which chlorophyll is energized by light and in turn energizes a molecule of ATP (*see under* METABOLISM), and the "dark reactions" in which the energized ATP acts on the carbon dioxide and water to form carbohydrates, "resetting" both itself and the chlorophyll molecule back to their original, lower-energy levels. For convenience, the total set of reactions is usually summarized in a single chemical equation:

$$6CO_2+6H_2O+\mathit{light} \rightarrow C_6H_{12}O_6+6O_2$$

phototropism in botany, the alteration of the GROWTH FORM of a plant in response to light. Most plants are *positively phototrophic;* that is, they tend to grow toward the light, elongating and bending their SHOOTS so that their leaves and flowers get full-light exposure. The phenomenon appears to be controlled by the movement of AUXINS from light to dark areas of the plant, so that the CELLS in the dark areas are stimulated to lengthen faster than those in the light areas. This *differential growth* causes a bending of the stem toward the light. In the sunflowers, phototropism is photoperiodic in nature (*see* PHOTOPERIODISM), with differential growth proceeding around the stem on a daily basis, causing the stem to twist so that the flower head literally follows the Sun across the heavens. Phototropism is also responsible for the often strikingly uniform nature of the CANOPY of a forest, as the branches of the trees spread out to fill every space at the canopy level where light may be available.

photovoltaic (PV) cells *See* SOLAR ENERGY.

phreatic zones in the structure of a river or stream, the area beneath the streambed containing groundwater. *Compare* with BENTHIC, HYPORHEIC, and RIPARIAN ZONES.

phreatophyte in the biological sciences, a plant that obtains its water from the saturated zone of the soil (*see* WATER TABLE) or from the CAPILLARY FRINGE just above the saturated zone, rather than obtaining it from soil moisture. Phreatophytes commonly grow along watercourses, where the saturated zone is close to the surface and their roots can reach it easily. They transpirate massive amounts of water (*see* TRANSPIRATION), and dense growths of them may significantly reduce the volume of water in a stream; hence, eradicating them is one means of increasing streamflow. Care must be taken, however, that the removal of phreatophytes does not increase water temperature (by eliminating shade), promote bank EROSION (by destroying the root network that holds the bank in place), and eliminate the COVER required by animals and birds that use the stream as a water source. Common phreatophytes include cottonwoods, willows, and tamarisk (salt cedar). *See also* RIPARIAN ZONE.

phyllite in geology, a dark, fine-grained, tightly layered rock resulting from the incomplete metamorphosis of SHALE into SCHIST (*see* METAMORPHIC ROCK). It lies between SLATE and schist on the metamorphic series. The surfaces of the individual layers in phyllite usually show a characteristic silky sheen due to the presence of large amounts of minute fragments of MICA.

phylum in the biological sciences, a major taxonomic division (*see* TAXONOMY), falling between the KINGDOM and the CLASS in the taxonomic order; that is, kingdoms are divided into phylums, which are further divided into classes. The term "phylum" is used principally for animals; the corresponding term for plants is *division.*

physiographic province in geology and physical geography, an area of the Earth's surface that is readily distinguishable from neighboring areas on the basis of rock type, landforms, and/or climate. A physiographic province does not need to be uniform throughout, but the complex of features in any one part of it should be similar to the complex in the other parts. For example, the Klamath Mountain province of California and Oregon is a disordered mix of GRANITES, PERIDOTITES, MARBLES, and other rock types. This mix is the primary feature differentiating it from the nearby Cascade province, which is virtually all composed of BASALT, while the Cascade province is in turn differentiated from the Basin and Range province—also predominantly basalt—by its multitude of landforms in contrast to the Basin and Range's regular alternation of HORSTS, GRABENS, and tilted fault block mountains and valleys. *Compare* BIOTIC PROVINCE.

phytoplankton PLANKTON that contain CHLOROPHYLL and are therefore plantlike in nature. Most phytoplankton are not true plants but class technically as MONERA or PROTISTA instead. The most prominent members of the group are the ALGAE and CYANOPHYTES ("blue-green algae"). Phytoplankton are extremely important from an environmental standpoint because they are the primary PRODUCERS in all aquatic ECOSYSTEMS, forming the basis for all waterborne FOOD CHAINS and providing, through the process of PHOTOSYNTHESIS, approximately 70% of the Earth's atmospheric OXYGEN. *Compare* ZOOPLANKTON.

phytotoxicity the ability of a chemical COMPOUND to kill or seriously harm plant life.

picocurie in chemistry and physics, a measure of RADIOACTIVITY, equivalent to one-trillionth of a CURIE (.000000000001, or 10^{-12}). A substance emitting one picocurie of radiation is undergoing radioactive decay at the rate of roughly one ATOM every 27 seconds.

picrite basalt in geology, a form of BASALT that contains more than 50% OLIVINE. It is generally found near the bottom of thick flows that have remained liquid long enough for the relatively heavy olivine to settle out.

pillow lava in geology, LAVA that has been extruded under water. Rapid cooling by the water causes the lava to solidify as a series of smooth, amorphous lumps roughly the size and shape of bed pillows. The presence of these pillows is one of the pieces of evidence used by geologists to trace the boundaries of ancient lakes, seas, and rivers.

Pinchot, Gifford (1865–1946) American forester and conservationist, born August 11, 1865 in Simsbury, Connecticut, and educated at Yale University (B.A., 1889) and through private studies in Europe with the British forester Sir Dietrich Brandis (1890–96). His work as a private forester on the Vanderbilt estate in North Carolina (1896–98) attracted widespread attention, and in 1898 he was appointed chief of the Division of Forestry of the U.S. Department of Agriculture. When Theodore Roosevelt became president in 1901 (*see* ROOSEVELT, THEODORE), Pinchot became one of his chief advisers. Working together against massive pressure from logging interests and a largely hostile Congress, the two men essentially created the modern NATIONAL FOREST SYSTEM. In 1903, while still serving as Chief of Forestry for the United States, he helped found a School of Forestry at Yale (later named for him); and when he was dismissed from the Forest Service by President William Howard Taft in 1910 for insubordination (he had criticized the Taft Administration's release of reserved forest lands in Alaska), he turned to teaching full time. In 1922 he was elected governor of Pennsylvania, serving two six-year terms (1923–1935). A devotee of the work of GEORGE PERKINS MARSH, Pinchot was strongly dedicated to finding ways of using resources that would benefit humans without harming the environment, and is usually credited with coining the term "conservation." He did not believe in WILDERNESS or parks, which led him to a celebrated confrontation with JOHN MUIR over the HETCH HETCHY project in Yosemite, but the two men worked closely on other issues. Pinchot died on October 4, 1946. The best account of his life is his own *Breaking New Ground,* published posthumously in 1947 and reprinted in 1989.

pine barren in ecology, a region of sandy, relatively flat soils where SPECIES of pine such as lodgepole (*Pinus contorta*) or jack (*P. banksiana*), which are normally PIONEER SPECIES, have become the dominant tree. In some regions, such as the New Jersey Pine Barrens, the pines form relatively dense STANDS of dwarfed trees. In other regions, the trees may be widely scattered, with grasses becoming dominant. Other pioneering trees, such as scrub oak, may be mixed with the pines. Cedar swamps (*see* SWAMP) are common in depressions. UNDERSTORY vegetation is adapted to the sandy soils and to the acidic DUFF characteristic of pines, and usually includes rare and unusual species and COMMUNITIES. Pine barrens are maintained in nature by a combination of the extremely sandy soils, which do not hold moisture well, and the common presence of WILDFIRES, which race through the barrens every few years and prevent natural SUCCESSION from continuing beyond the pioneer stage and can lead to dwarfism. Where fire suppression is aggressively practiced, the barrens tend to slowly convert to more "normal" vegetation types. *See* FIRE ECOLOGY; FIRE-MAINTAINED CLIMAX.

pinnate form in botany, a GROWTH FORM of plants that is characterized by leaf lobes arranged linearly along opposite sides of a central axis. The leaf lobes may be either portions of a single leaf, as in the oak (*simple pinnate form*), or separate leaflets arranged along opposite sides of a single stem, as in the walnut (*compound pinnate form*). *Compare* PALMATE FORM.

pioneer species in ecology, a SPECIES of living ORGANISM that is adapted to colonize barren areas such as bare rock, sand DUNES, cultivated fields, CLEARCUTS, and burned-over land. The term is often restricted to plants, although animals have their own pioneering patterns, and the very first organisms to colonize new lands are usually MONERA (bacteria and bluegreen algae), PROTISTA (algae, protozoans, etc.), and LICHENS. Pioneer species have a number of characteristics in common, including the ability to expand their POPULATIONS rapidly; high tolerance for extreme conditions of temperature, moisture, and soil and water chemistry; and the ability to manufacture most or all of their own food directly from ELEMENTS or INORGANIC COMPOUNDS, so that they do not depend on the presence of organic soil HORIZONS but can live well in MINERAL SOILS. As a rule they are adapted to life in bright sunlight, and will not reproduce in their own shade, so they will typically either be ANNUALS or, if PERENNIALS, will only produce a single generation on a given site before being replaced by the more shade-tolerant, less extreme-adapted species for which they have prepared the way. Typical pioneer species include COMPOSITES such as thistle, sunflower, and dandelion; fireweed; morning glory; arbor vitae; and lodgepole pine. *See also* SERE; SUCCESSION.

pirated stream in geology, a stream whose headwaters have been cut off by another stream, so that it becomes a TRIBUTARY (or occasionally, the MAINSTEM) of the new stream, leaving its old bed below the cutoff point either dry or carrying the much-reduced flow of what were once the original channel's tributaries. Piracy results when one stream erodes through the divide separating it from another whose BASE LEVEL is higher than the first. The water in the higher stream will then abandon its old bed to seek the lower level.

The pirating stream may erode through the divide either laterally (by widening its valley until it intersects a parallel neighboring valley) or through headward erosion (deepening and extending its channel at its source). *See also* BARBED TRIBUTARY.

pistil *See* FLOWER: *parts of a flower.*

pit and mound microtopography in undisturbed forests, the depressions made by the upended root structures of fallen trees. Pit and mound microtopography is characterized by depressed patches of bare earth where plants can grow readily, protected by natural berms and mulched and fertilized by decaying deadfall. In actively logged forests, trees are usually cut before they fall, thus eliminating the biological diversity supported by natural pits and mounds.

pitch (1) in engineering geology, the angle a slope makes with a horizontal plane. Pitch is equivalent to DIP, except that dip indicates a compass direction as well as deviation from the horizontal. Measurement of pitch is either in degrees (the size of the angle the slope makes with an imaginary horizontal plane intersecting it at the point of measurement) or in percent (the elevation difference between two points on a slope—the *rise*—divided by the horizontal distance between the points—the *run;* a 20% slope, for example, being one which rises 20 feet for every 100 feet). *Compare* STRIKE.

(2) in mountaineering, a portion of a climb that makes a logical unit, usually because it consists of a single type of terrain that requires the same climbing technique throughout. For example, a SCREE SLOPE with a rock face across the center of it would consist of three pitches: the scree below the rock, the rock, and the scree above the rock. Often one long pitch will be divided into two at some obvious point, such as a broad ledge, a change in rock texture, etc.

pitch in forestry and logging, a natural resin exuded by trees, especially CONIFERS. Pitch is apparently used by the tree principally as a means of sealing wounds to prevent the loss of sap and to shut out airborne MICROORGANISMS. A thick, sticky, aromatic liquid that hardens upon exposure to air into an amorphous solid, pitch is normally golden to dark brown in color, and is semitransparent or translucent. It is produced and transported in passageways within the wood of the tree known as *resin ducts.* Many conifers also have resin ducts in their needles. Cavities within the tree produced by internal breakage, insect attack, etc., fill up with it. Wood from a tree that has had to produce large amounts of pitch in response to environmental stress is often streaked or soaked with it; such wood is classed according to the amount of pitch in it as *light, medium, heavy,* or *massed.* Commercial uses of pitch include the manufacture of turpentine and of *rosin,* used in the manufacture of wood filler, paint, and plastics, and as a friction-producing substance to be rubbed on dancers' shoes, the bows of stringed instruments, etc.

pitcher plant in botany, a type of CARNIVOROUS PLANT that captures insects in a specialized leaf formed into the shape of a pitcher and filled with digestive juices. Like most carnivorous plants, pitcher plants usually grow in the acid waters of BOGS where obtaining NITROGEN by normal means is difficult; hence the adaptation to a carnivorous life style, which allows them to obtain this necessary NUTRIENT from the bodies of insects.

placer mine a mine working a *placer deposit,* that is, a deposit of MINERAL-bearing SAND, GRAVEL, or similar materials. Placers are the result of the EROSION of a mineral LODE by water that carries the eroded material away and concentrates it in GRAVEL BARS and other alluvial deposits (an *alluvial placer—see* ALLUVIUM) or in the beaches and SPITS of a large lake or ocean (a *beach placer*). Gold is the most common mineral found in placer deposits, although platinum, tin, gemstones, and a few other materials are occasionally mined in this manner. All techniques for placer mining involve digging up the gravels and other loose materials of the deposit and washing them in such a manner that the lighter GANGUE is swept away with the water while the heavier mineral-bearing materials remain behind. The simplest form of placer mining is gold panning. Nearly as simple is the "rocker" (or "riffle box", or "cradle"), consisting of a wooden box with bars ("riffle bars", or "riffles") across its bottom. A bucket of gravel is dumped in the upper end of the rocker, water is poured over it, and the box is rocked back and forth to wash the lighter gangue over the tops of the riffles while the mineral-bearing materials concentrate behind them. Considerably more destructive, from an environmental standpoint, are the practices known as "hydraulicking" and "dredging." In hydraulicking, a large nozzle (called a "giant") directs a powerful stream of water at a gravel bank, washing it down over a series of artificial gravel bars, usually in a convenient streambed, which act as a gigantic riffle box, sorting and concentrating the mineral-bearing materials from the artificially eroded hillside. Dredging utilizes a large suction dredge (*see* DREDGE) floating in a quiet, often artificially created, pool of water in the middle of a watercourse. It sucks up the gravels from the bed of the pool, washes them over large riffle boxes in its body, and deposits the gangue as piles of

TAILINGS behind it as it moves slowly along the watercourse. Both hydraulicking and dredging suspend considerable amounts of SILT and other PARTICULATES in a stream and leave its banks denuded of vegetation and therefore able to hold less water, contributing to downstream flooding. The tailings piles, in addition to being unsightly, are usually sterile, and will remain devoid of vegetation for many years, continuing to contribute to sedimentation and flooding problems downstream. Consequently, placer mining has become one of the more tightly controlled forms of mining, with hydraulicking outlawed and dredging severely regulated in most jurisdictions. *Compare* HARD ROCK MINE; STRIP MINE.

plagioclase in geology, a common rock-forming MINERAL, consisting of aluminum silicate with the addition in varying amounts of sodium, calcium, and extra aluminum and silicon. The type formula is $(Na,Ca)(Al,Si)AlSi_2O_8$. Plagioclase is one of the closely related series of minerals known collectively as FELDSPAR (*see* PLAGIOCLASE FELDSPAR).

plagioclase feldspar a form of FELDSPAR consisting mostly or entirely of PLAGIOCLASE. It can be identified in the field by the fine parallel lines (*striations*) on its faces observable with a hand lens.

plains a relatively level landform created by glaciation, drainage of seas and lakes, sediment deposition, and other long-term geological events. The GREAT PLAINS constitutes the billion-acre heartland of the United States. *See also* GRASSLAND BIOME.

plan control in Forest Service usage, an ordinance, regulation, or rule which may be used to legally implement a management plan. Plan controls include zoning ordinances, permit procedures, design and performance standards, administrative rules, user fees, and all other legal instruments that, when enforced, help obtain the plan's GOALS and OBJECTIVES.

plan element a phase or topic area of a plan, such as goal-setting, data-gathering, planning for open space, etc., which may be separated from the other elements of the plan and worked on as a unit. Plan elements are topic-specific rather than area-specific; they are divided from each other by function rather than by geographic boundaries. Forest Service usage recognizes a standard set of plan elements. These are (1) legal requirements and authorities; (2) objectives and targets; (3) management situation; (4) basic assumptions; (5) data collection; (6) land capability; (7) alternative situations; (8) plan selection; (9) functional planning; and (10) documentation. *See separate articles under most of these topics.*

plane survey a survey that treats the area being surveyed as a plane surface for mathematical purposes, with straight lines, triangles whose angles always add up to 180°, etc. Plane surveys are convenient because they simplify the mathematics of measuring and map making; however, since they do not take into account the curvature of the Earth, they are accurate only for small areas. *Compare* GEODETIC SURVEY.

plankter an individual plankton organism (*see* PLANKTON).

plankton in limnology and oceanography, those ORGANISMS that live suspended in bodies of water, drifting with the currents rather than swimming from point to point (*see* NEKTON) or lying on or in the bottom muck (*see* BENTHOS). As a rule, plankton are PROTISTA, MONERA, or microscopic or near-microscopic INVERTEBRATES. Some swim weakly in order to pursue PREY or to move upward or downward through the WATER COLUMN in response to light levels (*see* DIEL MOVEMENT); however, drifting is always the principal means by which they get from place to place. Some (*holoplankton*) remain in the planktonic form all their lives; others (*menoplankton*) are the larvae (*see* LARVA) of larger invertebrates. They are present in very large numbers in the PHOTIC ZONE, where they form the basis of all aquatic FOOD CHAINS and FOOD WEBS as well as being used directly for food by numerous larger animals, including the baleen whale. They are also responsible for generating roughly 70% of all atmospheric OXYGEN. Plankton are strongly influenced by water temperature and chemistry, allowing ocean and lake currents to be mapped by identifying the plankton SPECIES that are present at any given point. *See also* NANNOPLANKTON; NET PLANKTON; PHYTOPLANKTON; ZOOPLANKTON.

planning area in Forest Service usage, a large area of relatively uniform topography, vegetation, climate, wildlife, etc., for which uniform GOALS and OBJECTIVES can be set. Planning area boundaries cross district, forest, and regional boundaries (*see* FOREST SERVICE: *structure and function*), but often follow state or county boundaries for ease in coordination across agency boundaries. The concept is now obsolete. Area guides (*see* PLANNING AREA GUIDE) have been replaced by REGIONAL PLANS under the terms of the NATIONAL FOREST MANAGEMENT ACT.

Planning Area Guide in Forest Service usage, a document setting forth the GOALS, guidelines and OBJECTIVES that are to be followed during the formulation of UNIT PLANS. The term is obsolete. *See* PLANNING AREA; REGIONAL PLAN.

planning unit in Forest Service usage, a relatively small area consisting of roughly uniform topography, vegetation, wildlife, etc., for which detailed land-use plans ("UNIT PLANS") are to be prepared following the overall strategy set forth in a PLANNING AREA GUIDE. A planning unit is usually approximately the same size as a ranger district, but its boundaries conform more strictly to topographical features such as WATERSHEDS. The term is obsolete, unit plans having been replaced by FOREST PLANS under the terms of the NATIONAL FOREST MANAGEMENT ACT.

planosol a soil with a thick, fertile, well-developed A horizon (*see* SOIL: *the soil profile*) overlying a B horizon that contains large amounts of CLAY. The transition between the two horizons is generally abrupt. Because of the quality of their A horizons, planosols make fairly good agricultural soils, although the presence of the clay layer ("clay pan") means that they do not drain well. They develop in conditions of moderate rainfall, principally under STEPPES and SHRUB-STEPPES, where clays present in the upper portion of the soil may be leached downward to collect in the pan (*compare* CALICHE). The term is technically obsolete, having been replaced by the term *aqualf* (*see* ALFISOL); however, it is still in wide use.

plant in the biological sciences, any member of the plant KINGDOM of living things, characterized by multicellular development with tissue specialization, sexual reproduction with separate haploid and diploid generations (*see* ALTERNATION OF GENERATIONS), and cell walls (*see* CELL) made primarily of CELLULOSE. Nearly all are autotrophs (*see* AUTOTROPHISM), though some have lost their CHLOROPHYLL and become PARASITES or SAPROBES. The term "plant" technically includes the BRYOPHYTES and the VASCULAR PLANTS (ferns, conifers, and flowering plants) but not the ALGAE (kingdom PROTISTA), blue-green algae (kingdom MONERA) or fungi (kingdom Fungi: *see* FUNGUS).

plant louse *See* APHID.

plastics recycling the collection of waste plastic products for remanufacturing into useful products. Though plastics constitute nearly one-fourth the volume in landfills, so far only about 7% of new plastic products (by weight) are made from recycled materials. The reasons for this are the low cost of petroleum, from which plastic feedstocks are derived, and the high cost of segregating the many different types of plastics in the waste stream prior to remanufacturing. The problem of plastics in terms of waste management stems from their lack of biodegradability in landfills. Ordinary plastics persist for 200–400 years, and even the new biodegradable products can last several decades. Many experts believe that the use of plastics as throwaway containers should be severely curtailed. *See also* RECYCLING; SOLID WASTE.

plate tectonics in geology, theory of the Earth's crustal dynamics that explains the formation of large-scale features such as ocean basins, continents, and mountain ranges, as well as suggesting causes for EARTHQUAKES, VOLCANOES, and other TECTONIC ACTIVITIES. The theory states that the Earth's solid surface, the LITHOSPHERE, consists of seven large and perhaps twice that many small separate units, or *plates,* riding on convection currents in the hot liquid MANTLE beneath them. All major tectonic activities take place along the boundaries between the plates and are caused by the plates' motions in relation to each other. At some places the plates are pulling apart and a new crust is constantly being formed to fill in the gap. These places (such as the Mid-Atlantic Ridge) are known as *constructive zones,* and the pulling apart of the plates is called *seafloor spreading,* even when it is taking place on dry land such as in East Africa's Afar Triangle. At other places the plates are colliding. These collisions buckle up the crust to form mountains (such as the Himalayas, formed when the Indian subcontinent collided with the rest of Asia). Almost always one plate is forced under the other at these zones of collision, with the leading edge of the lower plate being drawn down (*subducted*) into the mantle and melted; hence such boundaries are known as *destructive zones.* Most volcanoes form along destructive zones as the subducting plate is melted and the resultant MAGMA seeks an opening to the surface. This is the reason, for example, for the presence of the long chain of volcanoes that lines the western edge of North America, including Mt. St. Helens, Lassen Peak, Mt. Katmai, and other recently active volcanic peaks. A third type of activity that takes place along plate boundaries is sideways slipping as one plate moves horizontally past another. Since this neither creates nor destroys crust, boundary zones of this type are known as *conservative zones.* Most major earthquakes, such as those along California's San Andreas Fault, take place in conservative zones as the plates rub against each other and are held by friction until enough energy builds up to move them by small jerks. Pieces of one plate may rub off and adhere to the other along a conservative boundary. These tiny "platelets" are known as TERRANES. A final piece in the plate tectonics puzzle is the presence of so-called "hot spots," which remain stable in relation to the Earth's interior. As the plates move slowly over them, the hot spots erupt as volcanoes in the plates' centers. Hot spots probably represent the convergence of the upward limbs of two or more of the convection currents that drive the plates, though some geologists theorize that they may result from ancient meteor impacts. They are responsible for—among other things—the Hawaiian island chain, Yellowstone National Park, Iceland, and California's Mammoth Crater.

HISTORY OF PLATE TECTONICS

Although the idea of *continental drift*—the breakup and slow separation of continents—was postulated as long ago as the 17th century by Sir Francis Bacon on the basis of the close "jigsaw-puzzle" fit of the coastline of the Americas to that of Europe and Africa, the first rigorous working out of the drifting-continents hypothesis was made by the German scientist Alfred Wegener in his 1915 book *Origin of the Continents and Oceans*. Wegener was called a crackpot and his work was ignored for nearly 50 years. In the early 1960s evidence began to accumulate in favor of Wegener's theories. This evidence included precise computer matching of the edges of the CONTINENTAL SHELVES of Europe and Africa and the Americas; the discovery of seafloor spreading along the Mid-Atlantic Ridge; the location of matching rock STRATA, fault patterns, and even glacial erosive features (*see* GLACIAL LANDFORMS) on opposite sides of the Atlantic; extensive seismic mapping, which proved that earthquakes cluster along the lines now thought to be the plate boundaries; and enough further evidence that the theory now seems thoroughly corroborated. Recent evidence suggests that the motion of the plates carrying the continents may be rhythmic: that they may separate, move outward, then reverse and move back together, on a roughly 250-million year cycle. If true, this could help explain phenomena dependent on massive rhythmic climate changes such as the ICE AGES and the cycles of mass EXTINCTION seen in the fossil record, including the extinction of the dinosaurs.

Plimsoll line (load line) line on a ship's hull indicating the amount of *freeboard* that it must legally maintain, that is, the amount of the ship's side that must remain above the waterline when it is fully loaded. A ship that is loaded so heavily that its waterline lies above its Plimsoll line is liable to be swamped in heavy seas. The Plimsoll line is actually a series of separate lines indicating legal freeboard for differing water conditions; it appears on the bow. From top to bottom, the lines generally indicate the load lines for tropical freshwater (TF), freshwater (F), tropical zone (T), summer conditions (S), winter conditions (W), and winter in the North Atlantic (WNA). Extra indications often include a deck line (that is, the location of the ship's deck) and a circle with a horizontal line through it that is placed on the summer load line. The line is named for Samuel Plimsoll (1824–1898), a member of the British Parliament who was primarily responsible for the passage of the Merchant Shipping Act of 1876, establishing the first shipping load restrictions. (*See* figure.)

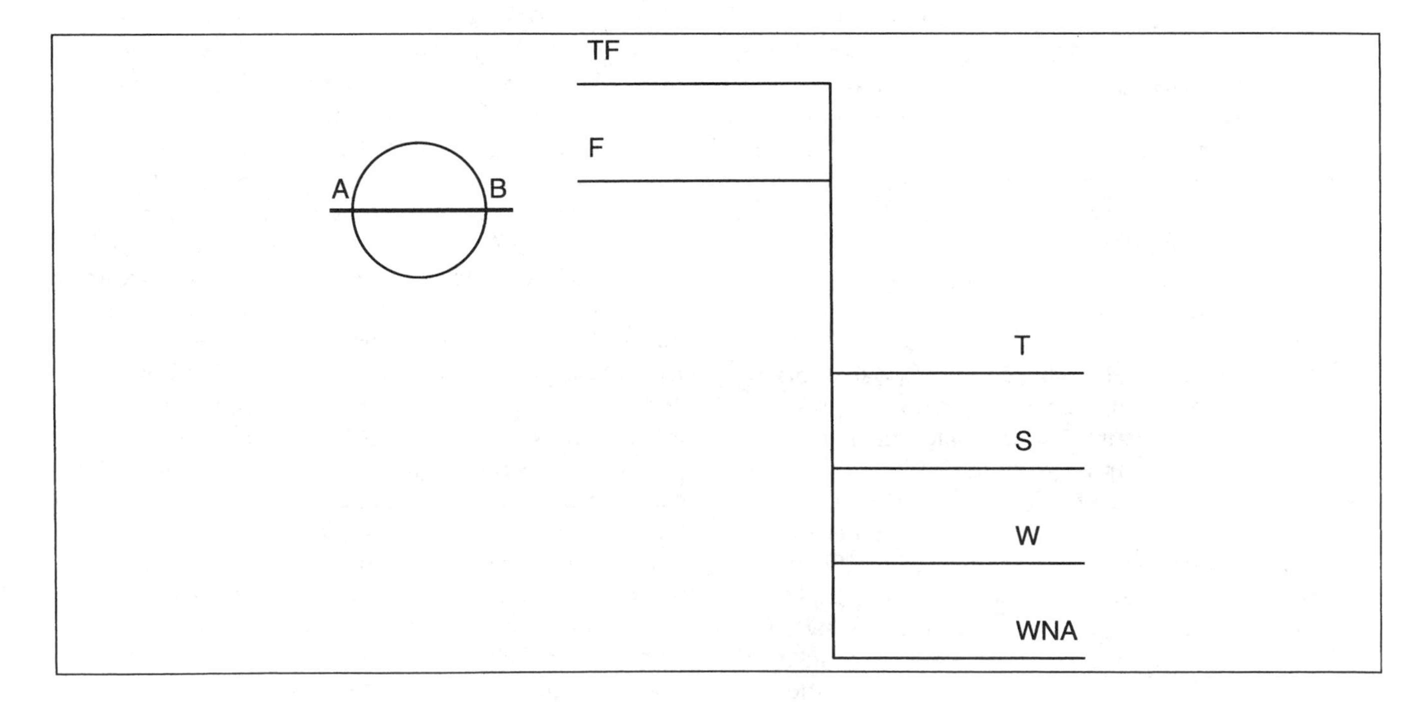

The Plimsoll line

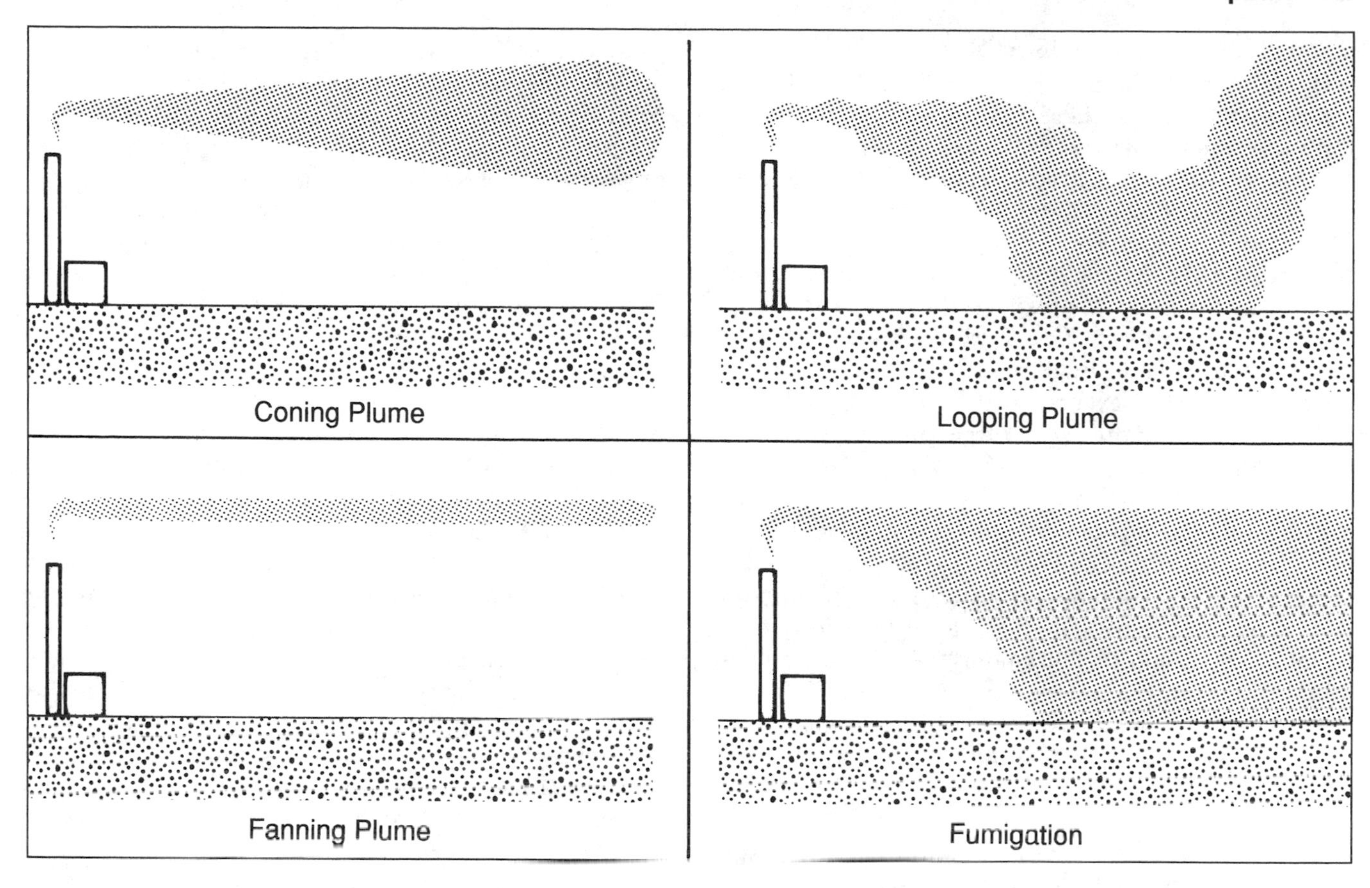

Types of plume

plume in pollution-control technology, the dispersion pattern made by the emissions from a pollution source as they spread out through the receiving medium (the water or air that the pollution is being poured into). The shape of a plume is determined by the interaction of three principal factors: (1) the location of the source; (2) the physical and chemical characteristics of the emitted material; and (3) the physical and chemical characteristics of the receiving medium. Of these three, the last is by far the most important. Thus, in the case of GROUNDWATER pollution, the shape of the plume is determined almost exclusively by the direction and rate of flow in the aquifer and by the amount of CAPILLARY ACTION on its fringes. Similarly, in a body of water, the shape of a pollutant plume will depend principally on current patterns, with some influence from the bouyancy of the pollutants and from the DIFFUSION rates of those pollutants that happen to be water-soluble. In air, bouyancy plays a relatively large part, influenced by such factors as the size of any PARTICULATES present and the density and temperature of any gases. Source location plays a significant role as well, due to the variability of conditions in different levels of the atmosphere. ("Source location" means here the *effective source,* that is, the height above ground at which the upward thrust of the pollutants as they leave the stack is overcome by gravity.) However, the principal factor is still the physics and chemistry of the receiving medium, that is, the air. Wind speed and direction are obvious influences. Less obvious, but still important, factors include TEMPERATURE, PRECIPITATION, ATMOSPHERIC PRESSURE, and ATMOSPHERIC STABILITY, as determined by the LAPSE RATE and the presence or absence of an INVERSION.

CLASSIFICATION OF PLUMES

Plumes in air are classified according to shape. The *coning plume,* the most common variety, simply spreads out downwind in the shape of a long, narrow cone with its apex at the effective source. The *looping plume,* which develops under the influence of a strong lapse rate, dips toward the ground and then climbs rapidly upward. Those times when the loop touches the ground can cause severe pollution hazards. The *fanning plume* is a coning plume that has been flattened out beneath a strong inversion so that it fans outward in a single horizontal layer. *Fumigation* takes place when a fanning plume spreads downward toward the ground in conditions of cold, still air. It represents the most hazardous conditions associated with air pollution.

THE VISIBLE PLUME

In some sources, the term "plume" is reserved for the *visible plume,* that is, those emissions from a stack that can be seen in the air. In most cases, the visible plume is composed primarily of water vapor and is not in itself hazardous. *See also* THERMAL POLLUTION.

pluton in geology, a body of IGNEOUS ROCK that has cooled beneath the surface. Strictly speaking, a pluton should consist of rock from a single source of MAGMA and be laid down in a single event or a closely related series of events, so that it is relatively uniform in composition. Very large plutons are called BATHOLITHS.

plutonium an artificially created radioactive element (*Pu*) that is extremely toxic. Plutonium was first created in 1940 by bombarding isotopes of uranium with deuteron atoms. The half-life of the most stable form of plutonium (Plutonium-244) is 80 million years. Plutonium-239, with a half-life of 25,000 years, is produced in large quantities by nuclear reactors and can be used as a nuclear fission fuel. *See also* NUCLEAR ENERGY.

pluvial climate in climatology, a rainy climate, especially, a formerly rainy climate in an area that now has a dry climate. Such *pluvial periods* generally correspond to periods of widespread glaciation (*see* ICE AGE).

pluvial lake in geology, a lake filled by rainfall and its associated RUNOFF rather than by GROUNDWATER and PERMANENT STREAMS. Pluvial lakes lie above the WATER TABLE and are generally separated from it by impermeable rock or clay. (*See* PERMEABILITY.) They are highly ephemeral, often appearing only during storms.

PMN short for *premanufacturing notification* (*see* TOXIC SUBSTANCES CONTROL ACT).

podzol in soil science, a strongly acidic soil with a well-developed A_2 horizon (*see* SOIL: *the soil profile*) consisting of a layer of white, gray, or light brown ash-like material with very low organic content sandwiched between a deep layer of mor humus (*see under* HUMUS) and a thick, dark B horizon containing a heavy accumulation of ORGANIC MATTER and MINERALS, especially iron, leached from the upper portion of the soil. Podzols are the characteristic soils of coniferous forests in cool to cold, humid regions. The term is now considered obsolete, all podzols having been reclassified as SPODOSOLS; however, it remains in widespread informal use. *See* SOIL: *classification of soils.*

podzolic soil a soil similar to a PODZOL but less acidic in nature, and with the characteristic A_2 horizon (*see* SOIL: *the soil profile*) less strongly developed. Podzolic soils are divided into *brown podzolics* and *gray-brown podzolics,* the former being less well-developed than the latter. Both types originate under deciduous or mixed deciduous/conifer forests under cool, moist conditions. The term is now considered obsolete, with some podzolics reclassified as INCEPTISOLS and others as SPODOSOLS. *See* SOIL: *classification of soils.*

poikilothermic organism in the biological sciences, an ORGANISM whose body temperature varies according to the temperature of the environment surrounding it. In popular terminology, poikilothermic organisms are "cold-blooded"; however, this term is misleading, as a so-called cold-blooded organism such as a reptile can have very hot blood if it happens to be lying on a rock in the sunshine. No organism is totally poikilothermic (even plants and MICROORGANISMS generate some heat through METABOLISM); however, all organisms except mammals and birds follow a generally poikilothermic strategy, reducing activity when cold and carefully controlling their exposure to the environment in order to take advantage as much as possible of sun, shade, and moisture as temperature controls. *Compare* HOMOTHERMALISM.

point source in pollution control, a source of POLLUTANTS that enter the environment at a single location, such as a sewage OUTFALL or a smokestack. A special type, the *mobile point source,* emits pollutants at a single location at any given time, but the location may vary with respect to the environment over time as the point source travels about. This type is particularly represented by automobile exhaust. Pollution from point sources is usually easier to control than that from NONPOINT SOURCES, since the source can be identified by following the PLUME backward and the pollutants are confined in a pipeline, smokestack, or similar conveyance before they are released into the environment, so that the entire waste stream can be diverted to a treatment plant. *Compare* AREA SOURCE.

poison any material with a detrimental effect on the METABOLISM of a living ORGANISM. "Poison" is a rather loosely defined category. Poisons may be TOXINS, irritants, or CARCINOGENS, or they may interfere with metabolism in other ways, such as preventing the absorption of food in the digestive tract or reducing production of important ENZYMES or hormones. The term is also used to refer to materials that interfere with complex chemical reactions of any nature, such as those involving a CATALYST (*see* CATALYTIC CONVERTER).

polar front in meteorology, a continuously existing weather FRONT that marks the location where cold air moving toward the equator from the poles meets warm air circulating over the middle LATITUDES. It may be thought of as the boundary between the central and the poleward HADLEY CELLS. The polar front moves poleward in the summer and equatorward in the winter, but remains continuously traceable all year around. Weather near it is almost always unsettled.

polar liquid in chemistry, a liquid in which the individual MOLECULES are electrically polarized, that is, one in which one end of each molecule is positively charged and the other end is negatively charged. Polarity arises in molecules with COVALENT BONDS when one of the molecule's ATOMS attracts the shared ELECTRONS more strongly than the other (or others). The electrons spend more of their orbital time in that end of the molecule, giving that end a slight negative charge due to the electrons' presence and leaving the other end with a slight positive charge due to their absence. Water is an example of a strongly polar liquid. The single OXYGEN atom in a water molecule forms the negatively charged apex of a "V," with each branch being formed by a positively charged HYDROGEN atom, a structure which is responsible for most of its unusual attributes (*see* WATER: *structure of water*). Polar liquids are important because of the ability of their electric charges to literally pull apart other polar molecules or ionic compounds, turning them into IONS—the process we know as DISSOCIATION. *See* SOLVENT; SOLUTION. *See also* DETERGENT; ELECTROLYTE.

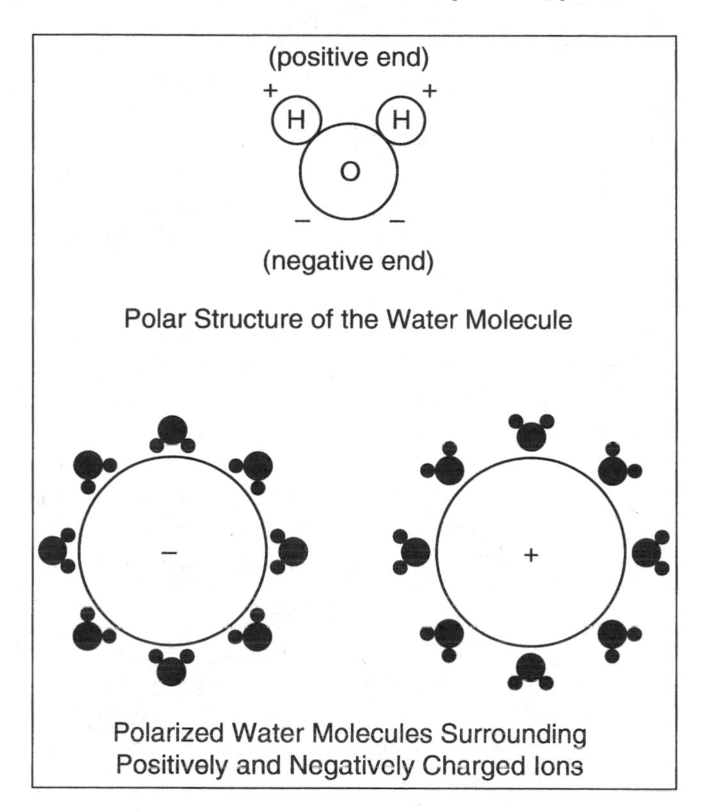

Polar liquid

polar region in geography, the region around either the North or the South Pole of the Earth. The outer limit of the polar region is generally considered to be the ISOTHERM corresponding to a warmest-month mean of 10°C; that is, the polar region consists of all lands for which the monthly mean temperature never rises about 10°C (50°F). Over land, this boundary corresponds roughly to the northern tree line (*see under* TIMBERLINE; there is no southern tree line due to the lack of land masses in the appropriate southern LATITUDES). The polar regions thus consist entirely of tundra (*see* TUNDRA BIOME), bare rock, and ice. The Sun never rises for much of the winter and, though it never sets for much of the summer, always remains low on the horizon, so that its rays strike obliquely, giving little warmth. PRECIPITATION—almost all of it in the form of SNOW—is low, usually less than 25 cm/yr (10 in/yr). Nevertheless, the polar climates are classified as humid, because the low air temperatures mean that even under conditions of very low ABSOLUTE HUMIDITY, RELATIVE HUMIDITY remains high. Bare ground, where it exists, is almost always underlain by PERMAFROST, meaning that in the warmer months the region is essentially one gigantic BOG. Despite these very harsh conditions, an astonishing number and variety of living ORGANISMS inhabit the polar regions, ranging from MOSSES and LICHENS to polar bears, musk oxen, and caribou. Since they are living very close to the limits of the conditions in which life can survive, these organisms are highly specialized, meaning that their life cycles are very easily disrupted. Hence, resource development in the polar regions needs to proceed very carefully to avoid wholesale extermination of POPULATIONS and perhaps of entire SPECIES. *See also* HOLARCTIC REGION; HOLARCTIC SPECIES; ALPINE ZONE.

pole-sized tree in forestry and logging, any tree between 5 inches and 10.9 inches DBH (*see* DIAMETER BREAST HEIGHT), that is, those trees that fall in the 6-, 8-, and 10-inch DIAMETER CLASSES. *Poletimber* consists of pole-sized trees with no more than 50% DEFECT, making them usable for lumber, although only a very small amount of lumber can be taken from each tree.

polishing pond (oxidation pond, lagoon) in sewage treatment technology, a shallow, unlined pond used as a SETTLING BASIN and biological treatment facility (*see* SECONDARY TREATMENT) for SEWAGE or industrial wastes. Polishing ponds contain AEROBIC BACTERIA as the decomposing agent (*see* DECOMPOSER) along with ALGAE that feed on the waste products of the bacteria and which in turn provide the bacteria with oxygen. The algae must be periodically harvested in order to prevent them from choking the pond. This is sometimes done by maintaining a POPULATION of brine shrimp in the water to feed on the algae, together with a population of pollution-tolerant fish to feed on the brine shrimp, thus constructing a complete ECOSYSTEM. Polishing ponds are typically rectangular flat-bottomed ponds 2 to 5 feet deep enclosed by earthen dikes. The long dimension of the rectangle should be no more than three times the length of the short dimension. BOD loadings (*see* BIOCHEMICAL OXYGEN DEMAND) must be kept light, in the neighborhood of 20 pounds of BOD per acre per day or less. Retention time within the pond is from 20 to 120 days, depending largely on the BOD loading.

pollen the "carrying case" utilized by the seed plants to enclose the male GAMETOPHYTE, or *sperm,* for transfer to the female gametophyte, or *egg.* The basic structure of all pollen grains is the same. It consists of an outer coating, or *exine* (composed of an extremely tough and resistant POLYMER called *sporopollenin*) surrounding an inner coating known as the *intine,* which in turn surrounds the sperm, usually two in number. The shape taken by this basic structure, however, varies greatly from FAMILY to family, from GENUS to genus, and even in many cases from SPECIES to species, so that it is often possible to completely identify a plant from its pollen alone. Pollen is also highly resistant to decay in the environment, so that it fossilizes easily, making it possible to determine the species makeup of ancient plant COMMUNITIES. The pollen of wind-pollinators such as GRASSES and CONIFERS is generally small, dry, and light; that of insect-pollinators such as the majority of flowering plants is often barbed or sticky to facilitate its carriage by the pollinating insect.

pollen records used by ecologists and other scientists to determine changes in plant life and therefore in climate and other environmental conditions over long stretches of geologic time. Pollen deposits can be found in lakebeds, glaciers, and various other strata.

pollutant a CONTAMINANT whose presence is damaging to the health of the environment. Pollutants are either substances that are useful in some context but have escaped from that context into other parts of the environment where they are harmful instead of useful (DDT, POLYCHLORINATED BIPHENYLS, PHOSPHATES, etc.) or they are waste products that have been inadequately disposed of or contained (DIOXIN, FLY ASH, human or animal FECES, etc.) *See* POLLUTION. *See also* CONVENTIONAL POLLUTANT; MICROCONTAMINANT; PARTICULATE; SUSPENDED SOLID; etc.

pollutant standard index (PSI) in air-pollution control, a measurement of air quality mandated by the 1977 amendments to the CLEAN AIR ACT in order to allow standardization of comparisons of POLLUTANT levels from community to community. The PSI is based on five so-called "criteria pollutants": sulfur dioxide (SO_2: *see* SULPHUR OXIDE), nitrogen dioxide (NO_2: *see* NITROGEN OXIDE), CARBON MONOXIDE (CO), OZONE (O_3), and TOTAL SUSPENDED PARTICULATES (TSP). Each of these pollutants is assigned a number from 0–100 based on its measured level taken as a percentage of the regulatory standard for health effects. For example, if the ambient level of CO (that is, the level in the AMBIENT AIR) is 79% of that judged to be harmful to human health, the CO number would be 79. The five numbers are then added together to obtain the PSI. PSI readings below 100 are classed as healthful; from 100 to 199, unhealthful; from 200 to 299, very unhealthful; and over 300, hazardous. The maximum possible PSI is 500. In this case, all five criteria pollutants would be at or above harmful levels.

pollution broadly, the contamination of some portion of the ENVIRONMENT by a POLLUTANT, especially one that is hazardous to living things or causes economic damage to the human community. The word covers an extraordinarily wide class of situations. *See,* e.g., AIR POLLUTION; WATER POLLUTION; THERMAL POLLUTION; NOISE POLLUTION; HAZARDOUS WASTE; NUCLEAR WASTE.

pollution trading the use of market mechanisms to encourage the reduction of air and water pollution. The general idea is that so long as overall emissions or effluent standards are met, nonpolluters or low polluters can sell "pollution credits" to those whose pollution loads are beyond legal limits. The concept was codified in federal law in 1990 with the CLEAN AIR ACT amendments which enabled the 110 most polluting power plants to buy and sell SO_2 (sulfur dioxide) emissions rights. Each plant was given a set number of annual emissions credits, which if not used could be sold or traded to other plants or banked. The ENVIRONMENTAL PROTECTION AGENCY (EPA) has estimated that the system reduced SO_2 emissions by 30% between 1994 and 1997. Based on this record, in 1997 the EPA proposed expanding the voluntary emissions trading system in eastern U.S. states to cover nitrogen oxides (NO_x). The pollution trading idea is now being tested by the EPA for water pollution under the CLEAN WATER ACT in various watersheds located in 10 U.S. states. The experiments provide the opportunity for pollution credit trades to be made between both point and nonpoint polluters. In general, the environmental community is divided on the long-term efficacy of pollution trading. Proponents claim that such trading is a practical approach that uses powerful market forces to achieve policy goals. Others fear that the trading mechanism will impede efforts to reach zero-pollution goals, which in the case of water is the objective of the law. A further, practical difficulty lies in the opportunity for cheating unless pollution trading schemes are closely monitored. Finally, there is an ethical and aesthetic component to the dispute, in that many believe that presently unpolluted airsheds or watersheds should remain that way, that polluters should not be allowed to degrade the environmental quality of pristine areas. *See also* NONPOINT POLLUTION.

polyaromatic hydrocarbon (polynuclear aromatic hydrocarbon, PAH) an aromatic hydrocarbon (*see* AROMATIC COMPOUND; HYDROCARBON) containing two or more BENZENE RINGS or other aromatic ring structures. Examples include naphthalene, anthracene, and benzo(a)pyrene. PAHs are environmentally persistent, often carcinogenic compounds (*see* CARCINOGEN) that are emitted from a wide variety of sources, including industrial oil and grease discharges, solid waste incinerators (both industrial and municipal) and the combustion of FOSSIL FUELS.

polybrominated biphenyl (PBB) in chemistry, any of a class of ORGANIC COMPOUNDS containing two or more bromine atoms attached to a bonded pair of BENZENE RINGS. PBBs have been used commercially as fire retardents. They are very similar in structure and properties to POLYCHLORINATED BIPHENYLS, and have similar—though generally less virulent—toxicological properties (*see* TOXICANT).

polychlorinated biphenyl (aroclor, PCB) in chemistry, any of a class of ORGANIC COMPOUNDS containing two or more CHLORINE atoms attached to a bonded pair of BENZENE RINGS. As a class, PCBs are odorless, colorless, highly viscous (*see* VISCOSITY) fluids that are very nearly chemically inert. They have excellent dielectric properties (*see* DIELECTRIC) and were once widely used as ASKAREL LIQUIDS to insulate electrical transformers. They were also utilized as hydraulic fluids due to their high viscosity and chemical and thermal stability. For a time, they were included in a common formula for newsprint ink. They have relatively low ACUTE TOXICITY; however, they are effective MUTAGENS and CARCINOGENS, and are highly lipophilic in nature (*see* LIPOPHILIC SUBSTANCE), meaning that they accumulate easily in the body fat of

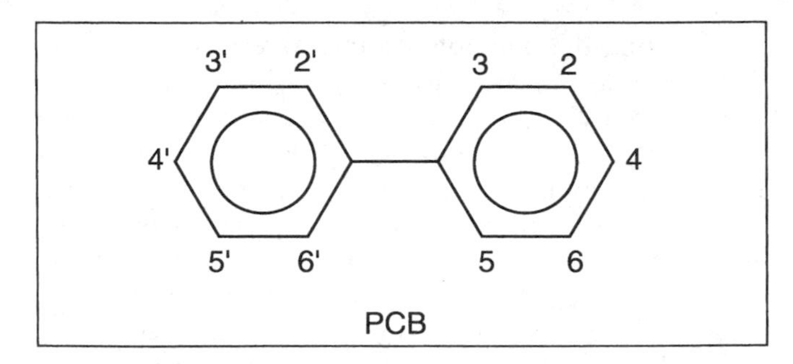

Structure of polychlorinated biphenyl

ORGANISMS and are readily passed up the FOOD CHAIN so that high enough concentrations to cause toxic reactions can build up rapidly (symptoms: gastrointestinal [digestive tract] disturbance, numbness in hands and feet, changes in blood chemistry). Their extreme stability allows them to persist unchanged in the environment and in living tissue essentially forever. This combination of harmful characteristics has forced the use of PCBs to be discontinued throughout most of the world. They have been banned in the United States since 1979 under the terms of the TOXIC SUBSTANCES CONTROL ACT of 1976. Maximum allowable concentrations in foods meant for human consumption: United States, 5 mg/kg; Canada, 2 mg/kg. *See also* AERIAL DEPOSITION.

polycyclic organic matter (POM) polycyclic COMPOUNDS derived from living ORGANISMS, that is, ORGANIC COMPOUNDS containing two or more ring structures (*see* AROMATIC COMPOUNDS; BENZENE RING), found in animal or plant waste products or in their decomposing remains. Common sources of these materials, which are often found in watercourses, include rotting leaves and twigs, pollen, flower petals (the colors of flowers are almost all polycyclic HYDROCARBONS), sawmill wastes, and the urine and FECES of wildlife and livestock. POMS combine with CHLORINE in chlorinated water supplies to form TRIHALOMETHANES; hence these compounds, not particularly dangerous in themselves, must be removed from water supplies as thoroughly as possible before chlorination.

polymer in chemistry, a large MOLECULE made up of a chain of small repeated units (*see* MONOMER). Polymers occur naturally in hair, feathers, woody tissue, and many other parts of ORGANISMS. They are synthesized to form plastics, synthetic rubbers, nonstick coatings for cookware, etc. Polymer chains often contain several thousand monomers. The exact number may vary from molecule to molecule of the same substance, as it is the chain structure itself, and the nature of the chained monomer, that is important in determining the compound's characteristics. *Addition polymers* are formed by simply linking individual monomers together; *condensation polymers* are formed by linking monomers in such a way that part of each monomer breaks off at the point of linkage, releasing two or more ATOMS at each link, which then combine to form a second substance—usually water—as a BYPRODUCT. *Copolymers* are polymers containing more than one type of monomer. They may be formed as either addition or condensation polymers. Copolymers often form complex two-dimensional webs rather than the one-dimensional chains typical of the single-monomer polymers.

polypedon in soil science, a soil body showing relatively uniform characteristics throughout, generally well delineated from other soil bodies. A polypedon is to a PEDON as a POPULATION is to an individual among living things; a swarm of gnats to a single gnat, a forest to a single tree, etc. Polypedons are functionally equivalent to soil series (*see* SEVENTH APPROXIMATION; GREAT SOIL GROUPS).

POM *See* POLYCYCLIC ORGANIC MATTER.

pome *See* FRUIT.

pond a body of standing water that is smaller than a LAKE. The line between "pond" and "lake" has proved difficult to draw with any exactness. The most useful criterion is probably to class all bodies of water small enough that a rainstorm will significantly change the water chemistry as ponds and those larger than this category as lakes, though this classification is far from perfect (Thoreau's Walden Pond would, under this system, be called a lake), and it breaks down completely when PLUVIAL LAKES are considered, as some of these more or less ephermeral bodies of water are very large indeed. The term "pond" is also used for a variety of artificial IMPOUNDMENTS created for industrial or community-health purposes (*see,* e.g., LOG POND; HOT POND; FACULTATIVE POND; POLISHING POND).

ponding (1) the accumulation of a fluid, such as water, in the surface irregularities of the material it is running across. Ponding of RUNOFF during a rainstorm increases the amount that soaks into the ground.

(2) (a) in sanitary engineering, the treatment of biological wastes by bacteriological action in a sewage LAGOON.

(2) (b) in sanitary engineering, the accumulation of water on the surface of a TRICKLING FILTER due to the clogging of the filter by excess bacteria (*see* BACTERIUM) or other insoluble materials.

poor rock in mining, an ORE in which the concentration of metal or other valuable material is too low to make mining profitable. The dividing line between LEAN ORE and poor rock is dependent upon the MINERAL involved, the state of processing technology, and the market for the mined product. Iron miners at the turn of the century, for example, considered "poor rock" any ore containing less than 50% iron. Today that figure has been lowered to 25%. *See also* TAILINGS; BENEFICIATING.

population (1) in the biological sciences, a group of individuals of a particular SPECIES living within a single geographical area. Generally speaking, the term "population" implies some degree of continuity and interaction; that is, all individuals in a given population have the ability in some sense to interact with other members of the population, but do not interact with members of other populations. The grizzly bears of Yellowstone National Park, for example, form a separate population from the grizzly bears of Glacier National Park; the white pine of Michigan forms a separate population from the white pine of Maine; and so forth. Populations often develop their own identifying characteristics: unique growth forms, coloration, behavior patterns, etc. In extreme cases, a population that is isolated for a long period of time may develop into a new species. The term is also used, with appropriate modifiers, for specific portions of a group of geographically related individuals. The *breeding population,* for example, includes all individuals within the general population who are of breeding age.

(2) the number of individuals of a specified type within a specified area. If the type of individual is not specified, it is assumed that "population" refers to humans. Thus, the *population* of the United States is 270 million (humans), while the *land bird population* of the United States is about 5.6 billion (land birds).

The Population Council international population-control organization founded in 1952 to provide support and technical expertise for research to help establish government policies to stabilize world population at a sustainable level. The council's 1999 budget of $69.1 million allows it to underwrite research studies, awards, fellowships, and local training, and to operate regional offices in five different nations (Egypt, India, Kenya, Mexico, and Senegal). Headquarters: One Dag Hammarskjold Plaza, New York, NY 10017. Phone: (212) 339-0500.

Population-Environment Balance a grassroots organization committed to educating the public concerning the negative effects of population growth on the environment. Founded in 1973, the organization advocates U.S.

population stabilization, responsible immigration policy, and increased funding for contraceptive research. Publications include *Balance Activist* and *Action Alerts*. Membership (1999): 10,000. Address: 2000 P Street NW, Suite 210, Washington, DC 20036. Phone: (202) 955-5700. Website: www.balance.org.

population explosion a rapid increase in the number of individuals of a specific type within a specified area, generally to the detriment of the ENVIRONMENT within which the explosion occurs. Population explosions are caused by a relaxation of controls over the birthrate and/or the death rate within a POPULATION, allowing the population to increase at or close to the rate of its BIOTIC POTENTIAL. Population explosions ultimately lead to population collapses as the CARRYING CAPACITY of the environment is exceeded (*see* KAIBAB PLATEAU). The speed of a population explosion is measured by its *doubling rate,* that is, the length of time that it will take the population to double in size. Very small changes in the percentage growth rate of a population can make enormous changes in the doubling rate. Human population worldwide is currently undergoing a population explosion in which the doubling rate is roughly 35 years, or about once per generation. It now stands at approximately 6 billion. This extraordinarily high figure results from a percentage growth rate of just 2% per year. *See also* IRRUPTION; DEMOGRAPHIC TRANSITION; DEMOGRAPHIC TRANSPOSITION.

Population Institute NONGOVERNMENTAL ORGANIZATION dedicated to achieving a more equitable balance between the world's population, environment, and resources. With members in 172 countries, it works with educational leaders, mass media, and policy makers through a network of leaders who volunteer to speak out about population issues. The institute sponsors World Population Awareness Week, observed each year in October with more than 3,000 events around the world; Contraceptive Empowerment Fund, which provides for reproductive health programs in some of the poorest communities of the world; and *POPLINE,* a bimonthly newspaper with a circulation of more than 65,000. Address: 107 Second Street NE, Washington, DC 20002. Phone: (202) 544-3300. Website: www.populationinstitute.org.

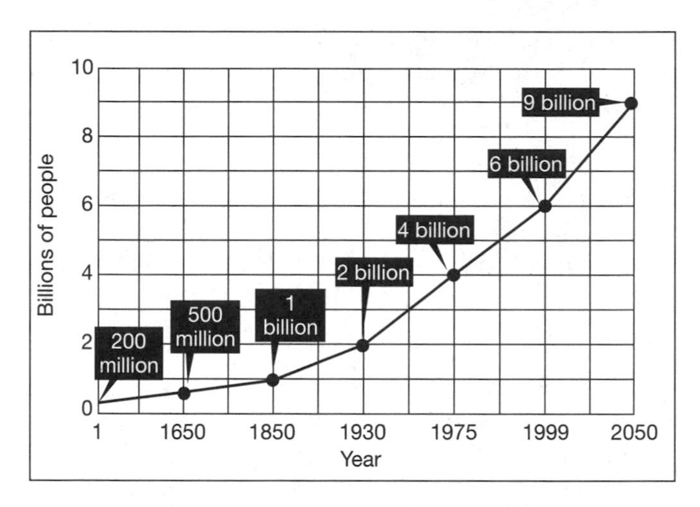

World population growth, A.D. 1–2050 ***(U.S. Census Bureau)***

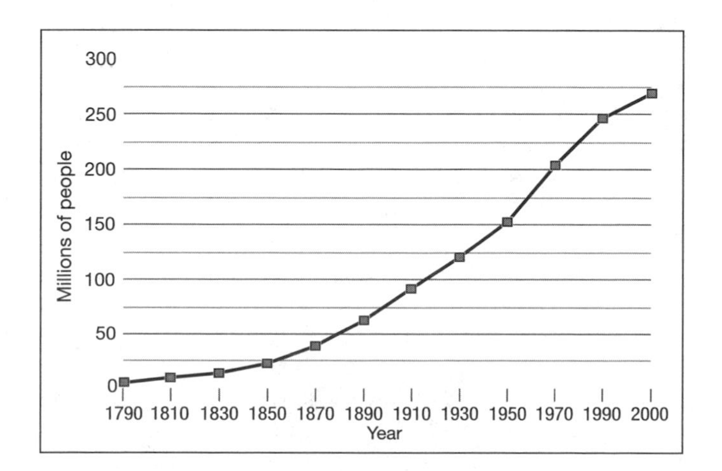

U.S. population, 1790–2000 ***(U.S. Census Bureau)***

Population Reference Bureau nonprofit educational organization founded in 1929 to compile, interpret, and distribute data on population dynamics both nationally and globally. Programs include policy studies, educational programs, and a substantial communications and publications program. Membership (1999): 4,000. Address: 1875 Connecticut Avenue NW, Suite 520, Washington, DC 20009. Phone: (202) 483-1100. Website: www.prb.org.

pork barrel in political terminology, a type of legislation that benefits a small number of states or legislative districts at the expense of the nation as a whole. Pork barrel legislation is attractive to legislators from those areas into which the benefits will flow because their constituents will obtain all of the benefits while paying only a small portion of the costs (which are borne nationwide). The advantages of this economic imbalance are often large enough to blind legislators (and local boosters) to the environmental problems of large public-works projects such as dams and freeways, a circumstance that accounts for the term's perjorative nature among environmentalists, though "pork barrel" is not necessarily synonymous with environmental degradation. *See also* LOGROLLING; OMNIBUS BILL; CHRISTMAS-TREE BILL; IRON TRIANGLE.

porosity in hydrology, a measure of the amount of water that an AQUIFER can hold when it is filled to capacity. Porosity depends upon the size and number of cavities ("pores") within the aquifer, and is independent of the connections which may or may not exist between these cavities. *Compare* PERMEABILITY.

porphyritic rock any IGNEOUS ROCK containing large crystals or rock grains in a fine-grained or glassy-textured matrix. *See also* PHENOCRYST.

post-climax stage in ecology, a successional stage (*see* SUCCESSION) beyond the CLIMATIC CLIMAX for a region, reached in local areas within the region where the MICROCLIMATE is wetter and/or warmer than the climate of the region as a whole, allowing the COMMUNITY to develop further than it normally would. The CLOUD FOREST is a fairly widespread example of a post-climax stage, consisting of a narrow elevation band within a mesic forest community (*see* MESIC SITE) which takes on rainforest characteristics (*see* RAIN FOREST BIOME) due to being frequently bathed in moist mists.

potable water water that is fit for human consumption. The term implies that the water is drinkable as well as safe, that is, that it is free of unpleasant odors, tastes, and colors, as well as containing no TOXINS, CARCINOGENS, MICROORGANISMS, or other health hazards. For standards regarding potability of public water supplies, *see* SAFE DRINKING WATER ACT.

potential yield in forestry, the maximum amount of timber that can be harvested annually from a given area under the principals of SUSTAINED YIELD. *Compare* PROGRAMMED ANNUAL HARVEST.

pour point in chemistry, the lowest temperature at which a given substance will flow (can be poured from a container) under a given set of conditions. The pour point is often somewhat higher than the FREEZING POINT, especially for highly viscous liquids (*see* VISCOSITY).

powder snow snow composed of very small, separate, crystalline flakes. Powder snow forms during cold snowfalls. It does not normally last long, as heat, SUBLIMATION, and the pressure of overlying snow all tend to consolidate the powder and fuse it together. In texture and behavior in the environment it is remarkably similar to sand, particularly in its ability to form "dunes" (snowdrifts). It is considered the best snow for skiing because skis will not normally stick to it due to its low temperature and compact crystal structure.

Powell, John Wesley (1834–1902) American ethnologist, geologist, explorer, and government administrator. Noted for his exploration of the Grand Canyon, Powell conducted geological surveys throughout the West, promoted conservation of the region (with a special emphasis on the lack of water supply for agriculture), and classified Indian languages within the area. After studying at Oberlin College and Wheaton College and serving in the Union Army during the Civil War, Powell became a professor of geology at Illinois Wesleyan University and then museum curator at Illinois Normal College (now Illinois State University). Beginning in 1867, he led a series of expeditions to the Green River and Colorado River canyons where he formulated the basic principles of structural geology. In 1879 he was appointed the first director of the U.S. Bureau of Ethnology, and he compiled and later published the first complete classification and distribution map of 58 Indian languages in the United States and Canada. He was the first director of the U.S. GEOLOGICAL SURVEY, from 1881 to 1892, and the first director of the Smithsonian Institution's Bureau of American Ethnology, from 1879 to 1902. Powell's 1879 government report, *Lands of the Arid Region of the United States* (available in a reprint edition), prophetically warned against the overdevelopment of western lands under the Homestead Act (1862). His concerns are now seen to have been well-founded, especially regarding the destruction of rangeland and the limitations of irrigation agriculture, but his recommendations were ignored. Powell also published *Explorations of the Colorado River of the West* (1875) and *An Introduction to the Study of Indian Languages* (1877). Wallace Stegner's *Beyond the Hundredth Meridian: John Wesley Powell and the Second Coming of the West* (1954) provides a definitive biography.

power administrations, federal originally a group of five agencies within the U.S. Department of Energy (*see* ENERGY, DEPARTMENT OF) charged with marketing and distributing the electricity generated by federally owned hydroelectric facilities. Most of these facilities are dams operated by the ARMY CORPS OF ENGINEERS and the BUREAU OF RECLAMATION; however, a few are operated by other agencies, especially within the area served by the Western Area Power Administration, which contains a large number of dams operated by the INTERNATIONAL BOUNDARY AND WATER COMMISSION. The power administrations construct and operate most high-voltage transmission lines in their regions, and coordinate the flow of power along the lines within their regions and among the several regions so that surpluses in one area may be used to offset deficiencies in other areas. They may have other duties as well. The Southeastern Power Administration, for example, participates in the planning and construction of generating facilities, while the Bonneville Power Administration is charged by the Northwest Power Planning Act of 1980 with promoting energy conservation, fish and wildlife enhancement, and the use of renewable resources as well as coordinating power transmission. All of the power administrations are legally required to sell power at the lowest possible cost "consistent with wise business practices"; most also are charged with giving preference to public utilities and cooperatives. Power is sold to other federal agencies, and occasionally to preferred customers, at cost. The five power administrations, their dates of formation, and their regions are: the Alaska Power Administration (1944: Alaska); the Bonneville Power Administration (1937: Idaho, Montana, Oregon, and Washington); the Southeastern Power Administration (1950: Alabama, Florida, Georgia, Kentucky, Mississippi, North Carolina, South Carolina, Tennessee, Virginia, and West Virginia); the Southwestern Power Administration (1943: Arkansas, Kansas, Louisiana, Missouri, Oklahoma, and Texas); and the Western Area Power Administration (1977: Arizona, California, Colorado, Iowa, Kansas, Minnesota, Montana, Nebraska, Nevada, New Mexico, North Dakota, South Dakota, Texas, Utah, and Wyoming). The ALASKA POWER ADMINISTRATION was privatized in 1995.

ppb an abbreviation for *parts per billion.*

ppm an abbreviation for *parts per million.*

prairie *See* GRASSLAND BIOME; GREAT PLAINS; BUFFALO COMMONS.

preattack plan (**presuppression plan**) in Forest Service terminology, a plan for the suppression of WILDFIRE, prepared as part of the forest planning process (*see* FOREST PLAN) and kept on file for quick reference if a fire breaks out. The preattack plan includes the organization of the fire management team; methods for financing fire suppression efforts; means of surveillance and monitoring fire progress; the training, mobilization, and dispatching of firefighting teams; fire weather forecasting and hazard ratings; and area-specific data such as the location of natural FIREBREAKS, water sources, sensitive resources, extreme FUEL HAZARDS, and so on.

precipitate in chemistry or physics, in the broadest sense, a material of one phase (solid, liquid, or gas) that forms in a dispersed manner throughout a material of a second (usually higher) phase as a result of some chemical or physical reaction. The most familiar precipitate in everyday life is the water droplets—clouds, mist or fog—that form when air saturated with water vapor cools down and can no longer hold all the vapor within it. The extra vapor condenses into liquid water at a constant rate throughout the cooling air mass, forming clouds (*see* PRECIPITATION). Precipitates in reactions are often solids that form within liquids, either due to the SUPERSATURATION of a SOLUTION (in which case part of the solute precipitates out) or as a result of a reaction between two dissolved substances in which a third, insoluble substance is formed. The solid precipitate may then be separated from the liquid by filtering, by centrifuging, or by simply allowing it to settle to the bottom. Occasionally, the precipitate and the material it precipitates within are in the same phase. Some metal ALLOYS, for example, are formed by the precipitation of solids within other solids.

precipitation in meteorology, rain, snow, sleet, hail, or any other weather event in which solid or liquid water falls out of the sky. Precipitation results from the condensation of water vapor from supersaturated air, forming tiny water droplets (or, if the air mass is cold enough, tiny particles of ice). These are carried by air currents until they coalesce with other water droplets or ice particles into drops, snowflakes, hailstones, etc., which are too big to be carried in the air any longer and which therefore fall to the ground. The amount and type of precipitation that falls on an area are extremely important in determining the makeup of the ecological COMMUNITY in that area (*see,* e.g., DESERT BIOME; RAIN FOREST BIOME). *See also* HYDROLOGIC CYCLE; CLIMATE; and separate articles on the various forms of precipitation (RAIN; SNOW; HAIL; etc.).

precommercial thinning in forestry, the removal of some of the trees from a dense STAND of UNMERCHANTABLE TIMBER in order to allow the remaining trees to reach merchantable size more rapidly, or to meet some other harvest goal such as improved production of poletimber (*see under* POLE-SIZE TREE) or wood pulp. *Compare* COMMERCIAL THINNING.

predator a living ORGANISM that captures and kills other living organisms in order to eat them. The organisms that are captured and killed are called PREY, and the act of capturing and killing them is known as *predation.* The word "capture" implies that both predator and prey are motile (*see* MOTILITY), and this is generally the case, though a number of examples of so-called "passive predation" are known in which the predator is stationary and only the prey is motile (the sea anenome, CARNIVOROUS PLANTS). Predator and prey almost always belong to the same KINGDOM: animals prey on other animals, PROTISTA prey on other protista, etc. (again, carnivorous plants are an exception to this rule, along with FILTER FEEDERS and those protista whose prey consists principally of bacteria [*see* BACTERIUM]). They are generally K-SELECTED SPECIES, and normally have longer lives and a higher degree of behavioral flexibility than their prey. From an ecological standpoint, predators are essential members of any COMMUNITY or ECOSYSTEM, serving as controls on prey numbers and helping to weed out less genetically fit individuals from the prey POPULATION by killing them before they have a chance to reproduce. *See also* PREDATOR CONTROL; PREDATOR-PREY RELATIONSHIP.

predator control the protection of LIVESTOCK from wild PREDATORS, such as coyotes, cougars, bears, bobcats, feral dogs (*see* FERAL ANIMAL), etc. Predator control may be divided into two types, *non-lethal* and *lethal*. Non-lethal methods are relatively noncontroversial. They include live-trapping and relocating, fencing, guard dogs, and the use of repellants, either chemical (substances that cause the predator to lose its desire to eat the livestock) or physical (noisemakers, ultrasound generators, etc.) Lethal methods (that is, methods that kill the predator instead of merely chasing it away) are considerably more controversial. These include poisoning (*see* BAIT STATION; 1080; COYOTE GETTER); shooting, often from an airplane; *denning*, in which a denful of cubs or pups is located and destroyed; and the use of the leghold trap, which captures and holds the predator so that it may be easily shot or otherwise destroyed. Studies of intensive predator-control efforts using lethal methods have demonstrated that these efforts generally cost more to operate than they prevent in livestock losses, and that poisoning efforts in particular sometimes actually increase livestock losses by poisoning rodents along with the predators, to the point that the predators must increase their dependence on livestock to make up for the lack of rodent prey. On the other hand, lethal control methods that are targeted directly at individual livestock-killing animals have a reasonable place in livestock management. These include the use of dogs to track predators from a kill; the equipping of sheepherders with rifles for the destruction of animals directly observed in livestock-killing activity; and the use of poisoned collars on livestock to poison animals that attack them, provided that the poison is one that does not persist in the environment or build up in the FOOD CHAIN and that the collars are only placed on animals that are actually at risk (that is, in an area where predator depradations on livestock have already been demonstrated). *See also* PREDATOR-PREY RELATIONSHIP.

predator-prey relationship in ecology, any of the set of relationships that exist between a PREDATOR and its preferred PREY. Predator-prey relationships are often complex interdependencies that tie the predator and prey POPULATIONS closely together in a dynamic equilibrium, so that changes in the status of one tend to change the status of the other in the same direction. For example, a reduction in the number of ladybugs—a principal predator on APHIDS—will initially lead to a POPULATION EXPLOSION of aphids, but the resulting defoliation of the aphids' food plants will eventually cause the aphid population to crash, so that the reduction of ladybug numbers ultimately leads to a reduction of aphid numbers as well. *See also* PREDATOR CONTROL; BALANCE OF NATURE; KAIBAB PLATEAU.

preferred alternative in environmental law, an action that an agency proposes to carry out, expressed as one of a range of alternatives in an ENVIRONMENTAL IMPACT STATEMENT.

prelogging in forestry and logging, the removal of downed timber, UNDERSTORY vegetation, etc., which might cause damage to felled trees during a logging operation. Prelogging can be carried out for profit if the downed material is sound and/or the understory vegetation is large enough to be commercially valuable for wood chips, poles, firewood, or some similar use.

preprocessing *See* BENEFICIATING.

prescribed burn in forestry, a fire set on purpose under precisely controlled ("prescribed") conditions in order to achieve some desired silvicultural goal. A prescribed burn differs from a CONTROLLED BURN in being part of an overall silvicultural management plan rather than simply being set to reduce a FUEL HAZARD, although fuel-hazard reduction may be its principal (or even its only) goal. Besides hazard reduction, prescribed burning may be used to prepare seedbeds for the planting of new trees, to remove competing vegetation, to improve BROWSE for wildlife and livestock, or to allow for the reproduction of FIRE-DEPENDENT SPECIES. It is always done under carefully managed conditions in which both the intensity of the fire and the acreage burned can be controlled. Even so, prescribed and controlled burns can become forest fires in themselves, such as the fire that destroyed almost 50,000 acres of forest land and more than 200 homes in Los Alamos, New Mexico, in 2000. The fire originated as a controlled burn by the National Park Service to clear 20 areas of brush in a remote area of Bandelier National Monument.

prescribed cut *See* ALLOWABLE CUT.

prescription in forestry, *see* SILVICULTURAL PRESCRIPTION; HARVEST PRESCRIPTION.

preservationism an approach to protecting publicly valued lands and landscapes that relies on acquisition and rigorous management of land resources for wildlife, natural beauty, ecological integrity, recreation, and similar purposes. In policy debates, "preservation" is often contrasted with "conservation," described by some as a more flexible land management approach which implies economic use of the land. For example, National Parks are "preserved," whereas National Forests are "conserved," since they are open to logging, mining, and other uses. The distinction between the two concepts was most clearly drawn in the late 19th and early 20th centuries by JOHN MUIR and GIFFORD PINCHOT. Muir, founder of the SIERRA CLUB, believed that outstanding natural areas should be preserved for their own sake as an ethical proposition, and Pinchot, the father of modern forestry in the United States, believed that forests could be cut commercially and the land preserved at the same time. *See also* CONSERVATION MOVEMENT.

preservation of character in landscape management, management of a visual landscape so that only ecological changes occur. Preservation of character is the most restrictive visual-alteration class, and is generally reserved for WILDERNESS AREAS, NATIONAL PARKS, and similar preserves, although it may be called for in the management of some SPECIAL INTEREST AREAS. *Compare* RETENTION OF CHARACTER. *See also* VISUAL RESOURCE MANAGEMENT SYSTEM.

presumed test *See under* COLIFORM COUNT.

presuppression plan *See* PREATTACK PLAN.

pretreatment in waste-treatment technology, treatment of waterborne wastes before they enter a SEWAGE TREATMENT PLANT in order to remove materials that might create problems for the plant, such as excessive grease, fat and lard; highly corrosive substances; metal slivers; etc. Municipal wastes are usually pretreated as they go into the sewage plant. This pretreatment almost always includes bar screens to remove large floating debris such as logs or tires, and may also include a *skimming tank* to remove floating oils and greases and a GRIT CHAMBER to remove sand and other coarse settleable materials. Industrial wastes are generally pretreated at the source, before release to the sewers. This industrial pretreatment should remove most toxic chemicals (especially those requiring unusual treatment procedures or those that would kill the ORGANISMS used for biological treatment; *see* SECONDARY TREATMENT) and should compact the wastes into a smaller stream, as well as meeting regular municipal pretreatment requirements. Industrial pretreatment is subject to federal standards (PRETREATMENT STANDARDS, or PTS) under the terms of the CLEAN WATER ACT. These standards are less strict for retrofitting older plants with pretreatment facilities ("pretreatment of existing sources," or PTES) than they are for new plants ("pretreatment of new sources," or PTNS).

pretreatment standards in pollution control, standards established for the removal of POLLUTANTS from industrial wastewaters before the wastewaters may be discharged into municipal sewer systems. Pretreatment standards are set primarily to prevent damage to municipal sewage treatment plants, especially by toxic compounds that may kill the bacteria in a SECONDARY TREATMENT FACILITY, and to halt the discharge of pollutants that will pass through a municipal treatment plant unchanged and thus will pollute the RECEIVING WATERS. Standards may also be imposed to prevent the inclusion of contaminants in the SLUDGE produced by the municipal plant, or to reduce health hazards to sewage system employees. Guidelines for setting pretreatment standards are provided by the ENVIRONMENTAL PROTECTION AGENCY under the terms of sec. 307 of the CLEAN WATER ACT. They may be set either for specific pollutants (especially the PRIORITY POLLUTANTS) or for specific types of industry (that is, specific types of pretreatment plants may be required to be installed by users of specific types of industrial processes). New source standards may differ from retrofit standards. The actual setting and enforcement of pretreatment standards for a given sewage system is up to the municipality that operates the system, as long as the system itself continues to operate within the EFFLUENT STANDARDS established by the Clean Water Act.

prevailing wind a wind that blows regularly from the same direction. The prevailing wind for any given location is a function of both topography and meteorology, that is, it is affected both by the presence of LANDFORMS such as mountains that block or deflect winds and large bodies of water that cause differential heating and cooling, and by the paths of storms caused by regular circulation patterns within the atmosphere. It may strongly affect vegetation patterns and the GROWTH FORM taken by trees and other large plants.

prevention of significant deterioration (PSD) in pollution control, program established under sec. 107 of the CLEAN AIR ACT to maintain current levels of air quality in those areas where the air is already cleaner than required by the NATIONAL AMBIENT AIR QUALITY STANDARDS (NAASQs), or where air quality has not been adequately measured. PSD standards set budget ceilings, or *increment standards,* for each POLLUTANT covered by the Clean Air Act. No discharge is allowed that would exceed these increment standards. The increment standards themselves are set according to the AIR QUALITY CLASS each airshed covered by the PSD program is classified into, with the least deterioration permitted in Class I airsheds such as those surrounding large National Parks. Industries wishing to locate in PSD areas must obtain a permit from the state (or from the ENVIRONMENTAL PROTECTION AGENCY in those states that have not filed adequate STATE IMPLEMENTATION PLANS). These permits are issued only if the industry may be certified as using the BEST AVAILABLE CONTROL TECHNOLOGY for the pollutants it emits and if the overall level of pollutants in the airshed will remain below the increment standards for its air quality class.

prey any living ORGANISM that is captured and killed by another living organism in order to serve as a food source. The word "captured" implies mobility, that is, the prey must be capable of taking evasive action to avoid capture, a qualification that eliminates the plants eaten by HERBIVORES from the definition. (Some ecologists would prefer that this distinction not be made.) Generally speaking, prey SPECIES and their PREDATORS belong to the same KINGDOM: animals are preyed upon by other animals, PROTISTA are preyed upon by other protista, and so forth (but see the discussion of exceptions under PREDATOR). Prey are usually R-SELECTED SPECIES. They are behaviorally less complex than predators, and they lead shorter lives, circumstances that they compensate for through extreme fecundity. *See also* PREDATOR-PREY RELATIONSHIPS.

Price-Anderson Act federal (U.S.) legislation to limit the financial liability of builders and operators of nuclear power plants (*see* NUCLEAR ENERGY), signed into law by President Dwight Eisenhower on September 2, 1957. Designed to promote nuclear power by improving its insurability, the Price-Anderson Act—actually an amendment to the Atomic Energy Act of 1954—established an absolute upper limit of $560 million on liability for damages resulting from an accident at any commercial nuclear reactor, including property damages, medical expenses, loss of life, and punitive and "mental suffering" damage awards. A "no-fault" clause added in 1966 provides that following an "extraordinary nuclear occurrence" (ENO), all claims for damages will be consolidated into a single court, and that the companies responsible for the ENO would assume the full $560 million liability. Otherwise, $160 million was to be provided by insurance policies sold to the utilities and the remainder comes from a financial pool established by the utilities themselves, for which they were assessed $50 million for each separate plant that they operate. In 1988 Price-Anderson was amended to increase the liability cap to $7 billion for severe accidents, more than 12 times higher than the original act had provided. In view of the nuclear accidents and near-accidents in the United States, the cost and difficulty of waste disposal, and a host of other problems, the Price-Anderson Act has provided a subsidy to the nuclear industry that many environmentalists believe is unwarranted. *See also* NUCLEAR WASTE AND CLEANUP.

primary growth in botany, an increase in height (or length) of a plant SHOOT or root. Some thickening may also take place, but is incidental to the process of lengthening. Primary growth is controlled by the *apical meristem,* a center of CELL division located in the growing tip of the shoot or root. It is called "primary" because it is always the first type of growth to occur when a seed sprouts, as apical meristems for root and shoot development are present in the plant embryo. *Compare* SECONDARY GROWTH.

primary productivity in limnology, the amount of new BIOMASS created within a body of water during a given period of time. More technically, primary productivity is defined as the amount of INORGANIC CARBON converted to ORGANIC CARBON during the specified period, usually expressed as $kg/m^2/day$. Primary productivity depends entirely on PHOTOSYNTHESIS (but *see* CHEMOSYNTHESIS); therefore, it is exclusively a function of plants and PHYTOPLANKTON, and takes place only in the PHOTIC ZONE. In small bodies of water, much of the primary productivity is carried on by MACROPHYTES. In large bodies of water, the burden shifts to phytoplankton due to the much greater proportion of deepwater areas where macrophytes cannot grow. *See also* TROPHIC LEVEL.

primary succession the establishment of natural plant life in areas that had not previously supported any life, such as areas covered by lava flows or newly exposed substrates in glaciated areas. *Compare* SECONDARY SUCCESSION; *see also* SUCCESSION.

primary treatment in sanitary engineering, the treatment of SEWAGE or other waterborne wastes to remove floating and suspended solids. Primary treatment is entirely physical; that is, it depends upon physical processes (screening, settling, etc.) rather than biological or chemical processes to accomplish its task. Most primary-treatment facilities consist of three stages: a screen made of metal bars set 1 inch apart to catch large material; a GRIT CHAMBER to remove sand, metal fragments, and other non-organic solids; and a SETTLING BASIN (also known as a *primary clarifier*) to remove the remaining organic solids (*see* ORGANIC MATTER). An optional fourth stage, a mechanical chopper or *comminutor,* may be placed between the grit chamber and the settling basin to chop the solids that have passed through the screen into fragments 1/8 inch or smaller in diameter. Primary treatment removes up to 60% of the solids, and up to 40% of the organic wastes, from the WASTE STREAM. The accumulated solids, known as SLUDGE, must be periodically or continuously removed and disposed of, a process that almost always requires additional treatment. The cleansed EFFLUENT may either be released into the environment (increasingly rare) or channeled into a secondary or tertiary treatment facility, depending upon the treatment standards required (*see* SECONDARY TREATMENT; TERTIARY TREATMENT). *See also* PRETREATMENT; SEWAGE TREATMENT PLANT.

primary wave *See* P-WAVE.

prime farmland a generalized description of agricultural land with the best soils and other attributes for producing food, feed, forage, fiber, and oilseed crops. Consisting of land in the USDA's soil capability classes I–III, prime farmland has an adequate water supply, is neither overly acid or alkaline, is free of rocks, and is not overly susceptible to soil erosion. Of the 931 million acres of land in farm use (1997), about 35% of it (325 million acres) is prime farmland, most of it in states between the 100th meridian and the Appalachians. Prime farmland is also commonly found around urban areas, causing concern that some of the best soils are lost to URBAN SPRAWL. *See also* AGRICULTURAL LAND, U.S.

Primitive Area a land management category established by the FOREST SERVICE in 1929 to maintain large tracts of land on the NATIONAL FORESTS in a primitive, unspoiled state for recreational purposes. Primitive Areas were similar to WILDERNESS AREAS, except that limited timber harvesting was allowed and motorized vehicles could be used within the boundaries of the area for administrative purposes. The category was eliminated by the WILDERNESS ACT of 1964, which provided that all Primitive Areas in existence at that time would either be reclassified as Wilderness Areas or would be released for general forest uses by September 1974.

primitive recreation in land-use planning, all unmotorized types of recreation associated with unroaded land. The category includes hiking, backpacking, canoeing, fishing, hunting, and cross-country skiing. It does not include motorized camping or the use of OFF-ROAD VEHICLES. Primitive recreation is the only form of recreation practicable in WILDERNESS AREAS, but it is not limited to such areas. Any piece of unroaded land can be used. *Compare* DISPERSED RECREATION; DEVELOPED RECREATION.

prior appropriation (Colorado rights) in water law, a system of WATER RIGHTS that allocates scarce water resources by priority of time; that is, those who have obtained water rights recently ("junior rights") are not allowed to use any water until all those who obtained rights earlier ("senior rights") have used the full amount of water specified in their right. The doctrine of prior appropriation grew out of the so-called "miners' laws" of the mid-19th century western United States where, because of the scarcity of streams, RIPARIAN DOCTRINE was unworkable. It has since been refined and in some cases drastically limited by statute, but its principals remain unchanged. Under prior appropriation, water is considered to be publicly held. The right to use that water, however, is private property, and may within limits be treated as its owner sees fit. Thus, water may be transferred for use far from a watercourse, may be used for a purpose other than that for which it was originally obtained, and (in some states) may be bought and sold. The right is attached to the land on which it is used, and when that land is sold the right passes to the purchaser, losing no seniority in the transaction. There are, however, certain limitations on the obtaining and holding of rights. The water must be put to BENEFICIAL USE, usually as defined by statute. (Uses commonly defined as "beneficial" include agriculture, mining, industrial development, domestic supply, and usually recreation and the maintenance of fish and wildlife POPULATIONS.) The applicant for the right must also satisfy a *test of intent.* For instance, if water is diverted for the purpose of draining a MARSH, that diversion cannot be used to establish priority of right if the water is later used for IRRIGATION; the date used to establish the irrigation right's seniority will be the date on which it first started to be used, not when it first started to be diverted. Most prior-appropriation states require the water to actually be diverted from the watercourse in which it naturally occurred before seniority can be established, although the principal of INSTREAM RIGHTS—the beneficial use of water as it flows naturally in the stream—has widely eroded that requirement in recent years. Rights that are not used for a specified period of time—usually five years—may be declared abandoned, causing them to lose all seniority. A right may not normally be obtained or held for more than the amount of water commonly recognized as adequate for the use to which it will be put (*see* DUTY OF WATER). Finally, prior-appropriation rights are usually subject to some form of preferential-rights restrictions as well. Types of use are legislatively categorized from "most preferred" (e.g., municipal water supply) to "least preferred" (e.g., recreation), allowing those in the most preferred categories to condemn and purchase the rights of those in less-preferred categories and allowing distinctions to be made between right-holders whose priorities date to the same time but who are using their rights for different purposes. A sensible and equitable means of distributing scarce water when users were few, all rights were relatively young, and all uses were similar in importance, prior appropriation has come under increasing attack during recent years because of its tendency to promote water waste by senior users, who are encouraged to continue to grow water-hungry crops and engage in wasteful irrigation practices in order to conserve the size of their right by the fact that any part of it they do not "use" for a given statutory period may be declared abandoned and taken from them. *See also* HYBRID RIGHTS.

priority pollutant in environmental law, a water POLLUTANT listed by the ENVIRONMENTAL PROTECTION AGENCY under the terms of the 1977 amendments to the CLEAN WATER ACT. Priority pollutants are subject to federal discharge limitations based on BEST AVAILABLE TECHNOLOGY standards. The list is updated regularly, with the updates published in the CODE OF FEDERAL REGULATIONS.

private sector in economics, the area of the economy encompassing all transactions in which a government body is not directly involved. Private sector transactions may be regulated by government bodies, but these government bodies are not directly involved in the sale, purchase, use, construction, or other contracted aspects of the transaction. For example, a farmer who subdivides land for sale to individual homeowners is engaged in a private-sector transaction, but that transaction is subject to zoning and other land use laws and regulations, as well as to consumer-protection laws and laws regarding the recording of deeds.

process water water that is being utilized within an industrial plant, that is, water that is part of a process used to manufacture something. Strictly speaking, the term "process water" refers only to water used directly for an industrial process, that is, as a raw material, cleansing agent, ELECTROLYTE, source of steam, etc. However, the term is sometimes applied to industrial cooling water as well. *Compare* FEED WATER.

producer in ecology, an ORGANISM that converts INORGANIC CARBON to ORGANIC CARBON, generally through PHOTOSYNTHESIS, although there are a few bacteria (*see* BACTERIUM) that practice CHEMOSYNTHESIS instead. ALGAE, photosynthetic bacteria, and most plants are classified as producers, which occupy the lowest level on the ECOLOGICAL PYRAMID and are depended upon by all other organisms to produce the complex MOLECULES necessary to support life. *Compare* CONSUMER. *See also* AUTOTROPHISM; PRIMARY PRODUCTIVITY.

productivity of an ecosystem, *see* PRIMARY PRODUCTIVITY.

product water in water-treatment terminology, *see* FINISHED WATER.

proglacial lake in geology, a lake formed adjacent to an active GLACIER and filled primarily by the glacier's meltwaters. One or more banks of a proglacial lake are always formed of glacial ice. Proglacial lakes range in size from tiny ponds to immense bodies of water. The Great Lakes, for example, were originally formed as proglacial lakes of the CONTINENTAL GLACIER.

programmed annual harvest in forestry, the amount of timber scheduled to be cut from a piece of COMMERCIAL FOREST LAND in any given year. The programmed annual harvest may be greater or less than the ALLOWABLE SALE QUANTITY, depending upon the management strategy being followed; however, the average programmed annual harvest on any given plot of land cannot exceed the allowable sale quantity over the life of the applicable FOREST PLAN, usually a period of 10 years or less.

project economic costs in BENEFIT/COST ANALYSIS, the costs directly incurred by the construction and operation of a project, such as the costs of materials, the salaries of construction workers and operators, the annual maintenance costs, etc. *Compare* INDUCED COSTS.

prokaryote in the biological sciences, a living ORGANISM made up of simplified CELLS lacking a true nucleus and containing few organelles and no mitochondria (*see* CELL: *structure of cells*). Those prokaryotes that contain CHLOROPHYLL do not confine it to CHLOROPLASTS, but spread it throughout the cell. The prokaryotes make up a separate KINGDOM of living things, the MONERA. It is generally thought that the EUKARYOTES evolved from the prokaryotes, and it seems fairly certain that the mitochondria and chloroplasts within eukaryotic cells evolved from captive prokaryotic cells that entered into SYMBIOSIS with their hosts. Prokaryotes occasionally form colonies in which some members of the colony take on specialized tasks such as food capture or attachment to the SUBSTRATE, but no true multicellular prokaryotic organisms are known. The group includes (and is limited to) the bacteria and bluegreen algae (*see* BACTERIUM; CYANOPHYTE).

property rights a constitutional issue related to the implementation of some environmental policies. A part of the Fifth Amendment of the Constitution, which deals mainly with proscriptions against self-incrimination, reads "nor shall private property be taken for public use without just compensation." Known as the "takings clause," this passage is used by those opposing land-use and other governmentally-mandated controls to demand payment for the imposition of regulations, thus frustrating public policies undertaken for public benefit. Commonly, property-rights lawsuits are brought by owners against governments for large-lot zoning or for restrictions on the use of land to protect endangered species, to preserve wetlands, and the like. Despite the clause, the courts have allowed fairly wide latitude to governments to control the uses of land under the "general welfare" doctrine (Article I, Section 8). Property rights are also a concern for those interested in the application of a LAND ETHIC, and for religious bodies that wish to engage in the development of environmental policies. In the Hebrew Bible, absolute property rights are constrained by religious law in the Pentateuch, and are taken up in Psalms, Isaiah, Job, and elsewhere. *See also* LAND USE CONSTRAINTS; LAND USE PLANNING; RELIGION AND THE ENVIRONMENT.

protective resemblance *See* MIMICRY.

protein one of a large group of complex organic molecules (*see* ORGANIC COMPOUND; MOLECULE) that are one of the primary components of living tissue. Proteins are composed almost exclusively of AMINO ACIDS, although some (the *conjugated proteins*) have other molecular groups attached to them. A single ORGANISM may be composed of thousands or even hundreds of thousands of different types of protein, each doing a separate, highly specific job: building bones (*karotins*), forming muscles (*actomyosins*), assisting in the digestion and metabolism of food (*enzymes*), carrying chemically coded messages through the bloodstream and the lymphatic system (*hormones*), and so forth. Because of this high specificity, proteins can also be dangerous. A protein in the wrong spot can cause a reaction ranging from mild irritation to the death of the invaded organism. *See* ALLERGEN; TOXIN.

STRUCTURE

The basic (*primary*) structure of a protein molecule is a *polypeptide chain,* which consists of a group of amino acids joined end-to-end by the bonding of the amine GROUP from one amino acid to the carboxyl group of the next, a connection known as the *peptide bond.* Polypeptide chains may be 1,000 or more amino acids in length. Proteins also show *secondary structure,* a term relating to the physical shape taken by the polypeptide chains. This is mostly commonly a spiral (the *alpha helix*), but it may also consist of two or more chains lying side by side and forming a flat structure known as a *beta structure* or *pleated sheet,* or a complex braid called a *triple helix* or *collagen.* Finally, both single and multiple-chain proteins may show *tertiary structure,* curling back upon themselves in tightly constricted lumps known as *globular proteins.* Globular proteins are the most biologically active forms of these molecules.

Protista in biology, one of the five KINGDOMS of living things (the other four: plants, animals, fungi [*see* FUNGUS], and MONERA [bacteria and blue green algae]). The Protista are primarily one-celled ORGANISMS, although some multicellular forms are classified here for want of a better place to put them. All are EUKARYOTES, that is, have large, complex cells with true nucleii. As defined by most authorities, the kingdom includes all PROTOZOANS, most ALGAE, and (usually) the so-called SLIME MOLDS.

proton *See* ATOM: *structure of the atom.*

protozoan any of numerous one-celled heterotrophic, motile ORGANISMS (*see* HETEROTROPH; MOTILITY) commonly found in bodies of water and in the soil, including AMOEBAS, most FLAGELLATES, paramecia (*see* PARAMECIUM) and others. Once considered primitive animals (hence the name), protozoans are now classed with the ALGAE in a separate KINGDOM (*see* PROTISTA). A few are parasitic in humans and other animals, causing (among other diseases) African sleeping sickness and AMOEBIC DISENTERY (*see* PARASITE). Most, however, function as primary consumers (*see* CONSUMER), making them an important part of the FOOD CHAIN, especially in aquatic ECOSYSTEMS, where they make up a large portion of the PLANKTON. *See also* MICROORGANISM.

province in geography, see PHYSIOGRAPHIC PROVINCE.

pseudokarst in geology, a type of topography that closely resembles KARST TOPOGRAPHY but is developed in BASALT rather than in LIMESTONE. Pseudokarst results from the formation of LAVA TUBES in active

PAHOEHOE flows, with subsequent ceiling collapse forming SINKS, *blind valleys* (small valleys whose bottom ends are closed by a rock wall; the stream forming the valley escapes through a cavern at the base of the wall), and other features normally associated with solutional caverns in limestone. Examples of pseudokarst are found in active volcanic regions worldwide. In the United States, the best examples are in Hawaii, northern California, central Oregon, and southwestern Washington.

psi abbreviation for POUNDS PER SQUARE INCH.

PSI *See* POLLUTANT STANDARD INDEX.

psia abbreviation for pounds per square inch, absolute pressure (*see* ABSOLUTE PRESSURE).

psig abbreviation for pounds per square inch, gauge pressure (*see* GAUGE PRESSURE).

pteridophyte in botany, any VASCULAR PLANT that produces spores instead of seeds, including the horsetails, club mosses, whisk ferns, and true ferns. The presence of vascular (liquid-conducting) tissue and the development of a well-differentiated structure including roots, stems, and leaflike appendages (known as *fronds*) show the close relationship of the pteridophytes to the seed plants. However, they lack both flowers and cones, and though their haploid generation, like that of seed plants, is much smaller and less well developed than their diploid generation (*see* ALTERNATION OF GENERATIONS), it develops as a separate, independent plant rather than being parasitic on the more highly developed diploid plant. *Compare* SPERMATOPHYTE; BRYOPHYTE.

public domain in the United States, federal land that has always been in public ownership and that has not been specifically designated for use as a NATIONAL FOREST, NATIONAL PARK, WILDLIFE REFUGE, military reservation, etc. The public domain is managed by the BUREAU OF LAND MANAGEMENT under the terms of the FEDERAL LAND POLICY AND MANAGEMENT ACT OF 1976, the TAYLOR GRAZING ACT OF 1934, and various other laws. It currently constitutes about 338.6 million acres, almost all in the western states. Most of this land is managed for timber production and livestock grazing, with some set aside for WILDERNESS AREAS, RESEARCH NATURAL AREAS, recreation areas, and similar uses.

HISTORY OF THE PUBLIC DOMAIN

The first public domain consisted of the western lands ceded to the federal government by the original 13 colonies as part of the terms of the Articles of Confederation in 1780. These lands comprised most of the area between the Mississippi River and the Appalachian Divide, a total of approximately 236 million acres. The Louisiana Purchase of 1803 added another 530 million acres, and various purchases and treaties over the next century added the rest of the western states (except Texas) and Alaska. At one time or another, more than 1.8 billion acres of land has been part of the public domain. Much of this was sold to raise revenue, was donated to states to raise money for education or to corporations and individuals as rewards for services rendered (*see,* e.g., RAILROAD GRANTS; CHECKERBOARD OWNERSHIP), or was parceled out to individuals through the HOMESTEAD ACT and other legislation designed to encourage settlement of the west. During the first half of the 20th century, federal policy toward the public domain gradually changed from disposal to stewardship, and today the only public-domain lands sold are those which have for one reason or another been declared surplus—a very small amount (less than 100,000 acres) each year. *See also* PUBLIC LAND; ACQUIRED LANDS; PUBLIC LAND OFFICE; PUBLIC LAND LAW REVIEW COMMISSION.

Public Employees for Environmental Responsibility (PEER) an association of government professionals such as biologists, scientists, resource managers, or law enforcement officials. PEER is dedicated to the conservation of the environment of the United States through sustainable management of public resources, environmental ethics, and professional integrity. It publishes *PEEReview* and various issue papers. Address: 2001 S Street NW, Suite 570, Washington, DC 20009. Phone: (202) 265-7337.

Public Health Service agency within the Department of Health and Human Services (*see* HEALTH AND HUMAN SERVICES, DEPARTMENT OF) charged with maintaining the physical, mental and environmental health of American citizens. Created by Congress on July 1, 1944, the service is headed by the assistant secretary for health and human services. He is assisted by five deputy assistant secretaries and by the surgeon general of the United States. Among the environmentally related programs of the Public Health Service are the National Center for Disease Statistics, which keeps track of all statistics relating to diseases, both contagious and environmentally caused; the National Institute of Environmental Health Sciences and the National Cancer Institute, which initiate and support research programs on the effects of environmental factors on human health, especially as those factors result from pollution and other human activities; the AGENCY FOR TOXIC SUBSTANCES AND DISEASE REGISTRY; the CENTERS FOR DISEASE CONTROL; and the FOOD AND

DRUG ADMINISTRATION. Headquarters: 200 Independence Avenue SW, Washington, DC 20201. Phone: (202) 619-0257. Website: http://phs.os.dhhs.gov/phs.

public land in the United States, any land under the jurisdiction of a government body, including federal land, state land, and county and municipal holdings. There are approximately 564 million acres of land owned by the federal government, about 25% of the total U.S. land area (2,271 million acres). The largest federal landowner is the Department of the Interior (*see* INTERIOR, DEPARTMENT OF). The second largest is the Department of Agriculture (*see* AGRICULTURE, DEPARTMENT OF), with almost all of its lands managed by the FOREST SERVICE. Other public lands are managed principally by the Department of Defense and by the General Services Administration, which has management jurisdiction over post offices and other federally owned buildings. *See also* PUBLIC DOMAIN; ACQUIRED LANDS; and entries for the various types of federally-owned land (NATIONAL PARK; NATIONAL FOREST; WILDLIFE REFUGE; WILDERNESS AREA; etc.).

Public Land Law Review Commission federal study commission established by Congress in 1964 to study the laws and regulations pertaining to the public lands of the United States and to make recommendations for altering and updating them. The commission's controversial final report, *One-Third of the Nation's Lands,* was submitted to President Richard Nixon in 1970. Criticized by both environmentalists and developers, it nevertheless led to a number of important and positive changes in federal land management policies, including (ultimately) the FEDERAL LAND POLICY AND MANAGEMENT ACT OF 1976. *See also* BUREAU OF LAND MANAGEMENT: *History.*

pueblo rights in water law, WATER RIGHTS held by certain cities and Indian pueblos in the American southwest that date back to Spanish laws made as early as the 16th century to protect the rights of existing pueblos and villages. Pueblo rights allow the public body holding them to tap all surface and groundwater sources within its boundaries to meet all of its needs, including those for further development and increased per capita use, even to the point of completely using them up. They have been held by both federal and state courts to take precedence over all rights established afterward. Pueblo rights can only be held by a city or pueblo that can trace its existence in an unbroken line back to the original holder of the rights. This includes Santa Fe (NM), Los Angeles (CA), and several other cities in New Mexico and California as well as various New Mexican Indian pueblos and, by implication, similar continuously-existing communities in Arizona, Colorado, Nevada, Utah, and Texas.

pulp the basic ingredient from which inexpensive papers are made. Pulp is made by grinding wood which is then boiled, treated with various acids and chemical compounds, drained, and pressed by rollers. Pulp and paper mills have been notorious air and water polluters in the past but are now regulated by the CLEAN AIR ACT, Water Pollution Act, and other statutes. *See also* AIR POLLUTION; WATER POLLUTION.

pumice in geology, a type of PYROCLASTIC ROCK formed of RHYOLITE that has been made frothy by mixing with air, usually during explosive ejection from a volcanic vent. The air remains trapped in the solidified rock as small bubbles called VESICLES. Because of the trapped air, pumice is extremely light in weight, and will often float if tossed into water. It is usually soft and structurally weak, and erodes rapidly. Pumice deposits are sometimes mined for use as scouring material or cement filler.

pumped storage a type of hydroelectric generating facility (*see* HYDROELECTRIC POWER) in which idle capacity from other electric power generating plants is used during times of low demand to pump water into an elevated storage RESERVOIR so that it may be released to run through turbines and generate electricity during times of PEAK LOAD. The same equipment is generally used both to pump the water and to generate electricity. An electric motor drives a pump to fill the reservoir. When water is released from the reservoir, the pump acts as a hydraulic turbine and the motor acts as a generator. Pumped storage facilities have the effect of smoothing out the demand curve so that the load on the primary facilities becomes more equal around the clock, allowing them to be run more efficiently. It also reduces the total amount of power that must be produced by the primary facilities during times of peak load. However, the efficiency of the generating grid as a whole is actually reduced due to losses from friction in the PENSTOCK and from electrical resistance in the pump motor and transmission lines. A pumped-storage reservoir fluctuates from full to empty on a daily basis, and therefore cannot support aquatic life. These fluctuations give it an extreme "bathtub-ring" effect (*see* DRAWDOWN). The source from which the water is drawn for the facility may also suffer from fluctuations in level, depending upon the size relationship between the reservoir and the source. A proposal for a pumped storage generating plant at STORM KING mountain in the Hudson highlands north of New York City provided the basis for one of the most protracted environmental lawsuits on record.

pure stand in forestry, a STAND in which all the trees belong to a single SPECIES. Pure stands are considerably more susceptible to environmental hazards such as WINDTHROW, disease, and insect infestations than are MIXED STANDS, due to their similarity in genetic make-up and GROWTH FORM. For this reason, they are rare in nature, although they may sometimes grow up following a fire, landslide, or other natural disaster that clears a large growing site and gives it uniform characteristics throughout. Pure stands of desirable species such as white pine or DOUGLAS FIR are often the goal of commercial forestry (*see* MONOCULTURE).

put-and-take fishery in wildlife management, the stocking of a body of water with hatchery-raised sport fish so that fishermen may have something to catch. In a true put-and-take fishery, the fish being stocked cannot breed in the waters they are put into, so the population is maintained entirely by artificial means. Put-and-take fisheries are becoming increasingly common as natural sportfish stocks decline due to water pollution, EUTROPHICATION, invasion by TRASH FISH, ACID RAIN, etc.

P-wave in geology, one of several types of seismic (earth-propagated) waves created by an EARTHQUAKE. The P-wave is a *compressional wave,* that is, one that moves by alternate compressions and rarefactions, and is thus identical in function to a sound wave. It causes the rumbling or cracking sound generally associated with earthquakes, as it moves the ground alternately toward and away from the earthquake's epicenter. Moving at 5 to 9 km/sec, it is the first wave to arrive in earthquake-affected areas, hence the designation P-wave (for Primary wave). P waves are the principal cause of TSUNAMIS, which are also compressional in nature. *Compare* S-WAVE.

pyramid, ecological *See* ECOLOGICAL PYRAMID.

pyroclastic rock in geology, rock made of material that has been violently ejected from VOLCANOS. It is generally soft, poorly compacted, and easily eroded. Pyroclastic rock is often ejected in the form of *pyroclastic flows,* masses of solid particles mixed with superheated gases that move down mountainsides at speeds of up to 200 miles per hour (*see* NUÉE ARDENTE). Examples of pyroclastic rock include TUFF and PUMICE.

pyroforic liquid any liquid whose FLASH POINT is at or below the boiling point of water (100°C). Pyroforic liquids are highly flammable, making them extremely hazardous to transport and store.

pyrolysis the chemical decomposition of a COMPOUND by the application of enough heat to break the bonds holdings its MOLECULES together. Although the products may later recombine into other, simpler compounds, true combustion is not involved. Pyrolysis of combustible materials may be obtained by heating them in the absence of OXYGEN, and it is sometimes used as a means of destroying hazardous materials whose combustion products might be environmentally dangerous. It can also be used as a means of reducing the volume of SOLID WASTES before landfilling them (*see* SANITARY LANDFILL) or treating them to reclaim metals and other useful products (*see* RECYCLING). Pyrolysis of wood wastes can be used to produce synthetic PETROLEUM. In a wood fire, pyrolysis takes place in the layers of wood beneath the burning exterior layer where oxygen cannot reach. This is the process that produces charcoal. Pyrolized wood burns hotter and cleaner than nonpyrolized wood.

pyroxene any of a large group of rock-forming MINERALS, type formula $(Mg,Fe,Ca,Na)(Mg,Fe,Al,Si)Si_2(Al))_6$, commonly in BASALT and related IGNEOUS ROCKS, less commonly found in METAMORPHIC ROCKS (including, at times, marble). Pyroxene is the second mineral group of the BOWEN SERIES, forming after the crystallization of OLIVINE and before AMPHIBOLE. It is generally black to dark green in color, although lighter varieties exist (including jadeite, the most common variety of true jade). The most common variety, augite, is a glassy black mineral that gives GABBRO its characteristic dark color. HARDNESS: 5–6; SPECIFIC GRAVITY: 3.4–3.6.

quad in energy terminology, one quadrillion (10^{15}) BRITISH THERMAL UNITS, roughly equivalent to the energy content of 8 billion gallons of GASOLINE. Total U.S. energy consumption equals about 75 quads annually.

quadrat *See* DENSITY: *measurement of density.*

quagga mussel *See* MUSSEL.

quality of life in the context of environmentalism, a policy objective that relates to standards other than individual income or gross national product. Quality of life measurements include human health, a sense of well-being, community stability, human rights, sustainable ecosystems, natural beauty, and similar benefits that cannot easily be quantified.

quality standards requirements in environmental laws for the reduction of pollution. *See also* CLEAN AIR ACT; CLEAN WATER ACT.

quartz an extremely common rock-forming MINERAL, comprising 28% of the Earth's crust and second in abundance only to FELDSPAR. Unlike most minerals, quartz is composed of a single COMPOUND, silicon dioxide (SiO_2, also known as silica). The pure mineral is colorless and transparent; however, most deposits of it contain small amounts of impurities, coloring them in many different hues and giving rise to a bewildering variety of common names. Geologists classify quartz in three basic varieties, according to the type of crystallization that has occurred. *Crystalline quartz* is composed of visible crystals, hexagonal and pointed on one or both ends, usually found in masses and given such names as amethyst, smoky quartz, milky quartz, rose quartz, and cairngorm. The masses often appear to grow from a carpet of tiny crystals known as *drusy quartz. Cryptocrystalline quartz* is composed of crystals too small to be seen by the naked eye. Almost always brightly colored by impurities, it goes by such names as jasper, chalcedony, carnelian, onyx, agate, and flint. *Amorphous* or *acrystalline quartz* is quartz that has not crystallized due to the inclusion of small amounts of water as an impurity. It is usually known as opal. All varieties of quartz have the same SPECIFIC GRAVITY, 2.65, and all except opal have the same HARDNESS, 7 (opal's hardness is 5.5). It is among the most thermally stable materials known, expanding and contracting very little upon heating. It is also nearly inert chemically. It is the principal constituent of glass, and is used in electronic equipment (watches, radio and television tuners, etc.) due to its ability in the crystalline form to emit perfectly regular high-frequency electrical pulses when placed under pressure. Quartz veins in rock are often associated with the occurrence of metals, especially gold (*see* QUARTZ MINE). *See also* QUARTZITE; SANDSTONE; SAND.

quartzite (ganister) a METAMORPHIC ROCK consisting of SANDSTONE that has been subjected to so much heat and pressure that the cementing material between the grains has become as hard as the grains themselves. Quartzite can always be told from sandstone under a microscope by the fact that breaks in the rock go through the individual grains rather than around them. Normally white, quartzite may be colored by impurities or by the presence of tiny gemstones (garnets, amethysts, etc.) created in the cementing materials during the metamorphic process. The rock is composed almost entirely of QUARTZ, and is among the hardest, toughest, most durable and least reactive materials in existence. The term *quartzite* is also occasionally used for extremely tough sandstones in which the cementing materials, as well as the sand grains, is made entirely or almost entirely of quartz.

quartz mine a gold mine developed among QUARTZ VEINS in hard rock rather than in erosive debris. Quartz mines are, in general, the least environmentally hazardous form of mining. *Compare* PLACER MINE. *See also* HARD ROCK MINE; LODE.

quick-disconnect coupling in hazardous-waste management, a coupling between two pipes or hoses, or between a pipe or hose and a tank, which consists of a simple screw-on, bayonet, or similar connection only, without the presence of spring-loaded valves (a simple ball valve may be present in one or both sides of the coupling). Quick-disconnect couplings are lighter and less complex than DRY-DISCONNECT COUPLINGS, and require only a simple on/off operation, without turning a valve to begin product flow. Thus they are much easier to use. However, they are also much more subject to spills, and should not be used for hazardous liquids or gases.

race *See* SUBSPECIES.

Rachel Carson Council environmental study institute, founded in 1965 to further the work of the environmental scientist and writer RACHEL CARSON (1907–1964). Primarily a library of information on PESTICIDES and other environmental CONTAMINANTS, the council also conducts symposia and has a small publications program. Prior to 1979 it was known as the Rachel Carson Trust for the Living Environment. Address: 8940 Jones Mill Road, Chevy Chase, MD 20815. Phone: (301) 652-1877. Website: http://members.oal.com/rccouncil/ourpage.

radical (free radical) in chemistry, a functional GROUP that has broken away from a molecule in such a way that it (the radical) is left with one unpaired electron. Radicals are highly reactive, tending to bond to other substances almost immediately. *Compare* ION.

radioactive waste *See* NUCLEAR WASTE AND CLEANUP.

radioactivity the release of radiation (ALPHA PARTICLES, BETA PARTICLES, GAMMA RAYS) that accompanies the spontaneous transformation of one ELEMENT into another, or one ISOTOPE into another, through the breakdown of a portion of the atomic nucleus (*see* ATOM: *structure of the atom*), a process known as *radioactive decay.* The release of a beta particle converts one neutron in the nucleus into a proton, raising the ATOMIC NUMBER by one while keeping the ATOMIC WEIGHT the same. The release of an alpha particle tears two protons and two neutrons out of the nucleus, lowering the atomic number by two and the atomic weight by four. The release of these particles is usually accompanied by a release of energy in the form of heat (used to drive nuclear reactors; *see* NUCLEAR ENERGY) and gamma rays. About 40 naturally occurring elements and isotopes are known to undergo spontaneous radioactive decay; others have been created in the laboratory. The presence of substances undergoing spontaneous radioactive decay can also induce decay in other, normally nonradioactive substances by physically knocking protons or neutrons out of the nonradioactive substance's nucleus.

ENVIRONMENTAL EFFECTS OF RADIOACTIVITY

Radioactivity causes biological damage in the same way that it induces decay in nonradioactive substances—by physically knocking MOLECULES apart. If the damage is to a molecule of DNA or RNA, cancers and genetic MUTATIONS can result. If the damage is to another molecule within the CELL, the cell's functions can be disrupted and the cell can die. The death of cells can cause dysfunction or death of ORGANISMS. The amount of damage that will occur from a given dose of radiation depends upon both the intensity of the dose and the type of radiation involved (alpha particles, being heavier, cause greater damage than do beta particles). Thus, while radiation *intensity* may be expressed directly in *rads* (the amount of radiation that will cause absorption of 0.01 joules of energy per kilogram of mass) or *roentgens* (roughly equivalent to rads, but measured as a loss of energy in the radiation being measured as it passes through a specified volume of air), radiation *doses* are usually expressed in *rems* (short for *r*adiation *e*quivalent in *m*an), where one rem is the equivalent of the damage caused by one rad of gamma radiation. One-time exposure to 1,000 rems will kill a human; one-time exposure to 100 rems will usually cause radiation sickness (nausea, vomiting, malaise, some neural damage). Effects of very small doses remain in question. Though most scientists agree that there is no lower threshold to radiation damage, it is not known whether damage is proportionately higher, proportionately lower, or linear (that is, exactly proportional) at very small doses as opposed to large ones. Standards set by the ENVIRONMENTAL PROTECTION AGENCY assume the linear model. They are currently placed at 25 millirems (0.025 rems) per person per year, over and above the local BACKGROUND LEVEL.

radiolarian any member of the Radiolaria, a large, diverse, and taxonomically indistinct group of PROTOZOANS characterized by an internal "skeleton" made of chitinous material and (usually) an external "shell" composed of silica and containing numerous pores through which stiff pseudopods protrude to catch the ORGANISM'S food. The skeleton is generally radially symmetrical, and is often highly fanciful in design, with numerous spikes and barbs protruding from it. It may be as much as 10 millimeters across. Radiolarians are distinguished from the similar FORAMINIFERANS by the fact that their pseudopods remain separate from each other and do not form a protoplasmic net. They are exclusively marine and almost always members of the PLANKTON community (that is, they drift freely rather than living on or in the bottom ooze). A few colonial forms are known, but most are single-cell organisms. Dead radiolarians usually fall to the ocean floor, forming so-called "radiolarian ooze"—often the principal component of the ocean floor, especially at depths below 13,000 feet. Radiolarian ooze compacts into chalk and LIMESTONE. Once classified by biologists as a separate ORDER, the radiolarians are no longer considered a single taxonomic group, but are divided among several orders and CLASSES.

radiosonde in meteorology, a package of weather instruments sent aloft in a balloon to measure conditions in the upper ATMOSPHERE. A radiosonde contains, as a minimum, a thermometer, a hygrometer, (humidity gauge), a BAROMETER, and a radio transmitter to send data from the instruments back to earth. Since the device transmits data in a steady stream, it is not essential to recover it. Radiosondes have been in use regularly since the 1920s. Radiosondes designed to be tracked by radar (usually by the use of a balloon made of radar-reflecting material) are known as *rawinsondes*.

radon an inert, odorless, and colorless gas that is a naturally occurring breakdown product of uranium. Radon can enter houses and other buildings from the soil under foundations and small amounts may be released into the air from drinking water. The gas is second only to smoking as the leading cause of lung cancer, and the combination poses a serious health risk. Currently (2000) about 160,000 people die annually from lung cancer in the United States. Of these deaths, about 12% can be attributed to a combination of smoking and the inhalation of indoor radon. The gas can be found in most regions of the United States, with the northeastern, middle Atlantic, upper midwestern, and northern Rocky Mountain states having higher than average levels. Houses with large basements are most likely to accumulate significant amounts of radon, especially in northern areas where there is no winter ventilation. The gas enters the basement from the soil beneath through foundation cracks, cellar doors, and various openings and may collect in dangerous concentrations. Radon standards are being established by the ENVIRONMENTAL PROTECTION AGENCY (EPA), based on findings that as many as 4–5 million U.S. homes have some radon gas, and that 50,000 to 100,000 have dangerous levels. The agency recommends that homeowners use inexpensive radon testing devices to determine whether remedial action should be taken. In houses with leaky cellar floors over permeable soils, researchers have found that up to 70% of the radon entry rate into the living area can be reduced through simple basement ventilation. In addition, the EPA advises that homeowners seal up cracks, and in extreme cases install diversion piping under the cellar floor to exhaust the radon harmlessly into the atmosphere. *See also* INDOOR POLLUTION.

radwaste *See* NUCLEAR WASTE.

railroad land grants in American history, grants of federal land made to private corporations during the middle to late 19th century as an inducement to build railroads. The grants were typically made in square-mile blocks (SECTIONS) arranged in a checkerboard pattern on either side of the proposed right-of-way. If the corporation did not complete the railroad, the lands would be returned (*revested*) to the federal government. Much of the privately held timberland in the western United States today was originally obtained as railroad land grants. *See also* CHECKERBOARD OWNERSHIP; OREGON & CALIFORNIA LANDS.

Rails-to-Trails Conservancy coalition of trail users of various types, founded in 1985 to encourage the conversion of abandoned railroad rights-of-way into trails. The conservancy's eventual goal is a unified national trail system along old rights-of-way.

Members include hikers, runners, bicyclists, park officials, and anyone else interested in the nonmotorized use of trails. The group lobbies for simplified abandonment regulations and improved local notification procedures. Membership (1999): 80,000. Headquarters: 1100 17th Street NW, 10th Floor, Washington, DC 20036. Phone: (202) 331-9696. Website: www.railtrails.org.

rain liquid PRECIPITATION. Rainfall occurs when the tiny droplets of condensed moisture that form clouds join together to form drops large enough so that the pull of gravity on them will overcome the bouyant effect of air currents. This generally happens in one of two ways: either enough turbulence exists in the cloud so that the droplets bump into each other hard enough to overcome SURFACE TENSION and coalesce, or (more commonly) the presence of tiny bits of solid matter such as dust particles or ice crystals act as nuclei for raindrop growth, attracting the cloud droplets and coalescing them into drops through electrostatic force. Raindrops built on ice-crystal nuclei often actually form as snow-flakes or hailstones, melting later as they fall through warm air on their way to the earth. The minimum size for raindrops that will fall through still air is 200 MICRONS in diameter (0.2 millimeters, or 0.008 inches)—roughly 20 times the diameter (and 8,000 times the volume) of the average cloud droplet. *See also* CLIMATE; RAIN FOREST BIOME; RAIN DAY; RAIN SHADOW.

Rainbow Warrior *See under* GREENPEACE.

Rainforest Action Network nonprofit organization founded in 1985 that works nationally and internationally to protect rain forests and defend the rights of indigenous peoples by nonviolent means such as letter-writing campaigns, demonstrations, and boycotts. The organization also creates educational materials, teacher resources, and information for community organizers. Membership (1999): 30,000. Address: 221 Pine Street, Suite 500, San Francisco, CA 94104. Phone: (415) 398-4404. Website: www.ran.org.

rain forest biome in ecology, a widespread and important BIOME found in regions of heavy rainfall. Most rain forests are located in the tropics, with the largest expanses in the Amazon Basin of South America, the Congo Basin of Africa, the Caribbean coasts of Central and South America and their associated islands, and the islands and peninsulas of Southeast Asia; however, a few temperate-zone rain forests exist, chiefly in Australia and New Zealand and on the Olympic Peninsula of the State of Washington in North America. Wherever they are found, rain forests show two striking climatic similarities: the rainfall is extremely heavy and spread nearly uniformly throughout the year, with no pronounced wet and dry seasons, and the temperature variation from season to season is very small. In fact, in most tropical rain forests—and probably most temperate rain forests as well—the diurnal (day to night) temperature shift is greater than the shift in mean temperatures from summer to winter. This means that conditions for plant growth are both highly favorable and nearly uniform throughout the year, leading to an astonishing profusion of plant life. The rain forest is the richest biome on Earth, both in terms of annual BIOMASS growth and in terms of the variety of SPECIES present. These forests typically have a CANOPY developed in three layers, with a broken OVERSTORY of very tall trees sticking up through a nearly unbroken middle-layer canopy that forms a roof over an UNDERSTORY consisting of younger trees of the taller species plus a few shorter species that are adapted to life below the canopy through the use of extremely large leaves for photosynthesizing in the shade (*see* PHOTOSYNTHESIS) coupled with a spindly GROWTH FORM for small total mass relative to the leaf area. There are few HERBS and FORBS on the ground beneath the trees, but a great number of EPIPHYTES and EPIPHYLLS cling to the branches and leaves of the canopy, together with climbing and trailing vines. Breaks in the canopy, such as riverbanks or road cuts, give rise to *jungle*—a prolific growth of plant life that crowds the edge of the opening to take advantage of the light, making travel difficult or impossible (contrary to Hollywood's representation, jungle does *not* extend much beyond the edges of the openings it occupies). The soils of the rain forest are very poor, both because most MINERALS have been leached out by the heavy rains (*see* LATERITIC SOIL) and because most of the NUTRIENTS are tied up in the existing vegetation. Consequently, clearing rain forests for agricultural use is at best misguided. Nevertheless, such clearing is proceeding today at a rapid enough pace to qualify as a major global environmental problem. Current estimates suggest that the tropical rainforest is being destroyed at the rate of 42 million acres per year (about 4,800 acres per hour). This destruction creates immense EROSION problems and threatens numerous plants and animals with EXTINCTION, and is probably significantly affecting the Earth's OXYGEN production and (through its impact on TRANSPIRATION RATES and on planetary ALBEDO) the temperature distribution and overall climate of the Earth as well. *See also* CLOUD FOREST.

rain shadow in meteorology, an area on the lee side of a mountain range or prominent mountain peak, in which little rain falls. Rain shadows result from the forced uplift and convergence of moisture-bearing AIR MASSES as they cross over the mountains, which cause extra rainfall on the mountains themselves and effectively "wring the air mass dry" before it can reach the area beyond the mountains. The effect can be quite striking. Sequim, Washington, in the rain shadow of the Olympic Mountains, gets less than 15 inches of precipitation annually, though it is less than 45 miles from the Hoh River rain forests, which receive more than 200 inches annually. The rain shadow of the Rocky Mountains is the principal cause of the dry climate on the Great Plains. Its effects are felt as far as Iowa and Missouri, 500 miles away. *See also* CLIMATE.

ranch unit in federal land-management terminology, privately-owned RANGELAND operated and managed as a single economic unit. It may consist of several discontinuous parcels.

range (1) in the biological sciences, the territory within which a SPECIES is normally found. The term may also be used with qualifiers to limit it to a specific individual or POPULATION (*home range*), specific season (WINTER RANGE; SUMMER RANGE), specific activity (*breeding range; hunting range*), and so on. *Compare* HABITAT.

(2) in livestock management, short for RANGELAND.

rangeland in livestock management, land on which the dominant vegetation is made up of GRASSES, HERBS and FORBS suitable for consumption by grazing LIVESTOCK. Natural rangeland is a CLIMAX COMMUNITY whose boundaries are determined largely by rainfall and soil conditions. It corresponds roughly to the GRASSLAND BIOME. In other areas, rangeland may be artificially maintained through such practices as forest clearing, PRESCRIBED BURNS, and IRRIGATION. *See also* RANGE MANAGEMENT; RANGE TREATMENT.

range management the management of a piece of land for the primary purpose of grazing LIVESTOCK. The principal factor taken into account by range managers is CARRYING CAPACITY, that is, how many animals the range can support without sustaining damage. This is determined by the type and amount of FORAGE available, the ability of the soil to withstand trampling by livestock without either compacting or eroding, and the presence of an adequate water supply, both for the livestock and for the wildlife and plants with which the livestock must share the range. A second factor is use conflicts; rangeland may also be used for wildlife HABITAT, recreation, water supply or timber harvest, and these uses can be strongly impacted by stock grazing. Once carrying capacity is determined, a range management plan must be worked out. The plan may involve an increase in carrying capacity through range improvements (planting better forage crops, constructing ponds or other water-supply structures, fencing against competing wildlife such as deer or antelope, etc.). Livestock may be rotated to different areas, allowing portions of the range a chance to recover; soil-stabilization measures may be taken on the resting portion of the range. As the plan is carried out, monitoring must be done on a regular basis to make certain that carrying capacity is not being exceeded. Monitoring involves studying two factors, *condition* (the current state of the range) and *trend* (the direction the condition is headed, that is, whether the range is getting better or getting worse). These are normally determined by examining the range for indicators such as LICHEN LINES, BROWSE LINES, GULLIES, LAG GRAVEL, and changes in the plant COMMUNITY (*see* INDICATOR SPECIES). *See also* ANIMAL UNIT MONTH.

range treatment (grazing treatment) the modification of RANGELAND in order to make it more productive for LIVESTOCK grazing. Range treatment includes physical improvements such as artificial waterholes and fences as well as biological improvement such as PRESCRIBED BURNS, seeding, and the elimination of undesirable species through hand removal or the use of BIOLOGICAL CONTROLS or BIOCIDES. *See also* RANGE MANAGEMENT.

rapid a stretch of a river or stream that descends abruptly but contains little or no free-falling water. Rapids can have numerous causes, including abrupt shifts in stream course, FAULTS, landslides, LAVA FLOWS, and differential EROSION of the river's bed, but the most common cause is the presence of boulders, gravel, and other debris brought down a TRIBUTARY STREAM and deposited in the MAINSTEM, where they form a rude dam. River runners classify rapids according to their severity and the dangers they pose to boats. Class I rapids contain wide, clear channels and have little or no wave action. Class II rapids may show waves up to two feet high, but they also have clear channels and pose no danger. Class III rapids have waves up to three feet, and contain rocks and eddies that may pose moderate difficulty, while class IV rapids have waves up to five feet and can pose serious dangers. Class V rapids contain waves over five feet high and many rocks and drops, and pose extreme danger. Class VI should not be run at all. The classification of a rapid may change with the time of year and with alterations in the river's course due to erosion and DEPOSITION. *See also* WHITE WATER.

rapid sand filter in water-supply technology, a type of filter used to remove small PARTICULATES from the FEED WATER of a water-supply system. It can also remove many PATHOGENS down to the level of bacteria (*see* BACTERIUM). A rapid sand filter consists of a bed of SAND of uniform particle size resting on a bed of GRAVEL, which it turn rests on (or is built around) a grid of perforated pipe. Water is introduced above the sand and is pulled down through the filter by a combination of gravity and (often) suction within the perforated pipe. The particulates and pathogens are removed by a combination of *sedimentation* (that is, being deposited on the sand grains), *flocculation* (biological interactions in the spaces between grains, leading to the formation of masses of stuck-together CELLS that are too large to fit through the spaces), and simple sieving. Periodically, the filter must be cleansed of the accumulated particulates and floc. This is done by passing water backwards through the filter (*backwashing*) at a rapid enough rate that the sand grains are agitated, freeing the materials stuck to them. The wash water is led out through a separate drain rather than becoming part of the next feed water cycle.

raptor any bird belonging either to the ORDER Falconiformes (vultures, falcons, hawks, and eagles) or the order Strigiformes (owls). Raptors are CARNIVORES, with legs and feet adapted to grasping PREY and with short, heavy, hooked beaks for tearing flesh apart. Their food is largely rodents, small reptiles and other birds, which they stalk by sight, avoiding detection by their prey either by soaring at a great height over the hunting ground (Falconiformes) or by flying over it at night in almost complete silence (Strigiformes). They are one of the most effective natural checks on small animal POPULATION EXPLOSIONS, and are legally protected in most parts of the world due to their importance in guarding agricultural crops.

Raptor Research Foundation professional organization that coordinates the distribution of information on the biology and conservation of raptors and their habitats. The specific concerns of the foundation are behavior, captive breeding, population monitoring, habitat ecology, migration patterns, and rehabilitation of injured birds of prey. Among its publications are a quarterly scholarly journal, *Journal of Raptor Research,* and a newsletter, *Wingspan.* Membership (1999): 1,200. Address: 3948 Development Street, Boise, ID 83706. Website: http://biology.boisestate.edu/raptor.

RARE process *See* ROADLESS AREA REVIEW AND EVALUATION.

rare species a SPECIES whose numbers are small relative to its total BIOTIC POTENTIAL. This includes both those species whose RANGE is broad but which occur only in small, scattered POPULATIONS throughout the range, and those species whose population densities may be large but whose range is limited to a very small geographical area. *Locally rare* species are rare within the boundaries of a particular region; they may be abundant elsewhere. Some states have written definitions of rare species into law. For example, California law states that a species is rare if any of the following conditions hold: it is confined to a small, specialized HABITAT; it is nowhere abundant throughout its range; its range is small enough that any reduction of available habitat within the range might cause it to become endangered (*see* ENDANGERED SPECIES); or its numbers are dependent on wildlife management, so that changes in management techniques might cause it to become endangered. *See also* THREATENED SPECIES; ENDANGERED SPECIES ACT.

ravine a large gully or small valley. Ravines are generally steep-sided, V-shaped features that are formed near the headwaters of streams. They may or may not have water in them all year around, but their formation and most of their subsequent enlargement is dependent upon running water.

ray flower *See under* COMPOSITE.

RCRA *See* RESOURCE CONSERVATION AND RECOVERY ACT.

RCWP *See* RURAL CLEAN WATER PROGRAM.

reach a stretch of river having similar characteristics. A stretch of FLAT WATER between two rapids, for example, might be classed as a reach; so might a stretch with a great deal of WHITE WATER, as opposed to a stretch with little or no white water. The term is also used more technically, but less often, to refer to any stretch between two bends, within which the river does not substantially change course.

reaction in chemistry, a change in the chemical structure of two or more substances, occurring when the substances come into contact with one another. Two substances may simply combine to form a third (HYDROGEN and OXYGEN, for example, react by combining to form WATER, according to the equation $2H_2+0_2 \longrightarrow 2H_20$); or part of each MOLECULE of one substance may transfer to another substance, transforming the properties of both substances (calcium hydroxide and CARBON DIOXIDE react by exchanging ATOMS to form calcium carbonate and water according to the equation $Ca(OH)_2+CO_2 \longrightarrow CaCO_3+H_20$, a reaction that is responsible for the hardening of brick mortar). Reactions may take place spontaneously between two substances, or they may require special conditions, such as the presence of heat, pressure, or a chemical CATALYST. *See also* REAGENT.

reactive in waste-management terminology, any explosive material, that is, any material that is capable of undergoing an explosive reaction at normal temperatures and pressures. The category includes substances that explode on contact with electric sparks, those that explode on contact with water, and those that explode if subject to bumping or to other forms of mechanical shock, as well as those that may undergo spontaneous combustion (that is, generate internal heat through chemical means, without external disturbance) leading to an explosion.

reagent (reactive agent) in chemistry, a substance that reacts with one or more other substances (*see* REACTION). Technically, the term applies to all such substances; however, it is generally restricted by usage to those substances whose principal use is to cause desired chemical reactions, such as the chemicals used by analytical chemists to test whether or not a substance reacts in a certain way, or those used by industry during the manufacture of complex synthetic substances such as plastics.

reasonably available control technology (RACT) in pollution control, the level of cleanup effort required of existing sources of air POLLUTANTS in NONATTAINMENT AREAS. The terminology comes from Section 170 of the CLEAN AIR ACT, as amended in 1977. Since "reasonably available" is not otherwise defined by the act (except to state that both economic and technological factors may be taken into account), the term is open to considerable interpretation. In general, it may be taken to mean that some effort must be expended by *all* generators of air pollutants in nonattainment areas to achieve attainment, while in areas that are already in compliance with the NATIONAL AMBIENT AIR QUALITY STANDARDS no further effort needs to be made by existing sources. *Compare* BEST AVAILABLE CONTROL TECHNOLOGY.

receiving water in waste-disposal technology, the body of water into which wastes are dumped. The nature of the receiving water dictates to some degree the type and degree of treatment that should be received by the waste before dumping and the quantity of waste that may be dumped. Rapidly flowing streams, for example, can accept more wastes than slowly flowing streams of the same size. Outfalls into streams that are naturally turbid (*see* TURBIDITY) may themselves be more turbid than outfalls into clear streams. Regulatory law generally does not address these differences, however. *See* EFFLUENT STANDARDS.

recharge in groundwater management, water that flows into an AQUIFER to replace that which is drawn off by wells or which flows out in springs. The rate of recharge of an aquifer is controlled in nature by a number of factors, including the amount of the aquifer that is exposed to the surface (the "recharge area") rather than being blocked by impermeable materials (*see* IMPERMEABLE LAYER), the POROSITY and PERMEABILITY of the material the aquifer is made of, and the amount of rainfall in the recharge area (most natural recharge results from rainfall). Recharge can be artificially increased by injecting water into wells, by constructing RESERVOIRS over recharge areas so that stream water stored in the reservoirs will soak into the ground and enter the aquifer, or simply by breaking up impermeable ground barriers so that more of the rainfall that falls over the aquifer will soak into it. *See* CONJUNCTIVE USE.

reclamation in land management, the conversion of land that was formally unfit for growing crops into agricultural land by manipulating its water supply. The term encompasses both the draining of lakes and WETLANDS in order to grow crops on their beds, and the process of supplying water by means of IRRIGATION to lands where the rainfall is too sparse to grow crops under natural conditions. Among the most drastic of human activities in terms of its effect on the environment, reclamation represents a total disturbance of the ecology of the area upon which it is practiced, altering conditions in the soil from xeric or humic to mesic (*see* XERIC SITE; HUMIC GLEY; MESIC SITE) and in the process completely changing the COMMUNITY of plants and animals that can live there. There are currently approximately 70 million acres of cropland under reclamation in the United States. About 51 million acres are irrigated and about 43 million acres are drained. The two figures add up to considerably more than 70 million due to overlap between the two categories (many wetlands are irrigated after drainage). *See also* BUREAU OF RECLAMATION.

Reclamation, Bureau of *See* BUREAU OF RECLAMATION.

recreation land class in federal land-management terminology, any of six standard categories into which recreation land can be classified. The recreation land classes were developed in 1962 by the Outdoor Recreation Resources Review Commission of the BUREAU OF OUTDOOR RECREATION and were at one time broadly adopted throughout the federal government. Their use has since been largely supplanted by the RECREATION OPPORTUNITY SPECTRUM. The six categories were:

Class I. High-density recreation areas, developed and managed for mass use (urban parks and playgrounds).
Class II. General developed outdoor recreation areas (campgrounds, picnic areas, etc.).
Class III. Natural areas (general forest and park lands, including areas with roads).
Class IV. Unique natural areas (areas containing outstanding natural features of unusual scenic beauty or scientific importance, including designated SPECIAL INTEREST AREAS).
Class V. Primitive areas (large areas of undisturbed roadless land, including all lands suitable for classification in the national Wilderness Preservation System [*see* WILDERNESS AREA]).
Class VI. Historic and cultural sites (lands qualifying as ARCHAEOLOGICAL RESOURCE lands or HISTORIC RESOURCE lands).

Recreation Opportunity Spectrum (ROS) the classification of land according to a continuous spectrum of recreation types based on their degree of "primitiveness," developed by the FOREST SERVICE after 1976 and since broadly adopted by recreation planners in other agencies and in the private sector in place of the earlier RECREATION LAND CLASS designations. As with recreation land class planning, Recreation Opportunity Spectrum planning classifies each piece of land into one of six categories, or *classes.* The classes vary significantly between the two systems, however. The six ROS classes are:

Class 1: primitive. Class 1 areas are relatively large blocks of land on which little or no modification of the natural environment has taken place. They are managed for low-impact, non-motorized DISPERSED RECREATION.
Class 2: semi-primitive non-motorized. These lands differ from class 1 lands primarily in being somewhat smaller, with higher use. Motorized recreation is still not permitted, although administrative roads may be present.
Class 3: semi-primitive motorized. Class 3 lands are essentially identical to class 2 lands, except that motorized recreation is permitted.
Class 4: roaded natural and roaded modified. Class 4 lands still look substantially natural, but modifications are often evident. User density is moderate to high, and the users encounter each other frequently. DEVELOPED RECREATION sites occur with some regularity. Resource activities such as logging and mining may be common.
Class 5: rural. These lands show a mix of natural and heavily modified land types, with agriculture and other land-based economic activities highly evident. High-density developed recreation sites may be present. There are not many buildings, but the human presence is everywhere noticeable.
Class 6: urban. Urban recreation lands are manicured, extensively modified, and heavily used. They include city parks, most county parks, and similar developed recreation sites on state and federal land, especially within or near urban areas.

recreation resource in Forest Service terminology, any resource that contributes or can potentially contribute to the enjoyment of leisure time by a significant number of individuals. The category is somewhat difficult to set limits on because of the great variety inherent in the term "recreation." Included are swimming holes, rocks available for collection, scenic drives, off-road vehicle sites, hiking trails, etc. The recreation resource category often overlaps into others. OLD-GROWTH forest is both a recreation resource and a timber resource, fishing streams are both a recreation resource and a wildlife resource, and so on. *See also* RECREATION LAND CLASS; RECREATION OPPORTUNITY SPECTRUM.

recreational river in land management, a river or section of a river set aside by Congress for recreational purposes under the terms of the National WILD AND SCENIC RIVERS ACT. The recreational river is the least restrictive of the three river classifications established by the act. It is defined as a river segment "readily accessible by road or railroad" that "may have some development" along its shore and "may have undergone some IMPOUNDMENT or DIVERSION in the past." It must be free-flowing; however, small dams and impoundments that do not detract from the overall free-flowing nature of a stretch of river may be included. Management is primarily for recreational purposes, and may include intensive DEVELOPED RECREATION sites on the riverbanks. Extensive agriculture is permitted, and barns, homes, cabins, stores, and other buildings may continue to be built. Water quality in the river is expected to be good enough to support water-contact sports and to keep a flourishing and largely natural aquatic COMMUNITY healthy; however, rivers with poorer water quality may be designated under the act if it can be demonstrated that recovery of good water quality is possible using existing technology and regulations. *Compare* WILD RIVER; SCENIC RIVER.

recycling in waste management, the reuse of materials that would otherwise be discarded. The materials may be cleaned and reused directly (*reuse*); they may be used as raw materials to make different finished goods (*remanufacture*); or they may be burned to obtain energy (*refuse-derived fuel,* or *RDF*). Recycling may take place within industrial processes (*internal* or *under-roof recycling*), including the reuse of cooling water, the collection of and reforging of cuttings from metal milling operations, the burning of trim ends in sawmills to produce heat, and so forth. Alternatively, it may take place after a product is used and discarded by consumers (*external* or *waste stream recycling*). This is ordinarily more difficult to manage than internal recycling due to the difficulties involved in separating usable from nonusable materials in the waste stream. The rewards, however, are great. They include a reduction in garbage bulk of up to 80% (and a corresponding reduction in dependence on SANITARY LANDFILLS), conservation of scarce natural resources, decreased waste disposal costs, and often, improved energy efficiency. (Aluminum remanufacture, for example, utilizes less than 10% of the energy required to process the same amount of aluminum from bauxite ORE). External recycling is made simpler if consumers practice *source separation*—the sorting of trash of different types into separate containers as it is discarded. Common categories of source separation include newspapers, other paper and wood products, metals, glass, and plastics.

THE POLITICS OF RECYCLING

Recyling programs have been widely adopted by industry, municipalities, and civic organizations over the past 20 years. Today, there are nearly 9,000 curbside recycling pickup programs in U.S. towns and cities, serving slightly more than half the population. There are also some 3,300 facilities for composting yard trimmings, 9,000 recycling drop-off centers, and tens of thousands of workplace programs. With regard to municipal solid waste (MSW), EPA data reveals that as of 1999, 28% was recycled, up from less than 10% in 1980. Although most Americans (75%) favor recycling, some oppose mandatory MSW programs as an unwanted and unnecessary government intrusion. Some of the main complaints are that recycling is not cost effective, that there is no market for recycled materials, and that landfill areas are so plentiful and safe that recycling is unnecessary. Those operating successful MSW programs reject these criticisms, pointing to recycling programs in cities such as Seattle, Cincinnati, and Green Bay, Wisconsin, which have a lower per-ton recycling cost than the per-ton cost of garbage disposal. Regarding the market for recycled materials, the criticism is true for some items, but not for others. Recycled paper and aluminum not only produce revenues for municipal programs, but are also increasingly relied on by the aluminum and paper industries, which have developed facilities that, in fact, require recycled materials. On the question of safety, 250 of the 1,200 toxic waste sites identified by the EPA on their Superfund National Priority List are former municipal solid waste landfills. Moreover, virtually all landfills are sources of harmful gaseous emissions, including volatile organic chemicals and methane, a greenhouse gas. In addition, many of the older landfills leach toxic substances into groundwater supplies. Despite the controversy, all but five states have waste reduction and/or recycling goals of various kinds, with several of them setting quite high targets. New Jersey, for example, has reduced its solid waste stream by 65%. For the future, recycling experts hope that the concept of "product stewardship" will take hold in the manufacturing sector. The idea here is that manufacturers of problematic materials (such as batteries, computers, and other items with toxic components) will take responsibility for their recovery after use. Another new direction in recycling efforts is to make producers responsible for their packaging, compensating government agencies for the cost of disposal. *See also* RESOURCE RECOVERY; SOLID WASTE.

redd *See* SPAWNING GRAVEL.

redox in chemistry, term used to refer to the simultaneous occurrence of REDUCTION and OXIDATION reactions. Since reduction and oxidation always occur together, the decision as to whether to refer to a reaction as "reduction," "oxidation," or "redox" depends on whether the speaker's interest is primarily in the reaction's reduction product or process, primarily in its oxidation product or process, or equally in both.

red tide a marine ALGAL BLOOM in which the blooming species is a toxic red dinoflagellate, generally a member of one of the two genera *Gymnodidium or Gonyaulax* (*see* GENUS). Red tides may be severe enough to literally color the water red. They poison fish and shellfish, and may also poison humans, animals, and birds that eat the poisoned fish or shellfish. Their causes are poorly understood, but appear to be similar to the causes of other algal blooms: an increase in a LIMITING NUTRIENT, a warming of the water, and other general factors that go along with EUTROPHICATION, either natural or cultural. Like other eutrophication phenomena, red tides seem to thrive on pollution. The incidence of red tides in Florida has increased dramatically with the buildup of population along the coastline, and New England—long thought too far north to be affected—had a red tide in 1972.

reducing agent in chemistry, any substance that causes another substance to be chemically reduced (*see* REDUCTION).

reduction in chemistry, originally, the removal of OXYGEN from or the addition of HYDROGEN to a COMPOUND; now more broadly defined as any reaction in which a chemical substance (ELEMENT or compound) gains one or more ELECTRONS. Reduction and OXIDATION always proceed simultaneously. *See also* REDOX.

reforestation the process of returning cutover, burned over, or otherwise deforested land to a forested condition. The term is broadly general; it includes preparation of the growing site (including any necessary BRUSH CONVERSION), planting the trees (*see* RESTOCKING), and tending the growing plantation until it is an established forest. *See also* REGENERATION.

Refuse Act important early federal pollution-control legislation, passed by Congress as part of the Rivers and Harbors Act of 1899 and signed into law by President William McKinley on March 3, 1899. The act (Section 13 of the Rivers and Harbors Act) makes it illegal to discharge any refuse material of any kind, other than runoff from streets and liquid discharge from sewers, into the navigable waters of the United States without an express permit to do so from the ARMY CORPS OF ENGINEERS. A fine of up to $2,500 is mandated for violations, half of which is to go to any person or persons providing information leading to the conviction of the violator. Largely ignored for a number of decades following its passage, the Refuse Act began to be used as a weapon against industrial pollution in the late 1950s. A 1960 Supreme Court decision affirmed this use of the Act, determining that all materials other than municipal sewage were subject to it. Later Supreme Court decisions broadened the definition of "refuse" to include products as well as waste materials when those products were accidentally discharged into navigable waters, as in oil spills; and to encompass other forms of water-quality damage, such as THERMAL POLLUTION and LEACHATE from waste dumps near navigable waters. On December 21, 1970, President Richard Nixon ordered the Corps of Engineers to establish and enforce uniform national permit procedures under the Refuse Act. These procedures, known as RAPP (for *Refuse Act Permit Program*), became the basis for the permit process outlined in the CLEAN WATER ACT OF 1972, which passed the responsibility for enforcement from the Corps to the ENVIRONMENTAL PROTECTION AGENCY.

refuse-derived fuel *See under* RESOURCE RECOVERY.

regeneration (reproduction) in forestry, the young trees and SEEDLINGS that have become established on a plot of land. Regeneration differs from REFORESTATION in that reforestation implies converting bared land or brush fields back into forest, while regeneration encompasses both reforestation and the development of a new generation of young trees under an existing CANOPY. *Natural regeneration* is regeneration that has become established without human intervention. *Artificial regeneration* is regeneration that has been accomplished by purposeful human effort such as spreading seed or setting out seedlings. The term is used for the act (human or natural) of establishing young trees as well as for the young trees themselves. *See also* BRUSH CONVERSION; REGENERATION CUTTING; SOIL-SITE FACTORS; RESTOCKING.

regeneration cutting (regeneration harvest, removal regeneration cutting) in forestry, the removal of older timber from a growing site in order to assist the establishment of SEEDLINGS and young trees. A regeneration cutting is similar to a RELEASE CUTTING, except that release cuttings remove trees that are of little or no value commercially in order to speed the establishment of a commercially viable crop, while regeneration cuttings include both those done strictly for release and those in which the removed timber is itself commercially valuable. The term "regeneration cutting" is sometimes applied to the cutting of any timber past ROTATION AGE; however, this use should be avoided, as it is based strictly on economic criteria with no consideration for the forest as anything but a place to grow FIBER as rapidly as possible.

regional benefits in environmental economics, the benefits accruing from a project to the region where the project is being carried out. Regional benefits include both primary and SECONDARY BENEFITS to the region. They do not include benefits that may accrue outside of the region, such as increased jobs from constructing turbines in Tennessee for a dam being built in California. The most common problem in determining the size of regional benefits lies in determining what is to be included in the region, which may be defined as a city, county, or multi-county area, a watershed or part of a watershed, a market area, etc., depending upon the type and scope of the project under consideration.

Regional Forester in Forest Service terminology, the chief administrative officer of a Forest Service Region. *See* FOREST SERVICE: *structure and function.*

Regional Plan in Forest Service usage, a document setting the planning direction for a National Forest

Region (*see* FOREST SERVICE: *structure and function*). Regional Plans are based on resource GOALS and guidelines established by the National Headquarters of the Forest Service under the terms of the NATIONAL FOREST MANAGEMENT ACT (NFMA) and the FOREST AND RANGELAND RENEWABLE RESOURCES PLANNING ACT (RPA). Each plan interprets these goals and guidelines in the context of the special characteristics of the region (regions with high-quality timber land, for example, would interpret the timber management goal in a more production-oriented fashion than those with low-quality timber land). The Regional Plans, in turn, provide guidance for the FOREST PLANS. Mandated under the terms of the NFMA, Regional Plans supplant the now-obsolete PLANNING AREA GUIDES of the old unit planning process.

regional planning *See* LAND USE PLANNING.

Regional Supplement *See under* FOREST SERVICE MANUAL.

regolith in geology, the covering of loose or lightly consolidated material that overlies BEDROCK over most of the earth's surface, including SOIL; SCREE, TALUS, and similar loose rock material (known collectively as the *rock mantle*); glacial deposits such as TILL and DRIFT; waterborne deposits such as sand, gravel, and ALLUVIUM; and organic materials such as PEAT and HUMUS. *See also* REGOSOL.

regosol in geology or soil science, a soil developed principally from weathered MINERAL fragments such as rock grains, SAND, DRIFT, and ASH, with little or no ORGANIC MATTER present. HORIZONS are poorly developed or absent altogether. Regosols do not hold moisture, and make poor soils for plant growth. They develop mostly at high altitudes and/or latitudes, or in arid or semiarid regions. The term "regosol" is considered obsolete in a taxonomic sense, and the group has been divided among several other classifications (*see* e.g., ENTISOL; INCEPTISOL). However, it remains in general use among scientists as a descriptive term.

regulatory complex the broad range of government regulations and regulatory agencies at all levels (federal, state, and local) that may affect a decision or action.

reinjection well *See* DEEPWELL DISPOSAL.

reintroduction of species in wildlife biology, restoring locally extinct species to part of their original range. Under the 1982 amendments to the ENDANGERED SPECIES ACT (ESA), experimental populations of endangered species may be reintroduced, including predatory species such as wolves, grizzly bears, and mountain lions. The best-known of these experimental efforts concerned restoring the gray wolf to Yellowstone National Park and a wilderness area in central Idaho. Formerly a "top predator" that once helped keep the now-burgeoning elk populations in check, the wolf was purposely eliminated by the 1930s as part of national park policy. The reintroduction plan allowed local ranchers to kill wolves (150 were released) involved in predation of domestic stock, and ranchers could receive indemnification for stock loss from a special fund set up by the National Wildlife Federation into the bargain. Nevertheless, the reintroduction was fraught with controversy, with political conservatives and anti-environmentalists joining local ranchers to frustrate the program. Meanwhile, gray wolves that were beginning to trickle into northern Montana from Canada were expected to repopulate Yellowstone and nearby wilderness areas naturally, in which case the wolf would be fully protected under the Endangered Species Act, in contrast to the experimentally reintroduced animals which were not. In 1997 a federal district judge ruled the reintroduction program illegal because under the ESA's rules, the experimental population would effectively put native wolves at risk. The experimental animals could be shot, but not those arriving naturally, and it would be impossible to tell the difference. Conservationists believed the decision to be a specious, if not a disingenuous, way to frustrate the restoration program. They continue to press for a full-scale reintroduction of wolves not only in the northern Rockies, but also in the Southwest (Mexican wolf) and the Southeast (red wolf). Other introduced species include the swift fox in Montana, the Canadian lynx in Colorado and, most successfully, the peregrine falcon. After the falcon was reintroduced in several areas around the country beginning in 1972, it was removed from the endangered list in 1999. A less certain outcome is predicted for an effort to reintroduce the California condor, of which only 20 were living in the wild. Several reintroductions have taken place since 1992 in California, and in Arizona and Utah as well.

relative humidity in meteorology, the amount of moisture presently in the air relative to the total amount of moisture the same mass of air could conceivably hold at the present temperature. The relative humidity is found by dividing the air's current moisture content by its maximum potential content as read from a chart and then multiplying by 100 to convert the resulting ratio to a percentage. For example, if the temperature is 20°C (68°F), the maximum potential moisture content of air at sea level is 14g/kg. If the actual (measured) moisture content was 8g/kg, the relative humidity would be (8/14) x 100, or 57%. Relative humidity obviously increases as the amount of moisture in the air increases. Less obviously, it also increases as the temperature drops, since air is capable of holding less moisture at lower temperatures. *See also* DEW POINT.

release in forestry, *see* RELEASE CUTTING; RELEASE LANGUAGE.

release cutting in forestry, the removal of undesired trees, brush, grass, etc., so that desirable trees can grow better. The most common use of the term is in conjunction with the removal of a brush or hardwood OVERSTORY so that young CONIFERS that have developed beneath the overstory can get more sunlight and growing room, and so that they will no longer have to compete for water and soil minerals. However, removal of grass and FORBS from around young seedlings, or removal of genetically inferior trees from a dense STAND so that the remaining trees have more room and less competition (*see* PRE-COMMERCIAL THINNING; COMMERCIAL THINNING), also qualify as release cuttings. Release may also be done through the use of HERBICIDES instead of by cutting ("chemical release" instead of "hand release").

release language in environmental politics, language in a law that designates a particular area for timber harvest (as opposed to "wilderness language," which designates a particular area for wilderness). "Soft release" language simply designates an area for timber harvest; "hard release" contains one more more additional clauses designed to prevent future consideration of the area as wilderness (for example, language forbidding the Forest Service to consider wilderness for an area in its planning process; language forbidding lawsuits to be filed on timber harvest plans or individual sales; and so on). Release language an important factor in wilderness legislation in the controversy following completion of the RARE II process in 1979 (*see* ROADLESS AREA REVIEW AND EVALUATION).

relict species (relic species, relict vegetation) a SPECIES of plant that was once widespread in a region but is now found only in small, isolated POPULATIONS ("relict populations"). Relict species are generally part of a *relict community* that represents a former CLIMAX COMMUNITY for the region, now displaced by a different climax due to climatic change but able to hang on here and there because of the favorable MICROCLIMATE presented by a mountaintop, a protected ravine, etc. They may exist only as relicts (e.g., devils club, *Oplopanax horridum,* which is found only in the lowlands around Puget Sound and on one small island in Lake Superior, 2,000 miles away), or they may represent scattered outposts of a species or community that remains widespread elsewhere (e.g., the arctic/alpine vegetation found on the summits of a few mountains in North Carolina). In addition to being interesting for its own sake, relict vegetation is scientifically valuable because of the information it can provide about the former RANGES of plants and about the history of a region's climate.

reliction in law, the slow, imperceptible covering or uncovering of a lakebed by changes in the level of the lake's water over time. New shoreline uncovered by reliction belongs to the adjacent riparian landowner (*see* RIPARIAN LAND); new lakebed covered by reliction belongs to the owner of the adjacent portion of lakebed. If the covering-up or uncovering of lakebed lands takes place rapidly enough to be perceptible, property lines are unaffected, and the new shoreline (or lakebed) will belong to the same owner as it did before the change that made it shoreline (or lakebed) took place. *Compare* ACCRETION.

relief in geography or geology, the vertical characteristics of a landscape. The term may be used specifically to refer to the difference in elevation between the highest and lowest parts of the region being described (e.g., a landscape encompassing a river bottom 2,000 feet above sea level and a mountaintop 4,200 feet above sea level would have relief of 2,200 feet), or it may be used in a more general sense to indicate a landscape's overall topography (mountains, cliffs, and canyons are "high relief landscapes"; hills and plains are "low relief landscapes").

religion and the environment a movement in the late 20th/early 21st century in which the religious community in the United States has directly addressed contemporary environmental concerns. Scores of books have been published since the early 1990s on ecotheology and what is now called "creation spirituality." A dozen national ecumenical groups have been established, and most mainstream denominations have set up offices for the environment which distribute resource materials to local churches and synagogues providing advice on environmental work and worship. Scientific findings of the 1980s—global warming, mass species extinction, the ozone hole—have engendered the recent emphasis by theologians and religious leaders on the environmental crisis, although the relationship of religious bodies to the environment had been developing in a minor way since the 1960s. Perhaps the most important event in those years was a doctrinal gauntlet thrown down by Lynn White, a professor of medieval history, in his remarkable essay, "The Historical Roots of the Environmental Crisis," which appeared in the March 10, 1967, issue of

Science magazine. White identified the Judeo-Christian tradition of dualism (man apart from nature) and the associated biblical injunction that humans subdue the Earth and multiply as a primary cause of the escalating devastation of natural processes. "We shall continue to have a worsening ecologic crisis," he wrote, "until we reject the Christian axiom that nature has no reason for existence save to serve man." A few religious leaders were responding at the time to the wave of environmental interest with scattered Earth Day "preach-ins" and stewardship-oriented recycling, open space preservation, and beautification projects. But, as if to confirm White's thesis, the preach-ins quickly died out and the "greening" of American religion faded, especially during the 1980s, when environmental reform was successfully made a deeply divisive, partisan issue by an anti-environmental reaction to the spate of legislation enacted during the 1970s. Cynics explained that many ministers and rabbis, understanding that those businessmen who support the church financially are also those who oppose environmental laws, turned away from the ecological crisis as being significant to religious observance. Today, however, some serious theologians have begun to assert that dualism—which includes the anthropocentric view of humans in an exclusive relation to God—is not a necessary doctrine and may even be in error; that the church has shown "inadequate sensitivity to the cry of creation," as one official denominational report has put it. These sentiments were, in effect, forced on religious thinkers by the growing certitude of scientists since the late 1980s that global warming and rising seas, a thinning ozone layer, the loss of ecological function in the littoral and the depletion of fish and shellfish populations, the continued erosion of soil in an unsustainable industrialized agriculture, the extermination of species, the dying of trees of the temperate zone and the massive deforestation of the tropics, pose dire threats to the continued integrity of the Earth. In a statement subscribed to by a majority of the world's Nobel laureates in science, the Union of Concerned Scientists declared, "No more than one or a few decades remain before the chance to avert threats we now confront will be lost and the prospects for humanity immeasurably diminished." Heeding the scientific evidence, most mainstream American religious bodies now acknowledge that the crashing of global ecosystems is a transcendent moral question, not simply a concern of a vocal minority of "tree-huggers" and campus revolutionaries. Much of the impetus for this change came from an open letter to the religious community by Carl Sagan, Henry Kendall, Freeman Dyson, and two dozen other distinguished scientists stating that environmental reform was not merely a matter of technological adjustment but required a "vision of the sacred." The letter urged the religious organizations to provide moral leadership regarding the environmental crisis—indeed, to accomplish in a very short time a radical restructuring of energy use, agricultural practice, manufacturing technique, renewable resource exploitation, and regional planning. "We urgently appeal to the world religious community to commit, in word and deed, and as boldly as required, to preserve the environment of the Earth." In answer, the heads of major denominations, meeting in New York in 1991, actually took up the challenge. "The cause of environmental integrity and justice must occupy a position of utmost priority for people of faith," they asserted at the Religious Summit on the Environment. "Response to this issue can and must cross traditional religious and political lines. It has the potential to renew religious life." At its best, the recent ecological recasting of religious thought, scholarship, and denominational activity to meet the environmental crisis has been, so far, remarkably ecumenical and remarkably forthright and affirmative, as the writings of John Cobb, Daniel Swartz, Paul Santmire, John F. Haught, Sallie McFague, Matthew Fox, Larry Rasmussen, Rosemary Radford Ruether, and Thomas Berry (among many others) have shown. In Christian terms the new approach is summed up by John Haught, Distinguished Professor of Theology at Georgetown University and adviser to the Vatican on science and religion: "Concern for either local or global environmental welfare is not a very explicit part of the Christian tradition. Nevertheless . . . there is great promise for theological renewal in the ecologically ambiguous Christian traditions. In fact a rethinking of Christianity in terms of the environmental crisis is already under way. . . . An ecological theology is congruent both with contemporary science and the classic doctrine of the Trinity; a doctrine which renounces the idea that God exists only in isolated aseity. . . . By killing off the ecological richness of nature, we prohibit fresh creation, and in so doing we violate the mysterious future which faith knows by the name God. . . . For hope to survive, nature must thrive." At the end of the 1990s, according to the NATIONAL RELIGIOUS PARTNERSHIP FOR THE ENVIRONMENT, some 3,000 churches in the U.S. have active environmental programs.

remineralization the application of crushed rock dust to mimic the effect of glaciation in agricultural and forested areas of the Northern Hemisphere. Adherents of remineralization, of which there are more in Europe than in the United States, believe that massive remineralization will increase tree and plant growth in soils leached out by decades of ACID RAIN and therefore reduce climate change extremes stemming from the buildup of CO_2. *See also* FOREST DECLINE AND PATHOLOGY.

remote sensing the use of recording devices in aircraft or satellites, such as Landsat, to collect geographical and biological data from the Earth's surface. Products of remote sensing include aerial photographs, often in infrared film; multispectral imagery, using a wide range of the electromagnetic spectrum; radar data; and digital image recording of various kinds. Maps and statistical data can be extracted from the remote sensing information acquired by these means, which can be used to detect changes in forest cover, climate patterns, environmental impacts of urbanization, watershed dynamics, and the like. In 1995 the National Aeronautic and Space Administration (NASA), with several European and Japanese space agencies, inaugurated a 15-year "Mission to Planet Earth" program which will involve deploying the Earth Observing System (EOS), which consists of a series of satellites that will beam back remotely sensed data to land-based platforms around the world. The program is expected to foster a greater understanding of the complex dynamics of the world ecosystem, supplying data to scientists and national and international decision makers. *See also* GEOGRAPHICAL INFORMATION SYSTEM.

removal cutting in silviculture, the harvesting of SEED TREES after seedlings and young trees have become well established around them on a plot of land. *See* SHELTERWOOD SYSTEM. *See also* REGENERATION CUTTING.

removal regeneration cutting *See* REGENERATION CUTTING.

renewable energy energy produced by using a RENEWABLE RESOURCE. Renewable energy includes SOLAR ENERGY, HYDROELECTRIC POWER, HYDROGEN, and the burning of wood, METHANE, and similar fuels, as long as these fuels are produced as rapidly as they are used. *See also* ALTERNATIVE ENERGY SOURCE.

Renewable Natural Resources Foundation (RNRF) coalition of professional organizations founded in 1972 to share information concerning the management of renewable natural resources, including soil, water, wildlife, vegetation, and the atmosphere. Member groups are primarily societies of land managers and scientists. The RNRF provides centralized research services for these societies and promote interdisciplinary work among them. Address: 5430 Grosvenor Lane, Bethesda, MD 20814. Phone: (301)493-9101. Website:www.rnrf.org.

renewable resource a RESOURCE that is replaced, either by natural processes or by human activities such as agriculture, as rapidly as it is removed for use. The term "renewable resource" includes water; organic materials from living sources that may be grown and harvested, such as wood, wool, and cot-

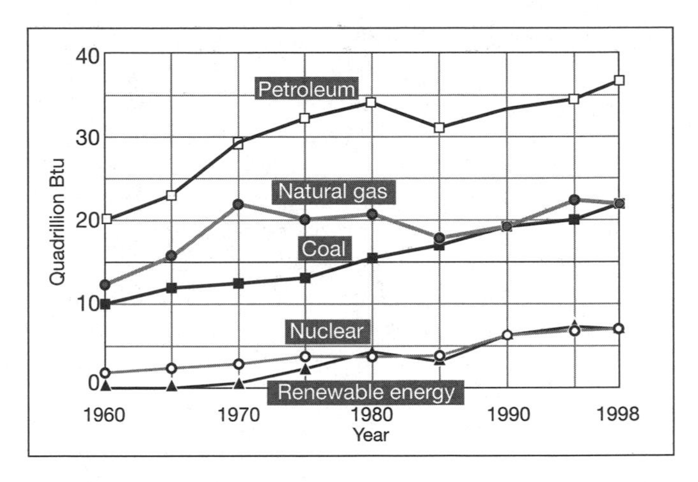

U.S. energy consumption, 1960–1998 ***(Energy Information Administration, U.S. Department of Energy,*** **Annual Energy Review, 1998**

ton; and inorganic forces such as wind, sunshine, and HYDROELECTRIC POWER. It does not include MINERALS or FOSSIL FUELS. Note that "renewable" does not mean "inexhaustible". Wood is renewable only so long as trees are cut no faster than they are grown; water is renewable only at the rate that streams and AQUIFERS are recharged by rainfall; and even sunshine is limited by cloud cover, nightfall, and the presence of topographic features, such as hills, which may block the sun for a large part of each day. For this reason, some economists prefer the term *flow resource* in place of *renewable resource* as a means of indicating that the availability of the resource depends on its replacement rate ("rate of flow") rather than its total volume. *Compare* NONRENEWABLE RESOURCE.

Renewable Resources Planning Act *See* FOREST AND RANGELAND RENEWABLE RESOURCES PLANNING ACT.

reproduction in forestry, *see* REGENERATION.

reproductive strategy in ecology, the complex of physiological and behavioral characteristics that a SPECIES has evolved to make certain that there are always enough individuals surviving into adulthood to perpetuate the species. There are two basic types of reproductive strategies in nature. The most common, known to ecologists as the "r-strategy," is simply to produce enough offspring that even if a high percentage are killed before reaching reproductive age there will be enough left to reproduce. The alternate, known as the "k-strategy," is for adults of a species to aggressively protect their offspring and train them in survival. Under this strategy, only a few offspring are produced, but a high percentage of them survive to adulthood. K-strategy and r-strategy, of course, represent the poles of a continuum, between which any combination of the two strategies can occur. *See also* K-SELECTED SPECIES; R-SELECTED SPECIES.

research natural area an area set aside by a public or private agency in order to protect a natural feature of particular scientific interest. "Natural feature" here generally means a plant COMMUNITY, though research natural areas may be set up to protect outstanding geological sites or CRITICAL HABITAT for rare or endangered animal SPECIES. A research natural area should be large enough to fully protect the community of interest; 300 acres is generally considered minimal, though smaller areas will be set aside if conditions require it. Management is directed toward the maintenance of the natural community that caused the area to be set aside. Timber harvest, mining, the grazing of domestic livestock, and other resource-intensive uses are forbidden, and recreation may be barred if the managing agency determines that it is interfering with research use of the area. Limited manipulation of the ECOSYSTEM is allowed if necessary to preserve the community. For example, if the area has been set aside to study a FIRE-MAINTAINED CLIMAX, the managers can use PRESCRIBED BURNS to make certain that the fire-maintained climax remains in existence. Both the BUREAU OF LAND MANAGEMENT and the FOREST SERVICE attempt as a matter of policy to maintain research natural areas for representative samples of all types of vegetation within their jurisdiction, as well as using them to protect RELICT SPECIES, ENDANGERED SPECIES, and rare or unusual communities (such as WETLANDS in predominantly dry regions). Establishment of these areas is an administrative decision, and is made at the discretion of the administering agency rather than by Congress.

reserved component in forestry, that portion of the TIMBER BASE that has been withdrawn from logging by laws or by administrative rules. The reserved component includes all lands that grow timber rapidly enough to allow harvest (generally, 20 cubic feet per acre per year; *see* SITE INDEX) but that are included within WILDERNESS AREAS, RESEARCH NATURAL AREAS, or similar preserves where logging is not allowed. These lands are not included when calculating the ALLOWABLE CUT. *Compare* STANDARD COMPONENT; SPECIAL COMPONENT; MARGINAL COMPONENT.

reservoir (1) in hydrology, *see* DAM.

(2) in general, a supply of some RESOURCE large enough to sustain a flow of the resource over time. Thus, a forest is a reservoir of timber; the atmosphere is a reservoir of OXYGEN; a POPULATION of living ORGANISMS is a reservoir of GENES. The term may also be used in a more restrictive sense to mean a closed container capable of holding some desired material and meting it out slowly, such as a tank used to supply oxygen to a welding torch or an aqualung.

resistivity the electrical resistance of a material over a standard distance, usually expressed in terms that combine resistance and distance, most commonly *ohm-centimeters*. Resistivity is used in environmental science principally to indicate how much a metal is likely to be subject to stray-current corrosion (*see* CORROSION: *types of corrosion*) if buried in different types of soil. In general, the higher the resistivity of the soil, the lower the incidence of corrosion. Typical values: poorly drained clays, 2,000 ohm-cm (low resistivity, high corrosion potential); well-drained gravels, 10,000 ohm-cm (high resistivity, low corrosion potential).

resource in economics, anything that can be used in such a way as to improve the welfare of the user. Note that this is an extremely broad definition. A tree is a resource to someone who wants to cut it down and build a house out of it, but it is equally a resource to someone who wants to merely sit and contemplate its beauty if this contemplation improves his welfare by making him feel better. It may also be a resource to a bird for nest-building or to a leaf-eating insect as a food source, as the definition makes no distinction between human and non-human "users." In order to make sense of the term, therefore, it is usually necessary to subdivide the category "resource" according to one of several standard schemes. One such division is by source: *natural resources* (unmodified resources given to us by the environment, sometimes termed *raw materials*); *capital resources* (items constructed or modified for use by humans); and *human resources* (human energy, thought, creativity, etc.). A second standard division of resources is by use: *commodities* (items that will be physically manipulated by the user) and *amenities* (items that are used to give pleasure, improve spiritual health, etc., without being touched or changed by the user). A third type of overall division is by availability: *renewable* or *flow resources* (those that are replenishable over time and whose availability is therefore determined by the rate at which they are used) and *nonrenewable* or *fund resources* (those that are not replenished and whose availability is therefore determined by the total amount present, the rate of use being essentially irrelevant). Because of this welter of confusing and overlapping terminology, it is necessary when talking about resources to make certain that everyone is using the same definitions and applying them to the same items in the same manner. *See also* RESOURCE ACTIVITY; RESOURCE BASE; RESOURCE

RECOVERY; RECYCLING; and separate articles under most terms *italicized* above.

resource activity generally, the actions necessary to obtain a NATURAL RESOURCE for use by humans, including mining, timber harvesting, well drilling, and so on. The term is usually limited to those situations involving the obtaining of commodities, but there is no real reason why it could not be used for amenities as well (*see amenity value*), in which case trail building, for example, would qualify as a resource activity.

resource base (1) the total amount of a given RESOURCE or set of resources available within a specific geographical area, both now and through the foreseeable future. The resource base is divided into categories (*reserves*) according to how much information is available concerning the supply. *Proved reserves* are those whose quantity has been measured; *probable reserves* are those that have been identified but not measured; and *possible reserves* are those whose existence has been inferred from data but have not yet been physically located. Beyond these three classes of *identified reserves* are the *undiscovered reserves*, including the *hypothetical* (those that theory says should be present but have no data to support them) and the *speculative* (those guessed at by various means, including statistical averages, cursory observation, and simple hunches). The resource base may also be divided into *economic* and *noneconomic* reserves, depending upon whether or not the resources in question may be recovered at a profit. The line between economic and noneconomic reserves, known as the *economic threshold*, varies considerably according to technology, availability of alternate supplies, and demand for the resource.

(2) in Forest Service usage, the resource base of a National Forest is the total amount of resources found within the boundaries of the forest, using the broadest possible definition of the term "resources."

Resource Conservation and Recovery Act (RCRA) federal legislation to regulate the disposal of HAZARDOUS WASTES, signed into law by President Gerald Ford on October 21, 1976. Couched as a series of amendments to the SOLID WASTE DISPOSAL ACT of 1965, the act sets up what has been called "cradle-to-grave" controls over the generation, handling, and disposal of hazardous wastes, including the setting of standards for containers, labels, handling procedures, facilities design, and record-keeping practices; the establishment of a manifest system for tracking all hazardous wastes from their point of origin to their point of ultimate disposal; and the assurance of the public's right to know the amounts and types of wastes handled and disposed of at each facility. Other provisions require the Administrator of the ENVIRONMENTAL PROTECTION AGENCY (EPA) to establish a list of hazardous materials to which the RCRA standards and requirements will apply, and to issue guidelines for the handling and disposal of these materials, taking into account public health, surface water and groundwater quality, air quality, safety, and esthetics; tighten the controls of the Solid Waste Disposal Act over open dumping; continue the authorization of federal funds and technical assistance for states and municipalities setting up hazardous-waste programs and facilities; create an Office of Solid Waste within the EPA; and require disclosure of the financial interests of all EPA officials concerned with the administration of solid waste programs, as they relate to companies that might be regulated under the provisions of the act. By the end of the 1990s, EPA and state regulatory agencies were required under the act to monitor more than 750,000 facilities creating hazardous waste and 15,000 waste haulers. To cope with the administrative difficulties implied by these numbers, the EPA is now striving to streamline its oversight efforts and make corrective actions required under the act more flexible. The agency has estimated that 5,000 facilities need corrective action, more than three times the number of abandoned toxic waste sites under the Superfund cleanup program. *See also* COMPREHENSIVE ENVIRONMENTAL RESPONSE, COMPENSATION AND LIABILITY ACT.

Resource Planning Act *See* FOREST AND RANGELAND RENEWABLE RESOURCES PLANNING ACT.

resource recovery strictly, the process of obtaining energy from waste materials, especially SOLID WASTES. Resource recovery may involve burning the wastes directly in incinerators designed to utilize the heat from the burning process for space heating, electrical generation, or other uses, or it may involve processing the wastes into solid fuels through separation of noncombustibles, compaction, PYROLISIS, and other techniques. These *refuse-derived fuels* (RDF) are functionally equivalent to COAL (though with only half the energy content), and may be transported and used in standard boilers. The term "resource recovery" is also sometimes used to indicate any form of RECYCLING in which the recycled product is in the form of raw materials rather than reusable manufactured goods. *See also* COGENERATION.

Resources for the Future (RFF) an independent, nonprofit research organization founded in 1952. Through the efforts of more than 40 researchers (economists, engineers, ecologists, city and regional planners, public policy analysts), RFF investigates areas such as climate change, electric utility restructuring, and sustainable forestry with special emphasis on BENEFIT/COST ANALYSIS and the use of market forces rather than regulation to achieve conservation objectives. Information gathered by RFF is distributed by various means, including congressional testimony, briefing government officials, publications, seminars, and workshops. Many RFF documents can be accessed on its website. Address: 1616 P Street NW, Washington, DC 20036. Phone: (202) 328-5000. Website: www.rff.org.

respiration in the biological sciences, the set of processes through which energy is obtained from sugars and other CARBOHYDRATES in the body of a living ORGANISM. Respiration includes both the processes taking place with the CELL (*cellular respiration,* or *internal respiration*) and the exchange of the HYDROGEN receptors and gaseous waste products involved in these processes with the organism's environment (*external respiration*). Internal respiration is essentially identical to catabolism (*see under* METABOLISM). External respiration generally involves the uptake of OXYGEN and the release of CARBON DIOXIDE. For one-celled organisms this is accomplished by diffusion through the cell wall. In animals it takes place in the lungs or gills and is the process we know as "breathing," and in plants it takes place in the leaves and is a part of the process known as TRANSPIRATION. In ANAEROBIC BACTERIA, which use SULFUR instead of oxygen as a hydrogen receptor, external respiration involves the exchange of sulfur and hydrogen sulfide (H_2S) with the environment.

restocking in forestry, the process of replanting cutover, burned over, or otherwise deforested land with new young trees. *See* REGENERATION; REFORESTATION; STOCKING LEVEL.

restoration ecology the study of the means by which disturbed ecosystems can be artificially restored to reestablish the viability of their inherent natural processes. Because so many crucial ecosystems have been destroyed around the world by industrial development, environmentalists are increasingly recognizing that it is not enough to save the few pristine areas that remain, but to figure out how to restore those that have been damaged. The favorite targets for restoration ecology are wetlands—salt marshes in estuarine areas, prairie potholes, and freshwater swamps and bogs. Other projects include rebuilding rivers and streams, restoring botanical diversity to forest lands, and reintroducing KEYSTONE SPECIES into damaged natural areas. One of the most ambitious restoration ecology projects is being undertaken in the FLORIDA EVERGLADES by the ARMY CORPS OF ENGINEERS. While the new techniques are promising, many environmentalists are concerned that the "technological fix" aspect of restoration ecology will suggest to policy makers in government and industry that the destruction of pristine areas is permissible since they can always be restored. At the same time, so much BIODIVERSITY has been lost that restoration ecology must be a necessary component of any comprehensive environmental policy.

restricted component in Forest Service terminology, that portion of the COMMERCIAL LAND BASE on which timber harvest is restricted by laws or regulations to a level below that which the land could yield under intensive management. The restricted component includes STREAMSIDE MANAGEMENT UNITS, SPECIAL INTEREST AREAS, and so forth. It excludes WILDERNESS AREAS, as they are not part of the commercial land base.

retention (1) in hydrology, the amount of water that is taken up by soil and vegetation during a rainfall. It is measured by taking the difference between the amount of rainfall and the amount of RUNOFF on a given plot of land.

(2) in landscape management, short for RETENTION OF CHARACTER.

retention of character in landscape management, the management of a VIEWSHED in such a way that management activities are not evident to the visitor. Only such activities may go on as will not be noticed when looking at the view, which remains completely natural in appearance. Retention of character is the second most restrictive category in the VISUAL RESOURCE MANAGEMENT SYSTEM, falling between PRESERVATION OF CHARACTER and PARTIAL RETENTION OF CHARACTER.

retention time in hydrology, the average amount of time a drop of water that enters a lake or pond will remain there before flowing out through the outlet stream. Retention time is calculated by dividing the rate of flow of the outlet stream into the volume of the body of water. It may also be determined by calculating the inflow (average annual precipitation multiplied by the area of the lake's WATERSHED) and dividing this figure into the water volume. Differences in retention time have a great effect on the susceptibility of different lakes or ponds to POLLUTION, those with long retention times being both considerably more sensitive to the presence of POLLUTANTS and considerably harder to clean up once they have become polluted.

return flow in irrigation, water applied to a crop that is not taken up by the crop but instead flows back into the river system or AQUIFER from which it was taken. Return flow represents excess water that is not needed by the plants. It generally carries a high load of SALTS, FERTILIZERS and other pollutants that it has picked up while passing through or over the soil of the field, and therefore represents a threat to the quality of water in the source. A certain amount is necessary, however, to prevent salt buildup in the irrigated field.

reverse osmosis (ultrafiltration) the process of removing dissolved POLLUTANTS from water by forcing it through a semipermeable membrane with holes large enough to pass the water MOLECULES but small enough to block passage of the (usually much larger) molecules of the pollutants. In reverse osmosis, hydraulic pressure is used to overcome osmotic pressure (*see* OSMOSIS), so that the water flows against the osmotic gradient, that is, moves through the membrane in such a way as to increase the difference in the concentration of pollutants on the two sides of the membrane rather than reducing it. The process consumes large amounts of energy, particularly as the holes in the semipermeable membrane approach closer and closer to the size of water molecules (to remove pollutants of smaller and smaller molecular size). However, it is in most cases less energy-consumptive than other methods of water purification (such as DISTILLATION or FREEZE DISTILLATION) and in some cases—particularly those involving volatile pollutants (*see* VOLATILITY; VOLATILE ORGANIC COMPOUNDS; VOLATILE SOLIDS)—it may be the only way in which complete purification can be assured.

rhizome in botany, an underground stem, generally growing horizontally. Short, thick rhizomes are used by plants to store food; longer, thin ones may be used as means of vegetative reproduction (*see* ASEXUAL REPRODUCTION). In either case, adventitious roots usually form on the underside of the rhizome, and adventitious leaves and stalks may form where the upper side of the rhizome is close to the surface (*see* ADVENTITIOUS GROWTH). *Compare* BULB; TUBER.

rhumb *See under* MERCATOR PROJECTION.

rhyolite in geology, a light-colored IGNEOUS ROCK, chemically identical to GRANITE but extrusive rather than intrusive in origin (*see* EXTRUSIVE ROCK; INTRUSIVE ROCK). It differs physically from granite in that it is *aphanitic,* that is, that its crystal structure is too small to be seen by the unaided eye.

ribonucleic acid *See* RNA.

Richter scale in geology, scale for measuring the magnitude of EARTHQUAKES, developed around 1940 at the California Institute of Technology by Charles F. Richter (1900–1987) and others. The Richter scale is based on SEISMOGRAPH readings, and is logarithmic in nature, with each increase of one unit in the scale representing a multiple of 10 in the size of the seismic wave, which corresponds to a multiple of 31.5 in the amount of energy released at the earthquake's epicenter. For example, an earthquake of magnitude 7 on the Richter scale would show a seismic wave 10 times as high as one of magnitude 6, and 100 (10 x 10) times as high as one of magnitude 5. The amount of energy released by the magnitude 7 quake would be 31.5 times as great as that released by the quake of magnitude 6, and 992 (31.5 x 31.5) times as great as that released by the quake of magnitude 5. A magnitude 6.35 quake releases roughly as much energy as the atomic bomb dropped on Hiroshima in 1945. The 1906 San Francisco earthquake was magnitude 8.3—the same as nearly 1,000 Hiroshima bombs. The largest earthquake ever recorded, the Iran quake of 1972, hit 9.5—roughly 315,000 Hiroshima bombs. *See also* EARTHQUAKE; MERCALLI SCALE.

ridge tillage *See* CONSERVATION TILLAGE.

riffle a small RAPID, generally one without waves or with only very small ones. Riffles are usually formed by gravel BARS, and contain no holes or large rocks. The water is normally only a few inches deep.

rift (1) in geology, the boundary between two separating continental plates (*see* PLATE TECTONICS; RIFT VALLEY).

(2) in oceanography, the thin sheet of water which runs up a beach after a wave has broken.

rift valley in geology, a valley formed along the boundary between two separating continental plates (*see* PLATE TECTONICS). As the plates separate, the land between them sinks downward as a large linear GRABEN. Rift valleys are usually steep-sided and flat-bottomed, and often contain large lakes. The best-known example is the Great Rift Valley of eastern Africa.

rift zone in oceanography, the area of a beach where RIFT occurs; that is, the area above the zone of breaking waves and below the zone of dry sand. The rift zone shifts up and down the beach with the tide. *See also* LITTORINA ZONE.

rill a very small stream, usually less than 15 centimeters across (6 inches). The term is used for small permanent streams emanating from springs, for EPHEMERAL STREAMS that develop on hillsides during and immediately after rainstorms, and for the numerous, tiny, extremely ephemeral streams developed by wave RIFT as it runs back into the sea from its high point on the beach. *See also* RILL CHANNEL; RILL EROSION.

rill erosion the EROSION of soil on a hillside by the formation of RILLS during a rainstorm. No single rill moves more than a few cubic centimeters of soil to form its channel. However, since thousands of rill channels may form, be erased, and reform during the course of a single rainstorm, the overall contribution of rill erosion to soil movement is often considerable. *See also* SHEET EROSION.

rimrock a vertical exposure of BEDROCK at the edge of a PLATEAU. Rimrock is characteristic of landforms in dry climates where EROSION is largely limited to MASS WASTAGE and the rounding effects of SHEET EROSION are not often observed.

Rio Earth Summit (U.N. Conference on Environment and Development) a 1992 United Nations conference held in Rio de Janiero on the 20th anniversary of the STOCKHOLM CONFERENCE ON ENVIRONMENT. Representatives of 178 nations attended, plus nearly 8,000 NONGOVERNMENTAL ORGANIZATIONS (NGOs) at an unofficial companion conference called the Global Forum. The Rio Earth Summit was meant to reconcile the twin goals of increasing economic growth in developing nations and the establishment of international agreements to deal with global environment crises. The delegates adopted a wide-ranging, nonbinding framework plan toward this end entitled "Agenda 21," an 800-page document calling for improved policies governing atmosphere, ocean pollution, soil and water conservation, toxic waste disposal, energy efficiency, population growth, and the transfer and financing of environmental technologies. In an associated action, the conference issued the "Rio Declaration on Environment and Development," also nonbinding, which gave priority to the economic needs of poorer countries as "the most environmentally vulnerable." To encourage follow-up action on Agenda 21, the conference created a high-level oversight group called the Commission on Sustainable Development. Two treaties were also adopted at Rio: the Framework Convention on Climate Change, which would ultimately lead to the KYOTO PROTOCOL, was intended to curb emissions of greenhouse gases considered responsible for GLOBAL WARMING; and the Biodiversity Convention treaty, which required participating nations to inventory plant and wildlife resources and develop plans to protect endangered species. The delegates also adopted a nonbinding Statement of Forest Principles, concerning the sustainable management of forest resources. While the Rio Earth Summit was, and still is, thought to have failed to produce any solid agreements that would truly meet the many formidable global environmental crises, its treaties and conventions and official and unofficial conclaves are still ramifying, most particularly in terms of the Kyoto accords and the potentially significant EARTH CHARTER as an exercise in global populist politics at the international level.

Rio + 5 Forum a meeting of 500 representatives of nongovernmental organizations that participated in the RIO EARTH SUMMIT. The forum was organized by Maurice Strong, a Canadian businessman and the prime mover of the Rio Earth Summit, and Mikhail Gorbachev, former USSR leader. As part of the forum, the independent Earth Charter Commission, which had been working separately on a declaration of principles, presented a benchmark version of its EARTH CHARTER, now thought to be a lasting contribution by the NGOs toward the role of the United Nations in fostering global environmental reform.

riparian doctrine in water law, a system of WATER RIGHTS based on the principal that the owner of land along a watercourse (*see* RIPARIAN LAND) should have the right to use the water flowing past his property. Based on principles laid down in the Justinian Code (sixth century A.D.), if not earlier, riparian doctrine in its purest form grants rights to the use of surface waters only to those who own property along a *watercourse,* defined as a stream, river, lake, or pond. The water may only be used on lands that border the stream; however, it may generally be used anywhere on those lands (some states require it to remain in the WATERSHED of the watercourse it is drawn from). Riparian rights are conveyed with the sale of the property, but lands subdivided from a riparian piece in such a way that they have no border on the water lose all rights to that water. The water must be shared with other riparian owners; hence, the use made of it may not degrade either the quality or the quantity of the water in the watercourse. In practice, this "pure" form of the riparian doctrine is generally hedged in by restrictions. The doctrine does not normally apply, for example, to artificial watercourses (canals and reservoirs) nor to "unreasonable" use. Municipalities that require water for their residents may acquire riparian rights through condemnation, even though they may not necessarily own waterfront property. Finally, most states that build their water law on the riparian doctrine have enacted some sort of a permit system, requiring riparian owners to obtain a permit from a state agency before they can exercise their water rights. These permits may spell out restrictions on the use of the water, may be terminated if abandoned or abused, and may (usually) be suspended during times of water shortage. The priority for suspension is generally least-necessary uses first, rather than the strict prior-in-time priority enforced under the doctrine of PRIOR APPROPRIATION. Twenty-nine states, almost all of them in the eastern half of the country, currently use some form of the riparian doctrine; in addition, nearly all states recognize the rights of riparian landowners to nonconsumptive uses such as fishing, boating, and the building of docks and wharves. *See also* HYBRID RIGHTS.

riparian land land along the edge of a stream or river. In ecology, riparian land is defined as land within the RIPARIAN ZONE. In law, the term is a little broader. For legal purposes, RIPARIAN LAND is any block of land under single ownership that borders or contains part of a permanent year-round stream and that lies totally within a single WATERSHED. For example, a lot 50 feet wide along a stream but extending 1,000 feet back from the stream would all be considered riparian land for purposes of law, provided that it was all within the stream's watershed; however, a lot whose border paralleled the stream for a mile but which remained 25 feet away the whole distance would not be considered riparian land. *See* RIPARIAN DOCTRINE.

riparian right *See* RIPARIAN DOCTRINE.

riparian vegetation plants that live in the area immediately bordering a river or stream. Strictly speaking, riparian vegetation is limited to PHREATOPHYTES, emergent MACROPHYTES, and those plants that form ASSOCIATIONS with phreatophytes and emergent macrophytes. Examples: cottonwood, rush, cattail. *See also* RIPARIAN ZONE; RIPARIAN LAND.

riparian zone the lands immediately bordering a stream or river. The riparian zone is generally considered to extend outward only as far as the plant COMMUNITY is dominated by RIPARIAN VEGETATION. *See also* RIPARIAN LAND.

riprap large, angular pieces of stone placed along a streambank or lake or ocean shoreline to prevent erosion. Riprap absorbs the mechanical force of wave or current action without moving, thereby preventing the soil behind it from moving as well. Properly installed riprap is composed of stones of several different sizes in order to fit together tightly and minimize the size of the spaces between stones. The bank it is laid on (not merely thrown on) should lie at an angle of 25 to 50% (14° to 27°). *See also* ARMOR STONE; EROSION: *control of erosion.*

risk assessment and management with reference to the environment, the use of risk analysis to promulgate environmental laws and regulations based on a comparison of risk-management approaches. Risk management professions seek to quantify risk levels to individuals, such as cancer risks from pesticide use, so that they can be balanced against potential benefits to society as a whole. The Environmental Protection Agency, the Food and Drug Administration, and the Occupational Health and Safety Administration, among others, all use sophisticated analytical methods to determine risks associated with various environmental policies.

River Network coordinates the dissemination of information resources and expertise in support of local groups engaged in protecting and restoring rivers and watersheds. The River Network has published several helpful guides for grassroots organizations. Headquarters: P.O. Box 8787, 520 SW 6th Avenue, Suite 1130, Portland, OR 97204. Phone: (503) 241-3506. Website: www.rivernetwork.org.

RNA (ribonucleic acid) any of a very large number of organic acids (*see* ORGANIC COMPOUND; ACID) found in the living CELL, where they are used to guide the construction of protein molecules (*see* PROTEIN; MOLECULE.) RNAs are very similar in structure to DNAs, differing in only three important ways: (1) the sugar used in their construction is ribose rather than deoxyribose, meaning that it has one more oxygen ATOM in its molecular structure; (2) the RNA molecule is a single, rather than a double, helix; and (3) where the nucleotide *thymine* would appear in DNA, RNA contains the nucleotide *uracil*—thymine with its attached methyl GROUP replaced by a single hydrogen atom. There are three distinct types of RNA in the cell. *Messenger RNA* (mRNA) forms along one of the two strands of a DNA molecule, forming a complementary copy of the MONOMER sequence within the strand of DNA (except, of course, for the minor differences in sugar and nucleotide structure indicated above). *Transfer RNA* (tRNA) also apparently forms along DNA strands. Its molecule is much shorter than that of mRNA. One end attaches to an AMINO ACID; the other end matches up to a portion of the mRNA according to a three-unit "code" known as a *codon,* the sequence of codons in the mRNA determining the sequence of the tRNA lined up along it and thus the sequence of the amino acids within the protein molecule being formed. The third type of RNA, *ribosomal RNA* (rRNA), is not well understood, but appears to be connected with the synthesis of the amino acids that go to make up the proteins. It forms small compact bodies known as *ribosomes* within the cell. Along with their apparent role in amino acid formation, the ribosomes seem to serve as "anchor points" for the mRNA and tRNA molecules during protein construction.

ENVIRONMENTAL SIGNIFICANCE OF RNA

RNA is the only means used by living ORGANISMS in forming proteins; thus damage to it destroys the body's ability to renew cells. This destruction can be caused by radiation (*see* RADIOACTIVITY) or by chemical MUTAGENS—the same mechanisms that cripple or destroy DNA. The RNA of all organisms uses the same set of codons, meaning that mRNA extracted from a BACTERIUM, for example, can be "read" by the tRNA of a human cell, providing powerful evidence for the interconnectedness of all living things through evolution, as well as making possible the science of genetic engineering. *See also* VIRUS.

road in federal agency usage, a route maintained by mechanical means for the regular use of motor vehicles. Those routes worn across the landscape by vehicular passage alone, and those routes that are no longer maintained and have become impassable to vehicles, do not qualify as roads and may not be used in the determination of ROADLESS AREA boundaries. *See also* WAY.

roadhead the end of a road, especially one penetrating into an otherwise roadless block of land. "Extending the roadhead" is the same as extending the road; that is, the roadhead is extended further back into the roadless area by building a new section of road beyond the former road's end. *Compare* TRAILHEAD.

roadless area loosely, any area without roads. More formally, as used by federal agencies, a roadless area is an area of 5,000 acres or more, or an area of less than 5,000 acres adjacent to a declared WILDERNESS AREA, or an island of any size, on which no motor vehicle routes exist that meet the federal definition of a ROAD. Roadless areas serve as the RESOURCE BASE from which wilderness areas are derived. They were very loosely protected by the WILDERNESS ACT OF 1964 and by the FEDERAL LAND POLICY AND MANAGEMENT ACT OF 1976, both of which required the study of roadless areas for their wilderness potential before any new roads could be built in them; however, this protection has largely been eroded by the completion of the ROADLESS AREA REVIEW AND EVALUATION process and by the addition by Congress of RELEASE LANGUAGE to most wilderness bills passed since 1978.

Roadless Area Review and Evaluation (RARE) process undertaken by the FOREST SERVICE to determine which ROADLESS AREAS on the NATIONAL FORESTS should be recommended for wilderness designation (*see* WILDERNESS AREA) and which should be retained in the COMMERCIAL LAND BASE. The process was actually undertaken twice. The original evaluation process, known today as RARE I, began in 1967 as an informal inventory of all roadless lands of 5,000 acres or more on the National Forest System. A directive from the Chief of the Forest Service in 1971 required that the inventory be completed, and recommendations as to which areas should be studied for wilderness be filed with his office, by June 30, 1972. A final ENVIRONMENTAL IMPACT STATEMENT based on these recommendations was released on October 15, 1973. Out of an inventoried base of 1,449 roadless areas totaling 55.9 million acres, 274 areas totaling 12.3 million acres were recommended for wilderness study.

FLAWS IN RARE I

As the Forest Service itself immediately admitted, there were serious flaws in RARE I. Due to improper boundaries and overly strict definitions of what should be considered "roadless," approximately 17 million acres of roadless land had been left out of the inventory. Areas that overlapped National Forest, Ranger District, or sometimes county or state boundaries had been broken up along these administrative lines and inventoried as several small units instead of one large one. Finally, the analysis used in choosing areas for wilderness study had been seriously flawed. It was based on a so-called "effectiveness index" (EI) calculated by multiplying the area's size in acres by its "quality index" (QI), which in turn had been calculated using a weighted set of estimates that also depended largely on size, thus multiplying a size factor by another size factor and making size alone the prime criterion for designating areas for wilderness study.

RARE II

In response to these criticisms, a second inventory, known as RARE II, was announced by Assistant Secretary of Agriculture Rupert Cutler in June 1977. The roadless lands were reinventoried; contiguous units on adjacent forests were treated as single roadless areas; and a new set of criteria, known as the Wilderness Attributes Rating System (WARS), was applied to the new inventory. The RARE II Final Environmental Impact Statement, released on January 4, 1979, identified 2,919 roadless areas totaling approximately 62 million acres. Of these, 1,981 areas totaling 15.1 million acres were recommended for "instant wilderness" designation, while 10.8 million acres were recommended for further study ("further planning"). The remainder was released to general forest uses.

AFTERMATH OF RARE II

Controversy continued to dog the process following the conclusion of RARE II, principally as charges that the decisions on most areas had been made politically, with little reference to the WARS ratings. The controversy resulted in the passage of several major wilderness bills by Congress during the next five years. A lawsuit, *California vs. Bergland,* was filed by the State of California against the Forest Service in the fall of 1979 challenging the RARE II inventory. As a result of this lawsuit, the 9th circuit court (California) declared the RARE II Environmental Impact Statement inadequate. On February 1, 1983, acting in response to this ruling, Assistant Secretary of Agriculture John Crowell (in an order that came to be called "RARE III") announced that all roadless areas would be returned to the forest planning process (*see* FOREST PLAN) for reevaluation on their merits. Since then, environmental organizations have actively pressed for the Forest Service to discontinue building new roads into roadless areas, based on calculations that such construction is not cost-effective and amounts only to a hefty subsidy to the forest products industry. In response, road building has been much reduced and in some areas "obliteration" projects have been undertaken in National Forests to eliminate dangerous and environmentally problematical roads. In 2000, President Clinton asked that no more new roads be built in roadless areas of the National Forests. *See also* RELEASE LANGUAGE.

roadside census *See under* STRIP CENSUS.

roches moutonnées in geology, hummocked bedrock resulting from glacial action. The name is French for "sheep rocks" and evidently refers to the appearance of the formations, which resemble masses of huddled sheep. Roches moutonnées are characteristic of hard, crystalline bedrock structures. They are smooth on the upslope side as a result of abrasion and rough on the downslope side as a result of plucking (*see* the description of both methods of rock shaping under GLACIAL LANDFORMS). *Compare* CRAG-AND-TAIL.

rock any consolidated, massive MINERAL material, "consolidated" in this case meaning not separable into individual grains or crystals by hand alone (without tools) and "massive" meaning containing numerous grains and/or crystals. Most rocks are MIXTURES of two or more minerals; however, a rock may be composed of a single mineral if its structure meets the definitions of "consolidated" and "massive" given above (examples: SANDSTONE, composed primarily or exclusively of QUARTZ; LIMESTONE, composed primarily or exclusively of calcite). Classification of rocks by geologists is done according to mineral content, grain or crystal size (*see,* e.g., PORPHYRITIC ROCK; CRYPTOCRYSTALLINE ROCK), and means of formation (*see* IGNEOUS ROCK; METAMORPHIC ROCK; SEDIMENTARY ROCK). *See also* BEDROCK; BASEMENT ROCK; REGOLITH.

Rockefeller, Laurance S. (1910–) American philanthropist and conservation leader. Educated at Princeton University, he participated in family aviation businesses including the founding of Eastern Airlines. Among his main areas of philanthropy are the preservation of parks and natural areas. Rockefeller donated the 5,000 acres to the U.S. government that became the Virgin Islands National Park. He has been director of the Jackson Hole Preserve and was responsible for the donation of 33,000 acres of land to the Grand Teton National Park. From 1969 to 1973 he was the director of the Citizens Advisory Committee on Environmental Quality and has also served on the board of the AMERICAN CONSERVATION ASSOCIATION.

rock mantle *See* REGOLITH.

rock phosphate *See* PHOSPHATE ROCK.

Rocky Flats a now-abandoned nuclear weapons facility located 16 miles from downtown Denver, Colorado, with some suburbs now within three miles of the plant. Rocky Flats, which manufactured plutonium components from 1952 to 1989 for the U.S. cold war arsenal, had a notorious history and still presents a daunting cleanup task. In 1969 a serious fire (one of several) broke out at the plant and might well have caused thousands of lingering deaths and illnesses from radiation exposure in the Denver metropolitan area had it not been contained. While no one died, the fire broke all previous records for U.S. industrial accidents in terms of the cost of materials and structures destroyed—nearly $71 million worth. In 1989, after decades of irresponsible management, the FBI raided the 6,500-acre site, charging the then-manager, a private corporation, with illegally dumping hazardous substances and knowingly contaminating nearby watercourses. After the raid the factory was closed, leaving behind some 14 tons of weapons grade plutonium, much of it in insecure containers. The cleanup program started in earnest in 1993 with a workforce of more than 4,000. By 2010 most of the major buildings will be demolished (about 100 of them) and all but 300 acres—some of which will remain toxic for thousands of years—are to be decontaminated. The construction of the WASTE ISOLATION PILOT PROJECT (WIPP) in Carlsbad, New Mexico, was in substantial part a response to the need for a place to store the toxic materials that had accumulated at Rocky Flats. The first shipment from Rocky Flats to Carlsbad, consisting of

26 drums of transuranic waste, took place in June 1999. Some 12,000 drums of transuranic waste slated for WIPP remain.

Rocky Mountain Arsenal Sixty-two square kilometer (24 square mile) military reservation on the northeast edge of Denver, Colorado, adjacent to Stapleton International Airport (now abandoned) and adjoining the South Platte River. The arsenal grounds and its surroundings have been considered one of the most polluted spots in the United States. Established during World War II to manufacture chemical-warfare agents and later modified to produce PESTICIDES and DEFOLIANTS as well, the arsenal has polluted at least 77 square kilometers (30 square miles) of the aquifer it sits on, threatening drinking water supplies for rural residences and the nearby city of Brighton. Crop and livestock damage was reported on adjacent farms beginning in 1951. Prior to 1957, all wastes on the site were dumped in unlined pits; after that, a number of other approaches were tried, none completely successfully. The most serious problems occurred in the early 1970s, when use of a 3,650-meter (12,000-foot) injection well on the site was linked to the occurrence of small earthquakes in the surrounding region. At about the same time, traces of 2,4-D were discovered in the site's groundwater: since the chemical had never been manufactured there, chemists concluded that it had been formed spontaneously on-site in the waste-disposal pits. In 1987 most of the arsenal's vast area was declared a Superfund site, under the COMPREHENSIVE ENVIRONMENTAL RESPONSE, COMPENSATION, AND LIABILITY ACT (CERCLA), with the U.S. Army (for chemical warfare agents) and Shell Oil (for pesticide manufacturing) being found financially responsible. In the recent history of the arsenal, the 1980s were characterized by a series of rancorous lawsuits involving various agencies of the federal and state governments and private industry. In 1986 in the midst of this confusion some two dozen bald eagles, then an endangered species, were discovered to have a winter roost in a stand of cottonwood trees on the arsenal grounds. This brought another legal player into the cleanup disputes, since under the ENDANGERED SPECIES ACT the Fish and Wildlife Service could conceivably take any action necessary to protect the eagles. At length it became clear that instead of a complicating factor, the eagles presented a way to focus the cleanup effort, for studies of the arsenal eagles revealed that they were, in fact, not so seriously contaminated by pesticides as might be expected. In fact, based on an examination of muscle and fat tissue, the arsenal eagles were healthier than those found at study sites that had been set up for comparison. A later survey revealed that the arsenal lands contained some 300 species of wildlife, including more than 6,000 mule deer and 275 white-tailed deer which were apparently thriving. In 1989 the National Wildlife Federation urged Congress to designate the arsenal as a National Wildlife Refuge, and in 1991 Colorado Representative Patricia Schroeder held hearings on the refuge proposal, which was passed in 1992 as the Rocky Mountain Arsenal National Wildlife Refuge Act. In 1996 a 14-year, $2 billion remediation program was begun, which will clean up some 17,000 acres of the arsenal's land. By 1999 some of the cleanup had already been completed, with buildings where nerve gas and mustard gas were made cleansed and demolished, and contaminated material buried in offsite landfills. A chemical dump site has been capped with crushed concrete from the runways of the former Stapleton airport; and two massive, double-lined landfill sites on arsenal grounds have been constructed to hold pesticide and nerve gas wastes. Work has also begun to restore about half of the refuge's 17,000 acres to the original short-grass prairie biome, with plantings of blue grama, buffalo, and western wheat grasses plus various native annual plants. While the Rocky Mountain Arsenal refuge will never be as pristine as others in the national wildlife refuge systems, the remediation effort is perhaps the most remarkable ever on a Superfund site once thought to be nearly impossible to clean up.

Rocky Mountain Institute a think tank best known for developing approaches to alternative energy use. Established in 1982 by physicist AMORY LOVINS, the institute seeks market-oriented solutions and promotes the transformation of corporations to embrace "natural capitalism" and end-use/least-cost approach. It has a staff of more than 45 full-time workers and an annual budget of nearly $5 million to provide basic research, education and outreach, and consulting services. *Rocky Mountain Institute Solutions* is its regularly published newsletter. Address: 1739 Snowmass Creek Road, Snowmass, CO 81654. Phone: (970) 927-3851. Website: www.rmi.org.

Rodale, Robert (1930–1990) American publisher. Rodale, son of J. I. Rodale, an advocate of ORGANIC GARDENING AND FARMING, developed his father's magazine publishing effort into a sizeable business involving many magazines, books, and other materials. Robert advanced the notion of organic farming to "regenerative agriculture," a combination of organic farming and sustainable agriculture in a community-based economic setting, through which a farmer could produce healthy crops at a profit, protect the environment, and improve the farmer's own natural resources utilizing careful stewardship. In 1971, after inheriting Rodale Press (publisher of *Organic Gardening* and *Prevention,* among other magazines), Robert established a 320-acre research facility in Pennsylvania to study regenerative agriculture and conduct comparative studies with conventional agriculture. In 1985 the U.S. Department of Agriculture agreed to station soil scientists at the Rodale facility and research collaboration also began with the land grant universities. The research facility grew to include three regional research locations networked with 34 farmers from the Mid-Atlantic states to the Great Plains, providing further testing for Rodale's regenerative farming practices. In 1979 Rodale Press expanded with *The New Farm,* a magazine sharing technical information and the fruits of Rodale Institute's research with the public. In the 1980s the Rodale Institute expanded into Tanzania, Senegal, Guatemala, and the Soviet Union. Robert Rodale was killed in an automobile accident in 1990 while in Moscow making arrangements for the publication of a Russian-language edition of *The New Farm.* His books include *Sane Living in a Mad World: A Guide to the Organic Way of Life* (1972); *The Best Health Ideas I Know; Including My Personal Plan for Living* (1974); *Our Next Frontier: A Personal Guide for Tomorrow's Lifestyle* (1981); and *Save Three Lives: A Plan for Famine Prevention* (1991).

rodent in the biological sciences, any mammal belonging to the ORDER Rodentia. The rodents are characterized by the presence of two pairs of long, sharp, constantly growing incisors (front teeth), one pair in the upper jaw and one in the lower, adapting the animals for gnawing hard food such as the stems of WOODY PLANTS. They have no canine teeth, having instead a gap between the incisors and the molars. Most rodents are small, although a few (beaver, capybara) achieve fairly large size. They breed prolifically (*see* R-SELECTED SPECIES) and have voracious appetites for plant material, and are important agricultural pests worldwide. They also serve as VECTORS for a number of diseases, including tularemia and bubonic plague. However, they also serve as the principal food source for coyotes, bobcats, RAPTORS, and other wild CARNIVORES, help aereate soil with their tunnels, and occasionally consume insect pests such as grasshoppers and locusts. Mice and voles spread the seeds of beneficial plants in their droppings; beavers build dams that help control EROSION on mountain streams. The Rodentia are the most numerous of all mammals. The order contains more than 30 FAMILIES, about 350 genera (*see* GENUS), and almost 5,000 different SPECIES.

rodenticide a PESTICIDE directed primarily against RODENTS. The most common rodenticides are the arsenicals (that is, compounds of ARSENIC), although other materials, notably zinc phosphate and other ZINC compounds, are often used as well. Since rodents are mammals, and all mammals share a number of common metabolic traits (*see* METABOLISM), rodenticides are often extremely toxic to other mammals, including humans. They must therefore always be used with extreme care.

Roosevelt, Theodore (1858–1919) American president and pioneering conservationist, born October 27, 1858, in New York City. An ardent naturalist and big-game hunter who had spent three years (1884–1886) ranching in the North Dakota badlands—where a NATIONAL PARK is now named for him—Roosevelt was governor of New York when he was elected vice president of the United States in 1900. He succeeded to the presidency on the death of William McKinley from an assassin's bullet on September 14, 1901. The most active environmentalist ever to achieve the presidency, Roosevelt was the chief moving force behind the ANTIQUITIES ACT OF 1906, which he used to set aside 18 NATIONAL MONUMENTS, including some (Grand Canyon, Olympic) that would later become NATIONAL PARKS. He also originated the concept of the NATIONAL WILDLIFE REFUGE and proclaimed 50 of them throughout the country (the first: Pelican Island, Florida, established March 14, 1903); created the BUREAU OF RECLAMATION; convened the National Conservation Commission of 1908, which provided the first thorough report on the state of the nation's natural resources; and (with GIFFORD PINCHOT) established the modern outlines of the NATIONAL FOREST SYSTEM. The parklands refuges and forests he proclaimed total approximately 230 million acres. He died January 6, 1919, at his Sagamore Hill estate at Oyster Bay, Long Island. *The Rise of Theodore Roosevelt* by Edmund Morris (1988) provides a comprehensive biography.

rootstock in botany, a short, vertical RHIZOME.

root wad in forestry, the mass of roots and soil that remains attached to the base of tree after the tree has blown over.

ROS *See* RECREATION OPPORTUNITY SPECTRUM.

rotation in forestry, one complete growth cycle in a managed forest. Rotations are generally measured from harvest to harvest; that is, one rotation equals the length of time from the harvest of one crop of trees on a given plot of land till the time the next crop on the same plot of land is ready for harvest. *See also* ROTATION AGE; SUSTAINED YIELD; AREA ROTATION.

rotation age in forestry, the age at which a tree is expected to be ready for harvest. Rotation age depends on a number of factors, including the SPECIES of tree involved, the SITE INDEX of the land on which it is growing, and the MANAGEMENT SITUATION for the growing site and for the forest as a whole. The simplest way of calculating rotation age, known as *cordwood rotation,* is simply to set it equal to CULMINATION OF MEAN ANNUAL INCREMENT for the given species and site. The other primary method, *sawtimber rotation,* is defined as the number of years required for timber to reach merchantable size, generally defined as 17–20 inch DBH (*see* DIAMETER BREAST HEIGHT). Either method of figuring rotation age can be affected by the practice of TIMBER STAND IMPROVEMENT techniques such as THINNING, RELEASE CUTTING, etc. *See also* ROTATION; SUSTAINED YIELD; AREA ROTATION.

rotenone a natural ketone, chemical formula $C_{23}H_{22}O_6$, derived from the roots of several species of LEGUMES, notably *Derris elliptica* (Japanese *Roten,* found in eastern Asia and in Australia) and used worldwide as an insecticide and fish poison. Rotenone is a colorless crystalline solid that is insoluble in water but dissolves readily in organic solvents (*see* ORGANIC COMPOUND; SOLVENT), which serve as the principal vehicle for its application. It decomposes rapidly upon exposure to light and OXYGEN, and is therefore relatively safe to use. Functionally, it is a respiratory inhibitor, paralyzing the respiratory system, and is thus much more toxic when inhaled than when ingested. Symptoms of mild poisoning include numbness of the mouth and nose, nausea and vomiting, and muscle tremors; heavier doses cause increasing degrees of breathing difficulty, which can eventually lead to death by asphyxiation. Chronic long-term exposure brings about changes in the fatty tissue or organs, especially the liver and kidneys. The substance is also a mild irritant, and may cause a rash on contact with skin. It is nevertheless among the safest and least environmentally hazardous pesticides in use. LD_{50} (rats): orally, 133 mg/kg; inhaled, 5 mg/kg.

rotifier in the biological sciences, any of numerous SPECIES of tiny to microscopic animals belonging to the PHYLUM Rotifera. Rotifers are aquatic, principally freshwater ORGANISMS. Most are benthic (*see* BENTHOS), but a few are members of the PLANKTON community. They are characterized by the presence of *coronas,* one or more circular (or roughly circular) groups of short filaments called *cilia,* which beat progressively around the circle, giving them the illusion of being rotating wheels; the coronas surround the animals' mouths and serve to drive food into them. Many species are *eutelate,* that is, every adult individual of the species is made of exactly the same number of CELLS. Most reproduce largely by PARTHENOGENESIS, and some reproduce exclusively in that manner. Males, where present, are much smaller than the females. Rotifers range in size from roughly .04 mm to 3 mm in size, with most clustered at about .1 mm. Their eggs have a tough, leathery coating, allowing them to survive long periods without water; thus these animals are among the most common FAUNA of VERNAL POOLS and other seasonal water bodies. A few live on the surface of MOSSES and LICHENS, becoming active only when the host plants are drenched with a surface film of water. The phylum contains about 100 genera (*see* GENUS) and 1,700–2,000 SPECIES.

round forty in logging, term used to describe a common practice used by timber thieves on the public lands in the 19th century. To cut a round forty, the thief purchased a 40-acre piece of PUBLIC DOMAIN timberland, and then "rounded" it by cutting all the timber around it as well as on it, sometimes for several miles in every direction.

roundwood in the forest products industry, the logs that become fuel, lumber, paper, and other wood products.

roundworm *See* NEMATODE.

RPA *See* FOREST AND RANGELAND RENEWABLE RESOURCES PLANNING ACT.

r-selected species in ecology, any SPECIES whose POPULATION size is controlled primarily by DENSITY-INDEPENDENT FACTORS. R-selected species tend to devote a high proportion of their energy to reproduction. They have numerous off-spring, only a few of which will themselves survive to reproductive age. R-selected animals typically lead short lives dominated by instinct rather than by learning. Examples include mice, frogs, most insects, and nearly all plants. *Compare* K-SELECTED SPECIES.

ruminant loosely, any mammalian HERBIVORE that "chews its cud"; that is, a mammal with a multi-chambered stomach in which the contents of the first chamber (the *rumen*) may be brought back to the mouth at will for more thorough chewing after it has been partially digested. The ruminant feeding system is an adaptation to a diet (grass, small woody plants) that is particularly rich in CELLULOSE. The animal bolts its food into the rumen while eating, then sits around chewing it later at leisure. This adaptation also allows the animal to eat more rapidly, reducing its time on the feeding range, where it is often especially susceptible to PREDATORS. The "true ruminants" (those with four distinct stomachs) include sheep, goats, cattle, deer, giraffe, bison, and antelope. Others characterized as ruminants include the camels (camel, dromedary, llama, alpaca) and the chevrotains, a group of Asian animals that resemble small antelopes with large hindquarters and appear to be closely related to the evolutionary ancestor of the deer. *See also* UNGULATE.

runner a RHIZOME lying above the ground. Runners are used as means of propagation by numerous plant species, including strawberries, morning glory, and a number of different grasses (*see* ASEXUAL REPRODUCTION).

runoff in hydrology, the sum of all water flowing out of a region over the surface of the Earth. *Channeled runoff* is water flowing in streams (*see* STREAMFLOW; DISCHARGE); *overland flow* is water flowing directly across the land, either in a broad, undifferentiated form (SHEETFLOW) or as a series of tiny streamlets without fixed courses (RILLS). Approximately one-third to one-sixth of all PRECIPITATION falling on an area leaves it as runoff. The remainder is returned to the air (*see* EVAPORATION; TRANSPIRATION; EVAPOTRANSPIRATION) or soaks into the ground (*see* INFILTRATION).

OVERLAND FLOW

Overland flow is by nature ephemeral; it occurs only during and immediately following precipitation. Its volume is highly dependent on the nature of the land surface. Precipitation falling on bare rock is almost all converted into overland flow, except for a small amount returned to the air as evaporation. At the other extreme, precipitation falling on well-vegetated soil is likely to create little or no overland flow due to the rapid taking-up of water by the plants and to the protection they afford the soil, so that it is not mechanically compacted and its infiltration rate remains high. Overland flow also depends partly on the nature of the precipitation itself. A heavy rainstorm can overwhelm the infiltration capacity of even the most thirsty soil, leading to overland flow of all precipitation that falls after the soil is saturated (*see* FIELD CAPACITY), whereas the same amount of rain falling as a light drizzle over a long period may all have time to soak into the soil or be taken up by plants.

CHANNELED RUNOFF

All overland flow eventually becomes channeled runoff. In addition, channeled runoff includes *channel precipitation* (precipitation falling directly on the surface of the water), *interflow* (water flowing through the soil, some of which will reach a stream instead of the WATER TABLE), and *baseflow* (water contributed to a stream as springs and seeps from the water table). (Note that this creates a fuzzy boundary between infiltration and runoff. Both interflow and baseflow result from infiltration. Infiltration may therefore best be thought of as deferred runoff, though this deferral may last, in extreme cases, several million years (*see* e.g., OGALLALA AQUIFER). Channeled runoff is thus steadier and more predictable than overland flow, though it may fluctuate fairly greatly from wet to dry seasons and as the result of particularly heavy storms (*see* FLOOD). *See also* HYDROLOGIC CYCLE; HYDROLOGIC EQUATION; EROSION; SOIL WATER.

runout height in geology, the height to which WAVES flow up a beach after breaking. Runout height marks the upper limit of the zone in which wave energy is expended on a coastline. The bottom of this zone is the HALF-WAVE DEPTH. *See also* RIFT.

Rural Clean Water Program a 10-year experimental program conducted by the U.S. Department of Agriculture and the Environmental Protection Agency to encourage farmers to undertake agricultural practices that would reduce pollution. Starting out as an amendment to the CLEAN WATER ACT in 1977, the program was originally seen as a public works funding program that would apply to private farms and ranches in rural watershed areas, as a parallel effort to the funding of municipal sewage districts in towns and cities. The program's budget was originally authorized at $600 million with a later authorization (in 1980) to $800 million. However, the program remained unimplemented, with the generous authorizations never appropriated. Instead, Congress approved a $70 million plan to test the program in 20 watersheds across the country, representative of the 600 or so rural U.S. watersheds with agricultural runoff problems. The 20 test watersheds dealt with the application of management practices that would slow pesticide and chemical fertilizer runoff, contain animal wastes, and reduce sedimentation from soil erosion. Participating farmers were offered up to 75% of the cost of installing management practices to meet water quality goals, and researchers carefully monitored the results. While valuable experience was gained from the experiment, the Rural Clean Water Program as such was not funded after the conclusion of the 10-year test period. However, financial support for some of the management practices was made available through regular appropriations to existing USDA programs concerned with water quality. *See* NONPOINT POLLUTION.

rut in the biological sciences, the period when a male animal is sexually active. Rut corresponds to ESTRUS in the female. In some animals it is controlled by a hormone cycle, as is estrus; in others, it is stimulated by the behavior and/or scent of a female "in heat," and is thus not dependably cyclic in nature.

sacrificial anode *See under* CORROSION.

Safe Drinking Water Act federal legislation designed to protect the quality of tap water and to prevent the degradation of AQUIFERS used as drinking-water supplies, signed into law by President Gerald Ford on December 16, 1974. With the exception of the GROUND-WATER sections, which deal with the regulation of INJECTION WELLS, the Safe Drinking Water Act is concerned exclusively with the quality of FINISHED WATER. It makes no attempt to regulate the quality of water going into a treatment works, only that of the water coming out of the consumers' taps. The act requires the ENVIRONMENTAL PROTECTION AGENCY (EPA) to publish national drinking water standards of two types: *primary standards,* which relate to materials (such as bacterial contamination and chemical toxins) that may be detrimental to human health, and *secondary standards,* which relate to materials that color water, create bad tastes or odors, or otherwise make it unpalatable. These standards are to be enforced by the states; however, if a state does not adequately enforce them, the EPA is authorized to compel the state to use its enforcement authority. All public water systems (defined as any water-supply system that pipes water to at least 15 separate connections) are subject to the authority of the act and must monitor their water quality regularly: annually for chemical contamination, monthly for bacterial contamination, and if the water supply comes from a surface source, daily for TURBIDITY. The portion of the act dealing with injection wells requires the states to set and enforce permit regulations for all underground injection of fluids (except for those directly connected with the production of oil) in such a manner that no aquifer used as a drinking water source will be contaminated. If an aquifer is determined to be the sole source of drinking water for a community or rural area, no federal money may be spent on projects that risk contaminating it.

safe yield in water technology, the maximum amount of water that a given water system can reliably deliver. For GROUNDWATER systems, long-term safe yield is equivalent to the rate of RECHARGE of the ACUIFER from which the system gets its water; short-term safe yield depends on the rate of flow of water through the aquifer (the PERCOLATION RATE). For systems dependent on surface water, long-term safe yield is equivalent to the BASEFLOW of the stream or streams tapped by the water system, and short-term safe yield depends upon the amount of storage available, both natural (lakes, large streams) and artificial (RESERVOIRS).

sag curve *See* OXYGEN SAG CURVE.

St. Croix International Waterway Commission established by the state of Maine and the Canadian province of New Brunswick to develop and oversee management policies for 110 miles of the St. Croix River system where it forms the Canadian/United States boundary. The commission's management plan has promulgated 22 international policies concerning equitable resource allocation, the establishment of local alliances for proactive planning, resource preservation, and appropriate development. Address: Box 610, Calais, ME 04619. Phone: (506) 466-7550. Website: www.asf.ca/OrgsNB/SCIWC.

saline intrusion the invasion of a freshwater AQUIFER by brackish or salty water. Saline intrusion happens in two principal situations. The most common cause is overpumping of wells in coastal aquifers, creating a CONE OF DEPRESSION that reaches below sea level and allows seawater to invade the aquifer by gravity flow. A second cause is the penetration of an IMPERMEABLE LAYER separating a freshwater aquifer from a saline aquifer and thereby allowing the two to mix. If the saline aquifer is under artesian pressure (*see* ARTESIAN WELL), this mixing can be fairly rapid. In either case, the water in the freshwater aquifer becomes degraded; hence, saline intrusion is classified as a form of POLLUTION.

salinization of soils *See* ALKALINIZATION OF SOILS.

Salmonella a small but widespread GENUS of pathogenic bacteria (*see* PATHOGEN; BACTERIUM), which act as the causative agents for several different diseases, among them typhoid fever and some varieties of food poisoning. *Salmonella* is a gram-negative (*see* GRAM STAIN) rod bacteria that is *motile,* that is, moves about in fluids under its own power. Ten SPECIES are known, at least seven of which are dangerous to humans. Some are extremely virulent: *S. enteritidis,* the species that causes salmonellosis (food poisoning), can cause symptoms in as little as eight hours and seldom has an INCUBATION TIME time longer than two days. It is spread by contaminated food and water, usually caused by contact with FECES; even microscopic bits of fecal matter may harbor the bacterium. There are about 50,000 cases of salmonellosis reported in the United States each year. (*See also* DANYS' VIRUS.)

salmonid in the biological sciences, any fish belonging to the FAMILY Salmonidae. The salmonids include the true trout (genus *Salmo,* including the brown trout, the steelhead and the Atlantic salmon); the char (genus *Salvelinus,* including the lake trout, the brook trout, and the dolly varden); and the salmon (genus *Onchorhynchus,* including the Pacific salmon, the chum, and the kokanee). (None of these species lists is complete.) Salmonids are generally large, powerful PREDATORS with a relatively high degree of intelligence; many are anadramous (*see* ANADRAMOUS SPECIES) and nearly all spawn in running water (*see* SPAWNING GRAVEL). They include most popular game species and some economically important commercial species, and are the family of fish most often and most successfully raised in hatcheries. Salmonids are notoriously selective about the temperature, chemical content and suspended sediment load of the water they spawn in, making preservation of salmonid spawning HABITAT one of the most important and difficult tasks in wildlife management.

salt in chemistry, a COMPOUND formed by the union of positive and negative IONS rather than complete ATOMS (although complex salts may have complete atoms in them as well). Salts may be formed in a number of ways. The two most common methods are (1) by the direct union of a METAL with a non-metal, and (2) by the neutralization of an ACID with a BASE. *Normal salts* are those in which neutralization is complete; that is, there are no leftover hydrogen (H+) or hydroxide ((OH)–) ions in the salt. An *acid salt* (also known as a *hydrogen salt*) contains some hydrogen ions; a *basic salt* (or *hydroxy salts*) contains some hydroxide ions. *Organic salts* are those that contain organic ions (*see* ORGANIC COMPOUND). They may be acidic, basic, or normal. *Soluble salts* (that is, those that dissolve readily in water or another SOLVENT; *see* SOLUTION) make excellent ELECTROLYTES due to their separation into charged ions on DISSOCIATION.

SALTS IN THE ENVIRONMENT

Salts are among the most common of all classes of compounds found in nature. Many (if not most) MINERALS are salts; all living ORGANISMS use dissolved salts of various types (principally common table salt, NaCl) to form the electrolytes necessary for life processes (blood is an electrolyte, as is the fluid found within living CELLS). Salts are therefore necessary for life. They class as POLLUTANTS in certain cases, however, due to their presence either in excessive quantities or in the wrong types. Living bodies must maintain a precise concentration of salt, known as the *electrolytic balance,* in their body fluids. The consumption of excess salt can overwhelm this balance and upset or destroy bodily functions. In addition, some salts are directly toxic. The principal source of salts as pollutants is irrigated agriculture (*see* IRRIGATION; RETURN FLOW).

salt marsh a MARSH developed along a seacoast and fed by salt water. Salt marshes are of two types, those that develop in ESTUARIES and those that develop in tidal LAGOONS. Those in estuaries undergo twice-daily shifts of water salinity as the tide moves in and out, flushing the marsh first with ocean water and then with river water, while those in tidal lagoons go through similar cycles of drowning and uncovering with the tides; hence, both types are inhabited by ORGANISMS able to survive in a broad range of conditions. The dominant vegetation is cord grass (Salinas); other common organisms are brown ALGAE, mussels, and PLANKTON. Salt marshes are floored with fine soil particles (SILTS and CLAYS) that adsorb NUTRIENTS, giving them extremely high productivity. An acre of salt marsh produces roughly 33 times the BIOMASS of an acre of GRASSLAND. They also serve as buffers against the wave action of ocean storms. The same sorptive qualities of their silt and clay floor that makes them highly productive, however, also makes them extremely vulnerable to pollution, as most MICROCONTAMINANTS also sorb well to these particles; and the protection they provide to shorelines makes them vulnerable to infilling, as development on the stable shore behind them "spills over" into the marsh.

Salton Sea large saline lake in southern California that appeared in the early 20th century as a result of an IRRIGATION works malfunction. The lake varies seasonally in size, but its average dimensions are approximately 51 kilometers (32 miles) long by 21 kilometers (13 miles) wide, with a surface area of about 960 square kilometers (275 square miles) and a maximum depth of 20.4 meters (67 feet). Its surface lies approximately 225 feet below sea level. The Salton Sea began forming in March 1905, when floods on the Colorado River washed out poorly constructed headgates on the four-year-old All-American Canal, a private project designed to irrigate California's Imperial Valley. The washout resulted in a flow of approximately 100,000 cubic feet per minute washing down the canal, which burst its banks and began flooding the valley. After many failed attempts, the flooding was finally stopped in February 1907: by that time the new lake was 28 meters deep (93 feet) and covered nearly 1,200 square kilometers (450 square miles). It has since stabilized at about 2/3 its original size. The Salton Sea's primary source of water is RUNOFF from irrigated lands in the Imperial Valley: hence, its SALINITY will continue to increase indefinitely.

saltwater intrusion *See* SALINE INTRUSION.

salvage cutting in forestry, the harvesting of dead or dying timber outside of the normal ROTATION in order to be able to take them while they still have some commercial value. Salvage cuttings are commonly done after WILDFIRES or insect infestations. They are exempt from many of the regulatory provisions of the NATIONAL FOREST MANAGEMENT ACT, including those governing ROTATION AGE, CLEARCUTS and BENEFIT/COST ANALYSIS (that is, the necessity of proving that benefits exceed costs). The timber manager must be prepared to prove that the salvage is necessary in order to gain these exemptions, however.

salvage logging amendment enacted in 1995 as a rider to an unrelated bill to permit the cutting of diseased or otherwise afflicted timber in areas where Forest Service timber sales would not otherwise be allowed. Environmentalists were concerned that the term "salvage" could be defined so broadly that healthy trees could be logged off with the dead and dying; that there is no ecological justification for such logging since healthy forests need dead trees to remain viable; and that salvage logging would encouraging additional road building. Distaste over the transparent political machinations of the salvage logging amendment was one of the factors leading to broad support for ZERO CUT in national forests and for other reforms as well, such as the Clinton administration initiative barring any new roads in ROADLESS AREAS.

sand in geology and soil science, a collection of loose ROCK or MINERAL particles of very small size (.05 to 2.0mm in diameter, or .002 to .16in.). The individual particles are known as *sand grains.* The mineral content of sand is not part of its definition; however, the vast majority of sand grains are composed of QUARTZ. Sand is subdivided by scientists into five classes based on size: *very coarse* (1.0–2.0mm); *coarse* (0.5–1.0mm); *medium* (0.25–0.5mm); *fine* (0.1–0.25mm); and *very fine* (0.05–0.1mm). Because sand grains are hard and angular, they do not lie tightly against each other, making sand one of the least cohesive of natural materials and providing it with a multitude of relatively large spaces (*pores*) between the grains, thus allowing water to drain easily through it. These values make the presence of some sand necessary in agricultural soils, in order to keep the soil loose enough for root growth and to prevent it from retaining too much water (which can literally drown roots). *See* SOIL: *classification of soils. See also* SANDSTONE; SPIT; DUNE; SHORE DRIFT.

sand dune *See* DUNE.

sandspit *See* SPIT.

sandstone in geology, a common, highly variable SEDIMENTARY ROCK composed of SAND grains cemented together by some noncrystalline mineral solid such as silica (*see* QUARTZ) or calcite (*see* LIMESTONE: *types of limestone*). Nearly all sandstones are composed of quartz sands, although this is not a necessary part of the definition. Sandstones vary considerably in color and hardness, the variance depending partly on the size and angularity of the included sand grains but mostly on the mineral content of the cementing material, which makes up to 30%–40% of the mass of the stone. They exhibit no CLEAVAGE PLANES and tend to be excellent building stones, provided they are hard enough to be structurally competent (that is, not subject to cracking or crumbling). Sandstones may be thought of as fossilized beaches or DUNES, and often show familiar depositional features such as ripple marks and cross-bedding, as well as being among the most fertile sources of fossilized plants and animals (*see* FOSSIL). Those in which the cementing agent is silica are light in color and extremely tough, and may be virtually 100% quartz. Other typical cementing agents are calcite (limestone) and HEMATITE (an iron mineral; it typically produces a brown or red sandstone with relatively low structural strength). *Arkose sandstone* is sandstone containing FELDSPAR grains in addition to the quartz. *Greywacke* contains small flakes of mica. *Flag* is sandstone that splits easily along BEDDING PLANES, making flat, tough stones that may be used for flooring. Sandstone is one of the few non-IGNEOUS ROCK types that form DIKES. These typically result when sand is blown into a rock crevice or fracture, remaining there long enough for the grains to become cemented together. *See also* QUARTZITE.

sandy loam a LOAM that contains between 50% and 70% sand. The sand content may be higher than this (up to 85%) if the remainder of the soil is composed of CLAY rather than SILT. *See* SOIL: *classification of soils.*

sanitary landfill a means of disposing of SOLID WASTES by compacting them and burying them in the earth, widely used by municipalities throughout the world. Wastes going to a sanitary landfill should first be sorted to remove recyclable materials and hazardous substances, both of which should be disposed of elsewhere (*see* RECYCLING; HAZARDOUS WASTE). At the landfill, the wastes are spread in an even, two-foot-thick layer down a sloping surface called the *working face* (maximum slope: 30%): when this layer is completed, heavy machinery is run over it to crush it and compact it to a thickness of about 6 inches. The compacted material then forms the next working face. At the end of each day's work, the compacted wastes are covered with 6 inches of compacted soil, the surface of which forms the next day's first working face. Each day's accumulation thus forms a separate unit, or *cell,* completely enclosed by earth. Cells accumulate horizontally, lying against each other like tipped dominoes When the entire area appropriated for the landfill has been covered, a foot-thick layer of earth is spread over all of them, and a new layer of cells, or *lift,* is begun on top of the first layer. When the entire operation is completed, a final sealing layer of earth at least 2 (preferably 4) feet thick is spread over the entire landfill, which then may be put to other uses.

CLASSIFICATION OF LANDFILLS

Landfills are classed according to the means by which they are fitted to the disposal site. There are three main types. The *area landfill* utilizes a level site and simply builds the first lift on the surface, creating a mound. The ultimate height of the mound, of course, depends upon the height of each lift and the number of lifts employed. The *trench landfill* also utilizes a level site, but digs a hole in it first and then refills so that the final covering of the landfill comes back to the original ground level. It is less unsightly than the area landfill and has the advantage of yielding its own covering material (the soil removed to make the trench can be used as cover), but it can only be used in areas where the soils are deep enough and well-drained enough so that the bottom of the trench remains above both BEDROCK and the WATER TABLE. The *slope/ramp landfill* is used on sloping ground. The toe of the slope is excavated to form a level area (the *ramp*) on which the first lift of cells are built, with each lift then becoming part of the next ramp (the excavated material is again used for covering). A fourth category, the *quarry landfill,* is sometimes recognized; it is simply a trench land-

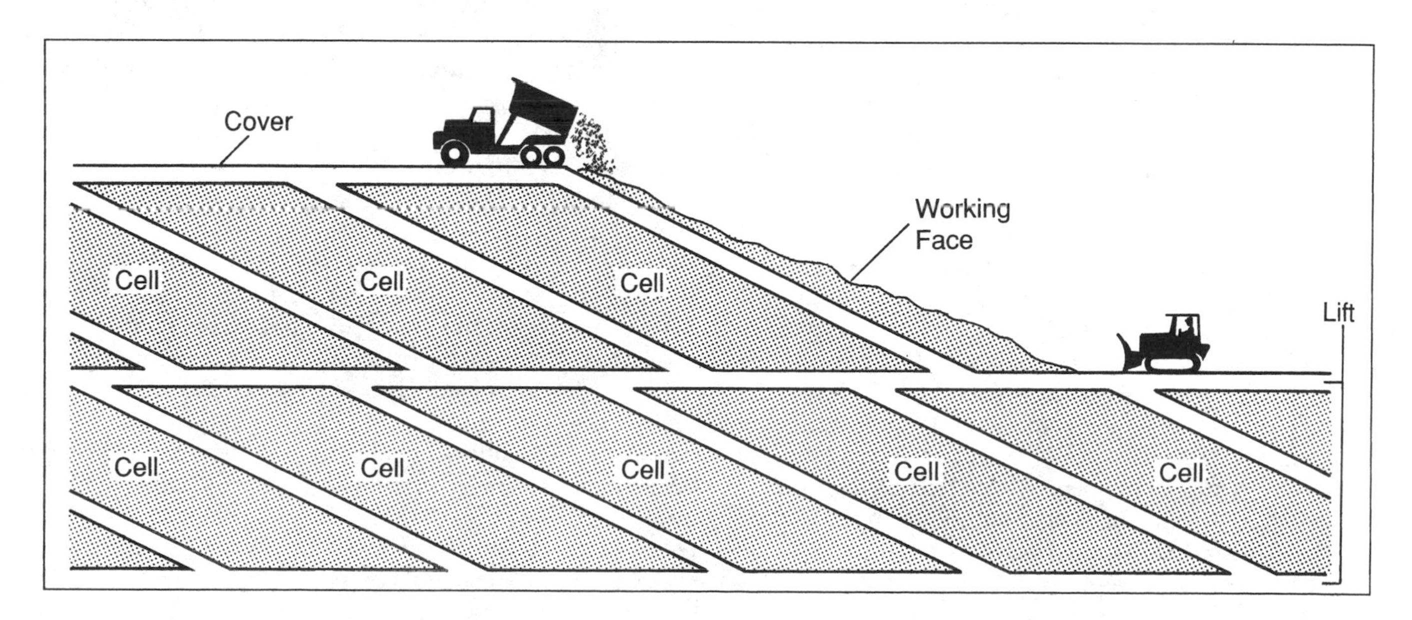

Sanitary landfill

fill built in a preexisting hole in the earth, such as an abandoned stone quarry or open-pit mine.

LANDFILLS AND THE ENVIRONMENT

Federal law and the laws of most states prohibit the disposal of HAZARDOUS MATERIALS in sanitary landfills (*see,* e.g., RESOURCE CONSERVATION AND RECOVERY ACT; SOLID WASTE DISPOSAL ACT; COMPREHENSIVE ENVIRONMENTAL RESPONSE, COMPENSATION AND LIABILITY ACT); nevertheless, the materials that go into these landfills are never entirely without hazard. Some hazardous materials are unavoidable, e.g., paint coverings, residues of furniture polish, drain cleaner, etc., in containers discarded as empty. Some are unrecognized as hazards until they are in place (transformers containing PCBs were once routinely landfilled; *see* POLYCHLORINATED BIPHENYL) and some hazardous materials are created within the landfill itself by chemical or biological action. Thus, the siting of sanitary landfills presents the same problems as the siting of hazardous waste dumps. LEACHATE must be prevented from reaching sensitive AQUIFERS (*see* AQUIFER SENSITIVITY; LE GRAND RATING SYSTEM), and vents must be provided at intervals within the fill to "bleed off" accumulating gases. Once full, a landfill cannot be used as a building site due to differential settling of the compacted material, which may tilt, twist, and crack foundations. Thus, it is most often pressed into service as parkland, and care must be taken to be certain that users of the parkland will not encounter pools of leachate that have been forced to the surface, accumulations of flammable gas, or other hazards. Finally, the size of landfill operations should be large enough that the need for siting new ones is reduced to a minimum, thus concentrating the environmental costs and keeping the impact on groundwater confined to a single aquifer or aquifer group. A standard rule of thumb is to provide 100 cubic yards of volume for each individual in the service area. At four cubic yards of compacted trash per individual per year, this should allow operations to continue at a single site for 25 years.

sap the liquid that carries NUTRIENTS throughout the body of a VASCULAR PLANT. Sap flows upward through the XYLEM and downward through the PHLOEM. The upward-trending sap ("raw sap") consists mostly of water with a few dissolved mineral IONS, while the downward-flowing sap ("elaborated sap") contains the CARBOHYDRATES and other foodstuffs produced through PHOTOSYNTHESIS in the plant's leaves. The circulation of sap is driven principally by evaporation from the leaf surfaces (*see* EVAPOTRANSPIRATION; GUTTATION).

sapling in forestry, a young tree. To qualify as a sapling in the United States, a tree must have a DBH (*see* DIAMETER BREAST HEIGHT) of at least 2 and no more than 4 inches. Trees within this size range are said to be "in the sapling stage." *Compare* SEEDLING.

saprobe in biology, any ORGANISM that obtains its NUTRIENTS from decaying ORGANIC MATTER, that is, from the bodies of other organisms that have died. Most saprobes are bacteria and fungi (*see* BACTERIUM; FUNGUS), though many PROTISTA and a few plants have adopted saprobic life-styles (animals that feed on the dead and decaying bodies of other animals are called SCAVENGERS rather than saprobes). Saprobes differ from PARASITES in that a parasite feeds on living organic matter, while a saprobe feeds on the remains of organisms that have died from other causes. They provide the principal means of breaking down previously living matter and recycling its nutrients for reuse by autotrophs (*see* AUTOTROPHISM) and are thus an extremely important part of the natural ecological cycle. The previous term for these organisms, *saprophyte,* is still widely seen. Its use should be avoided, due to its inaccuracy (the suffix *-phyte* means "plant," but most of these organisms are not plants). *See also* DECOMPOSER.

saprophyte *See* SAPROBE.

sapwood *See* XYLEM.

saturation threshold the concentration above which accumulation of a substance will no longer occur; for example, the concentration of a SOLUTION above which no further SOLUTE will dissolve. In environmental literature it is often seen in reference to BIOACCUMULATION, where it represents the upper limit beyond which an ORGANISM will no longer absorb and concentrate a given MICROCONTAMINANT. Not all microcontaminants show saturation thresholds (*see,* e.g., MIREX).

saturation zone *See water table.*

savanna *See under* GRASSLAND BIOME.

Save America's Forests coalition of local, regional, and national organizations founded in 1990 to work collectively on forest protection legislation, with a special emphasis on clearcutting. In 1995 the group campaigned unsuccessfully to repeal the SALVAGE LOGGING AMENDMENT (allowing clearcutting in the name of dead and diseased tree removal) passed by Congress, but were able to help defeat the so-called "son-of-salvage rider" in 1998. In 1999 the group proposed an "Act to Save America's Forests," supported by leading scientists, including E. O. WILSON and Jane Goodall. Membership (1999): 6,000. Address: 4 Library Court

SE, Washington, DC 20003. Phone: (202) 544-9219. Website: www.saveamericasforests.org. *See also* CLEARCUT.

Save the Whales wildlife conservation organization, founded in 1977 to lobby on behalf of the great whales and in opposition to commercial whaling. The group does not opposed subsistence whaling. Save the Whales is one of several whale-preservation groups that send representatives to the International Whaling Commission conference. It also engages in a substantial advertising campaign. Address: P.O. Box 2397, Venice, CA 90291. Website: www.savethewhales.org.

Save-the-Redwoods League (SRL) land-preservation organization, founded in 1918 to purchase significant stands of California redwoods in order to prevent them from being logged. The SRL works closely with national, state, and county park agencies, raising money by donation and subscription to purchase lands in the redwood forests to add to the park systems. It was instrumental in the establishment of Redwood National Park in 1968. Address: 114 Sansome Street, Room 605, San Francisco, CA 94104. Phone: (415) 362-2352. Website: www.savetheredwoods.org.

scaling diameter in logging, the diameter of a log as measured for the purpose of estimating the volume of wood the log contains. Scaling diameter is taken inside the bark at the small end of the log. The ruler or *scaling stick* is laid across the true center rather than the growth center. If the log is out of round, measurements are taken across the widest dimension and the narrowest dimension; the average of these two measurements is considered the scaling diameter. If for some reason the small end cannot be measured (for example, if it is broken instead of cut smooth), the large end is measured and then adjusted by subtracting a "standard taper" of 1 inch for every 10 feet of log length. All fractions of an inch are dropped. *See* LOG SCALING.

scarp in geology, generally, a sudden, steep drop of the Earth's surface, such as a line of cliffs or a soil CUTBANK; specifically, one formed during EARTHQUAKE activity. *See* FAULT; FAULT-LINE SCARP.

scat *See* FECES.

scavenger an animal or other motile organism (*see* MOTILITY; ORGANISM) that feeds on dead ORGANIC MATTER. A scavenger differs from a SAPROBE in that it consumes dead matter rather than merely absorbing nutrients from it. Once the matter is consumed the organism must move on to find more, hence the necessity for motility. Some scavengers are primarily or exclusively vegetarian (e.g., earthworms) and others are primarily or exclusively carnivorous. (e.g. vultures); however, most are opportunistic feeders, eating anything that comes their way as long as it once was alive and now is dead. They perform a valuable ecological function by consuming and thus removing carcasses that otherwise would disappear only through the slow and potentially pathogenic process (*see* PATHOGEN) of decomposition. *Compare* PREDATOR; DECOMPOSER.

Scenic America national organization concerned with preserving and enhancing the scenic character of America's communities and countryside. Scenic America, a successor organization to the "roadside council" movement, furnishes information and technical assistance on billboard and sign control to local affiliates and other groups. The organization also works on scenic byways, tree preservation, highway design, cellular tower siting, and other scenic conservation issues. Scenic America's six state affiliates include Kentucky, Michigan, Missouri, North Carolina, Ohio, and Texas. Among its publications are books, fact sheets, technical bulletins, and two periodicals, *Viewpoints* and *The Grassroots Advocate*. Headquarters: 801 Pennsylvania Avenue SE, Suite 300, Washington, DC 20003. Phone: (202) 543-6200. Website: www.scenic.org.

scenic area in Forest Service usage, *see* SPECIAL INTEREST AREA.

scenic climax in landscape management, the most spectacular scenery within a region, especially scenery striking enough to draw visitors to it for its scenic values alone. A scenic climax does not generally consist of a single feature, but of a group of features whose interrelationship is particularly striking. The term has both general and specific meanings. Yosemite Valley is *a* scenic climax of the United States (one of a number of areas containing striking scenery) and also *the* scenic climax of Yosemite National Park (the most scenic area within the park).

scenic river a river or section of a river designated as "scenic" under the terms of the National WILD AND SCENIC RIVERS ACT OF 1968. To qualify for designation as a Scenic River, a river or river segment must be free-flowing (or essentially free-flowing), must flow within a largely primitive or undeveloped WATERSHED, and must be substantially free of developments along its banks, although point access may be provided by roads. Designated Scenic Rivers may not be dammed, and mines and power-transmission corridors may not be located within one-quarter mile of their banks. The agency managing the river has the authority to condemn private RIPARIAN LAND along the river (*see* EMINENT DOMAIN) if such action is necessary to manage the river in conformity with the Wild and Scenic Rivers Act. *Compare* WILD RIVER; RECREATIONAL RIVER.

schist in geology, a relatively coarse-grained METAMORPHIC ROCK consisting of altered SLATE. It retains the easily separated parallel planes of the slate and its parent SHALE, but the planes are thicker and they have wavy, usually glistening surfaces (this surface condition is known as *schistocity*). Schist often contains a large amount of MICA created by the metamorphic process. Such schists are usually called *mica schists*. Schist grades metamorphically into *gneiss*. *See also* PHYLLITE.

Schumacher, E. F. (1911–1977) German-born British economist noted for his advocacy of "appropriate technology" for developing nations. After studying at Oxford, Schumacher served as an economic adviser to the British Control Commission in Germany from 1946 through 1950 and on the British National Coal Board from 1950 through 1970. He was president of the Soil Association, a British organic farming organization, and founded the Intermediate Technology Group to design technologies and tools appropriate to the capacity and traditions of developing nations who need them. His book, *Small Is Beautiful: A Study of Economics as Though People Mattered* (1973), translated into 15 languages, expressed his concern for the Third World, arguing that international assistance to poorer, less industrialized nations should always be given at the level of technological sophistication of the recipient, not that of the giver.

scientific method the means by which scientific information is adduced, gathered, and validated. According to British philosopher Karl Popper, an investigator must first form a concrete hypothesis based on all the available data and observations of a phenomenon. Second, he or she must try to prove the hypothesis false by collecting contrary data and observations and by making experiments. At length, if the effort to disprove the hypothesis fails, and the investigator has exhausted all conceivable ways to carry on further experiments to contradict it, then the hypothesis may be assumed to have undergone a rigorous application of the scientific method that either proves or disproves it.

SCLDF *See* SIERRA CLUB LEGAL DEFENSE FUND.

scree a collection of loose rock material lying on a mountain slope. Scree is smaller than TALUS, and generally lies higher on the mountain. It tends to accumulate in large quantities ("scree slopes") on the floors of COULIERS and other depressions on mountain faces. The term "scree" is sometimes also used by geologists in an inclusive sense to mean both scree and talus; that is, any portion of the REGOLITH that is made of rock fragments rather than of soil.

Scribner scale *See* LOG SCALING.

scrubber (stack scrubber, wet scrubber) in air pollution control, a device for removing POLLUTANTS from FLUE GASES by contacting them with water before they are released into the atmosphere. (Devices such as ELECTROSTATIC PRECIPITATORS and others that remove pollutants without the use of water are sometimes known as "dry scrubbers"; however, this term has no technical legitimacy.) Scrubbers wash PARTICULATES out of the gas stream, and remove soluble gases by dissolving them in the scrubber liquid. Those designed to remove SULFUR OXIDES often inject powdered LIMESTONE along with the water. The sulfur and limestone react in the presence of water to form gypsum ($CaSO_4 2H_2O$), which precipitates out and may be collected from the bottom of the scrubber.

DESIGN OF SCRUBBERS

Scrubbers come in a wide variety of different types. The flue gases may be passed through the water; alternatively, the water may be passed through the flue gases in the form of atomized droplets. *Spray towers* simply pass the gases upward past a series of nozzles that spray water through them. They are inexpensive to build and cheap to operate, but are relatively inefficient. *Cyclone scrubbers* induce a whirling motion by injecting the gases and the water droplets sideways into a circular chamber at high velocities. The resulting turbulence mixes the water and gases, assuring maximum contact, while throwing the heavier pollutants (and ultimately, the pollutant-filled water droplets) to the outside of the chamber by centrifugal force, where they can be efficiently collected. *Venturi scrubbers* inject the water droplets into the gas stream as it passes through a narrow opening. The constriction creates enough velocity to thoroughly entrain droplets (*see* ENTRAINMENT), and the turbulence beyond the opening provides the necessary mixing. *Packed-bed scrubbers* pass the flue gases through wetted frames packed with small bits of plastic cast in a variety of special shapes (saddles, slotted cones, etc.). These provide a large surface area on which the gases and the water can interact. Packed-bed scrubbers are further classified according to the relative direction of flow of the water and the gases, as *counterflow* (water and gases pass through the beds in opposite directions), *concurrent flow* (water and gases pass through the beds in the same direction) or *crossflow* (water and gases pass through the beds at right angles to each other). *Mobile-bed scrubbers* are packed-bed scrubbers in which the beds are made of plastic spheres packed loosely enough that the flow of gases and water keeps them in constant motion, thus reducing clogging of the pores in the beds. *Impingement-plate scrubbers* pass a flow of water over a plate with small holes in it. The flue gases pass through these holes and the water above them at high velocity, estraining spray from the water surface as they pass out of it and then striking against a baffle plate that thoroughly atomizes the water and gases together. (Note that the many varieties of scrubbers described here are only a representative sampling).

ADVANTAGES AND DISADVANTAGES OF SCRUBBERS

Scrubbers are among the most efficient methods of removing particulates and soluble gases from flue emissions, achieving removal rates of up to 95%. The materials collected by or created in the scrubber can often be recycled (gypsum, for example, can be used in the manufacture of wallboard). However, they are expensive to install and operate, are not always dependable, and may merely create a water-pollution problem rather than an air-pollution problem if the water used for the scrubbing process is not itself treated properly. Scrubber operation often creates a highly visible PLUME which, although it is composed almost completely of water vapor and is thus harmless, may pose a substantial public-relations problem.

SCS *See* SOIL CONSERVATION SERVICE.

sea breeze in meteorology, a wind that blows inland from an ocean, sea, or large lake onto a neighboring landmass. Sea breezes characteristically develop during the afternoon in warm weather. Rapid warming of the air over the land causes it to expand and rise, creating a localized zone of low pressure (*see* THERMAL LOW) into which the cooler, denser air above the sea flows. Air aloft moves from the land to the sea to compensate for the seaward-to-landward surface flow. *Compare* LAND BREEZE.

seafloor spreading *See* PLATE TECTONICS.

sea level *See* MEAN SEA LEVEL.

seamount in geology and oceanography, an isolated (or relatively isolated) underwater mountain peak that rises at least 500 meters (1,600 feet—some sources give 925 meters, or 3,000 feet) above the surrounding ocean floor. Seamounts are generally volcanic in origin. They tend to be arranged linearly along the margins of the Earth's crustal plates (*see* PLATE TECTONICS). By definition, seamounts remain totally submerged at all times. Those that don't, of course, are called islands or reefs. *See also* GUYOT.

sea stack a rock pillar or small island standing off a coastline and isolated from it by a narrow stretch of water or beach. Sea stacks are formed where the rock to seaward at the edge of a bluff is harder than the rock inland. WAVE EROSION removes the weaker landward rock, leaving the harder rock standing by itself. Common off all rocky seacosts, and also formed by large lakes, they are important nesting sites for seagulls, auks, and other water birds, and often contain large numbers of TIDEPOOLS on their lower slopes.

Seattle, Chief (1786–1866) Indian leader, diplomat, and orator. Seattle was chief of the Suquamish people of Puget Sound, near present-day Seattle, Washington. He is most famous for a speech delivered in January of 1854 during treaty negotiations with Washington territorial governor Isaac Stevens. Eyewitness accounts say that Chief Seattle spoke eloquently in his native language about his people and their coming displacement by new settlers. The essence of his speech, on the virtues of living in harmony with nature, has been embraced by many supporters of the environmental movement. Scholars are divided, however, on the authenticity of the many versions of his text currently in circulation. The speech was first published in an article by Dr. Henry A. Smith on October 29, 1887, in the *Seattle Sunday Star,* but it was not a verbatim transcript.

seawater intrusion *See* SALINE INTRUSION.

secchi disk a tool for measuring the TRANSPARENCY of a body of water through visual approximation. The secchi disk consists of a circular metal or plastic disk

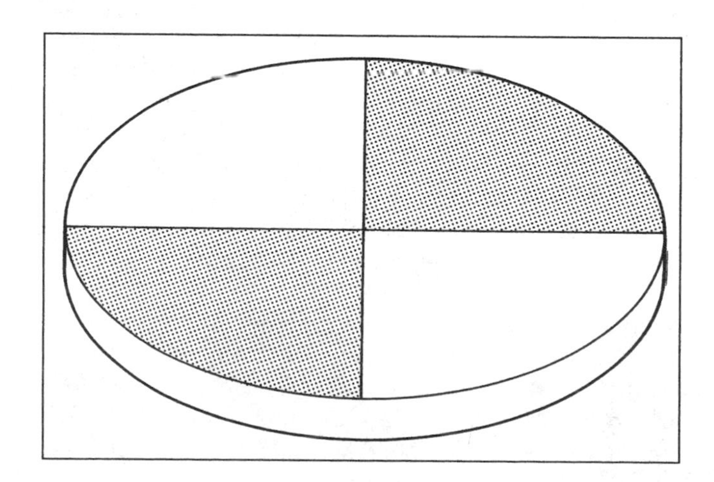

Secchi disk

12 inches in diameter with alternating black and white quadrants attached to the bottom of a weighted line marked off in depth measurements. The disk is allowed to sink until it just disappears. This depth is noted from the marks on the line, and the disk is slowly drawn up again until it just reappears. The average of the two depth measurements is called the *secchi depth* or *depth of visibility,* and is a reliable means of determining how far light will penetrate into a body of water—a useful index of its trophic state (*see* TROPHIC CLASSIFICATION SYSTEM). Secchi depths are nearly uniform throughout a given body of water, but vary greatly among different bodies of water, from as little as 15 centimeters (6 inches) to as much as 50 meters (160 feet).

secondary benefit in economics, a BENEFIT that accrues indirectly from an action instead of following as a direct result. Secondary benefits flow from primary (that is, direct) benefits. For example, a primary benefit from a construction project might be an increase in jobs; a secondary benefit flowing from that increase would be extra retail sales due to the increased buying power of the people who get the jobs. A primary benefit might be stabilizing the water supply; a secondary benefit flowing from that stability would be an increase in the economic development of the region served by the stabilized water supply. Secondary benefits are important in calculating BENEFIT/COST RATIOS for federal projects. They are generally the most controversial part of these calculations because there is no way to calculate them accurately and because they are often transfers from other parts of the economy rather than true growth factors (increases in retail sales in one region, for example, often come at the expense of sales in other regions). *See also* REGIONAL BENEFIT.

secondary clarifier *See under* SECONDARY TREATMENT.

secondary growth in botany, the thickening of plant SHOOTS or roots after lengthening ceases. Secondary growth originates in and is controlled by the *lateral meristems,* growth centers located in the plant's CAMBIUM. It is called "secondary" because the lateral meristems always develop after, rather than before, sprouting. Secondary growth takes place only in the WOODY PLANTS; it is not present in HERBS or in any of the BRYOPHYTES. *Compare* PRIMARY GROWTH.

secondary succession in ecology, SUCCESSION that takes place following a disturbance to the plant and animal COMMUNITY, as opposed to *primary succession,* which begins with bare ground such as new LAVA flow or an exposed lake bed. Secondary succession, also called *allogenic succession,* is typical of forests following logging, wildfire, or windstorms.

secondary treatment in sanitary engineering, treatment of sewage in order to lower its BOD (*see* BIOCHEMICAL OXYGEN DEMAND). It is called "secondary" because it is generally the second form of treatment given to the WASTE STREAM, following the physical removal of floating and suspended solids and of oil and grease (*see* PRIMARY TREATMENT). All forms of secondary treatment currently in use depend on biological action, usually the growth of bacteria (*see* BACTERIUM), to remove the BOD. The two most common methods are the ACTIVATED SLUDGE TANK and the TRICKLING FILTER. In most secondary-treatment facilities, effluent from BOD treatment is passed through a clarifier (the *secondary clarifier; see* SETTLING BASIN) designed to remove the BIOMASS developed during BOD removal. In the United States, a secondary-treatment facility must be able to remove at least 85% of the BOD in the waste stream to meet the standards specified in the CLEAN WATER ACT. A good facility can remove up to 95%. *See also* F/M RATIO; SEWAGE TREATMENT PLANT.

secondary wave *See* S-WAVE.

second growth in forestry, trees that have grown beyond the SAPLING stage but have not yet developed the characteristics of OLD GROWTH; that is, CULMINATION OF MEAN ANNUAL INCREMENT has not taken place and the forest has not developed typical old-growth plant and animal ASSOCIATIONS. The air in a second-growth forest is dryer that that in old growth, and the light level is higher, resulting in a very different FAUNA and FLORA. Second growth is typical of managed forests and of any natural area that has been through a large-scale growth disturbance such as an insect infestation or a forest fire. If left undisturbed, all second growth eventually becomes old growth.

second law of thermodynamics *See* THERMODYNAMICS, LAWS OF.

second-order lake in limnology, a lake in which overturn takes place at least once a year, that is, one in which the THERMOCLINE sinks to the bottom at some point during its annual cycle. The bottom waters of a second-order lake are warmed and reoxygenated at least once annually. As a general rule, second-order lakes in the temperate zones are those between 8 and 90 meters in depth (25 to 300 feet). *Compare* FIRST ORDER lake; THIRD ORDER LAKE.

section in surveying, one-thirty-sixth of a TOWNSHIP. A section is generally a square plot of land measuring one mile on each side and therefore containing precisely one square mile. However, surveying discrepancies in various parts of the country have blurred this definition, resulting in numerous sections larger or smaller than a square mile. *See also* SECTION LINE; TOWNSHIP AND RANGE SURVEYING SYSTEM.

section 404 permit in water politics, a permit from the federal government allowing DREDGE-AND-FILL activities in bodies of water or in WETLANDS. Section 404 permits are administered by the ARMY CORPS OF ENGINEERS under the provisions of section 404 of the CLEAN WATER ACT. The permits are required for all dredge-and-fill operations on all waterways or wetlands of any size (not just those definable as navigable; *see* NAVIGABLE STREAM) except for those operations that are incidental to other activities and do not substantially affect the water or wetland involved. A permit cannot be issued if wildlife HABITAT will be significantly affected or if water quality will be measurably degraded. The permitted activity must also conform to all other relevant statues, including the NATIONAL ENVIRONMENTAL POLICY ACT, the ENDANGERED SPECIES ACT, and the TOXIC SUBSTANCES CONTROL ACT. "Normal farming, silviculture, and ranching activities" are exempted provided the impact is minor and the purpose of the activity is not to change the use of the body of water or wetland involved.

section line in surveying, generally, one of a series of north-south or east-west lines marking off a map in SECTIONS. Every sixth section line is a TOWNSHIP LINE. Due to discrepancies in surveying that have since become codified into law, some section lines are at intervals other than one mile and/or lie at angles some distance from true north-south or east-west. Since surveys preceded settlement in most parts of the United States, property lines often lie along section lines. (*See*, e.g., HOMESTEAD ACT; RAILROAD LAND GRANTS; CHECKERBOARD OWNERSHIP). *See also* TOWNSHIP AND RANGE SURVEYING SYSTEM.

section tree *See* BEARING TREE.

sector in ecology, a division of an ASPECT. Sectors correspond roughly to months in the lunar or Gregorian calendars; however, their boundaries are chronologically fuzzy, depending on biological activity (budding, blooming, leaf fall, etc.) rather than on astronomical measurements.

secure landfill a SANITARY-LANDFILL designed and operated with extra safeguards so that it can legally be used for the disposal of HAZARDOUS WASTES. The design of a secure landfill must include the use of an IMPERMEABLE LINER beneath the entire landfill and as the capping material for each cell and lift. A collection and treatment system for LEACHATE must be installed. The site must be located so that it is not over a sensitive aquifer (*see* AQUIFER; AQUIFER SENSITIVITY; LE GRAND RATING SYSTEM), and MONITORING WELLS must be installed and be tested regularly to check for migration of hazardous materials into the local GROUNDWATER. These are minimum requirements. The best secure landfill operations double each of them, using dual impermeable liners (e.g., plastic sheeting covered by a layer of clay covered by more plastic sheeting covered by another layer of clay) with leachate collection systems beneath each liner and a monitoring system that includes not only groundwater monitoring but air monitoring and a sampling program for the leachate as it migrates through the landfill. No matter what procedures are taken, all secure landfills should be looked on as temporary expedients at best, as they do not deactivate hazardous materials but merely isolate them from the environment for a time. They are most successful in areas where net EVAPORATION exceeds net PRECIPITATION, reducing the amount of leachate that can develop at the site to minimum levels and thus significantly extending the effective isolation time.

sediment in geology, any loose material deposited on the Earth's surface at some point other than where it was formed. Sediments are classified into two broad categories. *Endogenous* (or *autochthonous*) sediments are those that are produced within a body of water and then sink out of it to the bottom. These include both chemical PRECIPITATES from IONS held in solution (rock salt, some calcites) and organic remains and BYPRODUCTS (the shells, bodies, and excrement of aquatic ORGANISMS living in the water). *Exogenous* (or *allochthonous*) sediments are those that are transported some distance from their point of origin before they are deposited. Most of these are the products of EROSION (SAND, SILT, GRAVEL, etc.), though some are organic products (HUMUS, POLLEN). Some exogenous sediments are transported through the air (*see* LOESS; AEOLIAN DEPOSIT) and some are transported by GLACIERS (*see* GLACIAL DRIFT), but the bulk of them are moved by water, either by waves and currents along a shoreline (*see* SHORE DRIFT) or by the flow of water downhill as SHEETFLOW, streams or rivers. Knowledge of the rate of sediment transport and deposition in a body of water is extremely important before the construction of GROINS, DAMS, LEVEES, and other water-manipulation projects, as both the useful life of the structure and its effects on lands downstream (or downcurrent) are determined largely by how rapidly to collects sediments. *See also* BED LOAD; SUSPENDED LOAD; DEPOSITION; TURBIDITY; SEDIMENTARY ROCK; SEDIMENT DELIVERY RATIO.

sedimentary rock rock formed from SEDIMENT DEPOSITS that have become consolidated, either by compression or by the deposition of some material (silica, calcite, etc.) that cements the individual sediment particles together. Sedimentary rocks characteristically show *bedding,* that is, contain a sequence of easily distinguishable layers that represent cyclical (often annual) deposition patterns in the original sediments. Examples: SANDSTONE, LIMESTONE, SHALE. *Compare* IGNEOUS ROCK; METAMORPHIC ROCK.

sediment delivery ratio in engineering geology, the ratio of the SEDIMENT entering a body of water over a year's time to the total amount of sediment eroded from the contributing WATERSHED over the same period. The sediment delivery ratio is affected by such factors as the amount of vegetation in the watershed, the steepness of its slopes, the INFILTRATION rate of it soils, and the proportion of the inflow to the body of water that reaches it overland as opposed to the proportion that reaches it as GROUNDWATER flow.

sediment tank *See* SETTLING BASIN.

seed leaf *See* COTYLEDON.

seedling in forestry, an infant tree, younger and smaller than a SAPLING. The term has no specific technical definition; however, as a general rule, it should not be applied to a tree more than five years old or larger than 2 inches in diameter at the base.

seed tree in forestry and logging, a tree left uncut during a timber harvest in order to provide a natural seed source for REGENERATION on the cutover land. Seed trees should ideally be selected for their superior growth vigor and form and for their ability to produce a good seed crop, so that the regeneration will consist of genetically superior SEEDLINGS in sufficient number to assure total REFORESTATION. *See also* SEED TREE CUTTING.

seed tree cutting in forestry and logging, a harvest method that leaves SEED TREES uncut in order to provide natural REGENERATION following the harvest. Once the regeneration is well established, the seed trees themselves may be harvested (the SEED TREE REMOVAL CUTTING). Three types of seed tree cutting are recognized by silviculturists. The *individual method* leaves a few seed trees—generally two to six per acre—widely scattered on what otherwise classes as a CLEARCUT. The *group method* leaves the same number of trees as the individual method, but concentrates them in a single group (or several groups) instead of scattering them over the CUT BLOCK. It has an advantage over the individual method in that the grouped seed trees are less subject to WINDTHROW, but a disadvantage in that the best seed trees usually are not located all in a group, so that selection strictly on the basis of genetic superiority cannot be done. The *reserved block method* is essentially the same as the group method, but leaves a larger group (or "block") or trees, generally by designating one or more cutblocks in a planned harvest to remain uncut until regeneration is established on the others. This method is essentially similar to, and grades into, a system of staggered PATCH CUTS. If properly designed and carried out, seed tree cuts can provide good regeneration of a site; however, their record is generally not good, usually because too few trees are left to provide good protection against windthrow and because it has proved difficult for logging operators to see the need for leaving the best trees as seed trees instead of leaving only UNMERCHANTABLE timber behind as seed source and harvesting and selling all of the superior STOCK. *Compare* SELECTIVE CUT; SHELTERWOOD SYSTEM.

seep a small SPRING from which water oozes rather than flows. Seeps generally cannot be utilized as water sources by humans and other large animals, though seepage over a wide enough area may produce a usable brook downslope.

seiche in hydrology, a tidelike change of level in a lake, bay, or other confined body of standing water resulting from an oscillation or "sloshing" of water from one end of the lake to the other end and back again. The seiche may be repeated several times, damping out slowly, like water sloshing in a bathtub or washbasin (which is, in fact, a miniature seiche). The most common type of seiche, the *wind seiche,* results from wind blowing across the water body and literally "piling up" water on the far shore. When the wind changes direction or dies, the piled-up water surges back, forming the seiche. In larger bodies of water a *pressure seiche* or *barometric seiche* may occur. This is caused by differences in ATMOSPHERIC PRESSURE from one end of the water body to the other, depressing the water level at the high-pressure end and raising it at the low-pressure end. When the pressure equalizes, the typical seiche oscillation and damping of levels takes place. Other possible, but not common, causes of seiches are underwater EARTHQUAKES, landslides and other forms of MASS WASTAGE that deposit material in a lake, and the calving of ICEBERGS. The size and oscillation period of a seiche depends upon the size and depth of the body of water and the shape of its basin. In large bodies like the Great Lakes, a seiche may reach as much as 15 feet in depth for the first surge and may oscillate noticeably for as much as a week. The change in water level is typically very rapid, and can be dangerous to recreationists along the shoreline.

seismic sea wave *See* TSUNAMI.

seismograph a device for measuring EARTHQUAKE tremors and similar vibrations in the Earth's crust (such as those created by explosions). The heart of a seismograph consists of a fixed frame, ideally resting on BEDROCK, within which a weight is suspended. The frame moves with the Earth, but the weight remains stationary (or nearly so) due to its inertia. The motion of the frame relative to the weight indicates the direction and scale of the Earth motion. The remainder of the instrument consists of mechanical or electronic devices for measuring and recording this motion difference, the most common of which is a steadily rotating drum of paper on which a pen traces a line whose peaks and dips graph the motion of the instrument (hence, the Earth) over time. *See also* RICHTER SCALE.

selective cut (selective logging) any method of timber harvest in which the logging crew cuts only scattered trees within a STAND rather than cutting the entire stand. Selective cutting is properly employed only in mixed-age stands. The trees of ROTATION AGE and older are cut, leaving the younger AGE CLASSES to continue growing. At the time of the next harvest cycle (usually 10 to 15 years later) some of the younger trees will have reached rotation age and will form the new harvest. REGENERATION is accomplished naturally. Selective cutting leaves the ground cover relatively undisturbed, maintains a closed CANOPY, and in general has a far less disturbing effect on the environment than any other method of harvest. It presents a number of practical difficulties, however, among them the numerous ENTRIES required; the greater care that must be taken in FELLING and YARDING; the difficulty of determining which trees have reached rotation age; the greater degree of WINDTHROW in the thinned-out stand; and the inability of many important commercial SPECIES, especially DOUGLAS FIR, to reproduce in their own shade, which means that the REPRODUCTION which comes up in a selectively cut forest is likely to be less-valuable species such as hemlock. Some of these difficulties may be oversome by practicing *diameter rotation* rather than age rotation (cut all trees above a standard diameter rather than above a standard age) and by modifying the classical "single-tree system" of selective harvest into a *group selection system,* in which trees are taken in scattered groups of eight or ten rather than as scattered individuals, providing small openings in the canopy for reproduction and leaving the remaining portion of the stand more thickly grown up and thus better protected against windthrow. *Compare* CLEARCUT; PATCH CUT; SEED TREE CUTTING; SHELTERWOOD SYSTEM. *See also* HIGH-GRADING.

selenium the 34th ELEMENT in the ATOMIC SERIES; atomic weight 78.96, chemical symbol Se. Selenium is classed as a *metalloid,* that is, a non-metal with some of the characteristics of a metal. It exists in three ALLOTROPIC FORMS, the most common of which is as gray or black hexagonal crystals (the others: red, semitransparent monoclinic crystals or a black or red-brown noncrystalline solid). In nature it is usually found in conjunction with sulfur, to which it is chemically related. Selenium is used in industry as a vulcanizing agent for rubber and as an additive to glass to improve its transparency or (in larger amounts) to give it a pinkish or reddish hue. Its rectifying ability (it passes alternating current in only one direction) and photoelectric characteristics (it conducts electricity approximately 1,000 times more efficiently in light than in darkness) make it an important component of consumer electronics goods and of solar cells (*see* SOLAR ENERGY). It is also one of the active agents in plain-paper office copiers. An important TRACE ELEMENT in the diets of warm-blooded animals (birds and mammals) at amounts of roughly 1 part per million, it is highly toxic in amounts that are only marginally larger, causing pallor, nervousness, skin irritation, gastrointestinal disturbance, liver dysfunction, and eventually death. It is listed as a PRIORITY POLLUTANT by the ENVIRONMENTAL PROTECTION AGENCY. *See also* SELENIUM ACCUMULATOR.

selenium accumulator a plant that bioaccumulates SELENIUM in its tissues if grown in selenium-rich soils (*see* BIOLOGICAL MAGNIFICATION). The rate of accumulation varies, but is commonly several hundred times the rate of selenium accumulation in other plants grown in the same soil. Selenium accumulators are poisonous to livestock or game animals grazing on them, causing sore feet, lethargy, and often death. They are most commonly found in arid or semiarid regions such as the western Great Plains. Example: milk vetch (*Astragalus* spp.).

sensible heat in meteorology, heat transferred by conduction and convection, as opposed to heat transferred by radiation. It is called sensible heat because it is the primary form of heat that affects the *sensible temperature*—the feel of hotness or coldness—of human skin. Sensible temperature depends on humidity, air speed, and other factors as well as the amount of heat present, and may vary considerably from the temperature as measured by a thermometer. *See also* BOWEN RATIO; WIND CHILL.

sepal one of the outermost parts of a flower. The sepals arise directly from the receptacle and form a ring (the *calyx*) just beneath the ring of petals. They are usually green and leaflike in appearance, but may often (as in many lilies and most orchids) resemble the petals in color, shape, and texture. They are evolutionarily derived from leaves, however, while the petals are evolutionarily derived from stamens. *See* FLOWER: *parts of a flower.*

septic tank a form of SEWAGE TREATMENT SYSTEM designed for use by a single building or small group of buildings rather than for centralized collection and treatment. A septic tank consists of a large rectangular tank, usually made of concrete, embedded in the ground. Sewage is led into the tank at one end; the solid matter is allowed to settle out; and both solid and liquid components are subjected to anaerobic digestion by bacteria resident in the tank (*see* ANAEROBIC BACTERIA; ANAEROBIC DIGESTER), so that both solids and BOD (*see* BIOCHEMICAL OXYGEN DEMAND) are removed. The septic tank thus serves as both a primary and a secondary treatment facility (*see* PRIMARY TREATMENT; SECONDARY TREATMENT). A DRAINFIELD designed to allow the effluent from the tank to trickle slowly into the soil, where it can be filtered and further decomposed before entering GROUNDWATER supplies or watercourses, completes the septic tank installation (*septic system*). Septic tanks should be employed only in soils with good PERCOLATION, and cannot be used in high-density developments because too may of them in one location can easily overwhelm the soil's capacity to assimilate the effluent. They should be sized to allow at least 24 hours of residence time for each day's incoming waste stream, plus storage room for up to three years' accumulation of digested matter (*see* SLUDGE) in the bottom of the tank (sometimes a two-chambered tank may be employed, with one chamber serving as a SETTLING BASIN and the other as a sludge digester and accumulator. In these cases, the digesting chamber needs to be sized for long term storage and the clarifying chamber for daily flow). A good rule of thumb is to make the total tank capacity equal to 22 times the expected daily flow; however, in no case should it be less than 500 gallons. To maintain optimum operating conditions, the tank must be emptied of sludge every few years. This is normally done by pumping the sludge into a tank truck, which then transports it to a municipal sludge treatment facility or (increasingly rarely) to a landfill. Septic tanks are very effective at eliminating suspended solids and BOD from domestic wastes, and can remove many pathogenic bacteria (*see* PATHOGEN; BACTERIUM). They are largely ineffectual against VIRUSES, however, and have no effect whatever on purely chemical POLLUTANTS such as HEAVY METALS and PCBs (*see* POLYCHLORINATED BIPHENYL), which are passed through unaltered. In fact, since the septic digestion process depends on biological activity, a septic tank can literally be poisoned by toxic compounds in the sewage stream entering it, resulting in a total loss of the ability of the system to treat BOD. Thus, septic tanks cannot be used for industrial installations or in any other situation where heavy chemical use is likely to occur.

sequence (rock sequence) in geology, the chronological order in which the layers of rock in a geologic FORMATION were laid down; also, any rock formation that shows such a sequence. *See also* SERIES.

seral species in ecology, any SPECIES that is characteristically part of a seral association (*see* SERE) rather than a CLIMAX COMMUNITY under the particular climate conditions in which it is found. A species may be seral over part of its RANGE and climatic over the rest. However, many seral species have growth characteristics that do not lend themselves well to being part of a climax community. In particular, most plants that have adapted to life as seral species cannot reproduce in their own shade and will therefore be replaced in the climax community by those which can. *Compare* PIONEER SPECIES.

seral stage *See* SERE.

sere in ecology, the series of slowly changing plant COMMUNITIES that occupy a site during the time between the establishment of the PIONEER SPECIES and the stabilization of the CLIMAX COMMUNITY. The sequence of events in a sere is fairly predictable for any given site, that is, given the geological location, the soil conditions, the climate, and the date of the last major disturbance, an ecologist can predict reasonably well what plants will make up the community currently occupying the site. Seres pass through recognizable stages (*seral stages*) in which one plant ASSOCIATION becomes temporarily dominant. On any given site, these stages will always occur in the same order unless they are interrupted by a disturbance such as fire or logging. Seral stages are almost never pure, but generally contain both old plants that are remnants of the previous stage and youthful plants that will be adults of the next stage. *See also* SUCCESSION.

series in geology, (1) a group of rock strata (*see* STRATUM) that occur in the same order throughout a given FORMATION. The series is characteristic of the formation, and allows the formation to be traced from one part of the landscape to another. A series in which all chronological layers are present is a *continuous* *series;* one in which one or more chronological gaps exist is a *discontinuous series* (*see* UNCONFORMITY). A series in which older rocks overlie younger ones is an *inverted series.* See also SEQUENCE.

(2) the order in which different minerals crystallize out of a molten MAGMA as it cools. *See* BOWEN SERIES.

serpentine in geology, a rock-forming MINERAL consisting of hydrated magnesium silicate, chemical formula $Mg_3Si_2O_5(OH)_4$. Serpentine rock is generally metamorphosed PERIDOTITE (*see* METAMORPHIC ROCK). It is usually found as a massive green to greenish-black amorphous rock, but it may form white, silky fibers (*see* ASBESTOS). Like peridotite, it often contains nickel and CHROMIUM, though rarely in commercial quantities. Soils weathered from serpentine present problems for plant growth, resulting in COMMUNITIES that differ considerably from those on nearby soils derived from other PARENT MATERIALS. The SPECIES that make up these communities are often endemics that cannot be found elsewhere (*see* ENDEMIC SPECIES). The individual plants are often stunted and/or widely spaced, with much bare rock and soil in evidence, giving rise to the term "serpentine barrens" or "serpentine desert." *See also* ULTRAMAFIC ROCK.

service in economics, a nonmaterial BENEFIT provided to a consumer. Like GOODS, services are economic entities that may be exchanged, paid for, or given. In general, services do not consume as many NATURAL RESOURCES as do goods, although most require an input of natural resources in the form of energy or specialized tools, and some (such as merchandising) require a flow of goods made from natural resources in order to function. Examples include design, teaching, maintenance and repair, transportation, etc.

sessile animal in biology, an animal that attaches itself permanently to a SUBSTRATE. Almost all sessile animals are marine ORGANISMS that feed on PLANKTON or small NEKTON species brought to them by the current, eliminating the need to move in order to feed and thus allowing the defensive and competitive advantages of attachment to come to the fore. Most are only sessile as adults; they move about as juveniles, thereby encouraging the colonization of new areas. Common examples include barnacles, coral, and sea anemones.

sessile growth in biology, a plant part such as a leaf, flower, or fruit that grows directly from the plant body instead of having a stem or stalk.

seston in limnology, all microscopic and near-microscopic material, both living and non-living, which is carried in suspension in a body of water (*see* SUSPENSION). The seston consists of the PLANKTON plus the TRIPTON. *See also* SUSPENDED SOLIDS.

settling basin (sediment tank, clarifier) in sewage treatment, a tank or basin in which the flow of SEWAGE is allowed to slow down far enough so that solid material in the sewage will settle out. Both the size of the basin and the rate of flow of the sewage through it are critical: in general, it should be large enough to allow a detention time of two hours (that is, water that flows into the tank will remain in it for two hours before flowing back out), and dimensioned so that this two-hour detention time never allows the velocity of water within the tank to exceed 4 feet per minute. In *horizontal-flow settling basins* (commonly used as primary clarifiers; *see* PRIMARY TREATMENT) the influent is introduced at one end and the clarified effluent is removed as overflow from the other end. In *vertical-flow settling basins* (commonly used as secondary clarifiers; *see* SECONDARY TREATMENT) the influent is introduced at the center of the tank and the effluent is removed as overflow all round the rim. In both types, the part of the rim where overflow occurs is usually lined with a "V-weir"—a metal plate with V-shaped notches cut in it—to make certain that the overflow is spread evenly along the rim. Mechanical scrapers and "flight mechanisms" are employed to remove the settled sludge from the bottom of the tank and to skim oil, grease, and other floating materials from the surface of the liquid in the tank. *See also* GRIT CHAMBER; SEWAGE TREATMENT PLANT.

seventh approximation in the earth sciences, a soil-classification scheme first proposed by scientists of the United States Department of Agriculture (USDA: *see* AGRICULTURE, DEPARTMENT OF) in 1960 and widely accepted since. The somewhat cryptic name refers to the method by which the system was developed. Taking the original USDA classification scheme as published in 1949 (*see* GREAT SOIL GROUPS) as a "first approximation," the system was refined through six succeeding approximations by circulating it among soil scientists throughout the United States, incorporating their suggestions, and recirculating the revised scheme. The seventh approximation was felt satisfactory for offering as a completed system. The scheme is officially known as the *National Cooperative Soil Survey.* It has also, but infrequently, been called *soil great groups* (to distinguish it from the old *great soil groups*).

METHOD

The classification of soils in the seventh approximation is based almost entirely on the presence or absence of one or more of a group of 12 easily recognized charac-

Soil Orders

Order	Characteristics	Diagnosis	World Area (%) (in mill. sq. mi.)
Entisol	recent	lack of any discernable horizons	6.5 (12.9%)
Inceptisol	young	horizons present but poorly developed	8.1 (16.1%)
Vertisol	inverted	vertical cracks; all horizons have high clay content	1.1 (2.2%)
Aridisol	arid climate	prominent calcic horizon	9.9 (19.7%)
Mollisol	soft, rich	prominent mollic horizon	4.6 (9.2%)
Spodosol	high ash content	prominent spodic horizon	2.8 (5.6%)
Alfisol	high clay content	prominent argillic horizon; ochric horizon	7.6 (15.1%)
Ultisol	heavily leached	all horizons heavily weathered; argillic horizon present	4.4 (8.8%)
Oxisol	high oxide content	prominent oxic horizon; forms laterite	4.8 (9.6%)
Histosol	high organic content	prominent histic horizon (peat)	0.4 (0.8%)

teristic HORIZONS, called *diagnostic horizons,* which bear names such as "argillic" (rich in clay), "spodic" (rich in ash), "oxic" (rich in oxides of iron and other metals), and so on. All soils are divided into 10 broad categories, or *orders,* based on their combination of diagnostic horizons. The orders are divided into *suborders* based mostly on climatic differences ("wet," "dry," "cool," etc.) and these in turn are separated into *great groups* on the basis of the presence or absence of some four dozen "formative elements," which include extra diagnostic horizons, extra climate elements, characteristic chemicals (such as MINERALS or SALTS), and prominent physical characteristics (such as blocky structure or excessive leaching). (The term "great groups" is somewhat unfortunate for this classification because it has no relationship whatever to the "great soil groups" of the older systems; these are more closely related to the new "orders.") The degree of resemblance of a specific soil to its great group definition places it in a *subgroup,* identified as "typic" (typical), "aquic" (wetter than normal), "xeric" (dryer than normal), and so on. The subgroup is further divided into *associations* (or *families*) on the basis of the one or more of the characteristics of the A horizon (mineral composition, texture, temperature or moisture gradient, etc.). Finally, the associations are divided into *series, types,* and *phases* according to similarities in profiles. These are given the name of some prominent geographical feature in the region where the soil was first identified. Series, which are functionally the same as POLYPEDONS, are geographically related soils that belong to the same association; types are members of a series divided on the basis of textural differences in the A horizon; and phases are soils of the same type that differ in depth, characteristics of the O horizon, etc., due to differences in local conditions (for example, hill soils may show a "steep phase" and a "level phase"). Series, type, and phase under the seventh approximation are usually the same as series, types and phases under the great soil group system, allowing soil maps prepared under the older system to continue to be useful. *Compare* UNIFIED SOIL CLASSIFICATION SYSTEM. *See also* SOIL: *classification of soils.*

sewage strictly, domestic wastewater, including human wastes (urine and fecal matter), washwater, EFFLUENT from garbage disposals, and anything else that is flushed down a toilet or washed down a drain. More loosely, the term "sewage" has come to mean anything carried in municipal sewers, including those industrial wastes that are delivered into the sewers rather than being treated and disposed of on-site. Sewage typically contains 100–300 parts per million of BOD (*see* BIOCHEMICAL OXYGEN DEMAND) and carries a high proportion of bacteria (*see* BACTERIUM), both living and dead. It may also carry MICROCONTAMINANTS and other toxic materials. *See also* SEWAGE TREATMENT PLANT; SEWER; SEWERAGE SYSTEM; BLACK WATER; GRAY WATER.

sewage treatment plant (STP) a facility for modifying SEWAGE to reduce its potential effects on the biological environment (*see* ENVIRONMENT; BIOSPHERE). The design of sewage treatment plants is highly variable depending upon the volume and character of the incoming WASTE STREAM and upon the characteristics required of the EFFLUENT. Therefore, these plants must normally be custom-designed, although prefabricated plants (*package plants*) are commonly used for smaller, nonindustrial communities and for specialized uses such as the PRETREATMENT of sewage from a subdivision before it passes it into the main sewage plant. All STPs use some form of PRIMARY TREATMENT, and most are also required to provide SECONDARY TREATMENT. (An exception is plants whose OUTFALLS lead directly into oceans or tidal rivers, and which can be shown not to adversely affect the quality of these RECEIVING WATERS; these may be allowed to discharge sewage following primary treatment only.) In addition, an increasing number of plants are now required to use TERTIARY TREATMENT, with the goal of achieving swimmable—if not drinkable—receiving waters in as much as of the nation as possible. Siting of these facilities is often extremely difficult. They should be at or near the bottom of the WATERSHED in which the SEWERAGE SYSTEM is located in order to allow for gravity flow of sewage (but *see* FORCED MAIN), and must normally be on a body of water large enough to absorb the effluent without damage. Usually they also must be located at some distance from residences because they qualify legally as a NUISANCE and thus reduce the value of nearby property (*see* NIMBY). Because of the siting difficulties, and the expense of construction of these often very complicated facilities, new STPs should be built large enough to accommodate at least 25 years, and preferably 50 years, of growth.

sewer a pipeline for carrying SEWAGE. Most sewers are made of concrete or clay pipes, although some are of cast iron, and current installations are increasingly using plastic. They are assembled into a *sewer system,* which forms the collection and transportation portion of a community waste-handling facility, or SEWERAGE SYSTEM. The smallest portion of the sewer system is the *connector,* which serves an individual house or business. The connectors feed into a *lateral,* which serves a single street or small neighborhood. Laterals are joined together by *mains,* serving large regions of a city. Larger systems may employ two types of main, the *branch* (serving several neighborhoods) and the *truck* (connecting a number of branches). Finally, a single *interceptor* collects all the sewage from the individual mains (or trunks) and carries it to the SEWAGE TREATMENT PLANT. Large plants may be served by more than one interceptor. Insofar as possible, sewer systems are designed so that sewage can flow to the treatment plant by gravity alone (but *see* FORCED MAIN). The rate of flow of the sewage should be between 2.5 ft/sec and 10 ft/sec. More than this, and grit carried in the sewage can physically abrade the pipes; less, and the sewage may become septic within the pipes, creating toxic gases and acids that may erode the sewer system. Sewers are designed to run only partially full and therefore tend to leak inward rather than outward, requiring excess DESIGN CAPACITY for water that enters the pipes from the WATER TABLE between the sewage source and the treatment plant (*see* INFILTRATION). The actual size of each portion of the system is determined by the size and characteristics of the population served. In general, laterals and connectors are sized to carry 400 gallons per capita per day (400 gcd), while mains, trunks, branches, and interceptors have a design capacity of 250 gcd. Connectors should be at least 4 inches in diameter. The minimum recommended size for all other portions of the system is 8 inches. *See also* COMBINED SEWER SYSTEM; STORM SEWER.

sewerage system the total system used by a community to dispose of its SEWAGE. A sewerage system consists of a sewer system (*see* SEWER), a SEWAGE TREATMENT PLANT, and some means of disposing of the EFFLUENT and SLUDGE from the sewage treatment plant. It may also include one or more pumping stations if the terrain does not allow the system to work by gravity flow alone, as where a city covers more than one WATERSHED or where it is so flat that sufficient slope cannot be built into the pipes. The actual design of the system will depend on the terrain, the type of sewage being handled (that is, the ratio of domestic wastewater to industrial wastes and the character of the industry producing wastes), and the standards the effluent must meet to avoid contaminating the RECEIVING WATERS.

shade-dependent species a plant SPECIES that cannot tolerate long exposure to direct sunlight. Shade-dependent species typically grow as UNDERSTORY plants in OLD-GROWTH forests and often cannot survive outside this setting. *Compare* SHADE-TOLERANT SPECIES; SHADE-INTOLERANT SPECIES.

shade-intolerant species a plant SPECIES that cannot grow in the shade. Shade-intolerant species are typically PIONEER SPECIES or species of early SERES in a successional sequence (*see* SUCCESSION). They are almost never CLIMAX SPECIES due to their inability to reproduce in their own shade. *Compare* SHADE-TOLERANT SPECIES; SHADE DEPENDENT SPECIES.

shade-tolerant species a plant SPECIES that can reproduce in the shade of other plants. Shade tolerance is a gradient. Some plants can tolerate only the open shade under an isolated tree, while others are able to grow and reproduce beneath the CANOPY of a dense STAND of CONIFERS. For trees, the most important point on the shade-tolerance gradient for any given species is the density of its own shade. Those species that are tolerant enough to reproduce in their own shade can perpetuate themselves permanently on a given site, thus becoming climax species (*see* CLIMAX COMMUNITY), while those that cannot reproduce in their shade are SERAL SPECIES that will be gradually replaced on a site by more tolerant species unless some site disturbance such as logging or fire takes place. *Compare* SHADE-DEPENDENT SPECIES; SHADE-INTOLERANT SPECIES.

shadscale desert a DESSERT in which the dominant vegetation is shadscale (*Atiplex confertifolia,* a sagebrushlike relative of the common garden beet), or a related species such as salt sage (*A. nuttallii*). Shadscale deserts form on soils that are heavy in SALTS, such as EVAPORITE BASINS. They are particularly common in the GREAT BASIN of the western United States.

shale a SEDIMENTARY ROCK made of CLAY that has been cemented together by calcium carbonate deposits (*see* LIMESTONE) or has simply been compressed into rock by its own weight or that of overlying STRATA. Shale is typically gray in color and shows fine parallel bedding, each bed indicating an annual deposit of new clay materials (*see* VARVE). The rock is soft and structurally weak and cannot be used for building, though it is often crumbled up and used as a constituent of cement. Shale containing calcium carbonate grades into MARL; shale containing sand and soil particles grades into siltstone and SANDSTONE. *See also* OIL SHALE.

shape factor in limnology, the ratio of the shoreline length of a lake to the circumference of a perfect circle having the same area as the surface of the lake. It is always greater than one, and increases as the shoreline becomes more irregular. The shape factor can be used as a rough index to the proportion of littoral (shoreline) HABITAT to open-water habitat in a lake and may therefore be used as a predictor of COMMUNITY balance within the lake environment. *Compare* VOLUME FACTOR.

shear wave *See* S-WAVE.

sheet erosion a type of EROSION in which a uniform (or nearly uniform) layer of the soil surface is removed simultaneously over a broad area. Sheet erosion results from SHEETFLOW. The phenomenon is not particularly common; however, RILL EROSION can lead to the same result if the rills constantly change course and direction on a smooth slope, and this much more common type of episodic (as opposed to simultaneous) surface removal is also sometimes called sheet erosion. *See also* RUNOFF.

sheetflow the generalized flow of water over the surface of the ground, not confined to channels. Sheetflow takes place only in heavy rains during which the INFILTRATION CAPACITY of the ground is exceeded so that the excess water must flow off. It carries this excess water into stream channels. If the soil is unprotected by plant cover, sheetflow usually carries numerous soil particles with it, a process known as SHEET EROSION. True sheetflow is relatively rare, as the unevenness of ground surfaces usually divides the flow into tiny rivulets. The constantly-changing courses of these rivulets, however, have the same effect on soil, plants, and receiving streams as does true sheetflow, and thus may be considered in the same category. *Compare* RILL.

shelter in ecology, protection from weather, PREDATORS, etc., adequate enough to ensure survival of an ORGANISM. Shelter is roughly synonymous with COVER. *See also* HABITAT.

shelterbelt (windbreak) long rows of trees, sometimes flanked by large shrubs, meant to reduce wind damage, soil erosion, and drifting snow in order to protect farmsteads and farm fields. During the 1930s shelterbelts were created on many farms under USDA programs in reaction to the soil loss sustained during the extended drought of the DUST BOWL years. During the 1970s, however, when farm size dramatically increased along with the size of farm machinery, most cropland shelterbelts were removed, since long passes for cultivating or harvesting were much more efficient than making frequent turns on 1930s-sized fields.

shelterwood system in logging and forestry, a timber-harvesting system designed to promote natural REGENERATION by leaving some of the OVERSTORY intact as a seed source. Shelterwood cutting is essentially the same as SEED TREE CUTTING, except that enough overstory trees are left to provided some shade for the growing site and to protect each other from WINDTHROW as well as to supply seed for the new generation of timber. The classical shelterwood cutting system involves three ENTRIES. The first entry, the *preparation cut,* is a light THINNING designed to let more light and rain through the overstory and to remove the brushy understory to prepare the forest floor as a seedbed by breaking up and decomposing the woody material and leaf LITTER on it (this cut is often skipped in contemporary practice). The second entry is the shelterwood cut itself. It aims for the removal of one-half the natural shade on the forest floor. The third entry, some years later, removes the remaining overstory to release the established regeneration (*see* RELEASE CUTTING). The shelterwood system leads to the same EVEN-AGE STAND composition as does clearcutting (*see* CLEARCUT), generally with considerably less harm to the soil and to watershed values. It is only marginally better aesthetically, however, and is much more expensive to accomplish; therefore, its use is generally limited to those places where soils and waters are at significant risk, or where frost damage to the seedlings or FROST HEAVE in the soil are potentially serious problems. *See* CLEARCUT; SELECTIVE CUT; NEW FORESTRY.

shoal area in limnology, the portion of a lake that is less than 3 meters deep. Rooted aquatic plants are limited to the shoal area, which is also the principal breeding and feeding ground for the lake's resident fish.

shoot in the biological sciences, generally, that portion of a plant that is above the surface of the ground. The term is also sometimes used to refer to a twig or a branch grown by a tree or other WOODY PLANT during the current growing season.

shore drift (littoral drift) in geology, the movement of SAND and other suspended matter laterally along the coastline of an ocean or large lake. Shore drift is carried by LONGSHORE CURRENTS, and is the prime source of building material for barrier islands, beaches, and spits. It can be seen from high vantage points as a broad light-colored band of inshore water, often several miles wide. Interruption of shore drift by ill-conceived engineering projects is one of the major causes of shoreline EROSION. *See also* BEACH NOURISHMENT.

short-day plant in botany, a plant that flowers only if the period of daylight to which it is exposed is shorter than a genetically determined "critical length." The critical length varies from SPECIES to species; within species, it is temperature-dependent, becoming longer at lower temperatures. Short-day plants bloom in the spring and the fall. *Compare* LONG-DAY PLANT; DAY-NEUTRAL PLANT. *See also* PHOTOPERIODISM.

short-term turbidity TURBIDITY caused by particles large enough to settle out of a sample of water if the water is left undisturbed for seven days. Short-term turbidity is usually formed by SAND, SOIL, and SILT particles suspended in the water. When caused by a one-time event such as a landslide, it may generally be expected to disappear rapidly enough to cause little environmental damage. However, if the source is an ongoing activity such as logging or roadbuilding, "short-term" turbidity may last a considerable length of time and cause severe disruption to aquatic ORGANISMS. *Compare* LONG-TERM TURBIDITY.

short ton a standard ton, that is, 2,000 pounds. *Compare* LONG TON.

shrub a woody plant less than 15 feet tall at maturity, with several main stems arising near ground level. Shrubs generally "put on girth" at a considerably slower rate than do trees, and are often adapted to more extreme climatic conditions. Some plants that normally grow as trees will become shrubs at the edges of their RANGES, for example at TIMBERLINE or at the edges of DESERTS; *see* KRUMMHOLTZ.

shrub-steppe a STEPPE with a scattered OVERSTORY of SHRUBS, usually sagebrush, rabbitbrush, juniper, mesquite, or similar arid-land SPECIES. The UNDERSTORY species are predominantly bunchgrasses (*see under* GRASS). Shrub-steppes form under drier soil conditions than those favoring a pure prairie type ("meadow steppe"). They are generally classed in the GRASSLAND BIOME, though they grade into desert on the drier sites (*see* DESERT BIOME).

sick building syndrome *See* INDOOR POLLUTION.

side notch in logging, an undercut extended around the side of a tree in order to induce it to fall into a desired LAY (*see* FEELING).

sierozem soil a soil type with a thin, light-colored A HORIZON (*see* SOIL: *the soil profile*) and a thicker B horizon that generally contains a high proportion of CLAY. The soil is low in ORGANIC MATTER, and is often underlain by an accumulation of calcic MINERALS (minerals high in calcium, usually in the form of calcium carbonate; *see* LIMESTONE) at the bottom of the B horizon. Sierozems develop under SHRUB-STEPPE communities in cool, arid climates. The term is now considered technically obsolete but is still in common use (*see* ARIDISOL).

Sierra Club environmental activist organization, founded in 1892 to protect and enjoy the mountain wilderness of the west. It has since expanded its interests to concern with all aspects of environmental protection on the planet, although wilderness preservation still ranks first on its agenda. Probably the best-known of all environmental organizations, the club concentrates principally on lobbying, presenting public input into the management decisions of federal agencies (either at hearings or through informal consultation) and increases public awareness of environmental threats. It also initiates (or cooperates with) a number of administrative appeals and lawsuits each year concerning actions by government bodies that it considers harmful to the environment. It has maintained an active outings program since its inception. Membership (1988): 401,000. Address: 730 Polk Street, San Francisco, CA 94109. Phone: (415) 776-2211. *See also* EARTH JUSTICE LEGAL DEFENSE FUND.

Sierra Club Legal Defense Fund (SCLDF) See EARTH JUSTICE LEGAL DEFENSE FUND.

Sikes Act federal legislation for the protection and conservation of wildlife on federal lands, signed into law by President Gerald Ford on October 18, 1974. Among other provisions, the act directs all federal land management agencies, including but not limited to the

FOREST SERVICE, the NATIONAL PARK SERVICE, and the BUREAU OF LAND MANAGEMENT, to develop and implement fish and wildlife conservation plans, protect and improve the HABITAT used by endangered or threatened species (*see* ENDANGERED SPECIES; THREATENED SPECIES) and cooperate with and provide assistance to the fish and wildlife and land-management agencies of the various states for wildlife protection activities.

sill in geology, a tabular rock-body—that is, one that is considerably thinner than it is broad or tall—that has intruded between the layers of a previously existing rock structure. Sills are similar to DIKES except that they are *concordant* (lie parallel to the prevailing rock structure instead of cutting across it).

silt a soil particle intermediate in size between CLAY and SAND. The actual size limitations for what are called silt particles vary somewhat according to which standard scale is being used. The most commonly-used scale in the United States is probably the USDA scale (United States Department of Agriculture), which designates as silt any particle between 0.002 and 0.05 millimeters in diameter. Other widely used standards include the WENTWORTH SCALE (silt: 1/256mm to 1/16, or roughly 0.004mm to 0.06), the MIT scale (Massachusetts Institute of Technology), 0.002 to 0.06mm; the USPRA scale (United States Public Roads Administration), 0.005 to 0.05; and the International Scale (International Society of Soil Scientists), 0.002 to 0.02mm. A soil is classified as a silt if 80% or more of its particles fall into the silt size classification. *See also* SILTATION.

siltation the deposition of SILT on the bed of a body of water. Siltation occurs whenever moving water slows down to the point that it can longer carry silt particles in SUSPENSION. The process results in the filling-up of lakes and reservoirs until they become alluvial plains (*see* ALLUVIUM). It is often very rapid: 40 miles of the 115-mile length of Lake Mead, formed in 1936 by the completion of Hoover Dam on the Colorado River, had already become silt plains by 1970, and the entire reservoir is expected to "silt in" within about 250 years.

silvicultural prescription in forestry, a description of the techniques ("silviculture treatments") to be used on a specific piece of land in order to grow and maintain a STAND of trees with a specific set of characteristics. A silvicultural prescription should include the techniques and the timing for the whole range of activities to be carried out on the treated site, including harvest, planting, BRUSH CONVERSION; THINNING, fertilizing removal of competing vegetation, etc. *Compare* HARVEST PRESCRIPTION.

single interval well *See under* MONITORING WELL.

singlejacking in logging, the harvest of timber by one individual working alone. Singlejacking is considered a dangerous practice and is strongly discouraged.

sink anything that collects significant quantities of some identifiable substance or type of energy. Thus a concrete wall can serve as a temporary *heat sink*, absorbing heat that falls upon it and reradiating it later. Similarly, a closed geological basin is a sink for the materials carried into it by water or wind; the ocean is a sink for the SALTS carried into it by rivers; etc.

sinkhole in geology, a hole in the ground caused by the collapse of an underground cavity, usually a portion of a cave. It may or may not allow access to the remainder of the cave. Sinkholes are characteristic both of KARST TOPOGRAPHY and of PSEUDOKARST. Their sides are usually vertical. They may be a few feet to several hundred feet in width and in depth. Very large "sinkholes" caused by the evacuation of a MAGMA CHAMBER within a VOLCANO and the subsequent collapse of the volcano's peak, as at Oregon's Crater Lake or Alaska's Katmai Volcano, are known as *collapse calderas* (*see under* CALDERA).

sinking agent any substance that will cause a floating POLLUTANT on the surface of a body of water (such as an oil slick) to sink below the surface. A sinking agent may either serve as a surfactant to mix the pollutant throughout the water instead of floating it on top (*see* WETTING AGENT), or it may bind chemically or physically with the pollutant to form a denser-than-water substance that will then sink to the bottom. Because they simply change the state of the pollutant rather than eliminating it, sinking agents are not generally recommended for pollution control; however, they have some limited uses in situations where the change of state may make cleanup easier, as for example in areas of heavy and rapid bottom SILTATION where the layer of pollutants carried to the bottom by the sinking agent will quickly be covered by a thick layer of SILT and thus removed from contact with the aquatic ECOSYSTEM.

sintering in chemistry and physics, the clumping together of a powdered material when it is heated to just below its melting point. Sintering forms the powder into irregular, porous lumps that remain fused when cooled. It is used in metallurgy to separate metals from their ORES. By heating a powdered ore to the point where the metal sinters, lumps of nearly pure metal may be formed, which may then be mechanically separated from the remaining ore powder. LEAN ORES may also be enriched in this manner (*see* BENEFICIATING). The process uses no chemicals and is thus less environmentally damaging than others, although the fact that the TAILINGS are in powdered form creates some disposal problems and can lead to air and water pollution.

SI *See* SITE INDEX.

SI system *See* METRIC SYSTEM.

site index (SI) in forestry, a number assigned to a piece of land (the "growing site") to indicate its ability to grow trees of a particular SPECIES. The site index is equivalent to the height a tree of the designated species can be expected to be at a standard age that varies with the species. For most timber species the standard age is 50 years, but for Ponderosa pine it is 80 years; for Douglas fir and other timber trees of the northern Pacific coast it is 100 years; for redwoods it has occasionally been set as high as 300 years; and for pulpwood plantations in the south, it is 35 years. The calculation of site index is done by measuring the average height of trees in a STAND, dividing by their average age, and multiplying by the standard age. For example, the site index of a stand of 70 year old Douglas firs 90 feet tall would be (90/70) x 100, or 128.5, indicating a relatively poor site. For loblolly pine on the East Coast, the same height and age figures would give a site index of 65, indicating a moderate site. In actual use, calculations are usually avoided by using a previously calculated graph of *site-index curves* on which plotting the age and height of a stand allows the index to be read directly.

SITE CLASS

In order to simplify the use of site indexes for planning purposes, they are usually lumped together into bands known as *site classes*. The boundaries of these classes and their total number varies according to the species being classed. There are generally five to seven, designated by Roman numerals, with the better sites given the lower numerals. For example, the site-classification scheme for Douglas fir (western Oregon and Washington) is:

Site I = SI 200 +
Site II = SI 170–199
Site III = SI 170–189
Site IV = SI 110–139
Site V = SI 80–109

A map of a forest by site class allows timber harvest and silvicultural activities to be focused where they are most appropriate and gives a fairly accurate preview both of the economic return from logging and the difficulty of REGENERATION on any given site. It is probably the single most important piece of information a forest manager can have.

site-specific factor in land-use planning, any PLAN ELEMENT, piece of data, physical constraint, etc., which operates on a specific site within a planning area but cannot be assumed to operate throughout the entire planning area. Soil type, for example, would be a sitespecific factor, whereas climate would not. *See also* SITE-SPECIFIC SPECIES.

site-specific species a SPECIES, especially a plant species, that depends on a narrow range of SITE-SPECIFIC FACTORS to survive and thus cannot live if removed from its original growing site. Site-specific species are generally ENDEMIC SPECIES with limited RANGES that are easily endangered and may be rendered extinct by man-caused disturbances such as logging mining, roadbuilding, etc.

Sixth Great Extinction *See* MASS EXTINCTION.

size class *See* DIAMETER CLASS.

skidding in logging, any form of YARDING in which the logs are dragged ("skidded") along the ground. Skidding can be highly destructive of plant and soil cover on the logging site, and can lead to severe EROSION; thus some form of HIGH-LEAD LOGGING is generally preferable from an environmental standpoint, especially on steep slopes. *See also* SKID ROAD.

skid road (skid trail) in logging, a route on the ground over which logs are skidded (*see* SKIDDING) from the point they are felled (*see* FELLING) to the LANDING. Skid roads should be laid out before logging begins, with attention paid to the topography and to network design so that only as many roads as are necessary are built and that they lay on the ground in such a way as to minimize EROSION. Experience shows roughly twice as much ground disturbance from haphazard "logger-choice" skidding as from a properly designed system of skid roads laid out in advance.

skyline lead in logging, a form of HIGH-LEAD LOGGING in which SPAR TREES are rigged at both ends of the high-lead cable, thus allowing logs to be lifted entirely off the ground during the yarding process. Skyline-lead yarding virtually eliminates the damage to soils that results from more conventional yarding practices and allows yarding to run more or less continuously, as the yarding cable is left in place over a pair of pulleys rather than being reeled in each time. It is somewhat more difficult to apply in OLD GROWTH

TIMBER than in SECOND-GROWTH, however, due both to the weight of the logs to be yarded and to the difficulty of establishing unobstructed cable routes.

SLAPP suit *See* ANTI-ENVIRONMENTALISM.

slash in logging, the woody material left on the ground after a logging operation. Slash includes limbs, tops, and other residue from BUCKING operations; uprooted stumps, pieces of bark, and severed or uprooted SHRUBS; and any other portions of WOODY PLANTS that are dead or dying as a result of the logging. The term is sometimes expanded to include windfalls and other natural woody debris, in which case slash resulting from a logging operation should be referred to as "logging slash." *See also* SLASH TREATMENT; FUEL; CULL.

slash-and-burn a form of agriculture widely practiced in tropical lands, involving the felling and burning of trees in order to produce an opening for crops that will be used for 1–3 years and then allowed to revert back to forest while another area is opened. Slash-and-burn fields generally give excellent crops the first year, with crop yields dropping roughly 30% the second year and 50% the third due to the loss of many NUTRIENTS through leaching by the heavy tropical rains and the removal of others in the harvested crops. This loss of productivity, coupled with the difficulty of keeping the field open as the forest attempts to reclaim it, are the factors that lead to the constant field relocation. Left to the forces of natural succession (*see* SUCCESSION), the fertility of the soil will recover in 30 to 50 years, at which time the slash-and-burn cycle can be repeated. Criticized in the past as "primitive" slash-and-burn is enjoying a minor renaissance today as the temporary loss of soil fertility that results from it is compared to the permanent loss of fertile soils to laterite formation resulting from large-scale agricultural clearing in the tropics (*see* LATERITIC SOIL).

slash treatment in logging and forestry, any method of handling SLASH to reduce fire danger and promote the recovery of a site after logging, including preparation for REFORESTATION. Ideally, slash treatment should be designed to return the NUTRIENTS in the slash to the soil on the site while eliminating FUEL BUILDUP. On federal forest land, the type of slash treatment to be used is generally designated in timber sale contracts. Typical methods include BROADCAST BURNS and YUM YARDING.

slate in geology, metamorphosed SHALE (*see* METAMORPHIC ROCK). Slate is an extremely hard, fine-grained rock that breaks easily along parallel CLEAVAGE PLANES, forming thin, flat sheets that are used as flooring and roof tiles. They were at one time widely utilized as blackboards and smaller erasable writing surfaces (hence the phrase "wipe the slate clean"). Slate deposits are usually impermeable (*see* IMPERMEABLE LAYER; PERMEABILITY), and form a barrier to GROUNDWATER flow.

sleet in meteorology, frozen rain. Sleet differs from HAIL in being simply frozen raindrops, rather than complex concentric structures. It is formed when falling rain encounters a layer of air next to the ground in which the temperature is below freezing (0°C:32°F) and freezes as it falls, reaching the earth as an ice pellet the same size as the raindrop that began the descent.

slickenside in geology, a polished rock face caused by EARTHQUAKE motion. Slickensides result when movement along a FAULT cracks a rock body and then rubs the two faces of the crack together as they are displaced relative to each other. They show STRIATIONS, and resemble faces polished by GLACIERS (*see* GLACIAL POLISH), except that the polishing is rarely as complete and the striations usually show small discontinuous steps, especially near their ends. Slickensides can be used to help determine the direction and magnitude of fault motion.

slime mold any ORGANISM belonging to the DIVISION Gymnomycota of the KINGDOM PROTISTA. The slime molds share characteristics with both plants and animals, and the difficulties that arose in deciding where to classify them are one of the principal reasons that biologists now classify living things into five kingdoms instead of two. The principal CLASS, the *plasmodial slime molds* (class Myxomycetes—approximately 540 SPECIES) spend most of their lives as AMOEBA-like masses of creeping protoplasm known as *plasmodia* (singular: plasmodium) that move slowly along the ground, engulfing and digesting BACTERIA, small FUNGI, and bits of dead and decaying ORGANIC MATTER. Under favorable circumstances, a single slime mold may grow to be as much as a meter across. Although the organism has many cell nuclei (*see* CELL: *structure of cells*), there are no interior cell walls; the entire plasmodium acts as a single giant multinucleate cell. When food and water conditions become unfavorable (and, for some species, when light conditions become favorable), the plasmodium stops moving and grows a whole forest of fungus-like spore-producing bodies (*sporangia*) on short stalks. These release spores that ultimately cross fertilize each other to create new plasmodia. The smaller class of slime molds, the *cellular slime molds* (class Acrasiomycetes—26 known species) also spend most of their lives as individual amoeba-like organisms. To reproduce, however, they gather together into a compact sluglike mass that produces a single sporangium similar to the multiple sporangia of the plasmodial slime molds. Both classes prefer cool, moist conditions such as may be found in the DUFF of a mature forest floor. Ecologically, they are DECOMPOSERS, and as such serve a valuable function in the life of a forest.

sludge (1) in geology and limnology, the mudlike muck on the bottom of a body of water, consisting of large amounts of ORGANIC MATTER mixed with particles of SILT, CLAY, and SAND and with enough water content to remain in a semiliquid state. *See also* SEDIMENT.

(2) in sanitary engineering, the solid material removed from the stream of SEWAGE by a SEWAGE TREATMENT PLANT. Sludge consists primarily of fecal matter (*see* FECES), but may also contain ground up food matter from garbage disposals; silt, sand, bits of leaves, etc. from COMBINED SEWER SYSTEMS; living MICROORGANISMS, especially ALGAE and bacteria (*see* BACTERIUM), that have entered the sewage at various points along its path and are now growing there; and chemical PRECIPITATES, especially if industrial OUTFALLS are connected to the sewer system in which the sludge originates. Sludge removed by PRIMARY TREATMENT is called *primary sludge;* that removed by SECONDARY TREATMENT is called *secondary sludge. See* CLARIFIER.

sludge worm *See* OLIGOCHAETE.

slump in geology and soil science, the downslope movement of soil or unconsolidated rock as a unit. In small slumps, such as the collapse of part of the edge of a CUTBANK, the unit that has slumped may break up on its way down the bank. In larger slumps, the slumped material usually remains together as a single unit in its new resting place. There is usually backward rotation of the soil mass involved during its slide down the hill, but it is recognizably the same piece of land as it was before the slump. Trees often continue to grow undisturbed in the new site, though their trunks may show a tilt into the hill. *Compare* EARTHFLOW.

slurry a mass of solid particles such as dust, rock fragments, etc., mixed with enough water so that it will flow like a liquid. The main body of a slurry is a thick SUSPENSION with enough mass so that particles too large to be suspended are nevertheless carried along by the drag of the slurry as a whole. Slurrys can be piped, pumped, tanked, and otherwise treated as liquids for transport purposes, and are an efficient means for moving materials (such as COAL), which are insoluble in water and which can be used in the form of the dust and small fragments that are obtained when the water is removed from the slurry at the destination point. Slurry spills from pipeline ruptures, valve failures, and so on, can cause excessive stream TURBIDITY but otherwise pose no special environmental problems, and should be treated as is any other spill of a potentially hazardous liquid (*see* HAZARDOUS WASTE).

small game in wildlife management, any GAME ANIMAL the size of a fox or smaller, including rabbits, raccoons, squirrels, opossums, and so on. GAME BIRDS are also usually included in the small-game category for management purposes. *Compare* BIG GAME ANIMAL.

smart growth *See* URBAN SPRAWL.

smog air pollution, especially air pollution that restricts visibility and is formed by the interaction of atmospheric gases with the waste products resulting from the burning of FOSSIL FUELS. There are two main types of smog. Ordinary ("classical") smog—a term coined in 1905 by Harold Des Voeux, a British physician—is a mixture of smoke and fog. Some COMPOUNDS in the smoke dissolve in the fog, turning the droplets of moisture that make up the fog into droplets of sulfuric ACID and other acids. This type of smog has been recognized as a health hazard since the middle of the 19th century. The other type of smog, *photochemical smog,* results from the release of gases, especially NITROGEN OXIDES (NO_x), in the exhausts of internal combustion engines. These gases combine with atmospheric gases in the presence of light to form a complex chemical "soup" with a characteristic brown color. Photochemical smog also dissolves in droplets of moisture to form acids, in this case principally nitric acid. It generally include significant quantities of OZONE. Both types of smog are serious health hazards, especially to those with chronic respiratory diseases such as EMPHYSEMA or ASTHMA, which are on the rise, especially in cities such as Los Angeles. In addition, smog kills vegetation, causes decay of stone, ironwork an other building materials, and contributes to the formation of ACID RAIN. *See also* AIR POLLUTION; PARTICULATES; AEROSOLS.

Smokey Bear a cartoon mascot of the U.S. Forest Service used to educate children (and adults) about the dangers of accidentally causing forest fires. Advertisements and signs depict Smokey asking that forest users break matches before discarding them, drown campfires, and generally be careful. The figure was devised in 1943 as part of a government educational program to counter the incendiary balloon bombs released by the Japanese forces that caused forest fires on the West Coast during the early years of World War II. The familiar slogan "Only You Can Prevent Forest Fires" was not introduced until after the war, via the broadcast voice of radio personality Jackson Weaver, who spoke the line into an empty wastepaper basket to achieve an authoritative, "bear-like" effect. Smokey became so popular that in 1952, after the development of Smokey toys and other commercial products, the U.S. Congress passed a law prohibiting the use of the likeness or the slogans by anyone except the Forest Service. The law was unenforced against merchandisers but was used to threaten environmental organizations that appropriated the Smokey cartoon for messages critical of Forest Service policies. In 1950, after a forest fire in southern New Mexico, a black bear cub found clinging to a charred tree was rescued by firefighters who, inevitably, nicknamed him Smokey Bear after the cartoon. The story caught the attention of national media and New Mexico officials shipped the bear cub to the National Zoo in Washington, D.C., as a "gift to the school children of America." The bear died in 1976 and was returned to New Mexico for burial. Despite his popularity with children, Smokey Bear has come in for a good deal of criticism from ecologists who point out that the slogan "Only YOU Can Prevent Forest Fires" gives the impression that careless visitors are a primary cause of fires, when in fact early Forest Service policies of fire suppression are more accurately the reason, since long-term suppression has allowed dangerous amounts of fuel wood to build up in forested areas, leading to disastrous "crown" fires that would ordinarily not be possible if occasional ground fires were allowed to clear forested areas of flammable debris. The Forest Service has since instituted PRESCRIBED BURNS but has not changed the slogan, and the bear's popularity goes on unabated. After more than half a century, Smokey Bear (not Smokey *the* Bear, and with an *e* in Smokey) still gets sackfuls of mail at a special address with a zip code shared by no one else: Smokey Bear Headquarters, Washington, D.C., 20252.

SMU *See* STREAMSIDE MANAGEMENT UNIT.

snag a dead tree that remains standing in place, especially one that has lost most or all of its leaves and/or branches. A "hard snag" is a snag whose outer surface remains sound; a "soft snag" is a snag whose outer surface has begun to decay; and a "whip" is a snag formed from a dead SAPLING. Snags are extremely important as wildlife HABITAT. Approximately 80 SPECIES of North American birds, and about half that number of mammals, utilize snags for shelter and as food sources (that is, they eat the insects which inhabit the snag), and POPULATIONS of these species can easily be driven into extinction by snag removal. For this reason, state and federal regulations require the leaving of all snags during logging operations except those that cause an imminent safety hazard to the loggers from breakage or those that will be a fire hazard during BROADCAST BURNS or other forms of PRESCRIBED BURN. A minimum of 2–10 snags per acre must be left, depending upon environmental conditions. If the required number cannot be left as natural snags, green LEAVE TREES ("green snags") may be designated that will be girdled (that is, have the bark removed in a complete circle around the trunk) following the logging operation to kill them and turn them into snags. A fresh snag created in this manner can be expected to last 30–50 years. Snags of all sizes, including whips, are useful to wildlife. However, some species (e.g., pileated woodpeckers) require large snags. Since these are also usable by those species that could use smaller snags, it is standard practice to choose relatively large trees (21 inches or more in DBH—*see* DIAMETER BREAST HEIGHT—at least 40 feet tall) as green snags.

snail darter *See under* TELLICO DAM.

sniff well *See* MONITORING WELL.

snow any form of PRECIPITATION that occurs as angular crystals of ICE. *Compare* RAIN; HAIL; SLEET. Snow crystals ("snowflakes") are generally hexagonal in form but may be square, irregular, or needlelike instead. They grow within cold clouds by a process of reverse SUBLIMATION, with water droplets passing from the gaseous state (water vapor) to the solid state (ice) as they come into contact with the growing flake. Usually a nucleus of some sort, such as a dust particle, is necessary to get the process going. When the flakes get heavy enough that turbulence within the cloud can no longer hold them up, they fall to earth as a snowstorm.

SNOW ON THE GROUND

Because of their angular nature, snowflakes do not lie tightly against each other when they first fall. Freshly fallen snow may contain as little as 10% water, the rest being air. Once on the ground, however, snow undergoes a complex series of processes that condense and "settle" it. Heat from the Sun above or the ground below can melt portions of the snow, allowing the resulting liquid water to run into some of the air spaces and refreeze. Movement of the snow surface by wind and other forces abrades it, breaking off the arms of the snowflakes and rounding them off so that they can lie closer together. Pressure on the lower layers of snow from the weight of the upper layers has a similar abrading and compacting effect. In addition, a significant rounding-off effect occurs through a process known as *destructive metamorphism,* resulting from the sublimation of water molecules from the tips of the snowflake's arms and their redeposit on the flake's central core. As a consequence of all these processes, the snow cover shrinks and firms up, becoming much more mechanically stable and increasing its water content dramatically, so that after two months' time the snow (now known as "old snow") may be as much as 80% to 90% water.

SNOW IN THE ENVIRONMENT

The most important function of snow, from an environmental standpoint, is its role as a water-storage device. In a typical temperate-zone climate of wet winters and dry summers, mountain snows that linger into the summer (the *snowpack*) provide meltwater that helps keep streams flowing and GROUNDWATER reservoirs recharged. It also serves as insulation during extremely cold weather, helping to keep plants, soils, and hibernating animals from being exposed to temperatures lower than 32°F (the bottom of the snow cover generally remains at this temperature year round.) Finally, snow's high ability to reflect incoming sunlight increases the Earth's ALBEDO for those areas in which it forms a relatively complete cover, a factor which can significantly alter weather patterns. *See also* SNOW SURVEY; WATERMELON SNOW; CORN SNOW; POWDER SNOW; FIRN; ICE; GLACIER; GRAUPEL.

snow survey a series of measurements taken of the snowpack (*see* SNOW: *snow in the environment*) to determine its depth and water content, with the purpose of predicting the amount of streamflow that will result when the snow melts. Snow surveys are usually made on a regular weekly or monthly basis throughout the cold months. For consistency, the measurements for a given region are always taken at the same series of spots (the *snow course*). Snow depths are reported as *equivalent depth,* that is the depth of the column of water that would result from the melting of a sample column of snow as tall as the measured snow depth.

SO in Forest Service terminology, abbreviation for Supervisor's Office, used to refer to the administrative headquarters of a National Forest.

Society for Conservation Biology professional organization committed to encouraging scientific study of the conservation and restoration of biological diversity. Members include resource managers, educators, government and private conservation workers, and students. The society's goals are to promote research; publish and distribute scientific, technical, and management materials; stimulate multidisciplinary communication and collaboration; educate the public; and recognize outstanding individuals and organizations in the field. It publishes the journal *Conservation Biology* and an online newsletter. Membership (1999): 5,700. Address: University of Washington, Box 351800, Seattle, WA 98195. Phone: (206) 616-4054. Website:http://conbio.rice.edu/scb.

Society of American Foresters (SAF) professional organization founded in 1900 to support scientific forest management in the United States. The SAF serves as the accrediting body for schools of forestry; publishes the *Journal of Forestry* and various other technical journals; and maintains numerous task forces ("working groups") on forestry-related issues. Membership (1999): 18,000. Address: 5400 Grosvenor Lane, Bethesda, MD 20814. Phone: (301) 897-8720. Website:www.safnet.org.

Society of Range Management professional society and conservation organization concerned with studying, conserving, managing, and sustaining rangeland resources. Founded in 1948, the society now has members in 48 countries, including many developing nations. The membership is made up of land managers, scientists, educators, and students. The society offers a professional accreditation program, sponsors college scholarships, and publishes the *Journal of Range Management* and *Rangelands*. Membership (1999): 4,000. Address: 1839 York Street, Denver, CO 80206. Phone: (303) 355-7070. Website:http://srm.org.

soft pesticide any PESTICIDE that breaks down readily in the environment, yielding harmless DECAY PRODUCTS.

softwood in forestry and logging, a CONIFER, that is, any tree belonging to the CLASS Gymnospermae (*see* GYMNOSPERM). Softwoods generally have softer wood than do HARDWOODS—hence the name—but this rule is not absolute. For example, the wood of alders (technically a hardwood) is considerably softer than the wood of hemlocks (technically a softwood).

SOHA short for spotted owl habitat area (*see* SPOTTED OWL MANAGEMENT AREA).

soil broadly, the REGOLITH, that is, any loose, inanimate material on the Earth's surface. More specifically, soil is usually defined as that part of the regolith that has been modified by weathering and by biological activity so that it is no longer either physically or chemically identical to its PARENT MATERIAL. An additional qualification added by agronomists, botanists, and foresters is that the soil must be capable of supporting plant life; hence these discipline usually consider "soil" to be identical to the SOLUM. Engineers, geologists, geographers, and land-use planners generally use the broader definition.

COMPOSITION OF SOIL

Soils vary immensely in the types of materials they are made of, due to the broad range of parent materials they may be derived from. However, nearly all soil materials may be placed into one or another of four broad categories. Three of these categories, making up 95% or more of the bulk of a typical soil, are mineral materials, classified by size. From the largest particle size to the smallest, these are SAND, SILT and CLAY (the size boundaries for these three categories have not been standardized; however, as a general rule, sand is more than 0.05mm in diameter, clay is less than 0.005mm in diameter, and silt is everything in between). The fourth category, HUMUS, is the organic component of the soil, and is composed of the remains and waste products of living ORGANISMS. It generally makes up 3%–5% of a soil, but may be much higher in wetland soils and scarce or nonexistent in desert soils.

STRUCTURE OF SOIL

The individual mineral grains and humus particles are rarely found separately in a soil; instead, they usually cling together into clumps of uniform texture and composition known as *aggregates* or *peds*. The shapes of the peds differ in a characteristic manner according to the mix of sand, silt, clay, humus, and moisture in the soil, and these characteristic shapes are referred to by soil scientists as the soil's *structure*. Four structural types are generally recognized, three of which are further subdivided into two or three subtypes each, making a total of nine structural categories. *Spheroidal soils* are those whose peds are small spheres. They include the relatively rare *single-grained soils* (in which the peds are individual mineral grains) as well as *granular soils* (nonporous spheres) and *crumbly soils* (porous spheres). *Blocklike soils* form rectangular cubes or blocks, and include the subcategories *blocky* (flat-faced, sharp-cornered blocks), *subangular blocky* (blocks with convex sides and rounded or obtruse-angled corners), and *massive* or *puddled* (large, irregular peds, or *clods*). *Platy soils* have peds that form thin, overlapping plates. Finally, *prismlike soils* form tall vertical columns. Two types are recognized: the *prismatic soils* (flat on top and with flat side faces) and *columnar soils* (rounded on top and often showing convex faces). Prismlike soils often break down into blocklike soils as the columns age and weather.

THE SOIL PROFILE

If allowed to develop long enough on one site, without disturbance, a soil will tend to differentiate from the top to the bottom into a series of horizontal bands of varying thicknesses known as *horizons*. The three principal horizons are labeled, from the top down, A, B, and C. The A horizon (also known as the *zone of elluviation*) is the portion of the soil from which minerals are leached and clay particles are removed by water trickling downward among the soil particles, while the B horizon (also known as the *zone of illuviation*) is the portion of the soil where the clay particles and leached minerals which have been removed from the A horizon are redeposited. The C horizon consists of unmodified or poorly modified parent materials. Above the A horizon there is often a thin layer of organic material—leaf litter, humus, and so on—which is known as the O horizon (or sometimes the A_0 horizon); below the C horizon lies the BEDROCK, sometimes referred to as the R (for *rock*) horizon. (Occasionally, a D horizon may be present in place of the C horizon, or between the C horizon and the R horizon. It consists of C-horizon type materials with a different mineral content from the horizons above it, and is generally a sign that the materials forming the solum have been transported into the area and overlain on top of a previous soil formed from local materials.) The boundaries between horizons are rarely abrupt; hence, each horizon is generally broken down further into subhorizons. Several different schemes exist for labeling these subhorizons. The most common uses subscripted numbers, as A_1, A_2, A_3; B_1, B_2, B_3; etc. In this scheme, the subscript 2 indicates the pure horizon, while 1 indicates a transitional zone to the horizon above, and 3 indicates a transitional zone to the horizon below. Another common scheme uses subscripted letters instead of numbers, with the letters indicating the mineral makeup of the subhorizon. A B_{ir} horizon is high in iron; a B_t is high in clay, etc. Transitional horizons in this scheme are often labeled with both letters, as AB, or A&B. The entire vertical structure of the soil from surface to bedrock, including all horizons and subhorizons (no matter how labeled), is known as the *soil profile*.

CLASSIFICATION OF SOILS

Numerous schemes have been developed for classifying soils into related categories. The simplest such scheme relies on the property known as *texture*, that is, on the relative proportions of sand, silt and clay found in a uniform sample of the soil. A mixture of 40% sand, 40% silt, and 20% clay is known as a LOAM. Mixtures at other proportions are given combined names (for example, increasing the proportion of sand in the mix grades a loam off to a *sandy loam*, a *loamy sand* and finally to a pure sand.) The textural classification scheme is usually represented as a triangle, with sand, silt, and clay at the three apeces and loam in the center. A more complex texture-related classification scheme, based on the soil's mechanical strength and compressibility, is used by civil engineers (*see* UNIFIED SOIL CLASSIFICATION SYSTEM). Agronomists, foresters, botanists, and others more concerned with the ability of a soil to grow plants than its ability to support structures use classification schemes based on soil profile as well as on texture. In the United States, the two most common systems are the "old" United States Department of Agriculture (USDA) scheme, formalized during the 1930s and 1940s and based on assumed genetic relationships between soil profiles (*see* GREAT SOIL GROUPS), and the "new" USDA scheme, published in 1960 and based on structural similarities rather than genetic criteria (*see* SEVENTH APPROXIMATION). *See also* PEDON; POLYPEDON; SOIL LOSS EQUATION.

Soil and Water Conservation Society professional organization promoting preservation, restoration, and wise management of soil, water, and related resources. The society, established in 1945 to support soil conservation programs of the USDA, now has chapters in the United States and Canada, conducts research and educational programs, and publishes books and an influential periodical, the *Journal of Soil and Water Conservation*. Membership (1999): 12,000. Address: 7515 NE Ankeny Road, Ankeny, IA 50021. Phone: (515) 289-2331. Website: www.swcs.org.

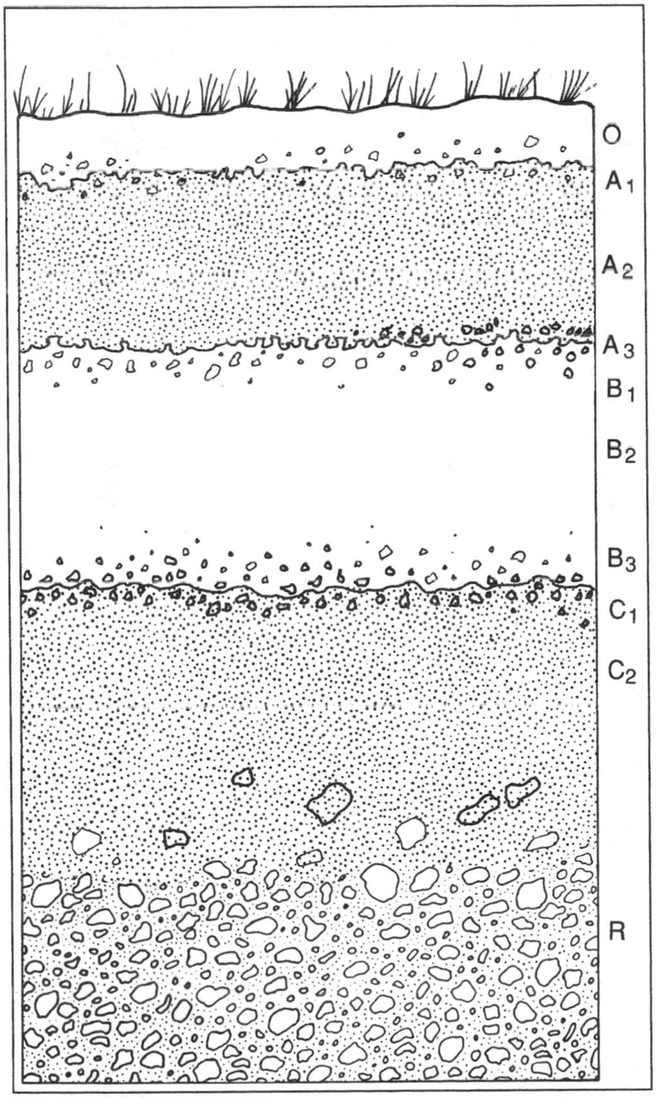

soil application of fertilizer, see under FERTILIZER.

soil conservation the adoption of various agricultural practices to reduce the wind and water EROSION of valuable soils. In general, topsoil in the United States is eroding 16 times faster that it can naturally form, a process that normally takes 500 years per inch. The loss of nutrients and organic matter can be made up for by the use of chemical fertilizers and animal waste, but not entirely. When all factors are taken into account, topsoil loss can reduce yields by 15 to 20%. Currently, topsoil loss stands at an annual rate of 1.9 million tons per year, down nearly 40% since 1982. This reduction was largely due to the increasing adoption of CONSERVATION TILLAGE and the removal of highly erodible cropland from cultivation under the CONSERVATION RESERVE PROGRAM. Soil conservation, although always perceived as necessary in farming practice, was not seen as a public issue requiring governmental attention in the United States until 1929, when Congress authorized the formation of soil conservation experiment stations which had been proposed by USDA soils scientist HUGH HAMMOND BENNETT. Beginning in 1933, under President Franklin Delano Roosevelt's New Deal, soil conservation projects were undertaken for the benefit of private farmland by the CIVILIAN CONSERVATION CORPS (CCC). The Soil Erosion Service was also established that year to provide farmers with equipment, seed, seedlings, and planning assistance for planting SHELTERBELTS and other purposes. In 1935 the Soil Conservation Service (now part of USDA's NATURAL RESOURCES CONSERVATION SERVICE) was established and was followed two years later with the Soil Conservation District program. Today, soil conservation programs of the U.S. government require only a fraction of the budget spent on them during the New Deal. In 1999 dollars, soil conservation on private land amounted to $5 billion in the 1930s. The budget is now $2.1 billion. Compared to soil losses in the DUST BOWL 1930s, soil erosion is much reduced, but soil conservationists are concerned that future reductions in erosion will be difficult to achieve with the slowing rate of adoption of conservation tillage techniques and the Conservation Reserve Program coming to a close.

The soil profile

Soil Conservation Service (SCS) agency within the U.S. Department of Agriculture (*see* AGRICULTURE, DEPARTMENT OF) charged with promoting agricultural practices that reduce or prevent soil EROSION. The agency also has responsibility for small-watershed development activities, including small FLOOD-control projects, POND construction, STREAM CHANNELIZATION and WETLANDS drainage, and is the lead federal agency for the control of NONPOINT POLLUTION from agricultural sources, especially increased SALT loads resulting from IRRIGATION. It administers no grant monies for these projects; however, it is the agency designated to provide technical expertise on soil and water conservation grants and small-watershed project grants administered by the Farmers Home Administration. In addition to these general activities, the SCS has been given several specific tasks, including leading the National Cooperative Soil Survey (which includes state and county soil agencies as well as the SCS); classifying and mapping farmlands in the United States; developing programs to reclaim abandoned COAL mines as agricultural land; running the federal SNOW SURVEY; and administering the Great Plains Conservation Program, a federal effort to stabilize agriculture in the Great Plains states through coordinated development and federal cost-sharing arrangements for conservation projects.

STRUCTURE AND FUNCTION

The SCS has no jurisdiction over any land; it operates instead through Soil Conservation Districts, which are established by farmers and ranchers on a cooperative basis under state and country regulations. The agency does not manage these districts but provides the technical expertise to assist in their management. The agency's own structure reflects the soil conservation district structure but it is not bound by it. Each state has a state director who supervises a number of field offices, called *area offices,* which in turn supervise work unit teams that are assigned to cover local areas ("work units") which usually correspond to the Soil Conservation Districts.

HISTORY

The Soil Conservation Service was established as the Soil Erosion Service of the Department of the Interior (*see* INTERIOR, DEPARTMENT OF THE) on August 25, 1933, under the terms of the National Industrial Recovery Act of 1933. In March 1935 it was moved to the Department of Agriculture, where it was renamed the Soil Conservation Service a month later (April 27, 1935) under the terms of the Soil Conservation Act of 1935. The first Soil Conservation Districts were established in 1936. There are now approximately 3,000 of these districts nationwide, covering more than 1 billion acres of farmland. Often criticized by environmentalists for its emphasis on structural means of flood control and its extensive efforts to drain and develop wetlands, the SCS nevertheless has been a major force for environmental protection in the 50 years of its existence, especially in terms of prevention of soil loss, retention of soil fertility, and reduction of SILTATION and other forms of agriculturally caused water pollution. Since 1977 it has been directed to consider scenic preservation and fish and wildlife HABITAT in the design of its projects, a pair of mandates that have contributed significantly to its care for the total environment. The SCS is now part of the NATURAL RESOURCES CONSERVATION SERVICE.

soil great groups *See* SEVENTH APPROXIMATION.

soil loss equation (universal soil loss equation) equation used by soil scientists to determine the amount of annual SOIL loss due to EROSION that can be expected under varying conditions and at varying locations. The equation is generally written

$$A=R\ K\ L\ S\ C\ P,$$

where R, the "rainfall factor," is the intensity of the average storm expected during the year multiplied by the total number of storms in an average year; K, the "soil factor," is a figure representing the erodibility of the particular soil being analyzed; L and S, the "slope factors," compare the downslope length (L) and gradient (S) of the plot of soil in question against a standard plot 22 meters long at a 9% gradient; C, the "crop factor," relates to the amount and type of vegetation present; and P, the "erosion-control factor," relates to agricultural practices, such as CONTOUR PLOWING, which may have been undertaken to slow down erosion. L and S are independent of location, and may be read (often together, as the "LS factor") from a universal table. The other factors are site-specific, and must be either experimentally determined or approximated from previous experimental work on closely related soils and under closely related conditions.

soil moisture tension (SMT) in soil science, a measure of the strength with which water is held in the soil by molecular binding and capillary action (*see* HYGROSCOPIC WATER; CAPILLARY WATER). It may be thought of as negative water pressure, in the sense that soil with a positive SMT will absorb water rather than draining it. Soils at FIELD CAPACITY have a low SMT. As the soil loses moisture to EVAPOTRANSPIRATION, the SMT increases. The same total amount of water in a soil will represent different SMTs depending upon the soil's composition and texture. Since water availability to plants depends on the SMT rather than on the total amount of water present, monitoring the moisture tension of a given soil plot is a far more efficient means of determining how much IRRIGATION water to use than is measuring its water content. Soil moisture tension is usually expressed in *bars,* where one bar = 100 kilopascals (100kPa). (One bar is equal to 1,000 millibars, the standard pressure unit of meteorology; *see* ATMOSPHERIC PRESSURE).

soil remnant in range management, an elevated "island" of soil protected by vegetation that has survived while the unprotected soil around it has been washed away by EROSION. Soil remnants are indicative of a badly deteriorating range. The original land surface can usually be traced from remnant to remnant, giving some clue as to how much erosion has taken place, although care must be taken to determine the actual surface of the soil remnant (which may be buried beneath erosion deposits from elsewhere on the slope). The soil of the remnants is usually TOPSOIL; the soil of the areas between is SUBSOIL, and is often DESERT PAVEMENT. The term "soil remnant" is also sometimes applied to the pedestals of soil that develop beneath pieces of LAG GRAVEL. This type of soil remnant is indicative of rapid, recent, and continuous erosion.

soil scarification disturbance of the SOIL or ground surface, especially when the disturbances are not of uniform dimensions and spacing. Soil scarification may be accidental, as when FELLING trees or YARDING logs in forestry, or it may be deliberate, as in the practice known as *chaining,* where a heavy chain dragged between two tractors churns up the soil surface and clears it of brush and small trees. Light scarification is often good for the soil, loosening and turning it over and helping to prepare it for planting. Heavy scarification, however, especially in the direction of soil drainage (that is, up and down a slope) can reduce productivity and drastically increase EROSION.

soil-site factors in forestry, the ENVIRONMENTAL FACTORS affecting tree growth, as opposed to *genetic factors* inherited from the tree's ancestral stock. Soil site factors include the effective depth of the SOIL (that is, the depth to which tree roots can penetrate, as determined by bedrock, the presence of CLAY subsoil [*hardpan*], availability of moisture, etc.), soil composition, SLOPE, ASPECT, CLIMATE (and MICROCLIMATE), and so on. *See also* EDAPHIC FACTORS; SITE-SPECIFIC FACTOR; SITE INDEX.

soil water water held in SOIL, SAND, or some similar loose material (*see* REGOLITH) at the earth's surface. The term generally refers only to water above the WATER TABLE; water below the water table is referred to as GROUNDWATER. *See also* GRAVITATIONAL WATER; HYGROSCOPIC WATER; CAPILLARY WATER.

sol in chemistry, any COLLOID in which the dispersed phase was originally a solid. The name "sol" by itself generally indicates a solid dispersed through a liquid (example: paint). Solids dispersed through other solids are called *solid sols* (example: sapphire), while solids dispersed through gases are called AEROSOLS. *Compare* GEL; EMULSION.

solar cell (photovoltaic cell) See SOLAR ENERGY.

solar collector *See* SOLAR ENERGY.

solar constant in meteorology, the amount of SOLAR ENERGY that regularly strikes the Earth's outer ATMOSPHERE. Measurements of the solar constant must be corrected both for variations in the angle at which the sun's rays strike the atmosphere and the annual variations in the distance between the Earth and the sun caused by eccentricities in the Earth's orbit. Technically, these corrections are made by defining the constant for a plane perpendicular to the Sun's rays located at the average (mean) distance from the Earth to the Sun. Corrected in this manner, the solar constant is equivalent to 1.94 gram-calories per square centimeter per minute, or about 380 million BRITISH THERMAL UNITS per square meter per year.

solar energy energy obtained from the Sun. The term is usually defined further to mean energy for human use derived directly from sunlight, as opposed to obtaining it through mediators such as the HYDROLOGIC CYCLE (HYDROELECTRIC POWER), plant growth (wood heat), differential atmospheric heating and cooling (power from winds and storms), and so on. The energy available directly from sunlight is immense. The earth as a whole receives roughly 2 calories of energy per square centimeter or surface each minute of daylight (*see* SOLAR CONSTANT), and even though much of that is reflected or diffused, the average *solar flux* (energy from the Sun) received on a square meter of land in the temperate latitudes is about 1,700 kilowatt-hours per year. The problem thus boils down to capturing this energy for human use. This is accomplished by solar collectors, of which there are four basic types. *Flat-plate collectors* consist of a flat surface, usually painted black to increase efficient heat absorption and radiation, covered by a sheet of glass mounted parallel to the flat surface and a few centimeters away. A fluid, usually water, is pumped through the space between the flat surface and the glass to absorb the heat and transfer it to the areas where it is needed. Flat-plate collectors are used in ACTIVE SOLAR DESIGNS. *Passive collectors,* used in PASSIVE SOLAR DESIGNS, consist primarily of expanses of glass windows oriented to face the Sun. Short-wave solar energy enters through the glass; is absorbed by surfaces within the room; and is reradiated by these surfaces as long-wave energy, which cannot pass back out through the glass and thus accumulates in the room's air space. The collector may be made more efficient by the addition of a *heat sink*, a dark-colored masonry wall or other structure with high SPECIFIC HEAT, which can absorb a considerable amount of energy and reradiate it slowly into the room (*see*, e.g., TROMBE WALL). A third type of collector, the *photovoltaic cell*, consists of a wafer of silica mounted directly beneath a glass plate. Solar radiation falling on the silica knocks electrons loose from some of its molecules, generating a slight electric current. A circular photovoltaic cell 8 centimeters in diameter (a little over 3 inches) generates roughly half a watt of electricity in direct sunlight. Photovoltaic cells are generally mounted in *banks* (arrays). A bank of 6,000 8-centimeter cells will supply enough electricity for the average home. Finally, *solar concentrators* consist of parabolic mirrors, which focus the rays falling on them onto a single point. The extremely high temperatures that result (as much as 3,000°C in large concentrators) can be used directly for industrial processes or to convert water to steam for electrical generation. Arrays of *tracking concentrators* geared to follow the Sun can be used to focus the sunlight on a central tower, where it can be used either for photovoltaic generation or for the creation of steam.

DESIGNING SOLAR INSTALLATIONS

The actual square footage required for a solar collector varies drastically with the type of collector and with the CLIMATE and LATITUDE of the site. Certain installation factors, however, are universal. The collector should lean north from the vertical (in the northern hemisphere) at an angle equal to the latitude, and should ideally face slightly west of due south, as the afternoon rays of the Sun spend less of their energy warming the atmosphere and thus retain slightly more of it to hit the collector. The designer must also take into account *solar access*—the amount of sunlight falling on the site unblocked by trees, hills, other buildings, etc. A concept that has proved particularly useful in determining solar access is the *solar envelope*, which may be defined as the space occupied by the shadow of a building, tree, or other object at any given time. For legal purposes, the time used is generally noon on the winter solstice (December 21). The height of a building's solar envelope above the ground at any point depends on the height and position of the Sun, the height and shape of the building, the slope of the ground, and the horizontal distance from the measuring point to the building wall.

USE OF SOLAR ENERGY

The construction of both active and passive solar collectors for new buildings is highly cost-effective. They can supply 30–40% of a home's space-heating needs, and virtually all of its water-heating needs, for a cost over and above normal construction amounting to about 7% of the price of the home. The typical payout time for balancing this additional cost through fuel savings is less than two years. Nevertheless, solar energy remains poorly utilized, partly due to the considerably higher costs involved in retrofitting an existing structure, but mostly because of misunderstandings concerning its efficiency and a de-emphasis of alternative energy research and assistance by the Reagan and Bush administrations during the 1980s. A study conducted in 1980, later quashed, indicated that solar energy would provide 20% of U.S. energy needs by 2000 if the then-current trends in the adoption of solar energy technologies were to continue. As it turned out, because of the withdrawal of government support and cheaper oil prices, only 5% of energy needs were met by solar energy in 2000. Nevertheless, technical innovations have continued, reducing the price of converting solar energy to electricity by half since 1990, which makes solar economical in some areas of the country. The cost of energy generated by photovoltaic cells remains high, however, although the National Renewable Energy Laboratory predicts that by 2020 the cost of PV cells will be only one-eighth its price in 2000. Given the technological advances, solar energy is, in the view of some experts, on its way back as a viable energy source.

solid waste technically, any solid material disposed of as no longer useful. In common usage the term has the somewhat more limited meaning of solid materials thrown out by householders—those materials referred to by engineers as *municipal solid waste* (MSW). Americans generate roughly 150 million tons of MSW each year, or about 3.4 pounds per person per day—approximately twice as much as the worldwide average among industrial nations. MSW is bulky and aesthetically repugnant, but it is nor normally particularly hazardous (although there are exceptions, such as discarded paint cans, medicine bottles, and disposable diapers). The underlying difficulty in disposing of these materials is thus the so-called NIMBY ("Not In My Back Yard") problem. Mayor Louie Welch of Houston, Texas, summed it up nicely in 1970: "Everybody wants us to pick up his garbage, but nobody wants us to put it down."

COMPOSITION OF SOLID WASTE

MSW consists of *garbage* (food scraps and other organic debris) and *rubbish* (everything that isn't garbage.) About 40% of the MSW from an average American city is paper and paper products; another 17% is yard wastes (grass clippings, dead leaves, etc.). Glass, metal, and food wastes are 8% to 10% each;

plastics average about 7%. Rubber, leather, textiles, wood, and miscellaneous wastes make up the remaining 10%.

METHODS OF SOLID WASTE DISPOSAL

There are currently four principal strategies in use to dispose of municipal solid waste. (A fifth method, dumping at sea, was prohibited by the Marine Protection, Research and Sanctuaries Act of 1972.) *Landfilling*, the most common option, consists of simply placing the material in a trash dump, or *tip*, and covering it with earth (*see* SANITARY LANDFILL). *Incineration* involves placing waste materials in a high-temperature incinerator, usually after separation of metals from the waste stream. This method achieves a 75%–85% reduction in the volume of the wastes, but produces stack emissions that include DIOXINS, FURANS, SULFUR OXIDES, FLY ASH, and other major contributors to air pollution. The ash that remains behind following incineration also contains concentrated volumes of dioxins and other hazardous materials. It is generally disposed of by landfilling. *RECYCLING* avoids the hazards of disposal by reusing the waste materials, but requires extra effort to separate the recyclable materials from the nonrecyclables and from each other, either in the home (*source separation*) or at the recycling facility. Recycling can only be used for those materials that are recyclable—currently, some 40% to 50% of the waste stream. The fourth method, *source reduction,* is actually not a means of disposal but a means of reducing the volume originally thrown out. It involves such strategies as increasing the durability of goods, reducing the size of packages, and so on, and is in little use in the United States. According to an EPA study, of the 217 million pounds of municipal solid waste generated in 1997, 28% was recovered. The remainder (156 million pounds) was landfilled. *See also* RECYCLING.

Solid Waste Disposal Act federal legislation to address the problems arising from the proliferation of improperly maintained waste dumps, signed into law by President Lyndon Johnson on October 20, 1965. The principal thrust of the act was to create a system of grants to municipalities for the creation and maintenance of adequate SOLID WASTE disposal facilities, including RECYCLING centers; however, it also mandated a national study by the Department of Health, Education and Welfare (*see* HEALTH AND HUMAN SERVICES, DEPARTMENT OF) concerning the locations of, and dangers posed by, the nation's HAZARDOUS WASTE dumps; banned open dumping; and directed the secretary of health, education, and welfare to authorize a study leading to the creation of federal guidelines for the disposal of hazardous waste. *See also* RESOURCE CONSERVATION AND RECOVERY ACT.

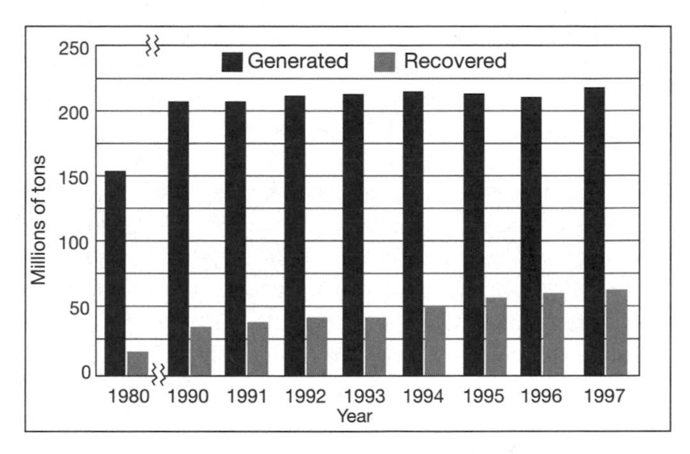

U.S. municipal solid waste—generation and recovery, 1980–1997 (millions of tons) ***(Characterization of Municipal Solid Waste in the United States: 1998. Franklin Associates, Ltd.)***

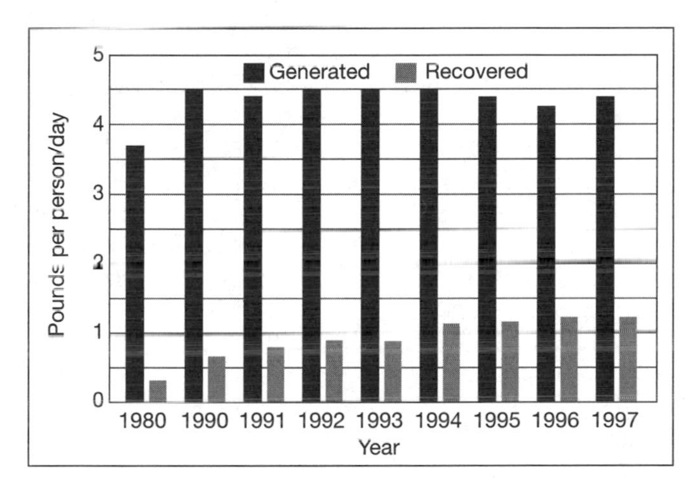

U.S. municipal solid waste—generation and recovery, 1980–1997 (pounds per person per day) (Characterization of Municipal Solid Waste in the United States: 1998. ***Franklin Associates, Ltd.)***

solifluction in the earth sciences, a slow downhill flow of water-saturated SOIL under the influence of gravity. Solifluction can take place in any environment where soil depth is limited and there is an adequate supply of water, but it is most common in regions of PERMAFROST due to the nearly unlimited water available to the soil as the winter frosts melt down to the permafrost layer each spring, coupled with the lubricating effect of the upper, semiliquid fringe of the permafrost itself. Its principal effect on the human environment is to make it difficult or impossible to locate solid foundations for structures such as buildings, highways, and pipelines.

solitude in Forest Service terminology, the state of being alone in a natural setting, where one may commune with one's surroundings undistracted by technology or by other humans. "Opportunities for solitude" defined in this rather restrictive manner are a major criterion used by the Forest Service in determining portions of the NATIONAL FORESTS to recommend for designation as WILDERNESS AREAS.

solonchak soil an INTRAZONAL SOIL that develops on poorly drained sites in arid or semiarid climates. The typical vegetation is SHRUB-STEPPE. Solonchak soils have thin, light-colored A HORIZONS (*see* SOIL: *the soil profile*), usually with SALT deposits on their surfaces. B horizons are lacking. The SUBSOIL is also light-colored. The soil as a whole contains very little ORGANIC MATTER. The term is now generally considered obsolete (*see* ARIDISOL) *Compare* SOLONETZ SOIL.

solonetz soil an INTRAZONAL soil that develops on moderately to well-drained sites in arid or semiarid climates. The typical vegetation is SHRUB-STEPPE. Solonetz soils superficially resemble SOLONCHAK SOILS, but the surface SALT deposits are lacking, and excavation will reveal a well-developed B HORIZON (*see* SOIL: *the soil profile*) with a dark color and a blocky or columnar texture. The lower portion of the B horizon usually contains concentrations of calcium and sodium salts, often cemented together; soil reaction as a whole is highly alkaline. The term is now generally considered obsolete (*see* ARIDISOL).

sols bruns acides a ZONAL SOIL that develops under conifer forests (*see* CONIFEROUS FOREST BIOME) in cool, moist climates. The A HORIZON (*see* SOIL: *the soil profile*) is thick and dark, with moderate to high organic content; the B horizon is also thick, with little or no CLAY present. Acidity is strong, and increases with depth. The soil is generally covered by a relatively thick layer of HUMUS. The term is now generally considered obsolete.

solubility in chemistry, the amount of one substance (the SOLUTE) that is able to dissolve in a second substance (the SOLVENT) under a specified set of conditions of temperature and pressure. Solubility is usually expressed as the number of GRAMS of solute that will dissolve in 100 grams of solvent. As a rough guide, the terms *strongly soluble, moderately soluble, slightly soluble,* and *insoluble* are used to characterize a substance's ability to dissolve. The solubility of gases generally decreases with increasing temperature; the solubility of liquids and solids generally increases with increasing temperature. The solubility of all substances increases with increasing pressure. Differences in solubility among different COMPOUNDS are among the most important of all chemical properties from an environmental standpoint, helping to govern processes ranging from differential EROSION to the transport of NUTRIENTS through the walls of CELLS in living tissue. *See also* SOLUTION.

solum in soil science, the upper portion of a soil, consisting of the O, A, and B HORIZONS (*see* SOIL: *the soil profile*). The solum is the portion of the soil that supports plant life; hence, agronomists and other agricultural scientists often use the terms *solum* and *soil* interchangeably.

solute in chemistry, a substance that has become dissolved in another substance (the SOLVENT). *See* SOLUTION; SOLUBILITY.

solution in chemistry, a special type of MIXTURE in which the mixing takes place on a molecular or submolecular level, that is, one in which the MOLECULES or IONS of one substance (the SOLUTE) are distributed evenly among the molecules of a second substance (the SOLVENT). Solutions are homogeneous (the ratio of solute to solvent remains the same throughout the solution) and stable (the solute will not settle out, no matter how long the solution sits, and it cannot be removed by a filter or a centrifuge). Nevertheless, solutions are not COMPOUNDS; the ratio of solute to solvent is not fixed, but can be continuously varied by adding more solute or more solvent to the solution, as long as the SOLUBILITY of the solute is not exceeded. A solute goes into solution in (is *dissolved by*) a solvent though a process called *solvation,* in which the individual molecules or ions of the solute are literally pulled off of the mass of the solute and surrounded by a swarm of molecules of the solvent, which are held against it by electrostatic charges. Solutions can occur between substances in any state (solid, liquid, or gas; some ALLOYS, for example, are solutions of a solid within another solid). However, the most common solutions are those in which the solvent is a liquid. Solutions in which the solvent is WATER, called *aqueous solutions,* are extremely important in natural processes (*see,* e.g., DISSOLVED OXYGEN; EROSION; NUTRIENT).

solvent in chemistry, a substance capable of dissolving another (*see* SOLUBILITY; SOLUTION). When standing alone the term usually refers to a substance, such as WATER or alcohol, which can dissolve a wide variety of other substances.

SOMU short for spotted owl management unit (*see* SPOTTED OWL MANAGEMENT AREA).

sonic boom the shock wave created by an aircraft flying faster than the speed of sound (roughly 760 miles per hour). Any object moving through the air compresses the air in front of it, creating a pressure wave that propagates forward at the speed of sound. This pressure wave sets the MOLECULES of the air moving in advance of the arrival of the object, aligning them so they will flow smoothly around it. If the object is moving faster than the speed of sound, however, it outruns the pressure wave it creates and is therefore constantly running into still air molecules, which must be abruptly disturbed to make way for the passage of the object. This abrupt disturbance is propagated outwards from the advancing object as a shock wave equivalent to that caused by a large explosion. The creation of sonic booms is constant as long as an aircraft is exceeding the speed of sound (not only at the moment that it passes through the "sound barrier"). It can be reduced somewhat by proper aircraft design, but it cannot be eliminated. Its effects are exactly the same as those of any other shock wave, including property damage, physiological damage to living ORGANISMS (for example, to human and animal ears), and psychological damage caused by the abrupt triggering of the "fight or flee" response. For these reasons, aircraft are generally forbidden by law to fly faster than the speed of sound when crossing populated areas.

sonoran zone *See under* LIFE ZONE.

sorbent any substance that will either absorb or adsorb another substance of group of substances (*see* ADSORPTION; ABSORPTION). Sorbents are widely used in pollution control, both on a routine basis to remove foreign substances (particularly ORGANIC COMPOUNDS) from drinking water and on an emergency basis to clean up after oil and chemical spills (*see,* e.g., ACTIVATED CARBON; IMBIBER BEADS).

South Coast Air Quality Management District (AQMD) the smog control agency for all or portions of Los Angeles, Orange, Riverside, and San Bernardino counties in California, where nearly 15 million people (about one-half California's population) live with some of the dirtiest air in the country. The agency is required to achieve and maintain healthful air quality through planning, regulation, monitoring, enforcement, improved technology, and public education. The poor air quality in the Los Angeles region is caused by a combination of 9 million polluting automobiles, industrial emissions, topography that impedes air movement, thermal INVERSION layers, and sunshine and warm temperatures that convert oxides of nitrogen to harmful tropospheric ozone. AQMD has the authority to control every air pollution source from large power plants and smoking vehicles to house paint and charcoal lighter fluid. Its $97 million annual budget comes from permit fees from polluting businesses and a surcharge added to vehicle registration fees. Overall, the AQMD has been seen as instrumental in the decrease of ozone levels, now less than half of what they were in the 1950s, plus a marked reduction of smog alerts. So-called Stage I alerts now occur less than half as often as they did 10 years ago, and the more severe Stage II alerts, which occurred 15 times a year on average in the 1980s, now seldom occur at all. Even so, the area covered by the AQMD must still reduce pollution drastically to meet federal air quality standards. Address: 21865 East Copley Drive, Diamond Bar, CA 91765. Phone: (909) 396-2000. Website: www.aqmd.gov.

Southeastern Power Administration *See* POWER ADMINISTRATIONS, FEDERAL.

Southwestern Power Administration *See* POWER ADMINISTRATIONS, FEDERAL.

SOx *See* SULFUR OXIDES.

spar tree in logging, a tree that supports the pulleys over which the cables of a high lead or SKYLINE LEAD yarding system (*see* YARDING; HIGH-LEAD LOGGING) are run. Spar trees are prepared by topping and limbing them (that is, removing the tops and limbs) to prevent the cables from snagging, and are usually stabilized through the use of support cables ("guy cables") running from the ground to a point midway or more up the tree. If a suitable tree does not exist near the LANDING to use as a spar tree, a *spar pole* may be used instead. This is a metal or wooden pole that is set upright beside the landing and held in place by guy cables. Spar poles have the advantage of portability, but are less stable—therefore, more dangerous—than spar trees.

spawn a large mass of eggs released during a single ovulation cycle. The borderline between a spawn and a "clutch" (a small group of eggs) is indistinct but is in the neighborhood of 30–50. (Most ORGANISMS that reproduce in clutches produce far fewer eggs than this, while most organisms that reproduce with spawns produce far more.) The term is used most commonly for the egg mass produced by fish, although most amphibians, some reptiles, and all insects and arachnids also produce spawns. *See also* SPAWNING BED; SPAWNING GRAVEL; SPAWNING RUN; SPAWNING STREAM; R-SELECTED SPECIES.

spawning bed the location where SPAWN is deposited, especially by aquatic ORGANISMS such as fish or sea turtles that tend to congregate and spawn in groups. Most SPECIES have a set of fairly restrictive requirements that must be met by the beds they spawn in. GRAVEL, SAND, or mud in which the eggs can be hidden is generally a prerequisite. Some species choose spawning beds only in running water; others, only in still water. Nearshore waters, offshore waters, or beaches may be used by various species. The maintenance of spawning beds of the proper type to be used by a given species is one of the most important regulators on the population size of that species; hence, activities that destroy spawning beds, such as DAM building, dredging (*see* DREDGE), or the fouling of bottom SEDIMENTS with POLLUTANTS, are among the most likely causes for endangerment or extinction of aquatic species (*see* ENDANGERED SPECIES; EXTINCT SPECIES). *See also* SPAWNING GRAVEL; SPAWNING RUN; SPAWNING STREAM.

spawning gravel a GRAVEL BAR used as a SPAWNING BED by aquatic animals, especially freshwater or anadramous fish. Though most fish excavate a small depression (known as a *redd*) in the spawning gravel in which to deposit their eggs, the advantage of the gravel is that such a depression is not really needed, the eggs being well protected by slipping into the spaces between the gravel pieces. For this reason, spawning gravels must be clean. Hence, any activity that increases SILTATION in a stream runs the risk of destroying the spawning gravels in the stream, thus depriving those SPECIES dependent on them of places to spawn and rendering them locally extinct.

spawning run a mass migration of aquatic ORGANISMS to a location where they will SPAWN together. Such runs may take place from offshore waters to nearshore waters, to favored beaches, or from large bodies of water up their tributary streams. A few SPECIES (e.g., the Atlantic eel) make spawning runs from freshwater or onshore shallows to deep water. *See also* ANADRAMOUS SPECIES; CATADRAMOUS SPECIES; SPAWNING BED; SPAWNING GRAVEL; SPAWNING STREAM.

spawning stream a stream used by an ANADRAMOUS SPECIES (e.g., salmon) as a location for spawning (*see* SPAWN). Most species have extremely strict requirements of temperature and transparency that must be met by their spawning streams. Warming the water by a few degrees (by cutting off the tree cover that would normally shade it; releasing waters warmed by industrial uses such as power-plant cooling; etc.) or increasing its TURBIDITY (by increasing SILT runoff; releasing NUTRIENTS into the water, which promote the growth of ALGAE and other MICROORGANISMS; etc.) can have as detrimental an effect on spawning success as would outright destruction of the SPAWNING BEDS. Many species, especially of salmon, return to the stream in which they were spawned to lay their own eggs, guided at least partially by the water chemistry; hence, changes in that chemistry can confuse the fish, and can lead to the same results as increasing heat or turbidity, even if the temperature and transparency of the water remain within limits normally tolerated by the species. *See also* SPAWNING GRAVEL.

SPDES *See* STATE POLLUTANT DISCHARGE ELIMINATION SYSTEM.

special component in Forest Service terminology, historically, COMMERCIAL FOREST LAND on which timber harvest is restricted due to conflict with other uses, such as recreation, WATERSHED protection, etc. Some harvest is allowed on special-component lands, but harvest plans must be carefully designed to avoid conflicts with the predominant use of the land. Therefore, the harvest volume is usually reduced considerably. The term was largely rendered obsolete by the NATIONAL FOREST MANAGEMENT ACT OF 1976. *Compare* STANDARD COMPONENT; MARGINAL COMPONENT; RESERVED COMPONENT.

special interest area in Forest Service terminology, an area of NATIONAL FOREST land that has been set aside by administrative action for recreational purposes. Special interest areas include scenic areas, BOTANICAL AREAS, ARCHAEOLOGICAL AREAS, GEOLOGICAL AREAS, and so forth. They do not include WILDERNESS AREAS (which are designated by Congress rather than by the Forest Service administration) or RESEARCH NATURAL AREAS (which are set aside under a separate set of regulations and often have state as well as federal laws governing them). They are designated at the regional level; that is, the local forest makes recommendations, but the actual setting aside of the land is done by the REGIONAL FORESTER. OFF-ROAD VEHICLE use is generally allowed in special interest areas (though it may be curtailed or eliminated if it damages the resource for which the area was set aside). Mining and timber harvest are allowed but restricted so that they do not interfere with the primary goal of recreation. Visual management is for RETENTION OR CHARACTER.

special interest group any group of individuals with a common interest in promoting or opposing a particular management activity, construction project, piece of legislation, etc. Special interest groups need not be formally organized. For example, hunters are considered a special interest group for all matters concerned with wildlife management, whether or not they belong to an active hunting organization. *See also* IRON TRIANGLE.

special use permit a permit issued by the FOREST SERVICE for the use of a designated portion of a NATIONAL FOREST for private purposes. The purpose may be either commercial (a ski area, a dude ranch, etc.) or noncommercial (a residence or vacation home, a water line, etc.). The land the permit applies to is known as the *permit area.* There are two types of permits: the *annual permit,* which must be renewed each year, and the *term permit,* which is for a length of time longer than a year. (The basic authority is for 30 years. However, flexibility in the law allows it to be as much as 50 or as little as two, and in practice most term permits run for 20 years, with ski areas usually doubling that.) Both term and annual permits may be used within a single permit area. For example, a ski area on National Forest land often has its permanent facilities (lodge and ski lifts) on land under term permit, while its ski runs are under annual permit. Permittees usually pay an annual fee based on a complicated formula involving the type of permit (annual or term), the area of land involved, and the relationship between the permittee's capital investment in the area and his gross proceeds from operating it. Annual permits are generally renewed automatically upon payment of the fee. Renewal of term permits requires review of the permit conditions but is usually also nearly automatic. Both types of permit may be revoked if the site is abandoned or if the permittee's use of it can be shown to be detrimental to the public interest.

species in the biological sciences, a group of living ORGANISMS that are physically capable of interbreeding with each other to produce fertile offspring, and which do not interbreed with (are *reproductively isolated from*) any other organisms. Individuals that are isolated from each other solely by distance and would reproduce if brought together are considered to be members of the same species; however, individuals that are physically capable of inter-breeding with each other but do not due to differences in HABITAT, breeding cycles (*see* ESTRUS; RUT), etc., are considered to belong to different species. *See* ALLOPATRIC SPECIES; SYMPATRIC SPECIES; SUBSPECIES. *See also* TAXONOMY.

species diversity *See* BIODIVERSITY.

species reintroduction *See* REINTRODUCTION OF SPECIES.

species-specific factor in the biological sciences, something that affects the distribution or survival rate of a particular SPECIES but does not affect other species that may be associated with it in the same HABITAT. Species-specific factors are generally genetically inherited traits such as tolerance for soil acidity, water requirements, NUTRIENT requirements, ability to withstand low temperatures, and so forth. *Compare* SITE-SPECIFIC FACTOR.

specific gravity the ratio of the density of a substance (that is, its mass per unit volume) to the density of a standard control substance. If the control substance is not specified, it is assumed to be either water at 4°C (for determining the specific gravity of liquids or solids) or air at 20°C (for determining the specific gravity of gases). The term is slowly being replaced in scientific literature by the equivalent term *relative density.*

specific heat the amount of heat necessary to change the temperature of a substance, that is, the amount of heat absorbed by the substance for each incremental rise in temperature. Specific heat varies from substance to substance, often dramatically. Water, for example, has a specific heat nearly three times as great as soil and rock, a difference that has a major effect on weather and climate (*see,* e.g., LAND BREEZE; SEA BREEZE; LAKE-EFFECT SNOW). The specific heat of a given substance is normally expressed as the amount of heat (in calories) necessary to raise one gram of the substance one degree Celsius. It may also be expressed in joules per thousand kilograms (J/Kkg).

spermatophyte any seed producing plant, including all ANGIOSPERMS (flowering plants) and GYMNOSPERMS (conifers and their associates), but not the ferns, mosses, liverworts, or hornworts. The term is not always recognized as a formal taxonomic classification (*see* TAXONOMY). *Compare* BRYOPHYTE; PTERIDOPHYTE.

spit in geology, a long, narrow ridge of SAND or other waterborne SEDIMENTS that extends into a body of water from its shoreline. Spits are formed where currents are deflected by irregularities in the shoreline. Deflection slows the current down, which reduces its ability to carry suspended materials, causing some of them to be deposited. Where currents usually flow in the same pattern (a river; a lake or ocean with predominant LONG-SHORE CURRENTS in one direction), the spit will develop in the direction of current flow and will generally be deflected back toward the shore at its outward end, forming a *hooked spit.* Where currents flow in opposing directions (for example, where two rivers meet), the spit will develop along the line of interaction between them. It will not develop a hook, and will be likely to extend straight out from the shoreline.

splitter in the biological sciences, one who tends to classify ORGANISMS with similar but slightly differing characteristics into difference SPECIES rather than refer to them as different RACES of the same species. *Compare* LUMPER.

spodosol any of a group of closely related soil types that develop from sandy PARENT MATERIALS under conditions of moderate to heavy rainfall. Spodosols generally have thin, light-colored A HORIZONS (*see* SOIL: *the soil profile*) with B horizons that contain a poorly mixed conglomeration of mineral and ORGANIC MATTER. There is an accumulation of aluminum and (usually) iron minerals, known as a *spodic layer,* low in the B horizon. These soils are strongly acidic and are low in fertility, making them very poor agricultural soils, although they often support rich forest communities. They commonly develop in coastal areas and in regions of relatively recent glacial activity, such as those areas of Canada and the northern United States that were covered by the CONTINENTAL GLACIER.

spoil the waste material from an excavation, such as that produced by a mine, a dredging operation, a building site, a road cut, etc. Spoil consists of SOIL, ROCK and SEDIMENTS that have no economic value at the time the excavation occurs but must be removed in the process of developing the excavated site for some economic use. Disposal of this material can be a serious environmental problem, both because of its sheer bulk and because it is often harmful in character. Mine spoil is generally acidic and high in MINERALS, many of which are toxic to plant life. It consists almost entirely of rock fragments, and therefore has low to nonexistent fertility. DREDGE spoil often contains a high proportion of POLLUTANTS that have become sorbed onto the sediment particles (*see* ADSORPTION; ABSORPTION; IN-PLACE POLLUTANTS). Spoil from construction projects is generally less hazardous than mine or dredge spoil, and can often be used for fill in other parts of the project or in nearby projects. *See also* TAILINGS; CONFINED DISPOSAL FACILITY.

sporophyte in the biological sciences, the diploid generation of an ORGANISM showing ALTERNATION OF GENERATIONS, that is, the generation whose CHROMOSOMES occur in matched (*homologous*) pairs in each CELL. Sporophytes produce spores rather than GAMETES. In the ANGIOSPERMS and GYMNOSPERMS, the sporophyte is the dominant generation, with the tiny GAMETOPHYTE parasitic upon it and only locatable by microscopic examination of the reproductive organs of the sporophyte.

Sport Fisheries and Wildlife, Bureau of *See* BUREAU OF SPORT FISHERIES AND WILDLIFE.

spotted owl management area ([SOMA] spotted owl habitat area [SOHA]) an area of OLD GROWTH forest on which timber harvest and road building have been banned in order to protect the HABITAT of the northern spotted owl (*Strix occidentalis*) and other old-growth dependent species. The spotted owl is a relatively large bird, ranging up to half a meter (19.5 inches) in length, with a wingspread that reaches up to 1.15 meters (45 inches). It inhabits dense growths of mature timber (*see* CULMINATION OF MEAN ANNUAL INCREMENT), where it feeds principally on rodents, supplementing its diet occasionally with small birds, reptiles, amphibians, and (rarely) large insects. Occasional sitings have been made in older SECOND GROWTH STANDS that have developed far enough to take on some of the characteristics of old growth. The owl is classified as rare or threatened over most of its range (*see* RARE SPECIES; THREATENED SPECIES), which includes the western slopes of the Sierra and Cascade mountain chains in California, Oregon, Washington, and British Columbia, and parts of the southern Rocky Mountains (the Rocky Mountain population is sometimes classed as a separate SUBSPECIES). It is considered an INDICATOR SPECIES for old growth; that is, where the owl exists, enough old growth probably remains to support most other old-growth dependent species. Guidelines for management of spotted owl habitat have been the subject of considerable controversy. Draft federal regulations call for the establishment of SOMAs of up to 2,200 acres of suitable habitat per nesting pair of owls: wherever possible, this acreage is to be drawn from special, marginal, or reserved lands (*see* SPECIAL COMPONENT; MARGINAL COMPONENT; RESERVED COMPONENT). SOMAs are to be drawn around all verified nesting sites, and in any case are to be established no more than 6 miles apart, with clusters of three being located within 12 miles of each other, wherever enough old growth remains to meet these requirements. Up to 1,000 acres of each SOMA will be withdrawn from the TIMBER BASE. The remainder will be included within calculations of the ALLOWABLE CUT, but will not be entered for at least 10 years following establishment of the SOMA. Court challenges to these regulations were filed by conservationists who felt they were not strict enough to prevent the bird's extinction. These challenges led to drastic reductions in the scheduled harvest on National Forest and Bureau of Land Management lands in the Pacific Northwest, creating a strong public outcry by supporters of the timber industry and leading to threats by Northwestern congressmen to introduce legislation repealing parts of the ENDANGERED SPECIES ACT, the FEDERAL LAND POLICY AND MANAGEMENT ACT, the NATIONAL FOREST MANAGEMENT ACT, and the MULTIPLE USE-SUSTAINED YIELD ACT. After some years of rancorous debate, the Clinton administration developed a forest plan in 1993 identifying habitat areas that could not be logged and areas that could be logged but only under special circumstances. Overall, the allowable cut in the restricted areas was reduced from 4.5 billion board feet to 1.2 billion board feet. To offset the economic impact of the plan, a $1.2 billion fund was set up for the retraining of displaced forest workers. In 1994 the plan was approved by the federal district court in that it did not violate the Endangered Species Act.

sprawl *See* URBAN SPRAWL.

spring in hydrology, a location where water flows out of the earth. Springs result from the intersection of the lower portion of an AQUIFER with the ground surface, allowing the water within the aquifer to flow out by gravity. The amount of flow depends both on the PERMEABILITY of the aquifer and on the amount of HEAD developed within the aquifer (that is, how far the highest portion of the WATER TABLE within the aquifer lies above the spring). The quality of the water issuing from a spring depends on the quality of the water within the aquifer. If the aquifer is polluted, the spring will be polluted; if the aquifer is developed from rocks or soils containing a high proportion of soluble minerals (*see* SOLUTION; MINERAL), the water will be mineralized. Springs supply most of the BASEFLOW of streams and often exist beneath the surface of lakes, where they help contribute to the lakes' water supply. *See also* HOT SPRING; SEEP.

squatter a person occupying land to which he or she has no legal claim.

stable substance in chemistry, a substance whose chemical composition does not alter, or which alters only very slowly, under normal environmental conditions. Once released into the environment, a stable substance will not break down (*decay* or *degrade*) into is components, but will retain its chemical identity virtually forever. *See also* PERSISTENT COMPOUND; HARD PESTICIDE.

stable system a group of ORGANISMS, objects, substances, etc., which interact with each other in such a way that the overall character of the group does not change over time. Changes can, and usually do, occur on an individual level. An OLD GROWTH forest is an example of a stable system, in this case a stable ECOSYSTEM. Although individual organisms within the ecosystem may die or migrate out of the area, they are replaced by others of the same SPECIES, keeping the species composition of the ecosystem constant. Similarly, a beach on which DEPOSITION exactly balances EROSION is a stable geologic system, and will remain the same size and in the same position even though the individual sand grains that make up the beach will change over time. All natural systems tend toward stability, and will achieve it if they are not disrupted for a long enough period of time. *See* SUCCESSION; CLIMAX COMMUNITY.

stade in glaciology, a period during which glacial deposits are being laid down (*see* TILL; OUTWASH DEPOSIT; MORAINE; KAME; ESKER; GLACIAL LANDFORM). Stades differ from nearly all other forms of geologic period in that they are variable in length. A single stade may represent a wide range of years, from three or four at the glacier's farthest extent to several thousand near its center of deposition. The time period between two separate stades is known as an INTERSTADIAL.

stage of a river, *see under* STREAMFLOW.

stage cut in forestry, the harvest of a timber STAND in two or more stages, usually beginning with the UNDERSTORY and moving up by height through the CANOPY. Properly designed stage cuts can reduce breakage of the felled timber, reduce soil EROSION, and promote natural REGENERATION of the cutover land, especially if the stages are separated in time by several years. *See also* SHELTERWOOD SYSTEM.

staggered setting in logging and forestry, the harvesting of CLEARCUTS that are relatively small and offset from each other so as to produce a mosaic of harvested and unharvested timber. Staggered settings improve the appearance of clearcuts, protect the remaining timber against WINDTHROW, and increase the chances that reseeding can take place from natural stock. *See also* PATCH CUT.

stagnant stand in forestry, a timber STAND in which growth is not occurring, or in which each year's growth is offset by the death of individual trees within the stand—in technical terms, when the NET ANNUAL INCREMENT of the stand is at or approaching zero. Stagnation in young stands usually results from trees growing too close together, so that they interfere with each other's growth. These stands can be encouraged to grow again by removing some of the trees (*see* RELEASE CUTTING; THINNING).

stamen *See* FLOWER: *parts of a flower.*

stand in forestry, a group of living trees of roughly homogenous species composition (*see* SPECIES): the borders of the stand are assumed to lie where the species composition changes. The size of a stand can thus be anywhere from a few trees to several hundred square miles of forest, although the FOREST SERVICE has occasionally used a working definition that requires a minimum size of one acre with a minimum STOCKING LEVEL of at least 10% throughout the area. Stands are classified by both species and AGE CLASS. *See*, e.g., MIXED STAND; EVEN-AGE STAND.

standard component in Forest Service terminology, historically, COMMERCIAL FOREST LAND on which there are no special restrictions on timber harvest. Standard component lands are those from which the bulk of a forest's timber harvest will normally come. The term was rendered obsolete by the NATIONAL FOREST MANAGEMENT ACT OF 1976, but is still in wide informal use. *Compare* SPECIAL COMPONENT; MARGINAL COMPONENT; RESTRICTED COMPONENT.

Standard Federal Region in politics, one of 10 regions established by the Intergovernmental Cooperation Act of 1968 to improve cooperation among federal agencies, their state counterparts, and the citizens, by consolidating the regional headquarters of all agencies in the same city within a region and standardizing the boundaries of the regions these headquarters serve. The law requires that each new agency established must conform its administrative structure to the Standard Federal Regions map, and that all previously established agencies that do not conform to these regions must be brought into conformity at any time that they reorganize. The 10 Standard Federal Regions, their headquarters cities, and their administrative boundaries are:

Region I (Boston: Massachusetts, Connecticut, Rhode Island, Vermont, New Hampshire, Maine);
Region II (New York City: New York, New Jersey, Virgin Islands, Puerto Rico);
Region III (Philadelphia: Pennsylvania, Delaware, Maryland, Virginia, West Virginia, District of Columbia);
Region IV (Atlanta: Kentucky, Tennessee, North Carolina, South Carolina, Georgia, Florida, Alabama, Mississippi);
Region V (Chicago: Illinois, Indiana, Ohio, Michigan, Wisconsin, Minnesota);
Region VI (Dallas-Fort Worth: Texas, Louisiana, Arkansas, Oklahoma, New Mexico);
Region VII (Kansas City: Missouri, Kansas, Iowa, Nebraska);
Region VIII (Denver: Colorado, Utah, Wyoming, Montana, North Dakota, South Dakota);
Region IX (San Francisco: California, Nevada, Arizona, Hawaii);
Region X (Seattle: Washington, Oregon, Idaho, Alaska).

standard plate count *See under* COLIFORM COUNT.

standard temperature and pressure (STP) in the sciences, a temperature of 0°C and a pressure of 760mm of mercury (101,325 Pascals, or 1013.25 millibars—sometimes referred to as "one atmosphere" because it is equivalent to the pressure normally exerted by the earth's ATMOSPHERE at sea level). STP measurements are used to report values which change with pressure and temperature changes, such as the volume of a gas. This allows the reported values to be directly compared.

stand cut in Forest Service usage, a large CLEARCUT. Stand cuts are so named because they generally harvest an entire STAND of trees in one operation. To be called a stand cut, a clearcut must cover at least 100 acres. Current regulations limit maximum size to 500 acres (the largest clearcut that can legally be made).

standing committee a permanent legislative committee with jurisdiction over a particular type of legislation, such as environmental bills, budgetary bills, foreign policy bills, etc. The personnel of the committee changes with time, but its structure and role remain constant. The number of standing committees in the federal legislature is limited by law to 37, 22 in the House and 15 in the Senate.

standing crop in the biological sciences, the BIOMASS of a particular SPECIES in a particular ENVIRONMENT at a particular point in time. The term is used not only in agriculture and forestry, where its meaning is fairly obvious (e.g., the standing crop of pine on a CUT BLOCK is the total mass of all living pine trees on the block: "standing" because living trees stand upright; "crop" because they are about to be harvested) but also in disciplines as diverse as wildlife biology (the standing crop of deer in a forest is the total mass of all living deer in the forest) and water quality control (the standing crop of ALGAE in a lake is the total mass of all living algae in the lake water).

standing cull in forestry and logging, a living tree that will not be harvested during a logging operation due to the presence of DEFECTS that will render it unmerchantable (*see* MERCHANTABLE TIMBER). Standing culls are usually marked before the logging operation begins. They are often converted to SNAGS following logging in order to meet wildlife objectives.

stand tending *See* INTERMEDIATE CUT.

State, Department of division of the federal (U.S.) government charged with the planning and carrying out of foreign policy. It is headed by the secretary of state, a cabinet-level official who reports directly to the president. The oldest continuing civilian institution of the federal government, it was created as the Committee for Secret Correspondence in 1775 (before independence), became the Department of Foreign Affairs in 1777, and was renamed the Department of State in 1789 under Secretary of State Thomas Jefferson. The department's environmental arm, the Bureau of Oceans and Environmental and Scientific

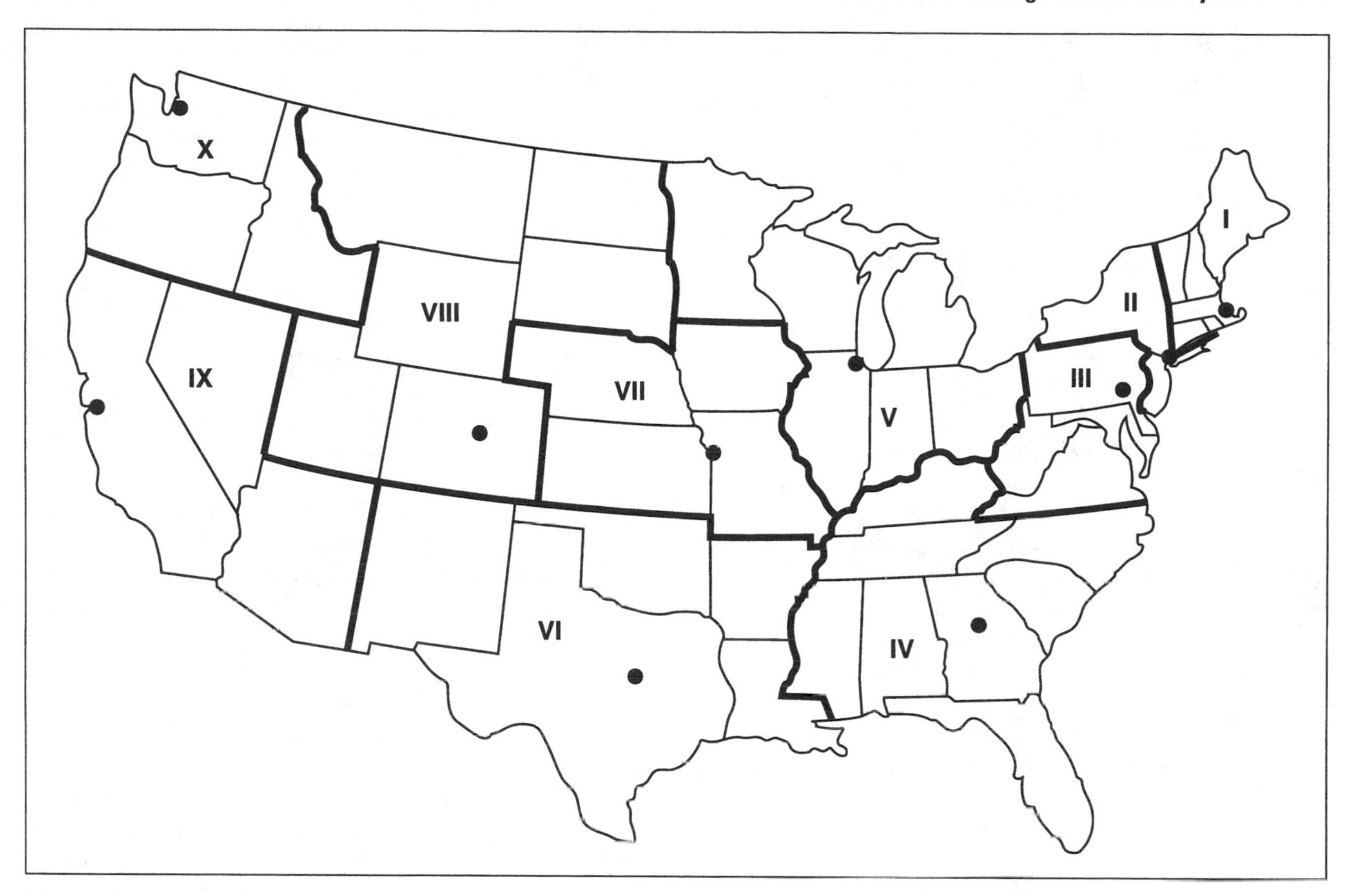

Standard federal regions

Affairs, has responsibility for the negotiation of treaties dealing with ocean pollution, atmospheric degradation, and other international environmental concerns, and with advising other agencies of the U.S. government when their actions might affect the international environment. Headquarters: 2201 C Street NW, Washington, D.C. 20520.

State Implementation Plan (SIP) in pollution control, a strategy for implementing the requirements of the CLEAR AIR ACT within the boundaries of a given state. State Implementation Plans are designed and administered by state and local governments under the oversight of the ENVIRONMENTAL PROTECTION AGENCY (EPA). A broad range of strategies is allowed, as long as these strategies conform to the requirements for implementation spelled out in the act. These requirements, found primarily in Section 110, deal primarily with the speed with which the plan will bring all areas of the state into conformity with the NATIONAL AMBIENT AIR QUALITY STANDARDS (NAAQSs). Strategies must be "continuous" rather than "intermittent", that is, must involve direct controls over source emissions rather than simply shutting down the source when the NAAQSs are exceeded. Dispersion (i.e., building taller stacks) may not be used as the sole control tool. REASONABLY AVAILABLE CONTROL TECHNOLOGY must be retrofitted to plants in NONATTAINMENT AREAS. Inspection and maintenance plans (I/M) for motor vehicle pollutant control systems may be required before the vehicles registered in nonattainment areas can be licensed. State Implementation Plans are submitted for approval to the Chief Administrator of the EPA, who may amend them if, in his judgment, they are not stringent enough to allow attainment of the NAAQSs within the timeline of the act. (He may not amend a plan for being too stringent; the act specifically allows states to set standards stricter than the NAAQSs.) States may legally refuse to design and enforce a State Implementation Plan; however, if a plan is not in force, the EPA will itself enforce the Clean Air Act regulations within the state. *See also* AIR QUALITY CLASS; BEST AVAILABLE CONTROL TECHNOLOGY.

State Pollutant Discharge Elimination System (SPDES) in pollution control, a system of state regulations designed to help the state meet the EFFLUENT STANDARDS of the CLEAN WATER ACT. SPDES regulations are similar to NATIONAL POLLUTANT DISCHARGE ELIMINATION SYSTEM (NPDES) regulations; however, the states may enact stricter standards than those in the federal legislation if they wish to do so. States with SPDES regulations in place that meet NPDES requirements may take over Clean Water Act enforcement from the ENVIRONMENTAL PROTECTION AGENCY. *See also* STATE WATER QUALITY MANAGEMENT PLAN.

State Water Quality Management Plan (SWQMP; 303 plan) in pollution control, a plan prepared by a state to bring the quality of its rivers, streams, and lakes into conformity with federal AMBIENT QUALITY STANDARDS. As spelled out in Section 303 of the CLEAN WATER ACT, SWQMPs are required to be the result of a "continuous planning process" (CPP) that evolves to meet changing conditions in the water bodies. States are allowed to designate particularly polluted stretches of rivers as "water quality limited segments" (WQLS). In these stretches, ambient quality standards do not need to be met, although the quality may not generally be allowed to deteriorate any further than it already has. SWQMPs are often combined with 208 PLANS into a single water-quality strategy document.

static level in hydrology, the level to which water will fill a well that is not being pumped. The static level corresponds to the WATER TABLE except in the case of ARTESIAN WELLS. In these, it stands some distance above the water table due to pressures within the AQUIFER tapped by the well. The static level is generally expressed in terms of distance below ground level (e.g., a static level of 33 feet is 33 feet below the surface of the ground). It is affected by CONES OF DEPRESSION as well as by changes in the regional water table, and is a more accurate measure of GROUNDWATER availability than can be obtained by strictly measuring the water table.

stationary source in pollution-control terminology, a source of air POLLUTANTS that is permanently fixed in one place. Stationary sources are further subdivided into POINT SOURCES (those from which the pollutants may be treated as if they were coming from a single point, such as smokestacks and other forms of industrial stacks) and AREA SOURCES (those that release pollutants diffusely over a broad area, such as construction projects, open HAZARDOUS-WASTE dumps, and commercial dry-cleaners). For laws and regulations governing stationary sources, *see* CLEAN AIR ACT; STATE IMPLEMENTATION PLAN; BEST AVAILABLE CONTROL TECHNOLOGY; LOWEST ACHIEVABLE EMISSION RATE; BUBBLE. *Compare* MOBILE SOURCE.

steppe a prairie, especially one dominated by short grasses and bunchgrass (*see* GRASS). Steppes develop on flat or rolling terrain in arid to semiarid climates (*see* ARID CLIMATE). Temperature does not appear to be a factor, although a CONTINENTAL CLIMATE seems to be necessary. Some authors use the word to denote only dry grasslands, while others use it for all grasslands. Extensive steppes are found in central Asia (where the term originated), central North America (the so-called "High Plains"), and southeastern South America (the Pampas). Smaller extents occur in East Africa, Australia, New Zealand, Madagascar, East Asia, and northwestern North America. *See also* SHRUB-STEPPE; GRASSLAND BIOME.

stewardship in the context of natural resource conservation, the prudent management of resources in order to maintain their ecological integrity while providing sustainable economic benefit. The notion of *stewardship* is biblical, with references in the Hebrew Bible concerning the vocation of the steward as the trusted servant of a landowner, an overseer of an estate or other valued possession (Genesis 43, 44; Chronicles 27, 28; Daniel 1, 16; Isaiah 22). In the Christian New Testament, the idea of stewardship is more metaphorical, wherein the steward, as God's disciple, has the responsibility of feeding and protecting those within the steward's charge but must be careful not to arrogate authority equal to that of the master (God the Creator). In the practical context of, say, managing forests, stewardship becomes the practice of caring for the resource and distributing its benefits but doing so in such a way that does not diminish the integrity of the forest, which is the gift of God (1 Corinthians 3, 4, 10; 1 Peter 4; Philippians 2; Titus 1). Despite its theological implications, the stewardship concept is seen by some philosophical ecologists (and not a few modern ecotheologians) as essentially self-contradictory, in that it refers to an anthropological economic function—i.e., resource use—rather than one involving a spiritual view of the sanctity of nature and the necessity for its protection for its own sake. *See also* RELIGION AND THE ENVIRONMENT.

stochastic change change that proceeds by random small steps, such as those introduced into a GENE POOL by MUTATION. Many natural processes (GENETIC DRIFT, BEACH NOURISHMENT, etc.) are best described as the sum of a series of stochastic changes.

stochastic observation in the sciences, gathering data by taking a series of observations at random points and times. Stochastic observation is particularly useful for monitoring slowly changing phenomena such as ecological processes or rapidly shifting but broadly uniform conditions such as those found in AIR MASSES.

stock (1) in agriculture, *see* LIVESTOCK.

(2) in the biological sciences, the small ancestral group from which a POPULATION is derived. The characteristics of the stock determine to a large extent the characteristics of the population. A stock of long-haired goats gives rise to a population of long-haired goats; a stock of cold-resistant tulips gives rise to a population of cold-resistant tulips; etc. The breeding of desired characteristics into domestic plants and animals by importing individuals with those characteristics and using them for cross-breeding is called IMPROVING THE STOCK.

(3) in geology, a small BATHOLITH, generally one with a surface area of less than 100 square kilometers (35 square miles).

Stockholm Conference on Environment officially the U.N. Conference on the Human Environment, which is called the Stockholm Conference because it was held in Stockholm, Sweden, in 1972. Bringing 113 nations together, this was the first of the United Nations' environmental conferences and has been a reference point for all conferences since. The conference created the U.N. Environment Programme (UNEP), an agency that grew in importance and which convened the RIO EARTH SUMMIT on the program's 20th anniversary. The 1972 conference also adopted a 26-point statement of environmental principles which has been guiding international discussions ever since.

stocking level in forestry, the ratio of the actual number of living trees on a plot of land to the number of living trees that should ideally be present in order to achieve maximum wood production. Land containing more trees than the ideal is said to be *overstocked* (*see* STAGNANT STAND); land containing less than the ideal is *understocked*. The number of trees that would be considered adequate stocking varies with the CAPABILITY of the land and with the age of the trees being considered, adequate stocking for SEEDLINGS being considerably higher than that for mature trees in order to allow for MORTALITY.

stocking rate *See under* RANGE MANAGEMENT.

stomata in botany, pores in the external surface of a plant, used by the plant for RESPIRATION. They may occur on all parts of the plant, but are normally concentrated heavily on the leaves. The stomata are designed to conserve moisture while allowing the free passage of OXYGEN and CARBON DIOXIDE. In addition, most plants have the ability to open and close them in response to external conditions of temperature and humidity, allowing fine control of the plant's water balance. Their pattern on the leaves is genetically determined, and may be used, especially in the CONIFERS, as a means of confirming SPECIES identification. *See also* TRANSPIRATION.

stool *See* FECES.

STORET in waste management, abbreviation for *storage and retrieval system.*

storm beach a deposit of SAND, beach PEBBLES, driftwood, etc., lying some distance above or inland from the normal beach, where it has been left by the high waves accompanying storms. The presence of well-developed storm beaches along a coastline is an indication that the coastline is subject to frequent and severe storms.

Storm King mountain on the west bank of the Hudson River at Cornwall, New York, approximately 65 kilometers (40 miles) above New York City. The mountain rises abruptly from the riverbank for more than 380 meters (1,250 feet); its summit is 408 meters above sea level (1,340 feet). In January 1963, the Consolidated Edison Power Company filed an application with the FEDERAL POWER COMMISSION (FPC) to build a PUMPED STORAGE facility at Storm King. The facility was to include a PENSTOCK 12 meters (40 feet) in diameter and three kilometers (10,000 feet) in length leading to a 240-acre reservoir with a capacity of 25,000 acre-feet at the head of Black Rock Hollow, between Whitehorse Mountain and Rattlesnake Hill southwest of Storm King, partially within the boundaries of Harvard University's Black Rock Forest Preserve. The associated two-megawatt power plant would have been at that time the largest in the world. The Storm King project was controversial from the beginning, not only because of its impact on the scenery but because of the threat it posed to the striped bass: 85% of the Hudson River POPULATION of this fish spawns in the vicinity of Storm King, and it was feared that the population could not survive the daily fluctuations in the river caused by filling and emptying the reservoir to operate the power plant (flows through the plant were estimated to be as high as 1 million cubic feet per minute). Nevertheless, a license was granted by the FPC in March 1965. In December 1965, the U.S. Court of Appeals remanded the license to the FPC, stating that "the renewed proceedings must include as a basic concern the preservation of natural beauty and of national historic shrines . . . the cost of a project is but one of several factors to be considered." The project continued to be before the FPC and in the courts for 15 years, until December 19, 1980, when a mediated agreement was announced between the power company and New York environmental groups in which the power company agreed to give up plans to build the Storm King project and donate project lands to the State of New York for park purposes and the environmental groups agreed to drop pending suits over COOLING TOWERS at three other power plants fronting the Hudson River.

storm runoff (1) in hydrology and water pollution control, overland runoff (*see* RUNOFF: *overland runoff*), especially that which occurs within urban areas and which must be handled by STORM SEWERS. Storm runoff is typically up to 100 times the flow of household and industrial SEWAGE carried by a city's SEWERAGE SYSTEM and therefore cannot be handled by the SEWAGE TREATMENT PLANT but must be passed around the plant. In the case of COMBINED SEWER SYSTEMS, sewage will be carried with the storm runoff around the plant and will be deposited, untreated, in the RECEIVING WATERS. Storm runoff itself can be heavily polluted, especially during the first few minutes of runoff (the *first flush*), as pollutants that have accumulated on the surface of the land are picked up by the water. For this reason, many municipalities make an attempt to catch and treat the first flush, letting the later, cleaner runoff go untreated. *See also* NONPOINT POLLUTION.

(2) in waste-management technology, the relatively uncontaminated RUNOFF that results from PRECIPITATION falling on those portions of a waste-treatment facility that are not directly involved in waste handling, such as building roofs, sidewalks, lawns, and so forth. Storm runoff requires no special handling, and is usually segregated from the contaminated runoff originating within the waste-handling areas of the facility and discharged directly into STORM SEWERS in order to avoid overloading the waste-treatment facility's CONTAINMENT SYSTEM.

storm sewer a SEWER designed to handle the RUNOFF that results from PRECIPITATION (rain, snow, hail, etc.), as opposed to one designed to handle domestic or industrial WASTEWATER. Storm sewers are usually considerably larger than the sewers that handle wastes. Their actual size for any given locality is determined by the size of the maximum storm event that can be expected for the local climate and by the INFILTRATION RATE of the local soil, which is in turn influenced by such factors as the vegetative cover of the surface, the amount of pavement and roof area, etc. Storm sewers are generally built in a network parallel to that of the wastewater sewers, but are emptied directly into the RECEIVING WATERS instead of going through the SEWAGE TREATMENT PLANT. The runoff enters them through screened openings in street curbs (*storm drains*) that are placed at least once a block. *See also* COMBINED SEWER SYSTEM.

STP *See* SEWAGE TREATMENT PLANT; STANDARD TEMPERATURE AND PRESSURE.

strata in geology, plural of STRATUM (*see also* STRATIGRAPHIC UNIT).

Strategic Lawsuits Against Public Participation *See* ANTI-ENVIRONMENTALISM.

Strategic Petroleum Reserve a U.S. government–owned stock of crude oil established in 1975 under the Energy Policy and Conservation Act. Undertaken to avoid repeating the shock to the U.S. economy of the OIL EMBARGO of the early 1970s and the later worldwide shortage in 1979, the reserve consists of five storage areas with a total capacity of 750 million barrels. Since 1990, reserve levels have hovered around 600 million barrels and have been drawn down from time to time to stabilize U.S. oil prices. The oil now stored in the Strategic Petroleum Reserve would last about 60 days if all foreign oil imports ceased and U.S. wells did not increase production. *See also* OIL RESERVES; ORGANIZATION OF PETROLEUM EXPORTING COUNTRIES.

stratification separation into horizontal layers. Surface water generally stratifies by temperature (*see* THERMAL STRATIFICATION), although it may sometimes stratify by mineral content instead. GROUNDWATER almost always stratifies by mineral content (*see* AQUIFER). GLACIERS and permanent snowfields stratify according to the annual accumulations of snow. Rocks also stratify according to time, but the time units involved are generally much longer than those for ice and snow (*see* STRATIGRAPHIC UNIT; STRATUM; BED). Living ORGANISMS within an ECOSYSTEM usually show *biostratification*. For example, within a body of water, some organisms will prefer the bottom (*see* BENTHOS); others will prefer the waters of middle depth (*see* NEKTON); and still others will remain close to the surface at all times (*see* NEUSTON).

stratigraphic unit in geology, a group of rock strata (*see* STRATUM) with some overall unifying feature that makes it useful for mapping FORMATIONS, correlating their ages, etc. *Biostratigraphic units* are unified by the COMMUNITIES of FOSSILS found in them.

Lithostratigraphic units are unified by MINERAL composition. *Chronostratigraphic units* are unified by their age of deposition, which may be identified through various means (*see* GEOLOGIC TIME).

stratigraphy the study of rock layers. *See* STRATIGRAPHIC UNIT; STRATUM; BED.

stratum (plural: **strata**) a single layer or set of closely related layers within a stratified rock FORMATION. In SEDIMENTARY ROCK, a stratum represents sediments laid down under a single, uniform set of conditions; in IGNEOUS ROCK, a stratum represents a single LAVA FLOW. Strata that may be apparent in METAMORPHIC ROCK are remnants of the stratification that existed within the igneous or sedimentary rock from which the metamorphic rock was derived, and are not particularly useful for dating purposes, although they may be used for mapping. *See also* STRATIGRAPHIC UNIT; BED; VARVE.

stray current corrosion *See under* CORROSION.

streak in geology, a deposit of color left behind by a ROCK or MINERAL when it is rubbed across a smooth hard surface. A rock or mineral of a given variety will normally show the same type of streak no matter what its external form or color may be; hence, the streak is a valuable tool in identifying rocks and minerals, and most geologists carry a small piece of hard ceramic of a neutral color in their field tools in order to perform the streak test under relatively controlled conditions.

stream channelization in engineering, modification of the channel of a stream in order to increase its rate of flow. Channelization generally involves the removal of obstructions such as rocks and GRAVEL BARS; the straightening of curves and MEANDERS; dredging to a uniform depth (*see* DREDGE); and (often) constructing streamside LEVEES and/or lining the modified channel with RIPRAP or concrete. By increasing the velocity of flow, channelization effectively increases the amount of water a stream can carry (that is, more water can pass through a given section of the stream during a given time), thus lowering flood levels and allowing the drainage of WETLANDS for agricultural purposes. However, the process has severe environmental effects, including the total destruction of the stream's BENTHOS and a consequent severe reduction in the amount of fish life that can be supported in the channelized section (usually by 80% or more); the loss of SPAWNING GRAVELS; increased EROSION and TURBIDITY due to the more rapid streamflow, and increased SILTATION downstream where the flow slows down again and drops its suspended materials; loss of WETLANDS and RIPARIAN VEGETATION; and severe aesthetic degradation. Channelized streams are also hydrologically unstable and require constant maintenance to prevent meanders, bars, and other natural conditions of unmodified stream channels from returning. For these reasons, channelization is not as popular today as it once was, and some channelized streams are being restored as much as possible to natural conditions. Most stream channelization in the United States has been carried out by the ARMY CORPS OF ENGINEERS (in their role of providing navigation improvements and flood control) and the SOIL CONSERVATION SERVICE (under the terms of the Watershed Protection and Flood Prevention Act of 1954). *See also* FLOOD: *methods of flood control.*

stream corridor *See* STREAMSIDE MANAGEMENT UNIT.

streamflow the amount of water flowing down a stream during a given time period. Streamflow depends on numerous factors, including the amount of PRECIPITATION in the stream's WATERSHED; the character of the channel (straight or crooked, deep or shallow, clear or obstructed, etc. [*see* STREAM CHANNELIZATION]); the ratio of precipitation to RUNOFF, which in turn is controlled by such factors as vegetation and INFILTRATION rate; and the amount of BASEFLOW provided by the AQUIFERS that the stream intersects. Measurements of streamflow are taken at *gauging stations,* which are located in areas where the channel shape is relatively uniform so that fluctuations in volume can be correlated strongly with fluctuations in water level. Gauging stations measure the depth of the water, which may be reported directly (as the stream's *stage*) or may be converted into a volume/time measurement (usually CUBIC FEET PER SECOND) from a *calibration curve* that is computed independently for each gauging station from the velocity and cross-sectional area of the stream at the station as measured (or calculated) at various stream stages.

streamside management unit (**[SMU] stream management zone [SMZ], stream corridor**) a long, narrow plot of land consisting of a stream and its associated lands, including both those lands that support RIPARIAN VEGETATION and those lands whose vegetation is a direct factor in controlling water conditions within the stream, such as trees that cool the water by providing shade or grasses whose root systems prevent SEDIMENTS from entering the stream. The boundary of an SMU is generally defined as a line parallel to the stream a set distance away. The distance varies with the type of stream and the amount of protection sought. Management practices within an SMU are designed to protect the stream's water quality and to keep its flow as steady as possible, and usually involve severe restrictions on mining, logging, and construction activities.

striation generally, one of a series of parallel scratches in a hard surface. In geology, the term usually refers to scratches left on glacial-polished bedrock (*see* GLACIAL POLISH; BEDROCK) by stones caught in the underside of GLACIERS. These scratches, which may be anywhere from a fraction of an inch to several feet deep, indicate the direction of the glacier's flow and (by their depth) the weight of the ice that once existed at that location. In mineralogy, "striations" refers to the microscopic parallel lines that mark some crystal faces (e.g., plagioclase feldspar; *see* FELDSPAR) as a result of the way in which the crystal grew.

strike the COMPASS BEARING of an imaginary horizontal line drawn across a sloping surface, such as a hillside or a tilted STRATUM of SEDIMENTARY ROCK. The strike is perpendicular to the DIP.

string bog a BOG developed on a slight slope, so that the water in the bog is able to flow very slowly in one direction. String bogs are usually long and narrow (hence the name). They tend to develop a regular pattern of alternating ridges and hollows (*flarks*) of PEAT at right angles to the direction of flow, probably caused by slow downhill creep of the watersoaked peat itself. The ridges may be dry and firm enough to support the growth of trees. The water chemistry and plant communities of string bogs are identical to those of other bogs.

strip in logging, the area of land on which a harvest operation is being carried out. A strip is essentially the same thing as a HARVEST UNIT, except that the harvest unit is looked at from the standpoint of silviculture, while the strip is looked at from the practical standpoint of the actual mechanical steps of the LOGGING SHOW. Before logging, the whole strip is usually examined by the lead feller (*see* BULL-OF-THE-WOODS) so that the sequence in which the trees are taken can be planned to avoid felling them onto each other (*see* LEAD). It is especially important to make the first cut ("open the strip") in the right location. On sloping ground, the opening is generally made at the bottom of the slope.

strip census a means of estimating the size of a POPULATION in a study region by walking a predetermined route (a TRANSECT) through the region, counting all sitings of the types of animals or plants you are interested in (the *quarry*), dividing by the area of the transect to get a unit count, and then multiplying that unit count by the size of the region as a whole. The area of the transect is found by multiplying its length by twice the average distance to the location where each individual quarry was first sited. In mathematical terms

$$P=AZ/2YX$$

where P=population size, A=area of the study region, Z=number of sitings, Y=average distance to each siting, and X=length of the transect. The strip census is fairly accurate for plants, but is inaccurate for animals due to individual differences in the *flushing distance*—the distance at which the quarry will flee from an approaching human, thereby making itself more easily seen. When used for wildlife populations, therefore, the estimate obtained from the strip census is usually multiplied by a correction factor based on previous experience with the SPECIES being counted. A variant of the strip census, the *roadside census*, involves driving slowly through an area and counting the individual quarry seen from the car. It is even less accurate than the strip census, but is sometimes useful for making a quick population estimate in order to see if more exact methods are warranted. In forestry, a strip census is referred to as a *strip cruise* (*see* TIMBER CRUISER).

strip cropping planting crops in alternating bands, such as corn and ryegrass, to reduce wind and water erosion and to maintain soil nutrients and organic matter. *See also* CONSERVATION TILLAGE; CONTOUR PLOWING.

strip mine (surface mine, open-pit mine) a mining technique in which the mined material, generally iron ORE, copper ore, or COAL, is obtained by removing any materials that lay over it (the OVERBURDEN) and piling them to the side to allow direct access to the mined material from above. The mined material is then simply dug out with power shovels. Strip mining is faster and cheaper than underground mining, uses fewer workers and is safer for those workers, and allows removal of up to 100% of the deposit (underground mining normally removes only about 50% of the deposit due to the need to leave pillars in place to support the roof). Unfortunately, the environmental hazards of the technique are considerable, including massive rock and soil disturbance; the destruction of vegetation by the highly acid SPOIL; pollution of rivers, streams, and lakes by RUNOFF leached through the spoil; the drying-up of SPRINGS and streams due to the interruption of AQUIFERS and other interferences with the WATER TABLE; and massive EROSION of the spoil slopes and

CUTBANKS, which are difficult to stabilize with vegetation because of their acidity. These effects may be mitigated somewhat by reclamation procedures (*see* SURFACE MINING CONTROL AND RECLAMATION ACT). *See also* LONGWALL MINE.

strontium 90 a toxic fallout product resulting from the explosion of nuclear devices. Strontium 90 is a salt that readily enters the food chain and has occurred in high quantities in the milk produced by cows subject to fallout from nuclear testing in Utah and elsewhere in the 1950s and early 1960s. In Utah the number of cases of childhood leukemia increased by 250% during the bomb-testing period. These findings were influential in the above-ground testing ban signed in 1963 by the United States and USSR.

structural basin a depression in the Earth's surface resulting from the underlying rock structure rather than from erosive activity. *See also* STRUCTURAL VALLEY.

structural valley in geology, a valley formed by the deformation of the Earth's surface through TECTONIC ACTIVITY rather than by the erosive power of a stream or GLACIER. It may be deepened, widened, or otherwise shaped by EROSION after its formation. A structural valley will usually be formed by a GRABEN or a SYNCLINE, although a few arise from other tectonic landforms. A valley created by erosion but aligned with structural features (e.g., a valley created by erosion along the inteface between a pair adjoining BATHOLITHS) is called a *structurally determined valley*. *See also* STRUCTURAL BASIN.

Student Conservation Association (SCA) offers conservation service opportunities, outdoor education, and career training for youth. SCA volunteers and interns provide more than 1 million hours of service annually in national parks, forests, refuges, and urban areas in all 50 states. They serve from four weeks to one year. Founded in 1957, the SCA has published numerous books. Recent publications include *Earth Work: A Resource Guide to Nationwide Green Jobs* (1994), *Lightly on the Land: The SCA Trail-Building and Maintenance Manual* (1996), and *Guide to Graduate Environmental Programs* (1997) as well as the monthly magazine *Earth Work* and *Earth Work Online*. Membership (1999): 24,000. Headquarters: Box 550, Charlestown, NH 03603. Phone: (603) 543-1700. Website: www.sca-inc.org.

subalpine parkland the region just below TIMBERLINE on a mountain slope, characterized by meadows and widely spaced groves of CONIFERS. Other common features include small streams and SPRINGS, TARNS, and rock outcrops. Wildflowers are abundant. Once considered a separate LIFE ZONE (*see* HUDSONIAN ZONE), subalpine parkland is currently treated by most ecologists as an ECOTONE between the TUNDRA BIOME and the CONIFEROUS FOREST BIOME.

subfamily in the biological sciences, a group of genera (*see* GENUS) that share a number of traits but are not distinct enough from other similar groups to warrant being classified as a separate FAMILY. *See also* TAXONOMY.

subgrade in engineering, the material on which a road is built. The subgrade consists of the material that is exposed after the roadbed has been graded and prepared but before any base material or pavement is laid down.

sublimation a process by which a solid changes directly to a gas without passing through a liquid stage in between. Sublimination takes place when the bonds joining MOLECULES into solid form are weak enough to allow some of those on the surface to escape. Most materials sublimate very slowly, but for some (i.e., CARBON DIOXIDE), the process is rapid enough that they cannot exist in the liquid state at all under normal ATMOSPHERIC PRESSURE. Sublimation is responsible for the odor of solids (we smell the escaped molecules) and for the slow disappearance of snow and ice from roads, hillsides, etc., even when temperatures remain below freezing. The term is also used for the reverse process, that is, the condensation of a vapor directly into a solid. *See also* EVAPORATION; CONDENSATION; VAPOR PRESSURE.

submerged macrophyte a plant, especially a VASCULAR PLANT, that is rooted on the bottom of a body of water and does not grow into the air, confining all its growth beneath the surface of the water. Submerged macrophytes are often classified as "nuisance" vegetation by recreational managers, as they snag fishing lines, disturb waders and swimmers, and can clog waterways thoroughly enough that boating is difficult or impossible. However, they also help anchor the bottom SEDIMENTS and provide feeding and breeding grounds for aquatic life. Submerged macrophytes are usually confined to depths above 9 meters (35 feet) due to their need for light and their intolerance for high water pressure. Examples: milfoil, pondweed. *Compare* EMERGENT MACROPHYTE; FLOATING MACROPHYTE.

submerged vegetation plants below the surface of the water. The term applies both to SUBMERGED MACROPHYTES and to EMERGENT MACROPHYTES and terrestrial plants that have been submerged by rising waters, as in a RESERVOIR.

subsoil The C HORIZON, that is, that portion of the soil that represents a transition to the PARENT MATERIAL. Subsoils have little organic content and are generally less cohesive mechanically than the upper layers of the soil. *Compare* TOPSOIL. *See also* SOIL: *the soil profile.*

subspecies (race) in biology, a subdivision of a SPECIES that shows distinct differences from the remainder of the species in coloring, shape, or behavior, but will nevertheless breed with individuals from the remainder of the species with little or no trouble. Subspecies usually develop in geographic isolation from each other, and represent separate POPULATIONS on their way to becoming separate species.

substrate (1) in the biological sciences, the substance upon which a living ORGANISM is growing. The substrate is usually nonliving, but may occasionally be another organism, as when barnacles attach themselves to a whale. The organism may obtain nutrients from the substrate (e.g., laboratory cultures of bacteria are grown on a substrate, such as agar, which can provide them with all the nutrients they need; *see* BACTERIUM), or it may use the substrate simply as a means of support (*see* SESSILE ANIMAL). The term *substrate* is also used by biochemists to refer to the chemical upon which an ENZYME acts within the body of a living organism.

(2) in geology, a layer of underlying rock, especially one on which the surface materials are resting uncomformably; *see* UNCONFORMITY; BEDROCK; COUNTRY ROCK.

subtidal algal zone in ecology, the zone directly below extreme low tide on an ocean shoreline, characterized by the growth of large brown ALGAE, particularly kelp of the genus *Laminaria*. It is sometimes called the *sublittoral zone* or the *Laminaria zone*. The outward boundary of this zone is at a depth of about 30 meters (100 feet).

suburbanization *See* URBAN SPRAWL.

succession in ecology, the gradual replacement of one type of COMMUNITY by another. Succession is a slow but continuous process, beginning with the invasion of a patch of open ground or a newly created body of water by PIONEER SPECIES and continuing through a series of recognizable stages known as SERES to the formation of a CLIMAX COMMUNITY in which the mix of SPECIES that form the community is no longer changing with successive generations. At this point succession ceases until a change in conditions allows it to begin again, either because a disturbance such as a windstorm or a landslide has removed the climax community and laid the ground bare again (*see* SECONDARY SUCCESSION) or because some factor such as CLIMATE, frequency of WILDFIRE, etc., has altered enough that the climax community can no longer maintain itself, allowing succession to begin once more. An understanding of succession leads to two principal conclusions. First, living communities are constantly in flux and change is inevitable. Even climax communities slowly evolve as climatic conditions change. Attempts to preserve some part of the landscape without change (seawalls, fire suppression, National Parks) are therefore doomed to failure. The only thing that can be preserved is the opportunity for nature to continue to change. Second, a climax (or near-climax) community is the result of thousands of years of succession and, once destroyed, cannot be recreated except by the same processes—if ever. Hence, human-induced changes in the ecology of a region are always permanent, at least on a human time scale. "Natural" conditions may again be obtained in relatively short order, but the community structure will be different and the species mix will ordinarily be far poorer, simply because the sere present is much closer to the pioneer stage than to the climax. The earliest scientific study of succession was undertaken by the American botanist and University of Chicago biology professor Henry Chandler Cowles (1869–1939) in the Indiana Dunes at the south end of Lake Michigan. His 1899 paper on the subject is now considered one of the foundations of the science of ecology. *See also* TROPHIC CLASSIFICATION SYSTEM.

succulent in botany, a plant in which the leaves and/or stems serve as water-storage devices. Succulent plant tissue is characteristically thick and fleshy, with a heavy outer covering: it is juicy when cut. The plants grow in areas where the water supply is episodic, as in DESERTS or in the crevices of rocks. Examples: cactus, stonecrop.

suitability of land, as defined by the CODE OF FEDERAL REGULATIONS, the appropriateness with which a given piece of land may be used for a given purpose, such as timber harvest, recreation, subdivision development, etc. Determination of suitability takes into account a number of factors, including the land's natural CAPABILITY; the ease with which management techniques may increase its ability to serve the stated purpose (e.g., can timber harvest be increased by THINNING or fertilization? Is space available for ski runs? etc.); potential conflicts between the stated purpose and other uses of the same land, both actual and potential; and so on. The former use of "suitability" as a synonym for "capability" is now considered obsolete.

suitable forest land in Forest Service terminology, land suitable to be used for growing timber. Determination of suitability takes into account not only the land's CAPABILITY (as expressed in the SITE INDEX), but also its ability to be harvested without harm to soil, water, or future timber crops; its administrative classification (that is, whether or not it has been withdrawn from harvest as part of a WILDERNESS AREA, SPOTTED OWL MANAGEMENT AREA, STREAMSIDE MANAGEMENT ZONE, or other restrictive classification); and its ability to be adequately restocked with thriving SEEDLINGS within five years of harvest. The total amount of suitable forest land on a National Forest is roughly synonymous with the TIMBER BASE. *See also* SUITABILITY OF LAND.

sulfur the 16th element of the ATOMIC SERIES, atomic weight 32.064, chemical symbol S. Sulfur is the 12th most common element, making up approximately 0.5% of the Earth's crust. Its chemistry is exceedingly complex. It can take the place of OXYGEN in most reactions (*see* OXIDATION). It also reacts *with* oxygen to form two separate compounds, SO_2 and SO_3 (*see* SULFUR OXIDE). In the free state it is a solid, S_8, whose molecules consist of eight sulfur atoms joined in a crown-shaped ring. In *rhombic sulfur,* the most stable of its several ALLOTROPIC FORMS, the S_8 molecules form yellow eight-sided crystals; in *monoclinic sulfur,* they form yellow needles. *Pharmaceutical sulfur,* formed by any of several methods, is a yellow powder made up of tiny rhombic S_8 crystals. Sulfur is odorless, tasteless, and has very low TOXICITY, although it may irritate sensitive tissues. It is insoluble in water (*see* SOLUBILITY). As a component of some PROTEINS, it is essential to the continuation of life. Its principal environmental hazards arise from its presence as an impurity in COAL and PETROLEUM deposits (*see* the discussion under SULFUR OXIDE).

sulfur oxide (SOx) general term for all COMPOUNDS of SULFUR and OXYGEN. Two stable sulfur oxides exist (11 others may be formed in the laboratory). *Sulfur dioxide* (SO_2), a heavy, colorless gas with a suffocating odor, is formed when sulfur is burned in air. It dissolves in water to form sulfurous acid (H_2SO_3). A highly irritating compound, it causes lung damage in humans and other animals and is toxic to plants in concentrations as low as 0.3 ppm. *Sulfur trioxide* (SO_3), a liquid, is formed by the combination of oxygen with sulfur dioxide under conditions of high heat. The heat necessary for the reaction may be lowered considerably in the presence of a CATALYST, such as platinum (found in automotive CATALYTIC CONVERTORS). Sulfur trioxide is noncorrosive and nonacidic when dry, but has an extremely high affinity for water, dissolving in it with explosive force to form sulfuric acid (H_2SO_4), among the most corrosive substances known. The heat of solution is high enough to boil the water, forming sulfuric acid vapor, a serious environmental hazard. Both sulfur dioxide and sulfur trioxide are formed during the combustion of COAL and PETROLEUM with sulfur present as an impurity. Their presence in the ATMOSPHERE from these sources is one of the chief causes of ACID RAIN. *See also* SMOG.

sullage *See* GRAY WATER.

sun belt in the United States, the tier of states lying below 37°N (*see* LATITUDE). These states, stretching from southern California to North Carolina, lie within the region of relatively high atmospheric pressure known as the *subtropical high* or *subtropical subsistence* (*see* HORSE LATITUDES; ATMOSPHERE: *dynamics of the atmosphere*) and thus usually experience more sunshine and less rain than the states further north. Because of its favorable climate, the sun belt attracts not only retirees, but industrial growth as well. New industries, many of them "high tech," have led to extensive URBAN SPRAWL throughout the region, with increases in developed land exceeding those of the Northeast and Midwest. For example, in the period 1950–90 urban "land consumption" in Philadelphia increased by 273 percent, in Atlanta by 973 percent.

sun cups a condition common in old SNOW or FIRN in which the snow surface develops a pattern of broad cup-shaped hollows up to a meter across and a few centimeters to half a meter or more in depth. The cause is not known, but appears to relate to differential SUBLIMATION caused by the angles at which sunlight is reflected off the snow's surface. Sun cups are especially common at high altitudes, where in the late summer they may have become so deep that the ridges between them break down, forming a field of closely spaced pillars (*nieve penitentes*) that can make mountain travel extremely difficult.

sunflecks the small patches of sun, often only a few centimeters across (1–2 inches), which penetrate the CANOPY of a forest and reach the forest floor. Sunflecks sweep across the UNDERSTORY vegetation as the day progresses, lighting various portions of it in turn, if only for a few moments, and thus allowing PHOTOSYNTHESIS to proceed. Without them, most understory plants could not survive. Shade tolerance may thus be thought of primarily as a plant's ability to utilize sunflecks efficiently (*see* SHADE-TOLERANT SPECIES).

sunspot a magnetic storm on the surface of the Sun, showing up as an irregular dark splotch. The cause is unknown. Sunspots release large quantities of elementary particles that interact with the Earth's magnetic field, forming the auroras (northern and southern lights). They also indirectly bring about an increase in the amount of carbon-14 (*see under* CARBON) in the ATMOSPHERE, so that ancient sunspot activity can be traced by the amount of carbon-14 incorporated into organic materials. Sunspot activity appears to follow a regular cycle, with maxima and minima occurring every 11 years. (Actually, the cycle is 22 years long, as the magnetic polarity of the sunspots reverses at each peak; one peak will be positive, the next negative, etc.) The correlation between sunspot activity and climate has been disputed; however, rainfall seems to reach a low point for much of the world during the high point of negative sunspot activity every 22 years, and there is strong evidence that the "little ice age"—a period of colder-than-normal winters and general glacial advance during the late 17th and early 18th centuries—corresponded with the near-total absence of the sunspot cycle during that period.

sunstroke in medicine, a rise in body temperature caused by exposure to too much heat, often in the form of intense sunlight. Symptoms include a flushed face, headache, nausea, dizziness, and muscular weakness. Most of these are caused by overexpansion of the capillary system as the body attempts to cool itself by circulating blood more rapidly. Sunstroke can usually be prevented by wearing a hat. A related condition, *heat exhaustion,* occurs as a result of profuse sweating. Its symptoms include dehydration and muscle cramps resulting from the loss of SALTS in the sweat. Treatment for both includes rest, physical cooling of the body (bathing in cold water is helpful), and—especially for heat exhaustion—drinking liquids and eating small quantities of salt.

supercar *See* HYBRID-ELECTRIC CAR.

supercooling in chemistry and physics, cooling a liquid in such a way that it remains in liquid form below the temperature where it would normally become a solid (*see* FUSION POINT). Supercooling represents a *metastable state,* that is, a state that is stable only so long as it is not disturbed, and a very small disturbance, such as jostling the liquid, can usually lead to crystallization. The tiny droplets of water in clouds often become supercooled due to their extremely small size. If conditions allow the droplets to grow, they will form ice crystals rather than larger droplets (this is the usual way in which hailstone formation is begun). So-called "amorphous solids" (i.e., noncrystalline solids), such as glass, are actually stabilized supercooled liquids, and most will form crystals if allowed to stand long enough. Class becomes brittle with age largely because of this crystallization process.

superfamily in biology, a loose grouping of related FAMILIES that is not distinct enough from other such groupings to be called an ORDER. *See* TAXONOMY.

Superfund *See* COMPREHENSIVE ENVIRONMENTAL RESPONSE, COMPENSATION AND LIABILITY ACT.

superphosphate *See under* PHOSPHATE.

supersaturation in the physical sciences, term referring to any situation in which the concentration of one substance in another is higher than the laws of physics would normally allow. Air is supersaturated with moisture if it is below the DEW POINT but condensation has not yet taken place. A SOLUTION is supersaturated if it contains more SOLUTE than the SOLVENT can normally hold without precipitating some of it out (*see* PRECIPITATION). Supersaturation occurs when the normal boundary conditions are exceeded by slow increments without any other disturbances. It is a highly unstable condition, and will revert rapidly to the more normal state if disturbed (for instance, a supersaturated solution will rapidly precipitate its excess solute if its container is bumped).

supertanker a very large tanker; that is, a very large ship designed to carry liquid cargo, specifically CRUDE OIL. The term is poorly defined, but generally refers to tanks with a DEADWEIGHT CAPACITY (dwc) of 100,000 tons or more. Supertankers of 200,000 to 300,000 tons are referred to technically as VLCC's (for Very Large Crude Carriers); those of more than 300,000 tons are called ULCC's (Ultra Large Crude Carriers). ULCC's have been built with capacities of as much as 550,000 tons: these ships are up to 1,500 feet long and 200 feet wide, with a draft of nearly 100 feet and with engines that generate as much as 45,000 horsepower. Supertankers are capable of moving vast quantities of oil at a relatively low cost, and are favored by the shipping industry for this reason. Their environmental costs, however, are high. These include the need to build and maintain deep-water ports (or else transfer their cargos of crude oil at sea, a hazardous operation that runs the risk of major spills); pollution from tank-cleaning operation and from ballast leaks; and the ever-present threat of accidents, made more likely by the awkward handling characteristics of these vessels, and their single-hull design. While virtually all merchant ships have double hulls, only 15% of oil tankers have them, which perpetuates the risk of disastrous oil spills such as that of the *EXXON VALDEZ* in Prince William

Sound in Alaska. On the positive side, they use considerably less fuel per ton of product carried than do smaller vessels, they are more stable in rough seas, and they are generally better equipped with computerized navigational gear and other accident-avoidance equipment. *Compare* BULK CARRIER. *See also* OIL SPILL.

Surface Mining Control and Reclamation Act of 1977 (SMCRA) federal (U.S.) legislation to control and mitigate the harmful environmental effects of strip-mining COAL (*see* STRIP MINE), signed into law by President Jimmy Carter on August 3, 1977. It does not affect the strip mining of materials other than coal. The SMCRA requires the setting of performance standards for environmentally sound means of coal strip mining, and for the reclamation of mined land after the coal is removed. Permits are required for all strip mine operations. To obtain a permit, an operator must file an acceptable reclamation plan and post a bond ensuring his/her ability to finance the reclamation. Certain lands may be declared unsuitable for strip mining; land placed in this category may not be mined at all. Permit fees, forfeited bonds, and fines for improper mining practices are placed in an Abandoned Mine Reclamation Fund established to finance the reclamation of mine pits created before the passage of the act. The act is enforced by the Office of Surface Mining Reclamation and Enforcement within the DEPARTMENT OF THE INTERIOR. Enforcement may be delegated to state governments if they pass laws and regulations as strict as, or stricter than, the federal laws. Technical assistance is provided by the office to those states that choose to take advantage of this provision. *See also* MOUNTAIN-TOP REMOVAL.

surface runoff *See* NONPOINT POLLUTION.

surface tension a combination of forces that causes the surface of a liquid to act as though it were a very thin elastic membrane. Surface tension is caused by an imbalance in the intermolecular bonds acting on the MOLECULES at the surface of the liquid. They are pulled downward and sideways by the molecules beneath and around them, but there is no counterbalancing upward pull because there are no molecules above them. As a consequence, all "extra" molecules at the surface are pulled downward into the liquid and the surface is as compact as possible. Surface tension varies from liquid to liquid. It is particularly strong in water, and is the force that holds up "water striders" and other aquatic insects that appear to walk on the water's surface. It is responsible or partly responsible for numerous phenomena, among them the spherical nature of water droplets, the rise of water in soil (*see* CAPILLARY ACTION) and in plants (*see* TRANSPIRATION), the splash when a stone is dropped into a pool, and the ability of "breathable" waterproof fabrics to pass water vapor but not liquid water. *See also* WETTING AGENT.

surfactant *See* WETTING AGENT.

survey plate a small metal plate marked with TOWNSHIP, RANGE and SECTION numbers and left as a field marker, usually nailed to a BEARING TREE. Survey plates normally have a standard township engraved on them. The surveyor punches a hole through the proper section corner and scratches township and range numbers in the areas provided for them.

survivorship the number of individuals in a POPULATION that remain alive after a given age. It is usually represented as a graph called a *survivorship curve,* where the horizontal (x-) axis represents ages at death and the vertical (y-) axis represents numbers of individuals. *Compare* AGE DISTRIBUTION. *Compare* BED LOAD; DISSOLVED LOAD. *See also* MORTALITY.

suspended load in geology, the SILT and CLAY particles carried in SUSPENSION by a stream. The size of the suspended load depends primarily on the character of the stream's WATERSHED—that is, the availability of silt and clay particles and the rate at which they wash into the stream—rather than on the velocity of the stream or the shape of its channel. Once in suspension, the suspended load tends to remain until the water stops moving altogether. *See also* EROSION; DEPOSITION.

suspended solid in wastewater treatment terminology, any CONTAMINANT in water that is carried in suspension rather than being dissolved (*see* SUSPENSION; SOLUTION). *Settleable solids* are suspended solids that will settle out if a water sample is allowed to sit undisturbed for two hours. *Nonsettleable solids* are those that remain in the water after the two-hour test period is over. Suspended solids are larger than DISSOLVED SOLIDS but smaller than COARSE SOLIDS. *See also* FLOATER; TOTAL SOLIDS; VOLATILE SOLIDS; FIXED SOLIDS.

suspension in chemistry, physics, geology, etc., a form of MIXTURE in which very fine particles of a solid (or liquid) are dispersed through the body of a liquid (or gas). The materials involved keep their own chemical and physical identities and will separate ("settle out") if allowed to stand undisturbed for a long enough period of time. *Compare* COLLOID; SOLUTION.

Susquehanna River Basin Commission charged with the protection and careful management of the Susquehanna River, the 16th-largest river in the United

States. The commission was established by a compact signed on December 24, 1970, between the federal government and the states of New York, Pennsylvania, and Maryland. Its principal concerns include flood control; development and use of surface and groundwater for municipal, agricultural, recreational, commercial and industrial purposes; protection and rehabilitation of fisheries and wetlands; safeguarding water quality and instream uses; and ensuring future river flow to the Chesapeake Bay. Address: 1721 North Front Street, Harrisburg, PA 17102. Phone: (717) 238-0422. Website: www.srbc.net.

sustainable agriculture the use of agricultural management practices that maintain soil quality and reduce the pollution of air and water by agricultural chemicals. Sustainable agriculture is usually considered a middle ground approach between large-scale industrial agricultural practice and organic farming. *See also* ORGANIC GARDENING AND FARMING.

sustainable development term applied to development policies that do not lead to diseconomies of scale associated with URBAN SPRAWL.

sustained yield in resource management, the harvest of timber or other RENEWABLE RESOURCES at a rate that may be maintained throughout the foreseeable future. The rate may be figured on either an annual or an episodic basis; that is, the resource may be used only at the rate at which it renews itself each year (e.g., cutting only a volume of timber equal to the MEAN ANNUAL INCREMENT; *see* NON-DECLINING EVEN FLOW) or it may be used all at once and then allowed to renew itself slowly (e.g., cutting all the timber on a block of land and then reseeding it and tending the growing forest until it contains the same volume of wood as when it was cut, at which time it may be cut again). Note that although the term is usually thought of as a timber management term (as in these examples), it applies equally well to other renewable resources, such as wildlife, water (*see* SAFE YIELD), FORAGE, etc. All federal agencies are required by law to practice sustained yield of all renewable resources under their jurisdiction. For the FOREST SERVICE, this requirement is contained in section 11(a) of the NATIONAL FOREST MANAGEMENT ACT. For the BUREAU OF LAND MANAGEMENT, the applicable law is section 102 of the FEDERAL LAND POLICY AND MANAGEMENT ACT. *See also* ALLOWABLE CUT; PROGRAMMED ANNUAL HARVEST.

swamp a type of WETLAND that supports woody growth (shrubs and trees) as well as GRASSES and other HERBS. The water table in a swamp is high, and there is generally standing water part or all of the year; however, PEAT does not develop and the soils are similar to those on nearby dryland sites. The trees in a swamp belong to water-tolerant genera (*see* GENUS) such as cedar (*Thuja*), willow (*Salix*), and bald cypress (*Taxodium*). *Compare* BOG; MARSH.

swash water that washes up a beach following the breaking of a wave. The height attained by swash (the *runout height*) is determined both by the height of the tide (or lake surface) and the energy of the wave. *See also* LITTORINA ZONE; RIFT.

S-wave (secondary wave) in seismology, a transverse ground wave caused by an EARTHQUAKE, that is, a ground wave that moves up and down or from side to side rather than alternately compressing and expanding the Earth. S-waves are so named because they travel more slowly than P-WAVES, and are thus the second set of waves to arrive at a SEISMOGRAPH following a nearby earthquake.

symbiosis (mutualism) in biology, a relationship between two ORGANISMS from which both derive some benefit. The organisms involved in such a relationship are called *symbionts* and the relationship itself is a *symbiotic relationship*. Symbiotic relationships may be very tight (as in the LICHENS, where the symbionts are so closely intertwined that they are usually treated by botanists as if they were a single organism) or rather loose (as in the case of birds that pick ticks and other ECTOPARASITES from the skins of large mammals; the birds obtain food and the mammals obtain relief from the parasites, but the two species remain fully independent and each may achieve the same benefit in other ways). Other examples of symbiotic relationships include bees and flowers, trees and MYCORRHYZA, and cattle and humans. *Compare* PARASITE; COMMENSAL SPECIES.

sympatric species two or more SPECIES that inhabit the same area. Sympatric species usually show more strongly divergent characteristics than do ALLOPATRIC SPECIES, that is, they specialize in certain areas in order to avoid COMPETITION. Two species of bird that are each capable of feeding on both fruit and insects, for example, may specialize in areas where they are sympatric so that one eats only fruit and the other eats only insects. This may eventually show up physiologically in different bill shapes and digestive tracts, creating separate RACES or even separate species out of sympatric and allopatric POPULATIONS of the same species. *See* CHARACTER DISPLACEMENT.

syncline in geology, a downward bow developed in strata (*see* STRATUM) that were originally laid horizontally. The ends of the strata involved in a syncline tilt upwards, while the center bulges downward. The determination as to whether bowed strata are synclines or ANTICLINES depend on the position of the strata when the bowing took place. A syncline may be tilted on its side or even inverted, but it will still be revealed as a syncline by the presence of strata laid down in the bow after the deformation (the forming of the syncline by TECTONIC activity such as volcanic action or EARTHQUAKES) but before the displacement (the tilting or inverting of the deformed formation).

synergy a combination of forces, actions, etc., in such a way that the combined result is greater than the sum of the results obtained when each works alone. Synergy is the phenomenon we refer to when we say "the whole is greater than the sum of the parts." Living systems are almost always synergistic in nature. The actions of even as simple an ORGANISM as a BACTERIUM cannot be predicted by analyzing its chemical makeup. The stability of ECOSYSTEMS results from synergistic relationships among organisms—one of the principal reasons that EXTINCTION, even of so-called "pest" species, is looked upon with such alarm by ecologists.

synfuels (synthetic fuels) substitutes for PETROLEUM and NATURAL GAS manufactured from other sources, such as COAL, SHALE, trash, or forestry and agricultural wastes. Not a new idea—Scotland was producing oil from shale in the 18th century and Germany made GASOLINE from COAL and ALCOHOL from potatoes to power its vehicles and aircraft during World War II—synfuels nevertheless enjoy a reputation as an exotic, futuristic and impractical idea, and efforts to begin manufacturing them on a large scale in the United States, including the formation by Congress in 1980 of a federal corporation (the U.S. Synthetic Fuels Corporation) to encourage their development, have not so far been particularly successful. Environmentally, synfuels show a mixture of good and bad traits. Coal and shale are usually obtained by STRIP MINING—a process that is inherently destructive to the environment—and processing these materials into fuels requires massive amounts of water, emits high levels of POLLUTANTS, and creates a considerable amount of highly acidic waste rock. On the other hand, METHANE and liquid fuels derived from cornstalks, forest SLASH, cattle manure, and other organic wastes have a lighter impact on the environment than conventional petroleum manufacture does, and have the additional advantage of being renewable. Synfuel derived from trash (refuse-derived fuel, or RDF: *see* RESOURCE RECOVERY) occupies a middle ground. It may help solve the municipal solid waste problem (*see* SOLID WASTE), and it is certainly renewable, but its manufacture releases large amounts of pollutants, and any serious effort at development will have to find ways to deal with this problem. There seems little doubt that synfuels of both the environmentally destructive and the environmentally benign varieties will play a larger role in the future than they do now. *See also* OIL SHALE.

systemic poison a POISON that is spread throughout the body of a living ORGANISM. The term is used both for poisons that act systematically (that is, that affect all parts of an organism) and for COMPOUNDS that spread systematically without harming an organism but happen to be poisonous to the organism's enemies. A plant treated with a systemic poison of this type will not itself be harmed but will kill the insects that normally feed on it. Systemic poisons are useful as PESTICIDES for ornamental plants but should not be used on food or fodder plants because they may harm humans and livestock as well as insects. Even for ornamentals, use should be carefully controlled. (Studies have shown, for example, that honey made from plants treated with systemics is usually contaminated with significant quantities of the systemic.) Examples: Demeton (Systox); OMPA (Pestox III).

T

tablemount *See* GUYOT.

taconite a type of iron ORE consisting of thin layers of iron-rich minerals such as HEMATITE alternating with slightly thicker layers of CHERT. Taconite is a low-grade ore (*see* LEAN ORE) containing between 20% and 35% iron, and must be concentrated (*see* BENEFICIATING) before it can be fed to a smelter. The product resulting from this concentration process is also generally known as taconite. Named for its superficial resemblance to the GRANITE found in the Taconic Mountains of New York and Vermont—but unrelated to that rock—taconite is found in upper Michigan and in Minnesota, where it forms the principal ores of the Marquette and Mesabi iron ranges. Approximately 90% of current U.S. steel production from domestic ores originates as taconite.

taiga *See* CONIFEROUS FOREST BIOME.

tailings the waste rock produced by a mine, smelter, or other MINERAL-extraction industry. Mine tailings consist of ORE that has too little mineral content to make processing it profitable (*see* LEAN ORE; POOR ROCK) plus the OVERBURDEN and COUNTRY ROCK that was removed to get at the ore. Tailings from processing operations consist of the pulverized matrix from which the product has been extracted. In form they are usually PEBBLE-sized rock chunks, but they may range in size from powder to large boulders. Tailings affect the environment in a number of ways. Dumped on the land, they alter WATERCOURSES and smother vegetation; rainwater leaching through them (*see* LEACHATE) may pick up toxic minerals. Tailings from uranium operations are radioactive; tailings from COAL operations are highly acidic and their leachate can destroy aquatic ECOSYSTEMS. Tailings from taconite beneficiating operations (*see* TACONITE; BENEFICIATING) dumped into Lake Superior during the 1950's and 1960's carried fibers of ASBESTOS, which made their way into drinking-water intakes as much as 60 miles away. *See also* TAILINGS DAM; TAILINGS POND; SPOIL.

tailings dam a DAM built to contain a pile of TAILINGS and its associated LEACHATE and keep them from affecting nearby WATERCOURSES.

tailings pond a body of water, usually man-made, in which TAILINGS from ore-processing operations are deposited. Tailings ponds are used as a means of containing fine tailings and, by keeping them underwater, preventing them from blowing away as dust. Their waters are normally heavily polluted and must be separated from nearby WATERCOURSES and from local AQUIFERS.

taking *See* EMINENT DOMAIN; PROPERTY RIGHTS.

Tallgrass Prairie Alliance environmental protection organization, founded in 1973 to preserve and protect the remaining remnants of the North American Tallgrass Prairies. Through a subsidiary, the Grassland Heritage Foundation, the alliance maintains a system of private preserves. It also works for federal and state park protection and for the creation of prairie preserves by other private parties. Prior to 1985 it was known as Save the Tallgrass Prairie. Membership (1988): 1,200. Address: 4101 W. 54th Terrace, Shawnee Mission, KS 66205. Phone: (913) 377-3326.

talus a sloping pile of rock fragments lying against the base of a cliff or steep slope. Talus consists of fist-sized or larger materials that have weathered from the cliff or slope, usually through FROST-WEDGING, and have been pulled to the bottom by gravity. If the talus is piled evenly along the base of the cliff, it is referred to as a *talus slope;* if it is piled in a triangular heap at the base of a COULOIR, it is referred to as *talus cone.* The piles thus formed are inherently unstable, and move slowly to level themselves, a process known as *talus creep.* If the talus is piling up slowly (erosion of the cliff or slope is slow), talus creep may be imperceptible on a human time scale; if its piling up rapidly, talus "creep" may be replaced by large-scale rapid movements of the talus known as *talus avalanches. Compare* SCREE.

tank farm a cluster of storage tanks, especially aboveground tanks. Tank farms for the storage of hazardous materials (that is: all materials but water) should include not only the tanks but some means of containing and cleaning up spills.

tannic acid (tannin) any large group of loosely related organic acids (*see* ORGANIC COMPOUND; ACID) formed in plant tissues, especially in leaves and bark. Different SPECIES of plants contain differing amounts of tannic acid. The highest percentage is found in oak GALLS, where it may comprise as much as 60% of the structure. Most tannic acids are ESTERS of glucose or other sugars, though some are derived from FLAVANOIDS instead. They are composed of CARBON, HYDROGEN, and OXYGEN (typical formula: $C_{27}H_{24}O_{18}$) in the form of a PHENOL-based POLYMER, and share a number of common properties, among them good solubility in water (*see* SOLUTION), the ability to form PRECIPITATES with many PROTEINS and with the SALTS of most metals, and a yellow-white color which darkens to deep brown on exposure to air and light. They contribute to the acidity of HUMUS and of BOG water, and are used by humans to tan leather and in the production of rubber and some inks. Their ability to bind metal salts makes them useful as an antidote for HEAVY METAL poisoning, although they must be used with care as they are themselves poisonous (LD_{50} [mice, orally], 6g/kg).

taper in forestry and logging, the gradual, usually uniform reduction in diameter from the base to the tip of a tree, especially a CONIFER; also, the gradual reduction in diameter of a log from the end that was originally nearer the bottom of the tree (the *butt end*) to the end that was nearer to the top. The average taper (*standard taper*) for large conifers is approximately 1 inch for every 8 feet. *See also* LOG SCALING; GIRARD FORM CLASS.

tape sampler in pollution control, a device for sampling PARTICULATE levels in the AMBIENT AIR at set intervals over a long period of time. Officially known as the American Iron and Steel Institute (AISI) Smoke and Haze Sampler, the tape sampler consists of a tube of a known diameter (usually 1 inch) through which air is forced at a fixed rate, commonly 0.25 cubic feet per minute. A tape of filter paper is mounted on reels across the end of the sampling tube so that all air that passes through the tube must also pass through the tape. Particulates in the air are deposited on the filter paper as a circular spot the size of the end of the tube. After a set time interval (generally 2–4 hours) the tape is reeled forward to a clean spot and the collection begins anew. The spots may be stamped with the date and time. The tape is analyzed by passing it over a light source and measuring the amount of light transmitted through the spot, producing a reading known as the *coefficient of haze* (COH), which may then be standardized by calculating the corresponding COH for a column of air 1,000 feet long (the "COH/1000"). Tape samplers have the advantage of being able to be run unattended for weeks or even months and still give an accurate reading of particulate levels on a daily (or even hourly) basis. *Compare* HIGH VOLUME SAMPLER.

taproot a large, deep central root. Taproots develop in GYMNOSPERMS and in DICOTYLEDONS; MONOCOTYLEDONS typically develop shallow, many-branched root systems instead (*see* FIBROUS ROOT SYSTEM). A *taproot system* consists of the taproot plus the secondary roots that branch from it. These secondary roots may be rudimentary (as in the carrot) or they may comprise most of the root system, with the taproot remaining relatively shallow and unimportant (as in the alder). Generally, plants with well-developed taproots are more difficult to uproot and can survive better in times of drought, whereas plants with poorly developed taproots or with fibrous root systems do a better job of anchoring the soil and preventing EROSION.

target (target goal) in Forest Service terminology, usually, a long-range OBJECTIVE. Like an objective, a target is clearly defined and is reachable by a series of concrete steps within a specified time frame; however, the time frame involved is in the range normally associated with GOALS rather than objectives. (An exception to this extended time frame is found in budgeting and performance rating, where the target time is normally one fiscal year). The term is also sometimes used simply as a synonym for objective.

target species the SPECIES of living ORGANISM toward which some management activity is directed. The term is most commonly used in relation to PESTICIDES. In this case, the target species is the species that the pesticide applier is attempting to kill. *Compare* NONTARGET SPECIES.

tarn a small mountain lake, especially one lying in a basin formed by glacial action. Tarns usually occupy CIRQUES, often above TIMBERLINE. They are almost always oligotrophic, although shallow ones may have mesotrophic or even eutrophic characteristics (*see* TROPHIC CLASSIFICATION SYSTEM). *See also* GLACIAL LANDFORMS.

taxis (plural: *taxes*), in biology, an involuntary movement by an ORGANISM in which the stimulus to make the movement and the control over the direction in which the movement takes place both come from outside the organism. The energy for the movement is supplied by the organism rather than by the stimulus (rolling down a hill does not count as a taxis). A taxis may involve the entire organism (for example, the AVOIDANCE behavior of MICROORGANISMS or the "pursuit" behavior of predatory microorganisms such as AMOEBAS, which is controlled by chemical stimuli released by the PREY) or it may involve only part of the organism (for example, the reflex that causes a human to pull his or her hand away from something hot).

taxol a cancer-fighting drug derived from the bark of the Pacific yew, a sparsely distributed tree of no commercial value as timber, found only in old growth forests of the Far West. Taxol was discovered in the late 1960s, but its therapeutic value in treating ovarian cancer was not understood until the late 1980s, after which demand for the drug exploded in order to deal with a cancer that was killing 12,000 women a year. To extract enough of the drug for the treatment of a single patient, the bark had to be stripped from three Pacific yews. Because much of the old growth, where the tree could be found, had already been CLEARCUT, the search for remaining trees to harvest for bark was intense, causing some fear that the species might become extinct. That the Pacific yew required the same old-growth environment as the endangered spotted owl, which by court order had caused all cutting to cease in old-growth forests, made the controversy especially vivid in that some chose to characterize the issue as whether owls or cancer victims should be saved. The problem was soon resolved, however, when drug companies and the forest products industry found that a hybrid form of the tree, which could be grown on tree farms, produced a semisynthetic version of taxol. An added benefit of the hybrid was that the entire tree could yield taxol, not just the bark. By 1993 the hybrid yews had grown enough for the first harvest, by the Weyerhaeuser Company. More recently, drug companies have been able to synthesize taxol in the laboratory, a significant development since the pressure to produce taxol was greatly increased with the finding that the drug was also effective in treating metastasized breast cancer in combination with standard chemotherapy regimes. Moreover, other studies have shown that taxol is effective in treating early-stage breast cancer as well. Since more than 180,000 women are diagnosed with breast cancer each year, the demand for taxol has increased by a whole order of magnitude since the substance was first used for ovarian cancer sufferers. In a larger context, the taxol discovery has given new urgency to the effort to discover natural-product drugs from threatened plants and trees in forested areas now being rapidly cleared, especially the tropical rain forests of Central and South America. *See also* SPOTTED OWL MANAGEMENT AREA.

taxonomy in the biological sciences, the systematic classification and naming of ORGANISMS. Taxonomy serves two purposes. First, it grants to every SPECIES of organism a name (the *scientific name*) that belongs only to that species and by which it will be recognized by scientists worldwide, thereby clarifying and improving communication about the species. (The need for this may be demonstrated by the common name *robin,* which refers to two unrelated birds, one found in Europe, the other in North America. The name "robin" therefore does not uniquely identify the North American species. The scientific name *Turdus migratorius,* however, does uniquely identify it.) Second, taxonomy demonstrates the patterns of relationships that exist among different groups of organisms, and therefore helps clarify evolutionary trends and mechanisms and show how the environment may shape an organism. A comparison among the scientific names of the Atlantic white cedar (*Chamaecyparis thyoides*), the Port Orford cedar (*Chamaecyparis lawsoniana*), and the Northern white cedar (*Thuja occidentalis*), for example, shows at a glance that the Port Orford and the Atlantic white are more closely related to each other than either is to the Northern white, though the common names and the proximity of ranges among the three species would indicate differently.

THE TAXONOMIC SYSTEM

Everyday language practices a rudimentary form of taxonomy. Depending on the broadness of the reference desired, for example, the same organism may be referred to as "an animal" (i.e., not a plant or a FUNGUS); a "mammal" (not a lizard, bird, fish, or insect); a "cat" (not a dog, kangaroo, or raccoon); or a "house cat" (not a lion or tiger). Scientific taxonomy works in much this same manner. Biologists divide the living world into a number of levels of classification, or *taxa* (singular: *taxon*). The broadest taxon is the KINGDOM. There are five kingdoms, and all living things belong to one, and only one, of these five. Each kingdom is divided into several phyla (*see* PHYLUM; in botany, this taxon is often referred to as a *division*); each phylum is divided into CLASSES, each class into ORDERS, and each order into FAMILIES. At the level of greatest detail are the GENUS and the SPECIES. These have special significance because the scientific name used to identify a particular species (the *binomial*) is made up of the genus name followed by the species name. This is a little like a family name being coupled to a given name. There may be several Browns in a room, and there may be several Johns in the same room, but there is likely to be only a single individual in the room referred to by the binomial "John Brown." Similarly, the species name *domesticus* is used in many genera, and the genus *Felis* includes several named species, but the binomial *Felis domesticus* refers only to a single species, the house cat.

HISTORY AND PRACTICE OF TAXONOMY

Although there were earlier attempts to rationalize the names of organisms in a systematic way, taxonomy is usually said to have begun with the Swedish scientist Carl von Linne (1707–1778), known more widely as Carolus Linnaeus. He was the first to utilize binomial names for organisms. Many refinements have since been made in his basic system, primarily by 19th century biologists following the concepts of Darwinian evolution. Today, taxonomy is recognized as a speciality within biology, and taxonomists normally work on nothing else but classifying and naming things. They place organisms into different taxa according to evolutionary relationships shown by evidence such as bone structure, reproductive organization, and microscopic similarities in tissue. The classifications that result are formalized by international agreement, so that all scientists everywhere in the world will use the same binomial to refer to the same species. The whole system is constantly undergoing review and revision. Names may change as new studies clarify and confirm, or disprove, suspected relationships.

Taylor Grazing Act federal legislation to protect the public RANGELANDS of the American west, signed into law by President Franklin Roosevelt on June 28, 1934. Essentially a repeal of the HOMESTEAD ACT, the Taylor Grazing Act authorized the secretary of the interior (*see* INTERIOR, DEPARTMENT OF THE) to withdraw large tracts of PUBLIC DOMAIN lands from settlement and occupancy. The withdrawn tracts, termed "grazing districts," were to be managed ". . . to regulate their occupancy and use, to preserve the land and its resources from destruction or unnecessary injury, [and] to provide for the orderly use, improvement, and development of the range. . . ." An advisory committee of ranchers and other livestock owners who used the rangelands was to be created for each grazing district. The original legislation limited the withdrawals to 80 million acres; this limitation was raised to 1,452 million acres in 1936, and eliminated altogether in 1954. The lands withdrawn under the Taylor Grazing Act are those known today as "BLM lands." The 59 grazing districts created under the act's terms have become the management districts of the BUREAU OF LAND MANAGEMENT, and the advisory committees have been expanded to include all users of the public lands rather than simply stockmen. *See also* FEDERAL LAND POLICY AND MANAGEMENT ACT OF 1976.

TCDD *See* DIOXIN.

TDS abbreviation for *total dissolved solids* (*see under* TOTAL SOLIDS).

technological fix in an environmental context, a phrase, usually pejorative, describing a solution to an

Sample Taxonomic Breakdowns for Several Species

	1.	2.	3.
kingdom	Animalia	Animalia	Plantae
phylum	Chordata	Arthropoda	Spermatophyta
class	Aves	Insecta	Angiospermae
order	Passeriformes	Diptera	Asterales
family	Sturnidae	Xylophagidae	Compositae
genus	*Sturnis*	*Musca*	*Taraxacum*
species	*vulgaris*	*domestica*	*officinale*
common name	starling	house fly	dandelion

environmental problem that is "merely" technological rather than systemic and comprehensive and that may have unforeseen adverse consequences. In agriculture, a technological fix often cited is the corporate development and marketing of genetically modified corn that contains a natural pesticide (Bt—*Bacillus thuringienses*) so that the need for spraying is reduced. While an environmental benefit may be seen in that chemical runoff and groundwater pollution is abated somewhat, the pollen from the modified corn was found to kill the larvae of the spectacular monarch butterfly. Some environmentalists would suggest that the planting of "Bt corn" was a technological fix, in that the best approach to controlling pests without pesticide spraying would be to adopt organic farming techniques to strengthen the resistance of corn to insect attack rather than genetically modifying it to exude a poison.

tectonic activity in geology, activity that results in movement or deformation of the Earth's crust. Tectonic activity includes EARTHQUAKES, volcanic intrusions (*see* INTRUSIVE ROCK; VOLCANO), seafloor spreading (*see* PLATE TECTONICS), ISOSTATIC REBOUND, and anything else that distorts bedrock structure. It generally results in the formation of large-scale landforms such as mountain ranges and STRUCTURAL VALLEYS. Rocks whose internal structure has been modified by tectonic activity are called *tectonites.*

tectonic basin in geology, a valley, lake basin, or similar depression in the Earth's surface that results from deformation of the local rock structure rather than from EROSION or DEPOSITION. The most common forms of tectonic basin are GRABENS and SYNCLINES. *See also* STRUCTURAL BASIN.

telecommunications towers *See* CELL TOWERS.

Teilhard de Chardin, Pierre (1881–1955) French Roman Catholic priest, geologist, paleontologist, and philosopher, who (against church teaching) interpreted evolution as a process involving both science and religion. Educated in Jesuit schools in France and England, Teilhard received a doctorate in paleontology from the Sorbonne in Paris in 1922. In 1918 he became a professor of geology at the Institut Catholique in Paris, but his teaching career was cut short because of his unorthodox ideas. His book *The Phenomenon of Man* (1955) detailed his theory of evolution as a movement toward the increasing complexity of organisms. Teilhard suggested that the emergence of man in the prehistory of the world brought forth a new dimension in evolution: that from the pre-human biosphere emerged the noosphere, a mind-layer (human consciousness) operating above and beyond the inhabitants of the Earth. From the noosphere increasingly complex social arrangements arise, leading to a higher (religious) consciousness. Ultimately, in Teilhard's system, the material and spiritual realms converge into a superconsciousness he calls the *Omega Point,* wherein God, the Omega, attracts organisms to himself and directs evolution to God's purposes. Aside from his philosophical and theological achievements, Teilhard was known for his contributions to geology and paleontology. After being released by the church from his teaching position in Paris, he worked in China for nearly 20 years and participated in the excavation that uncovered Peking man. In 1952 he moved to New York City to work with the Wenner Gren Foundation for Anthropological Research. Among his other published works (all published posthumously) are *Letters from a Traveler* (1956), *The Divine Milieu* (1957), and *The Future of Man* (1959).

Tellico Dam dam on the Little Tennessee River near Lenoir City in east-central Tennessee, which in the late 1970s became the site of a major test of the ENDANGERED SPECIES ACT. A project of the TENNESSEE VALLEY AUTHORITY, which first proposed it as early as 1936, Tellico Dam was authorized by Congress in 1966. Construction began the next year. In 1976, environmentalists sued to halt the project under the Endangered Species Act due to the presence in the project area of the entire remaining POPULATION of the snail darter, a 3-inch-long member of the perch family. On June 15, 1978, with the dam 98 percent complete, the U.S. Supreme Court ordered a halt to construction. When the Endangered Species Act was renewed by Congress on November 10, 1978, it carried an amendment authored by Senator Howard Baker (R-TN) creating an Endangered Species Committee (the so-called "God Committee") in the Executive Office of the President to rule on possible exemptions to the act on economic grounds. On January 23, 1979, at its first meeting, the committee voted unanimously to deny an exemption to Tellico, on the grounds that it would be more cost-effective to tear the dam down than to complete it. Baker and the local congressman, John Duncan, then pushed through an amendment to the 1979 public works appropriations bill (signed by President Jimmy Carter on September 25, 1979) specifically exempting Tellico from the Endangered Species Act and any other Federal legislation it was in violation of (including the NATIONAL ENVIRONMENTAL POLICY ACT, the CLEAN WATER ACT, the Federal Dam Safety Act, and the Historic Preservation Act). The floodgates were closed on November 29, 1979. An earthfill dam 1.2 kilometers (3/4 mile) in length and 39 meters (129 feet) high, Tellico has created a reservoir 53 kilometers (33 miles) long, with a surface area of

16,000 acres. Before the reservoir was filled, 700 snail darters were captured and transported to the nearby Hiwassee River to preserve the species.

temperate climate a CLIMATE marked by moderate average annual temperatures. There are two types: one in which temperature is moderate all year round (the *temperate oceanic climate*) and one in which the temperature is hot in the summer and cold in the winter (the *temperate continental climate*). The actual boundaries of these zones are not well pinned down, but climatologists generally consider the temperate climates to consist of all climates in which at least four months of the year, but no more than eight months of the year, have average temperatures exceeding 10°C (50°F). This is generally the region between 40° and 50° north and south latitude, although the local anomalies caused by the large landmasses in the northern hemisphere push the climate as far north as 70° over northern Norway and as far south as 30° over the Rocky Mountains and the Himalayas. *See also* TEMPERATE ZONE; TEMPERATE FOREST BIOME.

temperate deciduous forest *See* DECIDUOUS FOREST BIOME.

temperate forest biome a BIOME is recognized by some authorities that consist of the temperate phase of the DECIDUOUS FOREST BIOME, sometimes with the inclusion of the Pacific Coast phase of the CONIFEROUS FOREST BIOME.

temperate zone in geography, either of the two middle-latitude regions of the Earth's surface, extending from 23°27' north and south of the equator to 66°33' north and south, that is, the area between the Tropic of Cancer and the Arctic Circle in the northern hemisphere, and between the Tropic of Capricorn and the Antarctic Circle in the southern hemisphere. *See also* TEMPERATE CLIMATE; TEMPERATE FOREST BIOME.

temperature a measure of the "hotness" or "coldness" of a substance. Temperature is not dependent upon the *quantity* of heat present, but upon the *intensity* of that heat—not on the number of active MOLECULES the substance is composed of, but on the average level of activity of the molecules. A cup of boiling water is considerably hotter than a swimming pool full of water at 32°C (80°F), but it contains far less heat. One means of defining temperature is to classify it as *the ability of an object to transfer heat to another object or to its surroundings.* Heat will always transfer from an object at a higher temperature to an object at a lower temperature, regardless of the total heat present in each object. Two objects at the same temperature will not transfer heat to each other at all. When this situation exists, the objects are said to be in *thermal equilibrium.* Temperature is normally expressed by comparing the heat-transfer ability of an object to the heat-transfer ability of a standard, usually water at its freezing point (or, more accurately, at its TRIPLE POINT). *See also* ABSOLUTE TEMPERATURE.

1080 (compound 1080, sodium fluoracetate) a SALT of fluoroacetic acid, chemical formula $C_2H_2FNaO_2$, once widely used as a RODENTICIDE and predator poison (*see* PREDATOR CONTROL), especially in the American west. 1080 is odorless, colorless, tasteless, and highly lethal (LD_{50} [oral est.]: 2mg/kg). It is extremely stable in the environment and in animal METABOLISM, with pronounced secondary poisoning effects; that is, it will kill animals that feed on the carcasses of its victims. It is soluble in water and can contaminate watercourses and AQUIFERS. There is no known antidote. 1080 was normally used by injecting it into the carcass of a sheep, which was left out on the range as a "1080 bait station." Indiscriminate use of the compound led to the deaths of many NONTARGET SPECIES, and eventually to an EXECUTIVE ORDER by President Richard Nixon (EO 11643: February 2, 1972) banning the use of 1080, and all other predator-control poisons that cause secondary poisoning effects, on PUBLIC LANDS in the United States.

Tennessee-Tombigbee Waterway (Tenn-Tom) massive navigation project connecting the Tennessee River to the Tombigbee River in Mississippi and Alabama. Completed on December 11, 1984, the 375-kilometer (234-mile) waterway cost U.S. taxpayers $1.8 billion, making it the most expensive navigation project in history. Two hundred thirty-five million cubic meters (307 million cubic yards) of earth were displaced by the project—nearly half again as much as was moved to create the Panama Canal. The Tenn-Tom consists of three segments: the "river" segment, which consists of dredging, small navigation dams, and other channel improvements along 149 miles of the Tombigbee from Demopolis, Alabama, to Amory, Mississippi; the "lakes" segment, 46 miles of major navigation dams from Amory to the Tombigbee's headwaters; and the "divide" segment, 39 miles of canal (the "divide cut") connecting the lakes segment to Pickwick Reservoir on the Tennessee River at the Mississippi-Tennessee border. The channel is maintained throughout to a width of 85 meters (280 feet) and a depth of 2.75 meters (9 feet), with an overhead clearance of 17 meters (55 feet). Environmental costs included filling 51 valleys with dredge SPOIL, the destruction of 16,000 hectares (40,000 acres) of hardwood forest, and the removal of 6,900 hectares (17,000 acres) of farmland from production, as well as the lowering of the regional WATER TABLE along the divide cut and the elimination of the recreation, wildlife and water-quality benefits provided by the free-flowing stream prior to project construction. Benefits have primarily accrued to shippers in the form of lessened fuel costs and transit time for barge shipments to and from central Kentucky, though there has also been a small waterfowl benefit from the creation of 17,800 hectares (44,000 acres) of new lakes along the Mississippi FLYWAY. First proposed around 1870, the Tenn-Tom was authorized by Congress in 1946, but construction funds were not appropriated until 1971. During the 1970s it was a major focus of environmentalists' water project reform efforts. Suits were twice filed to halt construction, and in 1979 Senator Gaylord Nelson (D-WI) sponsored an unsuccessful bill to deauthorize the project.

Tennessee Valley Authority (TVA) independent federal (U.S.) agency charged with the development and management of the Tennessee River, its tributaries, and the land and resources of its drainage basin, an area of some 41,000 square miles covering most of the state of Tennessee and portions of Kentucky, Mississippi, Alabama, Georgia, North Carolina, and Virginia. The agency's extremely broad range of duties includes the maintenance of a 1,100-kilometer (650-mile), 2.7-meter (9-foot) deep shipping channel on the Tennessee MAINSTEM from Paducah, Kentucky, to Knoxville, Tennessee; the construction and operation of multipurpose dams (*see under* DAM) on the Tennessee and its tributaries, and the coordination of these dams with privately owned dams in the Tennessee basin and with dams operated by the ARMY CORPS OF ENGINEERS on the nearby Cumberland River system; the generation, transmission, and marketing of electrical power from its dams and from the coal-fired and nuclear steam-generation plants it also constructs and operates; watershed protection in the basin through forestry, reforestation, and soil conservation activities; recreation management; and the manufacture of nitrate fertilizer (*see under* FERTILIZER).

STRUCTURE AND FUNCTION

TVA is operated as a corporation that is wholly owned by the federal government. Its three-person board of directors is appointed by the president and confirmed by the Senate, and its operating budget is provided for by APPROPRIATIONS by Congress from the federal treasury, although the agency's electrical generation and marketing arm is made financially self-supporting by the sale of electricity to the 160-odd municipal utilities and corporations that have contracted with TVA as a wholesale supplier. Day-to-day operation is in the hands of a general manager appointed by the board: he/she oversees a staff of approximately 24,000 apportioned among a number of administrative divisions and six major program offices. These offices include an Office of Coal Gasification, an Office of Agricultural and Chemical Development, an Office of Power, an Office of Engineering Design and Construction, an Office of Economic and Community Development, and an Office of Natural Resources, which houses the Environmental Quality staff and is responsible for attempting to minimize the impact of the other five program offices on the environment and for making certain that TVA activities as a whole conform to federal laws on environmental protection. The Office of Natural Resources is also responsible for TVA's recreational developments, including the showcase Land Between The Lakes Regional Parkland lying between man-made reservoirs on the Tennessee and Cumberland Rivers just above their respective confluences with the Ohio.

HISTORY

The Tennessee Valley Authority was created by Congress on May 18, 1933. Its facilities at the time of its birth consisted of a single dam (Wilson) and two nitrate fertilizer plants, all near Muscle Shoals, Alabama. Controversial from the beginning, TVA survived a series of court challenges in the late 1930's that sought to have it declared unconstitutional and several similar challenges following World War II seeking to limit its scope to water resource development. Though from an environmental standpoint it is easy to argue with the agency's full-development bias, it is also true that TVA has provided an extraordinarily successful model for integrated resource management based on WATERSHED boundaries rather than on political borders. Headquarters: 400 West Summit Hill Drive, Knoxville, Tennessee 37902. Phone: (423) 632-2101. Website: www.tva.gov.

teratogen a substance or an ORGANISM that causes birth defects by affecting the development of the fetus rather than by altering its genetic makeup (*compare* MUTAGEN). Examples of teratogens include DIOXIN and the measles VIRUS.

terminal bud in the biological sciences, a plant bud located at the tip of a stem (as opposed to a *lateral bud,* which is located along the side of a stem). Terminal buds may develop into flowers, leaves or INFLORESCENCES. In forestry, the term *terminal bud* (or often just *terminal*) is usually used to refer specifically to the bud at the tip of the LEADER of a CONIFER, that is, to the uppermost bud on the tree.

terrain the physical aspects of a plot of land, especially as they affect the use of the land. Terrain is usually described in terms of elevation changes ("steep," "rolling," "flat," "jumbled," etc.) and/or vegetation ("forested," "grassy," "barren," etc.) The term "terrain" may also be used to mean a plot of land whose boundaries are seen in terms of changes in physical characteristics rather than as political lines. For example, the Rocky Mountains are a separate terrain (or "type of terrain") from the Great Plains. *Compare* TERRANE.

terpenes naturally occurring VOLATILE ORGANIC COMPOUNDS (VOCs) generated by trees. Terpenes provide the clean smell of coniferous woodlands and in the southern Appalachians contribute to the blue haze for which the Great Smoky Mountains and the Blue Ridge are named. In 1981 President Ronald Reagan stated that trees were more important causes of AIR POLLUTION than automobiles. While trees do emit hydrocarbons such as terpenes, the emissions are not air pollutants. *See also* ATMOSPHERE.

terrane in geology, a piece of landscape that is unified by some structural feature, such as all being part of the same FORMATION. Once considered obsolete, the term has had a revival in plate tectonic theory (*see* PLATE TECTONICS), where it is used to refer to a piece of the Earth's crust that has broken off of its original crustal plate and become attached to another. Most geologist now feel, for example, that all of North America west of the Cascade and Sierra Nevada mountain systems (that is, western California, Oregon, Washington and British Columbia, and nearly all of Alaska) is composed of many separate terranes which broke off the disintegrating Falleron Plate and plastered themselves onto ("became accreted to") the North American Plate over the last 50 million years.

territoriality the drive on the part of animals to protect a plot of land (the TERRITORY) against other members of its own SPECIES. Sometimes the territorial instinct is extended to other species as well. Territoriality is exhibited by nearly all animals, from insects to humans. The territory defended may belong to a single individual, a mated pair, or a group (such as a wolf pack). It may be small (as little as a square meter for some insects and reptiles) or large (as much as 1,500 square miles for some CANIDS). It may have fixed boundaries, or it may merely consist of all the space within a certain radius of an individual. Often both are seen. Humans, for example, may defend a fixed territory, but they also carry with them a roving territory, usually referred to as a *personal space,* which governs their distance from other humans in shared territory such as ticket lines and crowded beaches. Territorial defense is often rigidly ritualized, especially in the birds, which often mark their territorial boundaries by singing from conspicuous branches. Other animals may mark their territories by leaving deposits of scent (sometimes in the form of FECES, other times by rubbing special glands in the skin onto a boundary marker, or *scent post,* as when a domestic cat rubs its cheeks—the location of one of its sets of scent glands—against a chair). Animals that invade another individual's territory despite the markings are aggressively driven off. Ecologically, territoriality appears to serve primarily as a means of partitioning a scarce resource, such as FORAGE or nesting sites, with a minimum of actual fighting (if other individuals recognize and respect the territorial boundaries, no fighting takes place) therefore conserving energy. *See also* RANGE.

territory in behavioral ecology, the space occupied by an animal and defended against intrusion by others of the same SPECIES. Two general types of territory exist, the FEEDING TERRITORY (in which food-gathering activities take place) and the NESTING TERRITORY (in which the animal makes a home and rears its young). Some species may temporarily establish a third type, the MATING TERRITORY, in which courtship and mating occurs separated from the nest. The primary defensive mechanisms are warning—scent markings, display, territorial calls—although virtually all animals will fight, often to the death, if their territories are actually invaded by an aggressor. Nesting and mating territories are much more aggressively defended than are feeding territories, which will often (though not always!) be shared among several individuals. Size may range from 20 or 30 square miles (eagles, bears, cougars) down to a few square feet (insects, lizards, hummingbirds, penguins). Territoriality is important to a species because it tends to spread POPULATIONS of the species evenly through the available HABITAT and increase the survivability of each individual through its increased familiarity with the small piece of the environment it defends, enabling it to utilize all the food resources and to hide effectively from it enemies and PREDATORS. It has also given us birdsong, which is primarily a territorial call. *Compare* RANGE. *See also* TERRITORIALITY.

tertiary treatment (advanced treatment) treatment of SEWAGE or other waterborne wastes to a level beyond that reached by PRIMARY and SECONDARY TREATMENT. By definition, primary treatment is mechanical (settling, screening, etc.) and secondary treatment is biological (flocculation, digestion, etc.). In keeping with this division, tertiary treatment is often thought of as "chemical"; however, the term actually covers a wide variety of processes, only a few of which are strictly chemical in nature. The most common type of tertiary treatment is probably the use of a RAPID SAND FILTER or a *pebble filter* (a simple box or cage filled with PEBBLES through which the EFFLUENT is trickled) to clean up the output of the secondary treatment facility. Treatment with a chemical precipitant such as ferric chloride ($FeCl_3$) to remove PHOSPHORUS is also fairly common (*see* PRECIPITATE), as is chlorination of the waste stream to kill PATHOGENS. Other tertiary treatment processes that have been applied with varying degrees of success include filtration by ACTIVATED CARBON; REVERSE OSMOSIS; ION EXCHANGE treatment; and POLISHING PONDS. The actual choice of processes depends upon the problem being addressed. *See also* SEWAGE TREATMENT PLANT.

tetraethyl lead (TEL) an extremely poisonous viscous liquid used between 1926 and 1986 as an oxygenating additive to gasoline to reduce "ping" or "knock" in INTERNAL COMBUSTION ENGINES. TEL was discovered in 1854 by a German chemist, but no practical use was found for the compound until Thomas Midgely, a self-taught General Motors chemist, discovered in 1921 that when added to gasoline it increased the octane rating, thus reducing knock or ping. Not only could engines run more smoothly, but power could also be increased through higher compression rather than adding to displacement. By 1924 the product was given the brand name *Ethyl* by the Ethyl Gasoline Corporation, jointly owned by Standard Oil and General Motors, and was widely distributed, replacing ethanol (derived from vegetable oils) as an alternative additive with the same antiknock properties but without toxic effects. Although TEL was long known by health experts to be extremely poisonous, its manufacturers and their allies stubbornly denied that the compound was toxic. Meanwhile workers at TEL plants were dying from lead poisoning in some numbers. Children turned out to be extremely sensitive to leaded gas, suffering from lowered IQ and related disabilities, impaired hearing, and behavioral problems. Pregnant mothers poisoned by TEL could miscarry, or their offspring could suffer deformities or brain damage or both. Adults with elevated blood lead levels from TEL suffered from heart attacks, strokes, and premature death. Finally, in 1972, under the CLEAN AIR ACT, the ENVIRONMENTAL PROTECTION AGENCY (EPA) gave notice to manufacturers that TEL had to be phased out of gasoline products, and although contested in the courts, the ban held up. Despite attempts made by the Reagan administration to delay the phase-out, it was completed at last in 1986. Since then, TEL has continued to be marketed abroad, although the European Union banned leaded gasoline in 2000. *See also* GASOHOL; MBTE.

TFM (in full: **3-trifluoromethyl-4-nitrophenol**) an aromatic hydrocarbon compound (*see* AROMATIC COMPOUND; HYDROCARBON), chemical formula $C_7H_4F_3NO_3$, used primarily as a lampricide, that is, a pesticide targeted specifically at lampreys. TFM was discovered in 1958 as the result of a major research effort aimed at curbing the Great Lakes populations of the lamprey, a parasitic, eel-like fish which had entered the Lakes around 1922 via the St. Lawrence River and the Erie and Welland Canals and had proliferated there, reducing lake trout and whitefish populations in the Lakes by more than 90 percent. The chemical is selectively toxic to the lamprey: properly applied to streams that are the fish's spawning beds (*see* ANDRAMOUS FISH), it will kill up to 98 percent of the lamprey larvae without destroying other aquatic life, and for this reason it has become a staple tool of fisheries managers in the Great Lakes system, though there is increasing evidence that it has significant (though non-fatal) effects on other forms of life besides lampreys, and that the lamprey populations are becoming increasingly resistant to it and moving their spawning beds from streambed gravels to gravel bars in the open lakes.

thallium the 81st element in the atomic series; ATOMIC WEIGHT 204.37, chemical symbol Tl. One of the most toxic of the HEAVY METALS, thallium physically resembles LEAD. It is a soft, malleable, blue-gray metal with a low melting point. Chemically, it reacts in a manner similar to aluminum, forming COMPOUNDS with the same classes of substances. For instance, like aluminum, it reacts rapidly to OXYGEN on exposure to air, forming a tough oxide coating that then prevents further reaction. Some of its compounds react to infrared light, making them useful for the production of night-seeing devices and for so-called "electric eyes"; some other uses include the manufacture of optical glass, the production of low-temperature electrical switches (in amalgamation with MERCURY), and as a RODENTICIDE (in the form of its SALTS, especially thallium sulfate, Tl_2SO_4). Thallium and nearly all of its compounds are highly toxic. Symptoms of acute exposure include gastrointestinal (digestive system) distress, pain and tingling of hands and feet, coma, convulsions, and death. Chronic exposure to small amounts causes extreme weakness, pain in the hands and feet, and loss of hair.

thermal bar a vertical interface that develops between warm shoreward waters and cold open-lake waters in large, deep lakes. It is particularly well developed in the Great Lakes. The thermal bar forms as a result of the warming of shoreward shallows while the bulk of the lake remains cold due its great THERMAL MASS. It develops near the shore in the early spring and moves slowly outward as the weather warms, usually disappearing by mid-summer. Like the THERMOCLINE, which it resembles, the thermal bar prevents the mixing of warm and cold waters. In this case, the result is lack of circulation between open-lake and shore waters, resulting in a trapping of POLLUTANTS released from shore in the shoreward shallows, raising their apparent concentrations and multiplying their effects on aquatic life-forms, most of which also prefer the shallows.

thermal cover in wildlife management, COVER sought by animals as a protection against cold or heat rather than as a means of hiding themselves. Within dense stands of vegetation, the minimum air temperature is usually 3°F to 5°F warmer and the maximum air temperature is usually 3°F to 5°F cooler than the temperatures found in nearby open spaces.

thermal enrichment *See* THERMAL POLLUTION.

thermal inversion *See* INVERSION.

thermal low in meteorology, an area of low air pressure (*see* CYCLONE) caused by hot ground temperatures. Thermal lows are characteristic of dry, hot climates. The intense sunlight in these climates causes the ground surface to heat up. This in turn heats the air, causing it to rise, reducing the air pressure near the ground and creating an inward air flow from nearby AIR MASSES. Large-scale thermal lows are particularly common over the American southwest, where they cause most of the summer rainstorms. Rainfall in that region is roughly 10 times higher in the summer than it is in the spring or the fall.

thermal mass the ability of an object, or a body of liquid or gas, to absorb heat. Thermal mass is a function both of SPECIFIC HEAT and of the mass of the body involved. A large rock has a greater thermal mass than a small one; a body of water has a greater thermal mass than a body of rock the same size. The larger the thermal mass, the slower the temperature change for the same amount of input or release of energy. The very great thermal mass of the oceans, for example, is the reason that they remain at a nearly uniform temperature throughout the year.

thermal pollution (thermal enrichment) the artificial warming of a body of water (or sometimes, of air) to the point where the BIOTA inhabiting it are damaged. Thermal pollution usually results from the release of cooling waters, such as those used in power-generation equipment (*see* COOLING TOWER), into a river or lake. It may also result from the removal of streamside vegetation or the creation of shallow-water reservoirs along a watercourse. Sometimes thermal pollution acts directly on the biota. For example, it can raise the temperature of a stream beyond the rather narrow limits that many aquatic ORGANISMS require for breeding, thereby causing reproductive failure (trout and salmon are particularly sensitive to this). It can also encourage the growth of MICROORGANISMS, thus contributing to CULTURAL EUTROPHICATION. Its greatest effects, however, are on the SOLUBILITY of solids and gases in the water. A rise in temperature increases the solubility of solids, therefore raising the water's ability to carry pollutants in solution and effectively increasing its pollutant load. At the same time, a rise in temperature decreases the solubility of gases, causing DISSOLVED OXYGEN levels in the water to fall and increasing the chances that a body of water may become anoxic (*see* ANOXIA). As the dissolved oxygen levels are falling, BIOCHEMICAL OXYGEN DEMAND (BOD) is increasing, both because of the increase in the BIOMASS of algae, bacteria, and other heat-loving microorganisms and because this increase in heat speeds up the METABOLISM of all aquatic life. A rise of 10°C (18°F—by no means unheard of as the result of the return of power plant cooling water to the environment) is sufficient to double the rate of oxygen use by fish and other water-dwelling organisms. The overall effect of thermal pollution is thus to radically change—and almost always, to degrade—the ecology of the RECEIVING WATERS.

thermal stratification the separation of a liquid or a gas (or, occasionally, a group of living ORGANISMS) into horizontal layers according to temperature. The term is used by meteorologists to describe conditions in stagnant AIR MASSES (*see* INVERSION) and by ecologists to describe some aspects of the vertical distribution of insects and birds in the forest CANOPY. (These organism tend to choose their height above ground in part because of the differences in temperature at varying levels.) Its most common use in environmental science, however, is to describe seasonal changes in the circulation patterns of the water in lakes. During the summer, most lakes in TEMPERATE CLIMATES develop a three-layered thermal structure, with relatively warm water (the EPILIMNION) near the surface, relatively cold water (the HYPOLIMNION) in the depths, and a localized, narrow band (the THERMOCLINE) in between in which the temperature rapidly drops with depth. Early in the season, the thermocline is near the surface. As the waters warm up, the thermocline is driven lower and lower, decreasing the size of the hypolimnion and increasing the epilimnion. Eventually the thermocline disperses, either because it has reached the bottom of the lake or because cooling of the water with the onset of autumn has brought the epilimnion and hypolimnion into thermal equilibrium. This dispersal of the thermocline,

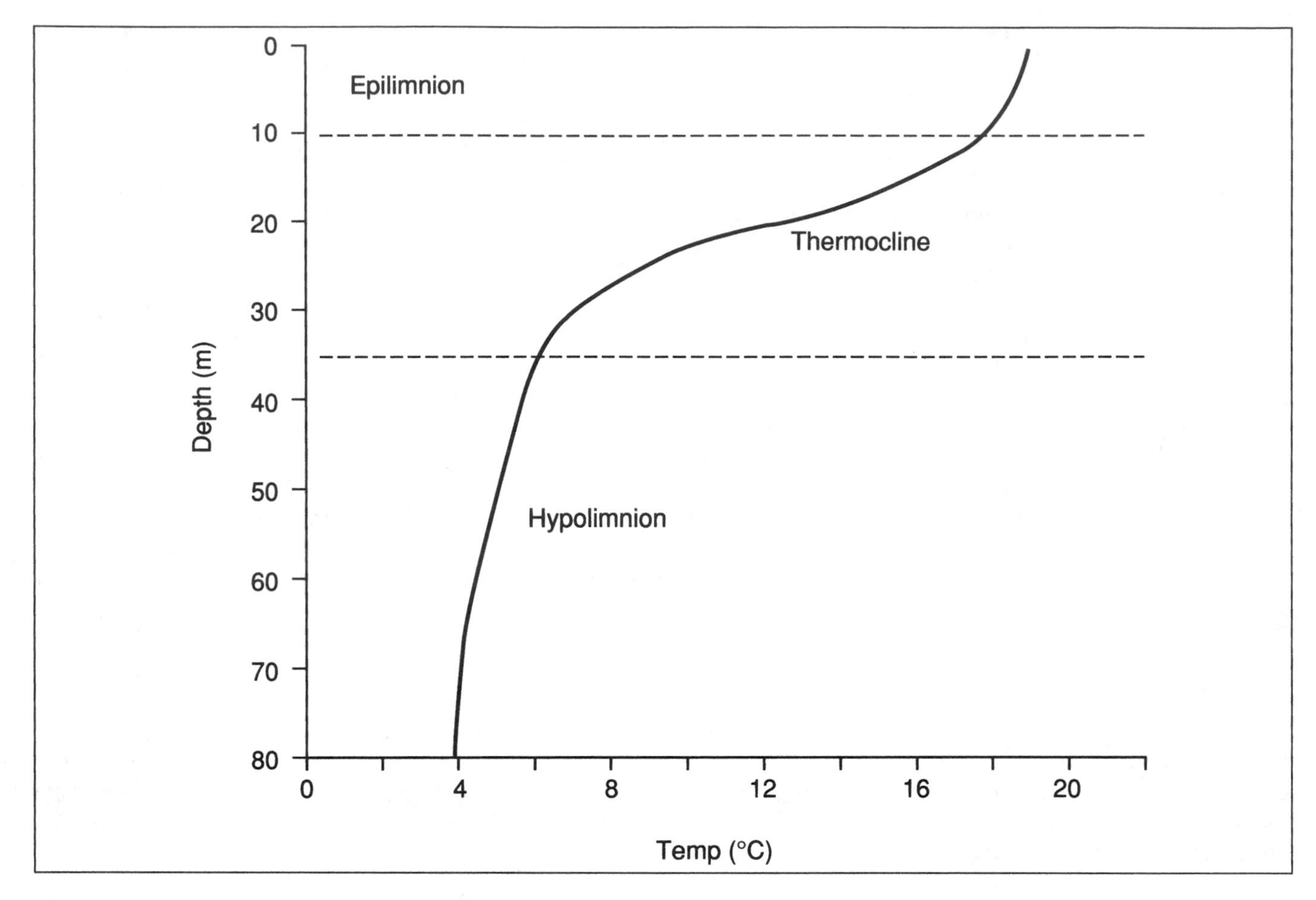

Thermal stratification (Seneca Lake, New York, August)

with consequent mixing of the waters all the way to the bottom, is known as the *fall overturn.* During the winter, when air temperatures fall below water temperatures, the water on the surface cools and sinks toward the bottom. However, if the temperature of the lake falls below 4°C (39°F) the surface water no longer sinks, due to the fact that water expands (and therefore becomes more bouyant) from 4°C down to its freezing point (0°C). When this happens, the water stratifies thermally again—this time with no thermocline and with the cold water (a very narrow layer) on top—and ice may form. This condition lasts until spring, when the air temperatures warm up again, warming the surface waters above 4° and reforming the thermocline (*spring overturn*). *See also* TROPHIC CLASSIFICATION SYSTEM.

thermocline (metalimnion) in limnology, the narrow band of water between the warm surface water of a lake (the EPILIMNION) and the colder water in its depths (the HYPOLIMNION). Temperatures in the thermocline drop rapidly with depth, at least 1°C per meter and often as much as 5°–7°C per meter. It acts as an effective barrier to windmixing—the primary means of water circulation in lakes—and therefore prevents any mixing of the epilimnion and hypolimnion, so that DISSOLVED OXYGEN (which enters the water from the surface) is not replenished in the hypolimnion, while NUTRIENTS (which enter the water from the bottom SEDIMENTS) are not replenished in the epilimnion. This dual "starvation," which continues as long as the thermocline is in existence, is the reason that large, deep lakes (which are more likely to develop a thermocline) are less productive than are small, shallow ones (*see* PRIMARY PRODUCTIVITY). *See also* THERMAL STRATIFICATION; TROPHIC CLASSIFICATION SYSTEM.

thermodynamics, laws of in the sciences, four basic principals governing the transfer and use of energy within physical systems—"physical systems" here meaning all systems composed of matter, that is, the universe or any subset of it. Elucidated during the late 19th and early 20th centuries, the laws of thermodynamics are of fundamental importance in understanding the functioning of ECOSYSTEMS and the physical and chemical laws which govern them.

THE FIRST LAW

Sometimes called the *law of conservation of energy,* the first law of thermodynamics states that *energy may neither be created nor destroyed;* in other words, that the net energy change of a system is always equal to the sum of the energy added to (or emitted by) the system and the work done on it (or by it). The first law is the reason the brakes of a car get hot; the energy of the car's forward motion is not destroyed when it is brought to a halt, but is transformed to heat at the brake linings and on the pavement.

THE SECOND LAW

Known as the *law of entropy,* the second law states that *all systems tend to move from order to disorder.* The amount of disorder in a system is called its *entropy;* thus this law may also be stated as *the entropy of the universe is always increasing.* In other words, the universe is running down. Entropy in a closed system tends toward equilibrium. Heat "flows" from a body at a higher temperature to a body at a lower temperature because this transfer brings the two bodies' entropy into equilibrium with each other.

THE THIRD LAW

The third law of thermodynamics is actually a minor corollary to the second. It states that *the entropy of a crystalline substance at a temperature of absolute zero is also zero.* Since the substance is ordered (crystals hold their molecules in a fixed matrix), and since it is not moving (at absolute zero, all molecular motion ceases), the order is perfect; there is no disorder. The third law is of interest principally to physicists engaged in measuring the absolute entropy of a system.

THE ZEROTH LAW

The so-called "zeroth law of thermodynamics" is actually the fourth, in terms of time. It is called the zeroth because it was realized, belatedly, that it undergirds the other three, which cannot function without it. Called the *law of equivalency of temperature,* it states that if two bodies are each at the same temperature as a third body, they are also at the same temperature as each other. Bodies at the same temperature do not necessarily contain the same amount of heat (*see* SPECIFIC HEAT), but their entropy is the same, and thus heat will not transfer between them.

THERMODYNAMICS AND THE ENVIRONMENT

It is difficult to underestimate the importance of the four laws of thermodynamics in understanding environmental systems. For example, since heat is the most random form of energy, one implication of the second law is that all systems produce waste heat: that it is not possible to create a cyclical process of converting heat to work and work back to heat without losing some of the heat along the way. The more mechanical and electrical energy we expend in doing work, the more we lose as heat—the real reason for the so-called "energy crisis." Living things appear to be increasing the ordering of the world (a process sometimes referred to as NEGENTROPY), but this is an illusion. In order to increase the order of a system, work must be done on it, and some of this work is inevitably lost as waste heat. METABOLISM is a heat-producing process for precisely that reason. The laws of thermodynamics are the inescapable bottom line by which the universe functions. The more energy we put into reordering the world for our own (human) priorities, the more we increase its total disorder—a problem that shows up as environmental degradation.

thermokarst a type of topography characterized by SINKHOLES and other collapse features that have developed due to the thawing of underground ice blocks. Thermokarst superficially resembles true KARST topography, but it is developed in soil rather than in rock, and there are no caverns or other solutional features. It is most commonly found on the arctic coastal plain, where it is associated with FROST HEAVE, PERMAFROST, PATTERNED GROUND, and other types of topography resulting from the freezing and thawing of GROUNDWATER.

thinning in forestry, the removal of some of the trees from a STAND in order to allow the others to grow better. Thinning may be done either before or after trees reach merchantable size (*see* PRECOMMERCIAL THINNING; COMMERCIAL THINNING; MERCHANTABLE TIMBER). In contrast to a SELECTIVE CUT, a thinning operation removes the inferior trees from a stand, leaving the best behind to grow faster, straighter, and/or taller (that is, "releasing" them from competition: *see* RELEASE CUTTING). Four general types of thinning are recognized. *Low thinning* (also called "German thinning") removes the UNDERSTORY trees, including those that have been suppressed and those that are simply younger than the dominant trees of the stand. It is classed as A, B, C, or D thinning depending upon how much material is removed. It seldom creates openings in the forest CANOPY. *Crown thinning* ("thinning from above"), by contrast, is expressly designed to create openings in the canopy by removing some of the dominant trees of the stand to give the others more room to grow. *Selective thinning* removes the best trees, rather than the worst trees, from a stand, to improve the growth potential of those left behind. It is essentially a type of harvest rather than a thinning operation, although it may be chosen in place of other types of thinning if the trees that will be released show good potential for growth. Finally, *free thinning* is a combination of the other three methods based on forest conditions at each location in the stand, and is the most common variety of thinning in natural stands or stands that otherwise get little attention from the forestor. The timing of a thinning, and the amount to be thinned, depend on the silvicultural result sought. Generally, earlier thinning produces greater volume but more TAPER and more KNOTS. The stand density sought will vary from species to species and site to site (*see* SITE INDEX), but is generally a 12-foot space between stem centers for CONIFERS on high-site land and 15 to 20 feet on the lower sites. *See also* CLEANING CUT.

third-order lake in limnology, a lake that does not develop a THERMOCLINE. Third-order lakes are generally warm and shallow—less than 8 meters (26 feet) in depth—and are much more likely to develop problems with EUTROPHICATION than are lakes of the first or second order (*see* FIRST ORDER LAKE; SECOND ORDER LAKE.) *See also* TROPHIC CLASSIFICATION SYSTEM.

thirst belt derogatory name for the American southwest derived from its need to import water to make up for its low rainfall. The term is used most often by residents of the states from which the water is taken. *See also* SUN BELT.

tholeiitite basalt a type of BASALT that is particularly poor in OLIVINE. Tholeiitite basalt (or *tholeiitic basalt,* or simply *tholeiitite*) is characteristic of ocean floors rather than of continental LAVA FLOWS, and is thought to be formed from so-called "primary magma," that is, MAGMA that originates from the Earth's MANTLE rather than from recycled crustal plates (*see* PLATE TECTONICS).

Thomas Report (Jack Ward Thomas Report) Report released on April 4, 1990, by a task force (the "Interagency Scientific Committee," or ICS) set up to review the status of the spotted owl (*see under* SPOTTED OWL MANAGEMENT AREA) in the OLD GROWTH forests of the Pacific Northwest. Chaired by biologist Jack Ward Thomas of the United States FOREST SERVICE and including representatives from the Forest Service, the BUREAU OF LAND MANAGEMENT and the FISH AND WILDLIFE SERVICE, the ICS concluded that the owl was threatened or endangered over most of its range (*see* THREATENED SPECIES; ENDANGERED SPECIES) and that spotted owl management areas were not adequate to keep it from EXTINCTION: committee members recommended instead the establishment of a system of *habitat conservation areas* (HCAs) of up to 50,000 acres, large enough to provide homes for 15 to 20 pairs of owls each. Travel corridors were to be left between the HCAs to allow genetic mixing of the entire owl population. In all, roughly 3 million acres of timberland was to be taken out of production, resulting in a net loss of 20% to 40% of the annual harvest in Oregon, Washington, and California, at an estimated cost of 10,000 to 60,000 jobs (most probable figure: 30,000). On June 22, 1990, following the recommendations of the report, the Fish and Wildlife Service officially listed the spotted owl as a threatened species. In the summer of 1990, Forest Service officials announced that future National Forest planning would comply with the goals of the Thomas Report. In 1994, after protracted negotiations and much controversy, a final plan to preserve some of the old-growth habitat for the spotted owl was approved by the federal district court that had banned all cutting until a plan could be developed.

Thoreau, Henry David (1817–1862) American writer and transcendentalist philosopher, born on July 12, 1817, in Concord, Massachusetts, and educated at Harvard University (B.A., 1837). For a short time he taught school in Concord, but he found the work unpleasant, and he left teaching within a year, supporting himself thereafter with his writing and a series of odd jobs including surveying and pencil-making. He also received a great deal of financial support from his friend and fellow transcendentalist Ralph Waldo Emerson. It was on Emerson's land at Walden Pond near Concord that Thoreau, determined to prove the value of living simply, built a small, spartan cabin and lived in it from 1845 to 1846. His book based on that experience, *Walden* (1854), remains the best articulation of the idea of living as close to nature as possible that has ever been published. Two other works, *A Week on the Concord and Merrimac Rivers* (1849) and *The Maine Woods* (published posthumously in 1864), approach *Walden* in their celebration of living a life in nature rather than off of it. Thoreau was also a passionate crusader against slavery and war, and his 1849 essay *On Civil Disobedience* is as important a document in the history of social activism as *Walden* is in the history of environmental consciousness. He died of "consumption" (probably tuberculosis) on May 6, 1862, at Emerson's home in Concord.

threatened species according to the ENDANGERED SPECIES ACT, any SPECIES that is likely to become an ENDANGERED SPECIES within the foreseeable future throughout all or a significant portion of its RANGE. The category includes RARE SPECIES, (i.e., those whose natural numbers are low), DEPLETED SPECIES (those whose populations have been reduced by human activities or natural disasters), and species endemic to small areas that are threatened with development (*see* ENDEMIC SPECIES). Lists of threatened species are prepared by the secretary of the interior (*see* INTERIOR, DEPARTMENT OF THE) and published in the FEDERAL REGISTER. Species on these lists receive essentially the same protection as those on lists of ENDANGERED SPECIES, although the goal of management is to keep them from becoming endangered rather than to keep them from becoming EXTINCT.

Three Mile Island nuclear power plant near Harrisburg, Pennsylvania, that on March 28, 1979, became the site of the worst U.S. commercial nuclear accident to date. Located on an island in the Susquehanna River approximately 10 kilometers (6 miles) below Harrisburg, the Three Mile Island plant consisted of two functioning pressurized water reactors (*see under* NUCLEAR ENERGY), each with its own containment building and set of cooling towers, with a combined capacity of approximately 1.7 megawatts. The accident took place in Unit II, which had been in operation for only three months, since December 30, 1978. At 4:00 A.M. on March 28 the feedwater pumps to the reactor shut down, apparently in response to a stuck pressure relief valve. The reactor failed to shut down simultaneously, and its core temperature began to climb. A malfunctioning pressure gauge led plant operators to turn off the emergency core cooling system manually, allowing water levels to drop in the containment building, exposing the top two feet of the reactor's core, which overheated drastically, reaching temperatures greater than 2,760°C (5,000°F) and melting more than half of the fuel rods. A 75-cubic-meter (850-cubic-foot) "bubble" of hydrogen gas collected in the top of the containment building. Radiation levels within the building reached 30,000 rads, high enough to kill a human after less than one minute's exposure. Releases of radioactive steam took place at least five times, raising radiation levels around the plant to between 10 and 30 times normal background levels and forcing many surrounding homes to be evacuated. The investigation into the accident determined that the plant had operated for the previous two weeks with its backup pumps inoperable, and that the containment building was not fully sealed—both violations of federal regulations. Unit I, shut down at the time of the accident to Unit II, was restarted on October 9, 1985.

thunder egg *See* GEODE.

TIC *See* TOTAL ION CONCENTRATION.

tidal power generation the use of strong tidal flows to drive electrical turbines. Tidal power plants are located in France and on the Bay of Fundy in Canada, where the daily fluctuations between high and low water can reach 43 feet.

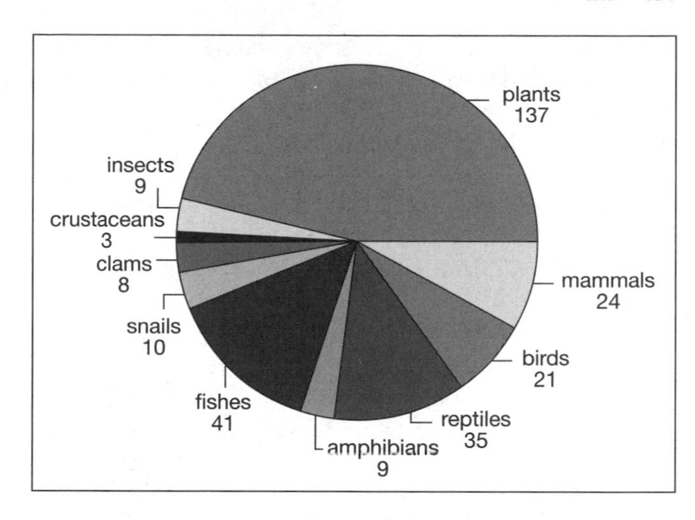

U.S. threatened species as of 1999 ***(U.S. Fish and Wildlife Service,*** **Endangered Species Technical Bulletin,** ***quarterly)***

tide flat a broad, flat stretch of land, generally made of SILTS and other fine soil particles, which lies between the elevations of high and low tide and is thus alternately covered and uncovered by water as the tide flows in and out. Tide flats generally develop in rivermouth bays, behind sea islands lying in the course of the prevailing SHORE DRIFT, and in other areas that are protected from waves and where an adequate sediment source exists. Warmed by the Sun and provided with a constantly replenished supply of water and NUTRIENTS by incoming tides, tide-flat muds become a rich organic "soup" in which numerous ORGANISMS thrive, and are among the most productive ECOSYSTEMS on Earth. Well-established tide flats generally develop into SALT MARSHES.

tidepool a depression in a SEA STACK or other coastal rock formation that lies between the elevations of high and low tide and thus catches and holds seawater between tides, while the area around it is temporarily dry. Since tidepool water is replenished every time the tide comes in, providing a new supply of OXYGEN and NUTRIENTS, tidepools can support a much greater BIOMASS than similarly sized pools that are isolated from tidal effects. ORGANISMS that inhabit tidepools (sea anemones, starfish, crabs, snails, mussels, barnacles, etc.) generally develop feeding patterns that reflect the tide cycle, coming out to feed as the tide comes in. These feeding patterns usually remain synchronized with the tides even if the organisms are removed from the tidepool and placed in a laboratory holding tank that is held at a constant water level.

tile field *See* DRAINFIELD.

till in geology, materials deposited directly by a GLACIER, as opposed to *fluvioglacial deposits,* which are deposited by the glacier's meltwaters. Till is unsorted, and includes materials from the size of BOULDERS down to CLAY particles. It is sometimes referred to as *boulder clay* for that reason. *Basal till* is material carried along beneath the glacier; *ablative till* is carried on the glacier's surface or pushed along in front of it. Landforms created by till tend to take on characteristic shapes depending upon their means of formation (*see* MORAINE; KAME). *See also* DRIFT.

timber originally, a log or pole used as structural support in a ship or building; hence, any wood strong enough and sound enough to be used for structural support, and therefore (the most common current usage) a tree or a STAND of trees seen as a source of structural wood ("standing timber").

timber base in Forest Service terminology, the inventory from which the ALLOWABLE CUT is determined. The term is poorly defined; however, it is generally taken to mean the total of all SUITABLE FOREST LAND, regardless of STOCKING LEVEL, in a given NATIONAL FOREST, Ranger District, Region, or similar Forest Service management unit.

timber beast a logger; also, by extension, a forestor primarily concerned with TIMBER management. The term is most commonly used by loggers and timber managers themselves.

timber cruiser a person employed to walk through a STAND of TIMBER and estimate its value and condition as a crop, a process known as "cruising the stand" or "cruising the timber." A timber cruiser notes the total timber volume of the stand, the number of trees and their average size and age, the SPECIES mix and total volume of each species, the type and condition of CONK, WINDTHROW, and other indicators of poor stand health, and any factors that might make harvesting or YARDING difficult or expensive, such as steep terrain or the presence of large boulders.

timberline the ECOTONE that marks the upper limits of tree growth on mountain slopes or the most poleward limit of tree growth in the Arctic or Antarctic regions. On mountain slopes, timberline is actually a broad belt, covering an elevation difference of as much as 500 meters (1,600 feet). At its bottom margin is the *forest line,* marking the upper elevation limit of the continuous, closed canopy forest (*see* CANOPY). Beyond the forest line, trees grow in scattered clusters in protected locations up to an elevation known as the *tree line,* which marks the upper limit of a tree's ability to grow upright. Species that are trees below the tree line often continue up the mountain as KRUMMHOLTZ to an elevation known as the *scrub line,* above which tree species can no longer survive at all. The land between the forest line and the tree line is the area generally referred to as timberline. The elevation of this area varies greatly with conditions. As a general rule, it is higher near the equator than it is near the poles, higher on inland mountains than on coastal mountains at the same LATITUDE, and higher on the south and west slopes of a MOUNTAIN RANGE than it is on the north and east slopes.

THE POLAR TREE LINE

At the northern and southern limits of tree growth in flat country (such as the arctic plain of North America) the forest stops suddenly at an abrupt border in which the forest line, tree line, and scrub line are for practical purposes one and the same. This *polar tree line,* which separates the taiga (*see* CONIFEROUS FOREST BIOME) from the tundra (*see* TUNDRA BIOME), is sometimes referred to as the northern (or southern) timberline. *See also* HUDSONIAN ZONE; SUBALPINE PARKLAND.

timber management compartment in Forest Service usage, an area of forest land with permanent, easily locatable boundaries, that may be treated as a unit for management purposes. The boundaries of a timber compartment are usually natural lines such as ridges or rivers, though properly surveyed, well-established artificial lines (such as a county or state line or the National Forest boundary) will do just as well. Timber management compartments are most useful for assigning planning personnel to such site-oriented tasks as data gathering and the planning of specific sales. Their boundaries should largely be ignored for resource-allocation planning such as the creation of an overall FOREST PLAN.

timbershed in forestry, the area from which a sawmill or a group of sawmills (such as those located in a particular community) draws its TIMBER. Timbershed boundaries depend primarily upon transportation difficulties. Logs will usually be transported to the easiest mill to reach, rather than the closest one. It is impossible to draw timbershed boundaries precisely, however, due to factors such as the need of a given mill for logs, which may outweigh transportation problems in bidding on the timber in a given tract of federal timberland.

timber stand improvement (TSI) in forestry, any silvicultural treatment (*see under* SILVICULTURAL PRESCRIPTION) designed to improve the timber yield from a STAND of trees. TSI procedures fall into three categories. The first category includes those management practices that improve volume growth, such as THINNING, fertilizing (*see* FERTILIZER), etc. In the second category are those practices that improve the quality of the wood laid down by the individual trees, such as pruning. The final category consists of practices that improve the stand composition (that is, increase the proportion of desired timber SPECIES over species that produce inferior wood). This includes brush suppression and other management practices designed to remove competing vegetation. Note that some TSI practices cross categories. Thinning is usually done to improve volume growth, but it can also greatly improve stand composition (by thinning undesirable species and weaker specimens of the desired species); REFORESTATION primarily increases volume production, but it is also a means of improving stand composition (because only good genetic stock of desirable timber species will generally be planted).

Times Beach a Missouri resort area where a DIOXIN emergency occurred in 1983. The unsurfaced dirt streets had been sprayed with dioxin-laden oil, producing soil contamination at 100 times the emergency level. The ENVIRONMENTAL PROTECTION AGENCY (EPA) evacuated the city and purchased properties so that they could be decontaminated.

TM in forestry, abbreviation for *timber management.*

TMA in forestry, abbreviation for *timber management assistance.*

Tokaimura nuclear disaster a nuclear core meltdown on September 30, 1999, in a nuclear-fueled processing plant in Tokaimura, Japan, about 60 miles northeast of Tokyo. The meltdown occurred after too much reactive uranium was added to a purification tank, exposing 35 persons to excessive doses of radiation. Three were seriously injured. With radiation levels jumping to 20 thousand times acceptable levels, 300,000 residents of the city were ordered to remain in their homes with windows closed and air conditioning off. Since Japan depends on its 51 nuclear reactors for some 30% of its electrical power needs, nuclear power authorities have been extremely safety-conscious. In this connection, the Tokaimura accident has suggested to many that even the most rigorous safety measures cannot completely guard against human error in the operation of nuclear facilities. *See also* CHERNOBYL; CHINA SYNDROME; NUCLEAR POWER; THREE MILE ISLAND.

tolerance in forestry, *see* SHADE-TOLERANT SPECIES.

toluene in chemistry, an aromatic hydrocarbon (*see* AROMATIC COMPOUND: HYDROCARBON), chemical formula $C_6H_5CH_3$, consisting of a BENZENE RING in which one of the HYDROGEN ATOMS has been replaced by a methyl GROUP (a CARBON atom with three hydrogen atoms attached to it). A colorless, volatile liquid (*see* VOLATILITY) with a pronounced "chemical" odor and a low FLASH POINT, toluene has many of the same properties as BENZENE but is considerably less toxic (*see* TOXICITY) and carcinogenic (*see* CARCINOGEN), making it useful as a benzene substitute in industrial processes. It is a good SOLVENT, especially for ORGANIC COMPOUNDS, and is used to make food extracts and in the production of some decaffeinated coffee. Nevertheless, its toxicity is high enough to cause it to be listed as a PRIORITY POLLUTANT by the ENVIRONMENTAL PROTECTION AGENCY. Symptoms (acute): headache, narcosis (sleepiness), and nausea; (chronic): the growth of abnormally large red cells in the blood (*macrocytic anemia*). LD_{50} (rats, orally): 7.532 ml/kg.

tonne the metric equivalent of the ton, equal in weight to 1,000 kilograms. It is equivalent to slightly under 2,205 pounds, roughly the same as a LONG TON and about 10% larger than a regular (short) ton.

top predator a PREDATOR at the top of a FOOD CHAIN or FOOD WEB. The top predator preys on ORGANISMS below it in the chain or web, but there is nothing that preys on the top predator. As a rule, top predators are large CARNIVORES with very slow birth rates and a long period of dependency for the young (*see* K-SELECTED SPECIES). Top predators are the key to POPULATION control for most PREY species. As the capstone of the ECOLOGICAL PYRAMID, they exert predation pressure on the prey, but are themselves controlled strictly by non-predation factors. They are also the organisms most at risk from environmental pollution, because they prey on the organisms at high levels on the food chain, allowing BIOLOGICAL MAGNIFICATION a maximum chance to concentrate pollutants in their tissues.

topographic map a map that shows the TOPOGRAPHY of a piece of land as well as (or instead of) the location of physical or political features such as roads, cities, lakes, jurisdictional boundaries, etc. The most common means of mapping is through the use of *contour lines,* lines connecting adjacent points of equal elevation on the mapped landscape. The elevation distance between neighboring lines, known as the *contour interval,* is constant for any specific map, and is normally given in the map legend. Contour lines that are close together indicate steep slopes; contour lines that are far apart indicate relatively flat areas. A practiced reader can gain a fairly complete picture of the local topography from a good contour map, which is why they are the maps of choice for most outdoorsmen, as well as being required for site planning prior to building construction, road construction, and similar activities.

topographic survey a survey in which measurements are taken and points are plotted in three dimensions (LATITUDE, LONGITUDE, and elevation above MEAN SEA LEVEL) rather than in just two (latitude and longitude). The result of a topographic survey is a TOPOGRAPHIC MAP. Most topographic survey work today is done by establishing reference points on the ground and using stereoscopic photographs taken by aircraft or satellites to map the terrain around the reference points.

topography the shape of a tract of land, especially in regards to elevation and RELIEF. For the most part, the words *topography* and TERRAIN are interchangeable; however, "topography" generally refers to a smaller area, and is more clearly restricted to simple physical description without regard to potential use by humans.

topsoil the upper layer or layers of the soil, containing a high percentage of ORGANIC MATTER and therefore having high fertility. The topsoil is usually defined as equivalent to the O and A HORIZONS (*see* SOIL: *the soil profile*). However, some authorities prefer to restrict it to the O horizon alone. Most soil animals (worms, pill bugs, etc.) are restricted to the topsoil, which also contains most or all of the soil's HUMUS. The presence of adequate topsoil in a plowed field should extend at least as deep as the ground will be turned by the plow.

tornado a type of intense storm characterized by the presence of a funnel-shaped cloud accompanied by extremely high winds. The precise cause of tornadoes is not known; however, they almost always arise from intense thunderstorms spawned along the edge of an advancing cold FRONT, where the temperature differential along the front is particularly high. They are among the most destructive forces in nature. Their winds, reaching speeds as high as 650 kilometers/hour (390 mph), are capable of driving straws through wooden planks and (in one documented case) bodily lifting and carrying objects as large as an 83-ton railway carriage with 117 passengers aboard, while the pressure differential between the interior and exterior of the funnel is sufficient to create a partial vacuum within the funnel, causing buildings within it to literally explode from the force of the air trapped within them. Tornados occur in all months of the year, but the greatest concentration of them is in the months of April through September, with May being the month during which they occur most frequently. Approximately 700 per year are reported in the United States, mostly in the central Midwest, where polar fronts moving southward from Canada collide with warm air from the Gulf of Mexico, creating great temperature differentials, and where the relatively flat terrain allows the storms room to develop and move about.

Torry Canyon the name of the ship involved in the first great ocean OIL SPILL. The *Canyon,* a 970-foot-long tanker with a capacity of 117,000 tons, hit a reef near the Scilly Isles off Land's End, Cornwall, England, on March 18, 1967. The impact broke the vessel in two, rupturing six of her 18 tanks and spilling approximately 34,000 tons of crude oil into the English Channel, where it fouled beaches and destroyed aquatic life in Cornwall and in Brittany, France, on the far side of the Channel. At the inquest into the accident, it was established that the Captain and First Mate had been arguing shortly before the ship ran aground, and it was assumed that clouded judgment from the tension of that argument was the primary factor in the navigational error which led to the grounding.

total dissolved solids *See under* TOTAL SOLIDS.

total ion concentration (TIC) in limnology, the number of IONS in a body of water, expressed as mass per unit volume, typically milligrams per liter (mg/l). Total ion concentration is not quite the same thing as total dissolved solids (*see under* TOTAL SOLIDS), because not all materials ionize when they dissolve. It can be measured indirectly by measuring the CONDUCTIVITY of a sample of the water, or directly by measuring the concentrations of the seven ions that make up over 99% of all the ions normally present in a body of water. These seven ions are sodium (Na^+), potassium (K^+), calcium (Ca^{++}), magnesium (Mg^{++}), chloride (Cl^-), sulfate (SO_{4++}, and bicarbonate (HCO_{3-}). TIC in natural lakes ranges from as low as 60 mg/l (in ultraoligotrophic lakes; *see* TROPHIC CLASSIFICATION SYSTEM) to 60,000 mg/l or more (in the alkaline evaporation lakes of the GREAT BASIN).

total solids (TS) in pollution control, the total load of solid POLLUTANTS in a water sample. Total solids are technically defined as the residue remaining following evaporation of the water sample in a crucible heated to 103°C (217°F). *Total dissolved solids* (TDS) are all solids dissolved in the water sample (*see* SOLUTION), *total suspended solids* (TSS) are all solids suspended in the sample but not dissolved (*see* SUSPENSION). Total dissolved solids may be separated from total suspended by means of a Gooch crucible, a clay vessel whose bottom is designed to serve as a filter. The sample is vacuum-drawn through the crucible, which retains the suspended solids while allowing the dissolved solids to pass. Analysis is usually gravimetric; that is, the crucible is simply weighed before and after the measuring operation, and the difference in the two weights is taken as the amount of solids present. *See also* FIXED SOLIDS; VOLATILE SOLIDS.

total suspended particulates (**TSP**) in pollution control, the total amount of PARTICULATE matter, both solid and liquid, suspended in the air at any given time. The NATIONAL AMBIENT AIR QUALITY STANDARDS call for a maximum of 75 micrograms/cubic meter TSP; the national average is currently around 60. Particulate emissions by U.S. industrial plants were cut in half between 1970 and 1980, dropping from 17.9 million metric tons (mmt) to 8.7 mmt. TSP measurements during that same period showed a drop of only 3% (61 micrograms/m^3 to 59 micrograms/m^3, lending some credence to industry's claim that most particulate emissions are from natural sources.

total suspended solids *See* TOTAL SOLIDS.

town a small city. Geographers generally class as a "town" any incorporated city with a central government and a population of between 2,500 and 10,000 (occasionally, 20,000) people. *Compare* VILLAGE; HAMLET.

township in surveying, a unit of land 6 miles square, laid out in relation to a rectangular grid of survey lines based on a predetermined MERIDIAN and BASELINE (*see* TOWNSHIP AND RANGE SURVEYING SYSTEM). A township contains 36 SECTIONS.

township and range surveying system (**TRSS, rectangular survey, government survey**) a system of land survey that divides a piece of land into square units aligned north-south and east-west. The basic unit is the TOWNSHIP, a square 6 miles on a side. The corners of townships are determined according to the intersection of a predetermined north-south MERIDIAN and east-west BASELINE, a location that is usually given an official name based on the meridian (e.g., the *Willamette Meridian* refers to the intersection of the meridian and baseline determined by the original survey of Portland, Oregon; it is used as the point of reference for nearly all surveys in western Oregon and Washington). The distance from the baseline is given in numbers of townships north or south, abbreviated *T (T2N: T41S),* while the distance from the meridian is given in numbers of townships east or west, a measurement referred to as the *range* and abbreviated *R (R14W: R21E).* Any given township can be specified precisely by its township number, range number, and meridian of reference; thus "T17N, R7E, Willamette Meridian" refers to the township located at the northwestern corner of Mt. Rainier National Park in the State of Washington. Townships are subdivided into SECTIONS, numbered from 1 to 36. The numbering begins in the northeast corner of the township and proceeds in a snakewise fashion, with nos. 1–6 lying along the northern border and numbered from right to left, nos. 7–12 directly below them but numbered from left to right, and so forth. The township and range surveying system was developed in response to the Ordinance of 1785 passed by the U.S. Congress. It replaced an older system, known as *metes and bounds,* based on physical features and preexisting property lines, which the early Congress felt was leading to too much confusion. The new system, applied by law to all newly-surveyed lands north and west of the Ohio River, had a profound impact on land settlement and development patterns in the American west (*see,* e.g., RAILROAD LAND GRANTS; CHECKERBOARD OWNERSHIP).

Township and Range Surveying System (diagram of a township)

6	5	4	3	2	1		
7	8	9	10	11	12		
18	17	16	15	14	13		
19	20	21	22	23	24		
30	29	28	27	26	25	NW	NE
31	32	33	34	35	36	qtr	qtr
						SW	SE
						qtr	qtr
township						section	

SECTIONS WITHIN A TOWNSHIP. Each section is one mile on a side with boundaries oriented due north-south and east-west. Portions of a section are labeled by *halves* and *quarters,* i.e., *Northwest Quarter; South Half of the Southwest Quarter,* etc.

toxaphene a PESTICIDE, approximate formula $C_{10}H_{10}Cl_8$, once widely used in the United States, now banned due to its contribution to TOXIC PRECIPITATION and its extreme TOXICITY to fish and other aquatic ORGANISMS. Toxaphene is not a COMPOUND but a MIXTURE, containing approximately 175 closely-related compounds of a type known as "ten-carbon chlorinated hydrocarbons"—that is, CHLORINATED HYDROCARBONS whose molecules contain 10 carbon atoms (*see* CARBON; ATOM) and varying numbers of CHLORINE and HYDROGEN atoms. The mixture is exactly reproducible, the same compounds being formed in the same proportions each time a series of key REACTIONS is performed. A yellow, waxlike substance with a low melting point, toxaphene is slightly soluble in water (*see* SOLUBILITY) and very soluble in most organic SOLVENTS. At the high point of its use (1976), approximately 33 million pounds of it were applied per year, mostly to cotton fields in the American southeast, where it proved effective against the boll weevil and other cotton pests. Use declined to 16 million pounds by 1981 and was cut off almost entirely by the ENVIRONMENTAL PROTECTION AGENCY on October 18, 1982 (effective 30 days later). The EPA's decree banned the sale and manufacture of the substance but allowed the continued use of current stockpiles. The impetus for the ban was the discovery of significant amounts of toxaphene in the waters of the Great Lakes, over 1,000 miles from the nearest area of extensive use. The compound can be absorbed through the skin; it acts on the central nervous system, causing tremors, disorientation, unconsciousness, and eventually death. LD_{50} (orally, dogs): 20 mg/kg.

toxicant a POISON, that is, any substance that causes the illness or death of an ORGANISM by disrupting its bodily functions. The word "toxicant" is a general term that covers both inorganic poisons (such as ARSENIC) and biologically-produced TOXINS. *Compare* MUTAGEN; TERATOGEN; CARCINOGEN. *See also* TOXICITY.

toxicity the ability of a substance to cause illness or death by disrupting bodily functions after it is ingested, inhaled, or absorbed through the skin. The toxicity of most substances varies with the path of entry into the body. A substance that is strongly toxic when inhaled, for example, may be only mildly toxic when ingested, and may not absorb through the skin at all. Toxicity is usually expressed in terms of the concentration of the substance that will kill 50% of those exposed to it for one hour (see LD_{50}, LC_{50}). *See also* TOXICANT; TOXIN.

toxic metals persistent pollutants that, because they can be neither created or destroyed, present special problems with regard to public health and environmental control policies. Among the most toxic are LEAD, MERCURY, CADMIUM, and ARSENIC.

toxic precipitation the presence of toxic substances (*see* TOXICANT) in rainfall, snowfall, and other forms of PRECIPITATION. These substances are typically present in extremely minute amounts, so that the rainfall itself is not hazardous. Nevertheless, the phenomenon is vitally important to the health of the total ENVIRONMENT, and is in fact the principal means by which most POLLUTANTS spread outside of inhabited areas. The presence of DDT and other PESTICIDES in the arctic and antarctic ECOSYSTEMS is a direct result of toxic precipitation, as is the presence of PESTICIDES, PCBs (*see* POLYCHLORINATED BIPHENYLS), and other manmade chemicals in the lakes of Isle Royale National Park, the Swiss Alps, and similar remote waters. The problem is not small. Studies have demonstrated that more than 90% of the POLYAROMATIC HYDROCARBONS found in the waters of Lake Superior have arrived via toxic precipitation, at the rate of as much as 163 metric tons per year.

MECHANICS OF TOXIC PRECIPITATION

Toxic substances reach the ATMOSPHERE by several pathways. Pesticides, of course, are usually sprayed directly into the atmosphere as droplets. As little as 12% of the sprayed pesticide may reach the target area, while the remainder may drift for some distance from the site. Many substances volatilize easily (*see* VOLATILITY) and are thus present in the air as vapors; others are suspended as PARTICULATES. Once in the air, they often travel great distances. In the case of substances such as PCB's and DDT, whose SOLUBILITY in water is small, they are washed out by the rain only a little at a time, allowing them to remain in circulation in the atmosphere long enough to reach remote areas such as the arctic and antarctic. Once they do fall from the air, however, they begin to be reconcentrated. Most of the toxic substances which fall over the entire WATERSHED of a lake will eventually be washed into the lake, concentrating them roughly by the ratio of the area of the watershed to the area of the lake. Once in the lake, they are subject to bioaccumulation through the FOOD CHAIN (*see* BIOLOGICAL MAGNIFICATION).

HISTORY

Toxic precipitation was first noted by the limnoligist Wayland Swain in 1978, during studies of the fish populations of Siskiwit Lake in Isle Royale National Park, Michigan. Numerous studies since have confirmed the scope and magnitude of the problem, and it has rapidly taken its place alongside ACID RAIN, depletion of the OZONE, the GREENHOUSE EFFECT, and the so-called "killer" SMOGS as one of the five principal human-caused problems of the ATMOSPHERE. *See also* AIR POLLUTION.

Toxic Release Inventory a database administered by the EPA providing data on AIR POLLUTION releases by individual plants and factories. Established by law in 1985, the TRI requires that 25,000 facilities report their releases annually for 300 different polluting compounds. Under the law, the TRI must be made available on computer disk and over the Internet, as well as in hard copy. Commercial users of the TRI include insurance companies, stockbrokers, developers, and planners. Activists also use the data to publicize local pollution problems. The website address is www.epa.gov/opptintr/tri.

Toxic Substances Control Act (TSCA) federal legislation to prevent public injury from the uncontrolled commercial or industrial use of toxic chemicals, signed into law by President Gerald Ford on October 11, 1976. The most important provision of the act is probably Section 5, which requires manufacturers to notify the ENVIRONMENTAL PROTECTION AGENCY (EPA) at least 90 days prior to the commercial introduction of a new chemical, either manufactured or imported. This *premanufacturing notification* (PMN) is designed to give the EPA a chance to assess the environmental dangers posed by the new substance and to prescribe special handling (including partial or complete bans) if necessary. Other important provisions allow the EPA to restrict or ban chemicals already in use if they are found to pose a substantial risk to the environment; provide authority for the agency to require manufacturers to test any substance for which insufficient safety data is available; and mandate the maintenance of a complete national inventory of all chemicals in commercial use, based on data on manufacturing, use and disposal that commercial chemical producers are required to report to the government. Finally, the act bans outright any further use of PCBs (*see* POLYCHLORINATED BIPHENYL) in the United States. Though it is a significant and potentially extremely powerful piece of legislation, the impact of the TSCSA has been somewhat muted by the profusion of loopholes (exempted from its provisions are eight categories of substances—foods, food additives, drugs, cosmetics, tobacco, pesticides, nuclear materials, and firearms and ammunition—which are regulated under other laws, plus any substance that the manufacturer can claim as a "trade secret") and by the overwhelming task of testing and regulating the approximately 65,000 chemical substances in regular commercial use.

toxic waste *See* HAZARDOUS WASTE.

toxin a poison, especially one produced by a living ORGANISM. In scientific usage the term is limited strictly to those substances—often PROTEINS or protein complexes—that are produced by one organism and are toxic to another (*see* TOXICITY). In popular usage, "toxin" has come to mean any ORGANIC COMPOUND, natural or man-made, which is toxic to living things. (In a sense, of course, man-made organic poisons fit the classic definition of "toxin" anyway: they are made by one living organism [humans] and are toxic to other organisms.) Toxins are usually classed by the type of tissue they act on (*see* CYTOTOXIN; ENTEROTOXIN; NEUROTOXIN). *See also* TOXICANT.

trace a faint trail, usually one which is established and maintained solely by the passage of animals and/or humans rather than by construction and maintenance techniques such as TREADWORK and BRUSHWORK. A trace is an indication of the route from point A to point B rather than being a means of improving ease of travel between the two points.

trace element (micronutrient) an ELEMENT required in very small amounts in the diet of a living ORGANISM. The typical dietary requirement for a trace element is less than 50 parts per million, with organisms usually showing a very limited range of tolerance to deviations from the required amount. Too much is often toxic (*see* TOXICITY), while too little can cause severe nutritional disorders. The metabolic use of these elements is varied, but most uses are related to the enzyme system (*see* ENZYME), either as a structural part of various enzymes or as CATALYSTS for their function in the body: a few form minor but essential structural components of specialized COMPOUNDS such as hemoglobin (animals) and chlorophyll (plants). Examples of trace elements include copper, selenium, manganese, cobalt, molybdenium, iodine, and iron. Examples of disorders stemming from their lack in the diet include goiter (iodine deficiency), anemia (iron, copper, or cobalt deficiency) and white-muscle disease (selenium deficiency). *See also* NUTRIENT.

traffic noise index (TNI) in noise-pollution control, the noise level from a roadway that is exceeded 10% of the time; that is, 10% of the sound reaching the observer from the roadway will be louder than the TNI and the remaining 90% will be at or below the index level. The measurement is normally given in dB(A) (*see* DECIBEL; A-WEIGHTED SOUND SCALE). *See also* NOISE POLLUTION LEVEL.

tragedy of the commons a metaphor concerning the overuse and potential destruction of ecosystems and natural resources by an exploding population. The phrase was coined by ecologist GARRETT HARDIN in a lecture at an American Society for the Advancement of Science meeting in 1968 and since then has entered the language of the environmental movement. In his lecture and in subsequent writings, Hardin points out that when a pasture is a public commons, all sheepherders have a right to use it. But if there is no system of restraint to keep from overusing it, any given sheepherder will risk the destruction of the common resource by taking it beyond its threshold capacity, based on the fear that his neighbor will gain an advantage by doing so sooner and thus destroying the resource for all. Harden likens the commons to the planetary ecosystem, which can also be ruined by unrestrained overpopulation and development. "In a crowded world of less than perfect human beings," he writes, "mutual ruin is inevitable if there are no controls. That is the tragedy of the commons." *See also* LAND ETHIC.

trailhead the intersection of a trail with a road, railroad, waterway, or other non-trail transportation corridor. Trailheads form the principal access points to WILDERNESS AREAS, and are usually the focus of at least minor development, in the form of parking areas, corrals and loading chutes for horses, signs, wilderness registers, and so on. *See also* TRAILHEAD SURVEY; ROADHEAD.

trailhead survey a poll of visitors passing through a TRAILHEAD. Since access to WILDERNESS AREAS and other roadless regions is primarily through trailheads, trailhead surveys usually provide the most accurate information on the use of these areas. Several types of trailhead survey are commonly utilized by recreation managers, including self-registration booths (where trail users "sign in", usually with some information about their expected length of stay and destination); automatic counters such as infrared beams ("electric eyes") or motion-picture cameras set to record one frame every 30 seconds; daily counts of parked cars; or direct interviews of visitors by management personnel.

trampling displacement a form of soil EROSION in which the motion of the eroded materials is caused by the feet of animals, including humans. Like most forms of erosion, trampling displacement takes place more often on bare ground than on ground with a good covering of vegetation, and more often on steep slopes than on gentle ones. The term implies a downslope motion of the displaced materials. Churning soil into dust or mud is not trampling displacement unless the soil or mud moves away from its original location. Overgrazed hillside meadows are the prime victims of trampling displacement, but it can also take place on overused trails, on SCREE SLOPES, and virtually anywhere where the soil is not fastened down firmly enough to resist foot traffic.

Trans-Alaskan Pipeline *See* ALASKA PIPELINE.

transcendentalism a nature-centered religious and philosophical movement that flourished in the mid-1800s in New England. RALPH WALDO EMERSON and HENRY DAVID THOREAU are today the best-known proponents of transcendentalism, which is so named for Kant's "transcendental" ideas—i.e., ideas beyond those that can be adduced through sensations of the material world. In general, members of the transcendental movement believed in the divinity of man and in nature's being "the oversoul," in Emerson's term. Transcendentalists also called for the rejection of traditional authority and borrowed spiritual precepts from eastern religions. Remnants of transcendentalism can be found in modern ecological theology and DEEP ECOLOGY. *See also* RELIGION AND THE ENVIRONMENT.

transect a one-time pass through an area along a predetermined course, usually a straight line, for the purpose of gathering data or establishing measurements. The act of making this pass is called *running a transect.* Transects are most commonly used as a form of preliminary study to determine whether or not a larger-scale study of the area is warranted. Among those who make the most use of the technique are botanists, wildlife biologists, range managers, and TIMBER CRUISERS. In each case, the purpose is to establish a preliminary census of the type, number, size, and condition of plants (animals, trees) present in an area.

transfer of development rights (TDR) See DEVELOPMENT RIGHTS.

transformer in ecology, an ORGANISM that can convert inorganic nitrogen compounds (*see* INORGANIC COMPOUND; NITROGEN) into the organic forms that other organisms can use for building PROTEINS. Transformers occupy a NICHE between DECOMPOSERS, which break down dead organic material so that its constituents can be reused, and PRODUCERS, which rebuild these materials into living tissue. Most transformers are MICROORGANISMS, commonly bacteria (*see* BACTERIUM).

transgenic crops *See* GENETIC ENGINEERING.

transition element (transition metal) in chemistry, an ELEMENT in which the ELECTRONS available for chemical bonding (*the valence electrons*) occur in more than one of the outer shells (*see* ATOM: *structure of the atom*). The transition elements, which are all METALS, share a number of characteristics, including multiple valence states (the ability to form COMPOUNDS with other elements in varying proportions), high reactivity (the ability to form compounds with many different elements), and the common presence of bright colors (reds, blues, greens, and yellows) as a characteristic of their compounds. They usually form ALLOYS easily. Transition elements that are close together in the PERIODIC TABLE show extremely similar properties due to the fact that their atoms have identical outer electron shells, differing only in one or two places in the inner shells. Sample transition elements include MERCURY, gold, copper, CHROMIUM, molybdenium, and tungsten.

Transition zone in the Merriam Life Zone system (*see* LIFE ZONE), the zone between the UPPER SONORAN ZONE (or Upper Austral zone) and the CANADIAN

ZONE. It is roughly equivalent to the TEMPERATE FOREST BIOME.

transparency of water, the ease with which light is able to penetrate a body of water. Since light is necessary for PHOTOSYNTHESIS, the transparency of a body of water has a great deal of bearing on how many plants and other autotrophs (*see* AUTOTROPHISM) it can support. The opposite is also true, however; the presence of large numbers of autotrophs, especially microscopic ones (*see* MICROORGANISM), can significantly lower the transparency of a body of water. Transparency is thus an excellent indicator of TROPHIC LEVEL, with oligotrophic bodies of water normally being much more transparent than eutrophic ones (*see* TROPHIC CLASSIFICATION SYSTEM). *See also* TURBIDITY; COMPENSATION DEPTH; SECCHI DISK.

transpiration the loss of water vapor to the air from the leaf surfaces or other parts of a plant. Transpiration results from the fact that the plant must keep some moist tissues exposed to the air at all times in order for carbon dioxide exchange to take place (*see* CARBON DIOXIDE; PHOTOSYNTHESIS). Water constantly evaporates from the exposed tissues. The amount of water lost in this manner can be immense. Under normal conditions, an acre of moderately sized HARDWOOD trees will transpirate more than 1 million gallons of water into the atmosphere in the course of a single year.

FUNCTIONS OF TRANSPIRATION

Along with its role in carbon-dioxide exchange, transpiration serves several other functions. It serves as the "pump" for raising water to the tops of trees, as the loss of water from the leaves creates a partial vacuum into which water flows from the XYLEM tubes, producing tension in the water columns within the tubes and literally pulling water up from the roots of the plant. It cools the leaves, and therefore the plant, in hot weather. On a larger scale, transpiration from forests cools the air and increases the HUMIDITY both within the forests and in areas downwind from them, thus serving both as an area-wide "air conditioner" and as a significant source of increased PRECIPITATION.

CONTROLS ON TRANSPIRATION

Approximately 90% of transpiration occurs through the STOMATA, the small pores on the surfaces of leaves and young stems. Thus, by closing its stomata, a plant may reduce transpiration to one-tenth its normal amount. Other important controls seen in plants include leaf size (the smaller the leaf area, the smaller the amount of transpiration) and the presence of leaf hairs (the hairier the leaf, the less wind and other evaporation-increasing air turbulence can reach the stomata). This is why plants in desert regions usually have small, hairy leaves. *See also* EVAPOTRANSPIRATION; RESPIRATION; WATER COLUMN.

Transportation, Department of (DOT) United States government agency created by Congress on October 15, 1966 and charged with overseeing all functions of the federal government relating to transportation except for those reserved to the military. A part of the EXECUTIVE BRANCH of the federal government, the department is headed by a cabinet-level officer (the Secretary of Transportation) and divided into 13 administrations. The Bureau of Transportation Statistics compiles, analyzes, and publishes transportation statistics; the U.S. Coast Guard ensures safety on American waterways; the Federal Aviation Administration oversees the safety of civilian aviation through issuance and enforcement of regulations; the Federal Highway Administration coordinates highway programs with the states; the Federal Motor Carrier Safety Administration ensures the safety of commercial motor vehicles; the Federal Railroad Administration promotes safe and environmentally friendly rail transportation; the Federal Transit Administration assists in the development and improvement of mass transit systems; the Maritime Administration maintains the nation's merchant marine; the National Highway Traffic Safety Administration works to reduce injury, death, and economic loss from motor vehicle crashes; the Research and Special Programs Administration oversees safe transportation of hazardous materials; the Saint Lawrence Seaway Development Corporation operates the waterway between the Great Lakes and the Atlantic Ocean; the Surface Transportation Board regulates interstate surface transportation, primarily trains; and the Transportation Administrative Services Center provides administrative and technical services for the DOT and other government agencies. The DOT has an annual budget of more than $54 million and nearly 100,000 employees nationwide. Headquarters: 400 Seventh Street SW, Washington, DC 20590. Phone: (202) 366-4000. Website: www.dot.gov.

transuranic elements those ELEMENTS lying higher than URANIUM in the ATOMIC SERIES. All transuranic elements are manmade. Most are radioactive and highly toxic. The group includes Neptunium (Np: atomic number 93), Plutonium (Pu: atomic number 94), Americium (Am: atomic number 95), Curium (Cm: atomic number 96), Berkelium (Bk: atomic number 97), Californium (Cf: atomic number 98), Einsteinium (Es: atomic number 99), Fermium (Fm: atomic number 100), Mendelevium (Md: atomic number 101), Nobelium (No: atomic number 102), and Lawrencium (Lr: atomic number 103). *See also* NUCLEAR WASTE.

traprock *See* BASALT.

trash fish in wildlife management, a SPECIES of fish which is not sought by either sport or commercial fishermen, either because it is not tasty enough or because it does not put up enough of a fight. Trash fish caught in commercial operations are generally ground up for fertilizer or made into pet food. The category includes numerous species. Those most prominent in freshwater are probably suckers, chubs, squawfish, shiners, and sculpins. A few species (carp, catfish, bluegill) are sometimes regarded as trash fish and sometimes not, depending both on their abundance in a given body of water and the attitude of the fisherman who catches them.

travel influence zone in Forest Service terminology, a visual management zone consisting of those lands immediately along an overland travel route that is likely to attract recreationists, plus those lands that form the visual backdrop for the travel route. "Overland travel routes" include roads, trails, railroads, tramways, and ski lifts. The travel influence zone classification is usually also extended to the lands around developed recreation sites, such as campgrounds and picnic areas, along these overland travel routes. Rivers and other waterborne transportation routes are not included in this category (*see* WATER INFLUENCE ZONE). Visual management in travel influence zones is weighted toward RETENTION, especially in the FOREGROUND. The MIDDLEGROUND and BACKGROUND may be managed as lower visual classes, depending on the amount and type of traffic along the travel route and on the distance between the travel route and the background landscape. *See also* VISUAL RESOURCE MANAGEMENT SYSTEM.

trawling a fishing method using huge, funnel-necked nets that are dragged along the ocean floor. Trawling causes serious impacts on ocean ecosystems by disturbing sea bottoms, thus disrupting communities of animals and plants that are essential to sea life. Some trawled areas are not expected to recover ecologically for hundreds of years. Trawling is thought to be a main cause of FISHERIES DECLINE and OCEAN POLLUTION. *See also* DRIFT-NET FISHING.

treadwork the construction or maintenance of a trail on the ground, as opposed to the flagging of the trail's route (*see* FLAG LINE) and the clearing of vegetation along it. Treadwork levels the trailbed (the *tread*), clears it of obstacles such as rocks and tree roots, and makes it visible and easy to follow. It may also include the use of surfacing materials such as bark chips, gravel, or asphalt. *Compare* BRUSHWORK.

tree hole succession a form of ecological SUCCESSION that is confined to water trapped in the cavities in trees or stumps. Tree hole succession is a characteristic of rain forests (*see* RAIN FOREST BIOME), CLOUD FORESTS, and other particularly damp environments. It may also occur during the wet season in moderately damp climates such as that of the eastern United States. A complete tiny ECOSYSTEM may develop in a tree hole, with PRODUCERS (ALGAE), first- and second-order CONSUMERS (PROTOZOANS, hydras, insect larvae [*see* LARVA], and even tadpoles), and DECOMPOSERS (bacteria and fungi; *see* BACTERIUM; FUNGUS), all actively functioning. Succession in new tree holes proceeds rapidly, and may be easily followed from the pioneer stage (the invasion of algae and other pioneering PROTISTA) through various seral stages (*see* SERE) to something approaching a CLIMAX COMMUNITY, with a rich and varied SPECIES composition. The speed at which tree hole succession takes place is a result of the amount of ORGANIC MATTER present in the tree-hole water from the constant decomposition of the outer surface of the tree, which provides a highly enriched eutrophic environment (*see* EUTROPHICATION) in which life can easily flourish.

tree line *See under* TIMBERLINE.

TreePeople started in 1973 when founder Andy Lipkis and his teenage friends planted trees to revitalize a dying forest. Over the years TreePeople has involved thousands of students and volunteers in urban forestry through education, training programs, and neighborhood renewal projects in southern California. TreePeople has published several manuals, including *A Planter's Guide to the Urban Forest* and *The Simple Act of Planting a Tree,* and offers *Seedling News,* a newsletter. Membership (1999): 20,000. Address: 12601 Mulholland Drive, Beverly Hills, CA 90210. Phone: (818) 753-4600.

trend (1) of a plant COMMUNITY, *see* RANGE MANAGEMENT.

(2) in geology, the compass bearing of the line of intersection between two faces, or between separate rock FORMATIONS, or between a rock outcrop and the surface of the ground. *Compare* DIP; STRIKE.

(3) in population biology, the direction POPULATION size is moving with time; that is, whether the population is growing, shrinking, or remaining stable.

trend count in the biological sciences, a series of POPULATION counts taken at regular intervals in order to ascertain the TREND of a study population. The length of the intervals depends on the type of ORGANISM being studied. A trend count for insect populations may have to be taken daily, whereas a trend count for the reproduction of CONIFERS in a CLEARCUT may be made only once every five or 10 years. Trend counts of mammal and bird populations are normally made annually. An example is the so-called "Christmas bird count" sponsored each December by the NATIONAL AUDUBON SOCIETY.

tributary a stream that empties into another, larger stream instead of into a lake or ocean. The tributary contributes its water to the larger stream (the MAINSTEM), hence the name. (Occasionally, the larger of two streams that join will be called the "tributary": the geography at the mouth of the Ohio River, for example, is such that the Mississippi is referred to as the mainstem and the Ohio the tributary, although the Ohio normally carries considerably more water).

CLASSIFICATION OF TRIBUTARIES

Tributaries are classified ("ordered") by geographers according to the number of sub-tributaries each one has. A stream with no tributaries is a *first-order stream;* a stream formed from two first-order streams is a *second-order stream;* a stream with at least one second-order stream flowing into it is a *third-order stream;* and so on. *See also* DRAINAGE DENSITY.

trichloroethane a CHLORINATED HYDROCARBON, chemical formula $C_2H_3Cl_3$, consisting of an ethane MOLECULE in which three of the six HYDROGEN ATOMS have been replaced by CHLORINE atoms. Trichloroethane exists in two isomeric forms (*see* ISOMER). In 1,1,1-trichloroethane (CH_3CCl_3), also known as *methylchloroform,* all three chlorine atoms are attached to the first of the two carbon atoms in the ethane molecule. In 1,1,2-trichloroethane ($CH_2ClCHCl_2$), or *vinyl trichloride,* two of the chlorine atoms are attached to the first carbon atom and one is attached to the second carbon atom. Both are colorless, nonflammable liquids that are good organic SOLVENTS and will not dissolve in, or mix well with, water. They are used as industrial solvents, primarily to clean cold-metal type and plastic molds. The two isomers have slightly different melting and boiling points and different odors. Both can irritate mucus membranes and cause drowsiness in high enough concentrations. In addition, vinyl trichloride is a CARCINOGEN, and has been listed as a PRIORITY POLLUTANT by the ENVIRONMENTAL PROTECTION AGENCY (methylchloroform is not currently listed).

trichloroethylene (TCE) a CHLORINATED HYDROCARBON, chemical formula C_2HCl_3, structural formula $CCl_2 = CHCl$, formed by substituting CHLORINE ATOMS for three of the four HYDROGEN atoms in the ethylene MOLECULE. A nonflammable, highly mobile liquid with a characteristic chloroform-like odor, TCE has been widely used as an industrial SOLVENT, especially in the rubber and paint and varnish industries, and for dry-cleaning clothing and degreasing metal parts. It has also seen some use as an anaesthetic and as a raw material for the manufacture of other ORGANIC COMPOUNDS, especially drugs. Moderately toxic (LD_{50} [rats, orally]:4.92 ml/kg; LC_{50} [rats, 4 hours]: 8,000 ppm), it causes ALCOHOL-like inebriation symptoms in small doses, acts as a narcotic in larger doses, and eventually leads to death from heart failure. TCE is also a known CARCINOGEN, and is listed as a PRIORITY POLLUTANT by the ENVIRONMENTAL PROTECTION AGENCY.

trichlorofluoromethane *See* CHLOROFLUOROMETHANE.

trichloromethane *See* CHLOROFORM.

trickling filter in pollution control, a widely used device for the SECONDARY TREATMENT of municipal SEWAGE, utilizing a bed of fist-sized rocks (3–4.5 in. in diameter). The wastewater to be treated is sprayed over the surface of the rock bed and trickles slowly down through it. The rocks rapidly develop a living film of ALGAE, fungi (*see* FUNGUS), and bacteria (*see* BACTERIUM) that utilize the NUTRIENTS from the wastewater, removing them from the water in the process and dramatically lowering its BOD (*see* BIOCHEMICAL OXYGEN DEMAND). The film also acts as a sorptive surface (*see* ADSORPTION; ABSORPTION) for organic MOLECULES and small particles, removing many of them along with the nutrients. It has been demonstrated that a trickling filter at peak operation is capable of removing approximately 98% of the VIRUSES in the waste stream. Trickling filters are usually round, with rotating arms to spread the sewage evenly across the surface of the rock bed. The bed itself has an optimal depth of 5 to 6 feet. The treated water is drawn off at the bottom, which is also laced with ducts to draw air through the rock bed, as most of the trickling filter ORGANISMS are aerobic (that is, depend upon OXYGEN [*see* AEROBIC BACTERIA; ANAEROBIC BACTERIA]) and the filter will quickly become foul-smelling and inoperative without adequate air circulation. Some trickling filters use molded plastic pieces in place of the rocks. These can be designed with considerably more surface area and air-circulation space than natural rocks, enabling construction of a variant of the trickling filter known as a *biological tower*—essentially a 10-meter (35-foot) deep trickling filter. Trickling filters (and biological towers) are often built in series, with the EFFLUENT from one filter becoming the FEED WATER to the next. The effluent may also be recirculated a second time through the same filter or filter series. *Compare* ACTIVATED SLUDGE.

trihalomethane (THM) any of a group of chemical COMPOUNDS consisting of a methane MOLECULE (CH_4) with three of its four HYDROGEN ATOMS replaced by HALOGENS (FLUORINE, CHLORINE, bromine, or iodine). The three halogens may be the same (three chlorine, $CHCl_3$; three bromine, $CHBr_3$; etc.) or may differ from each other (two bromine and one chlorine, $CHClBr_2$; two chlorine and one iodine, $CHICl_2$; etc.). Those in which the three are the same element are also known as *haloforms* (chloroform, bromoform, fluoroform, iodoform). The trihalomethanes have similar physical and toxicological properties, and are classed together for regulatory purposes. Current drinking water standards in the United States require them to be present in amounts of less than 100 parts per billion (ppb). *See also* POLYCYCLIC ORGANIC MATTER; CHLOROFORM.

triple point in chemistry and physics, a specific condition of temperature and pressure at which all three phases (solid, liquid and gas) of an ELEMENT or COMPOUND exist in equilibrium with each other. There is one, and only one, triple point for each element and compound; hence, this point is more useful for defining a substance than are the boiling and melting points, as these points vary with the ATMOSPHERIC PRESSURE on the substance.

triple superphosphate a concentrated form of superphosphate fertilizer (*see under* PHOSPHATE), made by treating PHOSPHATE ROCK with phosphoric acid (H_3PO_4) instead of sulfuric acid (H_2SO_4). Triple superphosphate has somewhat more than twice as much phosphorus as ordinary superphosphate and may therefore be applied to the soil in smaller amounts. Chemical formula: $Ca(H_2PO_4)_2$.

tripton in limnology, the nonliving suspended matter in a body of water. The tripton includes CLAY particles and other suspended SEDIMENTS; the excrement of aquatic ORGANISMS (excluding the living bacterial component, where present; *see* FECES); and the suspended bodies of dead PLANKTON and small NEKTON. The amount and composition of the tripton has a significant bearing on the trophic status of a body of water (*see* TROPHIC CLASSIFICATION SYSTEM) as well as on its color and TRANSPARENCY. *See also* SESTON.

Trombe wall a form of passive solar installation (*see* PASSIVE SOLAR DESIGN; SOLAR ENERGY) consisting of a masonry wall 6–18 inches thick with gaps at the top and the bottom. The wall faces the Sun, with the space in front of it enclosed by glass. Air heated by the Sun between the glass and the Trombe wall rises and passes into the room behind the wall through the gaps at the top. It is replaced by cool air drawn in from the room through the gaps at the bottom. The wall also absorbs heat which moves slowly through it, radiating into the room after the Sun goes down.

trophic classification system in limnology, a method of classifying standing bodies of water (lakes, bays, ponds, etc.) according to the number of living ORGANISMS they will support per unit area of water surface. The principal measurement upon which the trophic classification system is based is PRIMARY PRODUCTIVITY: loosely defined, the amount of organic growth that takes place per unit area over a given amount of time, usually a day. Primary productivity, in turn, is dependent upon two principal factors: the amount of NUTRIENTS entering the lake water and the amount of energy available in the water, both in the form of heat (warm lakes are more productive than cold ones) and light (shallow lakes, in which light can reach the lake bottom, are more productive than deep ones). Thus, a lake's trophic classification is dependent primarily on its depth, its geographical location (that is, whether it is in a cold or warm climate), and the amount of nutrients available in its WATERSHED and in the SEDIMENTS that have built up in its bed. Five trophic classifications are generally recognized by limnologists. An *oligotrophic* lake is one in which productivity is low; a *eutrophic* lake is one in which productivity is high. Extremely oligotrophic lakes are called *ultraoligotrophic;* extremely eutrophic lakes are called *hypereutrophic.* Lakes intermediate in character between eutrophic and oligotrophic are called *mesotrophic.* (A sixth category, *dystrophic,* is sometimes recognized. These are bodies of water that do not contain enough DISSOLVED OXYGEN to maintain life, generally because hypereutrophic conditions have caused such rapid productivity that all the oxygen has been used up.) The trophic classification system was developed by the biologist C. A. Weber in Germany around 1907 to describe the conditions of freshwater BOGS. It was first applied to lakes by the pioneer German limnologist Einar Naumann in 1919. Today it is considered among the most important tools of lake management. *See also* EUTROPHICATION.

trophic level in ecology, the level an ORGANISM occupies on the FOOD CHAIN. The first trophic level is occupied by the PRODUCERS, that is, the green plants and other autotrophs (*see* AUTOTROPHISM). The second level is made up of HERBIVORES ("first-order consumers"; *see* CONSUMER). All higher levels are filled with CARNIVORES. An organism may occupy more than one level. Thus, a human eating salmon garnished with lemon juice is feeding simultaneously on the second and fourth levels (second, because he or she is eating the lemon, a plant; fourth, because he or she is eating the salmon, itself a carnivore and therefore on at least the third trophic level). *See also* FOOD WEB; ECOLOGICAL PYRAMID.

tropical forest biome a BIOME recognized by some authorities that combines the tropical phases of the RAIN FOREST BIOME and the DECIDUOUS FOREST BIOME. It covers a broad enough range of vegetational types that calling it a single biome is probably inappropriate, although temperature conditions are relatively uniform throughout.

tropical scrub forest a forest of low, shrubby trees found in dry regions of the tropics. It is sometimes considered a separate BIOME. However, it is probably better treated as an ECOTONE between the tropical deciduous forest (*see* DECIDUOUS FOREST BIOME) and the tropical manifestation of the GRASSLAND BIOME ("savanna").

tropopause in meteorology, the boundary between the TROPOSPHERE and the stratosphere (*see* ATMOSPHERE: *structure of the atmosphere*). The tropopause is somewhat analogous to the THERMOCLINE in a body of water. Below it, the air is turbulent, and the temperature varies with the height above the Earth's surface; above it, the air is still, and the temperature remains relatively constant over a large vertical span. The tropopause is approximately 9 km (5.4 mi) high over the poles and 16 km (9.6 mi) high over the equator. It does not make a smooth transition between these elevations, but changes elevation in two abrupt steps, one above the HORSE LATITUDES and the other above the POLAR FRONT. Thus, the tropopause can be seen as the upper limbs of the HADLEY CELLS in the lower atmosphere. *See also* JET STREAM.

troposphere in meteorology, the lowest level of the ATMOSPHERE, reaching from the Earth's surface to the TROPOPAUSE, between 9 and 16 km high. Although the troposphere represents only about 1.2% of the depth of the atmosphere as a whole, it contains more than 75% of the atmospheric mass. Within the troposphere, temperatures decrease with height (*see* LAPSE RATE) and convection currents keep the air stirred up. Nearly all weather events (storms, winds, migrating high and low pressure cells, etc.) occur in the troposphere. The only exceptions are occasional thunderstorms whose upwelling air masses are powerful enough to breach the tropopause. *See also* ATMOSPHERE: *structure of the atmosphere*.

Trout Unlimited angler's organization, founded in 1959 to work for the preservation of cold water sport fisheries, especially trout, salmon, and steelhead. Although it does some congressional lobbying, the group emphasizes influencing agency regulations and making certain that field officers enforce them. It also provides volunteer design and labor for stream-enhancement and restoration work, primarily on federal land. Trout Unlimited has affiliates in Australia, Canada, and New Zealand. Membership (1999): 100,000. Headquarters: 1500 Wilson Boulevard, Suite 310, Arlington, VA 22209. Phone: (703) 522-0200. Website: www.tu.org.

true fir in forestry, a member of the GENUS *Abies*. The term is used to differentiate members of this genus from the DOUGLAS FIRS (genus *Pseudotsuga*).

trunk *See under* SEWER.

Trust for Public Land land preservation organization, founded in 1972 to purchase private lands for transfer to public agencies for use as parks and open space. The trust's also assists individuals and other groups who wish to find ways to purchase and preserve open space, and works with farmers, ranchers, and rural communities to keep agricultural land in production rather than turning it over to urban development. Through TPL's efforts, more than 1,400 areas in 45 states have been protected. Address: 116 New Montgomery Street, San Francisco, CA 94105. Phone: (415) 495-4014. Website: www.tpl.org.

TSI *See* TIMBER STAND IMPROVEMENT.

tsunami (seismic sea wave [incorrectly, **tidal wave**]) a large WAVE caused by an undersea EARTHQUAKE or other major TECTONIC ACTIVITY (such as a volcanic eruption). The exact mechanism of tsunami formation is unknown (and may differ from case to case), but most are probably the result of sudden vertical displacements of the sea floor along fault lines, producing disturbances in the water similar to those created when a stone is thrown into a pond, but on a vastly larger scale. The crests of the waves created by a tsunami are as much as 200 km (120 mi) apart, meaning that the HALF-WAVE depth is up to 100 km (60 mi). Therefore, these waves "feel bottom" even in the deepest part of the ocean. In mid-ocean they show up as a long swell 1 or 2 meters (3–7 feet) high, which is, however, moving extremely rapidly (up to 900 km/hr). As they approach land, they are constricted by the rising bottom and their energy is funneled by irregularities such as reefs, islands, and headlands, resulting in shorter, slower, but much higher wave crests—as high as 30 meters (100 feet), half of which will be above the normal ocean level and the other half below. Such large waves can sweep far inland, causing much loss of life and enormous property damage. Like ripples from a tossed stone, the gigantic ocean ripples of the tsunami consist of several waves and troughs that follow each other at intervals of roughly one hour, eventually damping out. Tsunamis are most common in the Pacific Ocean, where since 1946 a "tsunami warning network" of seismic stations and ocean-level gauges has successfully prevented most loss of life.

tuber in biology, a large, fleshy, underground food storage organ utilized by some plants. It may also assist the plant in vegetative propagation (*see* ASEXUAL REPRODUCTION). Tubers are stem tissue; they differ from CORMS in that they arise from underground branches of the stem (*stolons,* or *rhizomes*) rather than from the central stem itself. They are not part of the plant's root system. The most familiar tuber is undoubtedly the Irish, or white, potato. *Compare* BULB.

Tucson lawn in arid regions (*see* ARID CLIMATE), a lawn planted in native vegetation that is adapted to dry conditions and thus does not need to be watered regularly. Pioneered in Tucson, Arizona, Tucson lawns are a major water-saving strategy for municipalities. Replacing grass lawns with Tucson lawns can save several thousand gallons of water per household per week. (One hour of lawn watering uses approximately 600 gallons of water.)

tuff (tuffaceous rock) any of several varieties of rock formed from solidified pyroclastic flows (*see* PYROCLASTIC ROCK). Tuff usually consists of VOLCANIC ASH that has been compressed into rock by the weight of overlying materials. However, in some tuffs, known as *welded tuffs,* the materials comprising the tuff are not compressed but are welded together by their own heat. Welded tuffs are generally banded as if they had been laid down as STRATA. Tuff containing small bits of obsidian is known as *lapilli tuff.* If it contains larger rock fragments of various types, it is known as *tuff breccia.* These are usually soft, mechanically weak rocks, although some (especially the welded tuffs) approach the hardness of GRANITE.

tundra biome in the biological sciences, a major terrestrial BIOME of high mountain and polar regions, characterized by cold winters and cool summers; thin, NITROGEN-poor soils; and a near-total absence of trees. FOOD WEBS are short and simple. Plants are primarily low-lying PERENNIALS that reproduce by vegetative means (*see* ASEXUAL REPRODUCTION) to avoid the difficulties of setting seed in the short, cool summers. They tend to grown in clumps, with much bare ground in between. WOODY PLANTS are limited to SHRUBS, mostly along water-courses, although some of those "shrubs" are species that would be trees in more favorable growing conditions. MOSSES and LICHENS are common. Animal life, perhaps surprisingly, is fairly extensive. The tundra serves as the breeding ground for many WATERFOWL, and is home to numerous burrowing animals and to large nomadic herds of caribou and muskoxen, both of which feed principally on lichens and mosses. In polar tundra, PERMAFROST is usually present a short distance below the soil surface. In alpine regions, little permafrost exists, but rock layers lying close to the surface present similar barriers to water flow and the growth of plant roots. Precipitation on the polar tundra is light, averaging less than 25 centimeters (10 inches) per year. On the alpine tundra it is considerably heavier, but the rapid RUNOFF caused by the steep mountain slopes gives the two areas similar soil-moisture characteristics. In both locations, precipitation is primarily in the form of snow. Water features (bogs, soggy soil, and small ponds and brooks) are common during the short season when average temperatures lie above freezing.

DIVISIONS AND DISTRIBUTION

Tundra is usually divided into four types, or ZONES, which lie concentrically around the coldest regions (the poles and the mountain summits). *Mat tundra,* just beyond the treeline, consists of low-lying woody plants such as heath and willow; trees may be present in the form of KRUMMHOLZ. *Grass tundra* occupies the more fertile sites in the mat tundra zone and extends somewhat further north or up. It consists of hardy grasses, sedges and wildflowers. *Lichen-moss barrens* consisting of lichens and mosses lie above the grass tundra. Above them is *ice tundra,* where no plant life exists at all. Tundra covers about 20% of the Earth's land surface, primarily north of the Arctic Circle, where it forms vast treeless plains in northern North America and Eurasia. It is also found in scattered patches on high mountains in the temperate and tropical zones and around the edges of the Antarctic ice cap. Virtually all of the arctic tundra plants are circumpolar in distribution (*see* HOLARCTIC SPECIES). About 40% of the alpine plants are also found in the arctic, with the remainder often ENDEMICS that have developed in each alpine region as it became isolated from the others. *See also* ALPINE ZONE; PATTERNED GROUND.

turbidity the presence of suspended materials—principally SILTS, CLAYS, and MICROORGANISMS, living and dead—in a body of water. Turbidity caused by silts and clays (*inorganic turbidity*) is generally the result of wave action or stream EROSION acting on a previously deposited bed of these materials. Turbidity caused by living ORGANISMS (*organic turbidity*) commonly results from EUTROPHICATION. Both types of turbidity limit the depth to which light can penetrate the water (*see* TRANSPARENCY), lowering its productivity. In the case of organic turbidity, this serves as a negative feedback loop, decreasing the numbers of new organisms produced and therefore keeping the turbidity from rising beyond a certain range. Organic turbidity can be a health hazard due to the presence of PATHOGENS among the microorganisms causing the turbidity. Inorganic turbidity is aesthetically displeasing, and can conceal the presence of pathogens that cling to the silt or clay particles. Thus, turbidity must be removed during POTABLE WATER treatment. *See* SETTLING BASIN; RAPID SAND FILTER; JACKSON TURBIDITY UNIT.

turgor in biology, a condition of living tissue in which there is enough liquid in the tissue to create an outward pressure on the tissue walls, causing the tissue to become stiff. The term is most often used in connection with plant CELLS that have absorbed water by OSMOSIS until the pressure within them (the *turgor pressure*) exceeds the pressure around them, causing the cell wall to bulge outward under tension. Turgor is the principal means of support of non-woody plant tissue such as the stems of HERBS. Loss of turgor due to drought is the reason why plants "wilt." The term is also used to describe healthy tissue in animals, in which the blood pressure is adequate to keep the circulatory system full and therefore maintain proper elasticity in the tissue.

Turner, Frederick Jackson (1861–1932) American historian and educator noted for his thesis that the frontier has shaped the character of the American people. Educated at the University of Wisconsin and Johns Hopkins University, he taught from 1885 to 1910 at the University of Wisconsin and from 1910 to 1924 at Harvard University. His 1893 paper presented to the American Historical Association, "The Significance of the Frontier in American History," was a new and creative interpretation of U.S. history. Turner believed that it was the exploration and settlement of the western frontier, not the European influence of the initial colonists, that created the American ethos. Such ideas expressed a significant break with the thought of historians of his time and greatly shaped the thought and writing of later American historians. Turner retired to California in 1925 due to poor health and worked as a research associate at the Huntington Library in San Marino. His 1893 paper, together with other essays, was reprinted in *The Frontier in American History* (1920). His books include *The Rise of the New West* (1906) and *The Significance of Sections in American History* (1932), for which he was awarded the Pulitzer Prize for history. A biography, *Frederick Jackson Turner,* was published by historian Ralph A. Billington in 1973.

turnover of a lake, *see under* THERMOCLINE.

turnover time in ecology, the time required for one complete period of a BIOGEOCHEMICAL CYCLE; that is, the time it takes a sample MOLECULE of a given substance to be incorporated into an ORGANISM, cycled through the FOOD CHAIN, released back into the ENVIRONMENT through decomposition (*see* DECOMPOSER), and acted on by environmental forces (such as EROSION and DEPOSITION, stream transport, and TECTONIC ACTIVITY) so that it is ready to be taken up by an organism again. Turnover time varies greatly with conditions. In a highly eutrophic lake (*see* TROPHIC CLASSIFICATION SYSTEM) for example, the turnover time for PHOSPHORUS may be as little as one minute; however, if the phosphorus becomes incorporated into PHOSPHATE ROCK the turnover time may be on the order of several million years. Thus, turnover time serves as one measure of the productivity of an ECOSYSTEM.

TVA *See* TENNESSEE VALLEY AUTHORITY.

2,4-D (2,4-dichlorophenoxyacetic acid) a chlorinated hydrocarbon herbicide (*see* CHLORINATED HYDROCARBON; HERBICIDE), chemical formula $C_8H_6Cl_2O_3$, widely used as a lawn-weed killer and as a tool for BRUSH CONVERSION in forestry. It is strongly toxic to DICOTYLEDONS but has relatively low TOXICITY for other plants and for animals. 2,4-D consists of a BENZENE RING with an acetic acid "tail" containing an extra OXYGEN ATOM attached to the number 1 position on the ring and with CHLORINE atoms attached to the number 2 and number 4 positions. Insoluble in water (*see* SOLUBILITY), it is normally applied to vegetation by converting it to a soluble SALT (usually through the addition of sodium), dissolving it in water, and spraying it from a nozzle. Metabolic processes in both plants and animals (*see* METABOLISM) convert it back to the acid form. Ingested 2,4-D is excreted rapidly in the urine. It also breaks down readily in the environment, and thus poses no threat of BIOLOGICAL MAGNIFICATION or long-term persistence. Symptoms of ACUTE TOXICITY do not normally occur until doses reach levels 100 times or more of those normally used in vegetation management. CHRONIC TOXICITY requires long-term exposure (several weeks or more) at fairly high levels. The material is a TERATOGEN, but fetal damage does not occur until dosages reach nearly toxic levels for the mother. There is some evidence that it may be mildly mutagenic and carcinogenic (*see* MUTAGEN; CARCINOGEN), but results of tests for genetic damage and cancer have been largely inconclusive. 2,4-D has a strong medicinal odor and taste and is not likely to be ingested accidentally. For all these reasons, it is classed as among the safest of herbicides. Nevertheless, the possibility of cancer and mutations, however remote, requires that it be used only under careful restrictions. Symptoms (chronic and acute): gastrointestinal (digestive system) distress, eye and skin irritation, nerve damage, weakness and vertigo, and respiratory system failure. LD_{50} (rats, orally): 100 mg/kg. *See also* 2,4,5-T: DIOXIN.

2,4,5-T (2,4,5-trichlorophenoxyacetic acid) a chlorinated hydrocarbon herbicide (*see* CHLORINATED HYDROCARBON; HERBICIDE), chemical formula $C_8H_5Cl_3O_3$. A close relative of 2,4-D—it differs only in having one extra CHLORINE ATOM per MOLECULE—it has nearly identical effects and health hazards, and was once widely used, usually mixed with 2,4-D (e.g., in the DEFOLIANT known as Agent Orange used by the American military in Vietnam). However, it is always unavoidably contaminated during production by TCDD, one of the most potent forms of DIOXIN, and the impossibility of completely removing this CONTAMINANT makes 2,4,5-T much more dangerous to use than 2,4-D. For this reason, its registration was annulled by the ENVIRONMENTAL PROTECTION AGENCY in 1978, and it is no longer used in the United States.

208 plan (Areawide Waste Treatment Management Plan) in pollution control, one of a coordinated nationwide system of plans to protect water quality in streams and rivers, based on (and required by) Section 208 of the CLEAN WATER ACT. Originating with the 1972 amendments to the act, Section 208 requires each state to map water quality planning areas; designate a lead agency for planning within each area; and prepare a plan, to be filed with the ENVIRONMENTAL PROTECTION AGENCY, for controlling both POINT SOURCES and NONPOINT SOURCES within the planning area. The plan must contain adequate facilities for SEWAGE treatment and treatment of industrial discharges, as well as a specified set of BEST MANAGEMENT PRACTICES to control nonpoint pollution through land-use planning. NPDES permits (*see* NATIONAL POLLUTANT DISCHARGE ELIMINATION SYSTEM) may not be granted unless the applicant demonstrates compliance with the 208 plan. Federal funding was originally provided for up to 100% of the 208 planning and administrating process; however, this funding largely dried up in the early 1980s, and the plans, though still in force, have been largely subsumed into the STATE WATER QUALITY MANAGEMENT PLANS required by Section 303 of the act.

type conversion in forestry, the forced conversion of a growing site from one type of vegetation to another (*see,* e.g., BRUSH CONVERSION). Type conversion is usually a form of sped-up SUCCESSION (that is, the conversion proceeds from one SERE to another which is closer to the CLIMAX for the site); however, the site may be modified into a new successional pattern instead, as when IRRIGATION converts a tract of desert vegetation into cropland, lawn, or forest.

type genus *See under* FAMILY.

type location *See under* TYPE SPECIMEN.

type specimen in biology, the specimen (that is, the individual ORGANISM) used to describe a new SPECIES. The type specimen is ordinarily collected and placed in a museum, herbarium, or other location where it can be studied and referred to easily. The location it was collected from (that is, the place where the person who first described the species found the specimen he or she referred to for this description) is known as the *type location.* If the type specimen was collected by the original discoverer and designated by him or her as the type, it is known as a *holotype.* If it was collected by the original discoverer but designated as the type specimen by someone else, it is a *lectotype.* If all specimens collected by the discoverer have been destroyed, or if the discoverer collected no samples, a type specimen must not only be named but collected by someone else. This specimen, known as a *neotype,* should preferably be collected at the original type location.

typhoon *See* HURRICANE.

U

Udall, Stewart Lee (1920–) conservationist, legislator, and secretary of the interior. Born in 1920, Udall served in the U.S. Air Force during World War II, practiced law for several years, and served as a representative from Arizona in Congress, where he supported natural resource conservation. He was appointed secretary of the interior by President Kennedy and served in this position from 1961 to 1969. He worked to curb abuses and exploitation of public lands, helped create the WILDERNESS ACT OF 1964, greatly increased the NATIONAL PARK SYSTEM, and reorganized the BUREAU OF INDIAN AFFAIRS. He became a private consultant in 1969 for Overview, Inc. and as a private attorney helped develop the Radiation Exposure Compensation Act, which recognized the government's responsibility in the coverup of hazards of atomic testing and provided for compensation to victims. He is the author of a best-selling and influential history of the conservation movement, *The Quiet Crisis* (1963); *National Parks of America* (1966); *1976: Agenda for Tomorrow* (1968); and *The Myths of August: A Personal Exploration of Our Tragic Cold War Affair with the Atom* (1998).

ultisol a type of soil characteristic of warm, moist climates, consisting of a very thin A HORIZON (*see* SOIL: *the soil profile*) overlaying a thick, reddish brown B horizon containing a relatively high proportion of CLAY. Except in the A horizon, ORGANIC MATTER is nearly nonexistent. The soil is acidic (BASE SATURATION: less than 35%), and is low in most MINERALS due to the leaching action of the heavy rainfall along with the great age of the soil. Aluminum content is high due to the presence of the clays. Ultisols have very low natural fertility and make poor agricultural soils. Their natural cover is a forest consisting of trees with deep TAPROOTS that are able to reach mineral accumulations in the deep SUBSOIL and "pump" them up for the use of the shallow-rooted UNDERSTORY plants. Once these trees are cleared, productivity rapidly decreases unless the soils are heavily fertilized and treated with lime or other acid-neutralizing materials. Ultisols form the bulk of the soils in the so-called "red clay region" of the south-eastern United States.

ultrafiltration *See* REVERSE OSMOSIS.

ultramafic rock in geology, rock containing less than 45% silica; hence, little or no QUARTZ or FELDSPAR. They are commonly rich in iron and magnesium and weather to soils that are highly basic (*see* BASE). Most are igneous in origin (*see* IGNEOUS ROCK). Common examples: BASALT, SERPENTINE, PERIDOTITE.

ultraoligotrophic lake *See under* TROPHIC CLASSIFICATION SYSTEM.

ultraviolet (UV) radiation the emission of powerful electromagnetic light waves from the Sun and certain artificial sources such as welding arcs, fluorescent lamps, and lasers. Ultraviolet light is divided into the "A," "B," and "C" ranges, which correspond to the wavelengths. Wavelengths longer than UV-A are called violet light (hence *ultra*violet), and those shorter than UV-C are called X rays. The OZONE LAYER filters out almost all the short-wavelength UV-C, which is the most damaging to various organisms, including plants and people. About half the ultraviolet light in the B range, also harmful, is blocked by the ozone layer; but most of the relatively harmless rays of the A range strike the Earth. Excess exposure to UV radiation can injure the cellular structure of plants, animals, and other organisms, producing blindness and skin cancers and interfering with cellular division in eggs (*see* AMPHIBIAN DECLINES AND DEFORMITIES). In trees and plants, excess UV radiation in the B range can deform leaves and needles and inhibit growth, leading to weakness and increased susceptibility to the impacts of drought, cold, and pests and diseases (*see* FOREST DECLINE AND PATHOLOGY).

unconformity in geology, a line in a rock FORMATION separating a SERIES of older rocks from a series of considerably younger ones. An unconformity is a gap in the geologic record; it usually represents an ancient erosion surface (that is, an old ground surface that was leveled by EROSION, wearing away the intermediate-aged rock in the process, before young rock was deposited on top of it), but may instead be merely a gap in DEPOSITION. Geologists usually recognize three forms of unconformity. In a *disconformity,* the older and younger rocks lie in the same orientation. There is merely a chronological gap in the series. In an *angular unconformity,* the older rock STRATA lie at a different angle than the younger strata, indicating folding and/or faulting of the older rocks before the erosion and later deposition took place (*see* EARTHQUAKE; TECTONIC ACTIVITY). Finally, a *nonconformity* represents an erosional surface developed on GRANITE, SERPENTINE, or some other non-stratified rock, with stratified rock (BASALT, LIMESTONE, etc.) above it. Small-scale unconformities are sometimes distinguished from larger ones by referring to them as *paraconformities* (if the time gap is small) or *local unconformities* (if the area involved is small).

unconsolidated formation in geology, a FORMATION made of a soft, brittle rock, such as siltstone or MARL, that crumbles easily upon exposure to the environment. *See also* UNCONSOLIDATED MATERIAL.

unconsolidated material in geology and engineering, the REGOLITH, that is, the loose SOIL, SAND, and ROCK fragments that lie on top of the solid BEDROCK. *See also* UNCONSOLIDATED FORMATION.

understory in forestry, the plants that grow beneath the CANOPY of a living forest. Understory plants include HERBS, SHRUBS, and small trees (both those that are permanently small, such as the HARDWOODS that grow as understory trees beneath CONIFERS, and those that are small because they are young and have not yet grown into the canopy). Plants that can survive in the understory typically have large leaf surfaces to enable them to photosynthesize efficiently on the small amount of light that is able to penetrate the canopy (*see* PHOTOSYNTHESIS; SUNFLECKS). *Compare* OVERSTORY. *See also* SHADE-TOLERANT SPECIES.

ungulate any mammal with hooves. The ungulates are divided into two ORDERS, those with an odd number of toes (the Perissodactyla, including horses, rhinoceroses, and tapirs) and those with an even number of toes (the Artiodactyla, including cows, sheep, deer, antelope, bison, camels, pigs, goats, and hippopotamuses). Most ungulates are HERBIVORES and none are pure CARNIVORES. They are herd animals that usually occupy open spaces where their principal defense against PREDATORS is speed. Most have horns or antlers. The natural RANGE of the Perissodactyls is limited to the old world (Africa and Eurasia); native Artiodactyls are found on every continent except Australia. Domestic ungulates of both orders have, of course, been introduced everywhere. *See also* RUMINANT.

Unified Soil Classification System (USCS) a classification of soils based on their ability to support structures, developed by Arthur Casagrande for the ARMY CORPS OF ENGINEERS following World War II and since adopted as a standard for civil engineering work in the United States. The USCS divides soils into six categories according to particle size and/ or composition, giving each a letter designation: G (GRAVEL), S (SAND), M (fine sand and SILT), C (CLAY), O (organic materials other than PEAT) and P_t (peat). G and S soils are further categorized as W (well graded by size), P (poorly graded by size), or C (containing a significant clay fraction, which improves binding ability); M, C, O, and P_t soils are categorized as either L (low compressibility) or H (high compressibility). Thus, a GC soil is a gravelly clay; an ML soil is a silty soil that does not compress easily; and so on. Soil maps based on the USCS are usually very limited in scope, covering only the immediate environs of the construction site where the information is needed. *Compare* SEVENTH APPROXIMATION; GREAT SOIL GROUPS.

Unified Soil Classification System

Classification		Qualifiers	
G	gravelly soils	W	well graded and clean
S	sandy soils	C	high clay fraction
		P	poorly graded but fairly clean
M	very fine sand	L	low to medium compressibility low plasticity
C	inorganic clay		
O	organic silts	H	high compressibility and plasticity
P_t	peat		

uniformitarianism in geology, the concept that geologic change is generally gradual and that it always follows the same rules, allowing the causes of events in the distant past to be understood by extrapolating current processes such as EROSION, EARTHQUAKES, VOLCANIC ERUPTIONS (*see* VOLCANO; LAVA), etc. *Compare* CATASTROPHISM.

Union of Concerned Scientists (UCS) founded in 1969 by a group of faculty members and students at the Massachussetts Institute of Technology who were concerned about the misuse of science and technology in society. The organization calls for refocusing scientific research to solve environmental and social problems. UCS publishes books and papers on a variety of topics, including biotechnology, arms control, renewable energy sources, global warming, and transportation; a quarterly magazine, *Nucleus;* and a quarterly newsletter, *Earthwise.* Membership (1999): 70,000. Address: Two Brattle Square, Cambridge, MA 02238. Phone: (617) 547-5552. Website: www.ucsusa.org.

unique species in Forest Service usage, a SPECIES of plant or animal that is not endangered (*see* ENDANGERED SPECIES) but is of special interest due to its beauty, its unusual properties, its historical value, or its local rarity. The Alaska Cedar, for example, is not a unique species in Alaska, but is listed as a unique species in California because it is found in any two small STANDS within the boundaries of that state. Unique plant species are protected by the establishment of BOTANICAL AREAS or RESEARCH NATURAL AREAS wherever possible. Unique animal species are generally accorded the same sort of protection as THREATENED SPECIES during the planning process.

United Nations Conference on Environment and Development (UNCED) *See* RIO EARTH SUMMIT.

United Nations Educational, Scientific, and Cultural Organization (UNESCO) established in 1945 to encourage international collaboration in education, science, culture, and communication and therefore to advance global respect for justice, the rule of law, and human rights. Following World War II, UNESCO aided in the physical reconstruction of Europe's libraries and museums. Since 1950 it has organized projects in Latin America, Asia, and Africa, and in 1959 worked to preserve and restore monuments threatened by the ASWAN HIGH DAM in Egypt. The United States withdrew its membership in 1984, and Great Britain in 1985, charging UNESCO with budgetary extravagance and hostility to the press. Great Britain rejoined in 1997, but the United States is not currently a member. Address: 7, place de Fontenoy, 75352 Paris 07 SP, France. Phone: 33 1 45 10 00. Website: www.unesco.org.

United Nations Environment Programme (UNEP) the program of the United Nations charged with providing leadership on environmental issues. Divisions within UNEP are Environmental Information, Assessment and Early Warning; Environmental Policy Development and Law; Environmental Policy Implementation; Technology, Industry, and Economics; and Regional Cooperation and Representation. UNEP sponsored the 1992 RIO EARTH SUMMIT and other international meetings and organizes World Environment Day, the Environmental Sabbath, and the Clean Up the World Campaign. Regional Office for North America: 1818 H Street NW, Washington, DC 20433. Phone: (202) 458-9695. Website: www.unep.org.

United Nations Man and the Biosphere Program *See* INTERNATIONAL BIOSPHERE RESERVES.

United Nations World Health Organization (WHO) established in 1948 and headquartered in Geneva, Switzerland, and comprised of 191 member states. In 1981 WHO adopted a mission of "health for all by the year 2000" and has made notable progress against polio, cholera, leprosy, malaria, and tuberculosis. It operates public health programs in more than 100 countries providing immunization and sanitation instruction. The WHO has also drafted international quarantine requirements and sanitary recommendations to aid in preventing the spread of disease. Recently WHO has begun to research health problems related to environmental pollution. *The Riches of the Poor: A Journey Round the World Health Organization* (1988), by George Mikes, and *World Health Organization: A Brief Summary of Its Work* (1989), edited by Patricia Wood, take a closer look at the WHO. Address: 2 United Nations Plaza, DC-2 Building, Rooms 0956 to 0976, New York, NY 10017. Phone: (212) 963-4388. Website: www.who.int.

United States Agency for International Development (USAID) agency of the Department of State, founded in 1961 to consolidate nonmilitary foreign aid programs. USAID is responsible for administering economic support to more than 80 nations for agriculture, rural development, nutrition and health programs, population control, and market-related development. USAID also administers disaster relief such as the $767 million sent to postwar Bosnia-Herzegovina. Address: Ronald Reagan Building, Washington, DC 20523. Phone: (202) 712-4810. Website: www.info.usaid.gov. *See also* STATE, DEPARTMENT OF.

United States and Mexico International Boundary and Water Commission *See* INTERNATIONAL BOUNDARY AND WATER COMMISSION.

United States Congress the legislative branch of the U.S. government established in 1789 by Article 1 of the Constitution. It consists of two houses: the Senate, comprised of two senators elected from each state to serve six-year terms, and the House of Representatives, elected according the populations of the states and serving two-year terms. As the legislative branch, Congress is responsible for drafting bills which by their passage by majority vote in each house are sent to the president. If the president signs a bill within 10 days, it becomes law. If the bill is vetoed by the president, a two-thirds majority vote of each house is necessary to make the bill a law. Congress has supported the environmental movement by passage of laws such as the WILDERNESS ACT OF 1964, the NATIONAL TRAILS ACT OF 1968, the National WILD AND SCENIC RIVERS ACT (1968), the CLEAN AIR ACT (1963, amended in 1970 and 1977), the NATIONAL ENVIRONMENTAL POLICY ACT (1970), the CLEAN WATER ACT (1972), and the ENDANGERED SPECIES ACT (1973).

United States Green Building Council coalition founded in 1993 to promote the understanding, development, and implementation of GREEN BUILDING standards and design practices. Members include building product manufacturers, building owners and managers, financial and insurance firms, professional societies, architects, designers, engineers, contractors and builders, environmental groups, utilities, universities, research institutes, and government agencies. It publishes *Green Building Case Studies* and has developed the *Sustainable Building Technical Manual* and the Leadership in Energy and Environmental Design building rating system that provides a national standard for what constitutes a "green" building and will provide motivation to build "green." Membership (1999): 350. Address: 110 Sutter Street, Suite 410, San Francisco, CA 94104. Phone: (415) 445-9500. Website: www.usgbc.org.

United States Man and the Biosphere Program an agency of the U.S. Department of State operating in association with the United Nations' Man and the Biosphere Program. The U.S. program is organized as a consortium of federal agencies, including the Agency for International Development, U.S. Forest Service, National Oceanic and Atmospheric Administration, Bureau of Land Management, U.S. Geological Survey, Biological Resources Division of the Department of the Interior, National Park Service, Environmental Protection Agency, National Aeronautics and Space Administration, National Institutes of Health, National Science Foundation, Peace Corps, and Smithsonian Institution. In the United States, there are 99 BIOSPHERE RESERVES. Address: U.S. MAB Secretariat, OES/ETC/MAB SA-44C, First Floor, Dept. of State, Washington, DC 20522. Phone: (202) 776-8318. Website: www.mabnet.org. *See also* INTERNATIONAL BIOSPHERE RESERVES.

United States Public Interest Research Group (USPIRG) environmental and public interest research and advocacy organization operating in concert with state public interest research groups. Set up by RALPH NADER in 1983, the USPIRG researches, publishes, and distributes an annual "Congressional Scorecard," sponsors special projects research, and operates an internship program for students. Address: 218 D Street SE, Washington, DC 20003. Phone: (202) 546-9707. Website: http://pirg.org/uspirg/index.html.

United States Synthetic Fuels Corporation (USSFC) obsolete federal (U.S.) agency created by Congress on June 30, 1980, and charged with promoting the development of synthetic fuels (*see* SYNFUELS) by providing financial incentives to private industries to build commercial-scale synthetic fuels plants. The agency functioned as an investment banking corporation that was wholly owned by the federal government. In addition, it was required to report to Congress on progress in developing economically and environmentally sound means of producing synthetic fuels. The USSFC was discontinued by EXECUTIVE ORDER on April 18, 1986, and its functions were transferred to the secretary of the treasury.

unit plan *See* PLANNING UNIT.

unmerchantable timber standing timber that cannot be sold. The term includes standing CULLS (that is, trees whose growth form, size, or condition will not allow them to be used by a potential buyer); underaged trees; and *unmerchantable species,* those tree SPECIES for which there is no current market demand. *Compare* MERCHANTABLE TIMBER.

universal soil loss equation *See* SOIL LOSS EQUATION.

unsaturated zone (zone of aeration) in soil science, the portion of the soil lying above the WATER TABLE. The unsaturated zone includes three subzones. In the *zone of soil water,* at the surface of the soil, water exists as both water of adhesion and water of cohesion (*see* HYGROSCOPIC WATER). Below the zone of soil water is the *intermediate zone,* where water exists as water of adhesion only: the water in this subzone is not available for plant growth, and roots do not normally penetrate it. Beneath the intermediate zone and immediately above the water table lies the CAPILLARY FRINGE. *See also* SOIL WATER.

unstable liquid (reactive liquid) a liquid that may decompose and/or explode spontaneously under normal conditions of temperature and pressure, either alone or when combined with water. Unstable liquids generally remain stable if undisturbed, but are extremely sensitive to small shocks such as those that occur in normal handling, including pounding, bumping, jostling, or overheating.

upland disposal *See under* CONFINED DISPOSAL FACILITY.

upland game bird any bird, other than a WATERFOWL, that is managed for sport hunting, with open and closed seasons, BAG LIMITS, etc. (*see* GAME ANIMAL). Most upland game birds are GALLINACEOUS BIRDS (pheasant, quail, turkey, etc.) but some Columbidae (pigeons and doves) and a few members of other families, such as cranes, rails, and sandpipers, are usually included in this category.

Upper Colorado River Commission an administrative agency concerned with river conservation and management, with appointed commissioners from Colorado, New Mexico, Utah, and Wyoming, plus a commissioner representing the federal government who is appointed by the president. Address: 355 S. 400 East Street, Salt Lake City, UT 84111. Phone: (801) 531-1150.

upper sonoran zone in the Merriam Life Zones system (*see* LIFE ZONE), the western portion of the upper austral zone, corresponding approximately to the CHAPPARAL BIOME. Typical vegetation includes juniper, pinyon pine, scrub oak, and manzanita.

upwelling the upward movement of water in rivers, streams, ponds, lakes, and oceans from the bottom layer to the top. Most upwelling occurs with seasonal changes of the relative temperatures of the surface and bottom layers and, together with the opposite effect, *downwelling,* helps to distribute nutrients throughout the water column.

uranium the 92nd ELEMENT in the ATOMIC SERIES, ATOMIC WEIGHT 238.029, chemical symbol U. The heaviest naturally occurring element (that is, the one with the largest ATOMS), uranium is a silvery, lustrous METAL with a moderately high natural RADIOACTIVITY resulting from the fact that its nucleus is too large to be entirely stable. ISOTOPES with atomic weights from 227 to 240 have been produced artificially, but only three of these (U^{234}, U^{235}, and U^{238}) are known to exist outside the laboratory. Uranium combines readily with OXYGEN and will burn in air at a temperature of 170°C (338°F), forming several uranium oxides, primarily U_3O_6 (others: UO_2; UO_3; UO_4; U_2O_7). Due to this affinity for oxygen it is not found in the free state in nature. The metal is produced commercially from two primary ORES, pitchblende and uranite, which are both mixtures of U_3O_8 and UO_2; uranite has a higher UO_2 content than does pitchblende. Its only significant uses are as a fuel for nuclear reactors (*see* NUCLEAR ENERGY) and as an explosive in nuclear weapons.

PRODUCTION OF URANIUM

More than 99% of all natural uranium occurs in the form U^{238}, which, though radioactive, is not sufficiently so to power reactors or bombs: hence uranium must be "enriched" before use to increase its U^{234} and U^{235} content from the natural ratio of 1:140 up to about 1:30, or 3% (higher for weapons-grade material). Enrichment is a complicated procedure which begins with the digestion of the pitchblende and uranite ores in nitric acid, forming uranyl nitrate ($UO_2(NO_3)_2$), which is put through a series of REACTIONS whose end product is pure UO_2. This compound is treated with anhydrous hydrofluoric acid (HF) to produce uranium heptafluoride (UF_6), a GAS. The gas is passed through diffusion barriers which slow down the slightly larger $U^{238}F_6$, allowing the UF_6 formed from the other two isotopes to become concentrated. This enriched gas is then converted back to U_3O_8 and formed into rods or pellets to serve as nuclear fuel. The process poses several severe environmental hazards, including the potential releases of nitric and hydrofluoric acids into the environment, as well as the potential of escape of the highly toxic and radioactive uranium heptafluoride gas.

urban forestry the planting, maintenance, and conservation of trees and forests in metropolitan areas. As distinct from commercial forestry, urban forestry is concerned with the preservation of tree cover in cities and suburbs for environmental, aesthetic, and cultural reasons—a need that has become acute with post–World War II metropolitan growth and development. According to one study, metropolitan areas such as Seattle, Washington-Baltimore, and Atlanta have lost more than a third of their tree cover since 1970. According to AMERICAN FORESTS, a nonprofit organization, metropolitan areas should have an overall tree cover of 40% (30% in the arid Southwest), made up of a 15% cover in downtown commercial areas, 25% cover in urban residential areas, and 50% cover in suburban areas. The importance of urban forestry relates to the cooling effect of trees (*see* URBAN HEAT ISLAND), the mitigation of storm-water runoff, the reduction of ambient air pollution, and the aesthetic and cultural contribution of trees and woodlands to urban living. According to a 1985 analysis, a 50-year-old urban tree can save $73 in air conditioning costs and $75 in erosion and storm-water control costs, as well as provide $75 worth of wildlife shelter and produce $50 worth of air pollution control. Added up and compounded over the 50-year period, such a tree would be worth $57,151 to the people of the city. As these benefits have become better known, many cities and civic organizations have instituted tree planting campaigns in recent years to restore urban forests, and forestry schools and others are increasing their training of students in the specialized skills and knowledge required by urban forestry. *See also* GLOBAL RELEAF.

urban growth boundary in land use planning, a boundary delimiting urban services—transportation, garbage collection, water and gas mains, and the like—in order to control URBAN SPRAWL. *See also* OREGON LAND USE LAWS.

urban heat island in meteorology, a term used to describe the fact that temperatures within a city are usually several degrees warmer than those in the surrounding countryside. The urban heat island effect is most noticeable in large cities on clear, calm winter nights, when the temperature differential may be as much as 15°C (27°F); however, the effect exists to some degree in cities of all sizes and under all climate conditions. The principal cause of urban heat islands is the large amount of land surface that is covered with stone and stonelike materials, such as concrete buildings and pavement. These are efficient radiators that absorb solar energy and reradiate it as heat, usually continuing this reradiation long after the Sun goes down. Lesser, but still important, causes include the lack of standing water within cities due to efficient drainage systems, causing heat that might otherwise go to evaporating water MOLECULES to become available for warming the air and the buildings; the interruption of windflow by tall buildings, creating an "island of calm" in which heat can increase, and the complex heat-reflection patterns that develop among these tall buildings; and the presence of waste heat created by human activities such as space heating, power generation, and the operation of vehicles. The urban heat island reduces power-consumption needs for space heating and increases the number of frost-free days within the city, and to this extent it is beneficial. However, it also tends to create a localized temperature INVERSION over the city from which the POLLUTANTS produced by the city cannot escape. This phenomenon, known variously as a "haze hood" or a "dust dome," increases PARTICULATE levels within the city as much as 1,000 times over those in the surrounding countryside and is probably the major fac-

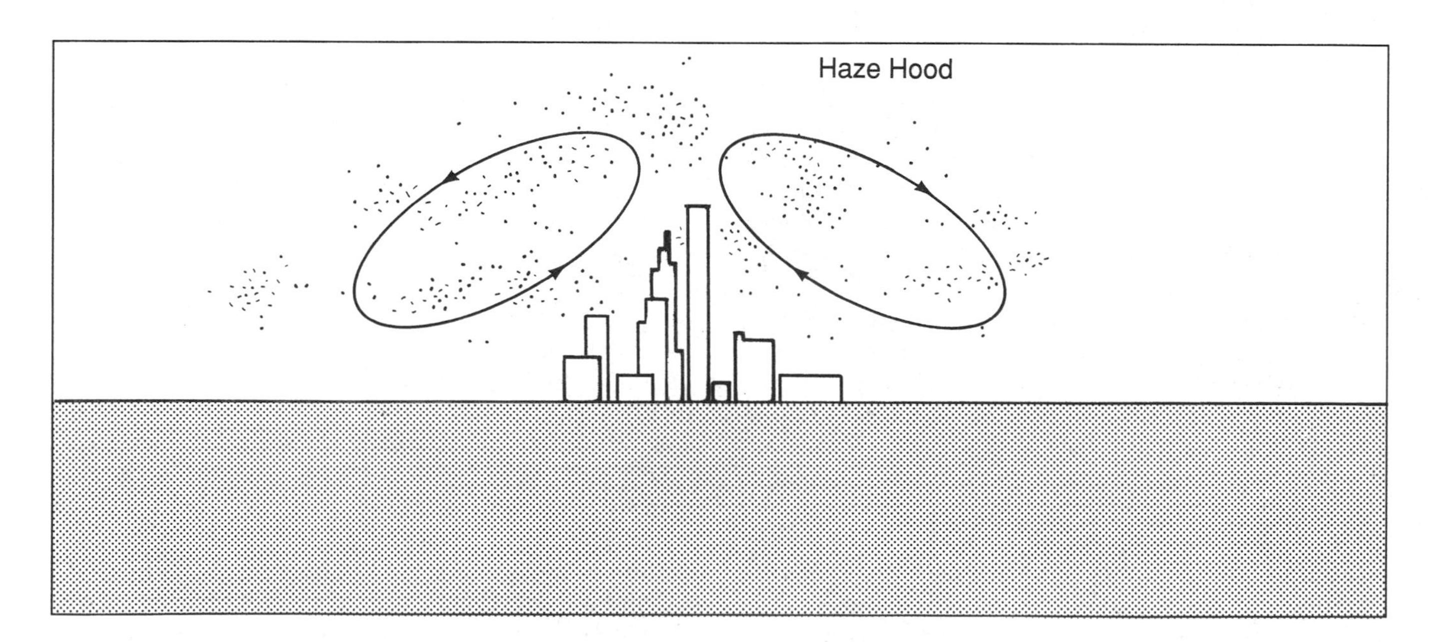

Urban heat island

tor in the so-called "urban precipitation effect," which causes rainfall within (and immediately downwind from) major cities to be as much as 10% higher than rainfall in the surrounding upwind areas.

Urban Mass Transportation Administration (UMTA) agency within the U.S. Department of Transportation (*see* TRANSPORTATION, DEPARTMENT OF) created by President Lyndon Johnson on July 1, 1968, under the terms of the Urban Mass Transportation Act of 1964. The UMTA is primarily a financing agency, providing grants and loans for the development and operation of MASS TRANSIT systems within urban areas. Public agencies at all non-federal levels (state, county, metropolitan) are eligible for UMTA assistance, which may come as outright grants, matching-fund grants, or loans for up to 80% of project costs (dependent upon eligibility guidelines). Private corporations may be eligible as well if they apply through a public agency. The agency also provides limited technical assistance toward the development of mass transit systems. Headquarters: 400 Seventh Street SW, Washington DC 20590.

urban renewal the demolition and rebuilding of center-city residential and commercial structures as well as those of other areas. Traditionally thought of as "slum clearance," urban renewal became a major public-works endeavor in American cities in the years following World War II. In hindsight, most urbanologists now believe that the effort was misguided and led to major infrastructure, environmental, and social problems. Along with slums, safe, pleasant low-income neighborhoods were demolished; and convention centers, municipal office buildings, and high-rise apartment blocks called "housing projects" were built in their place. While providing sanitary housing, the projects could never recapture the community coherence and safety of the destroyed neighborhoods, which were frequently destroyed in the name of renewal although the object was in fact to clear rights of way for urban sections of the interstate highway system. The highways encouraged "white flight" from inner cities and sprawl development in the suburbs and beyond. Many public housing projects, most famously the Pruitt-Igoe houses in St. Louis, became so crime-ridden and poorly maintained that they had to be demolished, only 20 years after they were built. The misadventure of urban renewal was thoroughly chronicled in Jane Jacob's classic book, *The Death and Life of Great American Cities* (1961). Reflecting Jacobs's views, current efforts to revitalize inner-city neighborhoods are now generally pursued on a smaller scale, and make an effort to preserve existing neighborhoods rather than destroying them. *See also* URBAN SPRAWL.

urban sprawl the dominant urban form of metropolitan America. Sprawling urban development has been a characteristic of some cities (London, for example) for centuries, but only since the end of World War II has it become a major national problem—in the United States as well as other countries. In the United States, urban sprawl—beyond the suburbanizing effect of streetcar and rail lines and parkways for automobiles—began with the way developers met the housing shortage of the late 1940s. Instead of building houses one at a time, or in small numbers and erecting apartment buildings, which had been the mode during the depression years and to a limited extent during the early 1940s, the intense pressure for low-cost family housing led developers to explore the possibility of mass housing on agricultural land around the urban fringe. In those places where rail and automobile commuting was possible (if time-consuming), hundreds of thousands of houses were built assembly-line fashion on former farmland far, but not too far, from center-city employment areas. In 1949 a returning veteran using the GI Bill could buy his family a two- or three-bedroom house for $10,000 or less with no down payment and a mortgage interest rate of 4%. In this way the age of sprawl began. In the late 1940s and early 1950s, urbanologists decried the middle-class flight to these far suburbs, but the economics and the social benefits of single-family housing, even with an hour-long commute, were powerful. Sprawl development was encouraged by "push" factors that persuaded young couples with children to escape the perceived unpleasantness of the city, and by the "pull" factors: grass, trees, safety, better schools. During the 1950s sprawl development slackened, for this was a decade characterized by three economic recessions. But by the early 1960s sprawl was dominant again, given a major economic expansion and the building of the interstate highway system which, together with the GI Bill renewed for Korean veterans, encouraged development even more distant from city cores than before. Suddenly a city like New York, with its five boroughs, had to be seen as a metropolitan area with 22, not five, urban and urbanizing counties. Then, during the 1970s and 1980s, the sprawl pattern of development went through a metamorphosis that would, for all intents and purposes, make it virtually unstoppable. This was the establishment of satellite, or "edge," cities around major metropolitan areas. Served by interstate highways and therefore able to draw from a wide, regional labor pool, new (or old) industries and corporations could be located on low-cost land away from city cores with very few competitive disadvantages, and many advantages. Suburban cities like White Plains, New York, or San Jose, California, became urban centers in their own right, ramifying sprawl to new orders of magnitude, since workers could now commute to their satellite city jobs from even further distances from the old city cores. As predicted by early urbanologists, metropolitan areas grew to urban regions hundreds of miles across, so that in effect whole sections of the country were blanketed with sprawl development, as in Los Angeles to San Diego, for instance, or Boston to Washington. After the intense period of the 1960s and 1970s, sprawl development abated somewhat during the 1980s and early 1990s, again due to an economic downturn. But it came back with a vengeance during the economic expansion of the mid-1990s as farmland was lost to development at the rate of more than 3 million acres a year. Metropolitan regions, while growing only moderately in population, grew enormously in land coverage, as single-family houses, now affordable to a larger segment of the population, covered vast acreages. Throughout what may be called the "era of sprawl," land use planners have warned that urban sprawl can exact terrible tolls on society. Because of its spreading pattern, commuting time and costs increase; "white flight" continues, impoverishing inner-city areas; pollution from automobile commuting becomes endemic, with scores of cities affected; the cost of services—fire, education, policing, water—increases as the radius of metropolitanization increases; natural beauty disappears; and overall environmental quality is degraded. Even so, those with a stake in the continuation of the sprawl pattern of development argue that sprawl enables more people to own their own homes, it encourages economic expansion, and it has a democratizing effect on society. While not denying these benefits, present-day environmentalists and land use planners offer an alternative vision that they call "smart growth," which emphasizes infill development, alternatives to automobile commuting (such as bike paths), and the preservation of cultural assets in the inner city to provide "pull" factors that will counter a continuing pattern of sprawl development. *See also* LAND USE PLANNING; ONE THOUSAND FRIENDS; OREGON LAND USE LAWS.

USACE *See* ARMY CORPS OF ENGINEERS.

USAID *See* UNITED STATES AGENCY FOR INTERNATIONAL DEVELOPMENT.

USDA *See* AGRICULTURE, DEPARTMENT OF.

USEPA *See* ENVIRONMENTAL PROTECTION AGENCY.

user fee a fee charged for the use of a public facility by a private party. Examples of user fees include road and bridge tolls, park entrance fees, fees charged for grazing permits on the public RANGE, fees charged for the passage of ships through locks and canals, and so forth. The purpose of user fees is not generally to raise revenue but to assure that the costs of building, operating, and maintaining a public facility fall most heavily on those who use the facility, thus eliminating one of the major causes of PORK BARREL politics—the ability to make the public at large pay for a project that will benefit only a small number of people.

USFS *See* FOREST SERVICE.

USGS *See* GEOLOGIC SURVEY.

vagility in the biological sciences, the capacity of an ORGANISM for sustained free movement. The term is usually used in a comparative sense. An organism that is more vagile than another can travel for longer periods and cover more territory. For example, AMOEBAS are more vagile than paramecia (*see* PARAMECIUM); birds are more vagile than mosquitoes; etc.

Valdez Principles a code of conduct for corporations developed by a coalition of environmental and allied organizations as a response to the ecological damage to Alaska's Prince William Sound caused by the *EXXON VALDEZ* oil spill. The Valdez Principles call on corporations to disclose any hazardous operation; provide annual audits on environmental aspects of their operations; minimize pollution; use renewable resources and conserve nonrenewable materials; preserve biodiversity; reduce and properly dispose of hazardous wastes; use energy-efficient processes; market safe products and services; restore damaged environments and compensate for human injury; include at least one environmental expert on the board of directors; and appoint a senior executive for environmental affairs.

valence in chemistry, the number of ELECTRONS that an ELEMENT or a GROUP will share with others in the process of forming COMPOUNDS. The sharable electrons are those in the outer, or *valence,* shell of each atom (*see* ATOM: *structure of the atom*). Atoms or groups that will share a single electron are called *monovalent;* those that will share two are called *divalent;* three, *trivalent;* etc. Elements such as the so-called "noble gases" (neon, argon, krypton, xenon, radon, and helium) that do not form compounds are called *zero valent.* Elements or groups that give up electrons while forming compounds are said to have *negative valence.* Those that accept electrons have *positive valence.* For a compound to be stable, its positive and negative valence totals must be equal. For example, in water (H_2O), the two hydrogen atoms each have a negative valence of one for a total negative valence of two, while the single oxygen atom has a positive valence of two, bringing the net valence to zero.

Valley of the Drums illegal waste dump near Brooks, Kentucky, found in 1979 to contain one of the highest concentrations of hazardous chemicals ever recorded. Officially called the "A. L. Taylor Drum Cleaning Service," the 23-acre disposal site lay approximately 15 miles south of the Louisville at the headwaters of Wilson Creek. The owner, Arthur L. Taylor, began dumping toxic chemicals on the site as early as 1967: he was repeatedly cited by the state for illegal dumping during the 1970s, but the citations were not followed up. Taylor died in November 1978, approximately six weeks before ENVIRONMENTAL PROTECTION AGENCY officials became aware of the site in the first week of January 1979. A count showed 17,051 toxics-containing barrels on the ground's surface, mostly piled in heaps on the front five acres of the property. Estimates of buried barrels ranged up to 200,000, although the actual number was probably only about a tenth of that. Concentrations of PHENOLS, PCBS, and 11 other toxic or carcinogenous chemicals (*see* TOXIC SUBSTANCE; CARCINOGEN) were found in Wilson Creek, at levels of up to 100,000 parts per billion. Surface RUNOFF contained BENZENE, HEAVY METALS, and other hazardous materials, and a total of 197 different toxic substances were found in the property's soil. The site was declared an environmental emergency by the EPA on March 2, 1979. Discovery of the Valley of the Drums helped focus national attention on the inadequacies of current hazardous-waste disposal practices and was, along with LOVE CANAL, one of the two primary moving forces behind the passage of the so-called "superfund" act of 1980: *see* COMPREHENSIVE ENVIRONMENTAL RESPONSE, COMPENSATION AND LIABILITY ACT.

valley train in geology, a glacial deposit consisting of SAND, GRAVEL, and other materials that have been washed out of a terminal moraine (*see under* MORAINE) by the meltwater stream of a GLACIER and deposited along the stream's course. A valley train may be thought of as an outwash plain (*see* OUTWASH DEPOSIT) confined to a streambed or to a narrow valley. *See also* GLACIAL LANDFORMS.

vapor pressure in chemistry and physics, the pressure at which the number of MOLECULES evaporating from a liquid or solid (*see* EVAPORATION) exactly balances the number of molecules being reincorporated into the liquid or solid, so that the liquid or solid is no longer losing volume. Vapor pressure varies dramatically with temperature, as well as varying widely among different classes of COMPOUNDS.

vapor recovery system in waste management, a collection system designed to trap and store hazardous vapors that escape from volatile liquids (*see* VOLATILITY) during handling or storage operations. Vapor recovery systems typically consist of pipe or hose lines leading from the points of likely vapor leakage (such as couplings on hoses used for transferring liquids from one tank to another) to the tank from which the liquid is being removed. The vapors collected by these pipes or hoses are used instead of air to replace the liquid volume in the tank as the liquid is withdrawn. In some systems a condenser and/or refrigeration unit is used to return the vapors to the liquid state and separate the air from them before returning them to the tank.

vapor well *See under* MONITORING WELL.

variety class in National Forest terminology, a classification of the visual landscape according to its presumed attractiveness to the viewer. Three variety classes are recognized. Class A landscapes ("Distinctive") are those with rugged landforms, a great deal of variety in vegetation types, and/or attractive water features. They stand out from the general terrain of the region, and are the areas most likely to attract sightseers. Class B landscapes ("Common") are those that are considered normal for the region; they may be rugged and varied, but they do not stand out as being particularly unusual when compared to other nearby areas. Class C landscapes ("Mineral") are those that have little or no variety and are of no particular interest to sightseers. Note that these are not absolute categories but are highly dependent upon the context within which a particular landscape is found. A piece of flat, unbroken forests that might be called Class C if it were located next to a mountain or canyon might easily be upgraded to Class A if it were surrounded by desert. *See also* VISUAL RESOURCE MANAGEMENT SYSTEM.

varve in geology, a layer of SEDIMENT, representing a single year's DEPOSITION, which is easily distinguishable from the layers above and below it. Varves are characteristic of cold freshwater lakes, especially those closely associated with GLACIERS (*see* PROGLACIAL LAKE). They appear as alternate bands of coarse and fine sediments, the coarse bands representing deposition during periods of heavy RUNOFF (that is, during the spring melt) while the fine bands represent the slower deposition during normal runoff. Each pair of bands, or "couplet" (that is, one fine band plus one coarse band) makes up a single varve. Varves are useful for dating LACUSTRINE DEPOSITS; however, care must be taken to distinguish them from similar layered deposits ("rhythmites") that have been put down during sporadic, non-annular events such as occasional floods.

vascular plant in biology, any plant SPECIES that conducts food and water within its body by means of vascular tissue—that is, XYLEM and PHLOEM. Because they contain long, stiff-walled conducting tubes ("veins"), vascular plant tissues are both an efficient means of liquid transport and a useful "prop" for the construction of tall plant bodies. Vascular plants are the only plants with true roots and are thus the only ones capable of colonizing dry sites. They are also the only plants with true leaves, which require the presence of vascular tissue to serve as veins. The category encompasses all of the so-called "green plants" except the BRYOPHYTES (mosses, horsetails, and ferns). *See also* ANGIOSPERM; GYMNOSPERM.

vector in epidemiology, the means by which a PATHOGEN is transferred from one HOST to another. Vectors are often other living ORGANISMS. For example, mosquitoes serve as the principal vector for *Plasmodium,* the parasitic PROTOZOAN that causes malaria in humans, picking it up by biting an infected individual and carrying it in their digestive tracts and saliva to the next individual they bite. Non-living agents such as wind or water may also serve as vectors. Infected water supplies, for example, are a prime vector for hepatitis. *See also* VECTOR CONTROL.

vector control in the health sciences, the control of disease by attacking the means of the disease's transmission (*see* VECTOR) rather than by attacking the disease ORGANISM itself. Though the term is a broad one, encompassing mechanical controls (air filtration, covering the mouth and nose during sneezes, etc.) as well as biological ones, it is usually associated most strongly with the control of carrier organisms such as mosquitoes, ticks, mites, lice, and fleas, and the primary responsibility of the public *vector control officer* hired by many counties and municipalities is to destroy these organisms or to prevent them from coming into contact with humans and with domestic animals. In the past, the principal means of doing this was through the use of PESTICIDES such as DDT. Although this approach is still used more often than is probably necessary, the emphasis today has shifted to mechanical and biological controls, including covering or draining the stagnant ponds in which mosquitoes breed; introducing organisms that feed on mosquito LARVAE or on adult mosquitoes; blocking access to buildings by rats and other carriers of fleas and lice; and the use of chemical repellents to prevent vector insects from biting.

vegetal composition the plant SPECIES that are found in a given COMMUNITY, together with some indications as to their distribution and frequency of occurrence within the community structure. Vegetal composition tends to be fairly uniform within a given community, although it may show variations from region to region. Changes in the composition of the plant cover may therefore normally be taken to represent changes in the community, which in turn usually indicate significant changes in one or more ENVIRONMENTAL FACTORS such as amount of moisture, amount of sunlight, soil type, etc.

vegetation type in forestry, a plant COMMUNITY characterized by its dominant type rather than by its dominant SPECIES: "coniferous forest," "grassland," "oak woodland," etc. Depending upon the circumstances, "vegetation type" can refer to anything from a single COMMUNITY to an entire BIOME or even a group of biomes. *See also* TYPE CONVERSION.

vegetative reproduction *See under* ASEXUAL REPRODUCTION.

vehicular way *See* WAY.

vein in mining, a sheetlike ORE BODY formed by the filling of a crack in a BATHOLITH or other body of (usually) IGNEOUS ROCK by MINERAL-rich ORE. Veins may be hundreds—or even thousands—of feet long and wide, but are generally only a few inches thick. They appear to be formed primarily during the slow cooling of a batholith. Since gold, silver and copper are not constituents of rock-forming minerals, they tend to concentrate in the molten center of the batholith as it cools. Eventually the hard exterior cracks under the pressure of gases also trapped within the molten center, and the materials from the center flow up and fill the cracks, crystallizing out primarily as QUARTZ (the last mineral to crystallize from a MAGMA; *see* BOWEN SERIES). The metals are carried along and become trapped in the quartz matrix. So-called "hard rock mining"—the mining of minerals directly from a rock body, as opposed to the location of them in stream sediments or other erosive deposits—almost always takes place along veins. *See also* LODE.

ventifact a stone whose appearance has been modified by the action of wind. Ventifacts usually result from the abrasive action of wind-driven SAND, although in a few cases they result from the wind's ability to carry away the weaker parts of a FORMATION (*see* YARDANG). Ventifacts commonly show surface polish; the surface may also be grooved and/or pitted. *Faceting* (the development of relatively flat faces with sharply angled boundaries between them) often occurs on the edge of a ventifact that faces the prevailing wind. Stones faceted in this way are referred to by a German term that combines the word "edge" (*kanter*) with a prefix denoting the number of edges (*ein, zwei, drei*), so that a "dreikanter," for example, is a ventifact with three edges, therefore with four facets. *See also* LAG GRAVEL.

vernal pool a small pond that exists only during the wet season, generally the period of spring RUNOFF, though it may form earlier and persist longer. It is always dry at least part of the year. Vernal pools result either from an annual elevation in the level of the general WATER TABLE or from the development of a temporary PERCHED WATER TABLE in thin soil on top of a layer of impermeable rock such as BASALT or GRANITE (*see* IMPERMEABLE LAYER). They can support a surprisingly complex and active COMMUNITY of ORGANISMS, including PERENNIAL PLANTS, hydras and ROTIFERS, insects, and even frogs and small fish—all of which are capable of estivating through the dry season (*see* ESTIVATION). The ecological structure of vernal pool communities is particularly fragile due to the environmental stress produced by the wide swing in water conditions between the pool stage and the dry-ground stage and they therefore require protection from disturbances such as OFF-ROAD VEHICLE use, grazing, etc.—even in the dry season—if they are to survive.

vertebrate any animal belonging to the SUBPHYLUM Vertebrata of the PHYLUM Chordata (*see* CHORDATE). The principal vertebrate characteristic is the backbone, a flexible bony structure consisting of a number of disc-shaped individual bones, the *vertebrae,* connected to each other and serving as both support for the animal's body and protection for its spinal chord. Other characteristics usually found in the vertebrates are a relatively large brain protected by a bony cranium ("skull"); four limbs (which may be developed variously as fins, legs, or wings); and well-developed circulatory and digestive systems, including a multi-chambered heart and a division of the digestive tract into esophagus, stomach, and intestine. Vertebrates are the most mobile and adaptable of all ORGANISMS, and have colonized the broadest range of ENVIRONMENTS. Individual vertebrate SPECIES can survive in a greater range of HABITATS than can individual species of any other phylum, with the possible exception of the ANGIOSPERMS. The group includes mammals, reptiles, birds, fish, and amphibians.

vertisol in soil science, a soil that develops deep vertical cracks during part of the year. The name *vertisol,* however, does not refer directly to the vertical cracks, but to the fact that the soil tends to invert itself with time, bringing lower layers up to the surface and at least partially mixing the HORIZONS. Vertisols develop in areas with CLAY-rich PARENT MATERIALS and a pronounced wet season/dry season climate regime. Their high clay content makes them swell when wet and shrink when dry, causing the cracks to develop. Surface materials break off the edges of the cracks and accumulate in the bottom of them, mixing the surface layers down into the SUBSOIL and creating a localized excess of materials that pushes subsoil materials toward the surface when it expands in the wet season. Mixing therefore takes place in both wet and dry seasons, making the soil strikingly homogeneous, although organic material still tends to be primarily concentrated in the upper part of the soil profile. Under natural conditions, vertisols usually support STEPPE or SHRUB-STEPPE vegetation. With the use of irrigation during the dry season they can be extremely productive agricultural soils because the slow vertical mixing they naturally undergo prevents TRACE ELEMENTS from leaching out of them. *See also* GRUMUSOL.

vesicle in general, any small opening entirely surrounded by solid materials. The vesicle may be empty (filled only with air), or it may be filled with liquid or with a solid of a different character from the surrounding materials. The term is most commonly used in geology, where it refers to the small sealed openings in volcanic rocks such as BASALT, and in microbiology, where it refers to a small vacuole (*see* CELL: *structure of cells*). *See also* AMPHIBOLE.

viewshed in visual management, the total landscape that can be seen from a given point. The MANAGEMENT CONSTRAINTS on a viewshed vary with the number of people who are likely to see the view and with the expectations they are likely to have for it. A viewshed within an area renowned for its scenery, such as a NATIONAL PARK, is likely to be managed differently from a viewshed in a less sensitive area such as a piece of flat forestland. Management constraints differ further with the distance of the various parts of the viewshed from the point of view (*see* FOREGROUND; MIDDLEGROUND; BACKGROUND). *See also* VISUAL RESOURCE MANAGEMENT SYSTEM.

village a small population center, with a size ranging from 250 persons to 2,500 persons. Villages are intermediate in size between HAMLETS and TOWNS. They are often incorporated (*see* MUNICIPALITY) and almost always have a school and several churches and stores. They can serve as trading centers for regions containing up to 10,000 people, but are almost always tributary to larger urban centers for most trade purposes, even when those larger centers are some distance away.

vinyl in chemistry, any COMPOUND containing the vinyl GROUP, C_2H_3Cl– (structural formula: CH_2=CH–). Because this group polymerizes easily (*see* POLYMER), vinyl compounds are important constituents of plastics and synthetic rubbers. The word "vinyl" in common usage has come to denote any of several plastics that have the look and feel of leather (most common: *Vinylite,* a Union Carbide trade-named fiber that is a mixture of polyvinyl chloride and polyvinyl acetate). As CHLORINATED HYDROCARBONS, many vinyls have some distinct environmental hazards. Most of these are associated with their manufacture rather than their use, although care should be taken where a vinyl compound is likely to be abraded into dust fine enough to be breathed, as in sanding, filing, or sawing most plastic compounds. *See also* VINYL CHLORIDE.

vinyl chloride a vinyl monomer (*see* VINYL; MONOMER), chemical formula C_2H_3Cl (structural formula: CH_2=CHCl), used as a refrigerant and in the manufacture of a variety of plastics, notably *polyvinyl chloride* (PVC), a vinyl chloride POLYMER. Vinyl chloride is a gas at room temperature, liquefying at –14°C (7°F). As a liquid, it is extremely volatile (*see* VOLATILITY), so much so that a small amount spilled on a section of bare skin may cool the skin enough through rapid evaporation that frostbite results. Breathing the vapors causes "vinyl chloride disease," a depression of the central nervous system similar to alcoholism. More serious is the material's carcinogenity (*see* CARCINOGEN). Chronic exposure to small amounts has been proved in both human and animal studies to cause liver cancer. For this reason, vinyl chloride is listed as a PRIORITY POLLUTANT by the ENVIRONMENTAL PROTECTION AGENCY, even though the chemical's SOLUBILITY in water is extremely low.

virion *See under* VIRUS.

virus any of a large class of self-reproducing chemicals with some of the attributes of life-forms. Though debate still exists on the matter, most biologists today do not consider viruses to be "alive" in any conventional sense, and a growing body of terminology referring to them reflects this distinction. Thus, individual viruses are not called "organisms" but "viral particles" (or *virions*); the stages of their existence are not referred to as "life" and "death" but "activity" and "inactivity." They nonetheless have an enormous impact on the living environment, causing most infectious diseases, affecting the genetic structure of MICROORGANISMS, and reducing plant growth vigor and yield on a nearly universal basis.

STRUCTURE AND FUNCTION OF VIRUSES

The simplest viral particles consist of a single MOLECULE of DNA or RNA, usually (but not always) with a thin coating of PROTEIN called a *capsid.* The capsid is generally formed of repeating subunits known as *capsomeres.* More complex viruses may contain extra fragments of DNA and/or RNA and (occasionally) a few simple LIPIDS or ENZYMES. Many are highly symmetrical in shape, forming icosohedral (20-sided) or rod-shaped crystals (all of the so-called "spherical" viruses, such as those that cause herpes, are actually icosohedral). Others may have a main structure (the "head") that is symmetrical and houses the DNA or RNA, with a complex protein-based substructure (the "tail") attached to it that functions as a means of transferring the viral DNA or RNA to a host CELL. Many viruses also carry an "envelope" of proteins and lipids picked up from the host cell. Their activity relates strictly to reproduction. They do not metabolize (*see* METABOLISM) or otherwise convert or utilize energy, do not feed, do not move about under their own power, and show no behavioral reactions. When brought into contact with a host cell of a type they can utilize, however, they penetrate it—shedding their envelope, if present—and "commandeer" the host cell's reproductive mechanism, using it to replicate the viral DNA or RNA instead of its own. In this way, thousands of new viral particles are produced within an infected cell, which then bursts, scattering the particles to infect new cells. Such retroviruses can be destroyed by the body's immune system, once it has learned to recognize them (either through a previous infection or through *vaccination,* which consists of introducing inactive viral particles into the body for the purpose of alerting the immune system to their characteristics). However, they cannot be destroyed by ANTIBIOTICS. Thus the only effective weapons against viral epidemics are immunization and mechanical barriers to the spread of the virus from person to person (or animal to animal, or plant to plant).

ORIGIN OF VIRUSES

Though it was once thought that viruses were leftover examples of the self-replicating molecules that were the evolutionary ancestors of all living things, this is now considered unlikely, due to the inability of viruses to reproduce by themselves. Current views hold that viruses have arisen from three sources: (1) escaped fragments of DNA and RNA from living organisms; (2) degenerate bacteria that have lost their cellular structure and most of their other attributes, except their DNA or RNA; and (3) MUTATIONS of previously existing viruses. The retrovirus (HIV) that causes AIDS is thought to have "jumped" from a primate monkey or ape to a human being in Africa during the 1950s and subsequently infected millions around the world. There is some evidence that the rapid economic development of previously uninhabited areas in the tropics, where viruses have lived in niches heretofore uninhabited by humans, may result in more "jumps" by other viruses that can cause epidemics for which there is no cure.

viscosity in chemistry and physics, the internal resistance of a fluid to movement. In liquids, viscosity shows up primarily in flowing speed. Molasses, for example, which is a highly viscous liquid, pours out of a container much more slowly than water, which is only moderately viscous. In gases, viscosity is most evident in the resistance a gas shows to objects moving through it. Air resistance—a result of the air's viscosity—is a major factor in automotive and aircraft efficiency, but is also the force that makes flight possible. Viscosity is caused partly by the friction between MOLECULES and partly by the polar bonds that develop between them (*see* POLAR LIQUID). It is measured by a factor known as the *coefficient of viscosity,* or the *viscosity index,* expressed in kg/m·sec and generally designated as Ñ. This index decreases with increasing temperature in liquids (where the principal factor in increasing viscosity is the polar bond) and increases with increasing temperature in gases (where the principal factor is friction).

visitor day (**[VD] recreational visitor day [RVD]**) in recreation management, the use of a recreation facility by one person for 12 hours. The term "person" here does not necessarily imply a single individual. One visitor day can be the result of one individual visiting a site for 12 hours, or two individuals visiting a site for six hours each, or four individuals visiting the site for three hours each, etc. A party of 24 visiting a site for 1/2 hour would be recorded as one visitor day; so would three people visiting the site at separate times that totaled 12 hours, even if those times were spread over several months.

visual quality objective in Forest Service terminology, the goal toward which visual management activities in a given VIEWSHED are directed, expressed in terms of the amount of alteration of the natural landscape that is to be permitted in the viewshed (*see* PRESERVATION OF CHARACTER; RETENTION; PARTIAL RETENTION OF CHARACTER; MODIFICATION OF CHARACTER; MAXIMUM MODIFICATION). All management activities of any type in a given viewshed, including timber harvest, road building, recreation developments, etc., are required to meet the visual quality objectives for that viewshed. *See also* VISUAL RESOURCE MANAGEMENT SYSTEM.

Visual Resource Management System (**[VRMS] Visual Management System [VMS]**) in Forest Service terminology, a systematic approach to classifying and managing landscapes according to their scenic quality, established in 1974 with the publication of Agricultural Handbook No. 462 (*National Forest Landscape Management*) and since adopted as policy throughout the NATIONAL FOREST SYSTEM. The Visual Resource Management System requires two inventories, one physical and the other sociological. The physical inventory begins by breaking the forest into *visual character types*—regions with essentially common physical characteristics (mountains, forested plains, grasslands, etc.). These are broken down into *visual character subtypes,* such as river valleys, rocky peaks, lake basins, and so on. Visual character subtypes are further broken down into *visual landscape character units,* which are essentially parts of the character subtype that fall into specific VIEWSHEDS, or areas that can be seen from a single spot. Finally, each visual landscape character unit is placed in one of three VARIETY CLASSES: class A ("distinctive": highly varied and striking features), class B ("common": distinctive, but not striking, features), or class C ("mineral": unvaried and uniform terrain). The sociological inventory is generally carried out concurrently with the physical inventory; it consists of classifying each viewshed into one of three *sensitivity levels* depending upon the amount and type of use it receives. Level 1 ("highest sensitivity") consists of viewsheds of high use where at least one-fourth of all users will be concerned about scenic quality. This includes areas visible from parks, WILDERNESS AREAS, wild river corridors, and major highways and cities. Level 2 ("average sensitivity") includes viewsheds of low to moderate use where at least one-fourth of all users will be concerned about scenic quality, and areas of high use where fewer than one-fourth of the viewers will have these concerns. Examples include most secondary forest roads and river corridors. Level 3 ("lowest sensitivity") is all other areas. The sociological inventory also divides the viewsheds themselves into FOREGROUND, MIDGROUND, and BACKGROUND areas. Following classification of viewsheds by variety class and sensitivity level, a *visual quality objective* (VQO) is established for each viewshed. The five permitted VQOs are preservation (P), retention (R), partial retention (PR), modification (M), and maximum modification (MM) (see separate articles on each VQO). VQOs are assigned according to a standard matrix, as follows in the table below. Once these visual quality objects are set, they become policy, and all management activities in the viewshed, including timber harvest and roadbuilding, are required to conform to them.

Visual Resource Management System

Variety Class	Sensitivity level 1			2			3
	fg	mg	bg	fg	mg	bg	
A	R	R	R	PR	PR	PR	PR
B	R	PR	PR	PR	M	M	M/MM
C	PR	PR	M	M	M	MM	MM

variety class:
A = distinctive
B = common
C = mineral

sensitivity level:
1 = highest sensitivity
2 = moderate sensitivity
3 = lowest sensitivity

fg = foreground mg = middleground bg = background
R = retention of character
PR = partial retention of character
M = modification of character
MM = maximum modification of character

volatile organic compound (VOC) in pollution control, an ORGANIC COMPOUND that evaporates easily. Technically, VOCs are usually defined as organic compounds with a VAPOR PRESSURE of 1,300 Pascals (13 millibars, or about 1% of ATMOSPHERIC PRESSURE at sea level). Compounds that fit this description are often lumped together for regulatory purposes due to their similar physical behavior in the ATMOSPHERE.

volatile solids in chemistry and physics, the portion of the TOTAL SOLIDS in a water sample that are driven off by heating the crucible in which the solids have been collected to 650°C (1200°F) for 30 minutes. Analysis of volatile solids is usually *gravimetric,* that is, the sample is simply weighed before and after the volatiles are driven off and the difference between the two weights is taken as the amount of volatile solids present in the sample. *Compare* FIXED SOLIDS.

volatility in chemistry and physics, the ease with which a solid or liquid may be vaporized, that is, the speed at which EVAPORATION or SUBLIMATION (known collectively as *volatilization*) takes place. Volatility is usually expressed in loose terms. A *volatile substance (volatile liquid;* VOLATILE SOLID*)* is a substance that evaporates or sublimates easily, such as ALCOHOL or camphor; a *non-volatile* (or *involatile*) *substance* is one that does not evaporate or sublimate in any noticeable way in the time it is observed, such as a rock or a piece of wood. For any substance, observed volatility will vary with changes in the temperature, the VAPOR PRESSURE, and the turbulence at the interface between the substance and the surrounding atmosphere. Hot windy days, for example, promote considerably more volatilization than cold still ones. Volatile substances are more environmentally hazardous than nonvolatile ones because the conversion to vapor allows the substance to be transported through the air, spreading it more completely through the ECOSYSTEM and allowing it to attack ORGANISMS through their respiratory systems, usually a more dangerous route than through the digestive tract.

volcanic ash fine ashlike material produced by volcanic eruptions (*see* VOLCANO). The term is usually applied only to pyroclastic materials (*see* PYROCLASTIC ROCK) less than 4 mm (1/10 inch) in diameter. Explosive volcanic eruptions, such as that of Mt. Pinatubo in 1991, Mt. St. Helens in 1980, or Krakatoa in 1883, produce large quantities of volcanic ash, considerably altering nearby landscapes and temporarily creating worldwide climate change by reducing the ability of the ATMOSPHERE to transmit sunlight. *See also* NUÉE ARDENTE.

volcano any location where MAGMA erupts from beneath the surface of the Earth, forming EXTRUSIVE ROCKS. Volcanos occur in a wide variety of types, and are classified in several different ways. The most basic classification divides those that erupt along long cracks in the ground from those that erupt from single spots or groups of spots. The first type, known as *fissure eruptions,* create vast plains of LAVA such as those found in the Columbia Basin of Oregon and Washington and in the region of Lake Superior, while the second type, known as *vent eruptions,* create the conical mountains we have come to know as "volcanos." Fissure eruptions are rare (the only known eruption of this type in historic times was near Laki, Iceland, in 1783); hence it is only the vent-type volcanos that have been categorized further.

VIOLENCE OF ERUPTION

Five types of volcanic eruption are recognized, depending upon the violence with which magmatic materials are ejected from the volcano's vent. This in turn seems to depend on the amount of material plugging the vent between eruptions and the amount of gases trapped within the magma as it rises up the vent's throat. The calmest, the *Hawaiian eruption,* merely sends streams of lava over the rim of the CRATER. *Strombolian eruptions* include the ejection of steam and small amounts of PYROCLASTIC MATERIALS. *Vulcanian eruptions* emit large amounts of pyroclastics and steam, and may toss volcanic bombs for long distances. *Plinian eruptions* are violent enough to tear portions of the cone away. *Peleean eruptions,* the most violent, are Plinian eruptions distinguished by the presence of a NUÉE ARDENTE—a fiery cloud of pyroclastics moving down the slope of the cone at a high rate of speed.

TYPE OF CONE

A volcano may be further categorized by the type of cone built around its vent. *Cinder cones* are made entirely of pyroclastics. They are not normally very high above the surrounding terrain. Their slopes lie at the ANGLE OF REPOSE, generally about 33 degrees. *Shield volcanos* are made entirely of lava flows, with little or no pyroclastics. They are broad and flat, with slopes generally at four to 12 degrees. *Stratovolcanos,* or *composite cones,* are the most common type. They are composed of intermixed layers of lava flows and pyroclastics, and their slopes may range anywhere from four degrees up to the angle of repose.

LOCATION

A final means of classifying volcanos is by their location on the planet, which in turn depends on the type of TECTONIC ACTIVITY that formed them. *Rift volcanos* are located along regions where continental plates are pulling apart (*see* PLATE TECTONICS). *Subduction volcanos* are formed at places where plates are colliding. Volcanos that develop in the centers of continental plates are called *hot-spot volcanos,* and apparently derive from stationary "hot spots" over which the plates are moving. Examples of this last type of volcano include the Hawaiian Islands, Mammoth Crater in California, and the great CALDERA that holds Yellowstone National Park.

THE LIVES OF VOLCANOS

Although a few volcanos are continuously active, most go through alternate spells of activity and dormancy. Generally, the more violent the eruptive type, the longer the time between eruptions. In an explosively active volcano, such as Mt. St. Helens in Washington, a major (Plinian or Peleean) eruption will be followed by a number of smaller eruptions tapering off over the next few years, following which the volcano becomes dormant again for several decades or centuries. Eventually activity will totally cease, and the volcano will become dead. At this point it will usually begin eroding rapidly, as the pyroclastics that make up major portions of most volcanos are not particularly stable. The final stage of a volcano's life is normally a hard, steep-sided rock structure known as a *volcanic neck* or *volcanic plug,* which is composed of lava that hardened under pressure in the throat of the vent.

volume factor in limnology, the ratio of the volume of water in a lake to the volume of a cone whose base is a circle with the same area as the lake's surface and whose height is equal to the lake's greatest depth. A circular lake with a conical bed would have a volume factor of 1. A lake with high proportion of shallow water (such as Lake Erie) will have a volume factor less than 1, while a lake with steep sides and a high proportion of deep water (such as Oregon's Crater Lake) will have a volume factor greater than 1. In general, lakes with high volume factors will have lower productivity (be home to fewer fish and other forms of aquatic life) than those with a low volume factor, due to the lower proportion of shallow water to breed in. *Compare* SHAPE FACTOR.

voluntary simplicity a way of life based on reducing one's environmental impact. Many books and magazine articles stress the need to make a personal gesture toward the preservation of nature and natural resources with practices such as riding a bicycle to work, living in a smaller house or apartment, reducing the purchase of material goods, and the like. Voluntary simplicity is a central feature of "intentional communities" made of those who wish to live closer to the land, but it can also be practiced by individuals and families simply by becoming conscious of the environmental implications of their economic activities. Voluntary simplicity, sometimes referred to as *frugality,* has also been taken up by ecotheologically inclined religious organizations wishing to counter CONSUMERISM and waste.

VRMS *See* VISUAL RESOURCE MANAGEMENT SYSTEM.

vug a large VESICLE, that is, an enclosed cavity, ranging in size from roughly a centimeter to perhaps a meter across, within a rock. Like vesicles, vugs form at the time of the rock's formation, usually through the trapping of air in a frothy LAVA. They often become lined with minerals, in which case they are known as GEODES. *See also* AMYGDULE.

wake a WAVE made by an object moving through the water. Wakes are of two types: *bow wakes,* which result from the displacement of water at the forward end of the object, and *stern wakes,* which result from the inflow of water to fill the space left by the end of the object as it moves forward. Bow wakes travel outward in a V-shaped pattern and may seriously impact nearby shorelines. The impact is often more severe than that of normal waves because bow wakes can strike the shore at any angle, whereas waves tend to come largely or completely from one direction, that of the PREVAILING WIND. Stern wakes do not propagate outward to any great extent. They are, however, highly turbulent, and those from large, deep-draft vessels can impact the bottom of a water channel in areas not normally reached by wind-caused surface waves.

Waldsterben the "forest death" that took place in Germany in the early 1980s. First in the Black Forest of Bavaria, and then in other parts of Germany, stands of silver fir, Norway spruce, and European beech began dying. The event was unprecedented, and was finally traced to AIR POLLUTION. In recent years stricter pollution laws enacted by the German government have alleviated the damage. *Waldsterben* was characterized by a simultaneous and rapid decline of forests involving both coniferous and deciduous trees, with a general thinning and change of color; the "casting" (i.e., dropping) of apparently healthy green leaves and shoots which littered the forest floor; leaves and shoots that had grown abnormally small and misshapen, and which were distributed along the stems in unnatural ways; the formation of nearly vertical or "hanging" branches in spruce and fir; and "distress" crops of cones and seeds, which occurred for three years in a row. The *Waldsterben* led both Canada and the United States to begin investigating the impacts of air pollution on North American forests. In fact, American scientists who visited Germany found, upon returning to the United States, that some of the same conditions had taken place at higher elevations in the Appalachians, in places subject to considerable air pollution from factories and power plants in the Ohio and Tennessee River valleys. *See also* FOREST DECLINE AND PATHOLOGY.

Wallace realms a division of the world into units ("realms") according to the characteristic types of animals found in each unit, developed by the biologist Alfred Russell Wallace in 1876 and still, with minor modifications, in broad use today. There are three realms, divided into five (sometimes six) *regions.* The first realm, *Notogaea,* consists of Australia, New Zealand, New Guinea, and their surrounding islands; it is conterminous with the *Australo-Papuan region.* Realm number two, *Neogaea,* consists of South America, Central America, and the Caribbean Islands, and is conterminous with the *Neotropical region.* The third and largest realm, *Arctogaea* (sometimes known as *Metagaea*), includes North America, Eurasia, and Africa, and is commonly broken into three regions: the *Oriental region* (India and Southeast Asia); the *Ethiopian region* (Africa south and east of the Atlas Mountains) and the *Holarctic region* (North America, Europe, northern Asia, and the Atlas Mountain region of Africa: *see* HOLARCTIC SPECIES). The Holarctic region is sometimes broken down further into the *Nearctic* (North America) and the *Palearctic* (Eurasia and North Africa). The most striking boundary in the Wallace Realms system, known as *Wallace's Line,* runs between the islands of Borneo and Lombok in

Indonesia. The FAUNA on one side of this narrow channel is clearly Australo-Papuan, and on the other side, it is just as clearly Oriental. Wallace's Line is now known to correspond almost exactly to the edge of the Asian CONTINENTAL SHELF.

Wallop Amendment language added to the CLEAN WATER ACT by Senator Malcolm Wallop (R-Wyoming) that states that the Act shall not be construed to limit the rights of the states to control WATER RIGHTS or to enter into interstate compacts for the transfer of water, except that "legitimate water quality measures" may limit such rights "incidentally." The amendment, passed as part of the Clean Water Act Amendments of 1977, appears in the act as Section (101)(g).

WAPA short for Western Area Power Administration (*see* POWER ADMINISTRATIONS, FEDERAL).

Ward Valley site of an ultimately victorious struggle by a coalition of American Indian tribes and conservation organizations to keep a 1,000-acre area of the Mojave Desert of California from being used as a nuclear waste facility. The Indian tribes emphasized that Ward Valley was a sacred site, while environmentalists warned that nuclear waste could seep into the Colorado River, which provides water to a population of 22 million in the Southwest. The Indians of the Fort Mojave, Chemehuevi, Colorado River, Cocopah, and Quechan tribes used the area for burial and ceremonial purposes, and described it as a "spirit path" to nearby Spirit Mountain, considered to be the birthplace of their ancestors. They were also concerned about the environmental impacts, with a special perspective: In the cosmology of the Mojave tribes, if the Colorado River dies the people will disappear as well. In view of the issues raised, the federal government refused to grant a permit to the private contractor wishing to build the facility until matters in dispute were resolved, a situation which did not appear likely. Finally, in 1999, after a four-year period of protest and many demonstrations against the project, a federal court threw out a last-ditch lawsuit filed by Governor Pete Wilson of California, who had asked that the U.S. Bureau of Land Management (BLM) simply transfer the land to the State of California so that California, not BLM, could approve the project, thereby allowing it to go forward under state auspices. When the decision was handed down, the Indians and the environmentalists declared a victory, at least for the moment. *See also* NUCLEAR WASTE AND CLEANUP; YUCCA MOUNTAIN; WASTE ISOLATION PILOT PROJECT.

warning coloration a conspicuous, easily recognizable color or pattern of colors carried by an animal (or occasionally, by a plant) in order to warn potential PREDATORS not to eat it. ORGANISMS displaying warning coloration are usually either unpalatable or actually poisonous, and a predator eating one member of the SPECIES is not likely to repeat the mistake with a second, similarly colored organism, thereby assuring the overall safety of the species. Warning coloration is usually a combination of either bright red or bright yellow with black. *See also* MULLERIAN MIMICRY.

Waste Isolation Pilot Project (WIPP) located in Carlsbad, New Mexico, the world's first underground depository for the permanent disposal of transuranic waste resulting from the research and production of nuclear weapons. Transuranic waste is midlevel in "hotness"—between the long-half-life waste slated for YUCCA MOUNTAIN in Nevada and the short-half-life wastes that were to have been shipped to WARD VALLEY in California. The material shipped to WIPP consists of contaminated soil, clothing, junked machinery, and other items that while not extremely dangerous to handle would nevertheless have a long radiation half-life: 24,000 years. A 2,150-foot deep cave bored into a stable salt formation at a cost of $2 billion, the WIPP is to receive plutonium-contaminated material from the Los Alamos Laboratory in New Mexico; from ROCKY FLATS, a nuclear weapons factory near Denver, Colorado; and from many other facilities where transuranic waste is presently stored. WIPP was first conceived in the late 1960s after a series of fires at Rocky Flats convinced the Department of Energy (DOE) that above-ground storage of nuclear waste so close to a major population center could not be continued. After a protracted period of examining alternative sites, Carlsbad was chosen and construction authorized by Congress in 1979. Immediately, protests began with special emphasis on the long-term stability of the WIPP site and the dangers inherent in transporting nuclear material by road from 10 different places across 22 states to southeastern New Mexico—a "mobile Chernobyl," critics charged. In response, the DOE sought to reassure the public via a $31 million public relations campaign over the next 20 years. At length, in 1998 WIPP received a permit from the ENVIRONMENTAL PROTECTION AGENCY (EPA) to begin operating, which is required to certify the safety of such facilities. The first of an estimated 17 shipments from Los Alamos arrived in March 1999. Shipments from other nuclear waste storage areas throughout the country will follow.

waste management and recycling *See* SOLID WASTE; SOLID WASTE DISPOSAL ACT.

waste stream in waste management, the total amount of waste produced by a society, or by some unit of society such as a household, city, or industrial plant. The size of the waste stream is usually expressed in terms of flow figures: so many kilograms, cubic meters, etc., per day.

wastewater in sanitary engineering, the water flowing in a SEWERAGE SYSTEM. Wastewater consists of SEWAGE, sullage (*see* GRAY WATER), industrial wastes, and STORM RUNOFF. *See also* SEWAGE TREATMENT PLANT.

water dihydrogen oxide (H_2O). Among the most familiar and ubiquitous of all chemicals, water is also among the most unusual, with a group of unique properties that place it in a class by itself and allow it to be, along with CARBON and sunshine, one of the three necessities for the existence of life. Water is the only common material that exists in all three phases of matter—solid, liquid, and gaseous—at earth-normal temperatures and pressures, a fact that drives the HYDROLOGIC CYCLE, without which all water would remain in the seas. It is among a very few materials that expand instead of contract as they freeze, accounting for the fact that ICE floats and keeping the oceans from freezing from the bottom up and staying that way for the rest of eternity. Its extremely high SURFACE TENSION causes CAPILLARY ACTION, allowing it to be held in the soil and utilized by plants as well as helping to drive the circulation of body fluids from CELL to cell. It is an excellent SOLVENT, making it ideal as a means of carrying NUTRIENTS around within the body. And its extremely small molecular size allows it to slip through tiny pores in membranes while the larger MOLECULES it carries in solution are filtered out and left behind—an extremely important phenomenon that is used by cells to sort and control the balance of chemicals within them (*see* OSMOSIS).

STRUCTURE OF WATER

Most of water's unique properties are the result of a single phenomenon, the *hydrogen bond*. The water molecule is highly polar (*see* POLAR LIQUID), with the pair of HYDROGEN ATOMS forming a negatively charged V at one end and the OXYGEN atom a positive pole at the other. Each oxygen atom is electrostatically attracted to one of the hydrogen atoms of a neighboring molecule. Each hydrogen atom is likewise attracted to an oxygen atom. The result is an interlocking two-dimensional web of atoms within which each atom is holding onto three others through electrostatic attraction—accounting both for water's high surface tension and for the fact that, although it is a relatively tiny molecule, it is a liquid at earth-normal temperatures. As water freezes, the two-dimensional bonds extend to three dimensions, causing it to expand its volume by approximately 10% (icebergs float with nine-tenths of their bulk beneath the surface because they are nine-tenths as dense as the water they float in). The formation of these three-dimensional bonds begins at about 4°C (39°F): hence water is at is densest at this temperature, meaning that the ocean basins, and the waters of lakes and streams beneath the winter ice, remain at 4°C and do not cool further (any water that does cool further becomes less dense and therefore rises to the surface, where it either warms up or cools down even more, forming surface ice).

DISTRIBUTION OF WATER

The earth contains about 344 million cubic miles of water. About 315 million cubic miles of that total (92%) is in the oceans. Another 19.4 million cubic miles (5.6%) is in the subcrustal zone of the planet and is unavailable for use by humans. 7.2 million cubic miles (2%) is locked up in the Greenland and Antarctic ICE CAPS and in mountain GLACIERS and sea ice; 2.3 million (0.7%) lies beneath the ground in AQUIFERS. The remaining 60,000 cubic miles—less than 0.02% of the total—is what we usually think of when we think of water resources. It includes water in the ATMOSPHERE (3,600 cubic miles), and water in the soil and in the bodies of plants and animals (3,400 cubic miles). The distribution of this tiny freshwater resource over the Earth's surface is highly erratic. Rainfall (atmospheric moisture) varies widely, from less than an inch a year in the Peruvian deserts to over 200 inches per year in places such as Hawaii, Southeast Asia, and the Washington and British Columbia coasts. About one-fifth of the 53,000 cubic miles found in lakes and streams is in a single lake, BAIKAL, in the Soviet Union; another one-fifth is in the Great Lakes system of North America. The remaining three-fifths is spread through the rest of the world. One of the major sociopolitical problems in natural resources is, and always has been, attempts by humans to remedy these inequalities in the natural distribution of water.

DRINKING WATER

To serve as drinking water, a water source must be free of PATHOGENS and TOXINS. It is also preferable that it be odorless and colorless and that it taste good. (Pure water has no taste, which is not considered good; most drinking water is flavored slightly by dissolved MINERALS.) In order to obtain these conditions, most societies today routinely treat their water supplies. A water-treatment plant for drinking water supplies is remarkably similar to a SEWAGE TREATMENT PLANT, which is not too surprising, as they have the same goal: to remove as much of the foreign matter in the water as possible. The first step is generally screening to remove large materials, followed by *flocculation*—treatment with a chemical, usually powdered alum, to cause SILT and CLAY materials suspended in the water to settle out. A clarifier, or SETTLING BASIN, removes the coagulated material (FLOC) from the flocculation process, as well as other settleable materials such as sand grains. A filter (or set of filters in sequence) removes the material that doesn't settle out, as well as capturing many pathogens and some toxins with particularly large molecules (*see* RAPID SAND FILTER; ACTIVATED CARBON). CHLORINE is usually added after the filtration process to remove any additional pathogens and to prevent the growth of bacteria (*see* BACTERIUM), ALGAE, and other potentially harmful ORGANISMS in the distribution system (pipes and mains). Finally, the treated water is led to a CLEAR WELL where it is stored for use in the distribution system as needed. The degree of treatment of water depends largely on the quality of the source. GROUNDWATER will generally receive only chlorination (and perhaps deionization: *see* ION; ION EXCHANGE), while water from highly eutrophic sources (*see* EUTROPHICATION) may undergo all of the above processes plus aeration, extra filtration, and the addition of chlorine at several different points of the cycle.

See also, e.g., CLEAN WATER ACT; SAFE DRINKING WATER ACT; POLLUTANT; WATER RIGHT; WATERSHED; WATER TABLE; POTABLE WATER; etc.

Water and Power Resource Service short-lived name change for the BUREAU OF RECLAMATION, in effect from November 6, 1979 to May 18, 1981.

water bar a small log placed diagonally across a trail, or a small sloping trench inscribed across a trail, to divert running water off the trail and onto the adjacent hillside. The log, where present, is seldom more than 10 cm (about 4 inches) in diameter. Water bars help prevent EROSION during heavy rains, when the trail might otherwise turn into a WATERCOURSE. On steep slopes they are placed at roughly 6-meter (20-foot) intervals.

water column (1) in botany, the continuous column of water in the XYLEM cells, reaching from the roots to the CROWN of a plant. The water column is under tension due to EVAPORATION of water from the leaves, and is always being pulled upward (*see* TRANSPIRATION). If cut, it will "snap," pulling away from the cut on both sides and leaving the xylem cells closest to the cut empty.

(2) in hydrology and limnology, the water within a lake, river, pond, sea, etc., as opposed to the water at the surface or in contact with the bed. If a cylinder is imagined beginning at bedrock and passing upward through the lake and into the air beyond, it can be seen as being composed of a series of column-shaped segments and flat "slices" that would include (in order upward) the sediment column, the lake bed, the water column, the water surface, and the air column. Material suspended "in the water column" is thus someplace between the surface and the bed.

watercourse any channel on the Earth's surface that carries water part or all of the year, or which once carried water. The term is somewhat more inclusive than *stream,* covering the banks as well as the stream itself and being applicable not only to running water but to dry channels that may carry water only once or twice in the span of a human lifetime.

waterfowl any member of the FAMILY Anatidae, including all ducks, geese, swans, and mergansers. The term should not be used to refer to birds of other families, even those (such as loons, grebes, gulls, and pelicans) that spend most or all of their lives on the water. Waterfowl are characterized by three-toed webbed feet; narrow, pointed wings; and broad, flat bills with serrated edges. Most have long, supple necks and short legs. Relatively large and heavy-bodied, they are nevertheless strong fliers. Most migrate long distances, usually in large flocks, over well-determined courses (*see* FLYWAY; MIGRATION CORRIDOR). They usually mate for life. Waterfowl are strongly dependent on the presence of WETLANDS. The size of local waterfowl populations is far more closely tied to the availability of wetlands than it is to factors such as hunting pressure, disease, and predation (*see* PREDATOR), making the preservation of wetlands the one absolute necessity for the continued presence of waterfowl in any given region. *See also* DUCK STAMP.

water influence zone in Forest Service terminology, the area adjacent to a lake, river, or other body of water that is likely to be used for recreation purposes. The water influence zone includes the banks of the body of water and the lands immediately adjacent to them plus those lands that can be seen from the water or from the banks. MANAGEMENT CONSTRAINTS are similar to those for a TRAVEL INFLUENCE ZONE, except that in addition to management for visual quality, the lands in a water influence zone must be managed to avoid degrading the quality or quantity of the water or of the fishery. This requires strict EROSION controls, controls over potentially polluting activities such as grazing and mining, and careful management of RIPARIAN VEGETATION.

watermelon snow SNOW turned pink or red by the presence of one of several SPECIES of ALGAE of the GENUS *Chlamydomonas,* a member of the so-called "green algae" that turns red at low temperatures. Watermelon snow tastes, as well as looks like, watermelon; however, eating it is not advisable, as the algae can cause diarrhea and other forms of gastrointestinal distress (*see* GASTROINTESTINAL DISEASE). It is characteristic of old, well-settled snow in the mountains in summer.

water meter a device for measuring the flow of water in a pipe, used most commonly as a means of determining the amount of water used by individual customers of a water system. Because they allow billing for water to accurately reflect its use, water meters encourage conservation; their installation has caused water use in homes to decline by as much as 75%. In addition, they enable water managers to discover local patterns in the use of their systems, thereby improving the ability to spot leaks, to plan water-system upgrades, and to target educational programs on water conservation to those areas where the largest waste is going on. Among the most important tools in the water manager's repertoire, they are nevertheless not universally used. Among large cities in the United States without universal residential water meters are Denver, Sacramento, and New York.

water of crystallization (water of hydration) water bound into the ionic or molecular structure (*see* ION; MOLECULE) of a solid. Water of crystallization keeps its identity as water without actually becoming chemically incorporated into the solid's molecules or ions. However, the water molecules attach physically to the molecules or ions of the solid in definite, fixed proportions, changing the structural relationships of these units to one another and (usually) enabling crystals to form. Two forms of water of crystallization, *water of coordination* and *anion water,* bind directly to the molecules or ions of the solid. A third type, *lattice water,* occupies a place in the crystal without directly binding to the crystallizing substance, usually by attaching itself to water of coordination or anion water by means of HYDROGEN BONDS. (A fourth type, *zeolitic water,* is sometimes recognized. It is not bound directly into the crystalline structure but occupies the spaces between crystals in microcrystalline solids.) Water of crystallization is responsible for such phenomena as the formation of crystals in sugar and washing soda and the "setting" of concrete. It is usually indicated in chemical formulas by the use of a dot at the end of the formula followed by the number of water molecules per formula unit of solid in the crystalline form, as, e.g., $CuSO_4 \cdot 5H_2O$ (copper sulfate crystals).

water pollution the presence in water of harmful foreign substances. Both parts of this definition must be met. That is, the material must be HARMFUL (must have the potential of damaging the health of humans or the environment or of interfering with some industrial process) and it must be *foreign* (pure sea water is not considered polluted, even though it is harmful to animals and plants that are adapted to freshwater). Water pollution is a severe problem; there are approximately 11,750 reported pollution incidents in the United States each year, involving an average of nearly 1,600 gallons each—a total POLLUTANT load in the nation's waterways of some 18.6 million gallons annually. (This does not count permitted discharges of pollutants, which total several times this amount. Permitted industrial discharges of BIOCHEMICAL OXYGEN DEMAND directly into the Great Lakes, for example—not counting discharges into tributaries or into municipal SEWERAGE SYSTEMS that then discharge into the Lakes—amount to approximately 100,000 tons per year.) The cost of U.S. water pollution abatement activities for all types of pollution, permitted and accidental, amounts to an average of about $10 billion annually.

For types of pollutants, *see* TOTAL SOLIDS; TURBIDITY; BIOCHEMICAL OXYGEN DEMAND; PATHOGEN; MICROCONTAMINANT. For legal remedies, *see* CLEAN WATER ACT; SAFE DRINKING WATER ACT; COMPREHENSIVE ENVIRONMENTAL RESPONSE, COMPENSATION, AND LIABILITY ACT; GREAT LAKES WATER QUALITY AGREEMENT OF 1978. *See also* ACID RAIN; TOXIC PRECIPITATION; POINT SOURCE; NONPOINT POLLUTION; ZERO DISCHARGE; EFFLUENT; SEWAGE TREATMENT PLANT; and many other entries in this encyclopedia.

Water Pollution Control Federation (WPCF) professional organization founded in 1928 to collect and disseminate information concerning the nature and control of domestic and industrial wastewater. A federation of some 64 state, regional, and international organizations, the WPCF includes engineers, public officials, chemists, equipment manufacturers, and others interested in water pollution control from a professional standpoint. Membership (1999): 36,000. Address: 601 Wythe Street, Alexandria, VA 22314. Phone: (703) 684–2400.

water power *See* HYDROPOWER.

water quality standard usually, a regulatory standard set under the terms of the CLEAN WATER ACT to control the effects of POLLUTANTS in bodies of water. Water quality standards are set by the states; however, the federal government has some enforcement powers over them, including the ability to require a state to set stricter standards if the standards it has submitted do not meet the requirements of the act. Three types of standards are recognized (*see* AMBIENT QUALITY STANDARDS; EFFLUENT STANDARDS; BEST MANAGEMENT PRACTICES). *See also* BEST AVAILABLE CONTROL

TECHNOLOGY; BEST CONVENTIONAL WASTE TREATMENT TECHNOLOGY; NATIONAL POLLUTION DISCHARGE ELIMINATION SYSTEM.

Water Resources Council obsolete federal coordinating body, created by Congress in 1965 to encourage unified planning for all water conservation and development activities in the United States by all parties—federal, state, local, and private. The council produced two key studies of American water resources, the First National Water Assessment (1968) and the Second National Water Assessment (1979), before being disbanded by President Ronald Reagan in 1982.

water rights in law, legal rights held by an individual, a corporation, or a public body for the use of water from a stream, lake, or AQUIFER. Water rights vary widely according to state law and in some cases predate current legal structures. Those of Hawaii, for example, are based on the laws of the ancient Kingdom of Hawaii; those of Louisiana are based on the French Civil Code; and those of a number of cities in the American southwest (including Los Angeles) are based on so-called PUEBLO RIGHTS, dating back to the Spanish conquest, that have been held to take precedence over state law. Most states, however, govern the rights to surface water supplies according to one or another of three basic systems: PRIOR APPROPRIATION, RIPARIAN DOCTRINE or HYBRID RIGHTS. Rights to GROUNDWATER and to diffused surface waters (rainfall and unconfined RUNOFF) are fragmentary and undeveloped in most states, although Arizona has a thorough groundwater-rights law based on a variation of prior appropriation, while Texas, Utah, and a few other states have specifically extended their systems of water rights to include diffused surface waters.

watershed (drainage basin, catchment area) the land from which a stream gets its water. A drop of rain that falls anywhere in the watershed of a given stream, and does not evaporate back into the air, will eventually end up in the stream. The edges of a watershed (*watershed divides*) are easy to define in mountainous country, where they correspond to ridges. Rain falling on a ridge separating two streams will obviously flow into the stream on the side on which it has fallen. Divides are a little more difficult to find in flat country, where they are often low, nearly undetectable rises that may interfinger with each other in nearly random patterns, as in the region around the Minnesota–South Dakota border, where an imperceptible change in slope in the middle of the prairie separates the watersheds of the Red River (tributary to the Arctic Ocean) and the Mississippi (tributary to the Gulf of Mexico; *see* CONTINENTAL DIVIDE). Knowledge of watershed boundaries is important in all forms of land-use planning because they define the limits of the impacts of activities that affect water quality or quantity, such as soil disturbance, withdrawal of water from streams, or pollution discharges, requiring planning for these activities to be done on a watershed-wide basis.

water table the upper edge of the saturated zone in the soil or in an AQUIFER. Beneath the water table, all spaces between soil particles or within the material forming the aquifer are full of water; above the water table, some of these spaces are empty, allowing water to trickle down through them from space to space. The water table determines the position of the STATIC LEVEL in wells and is closely related to the surface elevation of WETLANDS, LAKES, and PERMANENT STREAMS (which are in most cases simply places where the water table is higher than the land surface). It also determines to a large extent the sensitivity of lands to GROUNDWATER pollution. Those lands with high water tables are much more susceptible to pollution than those in which the water table is well below the ground surface. *See also* PERCHED WATER TABLE; CAPILLARY FRINGE; SOIL WATER.

waterway user fee *See* USER FEE.

water witch *See* DOWSER.

Watt, James Gaius (1938–) American political figure, born January 31, 1938, in Lusk, Wyoming, and educated at the University of Wyoming (B.S., 1960; J.D., 1962). After an early career in government, during which he served in roles ranging from Senatorial aide (to Alan Simpson, R-WY) to the directorship of the BUREAU OF OUTDOOR RECREATION, Watt left Washington in 1977 to become president and chief legal officer of the Mountain States Legal Foundation, a conservative public-interest law firm devoted to upholding private property rights, particularly as they are limited by environmental regulations. In 1980 he was named secretary of the interior (*see* INTERIOR, DEPARTMENT OF THE) by incoming President Ronald Reagan. Watt's blunt, confrontational style and strong antienvironmental bias made his tenure at Interior among the most controversial of any cabinet official of modern times. A strict "born-again" Christian, he saw little point to stewardship of NATURAL RESOURCES due to his belief in the imminence of the second coming of Christ, and therefore heavily promoted resource extraction, even within protected areas. Public opposition to his policies is given credit by most observers for the phenomenal membership growth of major environmental organizations such as the NATIONAL AUDUBON SOCIETY and the SIERRA CLUB during the 1980s. He was forced out of office in 1983 amid controversy over his insensitivity to women, the handicapped, and racial minority groups. In 1989 Watt was accused of influence peddling in connection with a federal housing subsidy program of the Department of Housing and Urban Development. Upon his pleading not guilty to 25 counts of perjury and obstruction of justice, the charge was reduced in 1996 to a single misdemeanor. Watt pleaded guilty to the reduced charge—of trying to influence a federal grand jury—and was sentenced to 500 hours of community service and a $5,000 fine.

wave (1) in physics, any periodic phenomenon that transfers energy from one point to another. Waves come in two broad types: *transverse waves,* in which the motion of the wave is perpendicular to the direction it is being transmitted (example: water waves, in which the water moves up and down while the energy is transferred horizontally; see definition (2), below); and *compressional waves,* in which the motion of the wave is in the same plane as the energy transmission (example: sound waves, in which the energy is transmitted by tiny pressure alterations known as *condensations* and *rarefactions* that cause air molecules to move backward and forward in the direction of the wave travel). In all forms of wave motion, only the energy is transmitted. The transmitting material is temporarily displaced to transmit the wave but returns to its original position when the wave has passed. The amount of energy transmitted by a wave is dependent upon the relationship between its *amplitude* (the distance the transmitting medium is displaced as the wave passes), its *frequency* (the number of displacements that occur in a given time period), and its *wavelength* (the distance from crest to crest or trough to trough, or from condensation to condensation or rarefaction to rarefaction, of the wave). The amount of energy increases with increasing amplitude and frequency and decreases with increasing wavelength, so that (for example) waves of higher frequency and shorter wavelength will always carry more energy than waves of the same amplitude but with lower frequency and longer wavelength. Examples of waves include radio waves, earthquake waves (*see* P-WAVE; S-WAVE), and solar radiation.

(2) in hydrology, a wave (definition (1), above) taking place in water. Waves in water are transverse waves that cause the water level to alternately raise and lower as the wave moves across the surface of the water body. When they strike the shore, the energy they carry is dissipated against the materials of the shoreline, causing EROSION. Waves "break" when the downward half of the up-and-down wave motion is interfered with by the bottom of the water body, causing the upper half of the wave to outrun the lower half (*see* HALF-WAVE DEPTH). Most waves are caused by wind (but *see* TSUNAMI; WAKE).

way (vehicular way) in Forest Service and Bureau of Land Management terminology, a track for vehicles across the landscape that was created and is maintained solely by the passage of the vehicles. A way does not constitute a road for planning purposes, and cannot be used to exclude an area from designation as "roadless."

way trail a trail that was created and is maintained solely by the passage of its users, such as the so-called "fisherman's trails" that often appear along a shoreline. Way trails are not normally shown on maps or inventoried as part of a trail system.

weather day-to-day or moment-to-moment changes in atmospheric conditions, such as rain, sunshine, wind, clouds, etc. Weather is what is happening in the ATMOSPHERE at any given moment, as opposed to CLIMATE, which is what happens in the atmosphere over a long period of time. The conditions within a storm are weather; the likelihood of storms happening during a particular month of the year is climate. See also different types of weather, e.g. CYCLONIC STORM; HURRICANE; ANTICYCLONE; etc.

weather modification the practice of inducing precipitation or other changes in specific areas for economic or public benefit. Techniques to modify weather were discovered in 1946 wherein snowfall was produced when clouds were "seeded" via crushed dry ice (CO_2 in solid form) dropped from aircraft into supercooled stratocumulus clouds. The following year, experiments in cloud seeding with particles of silver iodide also proved successful. Today, most weather modification projects use silver iodide because it can be released at low altitudes in dry, warm air, unlike dry ice, which must be released at high altitudes. Cloud seeding can be used to dispel fog as well as cause precipitation. Most often, weather modification techniques are used to benefit farmers needful of rain. In one application, the so-called hail busters in the upper midwestern and prairie province farm districts of the United States and Canada can reduce the devastating impact of frequent hailstorms on agricultural crops by changing large hailstones into smaller ones. In most weather modification projects, the silver iodide pellets are dropped from small aircraft, although land-based delivery systems using three-foot-long rockets are sometimes used. In the United States, weather modification efforts are constrained by the possibility of lawsuits pursued by private individuals or public authorities for "stealing rain" in areas beset by long-term drought.

web of life metaphor for the vast complexity of ecosystems and the manifold interconnections of plants and animals with one another and with their environment. *See* ECOLOGY.

weed cutting in forestry, a type of THINNING made in a young STAND, typically at the SAPLING stage, in which not only the inferior trees within the stand but also BRUSH, SHRUBS, small HARDWOODS and other forms of competing vegetation are removed. Weed cuttings combine TIMBER STAND IMPROVEMENT and vegetation management activities in a single ENTRY, reducing both costs and impacts on the ENVIRONMENT of the stand.

Weeks Law federal legislation expanding the authority of the FOREST SERVICE, signed into law by President William Howard Taft on March 1, 1911. The Weeks Law had three principal thrusts. Sections 1–5 established the authority of the Forest Service to enter into cooperative agreements with state agencies for the purposes of resource management and fire protection; sections 6–11 gave the service the right to add lands to the NATIONAL FOREST SYSTEM through purchase, exchange, or the exercise of the power of EMINENT DOMAIN. Finally, sections 12 and 13 made the Forest Service accountable to state laws wherever such laws did not directly conflict with federal legislation, and directed that 25% of all moneys received by any NATIONAL FOREST during any given fiscal year should go to the states and counties in which the Forest was located for the purpose of building and maintaining roads and schools. *See also* CLARKE-MCNARY ACT.

weighting assigning values to a list of items based on some preestablished set of priorities. Weighting is usually done by applying multipliers to estimated values. For example, if one has assigned a weight of 5 to wildlife values and 3 to watershed values, a plot of land ranked as a 6 (on a scale of 1 to 10) for water production and only as a 4 (on a scale of 1 to 10) for wildlife production would, after weighting, be given a rank of 18 (3 x 6) for watershed and 20 (5 x 4) for wildlife, suggesting that management of the area should be skewed toward wildlife production even thought the land was slightly better suited to use as a watershed. The process has some value as a decision-making tool, but its results should never be interpreted rigidly, as one is usually multiplying one set of estimates by another set of estimates and thereby standing a good chance of magnifying errors.

well cluster *See* MONITORING WELL.

Wentworth scale in engineering geology, a scale used to measure the size of rocks. The Wentworth scale is logarithmic to the base 2, that is, each size range is a multiple of two greater than the size range below it. Major boundaries lie at 1/256 mm (clay to silt), 1/16th mm (silt to sand), 2 mm (sand to pebbles), 16 mm (pebbles to cobbles), and 256 mm (cobbles to boulders). The divisions within these boundaries carry subtitles, as *fine, coarse, very fine, very coarse,* etc.

The Wentworth Scale

Size (mm)	Category: Subclass	Class
1/256		clay
1/128	very fine	
1/64	fine	silt
1/32	medium	
1/16	coarse	
1/8	very fine	
1/4	fine	
1/2	medium	sand
1	coarse	
2	very coarse	
4	small	
8	medium	pebbles
16	coarse	
32	small	
64	medium	cobbles
128	large	
256		boulders

Western Area Power Administration *See* POWER ADMINISTRATIONS, FEDERAL.

Western Wood Products Association (WWPA) coalition of forest products manufacturers, founded in 1964 to promote the forest products industry in the 12 western U.S. states and to offer technical and statistical services to its members and to the public. Address: 522 SW Fifth Avenue, Suite 500, Portland, OR 97204. Phone: (503) 224-3930. Website: www.wwpa.org.

wetland any area of land for which the WATER TABLE is at or near the surface of the land for a significant portion of the year. "Significant" in this context generally means that the land must be wet long enough to support the growth of water-dependent vegetation such as reeds, cattails, mosses, etc. (*see* MACROPHYTE). Lakes, streams, and other bodies of water that are deep enough and/or contain swift enough currents that rooted vegetation cannot take hold are generally excluded from the definition; however, a wetland may contain a substantial amount of standing water, including PONDS and slow-moving WATERCOURSES. Among the most productive of all ECOSYSTEMS, wetlands provide HABITAT for numerous beneficial SPECIES of plants and animals. They also generally improve the water quality of streams by filtering out suspended SEDIMENTS and any POLLUTANTS that may be sorbed to them (*see* ADSORPTION; ABSORPTION). *See also BOG; MARSH; SWAMP; ESTUARY.*

wet-sand margin *See* SWASH.

wetting ability in chemistry and physics, the attraction between the MOLECULES of a liquid and those of an adjacent solid. If this attraction is greater than the liquid's SURFACE TENSION, the liquid is said to "wet" the solid. Wetting ability determines the angle between the surface of a liquid and the container it is in. This angle is characteristic of the type of solid/liquid interface involved (examples: glass/mercury, 140°; water/steel, 90°; water/glass, 0°). The phenomenon is important in numerous natural and industrial processes, including cleansing (*see* WETTING AGENT; DETERGENT), CAPILLARY ACTION, the SINTERING of ores, and most soil-water phenomena, including SOIL MOISTURE TENSION, FIELD CAPACITY, and the depth of the CAPILLARY FRINGE.

wetting agent a substance that lowers the SURFACE TENSION of a fluid, thereby increasing its ability to cling to solid surfaces (*see* WETTING ABILITY).

wet well *See under* MONITORING WELL.

whip (1) in political terminology, an officer in a congressional CAUCUS who is designated to get word to the other caucus members when legislation that they are interested in comes up for a vote. Large caucuses designate several whips (a "whip system") who get the word to each other first so that it can be spread more rapidly to the other members.

(2) in forestry, a small SAPLING left standing in a CLEAR CUT.

whistle-blowers in government or private industry, those who bring unsafe, hazardous, and dishonest practices to public attention. In one notable case, a top engineering employee reported to his management in 1990 that the nuclear power plant fuel rods the company was involved in manufacturing were potentially dangerous. No action was taken by the firm, and the engineer filed a complaint with the Nuclear Regulatory Commission, whereupon he was fired by his employer. The engineer sued the company and its officers for $260 million for wrongful discharge, fraud, and racketeering, which he described as a cover-up that posed "a risk to the life and health of millions of people around the world." The suit was settled out of court. Although there are protections in place for employees of government agencies, private sector whistle-blowers have historically been subject to retaliation by employers. At the same time, some whistle-blowing can result in a reward, notably in cases of private contractors defrauding government agencies. In the 1980s and 1990s especially, whistle blowing became an important adjunct to the enforcement of environmental and public health laws. The Occupational Health and Safety Administration (OSHA), for example, receives some 3,000 complaints each year from whistle-blowers. At the state level, occupational safety programs receive an additional 1,450 complaints. Corporate reprisal against whistle-blowers often takes the form of job termination, which OSHA has moved to alleviate by strengthening the protection of whistle-blowers through special legislation, the Hazard Reporting Protection Act of 1999. Under the act, OSHA can require corporations to reinstate employees terminated for lodging a complaint and to provide them with lost wages and benefits as well as any legal costs incurred.

White, Lynn, Jr. (1907–1987) American historian noted for his condemnation of Western religions as a cause of the environmental crisis. Educated at Stanford University, Union Theological School, and Princeton University, White taught at Princeton and the University of California in Los Angeles. A specialist in medieval history, White argued that the Judeo-Christian perception of creation and, especially, the biblical decree that man should have dominion over the natural world, morally cleared the way for the rampant exploitation of the world's natural resources. His paper, "The Historical Roots of our Ecological Crisis," appeared in *Science* magazine in 1967 (Vol. 155:3767). Among his books are *Medieval Technology and Social Change* (1966); *Medieval Religion and Technology: Collected Essays* (1978); and (as co-author), *Life and Work in Medieval Europe: The Evolution of Medieval Economy from the Fifth to the Fifteenth Century* (1982). *See also* RELIGION AND THE ENVIRONMENT.

white water a set of rapids. A "small amount of white water" is one or two sets of small rapids in a REACH of mostly quiet river; a "white water river" is a river with many sets of rapids; etc. The white color of white water is caused by the ENTRAINMENT of many small bubbles of air in the water as it flows over the rapids. *Compare* FLAT WATER.

WHO *See* UNITED NATIONS WORLD HEALTH ORGANIZATION.

whole-body analysis an analysis of the amount of MICROCONTAMINANTS in the entire body of a large ORGANISM such as a fish, bird, or mammal, as opposed to an analysis of specific tissues such as the liver, lungs, or reproductive tract. The steps involved are: (1) *homogenization,* reducing the entire body of the organism to a well-mixed semi-liquid mash, usually with the aid of a tool similar to a home food processor; (2) *lipid extraction,* removing the body fats and oils (*see* LIPID) from the mash with the aid of precipitating agents (*see* PRECIPITATE) and organic SOLVENTS (this step is used only in those cases where the microcontaminants sought are highly LIPOPHILIC substances that will be concentrated almost exclusively in the fatty tissues); (3) *concentration,* reducing the total volume of the sample through the use of methods that will increase the ratios of microcontaminants to tissue in the sample by easily calculable amounts, thus making the contaminants easier to find while still being able to work back through the calculations to find the original concentrations; and finally, *analysis,* the actual determination of the types and amounts of contaminants present, using the general tools of analytic chemistry. Whole-body analysis is called for primarily in cases where general levels of BIOLOGICAL MAGNIFICATION in a FOOD CHAIN need to be determined. It is of little use to determine the effects of contamination on the organism being analyzed, as these effects vary greatly according to where in the organism's body the contaminant has become concentrated.

wholistic approach *See* HOLISTIC APPROACH.

Wild and Scenic Rivers Act federal legislation to protect selected rivers from dam construction and excessive commercial development, signed into law by President Lyndon Johnson on October 2, 1968. Declaring that "the established national policy of dam and other construction at appropriate sections of the rivers of the United States needs to be complemented by a policy that would preserve other selected rivers or sections thereof in their free-flowing condition" (Section (1)(b)), the act defines three classes of protected river (*see* WILD RIVER; SCENIC RIVER; RECREATIONAL RIVER) and spells out in considerable detail the management restrictions to be placed on these rivers and the process through which a river is to be considered for classification. A corridor of land on each side of a protected river is also protected: this land is to average no more than 320 acres per linear mile of river through the protected stretch. The rights of landowners within this corridor are maintained, subject to restrictions on types of development that would affect water quality or other qualities for which the river was designated a part of the Wild and Scenic Rivers System. Rivers under official study for inclusion in the system are temporarily given the same protection as rivers actually within the system. This "study protection" may last up to three years. Management plans for federal lands that include rivers eligible for Wild, Scenic, or Recreational status must take into account this eligibility. Eight so-called "instant rivers" and 27 "study rivers" were designated by the act. By the late 1990s, the system had grown to include 150 protected rivers and river segments, with a combined length of approximately 16,000 miles.

Wilderness Act of 1964 federal legislation creating the National Wilderness Preservation System (*see* WILDERNESS AREA), signed into law by President Lyndon Johnson on September 3, 1964. Declaring it the policy of Congress "to secure for the American people of present and future generations the benefits of an enduring resource of wilderness," the act outlines procedures for congressional declaration of Wilderness Areas on federal lands administered by the NATIONAL PARK SERVICE, the FISH AND WILDLIFE SERVICE, and the FOREST SERVICE, (the provisions of the act were extended to BUREAU OF LAND MANAGEMENT lands 12 years later; *see* FEDERAL LAND POLICY AND MANAGEMENT ACT OF 1976). Section (2)(c) defines what is meant by "wilderness":

> . . . an area where the earth and its community of life are untrammeled by man, where man himself is a visitor who does not remain. An area of wilderness is further defined to mean in this Act an area of undeveloped Federal land retaining is primeval character and influence, without permanent improvements or human habitation, which is protected and managed so as to preserve its natural conditions and which (1) generally appears to have been affected primarily by the forces of nature, with the imprint of man's work substantially unnoticeable; (2) has outstanding opportunities for solitude or a primitive and unconfined type of recreation; (3) has at least five thousand acres of land or is of sufficient size as to make practicable its preservation and use in an unimpaired condition; and (4) may also contain ecological, geological, or other features of scientific, educational, scenic, or historical vale.

Once established by Congress, a wilderness area cannot be used for timber harvest, and no new roads or permanent structures may be built within it. Motorized vehicles are prohibited except those operated in conjunction with established mining operations, and new mines may not be located (but *see*

ASPINALL AMENDMENT). Virtually all other uses are allowed, including hunting, grazing, nonmotorized recreation, and (with approval from the president of the United States) water resources development. Exceptions to the bans on timber harvest, roads and motorized vehicles are allowed for fighting fires and insect infestations. The act established 9.1 million areas of so-called "instant wilderness" in 54 different Wilderness Areas. By the end of the 1990s, the system of statutory wilderness areas totaled 103 million acres in 649 units. The U.S. Forest Service manages 400 units; the Bureau of Land Management, 134; the Fish and Wildlife Service, 71; and the National Park Service, 44.

HISTORY

The need for preservation of areas of wilderness was recognized by the Forest Service as early as 1924 with the establishment of the Gila Wilderness in New Mexico. The concept was formalized by the agency in 1929 with the promulgation of Administrative Regulation L-20, permitting the secretary of agriculture to set aside PRIMITIVE AREAS of 100,000 acres or more each. These "L-20 lands" were supplemented in 1939 by "U-regulation lands" under a pair of regulations (U-1 and U-2) allowing the administrative set-aside of "Wilderness Areas" 100,000 acres or more in size and "Wild Areas" of between 5,000 and 100,000 acres. A push for legislatively defined rather than administratively defined wilderness areas was begun shortly after World War II. It received impetus from certain Forest Service actions in that period, including the administrative declassification of some portions of existing Wilderness, notably the removal of the French Pete Valley in Oregon from the Three Sisters Wilderness Area. In 1956, the first wilderness legislation was introduced into Congress by Senator Hubert H. Humphrey of Minnesota and Congressman John Saylor of Pennsylvania. Over the next nine years, some 65 different wilderness bills were introduced, culminating in 1964 with the passage of the compromise bill that became the Wilderness Act. *See also* ROADLESS AREA REVIEW AND EVALUATION.

wilderness area an area set aside by Congress under the terms of the WILDERNESS ACT OF 1964 as "an area where the earth and its community of life are untrammeled by man, where man himself is a visitor who does not remain." Lands designated as wilderness must meet three basic criteria: (1) they must be PUBLIC LANDS; (2) they must be at least 5,000 acres in size, or must be large enough to be managed as wilderness (the second portion of this clause allows the inclusion of islands, or ISLAND ECOSYSTEMS, or less than 5,000 acres, provided there is no other barrier to their management as wilderness); and (3) they must be substantially unmodified by human activities other than trails and rustic campsites, although previously established water development projects, mines, and their associated roads may be allowed. NATIONAL FOREST, BUREAU OF LAND MANAGEMENT, NATIONAL PARK, and National WILDLIFE REFUGE lands are all available for wilderness classification. Currently, 100 million acres of federal wilderness exist, mostly on National Forest lands in the west. The term is also used by some states (e.g., New York, Michigan, California) to refer to areas of state land managed in the manner of federal wildernesses.

The Wilderness Society (TWS) environmental organization founded in 1935 to encourage the preservation of wilderness areas on the public land, and to monitor their management once they are created. One of the prime movers behind the WILDERNESS ACT OF 1964, TWS has since concentrated on grass-roots organizing in support of specific wilderness proposals on NATIONAL FOREST and BUREAU OF LAND MANAGEMENT lands and on the development of a national wilderness ethic, including the preservation of OLD GROWTH, WETLANDS, and other natural areas. Membership (1999): 300,000. Address: 900 17th Street NW, Washington, DC 20006. Phone: (202) 833-2300. Website: www.tws.org.

wilderness study area (WSA) an area set aside to be studied for possible inclusion in the national wilderness preservation system (*see* WILDERNESS AREA; WILDERNESS ACT OF 1964). Wilderness study areas may be designated either by Congress or by the agency responsible for managing the land involved. They are managed under the same criteria as wilderness areas until a final decision is made either to include the area (or part of the area) in the wilderness system or to release it for COMMODITY or other non-wilderness use. The establishment of wilderness study areas was implied in the Wilderness Act, and was made explicit in the Eastern Wilderness Act of 1975, which defined the concept and established the first legislated wilderness study areas. Administratively established wilderness study areas have been designated through the National Forest planning process (*see* FOREST PLAN) and through the wilderness review process established for BUREAU OF LAND MANAGEMENT lands by the FEDERAL LAND POLICY AND MANAGEMENT ACT OF 1976. *See also* ROADLESS AREA REVIEW AND EVALUATION; PRIMITIVE AREA.

Wilderness Watch environmental organization founded in 1969 to monitor and report on the disappearance of wild areas in the United States, especially those on federal lands. Primarily a data bank, this small but respected organization also publishes occasional position papers on federal land-use practices. Address: P.O. Box 9175, Missoula, MT 59807. Phone: (406) 542-2048. Website: www.wildernesswatch.org.

wildfire a fire burning out of control in the natural ENVIRONMENT. Dwellings and other structures may be consumed by a wildfire, but it primarily burns vegetation, living and dead. Wildfires may be set naturally by lighting, or they may be started accidentally by humans, either as the result of a CONTROLLED BURN getting out of control or as a result of a careless act such as discarding a burning cigarette or operating a chain saw without a spark arrestor. Roughly 2.5 to 3 million acres of land are burned over by wildfires in the United States in any given year. Of this total, 80% to 90% is the result of fires that are human-caused.

TYPES OF WILDFIRE

Fires are classed by the type of environment being burned (*grass fire, brush fire, forest fire*) and also by their intensity. The classifications used for fire intensity apply primarily to forest fires, which are generally thought of as the most destructive and dangerous forms of wildfire. Five different classes of intensity are usually recognized. *Spot fires* burn slowly, in highly localized areas (the term *spotting* is also used for fires which begin ahead of an advancing fire line due to sparks or flaming debris carried on the wind). *Ground fires* burn the HUMUS and the dry LITTER on the forest floor. *Surface fires* advance over the ground, burning only downed FUEL and undergrowth and having little effect on standing trees or brush, while *crown fires* burn standing trees and brush but may or may not burn the material on the ground. The most extreme form of wildfire is the *firestorm,* an intensely hot, rapidly moving column of flame that destroys everything in its path. High winds are generally required to begin a firestorm, but once started the fire's heat creates its own winds by convection. The column of smoke may rise to the stratosphere, much like the mushroom cloud from an atomic bomb, whose pattern it very much resembles. True firestorms, fortunately, are fairly rare. The great fires that swept Yellowstone National Park in the summer and fall of 1988, for example, were 80% ground and surface fires and 14% crown fires. In only 6% of the burned area—less than 1% of the park—were firestorm conditions approached.

WILDFIRE MANAGEMENT

Since fire is a natural part of most ECOSYSTEMS (*see,* e.g., FIRE-MAINTAINED CLIMAX; FIRE DEPENDENT SPECIES) the goal of wildfire management should be to control and direct, rather than to eliminate, fire. In WILDERNESS AREAS and other natural areas, including most NATIONAL PARKS, the best method of doing this may be to simply let all wildfires burn, monitoring their progress to make certain that they do not expand outside the boundaries of the natural area. A variation of this policy is to let all lightning-caused fires burn but to suppress those caused by humans. In the modified landscape of the managed forest, stricter control measures are usually required. These include the reduction or elimination of FUEL BUILDUP through such practices as YUM yarding and the chipping of logging debris for mulch; the use of PRESCRIBED BURNS at times when they can be limited to an area that needs to be treated with fire; and the construction of fuelbreaks (*see under* FIREBREAK) around particularly sensitive areas such as recreational developments or community WATERSHEDS. Fire fighting (*fire suppression*) remains a necessary tool when fires threaten sensitive areas or expand to a scale beyond that which might be expected in an unmodified ecosystem. The principal tools of fire suppression are the firebreak; the *backfire* (a deliberately set, controlled fire ignited in the path of the wildfire to burn off the fuel and thus prevent the wildfire from expanding any further in that direction by depriving it of anything to burn); and the use of fire retardant chemicals dropped from aircraft either in the path of the fire to coat the vegetation with unburnable material or directly on the fire to cool and smother it. (Fire retardants must be chosen that are not themselves environmentally harmful: a common retardant is BENTONITE clay. Plain water may also be used, and has the advantage in many areas of being available in lakes and rivers, from which it may be lifted in large buckets carried by helicopters.) A wildfire is said to be *contained* when firebreaks have been built completely around it. It is said to be *controlled* when there is no longer any danger that the fire will jump the firebreaks. The phase of suppression following control, known as *mop-up,* consists of hand-suppressing all remaining hot spots within the firebreaks. Normally a fire cannot be declared completely out until the fire area has been dampened by rain or snow.

Wild Horse Organized Assistance (WHOA!) wildlife preservation organization, founded in 1971 to prevent the abuse and extinction of feral horses and burros (*see* FERAL ANIMAL) on the RANGELANDS of the American west. WHOA! works to set aside wild horse and burro sanctuaries, especially on federal lands, and to influence federal and state laws regarding the treatment of wild horses. It also monitors range conditions and the activities of agencies and firms engaged in wild horse round-ups. Membership (1999): 12,000. Address: PO Box 555, Reno, NV 89504. Phone: (702) 323-5908.

wildlife benefits improvements in conditions for wild animals that result from a proposed action, such as a dam, a timber harvest, the establishment of a park or WILDERNESS AREA, etc. Wildlife benefits include such things as increased COVER or FORAGE, protection of CRITICAL HABITAT, increased dependability of a water supply, etc. A common means of measuring them is to estimate the number of additional HUNTER DAYS or ANGLER DAYS that the population of wild animals in the region will support following the carrying out of the proposed action.

Wildlife Conservation Society a membership organization organized by a consortium of New York City organizations and facilities to support international wildlife conservation. The consortium includes the Bronx Zoo, the New York Aquarium, the Central Park Wildlife Center, and the Prospect Park Wildlife Center. The society conducts wildlife and wildlands conservation projects in 53 countries. Membership (1999): 100,000. Address: 185th Street and Southern Boulevard, Bronx, NY 10460. Phone: (718) 220-5100. Website: www.wcs.org.

wildlife habitat *See* HABITAT.

Wildlife Management Institute (WMI) established in 1911 by sportsmen concerned about declining wildlife populations. Since then, WMI has become an international scientific and educational organization which fosters improved professional management of North American natural resources. The institute sponsors the Annual North American Wildlife and Natural Resources Conference; participates in the Cooperative Wildlife Research Unit Program at over 40 land-grant colleges and universities; testifies before House and Senate committees and subcommittees on wildlife-related legislative matters; and publishes authoritative books, information flyers, booklets and brochures. Address: 1101 14th Street NW, Suite 801, Washington, DC 20005. Phone: (202) 371-1808. Website: www.wildlifemgt.org/wmi.

wildlife refuge an area set aside and managed to provide HABITAT for wildlife. Hunting, trapping, and fishing (for sport or PREDATOR CONTROL) are generally allowed on refuges; if these activities are curtailed or forbidden, the area is designated not as a refuge but as a *wildlife sanctuary.* Timber harvest, grazing, and similar activities may take place on a refuge if they can be shown to be congruent with habitat maintenance. Most wildlife refuges are operated by the federal government. The majority of these contain large areas of formerly private lands that have been purchased with funds obtained through the sale of DUCK STAMPS, and are geared primarily to the maintenance of WATERFOWL populations. Day-to-day management of refuges is in the hands of the United States FISH AND WILDLIFE SERVICE. Federal refuge lands are available for classification as wilderness (*see* WILDERNESS AREA; WILDERNESS ACT OF 1964). In addition to these large areas of federal land, there are a number of state, local, and private wildlife refuges in the United States, some as small as suburban backyards, others—such as those owned and operated by the NATIONAL AUDUBON SOCIETY, THE NATURE CONSERVANCY, and DUCKS UNLIMITED—similar in scale to the federal refuges. On these lands, management restrictions are set by the local agency or private group that owns the land, except that they may not set restrictions that are looser than those in federal wildlife-management laws such as the ENDANGERED SPECIES ACT.

wildlife sanctuary *See under* WILDLIFE REFUGE.

Wildlife Society, The international educational and scientific organization of professionals and students. The society offers a program for certification as a professional wildlife biologist; sponsors annual conferences; and publishes *The Journal of Wildlife Management,* the *Wildlife Society Bulletin, Wildlife Monographs,* and other topical books. Membership (1999): 9,600. Address: 5410 Grosvenor Lane, Bethesda, MD 20814. Phone: (301) 897-9770. Website: www.wildlife.org.

wild river a river or section of a river designated as "wild" under the terms of the National WILD AND SCENIC RIVERS ACT OF 1968. The "wild" designation is the most restrictive of the three river classifications provided for by the act (*compare* SCENIC RIVER; RECREATIONAL RIVER). To qualify for designation as a wild river, a river or river segment must be free-flowing (or essentially free-flowing); must flow within a largely primitive or undeveloped WATERSHED; and must have no roads along its banks, access to the river being only by boat or by trail. Small developments, such as boat-in or walk-in lodges, may be allowed if their impact is small in relation to the total river ENVIRONMENT. All management restrictions for scenic rivers apply also to wild rivers; in addition, water quality must be good enough year-around to allow water-contact sports, such as swimming and boating, without health hazards to the participants.

Wilson, E. O. (1929–) Pulitzer Prize–winning American biologist and author, noted for his work in the sociobiology and species diversity. Born in 1929, Edward Wilson studied at the University of Alabama and Harvard University, where he joined the faculty in

1956. His early work was in entomology, especially ants, but he later extended behavior mechanisms found in social insects as well as humans. In his 1975 book *Sociobiology: The New Synthesis,* he argued that all human behavior is genetically based and by definition, selfish. Opponents countered that such theories were dangerous and could be utilized to excuse racism and eugenics. In recent years Wilson has been a champion of sustaining global BIODIVERSITY and has expressed his great concern about the long-term effects of species EXTINCTION. Wilson believes we may be witnessing the Sixth Great Extinction caused by mankind's destruction of and negative pressures on ecosystems worldwide. His books include two Pulitzer Prize winners, *On Human Nature* (1978) and *The Ants* (1990, with Bert Holldobler). Other books include *Biodiversity* (1988), *Edward O Wilson: A Life in Science* (1990), *The Diversity of Life* (1992), *Naturalist* (1994, autobiography), *Temple Wilderness: The Nature of Spirituality* (1996), *In Search of Nature* (1996), and *Consilience: The Unity of Knowledge* (1998). *See also* MASS EXTINCTION.

wilt point in the plant sciences, the point at which soil dryness causes a plant to wilt (that is, lose TURGOR) to the point that it can no longer recover when water is added. The wilt point is sometimes referred to as the *permanent wilting percentage;* however, this is a misnomer, since it is not the percentage of water in the soil but the SOIL MOISTURE TENSION that is critical for determining the wilt point. The actual wilt point for a given soil varies with the plant SPECIES and the atmospheric conditions (desert plants can extract water at much higher soil moisture tensions than can crop plants such as corn or wheat, and any condition that increases the EVAPOTRANSPIRATION rate of a plant, such as high temperatures or winds, will decrease the plant's ability to draw water out of the soil against a high soil moisture tension). Plants in middle-latitude climates with moderate rain commonly have a wilt point of around 15 atmospheres.

windbreak *See* SHELTERBELT.

wind chill factor in meteorology, the difference between the actual temperature and the sensible temperatures (*see* SENSIBLE HEAT) caused by the wind. Wind cools the body both by increasing EVAPORATION and by carrying away the thin layer of air close to the skin that the body has already warmed, replacing it with cooler air that needs to be re-warmed. The relationship among the various factors is complex. In general, however, the wind chill factor increases both with lower temperatures and with higher wind speed. The *effective temperature* (air temperature minus wind chill factor) is the temperature that determines the incidence of frostbite, HYPOTHERMIA, and other conditions brought on by cold.

Wind Chill Factor Conversion Table

Wind Speed (mph)										
Temp (F)	50	40	30	20	10	0	–10	–20	–30	–40
5	2	3	3	4	4	5	5	6	6	7
10	10	12	14	16	19	21	23	26	28	30
15	14	18	21	25	28	33	35	38	42	45
20	18	22	26	30	35	39	43	47	52	56
30	22	27	32	38	43	48	53	59	64	69
40	24	30	36	41	47	53	59	65	70	76

To find the effective temperature at a given wind speed and thermometer reading, find the column with the thermometer reading and the row with the wind speed, locate the point at which they intersect, and subtract the number found there (the wind chill factor) from the thermometer reading at the top of the column. Example: a wind speed of 30 and a thermometer reading of –10 yields a wind chill factor of 53, yielding an effective temperature of –63.

windlean in forestry, LEAN caused by a tree's reaction to the PREVAILING WIND in an area. Windlean is usually uniform throughout a STAND, and is more pronounced in isolated trees or in small stands than it is in large forests due to the protection a large group of trees is able to give to its individual members.

wind power power generated by the wind and used for human purposes. Since nearly all winds are generated by the sun, through differential heating and cooling of the earth's ATMOSPHERE, wind power is actually a form of SOLAR ENERGY; however, it is usually treated separately. Used since ancient times to pump water and (primarily in the Netherlands) to grind corn and wheat, wind power may also be used on a limited scale to generate electricity. Wind-generated power is clean and renewable; however, it suffers from unreliability due to the extremely variable strength of the wind over time at any given location. It may also cause visual pollution, in that the large windmills required to generate significant amounts of power are usually unsightly, and the best available winds are often in highly scenic locations such as ocean shores, hilltops, and mountain passes. For these reasons, wind power is not likely ever to become commercially significant, although small, single-home and single-business installations may be remarkably successful in reducing dependence on FOSSIL FUELS and other centrally generated power sources.

CAPTURING THE WIND

Wind power is generated through the use of a *windmill* consisting of a set of blades (the *fan*) attached to a central *rotor.* Power is led from the rotor to the generator by means of a gear train or a set of belts. Three basic types of windmill are in use. The *American (multiblade) windmill* consists of a fan with 10 or more fixed blades forming a circle 6 to 10 feet in diameter. A rudder, the *fantail vane,* keeps the fan pointed into the wind. At high wind speeds, the fantail vane collapses sideways against the fan, turning it across the wind rather than into it and thus preventing damage from too-rapid rotation. Partly because of its inability to use winds of higher speed, the American windmill is the least efficient type. The *fan-type windmill* consists of two, three, or four blades up to 100 feet or more in length. Their pitch is variable, so that they can maintain a rotational speed of 30 revolutions per minute with any windspeed from 15 to 75 miles per hour. At speeds over 75 mph the blades turn edge-on into the wind, stopping the mill. Like the American mill, the fan-type mill is faced into the wind by a fantail vane. Fan-type mills, often grouped together as "wind farms," are the mills used to generate commercial quantities of wind power. Finally, *S-rotor windmills* utilize a vertical rotor with a blade forming a spiral around it which looks, from a little distance, like the letter S. Since the spiral can catch the wind from any direction, the S-rotor mill does not require a fantail vane and can thus be made much smaller, lighter, and simpler. It is the preferred windmill for individual home and small-business installations.

AVAILABILITY OF WIND POWER

The power available from the wind is primarily dependent upon its speed. It is figured according to the formula

$$P=(D/2)V^3$$

where P = power, D = air density, and V = wind velocity. Doubling the wind speed in this equation multiplies the power by eight times—meaning that areas with intermittent gusty high-speed winds are actually better locations for windmills than are areas with constant winds of lower speeds. The density factor is relatively minor, but it does mean that more power is available on the seacoast than in the mountains, given otherwise equivalent conditions, because the air is denser at sea level. Mapping the United States from the standpoint of the average annual power available from the wind (rather than the average annual windspeed) indicates that there are relatively few locations where wind power can be a major factor in electrical generation. Among these are the northern and central high plains regions from the face of the Rockies east to the 100th meridian, the Pacific Northwest, and the New England coast. The best winds are located in southern Wyoming (average power 400 watts per square meter of blade surface); the worst are in southern Georgia and Alabama, southern Arizona, and the Los Angeles basis, each with an average power of 50 watts or less per square meter of blade surface.

windscour the removal of soil particles from bared areas by wind action. Wind selectively removes only the fine materials from newly bared areas; hence a characteristic MICROTOPOGRAPHY develops, with shallow basins ("wind-scoured depressions") separated by low ridges of soil, 5–15 cm (2–6 in) high on which plants maintain a foothold. The basins are normally floored with coarse, loose materials. Windscour is characteristic of overgrazed ranges in moderate to strongly windy locations, such as ridgetops. Overgrazing bares the soil in the first place, and the associated trampling breaks up the large soil particles into finer ones, allowing the wind to take them away (*see* TRAMPLING DISPLACEMENT). *See also* LAG GRAVEL; DESERT PAVEMENT.

windthrow in forestry, the uprooting of trees through the action of wind; also, a tree that has been uprooted in this manner. Windthrow is a major factor in natural forest SUCCESSION, both on a small scale (*see* GAP PHASE REPLACEMENT) and, as a result of particularly strong storms such as HURRICANES, on a forest-wide basis. Its effect differs significantly from that of WILDFIRE in that wildfire primarily affects the UNDERSTORY and often leaves the OVERSTORY intact, while windthrow removes overstory trees almost exclusively, with the effect of releasing the understory for increased growth rather than destroying it. Several factors affect a tree's susceptibility to windthrow. Old trees have weakened root systems due to their lowered overall vitality. At the same time, they are usually larger than average, and present more surface to the wind, a pair of factors that combine to make OVERMATURE TIMBER particularly vulnerable to windthrow. CONIFERS generally have less resistance to windthrow than do HARDWOODS due to their shallower root system. Some SPECIES, notably the DOUGLAS FIR and most spruces, are especially susceptible. Finally, timber harvest—particularly clearcutting (*see* CLEARCUT)—can leave those trees that are not harvested susceptible to windthrow because trees that are used to being buffered by other trees, and have developed their root systems accordingly, are suddenly exposed to the full force of the wind. This effect is often increased by poorly planned CUT BLOCK borders that funnel the winds into narrow openings, thus increasing both their speed and their turbulence and making the trees at the bottleneck extremely vulnerable. *See also* ROOT WAD.

winter range in wildlife management, the portion of an animal's RANGE that it uses during the winter. Winter range is generally much smaller than summer range, resulting in crowding. It is generally also lower in elevation, although this is not necessary (mountain goats, for example, often go higher in the mountains in the winter, rather than lower, in order to eat the vegetation exposed where the wind has blown the snow cover off of the ridges). Winter range is part of the CRITICAL HABITAT of any given SPECIES, and must be preserved in order to preserve the species.

WIPP *See* WASTE ISOLATION PILOT PROJECT.

wise-use movement a loosely structured association of organizations and individuals opposed to governmental ownership and control of natural resource lands. Participants in the movement consist of the owners of large acreages of rangelands and forest lands in the West; inholders of acreage within parks, forests, and other public lands; and those involved in far-right conservative politics. The term "wise use" that has been appropriated by the movement is meant to suggest that environmental laws, such as the ENDANGERED SPECIES ACT which can limit certain land uses and resource management practices, are extreme and that the unregulated environmental actions of private landowners would result in a "wiser" form of conservation.

wolf reintroduction *See* REINTRODUCTION OF SPECIES.

wood fiber *See* PULP.

woody plant any plant, such as a tree or a shrub, whose tissues include a significant amount of secondary xylem (*see* SECONDARY GROWTH; XYLEM). Woody plants are characterized by the presence of bark on their stems, which are usually stiffer than the stems of plants such as HERBS that show no secondary growth. All woody plants are PERENNIALS, and most are relatively long-lived. The structural strength of secondary xylem ("woody tissue") allows woody plants to reach much greater size than nonwoody plants, although not all of them take advantage of this ability. (Many heaths [Ericaceae], for example, are considerably less than tree size, and some are only a few inches tall.)

woolly adelgid an insect pest found in coniferous forests that in certain areas has greatly damaged balsam firs and hemlock trees. The adelgid has become a serious problem in recent years in forest areas with high levels of AIR POLLUTION. *See also* FOREST DECLINE AND PATHOLOGY.

working circle in Forest Service terminology, an area of the forest of relatively uniform TOPOGRAPHY and SPECIES mix that is large enough to support at least one forest-based industry (such as a logging firm). The working circle was traditionally employed in timber harvest planning. The ALLOWABLE CUT was calculated within each working circle, based on a single SILVICULTURAL PRESCRIPTION or a set of interrelated prescriptions, and the allowable cut for the forest as a whole was treated as the sum of the cuts for the various working circles within the forest. Ranger-district boundaries—and even forest boundaries—were not usually considered in defining the working circle, which was determined by a combination of silvicultural and economic considerations rather than administrative divisions. The concept is now considered obsolete, and the working-circle plan (the *working plan*) has been replaced by the UNIT PLAN and, more recently, by the FOREST PLAN. *See also* TIMBERSHED.

workplace pollution *See* INDOOR POLLUTION.

World Bank specialized agency of the United Nations formally organized in 1945. The bank makes loans to member nations or to private companies with a guarantee from a member nation. The loans are intended to encourage international trade, increase productivity, or reduce debt between nations. The World Bank is comprised of five closely associated institutions: the International Bank for Reconstruction and Development; the International Development Association; the International Finance Corporation; the Multilateral Investment Guarantee Agency; and the International Centre for Settlement of Investment Disputes. The bank, which is self-sustaining, provides almost $30 billion in loans annually, striving to help "each developing country onto a path of stable, sustainable, and equitable growth." To provide the capital for loans, the bank sells international bonds and receives funds from its member nations in an amount based on the relative strength of the members' economy. Although its policies have been subject to a great deal of criticism on environmental and social justice matters, the bank works to ensure that all projects protect natural resources, strengthen private businesses, support government infrastructure, provide for long-term financial planning, and contribute to a reduction in overall poverty by creating a stable social structure. Headquarters: 1818 H Street NW, Washington, DC 20433. Phone: (202) 477-1234. Website: www.worldbank.org.

World Climate Research Programme *See* INTERGOVERNMENTAL PANEL ON CLIMATE CHANGE; GLOBAL CHANGE.

World Environment Center environmental coordinating body, founded in 1974 to act as a central clearinghouse for information concerning environmental protection worldwide. Sponsored originally by the United Nations Environment Programme, the center promotes networking among environmental groups all over the planet, gathers information and makes it available, and sponsors symposia to increase dialogue among all those concerned with environmental problems. Address: 419 Park Avenue South, Suite 1800, New York, NY 10016. Phone: (212) 683-4700. Website: www.wec.org.

World Heritage Sites a program established in 1972 (*see* STOCKHOLM CONFERENCE ON ENVIRONMENT) and administered by UNESCO (UNITED NATIONS EDUCATIONAL, SCIENTIFIC, AND CULTURAL ORGANIZATION) that identifies and lists natural and cultural sites of worldwide significance. There are now 500 such sites, including the Grand Canyon in the United States, the Great Wall in China, and Timbuktu in Mali. Presently there are 18 World Heritage Sites in the United States.

World Heritage Sites in the United States (with date of designation)

Mesa Verde National Park, Colorado (1978)
Yellowstone National Park, Wyoming (1978)
Grand Canyon National Park, Arizona (1979)
Everglades National Park, Florida (1979)
Independence Hall, Philadelphia, Pennsylvania (1979)
Redwood National Park, California (1980)
Mammoth Cave National Park, Kentucky (1981)
Olympic National Park, Washington (1981)
Cahokia Mounds State Historic Site, Illinois (1982)
Great Smoky Mountains National Park, North Carolina–Tennessee (1983)
La Foraleza and San Juan Historic Site, Puerto Rico (1983)
Statue of Liberty, New York (1984)
Yosemite National Park, California (1984)
Chaco Culture National Historical Park, New Mexico (1987)
Monticello and University of Virginia, Virginia (1987)
Hawaii Volcanoes National Park, Hawaii (1987)
Pueblo of Taos, New Mexico (1992)
Carlsbad Caverns National Park, New Mexico (1995)

World Resources Institute a policy research center created in 1982 to provide information, ideas, and solutions for global environmental problems. Itsenvironmental concerns are to halt human-caused climate change, reverse ecosystem degradation, guarantee public access to information on natural resources and the environment, and encourage policies and practices that increase prosperity while conserving materials and reducing waste. The institute publishes a variety of books, case studies, and reports. Address: 1709 New York Avenue NW, Washington, DC 20006. Phone: (202) 638-6300. Website: www.wri.org.

World Scientists' Warning to Humanity a statement issued in 1992 and signed by 1,700 notable scientists, including a majority of all living Nobel laureates in science, warning "all humanity" that "a great change in our stewardship of the earth . . . is required, if vast human misery is to be avoided and our global home on this planet is not to be irretrievably mutilated." The warning states that only "one or a few decades remain before the chance to avert the threats we now confront will be lost, the prospects for humanity immeasurably diminished." A folder containing the text of the document and a selected list of signers is available from the UNION OF CONCERNED SCIENTISTS.

World Trade Organization (WTO) an international body set up in 1994 by the 120 nations of the "Uruguay Round" of the GENERAL AGREEMENT ON TARIFFS AND TRADE. The WTO has the status of a major international organization (such as the United Nations and the WORLD BANK) with the power to oversee and enforce trade agreements. Widely supported by internationalists, free-market economists, major transnational industries, and most governments, the WTO is nevertheless viewed with suspicion by human rights, environmental, and labor organizations in the United States. A coalition of such groups disrupted a WTO meeting in Seattle in 1999, calling attention to the potential for environmental and human rights abuses by large corporations, which would be the major beneficiaries of lowered trade barriers. *See also* NORTH AMERICAN FREE TRADE AGREEMENT.

Worldwatch Institute a research organization created to advise policy makers and the public about global

World Scientists' Warning to Humanity

INTRODUCTION Human beings and the natural world are on a collision course. Human activities inflict harsh and often irreversible damage on the environment and on critical resources. If not checked, many of our current practices put at serious risk the future that we wish for human society and the plant and animal kingdoms, and may so alter the living world that it will be unable to sustain life in the manner that we know. Fundamental changes are urgent if we are to avoid the collision our present course will bring about.

THE ENVIRONMENT The environment is suffering critical stress:
The Atmosphere Stratospheric ozone depletion threatens us with enhanced ultraviolet radiation at the earth's surface, which can be damaging or lethal to many life forms. Air pollution near ground level, and acid precipitation, are already causing widespread injury to humans, forests, and crops.
Water Resources Heedless exploitation of depletable ground water supplies endangers food production and other essential human systems. Heavy demands on the world's surface waters have resulted in serious shortages in some 80 countries, containing 40 percent of the world's population. Pollution of rivers, lakes, and ground water further limits the supply.
Oceans Destructive pressure on the oceans is severe, particularly in the coastal regions which produce most of the world's food fish. The total marine catch is now at or above the estimated maximum sustainable yield. Some fisheries have already shown signs of collapse. Rivers carrying heavy burdens of eroded soil into the seas also carry industrial, municipal, agricultural, and livestock waste—some of it toxic.
Soil Loss of soil productivity, which is causing extensive land abandonment, is a widespread by-product of current practices in agriculture and animal husbandry. Since 1945, 11 percent of the earth's vegetated surface has been degraded—an area larger than India and China combined—and per capita food production in many parts of the world is decreasing.
Forests Tropical rain forests, as well as tropical and temperate dry forests, are being destroyed rapidly. At present rates, some critical forest types will be gone in a few years, and most of the tropical rain forest will be gone before the end of the next century. With them will go large numbers of plant and animal species.
Living Species The irreversible loss of species, which by 2100 may reach one-third of all species now living, is especially serious. We are losing the potential they hold for providing medicinal and other benefits, and the contribution that genetic diversity of life forms gives to the robustness of the world's biological systems and to the astonishing beauty of the earth itself.

Much of this damage is irreversible on a scale of centuries, or permanent. Other processes appear to pose additional threats. Increasing levels of gases in the atmosphere from human activities, including carbon dioxide released from fossil fuel burning and from deforestation, may alter climate on a global scale. Predictions of global warming are still uncertain—with projected effects ranging from tolerable to very severe—but the potential risks are very great.

Our massive tempering with the world's interdependent web of life—coupled with the environmental damage inflicted by deforestation, species loss, and climate change—could trigger widespread adverse effects, including unpredictable collapses of critical biological systems whose interactions and dynamics we only imperfectly understand.

Uncertainty over the extent of these effects cannot excuse complacency or delay in facing the threats.

POPULATION The earth is finite. Its ability to absorb wastes and destructive effluent is finite. Its ability to provide food and energy is finite. Its ability to provide for growing numbers of people is finite. And we are fast approaching many of the earth's limits. Current economic practices which damage the environment, in both developed and underdeveloped nations, cannot be continued without the risk that vital global systems will be damaged beyond repair.

Pressures resulting from unrestrained population growth put demands on the natural world that can overwhelm any efforts to achieve a sustainable future. If we are to halt the destruction of our environment, we must accept limits to that growth. A World Bank estimate indicates that world population will not stabilize at less than 12.4 billion, while the United Nations concludes that the eventual total could reach 14 billion, a near tripling of today's 5.4 billion. But, even at this moment, one person in five lives in absolute poverty without enough to eat, and one in ten suffers serious malnutrition.

No more than one or a few decades remain before the chance to avert the threats we now confront will be lost and the prospects for humanity immeasurably diminished.

WARNING We the undersigned, senior members of the world's scientific community, hereby warn all humanity of what lies ahead. A great change in our stewardship of the earth and the life on it is required, if vast human misery is to be avoided and our global home on this planet is not to be irretrievably mutilated.

WHAT WE MUST DO Five inextricably linked areas must be addressed simultaneously:

1. **We must bring environmentally damaging activities under control to restore and protect the integrity of the earth's systems we depend on.** We must, for example, move away from fossil fuels to more benign, inexhaustible energy sources to cut greenhouse gas emissions and the pollution of our air and water. Priority must be given to the development of energy sources matched to Third World needs—small-scale and relatively easy to implement.

   We must halt deforestation, injury to and loss of agricultural land, and the loss of terrestrial and marine plant and animal species.
2. **We must manage resources crucial to human welfare more effectively.** We must give high priority to efficient use of energy, water, and other materials, including expansion of conservation and recycling.
3. **We must stabilize population. This will be possible only if all nations recognize that it requires improved social and economic conditions, and the adoption of effective, voluntary family planning.**
4. **We must reduce and eventually eliminate poverty.**
5. **We must ensure sexual equality, and guarantee women control over their own reproductive decisions.**

The developed nations are the largest polluters in the world today. They must greatly reduce their overconsumption, if we are to reduce pressures on resources and the global environment. The developed nations have the obligation to provide aid and support to developing nations, because only the developed nations have the financial resources and the technical skills for these tasks.

Acting on this recognition is not altruism, but enlightened self-interest: whether industrialized or not, we all have one lifeboat. No nation can escape from injury when global biological systems are damaged. No nation can escape from conflicts over increasingly scarce resources. In addition, environmental and economic instabilities will cause mass migrations with incalculable consequences for developed and undeveloped nations alike.

Developing nations must realize that environmental damage is one of the gravest threats they face, and that attempts to blunt it will be overwhelmed if their populations go unchecked. The greatest peril is to become trapped in spirals of environmental decline, poverty, and unrest, leading to social, economic, and environmental collapse.

Success in this global endeavor will require a great reduction in violence and war. Resources now devoted to the preparation and conduct of war—amounting to over $1 trillion annually—will be badly needed in the new tasks and should be diverted to the new challenges.

A new ethic is required—a new attitude towards discharging our responsibility for caring for ourselves and for the earth. We must recognize the earth's limited capacity to provide for us. We must recognize its fragility. We must no longer allow it to be ravaged. This ethic must motivate a great movement, convincing reluctant leaders and reluctant governments and reluctant peoples themselves to effect the needed changes.

The scientists issuing this warning hope that our message will reach and affect people everywhere. We need the help of many.

We require the help of the world community of scientists—natural, social, economic, and political.
We require the help of the world's business and industrial leaders.
We require the help of the world's religious leaders.
We require the help of the world's peoples.
We call on all to join us in this task.

SPONSORED BY THE UNION OF CONCERNED SCIENTISTS, TWO BRATTLE SQUARE CAMBRIDGE, MA 02238 environmental problems and the connection between the world economy and its environmental support systems. The institute, founded in 1974 by LESTER BROWN, publishes reports on single issues as well as *State of the World* and *Vital Signs*, widely distributed and influential books exploring such topics as global warming, soil erosion, water shortages, renewable energy, deforestation, population growth, transportation, and ocean pollution. Address: 1776 Massachusetts Avenue NW, Washington, DC 20036. Phone: (202) 452-1999. Website: www.worldwatch.org.

World Wildlife Fund international conservation organization, founded in 1961 to protect endangered and threatened SPECIES of animals and plants (*see* ENDANGERED SPECIES; THREATENED SPECIES) and their HABITATS in all of the nations of the world, and to assist in the creation of natural reserves. The fund has an enviable reputation for technical expertise and a record of working with agencies, corporations and governments to keep their wildlife activities on a biologically sound footing. WWF has helped create and safeguard over 450 nature reserves and national parks worldwide. Membership (1999): 1.2 million. Address: 1250 24th Street NW, Washington, DC 20037. Phone: (202) 293-4800. Website: www.worldwildlife.org.

Wright, Frank Lloyd (1867–1959) American architect noted for designs that are respectful of nature and the surrounding landscape. Educated in civil engineering at the University of Wisconsin, Wright joined the Chicago architectural firm of Adler and Sullivan in 1887 and started his own office in 1893. Wright's architectural philosophy was based on the idea that a building should be "organic," evolving out of its natural surroundings. Known as the Prairie style, his architecture stood in stark contrast to the ornate neoclassic and Victorian models of the late 19th and early 20th centuries. Wright's building materials were chosen for their natural textures and colors as well as their structural properties, with interiors in an open layout in which one room flows into another. Also an innovator in building technology, he was the first architect to use precast, reinforced concrete blocks, air conditioning, indirect lighting, and panel heating. His attention to design features such as passive solar heat gain from south-facing windows and natural airflow placed him well ahead of his time. In the 1930s Wright founded the Taliesin Fellowship to support his apprentices. It was headquartered first at his world-famous house in Wisconsin (named Taliesin, which means "shining brow" in Welsh) and later at Taliesin West in Arizona as "winter quarters." Both structures remain and are open to public view. The best of Wright's work may also be seen at Fallingwater, a private house built in 1937 now owned by the Western Pennsylvania Conservancy in Bear Run Nature Preserve near Mill Run, Pennsylvania. Among his best-known public buildings is the circular Guggenheim Museum in New York City, completed in 1959. Wright's books include *An Autobiography* (1932, revised in 1943); *An Organic Architecture* (1939); *Genius and the Mobocracy* (1949); *Natural House* (1954); and *An American Architecture* (1955). Robert Twombly explored the architect's life and work in his book *Frank Lloyd Wright: His Life and His Architecture* (1979).

WSA *See* WILDERNESS STUDY AREA.

xenolith in geology, a rock included within a FORMATION to which it does not belong; for example, a piece of PERIODOTITE contained within a BASALT flow or an isolated piece of basalt in a SEDIMENTARY ROCK such as LIMESTONE or shale. Xenoliths usually result from contamination of the COUNTRY ROCK during its formation by imported materials, although in EXTRUSIVE ROCKS such as solidified LAVA flows they may instead be a part of the formation through which the rock was extruded that was torn off and carried along with the flow. A special type of xenolith, the *xenocryst,* is a small crystal that resembles a PHENOCRYST but is composed of materials not normally found in the surrounding rock (*see,* e.g., AMYGDULE). *See also* INCLUSION.

xeric site a growing site for trees or other vegetation that is dryer than normal, either because not enough rain falls or because the rain that does fall does not remain in the soil. Extremely well-drained, sandy soils can provide xeric characteristics in an otherwise mesic environment (see MESIC SITE), as can south-facing slopes from which the moisture evaporates rapidly after it has fallen. Sites that receive plenty of moisture, but which receive it in a form that is difficult for plants to use, are also usually referred to as xeric: examples include high alpine or far northern sites where most of the precipitation is in the form of snow (*see* TUNDRA BIOME), SALT MARSHES, and the acid waters of BOGS. *See also* ARID CLIMATE; DESERT BIOME; XEROPHYTE.

xeriscaping a water-saving landscaping technique that is increasingly used in low-rainfall areas of the United States. (*Xeros* is a Greek word meaning "dry".) In some southwestern cities up to 70% of water use during summer months is applied to landscaping. In past years new residents and even city governments tended to design landscapes and gardens in rapidly expanding urban regions such as Phoenix or Albuquerque—with precipitation rates of less than 10 inches a year (Hartford, Connecticut, by contrast, receives 44 inches per year) to mimic those in wetter climates with which they were familiar. The result has been not only a massive waste of increasingly scarce water, but also the creation of humid microclimates and pollen levels that can create the very health problems many newcomers to the southwest have sought to escape. Accordingly, since the 1980s many city governments throughout the arid regions of the western United States have turned to xeriscaping to reduce urban water use and have encouraged property owners to do the same. Typically, xeriscaping does away with the familiar bluegrass lawn as a primary landscaping element, substituting a smaller patch of native buffalo and grama grasses, or no lawn at all. Large rocks are used to provide design interest, and gravel beds with few or no plants in them are frequent. Xeriscaped trees and shrubs tend to be native, and with deep roots to withstand long, dry periods. Flowers are also chosen for low water requirements and drought resistance. Landscape designers emphasize, however, that xeriscaping is not "zero-scaping," that quite lush gardens and municipal plantings are possible. A key to xeriscaping success is the use of DRIP IRRIGATION systems rather than watering by sprinklers. Drip systems, first developed for kibbutz irrigation in Israel, deliver water directly to plant roots rather than allowing it to evaporate. Today some 40 states promote xeriscaping in one form or another to conserve water, but the practice is more common in areas with 20 inches or less annual precipitation, which includes all or parts of states west of the 100th meridian.

xerophyte in biology, a plant that is adapted to growth on a site where it receives little usable water (*see* XERIC SITE). Adaptations displayed by xerophytes include water-storage tissues, as in the cacti and other SUCCULENTS; small, waxy leaves with deeply indented STOMATA, to reduce EVAPOTRANSPIRATION; the presence of leaf hairs, which act to lower down the flow of air across the leaf surface and therefore reduce the EVAPORATION rate; and deep root system that are able to follow a sinking WATER TABLE well down into the soil.

xylem in botany, the tissue system of a VASCULAR PLANT through which water and minerals are transported upward from the roots to the leaves. The driving force is the evaporation of water from the leaves (*see* TRANSPIRATION). The conducting CELLS of the xylem, the *tracheary elements,* consists of two types. The first type, the *tracheid,* is an elongated cell whose walls contain numerous places, called *pits,* where the cell wall is reduced to a thin membrane through which water can pass. These cells overlap within the xylem, and the pits are concentrated in the overlapping sections. The second type, the *vessel element,* contains actual holes, as well as pits, in its end walls. These cells are joined end-to-end in long tubes called *vessels.* The xylem of GYMNOSPERMS contains only tracheids; ANGIOSPERMS normally have both tracheids and vessel elements. In order to conduct water, both tracheids and vessel elements must be essentially empty; they depend on vertical and horizontal belts ("strands" and "rays") of complete cells, also contained within the xylem, to take over the functions normally accomplished by the protoplasts. Xylem formed at the growing tips of plants, known as *primary xylem,* is formed from the aprical meristem (*see under* PRIMARY GROWTH). Xylem laid down along the stem of a plant, known as *secondary xylem,* is formed from the inside of the CAMBIUM LAYER just inside the bark. Both forms of xylem are stiffened by the presence of a compound known as LIGNIN; however, secondary xylem contains much more lignin than does primary xylem, so that plants that produce secondary xylem are much more rigid than those that do not, and are able to grow to much greater heights (*see* WOODY PLANT).

yardang a rounded furrow or groove in the earth caused by wind EROSION. Yardangs tend to occur in groups and to occupy slightly raised places in otherwise flat terrain. They are oriented in the direction of the PREVAILING WIND. *See also* VENTIFACT.

yarding in logging and forestry, the transport of log from the site on which a tree was felled (*see* FELLING) to the LANDING where it is to be loaded onto a truck or train. Several different methods of yarding (*yarding systems*) are in use, and one of the most important aspects of planning for a timber harvest is to determine which system should be employed for each CUT BLOCK. The wrong choice may do irreparable damage to the soil through EROSION and compaction. (*Compaction* here refers to the mechanical compression of the soil from the weight of the yarding equipment and/or the logs; it decreases the pore space between the soil particles, making it difficult for plant roots to penetrate and reducing the soil's water-holding capacity, leading to greatly reduced plant growth and vitality.) Factors to be considered include the steepness and erodibility of the site; the ASPECT; the type of harvest method to be used (e.g., CLEARCUT, SHELTERWOOD SYSTEM, etc.); the distance from the farthest stump to the landing; and the character of the UNDERSTORY (that is, whether it is composed of WOODY PLANTS or HERBS, whether or not it will need to be cleared for REFORESTATION, etc.). For some methods of yarding, *see* SKIDDING; HIGH-LEAD LOGGING; SKYLINE LEAD.

yearling an animal in its second year of life; that is, one that has passed its first birthday but not yet reached its second birthday. The term is most commonly used for LIVESTOCK and for large game HERBIVORES such as deer and elk (a "yearling calf", a "yearling colt", a "yearling fawn", etc.).

Yellowstone National Park the world's first national park, set aside in 1872 by an act of Congress. Located in northwestern Wyoming Yellowstone is best known for its geysers (such as Old Faithful) and other geothermal features that can be found in the central section of the park, which is a flattish CALDERA 1,000 square miles in area. Yellowstone Lake (320 feet deep) lies within the caldera and feeds the river that gave Yellowstone its name, since it flows through a canyon of yellow stone. Around the caldera, mountains rise steeply, with the highest (Eagle Peak) at 11,358 feet. Yellowstone was first reported by a scout for the Lewis and Clark Expedition, John Colter, who claimed to have traversed the Yellowstone area sometime between 1806 and 1810. The most reliable early accounts of the area, however, were by Jim Bridger, a mountain man who explored Yellowstone in the 1830s. His stories of the geysers, giant birds (trumpeter swans), huge grizzly bears and wolves were widely disbelieved until a scientific expedition into Yellowstone was organized in 1869 that confirmed the region's astonishing attributes. Today millions visit the park, leading to many management difficulties, especially in terms of wildlife preservation. Grizzly bears became nearly extinct in the park, as did trumpeter swans, but both are now on the increase; and wolves have been reintroduced in a still-controversial program. A devastating fire swept through the park in 1988, although its ecological consequences are now thought to have been salutary, with a profusion of wildflowers and new growth on the burned-over land. A current management issue at the park is the increasing use of snowmobiles in the winter, which disturbs wildlife and produces serious noise pollution. With regard to long-range ecological planning, park managers now see Yellowstone as part of a larger region called "the Greater Yellowstone Ecosystem," which, at 21,000 square miles, is six times larger than the park itself and includes surrounding national forests, wildlife refuges, and private lands in three states. By planning at the ecosystem level, the protection of endangered and threatened species is much more likely to be successful. *See also* ENDANGERED SPECIES ACT; NATIONAL PARK SYSTEM.

Yosemite National Park another of the "crown jewels" in the National Park System. Yosemite, set aside in 1890 and located in the Sierra Nevada mountains of California, is one of the most popular national parks, with 3.7 million recreation visits a year. The central feature of the park is the incomparably beautiful Yosemite valley, a gardenlike enclave, carved by glaciers, within the rugged mountains surrounding it. Nine waterfalls descend into the valley, including 2,435-foot Yosemite Falls, highest in America. Yosemite Valley, described by THEODORE ROOSEVELT as the "most beautiful place on earth," has become so burdened by overuse, including severe air pollution from tourist automobiles, that park authorities have now devised a plan to eliminate private automobiles altogether. Aside from Yosemite Valley, the park contains a number of other scenic wonders, including giant redwoods at Mariposa Grove; Tuolumne Meadows, which JOHN MUIR described as a "series of flowery lawns"; and high-peak vistas, such as Glacier Point. The now-flooded HETCH HETCHY valley also lies within the park. Muir and his fledgling SIERRA CLUB fought to preserve the Hetch Hetchy, but in vain. Despite the fact that it rivaled Yosemite Valley in natural beauty, it became a reservoir in 1923. The protracted controversy, which started in 1901, led directly to the establishment of the NATIONAL PARK SERVICE to provide professional guardianship and interpretative programs for outstanding natural areas.

Yucca Mountain a nuclear waste storage project located 100 miles northwest of Las Vegas, Nevada. The Department of Energy began studies at the site in 1993, viewing it as a likely area for the permanent burial of nuclear material under the Nuclear Waste Policy Act. As an apparently stable geological formation, with no circulating groundwater, a deep mine in the mountain could, planners said, contain nuclear waste safely for thousands of years. The mine, bored into rock approximately 1,400 feet deep, would consist of a 100-mile network of tunnels that could eventually accommodate 70,000 tons of high-level waste (that is, "hot" nuclear material with a long half-life), after which it would be closed. The opening date was originally scheduled for 1998 but was postponed until 2010, and the site is still embroiled in controversy. Although more than $1.5 billion has been expended on the project, environmental scientists still challenge government estimates of the stability of the formation and worry as well about the leakage of nuclear waste into groundwater strata. Moreover, Yucca Mountain is a sacred site of the Shoshone Indians, who have asserted that the tribe has never ceded its land to the United States and have consistently refused any settlement for their territory. *See also* NUCLEAR WASTE AND CLEANUP; WARD VALLEY; WASTE ISOLATION PILOT PROJECT.

YUM in forestry and logging, an acronym for *yard unmerchantable material,* that is, transport tops and limbs, culls (*see* CULL MATERIAL), and all other cut material to the LANDING regardless of whether or not it can be sold. The principal purpose of YUM yarding is to reduce the fuel buildup on the site (*see* FUEL) and to prepare the site for rapid reforestation. It also encourages more complete utilization of the cut material, as the culls and SLASH, once yarded, can be hauled away for firewood, chip manufacture, or other uses that do not require merchantable-quality logs.

Z

zebra mussel *See* MUSSEL.

zero cut a forest-management policy option that would eliminate all further commercial logging in national forests and other federal lands. Bipartisan legislation was introduced in 1997 (the National Forest Protection and Restoration Act—NFPRA) to cancel existing timber sales in roadless areas; prohibit new timber sales and phase out existing contracts in two years; redirect forest industry subsidies to retrain loggers displaced by the zero-cut policy; support ecological restoration research in devastated CLEARCUT areas; and subsidize the development of nonwood alternatives to paper and construction materials. Fiercely opposed by the forest products industry and their allies in government, the zero cut option has only slowly gained adherents. By early 2000 the NFPRA had gained only 71 cosponsors in the House. Environmental organizations, some of them once opposed to zero cut as too radical, are now lining up behind the policy, as is a powerful coalition of churches and religious organizations—the Religious Campaign for Forest Conservation. It is likely that a form of the zero-cut option will effectively take place on federal lands, if not through legislation then through barring the development of new logging roads, which are required so that areas not yet cut over can be reached. *See also* SALVAGE LOGGING AMENDMENT.

zero discharge in pollution-control technology, the total elimination of all traces of a particular POLLUTANT from the EFFLUENT of a particular source. Zero-discharge standards are difficult to implement due to the greater and greater effort required to remove each additional increment of a pollutant as its level in the ENVIRONMENT approaches zero. At some point, the environmental gain from removing more of the substance usually will be offset by the environmental harm caused by generating the proportionately large amounts of energy or producing the machinery and chemical substances needed for the removal process. Thus, zero discharge standards are appropriate only for persistent toxic chemicals (*see* PERSISTENT COMPOUND; TOXIC SUBSTANCE) that are environmentally harmful in very low concentrations. They have been incorporated as GOALS into most pollution-control legislation since the passage of the CLEAN WATER ACT amendments of 1972, but have not actually been implemented as standards except in a few cases, notably those concerning persistent PESTICIDES such as TOXAPHENE, MIREX, and DDT. *See also* TOXIC SUBSTANCES CONTROL ACT; GREAT LAKES WATER QUALITY AGREEMENT.

zero emission in general, the absence of any discernible air pollutants from vehicles, factories, or other sources. Specifically, the term applies to the "zero emission requirement" in California law, enforced by the California Air Resources Board (CARB), mandating that 10% of the state's automobile fleet be composed of "zero emissions vehicles" by 2003. While the state has allowed partial credit to HYBRID-ELECTRIC CARS in meeting the requirement, essentially a true zero-emission vehicle would have to be fully powered by electricity or a HYDROGEN FUEL CELL. Because California comprises such a large proportion of the new-car market, its zero emission standard has provided an important impetus in reducing dependence on gasoline in order to reduce AIR POLLUTION. Critics of the zero emission requirement maintain that ELECTRIC VEHICLES (EVs) simply move the source of pollution from the tailpipe to an electric utility smokestack. While CARB admits that no vehicle is truly emission-free, research has shown that on a mile-for-mile basis, power plant pollutant emissions that result from charging EV batteries are 10 times lower that the pollutant emissions from a conventional vehicle itself. And in urban areas (where the cars are, but power plants are not), EVs reduce carbon monoxide by 99%, hydrocarbons by 98%, and smog-causing nitrogen oxides by 92%.

Zero Population Growth (ZPG) population-control organization, founded in 1968 to promote a stable world population through public education, media outreach, and the lobbying of world governments. It prepares resource and teaching materials; collects, collates and disseminates statistics; and bestows annual awards for service to population stabilization. Membership (1999): 56,000. Address: 1400 16th Street NW, Suite 320, Washington, DC 20036. Phone: (202) 332-2200. Website: www.zpg.org.

zeroth law of thermodynamics *See* THERMODYNAMICS, LAWS OF.

zinc the 30th ELEMENT of the ATOMIC SERIES, ATOMIC WEIGHT 65.38, chemical symbol Zn. A silvery, moderately reactive metal with a relatively low melting point (419.5°C, or 787°F), zinc is chemically related to CADMIUM and MERCURY. It reacts with OXYGEN and CARBON DIOXIDE in the presence of water vapor to form a thin, tough bluish-white nonreactive coating that protects it against further corrosion, and is thus used to protect other metals, particularly iron, from OXIDATION, a process known as *galvanization.* Combined with copper, it forms brass, a tough, resilient ALLOY that was known to the early civilizations in the Middle East. Environmentally, zinc is a TRACE ELEMENT that is necessary in the diets of both plants and animals. It is found in the structure of numerous ENZYMES, where it apparently serves as an activator of the enzyme's function. In plants, zinc deficiencies show up principally in leaf development. In animals, including humans, zinc deficiencies can cause fetal damage, dwarfism, loss of hair, and in extreme cases the development of open sores on the skin. In both plants and animals, zinc deficiencies also lead to retarded maturation rates. Too much zinc, however, is also dangerous. In humans, the symptoms of zinc poisoning resemble those of a severe case of influenza, including generalized aching, fever, nausea and vomiting, weakness, and frequent coughing. It is listed as a PRIORITY POLLUTANT by the ENVIRONMENTAL PROTECTION AGENCY.

zonal soil any soil with well-developed HORIZONS (*see* SOIL: *the soil profile*). The development of zonal soils is regulated almost completely by climate and time; therefore, they can only develop on well-drained, relatively flat locations where the BEDROCK is far enough below the surface that it has little or no influence on the soil structure. Most zonal soils are good agricultural soils. *Compare* AZONAL SOIL; INTRAZONAL SOIL. *See also* SOIL: *classification of soils.*

zone of aeration *See* UNSATURATED ZONE.

zone of saturation *See* WATER TABLE.

zoning the principal means by which municipalities control the use of land. *See also* LAND USE PLANNING; URBAN SPRAWL.

zoo a public park or garden devoted to the display and conservation of (mainly) exotic animals. In recent years zoos have begun to play an increasingly significant role in ecological education and the preservation of biodiversity. Such features as monkey houses, originally used for their entertainment value for humans, have been replaced with exhibits that re-create the actual ecosystems in which animals live. With regard to ENDANGERED SPECIES, zoos have developed breeding programs and help to reintroduce endangered species into the wild. Notable U.S. zoos are located in New York, Chicago, Cincinnati, Detroit, Philadelphia, and San Diego, although zoos in many other cities have quite remarkable educational and endangered species programs.

zooplankton in limnology and oceanography, a planktonic ORGANISM (*see* PLANKTON) that obtains its food by feeding on other organisms rather than by PHOTOSYNTHESIS or a related process. Zooplankton may be either microscopic animals or PROTISTA. Ecologically, they are classed as first-order consumers (*see* CONSUMER; TROPHIC LEVEL; ECOLOGICAL PYRAMID) that serve as a link in the FOOD CHAIN between the microscopic AUTOTROPHS such as ALGAE and larger aquatic animals such as fish and crustaceans. *Compare* PHYTOPLANKTON.

BIBLIOGRAPHY

Abramovitz, Janet N., and Ashley T. Mattoon. "The Ongoing Threat to the World's Forests." *USA Today,* September 1999.

Ackland, Len. "The Day They Almost Lost Denver." *The Bulletin of the Atomic Scientists,* July–August 1999.

Adams, Jim, and Jim Detjen. *Warning: Toxic Waste.* Louisville, Ky.: A Louisville Courier-Journal Special Report, 1979.

Agras, Jean, and Duane Chapman. "The Kyoto Protocol, CAFÉ Standards, and Gasoline Taxes." *Contemporary Economic Policy,* July 1999.

Alford, Ross A., and Stephen J. Richards. "Global Amphibian Declines: A Problem in Applied Ecology." *Annual Review of Ecology and Systematics,* 1999.

Allen, Arthur. "Prodigal Sun." *Mother Jones,* March–April 2000.

American Geological Institute. *Dictionary of Geological Terms.* New York: Doubleday, 1962.

Andelman, David A. "The New Farm Economics." *Management Review,* December 1997.

Anderson, Kristine F. "Georgia's Dolphin Quest." *Defenders,* September–October 1991.

Anderson, Walter Truett. "There's No Going Back to Nature." *Mother Jones,* September–October 1996.

Andren, Anders W., Steven J. Eisenreich, et al. *Atmospheric Loadings to the Great Lakes.* Windsor: International Joint Commission, 1977.

Andrewartha, H. G. *Introduction to the Study of Animal Populations.* Chicago: University of Chicago Press, 1961.

Annual Report—Committee on the Assessment of Human Health Effects of Great Lakes Water Quality. Windsor: International Joint Commission (annual).

Applegate Lake Environmental Impact Statement. Portland: Army Corps of Engineers, 1976.

Applewhite, Philip, and Sam Wilson. *Understanding Biology.* New York: Holt, Rinehart, and Winston, 1978.

Ashworth, William. *The Carson Factor.* New York: Hawthorn, 1979.

———. *Nor Any Drop to Drink.* New York: Summit, 1982.

———. *Under the Influence: Congress, Lobbies, and the American Pork Barrel System.* New York: Dutton, 1981.

Asner, Gregory P., et al. "The Decoupling of Terrestrial Carbon and Nitrogen Cycles." *BioScience,* April 1997.

Assessment of Airborne Contaminants in the Great Lakes Basin Ecosystem. Windsor: International Joint Commission, 1980.

Ayers, Harvard, et al., eds. *An Appalachian Tragedy: Air Pollution and Tree Death in the Eastern Forests of North America.* San Francisco: Sierra Club, 1998.

Bailey, Robert G. *Descriptions of the Ecoregions of the United States.* Washington, D.C.: USDA Forest Service, 1995.

Bailey, Ronald, ed. *Earth Report 2000.* Published for the Competitive Enterprise Institute. New York: McGraw-Hill, 2000.

Baker, Beth. "Marine Mammal Protection Increased." *BioScience,* May 1995.

———. "New Federal Task Force Tackles Amphibian Troubles." *BioScience,* May 1999.

Baker, Linda. "Feeling the Heat." *E Magazine,* May–June 1998.

Baldwin, Margaret L., Mary F. Burdette, et al. *Coping with Oil Spills.* Watertown, N.Y.: St. Lawrence-Eastern Ontario Commission, 1983.

Ballard, Robert D. *Exploring Our Living Planet.* Washington, D.C.: National Geographic Society, 1983.

Batisse, Michael. "Biosphere Reserves: A Challenge for Biodiversity Conservation and Regional Development." *Environment,* June 1997.

Beaulieu, John D., and Paul W. Hughes. *Land Use Geology of Central Jackson County, Oregon.* Salem: Oregon Department of Geology and Mineral Resources, 1977.

Beier, Ann, et al. "A New Approach to Runoff-State Coastal Nonpoint Pollution Control Programs." *Journal of Soil and Water Conservation,* March–April 1994.

Bennett, Charles F. *Man and Earth's Ecosystems.* New York: Wiley, 1975.

Benson, Jim. "The State of the Land." *Conservation Voices,* April–May 1999.

Beuter, John H., Norman K. Johnson, et al. *Timber for Oregon's Tomorrow.* Corvallis: Oregon State University Press, 1976.

"Biosphere Reserves: What, Where, and Why?" *Focus* (American Geographic Society), spring 1989.

Birkeland, Peter W. *Pedology, Weathering and Geomorphological Research.* New York: Oxford, 1974.

Blake, Jane Cappiello. "Early Winter Bird Watching: The National Audubon Society Christmas Bird Count." *Country Journal,* November–December 1996.

Blanchard, Nanette. "The Quietest War." *E Magazine,* March–April 1998.

Blankenship, Karl. "The Chesapeake: A Bay Legacy." *American Forests,* autumn 1998.

Blaustein, Andrew R., and David B. Wake. "The Puzzle of Declining Amphibian Populations." *Scientific American,* April 1995.

Bloomfield, Molly M. *Chemistry and the Living Organism.* New York: Wiley, 1980.

Bohlen, Curtis C. "Protecting the Coast." *BioScience,* April 1990.

Bolgiano, Chris. *The Appalachian Forest.* Mechanicsburg, Pa.: Stackpole, 1998.

Boraiko, Allen A. "The Pesticide Dilemma." *National Geographic Magazine,* February 1980.

Borenstein, Seth. "Flash, Reaction, Disaster: Radiation Leaks at Japanese Nuclear Site." *Kansas City Star,* October 1, 1999.

Bosch, Xavier. "Spain Lets Latin America 'Repay' Debts by Protecting Environment." *Nature,* September 17, 1998.

Boudreaux, Richard. "Central Asia: Shrinking of Aral Sea Leaves Central Asians Suffering." *Los Angeles Times,* December 26, 1996.

Boughey, Arthur S. *Man and the Environment.* New York: Macmillan, 1971.

Bourne, Joel. "The Organic Revolution." *Audubon,* March–April 1999.

Boyle, Robert H. "Bringing Back the Chesapeake." *Audubon,* May–June 1999.

Bracken, Frank A. "Remarks." *Land and Water Conservation Fund Amendments and Historic Sites Reform Act,* U.S. Senate Committee Print. Washington: U.S. Government Printing Office, 1993.

Bracken, Susan, Louise Doran Bickram, et al. *The Canadian Almanac and Directory.* Toronto: Copp Clark Pitman Ltd. (annual).

Brown, Lester R., et al. *State of the World 1999.* New York: Norton, 1999.

———. *State of the World 2000.* New York: Norton, 2000.

———. *Vital Signs 1999.* New York: Norton, 1999.

Buchsbaum, Ralph, and Mildred Buchsbaum. *Basic Ecology.* Pittsburgh: Boxwood Press, 1957.

"Buffalo Commons Metaphor Is About Change." Fargo (N.D.) *Forum,* January 10, 1999.

Bunyard, Peter. "Eradicating the Amazon Rainforests Will Wreak Havoc on Climate." *The Ecologist,* March–April, 1999.

"Buried in Garbage." *Washington Times,* January 27, 1986.

Byrnes, Patricia. "Legislative Outlook: Dreary." *Wilderness,* summer 1995.

Callenbach, Ernest. *Ecology: A Pocket Guide.* Berkeley: University of California Press, 1998.

Campbell, I. C. "A Critique of Assimilative Capacity." *Journal of the Water Pollution Control Federation,* May 1981.

Canby, Thomas Y. "Water: Our Most Precious Resource." *National Geographic Magazine,* August 1980.

Candler, Julie. "The Mandate for Alternative Fuels." *Nation's Business,* June 1994.

Canning, Kathie. "Effluent Trading: Making It Work." *Pollution Engineering,* February 1999.

Cannon, Carl M. "U.S. Plans Cleanup of Waters." Baltimore *Sun,* February 19, 1998.

Capra, Fritjof. *The Web of Life.* New York: Anchor, 1996.

Carey, Alan, and Sandy Carey. "Symbol of Success." *National Wildlife,* August–September 1994.

Carlson, Alvar W. *The Spanish-American Homeland: Four Centuries in New Mexico's Río Arriba.* Baltimore: Johns Hopkins, 1990.

Carpenter, Betsy. "Serving up a Safer Food Supply." *U.S. News & World Report,* August 5, 1996.

Carrel, Chris. "The Clean Water Cause." *Sierra,* May–June 1999.

Carson, Rachel. *Silent Spring.* New York: Houghton Mifflin, 1962.

Carstensen, Vernon, ed. *The Public Lands: Studies in the History of the Public Domain.* Madison: University of Wisconsin Press, 1968.

Chalabi, Fadhil J. "OPEC: An Obituary." *Foreign Policy,* winter 1997–98.

Chameides, W. L., et al. "Ozone Pollution in the Rural United States and the New NAAQS." *Science,* May 9, 1997.

Charman, Karen. "Force Feeding Genetically Engineered Foods." *PR Watch,* Fourth Quarter 1999.

Chepesiuk, Ron. "A Sea of Trouble?" *Bulletin of the Atomic Scientists,* October 1997.

Chinn, Harvey, and Caroline S. Blesoe. "Internet Access to Ecological Information: The U.S. LTER All-Site Bibliography Project." *BioScience,* January 1997.

Churchill, Kent T. *Standard and Guidelines—Pacific Northwest Regional Plan.* Portland: USDA–Forest Service, 1981.

Clark, John O. E., and Stella Stiegeler. *The Facts On File Dictionary of Earth Science.* New York: Facts On File, 2000.

Clark, John W., Warren Viessman Jr., et al. *Water Supply and Pollution Control.* 3d ed. New York: Harper & Row, 1977.

Clarke, Wendy Mitman. "Vanishing Night Skies." *National Parks,* July–August 1999.

The Clean Air Act: A Briefing Book for Members of Congress. Washington: National Clean Air Coalition, 1981.

Clement, Thomas M., Jr., Glen Lopez, et al. *Engineering a Victory for Our Environment.* San Francisco: Sierra Club Special Publications, 1973.

Clepper, Henry. *Professional Forestry in the United States.* Baltimore: Johns Hopkins University Press, 1971.

Coghlan, Andy. "A Jab for Trees." *New Scientist,* December 12, 1998.

Cohn, Jeffrey P. "A Makeover for Rocky Mountain Arsenal." *BioScience,* April 1999.

A Compilation of Federal Laws Relating to Conservation and Development of Our Nation's Fish and Wildlife Resources, Environmental Quality and Oceanography. Washington: U.S. Government Printing Office, 1977.

Conservation Directory. Washington, D.C.: National Wildlife Federation (annual).

Contamination of Ground Water by Toxic Organic Chemicals. Washington: Council on Environmental Quality, 1981.

Conway, Steve. *Timber Cutting Practices.* 2nd ed. San Francisco: Miller Freeman, 1973.

Coppleman, Peter. *Protecting Roadless Lands in the National Forest Planning Process.* Washington, D.C.: The Wilderness Society, 1985.

Coram, Robert. "Water Worlds." *Audubon,* May–June 1995.

Corchado, Alfred, and Laurence Iliff. "Crackdowns, Global Economy Rapidly Changing Lives in Border Region." *Dallas Morning News,* August 6, 1998.

Corman, Rena. *Air Pollution Primer.* New York: American Lung Association, 1978.

Costanza, Robert, et al. "The Value of the World's Ecosystem Services and Natural Capital." *Nature,* May 15, 1997.

Couret, Christina. "Solving the Problem of Cell Tower Placement." *American City and County,* September 1999.

Covey, Bill, ed. *Environmental Analysis—Northern California Planning Area Guide.* San Francisco: USDA–Forest Service, 1976.

Craddock, Ashley. "Faulty Rods." *Mother Jones,* May–June 1994.

Crump, Kenny S., and Harry A. Guess. *Drinking Water and Cancer: Review of Recent Findings and Assessment of Risks.* Washington, D.C.: Council on Environmental Quality, 1980.

Culver, Alicia. "New Global Campaign to Exterminate Dirty Dozen Pesticides." *Multinational Monitor,* September 1985.

Culver, Alicia, and Rose Marie Audette. "Danger's in the Well." *Environmental Action,* March 1985.

Cunningham, Randy Duke. "Migratory Bird Hunting and Conservation Stamp Promotion Act [Remarks]." *Congressional Record,* Daily ed., July 16, 1998.

Cunninghame, Brian, et al. *Brothers Grazing Management Program Environmental Impact Statement.* Prineville, Ore.: Bureau of Land Management, 1982.

Curtis, Helena. *Biology.* New York: Worth, 1968.

Cushman, John H., Jr. "U.S. Parks Face Pressure to Ban Off-Road Vehicles." *New York Times,* February 20, 2000.

Cutler, M. Rupert, Peter Kirby, et al. *A Conservationist's Guide to National Forest Planning.* San Francisco: Sierra Club Special Publications, 1981.

Daily, Gretchen C., ed. *Nature's Services: Societal Dependence on Natural Ecosystems.* Washington: Island Press, 1997.

Daintith, John, ed. *The Facts On File Dictionary of Chemistry.* 3rd ed. New York: Facts On File, 1999.

Daniel, Glenda, and Jerry Sullivan. *The North Woods: A Sierra Club Naturalist's Guide.* San Francisco: Sierra Club Books, 1981.

Danielson, John A. *Air Pollution Engineering Manual.* Cincinnati, Ohio: National Center for Air Pollution Control, 1967.

Darling, F. Fraser, and John P. Milton, eds. *Future Environments of North America.* Garden City, N.Y.: Natural History Press, 1966.

Darwin, Charles. *The Origin of Species.* 1859. Reprint, New York: Mentor, 1958.

Davis, George D., et al. *The Lake Baikal Region in the Twenty-first Century: A Model of Sustainable Development or Continued Degradation.* Wadhams, N.Y.: Davis Associates, 1993.

Davis, Mike. *Ecology of Fear: Los Angeles and the Imagination of Disaster.* New York: Henry Holt, 1998.

Dean, William, ed. *Terms of the Trade.* 2nd ed. Eugene, Ore.: Random Lengths Publications, 1984.

DeBach, Paul. *Biological Control by Natural Enemies.* London: Cambridge University Press, 1974.

Denison, Richard A., and John F. Ruston. "Recycling Is Not Garbage." *Technology Review,* October 1997.

Derr, Thomas Sieger. "Global Eco-logic." *First Things,* February 2000.

Detwyler, Thomas R., ed. *Man's Impact on Environment.* New York: McGraw-Hill, 1971.

Devall, Bill, ed. *Clearcut: The Tragedy of Industrial Forestry.* San Francisco: Sierra Club Books and Earth Island Press, 1993.

DiPrimio, Juan C., et al. "Climate," *Environment,* March 1992.

Dilworth, J. R. *Log Scaling and Timber Cruising.* Corvallis, Ore.: O.S.U. Book Stores, Inc., 1971.

Disher, Brandy E. "Dissolving Medical Waste." *Environmental Health Perspectives,* July 1996.

Dobson, Andy P., et al. "Hopes for the Future: Restoration Ecology and Conservation Biology." *Science,* July 25, 1997.

Doig, Ivan. "The Murky Annals of Clearcutting." *Pacific Search,* December 1975.

Dorfman, Robert and Nancy S., eds. *Economics of the Environment: Selected Readings.* New York: W. W. Norton, 1972.

Dorr, John A., and Donald F. Eschman. *Geology of Michigan.* Ann Arbor: University of Michigan Press, 1970.

Downwind: The Acid Rain Story. Ottawa: Environment Canada, 1982.

Doyle, Paul. "The New Superfund: Will It Work this Time?" *State Legislatures,* February 1987.

Drew, Douglas. "Fear in Unknown Quantities." *Student Lawyer,* March 1985.

Ebenreck, Sara. "Measuring the Value of Trees." *American Forests,* July–August 1988.

Eblen, Ruth A., and William R. Eblen, eds. *The Encyclopedia of the Environment.* Boston: Houghton Mifflin, 1994.

Eicher, Don L., and A. Lee McAlester. *History of the Earth.* Englewood Cliffs, N.J.: Prentice-Hall, 1980.

"Eight States Urge Repeal of Fuel Mandate." *Chemical Market Reporter,* January 21, 2000.

Eisenberg, Evan. *The Ecology of Eden.* New York: Knopf, 1998.

El-Gawhary, Karim. "Delta Blues: The Nile." In Richard Swift, et al. *New Internationalist* (U.K.), November 1995.

Elliott, Carolyn. "My Holiday in a Giant Greenhouse." *Focus* (U.K.), February 1998.

Ellison, Lincoln, A. R. Croft, et al. *Indicators of Condition and Trend on High-Range Watersheds.* Washington: USDA-Forest Service, 1951.

Emmel, Thomas C. *Introduction to Ecology and Population Biology.* New York: Norton, 1973.

Englund, Will. "Troubled Waters in the Deepest Lake in the World." Minneapolis *Star Tribune,* January 16, 1994.

Environmental Impact Statement Guidelines. Rev. ed. Seattle: Environmental Protection Agency, 1973.

Environmental Management Strategy for the Great Lakes System. Windsor, Ont.: International Joint Commission, 1978.

"EPA and CAM Reach Agreement on Clean Air Act Rules." *Chemical Marketing Reporter,* August 26, 1996.

Erickson, Deborah. "Sea Sick." *Scientific American,* August 1992.

Evaluation of Remedial Measures to Control Nonpoint Sources of Water Pollution in the Great Lakes. Windsor, Ont.: International Joint Commission, 1977.

"Ex-Reagan Official Pleads in HUD Case." *New York Times Current Events Edition,* March 15, 1995.

Ezell, Jeff. "OSHA Supports New Whistleblower Legislation to Strengthen Protections for Workers." *Job Safety and Health Quarterly,* spring 1999.

Fairley, Peter. "Congress Dumps Delaney Clause." *Chemical Week,* July 31, 1996.

"Federal Republic of Germany." Washington, D.C.: *State Dept. Background Notes,* May 1987.

Feldman, David Lewis. "Revisiting the Energy Crisis: How Far Have We Come?" *Environment,* May 1995.

Fell, Nolan, and Peter Liss. "Can Algae Cool the Planet?" *New Scientist,* August 21, 1993.

Ferejohn, John A. *Pork Barrel Politics: Rivers and Harbors Legislation, 1947–68.* Stanford: Stanford University Press, 1974.

Fetto, John. "Home on the Organic Range." *American Demographics,* August 1999.

"Fill It Up with Premium Unleaded Electricity, Please." *Electrical Design,* November 2, 1998.

Filler, Martin. "The Cultivator." *The New Republic,* November 15, 1999.

Findley, Roger W., and Daniel A. Farber. *Environmental Law.* St. Paul: West Publishing Company, 1983.

Fite, Katherine V., et al. "Evidence of Retinal Light Damage in *Rana cascadae*: A Declining Amphibian Species." *Copeia* 4 (American Society of Ichthyologists and Herpetologists), 1998.

Flannery, Maura C. "Biotechnology and Bioengineering," *The American Biology Teacher,* June 1998.

Fletcher, Colin. *The New Complete Walker.* New York: Knopf, 1977.

Fogarty, David. *Great Lakes Toxic Hotspots: A Citizen's Action Guide.* Chicago: Lake Michigan Federation, 1985.

Forbes, Reginald D. *Woodlands for Profit and Pleasure.* Washington: American Forestry Association, 1971.

Foster, Robert J. *Geology.* 2nd ed. Chicago: Merrill, 1971.

Foth, Henry D. *Fundamentals of Soil Science.* 6th ed. New York: Wiley, 1978.

Fox, Matthew. *Creation Spirituality.* San Francisco: Harper, 1991.

Francl, Terry, et al. *The Kyoto Protocol and U.S. Agriculture.* Chicago: The Heartland Institute, 1998.

Franklin, Jerry F., and C. T. Dyrness. *Natural Vegetation of Oregon and Washington.* Portland: USDA-Forest Service, 1973.

Fred C. Hart Associates. *Technology for the Storage of Hazardous Liquids.* Albany: New York Department of Environmental Conservation, 1983.

Freeman, Jack. "Five Years Later: How Important Is an 'Earth Charter'?" *Earth Times,* April 1–15, 1997.

Friedrich, Robert L., and Robert B. Blystone. "Internet Teaching Resources for Remote Sensing and GIS." *BioScience,* March 1998.

Fukuta, Norihiko. "Cloud Seeding Clears the Air." *Physics World,* May 1998.

Funk & Wagnalls New Encyclopedia. New York: Funk & Wagnalls, 1975.

The Future Is Abundant. Arlington, Wash.: Tilth, 1982.

Gangloff, Deborah. "Ten Years and Counting." *American Forests,* winter 1999.

Gardner, Gary. "From Oasis to Mirage: The Aquifers That Won't Replenish." *World Watch,* May–June 1995.

———. "Preserving Global Cropland." In Lester R. Brown, ed. *State of the World 1997.* New York: Norton, 1997.

Garrett, Wilbur F., ed. *Energy: Facing Up to the Problem, Getting Down to Solutions.* Washington: National Geographic Special Publications, 1981.

Gedzelman, Stanley D. "Mysteries in the Clouds." *Weatherwise,* June–July 1995.

Getches, David H. *Water Law.* St. Paul: West Publishing Company, 1984.

Glavin, Terry. "Sea Change." *Canadian Geographic,* May–June 1999.

Glenn, Jim. "The State of Garbage in America." *BioCycle,* May 1998.

Glick, Daniel. "Having Owls and Jobs, Too." *National Wildlife,* August–September 1995.

Glysson, Eugene A., James R. Packard, et al. *The Problem of Solid-Waste Disposal.* Ann Arbor: University of Michigan Press, 1972.

Goehring, Jan, and Dawn Levy. "A Move in the Right Direction—TEA-21." *State Legislatures,* September 1998.

Goetting, Ann. "Ecofeminism." *Journal of Comparative Family Studies,* spring 1996.

Goldfarb, Theodore D., ed. *Sources: Notable Selections in Environmental Studies.* Guilford, Conn.: Dushkin, 1997.

Gore, Al. *Earth in the Balance: Ecology and the Human Spirit.* Boston: Houghton Mifflin, 1992.

Gould, Stephen Jay. "A Humongous Fungus Among Us." *Natural History,* July 1992.

Graham, Frank, Jr. *The Audubon Ark: A History of the National Audubon Society.* New York: Knopf, 1990.

Graeub, Ralph. *The Petkau Effect: Nuclear Radiation, People and Trees.* New York: Four Walls Eight Windows, 1992.

Gray, Gary, et al. "Carbon Debt: We All Have One." *American Forests,* summer 1996.

Gray, Peter. *Encyclopedia of the Biological Sciences.* New York: Reinhold, 1961.

Great Lakes Diversions and Consumptive Uses. Windsor: International Joint Commission, 1981.

Greer, Douglas F. *Business, Government, and Society.* New York: Macmillan, 1983.

Griffin, Donald R. *Bird Migration.* New York: Dover, 1974.

Griffin, Susan. *Women and Nature.* New York: Harper & Row, 1978.

Gross, M. Grant. *Oceanography: A View of the Earth.* 2nd ed. Englewood Cliffs, N.J.: Prentice-Hall, 1977.

Grove, Noel. "Superships: Giants that Move the World's Oil." *National Geographic Magazine,* July 1978.

Hahn, Benjamin W., J. Douglas Post, et al. *National Forest Resource Management.* Stanford: Stanford Environmental Law Society, 1978.

Halliday, Tim, and W. Ronald Heyer. "The Case of the Vanishing Frogs." *Technology Review,* May–June 1997.

Harbour, Ron. "What's the Alternative Power?" *Automotive Industries,* October 1999.

Hardwick, F. Russell, and Charles M. Knobler. *Chemistry: Man and Matter.* Waltham, Mass.: Ginn, 1970.

Hair, Jay D. "Rampant Consumption Poses Resource Threat." *International Wildlife,* September–October 1994.

Hall, Bert S. "Lynn Townsend White, Jr. (1907–1987)." *Technology and Culture,* January 1989.

Hall, Douglas John. *The Steward: A Biblical Symbol Come of Age.* Grand Rapids, Mich.: Eerdmans, 1990.

Halliday, Tim, and W. Ronald Heyer. "The Case of the Vanishing Frogs." *Technology Review,* May–June 1997.

Hamilton, Martha M. "World Bank Sets Up Fund to Assist Pollution-Control Effort." *Washington Post,* January 19, 2000.

Hansen, Chad. "The Big Lie: Logging and Forest Fires." *Earth Island Journal,* spring 2000.

Harden, Garrett. "The Tragedy of the Commons." In *Exploring New Ethics for Survival: The Voyage of the Spaceship Beagle.* New York: Viking, 1972.

Harley, Kelly L. "In Fighting a Dam Disaster, They Helped Make History." *National Wildlife,* December 1997–January 1998.

Hattis, Dale. "Drawing the Line: Quantitative Criteria for Risk Management." *Environment,* July–August 1996.

Haught, John F. *Science and Religion: From Conflict to Conversation.* New York: Paulist Press, 1995.

Hawkin, Paul, et al. *Natural Capitalism: Creating the Next Industrial Revolution.* Boston: Little, Brown, 1999.

Hayhurst, Chris. "WIPP Lash." *E Magazine,* January–February 1998.

Helvarg, David. "Blue Frontiers." *Audubon,* May–June 1995.

———. *The War Against the Greens.* San Francisco: Sierra Club Books, 1994.

Hendee, John C., George H. Stankey, et al. *Wilderness Management.* Washington: USDA–Forest Service, 1978.

Herman, Marc. "Rain Check." *Mother Jones,* March–April 1998.

Hertsgaard, Mark. *Earth Odyssey: Around the World in Search of Our Environmental Future.* New York: Broadway Books, 1998.

Hess, Glenn. "House Commerce Committee Endorses a Limited Superfund Reform Measure." *Chemical Market Reporter,* October 18, 1999.

———. "World Community Moves Ahead to Protect the Ozone Layer." *Chemical Market Reporter,* December 13, 1999.

Heydt, Bruce. "Washing the Starts from the Sky." *Country Journal,* November–December 1993.

Heyes, Anthony G., and Catherine Liston-Heyes. "Subsidy to Nuclear Power Through Price-Anderson Liability Limit." *Contemporary Economic Policy,* January 1998.

Hickman, Cleveland P., Jr., Larry S. Roberts, et al. *Biology of Animals.* St. Louis: C. V. Mosby, 1982.

High Plains Associates. *High Plains Ogallala Aquifer Regional Study.* Washington, D.C.: U.S. Department of Commerce, undated (early 1980s).

Highsmith, Richard M., Jr., J. Granville Jensen, et al. *Conservation in the United States.* Chicago: Rand McNally, 1969.

Hill, John R. *The Indiana Dunes—Legacy of Sand.* Bloomington: Indiana Department of Natural Resources, 1974.

Hill, John W. *Chemistry for Changing Times.* Minneapolis: Burgess, 1972.

Hine, Robert, ed. *The Facts On File Dictionary of Biology.* 3rd ed. New York: Facts On File, 1999.

Hinman, Norman D. "The Benefits of Biofuels." *Solar Today,* July–August 1997.

Honey, Martha S. "Treading Lightly? Ecotourism's Impact on the Environment." *Environment,* June 1999.

Horrigan, Alice. "Affordable by Design." *E Magazine,* July–August 1997.

Howard, William W. "Protecting Dolphins." *International Wildlife,* March–April 1996.

Hoyt, Joseph Bixby. *Man and the Earth.* Englewood Cliffs, N.J.: Prentice–Hall, 1967.

Hunt, Lee O. *Field Practice of Silviculture.* 2nd ed. Revised. Roseburg, Ore.: Lee O. Hunt, 1971.

Imlioff, Karl, W. J. Muller, et al. *Disposal of Sewage and Other Waterborne Wastes.* Ann Arbor: Ann Arbor Science, 1972.

Jackson, John N., and Fred A. Addis. *The Welland Canals: A Comprehensive Guide.* St. Catharines, Ontario: The Welland Canals Foundations, 1982.

"James Watt Sentenced for Role in Blocking HUD Investigation." *Wall Street Journal,* March 13, 1996.

Johnson, Daniel M., Richard R. Peterson, et al. *Atlas of Oregon Lakes.* Corvallis: Oregon State University Press, 1985.

Johnson, Jeffrey. "Hot, Bothered, and Nowhere to Go." *Environmental Action,* January 1986.

Johnston, David. "Former Interior Secretary Avoids Trial with a Guilty Plea." *New York Times,* January 3, 1996.

———. "Ex-Interior Chief Is Indicted in Influence-peddling Case." *New York Times Current Events Edition,* February 23, 1995.

Jones, Samuel B., Jr., and Arlene F. Luchsinger. *Plant Systematics.* New York: McGraw-Hill, 1979.

Kaiser, Jocelyn. "Wiping the Slate Clean at Biosphere 2." *Science,* August 19, 1994.

Karwatka, Dennis. "UNESCO World Heritage Sites." *Tech Directions,* May–June 1998.

Kay, Jane Holtz. "Carsick Country." *Sierra,* July–August 1999.

Kehoe, Keiki. *Unavailable at Any Price: Nuclear Insurance.* Washington: Environmental Policy Center, 1980.

Kelly, Tom. *How Many More Lakes Have to Die?* Ottawa: Canada Today, 1981.

Kendeigh, S. Charles. *Animal Ecology.* Englewood Cliffs, N.J.: Prentice-Hall, 1961.

Kerchove, Rene de. *International Maritime Dictionary.* 2nd ed. Princeton: Van Nostrand, 1961.

Ketchum, Richard M. *The Secret Life of the Forest.* New York: American Heritage, 1970.

King, Jonathan, and Matt Rothman. *Troubled Water.* Emmaus, Pa.: Rodale, 1985.

Kitman, Jamie Lincoln. "The Secret History of Lead." *The Nation,* March 20, 2000.

Klonsky, Karen, and Laura Tourte. "Organic Agricultural Production in the United States: Debates and Directions." *American Journal of Agricultural Economics,* September–October, 1998.

Knopman, Debra S., et al. "Civic Governance: Tackling Tough Land-Use Problems with Innovative Governance." *Environment,* December 1999.

Koek, Karin F., Susan B. Martin, et al. *Encyclopedia of Associations.* 23rd ed. Detroit: Gale Research, 1988.

Kuchenberg, Tom. *Reflections in a Tarnished Mirror: The Use and Abuse of the Great Lakes.* Sturgeon Bay, Wisc.: Golden Glow Publishing, 1978.

Lacayo, Richard. "The Brawl Over Sprawl." *Time,* March 22, 1999.

LaDuke, Winona. "Reclaiming Our Native Earth." *Earth Island Journal,* spring 2000.

Lambert, Maurice B. *Volcanoes.* Vancouver, B.C.: Douglas & Mcintyre, 1978.

Lancaster, John. "Dolphin Deaths Along Gulf States Spark Federal, State Investigations." *Underwater USA,* July 1990.

Landau, Norman J., and Paul D. Rheingold. *The Environmental Law Handbook.* New York: Ballantine, 1971.

Lapedes, Daniel N., ed. *McGraw-Hill Dictionary of Scientific and Technical Terms.* New York: McGraw-Hill, 1974.

"Latin American Focus: Environmental Group Focuses on Latin America." *Barracada Internacional,* January 1994.

Lee, Gary. "It Comes Down to Corn or Natural Gas." *Washington Post National Weekly,* August 15–21, 1994.

Leggett, Jeremy, and Peg Stevenson. "Fiddling While the World Burns." *Greenpeace,* November–December 1990.

Lehman, H. Jane. "Energy Efficient Labels for Your Home?" *Consumer's Research Magazine,* December 1995.

Lehr, Paul F., R. Will Burnett, et al. *Weather: A Guide to Phenomena and Forecasts.* New York: Golden Press, 1965.

Lehtinen, Ulla. "Environmental Racism: The U.S. Nuclear Industry and Native Americans." *Ecologist,* March–April 1997.

Lemonick, Michael D. "Rocky Horror Show." *Time,* November 27, 1995.

Leopold, Aldo. *A Sand County Almanac.* New York: Oxford, 1949.

Lerdau, Manuel, et al. "Plant Production and Emission of Organic Compounds." *BioScience,* June 1997.

Liesman, Steve. "Inside the Race to Profit from Global Warming." *Wall Street Journal,* October 19, 1999.

Linden, Eugene. "The Tortured Land." *Time,* September 4, 1995.

———. "Chicken of the Sea?" *Time,* March 4, 1996.

Little, Charles E. *The Dying of the Trees.* New York: Viking-Penguin, 1997.

Longman, Phillip J. "Who Pays for Sprawl? Hidden Subsidies Fuel the Growth of the Suburban Fringe." *U.S. News and World Report,* April 27, 1998.

Lovelock, James. *The Ages of Gaia: A Biography of Our Living Earth.* New York: Norton, 1988.

Lugar, Richard, and Joseph R. Biden, Jr. "Swapping Debt for Nature." *Christian Science Monitor,* August 5, 1998.

Lundgren, Lawrence. *Environmental Geology.* Englewood Cliffs, N.J.: Prentice-Hall, 1986.

Lutgens, Frederick K., and Edward J. Tarbuck. *The Atmosphere: An Introduction to Meteorology.* Englewood Cliffs, N.J.: Prentice-Hall, 1982.

Macan, T. T. *Ponds and Lakes.* London: Allen and Unwin, 1973.

Mackay, Katurah. "President Issues a Green Agenda." *National Parks,* March–April 1999.

MacKinnon, Ian. "It's Not Easy Being Green." *New Age Journal,* September–October 1999.

Management and Control of Heavy Metals in the Environment. London: CEP Consultants Ltd., 1979.

Managing the Public Rangelands. Washington: Bureau of Land Management, 1979.

Mangelsdorf, Martha, and Karen Freiberg. *We're Running Out.* Wichita, Kans.: A Wichita Eagle and Beacon Special Publication, 1979.

Marcus, Mary Brophy. "Tracking a Cancer Cure." *U.S. News and World Report,* June 1, 1998.

Margaronis, Maria. "The Politics of Food: As Biotech 'Frankenfoods' are Stuffed Down Their Throats, Consumers Rebel." *The Nation,* December 27, 1999.

Margulies, Lynn, and Dorion Sagan. *What Is Life?* New York: Simon & Schuster, 1995.

Mariner, Wastl. *Mountain Rescue Techniques.* Innsbruck, Austria: Oestereichischer Alpenverein, 1963.

Marsh, William M., and Jeff Dozier. *Landscape: An Introduction to Physical Geography.* Reading, Mass.: Addison-Wesley, 1981.

Martin, Russell. *A Story That Stands Like a Dam.* New York: Holt, 1989.

Mastny, Lisa. "Ozone Hole Is Largest Ever." *World Watch,* January–February, 1999.

Matthews, Anne. *Where the Buffalo Roam.* New York: Grove Weidenfeld, 1992.

Maugh, Thomas H., II. "Studies Renew Anxiety About Fading Ozone," *Los Angeles Times,* February 2, 1986.

McClenahen, John S. "NAFTA Works: And Not Just by Moving Production to Mexico." *Industry Week,* January 10, 2000.

McGinn, Anne Platt. "Oceans Are on the Critical List." *USA Today,* January 2000.

McGraw-Hill Encyclopedia of Science and Technology. 6th ed. New York: McGraw-Hill, 1987.

McManus, Reed. "What Money Can Buy: Paying for New Parklands with the Land and Water Conservation Fund." *Sierra,* January–February 1998.

McNamee, Thomas. *The Grizzly Bear.* New York: Knopf, 1984.

McNulty, Tim. "The New Battleground: Global ReLeaf Fights Global Warming." *American Forests,* winter 1999.

Mehta, Pushpa S., et al. "Bhopal Tragedy's Health Effects: A Review of Methyl Isocyanate Toxicity." *Journal of the American Medical Association,* December 5, 1990.

Menotti, Victor. "Forest Destruction and Globalisation." *The Ecologist,* May–June 1999.

Merchant, Carolyn. *The Death of Nature: Women, Ecology, and the Scientific Revolution.* San Francisco: Harper & Row, 1980.

Middleton, Beth. *Wetland Restoration.* New York: Wiley, 1999.

Miller, G. Tyler, Jr. *Environmental Science.* Annotated Instructor's Edition. Pacific Grove, Calif.: Brooks/Cole, 2001.

———. *Sustaining the Earth: An Integrated Approach.* 4th ed. Pacific Grove, Calif.: Brooks/Cole Publishing Company, 1999.

Miller, George H. "Toxic Substances in the Environment: A Resource Handbook of Toxic Substances." Louisville, Ky.: Toxic Substances Task Force of Jefferson County, 1980.

Milner, Mark. "EuroEye: Chernobyl Legacy Haunts European Bank." *Guardian* (U.K.), May 3, 1997.

"Minimata: Mercury's Crippling Legacy." *Multinational Monitor,* April 1987.

Mintzer, Irving. "Adrift on Stormy Seas." *Global Change,* December 1997.

Moll, Gary, and Sara Ebenreck, eds. "Shading Our Cities: A Resource Guide for Urban and Community Forests." Washington: Island Press, 1989.

Molles, Manuel C., Jr. *Ecology: Concepts and Applications.* Boston: WCB/McGraw-Hill, 1999.

Monastersky, Richard. "The Plankton-Climate Connection." *Science News,* December 5, 1987.

"Mono Lake Saved." *National Wildlife,* June–July 1995.

Moran, Joseph M., Michael D. Morgan, et al. *An Introduction to Environmental Sciences.* Boston: Little, Brown, 1973.

Mostert, Noel. *Supership.* New York: Knopf, 1974.

Motavalli, Jim. "The Ties that Blind." *E Magazine,* March–April 1997.

———. "Founding Father: Gaylord Nelson on Earth Day's Past, Present, and Future." *E Magazine,* March–April 1995.

Motavalli, Jim, and Jennifer Bogo. "Wild Ideas: 12 Trends for the New Millennium." *E Magazine,* January–February 2000.

Motluck, Alison. "Deadlier than the Harpoon." *New Scientist,* July 1, 1995.

Mountaineering: The Freedom of the Hills. Seattle: The Mountaineers, 1960.

"Muck and Morals." *The Economist,* February 9, 1995.

Mukerjee, Madhusree. "Persistently Toxic: The Union Carbide Accident in Bhopal Continues to Harm." *Scientific American,* June 1995.

Mulvaney, Kieran. "1998—The International Year of the Ocean: A Sea of Troubles." *E Magazine,* January–February 1998.

Nash, Roderick Frazier. *The Rights of Nature: A History of Environmental Ethics.* Madison: University of Wisconsin Press, 1989.

National Park Service. *The National Parks Index 1993.* Washington: U.S. Government Printing Office, 1993.

The Nation's Water Resources, 1975–2000. Washington: U.S. Water Resources Council, 1978.

Nebergall, William H., Henry F. Holtzclaw, Jr., et al. *General Chemistry.* 6th ed. Lexington, Mass.: Heath, 1980.

"New Standards Proposed for Radon in Drinking Water and Indoor Air." *Journal of Environmental Health,* March 2000.

Nicol, John. "Japan's Nuclear Nightmare." *Maclean's,* October 11, 1999.

Nivola, Pietro S., and Robert W. Crandall. "The Extra Mile: Rethinking Energy Policy for Automotive Transportation." *Brookings Review,* winter 1995.

Norback, Craig T., and Judith C., eds. *Hazardous Chemicals On File.* New York: Facts On File, 1988.

Northwest Conservation and Electric Power Plan. Portland, Ore.: Northwest Power Planning Council, 1983.

Norton, Rob. "Owls, Trees, and Ovarian Cancer." *Fortune,* February 5, 1996.

A Nuclear Waste Primer. Washington: League of Women Voters, 1980.

"NWF Fighting to Keep Wolves in Yellowstone." *National Wildlife,* April–May 1998.

O'Brien & Gere Engineers, Inc. *Siting Manual for Storing Hazardous Substances.* 2nd ed. Albany: New York Department of Environmental Conservation, 1984.

Officer, Charles, and Jake Page. *Tales of the Earth: Paroxysms and Perturbations of the Blue Planet.* New York: Oxford, 1993.

Oliver, John F., and John J. Hidore. *Climatology: An Introduction.* Columbus, Ohio: Merrill, 1984.

O'Neil, Robert. "Hitting the Roads." *American City & County,* November 1999.

Oppenheimer, Michael, and Robert Boyle. *Dead Heat: The Race Against the Greenhouse Effect.* New York: Basic Books, 1990.

Orr, David. "Zero Cut on Public Lands." *Earth Island Journal,* summer 1997.

Ott, Herman E. "The Kyoto Protocol: Unfinished Business." *Environment,* July–August 1998.

Owen, Oliver S. *Natural Resource Conservation: An Ecological Approach.* New York: Macmillan, 1971.

Padgett, Tim. "Mexico: The Young and the Restless." *Time,* July 21, 1997.

Parker, Robert. "Dioxin Clarification." *Pacific Northwest Weed Topics,* September 1983.

Paterson, I. H., and Clarence W. Olmstead. *North America.* 7th ed. New York: Oxford, 1984.

Patton, Suzanne Zolfo. "What Price MTBE." *E Magazine,* July–August 1998.

Pawlick, Thomas. "What's Killing Canada's Sugar Maples?" *International Wildlife,* January 1985.

Pearce, David. "Auditing the Earth: The Value of the World's Ecosystem Services and Natural Capital." *Environment,* March 1998.

Pearce, Fred. "A Dirty Business." *New Scientist,* January 23, 1999.

Pearl, Richard M. *How to Know the Minerals and Rocks.* New York: McGraw-Hill, 1955.

Peattie, Donald Culross. *A Natural History of Trees of Eastern and Central North America.* Boston: Houghton Mifflin, 1991.

Penny, Timothy J. "The Last Farm Bill?" *The American Enterprise.* July–August 1995.

Petit, Charles W. "Polar Meltdown." *U.S. News & World Report,* February 28, 2000.

Petulla, Joseph M. *American Environmental History.* San Francisco: Boyd and Fraser, 1977.

Phillips, David. "Long-Distance Pollution Soils Our Arctic." *Canadian Geographic,* May–June 1995.

Phillips, Kathryn. "Prying into the Lives of Frogs." *National Wildlife,* October–November 1999.

Podar, Mahesh, and Richard M. Kashmanian. "Charting a New Course." *Forum for Applied Research and Public Policy,* fall 1998.

Pollack, Andrew. "We Can Engineer Nature. But Should We?" *New York Times,* February 6, 2000.

Pollack, Susan. "Holding the World at Bay." *Sierra,* May–June 1996.

Popper, Deborah E., and Frank J. Popper. "The Buffalo Commons: Metaphor or Method." *Geographical Review,* October 1999.

———. "The Great Plains: From Dust to Dust." *Planning,* December 1987.

Powell, John Wesley. *Lands of the Arid Region of the United States.* 1879. Reprinted, with an introduction by T. H. Watkins. Boston: Harvard, 1983.

Powledge, Fred. "Changing Chesapeake." *BioScience,* June 1996.

The Principal Laws Relating to Forest Service Activities. Washington, D.C.: USDA-Forest Service, 1983.

Pring, George A., and Karen A. Tomb. *License to Waste: Legal Barriers to Conservation and Efficient Use of Water in the West.* New York: Matthew Bender, 1979.

Purdum, Todd S. "This Time, Los Angeles May Lose Water War." *New York Times,* June 15, 1998.

Purdy, Jedediah S. "Rape of the Appalachians." *American Prospect,* November–December 1998.

Quinn, Rebecca. "Floodplain Management Insures Against Losses." *Forum for Applied Research and Public Policy,* fall 1996.

"Radon in Drinking Water Study Finds Health Risk Is Small." *Journal of Environmental Health,* April 1999.

Rampton, Sheldon, and John Stauber. "Silencing Spring: Corporate Propaganda and the Takeover of the Environmental Movement," in Richard Hofrichter, ed. *Reclaiming the Environmental Debate: The Politics of Health in a Toxic Culture.* Cambridge, Mass.: MIT, 2000.

Raven, Peter H., Ray F. Evert, et al. *Biology of Plants.* 2nd ed. New York: Worth, 1976.

Reese, April. "Bad Air Days." *E Magazine.* November–December 1999.

Report on Great Lakes Water Quality. Windsor: International Joint Commission (annual).

Report to Congress on the Nation's Renewable Resources. Washington: USDA-Forest Service, 1979.

"Research Project Focuses on Improving Battery Materials." *Electronic Design,* May 3, 1999.

Richman, Michael. "Clearing the Way for Clean Water." *Water Environment & Technology,* March 1997.

Ritchie, David, and Alexander Gates. *Encyclopedia of Earthquakes and Volcanoes, New Edition.* New York: Facts On File, 2000.

Robinson, Glen O. *The Forest Service: A Study in Public Land Management.* Baltimore: Johns Hopkins, 1975.

Rogue River National Forest Management Planning: The Planning Process. Medford, Ore.: USDA-Forest Service, 1980.

Rogue River National Forest Proposed Land and Resource Management Plan. Medford, Ore.: USDA-Forest Service, 1987.

Rogue River National Forest Vegetation Management Plan for Site Preparation and Conifer Release. Medford, Ore.: USDA-Forest Service, 1984.

Rossano, A. T., Jr. *Air Pollution Control Guidebook for Management.* Stamford, Conn.: Environmental Science Service Division, E.R.A. Inc., 1969.

Rothenberg, Al. "Stempel Charged Up on NiMH Batteries." *Ward's Auto World,* March 1998.

Roush, G. Jon. "Halls of Fame." *Wilderness,* fall 1995.

Roy, Sergei. "The Big Bang." *Moscow News,* July 11–17, 1996.

Runte, Alfred. *National Parks: The American Experience.* Rev. ed. Lincoln: University of Nebraska Press, 1987.

Russell, Dick. "Health Problems in the Health Care Industry." *Amicus Journal,* winter 2000.

———. "Vacuuming the Seas." *E Magazine,* July–August 1996.

Ruttner, Franz. *Fundamentals of Limnology.* Toronto: University of Toronto Press, 1953.

Saab, Saleem S. "Move Over Drugs, There's Something Cooler on the Black Market—Freon." *Dickinson Journal of International Law,* spring 1998.

Sale, Kirkpatrick. "Deep Ecology and Its Critics." *The Nation,* May 14, 1988.

Samarrai, Fariss. "Little Alga Has Big Place in Global Climate." *Sea Frontiers,* spring 1995.

Sampson, R. Neil. *Farmland or Wasteland: A Time to Choose.* Emmaus, Pa.: Rodale, 1981.

Schemnitz, Sanford D., ed. *Wildlife Management Techniques Manual.* Washington, D.C.: The Wildlife Society, 1980.

Schorr, Daniel. "Blowing the Whistle." *New Leader,* November 4–18, 1996.

Schultz, Marilyn Spigel, and Vivian Loeb Kasen. *Encyclopedia of Community Planning & Environmental Management.* New York: Facts On File, 1984.

Schwarz, Charles F., Edward C. Thor, et al. *Wildland Planning Glossary.* Berkeley: USDA-Forest Service, 1976.

Seideman, David. "An Unlikely Friend of Forests." *Audubon,* March 1998.

Semple, Alison. "Growth of a Green Network." *The Geographical Magazine,* February 1995.

Seneca, Joseph J., and Michael K. Taussig. *Environmental Economics.* Englewood Cliffs, N.J.: Prentice-Hall, Inc., 1974.

Shabecoff, Philip. "After Decades of Deception, A Time to Act." in Harvard Ayers et al., *An Appalachian Tragedy: Air Pollution and Tree Death in the Eastern Forests of North America.* San Francisco: Sierra Club Books, 1998.

———. *A Fierce Green Fire.* New York: Hill and Wang, 1993.

———. "Real Rio: Behind the Scenes at the '92 Earth Summit." *Buzzworm,* September–October 1992.

Sharpe, William E., and Joy R. Drohan, eds. *The Effects of Acidic Deposition on Aquatic Ecosystems in Pennsylvania.* University Park: Environmental Resources Research Institute, State University of Pennsylvania, 1999.

Shenon, Philip. "H.U.D. Aide Says Developer Paid Watt $300,000 for Minimal Work." *New York Times Current Events Edition,* May 9, 1989.

Shephard, Paul, and Daniel McKinley, eds. *The Subversive Science: Essays Toward an Ecology of Man.* New York: Houghton Mifflin, 1969.

Sheremata, Davis. "Shooting the Clouds." *Canadian Geographic,* July–August 1998.

Sheridan, David. *Hard Rock Mining on the Public Land.* Washington, D.C.: Council on Environmental Quality, 1977.

———. *Off-Road Vehicles on Public Land.* Washington, D.C.: Council on Environmental Quality, 1979.

Sibley, George. "Glen Canyon: Using a Dam to Heal a River." *High Country News,* July 22, 1996.

Sierra Club Air Quality Campaign. *Clean Air Factsheets.* San Francisco: Sierra Club, 1981.

Silverberg, Robert. *The Challenge of Climate.* New York: Meredith, 1969.

Silvicultural Systems for the Major Forest Types of the U.S. Washington: USDA-Forest Service, 1973.

Siskiyou National Forest Chetco-Grayback Planning Unit Environmental Impact Statement. Grants Pass, Ore.: USDA-Forest Service, 1979.

Sittig, Marshall. *Handbook of Toxic and Hazardous Chemicals and Carcinogens.* 2nd ed. Park Ridge, N.J.: Noyes, 1985.

Skolnicoff, Eugene B. "The Role of Science in Policy." *Environment,* June 1999.

Smith, Dan. "The Case for Greener Cities." *American Forests,* autumn 1999.

Smith, Mark. "Conservation Reserve Program Approaches Acreage Limit." *Agricultural Outlook,* June–July 1999.

Snyder, Jesse. "CARB: Hybrids Okay as ZEVs." *Ward's Auto World,* September 1995.

Sohn, Louis B., and Kristen Gustafson. *The Law of the Sea.* St. Paul: West Publishing Company, 1984.

Somma, Mark, and Sue Tolleson-Rinehart. "Tracking the Elusive Green Women: Sex, Environmentalism, and Feminism in the United States and Europe." *Political Research Quarterly,* March 1997.

Soroos, Marvin S. "The Odyssey of Arctic Haze." *Environment,* December 1992.

Spangenburg, Ray, and Diane Moser. "The Mysterious Ozone Hole." *Space World,* October 1986.

Speidel, David H., Lon C. Ruedisili, et al., eds. *Perspectives on Water: Uses and Abuses.* New York: Oxford, 1988.

Spretnak, Charlene. *States of Grace.* New York: HarperCollins, 1991.

Spurr, Stephen H., and Burton V. Barnes. *Forest Ecology.* 2nd ed. New York: Ronald Press, 1973.

Starr, Cecie, and Ralph Taggart. *Biology: The Unity and Diversity of Life.* 3d ed. Belmont, Calif.: Wadsworth, 1984.

"State Alliance Calls for a Phaseout of MTBE." *Chemical Market Reporter,* February 14, 2000.

Statistical Abstract of the United States. Washington: U.S. Department of Commerce, 1984.

Steadman, David W. ". . . And Live on Pigeon Pie." *New York State Conservationist,* April 1996.

Stevens, William K. "A Dam Open, the Grand Canyon Roars Again." *New York Times,* February 25, 1997.

Stick, David. *The Outer Banks of North Carolina.* Chapel Hill: University of North Carolina Press, 1958.

Stover, Dawn. "Inside Biosphere II." *Popular Science,* November 1990.

Stopa, Marsha. "Stempel-Iacocca Bicycles May Plug Into Europe, Asia." *Automotive News,* March 9, 1998.

Strohmeyer, John. *Extreme Conditions: Big Oil and the Transformation of Alaska.* New York: Simon & Schuster, 1993.

Suplee, Curt. "El Niño/La Niña: Nature's Vicious Cycle." *National Geographic,* March 1999.

Svitil, Kathy A. "Collapse of a Food Chain." *Discover,* July 1995.

"Take a Deep Breath." *Chemecology,* June–July 1994.

Tarbuck, Edward J., and Frederick K. Lutgens. *The Earth: An Introduction to Physical Geology.* Columbus, Ohio: Merrill, 1984.

Taylor, David A. "The Greening of the Carbon Trade." *Americas,* July–August 1999.

Terres, John K. *The Audubon Society Encyclopedia of North American Birds.* New York: Knopf, 1980.

Theodore, Louis, and Anthony J. Buonicore, eds. *Air Pollution Control Equipment.* Englewood Cliffs, N.J.: Prentice-Hall, 1982.

Thombury, William D. *Principles of Geomorphology.* 2nd ed. New York: Wiley, 1969.

Thompson, Dick. "Capitol Hill Meltdown." *Time,* August 9, 1999.

Tilden, Freeman. *The National Parks,* revised and expanded by Paul Schullery. New York: Knopf, 1986.

Tools for Measuring Your Forest. Oregon State University Extension Service, 1983.

Torrens, Kevin D., et al. "Watershed-Based Credit Trading: Are You Ready?" *Pollution Engineering,* May 1999.

Toxic Substances Control Programs in the Great Lakes Basin. Windsor: International Joint Commission, 1983.

Toxic Substances in the Great Lakes. Chicago: Environmental Protection Agency, 1980.

Trewartha, Glenn T., Arthur H. Robinson, et al. *Fundamentals of Physical Geography.* 2nd ed. New York: McGraw-Hill, 1968.

Trumbull, Mark. "Fisheries Crisis Stretches Across the Globe." *Christian Science Monitor,* July 6, 1994.

Turner, Frederick. "Oh, Wilderness." *Outside,* April 1997.

Twombly, Renee. "Lowering Water's Octane." *Environmental Health Perspectives,* October 1998.

Union of Concerned Scientists. *World Scientists' Warning to Humanity.* Cambridge, Mass.: Union of Concerned Scientists, 1997.

U.S. Department of Agriculture. *Agriculture Fact Book 1998.* Washington, D.C.: Office of Communications, U.S. Department of Agriculture, 1998.

———. *The 1996 Farm Bill.* Washington, D.C.: Office of Communications, U.S. Department of Agriculture, 1996.

U.S. Department of the Interior. *The National Parks Index 1993.* Washington, D.C.: U.S. Government Printing Office, 1993.

The United States Government Manual. Washington: Office of the Federal Register, U.S. General Services Administration (annual).

Upgren, Arthur R. "Night Blindness." *Amicus Journal,* winter 1996.

Vesiland, P. Aarne. *Environmental Pollution and Control.* Ann Arbor: Ann Arbor Science, 1975.

Wackernagel, Mathis, and William Rees. *Our Ecological Footprint: Reducing Human Impact on the Earth.* Gabriola Island, B.C.: New Society Publishers, 1996.

Wade, Beth. "Bringing Down the Dams." *American City and County,* June 1999.

Wagner, Richard H. *Environment and Man.* 2nd ed. New York: Norton, 1974.

Wallach, Lori, and Robert Naiman. "Four and a Half Years Later." *Ecologist,* May–June 1998.

Walth, Bret. *Fire and Eden's Gate: Tom McCall and the Oregon Story.* Portland: Oregon Historical Society Press, 1994.

Wang, Fan, and Ian C. Ward. "Multiple Radon Entry in a House with a Cellar." *Journal of the Air & Waste Management Association,* June 1999.

Water Quality/Water Rights. Sacramento: California Water Resources Control Board, 1981.

Watkins, T. H. *The Hungry Years.* New York: Holt, 1999.

Waves Against the Shore: An Erosion Manual. Chicago: Lake Michigan Federation, 1978.

Weart, Wally. "Rocky Flats Launch: First Shipment of Radioactive Waste Trucked to New Mexico." *Traffic World,* June 28, 1999.

Weisskopf, Michael. "In Minamata, the City of Death, It's Hard to Tell if Life Goes On." *Washington Post,* May 11, 1987.

Wheelwright, Jeff. "Condors: Back from the Brink." *Smithsonian,* May 1997.

White, J. F., ed. *Study of the Earth: Readings in Geological Science.* Englewood Cliffs, N.J.: Prentice-Hall, 1962.

White, Peter T. "The Fascinating World of Trash." *National Geographic Magazine,* April 1983.

Whitman, David. "See Forests Through Trees." *U.S. News & World Report,* October 25, 1999.

Wildavsky, Aaron. *The Politics of the Budgetary Process.* 3d ed. Boston: Little, Brown, 1979.

Wiley, John P., Jr. "The Gaia Hypothesis—That Life Creates the Conditions It Needs—Has Its Day in Court: The Jury Is Out." *Smithsonian,* May 1988.

Wiley, Karen B., and Steven L. Rhodes. "From Weapons to Wildlife: The Transformation of the Rocky Mountain Arsenal." *Environment,* June 1998.

Williams, Huntington, III. "Banking on the Future." *Nature Conservancy,* May–June, 1992.

Willoughby, L. G. *Freshwater Biology.* New York: Pica Press, 1977.

Wilson, E. O. *The Diversity of Life.* Cambridge, Mass.: Harvard, 1992.

Windholz, Martha, Susan Budavari, et al., eds. *The Merck Index.* 9th and 10th eds. Rahway, N.J.: Merck & Co., 1976 and 1983.

"WIPP Accepts First Shipment of Radioactive Waste." *Mining Engineering,* May 1999.

Wirth, Timothy E. "Global Climate Change." Remarks Before a U.S. Senate Committee on Environmental and Public Works Hearing. Washington: U.S. Government Printing Office, 1998.

Wistreich, George A., and Max D. Lechtman. *Microbiology.* 4th ed. New York: Macmillan, 1984.

Woodwell, George M., ed. *The Earth in Transition: Patterns and Processes of Biotic Impoverishment.* New York: Cambridge University Press, 1990.

World Resources Institute. *World Resources, 1996–97.* New York: Oxford, 1996.

Worster, Donald. *Dust Bowl: The Southern Plains in the 1930s.* New York: Oxford, 1979.

Wuerthner, George. "Why Healthy Forests Need Dead Trees." *Earth Island Journal,* fall 1995.

Wyckoff, Jerome. *Geology: Our Changing Earth Through the Ages.* New York: Golden Press, 1976.

Yablokov, Alexey V. "Chernobyl Postmortem." *Forum for Applied Research and Public Policy,* spring 1999.

Yoon, Carol Kaesuk. "Reassessing Ecological Risks of Genetically Altered Crops." *New York Times,* November 3, 1999.

Yule, John-David, ed. *Concise Encyclopedia of the Sciences.* New York: Facts On File, 1978.

Zahl, Paul A. "Where Would We Be Without Algae?" *National Geographic Magazine,* March 1974.

Zakin, Susan. *Coyotes and Town Dogs.* New York: Viking, 1993.

Zim, Herbert S., and Paul R. Shaffer. *Rocks and Minerals.* New York: Golden Press, 1957.

Zimmer, Carl. "The Value of a Free Lunch." *Discover,* January 1998.

Zimmerman, Tim. "If World War III Comes, Blame Fish." *U.S. News & World Report,* October 21, 1996.

INDEX

Page numbers in *italic* indicate illustrations.
Page numbers followed by *t* indicate tables and by *m* indicate maps.

B

D

I

J

M

P

S

X

Y

Z